Introductory
DC/AC Electronics
Third Edition

Nigel P. Cook

PRENTICE HALL
Englewood Cliffs, New Jersey   Columbus, Ohio

Library of Congress Cataloging-in-Publication Data
Cook, Nigel P.
Introductory DC/AC electronics / Nigel P. Cook. —3rd ed.
p. cm.
Includes index.
ISBN 0-13-213547-7
1. Electronics. I. Title.
TK7816.C65 1995
621.381—dc20 95-15889

Cover photo: Phil Matt
Editor: Dave Garza
Production Editor: Sheryl Glicker Langner
Project Management: Elm Street Publishing Services, Inc.
Cover Designer: Tom Mack
Production Manager: Pamela D. Bennett
Marketing Manager: Debbie Yarnell
Supplements Editor: Judith Casillo
Illustrations: Rolin Graphics Inc.

This book was set in Times Ten by The Clarinda Company and was printed and bound by R. R. Donnelley & Sons Company. The cover was printed by Phoenix Color Corp.

Printed in the United States of America
10 9 8 7 6 5 4 3 2 1

ISBN: 0-13-213547-7

Prentice-Hall International (UK) Limited, *London*
Prentice-Hall of Australia Pty. Limited, *Sydney*
Prentice-Hall of Canada, Inc., *Toronto*
Prentice-Hall Hispanoamericana, S. A., *Mexico*
Prentice-Hall of India Private Limited, *New Delhi*
Prentice-Hall of Japan, Inc., *Tokyo*
Simon & Schuster Asia Pte. Ltd., *Singapore*
Editora Prentice-Hall do Brasil, Ltda., *Rio de Janeiro*

To my loving wife, Dawn,
and beautiful children,
Candice and Jonathan,
whose love inspires me.

This book is dedicated to
all of the innovative, energetic,
and motivational instructors
who go out of their way to
prove that mathematics, science,
and technology are within
everyone's grasp and can be
made easy, interesting, and fun.

Preface

TO THE STUDENT

The early pioneers in electronics were intrigued by the mystery and wonder of a newly discovered science, whereas people today are attracted by its ability to lend its hand to any application and accomplish almost anything imaginable. If you analyze exactly how you feel at this stage, you will probably discover that you have mixed emotions about the journey ahead. On one hand, imagination, curiosity, and excitement are driving you on, while apprehension and reservations may be slowing you down. Your enthusiasm will overcome any indecision you have once you become actively involved in electronics and realize that it is as exciting as you ever expected it to be.

ORGANIZATION OF THE TEXTBOOK

This textbook has been divided into four basic parts. Chapters 1 through 5 introduce you to the world of electronics and the fundamentals of electricity. Chapters 6 through 10 cover direct current, or DC, circuits, Chapters 11 through 18 cover alternating current, or AC, circuits, and Chapters 19 through 23 cover semiconductor principles, devices, and circuits.

PART I The Fundamentals of Electricity
- Chapter 1 The World of Electronics
- Chapter 2 Voltage and Current
- Chapter 3 Resistance and Power
- Chapter 4 Resistors
- Chapter 5 Experimenting in the Lab

PART II Direct-Current Electronics
- Chapter 6 Direct Current (DC)
- Chapter 7 Series DC Circuits
- Chapter 8 Parallel DC Circuits
- Chapter 9 Series–Parallel DC Circuits
- Chapter 10 The Series–Parallel Analog Multimeter

PART III Alternating-Current Electronics
- Chapter 11 Alternating Current (AC)

Chapter 12 AC Test Equipment and Troubleshooting Techniques
Chapter 13 Capacitance and Capacitors
Chapter 14 Capacitive Circuits, Testing, and Applications
Chapter 15 Electromagnetism and Electromagnetic Induction
Chapter 16 Inductance and Inductors
Chapter 17 Transformers
Chapter 18 Resistive, Inductive, and Capacitive (*RLC*) Circuits

PART IV Semiconductor Devices and Circuits

Chapter 19 Semiconductor Principles
Chapter 20 Junction Diodes
Chapter 21 Bipolar Junction Transistors (BJTs)
Chapter 22 Field Effect Transistors (FETs)
Chapter 23 Operational Amplifiers

The material covered in this book has been logically divided and sequenced to provide a gradual progression from the known to the unknown, and from the simple to the complex.

DEVELOPMENT, CLASS TESTING, AND REVIEWING

The first phase of development was conducted in the classroom with students and instructors as critics. Each topic was class-tested by videotaping each lesson, and the results were then evaluated and implemented. This feedback was invaluable and enabled me to fine-tune my presentation of topics and instill understanding and confidence in the students.

The second phase of development was to forward a copy of the revised manuscript to several instructors at schools throughout the country. These technical and topical critiques helped to mold the text into a more accurate form.

The third and final phase was to class-test the final revised manuscript and then commission the last technical review in the final stages of production.

ILLUSTRATED TOUR OF TEXTBOOK FEATURES

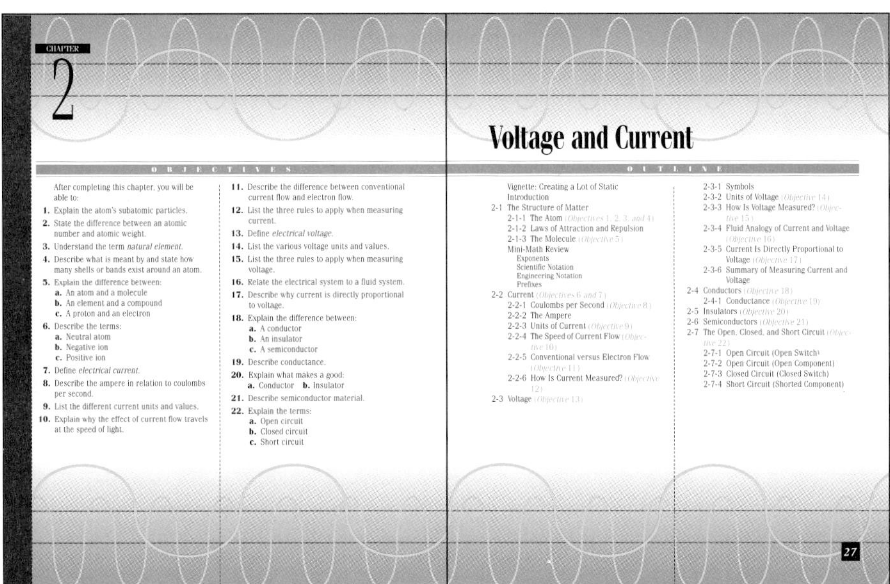

CHAPTER

2

Voltage and Current

OBJECTIVES

After completing this chapter, you will be able to:

1. Explain the atom's subatomic particles.
2. State the difference between an atomic number and atomic weight.
3. Understand the term *natural element*.
4. Describe what is meant by and state how many shells or bands exist around an atom.
5. Explain the difference between:
   a. An atom and a molecule
   b. An element and a compound
   c. A proton and an electron
6. Describe the terms:
   a. Neutral atom
   b. Negative ion
   c. Positive ion
7. Define *electrical current*.
8. Describe the ampere in relation to coulombs per second.
9. List the different current units and values.
10. Explain why the effect of current flow travels at the speed of light.
11. Describe the difference between conventional current flow and electron flow.
12. List the three rules to apply when measuring current.
13. Define *electrical voltage*.
14. List the various voltage units and values.
15. List the three rules to apply when measuring voltage.
16. Relate the electrical system to a fluid system.
17. Describe why current is directly proportional to voltage.
18. Explain the difference between:
    a. A conductor
    b. An insulator
    c. A semiconductor
19. Describe conductance.
20. Explain what makes a good:
    a. Conductor b. Insulator
21. Describe semiconductor material.
22. Explain the terms:
    a. Open circuit
    b. Closed circuit
    c. Short circuit

OUTLINE

Vignette: Creating a Lot of Static
Introduction
2-1 The Structure of Matter
2-1-1 The Atom *(Objectives 1, 2, 3, and 4)*
2-1-2 Laws of Attraction and Repulsion
2-1-3 The Molecule *(Objective 5)*
Mini-Math Review
Exponents
Scientific Notation
Engineering Notation
Prefixes
2-2 Current *(Objectives 6 and 7)*
2-2-1 Coulombs per Second *(Objective 8)*
2-2-2 The Ampere
2-2-3 Units of Current *(Objective 9)*
2-2-4 The Speed of Current Flow *(Objective 10)*
2-2-5 Conventional versus Electron Flow *(Objective 11)*
2-2-6 How Is Current Measured? *(Objective 12)*
2-3 Voltage *(Objective 13)*
2-3-1 Symbols
2-3-2 Units of Voltage *(Objective 14)*
2-3-3 How Is Voltage Measured? *(Objective 15)*
2-3-4 Fluid Analogy of Current and Voltage *(Objective 16)*
2-3-5 Current Is Directly Proportional to Voltage *(Objective 17)*
2-3-6 Summary of Measuring Current and Voltage
2-4 Conductors *(Objective 18)*
2-4-1 Conductance *(Objective 19)*
2-5 Insulators *(Objective 20)*
2-6 Semiconductors *(Objective 21)*
2-7 The Open, Closed, and Short Circuit *(Objective 22)*
2-7-1 Open Circuit (Open Switch)
2-7-2 Open Circuit (Open Component)
2-7-3 Closed Circuit (Closed Switch)
2-7-4 Short Circuit (Shorted Component)

27

Each chapter opening lists performance-based objectives, with each objective directly correlated to a chapter section on the following page.

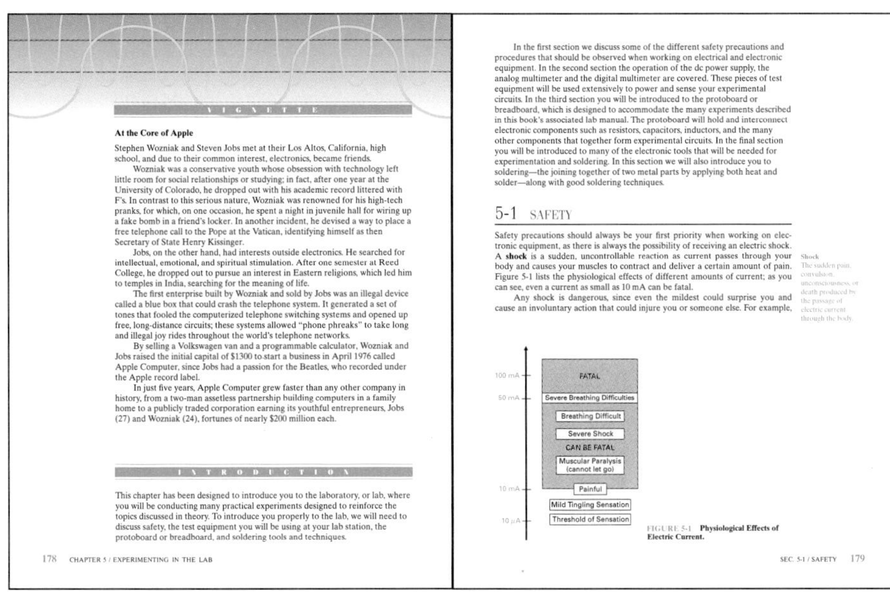

VIGNETTE

At the Core of Apple

Stephen Wozniak and Steven Jobs met at their Los Altos, California, high school, and due to their common interest, electronics, became friends.

Wozniak was a conservative youth whose obsession with technology left little room for social relationships or studying; in fact, after one year at the University of Colorado, he dropped out with his academic record littered with F's. In contrast to this serious nature, Wozniak was renowned for his high-tech pranks, for which, on one occasion, he spent a night in juvenile hall for wiring up a fake bomb in a friend's locker. In another incident, he devised a way to place a free telephone call to the Pope at the Vatican, identifying himself as then Secretary of State Henry Kissinger.

Jobs, on the other hand, had interests outside electronics. He searched for intellectual, emotional, and spiritual stimulation. After one semester at Reed College, he dropped out to pursue an interest in Eastern religions, which led him to temples in India, searching for the meaning of life.

The first enterprise built by Wozniak and sold by Jobs was an illegal device called a blue box that could crash the telephone system. It generated a set of tones that fooled the computerized telephone switching systems and opened up free, long-distance circuits; these systems allowed "phone phreaks" to take long and illegal joy rides throughout the world's telephone networks.

By selling a Volkswagen van and a programmable calculator, Wozniak and Jobs raised the initial capital of $1300 to start a business in April 1976 called Apple Computer, since Jobs had a passion for the Beatles, who recorded under the Apple record label.

In just five years, Apple Computer grew faster than any other company in history, from a two-man assetless partnership building computers in a family home to a publicly traded corporation earning its youthful entrepreneurs, Jobs (27) and Wozniak (24), fortunes of nearly $200 million each.

INTRODUCTION

This chapter has been designed to introduce you to the laboratory, or lab, where you will be conducting many practical experiments designed to reinforce the topics discussed in theory. To introduce you properly to the lab, we will need to discuss safety, the test equipment you will be using at your lab station, the protoboard or breadboard, and soldering tools and techniques.

178 CHAPTER 5 / EXPERIMENTING IN THE LAB

In the first section we discuss some of the different safety precautions and procedures that should be observed when working on electrical and electronic equipment. In the second section the operation of the dc power supply, the analog multimeter and the digital multimeter are covered. These pieces of test equipment will be used extensively to power and sense your experimental circuits. In the third section you will be introduced to the protoboard or breadboard, which is designed to accommodate the many experiments described in this book's associated lab manual. The protoboard will hold and interconnect electronic components such as resistors, capacitors, inductors, and the many other components that together form experimental circuits. In the final section you will be introduced to many of the electronic tools that will be needed for experimentation and soldering. In this section we will also introduce you to soldering—the joining together of two metal parts by applying both heat and solder—along with good soldering techniques.

5-1 SAFETY

Safety precautions should always be your first priority when working on electronic equipment, as there is always the possibility of receiving an electric shock. A **shock** is a sudden, uncontrollable reaction as current passes through your body and causes your muscles to contract and deliver a certain amount of pain. Figure 5-1 lists the physiological effects of different amounts of current; as you can see, even a current as small as 10 mA can be fatal.

Any shock is dangerous, since even the mildest could surprise you and cause an involuntary action that could injure you or someone else. For example,

Shock
The sudden pain, convulsion, unconsciousness, or death produced by the passage of electric current through the body.

FIGURE 5-1 **Physiological Effects of Electric Current.**

SEC. 5-1 / SAFETY 179

Motivational stories detail the people behind the electronics industry, and a conversational introduction reviews what has been previously covered and what is about to be covered.

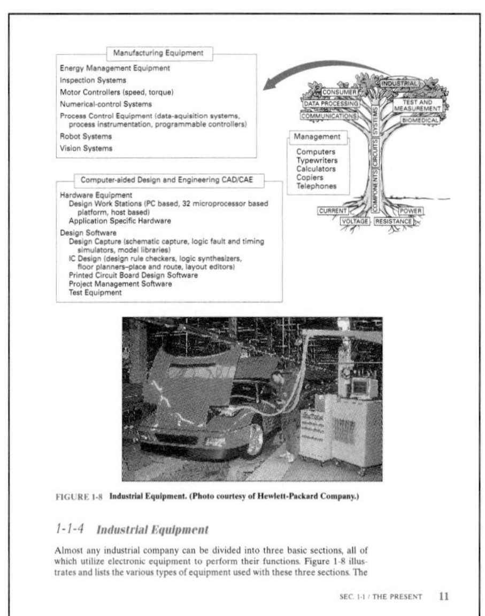

Manufacturing Equipment

Energy Management Equipment
Inspection Systems
Motor Controllers (speed, torque)
Numerical-control Systems
Process Control Equipment (data-aquisition systems, process instrumentation, programmable controllers)
Robot Systems
Vision Systems

Management

Computers
Typewriters
Calculators
Copiers
Telephones

Computer-aided Design and Engineering CAD/CAE

Hardware Equipment
Design Work Stations (PC based, 32 microprocessor based platform, host based)
Application Specific Hardware
Design Software
Design Capture (schematic capture, logic fault and timing simulators, model libraries)
IC Design (design rule checkers, logic synthesizers, floor planners–place and route, layout editors)
Printed Circuit Board Design Software
Project Management Software
Test Equipment

FIGURE 1-8 **Industrial Equipment. (Photo courtesy of Hewlett-Packard Company.)**

1-1-4 Industrial Equipment

Almost any industrial company can be divided into three basic sections, all of which utilize electronic equipment to perform their functions. Figure 1-8 illustrates and lists the various types of equipment used with these three sections. The

SEC. 1-1 / THE PRESENT 11

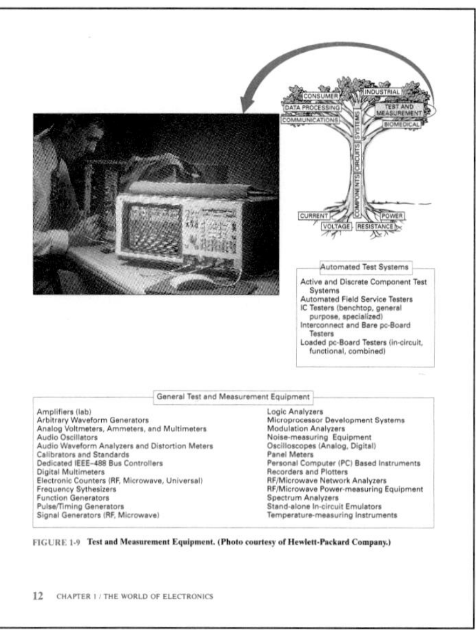

Automated Test Systems

Active and Discrete Component Test Systems
Automated Field Service Testers
IC Testers (benchtop, general purpose, specialized)
Interconnect and Bare pc-Board Testers
Loaded pc-Board Testers (in-circuit, functional, combined)

General Test and Measurement Equipment

Amplifiers (lab)
Arbitrary Waveform Generators
Analog Voltmeters, Ammeters, and Multimeters
Audio Oscillators
Audio Waveform Analyzers and Distortion Meters
Calibrators and Standards
Dedicated IEEE–488 Bus Controllers
Digital Multimeters
Electronic Counters (RF, Microwave, Universal)
Frequency Sythesizers
Function Generators
Pulse/Timing Generators
Signal Generators (RF, Microwave)
Logic Analyzers
Microprocessor Development Systems
Modulation Analyzers
Noise-measuring Equipment
Oscilloscopes (Analog, Digital)
Panel Meters
Personal Computer (PC) Based Instruments
Recorders and Plotters
RF/Microwave Network Analyzers
RF/Microwave Power-measuring Equipment
Spectrum Analyzers
Stand-alone In-circuit Emulators
Temperature-measuring Instruments

FIGURE 1-9 **Test and Measurement Equipment. (Photo courtesy of Hewlett-Packard Company.)**

12 CHAPTER 1 / THE WORLD OF ELECTRONICS

Chapter 1 includes extensive information about career opportunities for the electronics technician, providing interest in and motivation for the material which follows.

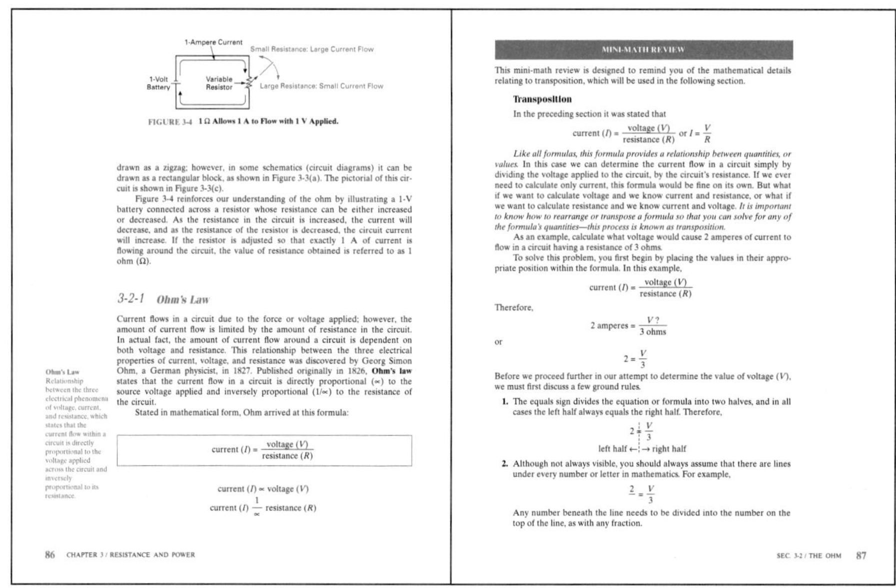

FIGURE 3-4 **1 Ω Allows 1 A to Flow with 1 V Applied.**

drawn as a zigzag; however, in some schematics (circuit diagrams) it can be drawn as a rectangular block, as shown in Figure 3-3(a). The pictorial of this circuit is shown in Figure 3-3(c).

Figure 3-4 reinforces our understanding of the ohm by illustrating a 1-V battery connected across a resistor whose resistance can be either increased or decreased. As the resistance in the circuit is increased, the current will decrease, and as the resistance of the resistor is decreased, the circuit current will increase. If the resistor is adjusted so that exactly 1 A of current is flowing around the circuit, the value of resistance obtained is referred to as 1 ohm (Ω).

3-2-1 Ohm's Law

Current flows in a circuit due to the force or voltage applied; however, the amount of current flow is limited by the amount of resistance in the circuit. In actual fact, the amount of current flow around a circuit is dependent on both voltage and resistance. This relationship between the three electrical properties of current, voltage, and resistance was discovered by Georg Simon Ohm, a German physicist, in 1827. Published originally in 1826, **Ohm's law** states that the current flow in a circuit is directly proportional (∝) to the source voltage applied and inversely proportional (1/∝) to the resistance of the circuit.

Ohm's Law Relationship between the three electrical phenomena of voltage, current, and resistance, which states that the current flow within a circuit is directly proportional to the voltage applied across the circuit and inversely proportional to its resistance.

Stated in mathematical form, Ohm arrived at this formula:

$$\text{current } (I) = \frac{\text{voltage } (V)}{\text{resistance } (R)}$$

$$\text{current } (I) \propto \text{voltage } (V)$$

$$\text{current } (I) \frac{1}{\propto} \text{resistance } (R)$$

86 CHAPTER 3 / RESISTANCE AND POWER

MINI-MATH REVIEW

This mini-math review is designed to remind you of the mathematical details relating to transposition, which will be used in the following section.

Transposition

In the preceding section it was stated that

$$\text{current } (I) = \frac{\text{voltage } (V)}{\text{resistance } (R)} \text{ or } I = \frac{V}{R}$$

Like all formulas, this formula provides a relationship between quantities, or values. In this case we can determine the current flow in a circuit simply by dividing the voltage applied to the circuit, by the circuit's resistance. If we ever need to calculate only current, this formula would be fine on its own. But what if we want to calculate voltage and we know current and resistance, or what if we want to calculate resistance and we know current and voltage. *It is important to know how to rearrange or transpose a formula so that you can solve for any of the formula's quantities—this process is known as transposition.*

As an example, calculate what voltage would cause 2 amperes of current to flow in a circuit having a resistance of 3 ohms.

To solve this problem, you first begin by placing the values in their appropriate position within the formula. In this example,

$$\text{current } (I) = \frac{\text{voltage } (V)}{\text{resistance } (R)}$$

Therefore,

$$2 \text{ amperes} = \frac{V\,?}{3 \text{ ohms}}$$

or

$$2 = \frac{V}{3}$$

Before we proceed further in our attempt to determine the value of voltage (V), we must first discuss a few ground rules.

1. The equals sign divides the equation or formula into two halves, and in all cases the left half always equals the right half. Therefore,

$$2 = \frac{V}{3}$$

left half ←¦→ right half

2. Although not always visible, you should always assume that there are lines under every number or letter in mathematics. For example,

$$\frac{2}{\ } = \frac{V}{3}$$

Any number beneath the line needs to be divided into the number on the top of the line, as with any fraction.

SEC. 3-2 / THE OHM 87

Mini-Math Reviews are included in appropriate places to review the necessary mathematics needed for the discussion at hand.

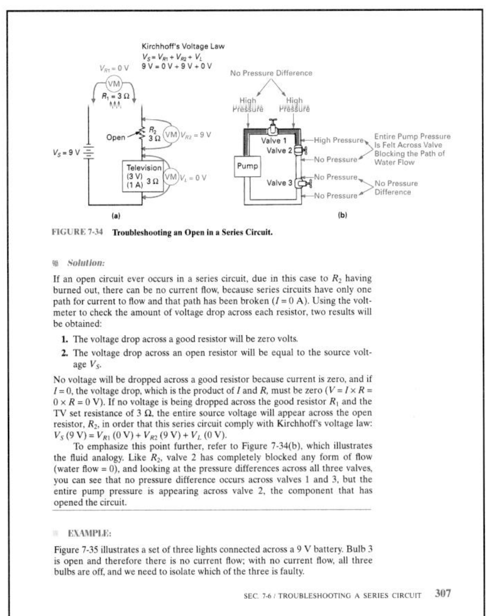

FIGURE 7-34 **Troubleshooting an Open in a Series Circuit.**

Solution:

If an open circuit ever occurs in a series circuit, due in this case to R_2 having burned out, there can be no current flow, because series circuits have only one path for current to flow and that path has been broken ($I = 0$ A). Using the voltmeter to check the amount of voltage drop across each resistor, two results will be obtained:

1. The voltage drop across a good resistor will be zero volts.
2. The voltage drop across an open resistor will be equal to the source voltage V_S.

No voltage will be dropped across a good resistor because current is zero, and if $I = 0$, the voltage drop, which is the product of I and R, must be zero ($V = I \times R = 0 \times R = 0$ V). If no voltage is being dropped across the good resistor R_1 and the TV set resistance of 3 Ω, the entire source voltage will appear across the open resistor, R_2, in order that this series circuit comply with Kirchhoff's voltage law: V_S (9 V) = V_{R1} (0 V) + V_{R2} (9 V) + V_L (0 V).

To emphasize this point further, refer to Figure 7-34(b), which illustrates the fluid analogy. Like R_2, valve 2 has completely blocked any form of flow (water flow = 0), and looking at the pressure differences across all three valves, you can see that no pressure difference occurs across valves 1 and 3, but the entire pump pressure is appearing across valve 2, the component that has opened the circuit.

EXAMPLE:

Figure 7-35 illustrates a set of three lights connected across a 9 V battery. Bulb 3 is open and therefore there is no current flow; with no current flow, all three bulbs are off, and we need to isolate which of the three is faulty.

SEC. 7-6 / TROUBLESHOOTING A SERIES CIRCUIT 307

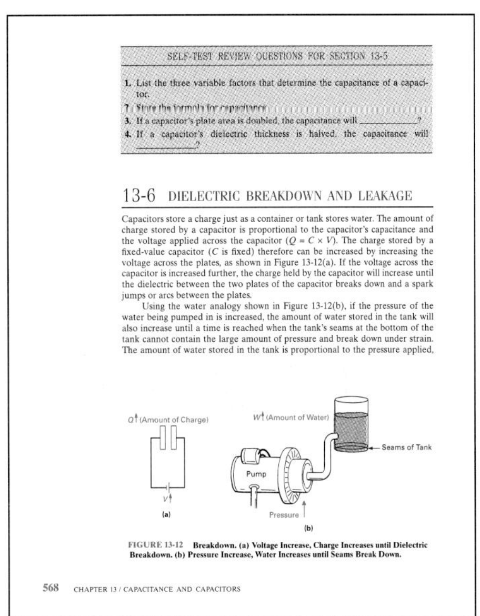

SELF-TEST REVIEW QUESTIONS FOR SECTION 13-5

1. List the three variable factors that determine the capacitance of a capacitor.
2. State the formula for capacitance.
3. If a capacitor's plate area is doubled, the capacitance will ________?
4. If a capacitor's dielectric thickness is halved, the capacitance will ________?

13-6 DIELECTRIC BREAKDOWN AND LEAKAGE

Capacitors store a charge just as a container or tank stores water. The amount of charge stored by a capacitor is proportional to the capacitor's capacitance and the voltage applied across the capacitor ($Q = C \times V$). The charge stored by a fixed-value capacitor (C is fixed) therefore can be increased by increasing the voltage across the plates, as shown in Figure 13-12(a). If the voltage across the capacitor is increased further, the charge held by the capacitor will increase until the dielectric between the two plates of the capacitor breaks down and a spark jumps or arcs between the plates.

Using the water analogy shown in Figure 13-12(b), if the pressure of the water being pumped in is increased, the amount of water stored in the tank will also increase until a time is reached when the tank's seams at the bottom of the tank cannot contain the large amount of pressure and break down under strain. The amount of water stored in the tank is proportional to the pressure applied,

FIGURE 13-12 **Breakdown. (a) Voltage Increase, Charge Increases until Dielectric Breakdown. (b) Pressure Increase, Water Increases until Seams Break Down.**

568 CHAPTER 13 / CAPACITANCE AND CAPACITORS

Concept analogies are used throughout the text to take the student from the known to the unknown.

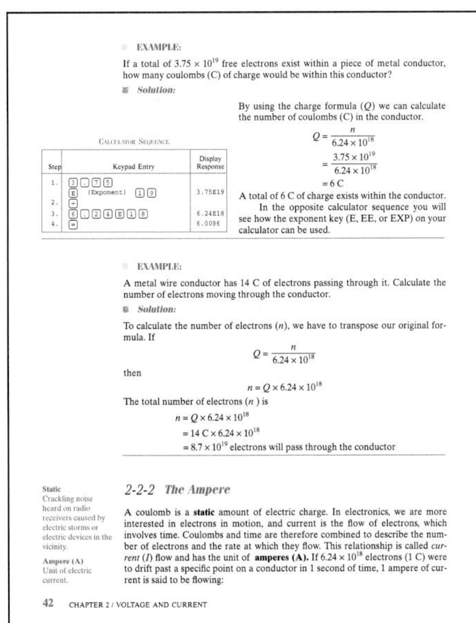

EXAMPLE:

If a total of 3.75×10^{19} free electrons exist within a piece of metal conductor, how many coulombs (C) of charge would be within this conductor?

Solution:

By using the charge formula (Q) we can calculate the number of coulombs (C) in the conductor.

$$Q = \frac{n}{6.24 \times 10^{18}} = \frac{3.75 \times 10^{19}}{6.24 \times 10^{18}} = 6 \text{ C}$$

A total of 6 C of charge exists within the conductor.

In the opposite calculator sequence you will see how the exponent key (E, EE, or EXP) on your calculator can be used.

CALCULATOR SEQUENCE

Step	Keypad Entry	Display Response
1.	3 . 7 5 E (Exponent) 1 9	3.75E19
2.	÷	
3.	6 . 2 4 E 1 8	6.24E18
4.	=	6.0096

EXAMPLE:

A metal wire conductor has 14 C of electrons passing through it. Calculate the number of electrons moving through the conductor.

Solution:

To calculate the number of electrons (n), we have to transpose our original formula. If

$$Q = \frac{n}{6.24 \times 10^{18}}$$

then

$$n = Q \times 6.24 \times 10^{18}$$

The total number of electrons (n) is

$$n = Q \times 6.24 \times 10^{18}$$
$$= 14 \text{ C} \times 6.24 \times 10^{18}$$
$$= 8.7 \times 10^{19} \text{ electrons will pass through the conductor}$$

2-2-2 *The Ampere*

Static
Crackling noise heard on radio receivers caused by electric storms or electric devices in the vicinity.

Ampere (A)
Unit of electric current.

A coulomb is a **static** amount of electric charge. In electronics, we are more interested in electrons in motion, and current is the flow of electrons, which involves time. Coulombs and time are therefore combined to describe the number of electrons and the rate at which they flow. This relationship is called *current* (I) flow and has the unit of **amperes (A).** If 6.24×10^{18} electrons (1 C) were to drift past a specific point on a conductor in 1 second of time, 1 ampere of current is said to be flowing:

42 CHAPTER 2 / VOLTAGE AND CURRENT

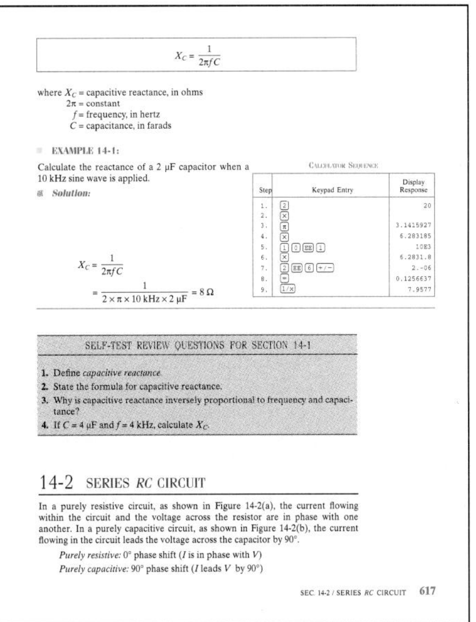

$$X_C = \frac{1}{2\pi f C}$$

where X_C = capacitive reactance, in ohms
2π = constant
f = frequency, in hertz
C = capacitance, in farads

EXAMPLE 14-1:

Calculate the reactance of a 2 μF capacitor when a 10 kHz sine wave is applied.

Solution:

$$X_C = \frac{1}{2\pi f C} = \frac{1}{2 \times \pi \times 10 \text{ kHz} \times 2 \text{ μF}} = 8 \, \Omega$$

CALCULATOR SEQUENCE

Step	Keypad Entry	Display Response
1.	2	20
2.	×	
3.	π	3.1415927
4.	×	6.283185
5.	1 0 EE 1	10E3
6.	×	6.2831.8
7.	2 EE 6 +/−	2.-06
8.	=	0.1256637
9.	1/x	7.9577

SELF-TEST REVIEW QUESTIONS FOR SECTION 14-1

1. Define *capacitive reactance.*
2. State the formula for capacitive reactance.
3. Why is capacitive reactance inversely proportional to frequency and capacitance?
4. If C = 4 μF and f = 4 kHz, calculate X_C.

14-2 SERIES *RC* CIRCUIT

In a purely resistive circuit, as shown in Figure 14-2(a), the current flowing within the circuit and the voltage across the resistor are in phase with one another. In a purely capacitive circuit, as shown in Figure 14-2(b), the current flowing in the circuit leads the voltage across the capacitor by 90°.

Purely resistive: 0° phase shift (I is in phase with V)
Purely capacitive: 90° phase shift (I leads V by 90°)

SEC. 14-2 / SERIES *RC* CIRCUIT 617

Calculator sequences are included in all examples that introduce new mathematical procedures, and a running glossary defines all new terms in the margin of the text.

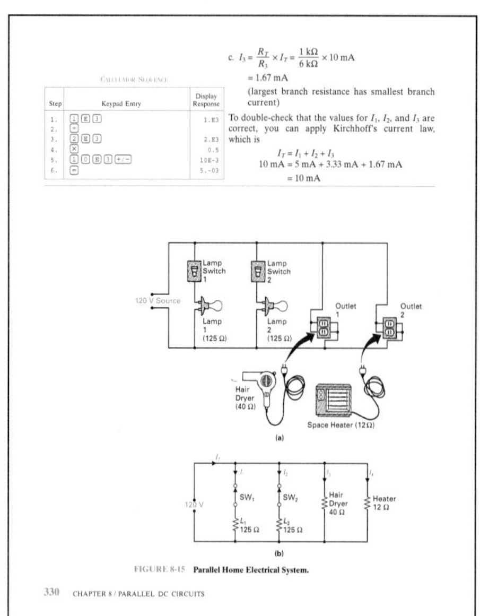

c. $I_3 = \frac{R_T}{R_3} \times I_T = \frac{1\ \text{k}\Omega}{6\ \text{k}\Omega} \times 10\ \text{mA}$

$= 1.67\ \text{mA}$

(largest branch resistance has smallest branch current)

CALCULATOR SEQUENCE

Step	Keypad Entry	Display Response
1.	1 E 3	1.E3
2.	÷	
3.	2 E 3	2.E3
4.	×	0.5
5.	1 0 E 3 +/−	10E-3
6.	=	5.-03

To double-check that the values for I_1, I_2, and I_3 are correct, you can apply Kirchhoff's current law, which is

$$I_T = I_1 + I_2 + I_3$$
$$10\ \text{mA} = 5\ \text{mA} + 3.33\ \text{mA} + 1.67\ \text{mA}$$
$$= 10\ \text{mA}$$

FIGURE 8-15 **Parallel Home Electrical System.**

330 CHAPTER 8 / PARALLEL DC CIRCUITS

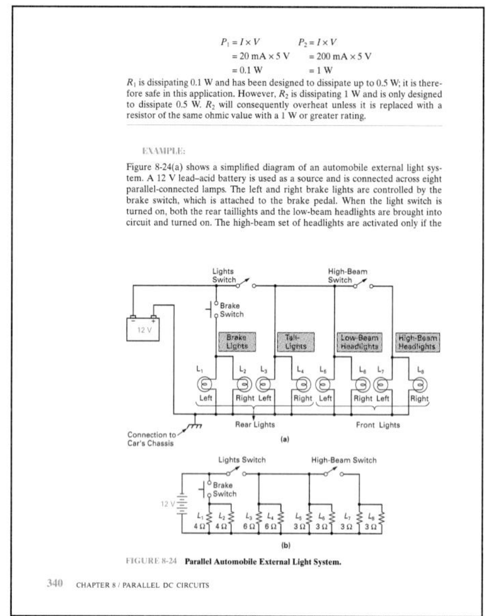

$$P_1 = I \times V \qquad P_2 = I \times V$$
$$= 20\ \text{mA} \times 5\ \text{V} \qquad = 200\ \text{mA} \times 5\ \text{V}$$
$$= 0.1\ \text{W} \qquad = 1\ \text{W}$$

R_1 is dissipating 0.1 W and has been designed to dissipate up to 0.5 W; it is therefore safe in this application. However, R_2 is dissipating 1 W and is only designed to dissipate 0.5 W. R_2 will consequently overheat unless it is replaced with a resistor of the same ohmic value with a 1 W or greater rating.

EXAMPLE:

Figure 8-24(a) shows a simplified diagram of an automobile external light system. A 12 V lead–acid battery is used as a source and is connected across eight parallel-connected lamps. The left and right brake lights are controlled by the brake switch, which is attached to the brake pedal. When the light switch is turned on, both the rear taillights and the low-beam headlights are brought into circuit and turned on. The high-beam set of headlights are activated only if the

FIGURE 8-24 **Parallel Automobile External Light System.**

340 CHAPTER 8 / PARALLEL DC CIRCUITS

Actual circuit examples are included to help the student connect a topic to the real world.

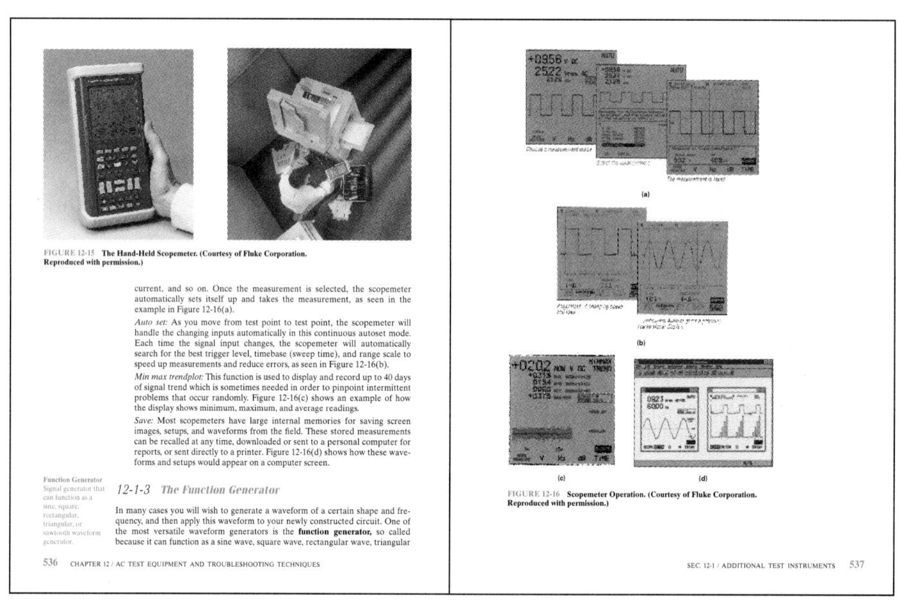

FIGURE 12-15 **The Hand-Held Scopemeter. (Courtesy of Fluke Corporation. Reproduced with permission.)**

current, and so on. Once the measurement is selected, the scopemeter automatically sets itself up and takes the measurement, as seen in the example in Figure 12-16(a).

Auto set: As you move from test point to test point, the scopemeter will handle the changing inputs automatically in this continuous autoset mode. Each time the signal input changes, the scopemeter will automatically search for the best trigger level, timebase (sweep time), and range scale to speed up measurements and reduce errors, as seen in Figure 12-16(b).

Min max trendplot: This function is used to display and record up to 40 days of signal trend which is sometimes needed in order to pinpoint intermittent problems that occur randomly. Figure 12-16(c) shows an example of how the display shows minimum, maximum, and average readings.

Save: Most scopemeters have large internal memories for saving screen images, setups, and waveforms from the field. These stored measurements can be recalled at any time, downloaded or sent to a personal computer for reports, or sent directly to a printer. Figure 12-16(d) shows how these waveforms and setups would appear on a computer screen.

Function Generator Signal generator that can function as a sine, square, rectangular, triangular, or sawtooth waveform generator.

12-1-3 The Function Generator

In many cases you will wish to generate a waveform of a certain shape and frequency, and then apply this waveform to your newly constructed circuit. One of the most versatile waveform generators is the **function generator,** so called because it can function as a sine wave, square wave, rectangular wave, triangular

536 CHAPTER 12 / AC TEST EQUIPMENT AND TROUBLESHOOTING TECHNIQUES

FIGURE 12-16 **Scopemeter Operation. (Courtesy of Fluke Corporation. Reproduced with permission.)**

SEC. 12-1 / ADDITIONAL TEST INSTRUMENTS 537

A strong testing, test equipment, and troubleshooting emphasis prepares the student for the working world.

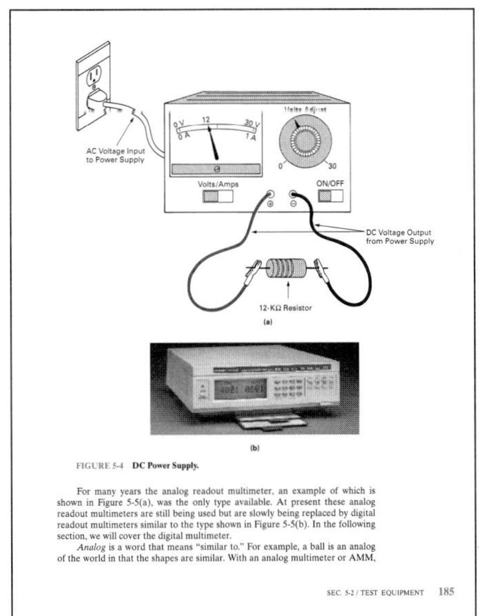

FIGURE 5-4 **DC Power Supply.**

For many years the analog readout multimeter, an example of which is shown in Figure 5-5(a), was the only type available. At present these analog readout multimeters are still being used but are slowly being replaced by digital readout multimeters similar to the type shown in Figure 5-5(b). In the following section, we will cover the digital multimeter.

Analog is a word that means "similar to." For example, a ball is an analog of the world in that the shapes are similar. With an analog multimeter or AMM,

SEC. 5-2 / TEST EQUIPMENT 185

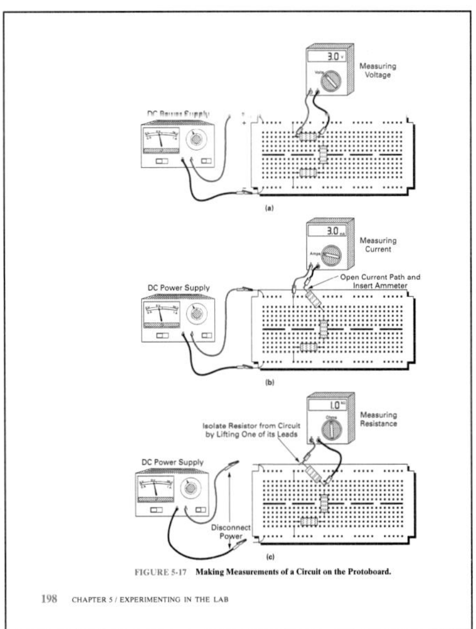

FIGURE 5-17 **Making Measurements of a Circuit on the Protoboard.**

198 CHAPTER 5 / EXPERIMENTING IN THE LAB

Chapter 5 introduces the student to experimenting in the lab by covering safety, test equipment, protoboards, and soldering tools and techniques.

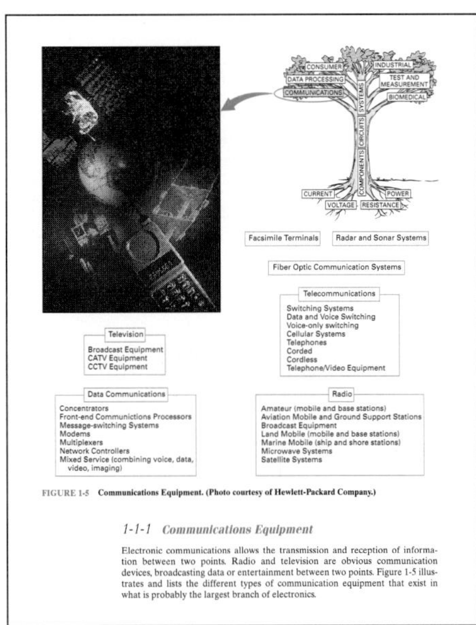

FIGURE 1-5 **Communications Equipment. (Photo courtesy of Hewlett-Packard Company.)**

1-1-1 Communications Equipment

Electronic communications allows the transmission and reception of information between two points. Radio and television are obvious communication devices, broadcasting data or entertainment between two points. Figure 1-5 illustrates and lists the different types of communication equipment that exist in what is probably the largest branch of electronics.

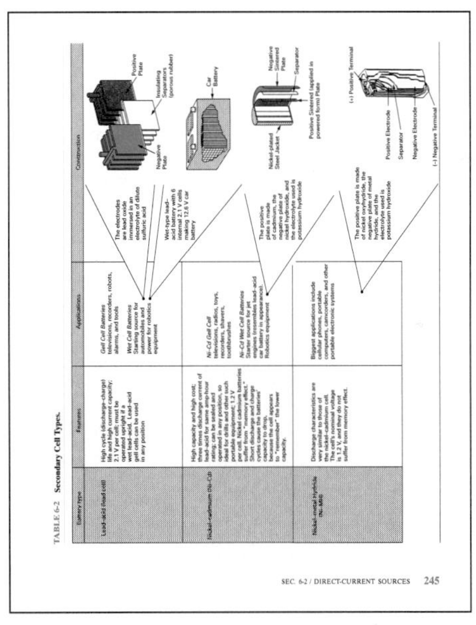
TABLE 6-2 **Secondary Cell Types.**

Battery type	Features	Applications	Construction
Lead-acid (lead cell)	High cycle (discharge-charge) life and high current capacity; 2.1 V per cell; must be operated upright if a wet lead-acid. Lead-acid gell cells can be used in any position	*Gell Cell Batteries* televisions, recorders, robots, alarms, and tools *Wet Cell Batteries* Starting source for automobiles and power for robotics equipment	The electrodes are lead oxide immersed in an electrolyte of dilute sulfuric acid. Wet-type lead-acid battery with 6 internal 2.1 V cells making 12.6 V car battery
Nickel-cadmium (Ni-Cd)	High capacity and high cost; three times discharge current of lead-acid for same amp-hour rating; can be sealed and operated in any position, so ideal for drills and other such portable equipment; 1.2 V per cell. Nickel cadmium batteries suffer from "memory effect." Short discharge and charge cycles cause the batteries' capacity to drop, because the cell appears to "remember" the lower capacity.	*Ni-Cd Gell Cell* televisions, radios, toys, recorders, shavers, toothbrushes *Ni-Cd Wet Cell Batteries* Starter source for jet engines (resembles lead-acid car battery in appearance). Robotics equipment	The positive plate is made of cadmium, the negative plate of nickel hydroxide, and the electrolyte used is potassium hydroxide
Nickel-metal Hydride (Ni-MH)	Discharge characteristics are very similar to those of the nickel-cadmium cell. The cell's nominal voltage is 1.2 V, and they do not suffer from memory effect.	Biggest applications include cellular phones, portable computers, camcorders, and other portable electronic systems	The positive plate is made of nickel oxyhydride, the negative plate of metal hydride, and the electrolyte used is potassium hydroxide

SEC. 6-2 / DIRECT-CURRENT SOURCES 245

Unique diagrams and tables, coupled with the textbook's conversational writing style, make even traditionally difficult topics easily accessible.

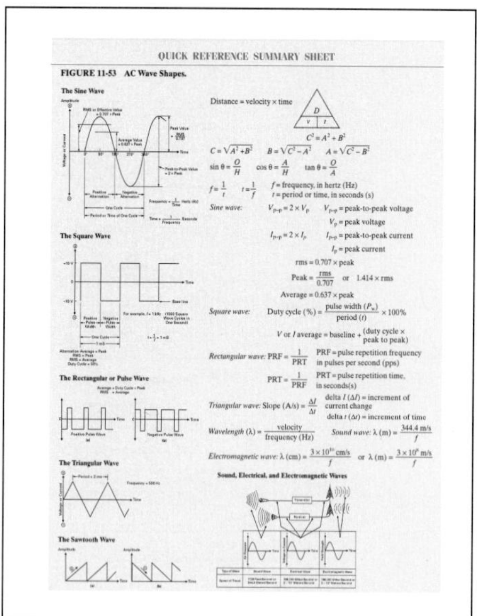

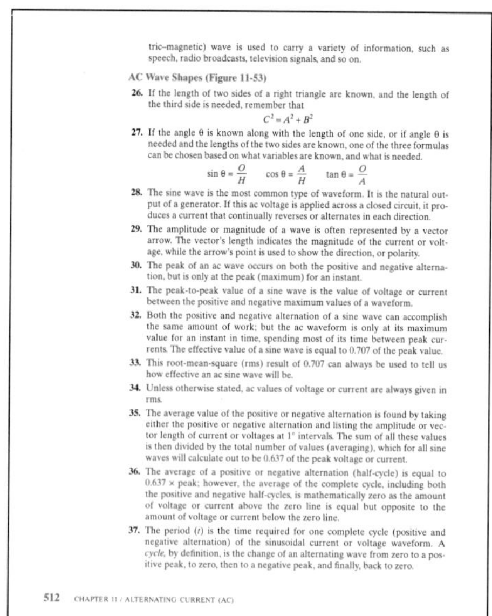

tric–magnetic) wave is used to carry a variety of information, such as speech, radio broadcasts, television signals, and so on.

AC Wave Shapes (Figure 11-53)

26. If the length of two sides of a right triangle are known, and the length of the third side is needed, remember that

$$C^2 = A^2 + B^2$$

27. If the angle θ is known along with the length of one side, or if angle θ is needed and the lengths of the two sides are known, one of the three formulas can be chosen based on what variables are known, and what is needed.

$$\sin\theta = \frac{O}{H} \qquad \cos\theta = \frac{A}{H} \qquad \tan\theta = \frac{O}{A}$$

28. The sine wave is the most common type of waveform. It is the natural output of a generator. If this ac voltage is applied across a closed circuit, it produces a current that continually reverses or alternates in each direction.
29. The amplitude or magnitude of a wave is often represented by a vector arrow. The vector's length indicates the magnitude of the current or voltage, while the arrow's point is used to show the direction, or polarity.
30. The peak of an ac wave occurs on both the positive and negative alternation, but is only at the peak (maximum) for an instant.
31. The peak-to-peak value of a sine wave is the value of voltage or current between the positive and negative maximum values of a waveform.
32. Both the positive and negative alternation of a sine wave can accomplish the same amount of work; but the ac waveform is only at its maximum value for an instant in time, spending most of its time between peak currents. The effective value of a sine wave is equal to 0.707 of the peak value.
33. This root-mean-square (rms) result of 0.707 can always be used to tell us how effective an ac sine wave will be.
34. Unless otherwise stated, ac values of voltage or current are always given in rms.
35. The average value of the positive or negative alternation is found by taking either the positive or negative alternation and listing the amplitude or vector length of current or voltages at 1° intervals. The sum of all these values is then divided by the total number of values (averaging), which for all sine waves will calculate out to be 0.637 of the peak voltage or current.
36. The average of a positive or negative alternation (half-cycle) is equal to 0.637 × peak; however, the average of the complete cycle, including both the positive and negative half-cycles, is mathematically zero as the amount of voltage or current above the zero line is equal but opposite to the amount of voltage or current below the zero line.
37. The period (t) is the time required for one complete cycle (positive and negative alternation) of the sinusoidal current or voltage waveform. A *cycle*, by definition, is the change of an alternating wave from zero to a positive peak, to zero, then to a negative peak, and finally, back to zero.

512 CHAPTER 11 / ALTERNATING CURRENT (AC)

Quick Reference Summary Sheets within the end-of-chapter summary provide a visual reference to all key facts, symbols, circuits, and formulas.

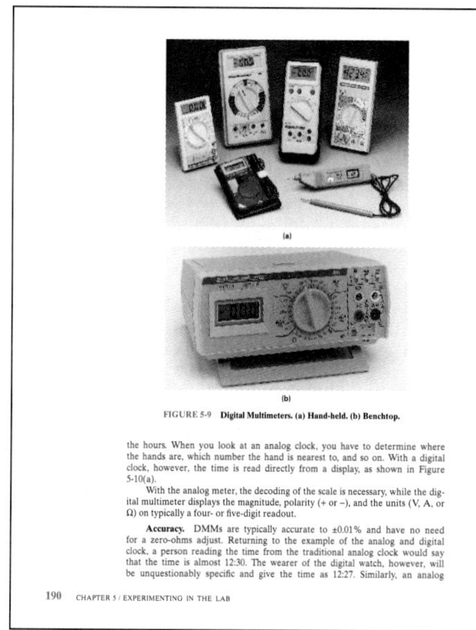

(a)

(b)

FIGURE 5-9 **Digital Multimeters. (a) Hand-held. (b) Benchtop.**

the hours. When you look at an analog clock, you have to determine where the hands are, which number the hand is nearest to, and so on. With a digital clock, however, the time is read directly from a display, as shown in Figure 5-10(a).

With the analog meter, the decoding of the scale is necessary, while the digital multimeter displays the magnitude, polarity (+ or −), and the units (V, A, or Ω) on typically a four- or five-digit readout.

Accuracy. DMMs are typically accurate to ±0.01% and have no need for a zero-ohms adjust. Returning to the example of the analog and digital clock, a person reading the time from the traditional analog clock would say that the time is almost 12:30. The wearer of the digital watch, however, will be unquestionably specific and give the time as 12:27. Similarly, an analog

190 CHAPTER 5 / EXPERIMENTING IN THE LAB

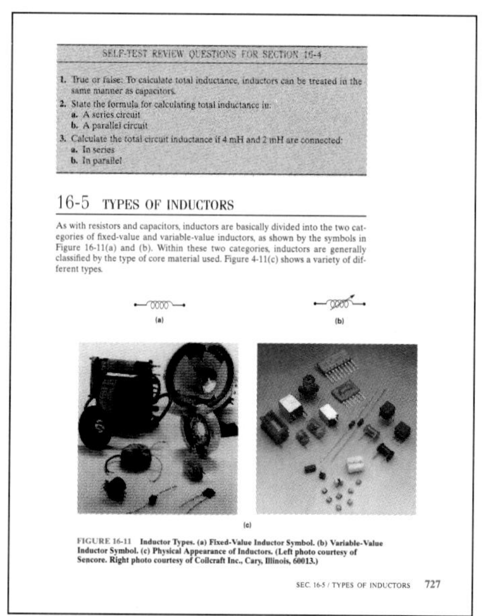

SELF-TEST REVIEW QUESTIONS FOR SECTION 16-4

1. True or false: To calculate total inductance, inductors can be treated in the same manner as capacitors.
2. State the formula for calculating total inductance in:
   a. A series circuit
   b. A parallel circuit
3. Calculate the total circuit inductance if 4 mH and 2 mH are connected:
   a. In series
   b. In parallel

16-5 TYPES OF INDUCTORS

As with resistors and capacitors, inductors are basically divided into the two categories of fixed-value and variable-value inductors, as shown by the symbols in Figure 16-11(a) and (b). Within these two categories, inductors are generally classified by the type of core material used. Figure 4-11(c) shows a variety of different types.

(a) (b)

(c)

FIGURE 16-11 **Inductor Types. (a) Fixed-Value Inductor Symbol. (b) Variable-Value Inductor Symbol. (c) Physical Appearance of Inductors. (Left photo courtesy of Sencore. Right photo courtesy of Coilcraft Inc., Cary, Illinois, 60013.)**

SEC. 16-5 / TYPES OF INDUCTORS 727

Many new photos have been included to help student recognition.

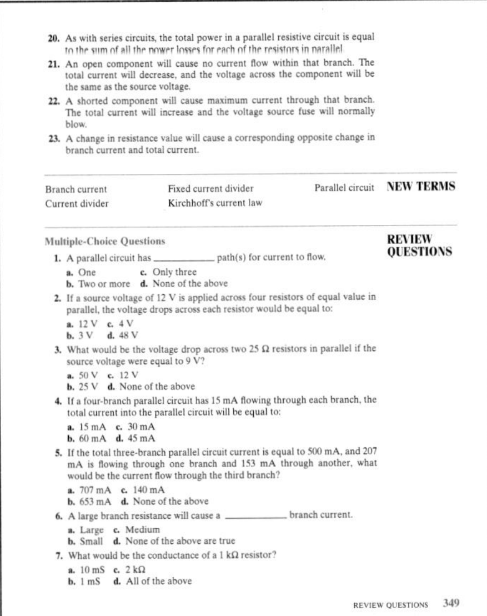

20. As with series circuits, the total power in a parallel resistive circuit is equal to the sum of all the power losses for each of the resistors in parallel.
21. An open component will cause no current flow within that branch. The total current will decrease, and the voltage across the component will be the same as the source voltage.
22. A shorted component will cause maximum current through that branch. The total current will increase and the voltage source fuse will normally blow.
23. A change in resistance value will cause a corresponding opposite change in branch current and total current.

NEW TERMS

Branch current · Current divider · Fixed current divider · Kirchhoff's current law · Parallel circuit

REVIEW QUESTIONS

Multiple-Choice Questions

1. A parallel circuit has ________ path(s) for current to flow.
   a. One c. Only three
   b. Two or more d. None of the above
2. If a source voltage of 12 V is applied across four resistors of equal value in parallel, the voltage drops across each resistor would be equal to:
   a. 12 V c. 4 V
   b. 3 V d. 48 V
3. What would be the voltage drop across two 25 Ω resistors in parallel if the source voltage were equal to 9 V?
   a. 50 V c. 12 V
   b. 25 V d. None of the above
4. If a four-branch parallel circuit has 15 mA flowing through each branch, the total current into the parallel circuit will be equal to:
   a. 15 mA c. 30 mA
   b. 60 mA d. 45 mA
5. If the total three-branch parallel circuit current is equal to 500 mA, and 207 mA is flowing through one branch and 153 mA through another, what would be the current flow through the third branch?
   a. 707 mA c. 140 mA
   b. 653 mA d. None of the above
6. A large branch resistance will cause a ________ branch current.
   a. Large c. Medium
   b. Small d. None of the above are true
7. What would be the conductance of a 1 kΩ resistor?
   a. 10 mS c. 2 kΩ
   b. 1 mS d. All of the above

REVIEW QUESTIONS 349

8. If only two resistors are connected in parallel, the total resistance equals:
   a. The sum of the resistance values
   b. Three times the value of one resistor
   c. The product over the sum
   d. All of the above
9. If resistors of equal value are connected in parallel, the total resistance can be calculated by:
   a. One resistor value divided by the number of parallel resistors
   b. The sum of the resistor values
   c. The number of parallel resistors divided by one resistor value
   d. All of the above could be true.
10. The total power in a parallel circuit is equal to the:
   a. Product of total current and total voltage
   b. Reciprocal of the individual power losses
   c. Sum of the individual power losses
   d. Both (a) and (b)
   e. Both (a) and (c)

Essay Questions

11. Describe the difference between a series and a parallel circuit. (8-1)
12. Explain and state mathematically the situation regarding voltage in a parallel circuit. (8-2)
13. State Kirchhoff's current law for parallel circuits. (8-3)
14. What is the current-divider formula? (8-3)
15. List the formulas for calculating the following total resistances: (8-4)
   a. Two resistors of different values
   b. More than two resistors of different values
   c. Equal-value resistors
16. Describe the relationship between branch current and branch resistance. (8-3)
17. Briefly describe total and individual power in a parallel resistive circuit. (8-5)
18. Discuss troubleshooting parallel circuits as applied to: (8-6)
   a. A shorted component b. An open component
19. Explain why parallel circuits have a smaller total resistance and larger total current than series circuits. (8-3)
20. Briefly describe why Kirchhoff's voltage law applies to series circuits and why Kirchhoff's current law relates to parallel circuits. (8-3)

Practice Problems

21. Calculate the total resistance of four 30 kΩ resistors in parallel.
22. Find the total resistance for each of the following parallel circuits:
   a. 330 Ω and 560 Ω c. 2.2 MΩ, 3 kΩ, and 220 Ω
   b. 47 kΩ, 33 kΩ, and 22 kΩ

350 CHAPTER 8 / PARALLEL DC CIRCUITS

An extensive end-of-chapter test bank tests the student's understanding with multiple-choice, essay, and practice problems and troubleshooting questions.

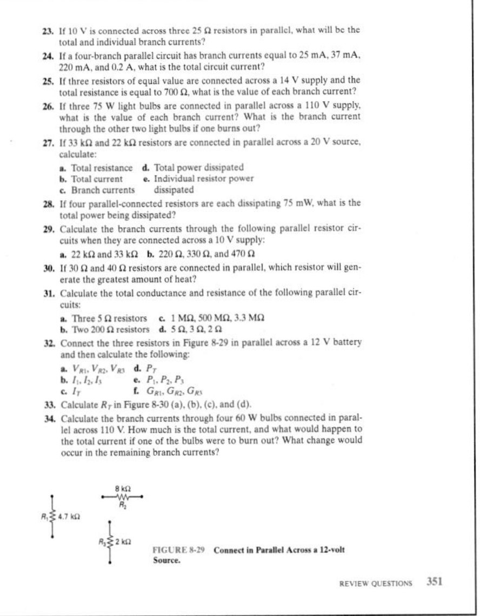

23. If 10 V is connected across three 25 Ω resistors in parallel, what will be the total and individual branch currents?
24. If a four-branch parallel circuit has branch currents equal to 25 mA, 37 mA, 220 mA, and 0.2 A, what is the total circuit current?
25. If three resistors of equal value are connected across a 14 V supply and the total resistance is equal to 700 Ω, what is the value of each branch current?
26. If three 75 W light bulbs are connected in parallel across a 110 V supply, what is the value of each branch current? What is the branch current through the other two light bulbs if one burns out?
27. If 33 kΩ and 22 kΩ resistors are connected in parallel across a 20 V source, calculate:
   a. Total resistance d. Total power dissipated
   b. Total current e. Individual resistor power dissipated
   c. Branch currents
28. If four parallel-connected resistors are each dissipating 75 mW, what is the total power being dissipated?
29. Calculate the branch currents through the following parallel resistor circuits when they are connected across a 10 V supply:
   a. 22 kΩ and 33 kΩ b. 220 Ω, 330 Ω, and 470 Ω
30. If 30 Ω and 40 Ω resistors are connected in parallel, which resistor will generate the greatest amount of heat?
31. Calculate the total conductance and resistance of the following parallel circuits:
   a. Three 5 Ω resistors c. 1 MΩ, 500 MΩ, 3.3 MΩ
   b. Two 200 Ω resistors d. 5 Ω, 3 Ω, 2 Ω
32. Connect the three resistors in Figure 8-29 in parallel across a 12 V battery and then calculate the following:
   a. V_{R1}, V_{R2}, V_{R3} d. P_T
   b. I_1, I_2, I_3 e. P_1, P_2, P_3
   c. I_T f. G_{R1}, G_{R2}, G_{R3}
33. Calculate R_T in Figure 8-30 (a), (b), (c), and (d).
34. Calculate the branch currents through four 60 W bulbs connected in parallel across 110 V. How much is the total current, and what would happen to the total current if one of the bulbs were to burn out? What change would occur in the remaining branch currents?

FIGURE 8-29 **Connect in Parallel Across a 12-volt Source.**

REVIEW QUESTIONS 351

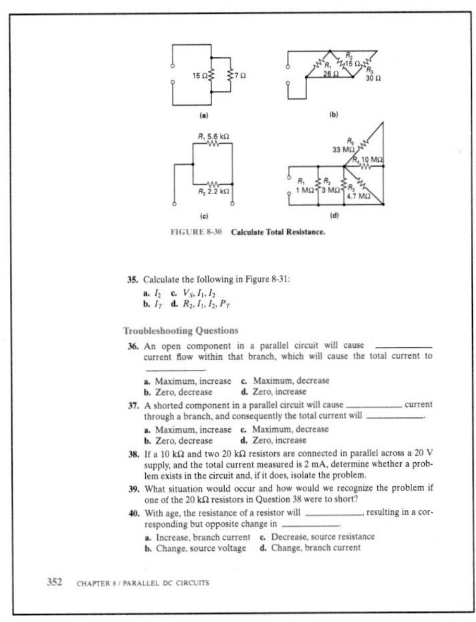

FIGURE 8-30 **Calculate Total Resistance.**

35. Calculate the following in Figure 8-31:
   a. I_2 c. V_S, I_1, I_2
   b. I_T d. R_2, I_1, I_2, P_T

Troubleshooting Questions

36. An open component in a parallel circuit will cause ________ current flow within that branch, which will cause the total current to ________.
   a. Maximum, increase c. Maximum, decrease
   b. Zero, decrease d. Zero, increase
37. A shorted component in a parallel circuit will cause ________ current through a branch, and consequently the total current will ________.
   a. Maximum, increase c. Maximum, decrease
   b. Zero, decrease d. Zero, increase
38. If a 10 kΩ and two 20 kΩ resistors are connected in parallel across a 20 V supply, and the total current measured is 2 mA, determine whether a problem exists in the circuit and, if it does, isolate the problem.
39. What situation would occur and how would we recognize the problem if one of the 20 kΩ resistors in Question 38 were to short?
40. With age, the resistance of a resistor will ________, resulting in a corresponding but opposite change in ________.
   a. Increase, branch current c. Decrease, source resistance
   b. Change, source voltage d. Change, branch current

352 CHAPTER 8 / PARALLEL DC CIRCUITS

TO THE INSTRUCTOR

Many people have asked me if there is one single factor involved with student retention. My answer is that whether a student enjoys electronics or hates it, and therefore stays in the program or leaves, is directly dependent on the instructor. I'm sure we have all seen the power we possess over a class of students. If this is used in a positive way, we can retain students despite any obstacles.

The responsibility of being the single most important factor in a student's education can be intimidating, unless we feel that we are fully supported. This support comes in the form of:

- a good textbook and set of teaching aids
- a good lab manual and set of test equipment
- a good course curriculum, and
- some teaching tricks of the trade

TEXTBOOK, LAB MANUAL, AND TEACHING AIDS

In all introductory texts, readability and relativity should be prime concerns to provide a complete understanding of the topics presented. However, in the pursuit of simplicity, boredom is often the price to be paid. Students are more involved in the learning process when they understand how the subject relates to their personal needs, career opportunities, and interests. Consequently, logical, current, and dynamic avenues of approach are needed to generate student enthusiasm, interest, and persistence.

This text has been specifically designed around the above philosophies, with a **complete picture introduction, real-life vignettes** at the beginning of each chapter, the use of **concept analogies** to take the student from the known to the unknown, a constant reference to **actual applications,** and a color insert that describes the many **career opportunities** available in the area of electronics.

The objective of this textbook, and its associated lab manual, is to provide a student with a solid foundation in the theoretical and practical knowledge needed to become an electronics technician. By definition, **an electronics technician must be able to diagnose, isolate, and repair electrical and electronic circuit and system malfunctions.** In order to achieve this goal, he or she must have a thorough knowledge of electronics, test equipment, and troubleshooting techniques. This text has been designed specifically with this objective.

Throughout this text I have also tried to provide to the student the basis for gaining *intelligence* along with an education. All too often education is mistaken for intelligence. Education is simply the acquisition of information. For example, just because I have acquired a first-rate set of woodworking tools does not make me a good carpenter. To be a good carpenter, I must first develop carpentry skills. There are many elements in that nebulous, indefinable quality we call "intelligence"; however, generally it falls into our understanding of **analogies,** which measure our ability to see relationships, and **mathematics, reasoning,** and **logic,** which measure our ability to think logically and make use of the facts we know. Therefore, instead of simply stating a formula and requiring a student to memorize it, which would be simply supplying an education, I have introduced the student to the quantities involved, compared them to something they know, and then studied the relationships to build up a formula in the same way as the pioneers. These analogies will develop a student's ability to see relationships while the study of relationships will develop good reasoning and logic skills, and therefore instill "intelligence." Although this takes a little more time in the early stages, the rewards and ease of understanding in the later stages will compensate for the initial effort.

With the lab manual, Gary Lancaster, Hugh Scriven, and I have tried to create **practical** experiments that inspire and challenge students. Our experiments were purposefully organized to work in conjunction with the textbook and translate all of its theoretical material into practical experience.

Along with the textbook and lab manual you will find a good set of teaching aids that will support your teaching effort.

ELECTRONIC TECHNICIAN TRAINING: INDUSTRY FEEDBACK

To be competitive, "electronic manufacturing companies" and "electronic servicing companies" must make a choice: either they compete with foreign nations on their terms with *lower wages, increased hours,* and *demoralized workers,* or they restructure the work organization and upgrade worker skills for a *high-skill, high-performance organization.* The high school graduate and experienced worker are therefore faced with the same simple choice: *high skills or low wages*. Since "high skills" would be the most obvious choice, the next question is, what skills should the electronics technician have to be competitive in today's and tomorrow's workforce? Posing this question to advisory board members and electronic panels from more than 40 electronic companies, the following feedback was obtained.

1. *Training must provide a core of knowledge for an electronic technician.* Training must produce an electronic technician. By definition, an electronic technician is a person having a complete understanding of the operation, application, and testing of electronic devices, circuits, and systems. Table 1 indicates in priority which training skills industry felt were most useful.

TABLE 1 Areas of Training Most Useful for Job Performance

ELECTRONIC MANUFACTURERS	ELECTRONIC SERVICING
1. Troubleshooting Techniques	1. Customer Relations
2. Use of Test Equipment	2. Troubleshooting Techniques
3. Circuit Recognition and Analysis	3. Use of Test Equipment
4. Adaptability and Teamwork Skills	4. Adaptability and Teamwork Skills
5. Customer Relations	5. Circuit Recognition and Analysis

TABLE 2 Test Equipment Most Useful for Job Performance

1. Digital Multimeter	5. Power Supply
2. Oscilloscope	6. Logic Analyzer
3. Logic Probe and Pulser	7. Signal Generators
4. Automatic Test Equipment	

Table 1 lists test equipment and troubleshooting equipment rate as high priorities. Since test equipment is a technician's tool of the trade, Table 2 lists, in order, the test equipment that industry felt should have the most emphasis.

2. *Technician must be trained to adapt to new technologies.* Training should supply "intelligence along with an education." By definition, education is the acquisition of information and an indication of the ability to make such acquisitions, and intelligence is our ability to think logically and make use of facts we know. This is dependent on our mathematical, logic, and reasoning skills, and our ability to continually learn and see relationships. Concentrating on math skills, logic and reasoning skills, learning skills, and relationship skills develops intelligence, which in turn develops better adaptability and organizational and problem solving skills. Success in these areas will ensure career advancement. To address each of these skills in a little more detail:
   a. *Math Skills:* As a tool, math should be integrated into appropriate position and, once taught, should be instantly applied so as to demonstrate need.
   b. *Logic and Reasoning Skills:* To give only an education, a student would be told to "memorize a formula." To give an education along with intelligence, training should explain why certain quantities are included in a formula and why they are either proportional or inversely proportional. In short, work first on concept comprehension.
   c. *Learning Skills:* The ability to adapt to new technology is directly dependent on an employee's ability to study and learn independently. Reading comprehension is therefore a necessity. To encourage this skill, training must be conversational and interactive, not dry and cold.
   d. *Relationship Skills:* Use analogies to go from the known to the unknown. Learning, retention, and comprehension are enhanced when parallels are explored, and this will in turn develop organizational and relationship skills.
3. *Training must develop better communication, teamwork, and customer service skills.* Industry's drive for high quality and reduced cost has led to the integration of electronic system design, manufacture, and service. There is a need, therefore, for increased teamwork at all levels of personnel, and technicians must have good communication and interpersonal skills. From the standpoint

of customer service skills, training should address how to work under pressure with customers on-site, dress professionally, and explain technical topics to nontechnical people. This can be achieved by having students constantly practice communicating their science (written and spoken).

4. *Training should educate students so that they have a better understanding of business practices, including manufacturing and production processes, and technical and business data.* Training should include tours of electronic companies and details on development procedures, job titles and responsibilities (technical and nontechnical), work ethics, environments, and so on.
5. *The electronic technician's role is evolving due to the complexity of electronic systems, the increase in computer-driven systems, and the economics of unit replacement versus component replacement.* To address these points:
   - **a.** Since computer work stations are becoming an increasingly important tool, training must provide a complete understanding of computer applications in industry.
   - **b.** As equipment becomes more complex, analytical and troubleshooting skills are critical for technicians. However, since yesterday's system is today's component, troubleshooting is moving away from a focus on discrete components to a subsystem and unit (functional) level of maintenance.
   - **c.** Since nearly all design, manufacturing, and service equipment are programmable, the ratio of a technician's involvement in hardware versus software has changed:

     Late 1980s: 80 percent Hardware, 20 percent Software
     Present: 60 percent Hardware, 40 percent Software

     Training should therefore include an introduction to computers and their applications at an early stage in the program. Technicians must be "application software proficient," and, therefore, the following should be covered: DOS, Windows, UNIX, Word Processing, Databases, Spreadsheets, Circuit Simulation Software, Basic Programming (such as C), and Networks.

Most employers agreed that the electronic technician's role is continually evolving to keep pace with the ever-changing electronics industry. As educators, we are also having to continually adapt to the needs of industry so that our graduates have the skills for both career and advancement.

TEACHING TECHNIQUES

Over the past 15 years I have taught thousands of students all aspects of electronics. In this time I have acquired a considerable amount of feedback regarding students' needs and responses to their introduction to the science of electricity and electronics. In this section I would like to share some of my teaching techniques that have enabled me to obtain a high level of student retention.

1. Using concept analogies wherever possible to take the student from the known to the unknown.

2. Keeping the objective "to produce an electrical/electronics technician" in mind at all times, and therefore concentrating on delivering a complete knowledge of electricity, electronics, test equipment, and troubleshooting techniques.
3. Explaining all topics with analogies, reasoning, and logic first, and reinforcing the topic with mathematics second. This provides for giving the students intelligence along with an education.
4. Assuming no prior knowledge of electricity, electronics, or technical mathematics. Starting with the complete picture or world of electronics, and including mini-math reviews where they are needed.
5. Encouraging student enthusiasm, interest, and persistence by including real-life vignettes of entrepreneurs, actual applications, and a career opportunity tour.

In addition, there are a few other teaching methods that have yielded a very good response in my classes, and so I will mention them to you for your consideration.

6. Application Presentations: These 20-minute presentations, generally on the first day of each week, describe and demonstrate systems such as a Jacob's Ladder, laser, computer system, a variety of application software, robots, test equipment, and so on. These presentations seem to generate a lot of student enthusiasm since they demonstrate the exciting end objective of their studies.

I have also turned the tables and had students give presentations on any electrical or electronic equipment. Although the topics seem to be limited to video games, remote controlled toys, and car stereo systems, the presentations have helped develop necessary student communication skills.

7. Schematic Descriptions and Troubleshooting: Once a certain knowledge of electricity, electronics, and test equipment has been mastered, I begin to apply this knowledge to actual system schematics. To cover a circuit fully, we follow four steps.
   a. First we study the purpose of a circuit, its inputs and outputs, and its schematic diagram and circuit description.
   b. Second, we compare the schematic to the actual printed circuit board so that we can identify the components.
   c. Next we study the troubleshooting guides or charts and analyze the process of diagnosing and isolating a problem.
   d. The final step is for me to introduce a problem and then have a team of normally two students diagnose, isolate, and repair the problem.
8. First Approximations of a Subject: It was philosopher René Descartes who first stated that to solve any problem you should start with the simple and then proceed to the complex. For Descartes there were three approximations; the **first approximation** was the simplest, the **second approximation** contained more detail, and the **third approximation** was the complex. I have applied this method to my teaching of all topics of electricity and electronics and have noticed a dramatic improvement in student comprehension. To explain this in a little more detail, a first approximation is a general descrip-

tion of a subject in which the key points and purposes are outlined. Following this complete picture description, we step through the details of the chapter, which is a second approximation of the subject. In keeping with the main objective, only the first and second approximations of a subject are needed for a technician since they cover purpose, construction, symbol, operation, characteristics, applications, testing, and troubleshooting. The third approximation of a subject covers design and therefore is only covered by engineering students.

These first approximations give the student a view of the complete picture, instead of having to wait until we finish all of the pieces in the chapter and then trying to connect them all together.

Like me, you may wish to develop first approximation presentations for all topics, especially the more difficult to grasp subjects, such as series–parallel circuits, capacitance, inductance, transistors, and so on.

If you should have any other ideas that you have found in your experience would assist other instructors, please send them to me in care of Prentice Hall.

INSTRUCTOR MOTIVATION: "THE NUTTY PROFESSOR"

I have been teaching electronics for nearly 15 years. In that time, teaching has given me many memorable and rewarding experiences. In contrast, it has been the source of many frustrating moments which have nearly brought me to the brink of madness.

As you are probably already aware, our chosen profession will most often work us day and night and most likely not make us a fortune. The key question is, therefore, why do we do it? The answer is simple. It comes from the personal interactions between the instructor and his or her students. It comes in the form of many intangibles, such as the delight of bringing a drink of knowledge to those who thirst for it, the challenge of the unknown and searching for answers, the pride in turning confusion and bewilderment into understanding and comprehension, and from turning an unskilled person into a working professional. There are mutual benefits to be reaped just from being involved with the learning process. The exchange of knowledge develops for both of us a better understanding of the subject and strong communicative skills, producing a student who is effective at work and in society as a whole.

Teaching is extremely demanding and often very frustrating, but the rewards well compensate the effort. It is a profession seldom understood except by those who really enjoy being involved with life. This vocation may be a "Mad, Mad, Mad World," but it's *my* world.

ACKNOWLEDGMENTS

A special thank-you goes to the professional, creative, and friendly people at Prentice Hall, namely: my editor, David Garza; my marketing manager, Debbie Yarnell; my sales manager, Todd Rossell; my production editors, Sheryl

Langner and Martha Beyerlein; and Judy Casillo, Maria Klimek, and Sylvia Huning.

A special thank-you goes to the husband and wife team of Steve and Marilyn Howe, who worked tirelessly and in a timely fashion to ensure a high level of quality and accuracy.

My appreciation and thanks are extended also to the following instructors who have reviewed and contributed greatly to the development of this textbook: Bob L. Bixler, Austin Community College; James C. Graves, Jr., Indian River Community College; John Hamilton, Spartan School of Aeronautics; William Jenkins, Electronics Institute, Kansas City, Missouri; Gary Lancaster, National Education Center; Leo R. Majewski, Dunwoody Institute; and Marcus S. Rasco, DeVry, Dallas.

Nigel P. Cook

Contents

PART I
The Fundamentals of Electricity *2*

Chapter 14

Chapter 15

Chapter 16

Introductory
DC/AC Electronics

CHAPTER 1

OBJECTIVES

After completing this chapter, you will be able to:

1. State the four basic phenomena on which electronics is based.
2. List some of the components and circuits that exist within electronic equipment.
3. List the seven branches of electronics.
4. Describe the types of equipment that can be found in each of the different branches or fields of electronics.
5. List the persons and companies largely responsible for the progression of electronics.
6. Explain how the unit names of many electrical quantities are derived.
7. Explain the process and people involved in the development of a product.

The World of Electronics

OUTLINE

VIGNETTE

Competitors Beware

During his years as a student at the University of Utah, Nolan Bushnell enjoyed playing the ancient Chinese game called GO. In this game, a word is used frequently throughout as a warning to your opponents that they are in jeopardy of losing with your next move.

With an initial investment of $500, Bushnell launched the video game industry in 1972, developing the first coin-operated Ping-Pong game, Pong. After almost an overnight success, Bushnell sold his company four years later for $15 million; the company was named after the polite Chinese term he learned in college, and as a warning to the competition, ATARI.

Following ATARI, Bushnell stepped up his ambition, taking high-tech to the masses through his Chuck E. Cheese's Pizza Time Theatres.

INTRODUCTION

In most cases, it is easier to build a jigsaw puzzle when you can see the completed picture on the front of the box. The same is true whenever anyone is trying to learn anything new, especially a science that contains many small pieces. In this chapter, you will be introduced to the "world of electricity and electronics." This complete picture introduction will help you see how all of the pieces in the electronics industry fit together.

1-1 THE PRESENT

At present you have probably heard mention of current, voltage, resistance, power, and some of the components, circuits, and systems used in electronics. Figure 1-1 shows how all of these items connect by comparing their relationship to the different parts of a tree. Working from the bottom up, you can see that the foundation of electronics rests on four basic roots or phenomena, known as current, voltage, resistance, and power. These are all related to one another, as can be seen by the formula circle in Figure 1-2.

Moving up the tree, the next block is electronic components, which were developed to manipulate or control voltage and current. Referring to Figure 1-3, you will see a listing of many electronic components, some names of which you may recognize.

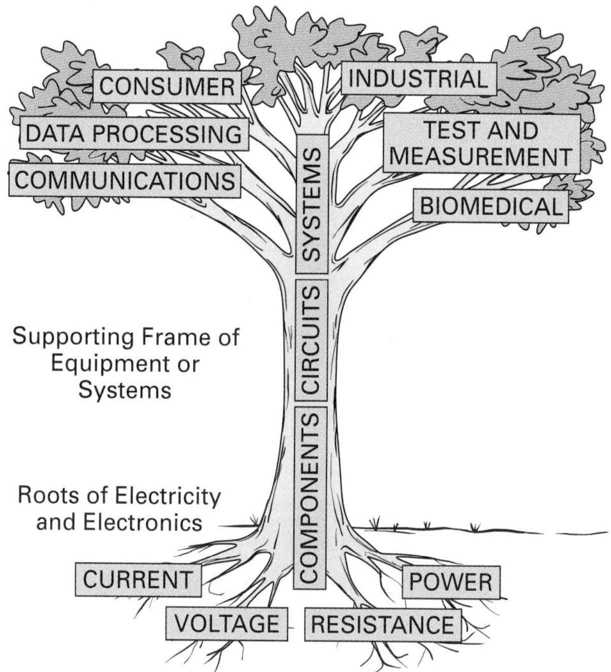

FIGURE 1-1 **The Tree of Electronics.**

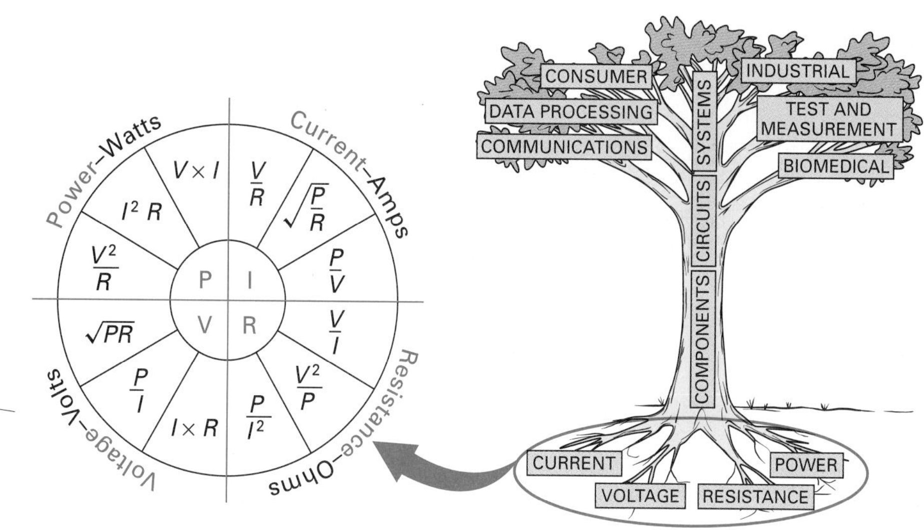

FIGURE 1-2 **The Roots of Electronics.**

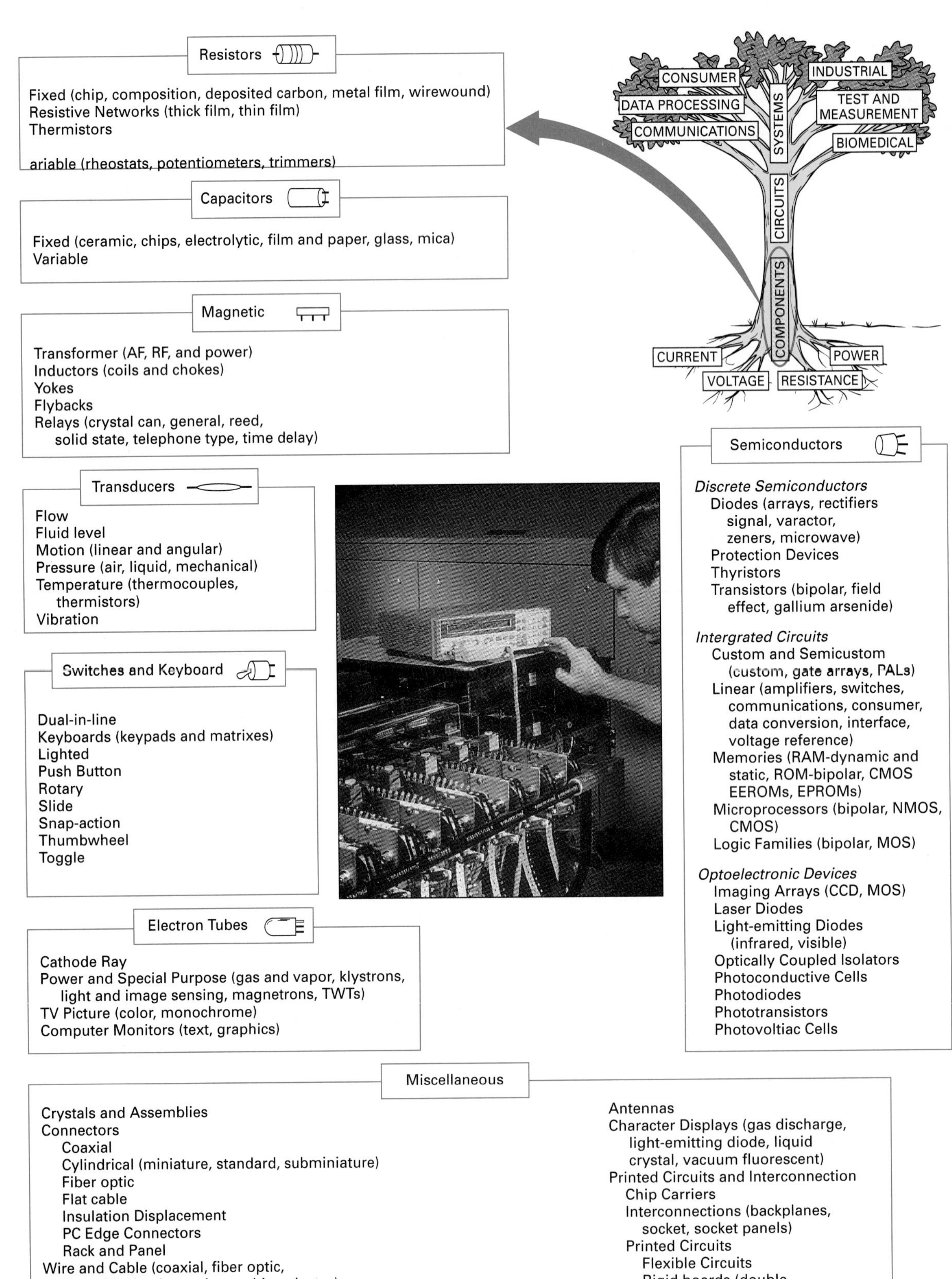

FIGURE 1-3 **Components. (Photo courtesy of Hewlett-Packard Company.)**

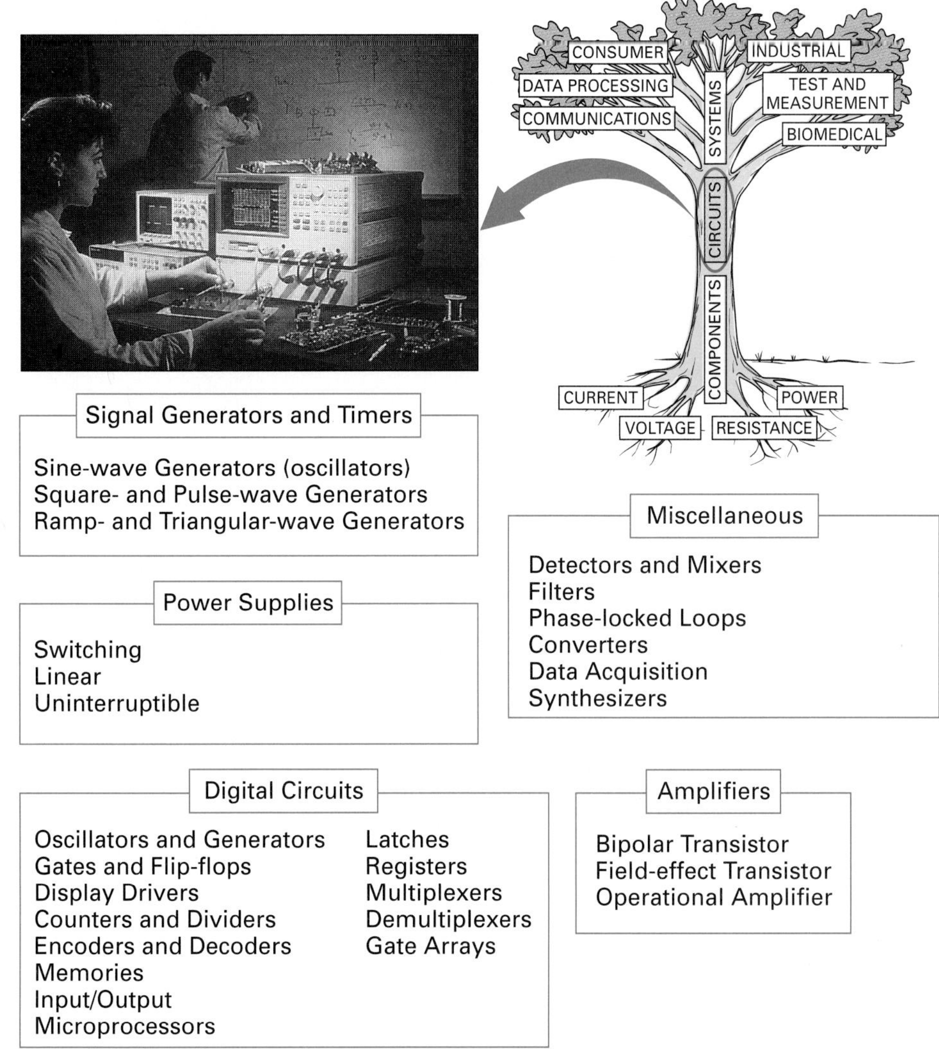

FIGURE 1-4 Circuits. (Photo courtesy of Hewlett-Packard Company.)

When a group of components is used in conjunction with one another, it forms a circuit, examples of which can be seen in Figure 1-4.

Just as components are the building blocks for circuits, circuits are in turn the building blocks for electronic systems or equipment, which can be categorized into one of six basic groups. Let us now examine these six different branches in more detail.

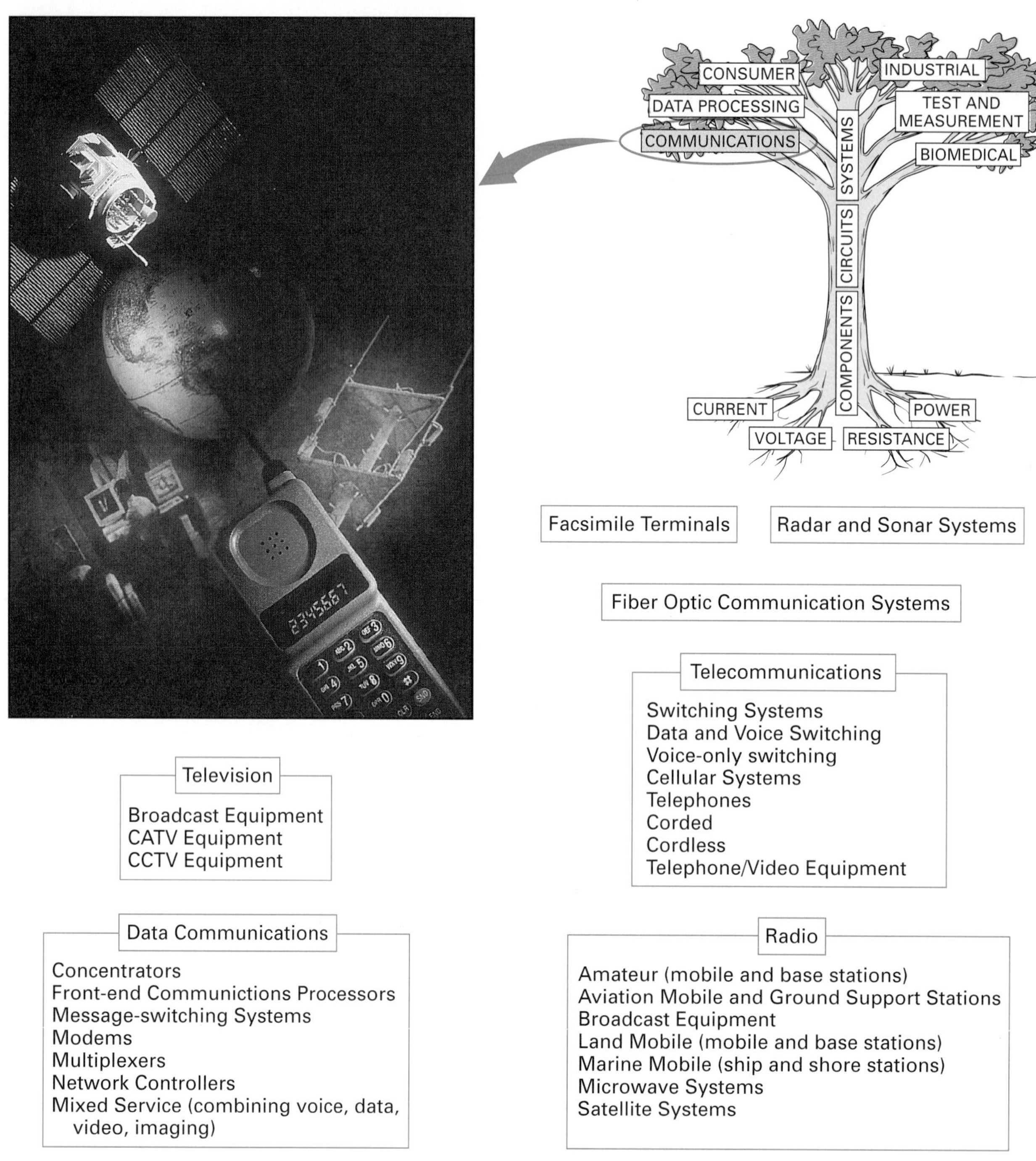

FIGURE 1-5 Communications Equipment. (Photo courtesy of Hewlett-Packard Company.)

1-1-1 Communications Equipment

Electronic communications allows the transmission and reception of information between two points. Radio and television are obvious communication devices, broadcasting data or entertainment between two points. Figure 1-5 illustrates and lists the different types of communication equipment that exist in what is probably the largest branch of electronics.

1-1-2 *Data Processing Equipment*

The computer is proving to be one of the most useful of all systems. Its ability to process, store, and manipulate large groups of information at an extremely fast rate makes it ideal for almost any and every application. Systems vary in complexity and capability, ranging from the Cray supercomputer to the home personal computer. The applications of word processing, record keeping, inventory, analysis, and accounting are but a few reasons why data processing systems, illustrated and listed in Figure 1-6, are used extensively.

CONSUMER
INDUSTRIAL
DATA PROCESSING
TEST AND MEASUREMENT
COMMUNICATIONS
BIOMEDICAL
SYSTEMS
CIRCUITS
COMPONENTS
CURRENT
POWER
VOLTAGE
RESISTANCE

Data Terminals

CRT Terminals
- ASCII Terminals
- Graphics Terminals (color, monochrome)

Remote Batch Job Entry Terminals

Computer Systems

Microcomputers and Supermicrocomputers
Minicomputers (personal computers) and Superminicomputers (technical workstations, multiuser)
Mainframe Computers
Supercomputers

I/O Peripherals

Computer Microfilm
Digitizers
Graphics Tablets
Light Pens
Trackball and Mice
Optical Scanning Devices
Plotters
Printers
- Impact
- Nonimpact (laser, thermal, electrostatic, inkjet)

Data Storage Devices

Fixed Disk (14, 8, $5\frac{1}{4}$, and $3\frac{1}{2}$ in.)
Flexible Disk (8, $5\frac{1}{4}$, and $3\frac{1}{2}$ in.)
Optical Disk Drives (read-only, write once, erasable)
Cassette
Cartridge Magnetic Tape ($\frac{1}{4}$ in.)
Cartridge Tape Drives ($\frac{1}{2}$ in.)
Reel-type Magnetic Tape Drives

FIGURE 1-6 **Data Processing Equipment. (Photo courtesy of Hewlett-Packard Company.)**

1-1-3 Consumer Equipment

Figure 1-7 illustrates and lists some of the current consumer electronic equipment available. From the smart computer-controlled automobiles, which provide navigational information and monitor engine functions and braking, to the compact disk players, video camcorders, satellite TV receivers, and wide-screen stereo TVs, this branch of electronics provides us with entertainment, information, safety, and, in the case of the pacemaker, life.

Audio Equipment

Car
Stereo Equipment
- Compact Systems (miniature components)
- Components (speakers, amps, turntables, tuners, tape decks)

Phonographs and Radio Phonographs
Radios (table, clock, portable)
Tape Players/Recorders
Compact Disk Players
Digital Tape Players

Video Equipment

TV Receiver (color, monochrome)
Projection TV Receivers
Video Cassette Recorders (VCRs)
Video Disk Players
Camcorders (8 mm, $\frac{1}{2}$ in.)
Home Satellite Receivers

Personal

Calculators, Cameras, Watches
Telephone Answering Equipment
Personal Computers
Microwave Ovens
Musical Equipment and Instruments
Pacemakers and Hearing Aids
Alarms and Smoke Detectors

Automobile Electronics

Dashboard
Engine Monitoring and Analysis
Computer Navigation Systems
Alarms
Telephones

FIGURE 1-7 Consumer Electronics Equipment. (Photo courtesy of Hewlett-Packard Company.)

Manufacturing Equipment

- Energy Management Equipment
- Inspection Systems
- Motor Controllers (speed, torque)
- Numerical-control Systems
- Process Control Equipment (data-aquisition systems, process instrumentation, programmable controllers)
- Robot Systems
- Vision Systems

Computer-aided Design and Engineering CAD/CAE

- Hardware Equipment
  - Design Work Stations (PC based, 32 microprocessor based platform, host based)
  - Application Specific Hardware
- Design Software
  - Design Capture (schematic capture, logic fault and timing simulators, model libraries)
  - IC Design (design rule checkers, logic synthesizers, floor planners–place and route, layout editors)
  - Printed Circuit Board Design Software
  - Project Management Software
  - Test Equipment

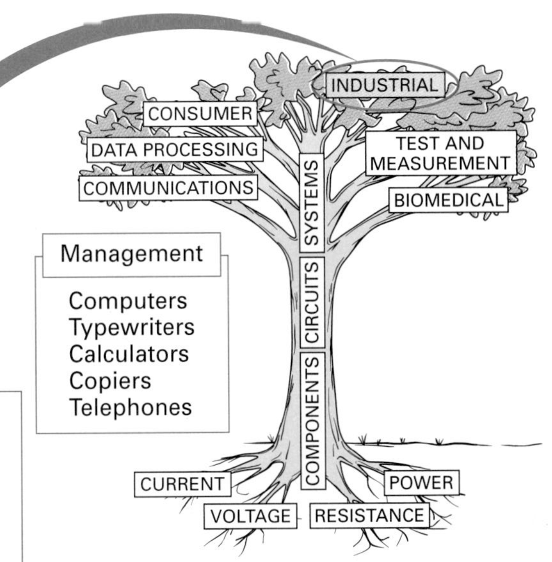

FIGURE 1-8 **Industrial Equipment. (Photo courtesy of Hewlett-Packard Company.)**

1-1-4 *Industrial Equipment*

Almost any industrial company can be divided into three basic sections, all of which utilize electronic equipment to perform their functions. Figure 1-8 illustrates and lists the various types of equipment used with these three sections. The

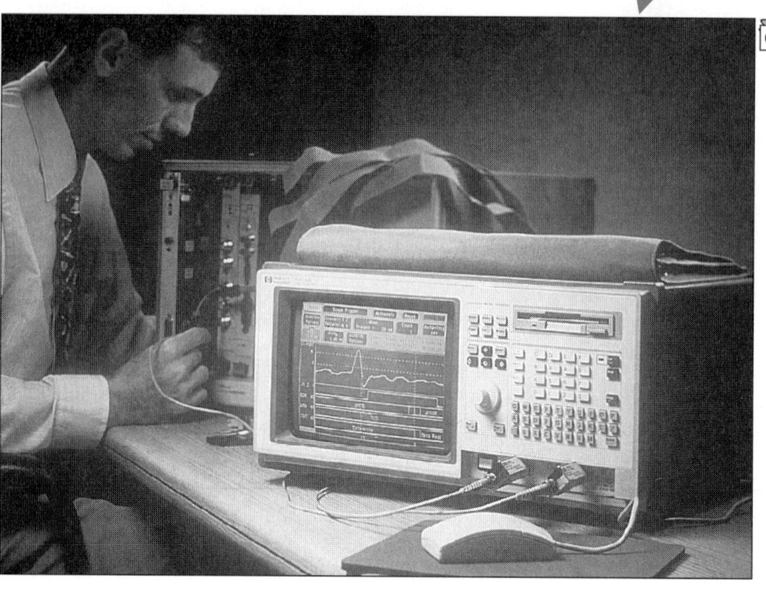

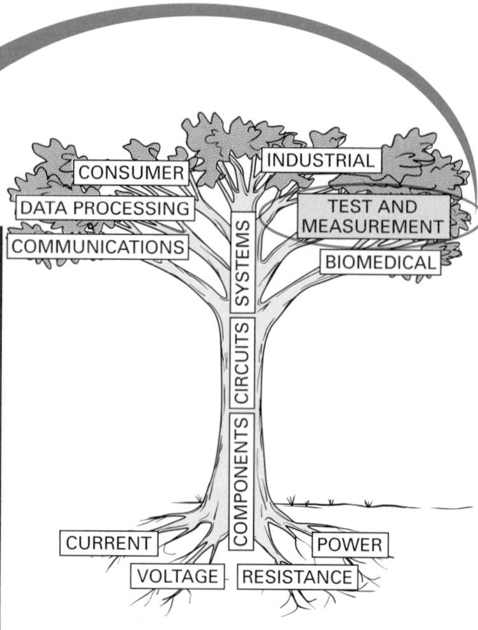

Automated Test Systems

- Active and Discrete Component Test Systems
- Automated Field Service Testers
- IC Testers (benchtop, general purpose, specialized)
- Interconnect and Bare pc-Board Testers
- Loaded pc-Board Testers (in-circuit, functional, combined)

General Test and Measurement Equipment

- Amplifiers (lab)
- Arbitrary Waveform Generators
- Analog Voltmeters, Ammeters, and Multimeters
- Audio Oscillators
- Audio Waveform Analyzers and Distortion Meters
- Calibrators and Standards
- Dedicated IEEE–488 Bus Controllers
- Digital Multimeters
- Electronic Counters (RF, Microwave, Universal)
- Frequency Sythesizers
- Function Generators
- Pulse/Timing Generators
- Signal Generators (RF, Microwave)
- Logic Analyzers
- Microprocessor Development Systems
- Modulation Analyzers
- Noise-measuring Equipment
- Oscilloscopes (Analog, Digital)
- Panel Meters
- Personal Computer (PC) Based Instruments
- Recorders and Plotters
- RF/Microwave Network Analyzers
- RF/Microwave Power-measuring Equipment
- Spectrum Analyzers
- Stand-alone In-circuit Emulators
- Temperature-measuring Instruments

FIGURE 1-9 **Test and Measurement Equipment. (Photo courtesy of Hewlett-Packard Company.)**

manufacturing section will typically use power, motor, and process control equipment, along with automatic insertion, inspection, and vision systems, for the fabrication of a product. The engineering section uses computers and test equipment for the design and testing of a product, while the management section uses electronic equipment such as computers, copiers, telephones, and so on.

1-1-5 *Test and Measurement Equipment*

The rising complexity of electronic components, circuits, and equipment is causing a demand for sophisticated automatic test equipment for both the manufacturer and customer to test their products. Figure 1-9 illustrates and lists some of the test and measurement equipment which can be basically classified as either stand-alone and/or computer-controlled test instruments.

1-1-6 *Biomedical Electronic Equipment*

Electronic equipment is used more and more within the biological and medical fields. Figure 1-10 illustrates and lists some of the medical equipment, which can be categorized simply as being either patient care or diagnostic equipment. In the operating room, the endoscope, which is an instrument used to examine the interior of a canal or hollow organ, and the laser, which is used to coagulate, cut, or vaporize tissue with extremely intense light, both reduce the use of invasive surgery. A large amount of monitoring equipment is used both in and out of operating rooms, and this equipment consists of generally large computer-controlled systems that can have a variety of modules inserted (based on the application) to monitor, on a continuous basis, body temperature, blood pressure, pulse rate, and so on. In the diagnostic group of equipment, the clinical laboratory test results are used as diagnostic tools. With the advances in automation and computerized information systems, multiple tests can be carried out at increased speeds. Diagnostic imaging, in which a computer constructs an image of a cross-sectional plane of the body, is probably one of the most interesting equipment areas.

Electronics
Science related to the behavior of electrons in devices.

Electrical Components
Components, circuits, and systems that manage the flow of *power.*

Electronic Components
Components, circuits, and systems that manage the flow of *information.*

Electricity
Science that states certain particles possess a force field, which with electrons is negative and with protons is positive. Electricity can be divided into two groups: static and dynamic. Static electricity deals with charges at rest, while dynamic electricity deals with charges in motion.

1-1-7 *Electrical and Electronic Components, Circuits, and Systems*

By definition, **electronics** is the branch of technology or science that deals with the use of **components** to control the flow of **electricity** in a vacuum, gas, liquid, semiconductor, conductor, or superconductor. Both electrical and electronic components, circuits, and systems control electron flow. However, their applications are distinctly different.

To properly manage power, electrical devices must perform such functions as *generating, distributing,* and *converting* electrical power.

To properly manage information, electronic devices must perform such functions as *generating, sensing, storing, retrieving, amplifying, transmitting, receiving,* and *displaying* information.

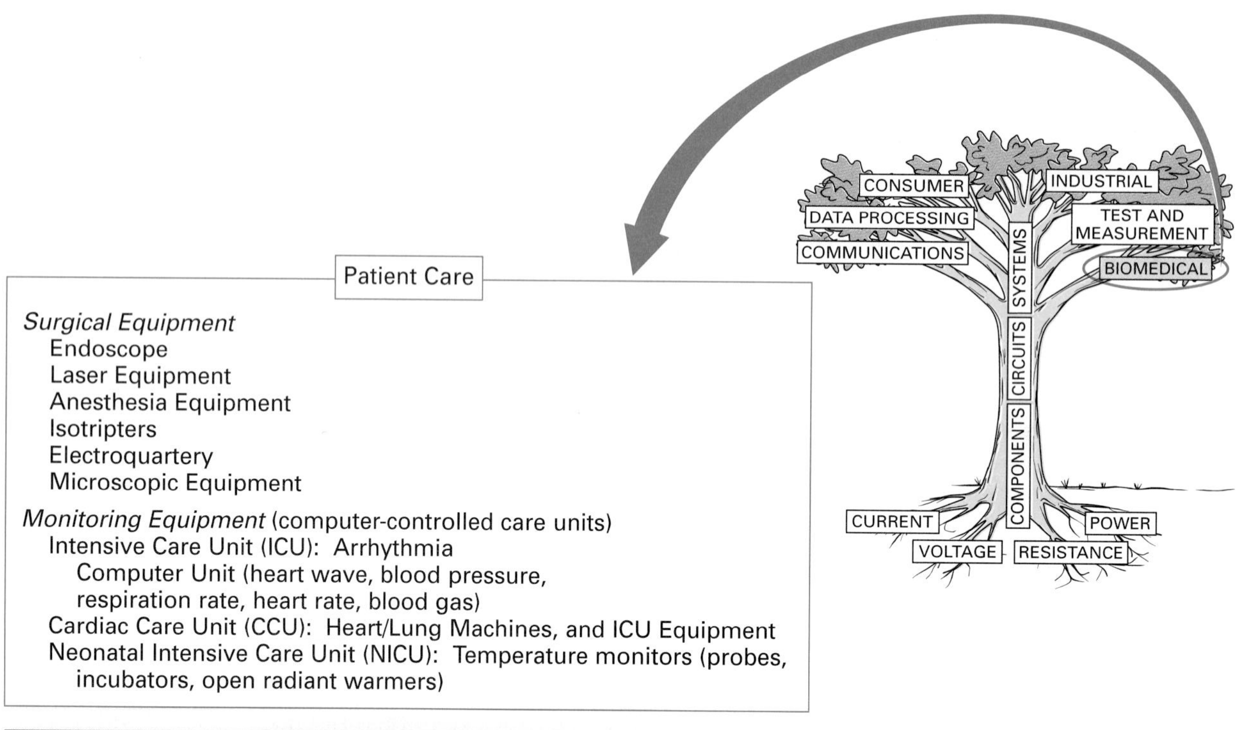

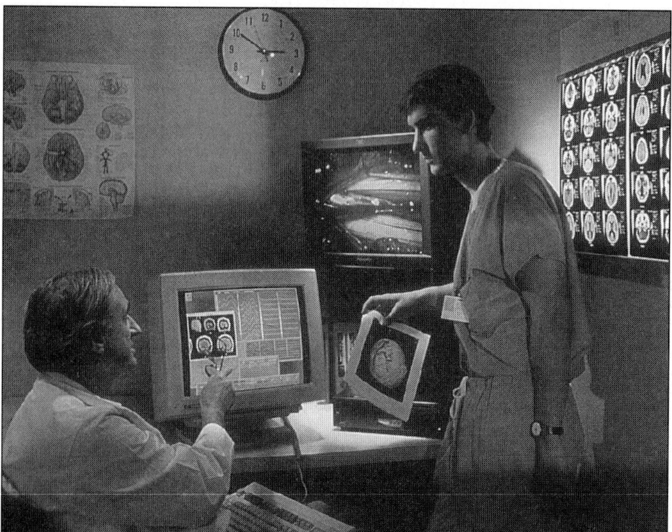

Diagnostic

Diagnosis Equipment
- X-Ray (computed tomography)
- Magnetic Resonance Imaging (MRI)
- Diagnostic Sounder
- Electrocardiograph (EKG)
- Electromyograph
- Electroencephalograph (EEG)
- Coagulograph
- Ultrasound (computed sonography)
- Nuclear Medicine (isotopes, spectroscopy)

Clinical Laboratory
- Automated Clinical Analyzers
- Centrifuge Incubators
- Cell Counters

FIGURE 1-10 **Biomedical Equipment. (Photo courtesy of Hewlett-Packard Company.)**

Some systems are designed specifically to manage the flow of power and therefore are only electrical, while other systems are designed to manage both power and information. For example, a television contains electrical components and circuits that manage the flow of electrical power from the wall outlet and electronic components and circuits that manage the flow of information or TV signals from the antenna or cable. The electrical circuits are needed since they supply power to the electronic circuits which in turn manage the flow of audio (sound) and video (picture) information signals.

SELF-TEST REVIEW QUESTIONS FOR SECTION 1-1

1. Name the four basic roots or phenomena on which the foundation of electronics rests.

2. List three examples of electronic components.

3. Name two circuits.

4. In which branch of electronics would radar be categorized?

1-2 THE PAST

Our way of life has undergone enormous change since the unveiling of electricity and electronics. Each of us, with just a little thought, can easily envisage how difficult it would be to manage in a world without electronics. The people responsible for the discovery of this science are discussed in this section. Figure 1-11 illustrates how a section of the tree of electronics has been extracted so that the rings of time can be viewed. These rings represent the age of the tree, with one ring being equal to approximately one year, and in this illustration the rings are used to summarize the progression of electricity and electronics.

Our history begins in 1600 with an English physician, William Gilbert (1544–1603), who documented many years of research and experiments on magnets and magnetic bodies such as amber and loadstones. Probably his most important discovery was that when rubbed with a cloth, amber would attract lightweight objects. In 1601 he was appointed physician to Queen Elizabeth I at a salary of $150 a year. He was the first to believe that the earth was nothing but a large magnet and that its magnetic field causes a needle to align itself between north and south.

Stephen Gray (1693–1736), also an Englishman, discovered that certain substances would conduct electricity. This lead was picked up in 1730 by Charles du Fay, a French experimenter, who believed that there were two types of electricity, which he called *vitreous* and *resinous* electricity.

One of the best known and most admired men in the latter half of the eighteenth century was the American Benjamin Franklin (1706–1790). He is reknowned for his kite-in-a-storm experiment, which proved that lightning is electricity. Franklin also discovered that there was only one type of electricity—that the two previously believed types were simply two characteristics of electricity. Vitreous was renamed *positive charge* and resinous was called *negative charge,* terms that were invented by Franklin, along with the terms *battery* and *conductor,* which are still used today.

The unit of electrical charge is named the coulomb, in honor of the French physicist Charles A. de Coulomb (1736–1806), who developed the laws of attraction and repulsion between charged bodies.

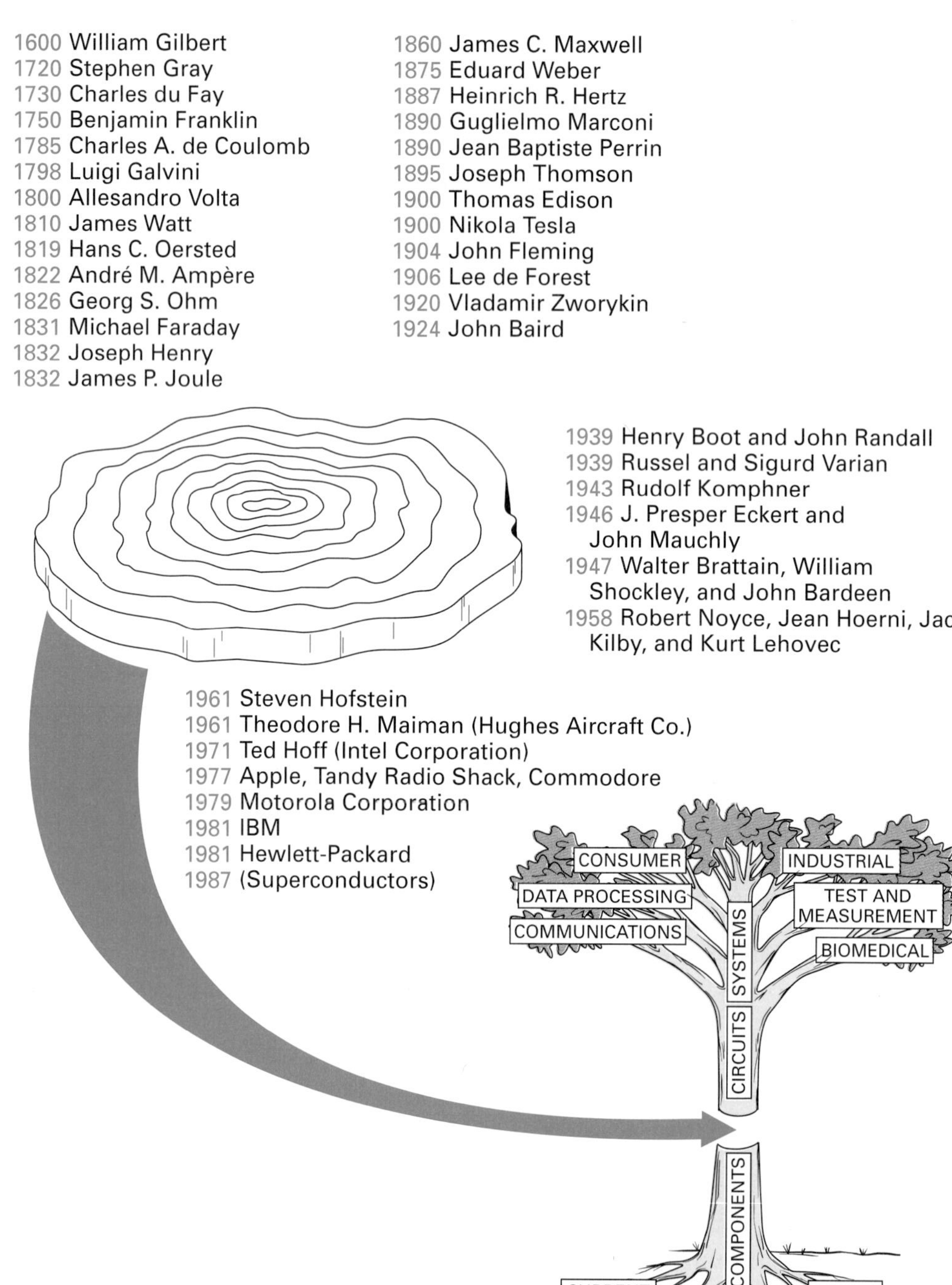

FIGURE 1-11 History of Electricity and Electronics.

The galvanometer, which is used to measure electrical current, was named after Luigi Galvani (1737–1798), who conducted many experiments with electrical current, or, as it was known at that time, "galvanism."

The unit of voltage, the volt, was named in honor of Alessandro Volta (1745–1827), an Italian physicist who is famous for his invention of the electric battery. In 1801, he was called to Paris by Napoleon to show his experiment on the generation of electric current.

The unit of power is the watt in honor of a Scottish engineer and inventor, James Watt (1736–1819), for his advances in the field of science.

In 1819, a Danish physicist, Hans C. Oersted (1777–1851), accidentally discovered an interesting phenomenon. Placing a compass near a current-carrying conductor, he noticed that the needle of the compass pointed to the conductor rather than north. He was quick to realize that electricity and magnetism were related, and in honor of his work, the unit oersted was adopted for the unit of magnetic field strength.

The unit of electrical current is the ampere, named in honor of André M. Ampère (1775–1836), a French physicist who pioneered in the study of electromagnetism. After hearing of Oersted's discoveries, he conducted further experiments and discovered that two current-carrying conductors would attract and repel one another, just like two magnets.

Ohm's law, the best known law in electrical circuits, was formulated by Georg S. Ohm (1787–1854), a German physicist. His law was so coldly received that his feelings were hurt and he resigned his teaching post. When his law was finally recognized, he was reinstated. In honor of his accomplishments, the unit of resistance is called the ohm.

In 1831, Michael Faraday (1791–1867), an English physicist, explored further Oersted's discovery of electromagnetism and discovered that a magnetic field could be used to produce electric current. These findings are today referred to as Faraday's laws of electromagnetic induction. A German-born scientist working in Russia extended Faraday's findings and found that the current induced in a conductor is such that it opposes the change in the magnetic field producing it. This is known today as Lenz's law, in honor of Heinrich F. E. Lenz. Michael Faraday also investigated static electricity and the lines of electric force, and it is in acknowledgement of his work in this area that the unit of capacitance is named the farad.

Joseph Henry (1797–1878), an American physicist, also conducted extensive studies into electromagnetism. Henry was the first to insulate the magnetic coil of wire and developed coils for telegraphy and motors. In recognition of his discovery of self-induction in 1832, the unit of inductance is called the henry.

James P. Joule (1818–1889), an English physicist and self-taught scientist, conducted extensive research into the relationships between electrical, chemical, and mechanical effects, which led him to the discovery that one energy form can be converted into another. For his achievements, his name was given to the unit of energy, the joule.

As a small boy, James C. Maxwell (1831–1879) was persistently inquisitive. He built many scientific toys before he was 8. At the age of 14 he wrote a paper on how to construct oval curves, and at 18 two of his papers were published. The supreme achievement of this Scottish physicist however was to translate Faraday's experiments into mathematical notation. This set of mathematical equations, known as Maxwell's equations, shows the relationship between electricity and magnetism.

Eduard W. Weber (1804–1891), a German physicist, made enduring contributions to the modern system of electrical units, and magnetic flux is measured in webers in honor of his work.

Heinrich R. Hertz (1857–1894), a German physicist, was the first to demonstrate the production and reception of electromagnetic (radio) waves. In honor of his work in this field, the unit of frequency is called the hertz.

Studying the experiments of Maxwell and Hertz, Guglielmo Marconi (1874–1937) invented a practical system of telegraphy communication. In an evolutionary process, Marconi extended his distance of communication from 1½ miles in 1896 to 6000 miles in 1902. In September 1899, Marconi equipped two U.S. ships with equipment and used them in the Atlantic Ocean to transmit to America the progress of the America's Cup yacht race.

The electron was first discovered by Jean B. Perrin (1870–1942), a French physicist who was awarded the Nobel prize for physics. Perrin discovered that cathode rays consisted of negatively charged particles, and these particles, which later became known as electrons, were measured by an English physicist, Joseph Thomson (1846–1914).

Thomas Edison (1847–1931) was a self-educated inventor best known for his development of the phonograph and the incandescent lamp. The first version of the phonograph cost $18 and was turned with a hand crank. A decade later it was motor driven, with cylindrical wax and then disk-type records. In 1879, after $40,000 worth of fruitless experiments, he succeeded in developing an incandescent lamp that consisted of a loop of carbonized cotton thread that glowed in a vacuum for 40 hours. In 50 years, he took out 1033 patents. Nikola Tesla (1856–1943) was the inventor of the induction motor and worked to improve power transmission. Working for Edison for a short time, they developed a hatred for one another that prompted Tesla to begin his own business. Edison promoted dc power distribution, while Tesla believed in ac, and eventually Tesla's reasoning was adopted worldwide. In 1912, they were nominated together for the Nobel prize in physics, but Tesla would have nothing to do with Edison, so the prize went to a third party. A theory of Tesla's, which up to this time has not proven possible, is the wireless transmission of electrical power by high-energy electromagnetic or radiant beams.

In 1904 John A. Fleming, a British scientist, saw the value of an effect that was discovered by Edison but for which he saw no practical purpose. The "Edison effect" permitted Fleming to develop the "Fleming valve," which passes current in only one direction. Its operation made it the first device able to convert alternating current into direct current and to detect radio waves.

Although the Fleming valve was an advance, it could not amplify or boost a signal. The "audion," developed by U.S. inventor Lee de Forest (1873–1961), sparked an era known as "vacuum-tube electronics" that brought about transcontinental telephony in 1915, radio broadcasting in 1920, radar in 1936, and television between 1927 and 1946, because of this triode vacuum tube's ability to amplify small signals.

The father of television, Vladamir C. Zworykin, developed the first television picture tube, called the kinescope, in 1920.

John L. Baird (1888–1946) was a British inventor and television pioneer.

He was the first to transmit television over a distance. He reproduced objects in 1924, transmitted recognizable human faces in 1925, demonstrated the first true television in 1926, and in 1939 he developed television in natural color.

During World War II, there was a need for microwave-frequency vacuum tubes. British inventor Henry Boot developed the magnetron in 1939, and in the same year an American brother duo, Russel and Sigurd Varian, invented the klystron. In 1943, the traveling wave tube amplifier was invented by Rudolf Komphner, and up to this day these three microwave vacuum tubes are still used extensively.

In 1946, J. Presper Eckert and John Mauchly unveiled ENIAC, which used over 300,000 vacuum tubes. ENIAC, which is an acronym for "electronic numerical integrator and computer," was the first large-scale electronic digital computer.

In 1947 Walter Brattain, William Shockley, and John Bardeen sparked an even greater era, known as "solid-state electronics," with their invention of the transistor at Bell Laboratories. Transistorized equipment is smaller, cheaper, more reliable, more robust, and consumes less power than its vacuum-tube counterparts.

In 1958, Robert Noyce, Jean Hoerni, Jack Kilby, and Kurt Lehovec all took part in the development of the integrated circuit, which incorporated many transistors and other components on a small chip of semiconductor material.

In 1961, Steven Hofstein devised the field-effect transistor used in MOS (metal oxide semiconductor) integrated circuits, and in the same year Theodore H. Maiman, a scientist working at Hughes Aircraft Company, built the first operational laser using a synthetic ruby crystal.

In 1971, Ted Hoff of Intel Corporation designed a microprocessor, the 4004, that had all the basic parts of a central processor. Intel improved on the 4-bit 4004 microprocessor and unveiled an 8-bit microprocessor in 1974 that could add two numbers in 2.5 millionths of a second.

Three mass-market personal computers emerged in 1977: the Apple II, Radio Shack's TRS-80, and the Commodore PET.

In 1979, Motorola Corporation continued to advance computers by creating a powerful and versatile 16-bit microprocessor that could multiply two numbers in 3.2 billionths of a second.

IBM, who had up to this time dominated the big computer market, entered the personal computer market with the IBM PC in 1981. Also in 1981, Hewlett-Packard unveiled its 32-bit microprocessor to further advance the speed and power of computers.

In scientific laboratories around the world, thousands of scientists began working furiously to develop a new technology that at the beginning of 1987 seemed little more than science fiction. Practical applications of superconductors cannot be fully realized at this time, just as no one could foresee all of the uses for the transistor when it was invented 40 years earlier. In the future, however, it is believed that all power will be distributed over *superconductor cables* because of their low power losses, while all information signals will be distributed over *fiber optic cables* because of their small size and large capacity.

SELF-TEST REVIEW QUESTIONS FOR SECTION 1-2

Name the person who was honored by the use of his name for the unit of the following electrical properties:

1. Resistance **3.** Voltage
2. Current **4.** Power

1-3 THE FUTURE

So many things have happened in the past, so many things are happening at present, and so many things are under development to unfold in the future that it has become very difficult to keep tabs on the progression in all fields of electronics.

Electronics is an exciting and rewarding field to be involved in. The main purpose of this chapter is not only to give you an insight into the past and present, but also to give you the reasons why we begin our studies of electronics by discussing current, voltage, resistance, and power. These roots are necessary to understand components, as components need to be comprehended to understand circuits, and in turn circuits are necessary to understand the exciting, wide range of systems that exists in the different branches of electronics.

Your future in the electronics industry begins with this course in DC/AC electronics. To give you an idea of where you are going and what you will be covering, Figure 1-12 breaks up your study of electronics into four basic steps, which correlate to the basic blocks shown in the tree.

SELF-TEST REVIEW QUESTIONS FOR SECTION 1-3

Which two of the following have been predicted by the government as having the largest use in the future?

1. Vacuum tubes **3.** Robots
2. Lasers **4.** Computers

1-4 PEOPLE IN ELECTRONICS

In any industrial company, people are required to keep the industry moving, and an electronics company is not any different. The photo essay and associated storyboard demonstrate how a typical electronics company would develop a product; but more important, they show the people involved in this product development process, their job titles and responsibilities, and how each person plays a very important role. Use this photo essay as a career opportunities tour.

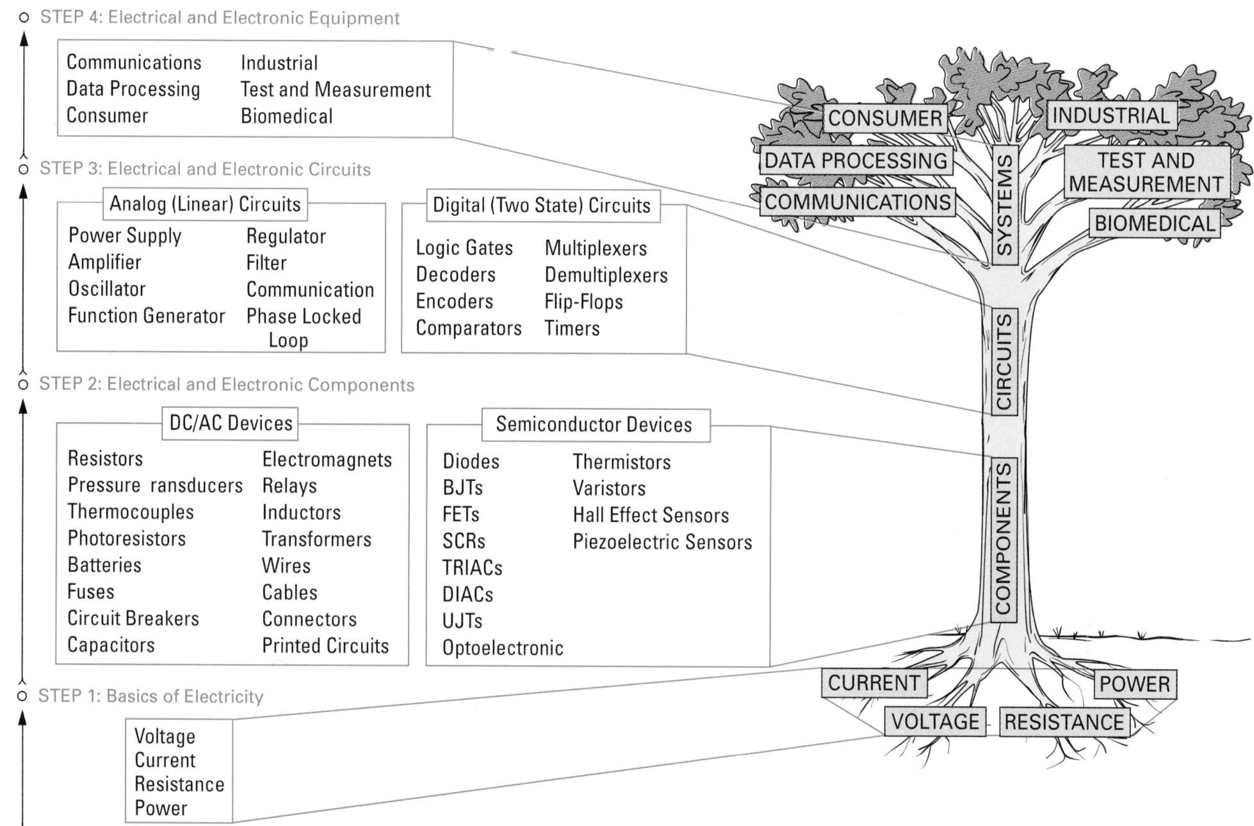

FIGURE 1-12 **The Steps Involved in Studying Electricity and Electronics.**

SUMMARY

1. Electronics is built on four basic roots or phenomena: current, voltage, resistance, and power.
2. Electronic components control the four basic roots or phenomena of electronics.
3. Groups of components are interconnected to perform certain functions, and these arrangements are known as circuits.
4. Just as components are the building blocks for circuits, circuits are in turn the building blocks for electronic equipment.
5. All electronic equipment can be categorized into one of six different branches or groups: communications, data processing, consumer, industrial, test and measurement, and biomedical.
6. The names for almost all the units of electricity and electronics are derived from early experimenters.
7. The four main quantities and units are:
   a. Current, which is measured in amperes or amps
   b. Voltage, which is measured in the unit of volts
   c. Resistance, which is measured in ohms
   d. Power, which is measured in watts

8. The engineer forms the concepts, designs, and modifications of all electronic circuits or systems.
9. Engineering technicians aid the engineer in the research and development of a product.
10. The technician is an expert in the diagnosis and repair of problems or malfunctions within electronic equipment.
11. The assembler is also an expert in his or her field, which is the fabrication of or manufacturing techniques used in the assembly of the final product.

REVIEW QUESTIONS

Multiple-Choice Questions

1. The unit of charge was named after:
   a. Alessandro Volta **c.** Georg Ohm
   b. Heinrich Hertz **d.** Charles Coulomb
2. ____________ developed the basic laws relating to current flow.
   a. André Ampère **c.** Michael Faraday
   b. Heinrich Hertz **d.** Intel Corporation
3. The first 16-bit microprocessor was developed by:
   a. Intel Corporation **c.** Texas Instruments
   b. Hewlett-Packard **d.** Motorola Corporation
4. The first operational laser was built by:
   a. Walter Brattain **c.** Theodore Maiman
   b. Guglielmo Marconi **d.** James Maxwell
5. The unit of frequency was named in honor of:
   a. Heinrich Hertz **c.** Georg Ohm
   b. Jean Perrin **d.** Greek experimenters
6. The person largely responsible for the era termed "vacuum-tube electronics" was:
   a. John Fleming **c.** Lee de Forest
   b. William Gilbert **d.** Benjamin Franklin
7. The carbon filament light bulb was invented by:
   a. Joseph Henry **c.** Thomas Edison
   b. James Joule **d.** Vladamir Zworykin
8. The persons responsible for the transistor and the subsequent "solid-state electronics" era were:
   a. Hertz and Marconi **c.** Noyce, Hoerni, Kilby, and Lehovec
   b. Eckert and Mauchly **d.** Brattain, Schockley, and Bardeen
9. The resistor, capacitor, and inductor are all examples of electronic circuits.
   a. True **b.** False
10. Amplifiers and oscillators are both examples of electronic ____________.
    a. Circuits **c.** Both (a) and (b)
    b. Components **d.** None of the above

Electronic Technicians in the Product Development Process

In the following 16 pages of this color insert you will be introduced to the many different types of "Electronic Technician." By definition, "An electronics technician must be able to diagnose, isolate, and repair electronic circuit and system malfunctions." In order to achieve this goal an electronic technician must have a thorough knowledge of electronics, test equipment and troubleshooting techniques.

In the basic block diagram below you can see the steps followed by a typical electronics company in the development of an electronic system. As you can see, these companies employ a variety of electronic technicians, all of which play a very important role in the development of an electronic system. In the following pages of this color insert we will examine these electronic technicians in more detail.

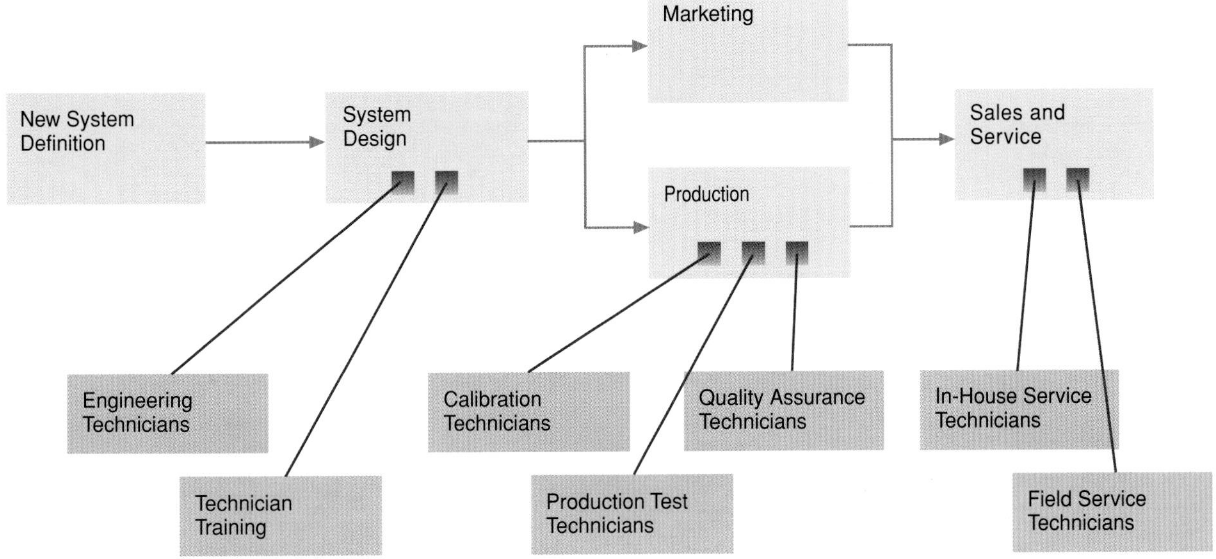

Development of an Electronic Product

The flow chart below expands on the basic block diagram on the previous page, showing the order from top to bottom, in which an electronic product is developed from conception to shipping. This photographic essay also shows the responsibilities of each person and how he or she fits into this process, giving you a clearer picture of the many career opportunities available.

ENGINEERS

The engineer is responsible for designing the electronic system and supervising its development. During this process, the engineer relies heavily on the engineering technician to assist in the prototyping of the design and the diagnosis and repair of problems.

TECHNICIANS

The technician is the expert in troubleshooting circuit and system malfunctions. Along with a thorough knowledge of all test equipment and how to use it to diagnose problems, the technician is also familiar with how to repair or replace faulty components. Referring to the flow chart, as you will see in the following photographs, technicians are involved in almost every step of product development.

ASSEMBLERS

The assembler is responsible for putting the electronic product together and is trained in the correct component handling and assembly techniques.

Product Conception
Product Definition
Design Breakdown
Hiring of Personnel
Hardware Design
Software Design
Mechanical Design
Parts Purchasing
Documentation
Breadboard Design
Prototype Layout, Construction, & Test
Build Pre-production Units
Quality Engineering Test
Stock Components
Design Test Systems
Technician Training
Chassis Fabrication
Component Insertion
Flow Soldering
Wire/Cable Hook-up
Final Assembly
Product Assembly
Product Test & Calibration
Quality Control Test
Product Release
Sales
Shipping
Customer
Customer Service

The first step in the process can be seen in this photograph. The key company managers meet to define the new product. This new product will have to meet the needs of the ever-expanding electronics industry. During this meeting, the new product's budget, target dates, and key features will be outlined.

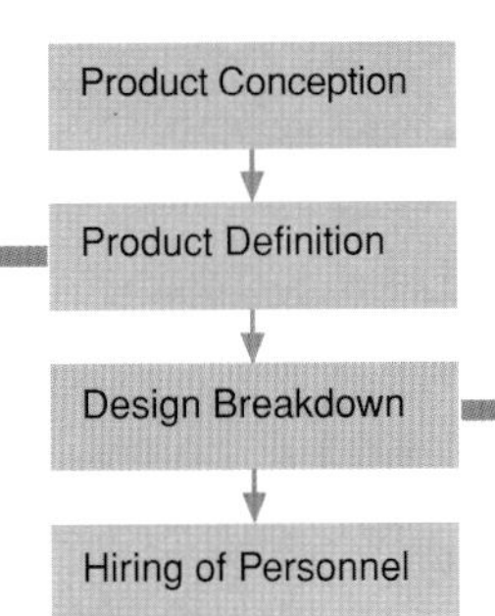

Once the product has been defined, it is the job of engineering to specify and design all major components of the new equipment. In this photograph, the design of the new product is being broken up into smaller tasks by an engineering manager and these task projects are being assigned to different engineers and engineering technicians.

New personnel may be needed by the company to develop and produce this new product. Here, a prospective employee is receiving a technical and personnel interview for a technician position.

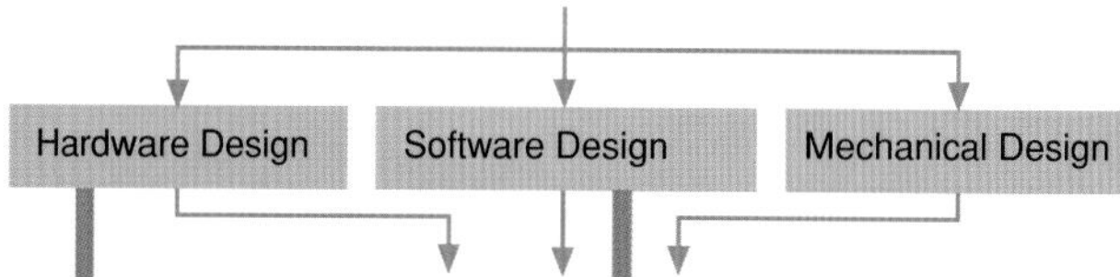

The hardware design engineer shown is entering the newly designed circuit, which consists of many interconnected electronic components, into a computer. The equipments hardware is the physical cables, components, and circuits you can see inside every piece of electronic equipment.

At the heart of many electronic products produced today is a computer controlling operation. The software design engineer is entering a list of instructions into the equipments computer memory, and it is this program of instructions that will control how the equipment operates.

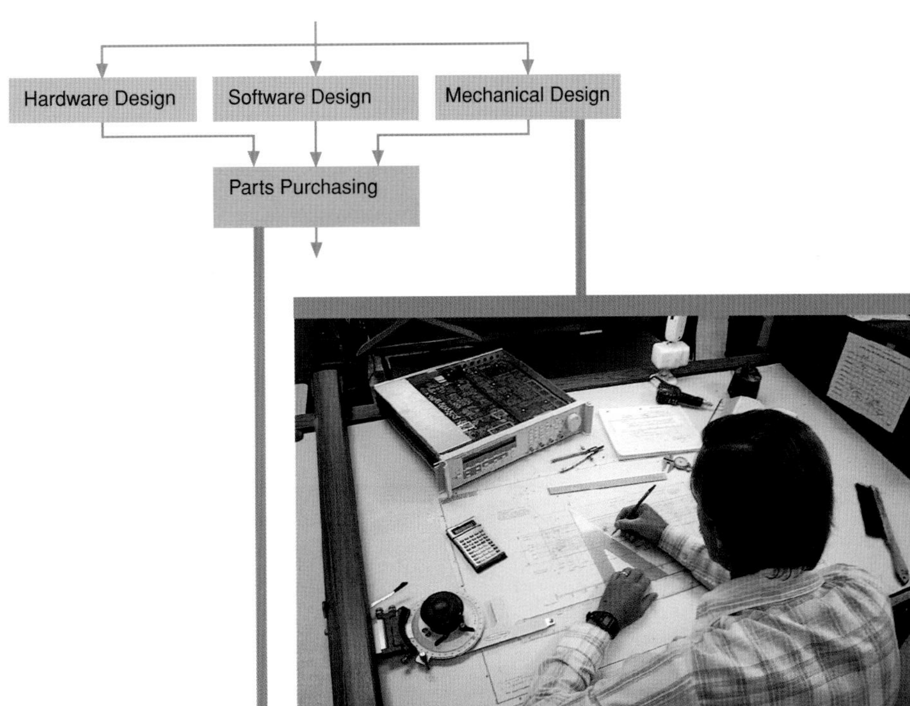

The software program resides within the equipment hardware, and both need to be encased inside a chassis or enclosure. In this photograph, a mechanical design engineer is designing the equipment housing and front panel.

Once the design is complete, the equipment parts have to be ordered so the product can be constructed. In this photograph, the engineering technician on the left is examining the variety of components available from the supplier, while the purchasing agent on the right is comparing costs.

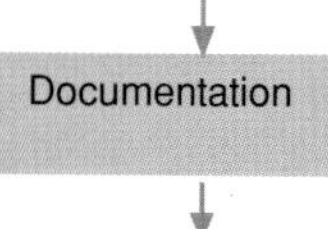

From the moment of conception to the final product, the unit is documented with mechanical drawings, schematic diagrams, and written instructions.

All documentation generated on a product is used to produce an operator's and maintenance manual that will accompany the finished unit when shipped. In this photograph, you can see a technical writer overseeing the drafting of some line art by an illustrator.

From sketches supplied by the engineers, a breadboard model of the design is constructed. The breadboard model is an experimental arrangement of a circuit in which the components are temporarily attached to a flat board. In this arrangement, the components can be tested to prove the feasibility of the circuit. A breadboard facilitates making easy changes when they are necessary. Here you can see an engineering technician breadboarding the design.

Breadboard Design

Prototype Layout, Construction, & Test

The engineer does not consider the final location of the components in constructing the breadboard model. At the next stage, however, a prototype working model completely representative of the final, mass-produced product is hand-assembled. The breadboard is replaced by a printed circuit board (PCB). In this scene, an engineering technician is producing a PCB layout from the design schematic diagrams.

The newly constructed prototype seen here is undergoing a complete evaluation of its mechanical and electrical form, design and performance.

An approved prototype will initiate the production of several pre-production units, which can be seen in this photograph. The pathways these units take through construction will be mapped and used for mass assembly of the new product.

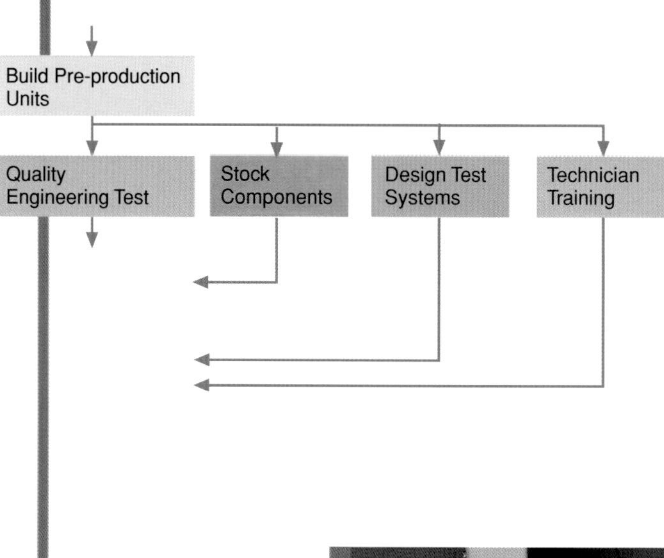

One of the pre-production units is taken through an extensive series of tests to determine whether it meets the standards listed. In this photograph, a quality assurance (QA) technician is evaluating the new product as it is put through an extensive series of tests.

The stockroom holds all of the raw materials needed to build products. In the foreground of this photograph, incoming components are being inspected and values verified, while in the background the parts kits needed for the many phases of assembly are being requisitioned.

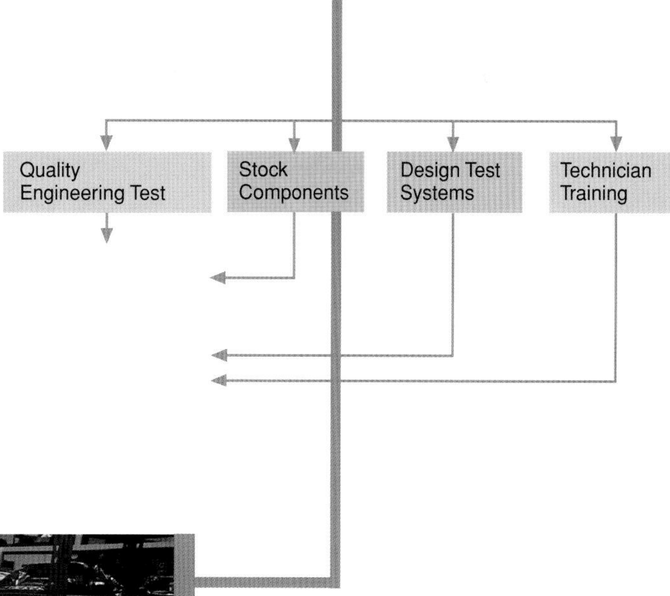

In this photograph, prepping of the many electronic components is taking place. In this process, wires, cables, and components are pre-cut to save time during the product assembly.

Once the assembled units come out of production, they will need to be tested. In this phase of the process, a test engineer is designing an automatic test system for the new product.

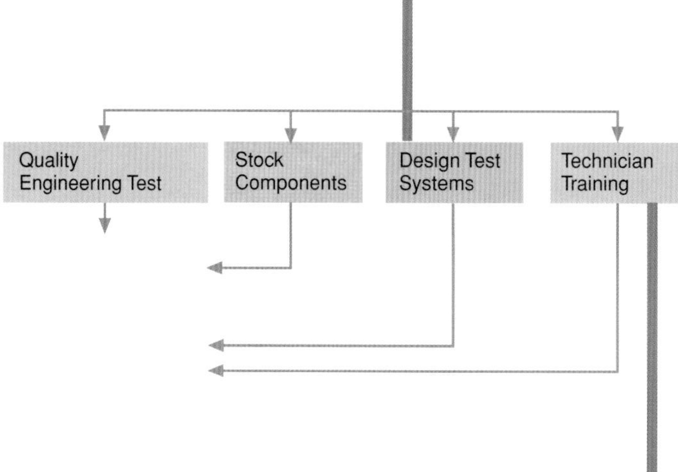

Training is a very important function. In this photograph, engineering technicians, production test technicians, quality control technicians, and customer service technicians are being taught the operation and component level maintenance of the new equipment.

The components which make up the unit are grouped into several areas where complex precision assembly takes place. This photograph of the machine shop shows the chassis or housing for the equipment being fabricated.

In this photograph, all of the electronic components are being inserted into their respective positions within the printed circuit boards.

Chassis Fabrication

Component Insertion

Flow Soldering

Wire/Cable Hook-up

Final Assembly

Product Assembly

Once the boards have been filled with components, they are run through a flow- or wave-soldering machine. This machine solders the components to the board by moving the printed circuit board over a flowing wave of molten solder in a solder bath.

The wires and cables that interconnect all of the separate boards and units within the equipment are added at this stage.

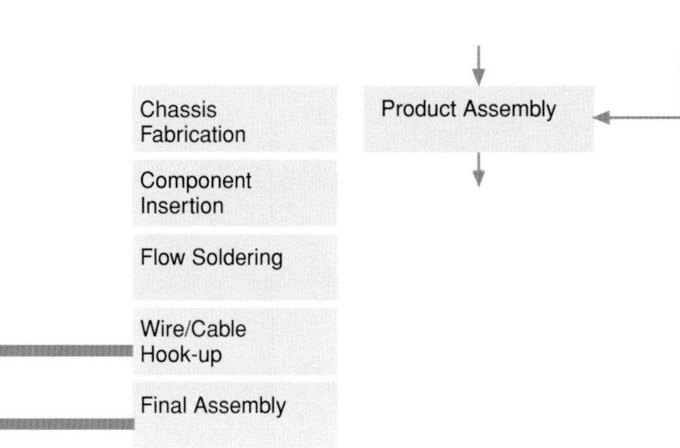

Final assembly of the product is taking place in this photograph. The equipment will have the remaining units inserted, and its front and rear panels will be mounted and connected.

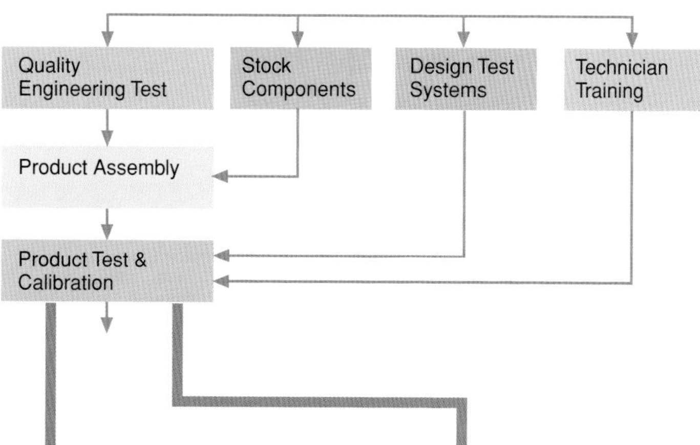

The more complex problems are handled by the production test technicians seen in this photograph. Once the system is fully operational, it is calibrated by a calibration technician.

After leaving assembly, the unit is hooked up to a test system and subjected to various testing procedures. The automatic test equipment (ATE) found at these stations perform many tests that would be too time-consuming for a technician to do manually.

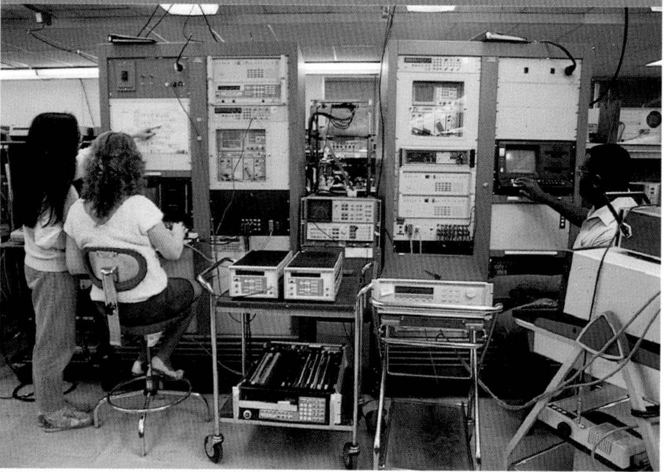

The quality control technicians seen in this photograph are performing the extensive series of electrical and mechanical final inspection tests. These will ensure that the performance standards listed in the unit's specifications are being met.

Quality Control Test

Product Release

From the time of product definition, the marketing personnel have been planning the advertising brochures and sales approach in preparation for the release of the product.

In this photograph, an applications engineer is making a sales call and will demonstrate to the customer the new product with all of its features and possible applications.

In the background of this photograph, you can see all of the equipment ready to be shipped. In the foreground, a delicate electronic unit is being carefully packaged to prevent damage during transit.

In this photograph, a customer is receiving instruction on the operation of the purchased unit.

Once the customer has received the electronic equipment, customer service provides assistance in maintenance and repair of the unit through direct in-house service or at service centers throughout the world. This photograph shows some in-house service technicians troubleshooting problems on returned units.

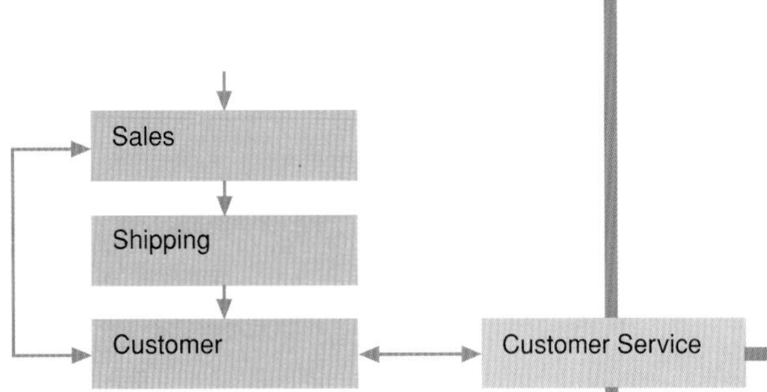

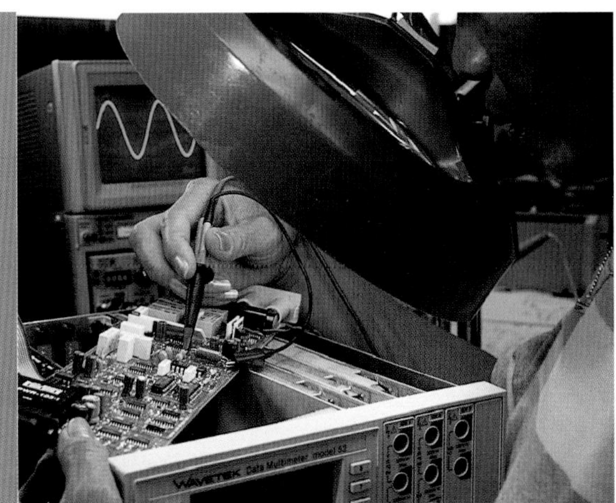

In this photograph, another in-house service technician is performing a test on a returned faulty unit containing the latest surface mount devices (SMD).

The field service technician seen here has been requested by the customer to make a service call on a malfunctioning unit that currently is under test.

11. What new technology has just recently been unveiled?

a. Microelectronics
b. Solid-state electronics
c. Superconductivity
d. Two of the above could be true

12. Radio equipment would be classified in what branch of electronics?

a. Data processing
b. Industrial
c. Biomedical
d. Test and measurement
e. None of the above

13. In which branch of electronics would the oscilloscope be found?

a. Data processing
b. Test and measurement
c. Military and government
d. Biomedical
e. None of the above

14. Fiber optic cables are used to carry ____________.

a. Power
b. Information
c. Both (a) and (b)
d. None of the above

15. Superconductor cables are highly desirable due to:

a. Their large size
b. Their low efficiency
c. Their low power losses
d. All of the above

Essay Questions

16. What are the three sections into which an industrial company can be divided? (1-1)
17. List the seven branches of electronic equipment. (1-1)
18. Give examples of five electronic components and three electronic circuits. (1-1)
19. Give examples of four pieces of equipment in each branch of electronics. (1-1)
20. List three attributes of the human body that medical electronic equipment can measure. (1-1-7)
21. List the four roots of electronics. (1-1)
22. Name the units for the following quantities: (1-1)

a. Current
b. Voltage
c. Resistance
d. Power

23. Who promoted the ac power distribution that is now used worldwide? (1-2)
24. What was the ENIAC? (1-2)
25. List two things for which Thomas Edison is best known. (1-2)
26. State the responsibilities of the following people: (1-4)

a. Engineer **b.** Technician **c.** Assembler

27. What is the difference between a marketing person and a salesperson? (1-4)
28. What tasks do the following people perform? (1-4)

a. Test engineer **b.** Calibration technician **c.** Sustaining engineer

29. At which stage in the product development process would an assembler get involved? (1-4)

30. What responsibilities do the following people have? (1-4)

a. Engineering/technical writer
b. Quality assurance technician
c. Customer service technician (in-house and field service)
d. Engineering technician

CHAPTER 2

OBJECTIVES

After completing this chapter, you will be able to:

1. Explain the atom's subatomic particles.

2. State the difference between an atomic number and atomic weight.

3. Understand the term *natural element.*

4. Describe what is meant by and state how many shells or bands exist around an atom.

5. Explain the difference between:
  - **a.** An atom and a molecule
  - **b.** An element and a compound
  - **c.** A proton and an electron

6. Describe the terms:
  - **a.** Neutral atom
  - **b.** Negative ion
  - **c.** Positive ion

7. Define *electrical current.*

8. Describe the ampere in relation to coulombs per second.

9. List the different current units and values.

10. Explain why the effect of current flow travels at the speed of light.

11. Describe the difference between conventional current flow and electron flow.

12. List the three rules to apply when measuring current.

13. Define *electrical voltage.*

14. List the various voltage units and values.

15. List the three rules to apply when measuring voltage.

16. Relate the electrical system to a fluid system.

17. Describe why current is directly proportional to voltage.

18. Explain the difference between:
  - **a.** A conductor
  - **b.** An insulator
  - **c.** A semiconductor

19. Describe conductance.

20. Explain what makes a good:
  a. Conductor **b.** Insulator

21. Describe semiconductor material.

22. Explain the terms:
  - **a.** Open circuit
  - **b.** Closed circuit
  - **c.** Short circuit

Voltage and Current

OUTLINE

VIGNETTE

Electron
Smallest subatomic particle of negative charge that orbits the nucleus of the atom.

Voltage (*V* or *E*)
Term used to designate electrical pressure or the force that causes current to flow.

Current (*I*)
Measured in amperes or amps, it is the flow of electrons through a conductor.

Resistance
Symbolized *R* and measured in ohms (Ω), it is the opposition to current flow with the dissipation of energy in the form of heat.

Power
Amount of energy converted by a component or circuit in a unit of time, normally seconds. It is measured in units of watts (joules/second).

Element
There are 107 different natural chemical substances or elements that exist on the earth and can be categorized as either being a gas, solid, or liquid.

Atom
Smallest particle of an element.

Creating a Lot of Static

Coulomb's law states that the force exerted between two electric charges is equal to the product of the charges, and is inversely proportional to the distance between them: $f = Q_1 \times Q_2/d^2$.

Ironically, Coulomb's law was not first discovered by Coulomb, but by Henry Cavendish, a wealthy scientist and philosopher. Cavendish did not publish his discovery, which he made several years before Coulomb discovered the law independently. James Clerk Maxwell published the scientific notebooks of Cavendish in 1879, describing his experiments and conclusions. However, about 100 years had passed, and Coulomb's name was firmly associated with the law. Many scientists demanded that the law be called Cavendish's law, while other scientists refused to change, stating that Coulomb was the discoverer because he made the law known promptly to the scientific community.

INTRODUCTION

If Chapter 1 were described as the big picture, Chapter 2 would be called the small picture. In this chapter we will examine the smallest and most significant part of electricity and electronics—the **electron.** By having a good understanding of the electron and the atom, we will be able to obtain a clearer understanding of our four basic electrical quantities: **voltage, current, resistance,** and **power.** In this chapter we will be discussing the basic building blocks of matter, the electrical quantities of voltage and current, and the difference between a conductor and insulator.

2-1 THE STRUCTURE OF MATTER

All of the matter on the earth and in the air surrounding the earth can be classified as being either a solid, liquid, or gas. A total of approximately 107 different natural elements exist in, on, and around the earth. An **element,** by definition, is a substance consisting of only one type of atom; in other words, every element has its own distinctive atom, which makes it different from all the other elements. This **atom** is the smallest particle into which an element can be divided without losing its identity, and a group of identical atoms is called an element, shown in Figure 2-1.

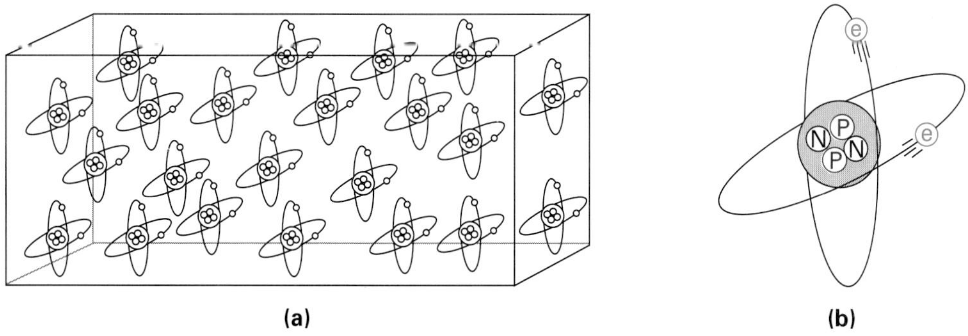

FIGURE 2-1 **(a) Element: Many Similar Atoms. (b) Atom: Smallest Unit.**

For the sake of discussion, let us take a small amount of either a solid, a liquid, or a gas and divide it into two pieces. Then we divide a resulting piece into two pieces, and keep repeating the process until we finally end up with a tiny remaining part. Viewing the part under the microscope, as shown in Figure 2-2, the substance can still be identified as the original element as it is still made up of many of the original solid, liquid or gas atoms. A small amount of gold, for example, the size of a pinpoint, will still contain several billion atoms. If the element subdivision is continued, however, a point will be reached at which a single atom will remain. Let us now analyze the atom in more detail.

2-1-1 *The Atom*

The word atom is a Greek word meaning a particle that is too small to be subdivided. At present, we cannot clearly see the atom; however, physicists and researchers do have the ability to record a picture as small as 12 billionths of an

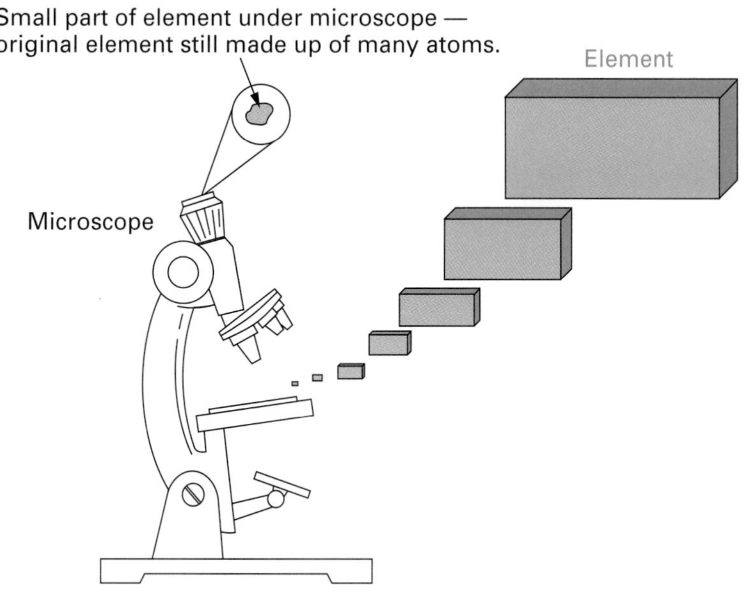

FIGURE 2-2 **Elements under the Microscope.**

inch (about the diameter of one atom), and this image displays the atom as a white fuzzy ball.

Subatomic Particles such as electrons, protons, and neutrons that are smaller than atoms.

In 1913, a Danish physicist, Neils Bohr, put forward a theory about the atom, and his basic model outlining the **subatomic** particles that make up the atom is still in use today and is illustrated in Figure 2-3. Bohr actually combined the ideas of Lord Rutherford's (1871–1937) nuclear atom with Max Planck's (1858–1947) and Albert Einstein's (1879–1955) quantum theory of radiation.

The three important particles of the atom are the *proton,* which has a positive charge, the *neutron,* which is neutral or has no charge, and the *electron,* which has a negative charge. Referring to Figure 2-3, you can see that the atom consists of a positively charged central mass called the *nucleus,* which is made up of protons and neutrons surrounded by a quantity of negatively charged orbiting electrons.

Atomic Number Number of positive charges or protons in the nucleus of an atom.

Table 2-1 lists the periodic table of the elements, in order of their atomic number. The **atomic number** of an atom describes the number of protons that exist within the nucleus.

The proton and the neutron are almost 2000 times heavier than the very small electron, so if we ignore the weight of the electron, we can use the fourth column in Table 2-1 (weight of an atom) to give us a clearer picture of the protons and neutrons within the atom's nucleus. For example, a hydrogen atom, shown in Figure 2-4(a), is the smallest of all atoms and has an atomic number of 1, which means that hydrogen has a one-proton nucleus. Helium, however [Figure 2-4(b)], is second on the table and has an atomic number of 2, indicating that two protons are within the nucleus. The **atomic weight** of helium, however, is 4, meaning that two protons and two neutrons make up the atom's nucleus.

Atomic Weight The relative weight of a neutral atom of an element, based on a neutral oxygen atom having an atomic weight of 16.

The number of neutrons within an atom's nucleus can therefore be calculated by subtracting the atomic number (protons) from the atomic weight (protons and neutrons). For example, Figure 2-5(a) illustrates a beryllium atom which has the following atomic number and weight.

Beryllium
Atomic number: 4 (protons)
Atomic weight: 9 (protons and neutrons)

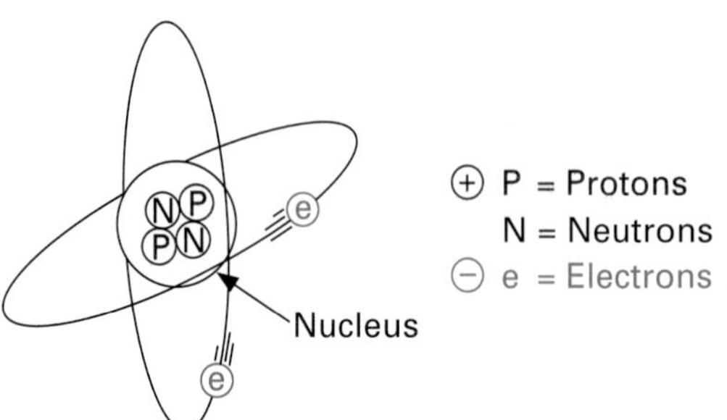

FIGURE 2-3 The Atom.

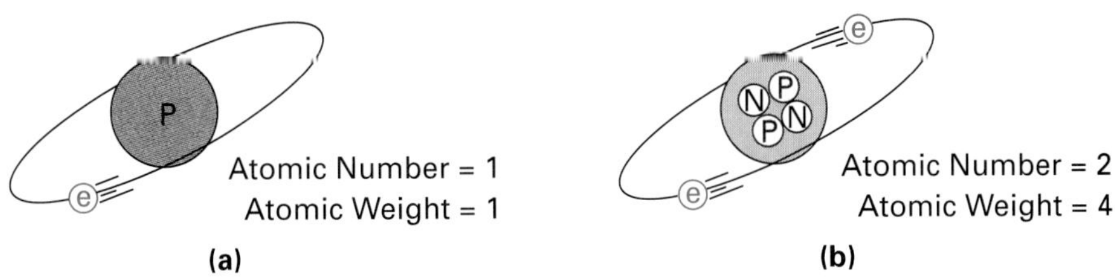

FIGURE 2-4 **(a) Hydrogen Atom. (b) Helium Atom.**

Subtracting the beryllium atom's weight from the beryllium atom number, we can determine the number of neutrons in the beryllium atom's nucleus, as shown in Figure 2-5(a).

In most instances, atoms like the beryllium atom will not be drawn in the three-dimensional way shown in Figure 2-5(a). Figure 2-5(b) shows how a beryllium atom could be more easily drawn as a two-dimensional figure.

A **neutral atom** or *balanced atom* is one that has an equal number of protons and orbiting electrons, so the net positive proton charge is equal but opposite to the net negative electron charge, resulting in a balanced or neutral state.

Neutral Atom
An atom in which the number of positive charges in the nucleus (protons) is equal to the number of negative charges (electrons) that surround the nucleus.

Beryllium

Atomic Number	Atomic Weight
4 Protons	4 Protons + 5 Neutrons

Number of Protons = 4

Number of Neutrons = 9 − 4 = 5

(a)

4 proton nuclei.

2 electrons in the first orbital path.

2 electrons in the second orbital path.

(b)

FIGURE 2-5 **Beryllium Atom.**

TABLE 2-1 Periodic Table of the Elements

ATOMIC NUMBER	ELEMENT NAME	SYMBOL	ATOMIC WEIGHT	ELECTRONS/SHELL							DISCOVERED	COMMENT
				K	L	M	N	O	P	Q		
1	Hydrogen	H	1.007	1							1766	Active gas
2	Helium	He	4.002	2							1895	Inert gas
3	Lithium	Li	6.941	2	1						1817	Solid
4	Beryllium	Be	9.01218	2	2						1798	Solid
5	Boron	B	10.81	2	3						1808	Solid
6	Carbon	C	12.011	2	4						Ancient	Semiconductor
7	Nitrogen	N	14.0067	2	5						1772	Gas
8	Oxygen	O	15.9994	2	6						1774	Gas
9	Fluorine	F	18.998403	2	7						1771	Active gas
10	Neon	Ne	20.179	2	8						1898	Inert gas
11	Sodium	Na	22.98977	2	8	1					1807	Solid
12	Magnesium	Mg	24.305	2	8	2					1755	Solid
13	Aluminum	Al	26.98154	2	8	3					1825	Metal conductor
14	Silicon	Si	28.0855	2	8	4					1823	Semiconductor
15	Phosphorus	P	30.97376	2	8	5					1669	Solid
16	Sulfur	S	32.06	2	8	6					Ancient	Solid
17	Chlorine	Cl	35.453	2	8	7					1774	Active gas
18	Argon	Ar	39.948	2	8	8					1894	Inert gas
19	Potassium	K	39.0983	2	8	8	1				1807	Solid
20	Calcium	Ca	40.08	2	8	8	2				1808	Solid
21	Scandium	Sc	44.9559	2	8	9	2				1879	Solid
22	Titanium	Ti	47.90	2	8	10	2				1791	Solid
23	Vanadium	V	50.9415	2	8	11	2				1831	Solid
24	Chromium	Cr	51.996	2	8	13	1				1798	Solid
25	Manganese	Mn	54.9380	2	8	13	2				1774	Solid
26	Iron	Fe	55.847	2	8	14	2				Ancient	Solid (magnetic)
27	Cobalt	Co	58.9332	2	8	15	2				1735	Solid
28	Nickel	Ni	58.70	2	8	16	2				1751	Solid
29	Copper	Cu	63.546	2	8	18	1				Ancient	Metal conductor
30	Zinc	Zn	65.38	2	8	18	3				1746	Solid
31	Gallium	Ga	69.72	2	8	18	4				1875	Liquid
32	Germanium	Ge	72.59	2	8	18	4				1886	Semiconductor
33	Arsenic	As	74.9216	2	8	18	5				1649	Solid
34	Selenium	Se	78.96	2	8	18	6				1818	Photosensitive
35	Bromine	Br	79.904	2	8	18	7				1898	Liquid
36	Krypton	Kr	83.80	2	8	18	8				1898	Inert gas
37	Rubidium	Rb	85.4678	2	8	18	8	1			1861	Solid
38	Strontium	Sr	87.62	2	8	18	8	2			1790	Solid

For example, Figure 2-6 illustrates a copper atom, which is the most commonly used metal in the field of electronics. It has an atomic number of 29, meaning that 29 protons and 29 electrons exist within the atom when it is in its neutral state.

Shells or Bands An orbital path containing a group of electrons that have a common energy level.

Orbiting electrons travel around the nucleus at varying distances from the nucleus, and these orbital paths are known as **shells or bands.** The orbital shell nearest the nucleus is referred to as the first or K shell. The second is known as

TABLE 2-1 (*continued*)

ATOMIC NUMBER[a]	ELEMENT NAME	SYMBOL	ATOMIC WEIGHT	ELECTRONS/SHELL K	L	M	N	O	P	Q	DISCOVERED	COMMENT
39	Yttrium	Y	88.9059	2	8	18	9	2			1843	Solid
40	Zirconium	Zr	91.22	2	8	18	10	2			1789	Solid
41	Niobium	Nb	92.9064	2	8	18	12	1			1801	Solid
42	Molybdenum	Mo	95.94	2	8	18	13	1			1781	Solid
43	Technetium	Tc	98.0	2	8	18	14	1			1937	Solid
44	Ruthenium	Ru	101.07	2	8	18	15	1			1844	Solid
45	Rhodium	Rh	102.9055	2	8	18	16	1			1803	Solid
46	Palladium	Pd	106.4	2	8	18	18				1803	Solid
47	Silver	Ag	107.868	2	8	18	18	1			Ancient	Metal conductor
48	Cadmium	Cd	112.41	2	8	18	18	2			1803	Solid
49	Indium	In	114.82	2	8	18	18	3			1863	Solid
50	Tin	Sn	118.69	2	8	18	18	4			Ancient	Solid
51	Antimony	Sb	121.75	2	8	18	18	5			Ancient	Solid
52	Tellurium	Te	127.60	2	8	18	18	6			1783	Solid
53	Iodine	I	126.9045	2	8	18	18	7			1811	Solid
54	Xenon	Xe	131.30	2	8	18	18	8			1898	Inert gas
55	Cesium	Cs	132.9054	2	8	18	18	8	1		1803	Liquid
56	Barium	Ba	137.33	2	8	18	18	8	2		1808	Solid
57	Lanthanum	La	138.9055	2	8	18	18	9	2		1839	Solid
72	Hafnium	Hf	178.49	2	8	18	32	10	2		1923	Solid
73	Tantalum	Ta	180.9479	2	8	18	32	11	2		1802	Solid
74	Tungsten	W	183.85	2	8	18	32	12	2		1783	Solid
75	Rhenium	Re	186.207	2	8	18	32	13	2		1925	Solid
76	Osmium	Os	190.2	2	8	18	32	14	2		1804	Solid
77	Iridium	Ir	192.22	2	8	18	32	15	2		1804	Solid
78	Platinum	Pt	195.09	2	8	18	32	16	2		1735	Solid
79	Gold	Au	196.9665	2	8	18	32	18	1		Ancient	Solid
80	Mercury	Hg	200.59	2	8	18	32	18	2		Ancient	Liquid
81	Thallium	Tl	204.37	2	8	18	32	18	3		1861	Solid
82	Lead	Pb	207.2	2	8	18	32	18	4		Ancient	Solid
83	Bismuth	Bi	208.9804	2	8	18	32	18	5		1753	Solid
84	Polonium	Po	209.0	2	8	18	32	18	6		1898	Solid
85	Astatine	At	210.0	2	8	18	32	18	7		1945	Solid
86	Radon	Rn	222.0	2	8	18	32	18	8		1900	Inert gas
87	Francium	Fr	223.0	2	8	18	32	18	8	1	1945	Liquid
88	Radium	Ra	226.0254	2	8	18	32	18	8	2	1898	Solid
89	Actinium	Ac	227.0278	2	8	18	32	18	9	2	1899	Solid

[a]Rare earth series 58–71 and 90–107 have been omitted

the L, the third is M, the fourth is N, the fifth is O, the sixth is P, and the seventh is referred to as the Q shell. There are seven shells available for electrons (K, L, M, N, O, P, and Q) around the nucleus, and each of these seven shells can only hold a certain number of electrons, as shown in Figure 2-7. The outermost electron-occupied shell is referred to as the **valence shell or ring,** and these electrons are termed *valence electrons.* In the case of the copper atom, a single valence electron exists in the valence N shell.

Valence Shell or Ring
Outermost shell formed by electrons.

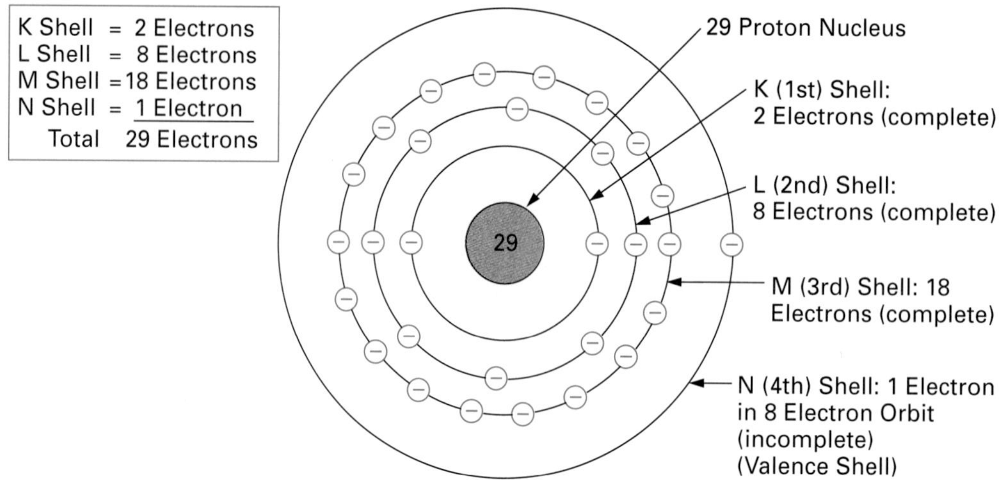

FIGURE 2-6 **Copper Atom.**

All matter exists in one of three states: solids, liquids, and gases. The atoms of a solid are fixed in relation to one another but vibrate in a back-and-forth motion, unlike liquid atoms, which can flow over each other. The atoms of a gas move rapidly in all directions and collide with one another. The far-right column of Table 2-1 indicates whether the element is a gas, a solid, or a liquid.

2-1-2 Laws of Attraction and Repulsion

For the sake of discussion and understanding, let us theoretically imagine that we are able to separate some positive and negative subatomic particles. Using these separated protons and electrons, let us carry out a few experiments, the results of which are illustrated in Figure 2-8. Studying Figure 2-8, you will notice that:

1. *Like charges* (positive and positive or negative and negative) repel one another.

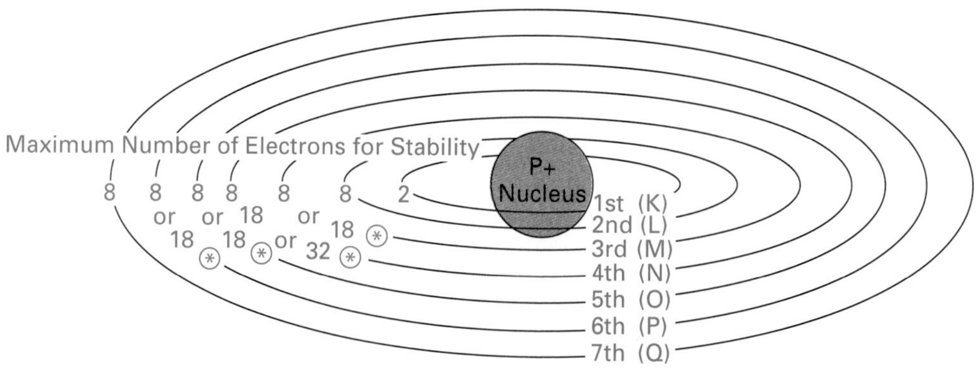

*The maximum number of electrons in these shells is dependent on the element's place ir the periodic table.

FIGURE 2-7 **Electrons and Shells.**

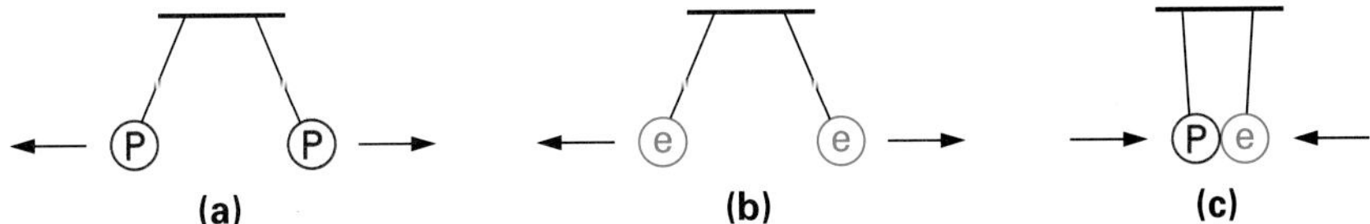

P = Proton (positive)
e = Electron (negative)

FIGURE 2-8 **Attraction and Repulsion. (a) Positive Repels Positive. (b) Negative Repels Negative. (c) Unlike Charges Attract.**

2. *Unlike charges* (positive and negative or negative and positive) attract one another.

Orbiting negative electrons are therefore attracted toward the positive nucleus, which leads us to the question of why the electrons do not fly into the atom's nucleus. The answer is that the orbiting electrons remain in their stable orbit due to two equal but opposite forces. The centrifugal outward force exerted on the electrons due to the orbit counteracts the attractive inward force trying to pull the electrons toward the nucleus due to the unlike charges.

Due to their distance from the nucleus, valence electrons are described as being loosely bound to the atom. The electrons can, therefore, easily be dislodged from their outer orbital shell by any external force, to become a **free electron.**

Free Electron
An electron that is not in any orbit around a nucleus.

2-1-3 The Molecule

An atom is the smallest unit of a natural element, or an element is a substance consisting of a large number of the same atom. Combinations of elements are known as **compounds,** and the smallest unit of a compound is called a **molecule,** just as the smallest unit of an element is an atom. Figure 2-9 summarizes how elements are made up of atoms and compounds are made up of molecules.

Compound
A material composed of united separate elements.

Molecule
Smallest particle of a compound that still retains its chemical characteristics.

Water is an example of a liquid compound in which the molecule (H_2O) is a combination of an explosive gas (hydrogen) and a very vital gas (oxygen). Table salt is another example of a compound; here the molecule is made up of a highly poisonous gas atom (chlorine) and a potentially explosive solid atom (sodium). These examples of compounds each contain atoms that, when alone, are both poisonous and explosive, yet when combined the resulting substance is as ordinary and basic as water and salt.

SELF-TEST REVIEW QUESTIONS FOR SECTION 2-1

1. Define the difference between an element and a compound.
2. Name the three subatomic particles that make up an atom.
3. What is the most commonly used metal in the field of electronics?
4. State the laws of attraction and repulsion.

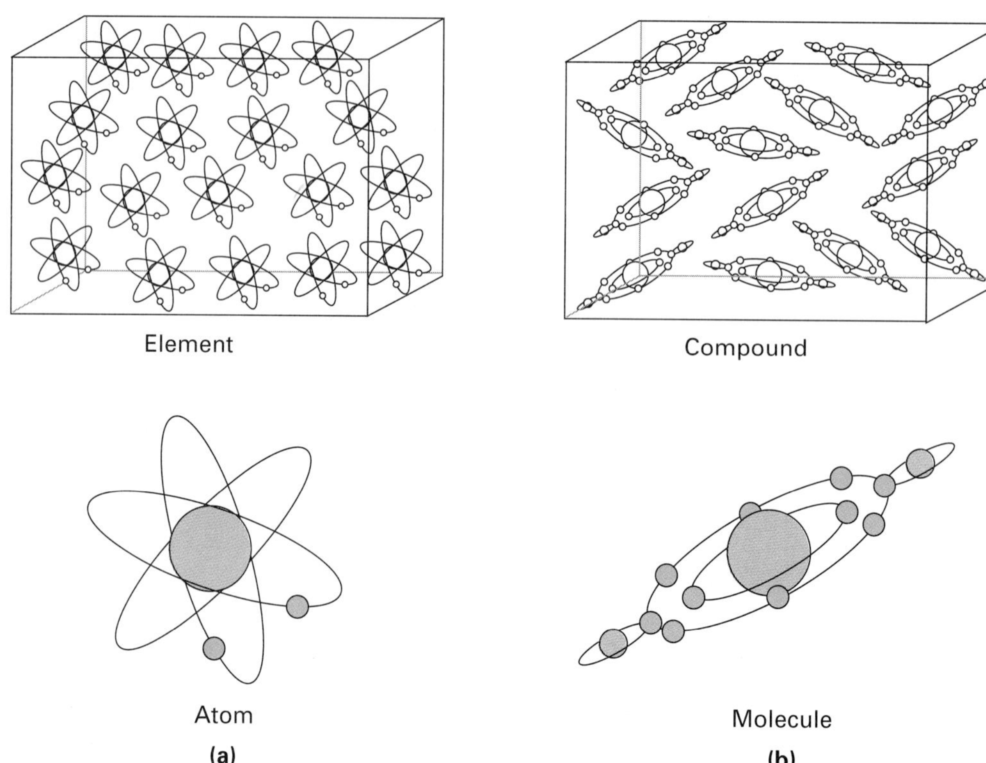

FIGURE 2-9 (a) An Element Is Made Up of Many Atoms. (b) A Compound Is Made Up of Many Molecules.

MINI-MATH REVIEW

This mini-math review is designed to remind you of the mathematical details relating to exponents, scientific notation, engineering notation, and prefixes, which will be used in the following section.

Exponents

Many of the values inserted into formulas contain numbers that have *exponents.* An exponent is a smaller number that appears to the right and slightly higher than another number; for example:

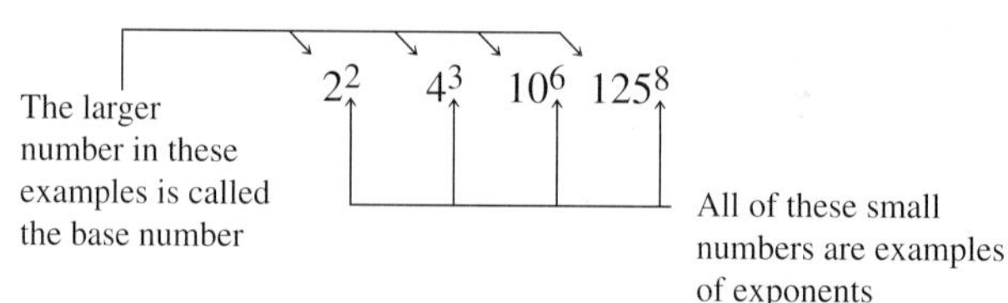

Exponents allow us to save space, as you can see in this example: $5 \times 5 \times 5 \times 5 = 5^4$. Here the exponent (^{4}) is used to indicate how many times its base (5)

is used as a *factor.* You can also see from this example that the other advantage is that *exponents are a sort of math shorthand.* It is much easier to write 5^4 than to write $5 \times 5 \times 5 \times 5$. Using your calculator, you will also find that *exponents allow you to work out problems more quickly.* For example, inputting "5 × 5 × 5 × 5" is a lot more time consuming than is simply pressing "5 y^x 4 =". The "y^x" key on your calculator allows you to enter in any base number (y) and any exponent (^{x}). No matter what method is used, however, the result in this example will always equal 625.

$$5 \times 5 \times 5 \times 5 = 5^4 = 625$$

As another example, what does 10^3 mean? It means that the base (10) is to be used as a factor three (^{3}) times; therefore,

$$10^3 = 10 \times 10 \times 10 = 1000$$

Scientific Notation

Many of the sciences, such as electronics, deal with numbers that contain a large number of zeros; for example: 14,000 and 0.000032. By using exponents we can eliminate the large number of zeros to obtain a shorthand version of the same number. This method is called *scientific notation* or *powers of 10.*

As an example, let us remove all of the zeros from the number 14,000 until we are left with simply 14. However, this number (14) is not equal to the original number (14,000) and therefore simply removing the zeros is not an accurate shorthand. Another number needs to be written along with the 14 to indicate what has been taken away—this is called a *multiplier.* The multiplier must indicate what you have to multiply 14 by to get back to 14,000; therefore,

$$14{,}000 = 14 \times 1000$$

As you know from our discussion on exponents, we can replace the 1000 with 10^3, since $1000 = 10 \times 10 \times 10$. Therefore, the scientific notation for 14,000 is 14×10^3. To convert 14×10^3 back to its original form (14,000), simply remember that each time a number is multiplied by 10, the decimal place is moved one position to the right. In this example 14 is multiplied by 10 three times (10^3 or $10 \times 10 \times 10$) and therefore the decimal point will have to be moved three positions to the right.

$$14 \times 10^3 = 14 \times 10 \times 10 \times 10 = 14.0.0.0. = 14{,}000$$

As another example, what is the scientific notation for the number 0.000032? If we once again remove all the zeros to obtain the number 32, we will again have to include a multiplier with 32 to indicate what 32 has to be multiplied by to return it to its original form. In this case 32 will have to be multiplied by $\frac{1}{1{,}000{,}000}$ (one millionth) to return it to 0.000032.

$$32 \times \frac{1}{1{,}000{,}000} = 0.000032$$

This can be verified because when you divide any number by 10, you move the decimal point one position to the left, and therefore to divide any number by 1,000,000, you simply move the decimal point six positions to the left.

$$32 \times \frac{1}{1{,}000{,}000} = \frac{32}{1{,}000{,}000} = \frac{32}{10 \times 10 \times 10 \times 10 \times 10 \times 10}$$

$$= 0.000032$$

Again, an exponent can be used in place of the $\frac{1}{1{,}000{,}000}$ multiplier. It will be

$$\frac{1}{1{,}000{,}000} = \frac{1}{10^6} = 0.000001 = 10^{-6}$$

Whenever you divide a number into 1, you get the *reciprocal* of that number. In this example, when you divide 1,000,000 into 1 you get 0.000001, which is equal to a negative exponent of 10^{-6}. This means that the multiplier 10^{-6} is indicating to move the decimal point back (to the left) by six places, and therefore

$$32 \times \frac{1}{1{,}000{,}000} = 32 \times 0.000001 = 32 \times 10^{-6} = 0.000032$$

Now that you know exactly what a multiplier means, you only have to remember these simple rules;

- A *negative exponent* tells you how many places *to the left* to move the decimal point.
- A *positive exponent* tells you how many places *to the right* to move the decimal point.

Remember one important point: a negative exponent does not indicate a negative number; it simply indicates a fraction. For example, "4×10^{-3} volts" means that 1 volt has been broken up into 1000 parts and we have 4 of those pieces, or $\frac{4}{1000}$.

To reinforce our understanding, let us try converting a few numbers to scientific notation.

Scientific Notation A floating-point system in which numbers are expressed as products consisting of a number between 1 and 10 multiplied by an appropriate power of 10.

$$230{,}000{,}000 = 230{,}000{,}000 = 23 \times 10^7$$

$$760{,}000 = 760{,}000 = 76 \times 10^4$$

$$0.0019 = 0.0019. = 19 \times 10^{-4}$$

$$0.00085 = 0.00085. = 85 \times 10^{-5}$$

Engineering Notation

Engineering Notation A floating-point system in which numbers are expressed as products consisting of a number that is greater than 1 multiplied by an appropriate power of 10 that is some multiple of three.

To be a little more specific, **scientific notation** is expressed as a base number between 1 and 10 with a multiplier that is a power of 10, for example, 3.2×10^{-5}. **Engineering notation** however is expressed as a base number that is greater than 1 with a multiplier that is a power of ten with an exponent that is some multiple of three, for example, 32×10^{-6}.

Prefixes

Although any power of 10 can be used (10^1, 10^2, 10^{-5}, 10^{-2}), most electrical or electronic values use exponents that are multiples of three (10^3, 10^6, 10^9, 10^{12} or 10^{-3}, 10^{-6}, 10^{-9}, 10^{-12}). These multiples have been given names and symbols, as shown below.

VALUE	NAME	SYMBOL		
10^{12} = 1,000,000,000,000	tera	T	↑	Multipliers that are greater than 1
10^{9} = 1,000,000,000	giga	G		
10^{6} = 1,000,000	mega	M		
10^{3} = 1,000	kilo	k		
10^{-3} = 1/1,000	milli	m		Multipliers that are less than 1 (fractions)
10^{-6} = 1/1,000,000	micro	μ		
10^{-9} = 1/1,000,000,000	nano	n		
10^{-12} = 1/1,000,000,000,000	pico	p	↓	

These multipliers are often called *prefixes*, because they precede units. For example, 25 kilovolts or 25 kV means 25×10^3 volts or 25,000 volts, or 50 milliamps (50 mA) means 50×10^{-3} amps or 50/1000ths of one amp.

2-2 CURRENT

The movement of electrons from one point to another is known as *electrical current.* Energy in the form of heat or light can cause an outer shell electron to be released from the valence shell of an atom. Once an electron is released, the atom is no longer electrically neutral and is called a **positive ion,** as it now has a net positive charge (more protons than electrons). The released electron tends to jump into a nearby atom, which will then have more electrons than protons and is referred to as a **negative ion.**

Positive Ion
Atom that has lost one or more of its electrons and therefore has more protons than electrons, resulting in a net positive charge.

Negative Ion
Atom that has more than the normal neutral amount of electrons.

Let us now take an example and see how electrons move from one point to another. Figure 2-10 illustrates a metal conductor connecting two charged objects. The metal conductor could be either gold, silver, or copper, but whichever it is, one common trait can be noted: The valence electrons in the outermost shell are very loosely bound and can easily be pulled from their parent atom.

The negative ions on the right in Figure 2-11 have more electrons than protons, while the positive ions on the left in Figure 2-11 have fewer electrons

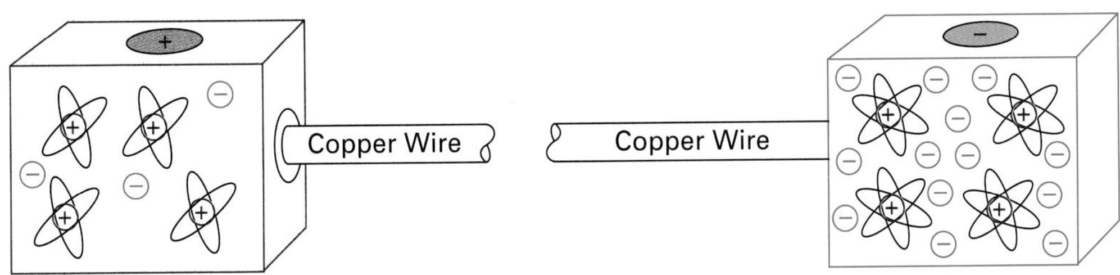

FIGURE 2-10 **Positive and Negative Charges.**

Positive Charge The charge that exists in a body which has fewer electrons than normal.

Negative Charge An electric charge which has more electrons than protons.

than protons and so display a **positive charge.** The strip of metal joining the two charges has its own atoms, which begin in the neutral condition. Let us now concentrate on one of the negative ions. In Figure 2-11(a), the extra electrons in the outer shells of the negative ions on the right side will feel the attraction of the positive ions on the left side and the repulsion of the outer negative ions, or **negative charge.** This will cause an electron in a negative ion to jump away from its parent atom's orbit and land in an adjacent atom to the left within the metal wire conductor, as shown in Figure 2-11(b). This adjacent atom now has an extra electron and is called a negative ion, while the initial parent negative ion becomes a neutral atom, which will now receive an elec-

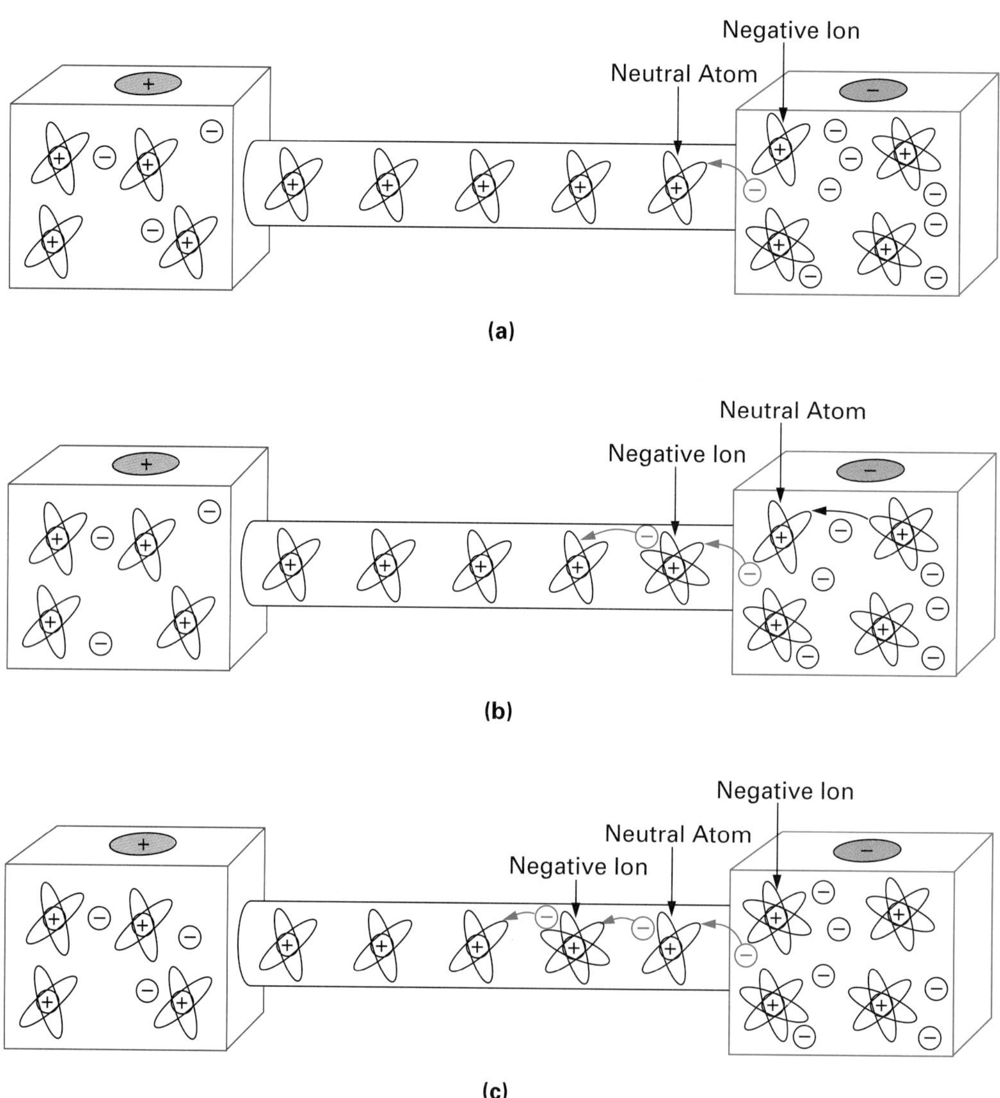

FIGURE 2-11 **Electron Migration Due to Forces of Positive Attraction and Negative Repulsion on Electrons.**

tron from one of the other negative ions, because their electrons are also feeling the attraction of the positive ions on the left side and the repulsion of the surrounding negative ions.

The electrons of the negative ion within the metal conductor feel the attraction of the positive ions, and eventually one of its electrons jumps to the left and into the adjacent atom, as shown in Figure 2-11(c). This continual movement to the left will produce a stream of electrons flowing from right to left. Millions upon millions of atoms within the conductor pass a continuous movement of billions upon billions of electrons from right to left. This electron flow is known as **electric current.**

Electric Current (*I*) Measured in amperes or amps, it is the flow of electrons through a conductor.

To summarize, we could say that as long as a force or pressure, produced by the positive charge and negative charge, exists it will cause electrons to flow from the negative to the positive terminal. The positive side has a deficiency of electrons and the negative side has an abundance, and so a continuous flow or migration of electrons takes place between the negative and positive terminal through our metal conducting wire. This electric current or electron flow is a measurable quantity, as will now be explained.

2-2-1 *Coulombs per Second*

There are 6.24×10^{18} electrons in 1 **coulomb of charge,** as illustrated in Figure 2-12. To calculate coulombs of charge (designated Q), we can use the formula

Coulomb of Charge Unit of electric charge. One coulomb equals 6.24×10^{18} electrons.

$$\text{charge, } Q = \frac{\text{total number of electrons } (n)}{6.24 \times 10^{18}}$$

where Q is the electric charge in coulombs.

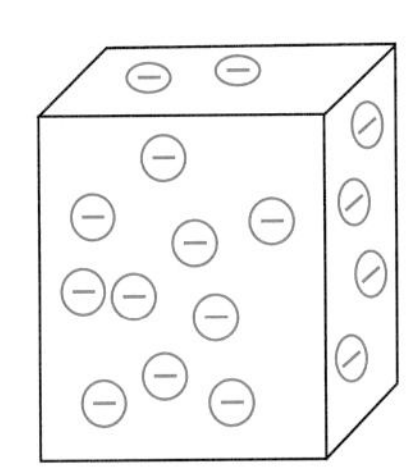

(a) 6.24×10^{18} Electrons = 1 Coulomb of Charge

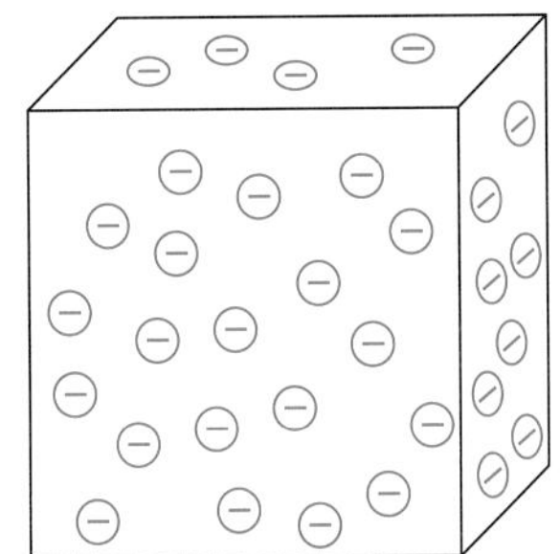

(b) 12.48×10^{18} Electrons = 2 Coulombs of Charge

FIGURE 2-12 **(a) 1 C of Charge. (b) 2 C of Charge.**

EXAMPLE:

If a total of 3.75×10^{19} free electrons exist within a piece of metal conductor, how many coulombs (C) of charge would be within this conductor?

Solution:

By using the charge formula (Q) we can calculate the number of coulombs (C) in the conductor.

$$Q = \frac{n}{6.24 \times 10^{18}}$$
$$= \frac{3.75 \times 10^{19}}{6.24 \times 10^{18}}$$
$$= 6\text{ C}$$

A total of 6 C of charge exists within the conductor.

In the opposite calculator sequence you will see how the exponent key (E, EE, or EXP) on your calculator can be used.

CALCULATOR SEQUENCE

Step	Keypad Entry	Display Response
1.	[3] [.] [7] [5] [E] (Exponent) [1] [9]	3.75E19
2.	[÷]	
3.	[6] [.] [2] [4] [E] [1] [8]	6.24E18
4.	[=]	6.0096

EXAMPLE:

A metal wire conductor has 14 C of electrons passing through it. Calculate the number of electrons moving through the conductor.

Solution:

To calculate the number of electrons (n), we have to transpose our original formula. If

$$Q = \frac{n}{6.24 \times 10^{18}}$$

then

$$n = Q \times 6.24 \times 10^{18}$$

The total number of electrons (n) is

$$n = Q \times 6.24 \times 10^{18}$$
$$= 14\text{ C} \times 6.24 \times 10^{18}$$
$$= 8.7 \times 10^{19}\text{ electrons will pass through the conductor}$$

2-2-2 The Ampere

Static
Crackling noise heard on radio receivers caused by electric storms or electric devices in the vicinity.

Ampere (A)
Unit of electric current.

A coulomb is a **static** amount of electric charge. In electronics, we are more interested in electrons in motion, and current is the flow of electrons, which involves time. Coulombs and time are therefore combined to describe the number of electrons and the rate at which they flow. This relationship is called *current* (I) flow and has the unit of **amperes (A).** If 6.24×10^{18} electrons (1 C) were to drift past a specific point on a conductor in 1 second of time, 1 ampere of current is said to be flowing:

$$\text{current } (I) = \frac{\text{coulombs } (Q)}{\text{time } (t)}$$

1 ampere = 1 coulomb per 1 second

$$1 \text{ A} = \frac{1 \text{ C}}{1 \text{ s}}$$

To summarize, we could say that 1 ampere equals a flow rate of 1 coulomb per second, and current is measured in amperes.

EXAMPLE:

If 5×10^{19} electrons pass a point in a conductor in 4 s, what is the amount of current flow in amperes?

Solution:

Current (I) is equal to Q/t. We must first convert electrons to coulombs.

$$Q = \frac{n}{6.24 \times 10^{18}}$$
$$= \frac{5 \times 10^{19}}{6.24 \times 10^{18}}$$
$$= 8 \text{ C}$$

Now, to calculate the amount of current, we use the formula

$$I = \frac{Q}{t}$$
$$= \frac{8 \text{ C}}{4 \text{ s}}$$
$$= 2 \text{ A}$$

This means that 2 A or 1.248×10^{19} electrons (2C) are passing a specific point in the conductor every second.

CALCULATOR SEQUENCE

Step	Keypad Entry	Display Response
1.	5 E (exponent) 1 9	5E19
2.	÷	
3.	6 . 2 4 E 1 8	6.24E18
4.	=	8.012
5.	÷	
6.	4	2.003
7.	=	2.003

2-2-3 Units of Current

Current within electronic equipment is normally a value in milliamperes or microamperes and very rarely exceeds 1 ampere. Table 2-2 lists all the prefixes related to current. For example, 1 milliampere is one-thousandth of an ampere, which means that if 1 ampere were divided into 1000 parts, 1 part of the 1000 parts would be flowing through the circuit.

TABLE 2-2 Current Units

NAME	SYMBOL	VALUE
Picoampere	pA	$10^{-12} = \frac{1}{1,000,000,000,000}$
Nanoampere	nA	$10^{-9} = \frac{1}{1,000,000,000}$
Microampere	μA	$10^{-6} = \frac{1}{1,000,000}$
Milliampere	mA	$10^{-3} = \frac{1}{1000}$
Ampere	A	$10^{0} = 1$
Kiloampere	kA	$10^{3} = 1000$
Megaampere	MA	$10^{6} = 1,000,000$
Gigaampere	GA	$10^{9} = 1,000,000,000$
Teraampere	TA	$10^{12} = 1,000,000,000,000$

EXAMPLE:

Convert the following:

a. 0.003 A = ______________ mA (milliamperes)
b. 0.07 mA = ______________ μA (microamperes)
c. 7333 mA = ______________ A (amperes)
d. 1275 μA = ______________ mA (milliamperes)

Solution:

a. 0.003A = ______________ mA. In this example, 0.003 A has to be converted so that it is represented in milliamperes (10^{-3} or 1/1000 of an ampere). The basic algebraic rule to be remembered is that both expressions on either side of the equals must be the same.

LEFT		RIGHT	
Base	Exponent	Base	Exponent
0.003×10^{0} =		______ $\times 10^{-3}$	

The exponent on the right in this example is going to be decreased 1000 times (10^{0} to 10^{-3}), so for the statement to balance the number on the right will have to be increased 1000 times; that is, the decimal point will have to be moved to the right three places (0.003 or 3). Therefore,

$$0.003 \times 10^{0} = 3 \times 10^{-3}$$

or

$$0.003 \text{ A} = 3 \times 10^{-3} \text{ A or } 3 \text{ mA}$$

b. 0.07 mA = ____________ μA. In this example the exponent is going from milliamperes to microamperes (10^{-3} to 10^{-6}) or 1000 times smaller, so the number must be made 1000 times greater.

0.070 or 70.0

Therefore, 0.07 mA = 70 μA.

c. 7333 mA = ____________ A. The exponent is going from milliamperes to amperes, increasing 1000 times, so the number must decrease 1000 times.

7333. or 7.333

Therefore, 7333 mA = 7.333 A.

d. 1275 μA = ____________ mA. The exponent is changing from microamperes to milliamperes, an increase of 1000 times, so the number must decrease by the same factor.

1275.0 or 1.275

Therefore, 1275 μA = 1.275 mA.

2-2-4 The Speed of Current Flow

Electrons will in fact move very slowly as they jump from parent to adjacent atom; however, the chain reaction occurs at the speed of light, which is 186,000 miles per second or 300,000,000 meters (m) per second (3×10^8 m/s). This chain reaction is best understood by using an analogy where we relate free electrons within a conductor to a string of Ping-Pong balls within a tube.

If one extra ball is inserted in one end, a ball will appear out of the other end almost instantly. Although each ball within the tube has only moved a small distance, the effect has traveled toward the end of the tube at the speed of light, as seen in Figure 2-13(a).

As far as electrons go, the first electron jump from parent to adjacent atom causes the next to jump, and the next, and next, and so on, as shown in Figure 2-13(b). Just as the first ball causes a chain reaction with the others, and even though the actual speed of the electrons is only a fraction of an inch per second, the effective velocity is at the speed of light.

2-2-5 Conventional versus Electron Flow

Electron Flow
A current produced by the movement of free electrons toward a positive terminal.

Conventional Current Flow
A current produced by the movement of positive charges toward a negative terminal.

Electrons drift from a negative to a positive charge, as illustrated in Figure 2-14. As already discussed, this current is known as **electron flow.**

In the eighteenth and nineteenth centuries, however, when very little was known about the atom, researchers believed wrongly that current was a flow of positive charges. Although this has now been proved incorrect, many texts still use **conventional current flow,** shown in Figure 2–15.

Whether conventional current flow or electron flow is used, the same answers to problems, measurements, or designs are obtained. The key point to remember is that direction is not important, but the amount of current flow is.

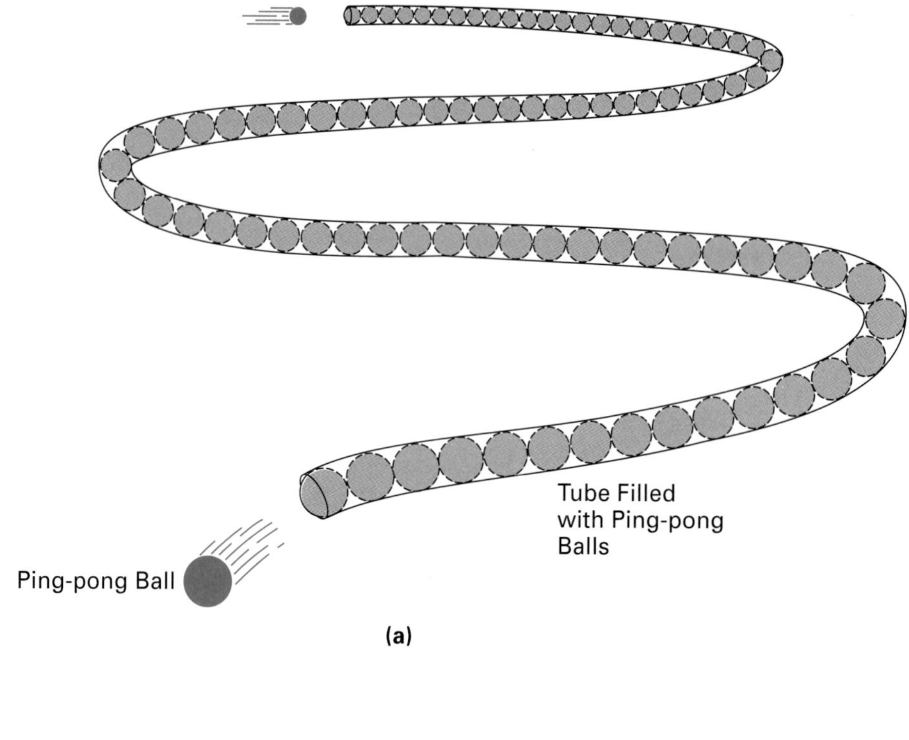

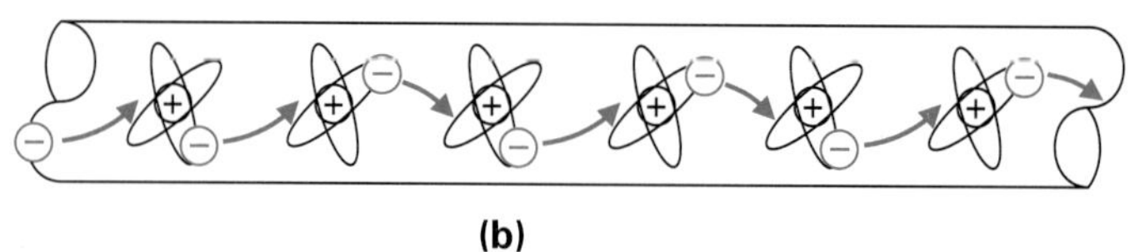

FIGURE 2-13 Chain Reaction. (a) Ping-Pong Ball Analogy. (b) Electrons.

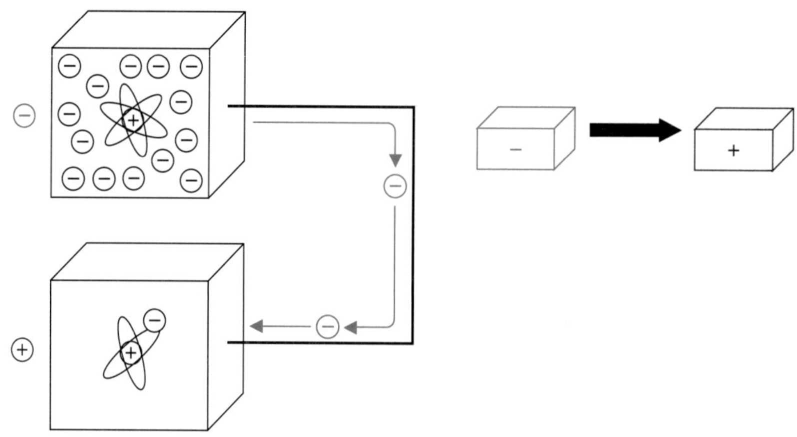

FIGURE 2-14 Electron Current Flow.

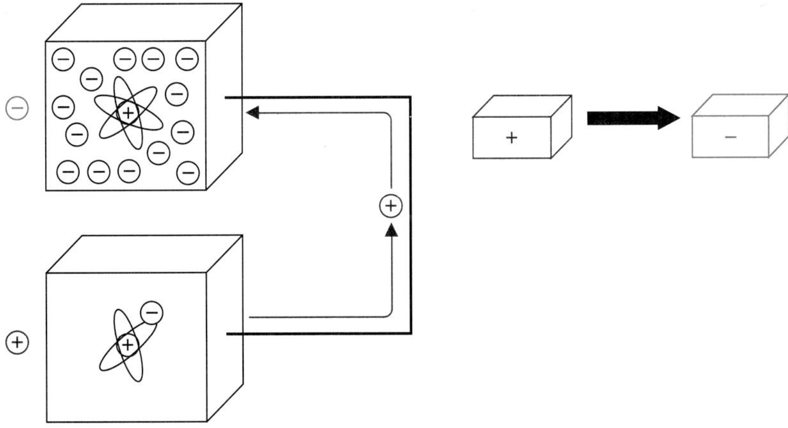

FIGURE 2-15 **Conventional Current Flow.**

Throughout this book we will be using electron flow so that we can relate back to the atom when necessary; if you wish to use conventional flow, just reverse the direction of the arrows. But to avoid confusion, be consistent with your choice of flow.

2-2-6 *How Is Current Measured?*

Ammeters (ampere meters) are used to measure the current flow within a circuit, as shown in Figure 2-16. In electronics, where current is generally small, the milliampere (mA) or microampere (μA) scale will generally be used, whereas in electrical high-current circuits, the meter will generally be used on an ampere scale.

Ammeter
Meter placed in the path of current flow to measure the amount.

Three rules must be remembered and applied when measuring current:

1. Always set the selector to the higher scale first (amperes) and then reduce as needed to milli- or microamperes, just in case a larger current than anticipated is within the circuit, in which case the meter could be damaged.

To explain this, let us take an example. If you were to weigh some potatoes on a set of scales and you were not aware of how much they weigh, the largest scale should initially be selected. For example, if you selected the 0 to 10 pound scale and the potatoes weighed 15 pounds, the meter needle would be forced violently to the right side and could damage the scale (this is known as *pegging* the meter). By always selecting the 0 to 50 pound scale and then stepping down to 0 to 40, 0 to 30, and 0 to 20 pounds, as needed, you would almost guarantee never pegging the meter. With the ammeter, we must apply the same philosophy and select a higher scale, then work down if needed.

2. The current meter must be connected so that the positive lead of the ammeter (red) is connected to the positive charge or polarity and the negative lead of the ammeter (black) is connected to the negative charge or

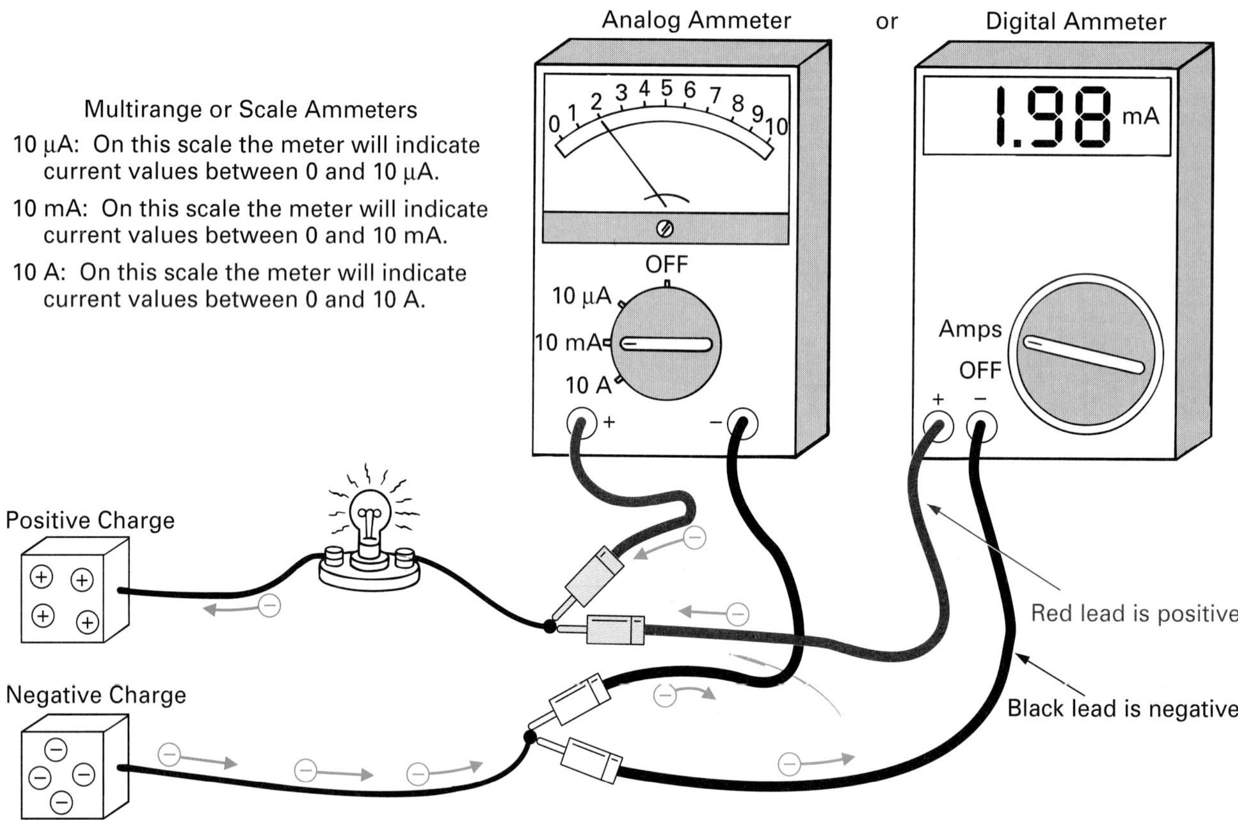

FIGURE 2-16 **Measuring Current with the Ammeter.**

polarity. Ammeters are sensitive to the polarity (positive or negative) of the charge.

3. If you wish to measure current flowing within a wire, the wire must be opened so that the ammeter measuring the current will be placed in the path of current flow.

SELF-TEST REVIEW QUESTIONS FOR SECTION 2-2

1. What is the unit of current?
2. Define current in relation to coulombs and time.
3. What is the difference between conventional and electron current flow?
4. List the three rules that should be applied when using the ammeter to measure current.

2-3 VOLTAGE

Voltage is the force or pressure exerted on electrons. Referring to Figure 2-17(a) and (b), you will notice two situations. Figure 2-17(a) shows highly concentrated positive and negative charges or potentials connected to one another by a copper wire. In this situation, a large potential difference or voltage is being applied across the copper atom's electrons. This force or voltage causes a large amount of copper atom electrons to move from right to left. Figure 2-17(b) illustrates a low concentration of positive and negative potentials, so a small voltage or pres-

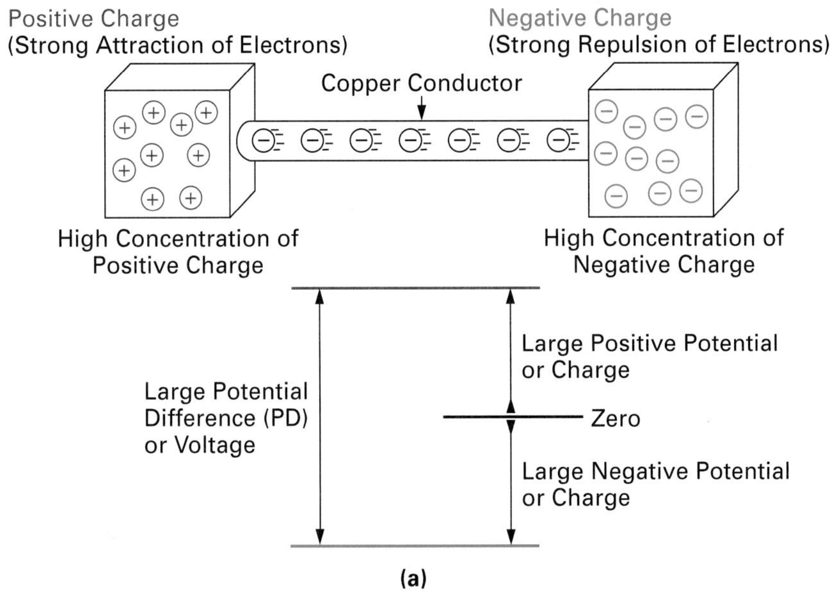

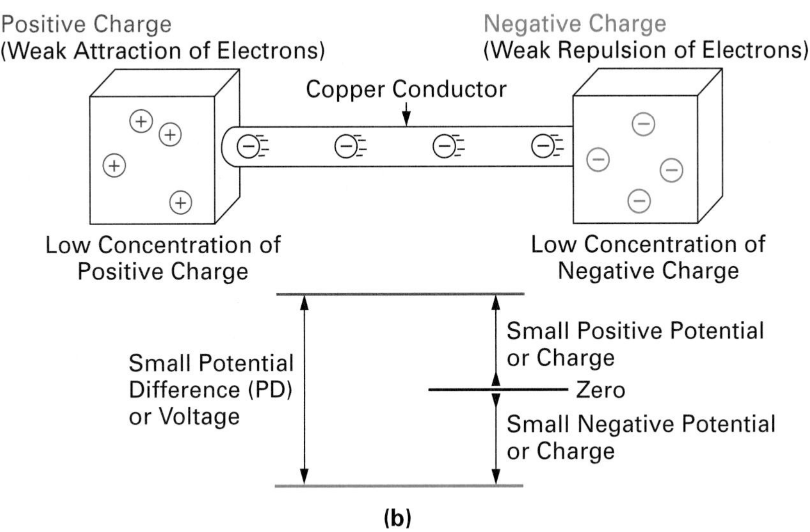

FIGURE 2-17 (a) Large Potential Difference or Voltage. (b) Small Potential Difference or Voltage.

sure is being applied across the conductor, causing a small amount of force, and therefore current, to move from right to left.

To summarize, then, we could say that a highly concentrated charge produces a high voltage, whereas a low concentrated charge produces a low voltage. Voltage is also appropriately known as the electron moving force or **electromotive force (emf),** and since two opposite potentials exist, one negative and one positive, the strength of the voltage can also be referred to as the amount of **potential difference (PD)** applied across the circuit. Referring back to Figure 2-17(a), we see that a large voltage, electromotive force, or potential difference exists across the copper conductor, while in Figure 2-17(b) a small voltage, potential difference, or electromotive force is exerted across the conductor.

Electromotive Force (emf)
Force that causes the motion of electrons due to a potential difference between two points.

Potential Difference (PD)
Voltage difference between two points, which will cause current to flow in a closed circuit.

Voltage is the force, pressure, potential difference (PD), or electromotive force (emf) that causes electron flow or current and is symbolized by italic uppercase *V*. The unit for voltage is the volt, symbolized by roman uppercase V. This can become a bit confusing. For example, when the voltage applied to a circuit equals 5 volts, the circuit notation would appear as

$$V = 5\text{ V}$$

You know the first *V* represents "voltage," not "volt," because 1 volt cannot equal 5 volts. To avoid confusion, some texts and circuits use *E*, symbolizing electromotive force, to represent voltage; for example,

$$E = 5\text{ V}$$

In this book, however, we maintain the original designation for voltage (*V*).

2-3-1 Symbols

Battery
Dc voltage source containing two or more cells that converts chemical energy into electrical energy.

A **battery,** illustrated in Figure 2-18, converts chemical energy into electrical energy. At the positive terminal of the battery, positive charges or ions (atoms with more protons than electrons) are present, and at the negative terminal, negative charges or ions (atoms with more electrons than protons) are available to supply electrons for current flow within a circuit. A battery, therefore, chemi-

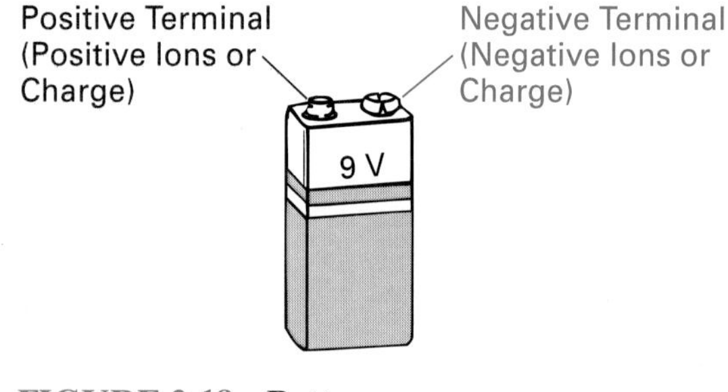

FIGURE 2-18 Battery.

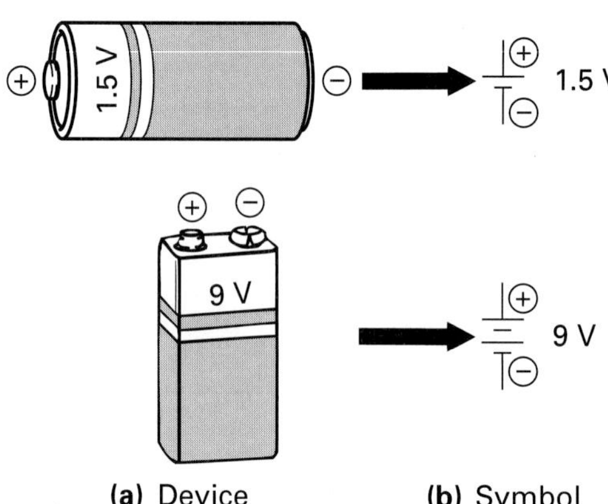

FIGURE 2-19 Batteries. (a) Pictorial. (b) Schematic Symbol.

cally generates negative and positive ions at its respective terminals. The symbol for the battery is illustrated in Figure 2-19.

In Figure 2-20(a) we have illustrated two other devices as they appear physically and in their electronic schematic symbols. The first new device is the light bulb, and the other is a piece of copper conductor wire with an alligator or crocodile clip on either end.

The 9-V battery shown in Figure 2-20(b) chemically generates positive and negative ions. The negative ions at the negative terminal force away the negative electrons, which are attracted by the positive charge or absence of electrons on the other terminal. Proceeding through the copper conductor wire, jumping

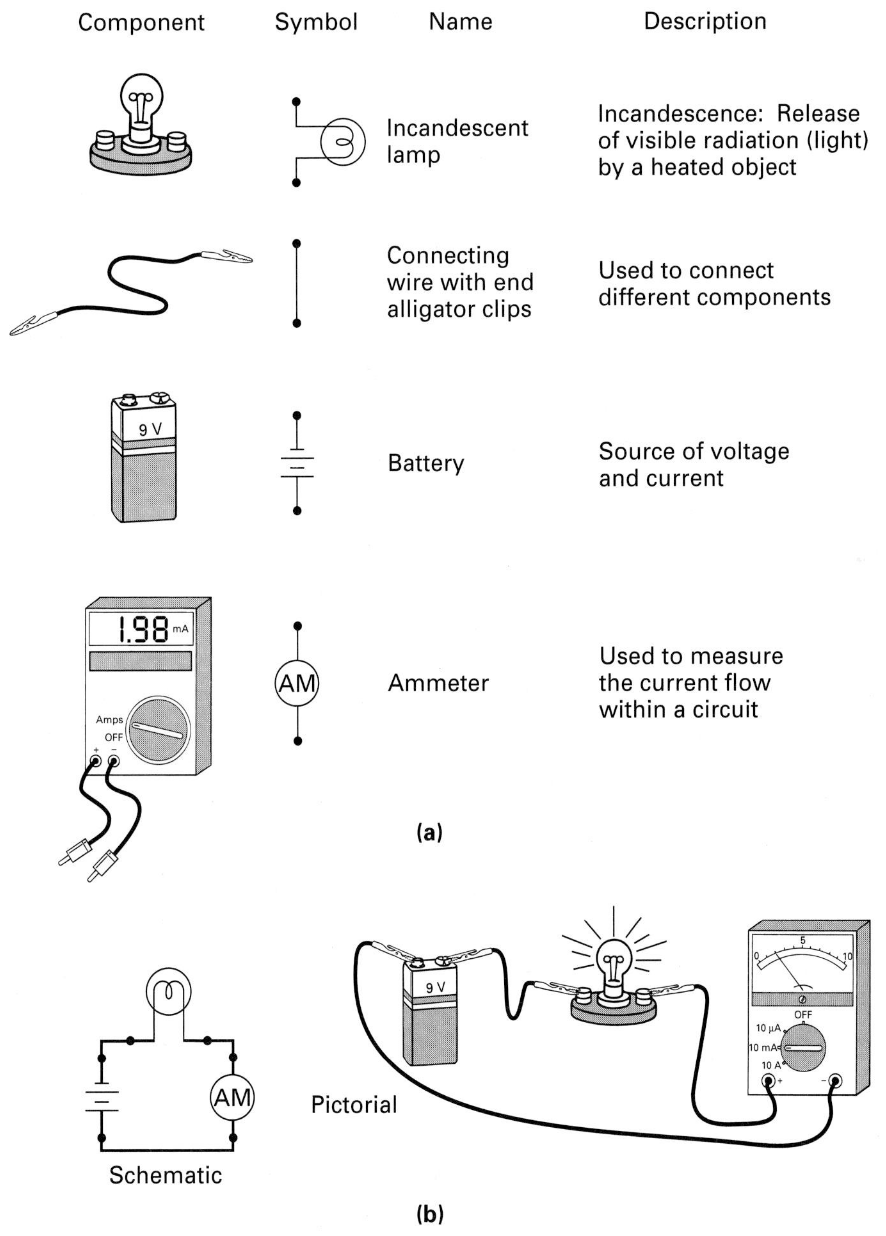

FIGURE 2-20 **(a) Components. (b) Example Circuit.**

from one atom to the next, they eventually reach the bulb. The electrons pass through the bulb, which glows due to this current passing through it. When emerging from the light bulb, the electrons travel through another connector cable and finally reach the positive terminal of the battery.

Studying Figure 2-20(b), you will notice two reasons why the circuit is drawn using symbols rather than illustrating the physical appearance:

1. A circuit with symbols may be drawn faster and more easily.
2. A circuit with symbols has less detail and clutter and is more easily comprehended because of having fewer distracting elements.

2-3-2 *Units of Voltage*

The unit for voltage is the volt (V). Voltage within electronic equipment is normally measured in volts, whereas heavy-duty industrial equipment normally requires high voltages that are generally measured in kilovolts (kV). Table 2-3 lists all the prefixes and values related to volts.

EXAMPLE:

Convert the following:

a. 3000 V = ____________ kV (kilovolts)
b. 0.14 V = ____________ mV (millivolts)
c. 1500 kV = ____________ MV (megavolts)

Solution:

a. 3000 V = 3 kV or 3×10^3 volts (exponent ↑ 1000, number ↓ 1000)
b. 0.14 V = 140 mV or 140 × 10-3 volt (exponent ↓ 1000, number ↑ 1000)
c. 1500 kV = 1.5 MV or 1.5 × 106 volts (exponent ↑ 1000, number ↓ 1000)

TABLE 2-3 Voltage Units

NAME	SYMBOL	VALUE
Picovolts	pV	$10^{-12} = \frac{1}{1,000,000,000,000}$
Nanovolts	nV	$10^{-9} = \frac{1}{1,000,000,000}$
Microvolts	μV	$10^{-6} = \frac{1}{1,000,000}$
Millivolts	mV	$10^{-3} = \frac{1}{1000}$
Volts	V	$10^0 = 1$
Kilovolts	kV	$10^3 = 1000$
Megavolts	MV	$10^6 = 1,000,000$
Gigavolts	GV	$10^9 = 1,000,000,000$
Teravolts	TV	$10^{12} = 1,000,000,000,000$

2-3-3 *How Is Voltage Measured?*

Voltmeters (voltage meters) are used to measure electrical pressure (voltage) anywhere in an electronic circuit. Figure 2-21 illustrates a voltmeter being used to measure a battery's voltage. A voltmeter is connected across a device, as illustrated in Figure 2-22(a) and (b), with the two leads of the voltmeter touching either side of the component, in this case a light bulb. Figure 2-22(a) indicates how to measure the voltage across light bulb 1 (L1), whereas Figure 2-22(b) indicates how to measure the voltage across light bulb 2 (L2).

Voltmeter
Instrument designed to measure the voltage or potential difference. Its scale can be graduated in kilovolts, volts, or millivolts.

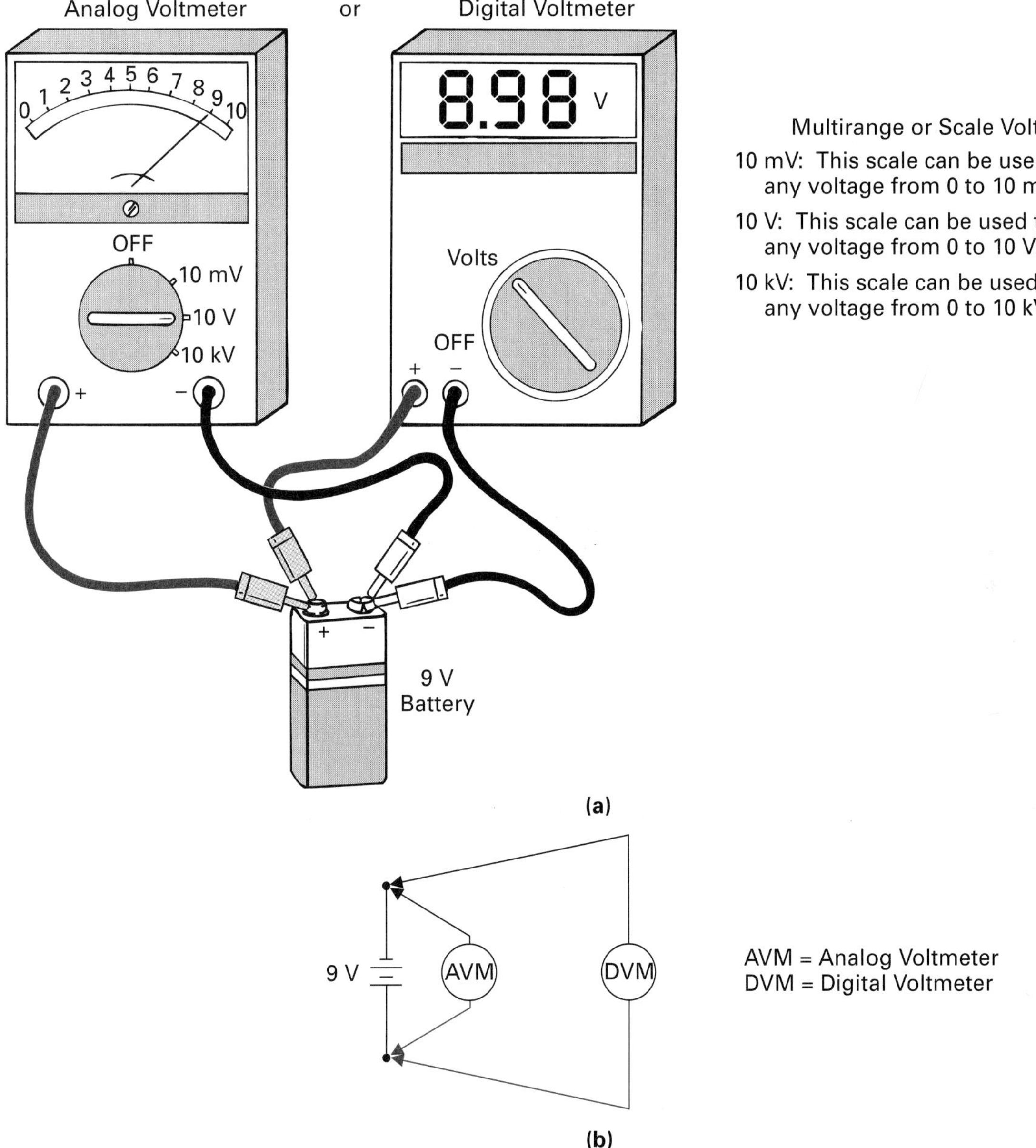

FIGURE 2-21 Using the Voltmeter to Measure Voltage. (a) Pictorial. (b) Schematic.

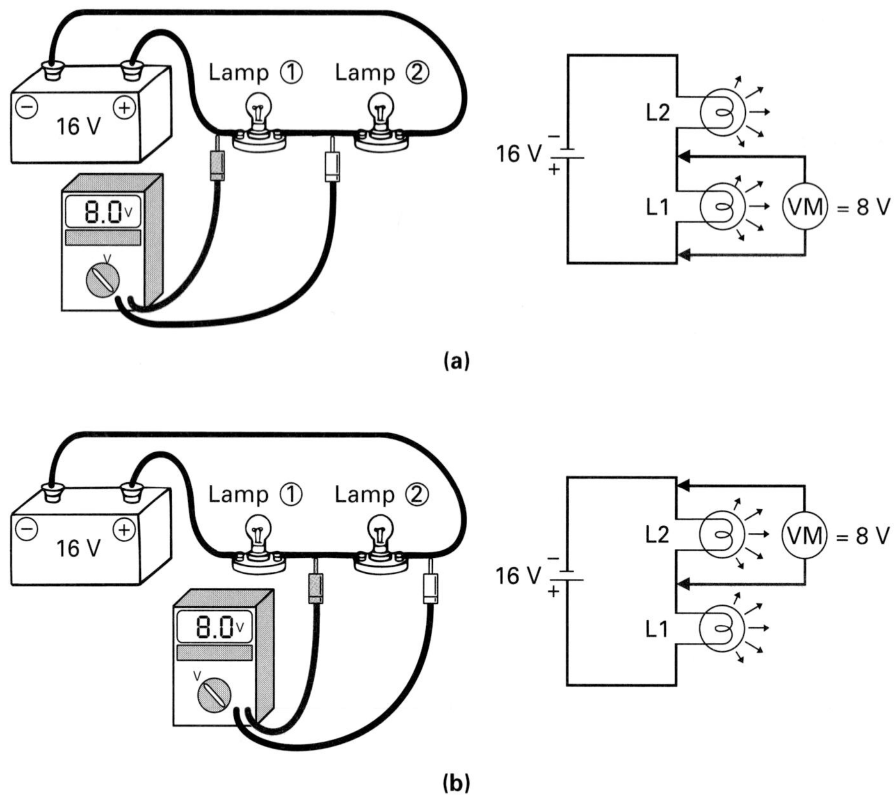

FIGURE 2-22 **Measuring Components across (a) Lamp 1 and (b) Lamp 2.**

Three techniques must be remembered and applied when measuring voltage:

1. To prevent meter damage, always set to the highest scale first (kV) and then reduce as needed to the volt or millivolt range, just in case a larger voltage than anticipated is within the circuit.
2. Always connect the positive lead (red) of the meter to the component lead that is closer to the positive side of the battery, and the negative lead (black) of the meter to the component lead that is closer to the negative side of the battery.
3. If you are measuring the voltage across a component, the voltmeter must be connected across the component and outside of the circuit.

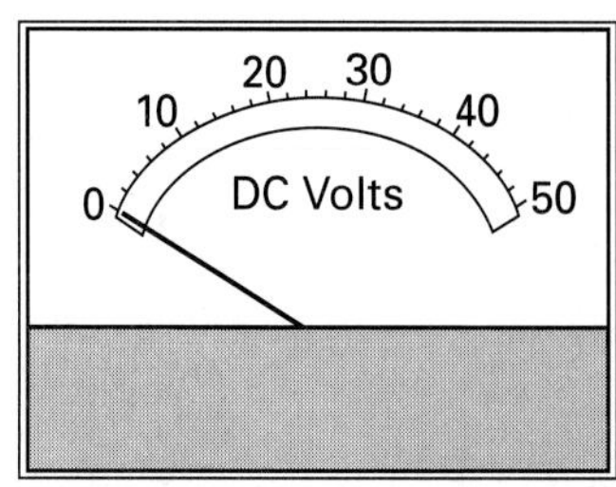

FIGURE 2-23 **Fixed-Range Voltmeter.**

In some applications, fixed-range voltmeters are used: for example, a voltage indicator within a car showing the condition of the battery (Figure 2-23). In this application, a fixed voltmeter of 0 to 15 V would be sufficient.

2-3-4 *Fluid Analogy of Current and Voltage*

In Figure 2-24(a), a system using a pump, pipes, and a waterwheel is being used to convert electrical energy into mechanical energy. The electrical energy is in the form of voltage, which is applied to the pump to send it into operation and cause water to flow. The pump generates:

1. A high pressure at the outlet port, which forces the water molecules out and into the system
2. A low pressure in the inlet port, which attracts the water molecules into the pump

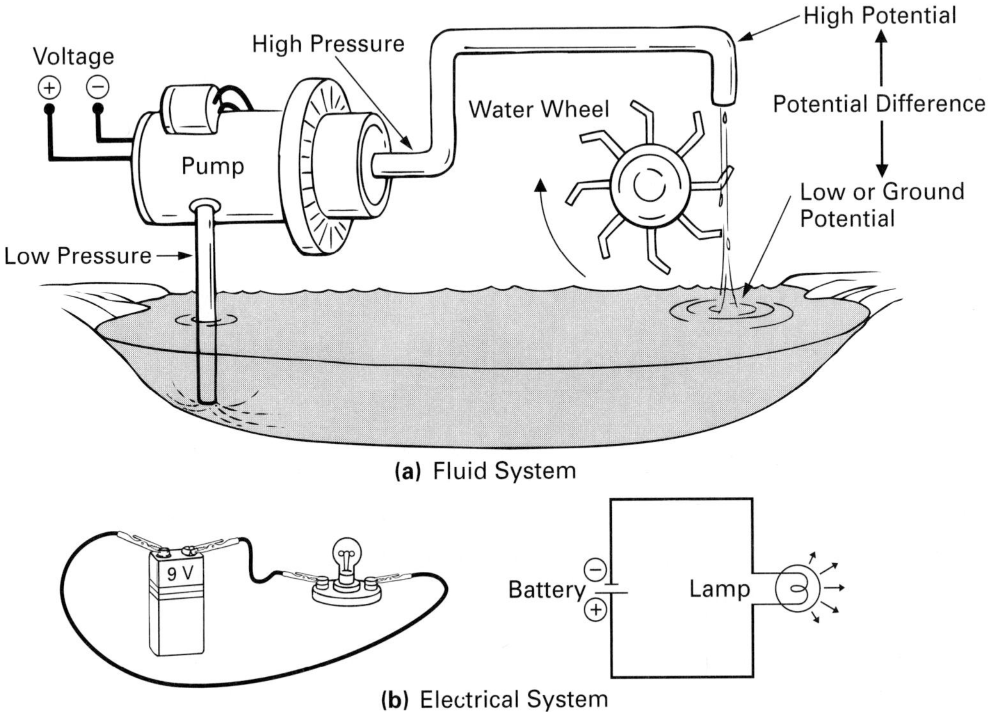

(a) Fluid System		(b) Electrical System
Pump generates pressure, which is the water moving force.	Equivalent to (≅)	Battery generates voltage, which is the electron moving force.
Water current flow.		Electron current flow.
High pressure or potential.		High voltage or potential.
Low pressure or potential.		Low voltage or potential.

FIGURE 2-24 Comparison between a Fluid System (a) and an Electrical System (b).

The water current flow is in the direction indicated, and the high pressure or potential within the piping will be used to drive the waterwheel around, producing mechanical energy. The remaining water is attracted into the pump due to the suction or low pressure existing on the inlet port. In fact, the amount of water entering the inlet port is the same as the amount of water leaving the outlet port. It can therefore be said that the water flow rate is the same throughout the circuit. The only changing element is the pressure felt at different points throughout the system.

In Figure 2-24(b), an electric circuit containing a battery, conductors, and a bulb is being used to convert electrical energy into light energy. The battery generates a voltage just as the pump generates pressure. This voltage causes electrons to move through conductors, just as pressure causes water molecules to move through the piping. The amount of water flow is dependent on the pump's pressure, and the amount of current or electron flow is dependent on the battery's voltage. Water flow through the wheel can be compared to current flow through the bulb. The high pressure is lost in turning the wheel and producing mechanical energy, just as voltage is lost in producing light energy out of the bulb. We cannot say that pressure or voltage flows: Pressure and voltage exist and cause water and current to flow, and it is this flow that is converted to mechanical energy in our fluid system and light energy in our electrical system. Voltage is the force of repulsion and attraction needed to cause current to flow through a circuit, and without this potential difference or pressure there cannot be current.

2-3-5 *Current Is Directly Proportional to Voltage*

Referring to the fluid system and then the electrical circuit in Figure 2-24, you can easily see that the flow is proportional to the pressure, which means: If the pump were to generate a greater pressure, a larger amount of water would flow through the system.

pressure $\uparrow$ water flow $\uparrow$

Similarly, if a larger voltage were applied to the electrical circuit, this larger electron moving force (emf) would cause more electrons or current to flow through the circuit, as shown in Figure 2-25.

Current (I) is therefore said to be *directly proportional* ($\propto$) to *voltage (V)*, as a voltage decrease ($V\downarrow$) results in a subsequent current decrease ($I\downarrow$), and similarly, a voltage increase ($V\uparrow$) causes a current increase ($I\uparrow$).

Current is directly proportional to voltage ($I \propto V$)

$V\downarrow$ causes a $I\downarrow$

$V\uparrow$ causes a $I\uparrow$

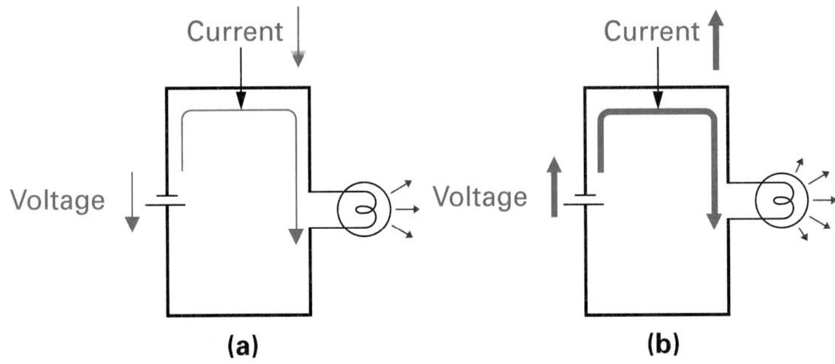

FIGURE 2-25 **(a) Small Voltage Produces a Small Current. (b) Large Voltage Produces a Large Current.**

As you can see from this example, directly **proportional** ($\propto$) is a phrase which means that one term will change in proportion, or in size, relative to another term.

Proportional
Having the same or constant ratio.

2-3-6 *Summary of Measuring Current and Voltage*

CURRENT (AMMETER)	*VOLTAGE (VOLTMETER)*
1. Always set the meter to the highest scale first.	
2. Ensure that the positive lead of the meter is connected closer to the positive side of the voltage supply (for example, battery), and the negative lead of the meter is connected closer to the negative side of the supply voltage.	
3. Always connect the ammeter in the path of current flow so that the electrons have to pass through the meter.	**3.** Always connect the voltmeter across the component so that the pressure or potential difference change across the component can be measured.

Figure 2-26 summarizes these measurements.

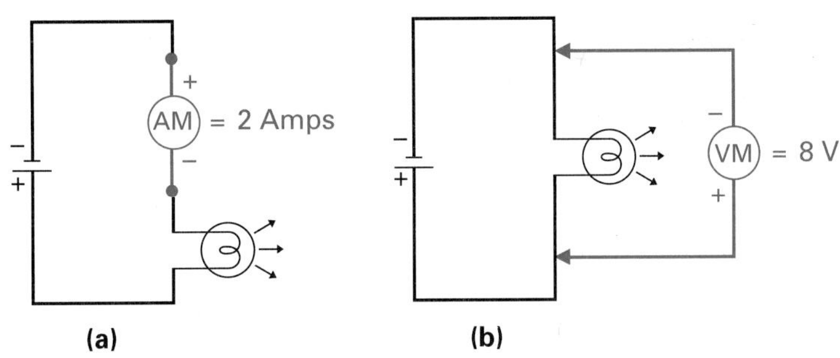

FIGURE 2-26 **Measurement of (a) Current and (b) Voltage.**

SELF-TEST REVIEW QUESTIONS FOR SECTION 2-3

1. What is the unit of voltage?
2. Convert 3 MV to kilovolts.
3. List the three rules that should be applied when using the voltmeter to measure voltage.
4. What is the relationship between current and voltage?

2-4 CONDUCTORS

A lightning bolt that splits or sets fire to a tree and the operation of your calculator are both electrical results achieved by the flow of electrons. The only difference is that your calculator's circuits control the flow of electrons, unlike the lightning bolt, which is the uncontrolled flow of electrons.

Materials that pass current easily, that is, offer little opposition (low resistance) to current, are called **conductors.** As mentioned previously, the atom has a maximum of seven orbital paths known as shells, which are named K, L, M, N, O, P, and Q, stepping out toward the outermost or valence shell.

Conductors
Length of wire whose properties are such that it will carry an electric current.

Conductors are materials or natural elements whose valence electrons can easily be removed from their parent atoms. They are therefore said to be sources of free electrons, and these free electrons provide us with circuit current. The precious metals of silver and gold are the best conductors. More specifically, a better conductor has:

1. Electrons in shells farthest away from the nucleus; these electrons feel very little nucleus attraction and can be broken away from their atom quite easily.
2. More electrons per atom.
3. An incomplete valence shell. This means that the valence shell does not have in it the maximum possible number of electrons. If the atom had its valence ring complete (full), there would be no holes (absence of an electron) in that shell, so no encouragement for adjacent atom electrons to jump from their parent atom into the next atom would exist, preventing the chain reaction known as current.

Economy must be considered when choosing a conductor. Large quantities of conductors using precious metals are obviously going to send the cost of equipment beyond reach. The conductor must also satisfy some physical requirements, in that we must be able to shape it into wires of different sizes and easily bend it to allow us to connect one circuit to the next.

Copper is the most commonly used conductor, as it meets the following three requirements:

1. It is a good source of electrons.
2. It is inexpensive.
3. It is physically pliable.

Aluminum is also a very popular conductor, and although it does not possess as many free electrons as copper, it has the two advantages of being less expensive and lighter than copper.

2-4-1 *Conductance*

Conductance is the measure of how good a conductor is at carrying current. Conductance (symbolized G) is equal to the reciprocal of resistance and is measured in the unit siemens (S):

Conductance (*G*) Measure of how well a circuit or path conducts or passes current. It is equal to the reciprocal of resistance.

$$\text{conductance } (G) = \frac{1}{\text{resistance } (R)}$$

This means that conductance is inversely proportional to resistance. For example, if the opposition to current flow (resistance) is low, the conductance is high and the material is said to have a good conductance.

$$\text{high conductance } G\uparrow = \frac{1}{R\downarrow \text{ (low resistance)}}$$

Conductance (G) values are given in siemens (S) and resistance (R) in ohms (Ω).

On the other hand, if the resistance of a conducting wire is high ($R\uparrow$), its conductance value is low ($G\downarrow$) and it is called a poor conductor. To summarize, we say that a good conductor has a high conductance value and a very small opposition or resistance to current flow.

EXAMPLE:

A household electric blanket offers 25 ohms of opposition or resistance against current flow. Calculate the conductance of the electric blanket's heating element.

Solution:

$$\text{conductance} = \frac{1}{\text{resistance}}$$

$$G = \frac{1}{R}$$

$$= \frac{1}{25 \text{ ohms}}$$

$$= 40 \text{ mS (millisiemens)}$$

EXAMPLE:

To further reinforce our understanding of a conductor, let us compare a good conductor (copper) to a poor conductor (carbon) as an example. Refer also to carbon and copper in the atomic periodic table (Table 2-1).

TABLE 2-4 Relative Conductivity of Conductors

MATERIAL	CONDUCTIVITY (RELATIVE)
Silver	1.000
Copper	0.945
Gold	0.652
Aluminum	0.575
Tungsten	0.297
Nichrome	0.015

Solution:

Carbon

1. The valence ring is the L or second shell, which feels a strong nucleus attractive force, discouraging the release of free electrons and thus of circuit current.
2. Only six electrons orbit the carbon atom, making it a poor source of electrons.
3. The valence shell (L, maximum eight electrons) is already half complete (four electrons), so few holes (only four) exist, discouraging the jumping of electrons from parent atoms into adjacent atom holes.

Copper

1. The valence ring is the N or fourth shell, which feels only a weak nucleus attractive force, encouraging the release of free electrons and thus of circuit current.
2. A total of 29 electrons orbit the nucleus, making it a good source of electrons.
3. The valence shell (N, maximum 32 electrons) is incomplete, as only one electron occupies it when neutral, so 31 holes exist, encouraging the jumping of electrons from parent atoms into adjacent atom holes.

To express how good or poor a conductor is, we must specify a reference point. The reference point we use is the best conductor, silver, which has a conductivity value of 1.0. Table 2-4 lists other conductors and their relative conductivity values with respect to the best, silver. The formula for calculating relative conductivity is

$$\text{Relative Conductivity} = \frac{\text{Conductor's Relative Conductivity}}{\text{Reference Conductor's Relative Conductivity}}$$

EXAMPLE:

What is the relative conductivity of tungsten if copper is used as the reference conductor?

Solution:

$$\text{tungsten} = 0.297$$
$$\text{copper} = 0.945$$
$$\text{relative conductivity} = \frac{\text{conductivity of conductor}}{\text{conductivity of reference}} = \frac{0.297}{0.945} = 0.314$$

CALCULATOR SEQUENCE

Step	Keypad Entry	Display Response
1.	[1] [6] [E] (exponent) [3]	16E3
2.	[÷]	
3.	[2] [E] [6] (1000 + K = E6)	2E6
4.	[=]	0.008

EXAMPLE:

What is the relative conductivity of silver if copper is used as the reference?

Solution:

$$\text{relative conductivity} = \frac{\text{silver}}{\text{copper}} = \frac{1.000}{0.945} = 1.058$$

SELF-TEST REVIEW QUESTIONS FOR SECTION 2-4

1. True or false: A conductor is a material used to block the flow of current.
2. List the three atomic properties that make a better conductor.
3. What is the most commonly used conductor in the field of electronics?
4. Calculate the conductance of a 35 ohm heater element.

2-5 INSULATORS

Materials that are used to block current, that is, offer high resistance or opposition to current flow, are called **insulators.** Just as certain materials permit the easy flow of current and so have good conductivity, certain materials allow small to almost no amount of free electrons to flow. These materials are known as insulators. Insulators can, with sufficient pressure or voltage applied across them, "break down" and conduct current; that is, the voltage, known as the **breakdown voltage,** must be great enough to dislodge the electrons from their close orbital shells (K, L shells) and send them off as free electrons.

Insulators
Materials that have few electrons per atom and those electrons are close to the nucleus and cannot be easily removed.

Breakdown Voltage
The voltage at which breakdown of an insulator occurs.

Dielectric Strength The maximum potential a material can withstand without rupture.

A good insulator or dielectric should have the maximum possible resistance and conduct no current at all. To express how good or poor an insulator is, we list a voltage that when applied across one-thousandth of an inch of this insulator material, will cause it to break down and conduct a large current. This measure of an insulator is known as its **dielectric strength.** Table 2-5 lists some of the more popular insulators and the value of kilovolts that will cause a centimeter of insulator to break down. From Table 2-5 we say that if 1 centimeter of paper is connected to a variable voltage source, a voltage of 500 kilovolts is needed to break down the paper and cause current to flow.

To calculate the dielectric thickness needed therefore to withstand a certain voltage, we can use the following formula.

$$\text{dielectric thickness} = \frac{\text{voltage to insulate}}{\text{insulator's breakdown voltage}}$$

EXAMPLE:

What thickness of mica would be needed to withstand 16,000 V?

Solution:

CALCULATOR SEQUENCE

Step	Keypad Entry	Display Response
1.	0 . 2 9 7	0.297
2.	÷	
3.	0 . 9 4 5	0.945
4.	=	0.314

$$\text{mica strength} = 2000 \text{ kV/cm}$$

$$\text{dielectric thickness} = \frac{16{,}000 \text{ V}}{2000 \text{ kV}} = 0.008 \text{ cm}$$

TABLE 2-5 Breakdown Voltages of Certain Insulators

MATERIAL	BREAKDOWN STRENGTH (kV/cm)
Mica	2000
Glass	900
Teflon	600
Paper	500
Rubber	275
Bakelite	151
Oil	145
Porcelain	70
Air	30

EXAMPLE:

What maximum voltage could 1 mm of air withstand?

Solution:

There are 10 mm in 1 cm. If air can withstand 30,000 V/cm, it can withstand 3000 V/mm.

SELF-TEST REVIEW QUESTIONS FOR SECTION 2-5

1. True or false: An insulator is a material used to block the flow of current.
2. What is considered to be the best insulator material?
3. Define *breakdown voltage.*
4. Would the conductance figure of a good insulator be large or small?

2-6 SEMICONDUCTORS

Certain materials are neither insulators (high resistance to current) nor conductors (low resistance to current), but in fact fall between the two, as seen in Figure 2-27. These materials are known as **semiconductors** because they conduct less than a metal conductor, but more than an insulator. Silicon is the most commonly used semiconductor for components such as transistors and integrated

Semiconductors Conductor with a resistivity figure between that of an insulator and a metal.

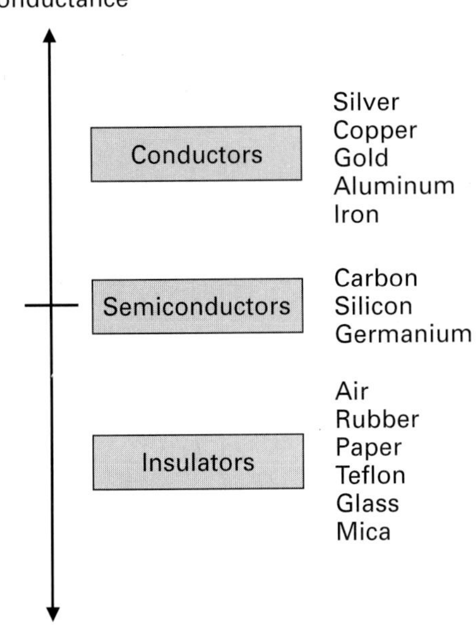

FIGURE 2-27 **Elements and Compounds Used in Electronics.**

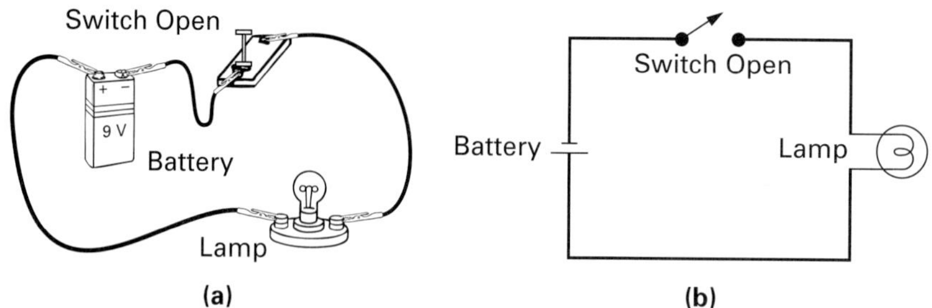

FIGURE 2-28 An Open Switch Causing an Open Circuit. (a) Pictorial. (b) Schematic.

circuits. Semiconductor materials, devices, and circuits are covered in more detail in chapters 19 through 29.

2-7 THE OPEN, CLOSED, AND SHORT CIRCUIT

A simple circuit using a battery, conductors, light bulb, and a switch is shown in Figure 2-28. A new component, the switch, is illustrated in Figure 2-28(a), and the schematic symbol is shown in Figure 2-28(b).

2-7-1 *Open Circuit (Open Switch)*

Open Circuit
Break in the path of current flow.

An opened switch (Figure 2-28) produces an "open" in the circuit, which prevents current flow as the extra electrons in the negative terminal of the battery cannot feel the attraction of the positive terminal due to the break in the path. The opened switch has produced an **open circuit.**

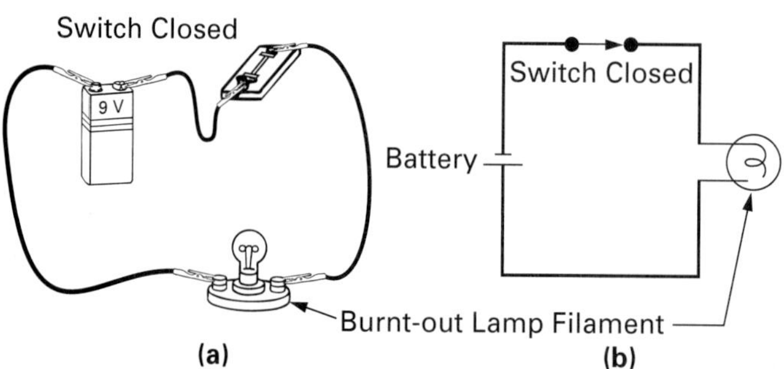

FIGURE 2-29 An Open Lamp Filament Causing an Open Circuit. (a) Pictorial. (b) Schematic.

2-7-2 Open Circuit (Open Component)

If the switch is now closed so as to make a complete path in the circuit, an open could still exist due to the failure of one of the components. For example, in Figure 2-29, the light bulb filament has burned out; in so doing, it creates an open in the circuit and again there will be no current flow. An open component therefore can also produce an open circuit.

2-7-3 Closed Circuit (Closed Switch)

A closed switch produces a **closed circuit** since current now has a complete path from the negative to positive terminal, as seen in Figure 2-30. The closed switch produces a closed circuit.

Closed Circuit
Circuit having a complete path for current to flow.

2-7-4 Short Circuit (Shorted Component)

A **short circuit** normally occurs when one point is accidentally connected to another. Figure 2-31(a) illustrates the physical appearance of the circuit and how the accident occurred. A set of pliers was accidentally laid across the two contacts connecting the light bulb. Figure 2-31(b) shows the schematic illustration with the effect of the short across the light bulb drawn in. All the current will flow through the metal of the pliers, which offers no resistance or opposition to the current flow; since the light bulb does have some resistance or opposition, almost no current will flow through it, and therefore no light will be produced.

Short Circuit
Also called a short; it is a low-resistance connection between two points in a circuit, typically causing a large amount of current flow.

SELF-TEST REVIEW QUESTIONS FOR SECTIONS 2-6 AND 2-7

1. What is a semiconductor?
2. What is the name of the most commonly used semiconductor material?
3. Does current flow through an open circuit?
4. Describe the difference between a closed circuit and a short circuit.

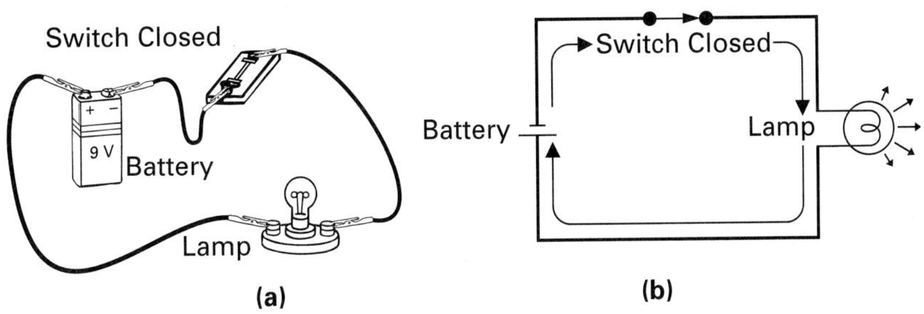

FIGURE 2-30 **Closed Switch Causing a Closed Circuit. (a) Pictorial. (b) Schematic.**

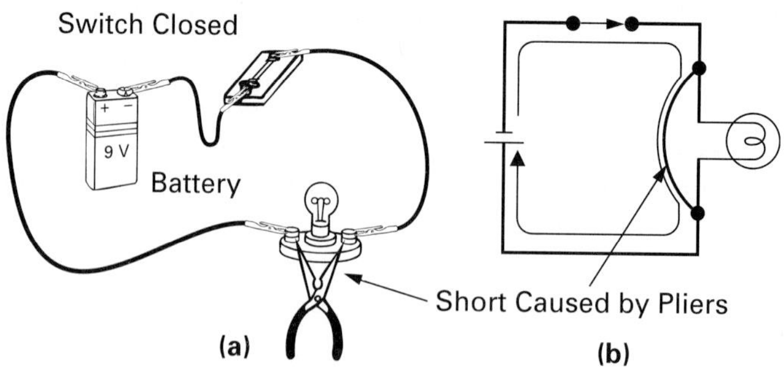

FIGURE 2-31 Pliers Causing a Short Circuit across a Lamp. (a) Pictorial. (b) Schematic.

SUMMARY

The Structure of Matter

1. All of the matter on the earth and in the air surrounding the earth can be classified as being either a solid, liquid, or gas. A total of approximately 107 different natural elements exist in, on, and around the earth.
2. An *element,* by definition, is a substance consisting of only one type of atom; in other words, every element has its own distinctive atom, which makes it different from all the other elements.
3. The *atom* is the smallest particle into which an element can be divided without losing its identity, and a group of identical atoms is called an element.
4. The word atom is a Greek word meaning a particle that is too small to be subdivided.
5. In 1913, a Danish physicist, Neils Bohr, put forward a theory about the atom, and his basic model outlining the *subatomic* particles that make up the atom is still in use today.
6. The three important particles of the atom are the *proton,* which has a positive charge, the *neutron,* which is neutral or has no charge, and the *electron,* which has a negative charge.
7. The atom consists of a positively charged central mass called the *nucleus,* which is made up of protons and neutrons surrounded by a quantity of negatively charged orbiting electrons.
8. The *atomic number* of an atom describes the number of protons that exist within the nucleus.
9. The proton and the neutron are almost 2000 times heavier than the very small electron.
10. A hydrogen atom is the smallest of all atoms and has an atomic number of 1, which means that hydrogen has a one-proton nucleus.
11. Helium is second on the table and has an atomic number of 2, indicating that two protons are within the nucleus. The *atomic weight* of helium, however, is 4, meaning that two protons and two neutrons make up the atom's nucleus.

12. The number of neutrons within an atom's nucleus can subsequently be calculated by subtracting the atomic number (protons) from the atomic weight (protons and neutrons).

13. A *neutral atom* or *balanced atom* is one that has an equal number of protons and orbiting electrons, so the net positive proton charge is equal but opposite to the net negative electron charge, resulting in a balanced or neutral state.

14. Orbiting electrons travel around the nucleus at varying distances from the nucleus, and these orbital paths are known as *shells or bands.* The orbital shell nearest the nucleus is referred to as the first or K shell. The second is known as the L, the third is M, the fourth is N, the fifth is O, the sixth is P, and the seventh is referred to as the Q shell. There are seven shells available for electrons (K, L, M, N, O, P, and Q) around the nucleus, and each of these seven shells can only hold a certain number of electrons. The outermost electron-occupied shell is referred to as the *valence shell or ring,* and these electrons are termed *valence electrons.*

15. All matter exists in one of three states: solids, liquids, and gases. The atoms of a solid are fixed in relation to one another but vibrate in a back-and-forth motion, unlike liquid atoms, which can flow over each other. The atoms of a gas move rapidly in all directions and collide with one another.

Laws of Attraction and Repulsion

16. *Like charges* (positive and positive or negative and negative) repel one another. *Unlike charges* (positive and negative or negative and positive) attract one another. Orbiting negative electrons are therefore attracted toward the positive nucleus, which leads us to the question of why the electrons do not fly into the atom's nucleus. The answer is that the orbiting electrons remain in their stable orbit due to equal but opposite forces. The centrifugal outward force exerted on the electrons due to the orbit counteracts the attractive inward force trying to pull the electrons toward the nucleus due to the unlike charges.

17. Due to their distance from the nucleus, valence electrons are described as being loosely bound to the atom. The electrons can, therefore, easily be dislodged from their outer orbital shell by any external force, to become a free electron.

The Molecule

18. An atom is the smallest unit of a natural element, or an element is a substance consisting of a large number of the same atom.

19. Combinations of elements are known as *compounds,* and the smallest unit of a compound is called a *molecule,* just as the smallest unit of an element is an atom.

20. Water is an example of a liquid compound in which the molecule (H_2O) is a combination of an explosive gas (hydrogen) and a very vital gas (oxygen). Table salt is another example of a compound; here the molecule is made up of a highly poisonous gas atom (chlorine) and a potentially explosive solid atom (sodium).

Exponents

21. Many of the values inserted into formulas contain numbers that have *exponents.* An exponent is a smaller number that appears to the right and slightly higher than another number.
22. *Exponents allow us to save space,* as you can see in this example:

$$5 \times 5 \times 5 \times 5 = 5^4$$

Here the exponent (^{4}) is used to indicate how many times its base (5) is used as a *factor.* You can also see from this example that the other advantage is that *exponents are a sort of math shorthand.* It is much easier to write 5^4 than to write $5 \times 5 \times 5 \times 5$. The "$y^x$" key on your calculator allows you to enter in any base number (y) and any exponent (^{x}).

Scientific Notation

23. Many of the sciences, such as electronics, deal with numbers that contain a large number of zeros. By using exponents we can eliminate the large number of zeros to obtain a shorthand version of the same number. This method is called *scientific notation* or *powers of 10.*
24. A *negative exponent* tells you how many places *to the left* to move the decimal point. A *positive exponent* tells you how many places *to the right* to move the decimal point. Remember one important point: that a negative exponent does not indicate a negative number; it simply indicates a fraction.

Engineering Notation

25. *Scientific notation* is expressed as a base number between 1 and 10 with an exponent that is a power of 10. *Engineering notation,* however, is expressed as a base number that is greater than 1 with an exponent that is some multiple of three.

Prefixes

26. Although any power of 10 can be used (10^1, 10^2, 10^{-5}, 10^{-2}), most electrical or electronic values use exponents that are multiples of three (10^3, 10^6, 10^9, 10^{12}, or 10^{-3}, 10^{-6}, 10^{-9}, 10^{-12}). These multiples have been given names and symbols, as shown below.

VALUE	NAME	SYMBOL	
10^{12} = 1,000,000,000,000	tera	T	Multipliers that are greater than 1
10^{9} = 1,000,000,000	giga	G	
10^{6} = 1,000,000	mega	M	
10^{3} = 1,000	kilo	k	
10^{-3} = 1/1,000	milli	m	Multipliers that are less than 1 (fractions)
10^{-6} = 1/1,000,000	micro	μ	
10^{-9} = 1/1,000,000,000	nano	n	
10^{-12} = 1/1,000,000,000,000	pico	p	

These multipliers are often called *prefixes,* because they precede units.

Current (Figure 2-32)

27. The movement of electrons from one point to another is known as *electrical current.*

28. Energy in the form of heat or light can cause an outer shell electron to be released from the valence shell of an atom. Once an electron is released, the atom is no longer electrically neutral and is called a *positive ion,* as it now has a net positive charge (more protons than electrons). The released electron tends to jump into a nearby atom, which will then have more electrons than protons and is referred to as a *negative ion.*

29. Millions upon millions of atoms within the conductor pass a continuous movement of billions upon billions of electrons from right to left. This electron flow is known as *electric current.*

30. As long as a force or pressure, produced by the positive charge and negative charge, exists it will cause electrons to flow from the negative to the positive terminal. The positive side has a deficiency of electrons and the negative side has an abundance, and so a continuous flow or migration of electrons takes place between the negative and positive terminal through our metal conducting wire. This electric current or electron flow is a measurable quantity.

31. There are 6.24×10^{18} electrons in 1 *coulomb of charge.*

32. A coulomb is a *static* amount of electric charge. In electronics, we are more interested in electrons in motion, and current is the flow of electrons, which involves time. Coulombs and time are therefore combined to describe the number of electrons and the rate at which they flow. This relationship is called *current (I)* flow and has the unit of *amperes* (A). If 6.24×10^{18} electrons (1 C) were to drift past a specific point on a conductor in 1 second of time, 1 ampere of current is said to be flowing.

33. Current within electronic equipment is normally a value in milliamperes or microamperes and very rarely exceeds 1 ampere.

QUICK REFERENCE SUMMARY SHEET

FIGURE 2-32 Current.

$$\text{Electric Charge } (Q) \text{ in coulombs } (C) = \frac{\text{total number of electrons } (n)}{6.24 \times 10^{18}}$$

$$\text{Electric Current } (I) \text{ in amperes } (A) = \frac{\text{coulombs } (Q)}{\text{time } (t)}$$

TABLE 2-2 Current Units

NAME	SYMBOL	VALUE
Picoampere	pA	$10^{-12} = \frac{1}{1,000,000,000,000}$
Nanoampere	nA	$10^{-9} = \frac{1}{1,000,000,000}$
Microampere	μA	$10^{-6} = \frac{1}{1,000,000}$
Milliampere	mA	$10^{-3} = \frac{1}{1000}$
Ampere	A	$10^{0} = 1$
Kiloampere	kA	$10^{3} = 1000$
Megaampere	MA	$10^{6} = 1,000,000$
Gigaampere	GA	$10^{9} = 1,000,000,000$
Teraampere	TA	$10^{12} = 1,000,000,000,000$

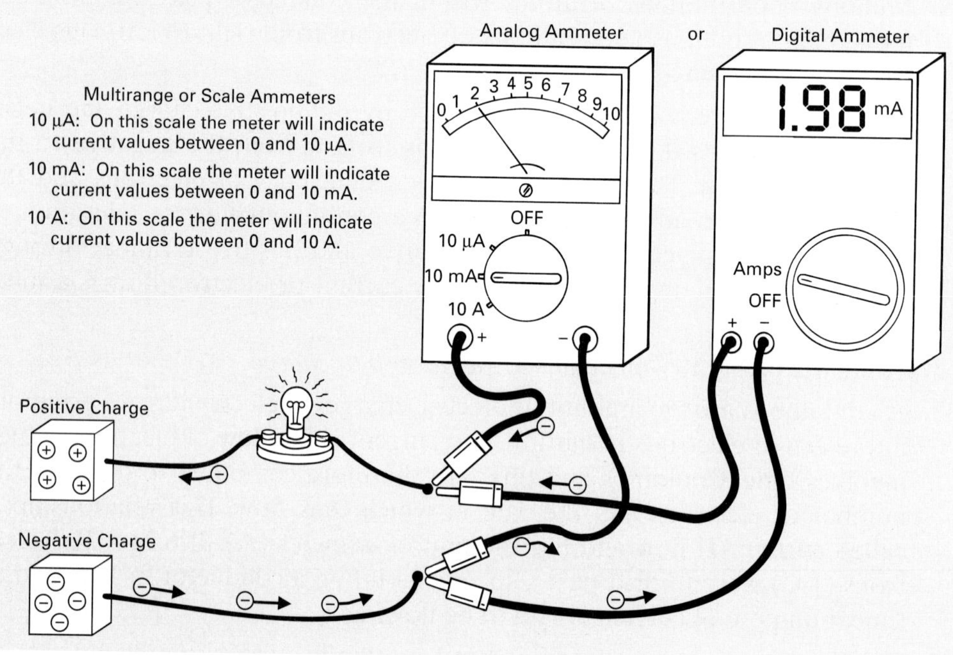

34. Electrons will in fact move very slowly as they jump from parent to adjacent atom; however, the chain reaction occurs at the speed of light, which is 186,000 miles per second or 300,000,000 meters (m) per second (3×10^8 m/s).

35. Electrons drift from a negative to a positive charge. This current is known as *electron flow*. In the eighteenth and nineteenth centuries, however, when very little was known about the atom, researchers believed wrongly that current was a flow of positive charges. Although this has now been proved incorrect many texts still use *conventional current flow*. Whether conventional current flow or electron flow is used, the same answers to problems, measurements, or designs are obtained. The key point to remember is that direction is not important, but the amount of current flow is.

36. *Ammeters* (ampere meters) are used to measure the current flow within a circuit. In electronics, where current is generally small, the milliampere (mA) or microampere (μA) scale will generally be used, whereas in electrical high-current circuits, the meter will generally be used on an ampere scale.

37. Three rules must be remembered and applied when measuring current:
    a. Always set the selector to the higher scale first (amperes) and then reduce as needed to milli- or microamperes, just in case a larger current than anticipated is within the circuit, in which case the meter could be damaged.
    b. The current meter must be connected so that the positive lead of the ammeter (red) is connected to the positive charge or polarity and the negative lead of the ammeter (black) is connected to the negative charge or polarity. Ammeters are sensitive to the polarity (positive or negative) of the charge.
    c. If you wish to measure current flowing within a wire, the wire must be opened so that the ammeter measuring the current will be placed in the path of current flow.

Voltage (Figure 2-33)

38. *Voltage* is the force or pressure exerted on electrons.

39. A highly concentrated charge produces a high voltage, whereas a low concentrated charge produces a low voltage. Voltage is also appropriately known as the electron moving force or *electromotive force (emf)*, and since two opposite potentials exist, one negative and one positive, the strength of the voltage can also be referred to as the amount of potential difference (PD) applied across the circuit.

40. Voltage is the force, pressure, *potential difference (PD)*, or electromotive force (emf) that causes electron flow or current and is symbolized by italic uppercase *V*. The unit for voltage is the volt, symbolized by roman uppercase V.

41. A *battery* converts chemical energy into electrical energy. At the positive terminal of the battery, positive charges or ions (atoms with more protons than electrons) are present, and at the negative terminal, negative charges

QUICK REFERENCE SUMMARY SHEET

FIGURE 2-33 Voltage.

TABLE 2-3 Voltage Units

NAME	SYMBOL	VALUE
Picovolts	pV	$10^{-12} = \frac{1}{1,000,000,000,000}$
Nanovolts	nV	$10^{-9} = \frac{1}{1,000,000,000}$
Microvolts	μV	$10^{-6} = \frac{1}{1,000,000}$
Millivolts	mV	$10^{-3} = \frac{1}{1000}$
Volts	V	$10^{0} = 1$
Kilovolts	kV	$10^{3} = 1000$
Megavolts	MV	$10^{6} = 1,000,000$
Gigavolts	GV	$10^{9} = 1,000,000,000$
Teravolts	TV	$10^{12} = 1,000,000,000,000$

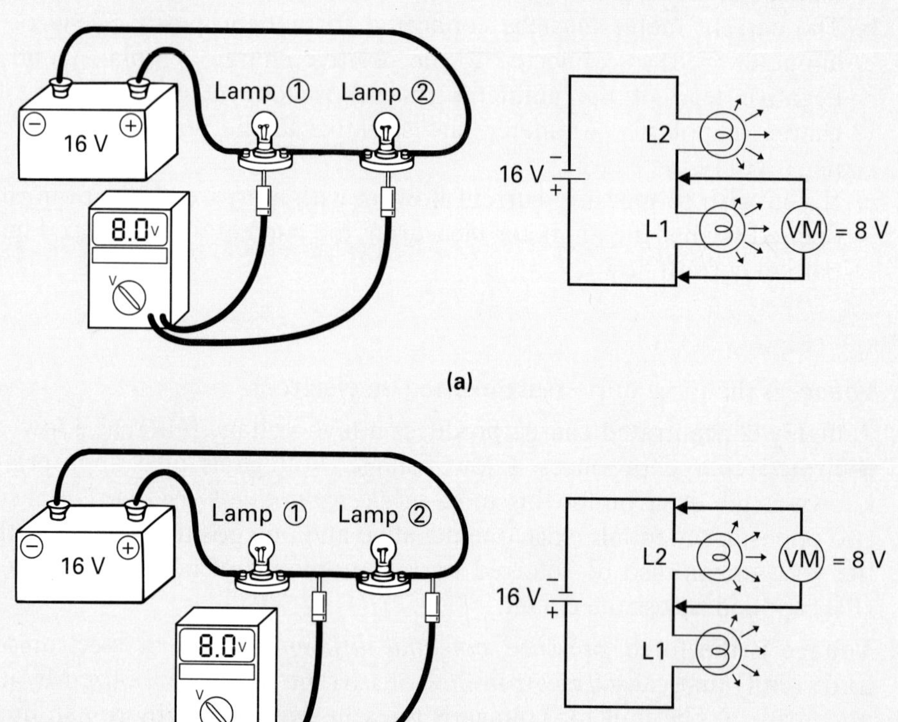

Current (*I*) is directly proportional (∝) to voltage (*V*).

$$I \propto V$$

or ions (atoms with more electrons than protons) are available to supply electrons for current flow within a circuit.

42. Voltage within electronic equipment is normally measured in volts, whereas heavy-duty industrial equipment normally requires high voltages that are generally measured in kilovolts (kV).
43. *Voltmeters* (voltage meters) are used to measure electrical pressure (voltage) anywhere in an electronic circuit. A voltmeter is connected across a device with the two leads of the voltmeter touching either side of the component.
44. Three techniques must be remembered and applied when measuring voltage:
    - **a.** To prevent meter damage, always set to the highest scale first (kV) and then reduce as needed to the volt or millivolt range, just in case a larger voltage than anticipated is within the circuit.
    - **b.** Always connect the positive lead (red) of the meter to the component lead that is closer to the positive side of the battery, and the negative lead (black) of the meter to the component lead that is closer to the negative side of the battery.
    - **c.** If you are measuring the voltage across a component, the voltmeter must be connected across the component and outside of the circuit.
45. If a larger voltage were applied to the electrical circuit, this larger electron moving force (emf) would cause more electrons or current to flow through the circuit. *Current (I)* is therefore said to be *directly proportional* ($\propto$) to *voltage (V),* as a voltage decrease ($V\downarrow$) results in a subsequent current decrease ($I\downarrow$), and similarly, a voltage increase ($V\uparrow$) causes a current increase ($I\uparrow$).
46. Directly *proportional* ($\propto$) is a phrase which means that one term will change in proportion, or in size, relative to another term.

Conductors and Insulators (Figure 2-34)

47. Materials that pass current easily, that is, offer little opposition (low resistance) to current, are called *conductors.* Conductors are materials or natural elements whose valence electrons can easily be removed from their parent atoms. The precious metals of silver and gold are the best conductors.
48. Copper is the most commonly used conductor, as it meets the following three requirements:
    - **a.** It is a good source of electrons.
    - **b.** It is inexpensive.
    - **c.** It is physically pliable.
49. *Conductance* is the measure of how good a conductor is at carrying current. Conductance (symbolized G) is equal to the reciprocal of resistance and is measured in the unit siemens (S): If the opposition to current flow (resistance) is low, the conductance is high and the material is said to have a good conductance. On the other hand, if the resistance of a conducting wire is high ($R\uparrow$), its conductance value is low ($G\downarrow$) and it is called a poor con-

QUICK REFERENCE SUMMARY SHEET

FIGURE 2-34 Conductors and Insulators.

Conductors

$$\text{Conductance } (G) = \frac{1}{\text{Resistance } (R)}$$

$$\text{Relative Conductivity} = \frac{\text{Conductor's Relative Conductivity}}{\text{Reference Conductor's Relative Conductivity}}$$

TABLE 2-4 Relative Conductivity of Conductors

MATERIAL	CONDUCTIVITY (RELATIVE)
Silver	1.000
Copper	0.945
Gold	0.652
Aluminum	0.575
Tungsten	0.297
Nichrome	0.015

Insulators

$$\text{Dielectric Thickness} = \frac{\text{Voltage to Insulate}}{\text{Insulator's Breakdown Voltage}}$$

TABLE 2-5 Breakdown Voltages of Certain Insulators

MATERIAL	BREAKDOWN STRENGTH (kV/cm)
Mica	2000
Glass	900
Teflon	600
Paper	500
Rubber	275
Bakelite	151
Oil	145
Porcelain	70
Air	30

ductor. To summarize, we say that a good conductor has a high conductance value and a very small opposition or resistance to current flow.

50. To express how good or poor a conductor is, we must specify a reference point. The reference point we use is the best conductor, silver, which has a conductivity value of 1.0.

51. Materials that are used to block current, that is, offer high resistance or opposition to current flow, are called *insulators.* Just as certain materials permit the easy flow of current and so have good conductivity, certain materials allow small to almost no amount of free electrons to flow. Insulators can, with sufficient pressure or voltage applied across them, "break down" and conduct current. A good insulator or dielectric should have the maximum possible resistance and conduct no current at all. To express how good or poor an insulator is, we list a voltage that when applied across one-thousandth of an inch of this insulator material, will cause it to break down and conduct a large current. This measure of an insulator is known as its *dielectric strength.*

Semiconductors

52. Certain materials are neither insulators (high resistance to current) nor conductors (low resistance to current), but in fact fall between the two. These materials are known as *semiconductors* because they conduct less than a metal conductor, but more than an insulator. Silicon is the most commonly used semiconductor for components such as transistors and integrated circuits.

The Open, Closed, and Short Circuit

53. An opened switch produces an "open" in the circuit, which prevents current flow as the extra electrons in the negative terminal of the battery cannot feel the attraction of the positive terminal due to the break in the path. The opened switch has produced an *open circuit.* If the switch is now closed so as to make a complete path in the circuit, an open could still exist due to the failure of one of the components. An open component can also produce an open circuit. A closed switch produces a *closed circuit* since current now has a complete path from the negative to positive terminal. The closed switch produces a closed circuit.
54. A *short circuit* normally occurs when one point is accidentally connected to another.

NEW TERMS

Ammeter
Ampere
Atom
Battery
Breakdown voltage
Charge
Closed circuit
Compound
Conductance
Conductor
Conventional current flow
Coulomb
Current
Dielectric strength
Electromotive force
Electron
Electron current flow
Element
Insulator
Ion
Molecule
Negative
Neutron

Nucleus	Proton	Siemens
Open circuit	Relative conductivity	Subatomic
Polarity	Resistance	Switch
Positive	Semiconductor	Voltage
Potential difference	Short circuit	Voltmeter

REVIEW QUESTIONS

Multiple-Choice Questions

1. Hydrogen is a:
   a. Gas **b.** Solid **c.** Liquid **d.** Other
2. Which of the following is a liquid?
   a. Calcium **b.** Magnesium **c.** Mercury **d.** Helium
3. The atomic number of an atom describes:
   a. The number of neutrons **c.** The number of nuclei
   b. The number of electrons **d.** The number of protons
4. The most commonly used metal in the field of electronics is:
   a. Silver **b.** Copper **c.** Mica **d.** Gold
5. The smallest unit of an element is:
   a. A compound **b.** An atom **c.** A molecule **d.** A proton
6. The smallest unit of a compound is:
   a. An element **b.** A neutron **c.** An electron **d.** A molecule
7. A water molecule is made up of:
   a. 2 parts hydrogen and 1 part oxygen
   b. Chlorine and sodium
   c. 1 part oxygen and 2 parts sodium
   d. 3 parts chlorine and 1 part hydrogen
8. A negative ion has:
   a. More protons than electrons **c.** More neutrons than protons
   b. More electrons than protons **d.** More neutrons than electrons
9. A positive ion has:
   a. Lost some of its electrons **c.** Lost neutrons
   b. Gained extra protons **d.** Gained more electrons
10. One coulomb of charge is equal to:
    a. 6.24×10^{18} electrons **c.** 6.24×10^{8} electrons
    b. 1018×10^{12} electrons **d.** 6.24×10^{81} electrons
11. If 14 C of charge passes by a point in 7 seconds, the current flow is said to be:
    a. 2 A **b.** 98 A **c.** 21 A **d.** 7 A
12. How many electrons are there within 16 C of charge?
    a. 9.98×10^{19} **b.** 14 **c.** 16 **d.** 10.73×10^{19}

13. Current is measured in:

a. Volts b. Coulombs/second c. Ohms d. Siemens

14. Voltage is measured in units of:

a. Amperes b. Ohms c. Siemens d. Volts

15. Another word used to describe voltage is:

a. Potential difference c. Electromotive force (emf)
b. Pressure d. All of the above

16. Conductors offer a ____________ resistance to current flow.

a. High b. Low c. Medium d. Maximum

17. Conductors have:

a. Electrons in shells farthest away from the nucleus
b. Relatively more electrons per atom
c. An incomplete valence shell
d. All of the above
e. None of the above

18. Conductance is the measure of how good a conductor is at passing current, and is measured in:

a. Siemens b. Volts c. Current d. Ohms

19. Insulators have:

a. Electrons close to the nucleus d. All of the above
b. Relatively few electrons per atom e. None of the above
c. An almost complete valence shell

20. An open circuit will cause:

a. No current flow c. A break in the circuit
b. Maximum current flow d. Both (a) and (c)

Essay Questions

21. Describe the three factors that make a good conductor. (2-4)
22. State the three rules that should be remembered and applied when measuring: (2-3-6)

a. Current b. Voltage

23. Why is current directly proportional to voltage? (2-3-5)
24. List the four most commonly used current units and their values in terms of the basic unit. (2-2-3)
25. What is the speed of light in (a) miles and (b) meters per second? (2-2-4)
26. List the four most commonly used voltage units and their values in terms of the basic unit. (2-3-2)
27. Describe what is meant by: (2-7)

a. An open circuit b. A closed circuit c. A short circuit

28. What is the difference between conventional and electron current flow? (2-2-5)

29. Give the unit and symbol for the following:

a. Voltage (V) is measured in ____________ (__). (2-3).
b. Current (I) is measured in ____________ (__). (2-2-2).
c. Conductance (G) is measured in ____________ (__). (2-4-1)
d. Resistance (R) is measured in ____________ (__). (2-4-1)

30. In relation to the structure of matter and the atom, describe: (2-1)

a. The atom's subatomic particles
b. An element (2-1)
c. A compound (2-1-3)
d. A molecule (2-1-3)
e. A neutral atom (2-1-1)
f. A positive ion (2-2)
g. A negative ion (2-2)

Practice Problems

31. What is the value of conductance in siemens for a 100 ohm resistor?

32. Calculate the total number of electrons in 6.5 C of charge.

33. Calculate the amount of current in amperes passing through a conductor if 3 C of charge passes by a point in 4 s.

34. Convert the following:

a. 0.014 A = ____________ mA
b. 1374 A = ____________ kA
c. 0.776 μA = ____________ nA
d. 0.91 mA = ____________ μA

35. Convert the following:

a. 1473 mV = ____________ V
b. 7143 V = ____________ kV
c. 0.139 kV = ____________ V
d. 0.390 MV = ____________ kV

36. What is the relative conductivity of copper if silver is used as the reference conductor?

37. What minimum thickness of porcelain will withstand 24,000 V?

38. To insulate a circuit from 10 V, what insulator thickness would be needed if the insulator is rated at 750 kV/cm?

39. Convert the following:

a. 2000 kV/cm to ____________ meters
b. 250 kV/cm to ____________ mm

40. What maximum voltage may be placed across 35 mm of mica without it breaking down?

CHAPTER

3

OBJECTIVES

After completing this chapter, you will be able to:

1. Define *resistance* and *ohm.*
2. Explain Ohm's law and its application.
3. Describe why:
   - **a.** Current is proportional to voltage.
   - **b.** Current is inversely proportional to resistance.
4. Explain what factors affect the resistance of conductors.
5. Explain superconductivity and its advantages.
6. List some of the different types of conductors and their connectors.
7. Describe and define the terms *energy, work,* and *power.*
8. Describe how the wattmeter can be used to measure power.
9. Explain what is meant by the term *kilowatt-hour.*

Resistance and Power

O U T L I N E

VIGNETTE

Genius of Chippewa Falls

In 1960, Seymour R. Cray, a young vice-president of engineering for Control Data Corporation, informed president William Norris that in order to build the world's most powerful computer he would need a small research lab built near his home. Norris would have shown any other employee the door, but Cray was his greatest asset, so in 1962 Cray moved into his lab, staffed by 34 people and nestled in the woods near his home overlooking the Chippewa River in Wisconsin. Eighteen months later the press was invited to view the 14- by 6-foot 6600 supercomputer that could execute 3 million instructions per second and contained 80 miles of circuitry and 350,000 transistors, which were so densely packed that a refrigeration cooling unit was needed due to the lack of airflow.

Cray left Control Data in 1972 and founded his own company, Cray Research. Four years later the $8.8 million Cray-1 scientific supercomputer outstripped the competition. It included some revolutionary design features, one of which is that since electronic signals cannot travel faster than the speed of light (1 foot per billionth of a second) the wire length should be kept as short as possible, because the longer the wire the longer it takes for a message to travel from one end to the other. With this in mind, Cray made sure that none of the supercomputer's conducting wires exceeded 4 feet in length.

In the summer of 1985, the Cray-2, Seymour Cray's latest design, was installed at Lawrence Livermore Laboratory. The Cray-2 was 12 times faster than the Cray-1, and its densely packed circuits are encased in clear Plexiglas and submerged in a bath of liquid coolant. The 60-year-old genius has moved on from his latest triumph, nicknamed "Bubbles," and is working on another revolution in the supercomputer field, because for Seymour Cray a triumph is merely a point of departure.

INTRODUCTION

Voltage, current, resistance, and power are the four basic concepts of prime importance in our study of electronics. In Chapter 2, voltage and current were introduced, and in this chapter we analyze resistance and power.

3-1 WHAT IS RESISTANCE?

Resistance is the opposition to current flow accompanied by the dissipation of heat. In Figures 3-1 and 3-2, we have again used the fluid analogy to explain the concept of resistance. In Figure 3-1(a), a valve has been opened almost completely, so a very small opposition to the water flow exists within the pipe. This small or low resistance within the pipe will not offer much opposition to water flow, so a large amount of water will flow through the pipe and gush from the outlet.

In Figure 3-1(b), a small **resistance** (which is symbolized by the zigzag line) has been placed in the circuit, providing very little resistance to the passage of current flow. This low resistance or small opposition will therefore allow a large amount of current to flow through the conductor, as illustrated by the heavy line. In Figure 3-2(a), the valve is almost completely closed, resulting in a high resistance or opposition to water flow, so only a trickle of water passes through the pipe and out from the outlet. In Figure 3-2(b), a large value of resistance has been placed in the circuit, causing a large opposition to the passage of current. This large resistance allows only a small amount of current to flow through the conductor, as illustrated by the light line.

Resistance
Symbolized *R* and measured in ohms (Ω), it is the opposition to current flow with the dissipation of energy in the form of heat.

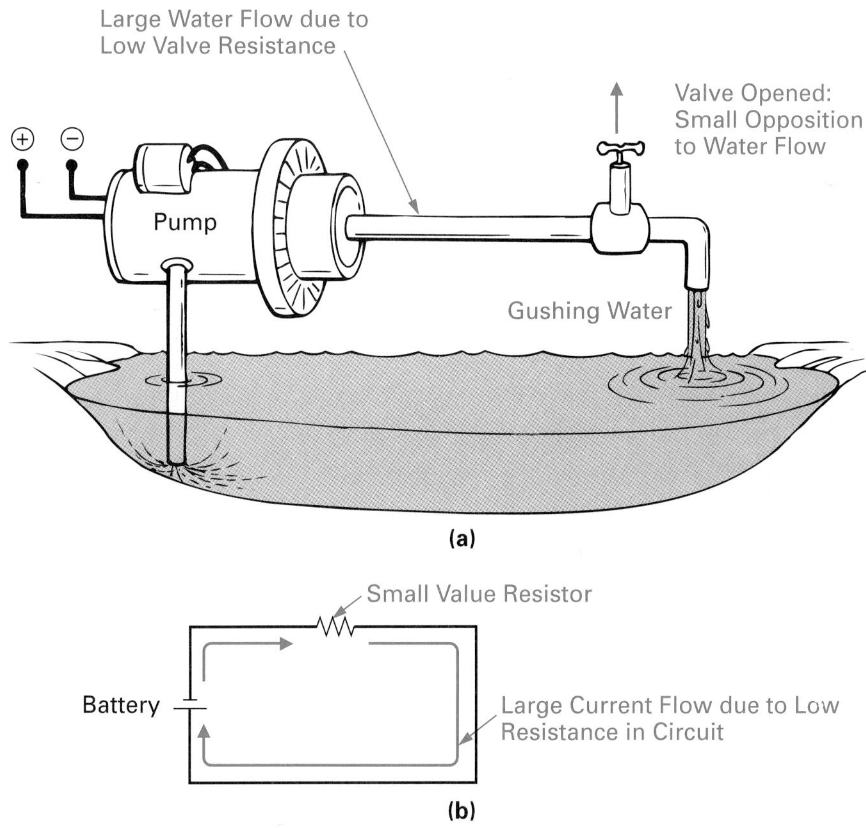

FIGURE 3-1 **Low Resistance. (a) Fluid System. (b) Electrical Circuit.**

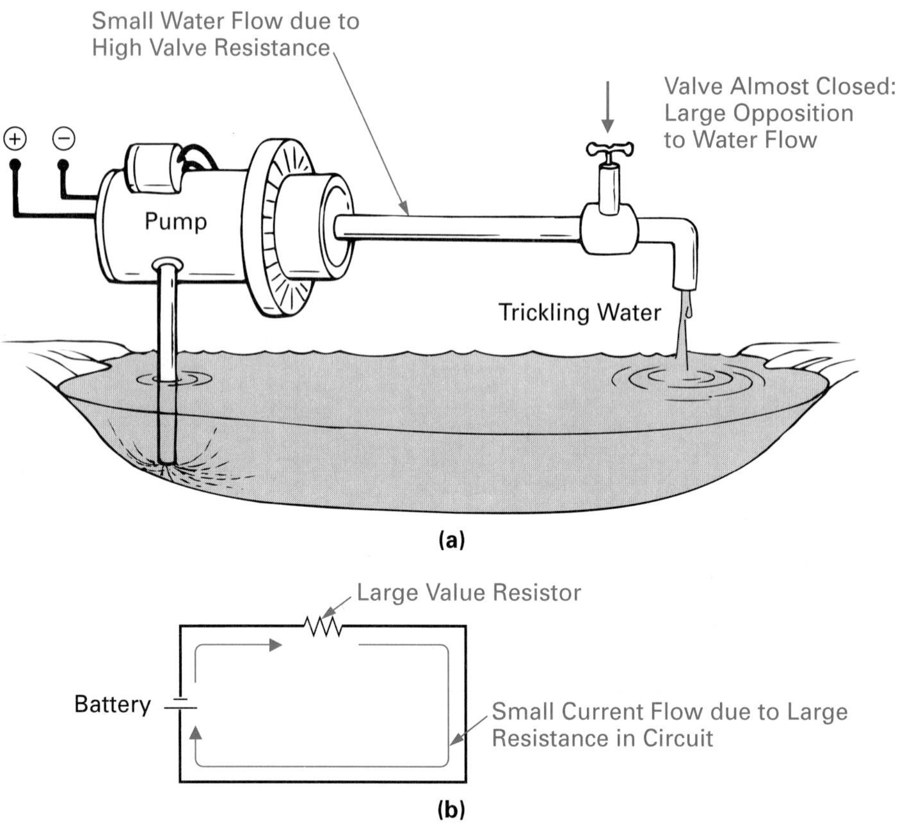

FIGURE 3-2 **High Resistance. (a) Fluid System. (b) Electrical Circuit.**

In both examples of low and high resistance, you may have noticed that resistance and current are inversely proportional ($1/\propto$) to one another; thus, if resistance is high, the current is low, and if resistance is low, the current is high.

Resistance (R) is **Inversely proportional** $\left(\frac{1}{\propto}\right)$ to **Current (I)**

$$R \quad \frac{1}{\propto} \quad I$$

If resistance is increased by some value, current will be decreased by the same value. For example, if resistance is doubled, current is halved (assuming a constant voltage).

$$R\uparrow \frac{1}{\propto} I\downarrow$$

In Figures 3-1 and 3-2, the fluid analogy has been used alongside the electrical circuit to help you understand the idea of low and high resistance. In between low and high resistance exist many different values of resistance, and we now need to analyze resistance further to be able to define clearly exactly how much resistance exists within a circuit, and not just say that resistance is generally low or high.

SELF-TEST REVIEW QUESTIONS FOR SECTION 3-1

1. What is resistance?
2. What is the difference between low and high resistance?
3. What is the relationship between current flow and resistance?
4. Would a very high resistance have a small or a large conductance figure?

3-2 THE OHM

Current is measured in amperes, voltage is measured in volts, and resistance is measured in **ohms,** in honor of Georg Simon Ohm and his work with current, voltage, and resistance. The larger the resistance, the larger the value of ohms and the more the resistor will oppose current flow. The ohm is given the symbol Ω, which is the Greek capital letter omega. By definition, 1 ohm is the value of resistance that will allow 1 ampere of current to flow through a circuit when a voltage of 1 volt is applied, as shown in Figure 3-3(b), where the resistor is

Ohm
Unit of resistance, symbolized by the Greek capital letter omega (Ω).

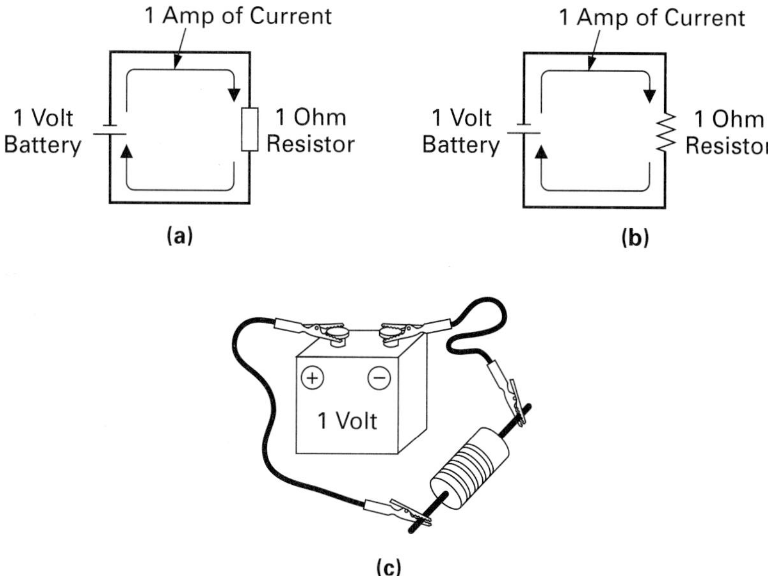

FIGURE 3-3 **One Ohm. (a) New and (b) Old Resistor Symbols. (c) Pictorial.**

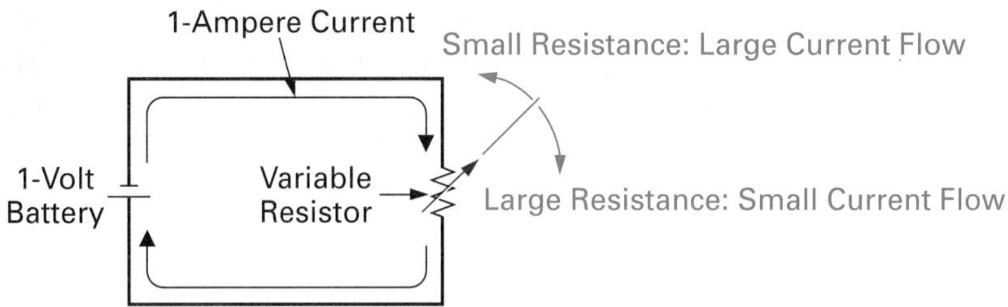

FIGURE 3-4 **1 Ω Allows 1 A to Flow with 1 V Applied.**

drawn as a zigzag; however, in some schematics (circuit diagrams) it can be drawn as a rectangular block, as shown in Figure 3-3(a). The pictorial of this circuit is shown in Figure 3-3(c).

Figure 3-4 reinforces our understanding of the ohm by illustrating a 1-V battery connected across a resistor whose resistance can be either increased or decreased. As the resistance in the circuit is increased, the current will decrease, and as the resistance of the resistor is decreased, the circuit current will increase. If the resistor is adjusted so that exactly 1 A of current is flowing around the circuit, the value of resistance obtained is referred to as 1 ohm (Ω).

3-2-1 *Ohm's Law*

Ohm's Law Relationship between the three electrical phenomena of voltage, current, and resistance, which states that the current flow within a circuit is directly proportional to the voltage applied across the circuit and inversely proportional to its resistance.

Current flows in a circuit due to the force or voltage applied; however, the amount of current flow is limited by the amount of resistance in the circuit. In actual fact, the amount of current flow around a circuit is dependent on both voltage and resistance. This relationship between the three electrical properties of current, voltage, and resistance was discovered by Georg Simon Ohm, a German physicist, in 1827. Published originally in 1826, **Ohm's law** states that the current flow in a circuit is directly proportional (∝) to the source voltage applied and inversely proportional (1/∝) to the resistance of the circuit.

Stated in mathematical form, Ohm arrived at this formula:

$$\text{current } (I) = \frac{\text{voltage } (V)}{\text{resistance } (R)}$$

$$\text{current } (I) \propto \text{voltage } (V)$$

$$\text{current } (I) \frac{1}{\propto} \text{resistance } (R)$$

MINI-MATH REVIEW

This mini-math review is designed to remind you of the mathematical details relating to transposition, which will be used in the following section.

Transposition

In the preceding section it was stated that

$$\text{current } (I) = \frac{\text{voltage } (V)}{\text{resistance } (R)} \text{ or } I = \frac{V}{R}$$

Like all formulas, this formula provides a relationship between quantities, or values. In this case we can determine the current flow in a circuit simply by dividing the voltage applied to the circuit, by the circuit's resistance. If we ever need to calculate only current, this formula would be fine on its own. But what if we want to calculate voltage and we know current and resistance, or what if we want to calculate resistance and we know current and voltage. *It is important to know how to rearrange or transpose a formula so that you can solve for any of the formula's quantities—this process is known as transposition.*

As an example, calculate what voltage would cause 2 amperes of current to flow in a circuit having a resistance of 3 ohms.

To solve this problem, you first begin by placing the values in their appropriate position within the formula. In this example,

$$\text{current } (I) = \frac{\text{voltage } (V)}{\text{resistance } (R)}$$

Therefore,

$$2 \text{ amperes} = \frac{V\,?}{3 \text{ ohms}}$$

or

$$2 = \frac{V}{3}$$

Before we proceed further in our attempt to determine the value of voltage (V), we must first discuss a few ground rules.

1. The equals sign divides the equation or formula into two halves, and in all cases the left half always equals the right half. Therefore,

   $$2 = \frac{V}{3}$$

   left half ←¦→ right half

2. Although not always visible, you should always assume that there are lines under every number or letter in mathematics. For example,

   $$\frac{2}{} = \frac{V}{3}$$

   Any number beneath the line needs to be divided into the number on the top of the line, as with any fraction.

Now that the basics have been covered, we can transpose the formula. To achieve this operation we use basic algebra and follow two steps:

First step: Move the unknown quantity so that it is above the line.

Second step: Move the unknown quantity so that it stands by itself on either side of the equals sign.

Looking at the first step and our example, you can see that "V" does sit above the line:

$$\frac{2}{} = \frac{V}{3}$$

therefore, we can move onto the second step.

Looking at the second step and our example, you can see that "V" does not stand by itself on either side of the equals sign. To satisfy this step, we must somehow move the "3" under the "V" away from the right side of the equation. *If you do exactly the same thing to both sides of the equation or formula, nothing is changed.* This means that you can add to, subtract from, multiply, or divide each side of the equals by the same number and nothing is changed. To move the "3" in our example away from the right side of the equation, we must do something to both sides of the equation to make it cancel. The solution is to multiply both sides by "3," as follows:

$$\frac{3 \times 2}{} = \frac{V \times 3}{3}$$

Since any number divided by the same number equals 1 ($3 \div 3 = 1$) and $V \times 1 = V$, the second step in our transposition process is achieved with "V" isolated on the right side of the equals sign. To summarize, the process was as follows:

$$2 = \frac{V}{3} \qquad \text{(solve for } V = ?)$$

$$3 \times 2 = \frac{V \times 3}{3} \qquad \text{(multiply both sides by 3)}$$

$$3 \times 2 = \frac{V \times 3}{3} \qquad \text{(cancel the 3's)}$$

$$3 \times 2 = V \times 1 \qquad (V \times 1 = V)$$

$$3 \times 2 = V$$

or

$$V = 3 \times 2$$

In this example we can see that $V = 3 \times 2 = 6$ V. This means that when a voltage of 6 volts is applied to a circuit containing 3 ohms of resistance, a current of 2 amperes will flow.

Let's return to our original problem, which was: How can you calculate voltage ($V = ?$) if I and R are known, and how do you calculate resistance ($R = ?$) if V and I are known? You can now see that all you have to do is transpose the original formula, as follows:

$$\text{current } (I) = \frac{\text{voltage } (V)}{\text{resistance } (R)}$$

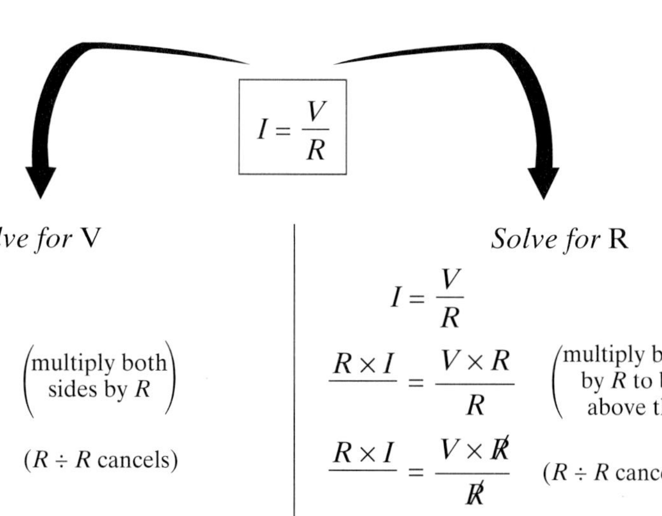

$$I = \frac{V}{R}$$

Solve for V

$$I = \frac{V}{R}$$

$$\frac{R \times I}{} = \frac{V \times R}{R} \quad \left(\text{multiply both sides by } R\right)$$

$$\frac{R \times I}{} = \frac{V \times \not{R}}{\not{R}} \quad (R \div R \text{ cancels})$$

$$\frac{R \times I}{} = \frac{V}{}$$

$$\boxed{V = I \times R}$$

Solve for R

$$I = \frac{V}{R}$$

$$\frac{R \times I}{} = \frac{V \times R}{R} \quad \left(\text{multiply both sides by } R \text{ to bring } R \text{ above the line}\right)$$

$$\frac{R \times I}{} = \frac{V \times \not{R}}{\not{R}} \quad (R \div R \text{ cancels})$$

$$\frac{R \times I}{I} = \frac{V}{I} \quad \left(\text{divide both sides by } I \text{ to isolate } R\right)$$

$$\frac{R \times \not{I}}{\not{I}} = \frac{V}{I} \quad (I \div I \text{ cancels})$$

$$\boxed{R = \frac{V}{I}}$$

To test if all these three formulas are correct, we can plug into each the values of the preceding example, as follows:

$$I = \frac{V}{R} \qquad V = I \times R \qquad R = \frac{V}{I}$$

$$2\text{ A} = \frac{6\text{ V}}{3\ \Omega} \qquad 6\text{ V} = 2\text{ A} \times 3\ \Omega \qquad 3\ \Omega = \frac{6\text{ V}}{2\text{ A}}$$

As you can see from this example, the transposition has been successful.

3-2-2 The Ohm's Law Triangle

The three forms of Ohm's law can be summarized by placing the three properties within a triangle for easy memory recall, as illustrated in Figure 3-5.

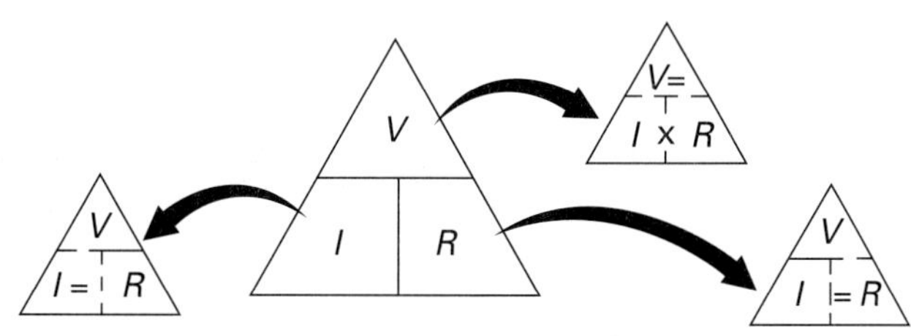

FIGURE 3-5 Ohm's Law Triangle.

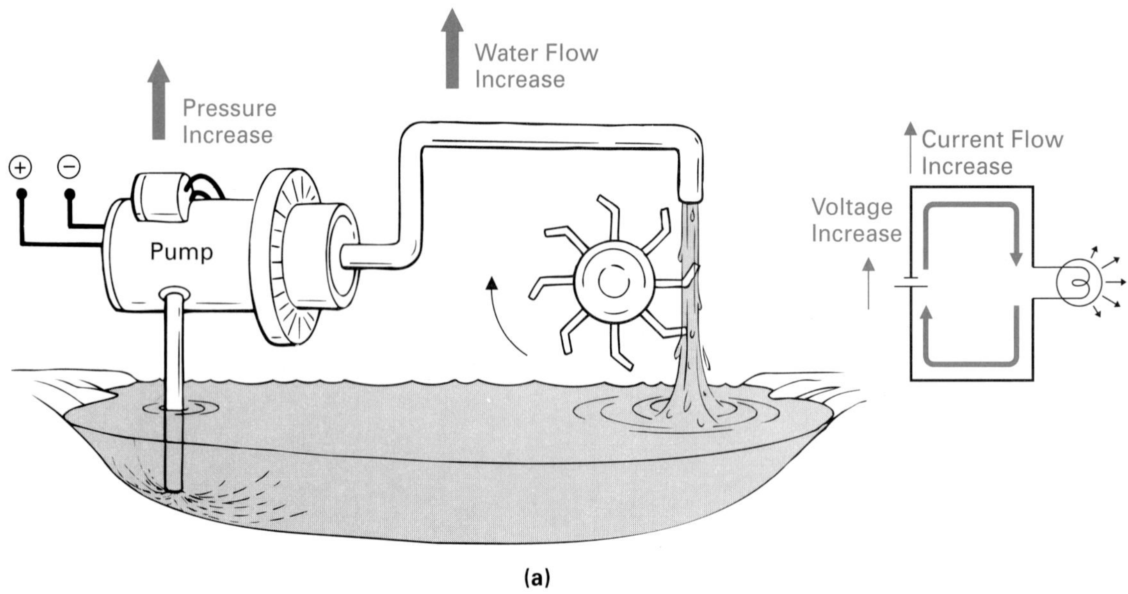

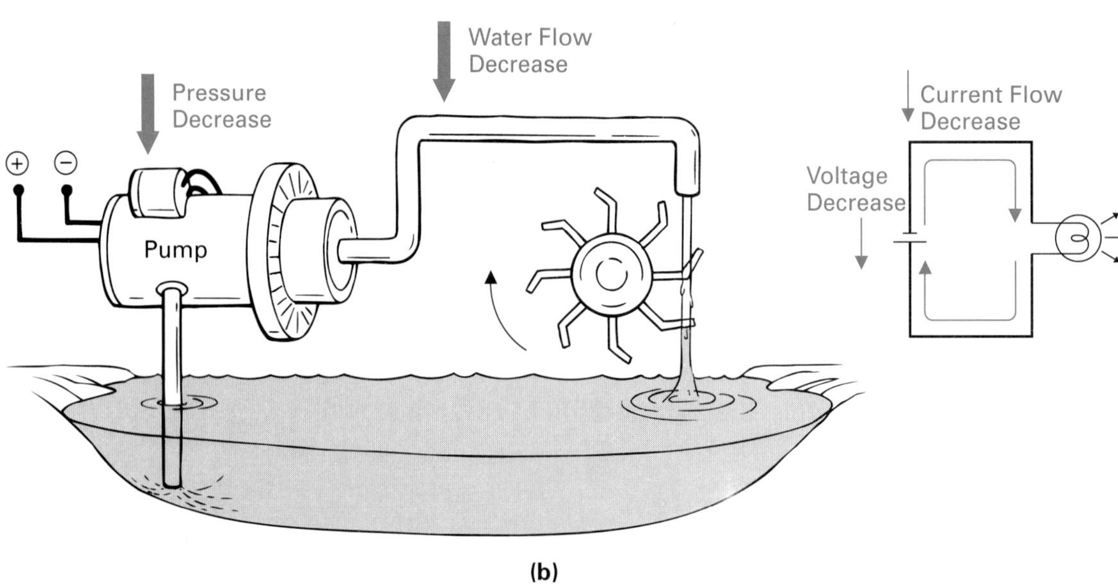

FIGURE 3-6 Current Flow Is Proportional to Voltage. (a) Pressure or Voltage Increase Causes a Water or Current Flow Increase (V↑, I ↑). (b) Pressure or Voltage Decrease Results in a Water or Current Flow Decrease (V↓, I ↓).

$$I = \frac{V}{R} \qquad V = I \times R \qquad R = \frac{V}{I}$$

3-2-3 *Current Is Proportional to Voltage* (I ∝ V)

Referring to Figure 3-6(a), you can see that an increase in the pump pressure will result in an increase in water flow. Similarly, an increase in voltage (electron moving force) will exert a greater pressure on the circuit electrons and cause an increase in current flow. This point is reinforced with Ohm's law:

$$I\uparrow = \frac{V\uparrow}{R}$$

EXAMPLE:

Referring to Figure 3-7, if resistance remains constant at 1 Ω and the applied voltage equals 2 V, what is the value of current flowing through the circuit?

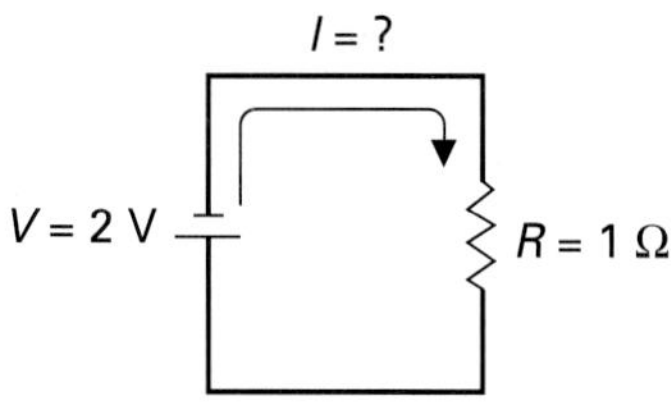

FIGURE 3-7

Solution:

$$\text{current } (I) = \frac{\text{voltage } (V)}{\text{resistance } (R)} = \frac{2\ V}{1\ \Omega} = 2\ \text{A}$$

CALCULATOR SEQUENCE

Step	Keypad Entry	Display Response
1.	2	2
2.	÷	
3.	1	1
4.	=	2

EXAMPLE:

Referring to Figure 3-8, if the voltage from Example 3-1 is now doubled to 4 V, what would be the change in current?

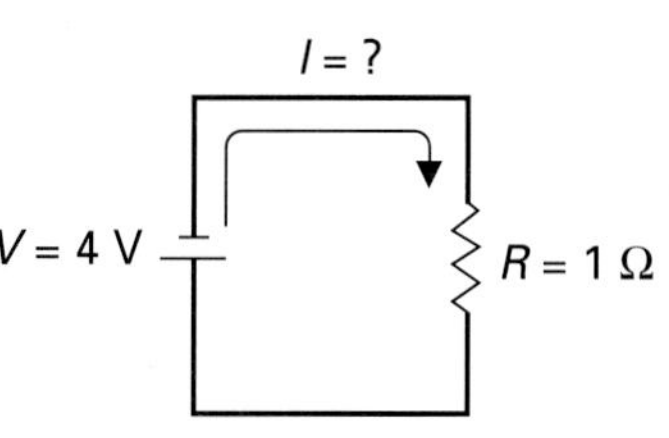

FIGURE 3-8

Solution:

$$\text{current } (I) = \frac{\text{voltage } (V)}{\text{resistance } (R)}$$

$$= \frac{4\text{ V}}{1\ \Omega}$$

$$= 4\text{ A}$$

On the other hand, a decrease in pump pressure or battery voltage will result in a small water or electron force, causing a decrease in current flow, as seen in Figure 3-6(b). This is again reinforced by Ohm's law.

$$I\downarrow = \frac{V\downarrow}{R}$$

Let's once again take an example to prove the point mathematically.

EXAMPLE:

If the circuit resistance equals 2 Ω and the applied voltage equals 8 V, what would be the circuit current?

Solution:

$$\text{current } (I) = \frac{\text{voltage } (V)}{\text{resistance } (R)}$$

$$= \frac{8\text{ V}}{2\ \Omega}$$

$$= 4\text{ A}$$

EXAMPLE:

If the voltage from Example 3-3 is now halved to 4 V, what would be the change in circuit current?

Solution:

$$\text{current } (I) = \frac{\text{voltage } (V)}{\text{resistance } (R)}$$

$$= \frac{4\text{ V}}{2\ \Omega}$$

$$= 2\text{ A}$$

In conclusion, we can say that if resistance were to remain constant and the voltage were to double, the current within the circuit would also double. Similarly, if the voltage were halved, the current would also halve, proving that current and voltage increase or decrease by the same percentage for the same value of resistance, which makes them directly proportional to one another.

3-2-4 *Current Is Inversely Proportional to Resistance* $\left(I \frac{1}{\propto} R\right)$

With the valve opened, as shown in Figure 3-9(a), a small opposition to the water flow will result, so a large water flow exists in a system with low resistance. Similarly, with the electronic circuit a small resistance ($R\downarrow$) allows a large current ($I\uparrow$) flow around the circuit.

Consequently, in both the fluid system and electronic circuit, a decrease in resistance causes an increase in current flow. Ohm's law reinforces this point.

$$\text{(resistance) } R\downarrow = \frac{V}{I\uparrow} \text{ (current)}$$

To prove this mathematically, let's take an example.

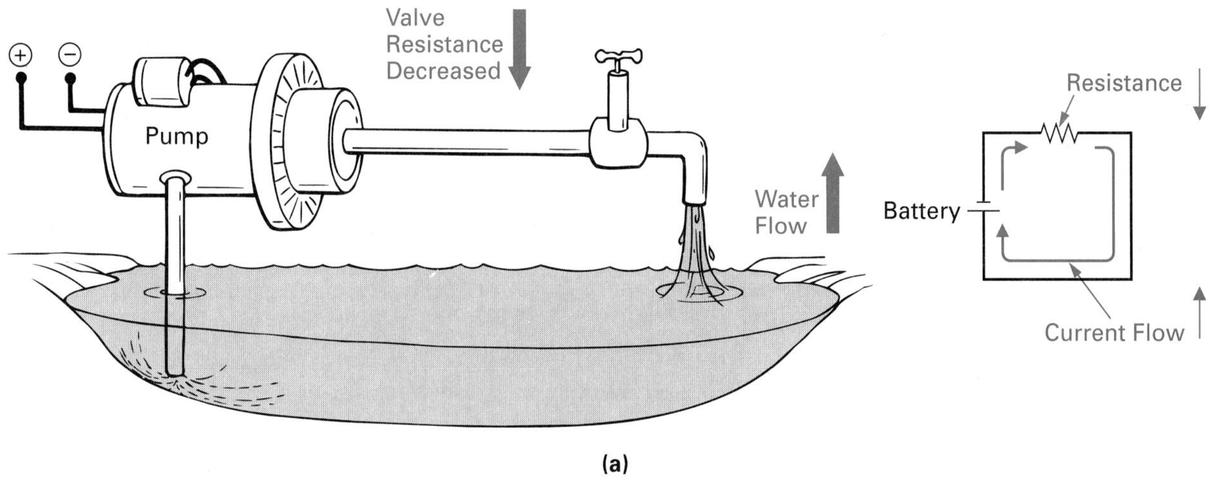

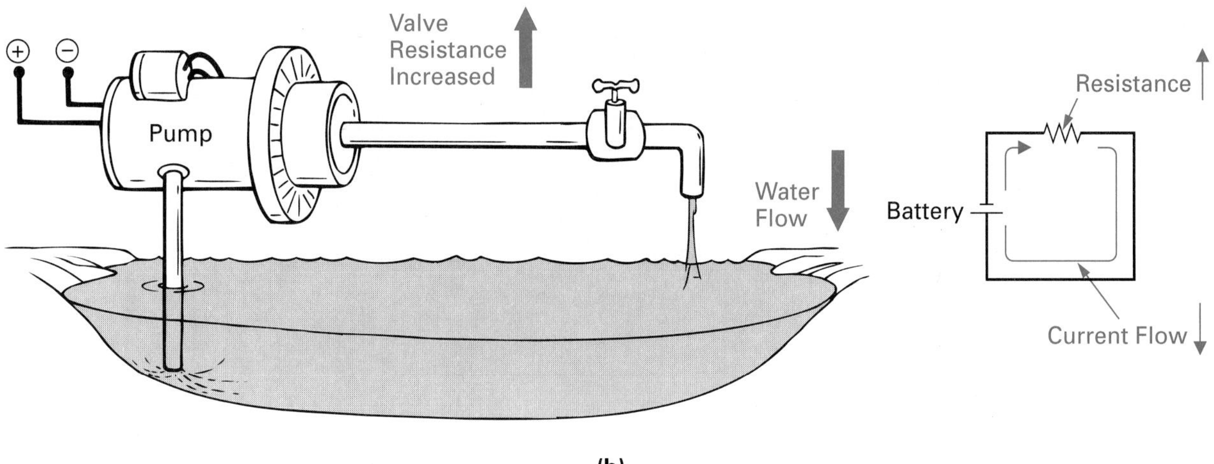

FIGURE 3-9 Current Is Inversely Proportional to Resistance. (a) Resistance Decrease Causes a Water or Current Flow Increase ($R\downarrow$, $I\uparrow$). (b) Resistance Increase Causes a Water or Current Flow Decrease ($R\uparrow$, $I\downarrow$).

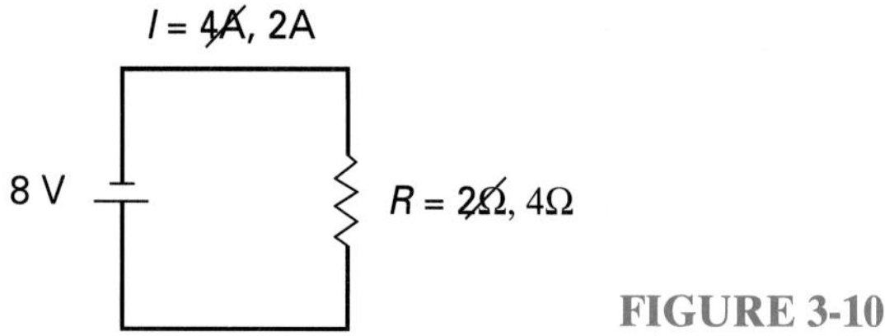

FIGURE 3-10

EXAMPLE:

If the applied circuit voltage is 8 V and the circuit resistance equals 2 Ω, what is the total amount of current flow?

Solution:

$$\text{current } (I) = \frac{\text{voltage } (V)}{\text{resistance } (R)} = \frac{8\text{ V}}{2\ \Omega} = 4\text{ A}$$

On the other hand, in Figure 3-9(b) the valve has been almost completely closed, offering a large opposition to the water flow; consequently, a small water flow results in a system with high resistance. With the electronic circuit in Figure 3-10, a large circuit resistance ($R\uparrow$) permits only a small current ($I\downarrow$) to flow around the circuit.

Ohm's law reinforces the illustration in Figure 3-9(b) by showing that a large resistance or opposition causes a small water or current flow.

$$\text{(resistance) } R\uparrow = \frac{V}{I\downarrow} \text{ (current)}$$

To prove this mathematically, let's take an example.

EXAMPLE:

If the applied voltage is still equal to 8 V and the circuit resistance is now 4 Ω, calculate the value of current flow through the circuit.

Solution:

$$\text{current } (I) = \frac{\text{voltage } (V)}{\text{resistance } (R)} = \frac{8\text{ V}}{4\ \Omega} = 2\text{ A}$$

In conclusion, we can say that if voltage were to remain constant and the resistance were to double, the current within the circuit would be halved. On the other hand, if the circuit resistance were halved, the circuit current would double, confirming that current is inversely proportional to resistance.

SELF-TEST REVIEW QUESTIONS FOR SECTION 3-2

1. Define 1 ohm in relation to current and voltage.
2. Calculate I if $V = 24$ V and $R = 6\ \Omega$.
3. What is the Ohm's law triangle?
4. What is the relationship between current and voltage; between current and resistance?
5. Calculate V if $I = 25$mA and $R = 1\ \text{k}\Omega$.
6. Calculate R if $V = 12$ V and $I = 100\ \mu$A.

3-3 CONDUCTORS AND THEIR RESISTANCE

Resistors are normally made out of materials that cause an opposition to current flow. Conductors, on the other hand, are not meant to offer any resistance or opposition to current flow. This, however, is not always the case since some are good conductors having a low resistance, while others are poor conductors having a high resistance. Conductance is the measure of how good a conductor is, and even the best conductors have some value of resistance. Up until this time, we have determined a conductor's resistance based on the circuit's *electrical characteristics,*

$$R = \frac{V}{I}$$

With this formula we could determine the conductor's resistance based on the voltage applied and circuit current. Another way to determine resistance is to examine the *physical factors* of a conductor.

The total resistance of a typical conductor, as shown in Figure 3-11, is determined by four main physical factors:

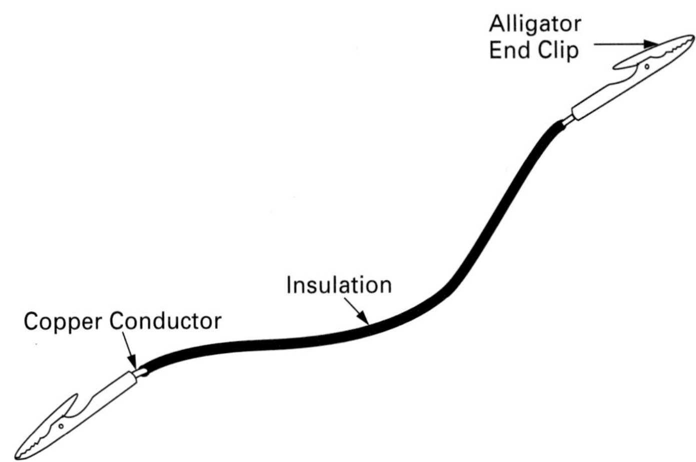

FIGURE 3-11 Copper Conductor.

1. The type of conducting material used for the flow of current
2. The conductor's cross-sectional area
3. The total length of the conductor
4. The temperature of the conductor

Let's now address each of these topics separately and discuss the reason why each will vary a conductor's resistance.

3-3-1 *Conducting Material*

In our discussion on conductors, we said that copper, for instance, is a better conducting material than carbon as copper had more free electrons ready for current flow; and as we have stated, a good conductor has a large conductance figure and a small resistance value. Carbon, however, has a smaller amount of free electrons than copper, resulting in a small conductance value and a high resistance figure.

3-3-2 *Conductor's Cross-Sectional Area*

The resistance of a conductor is inversely proportional to the conductor's cross-sectional area, which means: The greater the cross-sectional area, the lower the resistance [Figure 3-12(a)], and similarly, the smaller the cross-sectional area, the higher the resistance [Figure 3-12(b)].

We can relate this to our fluid analogy, shown in Figure 3-13. A larger pipe or conductor results in a larger water flow or current due to the lower resistance. On the other hand, the smaller the pipe or conductor, the less the amount of water or electron flow, due to the larger resistance (Figure 3-14).

We need a unit of measure to compare one thickness of wire conductor to another. Figure 3-15 illustrates a conductor with a diameter of 0.001 inch ($\frac{1}{1000}$ of an inch). This is termed 1 **mil,** and the diameters of all conductors are measured in mils.

Mil
One thousandth of an inch (0.001 in.).

As nearly all conductors are circular in cross section, the area of a conductor is measured in **circular mils** (cmil), which can be calculated by squaring the diameter.

Circular Mil (cmil)
A unit of area equal to the area of a circle whose diameter is 1 mil.

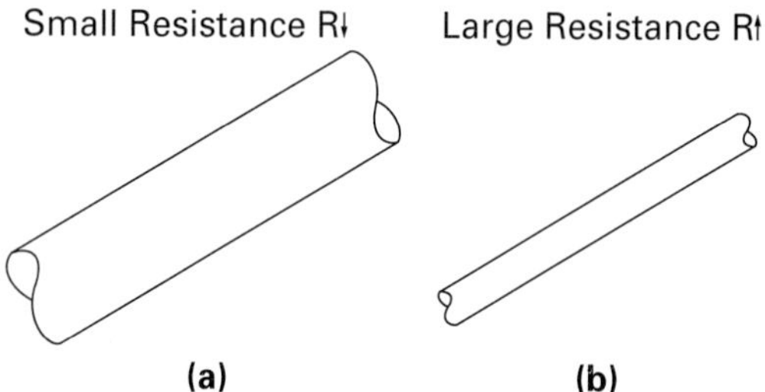

FIGURE 3-12 **Conduction. (a) Large Cross-Sectional Area. (b) Small Cross-Sectional Area.**

$$\text{circular mils (cmil)} = \text{diameter}^2$$

EXAMPLE:

If a conductor has a diameter of 71.96 mils, what is the circular mil area?

Solution:

$$\begin{aligned} \text{cmil} &= \text{diameter}^2 \\ &= 71.96^2 \\ &= 5178 \text{ cmil} \end{aligned}$$

CALCULATOR SEQUENCE

Step	Keypad Entry	Display Response
1.	7 1 . 9 6	71.96
2.	X^2	5178.24

3-3-3 *Conductor Length*

Increasing the length of the conductor used increases the amount of resistance within the circuit. For example, if a conductor has a resistance of 1 ohm for every 10 feet of conductor, then 30 feet of conductor are going to triple the conductor's resistance within the circuit to 3 ohms.

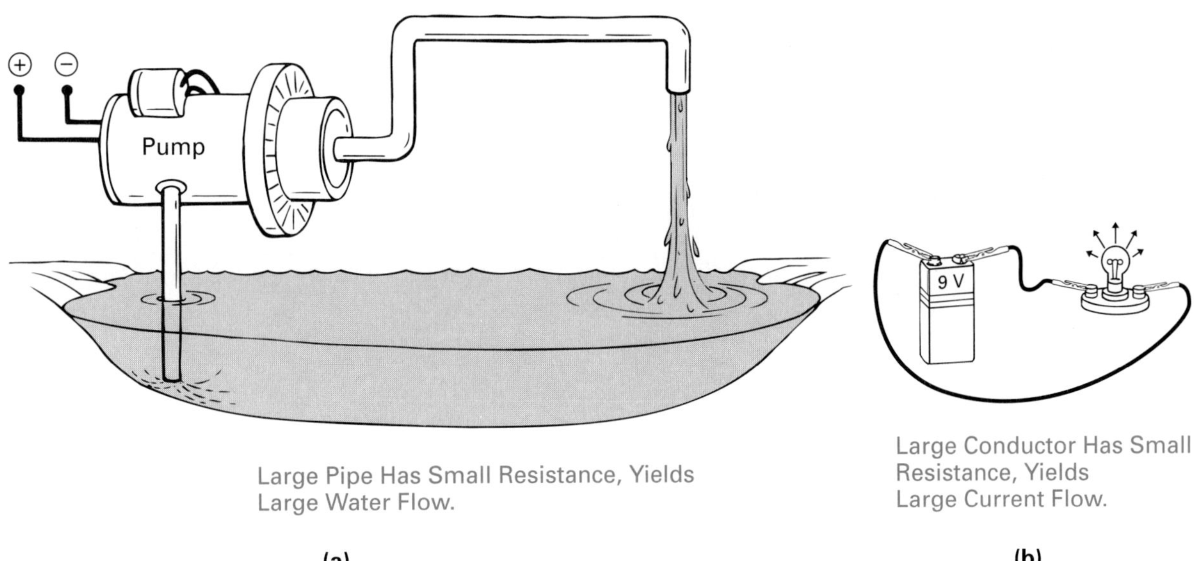

FIGURE 3-13 Large Cross-Sectional Area Causes Low Resistance. (a) Large Pipe (Small Resistance) Yields Large Water Flow. (b) Large Conductor (Small Resistance) Yields Large Current Flow.

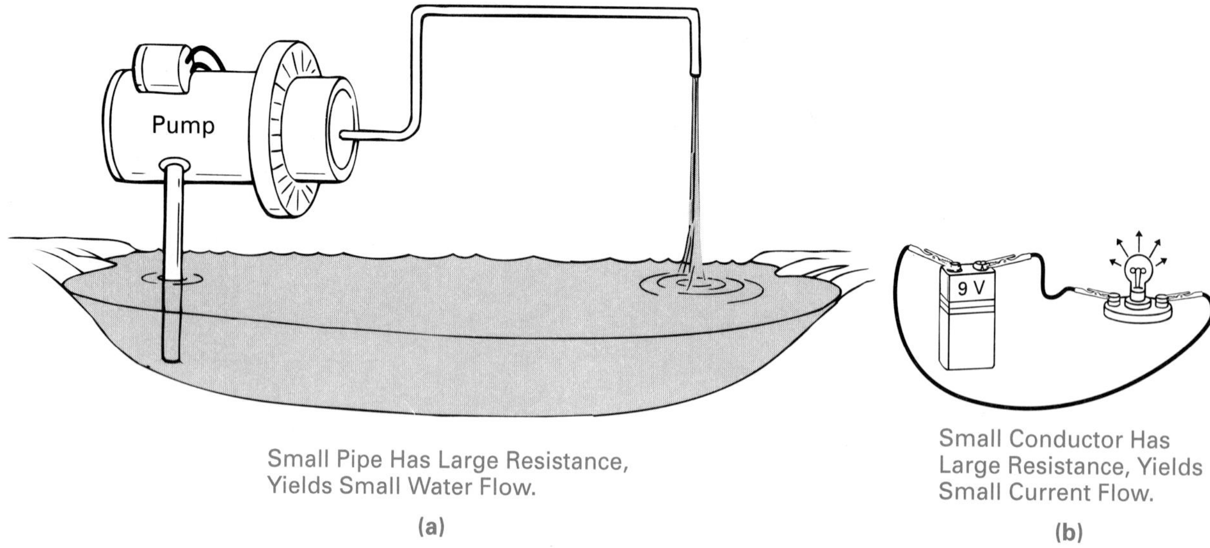

FIGURE 3-14 Small Cross-Sectional Area Causes High Resistance. (a) Small Pipe (Large Resistance) Yields Small Water Flow. (b) Small Conductor (Large Resistance) Yields Small Current Flow.

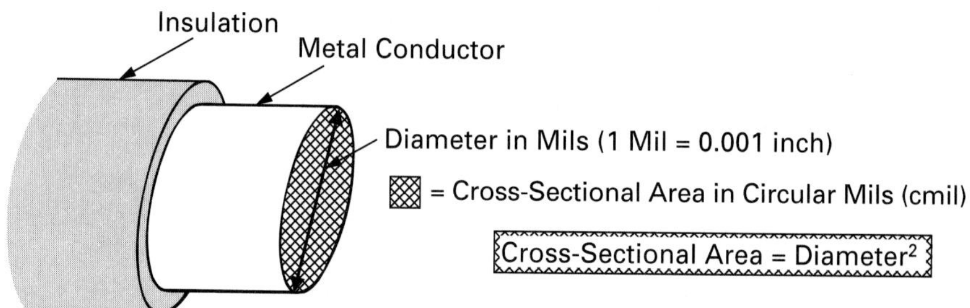

FIGURE 3-15 Measuring a Conductor's Cross-Sectional Area.

3-3-4 *Physical Resistance Formula*

By combining all of these physical factors, we can arrive at a formula for resistance. The resistance of any conductor is calculated by use of the following formula:

$$R = \frac{\rho \times l}{a}$$

where R = resistance of conductor, in ohms
ρ = resistivity of conducting material
l = length of conductor, in feet
a = area of conductor, in circular mils

TABLE 3-1 **Material Resistivity**

MATERIAL	RESISTIVITY (cmil/ft) IN OHMS
Silver	9.9
Copper	10.7
Gold	16.7
Aluminum	17.0
Tungsten	33.2
Zinc	37.4
Brass	42.0
Nickel	47.0
Platinum	60.2
Iron	70.0

Resistivity, by definition, is the resistance (in ohms) that a certain length of material (in centimeters) will offer to the flow of current. Table 3-1 lists the resistivity of the more commonly used conductors.

Resistivity
Measure of a material's resistance to current flow.

In summary, therefore, resistance can be calculated based either on its physical factors or on electrical performance.

$$\frac{\rho \times l}{a} = R = \frac{V}{I}$$

Physical $\Longleftrightarrow$ Electrical

EXAMPLE:

Calculate the resistance of 333 ft of copper conductor with a conductor area of 3257 cmil.

Solution:

$$R = \frac{\rho \times l}{a}$$

$$= \frac{10.7 \times 333}{3257}$$

$$= 1.09\ \Omega$$

CALCULATOR SEQUENCE

Step	Keypad Entry	Display Response
1.	[1] [0] [.] [7]	10.7
2.	[×]	
3.	[3] [3] [3]	333
4.	[÷]	3563.1
5.	[3] [2] [5] [7]	3257
6.	[=]	1.09

EXAMPLE:

Calculate the resistance of 1274 ft of aluminum conductor with a diameter of 86.3 mils.

■ ***Solution:***

$$R = \frac{\rho \times l}{a}$$

If the diameter is equal to 86.3 mils, the circular mil area equals d^2:

$$86.3^2 = 7447.7 \text{ cmil}$$

Referring to Table 3-1 you can see that the resistivity of aluminum is 17.0. Therefore,

$$R = \frac{17 \times 1274}{7447.7}$$
$$= 2.9\ \Omega$$

3-3-5 *Temperature Effects on Conductors*

When heat is applied to a conductor, the atoms within the conductor convert this thermal energy into another form of energy, in this case mechanical energy or movement. These random moving atoms cause collisions between the directed electrons (current flow) and the adjacent atoms, resulting in an opposition to the current flow (resistance).

Positive Temperature Coefficient of Resistance
The rate of increase in resistance relative to an increase in temperature.

Metallic conductors are said to have a **positive temperature coefficient of resistance** (+ Temp. Coe. of R), because the greater the heat applied to the conductor, the greater the atom movement, causing more collisions of atoms to occur, and consequently the greater the conductor's resistance.

heat ↑ resistance ↑

3-3-6 *Maximum Conductor Current*

Any time that current flows through any conductor, a certain resistance or opposition is inherent in that conductor. This resistance will convert current to heat, and the heat further increases the conductor's resistance, causing more heat to

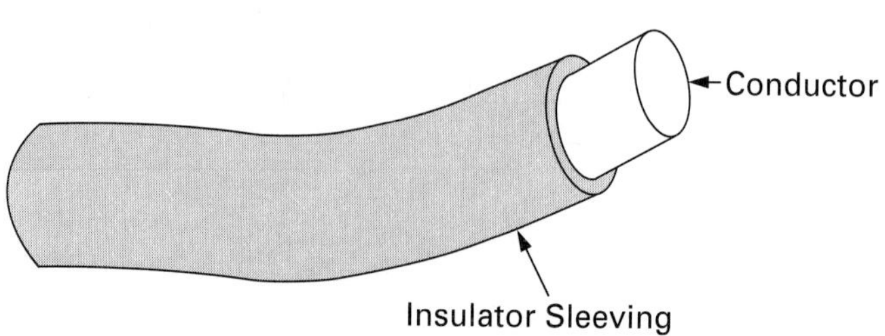

FIGURE 3-16 Conductor with Insulator.

be generated due to the opposition. Consequently, a conductor must be chosen carefully for each application so that it can carry the current without developing excessive heat. This is achieved by selecting a conductor with a greater cross-sectional area to decrease its resistance.

Conducting wires are normally covered with a plastic or rubber type of material, known as *insulation,* as shown in Figure 3-16. This insulation is used to protect the users and technicians from electrical shock and also to keep the conductor from physically contacting other conductors within the equipment. If the current through the conductor is too high, this insulation will burn due to the heat and may cause a fire hazard. The National Fire Protection Association has developed a set of standards known as the **American Wire Gauge** for all copper conductors, which lists their diameter, area, resistance, and maximum safe current in amperes. This table is given in Table 3-2. A rough guide for measuring wire size is shown in Figure 3-17.

American Wire Gauge
American wire gauge (AWG) is a system of numerical designations of wire sizes, with the first being 0000 (the largest size) and then going to 000, 00, 0, 1, 2, 3, and so on up to the smallest sizes of 40 and above.

TABLE 3-2 The American Wire Gauge (AWG) for Copper Conductor

AWG NUMBER[a]	DIAMETER (mils)	MAXIMUM CURRENT (A)	Ω/1000 ft
0000	460.0	230	0.0490
000	409.6	200	0.0618
00	364.8	175	0.0780
0	324.9	150	0.0983
1	289.3	130	0.1240
2	257.6	115	0.1563
3	229.4	100	0.1970
4	204.3	85	0.2485
5	181.9	75	0.3133
6	162.0	65	0.3951
7	144.3	55	0.4982
8	128.5	45	0.6282
9	114.4	40	0.7921
10	101.9	30	0.9981
11	90.74	25	1.260
12	80.81	20	1.588
13	71.96	17	2.003
14	64.08	15	2.525
15	57.07		3.184
16	50.82	6	4.016
17	45.26	Wires	5.064
18	40.30	of this 3	6.385
19	35.89	size	8.051
20	31.96	have	10.15
22	25.35	current	16.14
26	15.94	measured	40.81
30	10.03	in mA	103.21
40	3.145		1049.0

[a]The larger the AWG number, the smaller the size of the conductor.

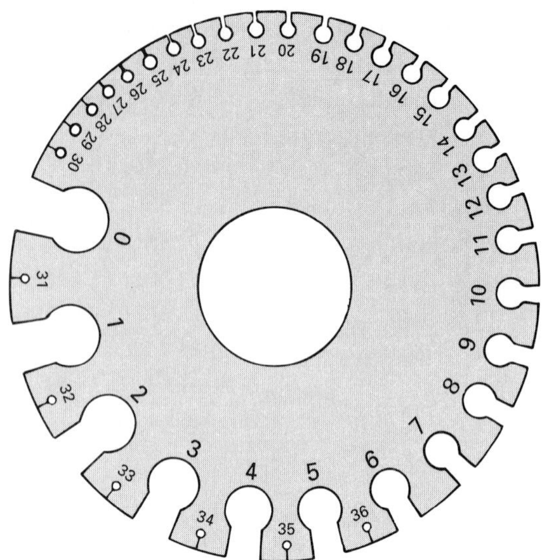

FIGURE 3-17 **Wire Gauge Size.**

3-3-7 Superconductivity

Superconductor
Metal such as lead or niobium that, when cooled to within a few degrees of absolute zero, can conduct current with no electrical resistance.

Conductors have a positive temperature coefficient, which means that if temperature increases, so does resistance; but what happens if the temperature is decreased? In 1911, a Dutch physicist, Heike Onnes, discovered that mercury (a liquid conductor) lost its resistance to electrical current when the temperature was decreased to –459.7 Fahrenheit (0 on the Kelvin temperature scale). Mercury actually became a **superconductor,** allowing a supercurrent to flow and not encounter any resistance and therefore not generate any heat. (Heat dissipated by any resistance can be calculated by the power formula $P = I^2 \times R$. If $R = 0$, the power dissipated by the conductor is zero watts.)

In the spring of 1986, two IBM scientists discovered that a conductor compound made up of barium, lanthanum, copper, and oxygen would superconduct (have no resistance to current flow) at –406°F. Since then other scientists have increased the temperature to a point that now conductor compounds can be made to superconduct at –300°F. Some scientific projects that need to make use of the advantages of superconductivity immerse these conductor compounds in a bath of liquid nitrogen so that they can achieve superconductivity. Liquid nitrogen, however, is difficult to handle, so the real scientific achievement will be to find a conductor compound that will superconduct at room temperature.

When discussing the AWG table (Table 3-2), a maximum value of current was stated for a given thickness of wire. This is because a large current will generate a large amount of heat if the resistance is too large, so to decrease the resistance and therefore heat generated ($P\uparrow = I^2R\uparrow$), a thicker conductor with a lower resistance is used. Superconductivity allows a standard 1000-A conductor, which would be approximately 2 in. thick, to be replaced by a superconductor about the thickness of a human hair. Not only will conductors be reduced dramatically in size, but the increased efficiency (as no power is being wasted in the form of heat) will result in high energy savings.

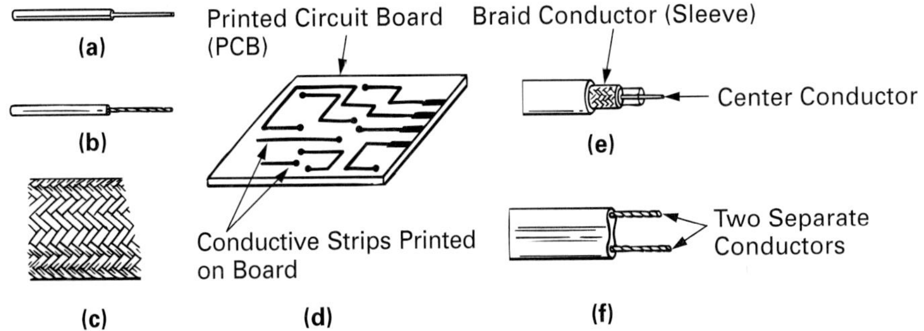

FIGURE 3-18 **Wires and Cables. (a) Solid Wire. (b) Stranded Wire. (c) Braided Wire. (d) Printed Wire. (e) Coaxial Cable. (f) Twin Lead.**

3-3-8 *Conductor and Connector Types*

A cable is made up of two or more wires. Figure 3-18 illustrates the different types of **wires** and **cables.** In Figure 3-18(a), (b), and (c), only one conductor exists within the insulation, so they are classified as wires, whereas the cables seen in Figure 3-18(e) and (f) have two conductors. The coaxial and twin-lead cables are most commonly used to connect TV signals into television sets. Figure 3-18(d) illustrates how printed conducting copper, silver, or gold paths exist on a plastic insulating board and are used to connect components such as resistors that are mounted on the other side.

Wire
Single solid or stranded group of conductors having a low resistance to current flow.

Cable
Group of two or more insulated wires.

Wires and cables have to connect from one point to another. Some are soldered directly, whereas others are attached to plugs that plug into sockets. A sample of connectors is shown in Figure 3-19.

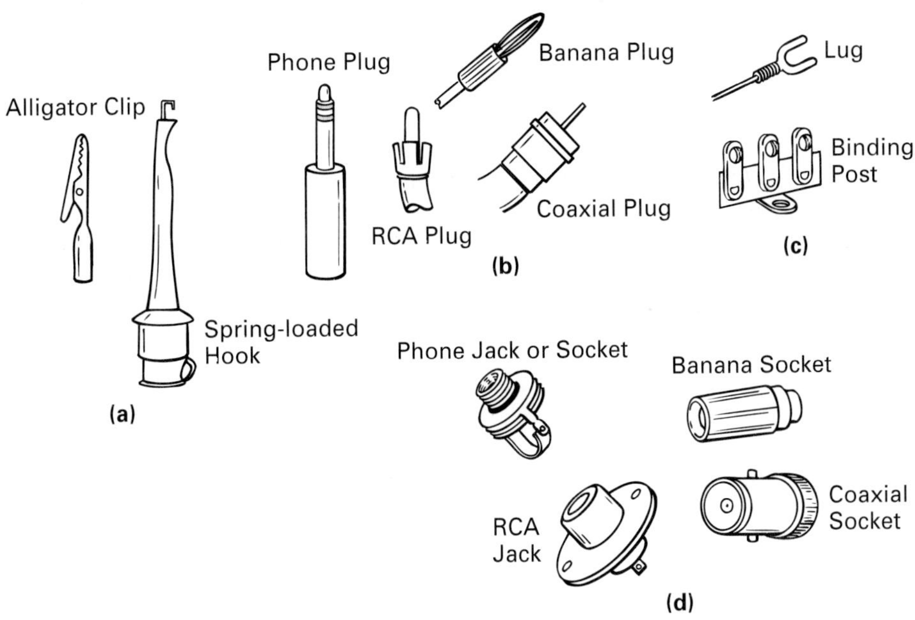

FIGURE 3-19 **Connectors. (a) Temporary Connectors. (b) Plugs. (c) Lug and Binding Post. (d) Sockets.**

SELF-TEST REVIEW QUESTIONS FOR SECTION 3-3

1. List the four factors that determine the total resistance of a conductor.
2. What is the relationship between the resistance of a conductor and its cross-sectional area, length, and resistivity?
3. True or false: Conductors are said to have a positive temperature coefficient.
4. True or false: The smaller the AWG number, the larger the size of the conductors.

3-4 ENERGY, WORK, AND POWER

Energy
Capacity to do work.

Work
Work is done anytime energy is transformed from one type to another, and the amount of work done is dependent on the amount of energy transformed.

Joule
The unit of work and energy.

Potential Energy
Energy that has the potential to do work because of its position relative to others.

Kinetic Energy
Energy associated with motion.

The sun provides us with a consistent supply of energy in the form of light. Coal and oil are fossilized vegetation that grew, among other things, due to the sun, and are examples of **energy** that the earth has stored for millions of years. It can be said, then, that all energy begins from the sun. On the earth, energy is not created or destroyed; it is merely transformed from one form to another. The transforming of energy from one form to another is called **work.** The greater the energy transformed, the more work that is done.

The six basic forms of energy are light, heat, magnetic, chemical, electrical, and mechanical energy. The unit for energy is the **joule** (J). "Potential" (position) and "kinetic" (motion) are two terms used when describing energy. A cart on top of a hill has **potential** (position) **energy,** while a cart rolling down a hill has **kinetic** (motion) **energy.** Potential and kinetic energy are best described by looking at the example of a swinging pendulum, as seen in Figure 3-20. When the pendulum is in its upmost position [Figure 3-20(a)], it has potential energy, due to its position relative to the resting position, yet it has no motion and so no kinetic energy. When the pendulum is in the position shown in Figure 3-20(b), it has no potential energy, yet it has motion or kinetic energy. In between these two points, the pendulum possesses a combination of both potential and kinetic energy, as shown in Figure 3-20(c).

Looking at Figure 3-21, let us try to summarize our discussion so far on energy and work. Chemical energy within the battery is converted to electrical energy when the electrons are attracted and repelled and set in motion. To use

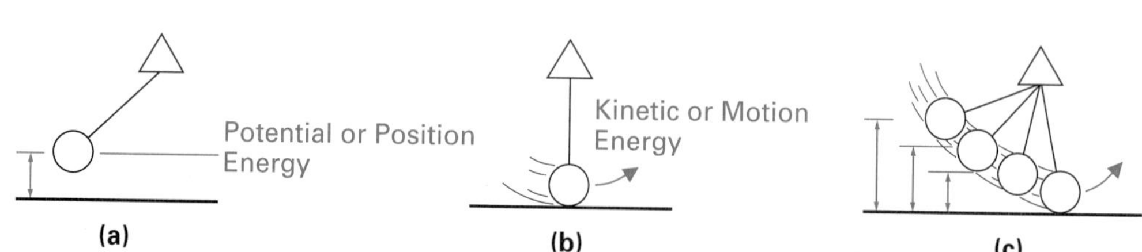

FIGURE 3-20 Pendulum. (a) Potential Energy. (b) Kinetic Energy. (c) Potential and Kinetic Energy.

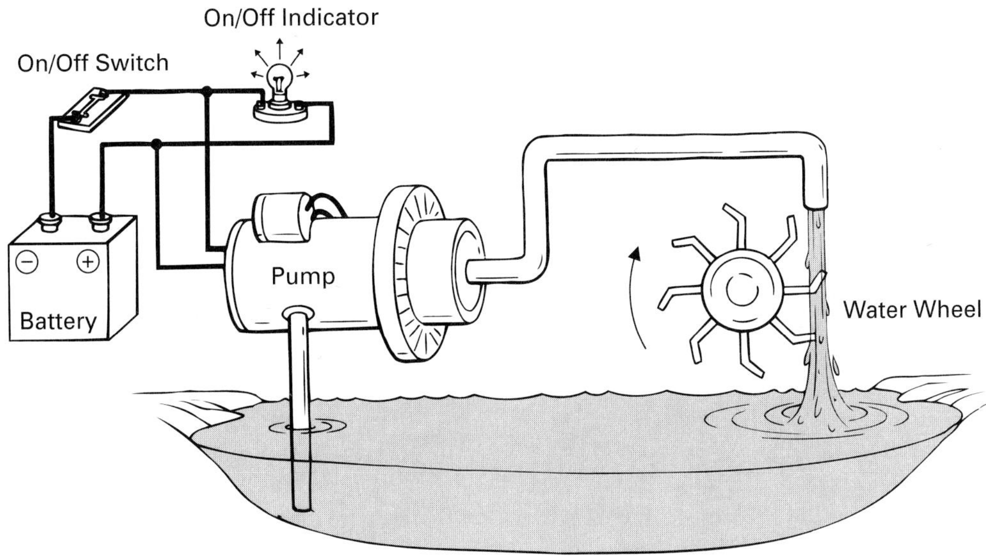

FIGURE 3-21 **Energy Transfer.**

the two terms, we could say that the battery has the potential energy to set electrons in motion, and these moving electrons are said to possess kinetic energy. This electrical energy drives two devices:

1. The light bulb, which converts electrical energy into light and heat energy
2. The pump, which uses the electrical energy to produce mechanical energy within the pump

The pump has the potential energy to cause water flow or kinetic energy, just as the battery has the potential energy to cause kinetic energy in the form of electron flow. The kinetic energy within the water flow is finally used to turn the waterwheel.

Work is being done everytime one form of energy is transformed to another. A device that converts one form of energy to another is called a **transducer;** in Figure 3-21, the battery, light bulb, and pump were all examples of transducers doing work at each stage of conversion.

Transducer
Any device that converts energy from one form to another.

The amount of work done is equal to the amount of energy transformed, and in both cases an equal amount of energy was transformed, so an equal amount of work was done. Energy and work have the same symbol (W), the same formula, and the same unit (the joule). Energy is merely the capacity, potential, or ability to do work, and work is done when a transformation of the potential, capacity, or ability takes place.

EXAMPLE:

One person walks around a track and takes 50 minutes, while another person runs around the track and takes 50 seconds. Both were full of energy before they walked or ran around the track, and during their travels around the track they converted the chemical energy within their bodies into the mechanical energy of movement.

a. Who exerted the most energy?

b. Who did the most work?

Solution:

Both exerted the same amount of energy. The runner exerted all his energy (for example, 100 J) in the short time of 50 seconds, while the walker spaced his energy (100 J) over 50 minutes. They both did the same amount of work. So the only difference between the runner and the walker is time.

3-4-1 Power

Power
Amount of energy converted by a component or circuit in a unit of time, normally seconds. It is measured in units of watts (joules/second).

Watt (W)
Unit of electric power required to do work at a rate of 1 joule/second. One watt of power is expended when 1 ampere of direct current flows through a resistance of 1 ohm.

Power (P) is the rate at which work is performed and is given the unit of **watt** (W), which is joules per second. Thus power involves a time factor.

Returning to our two persons walking and running around the track, we could say that the number of joules of energy exerted in 1 second by the runner was far greater than the number of joules of energy exerted in 1 second by the walker, although the total energy exerted by both persons around the entire track was equal and therefore the same amount of work was done. Even though the same amount of energy was used, and therefore the same amount of work was done by the runner and the walker, the output power of each was different. The runner exerted a large value of joules/second or watts (high power output) in a short space of time, while the walker exerted only a small value of joules/second or watts (low power output) over a longer period of time.

Whether discussing a runner, walker, electric motor, heater, refrigerator, light bulb, or compact disk player—power is power. The output power, or power ratings, of electrical, electronic, or mechanical devices can be expressed in watts and describes the number of joules of energy converted every second. The output power of rotating machines is given in the unit *horsepower* (hp), the output power of heaters is given in the unit *British thermal units per hour* (Btu/h), and the output power of cooling units is given in the unit *ton of refrigeration.* Despite the different names, they can all be expressed quite simply in the unit of watts. The conversions are as follows:

$$1 \text{ horsepower (hp)} = 746 \text{ W}$$

$$1 \text{ British thermal unit per hour (Btu/h)} = 0.293 \text{ W}$$

$$1 \text{ ton of refrigeration} = 3.52 \text{ kW (3520 W)}$$

Now we have an understanding of power, work, and energy. Let's reinforce our knowledge by introducing the energy formula with some mathematical problems related to electronics.

3-4-2 Calculating Energy

The amount of energy stored (W) is dependent on the coulombs of charge stored (Q) and the voltage (V).

$$W = Q \times V$$

where W = energy stored, in joules (J)
Q = coulombs of charge (1 coulomb = 6.24×10^{18} electrons)
V = voltage, in volts (V)

If you consider a battery as an example, you can probably better understand this formula. The battery's energy stored is dependent on how many coulombs of electrons it holds (current) and how much electrical pressure it is able to apply to these electrons (voltage).

EXAMPLE:

If a 1-V battery can store 6.24×10^{18} electrons, how much energy is the battery said to have?

Solution:

$$\begin{aligned} W &= Q \times V \\ &= 1\ \text{C} \times 1\ \text{V} \\ &= 1\ \text{J of energy} \end{aligned}$$

EXAMPLE:

How many coulombs of electrons would a 9-V battery have to store to have 63 J of energy?

Solution:

If $W = Q \times V$, then

$$Q = \frac{W}{V}$$

$$\text{coulombs of electrons } (Q) = \frac{\text{energy in joules } (W)}{\text{battery voltage } (V)}$$

$$= \frac{63\ \text{J}}{9\ \text{V}}$$

$$= 7\ \text{C of electrons}$$

or

$$7 \times 6.24 \times 10^{18} = 4.36 \times 10^{19}\ \text{electrons}$$

CALCULATOR SEQUENCE

Step	Keypad Entry	Display Response
1.	6 3	63
2.	÷	
3.	9	9
4.	=	7
5.	×	7
6.	6 . 2 4 E 1 8	6.24E18
7.	=	4.36E19

3-4-3 *Calculating Power*

Power, in relation to electricity and electronics, is the rate (t) at which electric energy (W) is converted into some other form. A power formula can therefore be derived as follows:

$$P = \frac{W}{t}$$

where P = power, in watts
W = energy, in joules
t = time, in seconds

Since $W = Q \times V$, we can substitute W in the formula above, to obtain

$$P = \frac{W}{t} = \frac{Q \times V}{t}$$

where Q = coulombs of charge
V = voltage, in volts

Since coulombs of charge $Q = I \times t$, we can substitute Q in the formula above, to obtain

$$P = \frac{Q \times V}{t} = \frac{(I \times t) \times V}{t}$$

By canceling t, we arrive at a final formula for power:

$$P = I \times V$$

where P = power, in watts (W)
I = current, in amperes (A)
V = voltage, in volts (V)

This formula states that the amount of power delivered to a device is dependent on the electrical pressure or voltage applied across the device and the current flowing through the device.

EXAMPLE:

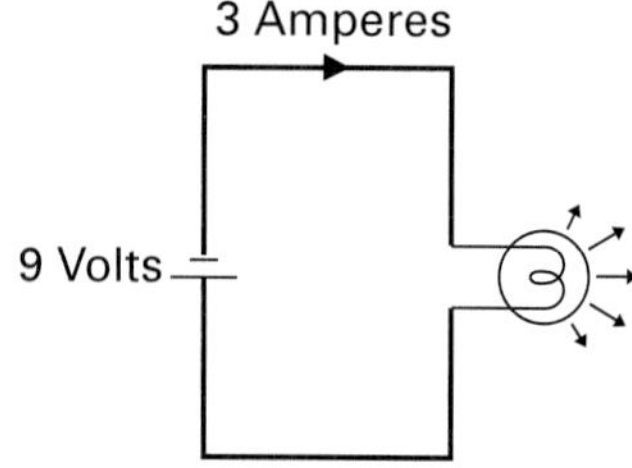

FIGURE 3-22 Calculating Power.

Power, in relation to electronics, is the rate at which electric energy is converted into some other form. In our example (Figure 3-22) it will be transformed from electric energy into light and heat energy by the light bulb. Power has the unit of watts, which is the number of joules of energy transformed per second (J/s). If 27 J of electric energy is being transformed into light and heat per second, how many watts of power does the light bulb convert?

■ ***Solution:***

$$\text{power} = \frac{\text{joules}}{\text{second}} = \frac{27\text{ J}}{1\text{ s}} = 27\text{ W}$$

The power output in Example 3-13 could have easily been calculated by merely multiplying current by voltage to arrive at the same result.

$$\text{power} = I \times V = 3\text{ A} \times 9\text{ V} = 27\text{ W}$$

Therefore, we could say that the light bulb dissipates 27 W, or 27 J per second.

Looking at the power formula $P = I \times V$, we can say that 1 watt of power is expended when 1 ampere of current flows through a circuit that has 1 volt applied or when 1 ampere flows through 1 ohm.

Like Ohm's law, we can transpose the power formula as follows:

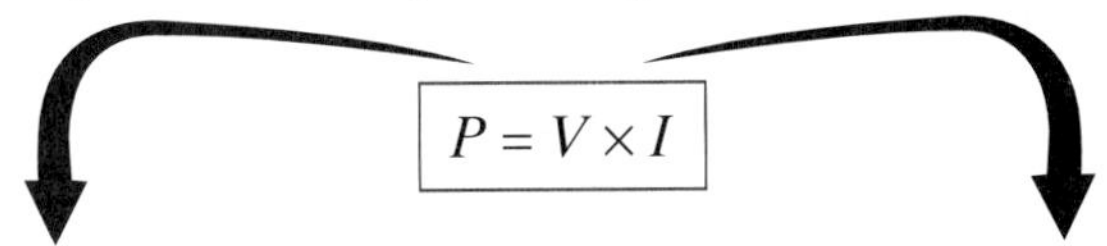

$$P = V \times I$$

Solve for V

$$P = V \times I$$

$$\frac{P}{I} = \frac{V \times I}{I} \quad \left(\text{divide both sides by } I\right)$$

$$\frac{P}{I} = \frac{V \times I}{I} \quad \left(I \div I = 1\text{, and } V \times 1 = V\right)$$

$$\frac{P}{I} = V$$

$$\boxed{V = \frac{P}{I}}$$

This formula can be used to calculate V when P and I are known

Solve for I

$$P = V \times I$$

$$\frac{P}{V} = \frac{V \times I}{V} \quad \left(\text{divide both sides by } V\right)$$

$$\frac{P}{V} = \frac{V \times I}{V} \quad \left(V \div V = 1\text{, and } I \times 1 = I\right)$$

$$\frac{P}{V} = I$$

$$\boxed{I = \frac{P}{V}}$$

This formula can be used to calculate I when P and V are known

MINI-MATH REVIEW

This mini-math review is designed to remind you of the mathematical details relating to substitution, which will be used in following sections.

Substitution

In the previous section we stated that $P = V \times I$, and by transposition we arrived at $V = P/I$ and $I = P/V$. When trying to calculate power we may not always have the values of V and I available. For example, we may know only I and R, or V and R. **Substitution** enables us to obtain alternative power formulas by replacing or substituting one mathematical term with an equivalent mathematical term. For example, since $I = V/R$, we could substitute I in any formula with V/R. Similarly, since $V = I \times R$, we could substitute V in any formula with $I \times R$. The following shows how we can substitute terms in the $P = V \times I$ formula to arrive at alternative power formulas for wattage calculations.

$$P = V \times I$$

$P = V \times I$ (substitute I with V/R)

$P = V \times \frac{V}{R}$ ($V \times V = V^2$)

$$P = \frac{V^2}{R}$$

This formula can be used to calculate P when V and R are known

$P = V \times I$ (substitute V with $I \times R$)

$P = (I \times R) \times I$ ($I \times I = I^2$)

$$P = I^2 \times R$$

This formula can be used to calculate P when I and R are known

EXAMPLE:

In Figure 3-23 a 12-V battery is connected across a 36-Ω resistor. How much power does the resistor dissipate?

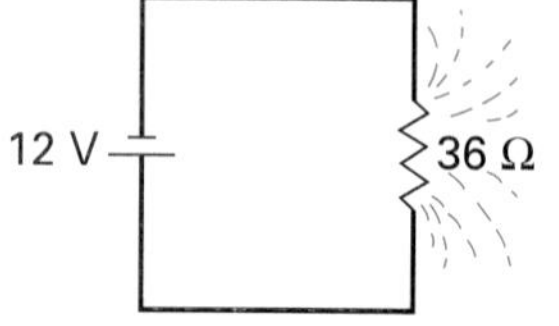

FIGURE 3-23

Solution:

CALCULATOR SEQUENCE

Step	Keypad Entry	Display Response
1.	[1] [2]	12
2.	[X²] (square key)	144
3.	[÷]	
4.	[3] [6]	36
5.	[=]	4

$$\text{power} = \frac{\text{voltage}^2}{\text{resistance}}$$

$$= \frac{12\ V^2}{36\ \Omega} = \frac{144}{36}$$

$$= 4\ W$$

Four joules of energy per second is being dissipated.

3-4-4 *Measuring Power*

Electrical power is measured with a wattmeter. The wattmeter shown in Figure 3-24 has four terminals, which must be connected correctly, as shown in Figure 3-25; in this example we are measuring the amount of power being supplied to a radio.

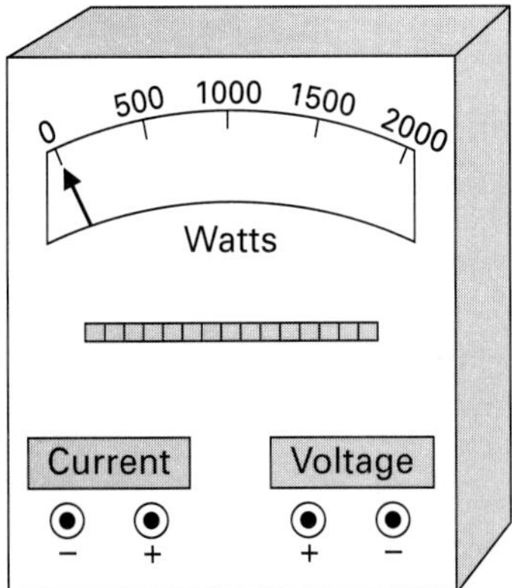

FIGURE 3-24 **Wattmeter.**

The current terminals of the wattmeter are connected so that the current passing from the battery to the radio has to pass through the wattmeter's current terminals so that the wattmeter can sense current, while the voltage terminals of the wattmeter are connected across the radio so as to sense the radio's voltage supply. The **wattmeter** is therefore like an ammeter and voltmeter in one package, and through a multiplication process ($P = I \times V$), we can obtain a direct reading of the amount of power being supplied to the radio.

Wattmeter
A meter used to measure electric power in watts.

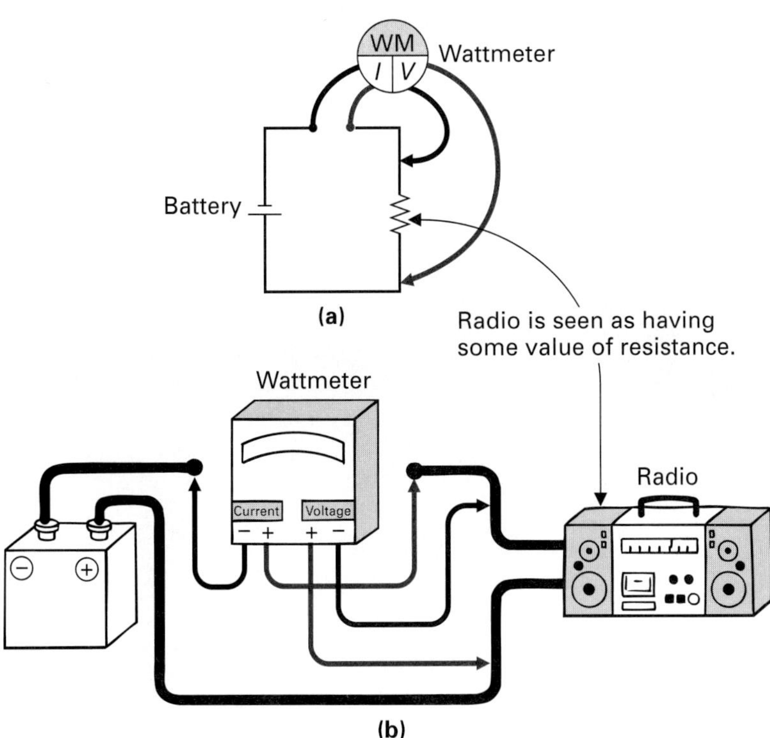

FIGURE 3-25 **Wattmeter Connections or Hookup. (a) Schematic Symbol. (b) Pictorial.**

3-4-5 The Kilowatt-Hour

Kilowatt-hour 1000 watts for 1 hour.

Kilowatt-hour meter A meter used by electric companies to measure a customer's electric power use in kilowatt-hours.

You and I pay for our electric energy in a unit called the **kilowatt-hour** (kWh). The **kilowatt-hour meter,** shown in Figure 3-26, measures how many kilowatt-hours are consumed, and the electric company then charges accordingly. Since power is the rate at which energy is used, if we multiply power and time, we can calculate how much energy has been consumed.

$$\text{energy consumed } (W) = \text{power } (P) \times \text{time } (t)$$

Energy is normally expressed in joules; however, since this formula uses the product of power (in watts) and time (in seconds or hours) we can use one of three other units of energy: the watt-second (Ws), watt-hour (Wh), or kilowatt-hour (kWh). The kilowatt-hour is most commonly used by electric companies. A *kilowatt-hour* of energy is consumed when you use 1000 watts (1 kW) of power in 1 hour.

$$\text{energy consumed (kWh)} = \text{power (kW)} \times \text{time (h)}$$

FIGURE 3-26 **Kilowatt-hour Meter Being Tested for Accuracy by a Journeyman Meter Tester. (Courtesy of Fred Vaughn, San Diego Gas and Electric.)**

FIGURE 3-26 *(cont.)* **Digital kilowatt-hour meter. (© Tim Stahl 1992, Courtesy of Fred Vaughn, San Diego Gas and Electric.)**

EXAMPLE:

If a 100-W light bulb is left on for 10 hours, how many kilowatt-hours will we be charged for?

Solution:

$$\begin{aligned}\text{power consumed (kWh)} &= \text{power (kW)} \times \text{time (hours)}\\ &= 0.1\text{ kW} \times 10\text{ hours} \quad (100\text{ W} = 0.1\text{ kW})\\ &= 1\text{ kWh}\end{aligned}$$

CALCULATOR SEQUENCE

Step	Keypad Entry	Display Response
1.	[0] [.] [1] [E] [3]	0.1E3
2.	[×]	
3.	[1] [0]	10
4.	[=]	1E3

EXAMPLE:

Figure 3-27 illustrates a typical household heater and an equivalent electrical circuit. The heater has a resistance of 7 Ω and the electric company is charging 6 cents/kWh. Calculate:

a. The energy consumed by the heater
b. The cost of running the heater for 7 hours

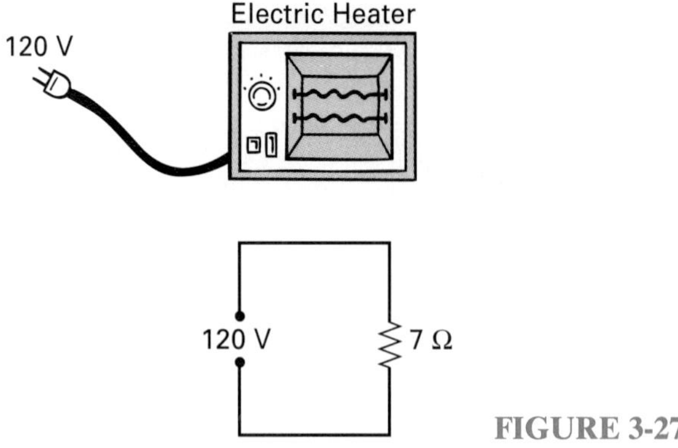

FIGURE 3-27

■ ***Solution:***

a. Power $(P) = \frac{V^2}{R} = \frac{120^2}{7}$

$= \frac{14400 \text{ V}}{7} = \frac{14.4 \text{ kV}}{7}$

$= 2057 \text{ W } (2 \text{ kW})$

b. Energy consumed = power (kW) × time (hours)

$= 2.057 \times 7 \text{ h}$

$= 14.399 \text{ kWh}$

Cost = kWh × rate = 14.399 × 6 cents

= 86 cents

SELF-TEST REVIEW QUESTIONS FOR SECTION 3-4

1. List the six basic forms of energy.
2. What is the difference between energy, work, and power?
3. List the formulas for calculating energy and power.
4. What is 1 kilowatt-hour of energy?

SUMMARY

Resistance (Figure 3-28)

1. Resistance is the opposition to current flow accompanied by the dissipation of heat.

2. Resistance and current are inversely proportional (1/∝) to one another; thus, if resistance is high, the current is low, and if resistance is low, the current is high. If resistance is increased by some value, current will be decreased by the same value. For example, if resistance is doubled, current is halved (assuming a constant voltage).

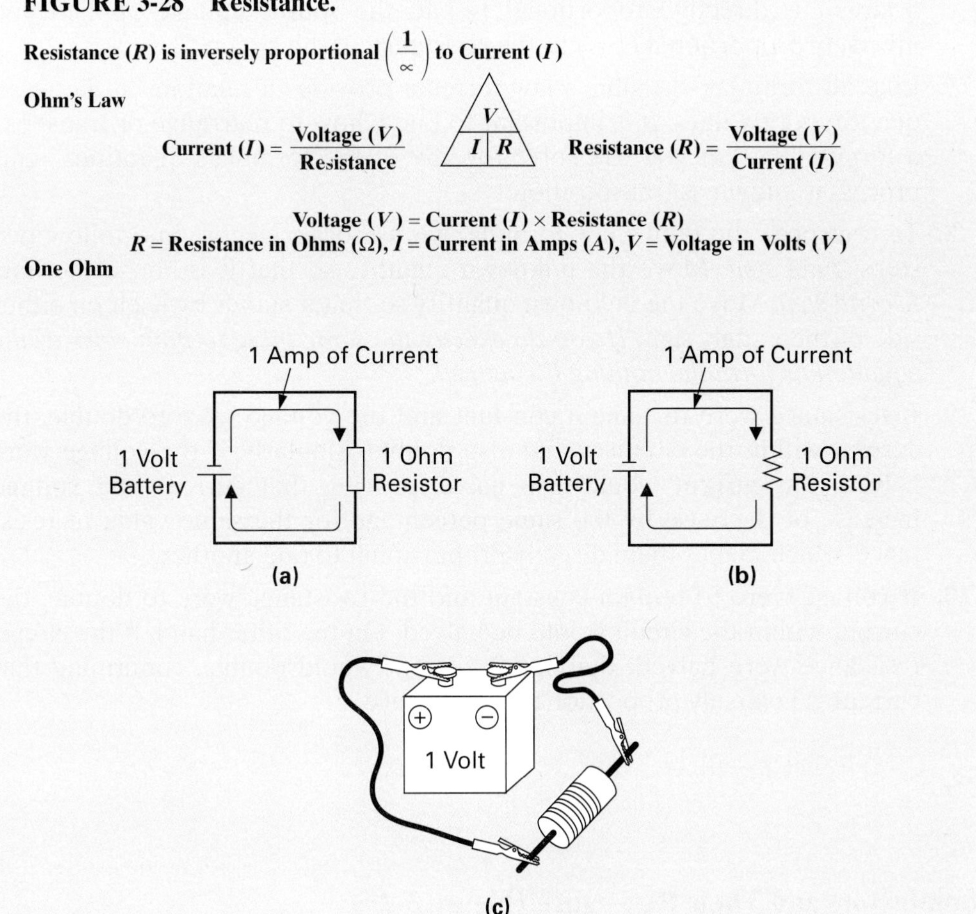

3. Current is measured in amperes, voltage is measured in volts, and resistance is measured in *ohms,* in honor of Georg Simon Ohm and his work with current, voltage, and resistance.
4. The larger the resistance, the larger the value of ohms and the more the resistor will oppose current flow.
5. The ohm is given the symbol Ω, which is the Greek capital letter omega. By definition, 1 ohm is the value of resistance that will allow 1 ampere of current to flow through a circuit when a voltage of 1 volt is applied. If a resistor is adjusted so that exactly 1 A of current is flowing around the circuit, the value of resistance obtained is referred to as 1 ohm (Ω).
6. Current flows in a circuit due to the force or voltage applied; however, the amount of current flow is limited by the amount of resistance in the circuit. In actual fact, the amount of current flow around a circuit is dependent on both voltage and resistance. This relationship between the three electrical

properties of current, voltage, and resistance was discovered by Georg Simon Ohm, a German physicist, in 1827.

Published originally in 1826, *Ohm's law* states that the current flow in a circuit is directly proportional ($\propto$) to the source voltage applied and inversely proportional ($1/\propto$) to the resistance of the circuit.

7. Like all formulas, the ohm's law formula provides a relationship between quantities, or values. It is important to know how to rearrange or transpose a formula so that you can solve for any of the formula's quantities—this process is known as transposition.
8. To transpose the ohm's law formula, we use basic algebra and follow two steps: *First step:* Move the unknown quantity so that it is above the line. *Second step:* Move the unknown quantity so that it stands by itself on either side of the equals sign. *If you do exactly the same thing to both sides of the equation or formula, nothing is changed.*
9. If resistance were to remain constant and the voltage were to double, the current within the circuit would also double. Similarly, if the voltage were halved, the current would also halve, proving that current and voltage increase or decrease by the same percentage for the same value of resistance, which makes them directly proportional to one another.
10. If voltage were to remain constant and the resistance were to double, the current within the circuit would be halved. On the other hand, if the circuit resistance were halved, the circuit current would double, confirming that current is inversely proportional to resistance.

Conductors and Their Resistance (Figure 3-29)

11. Resistors are normally made out of materials that cause an opposition to current flow. Conductance is the measure of how good a conductor is, and even the best conductors have some value of resistance.
12. Ohm's law can be used to determine the conductor's resistance based on the voltage applied and circuit current. Another way to determine resistance is to examine the *physical factors* of a conductor. The total resistance of a typical conductor is determined by four main physical factors:
    - **a.** The type of conducting material used for the flow of current
    - **b.** The conductor's cross-sectional area
    - **c.** The total length of the conductor
    - **d.** The temperature of the conductor
13. The resistance of a conductor is inversely proportional to the conductor's cross-sectional area. The greater the cross-sectional area, the lower the resistance, and similarly, the smaller the cross-sectional area, the higher the resistance.
14. As nearly all conductors are circular in cross section, the area of a conductor is measured in *circular mils* (cmil).

15. Increasing the length of the conductor used increases the amount of resistance within the circuit.

16. By combining all of a conductor's physical factors, we can arrive at a formula for resistance.

17. *Resistivity,* by definition, is the resistance (in ohms) that a certain length of material (in centimeters) will offer to the flow of current. In summary, therefore, resistance can be calculated based on either its physical factors or on electrical performance.

18. When heat is applied to a conductor, the atoms within the conductor convert this thermal energy into another form of energy, in this case mechanical energy or movement. These random moving atoms cause collisions between the directed electrons (current flow) and the adjacent atoms, resulting in an opposition to the current flow (resistance).

19. Metallic conductors are said to have a *positive temperature coefficient of resistance* (+ Temp. Coe. of R), because the greater the heat applied to the conductor, the greater the atom movement, causing more collisions of atoms to occur, and consequently the greater the conductor's resistance.

20. Any time that current flows through any conductor, a certain resistance or opposition is inherent in that conductor. This resistance will convert current to heat, and the heat further increases the conductor's resistance, causing more heat to be generated due to the opposition. Consequently, a conductor must be chosen carefully for each application so that it can carry the current without developing excessive heat. This is achieved by selecting a conductor with a greater cross-sectional area to decrease its resistance.

21. Conducting wires are normally covered with a plastic or rubber type of material, known as *insulation.* This insulation is used to protect the users and technicians from electrical shock and also to keep the conductor from physically contacting other conductors within the equipment. If the current through the conductor is too high, this insulation will burn due to the heat and may cause a fire hazard. The National Fire Protection Association has developed a set of standards known as the *American Wire Gauge* for all copper conductors, which lists their diameter, area, resistance, and maximum safe current in amperes.

22. Conductors have a positive temperature coefficient, which means that if temperature increases, so does resistance; but what happens if the temperature is decreased? In 1911, a Dutch physicist, Heike Onnes, discovered that mercury (a liquid conductor) lost its resistance to electrical current when the temperature was decreased to –459.7 Fahrenheit (0 on the Kelvin temperature scale). Mercury actually became a *superconductor,* allowing a supercurrent to flow and not encounter any resistance and therefore not generate any heat.

23. In the spring of 1986, two IBM scientists discovered that a conductor compound made up of barium, lanthanum, copper, and oxygen would superconduct (have no resistance to current flow) at –406° F.

Some scientific projects that need to make use of the advantages of superconductivity immerse these conductor compounds in a bath of liquid

FIGURE 3-29 Conductors and Their Resistance.

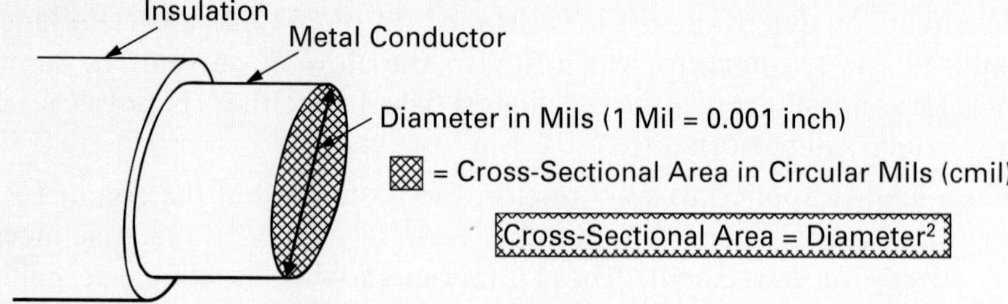

$$R = \frac{\rho \times l}{}$$

R = Resistance of Conductor in Ohms
ρ = Resistivity of Conducting Material
l = Length of Conductor in Feet
a = Area of Conductor in Circular Mils

TABLE 3-1 Material Resistivity

MATERIAL	RESISTIVITY (cmil/ft) IN OHMS
Silver	9.9
Copper	10.7
Gold	16.7
Aluminum	17.0
Tungsten	33.2
Zinc	37.4
Brass	42.0
Nickel	47.0
Platinum	60.2
Iron	70.0

Conductors have a positive temperature coefficient of resistance ($t \uparrow$ cause $R \uparrow$).

TABLE 3-2 The American Wire Gauge (AWG) for Copper Conductor

AWG NUMBER[a]	DIAMETER (mils)	MAXIMUM CURRENT (A)	Ω/1000 ft
0000	460.0	230	0.0490
000	409.6	200	0.0618
00	364.8	175	0.0780
0	324.9	150	0.0983
1	289.3	130	0.1240
2	257.6	115	0.1563
3	229.4	100	0.1970
4	204.3	85	0.2485
5	181.9	75	0.3133
6	162.0	65	0.3951
7	144.3	55	0.4982

FIGURE 3-29 ***(continued)*** **Conductors and Their Resistance.**

TABLE 3-2 **The American Wire Gauge (AWG) for Copper Conductor**

AWG NUMBER[a]	DIAMETER (mils)		MAXIMUM CURRENT (A)	Ω/1000 ft
8	128.5		45	0.6282
9	114.4		40	0.7921
10	101.9		30	0.9981
11	90.74		25	1.260
12	80.81		20	1.588
13	71.96		17	2.003
14	64.08		15	2.525
15	57.07			3.184
16	50.82		6	4.016
17	45.26	Wires		5.064
18	40.30	of this	3	6.385
19	35.89	size		8.051
20	31.96	have		10.15
22	25.35	current		16.14
26	15.94	measured		40.81
30	10.03	in mA		103.21
40	3.145			1049.0

[a]The larger the AWG number, the smaller the size of the conductor.

nitrogen so that they can achieve superconductivity. Liquid nitrogen, however, is difficult to handle, so the real scientific achievement will be to find a conductor compound that will superconduct at room temperature.

24. A cable is made up of two or more wires.

25. Wires and cables have to connect from one point to another. Some are soldered directly, whereas others are attached to plugs that plug into sockets.

Power (Figure 3-30)

26. The sun provides us with a consistent supply of energy in the form of light. Coal and oil are fossilized vegetation that grew, among other things, due to the sun, and are examples of *energy* that the earth has stored for millions of years. It can be said, then, that all energy begins from the sun.

On the earth, energy is not created or destroyed; it is merely transformed from one form to another. The transforming of energy from one form to another is called *work.* The greater the energy transformed, the more work that is done.

27. The six basic forms of energy are light, heat, magnetic, chemical, electrical, and mechanical energy. The unit for energy is the *joule* (J).

QUICK REFERENCE SUMMARY SHEET

FIGURE 3-30 Power.

$$W = Q \times V$$

W = Energy Stored in Joules (J)
Q = Coulombs of Charge
V = Voltage in Volts (V)

$$P = \frac{W}{t}$$

P = Power in Watts
W = Energy in Joules
t = Time in Seconds

$$P = I \times V$$

P = Power in Watts (W), I = Current in Amps (A), V = Voltage in Volts (V)

$$V = \frac{P}{I} \qquad I = \frac{P}{V}$$

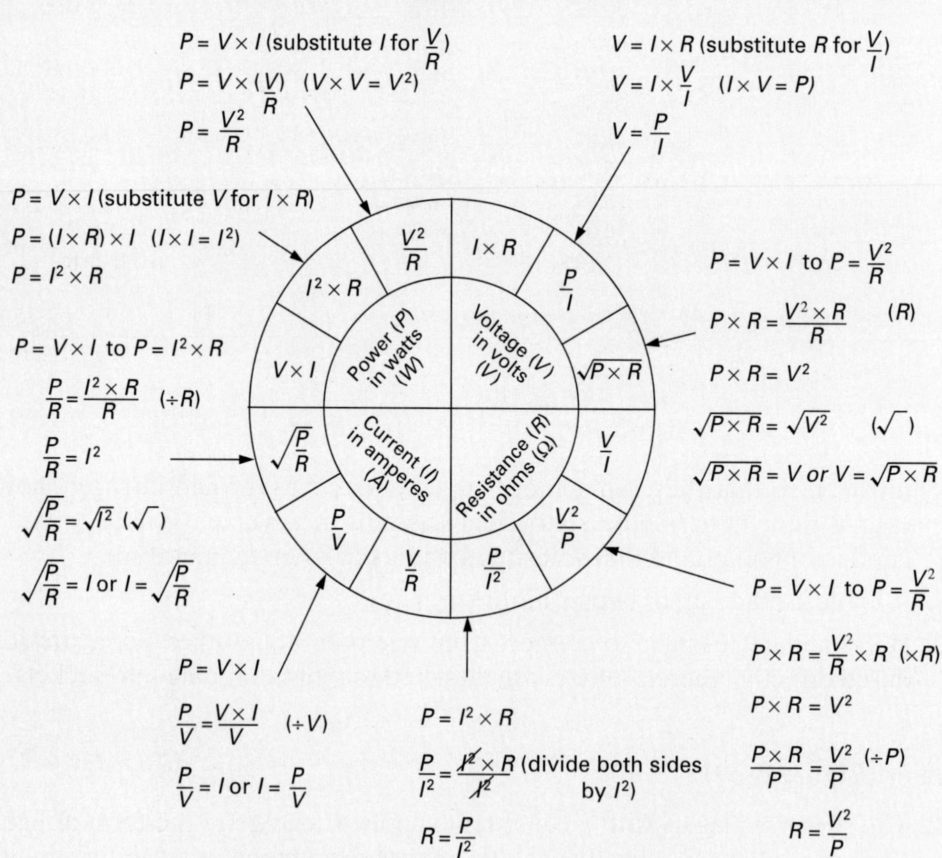

Energy Consumed (kWh) = Power (kW) × time (hours)

"Potential" (position) and "kinetic" (motion) are two terms used when describing energy. A cart on top of a hill has *potential* (position) *energy,* while a cart rolling down a hill has *kinetic* (motion) *energy.*

28. Chemical energy within the battery is converted to electrical energy when the electrons are attracted and repelled and set in motion. To use the two terms, we could say that the battery has the potential energy to set electrons in motion, and these moving electrons are said to possess kinetic energy.

29. Work is being done everytime one form of energy is transformed to another. A device that converts one form of energy to another is called a *transducer.*

30. Energy and work have the same symbol (W), the same formula, and the same unit (the joule). Energy is merely the capacity, potential, or ability to do work, and work is done when a transformation of the potential, capacity, or ability takes place.

31. *Power* (P) is the rate at which work is performed and is given the unit of *watt* (W), which is joules per second. Thus power involves a time factor.

32. Whether discussing a runner, walker, electric motor, heater, refrigerator, light bulb, or compact disk player—power is power. The output power, or power ratings, of electrical, electronic, or mechanical devices can be expressed in watts and describes the number of joules of energy converted every second.

The output power of rotating machines is given in the unit *horsepower* (hp), the output power of heaters is given in the unit *British thermal units per hour* (Btu/h), and the output power of cooling units is given in the unit *ton of refrigeration.* Despite the different names, they can all be expressed quite simply in the unit of watts.

33. The amount of energy stored (W) is dependent on the coulombs of charge stored (Q) and the voltage (V). The battery's energy stored is dependent on how many coulombs of electrons it holds (current) and how much electrical pressure it is able to apply to these electrons (voltage).

34. Power, in relation to electricity and electronics, is the rate (t) at which electric energy (W) is converted into some other form.

35. The $P = I \times V$ formula states that the amount of power delivered to a device is dependent on the electrical pressure or voltage applied across the device and the current flowing through the device. Looking at the power formula $P = I \times V$, we can say that 1 watt of power is expended when 1 ampere of current flows through a circuit that has 1 volt applied or when 1 ampere flows through 1 ohm.

36. When trying to calculate power we may not always have the values of V and I available. For example, we may know only I and R, or V and R. *Substitution* enables us to obtain alternative power formulas by replacing or substituting one mathematical term with an equivalent mathematical term. For example, since $I = V/R$, we could substitute I in any formula with V/R. Similarly, since $V = I \times R$, we could substitute V in any formula with $I \times R$.

37. Electrical power is measured with a wattmeter. The wattmeter has four terminals, which must be connected correctly. The current terminals of the wattmeter are connected so that the wattmeter's current terminals can sense current, while the voltage terminals of the wattmeter are connected to sense the voltage supply. The *wattmeter* is therefore like an ammeter and voltmeter in one package, and through a multiplication process ($P = I \times V$), we can obtain a direct reading of the amount of power being supplied.
38. You and I pay for our electric energy in a unit called the *kilowatt-hour* (kWh). The *kilowatt-hour meter* measures how many kilowatt-hours are consumed, and the electric company then charges accordingly. Since power is the rate at which energy is used, if we multiply power and time, we can calculate how much energy has been consumed.
39. The kilowatt-hour is most commonly used by electric companies. A *kilowatt-hour* of energy is consumed when you use 1000 watts (1 kW) of power in 1 hour.

NEW TERMS

Alligator clip
American Wire Gauge (AWG)
Binding post
Cable
Circular mil
Connector
Energy
Joule
Kelvin
Kilowatt hour
Kilowatt-hour meter
Kinetic Energy
Lug
Mil
Ohm
Phone
Plug
Positive Temperature Coefficient of Resistance
Potential Energy
Power
RCA plug and jack
Resistance
Resistivity
Socket
Superconductor
Temperature Coefficient of Resistance
Transducer
Watt
Wattmeter
Wire
Work

REVIEW QUESTIONS

Multiple-Choice Questions

1. Resistance is measured in:
   a. Ohms **c.** Amperes
   b. Volts **d.** Siemens
2. Current is proportional to:
   a. Resistance **c.** Both (a) and (b)
   b. Voltage **d.** None of the above
3. Current is inversely proportional to:
   a. Resistance **c.** Both (a) and (b)
   b. Voltage **d.** None of the above

4. If the applied voltage is equal to 15 V and the circuit resistance equals 5 Ω, the total circuit current would be equal to:

   a. 4 A b. 5 A c. 3 A d. 75 A

5. Calculate the applied voltage if 3 mA flows through a circuit resistance of 25 kΩ.

   a. 63 mV b. 25 V c. 77 μV d. 75 V

6. Energy is measured in:

   a. Volts b. Joules c. Amperes d. Watts

7. Chemical energy within a battery is converted into:

   a. Electrical energy c. Magnetic energy
   b. Mechanical energy d. Heat energy

8. The water pump has the potential to cause water flow, just as the battery has the potential to cause:

   a. Voltage c. Current
   b. Electron flow d. Both (b) and (c)

9. Work is measured in:

   a. Joules c. Amperes
   b. Volts d. Watts

10. The device that converts one energy form to another is called a:

    a. Transformer c. Transit
    b. Transducer d. Transistor

11. Power is the rate at which energy is transformed and is measured in:

    a. Joules c. Volts
    b. Watts d. Amperes

12. Power is measured by using a (an):

    a. Ammeter c. Ohmmeter
    b. Voltmeter d. Wattmeter

13. A good conductor has a:

    a. Large conductance figure c. Both (a) and (b)
    b. Small resistance figure d. None of the above

14. The resistance of a conductor is:

    a. Proportional to the length of the conductor
    b. Inversely proportional to the area of the conductor
    c. Both (a) and (b)
    d. None of the above

15. AWG is an abbreviation for:

    a. Alternate Wire Gauge c. American Wave Guide
    b. Alternating Wire Gauge d. American Wire Gauge

Essay Questions

16. What is resistance? (3-1)

17. Briefly describe why:

a. Current is proportional to voltages. (3-2-3)
b. Current is inversely proportional to resistance. (3-2-4)

18. State Ohm's law. (3-2-1)

19. List the three forms of Ohm's law. (3-2-2)

20. List the six basic forms of energy. (3-4)

21. Briefly describe the following terms:

a. Potential energy (3-4)
b. Kinetic energy (3-4)

22. What is a transducer? (3-4)

23. Define power. (3-4-1)

24. Give three formulas for electric power. (3-4-3)

25. Give the units for each of the following:

a. Energy **e.** Work
b. Power **f.** Current
c. Voltage **g.** Charge
d. Resistance **h.** Conductance

26. What is the difference between work and power? (3-4)

27. What instrument is used to measure electrical power? (3-4-4)

28. State the formula used by electric companies to determine the amount of power consumed? (3-4-5)

29. What is 1 kilowatt-hour? (3-4-5)

30. List the four factors that determine a conductor's resistance. (3-3)

31. What is a circular mil? (3-3-2)

32. Define the resistivity of a conducting material. (3-3-4)

33. Describe why conductors have a positive temperature coefficient of resistance. (3-3-5)

34. What is the purpose(s) of placing an insulating sheath over conducting wires? (3-3-6)

35. What is the American Wire Gauge? (3-3-6)

36. What is a superconductor? (3-3-7)

37. List some of the advantages of superconductivity. (3-3-7)

38. What is the difference between a wire and a cable? (3-3-8)

39. Give some examples of different wires and cables. (3-3-8)

40. List examples of different conductor connectors. (3-3-8)

Practice Problems

41. An electric heater with a resistance of 6 Ω is connected across a 120-V wall outlet.
    a. Calculate the current flow. (3-2)
    b. Draw the schematic diagram.
42. What source voltage would be needed to produce a current flow of 8 mA through a 16-kΩ resistor? (3-2)
43. If an electric toaster draws 10 A when connected to a power outlet of 120 V, what is its resistance? (3-2)
44. Calculate the power used in Problems 41, 42, and 43. (3-4-3)
45. Calculate the current flowing through the following light bulbs when they are connected across 120 V: (3-4-3)
    a. 300 W **b.** 100 W **c.** 60 W **d.** 25 W
46. If an electric company charges 9 cents/kWh, calculate the cost for each light bulb in Problem 45 if on for 10 hours. (3-4-5)
47. Indicate which of the following unit pairs is larger: (3-4)
    a. Millivolts or volts
    b. Microamperes or milliamperes
    c. Kilowatts or watts
    d. Kilohms or megohms
48. Calculate the resistance of 200 ft of copper having a diameter of 80 mils. (3-3-4)
49. What AWG size wire should be used to safely carry just over 15 A? (3-3-6)
50. Calculate the voltage dropped across 1000 ft of No. 4 copper conductor when a current of 7.5 A is flowing through it. (3-3-6)
51. Calculate the unknown resistance in a circuit when an ammeter indicates that a current of 12 mA is flowing and a voltmeter indicates 12 V. (3-2-4)
52. What battery voltage would use 1000 J of energy to move 40 C of charge through a circuit? (3-4-2)
53. Calculate the resistance of a light bulb that passes 500 mA of current when 120 V is applied. What is the bulb's wattage? (3-4)
54. Which of the following circuits has the largest resistance and which has the smallest? (3-4)
    a. $V = 120$ V, $I = 20$ mA **c.** $V = 9$ V, $I = 100$ μA
    b. $V = 12$ V, $I = 2$ A **d.** $V = 1.5$ V, $I = 4$ mA
55. Calculate the power dissipated in each circuit in Problem 54. (3-4-3)
56. How many watts are dissipated if 5000 J of energy is consumed in 25 s? (3-4-3)
57. Convert the following:
    a. 1000 W = ____________ kW
    b. 0.345 W = ____________ mW
    c. 1250×10^3 W = ____________ MW
    d. 0.00125 W = ____________ μW

58. What is the value of the resistor when a current of 4 A is causing 100 W to be dissipated? (3-4-3)

59. How many kilowatt-hours of energy are consumed in each of the following: (3-4-5)

a. 7500 W in 1 hour **c.** 127,000 W for half an hour
b. 25 W for 6 hours

60. What is the maximum output power of a 12-V, 300-mA power supply? (3-4-3)

CHAPTER

4

OBJECTIVES

After completing this chapter, you will be able to:

1. Describe the difference between a fixed- and a variable-value resistor.
2. Explain the differences between the six basic types of fixed-value resistors: carbon composition, carbon film, metal film, wirewound, metal oxide, and thick film.
3. Identify the different resistor wattage ratings, and their value and tolerance labelling methods.
4. Describe the SIP, DIP, and chip thick-film resistor packages.
5. Explain the following types of variable resistors:
   - **a.** Mechanically adjustable: rheostat and potentiometer.
   - **b.** Thermally adjustable: RTD, TFD, and thermistor.
   - **c.** Optically adjustable: photoresistor.
6. List the three rules to remember when measuring resistance with an ohmmeter.
7. Explain how a resistor's value and tolerance are printed on the body of the component either:
   - **a.** By color coding, or
   - **b.** Typographically
8. Describe the difference between a general-purpose and a precision resistor.
9. Explain why current through any resistance is always accompanied by heat.
10. State the purpose of the filament and ballast resistor.
11. Describe some of the more common resistor problems.

Resistors

OUTLINE

VIGNETTE

Catch the Wave

In the true tradition of the electronics industry, Joel Naive founded his company, Wavetek, in his garage in 1962. As the first of two employees, Joel concentrated on technology while his wife Kathy provided the business counsel. He worked on refining the design of a function generator and also on expanding its capabilities. Thus the Model 101 was born. The prototype was hand-carried to Los Angeles and was exhibited at the West Coast's Electronic Show (WESCON) in September 1962. Joel's total out-of-pocket expenses up to this point—including costs for the trip to LA and a bottle of Scotch for a friend who let Joel sleep on his couch—was $435. The 101 was a complete success. The product line was sold worldwide, and the company grew into an international organization, with annual sales of approximately $100 million.

In fact, Joel's approach to engineering is "simple." He starts each project by writing $E = IR$ on the chalkboard. Then he says, "Let's start with the basics." His approach to design is through practical experimentation in addition to theoretical mathematics.

Besides test and measurement, Joel has a passion for classic automobiles of the 1920s era, which he restores and enters in international rallies. This business he appropriately named Garage Number 2.

INTRODUCTION

Resistance would seem to be an undesirable effect, as it reduces current flow and wastes energy as it dissipates heat. Resistors, however, are probably used more than any other component in electronics, and in this chapter we discuss all the different types.

4-1 RESISTOR TYPES

Resistor
Component made of a material that opposes the flow of current and therefore has some value of resistance.

Conductors offer a certain small amount of resistance; however, in electronics this resistance is not normally enough, so additional resistance is needed to control the amount of current flow. The component used to supply this additional resistance is called a **resistor.**

There are two basic types of resistors: fixed value and variable value. The fixed-value resistor, examples of which can be seen in Figure 4-1, has a value of

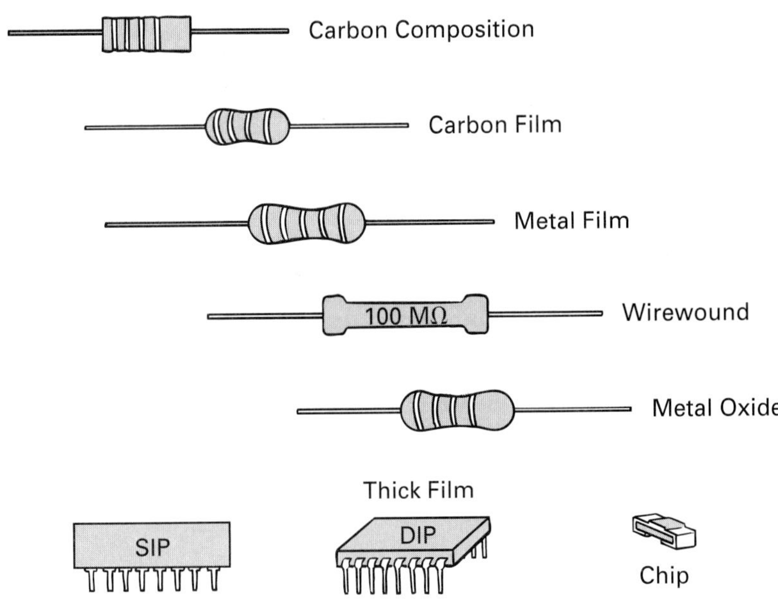

FIGURE 4-1 **Fixed-Value Resistors.**

resistance that cannot be changed and is the more common of the two. The variable-value resistor, on the other hand, has a range of values that can be selected generally by mechanically adjusting a control, as illustrated in Figure 4-2.

Let's discuss the different types of fixed- and variable-value resistors.

4-1-1 Fixed-Value Resistors

Fixed-Value Resistor
A resistor whose value cannot be changed.

Fixed-value resistors can be divided into six basic categories:

1. Carbon composition
2. Carbon film
3. Metal film
4. Wirewound
5. Metal oxide
6. Thick film

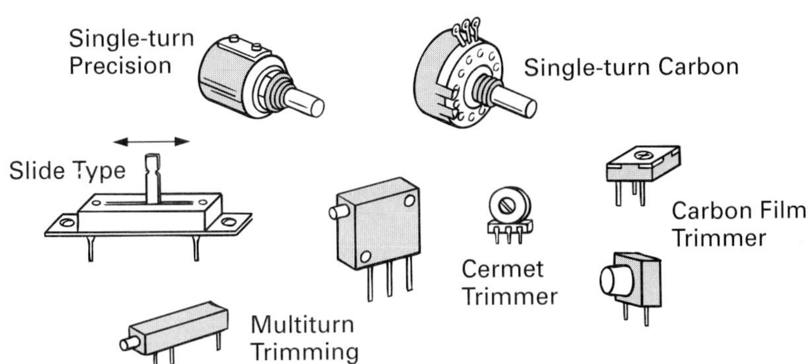

FIGURE 4-2 **Variable-Value Resistors.**

Carbon Composition Resistor Fixed resistor consisting of carbon particles mixed with a binder, which is molded and then baked. Also called a composition resistor.

Carbon Composition Resistors

This is the most common and least expensive type of fixed-value resistor, the appearance of which can be seen in Figure 4-3(a). It is constructed by placing a piece of resistive material, with embedded conductors at each end, within an insulating cylindrical molded case, as illustrated in Figure 4-3(b).

This resistive material is called carbon composition because powdered carbon and powdered insulator are bonded into a compound and used as the resistive material. By changing the ratio of powdered insulator to carbon, the value of resistance can be changed within the same area. For example, Figure 4-4 illustrates four fixed-value resistors, ranging from 2 Ω to 10 MΩ.

The 2-Ω resistor is exactly the same size as the 10-MΩ (10 million Ω) resistor. This is achieved by having more powdered carbon and less powdered insulator in the 2 Ω and less powdered carbon and more powdered insulator in the 10 MΩ. The color-coded rings or bands on the resistors in Figure 4-3 are a means of determining the value of resistance. This and other coding systems will be discussed later.

Dissipation Release of electrical energy in the form of heat.

Wattage Rating Maximum power a device can safely handle continuously.

The physical size of the resistors lets the user know how much power in the form of heat can be **dissipated,** as shown in Figure 4-5. As you already know, resistance is the opposition to current flow or electrons, and this opposition causes heat to be generated when current flows. The amount of heat dissipated per unit of time is measured in watts and each resistor has its own **wattage rating.** The bigger the resistor, the more heat that can be dissipated, so a large resistor could dissipate heat at a rate of 2 W, while a small resistor could dissipate heat at a rate of only ⅛ W. The key point is that heat should be dissipated faster than it is generated; if not, a larger resistor with a larger surface area to dissipate away the additional heat should be used so that the resistor will not burn up.

Tolerance Permissible deviation from a specified value, normally expressed as a percentage.

Another factor to consider when discussing resistors is their **tolerance.** Tolerance is the amount of deviation or error from the specified value. For example,

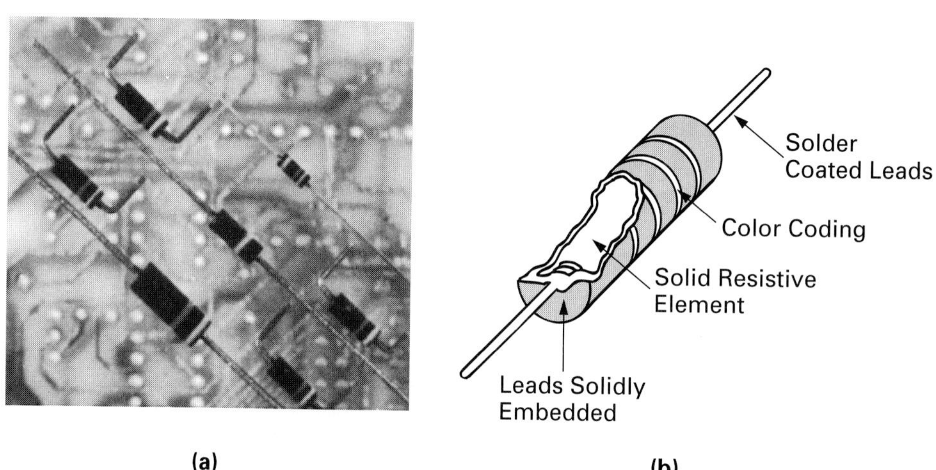

FIGURE 4-3 **Carbon Composition Resistors. (a) Appearance. (b) Construction. (Courtesy of Stackpole Electronics, Inc.)**

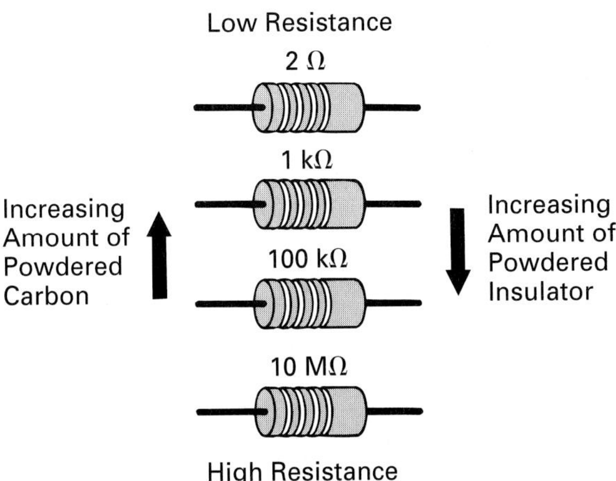

FIGURE 4-4 Carbon Composition Resistor Ratios.

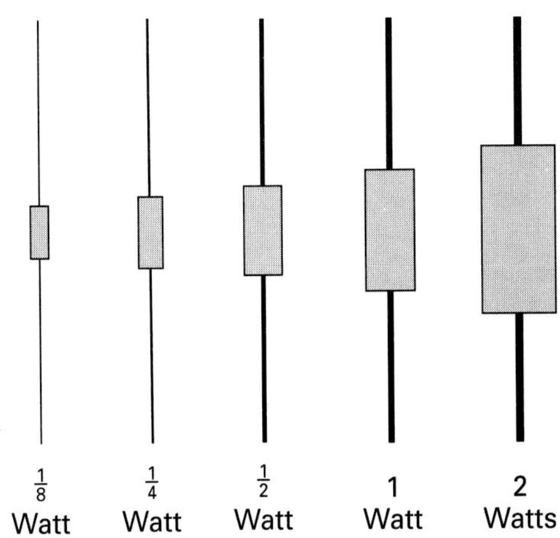

FIGURE 4-5 Resistor Wattage Rating Guide. All resistor silhouettes are drawn to scale.

a 1000-Ω (1-kΩ) resistor with a ±10% (plus and minus 10%) tolerance when manufactured could have a resistance anywhere between 900 and 1100 Ω.

$$\pm 10\% \text{ of } 1000 = 100$$

$$\underset{900}{\underset{\downarrow}{10\% -}} \boxed{1000} \underset{1100}{\underset{\downarrow}{+10\%}}$$

This means that two identically marked resistors when measured could be from 900 to 1100 Ω, a difference of 200 Ω. In some applications, this may be acceptable, although in others, where high precision is required, this deviation could be too large, so more expensive, smaller tolerance resistors are used.

EXAMPLE:

Calculate the amount of deviation of the following resistors:

a. 2.2 kΩ ± 10%
b. 5 MΩ ± 2%
c. 3 Ω ± 1%

Solution:

a. 10% of 2.2 kΩ = 220 Ω. For + 10%, the value is

$$2200 + 220\ \Omega = 2420\ \Omega$$
$$= 2.42\ \text{k}\Omega$$

For −10%, the value is

$$2200 - 220\ \Omega = 1980\ \Omega$$
$$= 1.98\ \text{k}\Omega$$

The resistor will measure anywhere from 1.98 kΩ to 2.42 kΩ.

CALCULATOR SEQUENCE

Step	Keypad Entry	Display Response
1.	1 0	10
2.	%	10
3.	×	0.10
4.	2 . 2 E 3	2.2E3
5.	=	220

b. 2% of 5 MΩ = 100 kΩ

$$5\ \text{M}\Omega + 100\ \text{k}\Omega = 5.1\ \text{M}\Omega$$
$$5\ \text{M}\Omega - 100\ \text{k}\Omega = 4.9\ \text{M}\Omega$$
$$\text{Deviation} = 4.9\ \text{M}\Omega \text{ to } 5.1\ \text{M}\Omega$$

c. 1% of 3 Ω = 0.03 Ω or 30 milliohms (mΩ).

$$3\ \Omega + 0.03\ \Omega = 3.03\ \Omega$$
$$3\ \Omega - 0.03\ \Omega = 2.97\ \Omega$$
$$\text{Deviation} = 2.97\ \Omega \text{ to } 3.03\ \Omega$$

Carbon Film Resistors

Carbon Film Resistor
Thin carbon film deposited on a ceramic form to create resistance.

Substrate
The mechanical insulating support on which a device is fabricated.

Figure 4-6(a) illustrates the physical appearance of **carbon film resistors.** They are constructed, as shown in Figure 4-6(b), by first depositing a thin layer or film of resistive material (a blend of carbon and insulator) on a ceramic (insulating) **substrate.** The film is then cut to form a helix or spiral. A greater ratio of carbon to insulator will achieve a low-resistance helix; on the other hand, a greater ratio of insulator to carbon will create a higher-resistance helix. Carbon film resistors have smaller tolerance figures (±5% to ±2%), are more stable (maintain same resistance value over a wide range of temperatures), and have less internally generated noise (random small bursts of voltage) than do carbon composition resistors.

Metal Film Resistors

Metal Film Resistor
A resistor in which a film of a metal, metal oxide, or alloy is deposited on an insulating substrate.

Figure 4-7(a) illustrates the physical appearance of some typical **metal film resistors.** They are constructed by spraying a thin film of metal on a ceramic cylinder (or substrate) and then cutting the film to form a substrate, as shown in

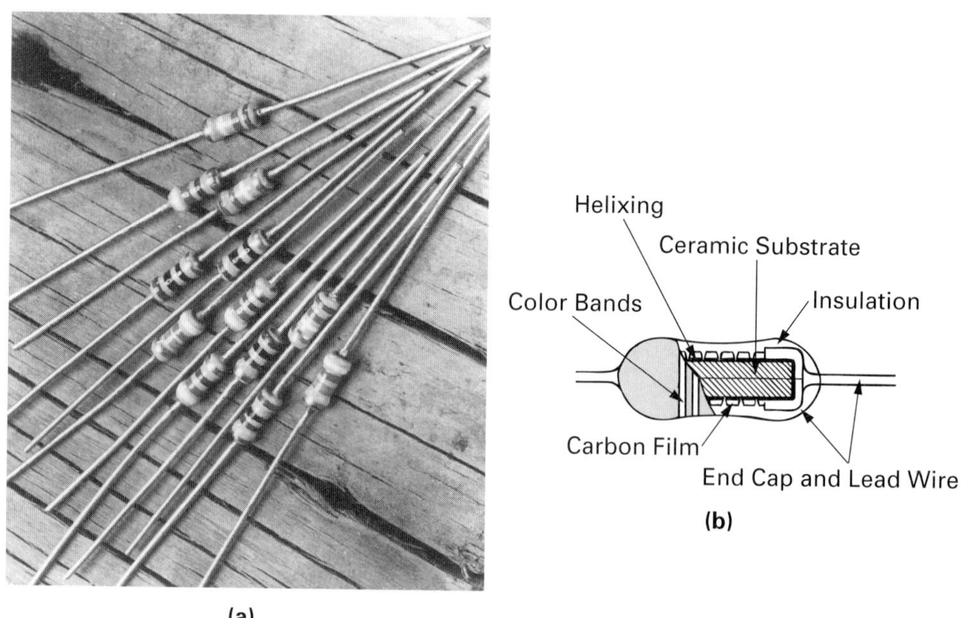

FIGURE 4-6 Carbon Film Resistors. (a) Appearance. (b) Construction. (Courtesy of Stackpole Electronics, Inc.)

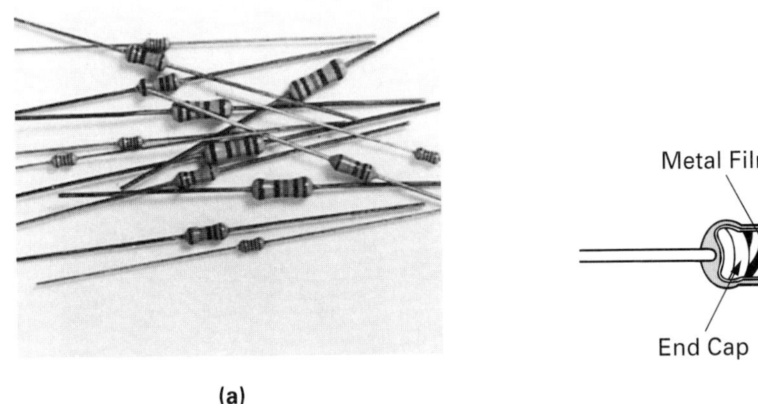

(a)

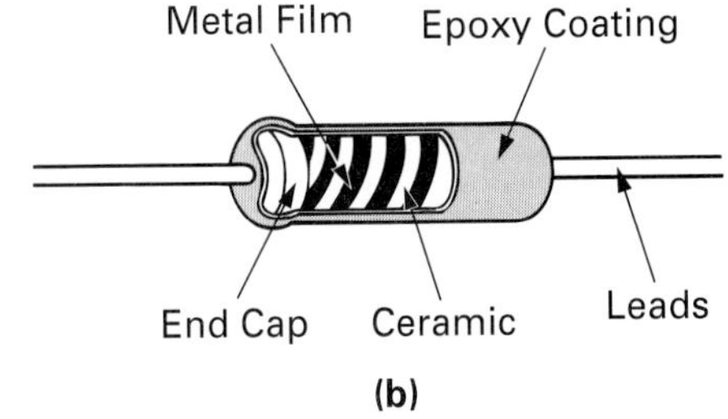

(b)

FIGURE 4-7 **Metal Film Resistors. (a) Appearance. (b) Construction. (Courtesy of Stackpole Electronics, Inc.)**

Figure 4-7(b). Metal film resistors have possibly the best tolerances commercially available of ±1% to ±0.1%. They also maintain a very stable resistance over a wide range of temperatures (good stability) and generate very little internal noise compared to any carbon resistor.

Wirewound Resistors

Figure 4-8(a) and (c) illustrate the appearance of different types of **wirewound resistors.** They are constructed, as can be seen in Figure 4-8(b), by wrapping a length of wire uniformly around a ceramic insulating core, with terminals making the connections at each end. The length and thickness of the wire are varied to change the resistance, which ranges from 1 Ω to 150 kΩ.

Wirewound Resistor
Resistor in which the resistive element is a length of high-resistance wire or ribbon, usually nichrome, wound onto an insulating form.

Since current flow is opposed by resistors and this opposition generates heat, the larger the physical size of the resistor, the greater the amount of heat that can be dissipated away and so the greater the current that can be passed through the resistor. Wirewound resistors are generally used in applications requiring low resistance values, which means that the current and therefore power dissipated are high ($R\downarrow = V/I\uparrow$, $P\uparrow = I^2\uparrow R$). Consequently, these resistors are designed to have large surface areas so that they can safely dissipate away the heat. The amount of heat that can be dissipated is measured in watts and is indicated on the resistor. A 10-W resistor can be used to dissipate 20 W if air is blown across its surface by a cooling fan or if it is immersed in a coolant. Wirewound resistors typically have good tolerances of ±1%; however, their large physical size and difficult manufacturing process make them very expensive.

Metal Oxide Resistors

Figure 4-9(a) illustrates the physical appearance of some **metal oxide resistors.** They are constructed, as can be seen in Figure 4-9(b), by depositing an oxide of a metal such as tin onto an insulating substrate. The ratio of oxide (insulator) to tin (conductor) will determine the resistor's resistance.

Metal Oxide Resistor
A metal film resistor in which an oxide of a metal (such as tin) is deposited as a film onto the substrate.

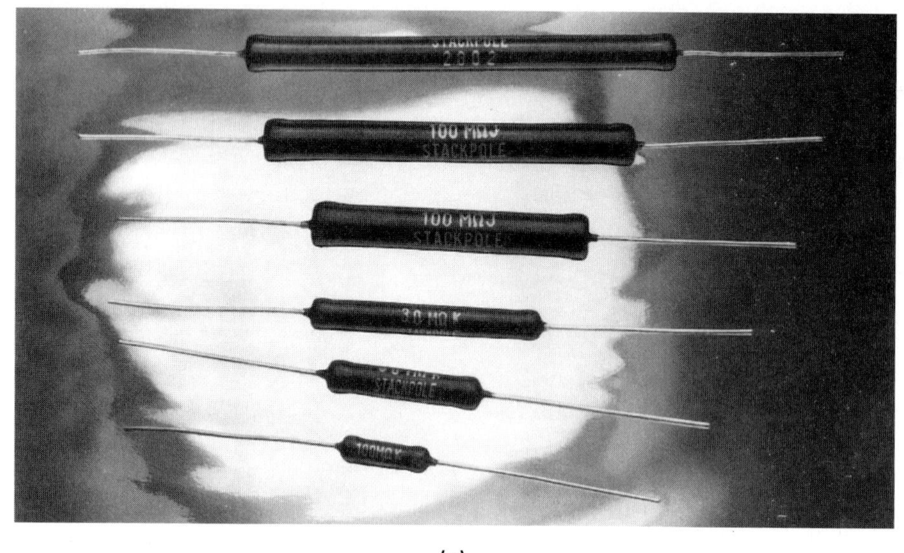

(a)

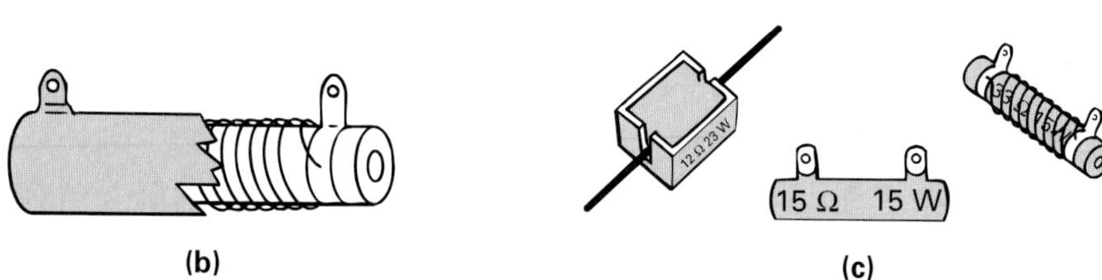

(b) (c)

FIGURE 4-8 Wirewound Resistors. (a) Appearance. (b) Construction. (c) Other Types. (Courtesy of Stackpole Electronics, Inc.)

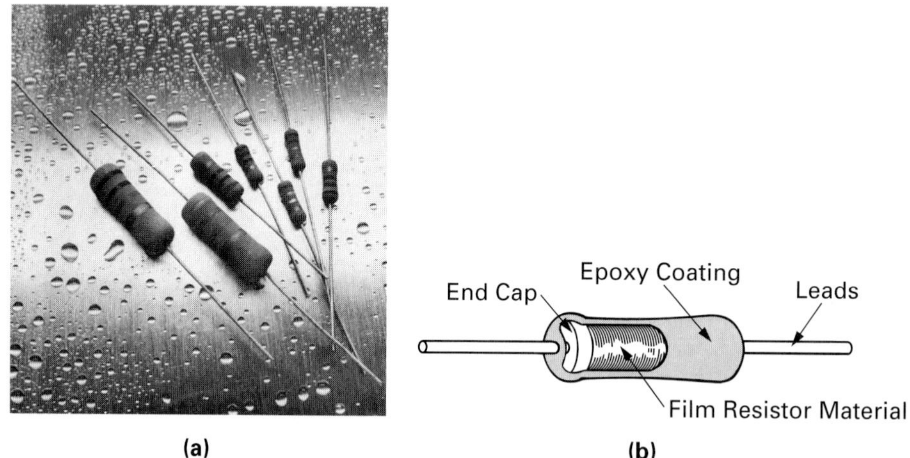

(a) (b)

FIGURE 4-9 Metal Oxide Resistors. (a) Appearance. (b) Construction. (Courtesy of Stackpole Electronics, Inc.)

Metal oxide resistors have excellent temperature stability, which is the ability of the resistor to maintain its value of resistance without change even when temperature is changed. There is a tendency when heat is increased for atoms to move, and in so doing they cause collisions, resulting in an opposition to current; so a temperature increase will cause a resistor's resistance to increase. This increase from the desired resistance is very small with metal oxide resistors.

Thick-Film Resistors

Figure 4-10 illustrates examples of **thick-film resistors.** Figure 4-10(a) and (b) shows the two different types of resistor networks, called SIPs and DIPs. The **single in-line package** is so called because all its lead connections are in a single line, whereas the **dual in-line package** has two lines of connecting pins. The chip resistors shown in Figure 4-10(c) are small thick-film resistors that are approximately the size of a pencil lead.

Thick-Film Resistors
Fixed-value resistor consisting of a thick-film resistive element made from metal particles and glass powder.

The SIP and DIP resistor networks are constructed by first screening on the internal conducting strip (silver) that connects the external pins to the resistive material, and then screening on the thick film of resistive paste (bismuth/ruthenate), the blend of which will determine resistance. The chip resistor uses the same resistive film paste material, which is deposited onto an insulating substrate with two end terminations and protected by a glass coat.

Single In-Line Package (SIP)
Package containing several electronic components (generally resistors) with a single row of external connecting pins.

The SIP and DIP resistor networks, once constructed, are trimmed by lasers to obtain close tolerances of typically ±2%. Resistance values ranging from 22 Ω to 2.2 MΩ are available, with a power rating of ½ W.

Dual In-Line Package (DIP)
Package that has two (dual) sets or lines of connecting pins.

The chip resistor shown in Figure 4-10(c) is commercially available with resistance values from 10 Ω to 3.3 MΩ, a ±2% tolerance, and a ⅛-W heat dissipation capability. They are ideally suited for applications requiring physically small sized resistors, as explained in the inset in Figure 4-10(c). The chip resistor is called a **surface mount technology** (SMT) device. The key advantage of SMT devices over "through-hole" devices is that a through-hole device needs both a hole in the printed circuit board (PCB) and a connecting pad around the hole. With the SMT device, no holes are needed since the package is soldered directly onto the surface of the PCB. Without the need for holes, pads can be placed closer together. This results in a considerable space saving, as you can see in the inset in Figure 4-10(c) by comparing the through-hole resistor to the surface mount chip resistor.

Surface Mount Technology
A method of installing tiny electronic components on the same side of a circuit board as the printed wiring pattern that interconnects them.

4-1-2 *Variable-Value Resistors*

In certain applications we may require a variation in resistance while in circuit: for example, in the volume adjustment on a radio or television. The component that achieves this is the variable resistor.

The resistance of **variable resistors** can be varied in one of three ways:

Variable Resistor
A resistor whose value can be changed.

1. Mechanically (user) adjustable:
   a. Rheostat
   b. Potentiometer

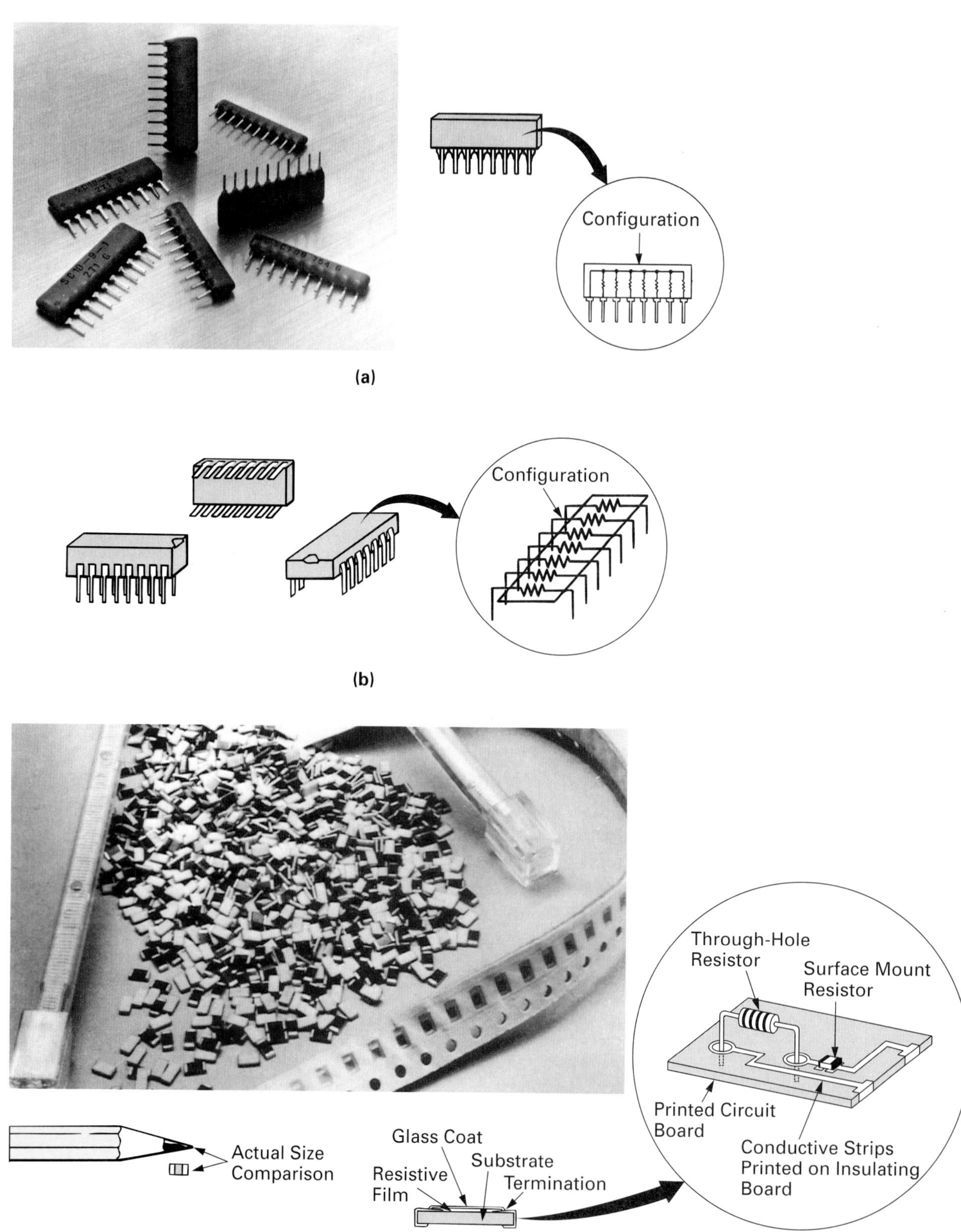

FIGURE 4-10 Thick-Film Resistors. (a) SIPs: Single In-Line Packages. (b) DIPs: Dual In-Line Packages. (c) Chips. (Courtesy of Stackpole Electronics, Inc.)

2. Thermally (heat) adjustable:
   a. Thermistor
   b. Resistive temperature detector (RTD)
   c. Thin-film detector (TFD)
3. Optically (light) adjustable:
   a. Photoresistor

Mechanically (User) Adjustable Variable Resistors

This category includes two types, both of which will cause a change in resistance when a shaft is rotated.

Rheostat (two terminals: A and B). Figure 4-11(a) shows the physical appearance of different rheostats, while Figure 4-11(b) illustrates schematic symbols. As can be seen in the construction of a circular rheostat in Figure 4-11(c), one terminal is connected to one side of the track and the other terminal of this two-terminal device is connected to a movable wiper. As the wiper is moved away from the end of the track with the terminal, the resistance between the stationary end terminal and the mobile wiper terminal increases. This is summarized in Figure 4-11(d), where the wiper has been moved down by a clockwise rotation of the shaft. Current would have to flow through a large resistance as it travels from one terminal to the other. On the other hand, as the wiper is moved closer to the end of the track connected to the terminal, the resistance decreases. This is summarized in Figure 4-11(e), which shows that as the wiper is moved up, as a result of turning the shaft counterclockwise, current will see only a small resistance between the two terminals.

Rheostat
Two-terminal variable resistor that, through mechanical turning of a shaft, can be used to vary its resistance and therefore its value of terminal to terminal current.

Rheostats come in many shapes and sizes, as can be seen in Figure 4-11(a). Some employ a straight-line motion to vary resistance, while others are classified as circular-motion rheostats. The resistive elements also vary; wirewound and carbon tracks are very popular. Cermet, which is a ceramic (insulator)–metal (conductor) mix the ratio of which can be used to produce different values of resistive tracks, is also used. A trimming rheostat is a miniature device used to change resistance by a small amount. Other circular-motion rheostats are available that require between two to ten turns to cover the full resistance range.

Potentiometer (three terminals: A, B, and C). Figure 4-12(a) illustrates the physical appearance of a variety of potentiometers, also called pots (slang), while Figure 4-12(b) shows its schematic symbol. You will probably notice that the difference between a rheostat and potentiometer is the number of terminals; the rheostat has two terminals and the potentiometer has three. With the rheostat, there were only two terminals and the resistance between the wiper and terminal varied as the wiper was adjusted. For the potentiometer illustrated in Figure 4-12(c), you can see that resistance can actually be measured across three separate combinations: between A and B (X), between B and C (Y), and between C and A (Z).

Potentiometer
Three-lead variable resistor that through mechanical turning of a shaft can be used to produce a variable voltage or potential.

The only difference between the rheostat and the potentiometer in construction is the connection of a third terminal to the other end of the resistive track, as can be seen in Figure 4-12(d), which shows the single-turn potentiometer. Also illustrated in this section of the figure is the construction of a multiturn

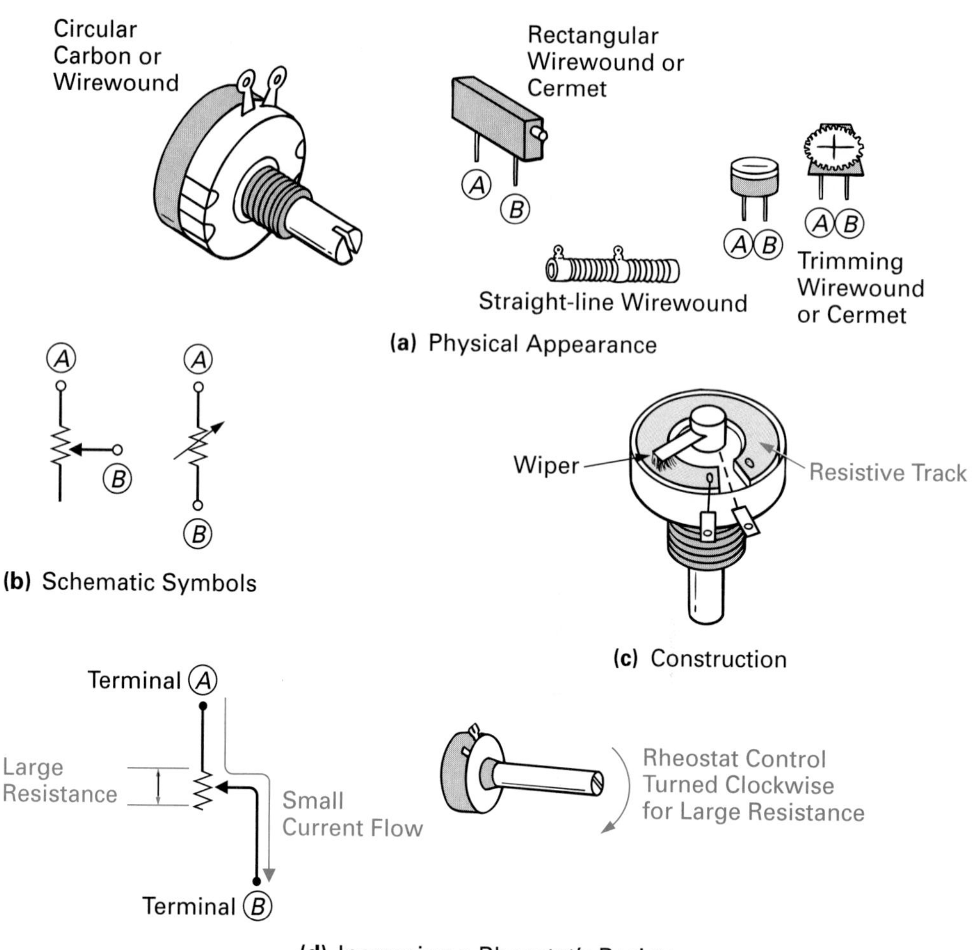

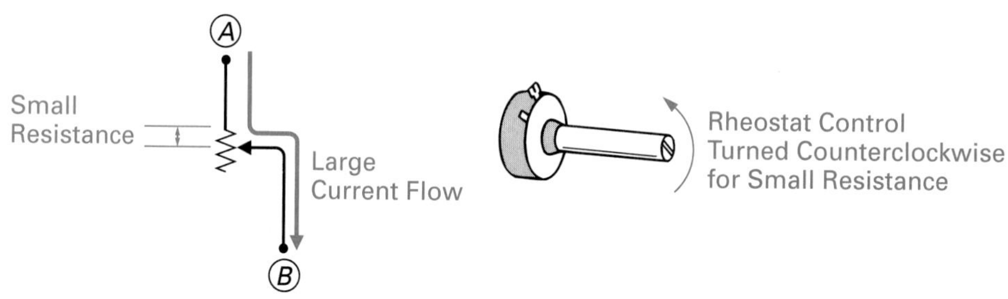

FIGURE 4-11 **Rheostat. (a) Physical Appearance. (b) Schematic Symbols. (c) Construction. (d) Increasing a Rheostat's Resistance. (e) Decreasing a Rheostat's Resistance.**

potentiometer in which a contact arm slides along a shaft and the resistive track is formed into a helix of 2 to 10 coils.

Figure 4-13(a) illustrates a 10-kΩ potentiometer. The resistance measured between terminals *A* and *C* will always be the same (10 kΩ) no matter where we put the wiper, because current still has to travel through the complete resistance between *A* and *C*, as illustrated in Figure 4-13(b) with a 10-kΩ potentiometer.

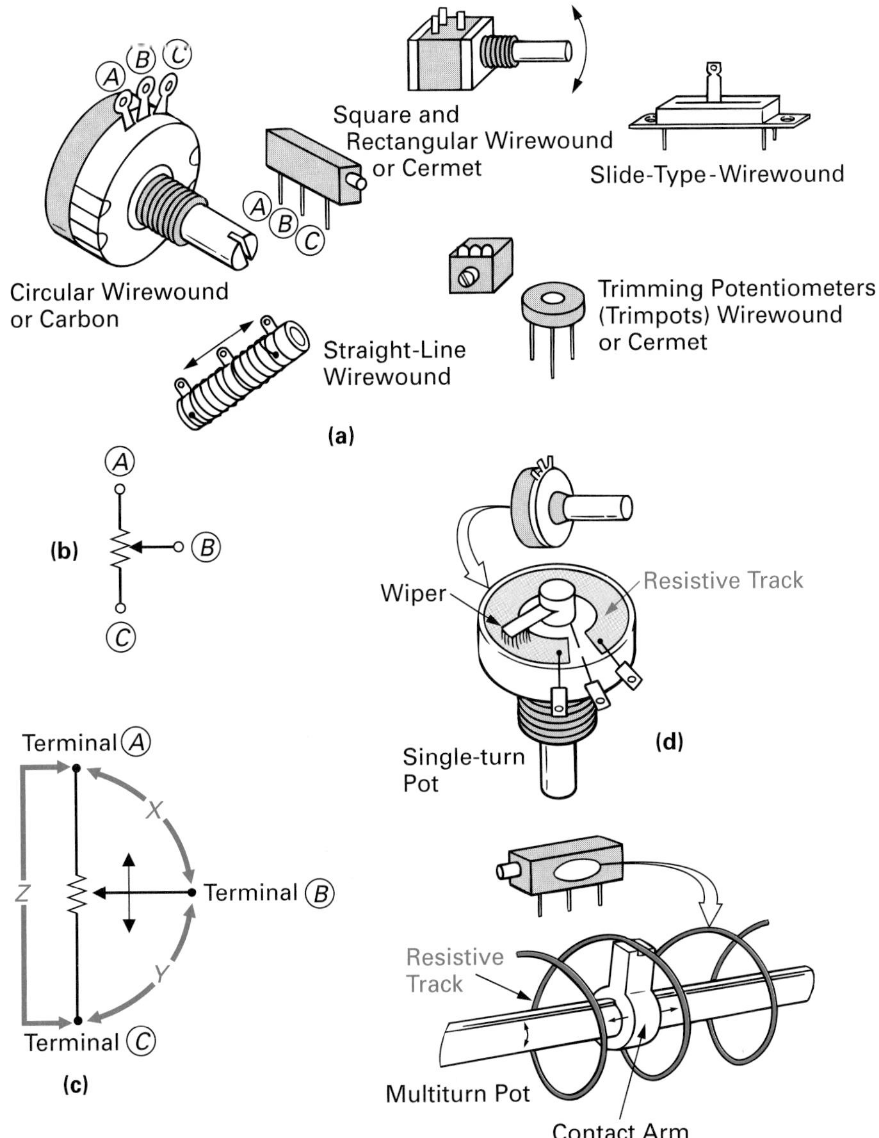

FIGURE 4-12 Potentiometer. (a) Physical Appearance. (b) Schematic Symbol. (c) Operation. (d) Construction.

The resistance between *A* and *B* (*X*) and *B* and *C* (*Y*) will, however, vary as the wiper's position is moved, as illustrated in Figure 4-14.

If the user mechanically turns the shaft in the clockwise direction, the resistance between *A* and *B* increases, while the resistance between *B* and *C* decreases. Similarly, if the user mechanically turns the shaft counterclockwise, a decrease occurs between *A* and *B* and there is a resulting increase in resistance between *B* and *C*. This point is summarized in Figure 4-15.

In some applications, you may see the symbol illustrated in Figure 4-16, where *B* is connected to *C* and only two terminals are hooked up by the user. In this situation, the potentiometer is being used as a rheostat.

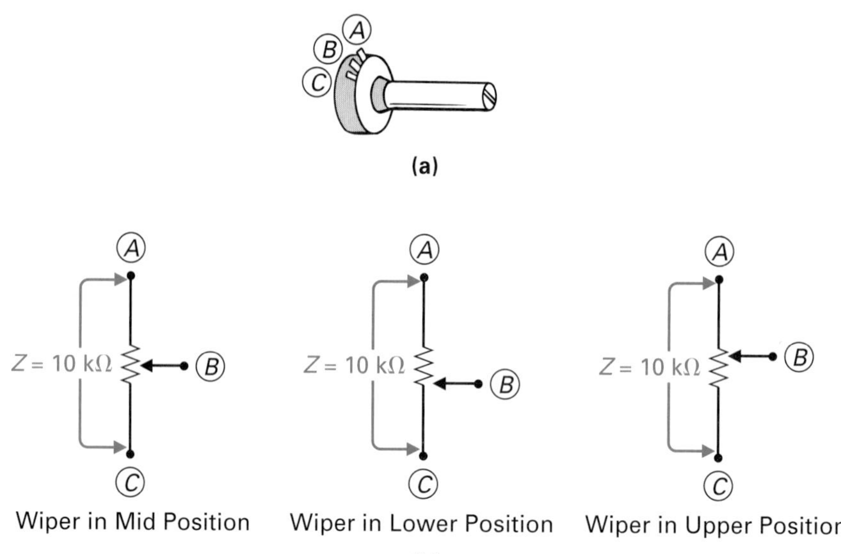

FIGURE 4-13 **A 10-kΩ Potentiometer. (a) Physical Appearance. (b) End-to-End Resistance of a Potentiometer Remains Constant.**

Linear Relationship between input and output in which the output varies in direct proportion to the input.

Tapered Nonuniform distribution of resistance per unit length throughout the element.

Like the rheostat, the potentiometer comes in many different shapes and sizes. Wirewound, carbon, and cermet resistive tracks, circular or straight-line motion, 2 to 10 multiturn, and other variations are available for different applications.

Whether rheostat or potentiometer, the resistive track can be classified as having either a **linear** or a **tapered** (nonlinear) resistance. In Figure 4-17(a) we have taken a 1-kΩ rheostat and illustrated the resistance value changes between *A* and *B* for a linear and a tapered one-turn rheostat.

The definition of linear is having an output that varies in direct proportion to the input. The input, in this case, is the user turning the shaft, and the output, as can be seen, is the linearly increasing resistance between *A* and *B*.

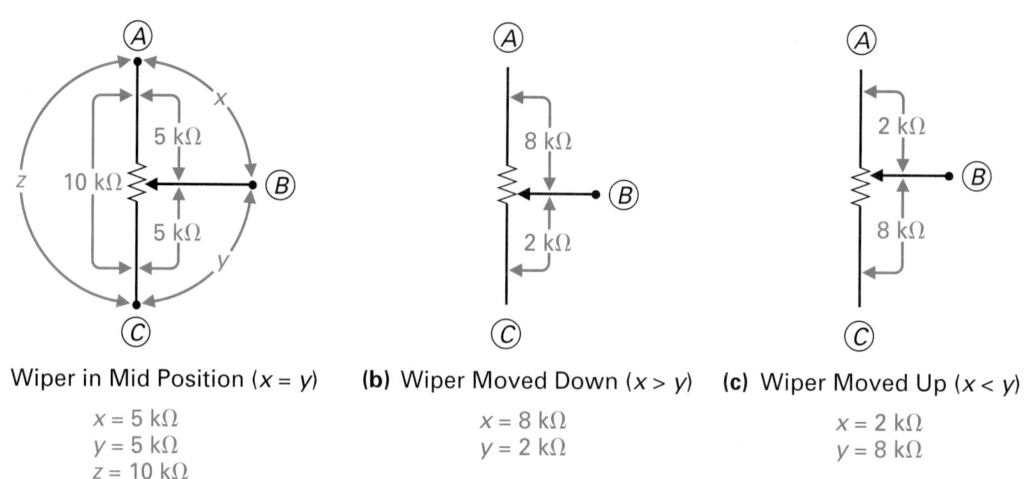

FIGURE 4-14 **Varying a Potentiometer's Resistance.**

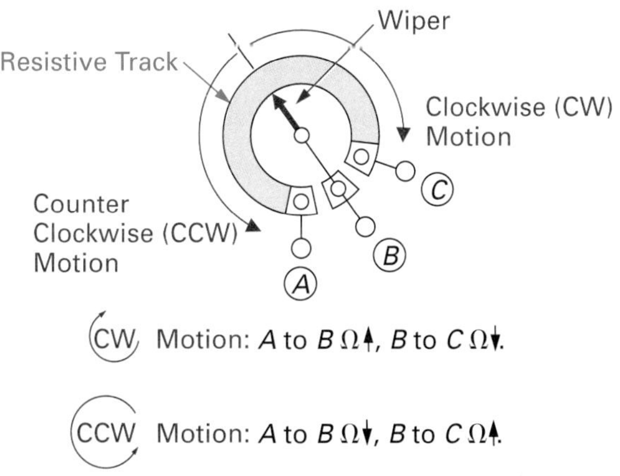

FIGURE 4-15 **Potentiometer's Operation.**

With a tapered rheostat or potentiometer, the resistance varies nonuniformly along its resistor element, sometimes being greater or less for equal shaft movement at various points along the resistance element, as shown in Figure 4-17(a). Figure 4-17(b) plots the position of the variable resistor's wiper against the resistance between the two output terminals. This graph effectively shows the difference between a linear increase and a nonlinear or tapered increase.

In Figure 4-18 you can see an application of a potentiometer as a volume control in a television set. By increasing or decreasing the potentiometer's resistance, the amount of current passing to the loudspeaker is varied and so is the volume.

Thermally (Heat) Adjustable Variable Resistors

When first discussing variable-value resistors, we talked about the rheostat and potentiometer, both of which have a mechanical input (the user turning the shaft) to produce a change in resistance. A **bolometer** is a device that, instead of changing its resistance when mechanical energy is applied, changes its resistance when heat energy is applied.

Bolometer
Device whose resistance changes when heated.

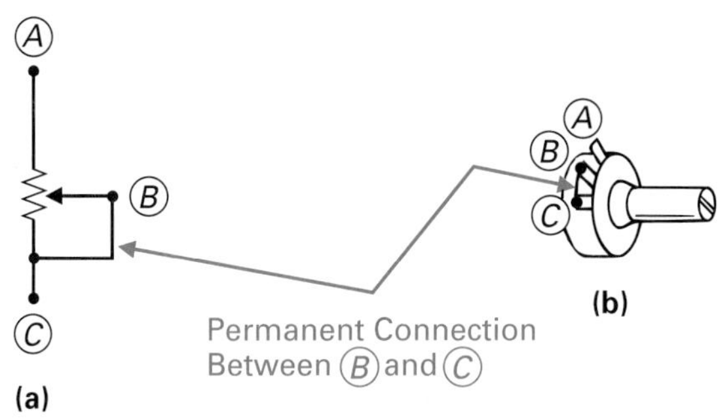

FIGURE 4-16 **The Potentiometer as a Rheostat. (a) Schematic Symbol. (b) Physical Appearance.**

	Fully CCW	$\frac{1}{4}$ CW	$\frac{1}{2}$ CW	$\frac{3}{4}$ CW	Fully CW
1 kΩ Rheostat (A, B)	A, B	A, B	A, B	A, B	A, B
Linear	0 Ω	250 Ω ($\frac{1}{4}$ of 1000 Ω)	500 Ω ($\frac{1}{2}$ of 1000 Ω)	750 Ω ($\frac{3}{4}$ of 1000 Ω)	1000 Ω
Tapered	0 Ω	350 Ω	625 Ω	900 Ω	1000 Ω
		These values have been arbitrarily chosen to illustrate a nonlinear change.			

(a)

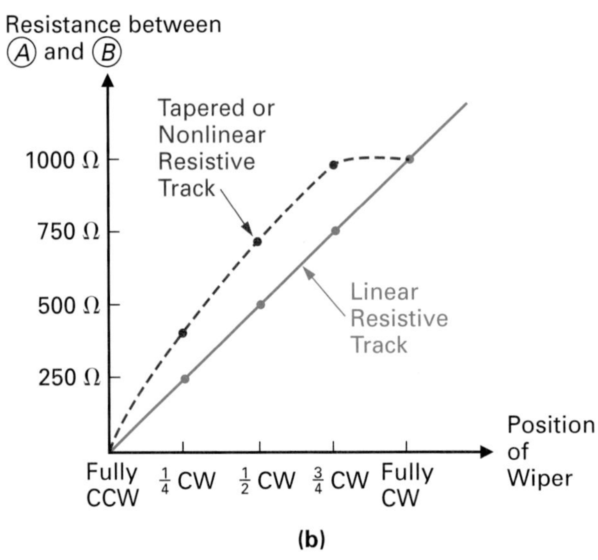

(b)

FIGURE 4-17 Linear versus Tapered Resistive Track.

Thermometry Relating to the measurement of temperature.

The measurement of temperature **(thermometry)** is probably the most common type of measurement used in industry today. Before discussing our three temperature sensors, let's first consider the four units of temperature measurement detailed in Table 4-1.

Commercially, temperature is normally expressed in degrees Celsius (°C) or degrees Fahrenheit (°F); however, although not commonly known, kelvin (K) and degrees Rankine (°R) are often used in industry, the kelvin being the international unit of temperature.

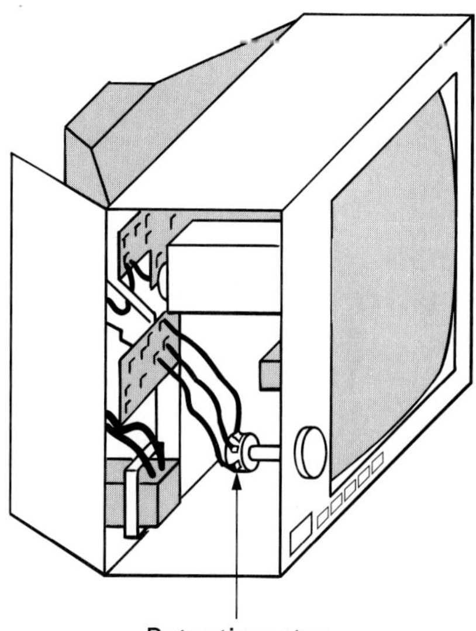

FIGURE 4-18 **Potentiometer as a Volume Control.**

Resistive Temperature Detector (RTD)
A temperature detector consisting of a fine coil of conducting wire (such as platinum) which will produce a relatively linear increase in resistance as temperature increases.

Thin-film Detector (TFD)
A temperature detector containing a thin layer of platinum, and used for very precise temperature readings.

There are basically three different types of temperature detectors, all of which are illustrated in Figure 4-19. The **resistive temperature detector** (RTD) and **thin-film detector** (TFD) shown in Figure 4-19(a) and (b) are both temperature sensors that contain a conducting material such as copper, nickel, or platinum and consequently have a positive temperature coefficient: resistance increases as temperature increases.

EXAMPLE:

Convert the following:

a. 74°F = ____________ °C
b. 45°C = ____________ °F
c. 25°C = ____________ K
d. 10°F = ____________ °R

TABLE 4-1 **Four Temperature Scales and Conversion Formulas**

	FAHRENHEIT	CELSIUS	KELVIN	RANKINE
Absolute zero	–459.69°F	–273.16°C	0 K	0°R
Melting point of ice (*x*)	32°F	0°C	273.16 K	491.69°R
	↓	↓	↓	↓
(Division between *x* and *y*)	(180°F)	(100°C)	(100 K)	(180°R)
	↓	↓	↓	↓
Boiling point of water (*y*)	212°F	100°C	373.16 K	671.69°R

$F = (^9/_5 \times C) + 32$
$C = {^5/_9} \times (F - 32)$
$R = F + 460$
$F = R - 460$
$K = C + 273$
$C = K - 273$

■ ***Solution:***

CALCULATOR SEQUENCE

Step	Keypad Entry	Display Response
1.	[7] [4]	74
2.	[−]	
3.	[3] [2]	32
4.	[=]	42
5.	[STO] (store result in memory)	
6.	[C/CE] (cancel display)	0
7.	[5]	5
8.	[÷]	
9.	[9]	9
10.	[=]	0.55555
11.	[×]	
12.	[RCL] (Recall value from memory)	42
13.	[=]	23.3

a. $C = \frac{5}{9} \times (F - 32)$

$= \frac{5}{9} \times (74 - 32)$

$= 0.555 \times 42 = 23.3°C$

b. $F = (\frac{9}{5} \times C) + 32$

$= (\frac{9}{5} \times 45) + 32$

$= (1.8 \times 45) + 32 = 113°F$

c. $K = C + 273$

$= 25 + 273 = 298\ K$

d. $R = F + 460$

$= 10 + 460 = 470°R$

Referring to the construction of the RTD in Figure 4-19(a), you can see that the sensing element consists of a coil of fine wire generally made of platinum, which gives a relatively linear increase in resistance as temperature increases, as indicated in the table in Figure 4-19(a).

The thin-film detector (TFD) shown in Figure 4-19(b) is constructed by placing a thin layer of platinum, for very precise temperature readings, on a ceramic substrate. Because of its small size, the TFD responds rapidly to temperature change and is ideally suited for surface temperature sensing.

Thermistor Temperature-sensitive semiconductor that has a negative temperature coefficient of resistance (as temperature increases, resistance decreases).

The **thermistor,** on the other hand, contains a semiconductor material that has a negative temperature coefficient: resistance decreases as temperature increases. A variety of thermistors can be seen in Figure 4-19(c). The thermistor is the most common type of temperature sensor and produces rapid and extremely large changes in resistance for very small changes in temperature, as shown in the associated table. The thermistor can be used as a temperature probe inside an oven. As the oven heats up, the thermistor's resistance decreases, and at a certain decreased resistance the thermistor turns off the oven. Once the temperature starts to drop, the thermistor's resistance increases, and this increase of resistance turns on the oven.

temperature ↑ thermistor's R ↓ oven off
temperature ↓ thermistor's R ↑ oven on

Optically (Light) Adjustable Variable Resistors

Photoresistor Also known as a photoconductive cell or light dependant resistor, it is a device whose resistance varies with the illumination of the cell.

Photo means illumination, and the **photoresistor** is a resistor that is photoconductive. This means that as the material is exposed to light, it will become more conductive and less resistive. Figure 4-20 illustrates the photoresistor, also called a light-dependent resistor (LDR), and its schematic symbol.

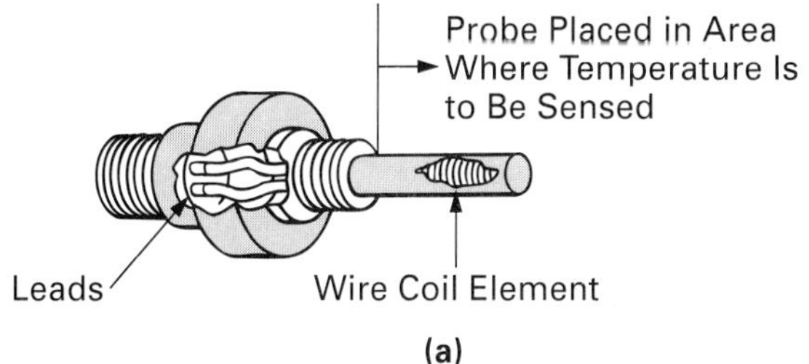

Platinum Winding			
°C	Ohms	°C	Ohms
−200	18.53	+200	175.84
−150	39.65	+250	194.08
−100	60.20	+300	212.03
−50	80.25	+350	229.69
±0	100.0	+400	247.06
+50	119.40	+450	264.14
+100	138.50	+500	280.93
+150	157.32	+550	297.16

(Conductors have a positive temperature coefficient of resistance—temp.↑, R↑)

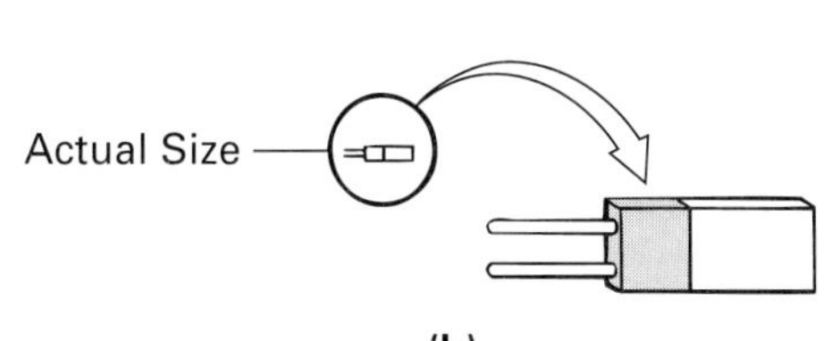

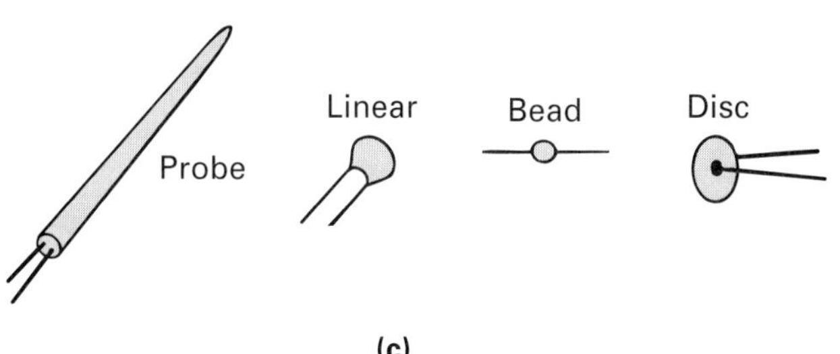

°C	Ohms
−50	100,000
0	7,500
+50	7,400
+100	100
+150	50
+200	27
+250	10
+300	7.5

(Semiconductors have a negative temperature coefficient of resistance—T↑, R↓)

FIGURE 4-19 **Temperature Sensors. (a) Resistive Temperature Detector (RTD). (b) Thin-Film Detector (TFD). (c) Thermistor.**

The photoresistor is a thin slice of photoconductive material whose resistance decreases as light is applied. The light energy is absorbed by the atoms within the photoconductive material, causing these atoms to release their valence electrons. This results in an increase in electrons, and therefore current passing through the photoresistor, so its resistance will have decreased. To summarize, we can say:

light energy ↑ conduction ↑ resistance ↓

Photoresistors can be used to turn on and off outdoor home security lights. During the day, the natural sunlight decreases the resistance of the photoresistor, and this low resistance is used to keep the lights off. At dusk, the sunlight is almost gone, and the photoresistor's resistance increases. This increased resistance is used to turn on the security lights.

sun up ↑ photoresistance ↓ security lights OFF
sun down ↓ photoresistance ↑ security lights ON

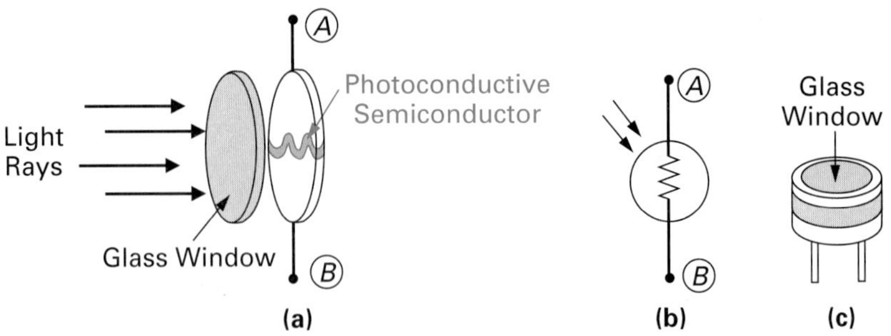

FIGURE 4-20 **Photoresistor. (a) Construction. (b) Schematic Symbol. (c) Physical Appearance.**

SELF-TEST REVIEW QUESTIONS FOR SECTION 4-1

1. List the six types of fixed-value resistors.
2. What is the difference between SIPs and DIPs?
3. Name the two types of mechanically adjustable variable resistors and state the difference between the two.
4. Describe a linear and a tapered potentiometer.
5. True or false: A thermistor has a negative temperature coefficient of resistance.
6. What name is given to the optically adjustable resistor?

4-2 HOW IS RESISTANCE MEASURED?

Ohmmeter
Measurement device used to measure electric resistance.

Resistance is measured with an **ohmmeter.** Figure 4-21(a) illustrates how a multirange analog ohmmeter and a digital ohmmeter can be used to measure the resistance of a resistor. The analog ohmmeter has four ranges. The simplest range has been selected ($R \times 1$), and on this scale the resistance value indicated can be read directly from the meter; in this example the value is 36 Ω. If we had selected the $R \times 10$ range and a value of 72 was obtained, then 72 would have to be multiplied by 10 ($72 \times 10 = 720$) to obtain the value of 720 Ω. If we were on the $R \times 100$ range and a reading of 15 was obtained, then $R \times 100 = 15 \times 100$, which equals 1.5 kΩ. The $R \times 1000$ scale, if selected, means that you would interpret 470 for example as $R \times 1000 = 470 \times 1000$, which equals 470 kΩ.

Three rules must be remembered and applied when measuring resistance.

1. Short the meter leads together and adjust the zero-ohms control so that the pointer is at 0 Ω. This calibrates the meter scale.
2. Turn off the circuit power and remove the component (if practicable) to be measured from circuit to protect the ohmmeter.

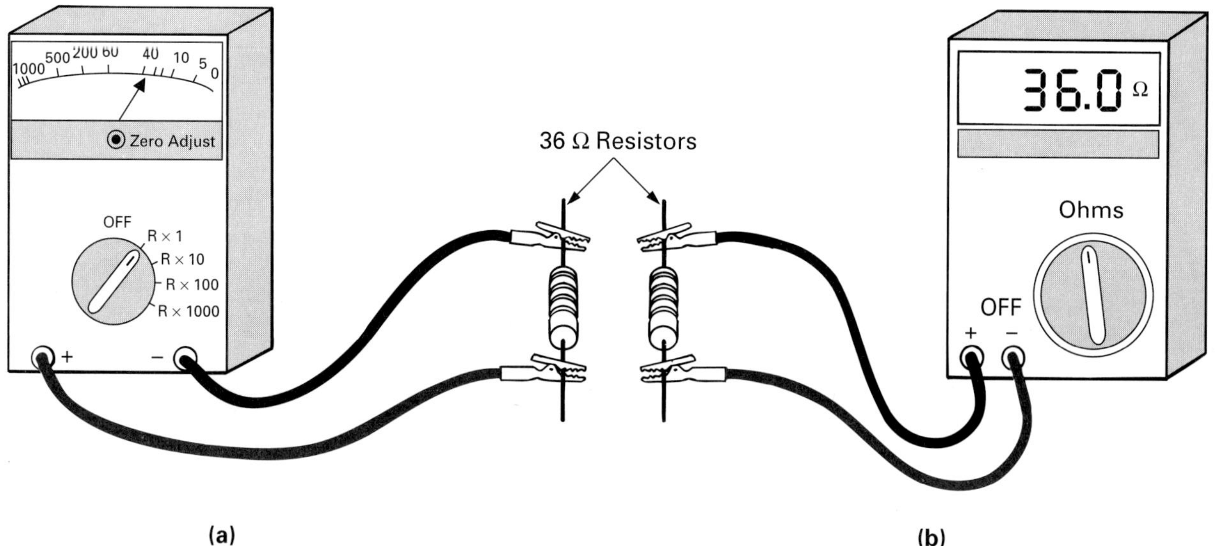

FIGURE 4-21 **Measuring Resistance.**

3. Connect the leads across the removed component, adjust the range scale until the pointer is approximately in the middle of the scale (for accuracy), and then multiply this value by the range scale selected.

With the digital auto-ranging ohmmeter shown in Figure 4-21(b), the measurement of resistance is a lot easier since you simply turn the selector to ohms and read off the reading from the display.

SELF-TEST REVIEW QUESTIONS FOR SECTION 4-2

1. Name the instrument used to measure resistance.
2. On the $R \times 1000$ range, how would a reading of 22 be interpreted?
3. On the $R \times 1$ range, how would a reading of 470 be interpreted?
4. List the three rules that should be applied when using the analog ohmmeter.

4-3 RESISTOR CODING

Using the ohmmeter, you can find the value of any resistor. However, manufacturers indicate the value and tolerance of resistors on the body of the component in one of two ways:

1. *Color code:* colored rings or bands
2. *Typographically:* printed alphanumerics (alphabet and numerals)

Examples of both of these can be seen in Figure 4-22.

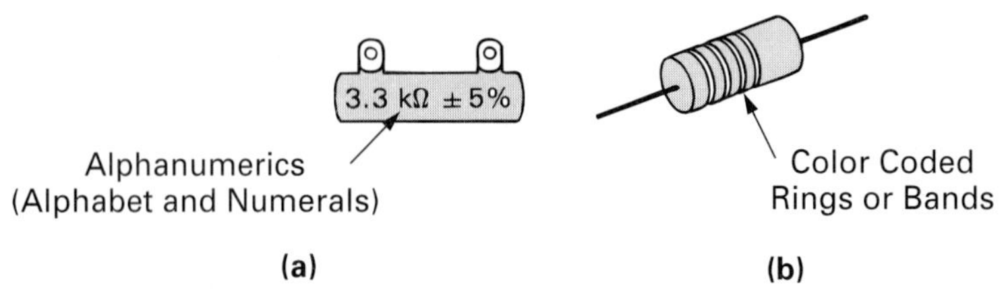

FIGURE 4-22 Resistor Value and Tolerance Indication. (a) Typographical. (b) Color Code.

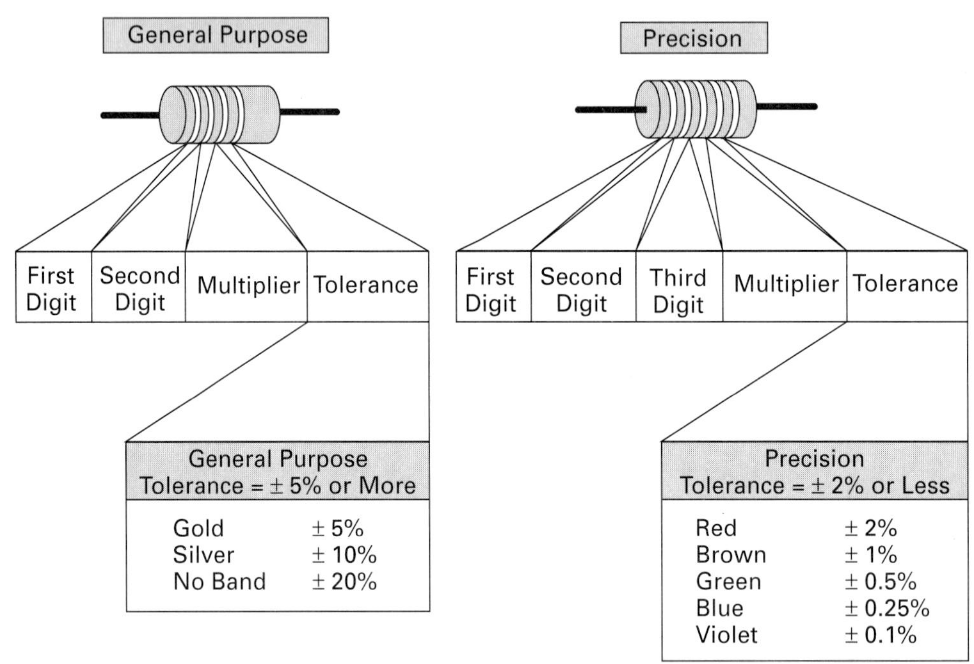

	Color	Digit Value	Multiplier		
Big	Black	0	1	One	1
Beautiful	Brown	1	10	One Zero	10
Roses	Red	2	100	Two Zeros	100
Occupy	Orange	3	1000	Three Zeros	1 k
Your	Yellow	4	10000	Four Zeros	10 k
Garden	Green	5	100000	Five Zeros	100 k
But	Blue	6	1000000	Six Zeros	1 M
Violets	Violet	7	10000000	Seven Zeros	10 M
Grow	Gray	8	–		
Wild	White	9	–		
So	Silver	–	10^{-2} or 0.01		1/100
Get some	Gold	–	10^{-1} or 0.1		1/10
Now	None	–			

FIGURE 4-23 General-Purpose and Precision Resistor Color Code.

4-3-1 *General-Purpose and Precision Color Code*

There are basically two different types of fixed resistors: general purpose and precision. Resistors with tolerances of ±2% or less are *precision resistors* and have five bands; resistors with tolerances of ±5% or greater have four bands and are referred to as *general-purpose resistors.* The color code and differences between precision and general-purpose resistors are illustrated in Figure 4-23.

When you pick up a resistor, notice that the bands are nearer one end; this end should be held in your left hand. If there are four bands on the resistor, follow the general-purpose resistor code; if five bands are present, adopt the precision resistor code.

General-Purpose Resistor Code

1. The first band on either a general-purpose or precision resistor can never be black, and it is the first digit of our number.
2. The second band indicates the second digit of the number.
3. The third band specifies the multiplier to be applied to the number, which ranges from × 1/100 to 10,000,000.
4. The fourth band describes the tolerance or deviation from the specified resistance, which is ±5% or greater.

Precision Resistor Code

1. The first band, like the general-purpose resistor, is never black and is the first digit of the three-digit number.
2. The second band provides the second digit.
3. The third band indicates the third and final digit of the number.
4. The fourth band specifies the multiplier to be applied to the number.
5. The fifth and final band indicates the tolerance figure of the precision resistor, which is always less than ±2%, which is why precision resistors are more expensive than general-purpose resistors.

EXAMPLE:

Figure 4-24 illustrates a ½-W resistor.

a. Is this a general or a precision resistor?
b. What is the resistor's value of resistance?
c. What tolerance does this resistor have, and what deviation plus and minus could occur to this value?

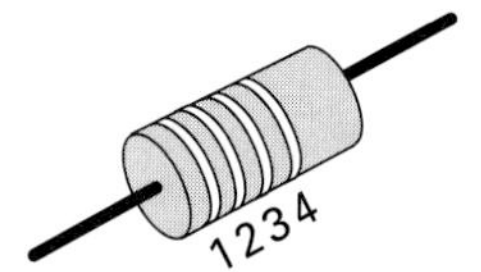

Band 1 = Green
Band 2 = Blue
Band 3 = Brown
Band 4 = Gold

FIGURE 4-24

Solution:

a. General purpose (four bands)
b. green blue × brown
   5 6 × 10 = 560 Ω

c. Tolerance band is gold, which is ±5%.

$$\text{deviation} = 5\% \text{ of } 560 = 28$$
$$560 + 28 = 588\ \Omega$$
$$560 - 28 = 532\ \Omega$$

The resistor could, when measured, be anywhere from 532 to 588 Ω.

EXAMPLE:

State the resistor's value and tolerance and whether it is general purpose or precision for the examples shown in Figure 4-25.

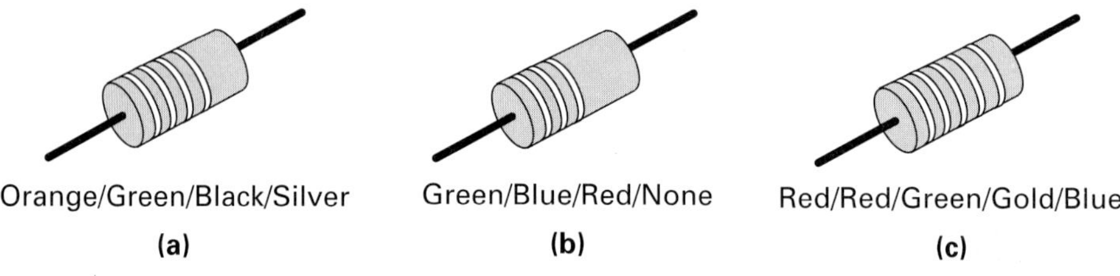

FIGURE 4-25

Solution:

a. orange green black silver (four bands = general purpose)
3 5 × 1 10% = 35 Ω ± 10%

b. green blue red none (four bands = general purpose)
5 6 × 100 20% = 5.6 kΩ ± 20%

c. red red green gold blue (five bands = precision)
2 2 5 × 0.1 0.25% = 22.5 Ω ± 0.25%

4-3-2 *Zero-Ohms Resistor*

Figure 4-26(a) illustrates the only resistor color code that you will probably have trouble with, the zero-ohms resistor, which has one black band. Zero ohms is equivalent to a straight piece of wire. You may ask yourself: Why do they manufacture a resistor of zero ohms since it really is not a resistor at all?

Printed circuit boards like the one shown in Figure 4-26(b) no longer have all their resistor and other components inserted by hand. *Automatic insertion equipment* replaces the human assembler and places the correct component in the correct position in a matter of seconds rather than hours. In some applications, a direct contact between two points is needed, in which case a piece of wire needs to be inserted, as shown in Figure 4-26(b). However, the resistor automatic insertion equipment will only handle resistors, not wire (called jumpers). In the past, manufacturers had to install jumpers manually after all the components were inserted. This caused enormous delays; but now the zero-ohms

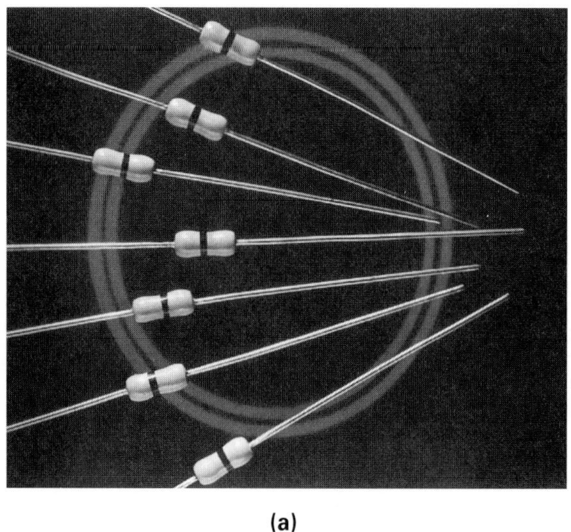

(a)

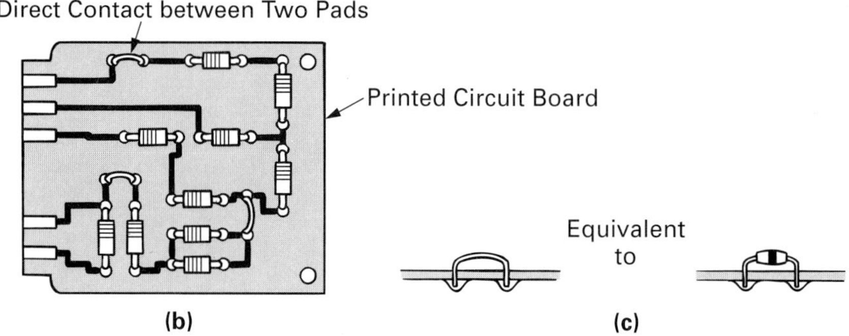

(b) (c)

FIGURE 4-26 **Zero-Ohms Resistor. (a) Appearance. (b) Printed Circuit Board. (c) Zero-Ohms Resistor as Jumper.**

ohms resistor can be installed automatically during the resistor insertion process and have exactly the same effect, as shown in Figure 4-26(c).

4-3-3 *Other Resistor Identification Methods*

If a resistor color code is not found, some other form of marking will always be made on the resistor. The larger fixed, wirewound resistors and nearly all variable resistors normally have the resistance value, tolerance and wattage printed on the resistor, as shown in Figure 4-27(a). SIP and DIP packages are normally coded with both letters and numerals, all of which have particular meaning, as shown in Figure 4-27(b). Chip resistors tend to be too small to have any intelligible form of marking and so will have to be measured with an ohmmeter. They are packaged, as previously shown in Figure 4-10(c), in either polythene bags, paper tape, or plastic magazines, and in all these cases the packaging will be used to specify values, tolerance, and wattage.

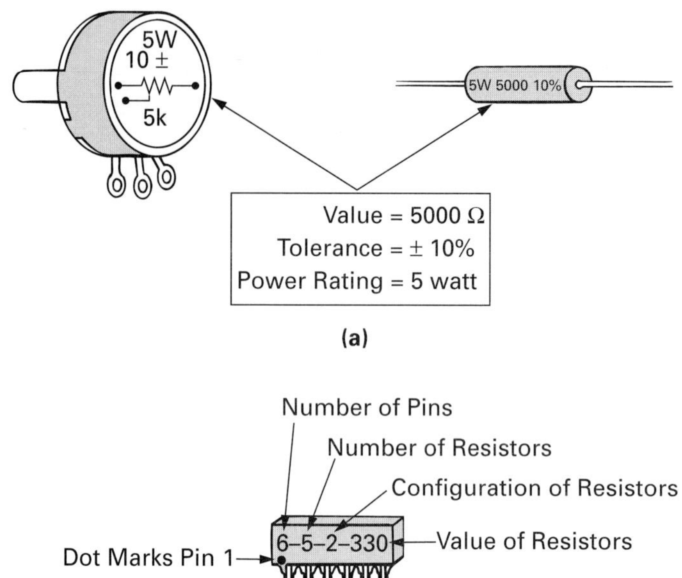

FIGURE 4-27 Typographical Value Indication. (a) Standard Variable and Fixed. (b) SIP and DIP.

SELF-TEST REVIEW QUESTIONS FOR SECTION 4-3

1. What tolerance differences occur between a general-purpose and a precision resistor?
2. A red/red/red/gold/blue resistor has what value and tolerance figure?
3. What color code would appear on a 4.7-kΩ, ±10% tolerance resistor?
4. Would a three-band resistor be considered a general-purpose type?

4-4 POWER DISSIPATION DUE TO RESISTANCE

The atoms of a resistor obstruct the flow of moving electrons (current), and this friction between the two causes heat to be generated. Current flow through any resistance therefore will always be accompanied by heat. In some instances, such as a heater or oven, the device has been designed specifically to generate heat. In other applications it may be necessary to include a resistor to reduce current flow, and the heat generated will be an unwanted side effect.

Any one of the three power formulas can be used to calculate the power dissipated by a resistance.

$$P = V \times I \qquad P = I^2 \times R \qquad P = \frac{V^2}{R}$$

However, since power is determined by the friction between current (I) and the circuit's resistance (R), the $P = I^2 \times R$ formula is more commonly used to calculate the heat generated.

EXAMPLE:

What wattage rating should a 33-Ω resistor have if the value of current is 100 mA?

Solution:

$$\begin{aligned} P &= I^2 \times R \\ &= (100\text{ mA}^2) \times 33\ \Omega \\ &= 0.33\text{ W} \end{aligned}$$

It would be safe therefore to use a ½-W (0.5-W) resistor.

EXAMPLE:

A coffee cup warming plate has a power rating of 120 V, 23 W. How much current is flowing through the heating element, and what is the heating element's resistance?

Solution:

$$P = V \times I \qquad I = \frac{P}{V}$$

$$I = \frac{23\text{ W}}{120\text{ V}} = 191.7\text{ mA}$$

$$R = \frac{V}{I} = \frac{120\text{ V}}{191.7\text{ mA}} = 626\ \Omega$$

EXAMPLE:

Calculate the current drawn by a 100-W light bulb if it is first used in the home (and therefore connected to a 120-V source) and then used in the car (and therefore connected across a 12-V source).

Solution:

$$P = V \times I \quad \text{therefore,} \quad I = \frac{P}{V}$$

In the home,

$$I = \frac{P}{V} = \frac{100\text{ W}}{120\text{ V}} = 0.83\text{ A}$$

In the car,

$$I = \frac{P}{V} = \frac{100\text{ W}}{12\text{ V}} = 8.33\text{ A}$$

This example brings out a very important point in relation to voltage, current, and power. In the home, only a small current needs to be supplied by the source when the applied voltage is large ($P = V\uparrow \times I\downarrow$); however, when the source voltage was small, as in the case of the car, a large current must be supplied by the source in order to deliver the same amount of power ($P = V\downarrow \times I\uparrow$).

4-5 FILAMENT RESISTOR

Filament Resistor
The resistor in a light bulb or electron tube.

Incandescent Lamp
An electric lamp that generates light when an electric current is passed through its filament of resistance, causing it to heat to incandescence.

Figure 4-28(a) illustrates the **filament resistor** within a glass bulb. This component is more commonly known as the household light bulb. The filament resistor is just a coil of wire that glows white hot when current is passed through it and, in so doing, dissipates both heat and light energy.

This **incandescent lamp** is an electric lamp in which electric current flowing through a filament of resistive material heats the filament until it glows and emits light. Figure 4-28(b) shows a test circuit for varying current through a small incandescent lamp. Potentiometer R_1 is used to vary the current flow through the lamp. The lamp is rated at 6 V, 60 mA. Using Ohm's law we can calculate the lamp's filament resistance at this rated voltage and current:

$$\text{filament resistance, } R = \frac{V}{I} = \frac{6\text{ V}}{60\text{ mA}} = 100\ \Omega$$

Cold Resistance
The resistance of a device when cold.

Hot Resistance
The resistance of a device when hot due to the generation of heat by electric current.

The filament material (tungsten, for example) is like all other conductors in that it has a positive temperature coefficient of resistance. Therefore, as the current through the filament increases, so does the temperature and so does the filament's resistance ($I\uparrow$, temperature $\uparrow$, $R\uparrow$). Consequently, when R_1's wiper is moved to the right so that it produces a high resistance ($R_1\uparrow$), the circuit current will be small ($I\downarrow$) and the lamp will glow dimly. Since the circuit current is small, the filament temperature will be small and so will the lamp's resistance ($I\downarrow$, temperature $\downarrow$, lamp resistance $\downarrow$). This small value of resistance is called the lamp's **cold resistance.** On the other hand, when R_1's wiper is moved to the left so that it produces a small resistance ($R\downarrow$), the circuit current will be large ($I\uparrow$), and the lamp will glow brightly. With the circuit current high, the filament temperature will be high and so will the lamp's resistance ($I\uparrow$, temperature $\uparrow$, lamp resistance $\uparrow$). This large value of resistance is called the lamp's **hot resistance.**

Figure 4-28(c) plots the filament voltage, which is being measured by the voltmeter, against the filament current, which is being measured by the ammeter. As you can see, an increase in current will cause a corresponding increase in filament resistance. Studying this graph, you may have also noticed that the lamp has been operated beyond its rated value of 6 V, 60 mA. Although the lamp can be operated beyond its rated value, at, for example 10 V, 80 mA, its life expectancy will be decreased dramatically—from several hundred hours to only a few hours.

Ballast Resistor
A resistor that increases in resistance when current increases. It can therefore maintain a constant current despite variations in line voltage.

A coil of wire similar to the one found in a light bulb, called a **ballast resistor,** is used to maintain a constant current despite variation in voltage. Since the coil of wire is a conductor and conductors have a positive temperature coefficient, an increase in voltage will cause a corresponding increase in current and therefore in the heat generated, which will result in an increase in the wire's resistance. This increase in resistance will decrease the initial current rise. Similarly, a decrease in voltage and therefore current ($I \propto V$) will result in a decrease in heat and wire resistance. This decrease in resistance will permit an increase in current to counteract the original decrease. Current is therefore regulated or maintained constant by the ballast resistor, despite variations in voltage.

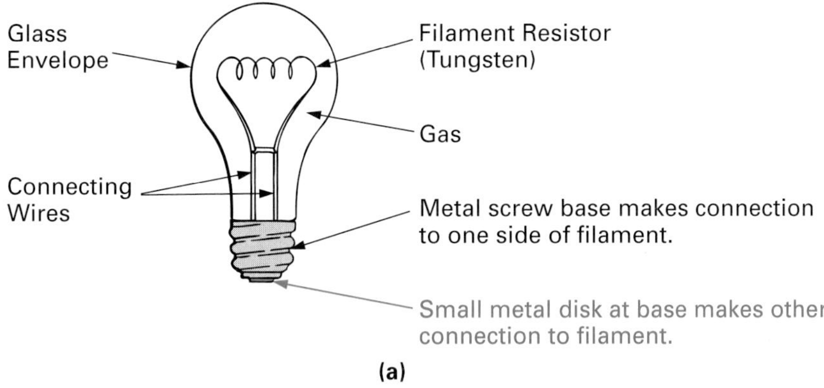

(a)

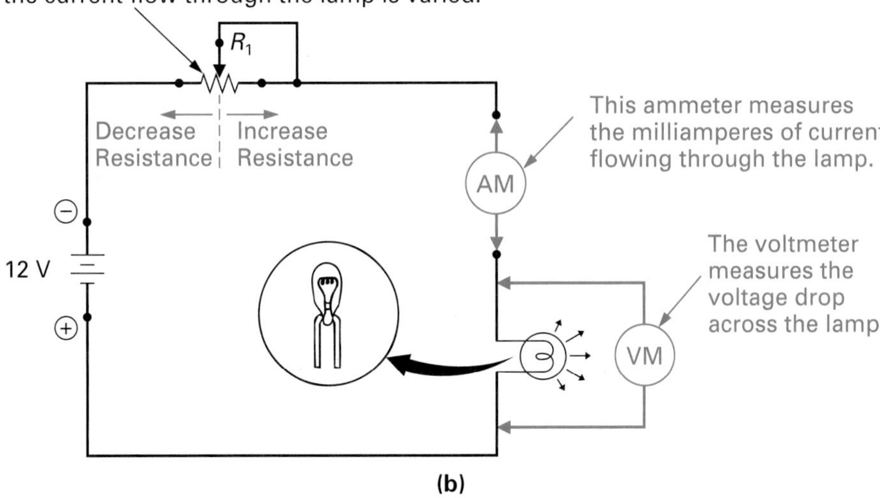

(b)

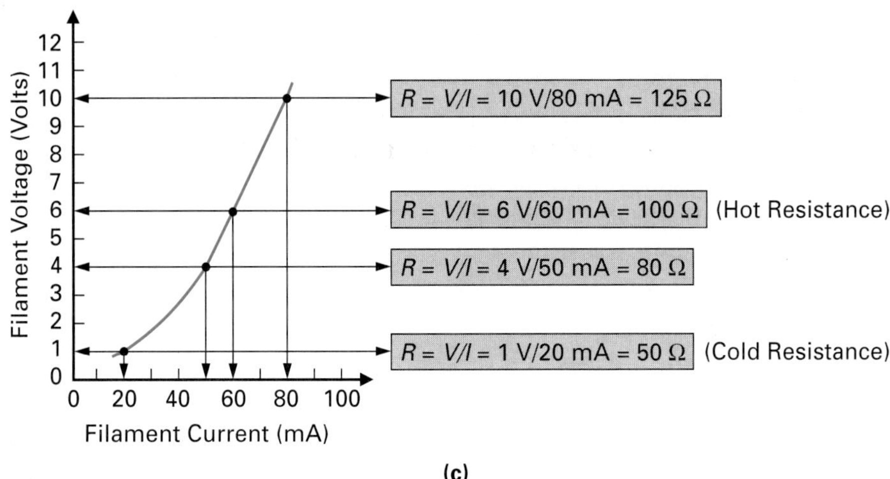

(c)

FIGURE 4-28 Filament Resistor.

SELF-TEST REVIEW QUESTIONS FOR SECTIONS 4-4 AND 4-5

1. Which formula is most commonly used to calculate the heat generated by a resistor?
2. An incandescent lamp has a ____________ (large/small) cold resistance and a ____________ (large/small) hot resistance.

4-6 TESTING RESISTORS

It is virtually impossible for resistors to short-circuit (no resistance or zero ohms) internally; however, they may be shorted out by some other device, providing a low-resistance path around the resistor. Generally, the resistor's internal resistive elements will begin to develop a higher resistance than its specified value (due to a partial internal open) or in some cases go completely open circuit (maximum resistance or infinite ohms).

Fixed-value resistors tend to go either partially or fully open as the resistive track begins to break down. Variable-value resistors have problems with the wiper making contact with the resistive track at all times. Faulty variable-value resistors in audio systems can normally be detected because they generate a scratchy noise whenever, for example, you adjust the volume or tone control.

The ohmmeter is the ideal instrument for verifying whether or not a resistor is functioning correctly. When checking a resistor's resistance, here are some points that should be remembered.

1. Always calibrate the meter first to zero. If the meter indicates 100 Ω when the two meter leads are shorted together (0 Ω), every reading you take will be in error by 100 ohms.
2. The ohmmeter has its own internal power source (a battery), so always turn off the circuit power and remove from the circuit the resistor to be measured. If this is not done, you will not only obtain inaccurate readings, but you can damage the ohmmeter.
3. The range scales on a digital multimeter often confuse people; for example, if a range is chosen so that the highest reading is 100 kΩ and a 1-MΩ resistor is measured, the meter will indicate an infinite-ohms reading and you may be misled into believing that you have found the problem (an open resistor). In the case of an analog ohmmeter, if the range scale multiplier is not applied, a misleading result is obtained. For example, if the $R \times 100$ range is selected, the resistance reading (R) must be multiplied by 100 to obtain the resistance.
4. Another point to keep in mind is tolerance. For example, if a suspected 1-kΩ (1000-Ω) resistor is measured with an ohmmeter and reads 1.2 kΩ (1200 Ω), it could be within tolerance if no tolerance band is present on the resistor's body (±20%). A 1-kΩ resistor with ±20% tolerance could measure anywhere between 800 and 1200 Ω

5. The ohmmeter's internal battery voltage is really too small to deliver an electrical shock; however, you should avoid touching the bare metal parts of the probes or resistor leads, as your body resistance of approximately 50 kΩ will affect your meter reading.

SELF-TEST REVIEW QUESTIONS FOR SECTION 4-6

1. What problem normally occurs with resistors? (open/shorts)
2. What instrument is used to verify a resistor's resistance?

SUMMARY

1. Conductors offer a certain small amount of resistance; however, in electronics this resistance is not normally enough, so additional resistance is needed to control the amount of current flow. The component used to supply this additional resistance is called a *resistor.*

   There are two basic types of resistors: fixed value and variable value. The fixed-value resistor has a value of resistance that cannot be changed and is the more common of the two. The variable-value resistor, on the other hand, has a range of values that can be selected generally by mechanically adjusting a control.

Fixed-Value Resistors (Figure 4-29)

2. *Fixed-value resistors* can be divided into six basic categories:
   - **a.** Carbon composition
   - **b.** Carbon film
   - **c.** Metal film
   - **d.** Wirewound
   - **e.** Metal oxide
   - **f.** Thick film
3. Carbon composition resistors are the most common and least expensive type of fixed-value resistor. It is constructed by placing a piece of resistive material, with embedded conductors at each end, within an insulating cylindrical molded case.

   This resistive material is called carbon composition because powdered carbon and powdered insulator are bonded into a compound and used as the resistive material. By changing the ratio of powdered insulator to carbon, the value of resistance can be changed within the same area.
4. The physical size of the resistors lets the user know how much power in the form of heat can be *dissipated.* As you already know, resistance is the opposition to current flow or electrons, and this opposition causes heat to be generated when current flows. The amount of heat dissipated per unit of time is measured in watts and each resistor has its own *wattage rating.* The bigger the resistor, the more heat that can be dissipated. The key point is that heat should be dissipated faster than it is generated; if not, a larger resistor with a larger surface area to dissipate away the additional heat should be used so that the resistor will not burn up.

QUICK REFERENCE SUMMARY SHEET

FIGURE 4-29 Fixed Value Resistors.

Type	Cost	Resistive Material	Advantages/Disadvantage
Carbon Composition 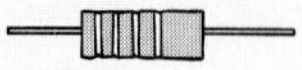	$	Powdered carbon and carbon insulator.	Low Cost but inherently noisy.
Carbon Film 	$$	A thin film of carbon and insulator.	Smaller tolerances (±5% to ±2%) and better temperature stability than carbon composition.
Metal Film 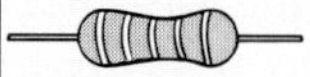	$$$	Thin metal film spiral on substrate.	Best tolerances (±1% to ±0.1%) and temperature stability; however, high cost.
Wirewound  100 MΩ	$$$$	Length of wire wrapped around ceramic core.	High power rating and good tolerance (±1%); however, very high cost.
Metal Oxide 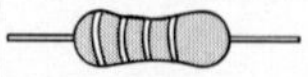	$$$	Oxide of a metal on an insulating substrate.	Has the best temperature stability; however, high cost.
Thick Film Networks Chips SIP DIP Chip	 $$$$ $$	Thick film of resistive paste on insulating substrate.	Small size makes them ideal for resistor networks (SIPs and DIPs) and chip resistors for SMT.

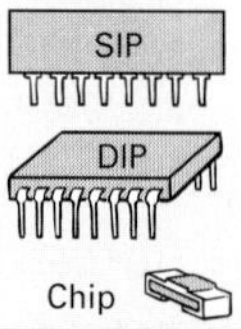

5. Another factor to consider when discussing resistors is their *tolerance.* Tolerance is the amount of deviation or error from the specified value. For example, a 1000-Ω (1-kΩ) resistor with a ± 10% (plus and minus 10%) tolerance when manufactured could have a resistance anywhere between 900 and 1100 Ω. In some applications where high precision is required, smaller tolerance resistors are used.
6. Carbon film resistors have a thin layer or film of resistive material (a blend of carbon and insulator) on a ceramic (insulating) substrate. The film is then cut to form a helix or spiral. Carbon film resistors have smaller tolerance figures (±5% to ±2%), are more stable (maintain same resistance value over a wide

range of temperatures), and have less internally generated noise (random small bursts of voltage) than do carbon composition resistors.

7. *Metal film resistors* are constructed by spraying a thin film of metal on a ceramic cylinder (or *substrate*) and then cutting the film to form a substrate. Metal film resistors have possibly the best tolerances commercially available of ±1% to ±0.1%. They also maintain a very stable resistance over a wide range of temperatures (good stability) and generate very little internal noise compared to any carbon resistor.

8. *Wirewound resistors* are constructed by wrapping a length of wire uniformly around a ceramic insulating core, with terminals making the connections at each end. The length and thickness of the wire are varied to change the resistance, which ranges from 1 Ω to 150 kΩ.

   Since current flow is opposed by resistors and this opposition generates heat, the larger the physical size of the resistor, the greater the amount of heat that can be dissipated away and so the greater the current that can be passed through the resistor.

   Wirewound resistors are generally used in applications requiring low resistance values, which means that the current and therefore power dissipated are high ($R\downarrow = V/I\uparrow$, $P\uparrow = I^2\uparrow R$). The amount of heat that can be dissipated is measured in watts and is indicated on the wirewound resistor.

9. The *metal oxide resistors* are constructed by depositing an oxide of a metal such as tin onto an insulating substrate. The ratio of oxide (insulator) to tin (conductor) will determine the resistor's resistance.

   Metal oxide resistors have excellent temperature stability, which is the ability of the resistor to maintain its value of resistance without change even when temperature is changed.

   There is a tendency when heat is increased for atoms to move, and in so doing they cause collisions, resulting in an opposition to current; so a temperature increase will cause a resistor's resistance to increase. This increase from the desired resistance is very small with metal oxide resistors.

10. The two different types of *thick-film resistor* networks are called SIPs and DIPs. The *single in-line package* is so called because all its lead connections are in a single line, whereas the *dual in-line package* has two lines of connecting pins. The chip resistors are small thick-film resistors that are approximately the size of a pencil lead.

    The SIP and DIP resistor networks are constructed by first screening on the internal conducting strip (silver) that connects the external pins to the resistive material, and then screening on the thick film of resistive paste (bismuth/ruthenate), the blend of which will determine resistance. The chip resistor uses the same resistive film paste material, which is deposited onto an insulating substrate with two end terminations and protected by a glass coat.

    The SIP and DIP resistor networks, once constructed, are trimmed by lasers to obtain close tolerances of typically ±2%. Resistance values ranging from 22 Ω to 2.2 MΩ are available, with a power rating of ½ W.

    The chip resistor is commercially available with resistance values from 10 Ω to 3.3 MΩ, a ±2% tolerance, and a ⅛-W heat dissipation capability.

Variable-Value Resistors

11. The resistance of *variable resistors* can be varied in one of three ways:

a. Mechanically (user) adjustable:

1) Rheostat

2) Potentiometer

b. Thermally (heat) adjustable:

1) Thermistor

2) Resistive temperature detector (RTD)

3) Thin-film detector (TFD)

c. Optically (light) adjustable:

1) Photoresistor

12. Mechanically (user) adjustable variable resistors (Figure 4-30) include two types, both of which will cause a change in resistance when a shaft is rotated. The *rheostat* is a two-terminal variable resistor that, through mechanical turning of a shaft, can be used to vary its resistance and therefore its value of terminal to terminal current.

13. Rheostats come in many shapes and sizes. Some employ a straight-line motion to vary resistance, while others are classified as circular-motion rheostats. The resistive elements also vary; wirewound and carbon tracks are very popular. Cermet, which is a ceramic (insulator)–metal (conductor) mix the ratio of which can be used to produce different values of resistive tracks, is also used. A trimming rheostat is a miniature device used to change resistance by a small amount. Other circular-motion rheostats are available that require between two to ten turns to cover the full resistance range.

14. The *potentiometer* is a three-lead variable resistor that through mechanical turning of a shaft can be used to produce a variable voltage or potential.

15. The difference between a rheostat and potentiometer is the number of terminals; the rheostat has two terminals and the potentiometer has three. With the rheostat, there were only two terminals and the resistance between the wiper and terminal varied as the wiper was adjusted. For the potentiometer, the resistance can actually be measured across three separate combinations: between A and B (X), between B and C (Y), and between C and A (Z).

The only difference between the rheostat and the potentiometer in construction is the connection of a third terminal to the other end of the resistive track.

16. Like the rheostat, the potentiometer comes in many different shapes and sizes. Wirewound, carbon, and cermet resistive tracks, circular or straight-line motion, 2 to 10 multiturn, and other variations are available for different applications.

17. Whether rheostat or potentiometer, the resistive track can be classified as having either a *linear* or a *tapered* (nonlinear) resistance.

The definition of linear is having an output that varies in direct proportion to the input. The input, in this case, is the user turning the shaft,

FIGURE 4-30 Mechanically (User) Adjustable Variable Resistor.

Rheostat

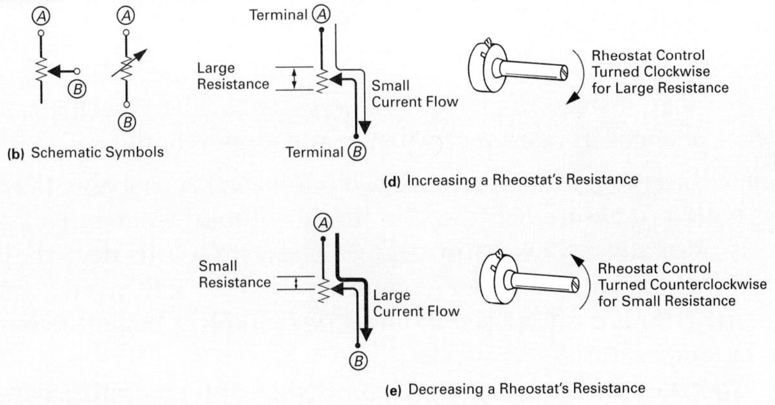

(A) (A) (B) (B)

(b) Schematic Symbols

Terminal (A)
Large Resistance
Small Current Flow
Terminal (B)
Rheostat Control Turned Clockwise for Large Resistance

(d) Increasing a Rheostat's Resistance

(A)
Small Resistance
Large Current Flow
(B)
Rheostat Control Turned Counterclockwise for Small Resistance

(e) Decreasing a Rheostat's Resistance

Potentiometer

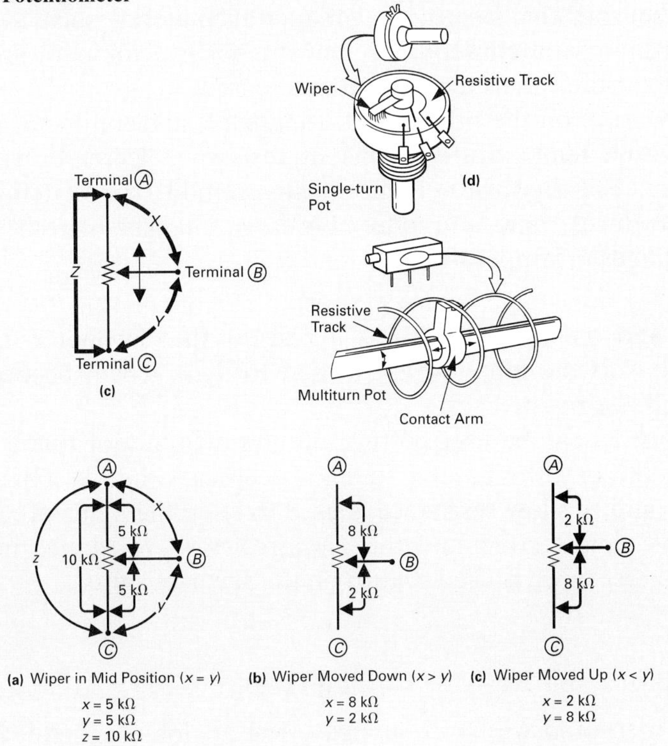

Wiper
Resistive Track
Single-turn Pot

(d)

Terminal (A)
X
Z
Terminal (B)
Y
Terminal (C)

(c)

Resistive Track
Multiturn Pot
Contact Arm

(A) 5 kΩ X Z 10 kΩ (B) 5 kΩ Y (C)

(a) Wiper in Mid Position ($x = y$)

$x = 5$ kΩ
$y = 5$ kΩ
$z = 10$ kΩ

(A) 8 kΩ (B) 2 kΩ (C)

(b) Wiper Moved Down ($x > y$)

$x = 8$ kΩ
$y = 2$ kΩ

(A) 2 kΩ (B) 8 kΩ (C)

(c) Wiper Moved Up ($x < y$)

$x = 2$ kΩ
$y = 8$ kΩ

Linear versus Tapered Resistive Tracks

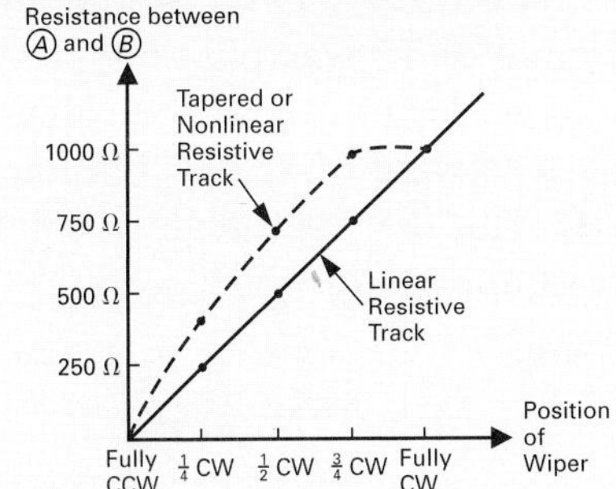

and the output, as can be seen, is the linearly increasing resistance between *A* and *B*.

With a tapered rheostat or potentiometer, the resistance varies nonuniformly along its resistor element, sometimes being greater or less for equal shaft movement at various points along the resistance element.

18. Thermally (heat) adjustable variable resistor (Figure 4-31). A *bolometer* is a device that, instead of changing its resistance when mechanical energy is applied, changes its resistance when heat energy is applied.

19. The measurement of temperature (*thermometry*) is probably the most common type of measurement used in industry today. Commercially, temperature is normally expressed in degrees Celsius (°C) or degrees Fahrenheit (°F); however, although not commonly known, kelvin (K) and degrees Rankine (°R) are often used in industry, the kelvin being the international unit of temperature.

There are basically three different types of temperature detectors. The *resistive temperature detector* (RTD) and *thin-film detector* (TFD) are both temperature sensors that contain a conducting material such as copper, nickel, or platinum and consequently have a positive temperature coefficient: resistance increases as temperature increases.

The *thermistor,* on the other hand, contains a semiconductor material that has a negative temperature coefficient: resistance decreases as temperature increases. The thermistor is the most common type of temperature sensor and produces rapid and extremely large changes in resistance for very small changes in temperature.

20. Optically (light) adjustable variable resistor (Figure 4-31). *Photo* means illumination, and the *photoresistor* is a resistor that is photoconductive. This means that as the material is exposed to light, it will become more conductive and less resistive.

Photoresistors can be used to turn on and off outdoor home security lights. During the day, the natural sunlight decreases the resistance of the photoresistor, and this low resistance is used to keep the lights off. At dusk, the sunlight is almost gone, and the photoresistor's resistance increases. This increased resistance is used to turn on the security lights.

Measuring Resistance and Resistor Coding (Figure 4-32)

21. Resistance is measured with an *ohmmeter.* The analog ohmmeter has four ranges. The simplest range has been selected ($R \times 1$), and on this scale the resistance value indicated can be read directly from the meter. If we had selected the $R \times 10$ range and a value of 72 was obtained, then 72 would have to be multiplied by 10 ($72 \times 10 = 720$) to obtain the value of 720 Ω. If we were on the $R \times 100$ range and a reading of 15 was obtained, then $R \times 100 = 15 \times 100$, which equals 1.5 kΩ. The $R \times 1000$ scale, if selected, means that you would interpret 470 for example as $R \times 1000 = 470 \times 1000$, which equals 470 kΩ.

FIGURE 4-31 Thermally and Optically Adjustable Variable Resistors.

Thermally (Heat) Adjustable Variable Resistors

	FAHRENHEIT	CELSIUS	KELVIN	RANKINE
Absolute zero	(180°F)	–273.16°C	0 K	0°R
Melting point of ice (x)	↓	0°C	273.16 K	491.69°R
	212°F	↓	↓	↓
(Division between x and y)		(100°C)	(100 K)	(180°R)
		↓	↓	↓
Boiling point of water (y)		100°C	373.16 K	671.69°R
–459.69°F				
32°F				

$F = (9/5 \times C) + 32$
$C = 5/9 \times (F - 32)$
$R = F + 460$
$F = R - 460$
$K = C + 273$
$C = K - 273$

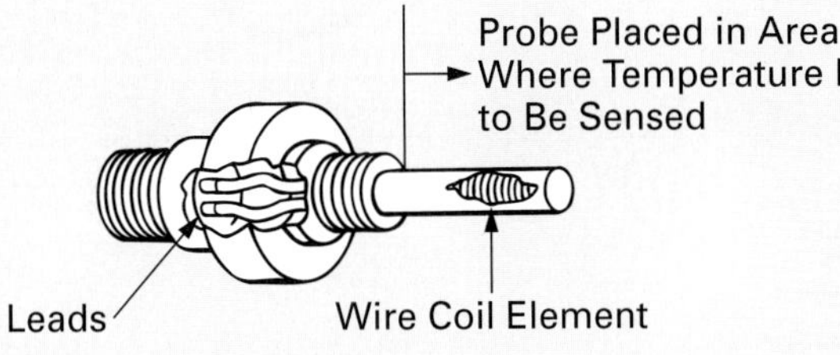

Platinum Winding			
°C	Ohms	°C	Ohms
–200	18.53	+200	175.84
–150	39.65	+250	194.08
–100	60.20	+300	212.03
–50	80.25	+350	229.69
±0	100.0	+400	247.06
+50	119.40	+450	264.14
+100	138.50	+500	280.93
+150	157.32	+550	297.16

(Conductors have a positive temperature coefficient of resistance—temp.↑, R↑)

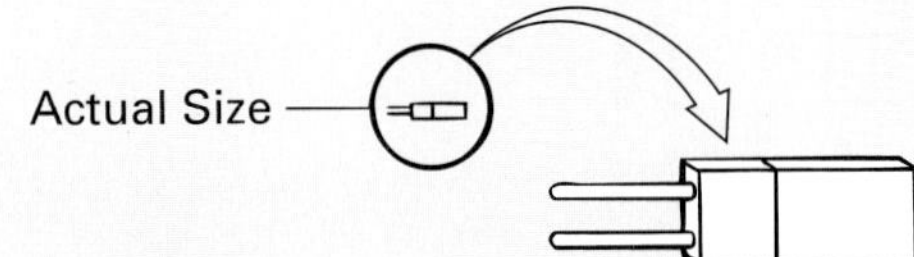

Probe
Linear
Bead
Disc

°C	Ohms
–50	100,000
0	7,500
+50	7,400
+100	100
+150	50
+200	27
+250	10
+300	7.5

(Semiconductors have a negative temperature coefficient of resistance—T↑, R↓)

Optically (Light) Adjustable Variable Resistor

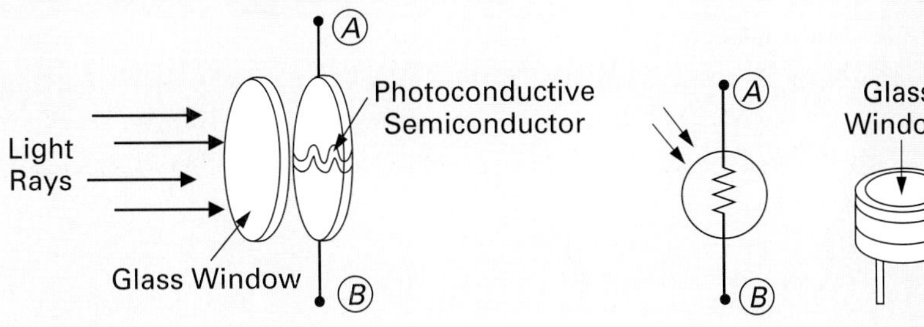

FIGURE 4-32 Measuring Resistance and Resistor Coding.

Measuring Resistance

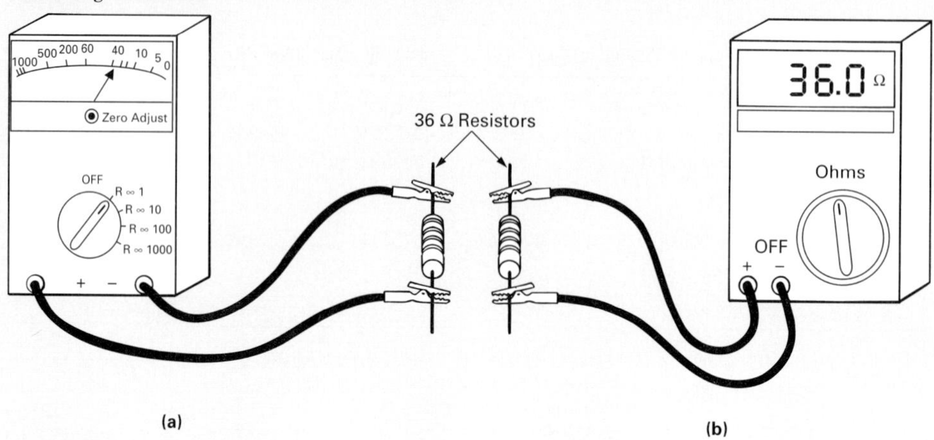

Resistor Color Code

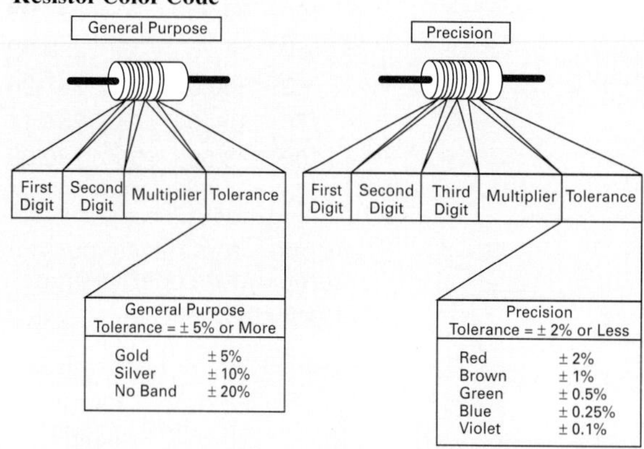

	Color	Digit Value	Multiplier		
Big	Black	0	1	One	1
Beautiful	Brown	1	10	One Zero	10
Roses	Red	2	100	Two Zeros	100
Occupy	Orange	3	1000	Three Zeros	1 k
Your	Yellow	4	10000	Four Zeros	10 k
Garden	Green	5	100000	Five Zeros	100 k
But	Blue	6	1000000	Six Zeros	1 M
Violets	Violet	7	10000000	Seven Zeros	10 M
Grow	Gray	8	–		
Wild	White	9	–		
So	Silver	–	10^{-2} or 0.01		1/100
Get some	Gold	–	10^{-1} or 0.1		1/10
Now	None	–			

Typographical Valve Indication

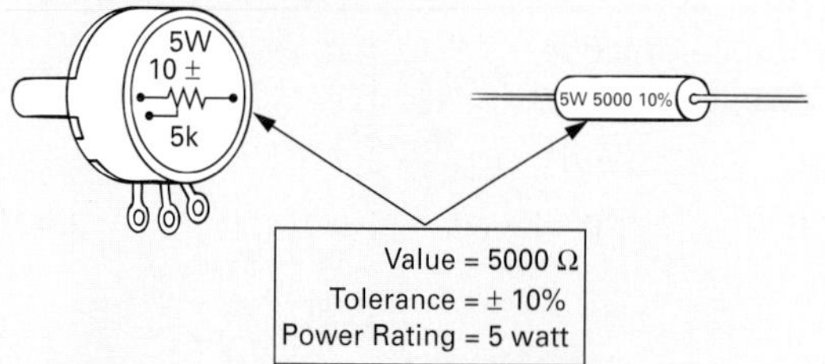

22. Three rules must be remembered and applied when measuring resistance:
    a. Short the meter leads together and adjust the zero-ohms control so that the pointer is at 0 Ω. This calibrates the meter scale.
    b. Turn off the circuit power and remove the component (if practicable) to be measured from circuit to protect the ohmmeter.
    c. Connect the leads across the removed component, adjust the range scale until the pointer is approximately in the middle of the scale (for accuracy), and then multiply this value by the range scale selected.
23. With the digital auto-ranging ohmmeter the measurement of resistance is a lot easier since you simply turn the selector to ohms and read off the reading from the display.
24. Using the ohmmeter, you can find the value of any resistor. However, manufacturers indicate the value and tolerance of resistors on the body of the component in one of two ways:
    a. *Color code:* colored rings or bands
    b. *Typographically:* printed alphanumerics (alphabet and numerals)
25. There are basically two different types of fixed resistors: general purpose and precision. Resistors with tolerances of ±2% or less are *precision resistors* and have five bands; resistors with tolerances of ±5% or greater have four bands and are referred to as *general-purpose resistors.*
26. For the general-purpose resistor code:
    a. The first band on either a general-purpose or precision resistor can never be black, and it is the first digit of our number.
    b. The second band indicates the second digit of the number.
    c. The third band specifies the multiplier to be applied to the number, which ranges from × $\frac{1}{100}$ to 10,000,000.
    d. The fourth band describes the tolerance or deviation from the specified resistance, which is ±5% or greater.
27. For the precision resistor code:
    a. The first band, like the general-purpose resistor, is never black and is the first digit of the three-digit number.
    b. The second band provides the second digit.
    c. The third band indicates the third and final digit of the number.
    d. The fourth band specifies the multiplier to be applied to the number.
    e. The fifth and final band indicates the tolerance figure of the precision resistor, which is always less than ±2%, which is why precision resistors are more expensive than general-purpose resistors.
28. Printed circuit boards no longer have all their resistor and other components inserted by hand. *Automatic insertion equipment* replaces the human assembler and places the correct component in the correct position in a matter of seconds rather than hours. The zero-ohms resistor can be installed automatically during the resistor insertion process and have exactly the same effect as a wire jumper.
29. If a resistor color code is not found, some other form of marking will always be made on the resistor. The larger fixed, wirewound resistors and

nearly all variable resistors normally have the resistance value, tolerance, and wattage printed on the resistor. SIP and DIP packages are normally coded with both letters and numerals, all of which have particular meaning. Chip resistors tend to be too small to have any intelligible form of marking and so will have to be measured with an ohmmeter. They are packaged in either polythene bags, paper tape, or plastic magazines, and in all these cases the packaging will be used to specify values, tolerance, and wattage.

Power Dissipation Due to Resistance

30. The atoms of a resistor obstruct the flow of moving electrons (current), and this friction between the two causes heat to be generated. Current flow through any resistance therefore will always be accompanied by heat. In some instances, such as a heater or oven, the device has been designed specifically to generate heat. In other applications it may be necessary to include a resistor to reduce current flow, and the heat generated will be an unwanted side effect.

31. Any one of the three power formulas can be used to calculate the power dissipated by a resistance. However, since power is determined by the friction between current (I) and the circuit's resistance (R), and $P = I^2 \times R$ formula is more commonly used to calculate the heat generated.

Filament Resistor

32. The *filament resistor* is more commonly known as the household light bulb. The filament resistor is just a coil of wire that glows white hot when current is passed through it and, in so doing, dissipates both heat and light energy.

This *incandescent lamp* is an electric lamp in which electric current flowing through a filament of resistive material heats the filament until it glows and emits light.

The filament material (tungsten, for example) is like all other conductors in that it has a positive temperature coefficient of resistance. Therefore, as the current through the filament increases, so does the temperature and so does the filament's resistance ($I\uparrow$, temperature $\uparrow$, $R\uparrow$).

33. A coil of wire similar to the one found in a light bulb, called a *ballast resistor,* is used to maintain a constant current despite variation in voltage. Since the coil of wire is a conductor and conductors have a positive temperature coefficient, an increase in voltage will cause a corresponding increase in current and therefore in the heat generated, which will result in an increase in the wire's resistance. This increase in resistance will decrease the initial current rise. Similarly, a decrease in voltage and therefore current ($I \propto V$) will result in a decrease in heat and wire resistance. This decrease in resistance will permit an increase in current to counteract the original decrease. Current is therefore regulated or maintained constant by the ballast resistor, despite variations in voltage.

Testing Resistors

34. It is virtually impossible for resistors to short-circuit internally; however, they may be shorted out by some other device, providing a low-resistance path around the resistor. Generally, the resistor's internal resistive elements will begin to develop a higher resistance than its specified value or in some cases go completely open circuit.

Fixed-value resistors tend to go either partially or fully open as the resistive track begins to break down. Variable-value resistors have problems with the wiper making contact with the resistive track at all times.

35. The ohmmeter is the ideal instrument for verifying whether or not a resistor is functioning correctly.

When checking a resistor's resistance, here are some points that should be remembered.

a. Always calibrate the meter first to zero.
b. The ohmmeter has its own internal power source (a battery), so always turn off the circuit power and remove from the circuit the resistor to be measured.
c. The range scales on a digital multimeter often confuse people; for example, if a range is chosen so that the highest reading is 100 kΩ and a 1-MΩ resistor is measured, the meter will indicate an infinite-ohms reading and you may be misled into believing that you have found the problem (an open resistor). In the case of an analog ohmmeter, if the range scale multiplier is not applied, a misleading result is obtained.
d. Another point to keep in mind is tolerance.
e. The ohmmeter's internal battery voltage is really too small to deliver an electrical shock; however, you should avoid touching the bare metal parts of the probes or resistor leads, as your body resistance of approximately 50 kΩ will affect your meter reading.

NEW TERMS

Alphanumeric
Ballast resistor
Carbon composition resistor
Carbon film resistor
Celsius (°C)
Cermet resistor
Chip resistor
Cold resistance
Color code
Dissipation
Dual in-line package (DIP)
Fahrenheit (°F)
Filament resistor
Fixed-value resistor
General-purpose resistor
Helix
Hot resistance
Incandescent lamp
Infinite (∞)
Jumper
Kelvin (K)
Linear
Metal film resistor
Metal oxide resistor

Ohmmeter
Photoresistor
Potentiometer
Precision resistor
Rankine (°R)
Resistive temperature detector (RTD)
Resistor
Rheostat
Single in-line package (SIP)
Substrate
Tapered
Temperature stability
Thermistor
Thermometry
Thick-film resistor
Thin-film detector (TFD)
Tolerance
Variable resistor
Wattage rating
Wirewound resistor

REVIEW QUESTIONS

Multiple-Choice Questions

1. The most common type of fixed-value resistor is the:
   a. Wirewound **c.** Film
   b. Carbon composition **d.** Metal oxide
2. A SIP package has:
   a. A single line of connectors **c.** No connectors
   b. A double line of connectors **d.** None of the above
3. The most commonly used mechanically adjustable variable-value resistor is the:
   a. Rheostat **c.** Potentiometer
   b. Thermistor **d.** Photoresistor
4. A thermistor has a negative temperature coefficient, which means that:
   a. As temperature increases, resistance increases.
   b. As temperature increases, conductance decreases.
   c. As temperature increases, resistance decreases.
   d. All of the above are true.
5. A ____________ -band resistor is called a general-purpose resistor, while a ____________ -band resistor is known as a precision resistor.
   a. Four, five **b.** Three, four **c.** One, four **d.** Two, seven
6. What would be the power dissipated by a 2-kΩ carbon composition resistor when a current of 20 mA is flowing through it?
   a. 40 W **c.** 1.25 W
   b. 0.8 W **d.** None of the above
7. What of the following is the most common type of temperature detector?
   a. RTD **c.** Thermistor
   b. TFD **d.** Barretter
8. A rheostat is a ____________ terminal device, while the potentiometer is a ____________ terminal device.
   a. 2, 3 **c.** 2, 4
   b. 1, 2 **d.** 3, 2

9. Which of the following is the unit of temperature in the International System of Units?

   a. Celsius
   b. Fahrenheit
   c. Kelvin
   d. Rankine

10. Which of the following temperature detectors has a positive temperature coefficient?

   a. RTD
   b. Thermistor
   c. TFD
   d. Both (a) and (b)
   e. Both (a) and (c)

11. Photoresistors are also called:

   a. TFDs
   b. LDRs
   c. RTDs
   d. TGFs

12. Photoresistors have:

   a. A positive temperature coefficient
   b. A negative temperature coefficient
   c. Neither (a) nor (b)

13. Resistance is measured with a (an):

   a. Wattmeter
   b. Milliammeter
   c. Ohmmeter
   d. A 100 meter

14. Which of the following is *not* true?

   a. Precision resistors have a tolerance of > 5%.
   b. General-purpose resistors can be recognized because they have either three or four bands.
   c. The fifth band indicates the tolerance of a precision resistor.
   d. The third band of a general-purpose resistor specifies the multiplier.

15. The first band on either a general-purpose or a precision resistor can never be:

   a. Brown
   b. Red or black stripe
   c. Black
   d. Red

16. What is the name of the resistor used to maintain a constant current despite variations in voltage by changing its resistance?

   a. RTD
   b. Thermistor
   c. Ballast resistor
   d. LDR

17. The term *infinite ohms* describes:

   a. A small finite resistance
   b. A resistance so large that a value cannot be placed on it
   c. A resistance between maximum and minimum
   d. None of the above

18. To calibrate an ohmmeter means:

   a. To make sure that it indicates 100 Ω when 0 Ω is being measured
   b. To zero the meter when the meter leads are touching
   c. To make sure that it indicates 0 Ω when the meter leads are apart
   d. Both (a) and (b)
   e. Both (a) and (c)

19. The typical problem with resistors is that they will:
   - **a.** Develop a partial internal open
   - **b.** Completely open circuit
   - **c.** Develop a higher resistance than specified
   - **d.** All of the above
   - **e.** Both (a) and (c)

20. A 10% tolerance, 2.7 MΩ carbon composition resistor measures 2.99 MΩ when checked with an ohmmeter. It is:
   - **a.** Within tolerance
   - **b.** Outside tolerance
   - **c.** Faulty
   - **d.** Both (a) and (c)
   - **e.** Both (a) and (b)

Essay Questions

21. Describe the difference between a fixed- and a variable-value resistor. (4-1)
22. List the six basic categories of fixed-value resistors. (4-1-1)
23. Explain why a resistor's size determines its wattage rating. (4-1-1)
24. Why are resistors given a tolerance figure, and which is best, a small or a large tolerance? (4-1-1)
25. Briefly describe the construction of:
   - **a.** Carbon composition fixed-value resistors (4-1-1)
   - **b.** Rheostats (4-1-2)
   - **c.** Potentiometers (4-1-2)
   - **d.** Multiturn precision potentiometers (4-1-2)
26. Draw the schematic symbol for a rheostat and potentiometer and describe the difference. Also show how a potentiometer can be used as a rheostat. (4-1-2)
27. Define linear and tapered resistive tracks. (4-1-2)
28. List the three rules that should be applied when using an ohmmeter to measure resistance. (4-2)
29. Describe the construction, operation, and temperature coefficient of the following: (4-1-2)
   - **a.** RTD
   - **b.** Photoresistor
   - **c.** Thermistor
   - **d.** TFD
30. List the four temperature scales in use today. (4-1-2)
31. Give the color code for the following resistor values: (4-3)
   - **a.** 1.2 MΩ, ± 10%
   - **b.** 10 Ω, ± 5%
   - **c.** 27 kΩ, ± 20%
   - **d.** 273 kΩ, ±0.5%
32. Give the resistor values for the following color codes: (4-3)
   - **a.** Orange, orange, black
   - **b.** Red, red, green, red, red
   - **c.** Brown, black, orange, gold
   - **d.** White, brown, brown, silver

33. For what application would the zero-ohms resistor be used? (4-3-2)
34. Describe the difference between a SIP and a DIP resistor package. (4-1-1)
35. What are the common problems encountered with fixed- and variable-value resistors? (4-6)

Practice Problems

36. If a 5.6-kΩ resistor has a tolerance of ±10%, what would be the allowable deviation in resistance above and below 5.6 kΩ? (4-3)
37. (a) If a current of 50 mA is flowing through a 10-kΩ, 25-W resistor, how much power is the resistor dissipating? (b) Can the current be increased, and if so by how much? (4-3)
38. What minimum wattage size could be used if a 12-Ω wirewound resistor were connected across a 12-V supply? (4-3)
39. Calculate the resistance deviation for the tolerances of the resistors listed in Problems 31 and 32. (4-1-1)
40. If a 1-kΩ rheostat has 50 V across it, calculate the different values of current if the rheostat is varied in 100-Ω steps. Referring to Figure 4-29, first calculate and insert the values into the table in Figure 4-33(a), and then plot these values in the graph in Figure 4-33(b).

Troubleshooting Questions

41. Briefly describe some of the problems that can occur with fixed and variable resistors. (4-6)

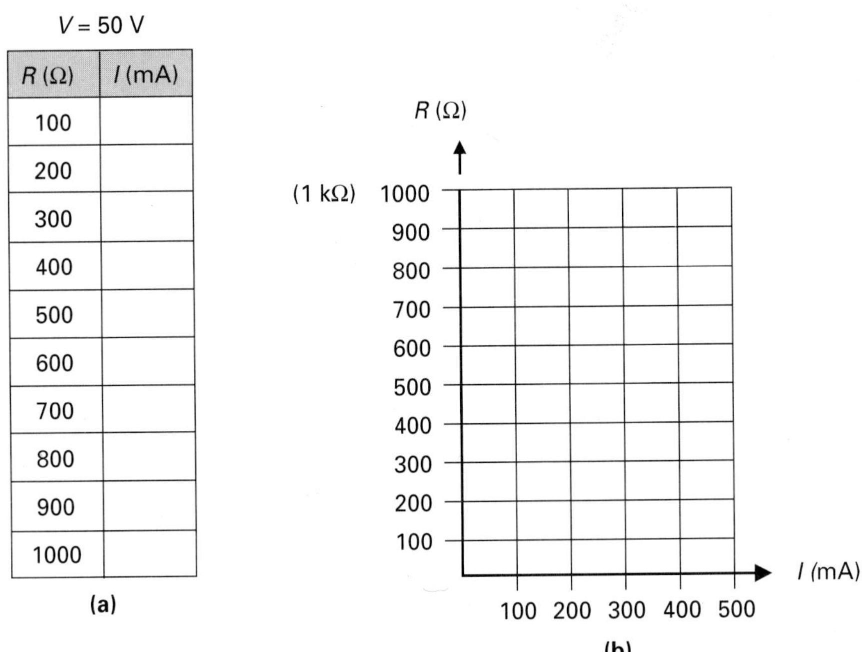

FIGURE 4-33 **Plotting a Rheostat's Current against Resistance.**

42. Describe why the tolerance of a resistor can make you think that it has a problem when in fact it does not. (4-6)

43. If a 20-kΩ, 1-W, ±10% tolerance resistor measures 20.39 kΩ, is it in or out of tolerance? (4-1-1)

44. If 33 μA is flowing through a carbon film 4.7-kΩ, ¼-W, ±5% resistor, will it burn up? (4-1-1)

45. What points should be remembered when using the ohmmeter to verify a resistor's value? (4-6)

CHAPTER

5

OBJECTIVES

After completing this chapter, you will be able to:

1. List the precautions that should be observed while troubleshooting electronic circuits or systems.
2. Define the term *body resistance.*
3. Briefly describe first aid, treatment, and resuscitation.
4. Explain the basic operation of the following test instruments:
   - **a.** The dc power supply
   - **b.** The analog multimeter
   - **c.** The digital multimeter
5. Describe the purpose and construction of the protoboard or breadboard.
6. Explain the purpose of most of the commonly used electronic tools, along with soldering principles and techniques.

Experimenting in the Lab

O U T L I N E

VIGNETTE

At the Core of Apple

Stephen Wozniak and Steven Jobs met at their Los Altos, California, high school, and due to their common interest, electronics, became friends.

Wozniak was a conservative youth whose obsession with technology left little room for social relationships or studying; in fact, after one year at the University of Colorado, he dropped out with his academic record littered with F's. In contrast to this serious nature, Wozniak was renowned for his high-tech pranks, for which, on one occasion, he spent a night in juvenile hall for wiring up a fake bomb in a friend's locker. In another incident, he devised a way to place a free telephone call to the Pope at the Vatican, identifying himself as then Secretary of State Henry Kissinger.

Jobs, on the other hand, had interests outside electronics. He searched for intellectual, emotional, and spiritual stimulation. After one semester at Reed College, he dropped out to pursue an interest in Eastern religions, which led him to temples in India, searching for the meaning of life.

The first enterprise built by Wozniak and sold by Jobs was an illegal device called a blue box that could crash the telephone system. It generated a set of tones that fooled the computerized telephone switching systems and opened up free, long-distance circuits; these systems allowed "phone phreaks" to take long and illegal joy rides throughout the world's telephone networks.

By selling a Volkswagen van and a programmable calculator, Wozniak and Jobs raised the initial capital of $1300 to start a business in April 1976 called Apple Computer, since Jobs had a passion for the Beatles, who recorded under the Apple record label.

In just five years, Apple Computer grew faster than any other company in history, from a two-man assetless partnership building computers in a family home to a publicly traded corporation earning its youthful entrepreneurs, Jobs (27) and Wozniak (24), fortunes of nearly $200 million each.

INTRODUCTION

This chapter has been designed to introduce you to the laboratory, or lab, where you will be conducting many practical experiments designed to reinforce the topics discussed in theory. To introduce you properly to the lab, we will need to discuss safety, the test equipment you will be using at your lab station, the protoboard or breadboard, and soldering tools and techniques.

In the first section we discuss some of the different safety precautions and procedures that should be observed when working on electrical and electronic equipment. In the second section the operation of the dc power supply, the analog multimeter and the digital multimeter are covered. These pieces of test equipment will be used extensively to power and sense your experimental circuits. In the third section you will be introduced to the protoboard or breadboard, which is designed to accommodate the many experiments described in this book's associated lab manual. The protoboard will hold and interconnect electronic components such as resistors, capacitors, inductors, and the many other components that together form experimental circuits. In the final section you will be introduced to many of the electronic tools that will be needed for experimentation and soldering. In this section we will also introduce you to soldering—the joining together of two metal parts by applying both heat and solder—along with good soldering techniques.

5-1 SAFETY

Safety precautions should always be your first priority when working on electronic equipment, as there is always the possibility of receiving an electric shock. A **shock** is a sudden, uncontrollable reaction as current passes through your body and causes your muscles to contract and deliver a certain amount of pain. Figure 5-1 lists the physiological effects of different amounts of current; as you can see, even a current as small as 10 mA can be fatal.

Shock
The sudden pain, convulsion, unconsciousness, or death produced by the passage of electric current through the body.

Any shock is dangerous, since even the mildest could surprise you and cause an involuntary action that could injure you or someone else. For example,

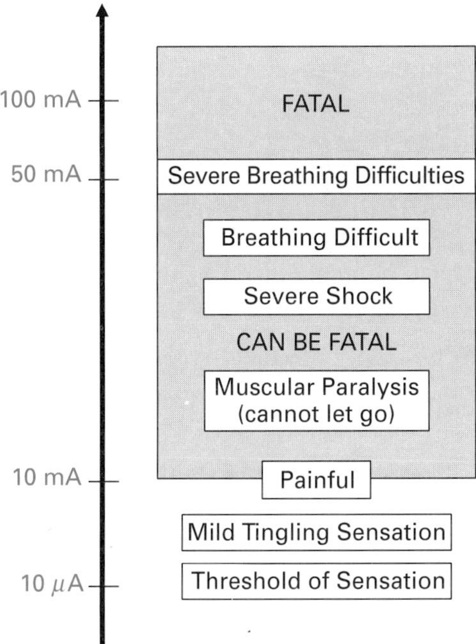

FIGURE 5-1 Physiological Effects of Electric Current.

a muscular spasm could throw you against a sharp object or move your arm to a point of higher voltage.

Body Resistance
The resistance of the human body.

5-1-1 *Body Resistance*

Voltage is the force that causes current to flow through a circuit. The amount of current is dependent on the circuit's resistance and the value of voltage ($I = V/R$). The greatest danger, therefore, is from high voltage points (more than 25 V). The resistance of your body also determines current, since a low resistance will cause a more dangerous high current. The human body has a resistance of about 10,000 to 50,000 Ω, depending on how good a contact you make with the **"live"** (power present) conductor. Skin resistance is generally quite high and will subsequently oppose current flow; however, your resistance is lowered if your skin is wet due to perspiration or if your skin has a cut or an abrasion.

Live
Term used to describe a circuit or piece of equipment that is ON and has current flow within it.

EXAMPLE:

Calculate the current through a body resistance of 10 kΩ if the body came in contact with 100 V. How much power is the body dissipating?

Solution:

$$I = \frac{V}{R} = \frac{100\text{ V}}{10\text{ k}\Omega} = 10\text{ mA}$$

$$P = V \times I = 100\text{ V} \times 10\text{ mA} = 1\text{ watt}$$

This value of current will be painful and possibly fatal.

Power is a point to consider. In the example, the body resistance was dissipating 1 watt. A power source may not be able to supply a fatal high power, and its output voltage will be pulled down due to the excessive load current through the body resistance. The current is then reduced to an amount that the power source can deliver.

5-1-2 *Precautions to Be Used When Troubleshooting*

The following procedures and precautions have been acquired from experienced technicians and should be applied whenever possible.

1. Except when absolutely necessary, do not work on electrical or electronic circuits or equipment when power is on.
2. When troubleshooting inside equipment, people tend to lean one hand on the chassis (the metal framework of the equipment) and hold the test lead or probe in the other hand. If the probing hand comes in contact with a high voltage, current will flow from one hand, through your chest (heart and lungs), and finally through the other hand to the chassis ground. To avoid this dangerous situation, always place the free hand in a pocket or behind your back while testing a piece of equipment with power on.

3. When troubleshooting equipment with power on, try to insulate yourself by not standing on a damp floor or leaning against any metal object and by wearing rubber-soled electrical safety shoes or standing on a rubber mat.
4. Electrolytic and other large-value capacitors (which will be covered in Chapters 13 and 14) can hold a dangerous voltage charge even after the equipment has been turned off. Be sure always to check if these capacitors are fully discharged by shorting the two terminals with a pair of *insulated* pliers or *insulated* test lead.
5. All tools should be well insulated. If not, it is not only dangerous to you, but probes that are not well insulated right down to the tip can cause short circuits between two points, resulting in additional circuit and equipment problems.
6. Always switch off the equipment and disconnect the power (since some equipment has power present even when off) before replacing any components. When removing components that get hot, such as resistors, allow enough time for cooling after the equipment has been turned off.
7. Necklaces, rings, and wristwatches have a low resistance and should be removed when working on equipment.
8. Inspect the equipment before working on it, and if it is in poor condition (frayed, cracked, or burned power cords, chipped plugs), turn off the equipment and replace these hazards.
9. Make sure that someone is present who can render assistance in the event of an emergency.
10. Make it a point to know the location of the power-off switch.
11. Cathode-ray tubes, which are the picture tubes within televisions and computer monitors, are highly evacuated and should therefore be handled with extreme care. If broken, the relatively high external pressure will cause an implosion (burst inward), which will result in the inward metal parts and glass fragments being expelled violently outward.

5-1-3 *First Aid, Treatment, Resuscitation*

Injury to Persons

Injury may be caused in a number of ways:

Shock. Electric shock is the effect produced on the body and in particular on the nerves by an electric current passing through it. Its magnitude depends on the strength of the current, which, in turn, depends on the voltage. Its effect varies according to the ohmic resistance of the body, which varies in different persons, and also according to the parts of the body between which the current flows (contact in the cardiac region can be particularly dangerous). It also depends on the current flow and on the surface resistance of the skin, which is much reduced when the skin is wet and is reduced to zero if the skin is penetrated.

Shock can be felt from voltages as low as 15 V, and at 20 to 25 V most people would experience pain. At about this voltage or a little higher, the victim may find himself unable to let go of the conductor and may suffer burning. It is believed that under certain conditions, death can be caused by voltages as low as 70, but generally the danger below 120 V ac is believed to be small (although not entirely negligible). Most serious and fatal accidents occur at the industrial 200 to 240 V ac and from 25 to 30 mA and over.

Injury can also be caused by a minor shock, not serious in itself but which has the effect of contracting the muscles sufficiently to result in a fall or other reaction.

Burns. Burns can be caused by the passage through the body of a heavy current if the body is in contact with a conductor, or by direct contact with an electrically heated surface. Burns can also be caused by the intense heat generated by the arcing from a short circuit. All cases of burns require immediate medical attention.

Explosion. An explosion can be caused by the ignition of flammable gases by a spark from an electric contact. In all cases where a flammable or ignitable atmosphere or vapor is present, special care is necessary.

Eye injuries. These can be caused by exposure to the strong ultraviolet rays of an electric arc. In these cases, the eyes may become inflamed and painful after a lapse of several hours, and there may be temporary loss of sight. Although very painful, the condition usually passes off within 24 hours. Lasers are also dangerous to the eyes due to their intensely concentrated beam, and therefore specially filtered protective glasses should be worn, as seen in Figure 5-2.

Precautions to protect the eyes must always be taken by wearing protective goggles when clipping leads or soldering. Permanent injury to the eyes can arise from the energy propagated by microwave equipment. No one should look

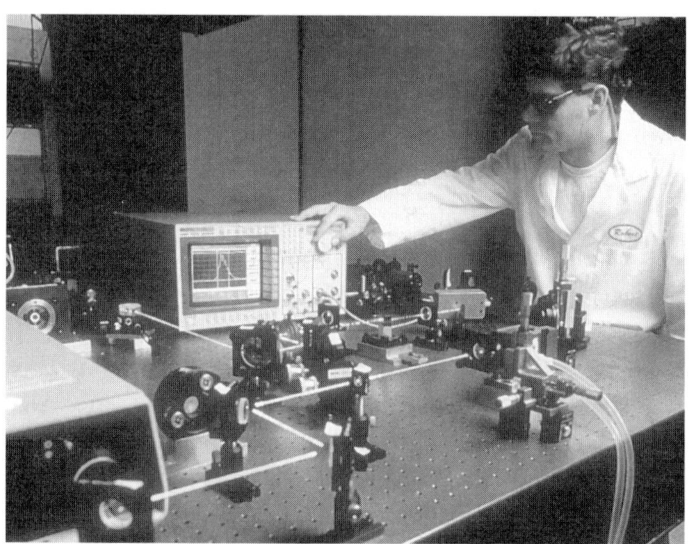

FIGURE 5-2 Protective Glasses for Laser Experimentation. (Photo courtesy of Hewlett-Packard Company.)

along a waveguide when it is on or examine a highly directional radiator at close distances.

Body injuries from microwave and radio-frequency equipment. The energy in microwave and radio-frequency equipment can damage the body, especially those parts with a low blood supply. The eyes are particularly vulnerable. The highest energy level to which operators should be subject is 1.0 mW/cm^2, and intensities exceeding 10 mW/cm^2 should always be avoided.

Resuscitation

You should familiarize yourself with the various methods of artificial respiration by contacting your local Red Cross for complete instruction. The mouth-to-mouth method of artificial respiration is the most effective of the resuscitation techniques. It is comparatively simple and produces the best and quickest results when done correctly.

Mouth-to-mouth resuscitation method. It is essential to begin artificial respiration without delay. *Do not touch the victim with your bare hands until the circuit is broken.* If this is not possible, *protect yourself* with dry insulating material and pull the victim clear of the conductor.

Step 1: Lay the patient on his or her back and, if on a slope, have the stomach slightly lower than the chest.

Step 2: Make a brief inspection of the mouth and throat to ensure that they are clear of obvious obstructions.

Step 3: Give the patient's head the maximum backward tilt so that the chin is prominent and the neck stretched to give a clear airway, as shown in Figure 5-3(a).

Step 4: Sealing off the patient's nose with your thumb and finger, open your mouth wide and make an airtight seal over the open patient's mouth and then blow, as shown in Figure 5-3(b). (New steps indicate using a plastic bag airway).

Step 5: After exhaling, turn your head to watch for chest movement, while inhaling deeply in readiness for blowing again, as shown in Figure 5-3(c).

Step 6: If the chest does not rise, check that the patient's mouth and throat are free of obstruction and that the head is tilted back as far as possible, and then blow again.

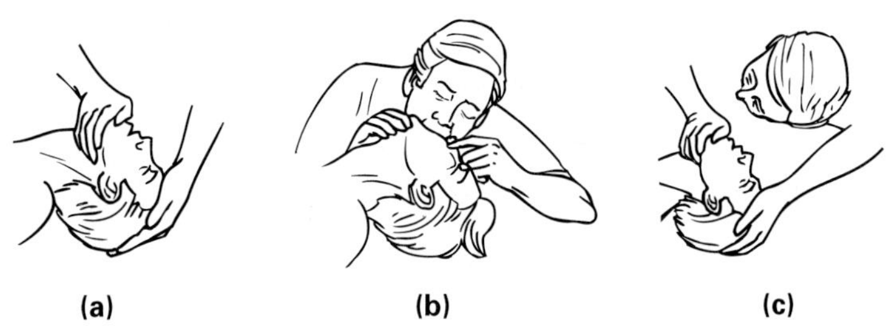

FIGURE 5-3 **Mouth-to-Mouth Resuscitation.**

SELF-TEST REVIEW QUESTIONS FOR SECTION 5-1

1. What is the approximate resistance of the human body?
2. List the precautions that should be taken when troubleshooting.

5-2 TEST EQUIPMENT

As a technician, your tools of the trade will be test equipment. In this section you will be introduced to the basic operation of the *dc power supply,* the *analog multimeter,* and the *digital multimeter* which you will be using in your experiments in the lab.

5-2-1 The DC Power Supply

DC Power Supply
A powerline, generator, battery, power pack, or other dc source of power for electronic equipment.

In the laboratory or repair shop, power can be obtained from a battery, but since ac power is so accessible at the wall outlet, a **dc power supply** is generally used to supply dc voltages. The dc power supply has a large advantage over the battery as a source of dc voltage in that it can quickly provide an accurate output voltage that can easily be varied by a control on the front panel, and it never runs down.

During your electronic studies you will use a dc power supply in the lab to provide different dc output voltages for various experiments. In Figure 5-4(a) you can see how a dc power supply can be used to supply 12 volts (12 V) to a 12-kilohm (12 kΩ) resistor. To explain the basic operation of this piece of equipment, the ON/OFF switch is used to turn on the power supply, while the VOLTS/AMPS switch is used to switch the meter between a voltmeter or ammeter. In the "volts" position, the meter indicates the voltage that is appearing across the output terminals, which in this example is 12 V. If the VOLTS/AMPS switch is placed in the "amps" position, the meter will now monitor the amount of current flowing out of the power supply, which in this example is 1 milliampere (1 mA). The VOLTS ADJUST control is used to either increase or decrease the output voltage. Figure 5-4(b) shows a photograph of a typical dc power supply.

Analog Multimeter
Piece of electronic test equipment that can perform multiple tasks in that it can be used to measure voltage, current, or resistance.

5-2-2 The Analog Multimeter (AMM)

The multimeter is a device for the measurement of current, voltage, and resistance. It is called a multimeter because it is like having three meters all rolled into one, in that it has

a. An ammeter for measuring current
b. A voltmeter for measuring voltage
c. An ohmmeter for measuring resistance

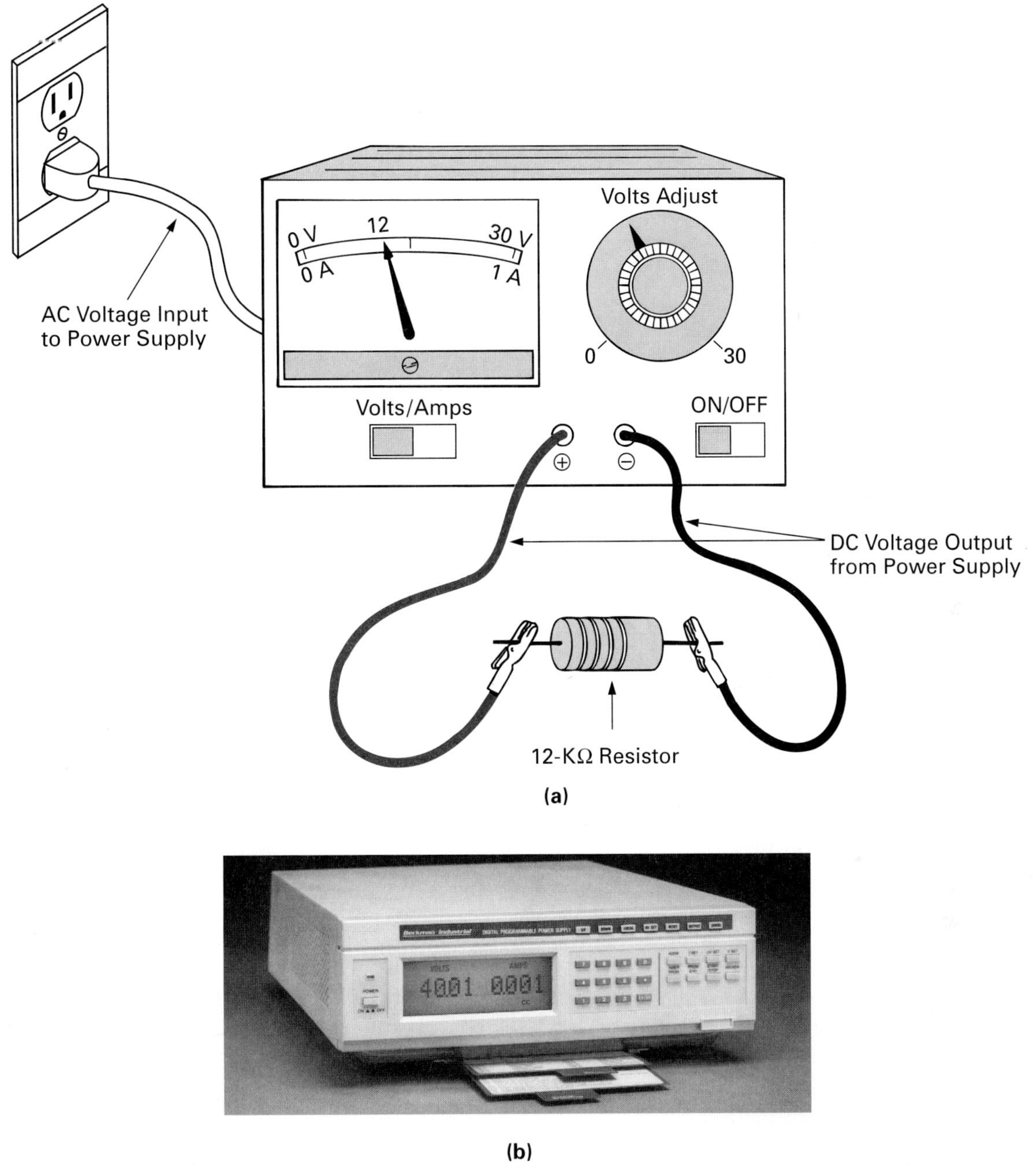

(a)

(b)

FIGURE 5-4 **DC Power Supply.**

For many years the analog readout multimeter, an example of which is shown in Figure 5-5(a), was the only type available. At present these analog readout multimeters are still being used but are slowly being replaced by digital readout multimeters similar to the type shown in Figure 5-5(b). In the following section, we will cover the digital multimeter.

Analog is a word that means "similar to." For example, a ball is an analog of the world in that the shapes are similar. With an analog multimeter or AMM,

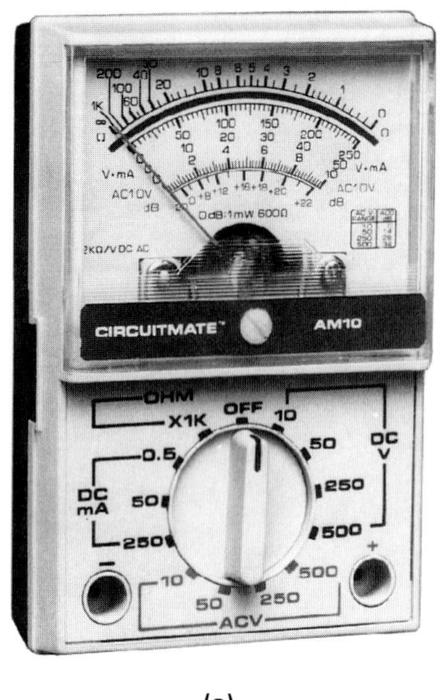

(a)

(b)

FIGURE 5-5 **Multimeter. (a) Analog Readout Multimeter. (b) Digital Readout Multimeter.**

the amount of pointer deflection across the scale is an analog (or similar to) the magnitude of the electrical property being measured. This type of indicator is therefore called an *analog display* as opposed to a *digital display,* which indicates the magnitude of the electrical property using decimal digits.

Let us now review the steps involved in measuring current, voltage, and resistance.

Measuring Current with the Analog Ammeter

Ammeter
Meter placed in the circuit path to measure the amount of current flow.

Ammeters (ampere meters) are used to measure the current flow within a circuit, as shown in Figure 5-6. In electronics, where current is generally small, the milliampere (mA) or microampere (μA) scale will generally be used, whereas in electrical high-current circuits, the meter will generally be used on an ampere scale.

Three rules must be remembered and applied when measuring current:

1. Always set the selector to the higher scale first (amperes), and then reduce as needed to milli- or microamperes, just in case a larger current than anticipated is within the circuit, in which case the meter could be damaged.
2. The current meter must be inserted in the circuit so that the positive lead of the ammeter (red) is connected to the positive charge or polarity, and the negative lead of the ammeter (black) is connected to the negative charge or polarity. Ammeters are polarity sensitive instruments.
3. If you wish to measure current flowing within a wire, the ammeter measuring the current must be inserted in the path of current flow.

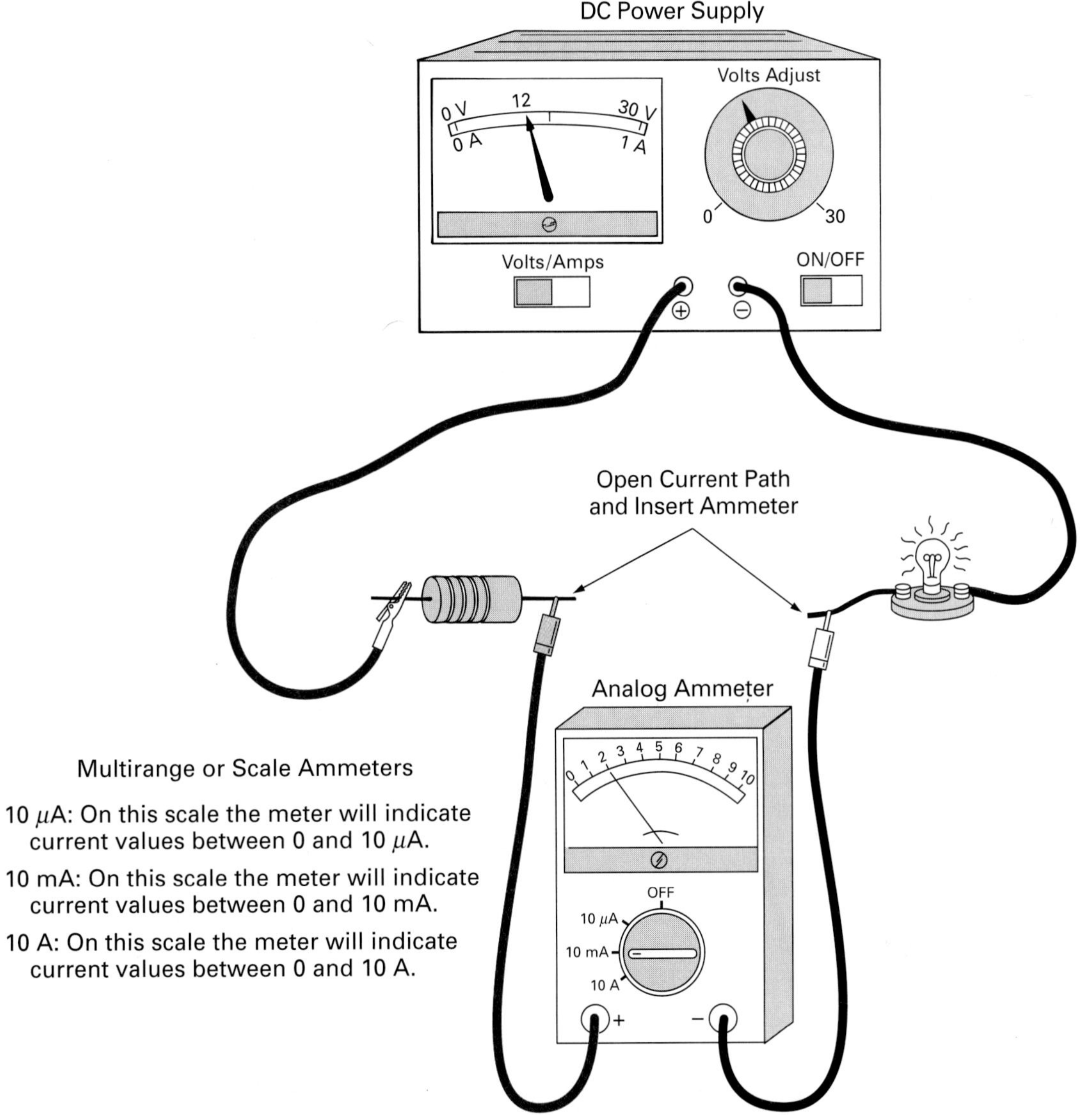

FIGURE 5-6 Measuring Current with the Analog Ammeter.

Measuring Voltage with the Analog Voltmeter

Voltmeters (voltage meters) are used to measure electrical pressure (voltage) anywhere in an electronic circuit. Figure 5-7 illustrates a voltmeter being used to measure the voltage drop across a resistor.

Voltmeter
Instrument designed to measure the voltage or potential difference. Its scale can be graduated in kilovolts, volts, or millivolts.

A voltmeter is connected across a device, as illustrated in Figure 5-7, with the two leads of the voltmeter touching either side of the component.

Three steps must be remembered and applied when measuring voltage:

1. Always set to the highest scale first (kV) and then reduce as needed to the volt and millivolt range, just in case a larger voltage than anticipated is within the circuit, in which case the meter could be damaged.

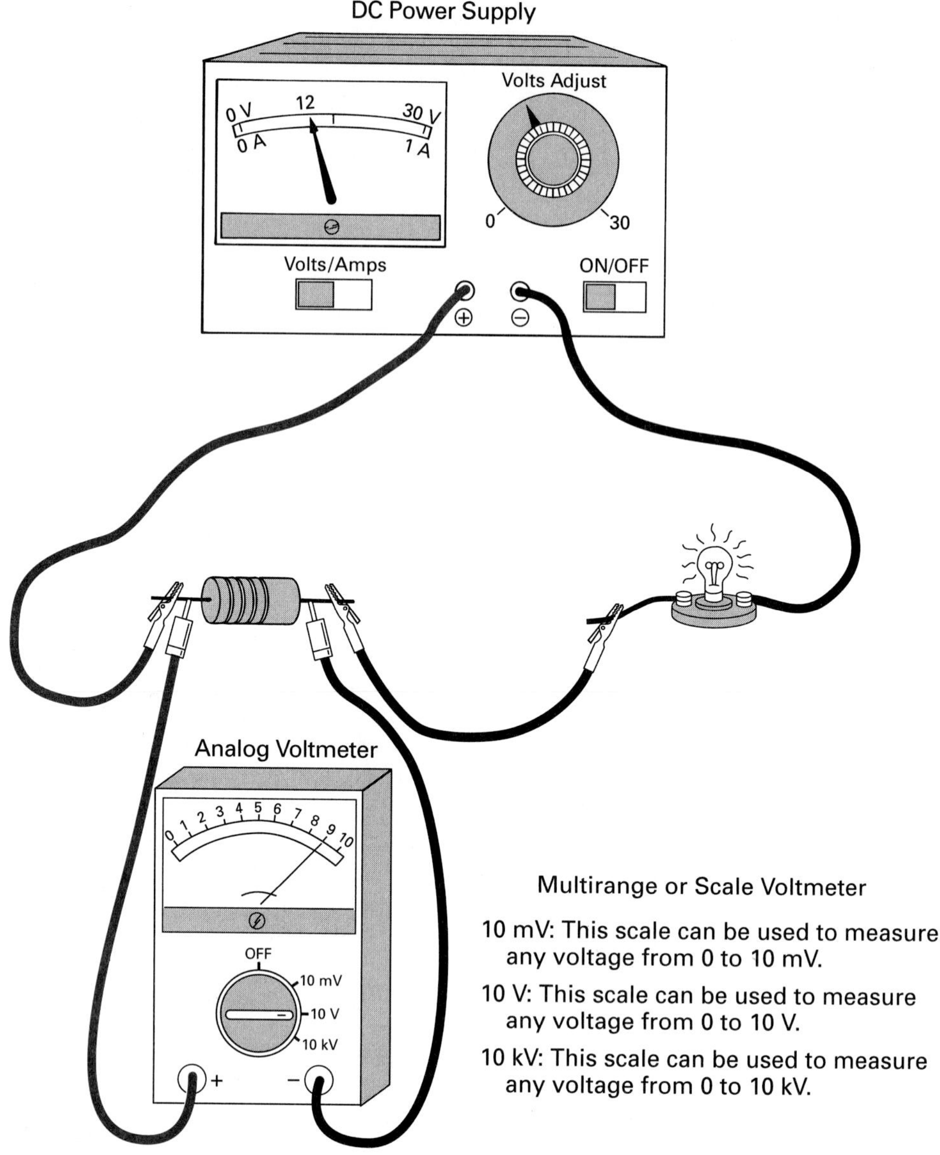

FIGURE 5-7 **Measuring Voltage with the Analog Voltmeter.**

2. Always connect the positive lead (red) of the meter to the component lead that is closer to the positive potential, and the negative lead (black) of the meter to the component lead that is closer to the negative potential. Voltmeters are also polarity sensitive instruments.
3. If you are measuring the voltage across a component, the voltmeter must be connected across the component and outside the circuit.

Measuring Resistance with the Analog Ohmmeter

Ohmmeter
Measurement device used to measure electric resistance.

Resistance is measured with an **ohmmeter.** Figure 5-8 illustrates a multirange ohmmeter with a resistor's resistance being measured. The meter has four ranges. The simplest range has been selected ($R \times 1$), and on this scale the resis-

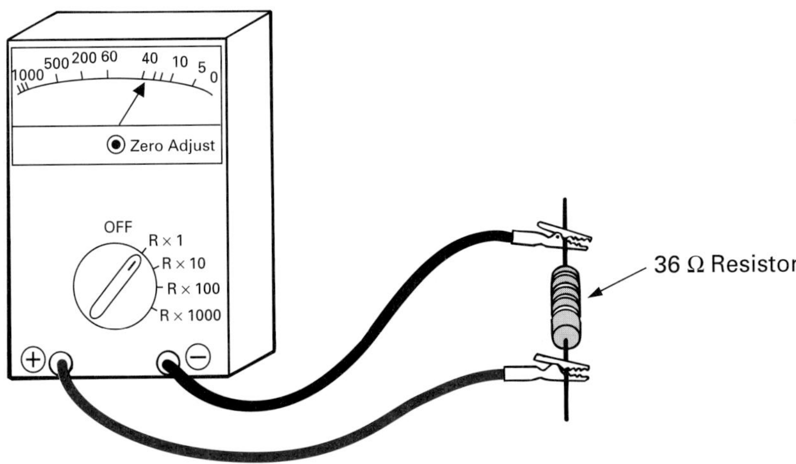

FIGURE 5-8 **Measuring Resistance with the Analog Ohmmeter.**

tance value indicated can be read directly from the meter; in this example the value is 36 Ω ($R \times 1 = 36 \times 1 = 36$ Ω). If we had selected the $R \times 10$ range, a meter reading of, for example, 71 would have to be multiplied by 10 ($71 \times 10 = 710$) to obtain the value of 710 Ω. If we were on the $R \times 100$ range, then a meter reading of 215 would be interpreted as $R \times 100 = 215 \times 100 = 21500$ or 21.5 kΩ. A meter reading of 14 on the $R \times 1000$ scale, if selected, means that you would interpret 14 as $R \times 1000 = 14 \times 1000$, which equals 14 kΩ.

Three rules must be remembered and applied when measuring resistance.

1. Short the meter leads together and adjust the zero-ohms control so the pointer is at 0 Ω. This calibrates the meter scale.
2. Turn off the power and remove the component (if practicable) to be measured from circuit to protect the ohmmeter.
3. Connect the leads across the component, adjust the range scale until the pointer is approximately in the middle of the scale (for accuracy), and then multiply this value by the range scale selected.

5-2-3 *The Digital Multimeter (DMM)*

The **digital multimeters** (DMM) illustrated in Figure 5-9 are gradually replacing the analog-type multimeter discussed previously because they are easier to read and have greater accuracy.

Digital Multimeter
Multimeter used to measure amperes, volts, and ohms and indicate the results on a digital readout display.

Analog versus Digital

To compare these two types of multimeters, let us examine their pros and cons in more detail.

Ease of reading. One of the greatest problems with an analog meter is the errors that occur due to the human factor when reading off the value from the many different scales. The thin analog needle against a calibrated scale is similar to the hands of a clock against the number scale indicating

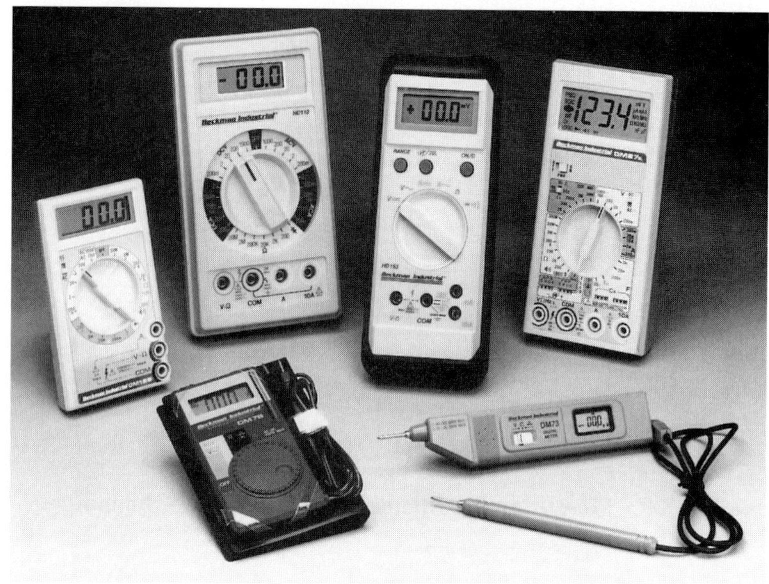

(a)

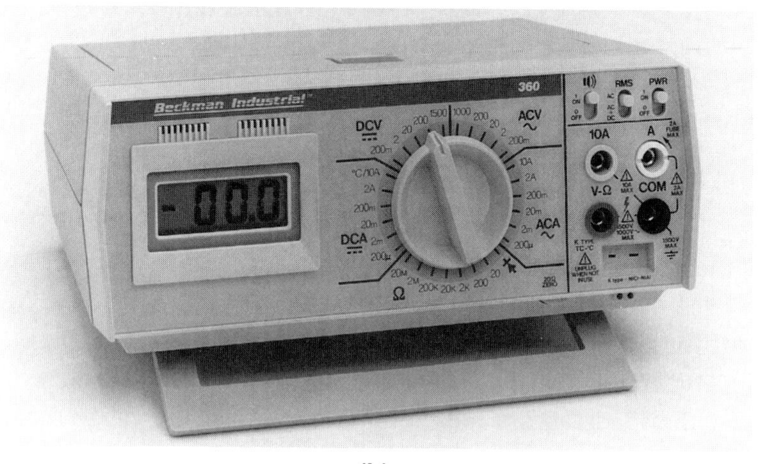

(b)

FIGURE 5-9 **Digital Multimeters. (a) Hand-held. (b) Benchtop.**

the hours. When you look at an analog clock, you have to determine where the hands are, which number the hand is nearest to, and so on. With a digital clock, however, the time is read directly from a display, as shown in Figure 5-10(a).

With the analog meter, the decoding of the scale is necessary, while the digital multimeter displays the magnitude, polarity (+ or −), and the units (V, A, or Ω) on typically a four- or five-digit readout.

Accuracy. DMMs are typically accurate to ±0.01% and have no need for a zero-ohms adjust. Returning to the example of the analog and digital clock, a person reading the time from the traditional analog clock would say that the time is almost 12:30. The wearer of the digital watch, however, will be unquestionably specific and give the time as 12:27. Similarly, an analog

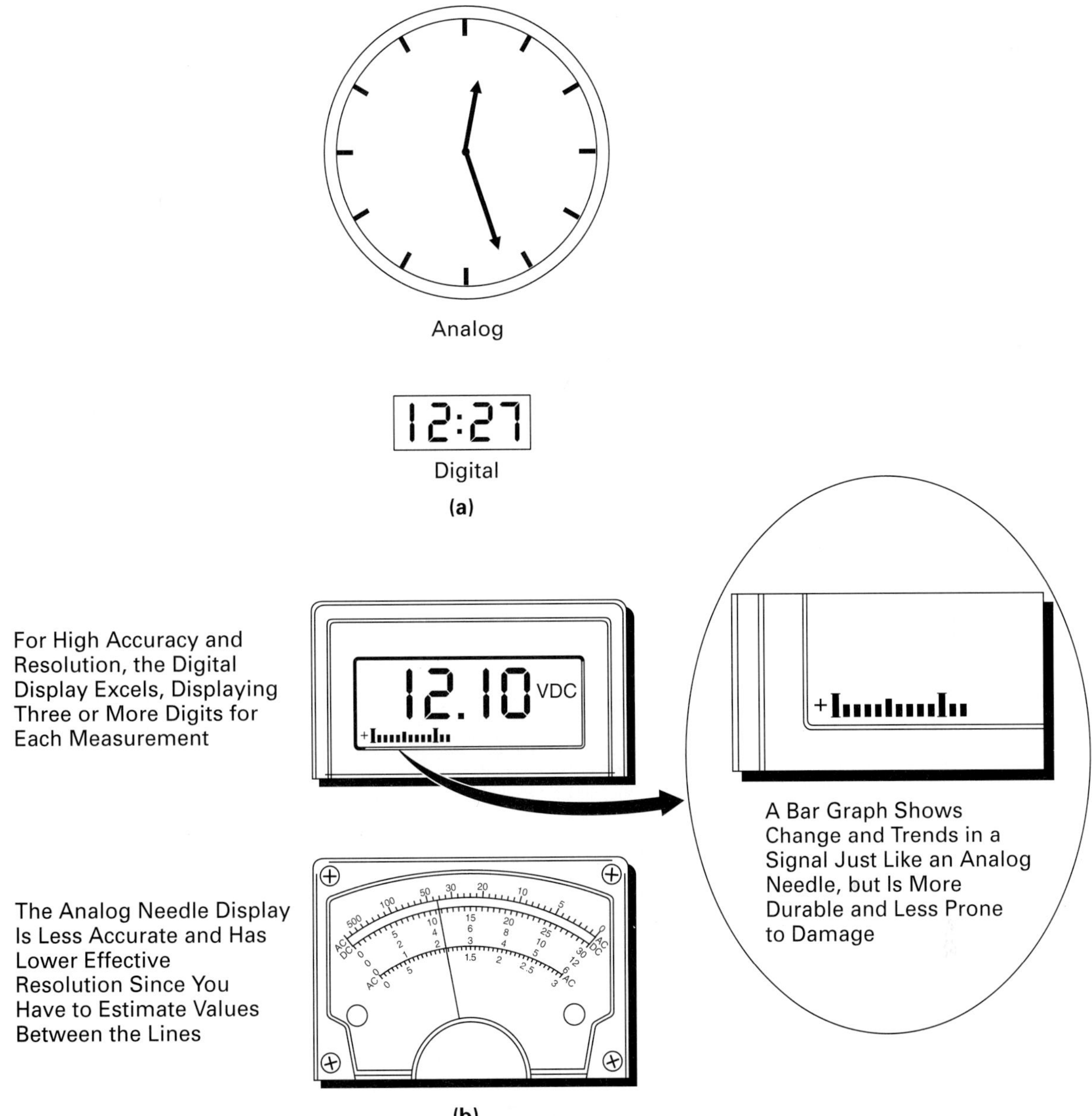

FIGURE 5-10 Analog versus Digital. (Courtesy of Fluke Corporation. Reproduced with permission.)

reading on a meter of about 7 V becomes 7.15 V with the far more accurate digital multimeter.

Digital meters have complicated internal circuitry which is why the digital readout meters have a more expensive price tag than the analog readout multimeters. Another disadvantage of the digital multimeter is its slow response to display the amount on the readout once it has been connected in circuit. To compensate for this disadvantage, most DMMs have a **bar graph display** below the digital readout, as shown in Figure 5-10(b), showing the magnitude of the measured quantity using more or less bars.

Bar Graph Display
A left-to-right or low-to-high set of bars that are turned on in successive order based on the magnitude they are meant to represent.

Using the Digital Multimeter

When troubleshooting electronic equipment, the multimeter becomes the technician's eyes, allowing him or her to see the situation and the problem within the circuit or system. Figure 5-11 illustrates a typical autoranging digital multimeter.

For the meter to be of any use, it must first be connected to the circuit or device to be tested. The two test leads for the meter, one of which is red and the other black, must be inserted into the correct meter lead jacks on the meter. The black lead is connected to the meter jack marked "c" or *common.* The red lead is connected to either the meter jack marked "300 mA or 10 A" (for amperes), or V–Ω (for volts or ohms).

Measuring current. Figure 5-12 shows the steps that should be followed when measuring current. The measurement of current is rarely performed when troubleshooting, as the circuit path has to be opened to insert the DMM in series with the current flow. However, if current is to be measured, the red lead is inserted into one of the ampere jacks.

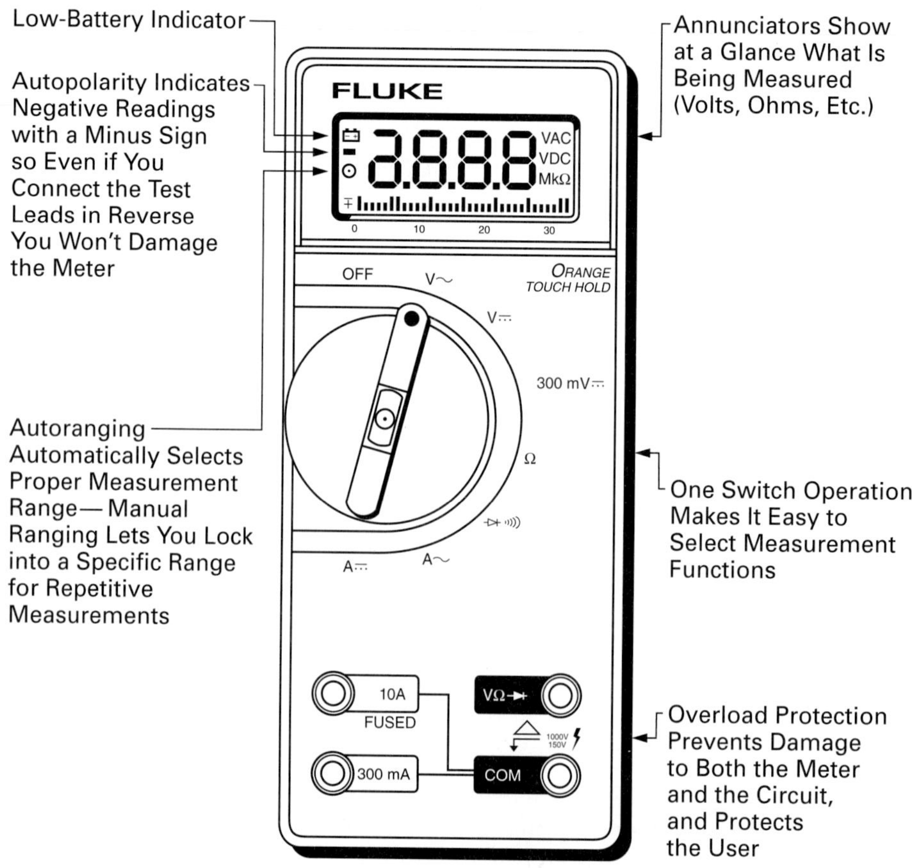

FIGURE 5-11 Typical Digital Multimeter. (Courtesy of Fluke Corporation. Reproduced with permission.)

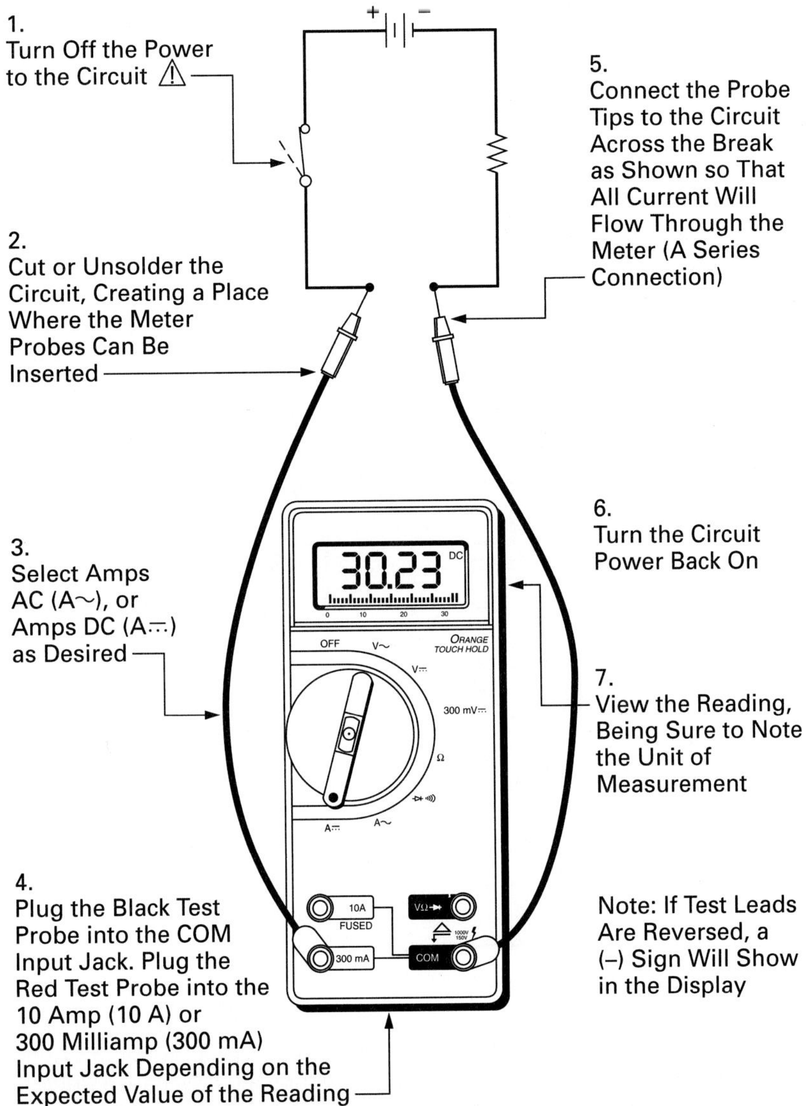

FIGURE 5-12 Measuring Current with a Digital Autoranging Multimeter (Courtesy of Fluke Corporation. Reproduced with permission.)

Measuring voltage. Figure 5-13 shows the steps that should be followed when measuring voltage. The measurement of voltage and resistance is where the DMM finds its greatest utilization. For voltage or resistance measurement, the red meter lead is inserted into the V–Ω (volt or ohms) meter jack.

Measuring resistance. Figure 5-14 shows the steps that should be followed when measuring resistance.

Remember that resistance measurements are carried out without power being applied to the component under test, and resistance values can vary by as

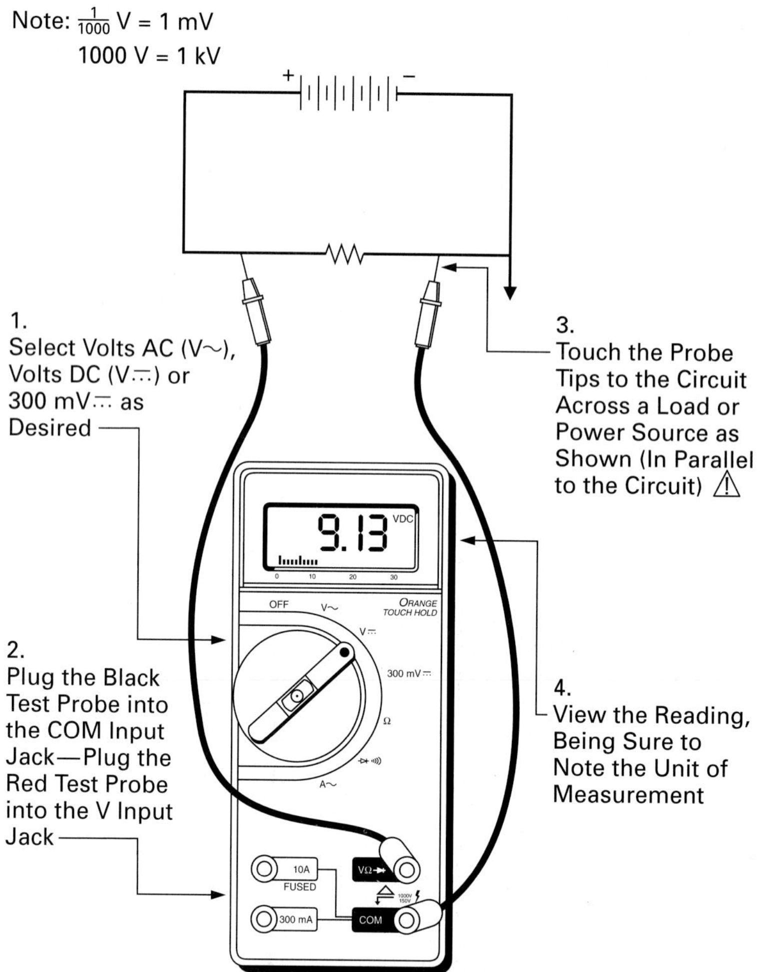

FIGURE 5-13 Measuring Voltage with a Digital Autoranging Multimeter (Courtesy of Fluke Corporation. Reproduced with permission.)

much as 20%, due to the tolerance of certain resistors; so do not be misled if your meter reading is slightly different from the color band value printed on the resistor. If a resistor's value is off and exceeds the tolerance, the resistor should be replaced.

A resistor will rarely short, but typically will open. If a resistor does open, the DMM display will flash on and off or display OL because the resistor has an infinite resistance (open circuit).

How to Make Resistance Measurements

Note: 1,000Ω = 1 kΩ

1,000,000Ω = 1 MΩ

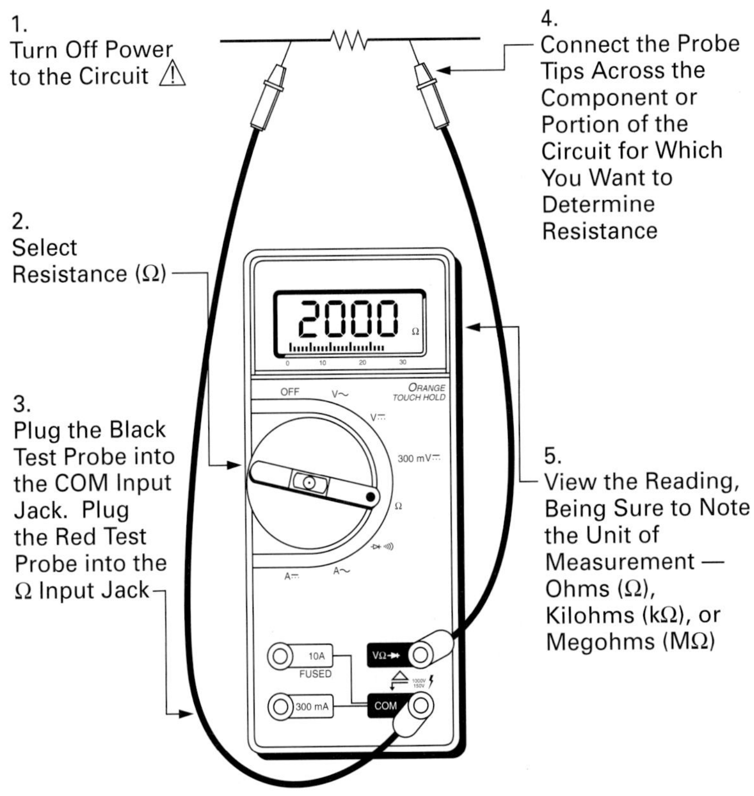

FIGURE 5-14 Measuring Resistance with a Digital Autoranging Multimeter (Courtesy of Fluke Corporation. Reproduced with permission.)

SELF-TEST REVIEW QUESTIONS FOR SECTION 5-2

1. What is the purpose of the dc power supply?
2. An ammeter is used to measure ____________, a voltmeter is used to measure ____________, and an ohmmeter is used to measure ____________.
3. What two key advantages does the DMM have over the AMM?
4. What is meant by the term "autoranging"?

5-3 PROTOBOARDS

Protoboard
An experimental arrangement of a circuit on a board. Also called breadboard.

The solderless prototyping board **(protoboard)** or breadboard is designed to accommodate the many experiments described in this textbook's associated lab manual. This protoboard will hold and interconnect resistors, capacitors, inductors, and many other components, as well as provide electrical power. Figure 5-15(a) shows an experimental circuit wired up on a protoboard.

Figure 5-15(b) shows the top view of a basic protoboard. As you can see by the cross section on the right side, electrical connector strips are within the protoboard. These conductive strips make a connection between the five hole groups. The bus strips have an electrical connector strip running from end to end, as shown in the cross section. They are usually connected to a power supply as seen in the example circuit in Figure 5-16(a). In this circuit, you can see that

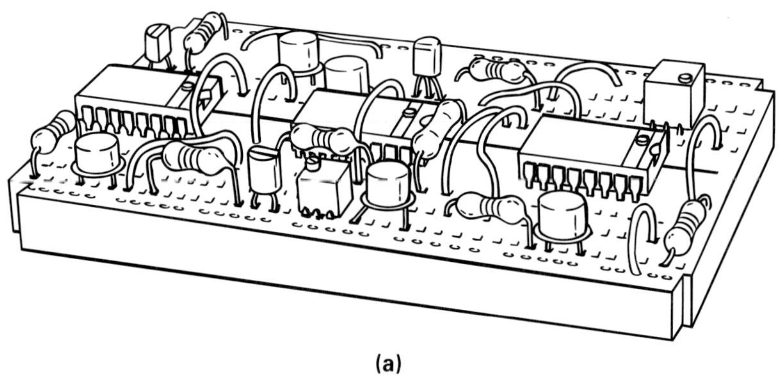

(a)

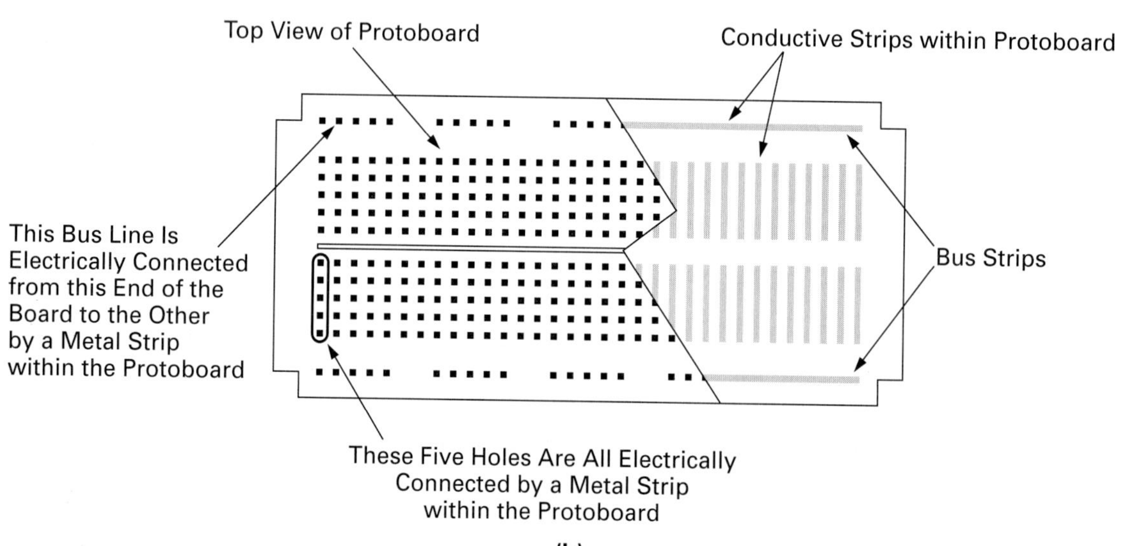

(b)

FIGURE 5-15 Experimenting with the Protoboard.

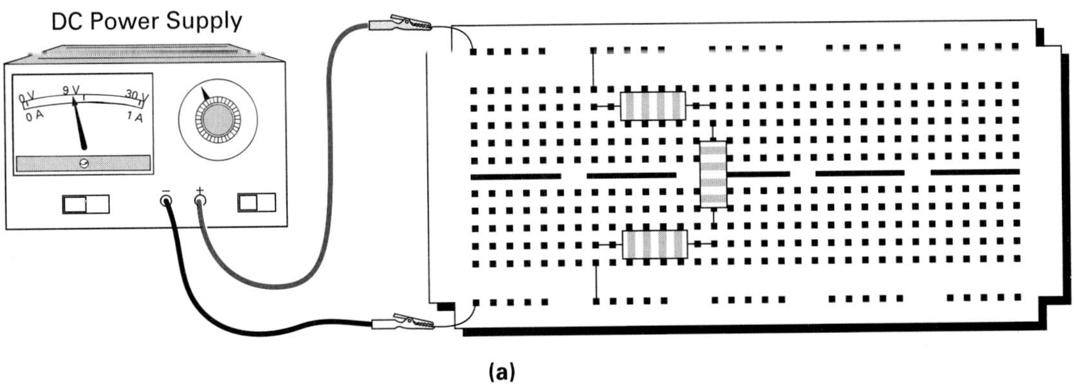

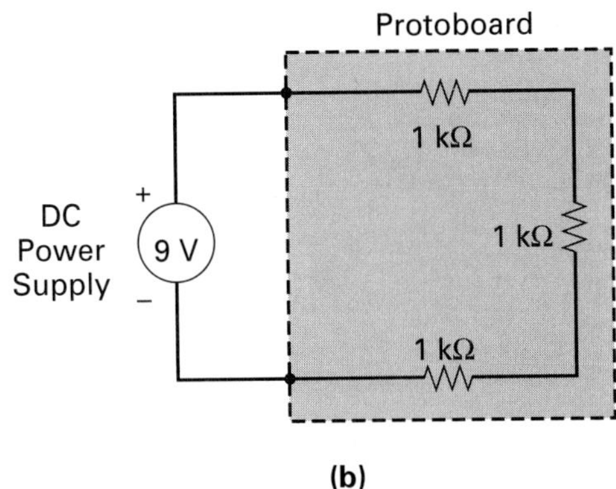

FIGURE 5-16 Constructing Circuits on the Protoboard.

the positive supply voltage is connected to the upper bus strip, while the negative supply (ground) is connected to the lower bus strip. These power supply "rails" can then be connected to a circuit formed on the protoboard with hookup wire, as shown in Figure 5-16(a). In this example, three resistors are connected end-to-end, as shown in the schematic diagram in Figure 5-16(b).

Figure 5-17 illustrates how the multimeter could be used to make voltage, current, and resistance measurements of a circuit constructed on the protoboard.

Other protoboards may vary slightly as far as layout, but you should be able to determine the pattern of conductive strips by making a few resistance checks with an ohmmeter.

SELF-TEST REVIEW QUESTIONS FOR SECTION 5-3

1. What is the purpose of the protoboard?
2. What is generally connected to the bus strips?

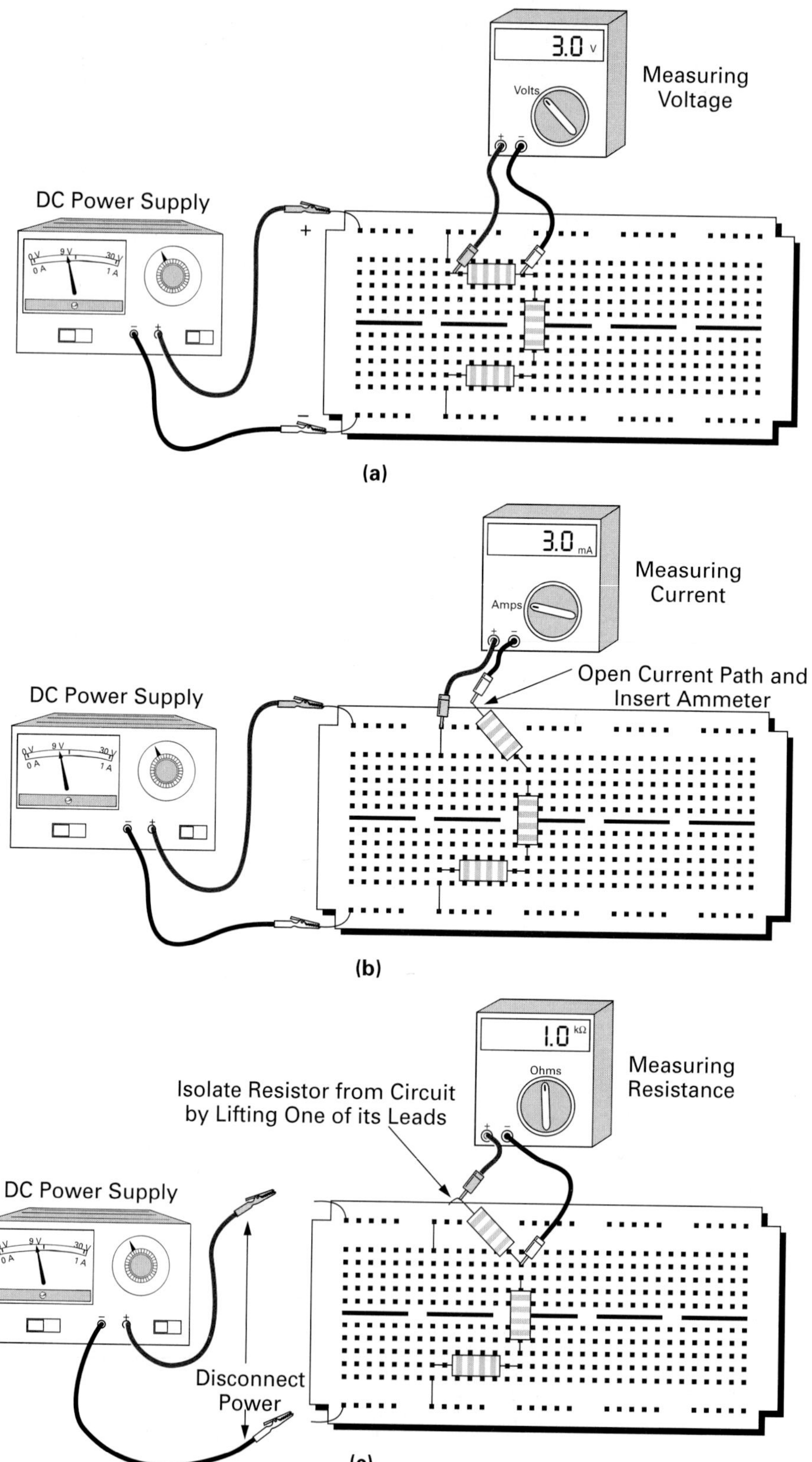

FIGURE 5-17 Making Measurements of a Circuit on the Protoboard.

5-4 SOLDERING TOOLS AND TECHNIQUES

Electronic components such as resistors, capacitors, diodes, transistors, and integrated circuits are combined to form circuits, which in turn are combined to form electronic systems. These components have their leads physically interconnected and then bonded with solder. Like the expression the "straw that broke the camel's back," one sloppy solder connection can cause an entire electronic system to fail.

5-4-1 *Soldering Tools*

Soldering is a skill that can be developed with practice and knowledge. A solder connection is the joining together of two metal parts by applying both heat and solder. The heat provided by the soldering iron is at a high enough temperature to melt the solder, making it a liquid that flows onto and slightly penetrates the two metal surfaces that need to be connected. Once the soldering iron and therefore the heat are removed, the solder cools. In so doing, it solidifies and bonds the two metal surfaces, producing a good *electrical and mechanical connection.* This is the purpose of soldering: to bond two metal surfaces together so that they are both electrically and mechanically connected.

Soldering
Process of joining two metallic surfaces to make an electrical contact by melting solder (usually tin and lead) across them.

You should at this time have a set of tools and toolbox with all the components needed for the experiments listed in the lab manual. Figure 5-18 illustrates the set of basic electronic tools that will be needed for soldering and experimentation. These tools include:

1. Solder wick: Used to remove solder from terminals.

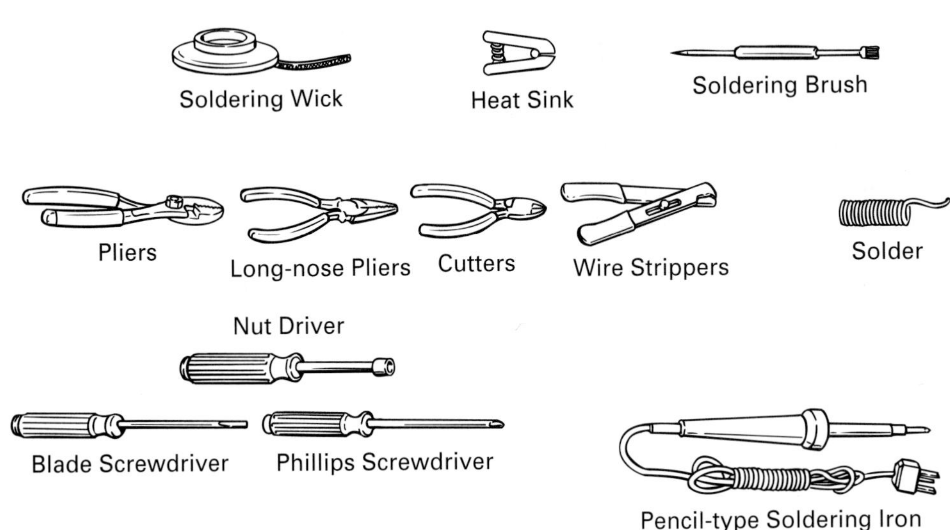

FIGURE 5-18 Basic Electronic Tools.

2. Heat sink: During the soldering process, this device is clamped onto the lead of especially heat-sensitive components to conduct the heat generated by the soldering iron away from the component.
3. Soldering brush: Used to clean off flux after soldering.
4. Pliers
5. Long-nose pliers: Used for gripping and bending; they are also known as needle-nose pliers.
6. Cutters: Also called dikes; these are available in many different sizes and are used to cut wires or cables.
7. Wire strippers: The strippers are designed to be adjustable or they have a variety of holes for stripping the insulating sleeve off a wire.
8. Solder: 60/40 rosin-core solder is most commonly used in electronics.
9. Nut driver
10. Blade screwdriver
11. Phillips screwdriver
12. Soldering iron

5-4-2 Wetting

Wetting
The coating of a contact surface.

Everytime you make a good electrical and mechanical connection, the solder will flow, when heated to its melting temperature, over the lead and terminal to be connected, as shown in Figure 5-19. The solder, in fact, actually penetrates the metals, and this embedding of the solder into the metal is called **wetting.** If the solder feathers out to a thin edge, good wetting is said to have occurred, and it is this that gives the connection its physical strength and electrical connection.

5-4-3 Solder and Flux

Solder
Metallic alloy that is used to join two metal surfaces.

Solder is a mixture of tin and lead, both of which have a low melting temperature with respect to other metals. This is necessary so that the soldering iron melts the solder and not the terminals or leads.

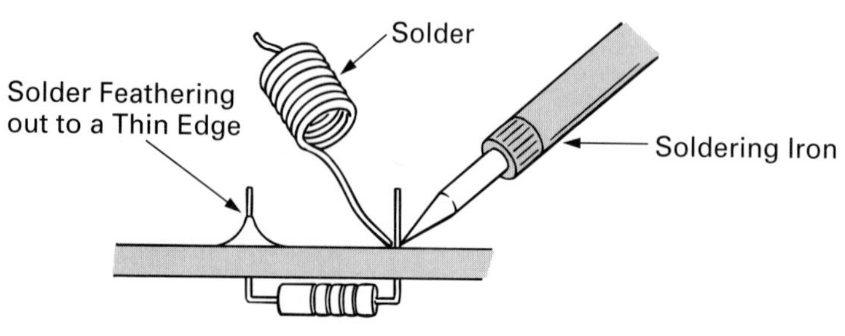

FIGURE 5-19 Wetting.

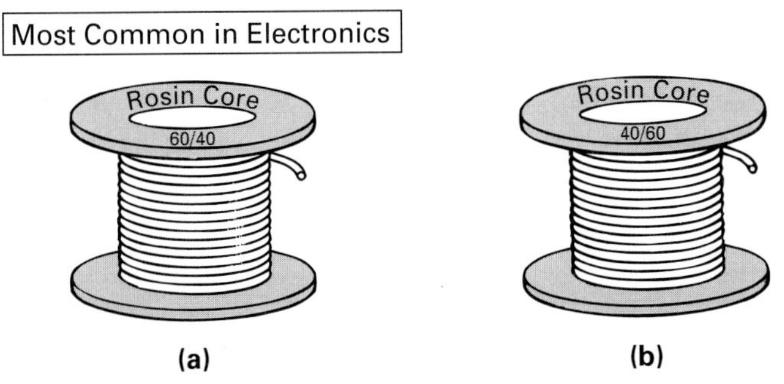

FIGURE 5-20 Tin/Lead Solder Ratios. (a) 60/40. (b) 40/60.

Different proportions of tin and lead are available to produce solder with different characteristics. For example, Figure 5-20(a) shows a 60/40 solder, which is a mixture of 60% tin and 40% lead, whereas Figure 5-20(b) shows a 40/60 solder that consists of 40% tin and 60% lead. The proportions of tin to lead determine the melting temperature of the solder; for example, 60/40 solder has a lower melting temperature than 40/60 solder. In electronics, 60/40 is most commonly used because of its low melting temperature, which means that a component lead or terminal will not have to be heated to a high temperature in order to make a connection. A 63/37 solder has an even lower melting temperature and is therefore even safer than 60/40 solder; however, it is more expensive.

Any lead or terminal is always exposed to the air, which forms an invisible insulating layer on the surface of leads, pins, terminals, and any other surface. A chemical substance is needed to remove this layer; otherwise, the solder would not be able to flow and stick to the metal contacts. **Flux** removes this invisible insulating oxide layer, and nearly all solder used in electronics contains the flux inside a type of solder tube, as shown in Figure 5-21. As the solder is applied to a heated connection, the flux will automatically remove any oxide. The two most common types of flux are rosin and acid. Acid-core solder is used only in sheet metal work and should never be used in electronics since it is highly corrosive. In electronics, you should only use rosin-core solder, and this is normally indicated as shown in Figure 5-20.

Flux
A material used to remove oxide films from the surfaces of metals in preparation for soldering.

A variety of solder diameters are available. The larger-diameter solder is used for terminals and large component leads and connections, whereas the

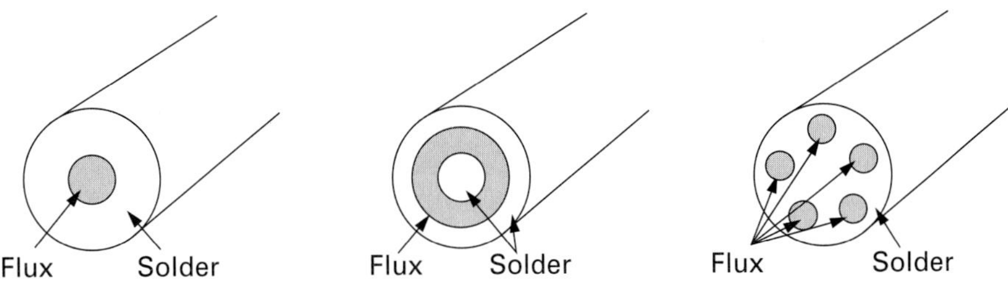

FIGURE 5-21 Rosin-Core Solder.

smaller-diameter solder is used for soldering terminals that are very close to one another on a printed circuit board, so the amount of solder being applied to the connection needs to be carefully controlled.

5-4-4 *Soldering Irons*

Soldering Iron Tool having an internal heating element that is used for heating a connection to melt solder.

Figure 5-22 illustrates the two basic types of **soldering irons.** They are rated in terms of wattage, which indicates the amount of power consumed. More important, the wattage rating of a soldering iron indicates the amount of heat it pro-

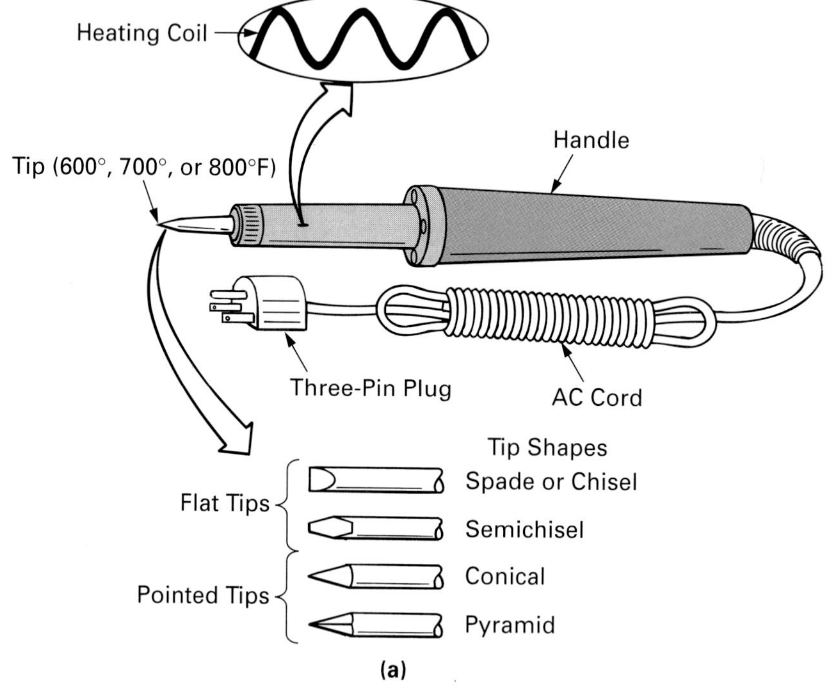

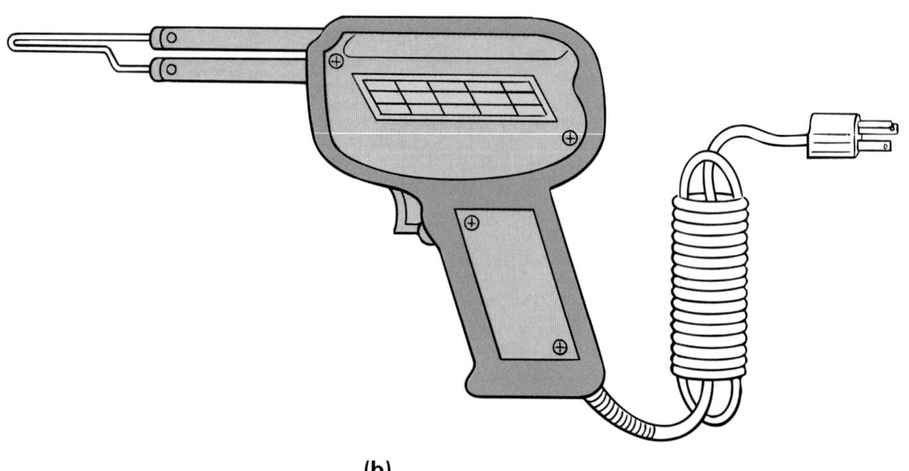

FIGURE 5-22 Types of Soldering Irons. (a) Pencil. (b) Gun.

duces. When an iron is applied to a connection, heat transfer takes place and heat is drained away from the iron to the metals to be connected; so a larger connection will need a larger-wattage soldering iron. A 25- to 60-W pencil iron is ideal for most electronic work.

For your safety and to protect voltage- and current-sensitive components, you should always use a soldering iron with a three-pin plug and three-wire ac power cord. The third ground wire will ground the iron's exposed metal areas and the tip to prevent electrical shock. This grounded tip will protect delicate MOS integrated circuits from leakage electricity and static charges. On the subject of safety, always turn off the equipment before soldering, as the grounded tip may cause a short circuit in the equipment and possible damage.

The tip of a soldering iron can sometimes be changed, and a different temperature tip will allow your iron to be used for different applications. The temperature of the tip selected should be governed by the size of the connection and the temperature sensitivity of the components. In general, a 700°F tip is ideal for most electronic applications; however, for delicate printed circuit boards, use a 600°F tip.

The tip of a soldering iron comes in a variety of shapes and sizes, as shown in Figure 5-22(a). The heat produced by the soldering iron heats the connection and melts the solder so that it flows over the connection. A tip shape should be chosen that will conduct the heat to the connection as quickly as possible, so as not to damage the component or PCB terminal. The tip shape that makes the best contact with the connection, and will therefore conduct the most heat, should be used.

As you regularly use your soldering iron, the tip will naturally accumulate dirt and oxide, and this contamination will reduce its effectiveness. The tip should be regularly cleaned by wiping it across a damp sponge, as shown in Figure 5-23(a). Soldering iron tips are generally made of copper, plated with either iron or nickel, so you should never clean a tip by filing the end, as this will probably remove the plating.

The tip of the soldering iron should always be **tinned** after it has been cleaned. Tinning the tip can be seen in Figure 5-23(b) and is achieved by applying a small layer of rosin-core solder to the tip. This will protect the tip from oxidation and increase the amount of heat transfer to the connection.

Tinned
Coated with a layer of tin or solder to prevent corrosion and simplify the soldering of connections.

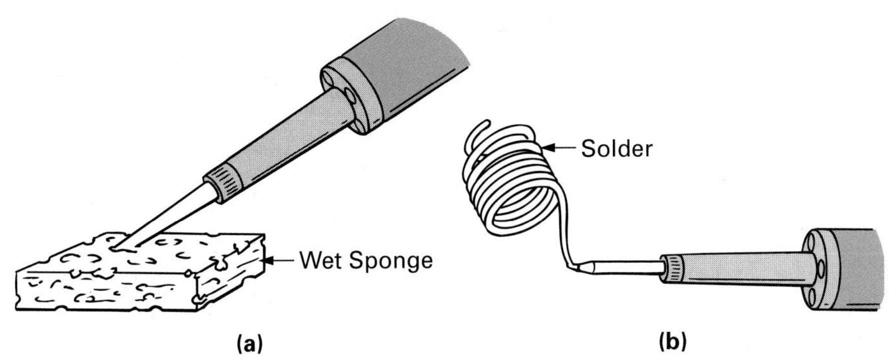

FIGURE 5-23 **Cleaning a Soldering Iron. (a) Cleaning the Tip. (b) Tinning the Tip.**

In summary, it is important to remember that *a soldering iron should not be used to melt the solder; its purpose is to heat the connection so that the solder will melt when it makes contact with the connection.*

5-4-5 Soldering Techniques

Before carrying out the six basic soldering steps, wires and components should be prepared. Wires should be wrapped around a terminal to give it a more solid physical support, as shown in Figure 5-24(a). Their insulation should be stripped back so as to leave a small gap. Too large a gap will expose the bare wire and possibly cause a short to another terminal, whereas too small a gap will cause the insulation to burn during soldering. Stranded wire should be tinned, as shown in Figure 5-24(b), and components should be mounted flat on the board, as shown in Figure 5-24(c).

Six-Step Soldering Procedure

To begin, always remember to wear safety goggles and a protective apron, and then proceed.

Step 1: Clean the tip on a damp sponge.

Step 2: Tin the tip if necessary.

Step 3: Heat the connection.

Step 4: Apply the solder to the opposite side of the connection.

Step 5: Leave the tip of the soldering iron on the connection only long enough to melt the solder.

Step 6: Remove the solder and then the iron; then let the melted solder cool and solidify undisturbed.

Once this procedure is complete, you should remove the excess leads with the cutters and remove the flux residue, because it collects dust and dirt that could produce electrical leakage paths. When inspecting your work, you should notice the following:

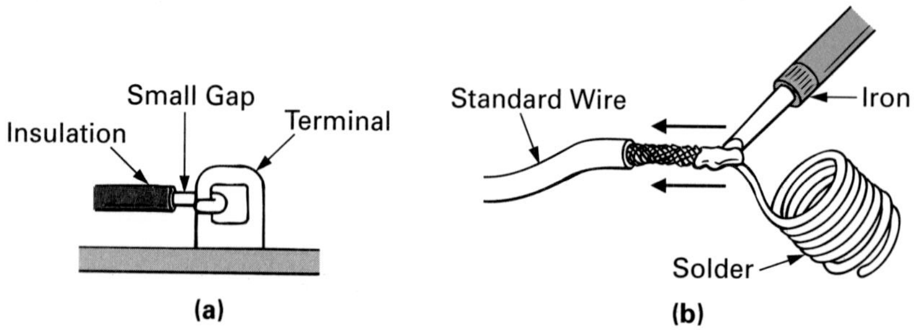

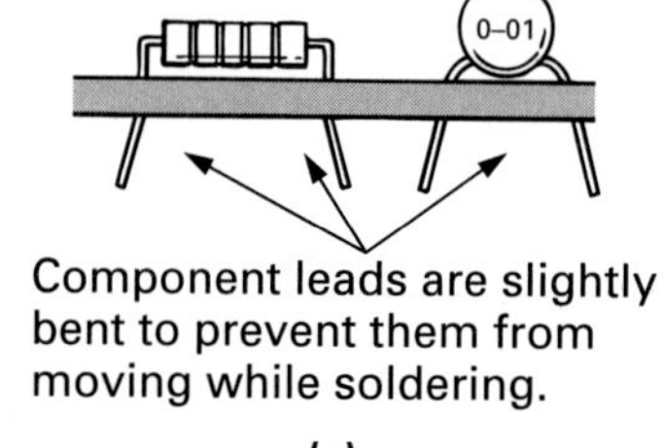

FIGURE 5-24 **Wire and Component Preparation.**

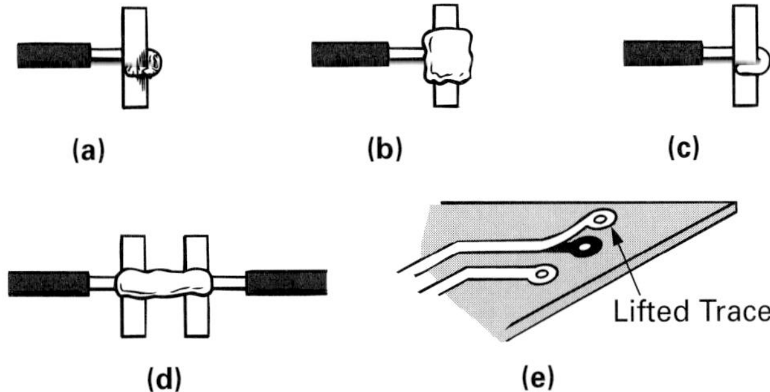

FIGURE 5-25 **Poor Solder Connections.**

1. The connection is *smooth and shiny.*
2. The solder should *feather out to a thin edge.*

Poor Solder Connections

1. *Cold solder joint:* This connection has a dull gray appearance; it makes a poor mechanical and electrical connection and is normally caused due to connector lead movement while the solder was cooling, or to insufficient heat [Figure 5-25(a)]. The cure is to resolder the connection.
2. *Excessive solder joint:* This is caused by too much solder being applied to the connection or by using solder of too large diameter [Figure 5-25(b)]. The cure is to desolder the excess with a wick.
3. *Insufficient solder joint:* Not enough solder was applied, so a poor bond exists between the lead and the terminal [Figure 5-25(c)]. The cure is to resolder as if it were a new joint.
4. *Solder bridge:* This occurs when adjacent terminals or traces on printed circuit boards are connected accidentally as a result of using too much solder [Figure 5-25(d)]. Most can be removed by desoldering the excess with a wick.
5. *Excessive heat:* On printed circuit boards, too much heat can lift the traces or track and ruin the entire board [Figure 5-25(e)].

5-4-6 Desoldering

If a faulty component has to be replaced or if a component or wire was soldered in an incorrect place, it will have to be desoldered. Figure 5-26 illustrates some of the tools used to desolder. The easiest way to remove a component is to remove all the solder and then disconnect the leads. All three tools shown in Figure 5-26 are designed to remove solder.

Use of the braided wick to remove solder can be seen in Figure 5-27(a). The braid contains many small wire strands, and when it is placed between the solder and the iron, the melted solder flows into the braid by capillary action. The solder-filled braid is then cut off and discarded.

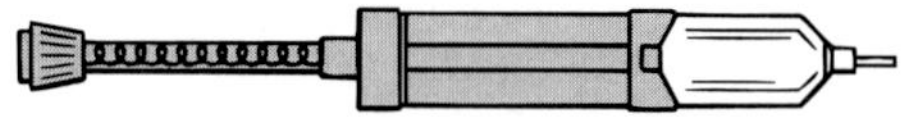

FIGURE 5-26 Desoldering Tools.

Desoldering The process of removing solder from a connection.

The **desoldering** bulb or spring-loaded plunger relies on suction rather than capillary action to remove the solder, as shown in Figure 5-27(b). The steps to follow are:

Step 1: Squeeze the desoldering bulb or cock the spring-loaded plunger.

Step 2: Melt the solder with the soldering iron.

Step 3: Remove the iron's tip.

Step 4: Quickly insert the tip of the bulb or plunger into the molten solder and then activate the suction.

When as much of the solder as possible has been removed, use the long-nose pliers to hold the lead and remove the component. It may be necessary to use the iron to heat the lead slightly to loosen up the residual solder so that the component lead can be removed with the pliers. Always be careful not to apply an excessive amount of heat or stress to either the component or the board.

5-4-7 *Safety*

Molten solder, like any other liquid, can splash or spill, and soldering irons are even hotter than the solder. Consequently, you should always wear protective safety glasses and an apron to protect yourself from burns. NEVER fling hot solder off an iron; always use the sponge.

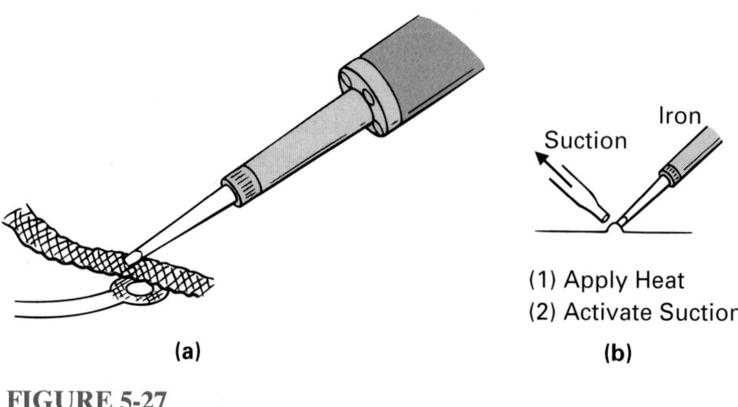

FIGURE 5-27

SELF-TEST REVIEW QUESTIONS FOR SECTION 5-4

1. A solder connection is the joining together of two metal parts by applying both ____________ and ____________.
2. In electronics a 60/40 solder is most commonly used and is composed of 60% ____________ and 40% ____________.
3. Describe the six-step soldering procedure.
4. What devices are available to remove solder?

SUMMARY

Safety

1. Safety precautions should always be your first priority when working on electronic equipment, as there is always the possibility of receiving an electric shock.
2. A shock is a sudden, uncontrollable reaction as current passes through your body and causes your muscles to contract and deliver a certain amount of pain.
3. Any shock is dangerous, since even the mildest could surprise you and cause an involuntary action that could injure yourself or someone else.
4. The human body has a resistance of about 10,000 to 50,000 Ω, depending on how good a contact you make with the "live" (power present) conductor.
5. Skin resistance is generally quite high and will subsequently oppose current flow; however, your resistance is lowered if your skin is wet due to perspiration or if your skin has a cut or an abrasion.

Precautions to Be Used When Troubleshooting

6. Except when absolutely necessary, do not work on electrical or electronic circuits or equipment when power is on.
7. To avoid dangerous situations, always place the free hand in a pocket or behind your back while testing a piece of equipment with power on.
8. When troubleshooting equipment with power on, try to insulate yourself.
9. Electrolytic and other large-value capacitors can hold a dangerous voltage charge even after the equipment has been turned off.
10. All tools should be well insulated.
11. Always switch off the equipment and disconnect the power before replacing any components.
12. Necklaces, rings, and wristwatches have a low resistance and should be removed when working on equipment.
13. Inspect the equipment before working on it, and if it is in poor condition turn off the equipment and replace these hazards.
14. Make sure that someone is present who can render assistance in the event of an emergency.

15. Make it a point to know the location of the power-off switch.
16. Cathode-ray tubes are highly evacuated and should therefore be handled with extreme care.

First Aid, Treatment, Resuscitation

17. Shock can be felt from voltages as low as 15 V, and at 20 to 25 V most people would experience pain. At about this voltage or a little higher, the victim may find himself unable to let go of the conductor and may suffer burning.
18. Injury can also be caused by a minor shock, not serious in itself but which has the effect of contracting the muscles sufficiently to result in a fall or other reaction.
19. Burns can be caused by the passage through the body of a heavy current.
20. An explosion can be caused by the ignition of flammable gases by a spark from an electric contact.
21. Eye injuries can be caused by exposure to the strong ultraviolet rays of an electric arc. Lasers are also dangerous to the eyes due to their intensely concentrated beam. Precautions to protect the eyes must always be taken by wearing protective goggles when clipping leads or soldering.
22. The energy in microwave and radio-frequency equipment can damage the body, especially those parts with a low blood supply.
23. You should familiarize yourself with the various methods of artificial respiration by contacting your local Red Cross for complete instruction.

The DC Power Supply

24. In the laboratory or repair shop, power can be obtained from a battery, but since ac power is so accessible, a dc power supply is generally used to supply dc voltages.
25. The power supply has a large advantage over the battery as a source of dc voltage in that it can quickly provide an accurate output voltage that can easily be varied by a control on the front panel, and it never runs down.
26. During your electronic studies you will use a dc power supply in the lab to provide different dc output voltages for various experiments.
27. The ON/OFF switch is used to turn on the power supply, while the VOLTS/AMPS switch is used to switch the meter between a voltmeter or ammeter. In the volts position, the meter indicates the voltage that is appearing across the output terminals. If the VOLT/AMPS switch is placed in the amps position, the meter will now monitor the amount of current flowing out of the power supply. The VOLTS ADJUST control is used to either increase or decrease the output voltage.

The Analog Multimeter (AMM) (Figure 5-28)

28. The multimeter is a device for the measurement of current, voltage, and resistance.

FIGURE 5-28 Making Measurements with the Analog Multimeter (AMM).

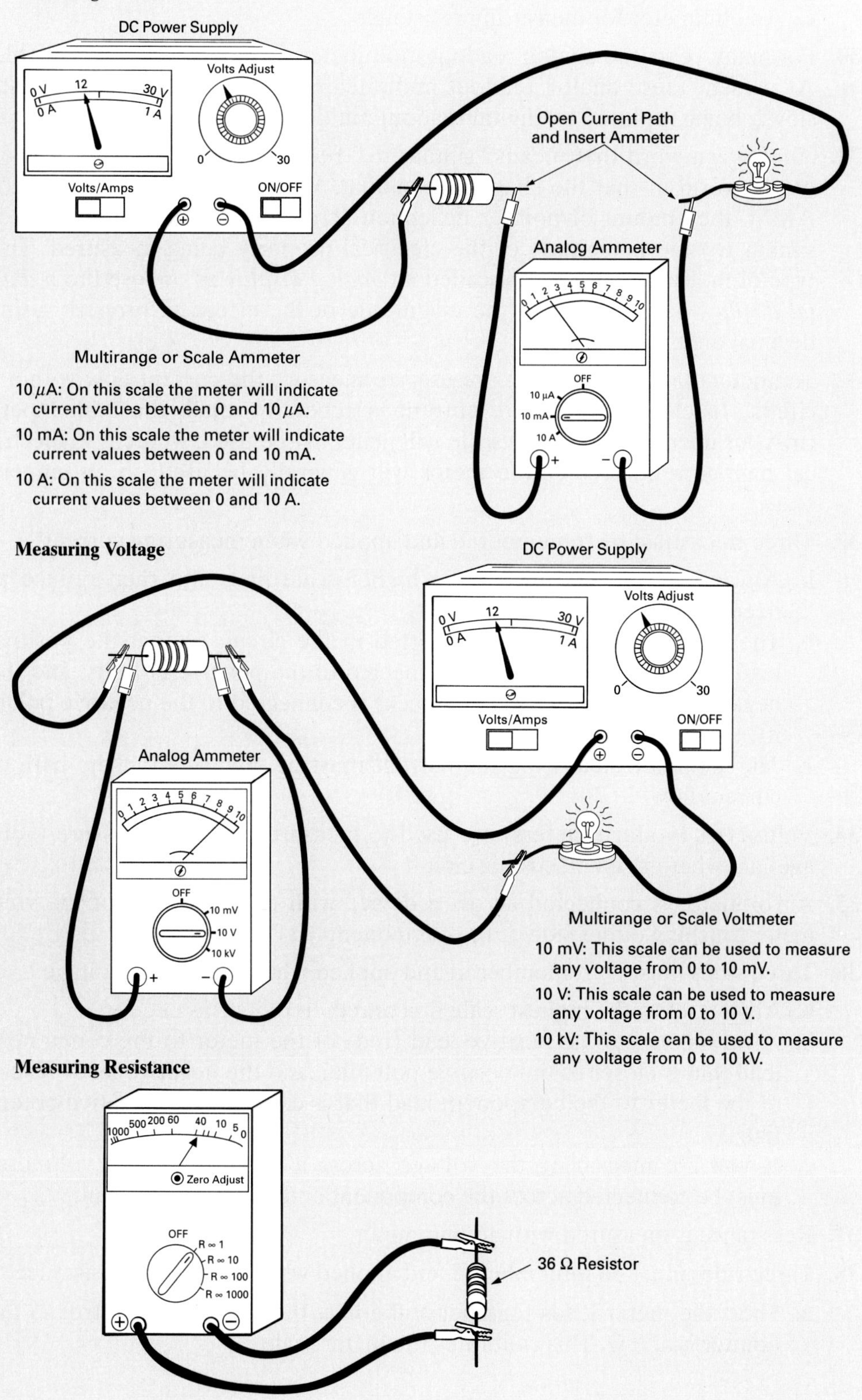

29. It is called a multimeter because it is like having three meters all rolled into one, in that it has
   a. An ammeter for measuring current
   b. A voltmeter for measuring voltage
   c. An ohmmeter for measuring resistance
30. For many years the analog readout multimeter was the only type available. At present these analog readout multimeters are still being used but are slowly being replaced by digital readout multimeters.
31. *Analog* is a word that means "similar to." For example, a ball is an analog of the world in that the shapes are similar. With an analog multimeter or AMM, the amount of pointer deflection across the scale is an analog (or similar to) the magnitude of the electrical property being measured. This type of indicator is therefore called an *analog display* as opposed to a *digital display,* which indicates the magnitude of the electrical property using decimal digits.
32. Ammeters (ampere meters) are used to measure the current flow within a circuit. In electronics, where current is generally small, the milliampere (mA) or microampere (μA) scale will generally be used, whereas in electrical high-current circuits, the meter will generally be used on an ampere scale.
33. Three rules must be remembered and applied when measuring current:
   a. Always set the selector to the higher scale first, and then reduce as needed.
   b. The current meter must be inserted in the circuit so that the positive lead of the ammeter (red) is connected to the positive polarity, and the negative lead of the ammeter (black) is connected to the negative polarity.
   c. The ammeter measuring the current must be connected in the path of current flow.
34. Voltmeters (voltage meters) are used to measure electrical pressure (voltage) anywhere in an electronic circuit.
35. A voltmeter is connected across a device with the two leads of the voltmeter touching either side of the component.
36. Three steps must be remembered and applied when measuring voltage:
   a. Always set to the highest scale first and then reduce as needed.
   b. Always connect the positive lead (red) of the meter to the component lead that is closer to the positive potential, and the negative lead (black) of the meter to the component lead that is closer to the negative potential.
   c. If you are measuring the voltage across a component, the voltmeter must be connected across the component and outside the circuit.
37. Resistance is measured with an ohmmeter.
38. Three rules must be remembered and applied when measuring resistance.
   a. Short the meter leads together and adjust the zero-ohms control so the pointer is at 0 Ω. This calibrates the meter scale.

b. Turn off the power and remove the component (if practicable) to be measured from circuit to protect the ohmmeter.
c. Connect the leads across the component, adjust the range scale until the pointer is approximately in the middle of the scale (for accuracy), and then multiply this value by the range scale selected.

The Digital Multimeter (DMM) (Figure 5-29)

39. The digital multimeters (DMM) are gradually replacing the analog-type multimeter because they are easier to read and have greater accuracy.

40. One of the greatest problems with an analog meter is the errors that occur due to the human factor when reading off the value from the many different scales.

41. When troubleshooting electronic equipment, the multimeter becomes the technician's eyes, allowing him or her to see the situation and the problem within the circuit or system.

42. The two test leads for the meter, one of which is red and the other black, must be inserted into the correct meter lead jacks on the meter. The black lead is connected to the meter jack marked "c" or *common.* The red lead is connected to the meter jack marked A for amperes or V-Ω for volts or ohms.

43. The solderless prototyping board (protoboard) or breadboard is designed to accommodate experiments. This protoboard will hold and interconnect resistors, capacitors, inductors, and many other components, as well as provide electrical power.

Soldering Tools and Techniques

44. Electronic components such as resistors, capacitors, diodes, transistors, and integrated circuits are combined to form circuits, which in turn are combined to form electronic systems. These components have their leads physically interconnected and then bonded with solder.

45. A solder connection is the joining together of two metal parts by applying both heat and solder.

46. The heat provided by the soldering iron is at a high enough temperature to melt the solder, making it a liquid that flows onto and slightly penetrates the two metal surfaces that need to be connected.

47. Once the soldering iron and therefore the heat are removed, the solder cools. In so doing, it solidifies and bonds the two metal surfaces, producing a good *electrical and mechanical connection*.

48. The solder actually penetrates the metals, and this embedding of the solder into the metal is called wetting.

49. Solder is a mixture of tin and lead, both of which have a low melting temperature with respect to other metals. This is necessary so that the soldering iron melts the solder and not the terminals or leads.

50. Different proportions of tin and lead are available to produce solder with different characteristics. A 60/40 solder is a mixture of 60% tin and 40% lead.

QUICK REFERENCE SUMMARY SHEET

FIGURE 5–29 Making Measurements with the Digital Multimeter (DMM).

Measuring Current

Measuring Voltage

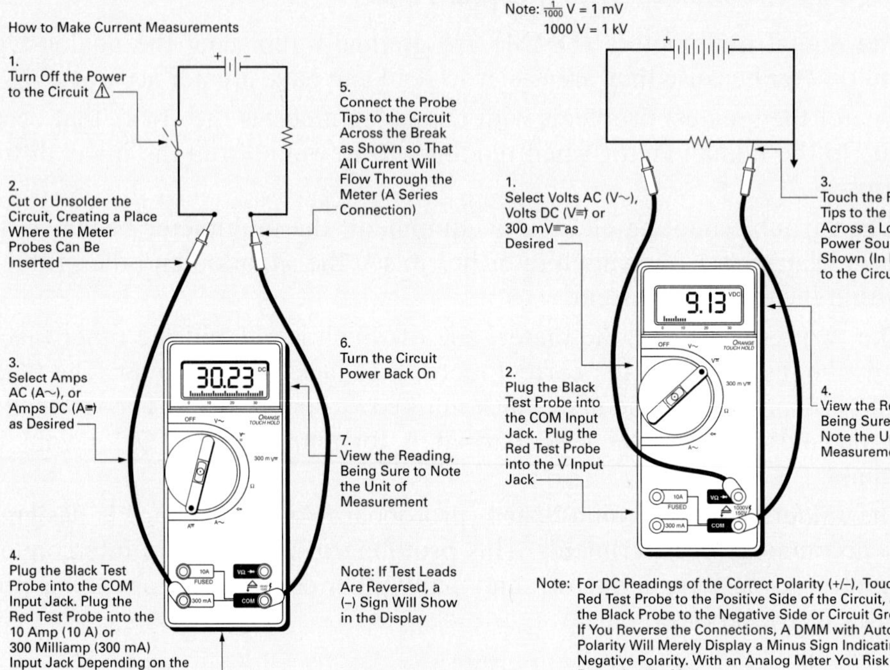

Measuring Resistance

Note: 1,000Ω = 1 kΩ
1,000,000Ω = 1 MΩ

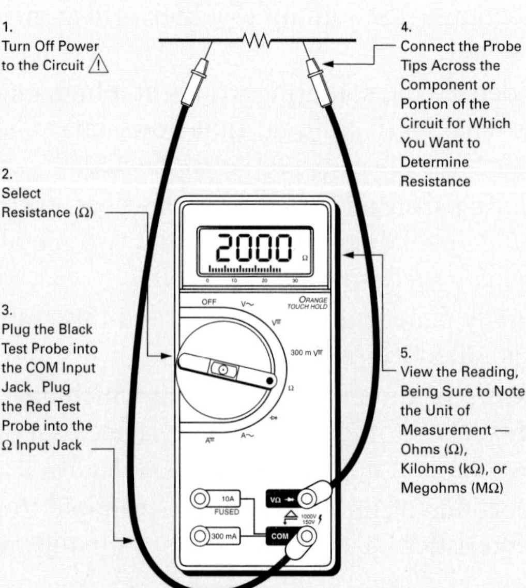

51. The proportions of tin to lead determine the melting temperature of the solder.

52. In electronics, 60/40 is most commonly used because of its low melting temperature, which means that a component lead or terminal will not have to be heated to a high temperature in order to make a connection.

53. A 63/37 solder has an even lower melting temperature is therefore even safer than 60/40 solder; however, it is more expensive.

54. Any lead or terminal is always exposed to the air, which forms an invisible insulating layer on the surface of leads, pins, terminals, and any other surface. A chemical substance is needed to remove this layer; otherwise, the solder would not be able to flow and stick to the metal contacts. Flux removes this invisible insulating oxide layer, and nearly all solder used in electronics contains the flux inside a type of solder tube.

55. The two most common types of flux are rosin and acid. Acid-core solder is used only in sheet metal work and should never be used in electronics since it is highly corrosive. In electronics, you should only use rosin-core solder.

56. A variety of solder diameters is available. The larger-diameter solder is used for terminals and large component leads and connections, whereas the smaller-diameter solder is used for soldering terminals that are very close to one another on a printed circuit board, so the amount of solder being applied to the connection needs to be carefully controlled.

57. Soldering irons are rated in terms of wattage, which indicates the amount of power consumed. More important, the wattage rating of a soldering iron indicates the amount of heat it produces. When an iron is applied to a connection, heat transfer takes place and heat is drained away from the iron to the metals to be connected; so a larger connection will need a larger-wattage soldering iron. A 25- to 60-W pencil iron is ideal for most electronic work.

58. For your safety and to protect voltage- and current-sensitive components, you should always use a soldering iron with a three-pin plug and three-wire ac power cord.

59. Always turn off the equipment before soldering, as the grounded tip may cause a short circuit in the equipment and possible damage.

60. The tip of a soldering iron can sometimes be changed, and a different temperature tip will allow your iron to be used for different applications.

61. In general, a 700°F tip is ideal for most electronic applications; however, for delicate printed circuit boards, use a 600°F tip.

62. The tip of a soldering iron comes in a variety of shapes and sizes. The tip shape that makes the best contact with the connection, and will therefore conduct the most heat, should be used.

63. As you regularly use your soldering iron, the tip will naturally accumulate dirt and oxide, and this contamination will reduce its effectiveness. The tip should be regularly cleaned by wiping it across a damp sponge.

64. The tip of the soldering iron should always be tinned after it has been cleaned. Tinning the tip will protect the tip from oxidation and increase the amount of heat transfer to the connection.

65. It is important to remember that *a soldering iron should not be used to melt the solder; its purpose is to heat the connection so that the solder will melt when it makes contact with the connection.*

66. Wires should be wrapped around a terminal to give it a more solid physical support. Their insulation should be stripped back so as to leave a small gap. Too large a gap will expose the bare wire and possibly cause a short to another terminal, whereas too small a gap will cause the insulation to burn during soldering. Stranded wire should be tinned, and components should be mounted flat on the board.

67. Always remember to wear safety goggles and a protective apron, and then proceed with the six step soldering procedure, which is:

Step 1 Clean the tip on a damp sponge.

Step 2 Tin the tip if necessary.

Step 3 Heat the connection.

Step 4 Apply the solder to the opposite side of the connection.

Step 5 Leave the tip of the soldering iron on the connection only long enough to melt the solder.

Step 6 Remove the solder and then the iron; then let the melted solder cool and solidify undisturbed.

Once this procedure is complete, you should remove the excess leads with the cutters and remove the flux residue, because it collects dust and dirt that could produce electrical leakage paths.

68. When inspecting your work, you should notice the following:

a. The connection is *smooth and shiny.*
b. The solder should *feather out to a thin edge.*

69. A *cold solder joint* has a dull gray appearance; it makes a poor mechanical and electrical connection and is normally caused due to connector lead movement while the solder was cooling, or to insufficient heat. The cure is to resolder the connection.

70. An *excessive solder joint* is caused by too much solder being applied to the connection or by using solder of too large diameter. The cure is to desolder the excess with a wick.

71. An *insufficient solder joint* happens because not enough solder was applied so a poor bond exists between the lead and the terminal. The cure is to resolder as if it were a new joint.

72. A *solder bridge* occurs when adjacent terminals or traces on printed circuit boards are connected accidentally as a result of using too much solder. Most can be removed by desoldering the excess with a wick.

73. On printed circuit boards, too much heat can lift the traces or track and ruin the entire board.

74. If a faulty component has to be replaced or if a component or wire was soldered in an incorrect place, it will have to be desoldered.
75. The easiest way to remove a component is to remove all the solder and then disconnect the leads.
76. The braided wick contains many small wire strands, and when it is placed between the solder and the iron, the melted solder flows into the braid by capillary action. The solder-filled braid is then cut off and discarded.
77. The desoldering bulb or spring-loaded plunger relies on suction rather than capillary action to remove the solder.
78. The steps to follow to desolder are:

    Step 1 Squeeze the desoldering bulb or cock the spring-loaded plunger.

    Step 2 Melt the solder with the soldering iron.

    Step 3 Remove the iron's tip.

    Step 4 Quickly insert the tip of the bulb or plunger into the molten solder and then activate the suction.

79. When as much of the solder as possible has been removed, use the long-nose pliers to hold the lead and remove the component.
80. It may be necessary to use the iron to heat the lead slightly to loosen up the residual solder so that the component lead can be removed with the pliers.
81. Always be careful not to apply an excessive amount of heat or stress to either the component or the board.
82. Molten solder, like any other liquid, can splash or spill, and soldering irons are even hotter than the solder.

NEW TERMS

Ammeter
Analog multimeter
Autoranging
Body resistance
Breadboard
Calibration
Dc power supply
Desoldering
Digital multimeter
Flux
Function generator
Live
Ohmmeter
Oscilloscope
Overload
Protoboard
Shock
Solder
Soldering
Soldering iron
Triggering
Voltmeter
Wetting

REVIEW QUESTIONS

Multiple-Choice Questions

1. What would be the current through a body resistance of 20 kΩ if it were to come in contact with 24 V?

   a. 12 mA **c.** 480 kA
   b. 1.2 mA **d.** 833.3 A

2. Could the current flow in Question 1 be considered fatal?

   a. Yes **b.** No

3. Which test instrument is used to measure current, voltage, and resistance?

   a. A dc power supply **c.** A function generator
   b. A multimeter **d.** An oscilloscope

4. Which instrument is designed to deliver a dc voltage for experimentation?

   a. A dc power supply **c.** A function generator
   b. A multimeter **d.** An oscilloscope

5. In which modes will a multimeter use its internal battery?

   a. Voltage **c.** Resistance
   b. Current **d.** None of the above

6. ____________ is needed within solder to remove an invisible insulating layer so that the solder can flow and stick to the metal contacts.

   a. Wetting **c.** Tin
   b. Flux **d.** Lead

7. The embedding of solder into the metal contacts is called:

   a. Wetting **c.** Tinning
   b. Flux **d.** None of the above

8. ____________ the tip of a soldering iron after it has been cleaned will protect it from oxidation and increase the amount of heat transfer to the connection.

   a. Fluxing **c.** Tinning
   b. Wetting **d.** None of the above

9. What tool can be used to desolder a connection?

   a. A bulb **c.** A spring-loaded plunger
   b. A braid **d.** All of the above

10. What is 60/40 rosin?

   a. A rosin-core 60% lead, 40% tin solder
   b. A 60% rosin, 40% lead solder
   c. A 60% tin, 40% lead solder with a rosin core
   d. All of the above

Essay Questions

11. What is the resistance of a human body? (5-1-1)
12. Why is it important for electronic technicians to understand resuscitation? (5-1-3)
13. What advantages does the dc power supply have over a battery as a voltage source? (5-2-1)
14. Choosing either an analog multimeter or digital multimeter, briefly describe the steps involved in measuring: (5-2-2)

   a. Current **b.** Voltage **c.** Resistance

15. What advantages does the DMM have over the AMM? (5-2-3)
16. Briefly describe the construction and function of a protoboard. (5-3)

17. Define the following soldering names and terms: (5-4)

    a. Wetting
    b. Rosin 60/40
    c. Tinning
    d. Desoldering

18. List the six-step soldering procedure. (5-4-5)
19. How should a good solder connection appear? (5-4-5)
20. List the four-step desoldering procedure. (5-4-6)

Practice Problems

21. If an analog ammeter indicates 15 on the 300 mA range, what value of current is present?
22. If an analog ohmmeter displays 47 on the $R \times 1000$ range, what is the resistance being measured?
23. If a digital voltmeter displays 200 on the 500 mV range, what is the value of voltage being measured?
24. If a DMM displays "OL," what does it indicate?
25. What current is flowing through the heating coil of a 60 W, 120 V pencil soldering iron?

CHAPTER 6

OBJECTIVES

After completing this chapter, you will be able to:

1. Describe how direct current can be generated:
   - **a.** Mechanically with friction and pressure
   - **b.** Thermally with a thermocouple
   - **c.** Optically with a photocell
   - **d.** Magnetically with a generator
   - **e.** Chemically with a battery
   - **f.** Electrically with a power supply
2. Understand the nature of magnetism and why certain materials will magnetize and certain materials will not magnetize.
3. Describe the magnetic field and the rules behind magnetic attraction and repulsion.
4. Explain how magnetic energy can be used to generate direct current.
5. State the differences between the three artificial magnets.
6. Describe the operation of both a primary and a secondary battery.
7. State the difference between a primary and a secondary cell.
8. Describe the features, applications, and construction of all the popular primary and secondary cells.
9. Explain the advantages of connecting batteries in series and parallel.
10. Define what is meant by a battery's:
    - **a.** Internal resistance
    - **b.** Maximum power transfer
    - **c.** Optimum power transfer
11. Describe the operation and use of various types of fuses, circuit breakers, and switches.

Direct Current (DC)

OUTLINE

VIGNETTE

The First Computer Bug

Mathematician Grace Murray Hopper, an extremely feisty and independent U.S. naval officer, was assigned to the Bureau of Ordnance Computation Project at Harvard during World War II. As Hopper recalled, "We were not programmers in those days, the word had not yet come over from England. We were 'coders,'" and with her colleagues she was assigned to compute ballistic firing tables on the Harvard Mark 1 computer. In carrying out this task, Hopper developed programming method fundamentals that are still in use.

Hopper is also credited, on a less important note, with creating a term frequently used today falling under the category of computer jargon. During the hot summer of 1945, the computer developed a mysterious problem. Upon investigation, Hopper discovered that a moth had somehow strayed into the computer and prevented the operation of one of the thousands of electromechanical relay switches. In her usual meticulous manner, Hopper removed the remains and taped and entered it into the logbook. In her own words, "From then on, when an officer came in to ask if we were accomplishing anything, we told him we were 'debugging' the computer," a term that is still used to describe the process of finding problems in a computer program.

INTRODUCTION

Direct current, abbreviated dc, is a flow of continuous current in only one direction and it is what we have been dealing with up to this point. A dc current is produced when a dc voltage source, such as a battery, is connected across a closed circuit. The requirement for a dc voltage source is that its output voltage polarity remain constant, and the fixed polarity of the voltage source will produce a unidirectional (one direction) current through a dc circuit. In this chapter, we will see how a dc voltage can be generated using several different methods, and we will examine fuses, circuit breakers, and switches.

6-1 SOURCE AND LOAD

Figure 6-1(a) illustrates a dc circuit with a battery connected across a light bulb. In this circuit, the potential energy of the battery produces kinetic energy or current that is used to produce light energy from the bulb. The bat-

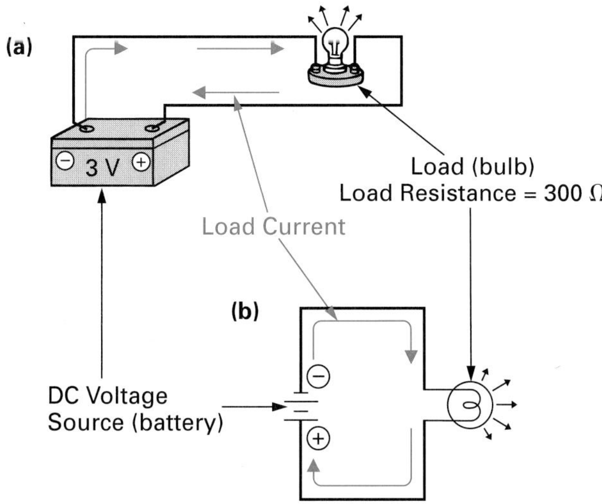

FIGURE 6-1 **Load on a Voltage Source. (a) Pictorial. (b) Schematic.**

tery is therefore the **source** in this circuit and the bulb is called a **load,** which by definition is a device that absorbs the energy being supplied and converts it into the desired form. The **load resistance** will determine how hard the voltage source has to work. For example, the bulb filament has a resistance of 300 Ω. The connecting wires have a combined resistance of 0.02 Ω (20 mΩ), and since this value is so small compared to the filament's resistance, it will be ignored, as it will have no noticeable effect. Consequently, a 300-Ω load resistance will permit 10 mA of **load current** to flow when a 3 V source is connected across the dc circuit ($I = V/R = 10$ mA). A change in filament resistance will result in a change in current; so if the load resistance is increased, the load current decreases ($I \downarrow = V/R \uparrow$), and conversely, if the load resistance is decreased, the load current increases ($I \uparrow = V/R \downarrow$) for a constant voltage source.

Source
Device that supplies the signal power or electric energy to a load.

Load
A component, circuit, or piece of equipment connected to the source. The load resistance will determine the load current.

Load Resistance
The resistance of the load.

Load Current
The current that is present in the load.

Figure 6-2 illustrates a heater, light bulb, motor, computer, robot, television, and microwave oven. All these and all other devices or pieces of equipment are connected to some form of supply (for example, a battery). The supply does not see all the circuitry and internal workings of the equipment; it simply sees the whole device or piece of equipment as some value of resistance. The resistance of the equipment and the supply voltage determine the value of current flow from the supply to the device or equipment. The battery or any other current source sees these devices or pieces of equipment as having some value of resistance, which is referred to as the device's *load resistance.*

EXAMPLE:

A component or piece of equipment offers a 390-Ω load resistance in a 9-V circuit. What current will flow in this circuit?

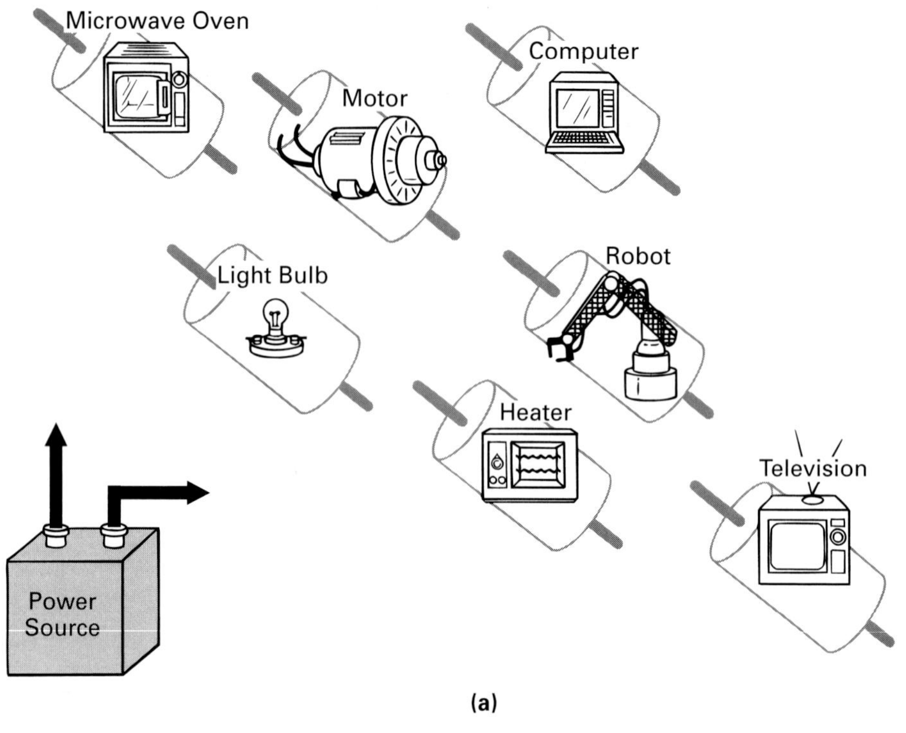

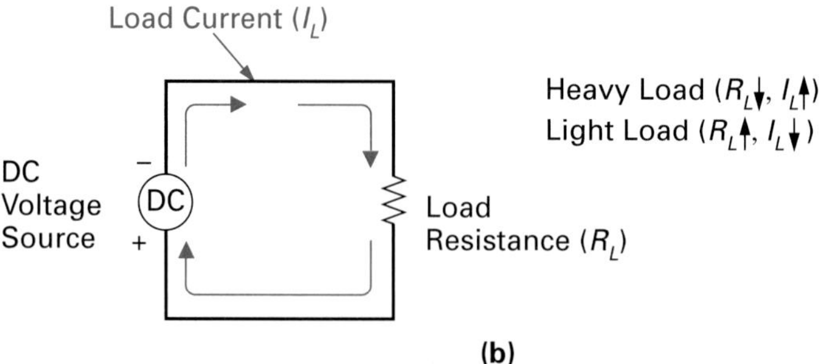

FIGURE 6-2 (a) Load Resistance of Equipment. (b) Schematic Symbols and Terms.

Solution:

$$\text{current } (I) = \frac{\text{voltage } (V)}{\text{resistance } (R)}$$

$$= \frac{9\text{ V}}{390\ \Omega}$$

$$= 23\text{ mA}$$

In summary, the phrase *load resistance* describes the device or equipment's circuit resistance, whereas the phrase *load current* describes the amount of current drawn by the device or equipment. A device that causes a large load current to flow, due to its small load resistance, is called a large or heavy load because it is heavily loading down or working the supply or battery. On the other hand, a device that causes a small load current to flow, due to its large load resistance, is referred to as a small or light load because it is only lightly loading down or working the supply or battery.

Load current and load resistance are therefore inversely proportional to one another in that a large load resistance ($R_L\uparrow$) causes a small load current ($I_L \downarrow$), referred to as a light load, and vice versa.

SELF-TEST REVIEW QUESTIONS FOR SECTION 6-1

1. Define *load resistance, load current, dc voltage source,* and *dc current.*
2. True or false: A large load current will flow when a large load resistance is connected across the supply.
3. What is the relationship between load resistance and load current?
4. Describe a heavy load resistance.

6-2 DIRECT-CURRENT SOURCES

The six basic forms of energy are mechanical, heat, light, magnetic, chemical, and electrical. To produce direct current (electrical energy), one of the other forms of energy must be used as a source and then converted or transformed into electrical energy in the form of a dc voltage that can be used to produce direct current.

6-2-1 *Mechanically Generated DC: Mechanical ⟶ Electrical*

When initially discussing current, we said that a positive charge was a material that has a deficiency or lack of electrons, while a negative charge has an excess of electrons. These positive and negative charges are known as *static* (stationary) *charges* or *static electricity.* Attraction and repulsion occur with positive and negative charges, as reviewed in Figure 6-3.

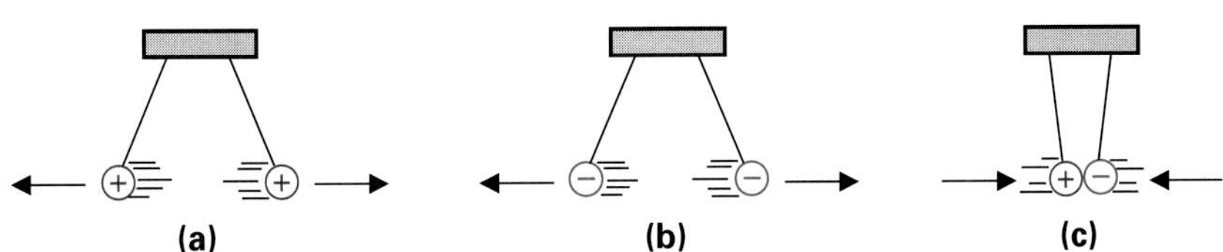

FIGURE 6-3 Attraction and Repulsion. (a) Positive Repels Positive. (b) Negative Repels Negative. (c) Unlike Charges Attract.

Friction

Friction
The rubbing or resistance to relative motion between two bodies in contact.

Friction is a form of mechanical energy, and Figure 6-4 illustrates how the electrons in a glass rod (insulator) can be forced out of their orbits and captured by another material, such as wool, by the work of rubbing. A static negative charge now exists in the wool, while a static positive charge exists in the glass. These mechanically generated static charges can be discharged in one of three ways, producing electrical energy, as illustrated in Figure 6-5.

Figure 6-5(a) illustrates the discharge of a static charge by placing a conductor between the two charges and allowing electrons to flow from one material to the other. Once a state of equilibrium is achieved, the two static charges will have neutralized and no further current will flow. If the connection has a small resistance, the discharge current will be brief and intense; on the other hand, if the connection has a large resistance, the discharge current will be prolonged and less intense.

Figure 6-5(b) illustrates the discharge of a static charge by contact. You may have noticed that your body can, through friction, build up a static charge that is discharged by contact once you touch another oppositely charged or neutralized material, giving you a shock.

Figure 6-5(c) illustrates how large static charges will discharge through the air, causing an arc between the two static charges known as lightning. Moving clouds in a storm cause friction between the molecules in the air. The buildup of a natural static charge in a cloud, if large enough, can cause a discharge or release of the electrical energy to the ground.

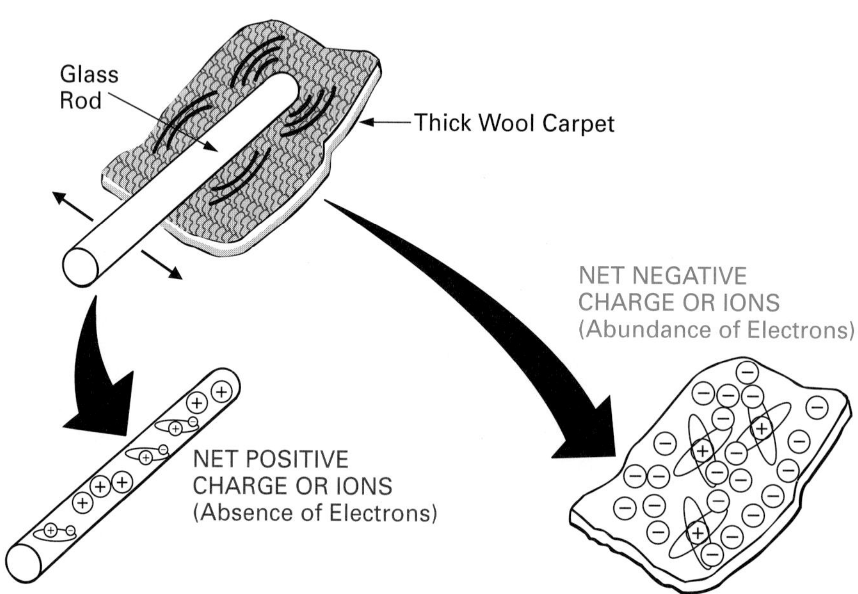

FIGURE 6-4 **Using Friction to Generate Static Electricity: Mechanical Energy to Electrical Energy.**

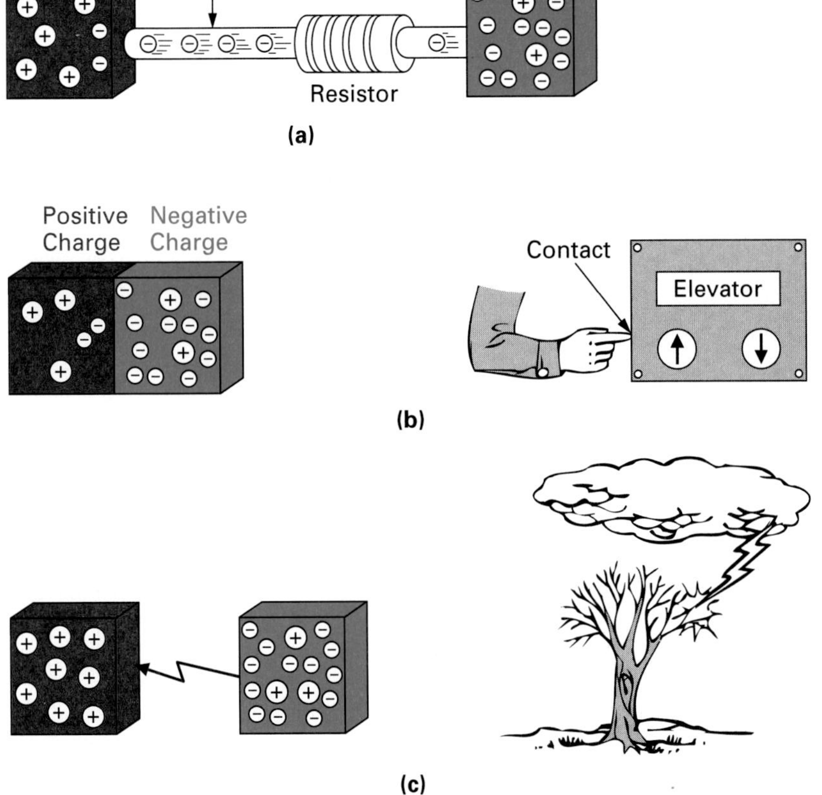

FIGURE 6-5 **Static Discharge. (a) Connection Static Discharge. (b) Contact Static Discharge. (c) Arc Static Discharge.**

Pressure

Mechanical energy in the form of **pressure** can also be used to generate electrical energy. Quartz and Rochelle salts are solid compounds that will produce electricity when a pressure is applied, as shown in Figure 6-6. Quartz is a natural or artificially grown crystal composed of silicon dioxide, from which thin slabs or plates are carefully cut and ground. Rochelle salt is a crystal of sodium potassium tartrate. They are both piezoelectric crystals. **Piezoelectric effect** is the generation of a voltage between the opposite faces of a crystal as a result of pressure being applied. This effect is made use of in most modern cigarette lighters. A voltage is produced between the opposite sides of a crystal (usually a crystalline lead compound) when it is either struck, pressured, or twisted. The resulting voltage is discharged across a gap to ignite the gas, which is released simultaneously by the operating button. The flame then burns for as long as the button is depressed. Piezoelectric effect also describes the reverse of what has just been discussed, as it was discovered that crystals would not only generate a voltage when mechanical force was applied, but would generate mechanical force when a voltage was applied.

Pressure
The application of force to something by something else in direct contact with it.

Piezoelectric Effect
The generation of a voltage between the opposite sides of a piezoelectric crystal as a result of pressure or twisting. Also, the reverse effect in which the application of voltage to opposite sides causes deformation to occur at the frequency of the applied voltage.

Figure 6-7(a) illustrates the physical appearance of a piezoresistive

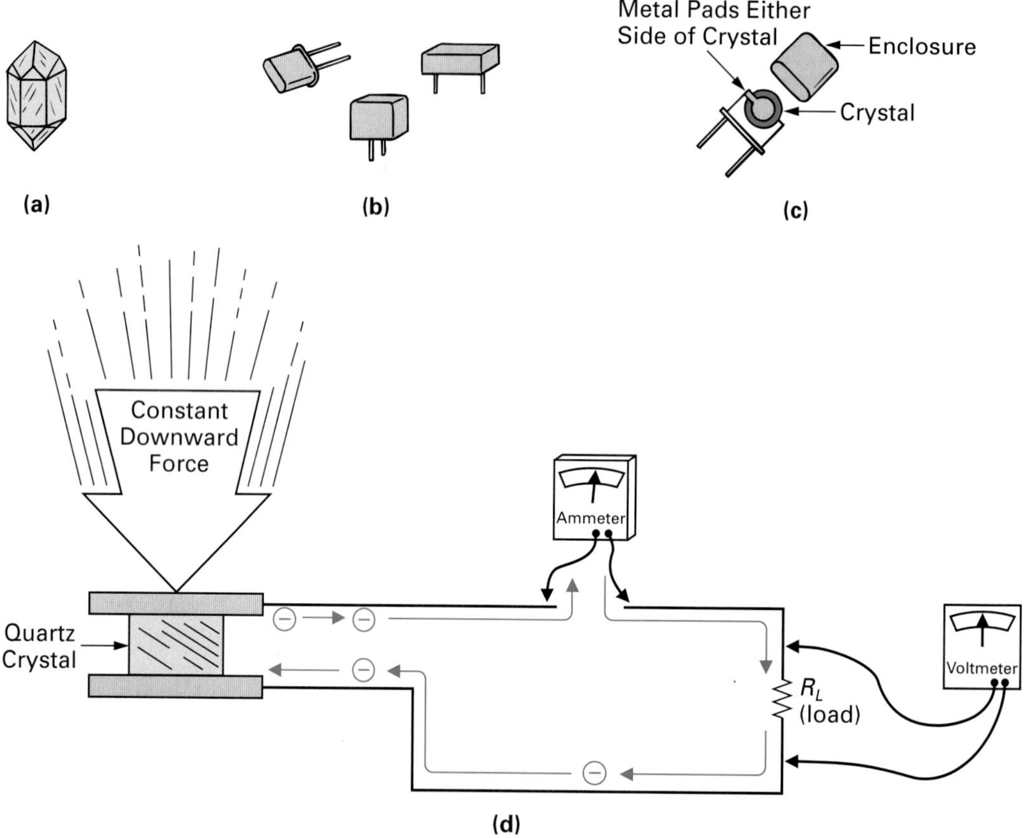

FIGURE 6-6 Direct Current from Pressure Using Crystals. (a) Quartz Crystal. (b) Physical Appearance. (c) Construction. (d) Operation.

diaphragm pressure sensor. Piezoresistance of a semiconductor is described as a change in resistance due to a change in the applied strain, and therefore semiconductor resistors can be used as pressure sensors. Four piezoresistors [Figure 6-7(b)] are buried in the surface of a thin, circular (silicon) diaphragm that has two pads connected to it making the connection. When pressure is applied, a stress or strain is applied to the diaphragm and therefore to the buried piezoresistors. The resistance between the pads will change depending on the amount of pressure applied to the diaphragm, and consequently a change in pressure (mechanical input) will cause a corresponding change in resistance (electrical output). Unlike quartz, which generates a voltage, the piezoresistive sensor needs an excitation voltage applied, and its value will determine the value of output voltage. Figure 6-7(c) illustrates the operation of this pressure transducer, which could be used, for example, to measure either the cooling system, hydraulic transmission, or fuel injection pressure in an automobile, which all vary in the 0 to 100 psi (pounds per square inch) range. The piezoresistive transducer would receive its excitation voltage from the car's battery (12 V) and produce a 0- to 100-mV output voltage, which would be sent to the car's computer, where the data would be analyzed and, if necessary, acted upon, as shown in Figure 6-7(d).

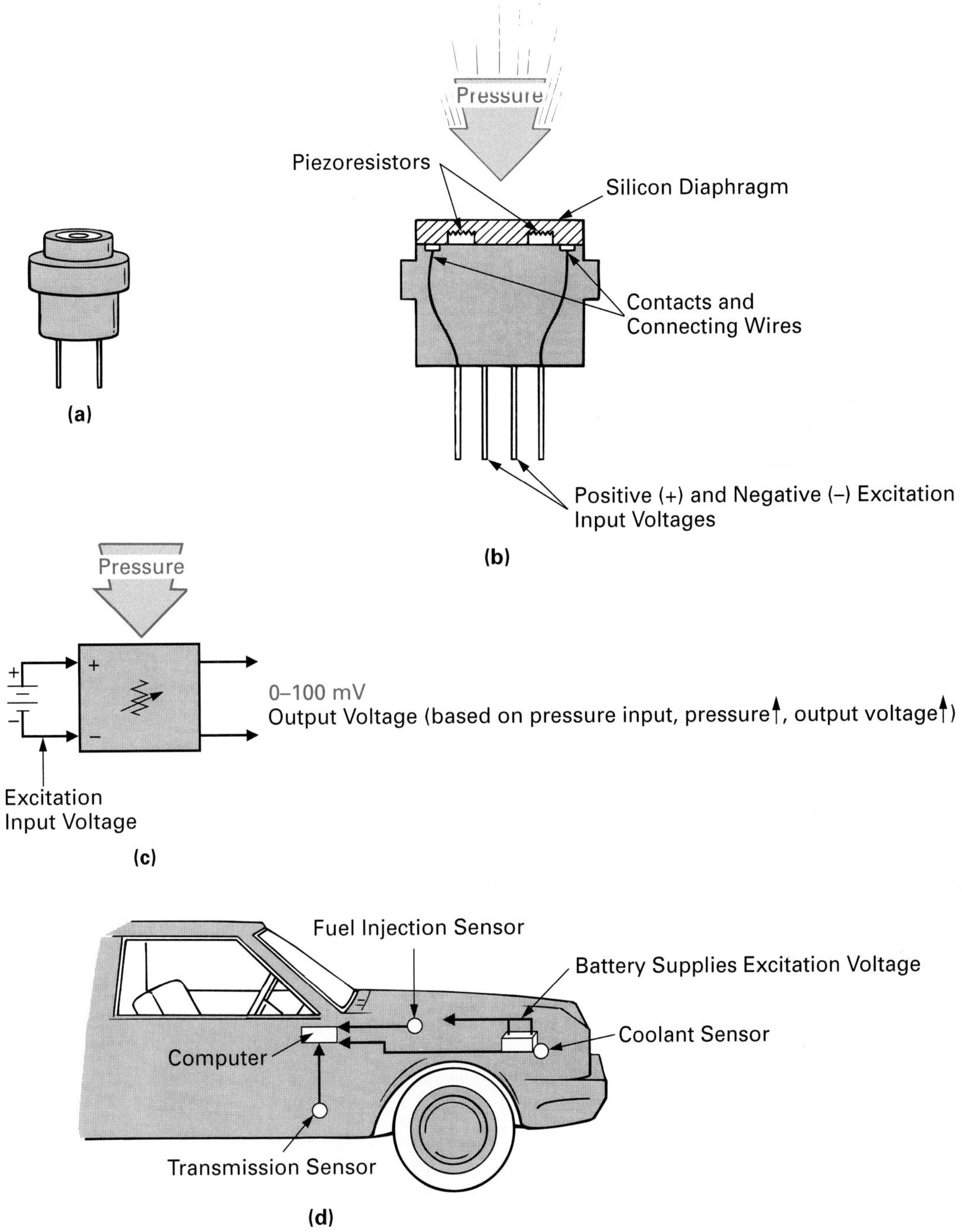

Pressure

Piezoresistors

Silicon Diaphragm

Contacts and
Connecting Wires

Positive (+) and Negative (–) Excitation
Input Voltages

(a)

(b)

Pressure

0–100 mV
Output Voltage (based on pressure input, pressure↑, output voltage↑)

Excitation
Input Voltage

(c)

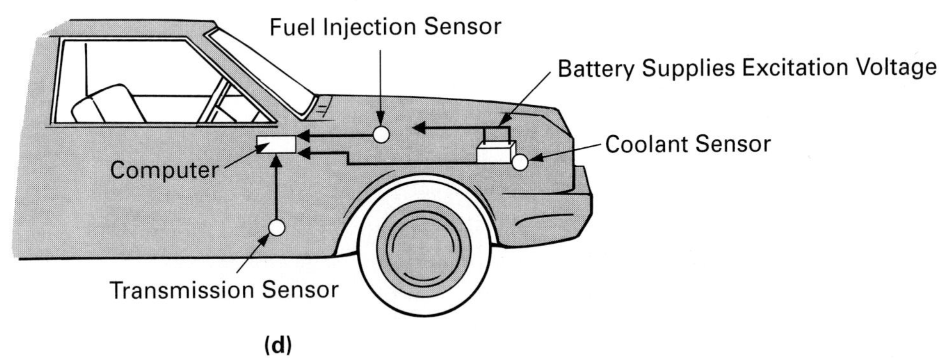

(d)

FIGURE 6-7 Direct Current from Pressure Using Piezoresistive Sensors. (a) Physical Appearance. (b) Construction. (c) Operation. (d) Application.

SELF-TEST REVIEW QUESTIONS FOR SECTION 6-2-1

1. How is friction used to generate dc?
2. How is pressure used to generate dc?

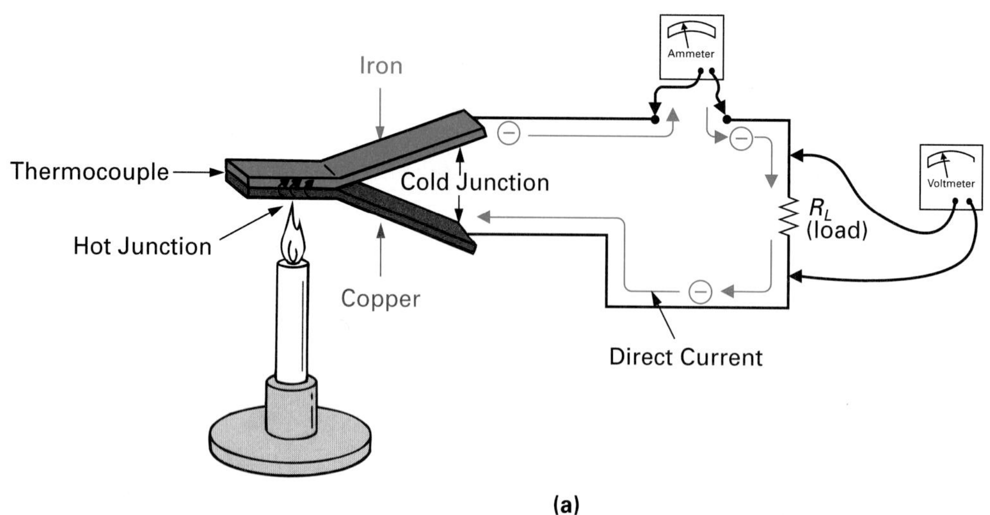

(a)

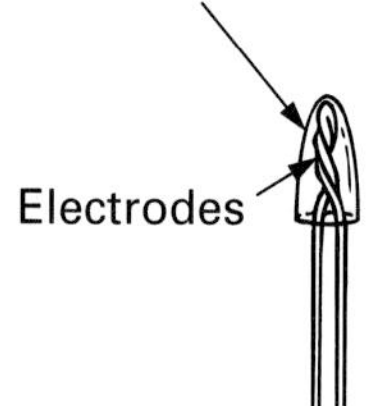

Two Metals	Temperature Range (°F)	Voltage Generated for Temperature Range (mV)
Iron vs. Copper–Nickel	32 to 1382	0 to 42.3
Nickel–Chromium vs. Nickel–Aluminum	–328 to 2282	–5.9 to 50.6
Platinum–6% Rhodium vs. Platinum–30% Rhodium	32 to 3092	0 to 12.4

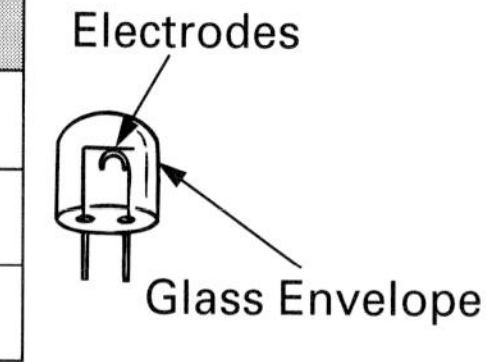

(b)

32.0°F
Sensor
Temperature Indicator

(c)

Water Heater
Shut-Off Valve
Thermocouple
Gas Inlet
Pilot Light

(d)

FIGURE 6-8 Direct Current from Heat Using the Thermocouple. (a) Operation. (b) Typical Thermocouples. (c) Indicating Temperature. (d) Water Heater Application.

6-2-2 Thermally Generated DC: Heat ⟶ Electrical

Heat can be used to generate an electric charge, as illustrated in Figure 6-8(a). When two dissimilar metals, such as copper and iron, are welded together and heated, an electrical charge is produced. This junction is called a **thermocouple,** and the heat causes the release of electrons from their parent atoms, resulting in a meter deflection indicating the generation of a charge. The size of the charge is proportional to the temperature difference between the two metals. Figure 6-8(b) lists some typical thermocouples and their voltages generated when heat is applied, and shows the physical appearance of two types.

Thermocouple
Temperature transducer consisting of two dissimilar metals welded together at one end to form a junction that when heated will generate a voltage.

Thermocouples like the one in Figure 6-8(c) are normally used to indicate temperature on a display calibrated in degrees. The thermocouple seen in Figure 6-8(d) is being heated by the gas pilot light, and the electrical energy generated will allow a valve to open and let gas through to the water heater. If, by accident, the pilot is extinguished by a gust of wind, the thermocouple will not receive any more heat and therefore will not generate electrical energy, which will consequently close the electrically operated valve and prevent a large explosion.

SELF-TEST REVIEW QUESTIONS FOR SECTION 6-2-2

1. What is the name given to the device used to convert heat to dc?
2. Name an application in which the transducer in Question 1 could be used.

6-2-3 Optically Generated DC: Light ⟶ Electrical

A **photovoltaic cell,** shown in Figure 6-9(a), also makes use of two dissimilar metals or semiconductor materials in its transformation of light to electrical energy. Figure 6-9(b) illustrates the construction and operation of a photoelectric cell, also called a **solar cell.**

Photovoltaic Cell
A solid-state device that generates electricity when light is applied. Also called a solar cell.

Solar Cell
Photovoltaic cell that converts light into electric energy. They are especially useful as a power source for space vehicles.

Photovoltaic Action
The development of a voltage across the junction of two dissimilar materials when light is applied.

A light-sensitive metal or semiconductor is placed behind a transparent piece of dissimilar metal. When the light illuminates the light-sensitive material, a charge is generated, causing current to flow and the meter to indicate current flow. This phenomenon, known as **photovoltaic action,** finds application as a light meter for photographic purposes and as an electrical power source for use by the satellite shown in Figure 6-9(c).

Electric power companies make use of solar cells to convert sunlight into electrical energy for the consumer. Out in space the solar cell may reach 75% efficiency, whereas on the earth they are only about 20% efficient, as the sun's rays are inhibited by the earth's atmosphere. The power obtained is on average about 100 milliwatts per square centimeter. Solar panels can also be found on calculators and emergency freeway telephones. Figure 6-10 shows how a satellite is generally covered with solar panels so that it can generate power to drive its electronic communication circuits.

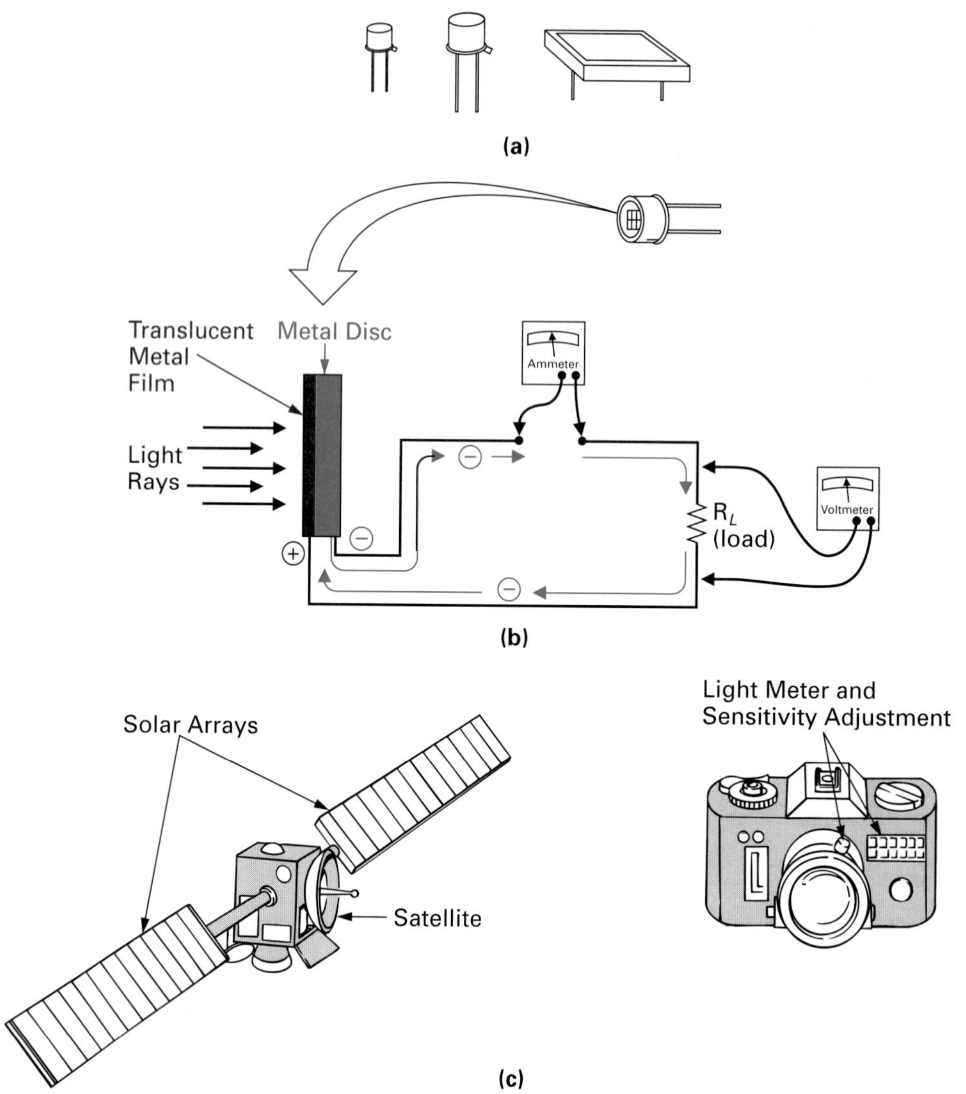

FIGURE 6-9 Direct Current from Light Using the Solar Cell. (a) Physical Appearance. (b) Construction and Operation. (c) Applications.

SELF-TEST REVIEW QUESTIONS FOR SECTION 6-2-3

1. What is the name of the component used to convert light energy to dc?

2. Name an application in which the transducer in Question 1 could be used.

6-2-4 *Magnetically Generated DC: Magnetic ⟶ Electrical*

Before discussing how magnetism can be used to generate electrical energy, let us first discuss the principles of magnetism.

FIGURE 6-10 **Solar Panels. (Photo courtesy of Hewlett-Packard Company.)**

The Nature of Magnetism

In the ancient city of Magnesia in Asia Minor, stones composed of iron oxide were found to have the strange ability to attract and repel one another. It was also noted that when suspended from a string, these stones would always point in only one direction. Historians are not sure whether the Chinese or Europeans first began using these "leading stones," or **lodestones** as they are now known, for long-distance land and sea travel, but the original compass was probably a lodestone suspended from a string. Later it was found that a metal needle could be magnetized by stroking it across a lodestone, and so the next form of compass consisted of a magnetized needle attached to a piece of wood floating in a bowl of water; hence the origin of the term *needle of a compass.*

Lodestone
A magnetite stone possessing magnetic polarity.

The lodestone shown in Figure 6-11(a) is an example of a natural magnet, unlike the needle, which is called an artificial magnet because it must be artificially magnetized in order to possess a magnetic field, but they both have the ability to point in only one direction by interacting with the earth's natural magnetic field, as shown in Figure 6-11(b).

Magnetism was recorded by the Greeks as early as 800 B.C., but it was not until 1269 that Petrus Peregrinus de Maricout found that the invisible magnetic lines produced by lodestones intersected at two points, which he called *poles.* In 1600, William Gilbert discovered that the magnetic field of a lodestone was lost when it was heated, but reappeared when the lodestone cooled. In 1819, Hans Christian Oersted discovered that any current-carrying conductor generated a

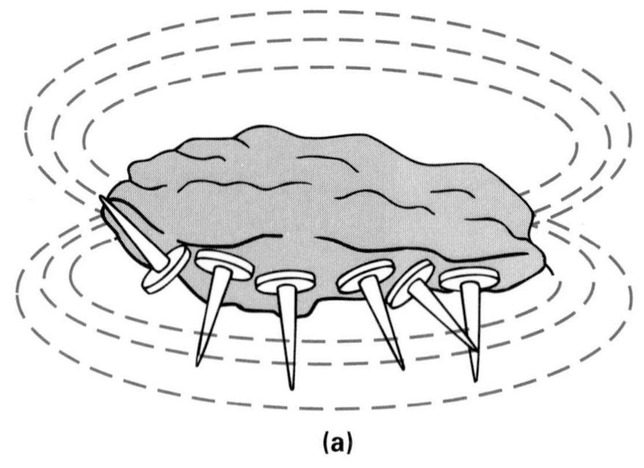

(a)

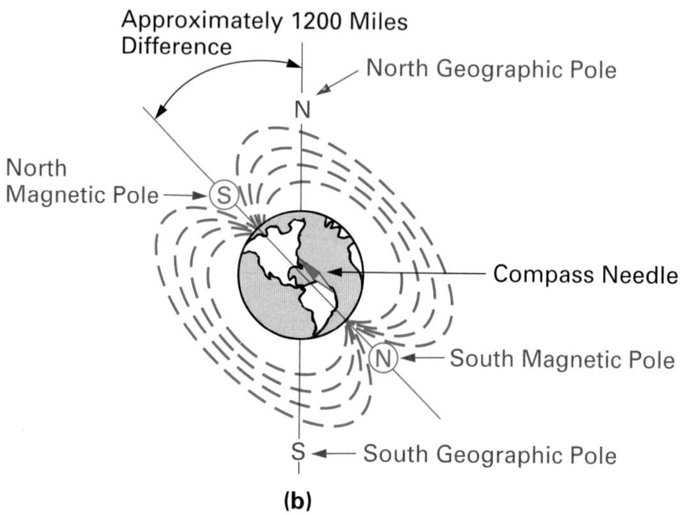

(b)

FIGURE 6-11 Natural Magnets. (a) Lodestone. (b) Earth.

magnetic field, and this interrelationship between electricity and magnetism became known as **electromagnetism.**

Electromagnetism Relates to the magnetic field generated around a conductor when current is passed through it.

In 1838, a German mathematician and physicist, Johann Gauss, published a paper on the earth's natural magnetic field, and in honor of his contribution the flux density of a magnetic field is measured in *gauss.* A magnetic flux density of 1 gauss (G) is produced 1 centimeter away from a wire carrying a current of 5 A. To give some examples, the field strength of a magnetized needle may only be a fraction of a gauss, whereas a lodestone 1 inch long may have a strength of 100 G. The earth's magnetic field has a magnetic field strength of 0.3 G at the equator and 0.7 G at either pole. Although this sounds small, a simple experiment can be conducted to prove how influential this invisible magnetic field is. If a screwdriver is positioned so that it points north and the screwdriver is tapped with a hammer several times, the screwdriver will actually become magnetized by the earth's magnetic field, as illustrated in Figure 6-12(a). The screwdriver is in fact an artificial magnet that was magnetized by a natural magnet (the earth). The atoms within the steel were originally misaligned. However, after being tapped and influenced by the earth's magnetic field, the atoms became aligned to reinforce one another, and will serve as a simple magnet as shown in Figure 6-12(b).

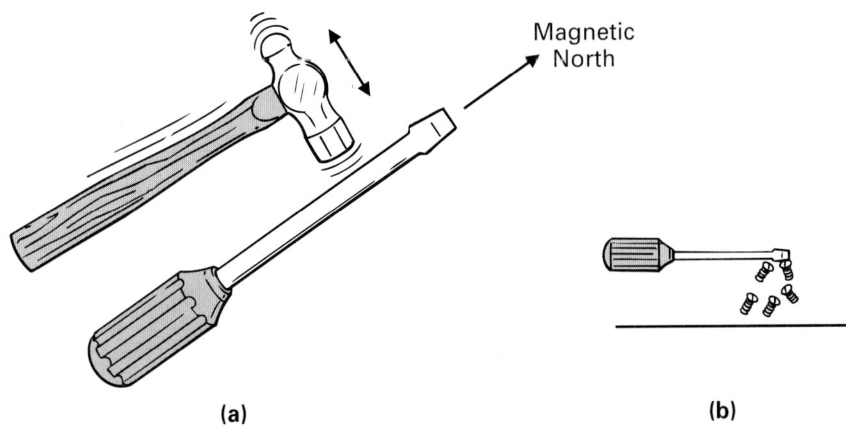

FIGURE 6-12 **Magnetizing a Screwdriver. (a) Method. (b) Result.**

Magnetism and the Atom

As you are already aware, an atom consists of electrons spinning around a nucleus; some atoms have few, while others have many electrons orbiting the atom's nucleus. A question arises when we find out that we can magnetize a screwdriver made of iron or nickel but cannot magnetize a screwdriver made of copper or tungsten. Why will certain materials magnetize whereas others will not? The answer is found in our smallest subatomic particle, the electron. Each electron is in fact a miniature magnet spinning around the nucleus. An electron at rest will not generate a magnetic field; however, orbiting electrons produce a magnetic field because of their high orbital velocity. Referring to your atomic periodic table of the elements (Table 2-1), you will notice that tungsten, for example, has 74 electrons (will not magnetize) and iron has 26 electrons (will magnetize). With tungsten, 37 electrons rotate around the nucleus in the clockwise direction, while 37 electrons orbit in the counterclockwise direction. The clockwise magnetic field will cancel with the counterclockwise magnetic field, so the net magnetic field is zero, which is why tungsten cannot be magnetized.

Iron has fewer electrons (26); however, 10 of the iron electrons spin clockwise and 16 spin counterclockwise, producing a net counterclockwise magnetic field of 6 electrons, which is why iron can be magnetized. If approximately 100 trillion atoms of the same magnetic polarity are all arranged in one direction, they are called a **domain.** Domains within unmagnetized materials are all misaligned, as shown in Figure 6-13(a), while aligned domains produce magnetized materials, shown in Figure 6-13(b).

Domain
Also known as a magnetic domain, it is a movable magnetized area in a magnetic material.

The Magnetic Field

The magnetic field produced by a magnet is an invisible force concentrated at two opposite points called the **magnetic poles,** shown in Figure 6-14. The north pole of the magnet, is the pole that would point south toward the earth's south magnetic pole if the magnet were free to rotate. The south pole of the magnet would point in a northerly direction if free to rotate. As you can see by again referring to Figure 6-14, the magnetic field is made up of invisible **flux lines** that leave the north pole and return at the south.

Magnetic Poles
Points of a magnet from which magnetic lines of force leave (north pole) and arrive (south pole).

Magnetic Flux
The magnetic lines of force produced by a magnet.

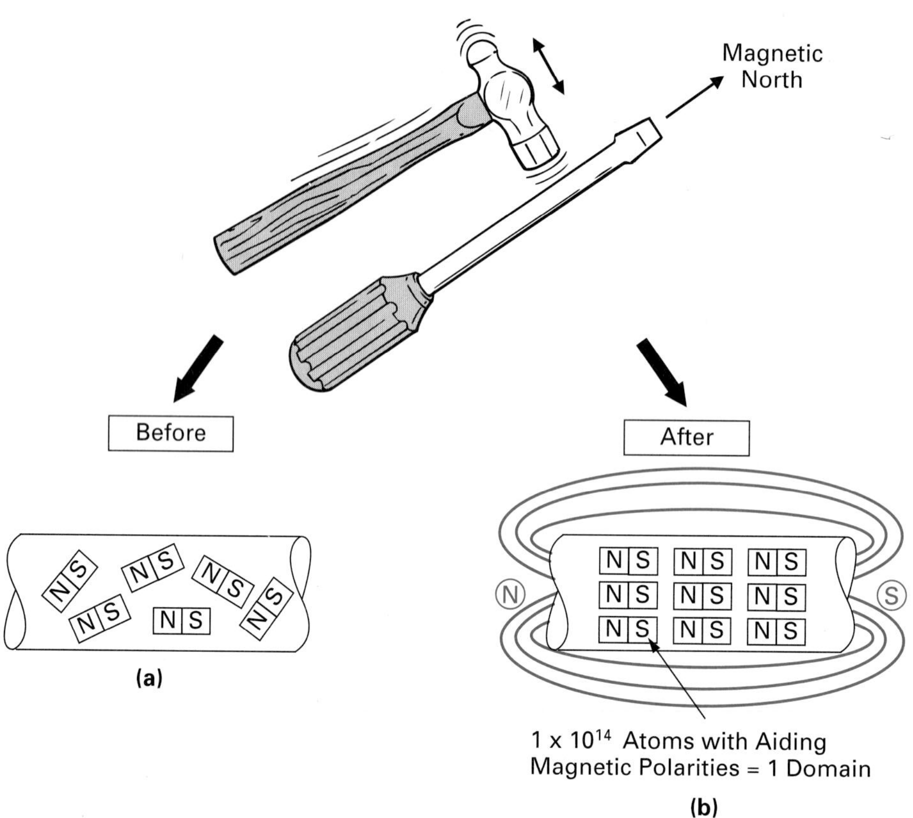

FIGURE 6-13 **(a) Misaligned Domains. (b) Aligned Domains.**

Attraction and Repulsion

When analyzing magnetic lines of flux within the magnetic field, it is noted that the law of attraction and repulsion could be applied to magnetic poles, just as it can be to positive and negative charges. This is summarized in Figure 6-15. A north against a north pole and south against a south pole will cause a force of repulsion between the two like poles [Figure 6-15(a)], while a south pole against

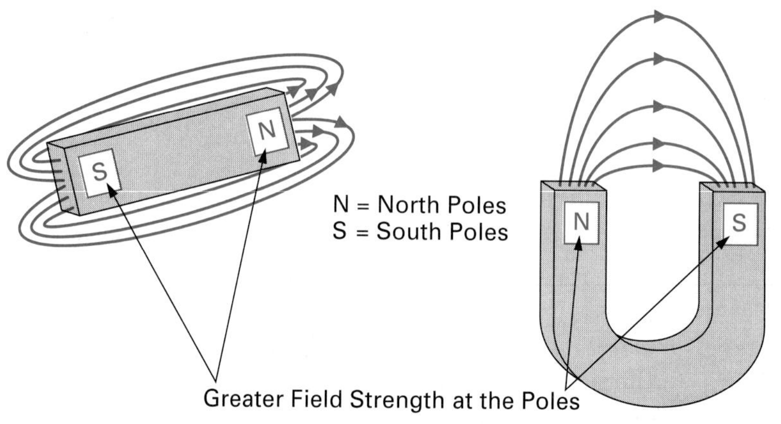

FIGURE 6-14 **Magnetic Field (Made Visible for Demonstration).**

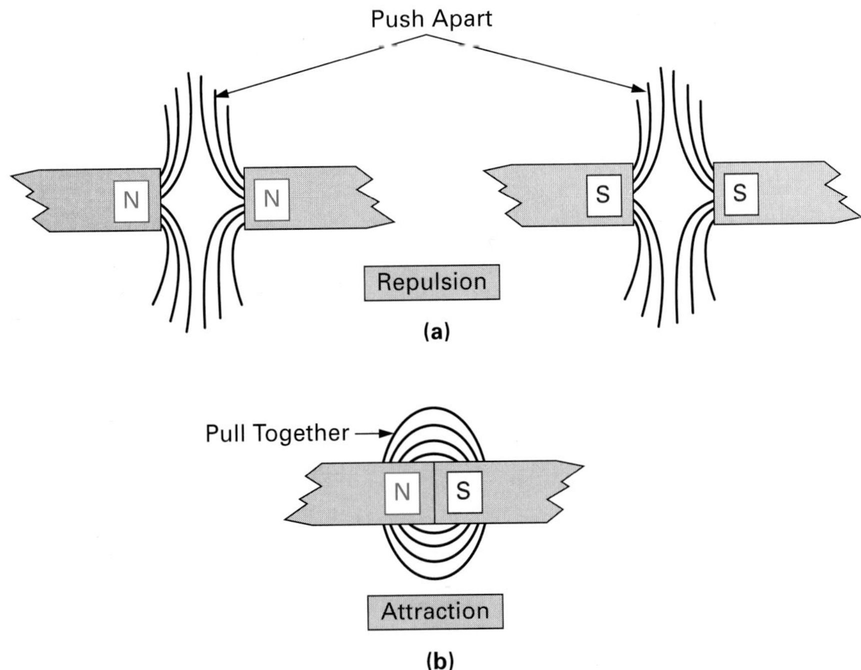

FIGURE 6-15 Attraction and Repulsion. (a) Like Poles Repel Each Other. (b) Unlike Poles Attract Each Other.

a north pole will result in a force of attraction between the two unlike poles [Figure 6-15(b)].

Artificial Magnets

There are three types of artificial magnets: (1) permanent magnets, (2) temporary magnets, and (3) electromagnets.

Permanent magnets. A permanent magnet is a piece of hardened steel or other special alloy that, when magnetized by the presence of a magnetic field, will retain its magnetism indefinitely (high retentivity), even when the material is removed from the magnetizing force or the magnetizing force is no longer present, as shown in Figure 6-16.

Temporary magnets. A temporary magnet is a piece of soft iron that will be magnetized while in the presence of a magnetizing field, but will be demagnetized the moment the soft iron is taken away from the magnetizing force or the magnetizing force is no longer present, as shown in Figure 6-17. Soft iron possesses low retentivity, which means it will not retain a magnetic field.

Electromagnets. An electromagnet is a temporary magnet that consists of an insulated wire wrapped around a soft-iron cylinder (core) to form a coil, as shown in Figure 6-18(a). A magnetic field is produced only while current flows through the coil, as shown in Figure 6-18(b). If a larger current is passed through the coil, a corresponding stronger magnetic field is generated, as shown in Figure 6-18(c).

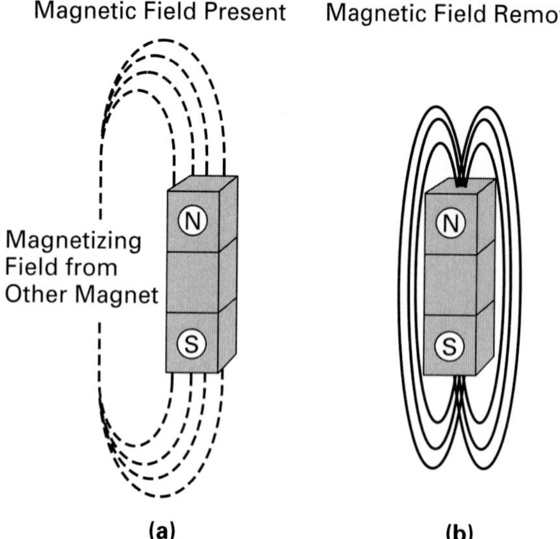

FIGURE 6-16 Permanent Magnets.
(a) Permanent magnet is magnetized when a magnetizing field is present (domains aligned).
(b) Permanent magnet retains the magnetic energy even when the magnetizing field is no longer present (domains remain aligned).

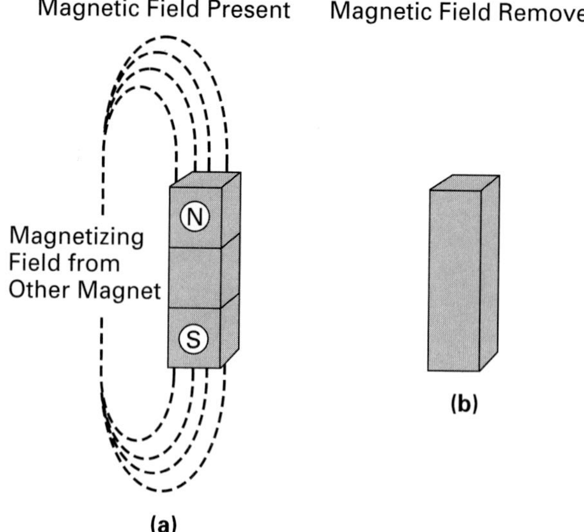

FIGURE 6-17 Temporary Magnets.
(a) Temporary magnet is magnetized when the magnetizing field is present (domains aligned).
(b) Temporary magnet loses magnetic energy when the magnetizing field is removed (domains no longer aligned).

Converting Magnetic Energy into Electrical Energy

A magnet like the one in Figure 6-19(a) can be used to generate electrical current by movement of the magnet past a conductor, as shown in Figure 6-19(b). When the magnet is stationary, no voltage is induced into the wire and therefore the light bulb is OFF. If the magnet is moved so that the magnetic lines of force cut through the wire conductor, a voltage is induced in the conductor, and this voltage produces a current in the circuit causing the light bulb to emit light. To achieve an energy conversion from magnetic to electric, you can either move the magnet past the wire or move the wire past the magnet.

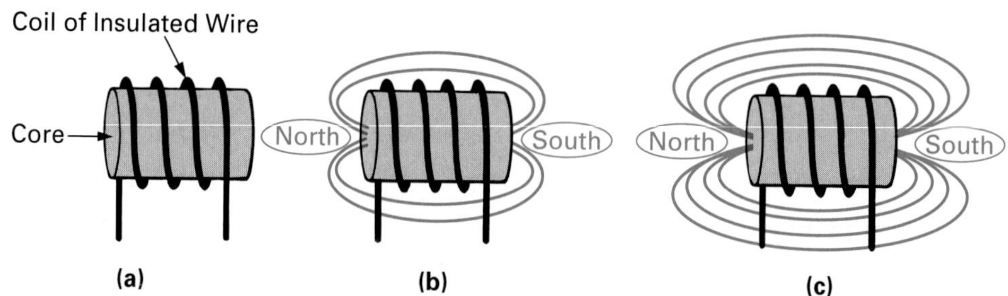

FIGURE 6-18 Electromagnets. (a) No Current, No Magnetic Field. (b) Small Current, Small Magnetic Field. (c) Large Current, Large Magnetic Field.

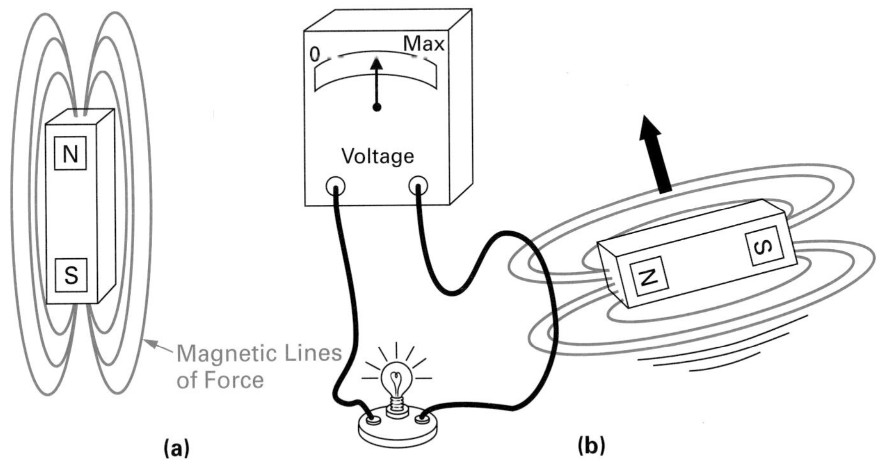

FIGURE 6-19 **Direct Current from Magnetism. (a) Stationary Magnet. (b) Moving Magnet.**

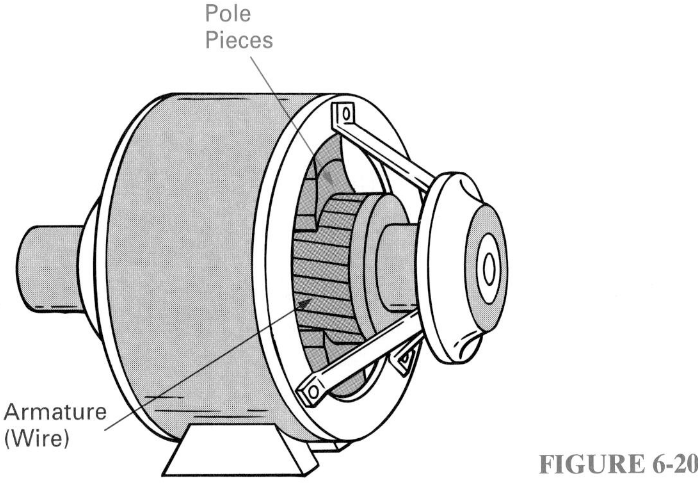

FIGURE 6-20 **Generator.**

To produce a continuous supply of electric current, the magnet or wire must be constantly in motion. The device that achieves this is the dc electric generator, illustrated in Figure 6-20.

SELF-TEST REVIEW QUESTIONS FOR SECTION 6-2-4

1. What is the name given to the natural magnetic rock?
2. How do the laws of attraction and repulsion apply to magnetism?
3. List the three types of artificial magnets.
4. What device is used to convert magnetic energy into electrical energy?

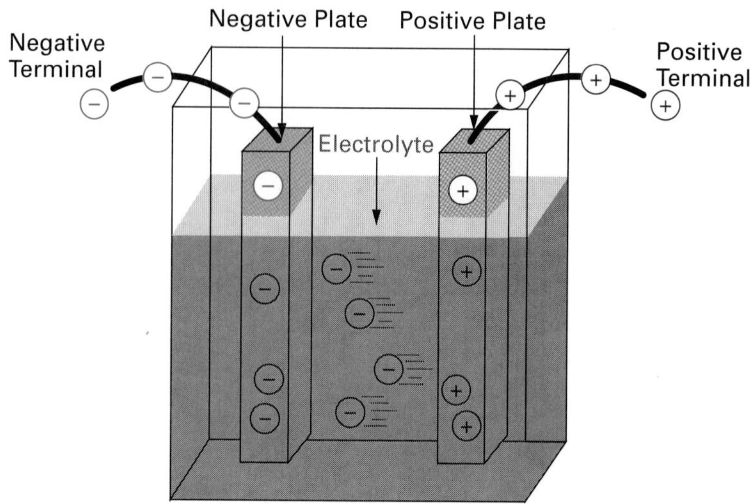

FIGURE 6-21 **Basic Battery.**

6-2-5 *Chemically Generated DC: Chemical ⟶ Electrical*

Voltaic Cell
Primary cell having two unlike metal electrodes immersed in a solution that chemically interacts with the plates to produce an emf.

The chemical voltage source (the battery) is called the **voltaic cell.** Its principle of operation was discovered by Alessandro Volta, an Italian physicist, in 1800.

A car uses a lead–acid battery, a portable radio has a carbon–zinc or alkaline battery, and, in some instances, you may be using the rechargeable nickel–cadmium batteries. No matter what type of battery is used, they all produce voltage by the same basic principle and all have three basic components, shown in Figure 6-21: (1) a negative plate, (2) a positive plate, and (3) an electrolyte.

Battery Operation

Electrolyte
Electrically conducting liquid (wet) or paste (dry).

Two dissimilar, separated metal plates such as copper and zinc are placed within a container that is filled with a liquid known as the **electrolyte,** usually an acid. A chemical reaction causes electrons to be repelled from one plate and attracted to the other, passing through the electrolyte. A large number of electrons collect on one plate (negative plate), while an absence or deficiency of electrons exists on the opposite, positive plate. The electrolyte acts on the two plates and transforms chemical energy into electrical energy, which can be taken from the cell at its two output terminals as an electrical current flow. If nothing is connected across the battery, as shown in Figure 6-22(a), a chemical reaction between the electrolyte and the negative electrode produces free electrons that travel from atom to atom and are then held in the negative electrode.

In Figure 6-22(b), a light bulb has been connected between the negative and positive electrodes. The mutual repulsion of the free electrons at the nega-

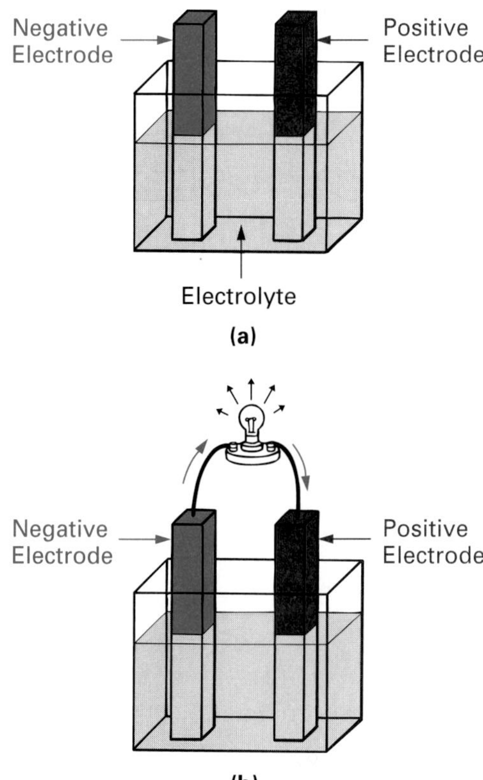

FIGURE 6-22 Battery Operation.

tive electrode combined with the attraction of the positive electrode results in a migration of free electrons (current flow) through the light bulb, causing its filament to produce light.

Primary Cells

The chemical reaction within the cell will actually dissolve the negative plate until eventually it will be eaten away completely. This discharging process also results in hydrogen gas bubbles forming around the positive plate, causing a resistance between the two plates known as the battery's internal resistance to increase (called *polarization*). To counteract this problem, all dry cells have a chemical within them known as a depolarizer agent, which reduces the buildup of these gas bubbles. With time, however, the depolarizer's effectiveness will be reduced, and the battery's internal resistance will increase as the battery reaches a completely discharged condition. These types of batteries are known as **primary** (first time, last time) **cells** because they discharge once and must then be discarded. Almost all primary cells have their electrolyte in paste form which is why they are also referred to as **dry cells.** Wet cells have an electrolyte that is in liquid form.

Primary Cell
Cell that produces electrical energy out through an internal electrochemical action; once discharged, it cannot be reused.

Dry Cell
Dc voltage-generating chemical cell using a nonliquid (paste) type of electrolyte.

Primary cell types. Figure 6-23 illustrates some of the various types of primary cells on the market today. The shelf life of a battery is the length of time a battery can remain on the shelf (in storage) and still retain its usability. Most pri-

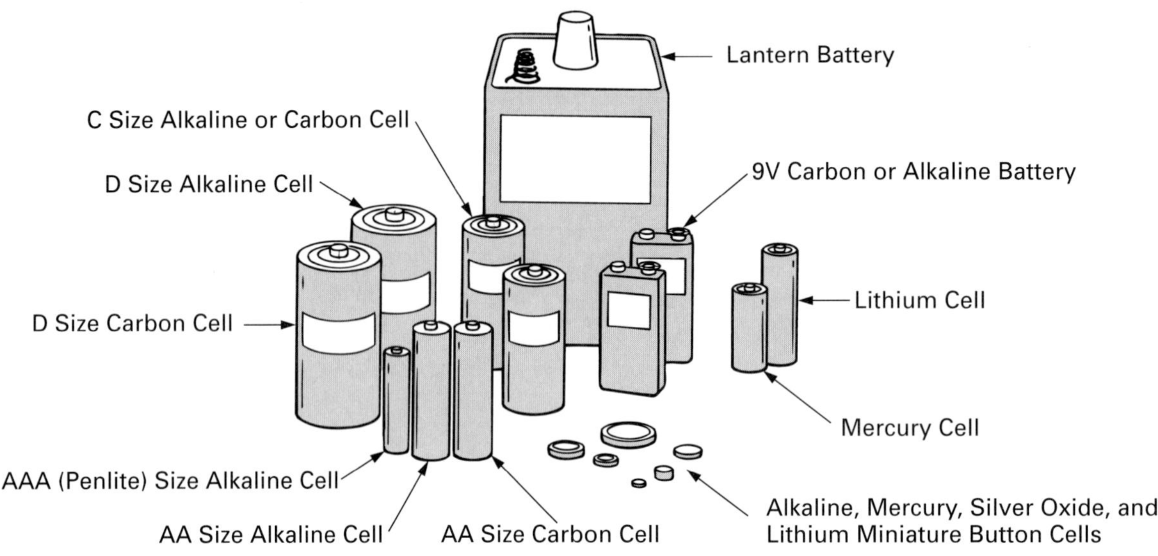

FIGURE 6-23 Primary Rectangular, Cylindrical, and Button Cells and Batteries.

mary cells will deteriorate in a three-year period to approximately 80% of their original capacity.

Of all the types of primary cells, five dominate the market as far as applications and sales are concerned. These are the carbon–zinc, alkaline–manganese, mercury, silver oxide, and lithium, all of which are described and illustrated in Table 6-1. The cell voltage in Table 6-1 describes the voltage produced by one cell (one set of plates). If two or more sets of plates are installed in one package, the component is called a battery.

Secondary Cells

Secondary Cells
An electrolytic cell for generating electric energy. Once discharged the cell may be restored or recharged by sending an electric current through the cell in the opposite direction to that of the discharge current.

Secondary cells operate on the same basic principle as that of primary cells. However, in this case the plates are not eaten away or dissolved; they only undergo a chemical change during discharge. Once discharged, the secondary cell can have the chemical change that occurred during discharge reversed by recharging, resulting once more in a fully charged cell. Restoring a secondary cell to the charged condition is achieved by sending a current through the cell in the direction opposite to that of the discharge current, as illustrated in Figure 6-24.

In Figure 6-24(a), we have first connected a light bulb across the battery, and free electrons are being supplied by the negative electrode and moving through the light bulb or load and back to the positive electrode.

After a certain amount of time, any secondary cell will run down or will have discharged to such an extent that no usable value of current can be produced. The surfaces of the plates are changed, and the electrolyte lacks the necessary chemicals.

In Figure 6-24(b), during the recharging process, we use a battery charger, which reverses the chemical process of discharge by forcing electrons back into the cell and restoring the battery to its charged condition. The battery charger

TABLE 6-1 Primary Cell Types

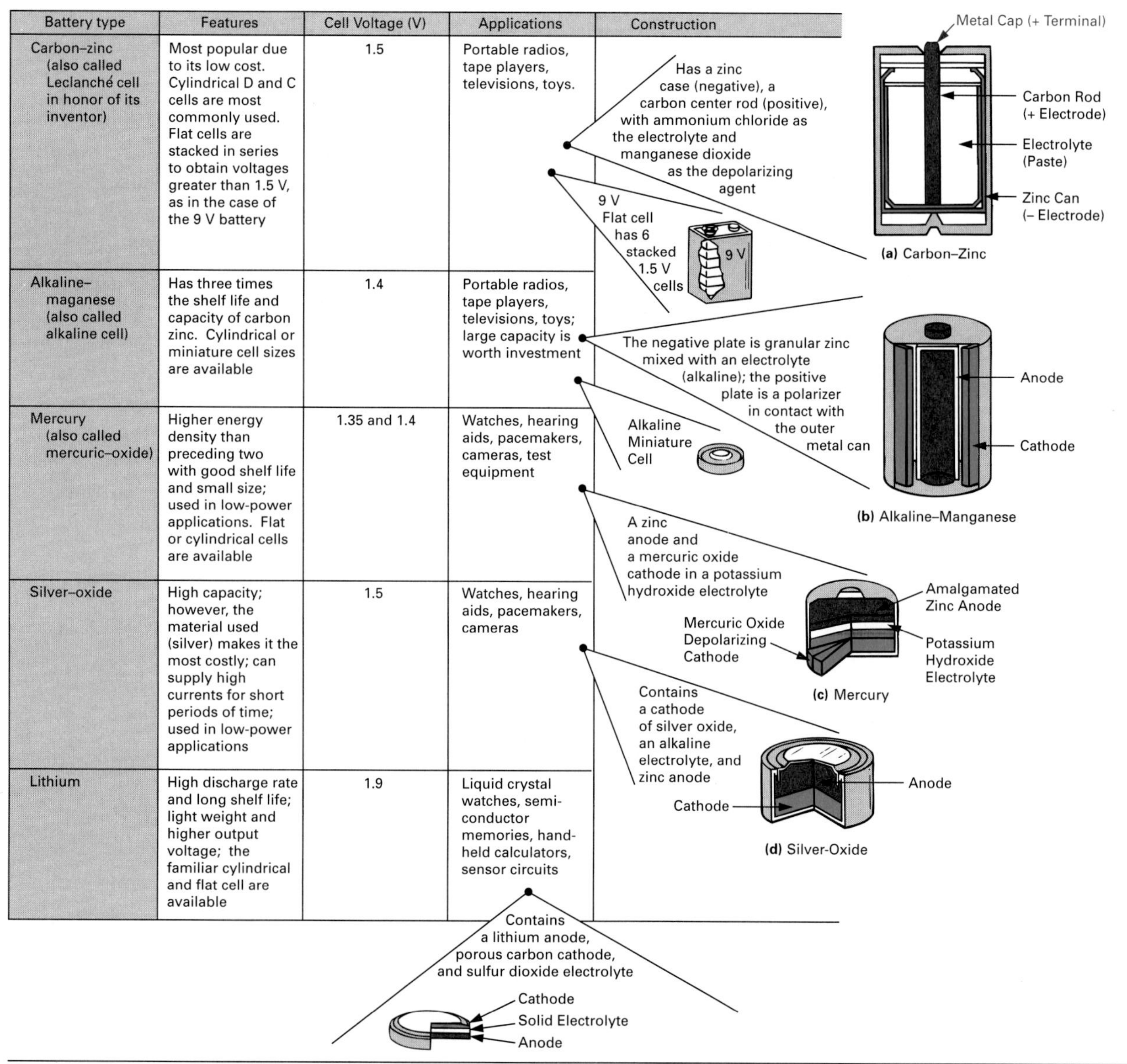

Battery type	Features	Cell Voltage (V)	Applications	Construction
Carbon–zinc (also called Leclanché cell in honor of its inventor)	Most popular due to its low cost. Cylindrical D and C cells are most commonly used. Flat cells are stacked in series to obtain voltages greater than 1.5 V, as in the case of the 9 V battery	1.5	Portable radios, tape players, televisions, toys.	Has a zinc case (negative), a carbon center rod (positive), with ammonium chloride as the electrolyte and manganese dioxide as the depolarizing agent. 9 V Flat cell has 6 stacked 1.5 V cells
Alkaline–maganese (also called alkaline cell)	Has three times the shelf life and capacity of carbon zinc. Cylindrical or miniature cell sizes are available	1.4	Portable radios, tape players, televisions, toys; large capacity is worth investment	The negative plate is granular zinc mixed with an electrolyte (alkaline); the positive plate is a polarizer in contact with the outer metal can. Alkaline Miniature Cell
Mercury (also called mercuric-oxide)	Higher energy density than preceding two with good shelf life and small size; used in low-power applications. Flat or cylindrical cells are available	1.35 and 1.4	Watches, hearing aids, pacemakers, cameras, test equipment	A zinc anode and a mercuric oxide cathode in a potassium hydroxide electrolyte
Silver-oxide	High capacity; however, the material used (silver) makes it the most costly; can supply high currents for short periods of time; used in low-power applications	1.5	Watches, hearing aids, pacemakers, cameras	Contains a cathode of silver oxide, an alkaline electrolyte, and zinc anode
Lithium	High discharge rate and long shelf life; light weight and higher output voltage; the familiar cylindrical and flat cell are available	1.9	Liquid crystal watches, semiconductor memories, hand-held calculators, sensor circuits	Contains a lithium anode, porous carbon cathode, and sulfur dioxide electrolyte

voltage is normally set to about 115% of the battery voltage. (For example, a battery charger connected across a 12-V battery would be set to 115% of 12 V, or 13.8 V.) If this voltage is set too high, an excessive current can result, causing the battery to overheat.

Secondary cells are often referred to as **wet cells** because the electrolyte is not normally in paste form (dry) as in the primary cell, but is in liquid (wet) form and is free to move and flow within the container. Figure 6-25(a) gives more information on the discharge process, and Figure 6-25(b) on the charge process, that occur to the plates and electrolyte of a lead–acid secondary cell.

Wet Cell
Cell using a liquid electrolyte.

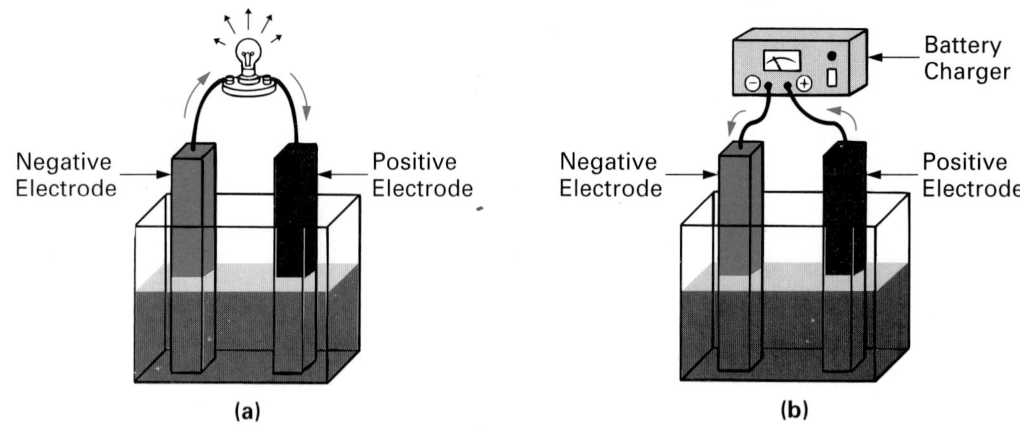

FIGURE 6-24 **Battery Operation. (a) Discharge. (b) Charge.**

Secondary cell types. Figure 6-26 illustrates the physical appearance of some of the types of secondary cells on the market today. The lead–acid and nickel–cadmium are the two most popular and are illustrated and described in Table 6-2 (p. 245). The two other types of secondary cells are the silver–zinc and silver–cadmium. Both have a large silver content and are therefore quite costly; however, they have the highest energy of all secondary cells and are used in specialized applications such as videotape recorders, portable televisions, and missiles.

Ampere-hours
If you multiply the amount of current, in amperes, that can be supplied for a given time frame in hours, a value of ampere-hours is obtained.

Secondary cell capacity. Capacity (C) is measured by the amount of **ampere-hours** (Ah) a battery can supply during discharge. If, for example, a battery has a discharge capacity of 10 ampere-hours, the battery could supply 1 ampere for 10 hours, 10 amperes for 1 hour, 5 amperes for 2 hours, and so on, during discharge. Automobile batteries (12 V) will typically have an ampere-hour rating of between 100 and 300 Ah.

Ampere-hour units are actually specifying the coulombs of charge in the battery. If, for example, a lead–acid car battery was rated at 150 Ah, this value would have to be converted to ampere-seconds to determine coulombs of charge, as 1 ampere-second (As) is equal to 1 coulomb. Ampere-hours are easily converted to ampere-seconds simply by multiplying ampere-hours by 3600, which is the number of seconds in 1 hour. In our example, a 150 Ah battery will have a charge of 54×10^4 coulombs.

EXAMPLE:

How many coulombs of charge does a 3.8-Ah lead–acid gel battery store?

Solution:

A 3.8-Ah battery can supply 3.8 A in 1 hour. This would be equivalent to 13,680 A in a second (3.8×3600). Since current is measured in coulombs/second, this battery when fully charged can store 13.68×10^3 C.

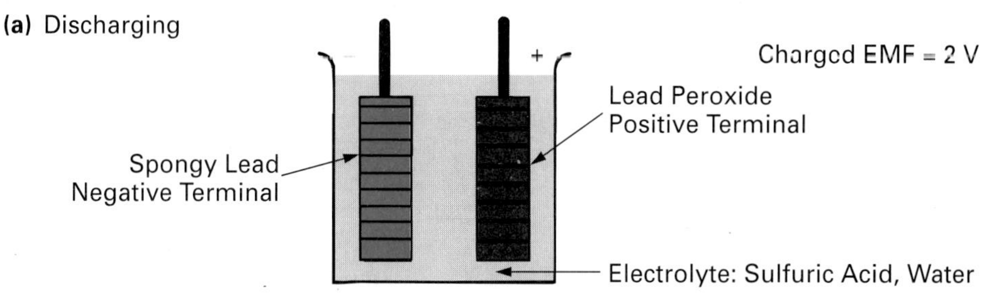

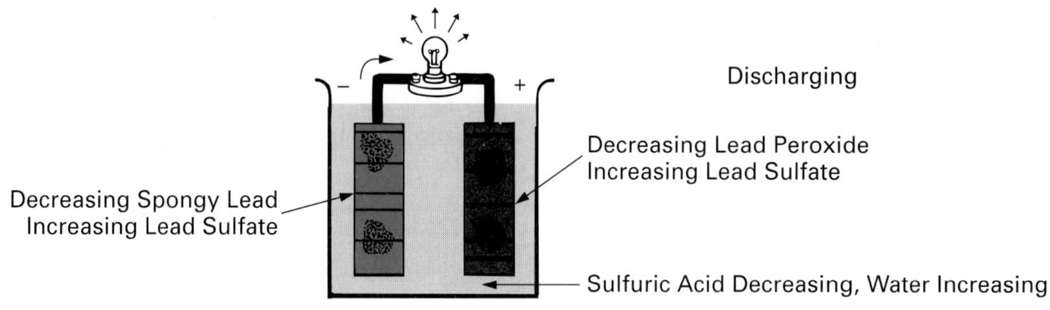

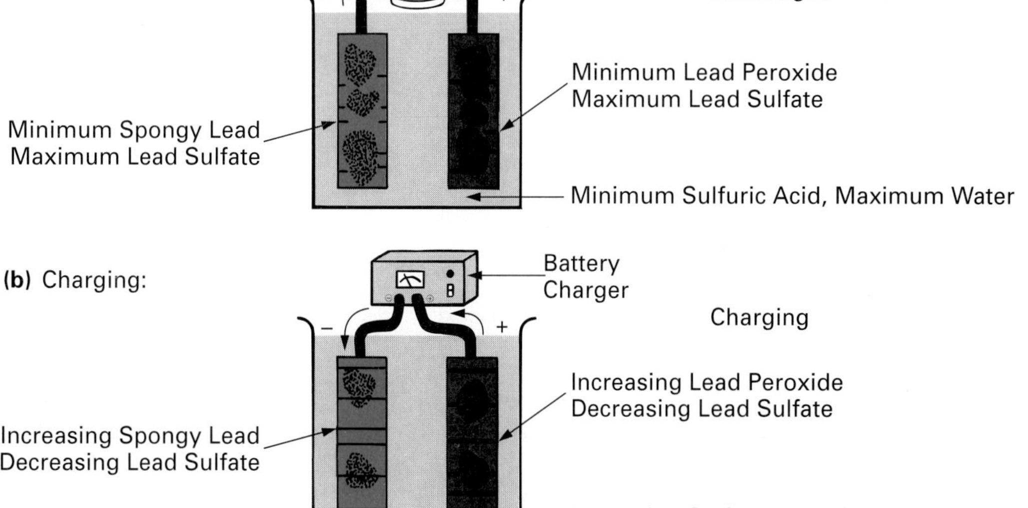

FIGURE 6-25 The Chemical Change of a Lead–Acid Cell During (a) Discharge and (b) Charge.

EXAMPLE:

A Ni-Cd battery is rated at 300 Ah and is fully discharged.

a. How many coulombs of charge must the battery charger put into the battery to restore it to full charge?

b. If the charger is supplying a charging current of 3 A, how long will it take the battery to fully charge?

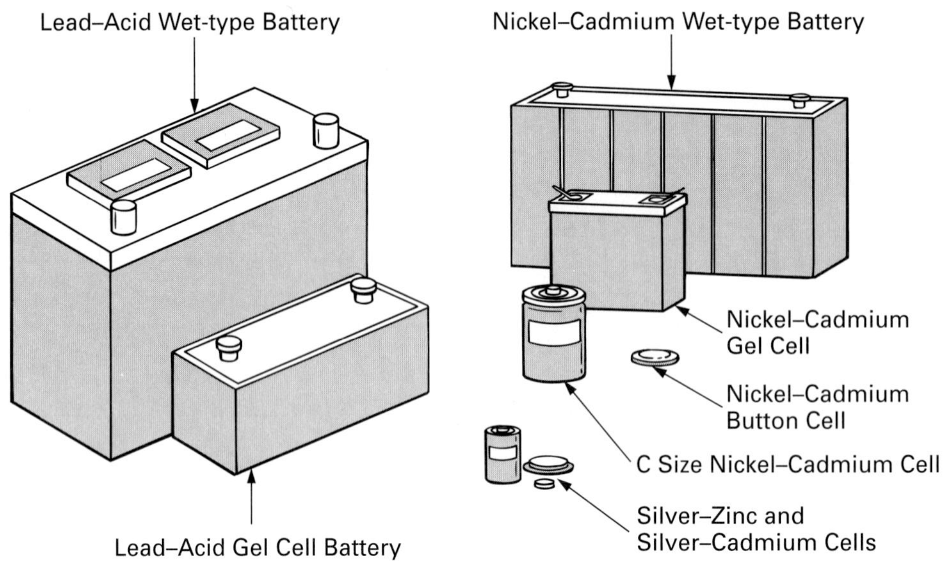

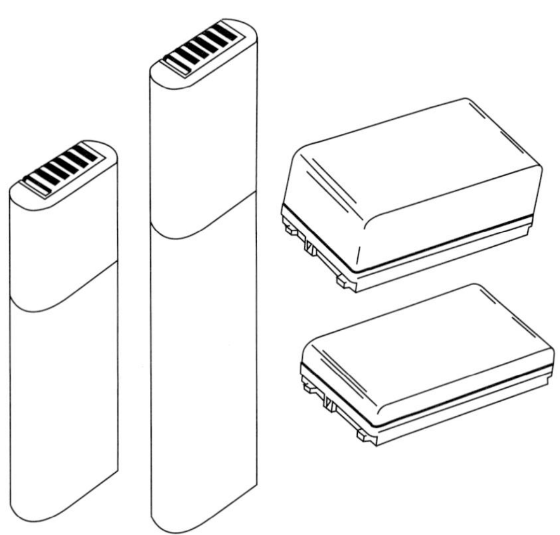

FIGURE 6-26 Secondary Rectangular, Cylindrical, and Button Cells and Batteries.

c. Once fully charged, the battery is connected across a load that is pulling a current of 30 A. How long will it take until the battery is fully discharged?

Solution:

a. The same amount that was taken out: 300 Ah $\times$ 3600 = 1080 $\times$ 10^3 C

b. $\dfrac{300 \text{ Ah}}{3 \text{ A}} = 100$ hours until charged

c. $\dfrac{300 \text{ Ah}}{30 \text{ A}} = 10$ hours until discharged

TABLE 6-2 Secondary Cell Types.

Battery type	Features	Applications	Construction
Lead–acid (lead cell)	High cycle (discharge–charge) life and high current capacity; 2.1 V per cell; must be operated upright if a wet lead–acid. Lead–acid gell cells can be used in any position	*Gell Cell Batteries* televisions, recorders, robots, alarms, and tools *Wet Cell Batteries* Starting source for automobiles and power for robotics equipment	The electrodes are lead oxide immersed in an electrolyte of dilute sulfuric acid 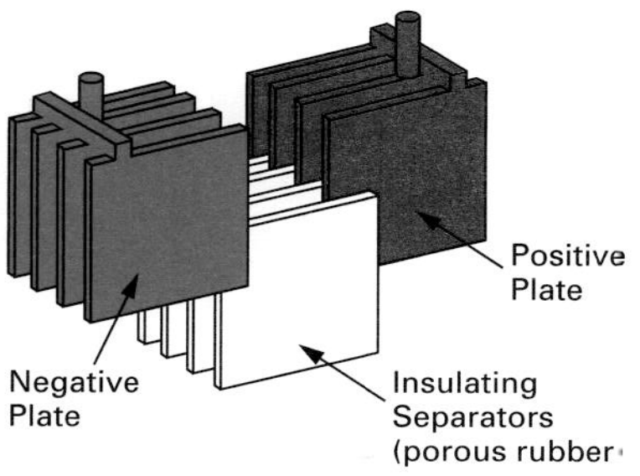 (Negative Plate; Positive Plate; Insulating Separators (porous rubber)) Wet-type lead–acid battery with 6 internal 2.1 V cells making 12.6 V car battery (Car Battery) 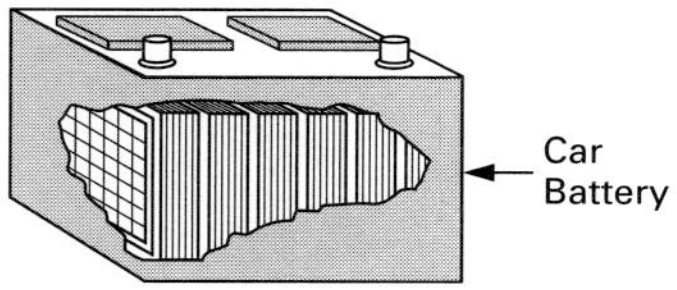
Nickel–cadmium (Ni–Cd)	High capacity and high cost; three times discharge current of lead–acid for same amp-hour rating; can be sealed and operated in any position, so ideal for drills and other such portable equipment; 1.2 V per cell. Nickel cadmium batteries suffer from "memory effect." Short discharge and charge cycles cause the batteries' capacity to drop, because the cell appears to "remember" the lower capacity.	*Ni–Cd Gell Cell* televisions, radios, toys, recorders, shavers, toothbrushes *Ni–Cd Wet Cell Batteries* Starter source for jet engines (resembles lead–acid car battery in appearance). Robotics equipment	The positive plate is made of cadmium, the negative plate of nickel hydroxide, and the electrolyte used is potassium hydroxide 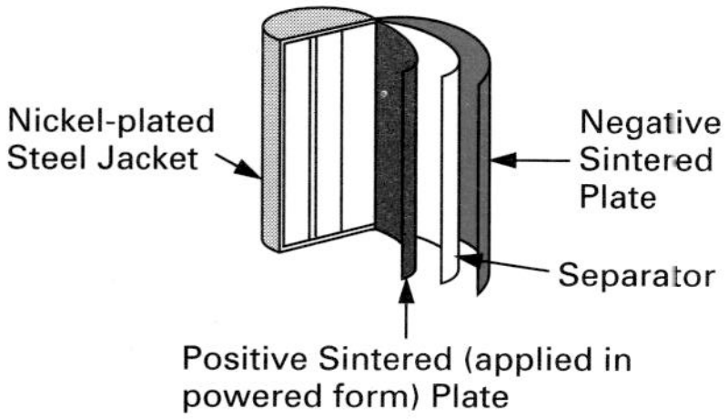 (Nickel-plated Steel Jacket; Negative Sintered Plate; Separator; Positive Sintered (applied in powered form) Plate)
Nickel–metal Hydride (Ni–MH)	Discharge characteristics are very similar to those of the nickel–cadmium cell. The cell's nominal voltage is 1.2 V, and they do not suffer from memory effect.	Biggest applications include cellular phones, portable computers, camcorders, and other portable electronic systems	The positive plate is made of nickel oxyhydride, the negative plate of metal hydride, and the electrolyte used is potassium hydroxide 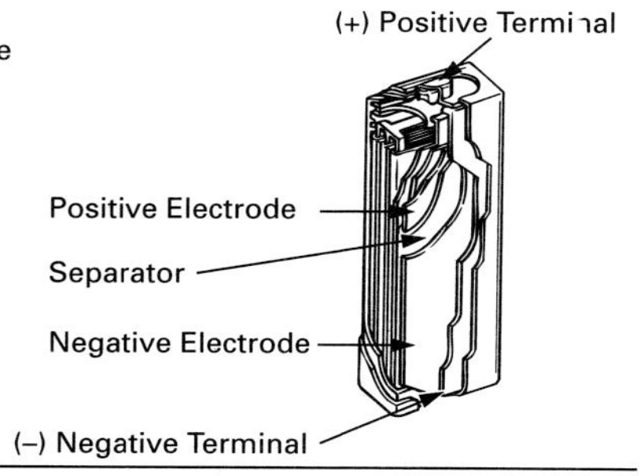 ((+) Positive Terminal; Positive Electrode; Separator; Negative Electrode; (–) Negative Terminal)

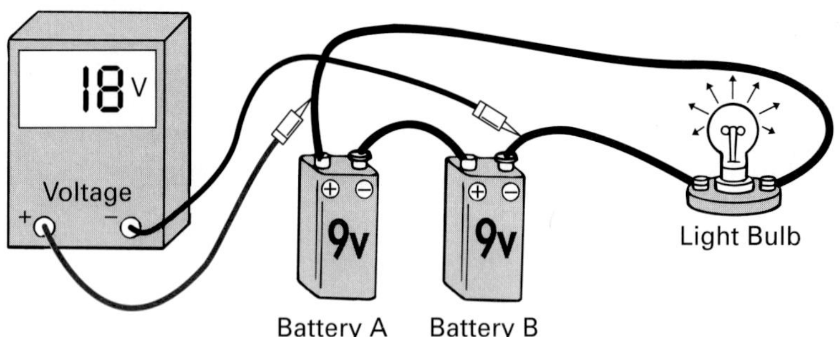

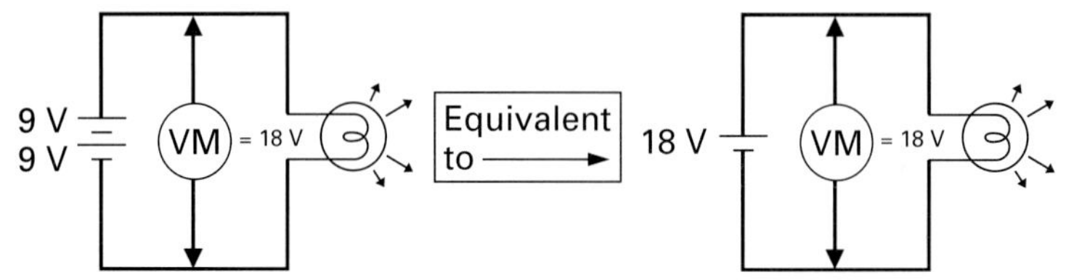

FIGURE 6-27 Batteries in Series.

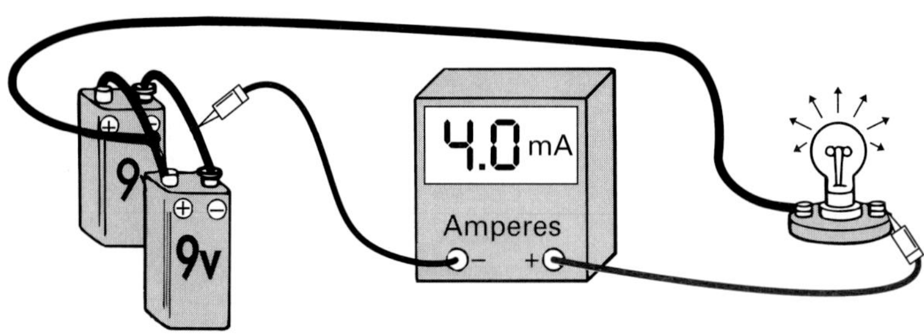

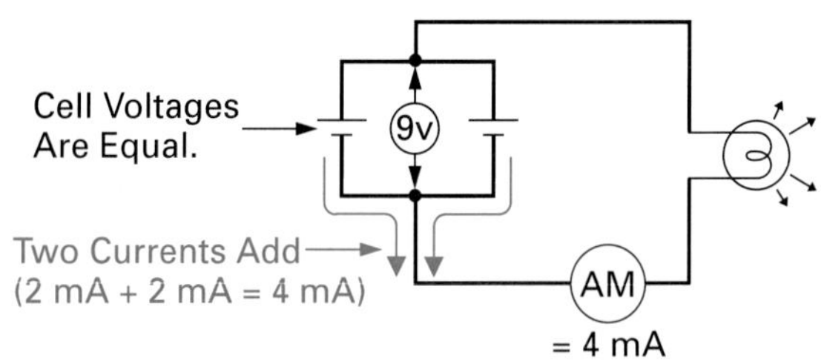

FIGURE 6-28 Batteries in Parallel.

Batteries in Series and Parallel

Batteries are often connected together to gain a higher voltage or current than can be obtained from one cell. Let's first analyze what can be obtained by connecting two cells in series, as shown in Figure 6-27.

When batteries are connected in series, the negative terminal of A is connected to the positive terminal of B. The total voltage across the bulb will be the sum of the two cell voltages, 18 V.

When batteries are connected in parallel, the negative terminal connects to the other negative terminal and the positive terminal connects to the positive. As can be seen in Figure 6-28, the voltage remains the same as for one cell; however, the current demand is now shared, and so the combined parallel batteries are able to supply a higher value of current compared to one battery. Each cell therefore provides half the total current flow through the load.

Internal Resistance

Figure 6-29 illustrates a typical lead–acid secondary cell on the left and its symbol on the right. When a circuit or piece of equipment is connected across the battery, as shown in Figure 6-29(a), the circuit or piece of equipment can be represented by its equivalent value of resistance, called the *load resistance* (R_L). In Figure 6-29(a) the switch is open, so the battery is not loaded and no load cur-

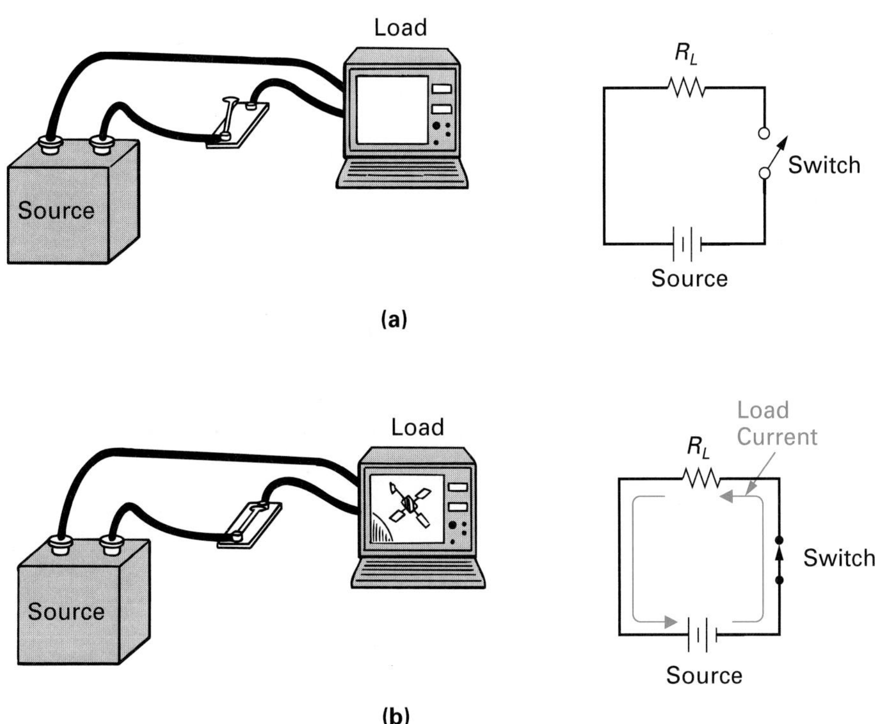

FIGURE 6-29 **Loading a Battery.**

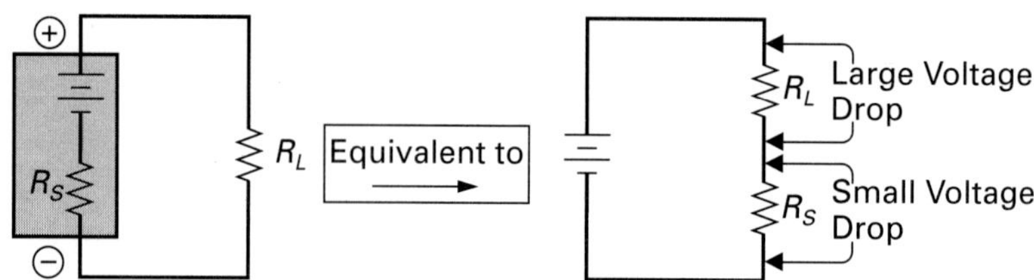

FIGURE 6-30 Source and Load Resistance Divide the Voltage.

rent flows; in Figure 6-29(b) the battery has a completed current path, as the switch is now closed, so a load current will flow, the value of which depends on the load resistance and battery voltage.

The battery just discussed is known as an *ideal voltage source;* however, in reality, there is no such thing as an ideal voltage source. Batteries or any other type of voltage source are not 100% efficient; they all possess some form of internal resistance. This internal resistance is represented by a resistor connected in series with the battery symbol as seen in Figure 6-30. As you can see in the schematic circuit illustrated in this figure, R_L and R_S together form the total resistive load.

Very little voltage will be dropped across R_S, as it is normally small with respect to R_L. Internal inefficiencies of batteries and all voltage sources must always be kept small because a large R_S would drop a greater amount of the voltage, resulting in less output voltage to the load and therefore a waste of electrical power. For example, a lead–acid cell will typically have an internal resistance of 0.01 Ω (10 mΩ).

Maximum Power Transfer
A theorem that states maximum power will be transferred from source to load when the source resistance is equal to the load resistance.

Maximum Power Transfer

Maximum power will be delivered to the load when the resistance of the load (R_L) is equal to the resistance of the source (R_S). The best way to see if this theorem is correct is to apply it to a series of examples and then make a comparison.

EXAMPLE:

Figure 6-31(a) illustrates a 10-V battery with a 5-Ω internal resistance connected across a load. The load in this case is a light bulb, which has a load resistance of 1 Ω.

$$\text{circuit current, } I = \frac{V}{R} = \frac{10\text{ V}}{R_S + R_L} = \frac{10\text{ V}}{6\ \Omega} = 1.66\text{ A}$$

The power supplied to the load is $P = I^2 \times R = 1.66^2 \times 1\ \Omega = 2.8\text{ W}$.

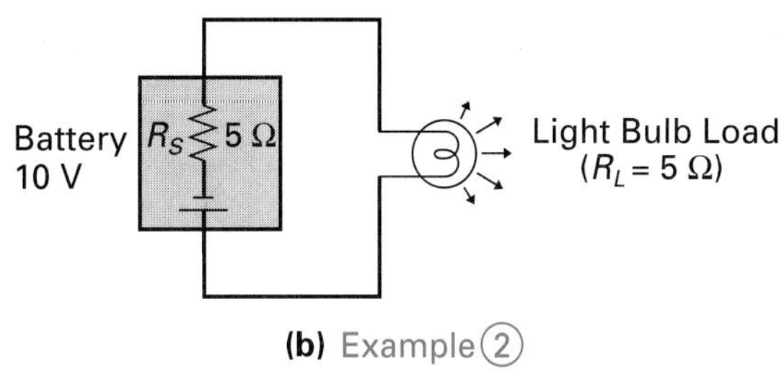

(b) Example ②

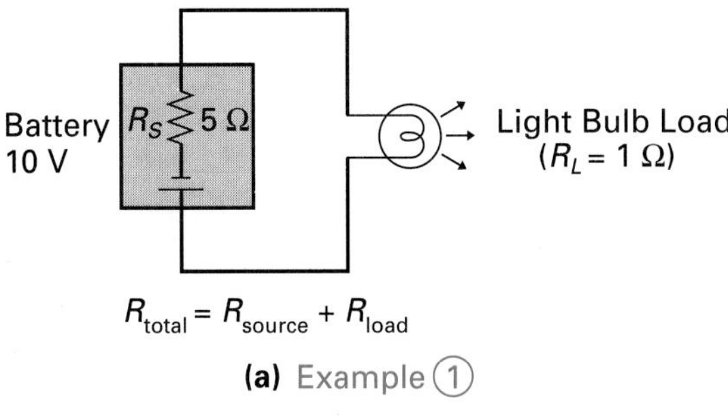

(a) Example ①

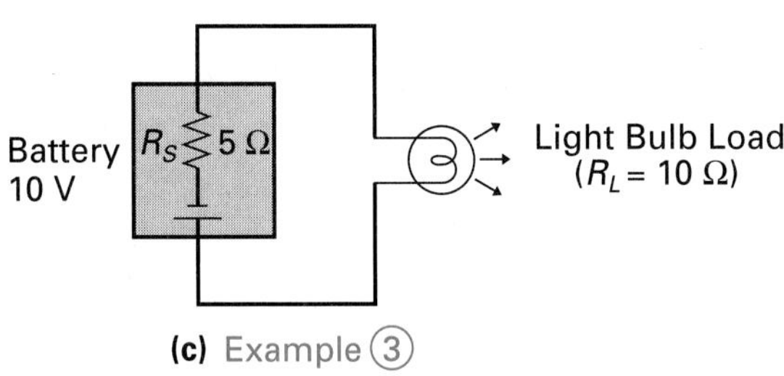

(c) Example ③

Power Supplied to Load (Watts)
5 W
4 W
3 W
2 W
1 W
1 Ω 5 Ω 10 Ω 15 Ω
Different Bulb Load Resistances (R_L)

(d) Curve

FIGURE 6-31 Maximum Power Transfer.

EXAMPLE:

Figure 6-31(b) illustrates the same battery and R_S connected across a 5-Ω light bulb.

$$\begin{aligned} I &= \frac{V}{R} \\ &= \frac{10\text{ V}}{R_S + R_L} \\ &= \frac{10\text{ V}}{10\ \Omega} \\ &= 1\text{ A} \end{aligned}$$

Power supplied is $P = I^2 \times R = 1^2 \times 5\ \Omega = 5\text{ W}$.

EXAMPLE:

Figure 6-31(c) illustrates the same battery again, and in this case a 10-Ω light bulb is connected in the circuit.

$$I = \frac{V}{R}$$
$$= \frac{10 \text{ V}}{R_S + R_L}$$
$$= \frac{10 \text{ V}}{15 \ \Omega}$$
$$= 0.67 \text{ A}$$

Thus $P = I^2 \times R = 0.67^2 \times 10 \ \Omega = 4.5$ W.

As can be seen by the graph in Figure 6-31(d), which plots the power supplied to the load against load resistance, maximum power is delivered to the load (5 W) when the load resistance is equal to the source resistance.

The maximum power transfer condition is only used in special cases such as the automobile starter, where the load resistance remains constant. In most other cases, where load resistance can vary over a range of values, circuits are designed with a load resistance that will cause the best amount of power to be delivered. This is known as **optimum power transfer.**

Optimum Power Transfer
Since the ideal maximum power transfer conditions cannot always be achieved, most designers try to achieve optimum power transfer and have the source resistance and load resistance as close in value as possible.

SELF-TEST REVIEW QUESTIONS FOR SECTION 6-2-5

1. List the three basic components in a battery.
2. List all the various primary cell types.
3. List the various types of secondary cells available.
4. True or false: Two 1.5-V batteries in series would produce a total voltage of 1.5 V.
5. How does a battery's internal resistance relate to the maximum power transfer theorem?
6. Why do batteries have an internal resistance?

6-2-6 *Electrically Generated DC*

Up until now we have only discussed electrical energy in the form of direct current (dc); however, as we will discover in Chapter 11, electrical energy is more readily available in the form of alternating current (ac); in fact, that is what arrives at the wall receptacle at your home and workplace. We can use a dc power supply, like the one seen in Figure 6-32, to convert the ac electrical energy at the wall outlet to dc electrical energy for experiments.

Power in the laboratory or repair shop can be obtained from a battery, but since ac power is so accessible, dc power supplies are used. The power supply has advantages over the battery as a source of voltage in that it can quickly provide an accurate voltage that can easily be varied by a control on the front panel, and it never runs down.

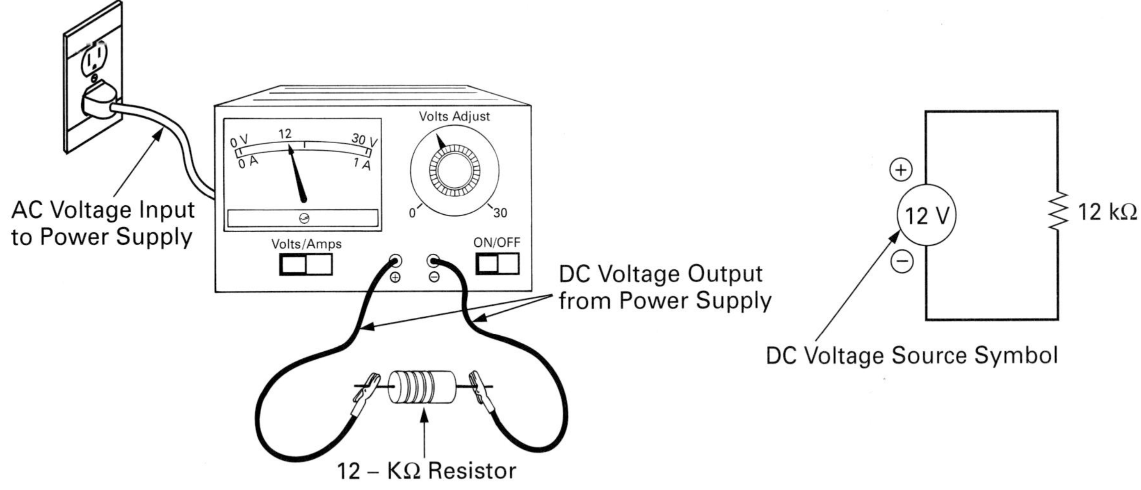

FIGURE 6-32 **Direct Current from Alternating Current.**

During your electronic studies, you will use a power supply in the lab to provide different voltages for various experiments. In Figure 6-32 you can see that the power supply is being used to supply 12 V to a 12 kΩ resistor.

To review the operation of the dc power supply, which was discussed in the previous chapter, the ON/OFF switch is used to turn on or off the power supply. The VOLTS/AMPS switch is used to switch the meter between a voltmeter or an ammeter. In the VOLTS position, the meter indicates whatever voltage has been set by the voltage adjustment control, in this example, 12 V. If the VOLTS/AMPS switch is placed in the AMPS position, the meter will now monitor the amount of current flowing out of the power supply, which in this example is 1 mA.

SELF-TEST REVIEW QUESTIONS FOR SECTION 6-2-6

1. An electric company supplies what form of electrical energy?
2. True or false: A dc power supply converts dc electrical energy into ac.
3. True or false: A volts/amperes switch on a typical power supply will determine whether the output is voltage or current.
4. What advantages does the dc power supply have over a battery?

6-3 EQUIPMENT PROTECTION

Current must be monitored and not be allowed to exceed a safe level so as to protect users from shock, to protect the equipment from damage, and to prevent fire hazards. There are two basic types of protective devices: fuses and circuit breakers. Let's now discuss each of these in more detail.

6-3-1 Fuses

Fuse
This circuit- or equipment-protecting device consists of a short, thin piece of wire that melts and breaks the current path if the current exceeds a rated damaging level.

A **fuse** is a type of metal resistor that consists of a wire link or element of low-melting point inside a casing, as illustrated in Figure 6-33. Fuses have a current rating that indicates the maximum amount of current they will allow to pass through the fuse to the equipment. Figure 6-34, for example, illustrates a battery supplying power to a radio with a 2-A fuse connected between the two. When current passes through the fuse, some of the electrical energy is transformed into heat. If the current through the thin metal element exceeds the current rating of this fuse, in our example 2-A, the excessive current will create enough heat to melt the element and "open" or "blow" the fuse, thus disconnecting the radio and protecting it from damage.

One important point to remember is that if the current increases to a damaging level, the fuse will open and protect the radio. This implies that something went wrong with the battery and it started to supply too much current. This is almost never the case. What actually happens is that, as with all equipment, eventually something internally breaks down. This can cause the overall load resistance of the piece of equipment to increase or decrease. An increase in load resistance ($R_L\uparrow$) means that the battery sees a higher resistance in the current path and therefore a small current would flow from the battery ($I \downarrow$), and the user would be aware of the problem due to the nonoperation of the equipment. If an internal equipment breakdown causes the equipment's load resistance to decrease ($R_L\downarrow$), the battery will see a smaller circuit resistance and will supply a heavier circuit current, ($I\uparrow$), which could severely damage the equipment before the user had time to turn it off. Fuse protection is needed to disconnect the current automatically in this situation, to protect the equipment from damage.

Fuse elements come in various shapes and sizes so as to produce either quick heating and then melting (fast blow) or delayed heating and then melting (slow blow), as illustrated in Figure 6-35. The reason for the variety is based in the application differences. When turned on, some pieces of equipment will be of such low resistance that a short momentary current surge, sometimes in the region of four times the fuse's current rating, will result in the first couple of seconds. If a fast-blow fuse were placed in the circuit, it would blow at the instant the equipment was turned on, even though the equipment did not need to be protected; it merely needs a large amount of current initially to start up. A slow-blow fuse would be ideal in this application, as it would permit the initial heavy current and would begin to heat up and yet not blow due to the delay; then the

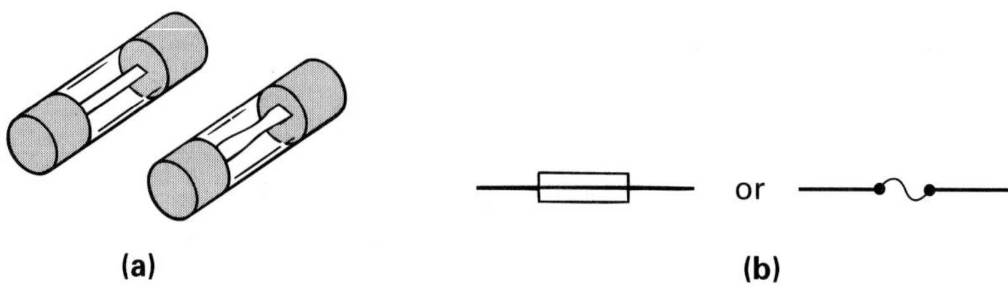

FIGURE 6-33 Fuse. (a) Physical Appearance. (b) Schematic Symbol.

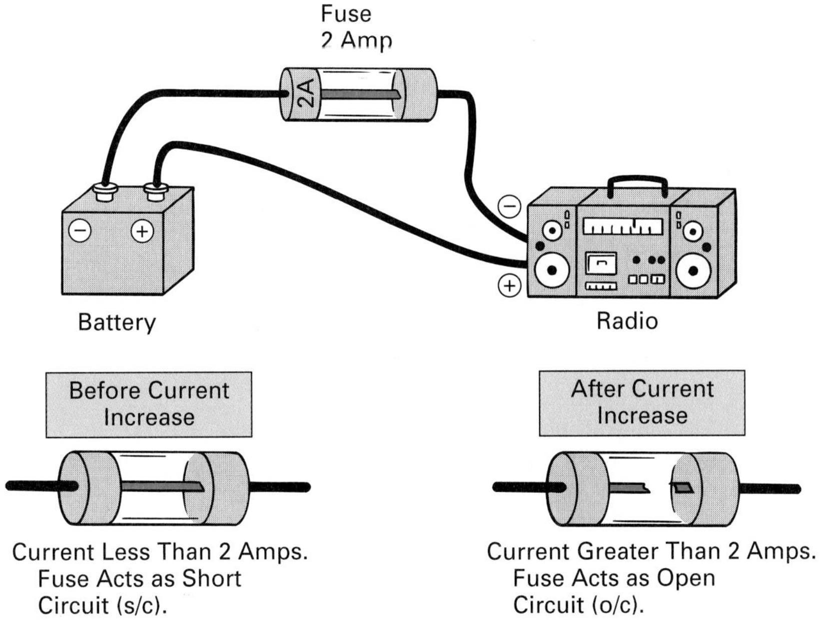

FIGURE 6-34 **Application of a Fuse.**

surge would end, the current would decrease, and the fuse would still be intact (some slow-blow fuses will allow a 400% overload current to flow for a few seconds).

When on, some equipment cannot take any increase in current; otherwise, damage will occur. The slow blow would not be good in this application, as it would allow an increase of current to pass to the equipment for too long a

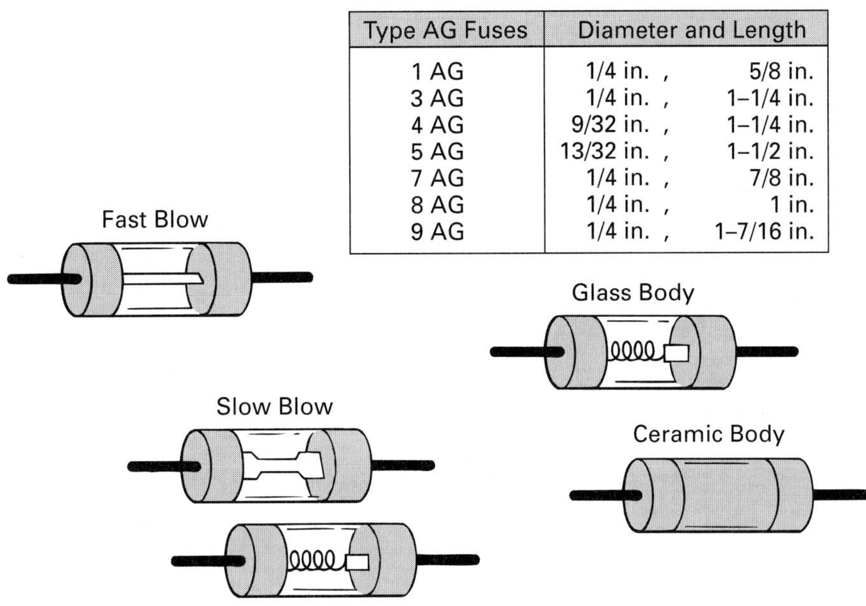

Type AG Fuses	Diameter and Length	
1 AG	1/4 in. ,	5/8 in.
3 AG	1/4 in. ,	1–1/4 in.
4 AG	9/32 in. ,	1–1/4 in.
5 AG	13/32 in. ,	1–1/2 in.
7 AG	1/4 in. ,	7/8 in.
8 AG	1/4 in. ,	1 in.
9 AG	1/4 in. ,	1–7/16 in.

FIGURE 6-35 **Fuse Types, Sizings, and Casings.**

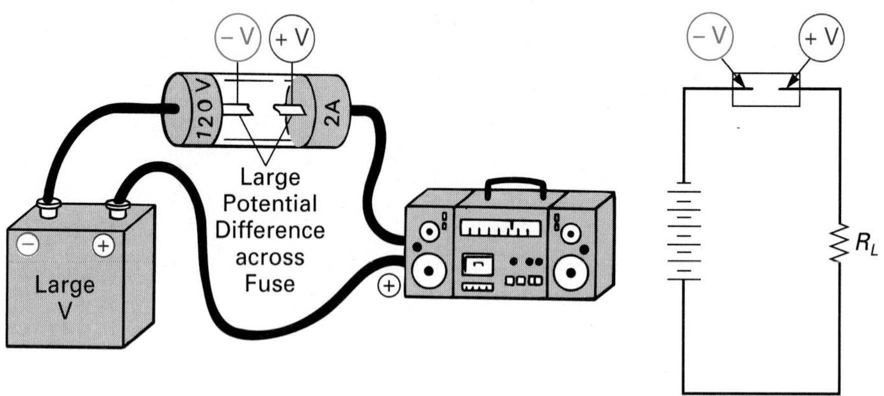

FIGURE 6-36 **Voltage Rating of Fuses.**

period of time, whereas the fast blow would disconnect the current instantly if the current rating of the fuse were exceeded, consequently preventing equipment damage. The automobile was the first application for a fuse in a glass holder, and a size standard named AG (automobile glass) was established (see Figure 6-35).

Fuses also have a voltage rating that indicates the maximum circuit voltage that can be applied across the fuse by the circuit in which the fuse resides. This rating, which is important after the fuse has blown, prevents arcing across the blown fuse contacts, as illustrated in Figure 6-36. Once the fuse has blown, the circuit's positive and negative voltages are now connected across the fuse contacts; and if the voltage is too great, an arc can jump across the gap, causing a sudden surge of current, damaging the equipment connected.

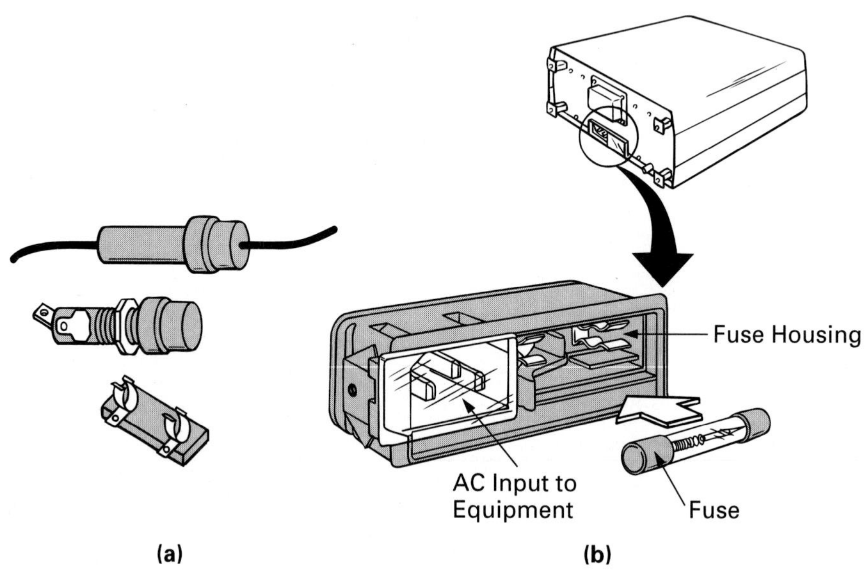

FIGURE 6-37 **Fuse Holders. (a) Holders. (b) Fuse and Holder in Equipment.**

Fuses are mounted within fuse holders and normally placed at the back of the equipment for easy access. Figure 6-37 illustrates fuse holders and how they are installed in a piece of test equipment. When replacing fuses, you must be sure that the power is off, because a potential is across it, and that a fuse with the correct current and voltage rating is used. Since it is merely a piece of wire, a good fuse should have a resistance of 0 Ω; that is, no voltage drop can be measured across it. A blown or burned-out fuse will, when removed, read infinite ohms, and when in its holder and power on, will have the full applied voltage across its two terminals.

6-3-2 *Circuit Breakers*

In many appliances and in a home electrical system, **circuit breakers** are used in place of fuses to prevent damaging current. A circuit breaker can open a current-carrying circuit without damaging itself, then be manually reset and used repeatedly, unlike a fuse, which must be replaced when it blows. To summarize, we can say that a circuit breaker is a reusable fuse. Figure 6-38 illustrates a circuit breaker's symbol, appearance, and typical application.

Circuit Breaker
Reusable fuse. This device will open a current-carrying path without damaging itself once the current value exceeds its maximum current rating.

There are three types of circuit breakers: (1) thermal type, (2) magnetic type, and (3) thermomagnetic type. Let's now discuss each of these in more detail.

Thermal Circuit Breakers

The operation of this type of circuit breaker depends on temperature expansion due to electrical heating. Figure 6-39 illustrates the construction of a thermal circuit breaker. A U-shaped bimetallic (two-metal) strip is attached to the housing of the circuit breaker; it is composed of a layer of brass on one side and a layer of steel on the other. The current arrives at terminal *A*, enters the right side of the bimetallic U-shaped strip, leaves the left side, and travels to the upper contact, which is engaging the bottom contact, where it leaves and exits the circuit breaker from terminal *B*.

If current were to begin to increase beyond the circuit breaker's current rating, heating of the bimetallic strip would occur. As with all metals, heat causes the strip to expand. Some metals expand more than others, and this mea-

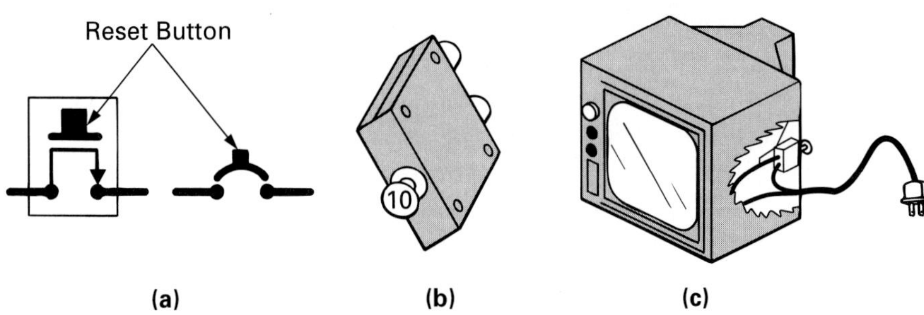

FIGURE 6-38 Circuit Breaker. (a) Symbol. (b) Appearance. (c) Typical Application.

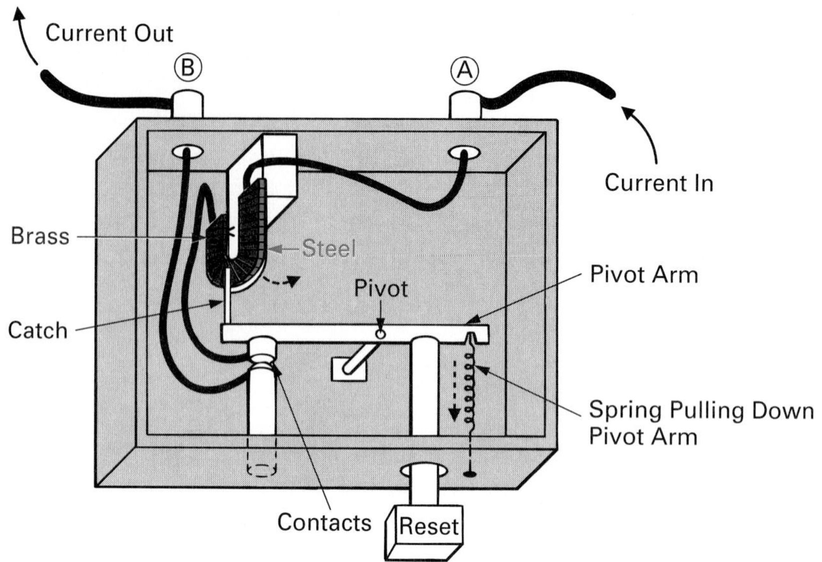

FIGURE 6-39 Thermal-Type Circuit Breaker.

surement is called the *thermal expansion coefficient.* In our situation, brass will expand more than steel, resulting in the lower end of the shaped bimetallic strip bending to the right. This allows a catch to release a pivoted arm and to be pulled down by the tension of a spring. This action lifts the left side of the arm and the attached top contact. The result is to separate the top contact surface from the bottom contact surface, thereby opening the current path and protecting the circuit from the excessive current. This is also known as *tripping* the breaker. The reset button must now be engaged to close the contacts; however, if the problem still exists, the breaker will just trip once more, due to the excessive current.

Magnetic Circuit Breakers

This type of circuit breaker depends on the response of an electromagnet to break the circuit for protection. Figure 6-40 illustrates the construction of a magnetic circuit breaker.

A small level of current flows through the coil of the electromagnet, providing a small amount of magnetic pull on the iron arm, pulling it to the left; however, this magnetic force cannot overcome the pull to the right being generated by the spring *A*. This safe value of current will therefore pass from the *A* terminal, through the coil to the top contact, out of the bottom contact, and then exit the breaker at terminal *B.*

If the current exceeds the current rating of the circuit breaker, an increase in current causes an increase in the current through the coil of the electromagnet, which generates a greater magnetic force on the vertical arm, which pulls the top half of the vertical arm to the left and the lower half below the pivot to the right. This releases the catch holding the horizontal arm, allowing spring *B* to pull the right side of the lower arm down and, consequently, open the contact and disconnect or trip the breaker. The reset button must now be engaged to

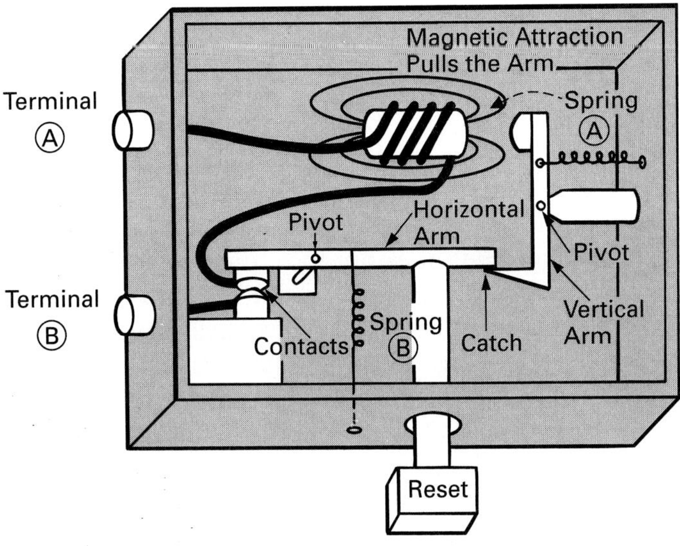

FIGURE 6-40 **Magnetic-Type Circuit Breaker.**

close the contacts; however, if the problem still exists, the breaker will trip once again, as the excessive circuit current still exists.

Thermomagnetic Circuit Breakers

Figure 6-41(a) illustrates the typical home- or residential-type thermomagnetic circuit breaker. The thermal circuit breaker is similar to a slow-blow fuse in that it is ideal for passing momentary surges without tripping because of the delay caused by heating of the bimetallic strip. The magnetic circuit breaker,

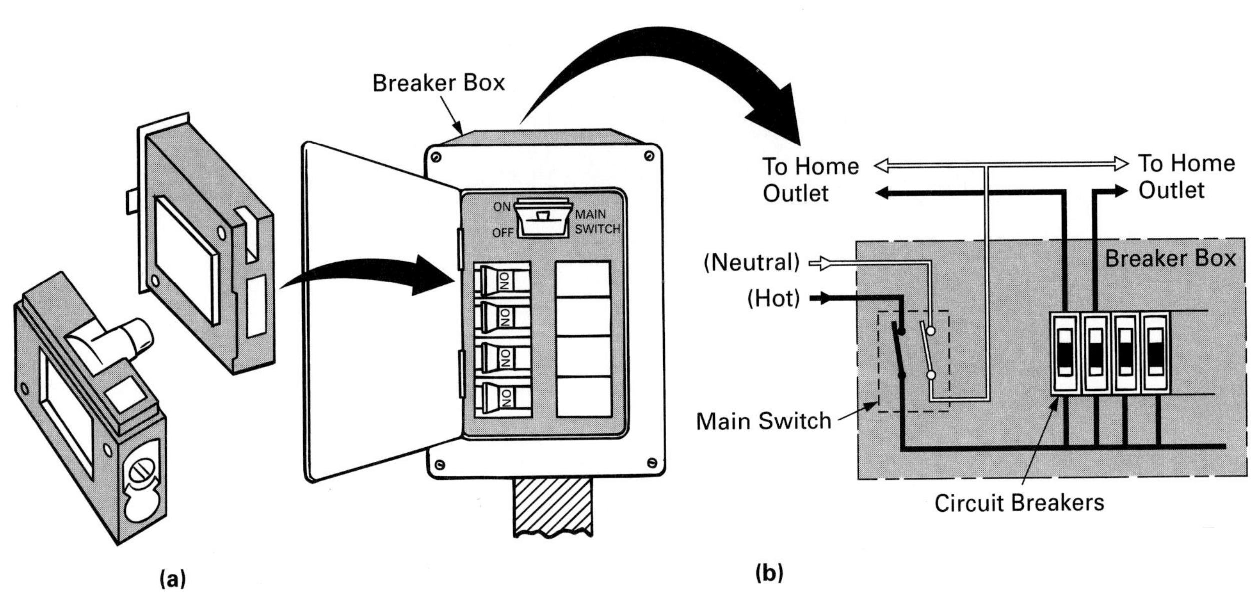

FIGURE 6-41 **Thermomagnetic-Type Circuit Breaker.**

however, is most similar to a fast-blow fuse in that it is tripped immediately when an increase of current occurs.

The thermomagnetic circuit breaker combines the advantages of both previously mentioned circuit breakers by incorporating both a bimetallic strip and an electromagnet as an actuating mechanism, and it operates as follows: For currents at and below the current rating, the circuit breaker is a short circuit and connects the current source to the load. For momentary overcurrent surges, the electromagnet is activated, but does not have enough force to trip the breaker. However, if overcurrent continues, the bimetallic strip will have heated and the combined forces of the bimetallic strip and electromagnet will trip the breaker and disconnect the current source from the load.

For a very large surge of current, the electromagnet receives enough current to trip the breaker independently of the bimetallic strip and disconnect the current source from the load. Figure 6-41(b) shows how a thermomagnetic circuit breaker is used in a typical residential circuit breaker cabinet.

6-4 SWITCHES

A *switch* is a device that completes (closes) or breaks (opens) the path of current, as seen in Figure 6-42. All mechanical switches can be classified into one of eight categories, illustrated in Figure 6-43. Many different variations of these eight different classifications exist. Figure 6-44 illustrates some of the different types on the market; but remember that they are all just basically a switch that merely opens or closes connections.

SELF-TEST REVIEW QUESTIONS FOR SECTIONS 6-3 AND 6-4

1. What do the current and voltage ratings of a fuse indicate?
2. What is the difference between a slow-blow and a fast-blow fuse?
3. List the three types of circuit breakers.
4. List the eight basic types of switches.

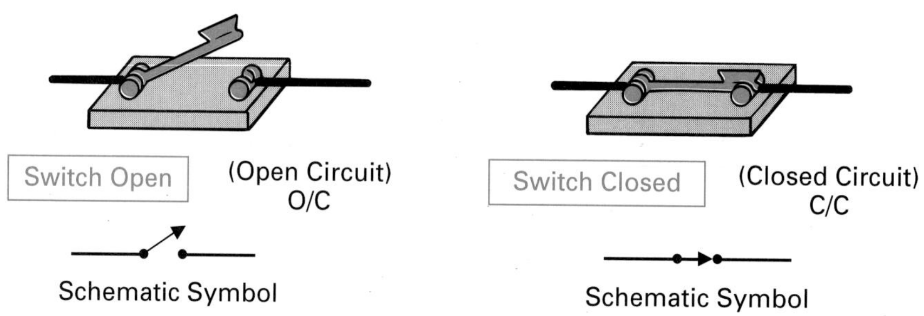

FIGURE 6-42 Open and Closed Switch.

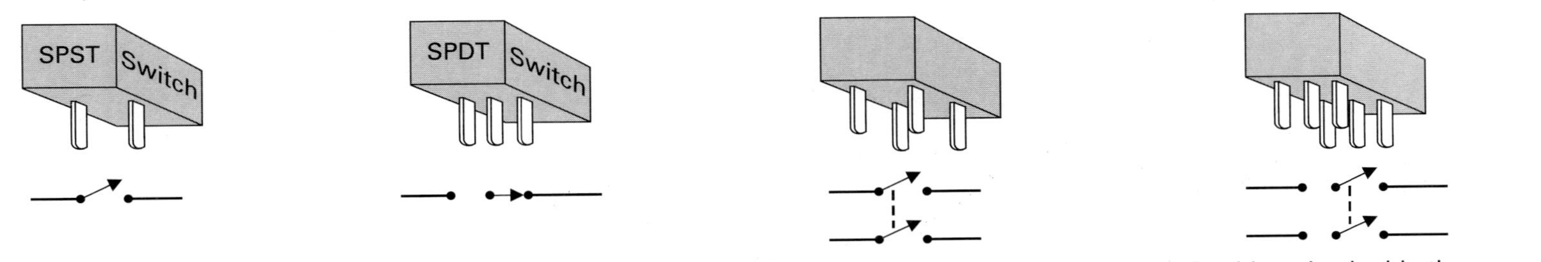

(a) Single-pole, single-throw (SPST). A two-terminal switch with only one pole or moving contact, which can be cast in one direction only (single throw).

(b) Single-pole, double-throw (SPDT). A three-terminal switch for connecting one terminal to either of two other terminals.

(c) Double-pole, single-throw (DPST). A four-terminal switch that is used to connect or disconnect two pairs of terminals simultaneously.

(d) Double-pole, double-throw (DPDT). A switch that has six terminals and is used to connect one pair of terminals to either of the other two pairs.

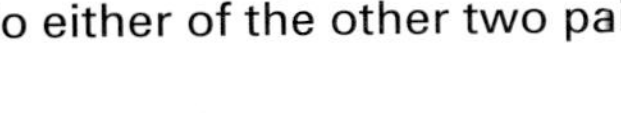

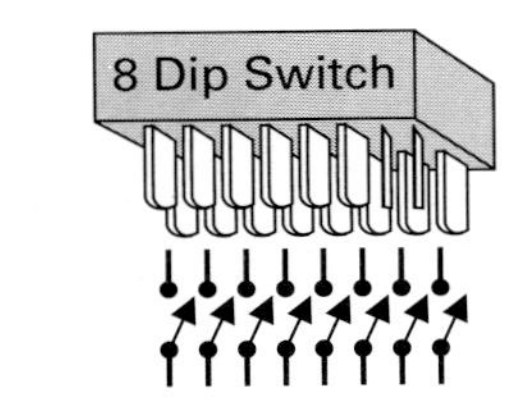

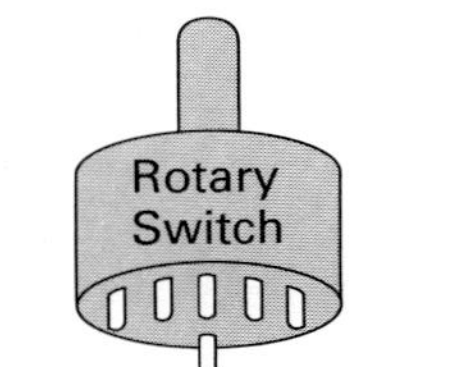

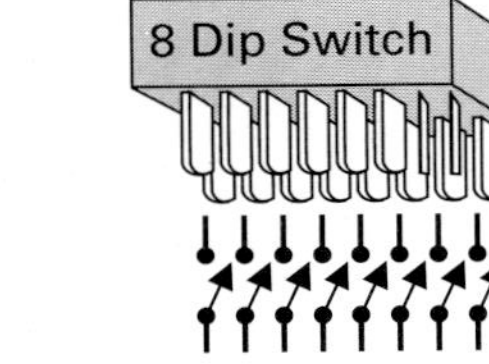

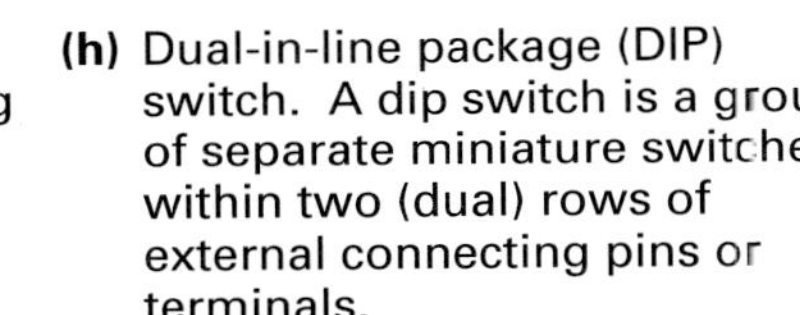

(e) Normally open push button (NOPB). A switch that will make contact and pass current when it is pressed.

(f) Normally closed push button (NCPB). A switch that normally makes contact and passes current, but disconnects or opens when pressed.

(g) Rotary. An electromechanical device that is capable of making (closing contacts) or breaking (opening contacts) in a circuit.

(h) Dual-in-line package (DIP) switch. A dip switch is a group of separate miniature switches within two (dual) rows of external connecting pins or terminals.

Single pole: one moving contact.
Double pole: two moving contacts.
Single throw: pole can be thrown or cast in only one direction.
Double throw: pole can be thrown or cast in two directions.

FIGURE 6-43 **Eight Basic Types of Switches.**

(a)

(b)

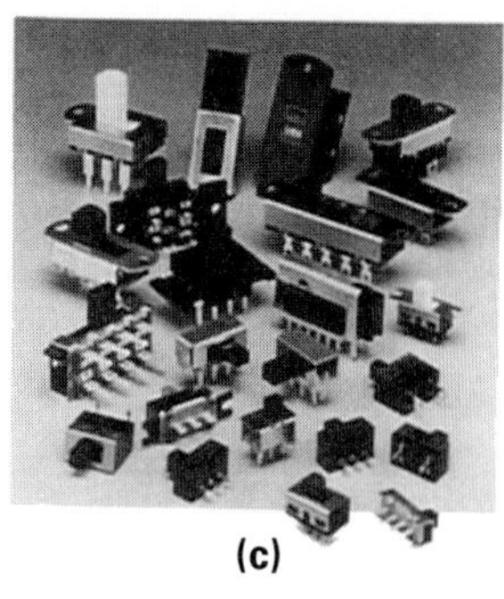
(c)

(d)

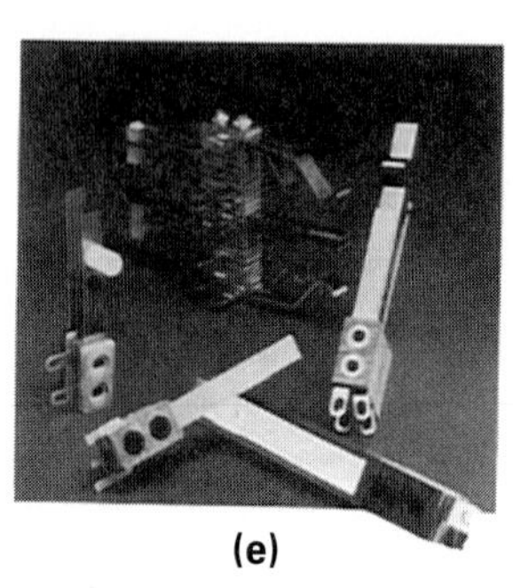
(e)

(f)

(g)

FIGURE 6-44 **Switches. (Courtesy of ITW Switches.)**

SUMMARY

Source and Load

1. The phrase *load resistance* describes the device or equipment's circuit resistance, whereas the phrase *load current* describes the amount of current drawn by the device or equipment.
2. A device that causes a large load current to flow, due to its small load resistance, is called a large or heavy load because it is heavily loading down or working the supply or battery. On the other hand, a device that causes a small load current to flow, due to its large load resistance, is referred to as a small or light load because it is only lightly loading down or working the supply or battery.
3. Load current and load resistance are therefore inversely proportional to one another in that a large load resistance ($R_L\uparrow$) causes a small load current ($I_L\downarrow$), referred to as a light load, and vice versa.
4. The six basic forms of energy are mechanical, heat, light, magnetic, chemical, and electrical. To produce direct current (electrical energy), one of the other forms of energy must be used as a source and then converted or transformed into electrical energy in the form of a dc voltage that can be used to produce direct current.

Mechanically Generated DC

5. Positive and negative charges are known as *static* (stationary) *charges* or *static electricity*. Attraction and repulsion occur with positive and negative charges.
6. Friction is a form of mechanical energy.
7. Electrons in a glass rod (insulator) can be forced out of their orbits and captured by another material, such as wool, by the work of rubbing. A static negative charge now exists in the wool, while a static positive charge exists in the glass. These mechanically generated static charges can be discharged in one of three ways, producing electrical energy.
8. Once a state of equilibrium is achieved, two static charges will be neutralized and no further current will flow. If the connection has a small resistance, the discharge current will be brief and intense; on the other hand, if the connection has a large resistance, the discharge current will be prolonged and less intense.
9. Moving clouds in a storm cause friction between the molecules in the air. The buildup of a natural static charge in a cloud, if large enough, can cause a discharge or release of the electrical energy to the ground.
10. Mechanical energy in the form of pressure can also be used to generate electrical energy.
11. Quartz and Rochelle salts are solid compounds that will produce electricity when a pressure is applied.
12. Quartz is a natural or artificially grown crystal composed of silicon dioxide, from which thin slabs or plates are carefully cut and ground. Rochelle salt is a crystal of sodium potassium tartrate. They are both piezoelectric crystals. Piezoelectric effect is the generation of a voltage between the opposite faces of a crystal as a result of pressure being applied.
13. A voltage is produced between the opposite sides of a crystal (usually a crystalline lead compound) when it is either struck, pressured, or twisted.
14. Piezoelectric effect also describes the reverse of what has just been discussed, as it was discovered that crystals would not only generate a voltage when mechanical force was applied, but would generate mechanical force when a voltage was applied.
15. Piezoresistance of a semiconductor is described as a change in resistance due to a change in the applied strain, and therefore semiconductor resistors can be used as pressure sensors.
16. Unlike quartz, which generates a voltage, the piezoresistive sensor needs an excitation voltage applied, and its value will determine the value of output voltage.

Thermally Generated DC

17. Heat can be used to generate an electric charge.
18. When two dissimilar metals, such as copper and iron, are welded together and heated, an electrical charge is produced. This junction is called a ther-

mocouple, and the heat causes the release of electrons from their parent atoms.

19. The size of the charge is proportional to the temperature difference between the two metals.
20. Thermocouples are normally used to indicate temperature on a display calibrated in degrees.

Optically Generated DC

21. A photovoltaic cell also makes use of two dissimilar metals or semiconductor materials in its transformation of light to electrical energy.
22. A photoelectric cell is also called a solar cell.
23. When the light illuminates the light-sensitive material, a charge is generated, causing current to flow.
24. This phenomenon, known as photovoltaic action, finds application as a light meter for photographic purposes and as an electrical power source for use by the satellite.
25. Electric power companies make use of solar cells to convert sunlight into electrical energy for the consumer.

Magnetically Generated DC

26. In the ancient city of Magnesia in Asia Minor, stones composed of iron oxide were found to have the strange ability to attract and repel one another. It was also noted that when suspended from a string, these stones would always point in only one direction.
27. Historians are not sure whether the Chinese or Europeans first began using these "leading stones," or lodestones as they are now known, for long-distance land and sea travel, but the original compass was probably a lodestone suspended from a string.
28. Later it was found that a metal needle could be magnetized by stroking it across a lodestone, and so the next form of compass consisted of a magnetized needle attached to a piece of wood floating in a bowl of water; hence the origin of the term *needle of a compass.*
29. The lodestone is an example of a natural magnet, unlike the needle, which is called an artificial magnet because it must be artificially magnetized in order to possess a magnetic field, but they both have the ability to point in only one direction by interacting with the earth's natural magnetic field.
30. Magnetism was recorded by the Greeks as early as 800 B.C., but it was not until 1269 that Petrus Peregrinus de Maricout found that the invisible magnetic lines produced by lodestones intersected at two points, which he called *poles.*
31. In 1600, William Gilbert discovered that the magnetic field of a lodestone was lost when it was heated, but reappeared when the lodestone cooled.
32. In 1819, Hans Christian Oersted discovered that any current-carrying conductor generated a magnetic field, and this interrelationship between electricity and magnetism became known as electromagnetism.

33. In 1838, a German mathematician and physicist, Johann Gauss, published a paper on the earth's natural magnetic field, and in honor of his contribution the flux density of a magnetic field is measured in *gauss.*

34. A magnetic flux density of 1 gauss (G) is produced 1 centimeter away from a wire carrying a current of 5 A. To give some examples, the field strength of a magnetized needle may only be a fraction of a gauss, whereas a lodestone 1 inch long may have a strength of 100 G. The earth's magnetic field has a magnetic field strength of 0.3 G at the equator and 0.7 G at either pole.

35. An atom consists of electrons spinning around a nucleus; some atoms have few, while others have many electrons orbiting the atom's nucleus. Each electron is in fact a miniature magnet spinning around the nucleus. An electron at rest will not generate a magnetic field; however, orbiting electrons produce a magnetic field because of their high orbital velocity.

36. Tungsten has 74 electrons (will not magnetize) and iron has 26 electrons (will magnetize). With tungsten, 37 electrons rotate around the nucleus in the clockwise direction, while 37 electrons orbit in the counterclockwise direction. The clockwise magnetic field will cancel with the counterclockwise magnetic field, so the net magnetic field is zero, which is why tungsten cannot be magnetized. Iron has fewer electrons (26); however, 10 of the iron electrons spin clockwise and 16 spin counterclockwise, producing a net counterclockwise magnetic field of 6 electrons, which is why iron can be magnetized.

37. If approximately 100 trillion atoms of the same magnetic polarity are all arranged in one direction, they are called a domain. Domains within unmagnetized materials are all misaligned while aligned domains produce magnetized materials.

38. The magnetic field produced by a magnet is an invisible force concentrated at two opposite points called the magnetic poles. The south pole of a magnet is the pole that would point north toward the earth's north magnetic pole if the magnet were free to rotate.

39. The magnetic field is made up of invisible flux lines that leave the north pole and return at the south.

40. The law of attraction and repulsion could be applied to magnetic poles, just as it can be to positive and negative charges. A north against a north pole and south against a south pole will cause a force of repulsion between the two like poles while a south pole against a north pole will result in a force of attraction between the two unlike poles.

41. There are three types of artificial magnets:

a. A permanent magnet is a piece of hardened steel or other special alloy that when magnetized by the presence of a magnetic field, will retain its magnetism indefinitely (high retentivity), even when the material is removed from the magnetizing force or the magnetizing force is no longer present.

b. A temporary magnet is a piece of soft iron that will be magnetized while in the presence of a magnetizing field, but will be demagnetized the

QUICK REFERENCE SUMMARY SHEET

FIGURE 6-45 Batteries.

Primary Cell Types

Battery type	Features	Cell Voltage (V)	Applications	Construction
Carbon-zinc (also called Leclanché cell in honor of its inventor)	Most popular due to its low cost. Cylindrical D and C cells are most commonly used. Flat cells are stacked in series to obtain voltages greater than 1.5 V, as in the case of the 9 V battery	1.5	Portable radios, tape players, televisions, toys.	Has a zinc case (negative), a carbon center rod (positive), with ammonium chloride as the electrolyte and manganese dioxide as the depolarizing agent. 9 V Flat cell has 6 stacked 1.5 V cells
Alkaline-maganese (also called alkaline cell)	Has three times the shelf life and capacity of carbon zinc. Cylindrical or miniature cell sizes are available	1.4	Portable radios, tape players, televisions, toys; large capacity is worth investment	The negative plate is granular zinc mixed with an electrolyte (alkaline); the positive plate is a polarizer in contact with the outer metal can
Mercury (also called mercuric-oxide)	Higher energy density than preceding two with good shelf life and small size; used in low-power applications. Flat or cylindrical cells are available	1.35 and 1.4	Watches, hearing aids, pacemakers, cameras, test equipment	A zinc anode and a mercuric oxide cathode in a potassium hydroxide electrolyte
Silver-oxide	High capacity; however, the material used (silver) makes it the most costly; can supply high currents for short periods of time; used in low-power applications	1.5	Watches, hearing aids, pacemakers, cameras	Contains a cathode of silver oxide, an alkaline electrolyte, and zinc anode
Lithium	High discharge rate and long shelf life; light weight and higher output voltage; the familiar cylindrical and flat cell are available	1.9	Liquid crystal watches, semi-conductor memories, hand-held calculators, sensor circuits	Contains a lithium anode, porous carbon cathode, and sulfur dioxide electrolyte

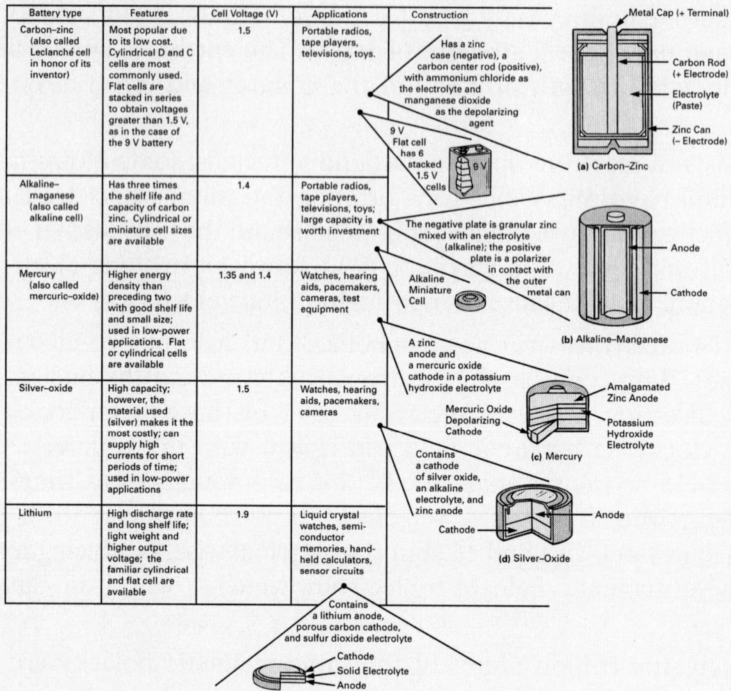

Secondary Cell Types

Battery type	Features	Applications	Construction
Lead-acid (lead cell)	High cycle (discharge-charge) life and high current capacity; 2.1 V per cell; must be operated upright if a wet lead-acid. Lead-acid gell cells can be used in any position	*Gell Cell Batteries* televisions, recorders, robots, alarms, and tools *Wet Cell Batteries* Starting source for automobiles and power for robotics equipment	The electrodes are lead oxide immersed in an electrolyte of dilute sulfuric acid Wet-type lead-acid battery with 6 internal 2.1 V cells making 12.6 V car battery
Nickel-cadmium (Ni-Cd)	High capacity and high cost; three times discharge current of lead-acid for same amp-hour rating; can be sealed and operated in any position, so ideal for drills and other such portable equipment; 1.2 V per cell. Nickel cadmium batteries suffer from "memory effect." Short discharge and charge cycles cause the batteries capacity to drop, because the cell appears to "remember" the lower capacity.	*Ni-Cd Gell Cell* televisions, radios, toys, recorders, shavers, toothbrushes *Ni-Cd Wet Cell Batteries* Starter source for jet engines (resembles lead-acid car battery in appearance). Robotics equipment	The positive plate is made of cadmium, the negative plate of nickel hydroxide, and the electrolyte used is potassium hydroxide
Nickel-metal Hydride (Ni-MH)	Discharge characteristics are very similar to those of the nickel-cadmium cell. The cell's nominal voltage is 1.2 V, and they do not suffer from memory effect.	Biggest applications include cellular phones, portable computers, camcorders, and other portable electronic systems	The positive plate is made of nickel oxyhydride, the negative plate of metal hydride, and the electrolyte used is potassium hydroxide

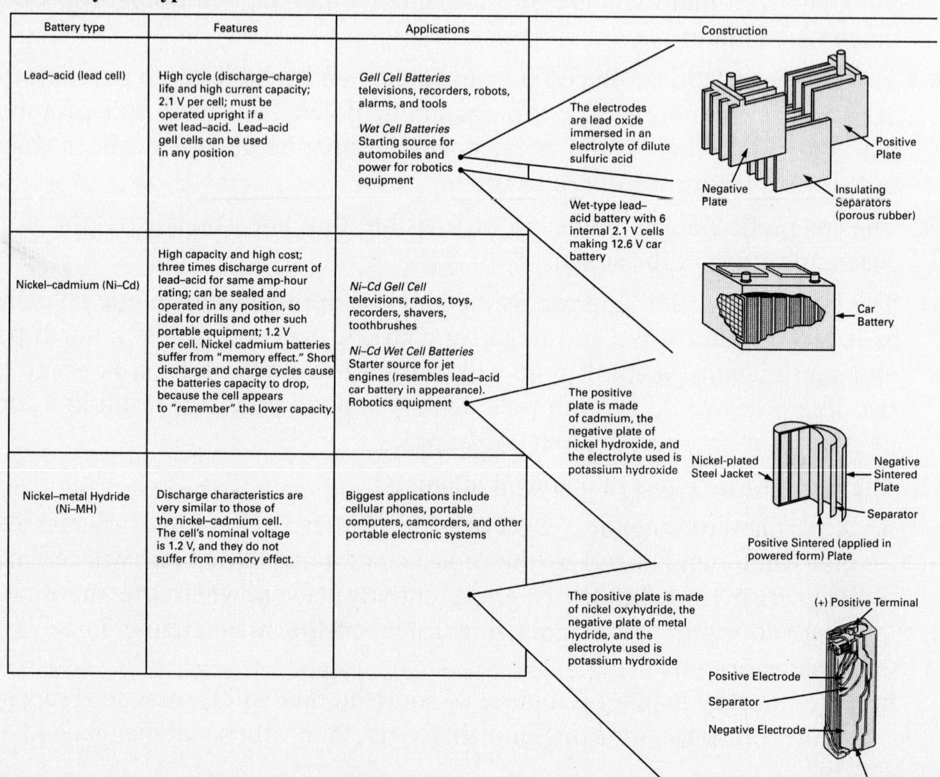

QUICK REFERENCE SUMMARY SHEET

FIGURE 6–45 (continued).

Identical cells

(a) In series:
$V_T = V_1 + V_2 + V_3 + \ldots$
(Total voltage is equal to the sum of all the cell voltages.)

(b) In parallel:
$V_T = V_1$ or V_2 or V_3 or . . .
(Total voltage is the same as for one cell voltage.)

moment the soft iron is taken away from the magnetizing force or the magnetizing force is no longer present. Soft iron possesses low retentivity which means it will not retain a magnetic field.

c. An electromagnet is a temporary magnet that consists of an insulated wire wrapped around a soft-iron cylinder (core) to form a coil. A magnetic field is produced only while current flows through the coil. If a larger current is passed through the coil, a corresponding stronger magnetic field is generated.

42. A magnet can be used to generate electrical current by movement of the magnet past a conductor. When the magnet is stationary, no current flows through the circuit, so the current meter indicates this zero current. If the magnet is moved so that the magnetic lines of force cross the wire conductor, a voltage is induced in the conductor.

43. To achieve an energy conversion from magnetic to electric, you can either move the magnet past the wire or move the wire past the magnet.

44. To produce a continuous supply of electric current, the magnet or wire must be constantly in motion. The device that achieves this is the dc electric generator.

Chemically Generated DC (Figure 6-45)

45. The chemical voltage source (the battery) is called the voltaic cell. Its principle of operation was discovered by Alessandro Volta, an Italian physicist, in 1800.

46. A battery has three basic components: (a) a negative plate, (b) a positive plate, and (c) an electrolyte.

47. Two dissimilar, separated metal plates such as copper and zinc are placed within a container that is filled with a liquid known as the electrolyte, usually an acid. A chemical reaction causes electrons to be repelled from one plate and attracted to the other, passing through the electrolyte. A large number of electrons collect on one plate (negative plate), while an absence or deficiency of electrons exists on the opposite, positive plate.

48. The electrolyte acts on the two plates and transforms chemical energy into electrical energy, which can be taken from the cell at its two output terminals as an electrical current flow.

49. The mutual repulsion of the free electrons at the negative electrode combined with the attraction of the positive electrode results in a migration of free electrons (current flow) through the light bulb, causing its filament to produce light.
50. The chemical reaction within the cell will actually dissolve the negative plate until eventually it will be eaten away completely.
51. These types of batteries are known as primary (first time, last time) cells because they discharge once and must then be discarded. Almost all primary cells have their electrolyte in paste form and are therefore referred to as dry cells, as opposed to wet cells, whose electrolyte is in liquid form.
52. The shelf life of a battery is the length of time a battery can remain on the shelf (in storage) and still retain its usability.
53. Of all the types of primary cells, five dominate the market as far as applications and sales are concerned. These are the carbon–zinc, alkaline–manganese, mercury, silver oxide, and lithium.
54. The cell voltage describes the voltage produced by one cell (one set of plates). If two or more sets of plates are installed in one package, the component is called a battery.
55. Secondary cells operate on the same basic principle as that of primary cells. However, in this case the plates are not eaten away or dissolved; they only undergo a chemical change during discharge. Once discharged, the secondary cell can have the chemical change that occurred during discharge reversed by recharging, resulting once more in a fully charged cell.
56. Restoring a secondary cell to the charged condition is achieved by sending a current through the cell in the direction opposite to that of the discharge current.
57. During the recharging process, we use a battery charger, which reverses the chemical process of discharge by forcing electrons back into the cell and restoring the battery to its charged condition. The battery charger voltage is normally set to about 115% of the battery voltage.
58. Secondary cells are often referred to as wet cells because the electrolyte is not normally in paste form (dry) as in the primary cell, but is in liquid (wet) form and is free to move and flow within the container. The lead–acid and nickel–cadmium are the two most popular. The two other types of secondary cells are the silver–zinc and silver–cadmium.
59. Capacity (C) is measured by the amount of ampere-hours (Ah) a battery can supply during discharge.
60. Ampere-hour units are actually specifying the coulombs of charge in the battery.
61. Batteries are often connected together to gain a higher voltage or current than can be obtained from one cell.
62. When batteries are connected in series, the negative terminal of battery A is connected to the positive terminal of battery B. The total voltage across the bulb will be the sum of the two cell voltages.

63. When batteries are connected in parallel, the negative terminal connects to the other negative terminal and the positive terminal connects to the positive. The voltage remains the same as for one cell; however, the current demand can now be shared.
64. Batteries or any other type of voltage source are not 100% efficient; they all possess some form of internal resistance. This internal resistance is represented by a resistor connected in series with the battery symbol.
65. Internal inefficiencies of batteries and all voltage sources must always be kept small because a large R_S would drop a greater amount of the voltage, resulting in less output voltage to the load and therefore a waste of electrical power.
66. Maximum power will be delivered to the load when the resistance of the load (R_L) is equal to the resistance of the source (R_S).
67. The maximum power transfer condition is only used in special cases such as the automobile starter, where the load resistance remains constant. In most other cases, where load resistance can vary over a range of values, circuits are designed with a load resistance that will cause the best amount of power to be delivered. This is known as optimum power transfer.

Electrically Generated DC

68. We can use ac electrical energy to generate dc electrical energy by use of a piece of equipment called a dc power supply.
69. The power supply has advantages over the battery as a source of voltage in that it can quickly provide an accurate voltage that can easily be varied by a control on the front panel, and it never runs down.

Fuses (Figure 6-46)

70. Current must be monitored and not be allowed to exceed a safe level so as to protect users from shock, to protect the equipment from damage, and to prevent fire hazards.
71. There are two basic types of protective devices: fuses and circuit breakers.
72. A fuse is a type of metal resistor that consists of a wire link or element of low-melting point inside a casing. Fuses have a current rating that indicates the maximum amount of current they will allow to pass through the fuse to the equipment.
73. When current passes through the fuse, some of the electrical energy is transformed into heat. If the current through the thin metal element exceeds the current rating of this fuse, the excessive current will create enough heat to melt the element and "open" or "blow" the fuse, thus disconnecting the equipment and protecting it from damage.
74. An increase in load resistance ($R_L\uparrow$) means that the battery sees a higher resistance in the current path and therefore a small current would flow from the battery ($I\downarrow$), and the user would be aware of the problem due to the nonoperation of the equipment.

QUICK REFERENCE SUMMARY SHEET

FIGURE 6-46 Fuses.

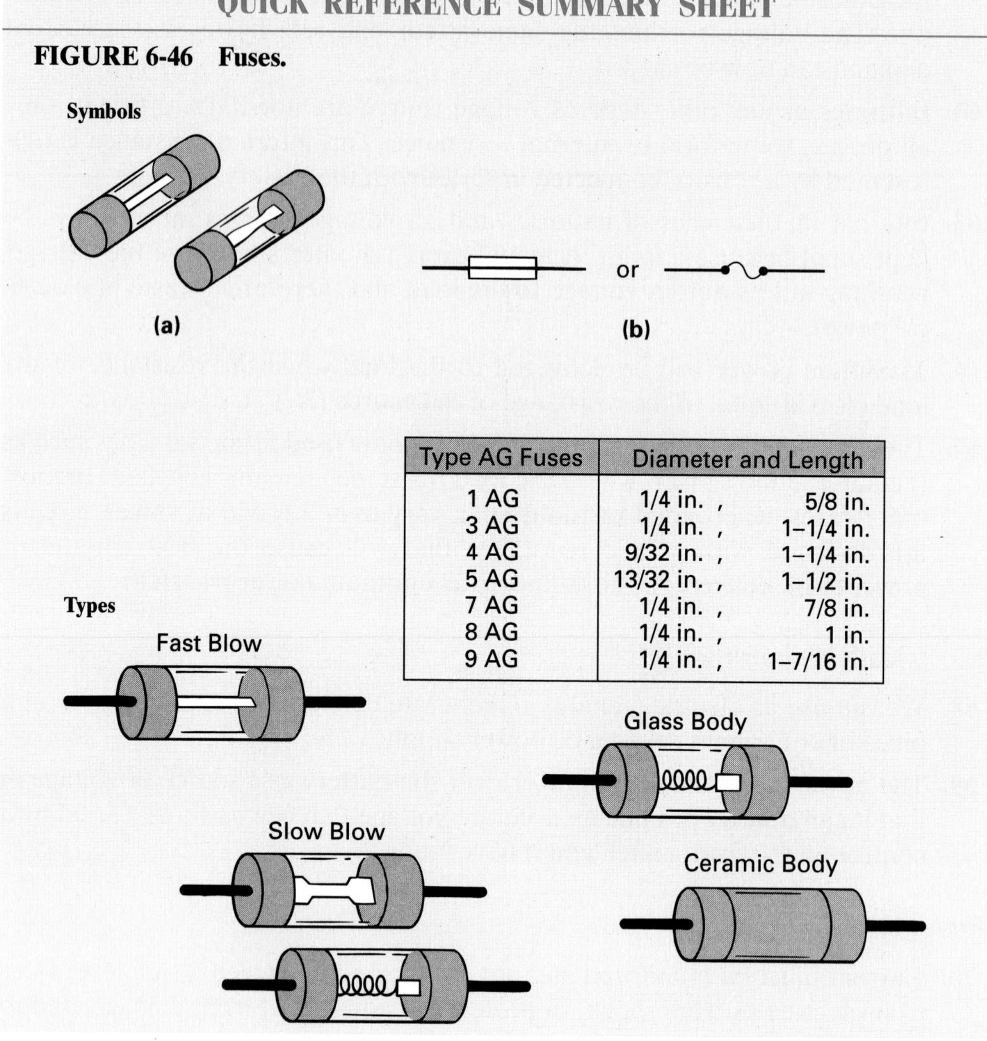

Type AG Fuses	Diameter and Length	
1 AG	1/4 in. ,	5/8 in.
3 AG	1/4 in. ,	1–1/4 in.
4 AG	9/32 in. ,	1–1/4 in.
5 AG	13/32 in. ,	1–1/2 in.
7 AG	1/4 in. ,	7/8 in.
8 AG	1/4 in. ,	1 in.
9 AG	1/4 in. ,	1–7/16 in.

75. If an internal equipment breakdown causes the equipment's load resistance to decrease ($R_L\downarrow$), the battery will see a smaller circuit resistance and will supply a heavier circuit current, ($I\uparrow$), which could severely damage the equipment before the user had time to turn it off. Fuse protection is needed to disconnect the current automatically in this situation, to protect the equipment from damage.

76. Fuse elements come in various shapes and sizes so as to produce either quick heating and then melting (fast blow) or delayed heating and then melting (slow blow).

77. Fuses also have a voltage rating that indicates the maximum circuit voltage that can be applied across the fuse by the circuit in which the fuse resides. This rating, which is important after the fuse has blown, prevents arcing across the blown fuse contacts.

78. Fuses are mounted within fuse holders and normally placed at the back of the equipment for easy access.

79. When replacing fuses, you must be sure that the power is off, because a potential is across it, and that a fuse with the correct current and voltage rating is used.

Circuit Breakers

80. In many appliances and in a home electrical system, circuit breakers are used in place of fuses to prevent damaging current.

81. A circuit breaker can open a current-carrying circuit without damaging itself, then be manually reset and used repeatedly, unlike a fuse, which must be replaced when it blows. There are three types of circuit breakers: (a) thermal type, (b) magnetic type, and (c) thermomagnetic type.

82. The operation of thermal circuit breakers depends on temperature expansion due to electrical heating. If current were to begin to increase beyond the circuit breaker's current rating, heating of the bimetallic strip would occur. The result is to separate the top contact surface from the bottom contact surface, thereby opening the current path and protecting the circuit from the excessive current. This is also known as *tripping* the breaker. The reset button must now be engaged to close the contacts.

83. Magnetic circuit breakers depend on the response of an electromagnet to break the circuit for protection. If the current exceeds the current rating of the circuit breaker, an increase in current causes an increase in the current through the coil of the electromagnet, which generates a greater magnetic force on the vertical arm, which opens the contact and disconnects or trips the breaker. The reset button must now be engaged to close the contacts.

84. The thermal circuit breaker is similar to a slow-blow fuse in that it is ideal for passing momentary surges without tripping because of the delay caused by heating of the bimetallic strip.

85. The magnetic circuit breaker, however, is most similar to a fast-blow fuse in that it is tripped immediately when an increase of current occurs.

86. The thermomagnetic circuit breaker combines the advantages of both previously mentioned circuit breakers by incorporating both a bimetallic strip and an electromagnet as an actuating mechanism. For currents at and below the current rating, the circuit breaker is a short circuit and connects the current source to the load. For momentary overcurrent surges, the electromagnet is activated, but does not have enough force to trip the breaker. However, if overcurrent continues, the bimetallic strip will have heated and the combined forces of the bimetallic strip and electromagnet will trip the breaker and disconnect the current source from the load. For a very large surge of current, the electromagnet receives enough current to trip the breaker independently of the bimetallic strip and disconnect the current source from the load.

FIGURE 6-47 Switches.

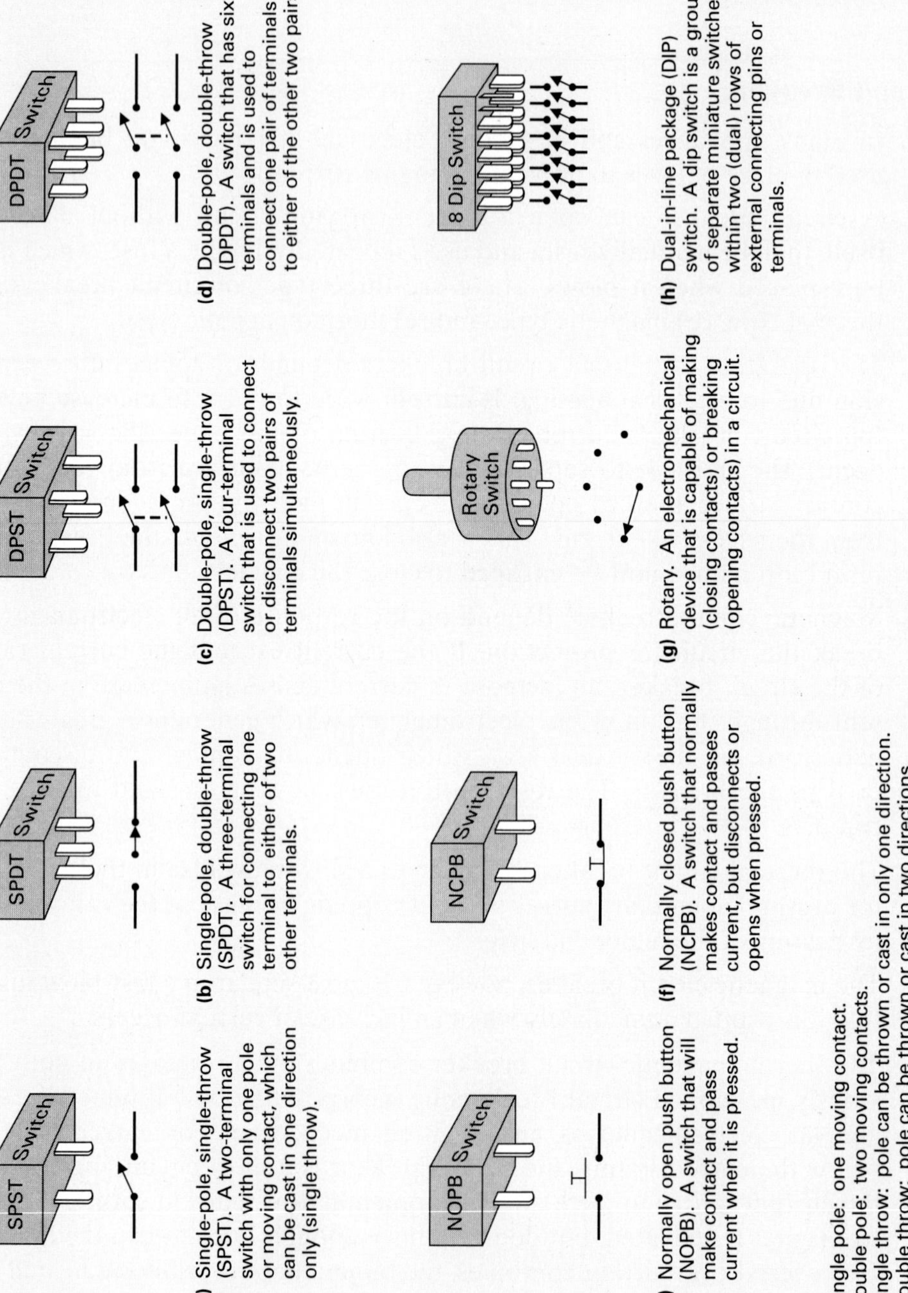

(a) Single-pole, single-throw (SPST). A two-terminal switch with only one pole or moving contact, which can be cast in one direction only (single throw).

(b) Single-pole, double-throw (SPDT). A three-terminal switch for connecting one terminal to either of two other terminals.

(c) Double-pole, single-throw (DPST). A four-terminal switch that is used to connect or disconnect two pairs of terminals simultaneously.

(d) Double-pole, double-throw (DPDT). A switch that has six terminals and is used to connect one pair of terminals to either of the other two pairs.

(e) Normally open push button (NOPB). A switch that will make contact and pass current when it is pressed.

(f) Normally closed push button (NCPB). A switch that normally makes contact and passes current, but disconnects or opens when pressed.

(g) Rotary. An electromechanical device that is capable of making (closing contacts) or breaking (opening contacts) in a circuit.

(h) Dual-in-line package (DIP) switch. A dip switch is a group of separate miniature switches within two (dual) rows of external connecting pins or terminals.

Single pole: one moving contact.
Double pole: two moving contacts.
Single throw: pole can be thrown or cast in only one direction.
Double throw: pole can be thrown or cast in two directions.

Switches (Figure 6-47)

87. A *switch* is a device that completes (closes) or breaks (opens) the path of current. All mechanical switches can be classified into one of eight categories, although many different variations of these eight different classifications exist.

NEW TERMS

Alkaline–manganese cell
Alternating current (ac)
Ampere-hour
Ampere-second
Artificial magnet
Battery
Battery charger
Blown fuse
Capacity
Carbon–zinc cell
Cell
Circuit breaker
Coil
Core
Dc generator
Depolarizer
Direct current (dc)
Discharge
Domain
Double-pole, double-throw (DPDT) switch
Double-pole, single-throw (DPST) switch
Dry cell
Dual in-line package (DIP) switch
Electrolyte
Electromagnet
Electromagnetism
Fast-blow fuse
Flux line
Friction
Fuse
Gauss
Generator
Lead–acid cell
Lithium cell
Load
Load current
Load resistance
Lodestone
Magnetic flux
Magnetic poles
Magnetic-type circuit breaker
Maximum power transfer
Mercuric oxide cell
Natural magnet
Nickel–cadmium cell
Normally closed pushbutton (NCPB) switch
Normally open push-button (NOPB) switch
Permanent magnet
Photocell
Photovoltaic action
Piezoelectric effect
Piezoresistance
Pole
Power supply
Pressure
Primary cell
Quartz
Rating
Retentivity
Rochelle salts
Rotary switch
Secondary cell
Shelf life
Silver–cadmium cell
Silver–oxide cell
Silver–zinc cell
Single-pole, double-throw (SPDT) switch
Single-pole, single-throw (SPST) switch
Slow-blow fuse
Solar cell
Solar panels
Solenoid
Source
Source current
Source resistance

Source voltage
Static charge
Supply
Supply voltage
Switch
Temporary magnet
Thermal-type circuit breaker
Thermocouple
Thermomagnetic-type circuit breaker
Voltaic cell
Wet cell

REVIEW QUESTIONS

Multiple-Choice Questions

1. Direct current is:

a. A reversing of current continually in a circuit
b. A flow of current in only one direction
c. Produced by an ac voltage
d. None of the above

2. If load resistance were doubled, load current will ____________ .

a. Halve **b.** Double **c.** Triple **d.** Remain the same

3. Quartz is a solid compound that will produce electricity when ____________ is applied.

a. Friction **b.** An electrolyte **c.** A magnetic field **d.** Pressure

4. Direct current is generated thermally by use of a:

a. Crystal **b.** Thermocouple **c.** Thermistor **d.** None of the above

5. An application of a photovoltaic cell would be:

a. A satellite power source
b. To turn on and off security lights
c. Both (a) and (b)
d. None of the above

6. The atomic theory of magnetism states that every ____________ in motion is a miniature magnet.

a. Electron **b.** Proton **c.** Neutron **d.** Photon

7. Magnetic flux lines ____________ the north pole and ____________ at the south pole.

a. Arrive, leave
b. Leave, arrive
c. Are never present, always present
d. None of the above

8. Like magnetic poles ____________, while unlike magnetic poles ____________ .

a. Attract, repel **b.** Repel, attract **c.** Repel, repel **d.** Attract, attract

9. Of the three artificial magnets, the ____________ holds its magnetic force even after the magnetizing force has been removed.

a. Permanent magnet
b. Temporary magnet
c. Electromagnet
d. Magnetite

10. Which type of battery can be used for only one discharge (and cannot be rejuvenated)?

a. A lead–acid cell
b. A secondary cell
c. A nickel–cadmium cell
d. A primary cell

11. A battery with a capacity of 12 ampere-hours could supply:

a. 12 A for 1 hour **c.** 6 A for 2 hours
b. 3 A for 4 hours **d.** All of the above

12. Batteries are normally connected in series to obtain a higher total:

a. Voltage **c.** Resistance
b. Current **d.** None of the above

13. The dc power supply converts:

a. ac to dc **c.** High dc to a low dc
b. dc to ac **d.** Low ac to a high ac

14. A fuse's current rating states the:

a. Maximum amount of current allowed to pass
b. Minimum amount of current allowed to pass
c. Maximum permissible circuit voltage
d. None of the above

15. A thermal-type circuit breaker is equivalent to a ____________ fuse, whereas a magnetic-type circuit breaker is equivalent to a ____________ fuse.

a. Slow-blow, slow-blow **c.** Slow-blow, fast-blow
b. Fast-blow, slow-blow **d.** Glass case, ceramic case

16. The type of circuit breaker found in the home is the:

a. Thermal type **c.** Magnetic type
b. Thermomagnetic type **d.** Carbon–zinc type

17. A single-pole, double-throw switch would have ____________ terminals.

a. Two **b.** Three **c.** Four **d.** Five

18. Maximum power will be transferred from the source to the load when the source resistance is equal to:

a. 25 Ω **b.** Load current **c.** Load resistance **d.** Source voltage

19. A switch will very simply ____________ (open) or ____________ (close) a path of current.

a. Break, make **b.** make, break **c.** s/c, o/c **d.** o/c, break

20. With switches, the word *pole* describes a:

a. Stationary contact **c.** Path of current
b. Moving contact **d.** None of the above

Essay Questions

21. Describe briefly how direct current can be generated:

a. Mechanically with pressure and friction (6-2-1) **d.** Magnetically (6-2-4)
b. Thermally (6-2-2) **e.** Chemically (6-2-5)
c. Optically (6-2-3) **f.** Electrically (6-2-6)

22. List the three artificial magnets (6-2-4)
23. Briefly describe the operation of a battery. (6-2-5)
24. Simply state the difference between a primary and a secondary cell. (6-2-5)

25. Describe what can be gained by connecting batteries in series and parallel. (6-2-5)
26. Describe a fuse's: (6-3-1)
    a. Current rating **b.** Voltage rating
27. Briefly describe the operation of the following types of circuit breakers: (6-3-2)
    a. Thermal type **b.** Magnetic type
28. List the eight different switch classifications. (6-4)
29. List the four popular primary cells. (6-2-5)
30. List the two popular secondary cells. (6-2-5)

Practice Problems

31. If a 50 W light bulb is connected across a 120 V source and the following fuses are available, which should be used to protect the circuit? (6-3-1)
    a. 0.5 A/120 V (fast blow) **c.** 1 A/12 V (slow blow)
    b. ¾ A/120 V (fast blow) **d.** 0.5 A/120 V (slow blow)
32. Draw a diagram and indicate the total source voltage of: (6-2-5)
    a. Six 1.5-V cells connected in series with one another
    b. Six 1.5-V cells connected in parallel with one another
33. Draw a diagram and indicate the polarities of a 12 V lead–acid battery being charged by a 15 V battery charger. If this battery is rated at 150 Ah: (6-2-5)
    a. How many coulombs of charge will be stored in the fully charged condition?
    b. How long will it take the battery to charge if a charging current of 5 A is flowing?
34. Show how to connect two 12 V batteries to increase (a) the voltage and (b) the current. (6-2-5)
35. If a 6 V nickel–cadmium battery discharges at a rate of 3.5 A in 4 hours:
    a. Calculate its ampere-hour and ampere-second rating.
    b. How long will it take the battery to discharge if a current of 2 A is drawn? (6-2-5)

CHAPTER

7

OBJECTIVES

After completing this chapter, you will be able to:

1. Describe a series circuit.
2. Identify series circuits.
3. Connect components so that they are in series with one another.
4. Describe why current remains the same throughout a series circuit.
5. Explain how to calculate total resistance in a series circuit.
6. Explain how Ohm's law can be applied to calculate current, voltage, and resistance.
7. Describe why the series circuit is known as a voltage divider.
8. Describe a fixed- and a variable-voltage divider.
9. Explain how to calculate power in a series circuit.
10. Describe how to troubleshoot and recognize:
    - **a.** An open component
    - **b.** A component value variation
    - **c.** A short circuit in a series circuit

Series DC Circuits

OUTLINE

VIGNETTE

Hitler's Computer Mistake

In the early days of World War II, German scientist Konrad Zuse, who designed and built the first general-purpose computer, proposed constructing a computer that would operate 1000 times faster than anything else at that time. He was going to redesign his Z3 computer, which was being used at that time to solve engineering problems in aircraft and missile design, and include vacuum tubes instead of electromechanical relay switches. This proposal was rejected by Hitler, who was not interested in this long-term, two-year project, as he was sure that the war was going to be, for him, a certain, quick victory. Due to Hitler's shortsightedness, this powerful computer, which could have been used to break British communication codes, was never developed. However, unknown to both Hitler and Zuse, the British code-breaking computer project, called Ultra, had highest priority and was moving rapidly toward completion.

INTRODUCTION

A series circuit, by definition, is the connecting of components end to end in a circuit to provide a single path for the current. This is true not only for resistors, but also for other components that can be connected in series; in all cases the components are connected in succession or strung together one after another so that only one path for current exists between the negative (–) and positive (+) terminals of the supply.

7-1 COMPONENTS IN SERIES

Series Circuit
Circuit in which the components are connected end to end so that current has only one path to follow throughout the circuit.

Figure 7-1 illustrates five examples of **series** resistive **circuits.** In all five examples, you will notice that the resistors are connected "in-line" with one another so that the current through the first resistor must pass through the second resistor, and the current through the second resistor must pass through the third, and so on.

EXAMPLE:

In Figure 7-2(a), seven resistors are laid out on a table top. With connecting wire, string all the resistors in series, starting at $R1$, and proceeding in numerical order through the resistors until reaching $R7$.

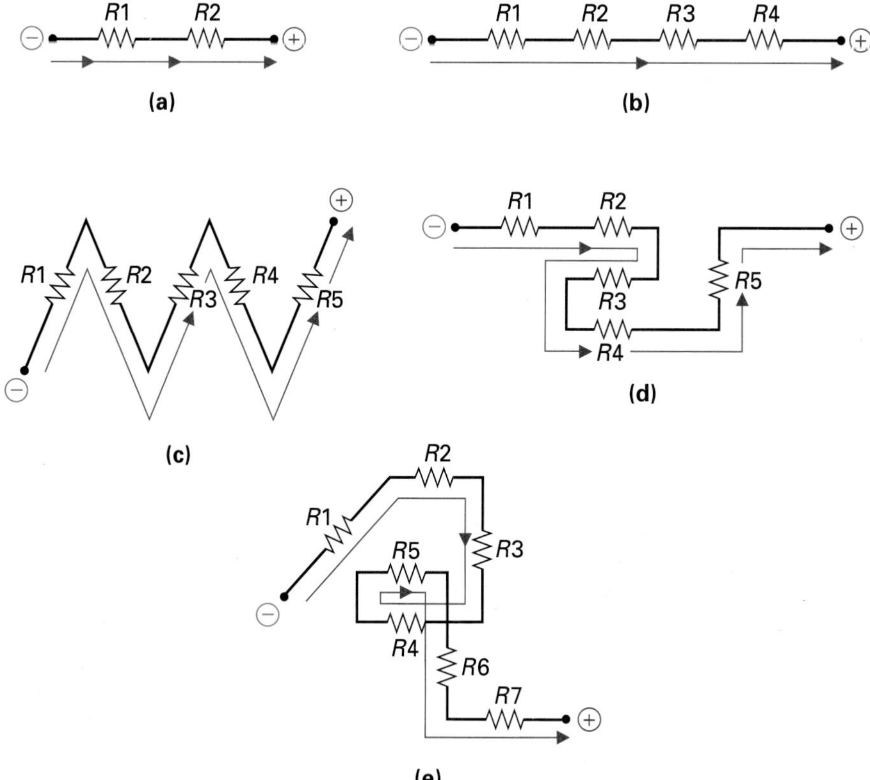

FIGURE 7-1 Five Series Resistive Circuits.

Solution:

In Figure 7-2(b), you can see that all the resistors are now connected in series, and the current has only one path to follow from negative to positive.

EXAMPLE:

Figure 7-3(a) shows four 1.5 V cells and three lamps. Using wires, connect all of the cells in series to create a 6 V battery source. Then connect all of the three lamps in series with one another, and finally, connect the 6 V battery source across the three-series-connected-lamp load.

Solution:

In Figure 7-3(b) you can see the final circuit containing a source, made up of four series-connected 1.5 V cells, and a load, consisting of three series-connected lamps. As explained in Chapter 6, when cells are connected in series, the total voltage (V_T) will be equal to the sum of all the cell voltages:

$$V_T = V_1 + V_2 + V_3 + V_4 = 1.5\text{ V} + 1.5\text{ V} + 1.5\text{ V} + 1.5\text{ V} = 6\text{ V}$$

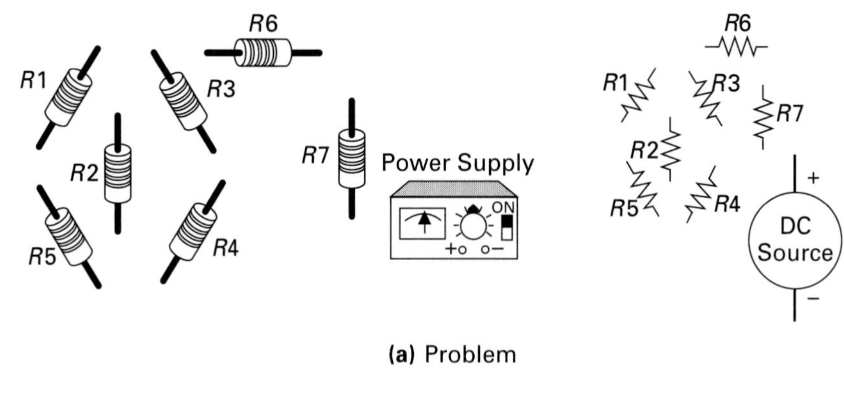

(a) Problem

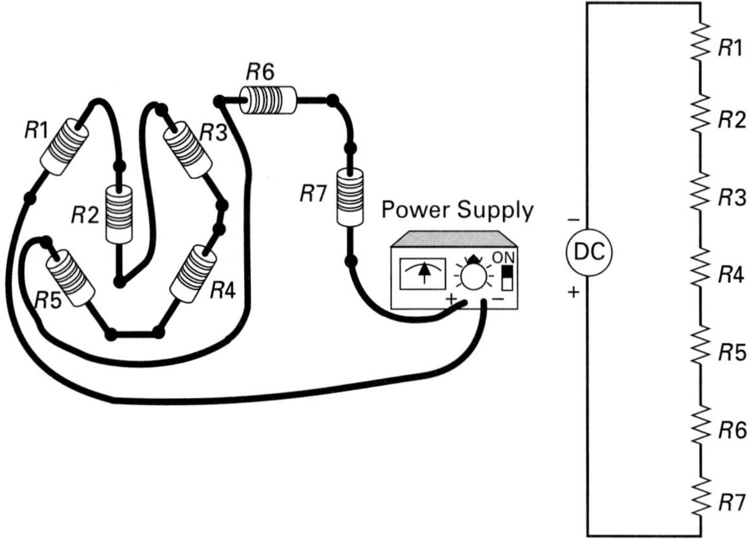

(b) Solution

FIGURE 7-2 Connecting Resistors in Series. (a) Problem. (b) Solution.

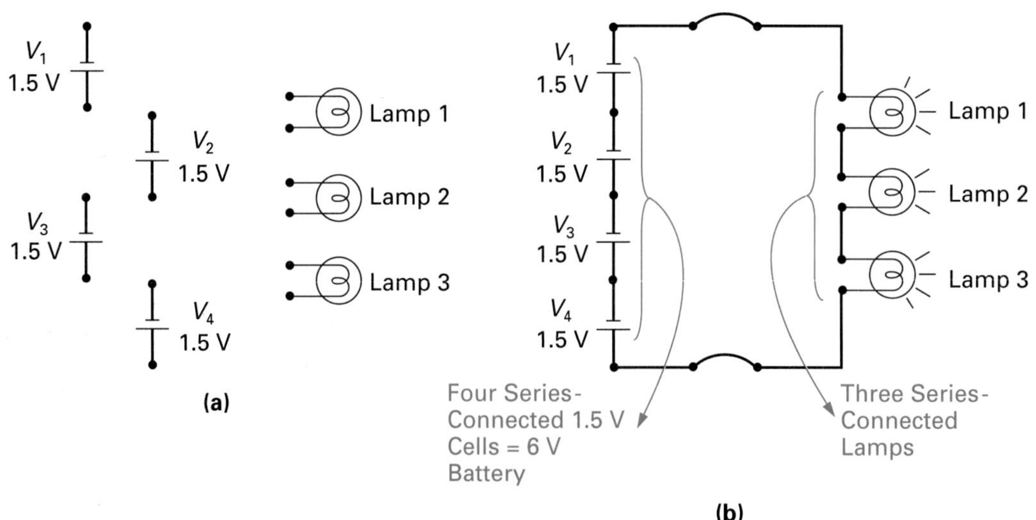

FIGURE 7-3 Series-Connected Cells and Lamps.

7-2 CURRENT IN A SERIES CIRCUIT

The current in a series circuit has only one path to follow and cannot divert in any other direction; consequently, the current through a series circuit is the same throughout that circuit.

Returning once again to the water analogy, you can see in Figure 7-4(a) that, if 2 gallons of water per second are being supplied by the pump, 2 gallons per second must be pulled into the pump; and if the rate at which water is leaving and arriving at the pump is the same, 2 gallons of water per second must be flowing throughout the circuit. The same value of water flow exists throughout a series-connected fluid system. If the values were adjusted to double the opposition to flow, then half, or 1 gallon per second, would be leaving the pump; but that same value of 1 gallon/second would be flowing throughout the system.

Similarly, with the electronic series circuit, shown in Figure 7-4(b), there is a total of 2 A leaving and 2 A arriving at the battery, so the same value of current exists throughout the series-connected electronic circuit. If the resistance of the circuit is doubled, then half, or 1 A of current, will leave the battery, but that same value of 1 A will flow throughout the entire circuit. Current is, therefore, exactly the same value at every point in a series circuit. This can be stated mathematically as

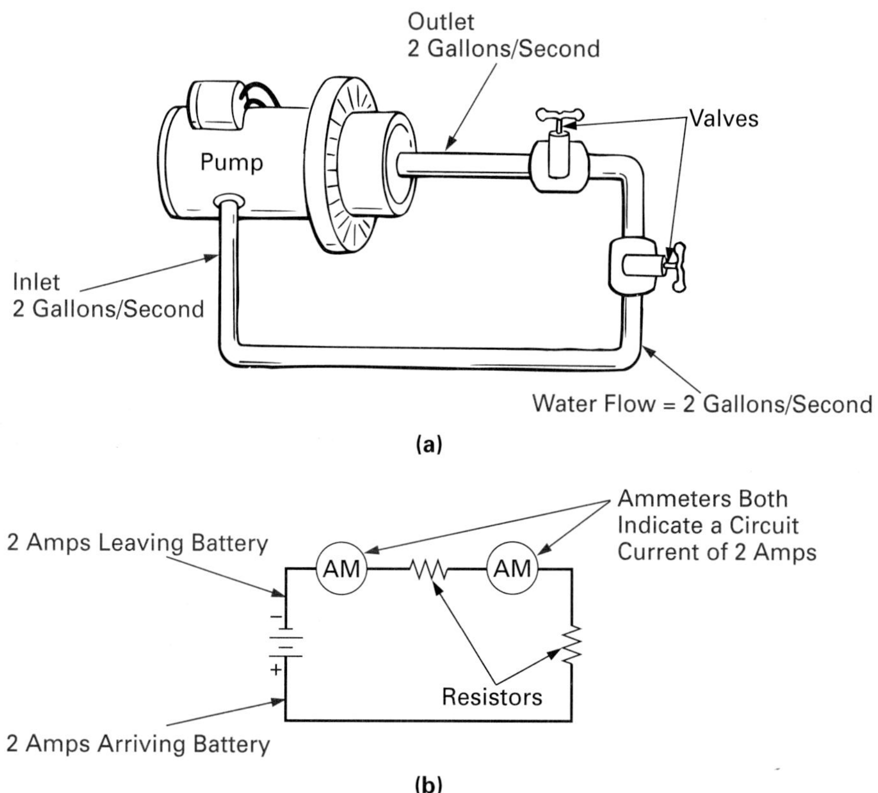

FIGURE 7-4 Series Circuit Current. (a) Fluid System. (b) Electric System.

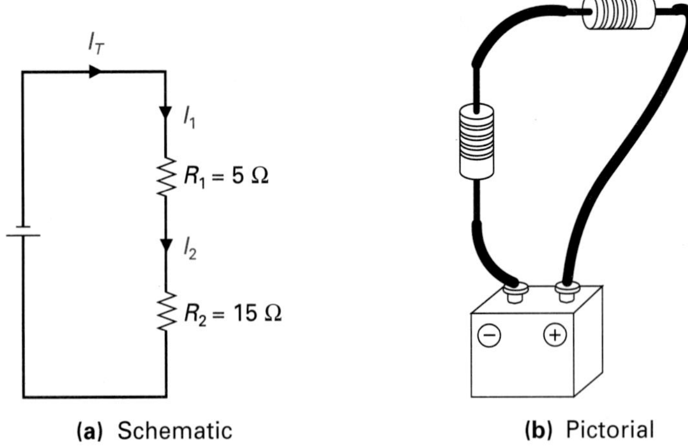

FIGURE 7-5 **(a) Schematic. (b) Pictorial.**

$$I_T = I_1 = I_2 = I_3 = \cdots$$

(Total current = current through R_1 = current through R_2 = current through R_3, and so on).

EXAMPLE:

In Figure 7-5, a total current (I_T) of 1 A is flowing out of a battery, through two end-to-end resistors R_1 and R_2, and returning to the battery. Calculate:

a. The current through R_1 (I_1)
b. The current through R_2 (I_2)

Solution:

Since R_1 and R_2 are connected in series, the current through both will be the same as the circuit current, which is equal to 1 A.

$$I_T = I_1 = I_2$$
$$1\ A = 1\ A = 1\ A$$

SELF-TEST REVIEW QUESTIONS FOR SECTIONS 7-1 AND 7-2

1. What is a series circuit?
2. What is the current flow through each of eight series-connected 8-Ω resistors if 8 A total current is flowing out of a battery?

7-3 RESISTANCE IN A SERIES CIRCUIT

Resistance is the opposition to current flow, and in a series circuit every resistor in series offers opposition to the current flow. In the water analogy of Figure 7-4, the total resistance or opposition to water flow is the sum of the two individual valve opposition values. Like the battery, the pump senses the total opposition in the circuit offered by all the valves or resistors, and the amount of current that flows is dependent on this resistance or opposition.

The total resistance in a series-connected electronic resistive circuit is thus equal to the sum of all the individual resistances, as shown in Figure 7-6(a)

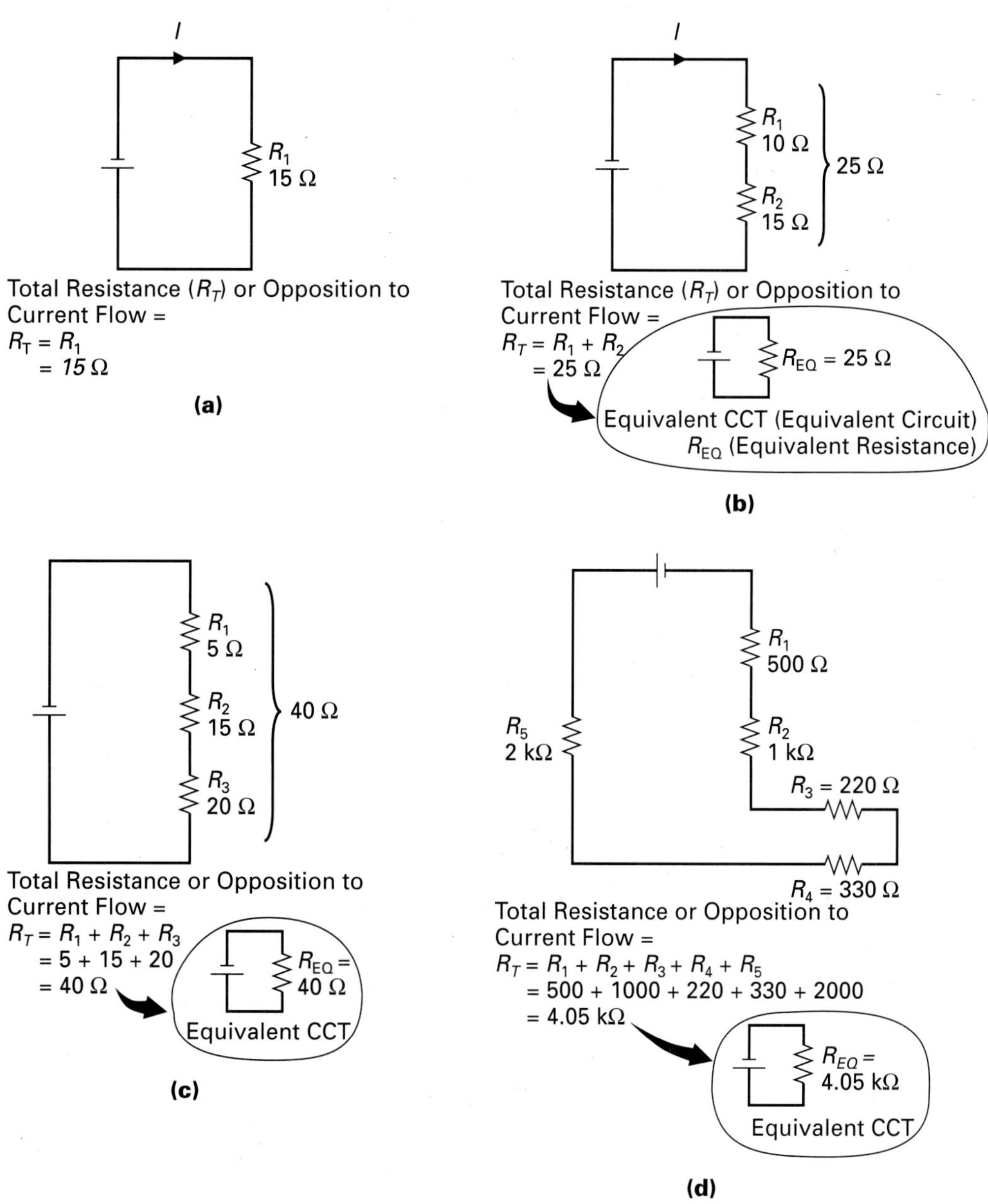

FIGURE 7-6 Total or Equivalent Resistance.

through (d). An equivalent circuit could therefore be drawn for each of the circuits in Figure 7-6(b), (c), and (d) with one resistor, of a value equal to the sum of all the series resistance values.

No matter how many resistors are connected in series, the total resistance or opposition to current flow is always equal to the sum of all the resistor values. This formula can be stated mathematically as

$$R_T = R_1 + R_2 + R_3 + \cdots$$

(Total resistance = value of R_1 + value of R_2 + value of R_3, and so on.)

Equivalent Resistance (R_{eq}) Total resistance of all the individual resistances in a circuit.

Total resistance (R_T) is the only opposition the battery can sense; it does not see the individual separate resistors, but one **equivalent resistance.** Based on its voltage and this total resistance, a value of current will be produced to flow through the circuit (Ohm's law, $I = V/R$).

EXAMPLE:

Referring to Figure 7-7, calculate

a. The circuit's total resistance
b. The current flowing through R_2

Solution:

a. $R_T = R_1 + R_2 + R_3 + R_4$
$= 25\ \Omega + 20\ \Omega + 33\ \Omega + 10\ \Omega$
$= 88\ \Omega$

b. $I_T = I_1 = I_2 = I_3 = I_4$. Therefore, $I_2 = I_T = 3$ A.

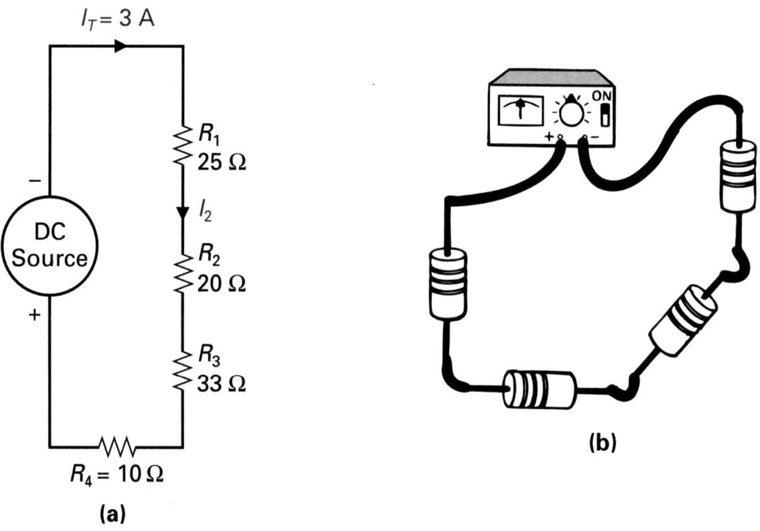

FIGURE 7-7 (a) Schematic. (b) Pictorial.

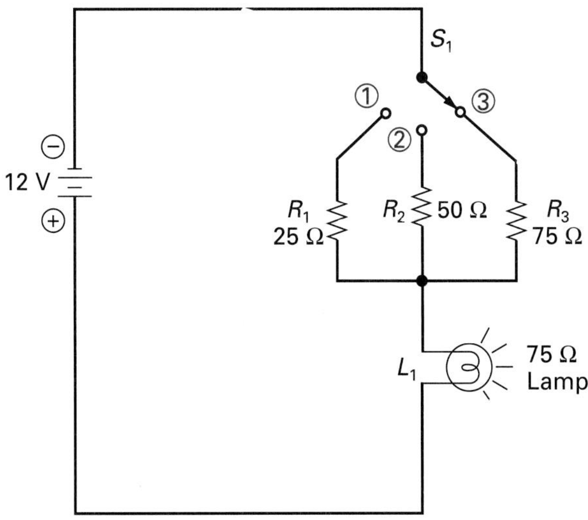

FIGURE 7-8 Three-Position Switch Controlling Lamp Brightness.

EXAMPLE:

Figure 7-8 shows how a single-pole three-position switch is being used to provide three different lamp brightness levels. In position ① R_1 is placed in series with the lamp, in position ② R_2 is placed in series with the lamp, and in position ③ R_3 is placed in series with the lamp. If the lamp has a resistance of 75 Ω, calculate the three values of current for each switch position.

Solution:

$$\text{Position 1: } R_T = R_1 + R_{\text{lamp}}$$

$$= 25\ \Omega + 75\ \Omega = 100\ \Omega$$

$$I_T = \frac{V_T}{R_T} + \frac{12\text{ V}}{100\ \Omega} = 120\text{ mA}$$

$$\text{Position 2: } R_T = R_2 + R_{\text{lamp}} = 50\ \Omega + 75\ \Omega = 125\ \Omega$$

$$I_T = \frac{V_T}{R_T} = \frac{12\text{ V}}{125\ \Omega} = 96\text{ mA}$$

$$\text{Position 3: } R_T = R_3 + R_{\text{lamp}} = 75\ \Omega + 75\ \Omega = 150\ \Omega$$

$$I_T = \frac{V_T}{R_T} = \frac{12\text{ V}}{150\ \Omega} = 80\text{ mA}$$

SELF-TEST REVIEW QUESTIONS FOR SECTION 7-3

1. State the total resistance formula for a series circuit.
2. Calculate R_T if $R_1 = 2\text{ k}\Omega$, $R_2 = 3\text{ k}\Omega$, and $R_3 = 4700\ \Omega$.

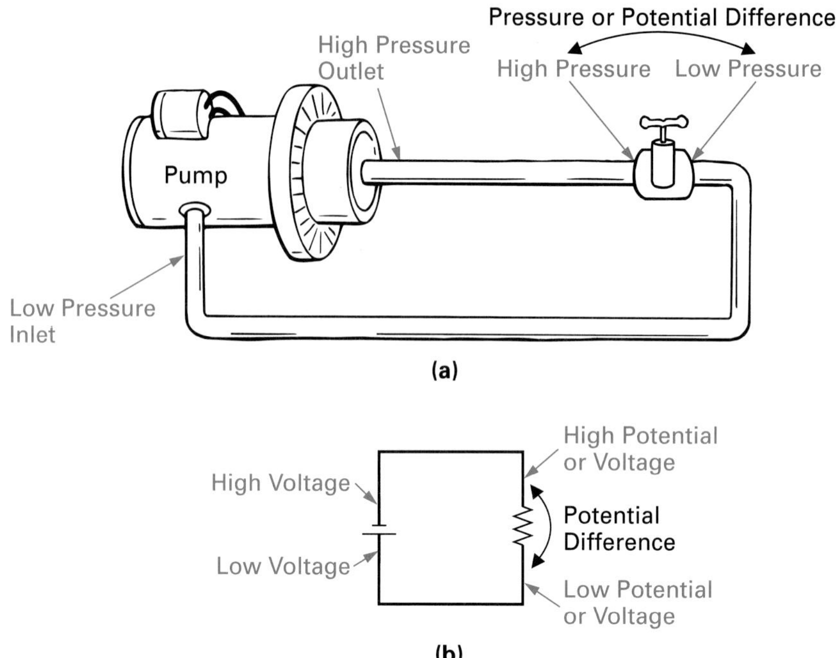

FIGURE 7-9 **Series Circuit Voltage. (a) Fluid Analogy of Potentional Difference. (b) Electrical Potential Difference.**

7-4 VOLTAGE IN A SERIES CIRCUIT

A potential difference or voltage drop will occur across each resistor in a series circuit when current is flowing. The amount of voltage drop is dependent on the value of the resistor and the amount of current flow.

Let's examine this idea of potential difference or voltage drop by returning, once more, to the water analogy. In Figure 7-9(a), you can see that the high pressure from the pump's outlet is present on the left side of the valve; however, on the right side of the valve the high pressure is no longer present. The high potential that exists on the left of the valve is not present on the right, so a potential or pressure difference is said to exist across the valve.

Similarly, with the electronic circuit in Figure 7-9(b), the battery produces a high voltage or potential that is present at the top of the resistor; however, the high voltage that exists at the top of the resistor is not present at the bottom. Therefore, a potential difference or voltage drop is said to occur across the resistor.

The voltage drop across resistors can be found by utilizing Ohm's law: $V = I \times R$.

EXAMPLE:

Referring to Figure 7-10, calculate:

a. Total resistance (R_T)
b. Amount of series current flowing throughout the circuit (I_T)

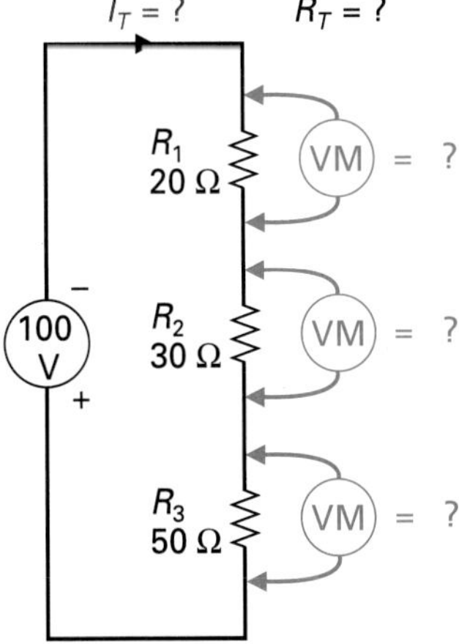

FIGURE 7-10 **Series Circuit Example.**

c. Voltage drop across R_1
d. Voltage drop across R_2
e. Voltage drop across R_3

Solution:

a. Total resistance $(R_T) = R_1 + R_2 + R_3$

$$= 20\ \Omega + 30\ \Omega + 50\ \Omega$$
$$= 100\ \Omega$$

b. Total current $(I_T) = \dfrac{V_{\text{Source}}}{R_T}$

$$= \frac{100\ \text{V}}{100\ \Omega}$$
$$= 1\ \text{A}$$

The same current will flow through the complete series circuit, so the current through R_1 will equal 1 A, the current through R_2 will equal 1 A, and the current through R_3 will equal 1 A.

c. Voltage across R_1 $(V_{R1}) = I_1 \times R_1$

$$= 1\ \text{A} \times 20\ \Omega$$
$$= 20\ \text{V}$$

d. Voltage across R_2 $(V_{R2}) = I_2 \times R_2$

$$= 1\ \text{A} \times 30\ \Omega$$
$$= 30\ \text{V}$$

e. Voltage across R_3 $(V_{R3}) = I_3 \times R_3$

$$= 1\ \text{A} \times 50\ \Omega$$
$$= 50\ \text{V}$$

Figure 7-11 shows the result of this example, and as you can see, the 20 Ω resistor drops 20 V, the 30 Ω resistor has 30 V across it, and the 50 Ω resistor has dropped 50 V. From this example, you will notice that the larger the resistor

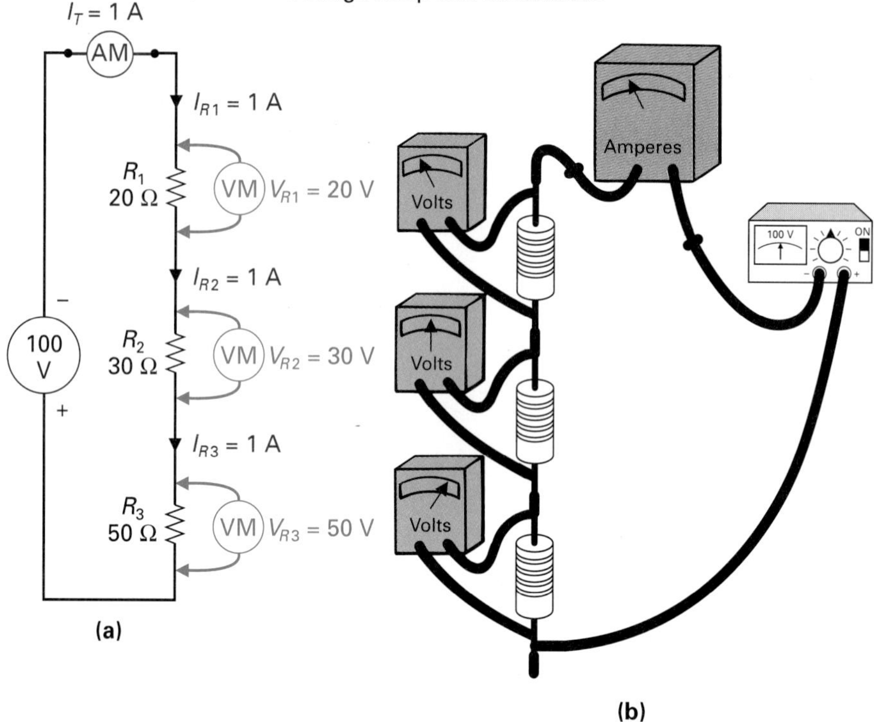

FIGURE 7-11 **(a) Schematic. (b) Pictorial.**

value, the larger the voltage drop. Resistance and voltage drops are consequently proportional to one another.

$$V_{\text{drop}} \uparrow = I(\text{constant}) \times R \uparrow$$

$$V_{\text{drop}} \downarrow = I(\text{constant}) \times R \downarrow$$

Another interesting point you may have noticed from Figure 7-11 is that, if you were to add up all the voltage drops around a series circuit, they would equal the source (V_S) applied:

$$\text{total voltage applied } (V_S \text{ or } V_T) = V_{R1} + V_{R2} + V_{R3} + \cdots$$

In the example, this is true, since

$$100 \text{ V} = 20 \text{ V} + 30 \text{ V} + 50 \text{ V}$$

$$100 \text{ V} = 100 \text{ V}$$

Kirchhoff's voltage law
The algebraic sum of the voltage drops in a closed path circuit is equal to the algebraic sum of the source voltage applied.

The series circuit has in fact divided up the applied voltage, and it appears proportionally across all the individual resistors. This characteristic was first observed by Gustav Kirchhoff in 1847; in honor of his discovery, this effect is known as **Kirchhoff's voltage law,** which states: The sum of the voltage drops in a series circuit is equal to the total voltage applied.

To summarize the effects of current, resistance, and voltage in a series circuit so far, we can say that:

1. The current in a series circuit has only one path to follow.
2. The value of current in a series circuit is the same throughout the entire circuit.
3. The total resistance in a series circuit is equal to the sum of all the resistances.
4. Resistance and voltage drops in a series circuit are proportional to one another, so a large resistance will have a large voltage drop and a small resistance will have a small voltage drop.
5. The sum of the voltage drops in a series circuit is equal to the total voltage applied.

EXAMPLE:

Calculate the voltage drop across the resistor R_1 in the circuit in Figure 7-12(a) for a resistance of 4 Ω or 2 Ω with a constant source of 4V.

Solution:

The voltage across R_1 can be calculated by using Ohm's law [Figure 7-12(b)].

$$V_{R1} = I_1 \times R_1$$
$$V_{R1} = 1\ \text{A} \times 4\ \Omega$$
$$= 4\ \text{V}$$

If the resistance is changed to 2 Ω, the current flow within the circuit would be equal to [Figure 7-12(c)]

$$I = \frac{V_S}{R} = \frac{4\ \text{V}}{2\ \Omega} = 2\ \text{A}$$

The voltage dropped across the resistor, however, would still be equal to

$$V_{R1} = I_1 \times R_1$$
$$= 2\ \text{A} \times 2\ \Omega$$
$$= 4\ \text{V}$$

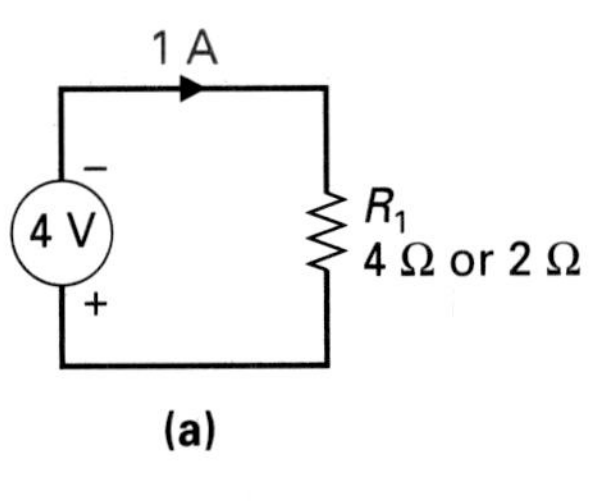

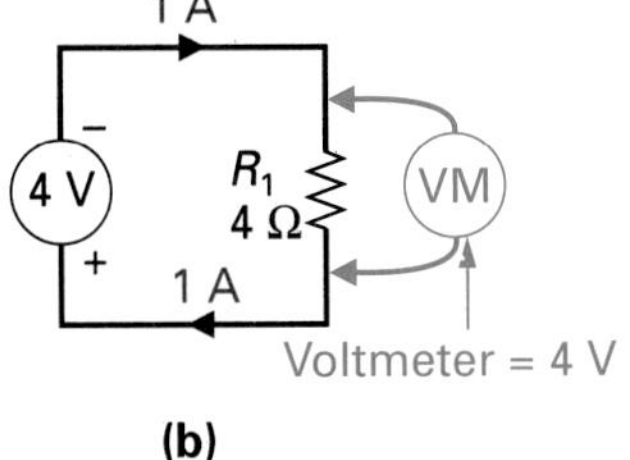

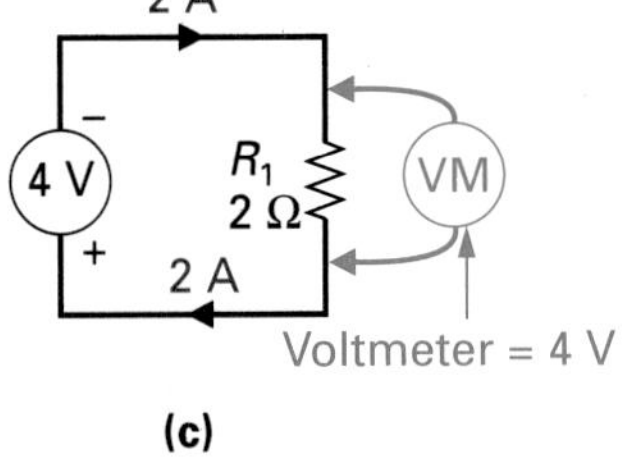

FIGURE 7-12 Single Resistor Circuit.

If only one resistor is connected in a series circuit, the entire applied voltage appears across this resistor. The amount of current flow is determined by the value of this single resistor and remains the same throughout the circuit.

EXAMPLE:

Referring to Figure 7-13(a), calculate:

a. Total circuit resistance
b. Value of current (I_T)

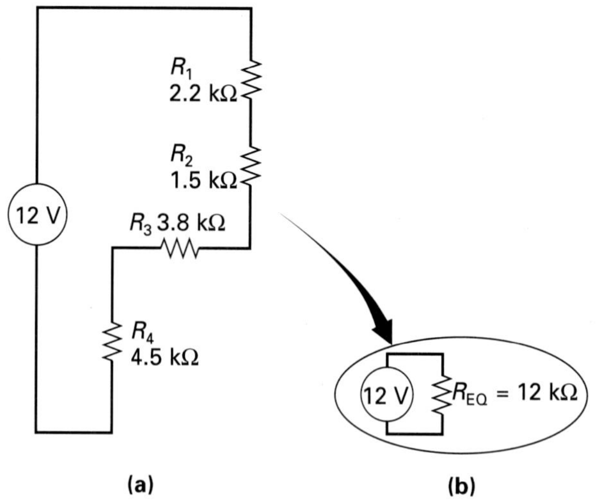

FIGURE 7-13 **Series Circuit Example.**

c. Voltage drop across each resistor
d. Then draw the circuit with respective voltages, resistances, and current inserted.

■ ***Solution:***

a. $R_T = R_1 + R_2 + R_3 + R_4$

$= 2.2\ \text{k}\Omega + 1.5\ \text{k}\Omega + 3.8\ \text{k}\Omega + 4.5\ \text{k}\Omega$

$= (2.2 \times 10^3) + (1.5 \times 10^3) + (3.8 \times 10^3) + (4.5 \times 10^3)$

$= 12\ \text{k}\Omega$ [Figure 7-13(b)]

b. $I_T = \dfrac{V_S}{R_T} = \dfrac{12\ \text{V}}{12\ \text{k}\Omega} = 1\ \text{mA}$

c. Voltage drop across each resistor:

$$V_{R1} = I_T \times R_1 = 1\ \text{mA} \times 2.2\ \text{k}\Omega = 2.2\ \text{V}$$

$$V_{R2} = I_T \times R_2 = 1\ \text{mA} \times 1.5\ \text{k}\Omega = 1.5\ \text{V}$$

$$V_{R3} = I_T \times R_3 = 1\ \text{mA} \times 3.8\ \text{k}\Omega = 3.8\ \text{V}$$

$$V_{R4} = I_T \times R_4 = 1\ \text{mA} \times 4.5\ \text{k}\Omega = 4.5\ \text{V}$$

d. See Figure 7-14.

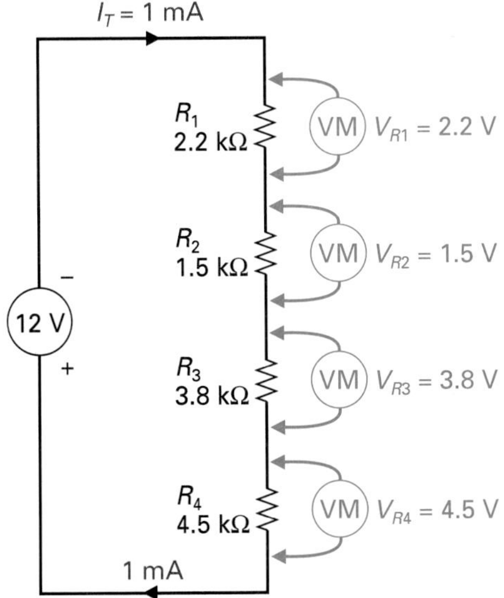

FIGURE 7-14 Series Circuit Example with All Values Inserted.

EXAMPLE:

Figure 7-15 shows a 12 V battery that has an internal source resistance (R_S or R_{int}) of 0.5 Ω (500 mΩ). If the battery was to supply its maximum safe current, which in this example is 2.5 A, what would be the output terminal voltage of the battery?

Solution:

If the battery was supplying its maximum safe current of 2.5 A, the output voltage (V_{out}) would equal the source voltage (V_S = 12 V) minus the voltage drop across the internal battery resistance (V_{Rint}). First let us calculate V_{Rint}:

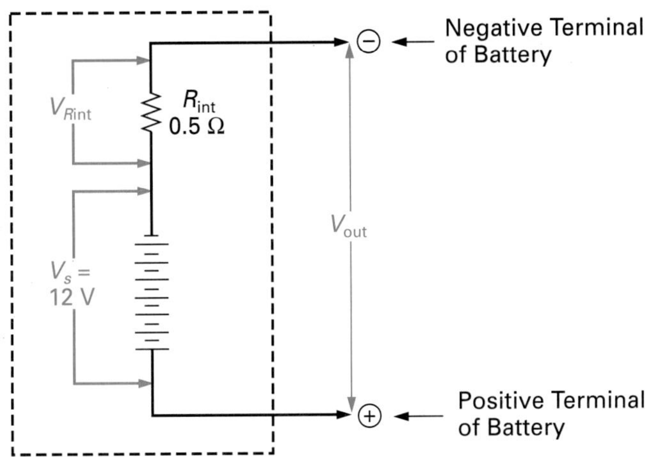

FIGURE 7-15 Voltage Drop across a Battery's Internal Series Resistance.

$$V_{Rint} = I \times R_{int}$$
$$= 2.5\ \text{A} \times 0.5\ \Omega = 1.25\ \text{V}$$

The output voltage therefore under full load will be

$$V_{out} = V_S - V_{Rint} = 12\ \text{V} - 1.25\ \text{V} = 10.75\ \text{V}$$

7-4-1 Fixed Voltage Divider

A series-connected circuit is often referred to as a *voltage-divider circuit.* The total voltage applied (V_T) or source voltage (V_S) is divided and dropped across the resistors in the series circuit. The amount of voltage dropped across a resistor is proportional to the value of resistance, and so a larger resistance causes a larger voltage drop across that resistor, while a smaller resistance causes a smaller voltage drop.

The voltage dropped across a resistor is normally a factor that needs to be calculated. The voltage-divider formula allows you to calculate the voltage drop across any resistor without having to work out current. This formula is stated as

$$V_X = \frac{R_X}{R_T} \times V_S$$

where V_X = voltage dropped across selected resistor
R_X = selected resistor's value
R_T = total series circuit resistance
V_S = source or applied voltage

Figure 7-16 illustrates a circuit from a previous problem. Normally, it would be necessary to:

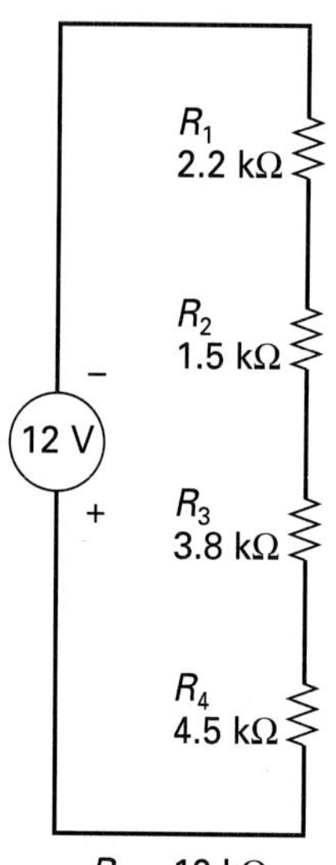

FIGURE 7-16 Series Circuit Example.

1. Calculate the total resistance by adding up all the resistance values.
2. Once we have the total resistance and source voltage (V_S), we could then calculate current.
3. Having calculated the current flowing through each resistor, we use the current value to calculate the voltage dropped across any one of the four resistors merely by multiplying current by the individual resistance value.

The voltage-divider formula allows us to bypass the last two steps. If we know total resistance, supply voltage, and the resistance value, we can calculate the voltage drop across the resistor without having to calculate steps 2 and 3. For example, what would be the voltage dropped across R_2 and R_4 in Figure 7-16? The voltage dropped across R_2 is

$$V_{R2} = \frac{R_2}{R_T} \times V_S$$
$$= \frac{1.5\text{ k}\Omega}{12\text{ k}\Omega} \times 12\text{ V}$$
$$= 1.5\text{ V}$$

CALCULATOR SEQUENCE

Step	Keypad Entry	Display Response
1.	1 . 5 E 3	1.5E3
2.	÷	
3.	1 2 E 3	12E3
4.	×	0.125
5.	1 2	12
6.	=	1.5

The voltage dropped across R_4 is

$$V_{R4} = \frac{R_4}{R_T} \times V_S$$
$$= \frac{4.5\text{ k}\Omega}{12\text{ k}\Omega} \times 12\text{ V}$$
$$= 4.5\text{ V}$$

The voltage-divider formula could also be used to find the voltage drop across multiple resistors. For example, what would be the voltage dropped across R_2 and R_3? (This is illustrated in Figure 7-17.) The voltage across $R_2 + R_3$ is

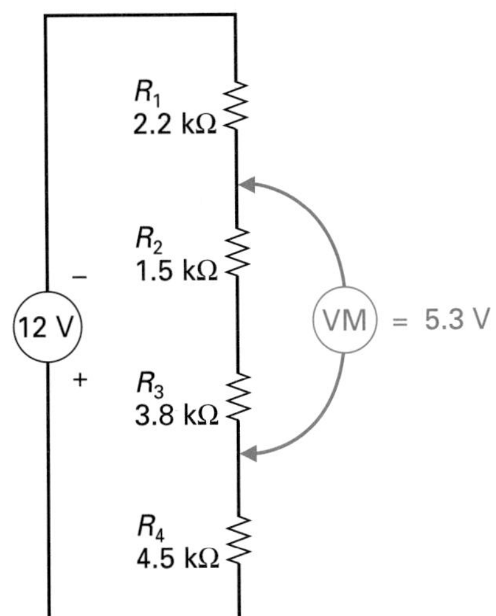

FIGURE 7-17 Series Circuit Example.

$$V_{R2} \text{ and } V_{R3} = \frac{R_2 + R_3}{R_T} \times V_S$$
$$= \frac{5.3\ \text{k}\Omega}{12\ \text{k}\Omega} \times 12\ \text{V}$$
$$= 5.3\ \text{V}$$

EXAMPLE:

Calculate the voltage drop across (see Figure 7-18):

a. R_1, R_2, and R_3 separately
b. R_2 and R_3 together
c. R_1, R_2, and R_3 together

Solution:

a. The voltage drop across a resistor is proportional to the resistance value. The total resistance (R_T) in this circuit is 100 Ω or 100% of R_T. R_1 is 20% of the total resistance, so 20% of the source voltage will appear across R_1. R_2 is 30% of the total resistance, so 30% of the source voltage will appear across R_2. R_3 is 50% of the total resistance, so 50% of the source voltage will appear across R_3. This was a very simple problem in which the figures worked out very neatly. The voltage-divider formula achieves the very same thing by calculating the ratio of the resistance value to the total resistance. This percentage is then multiplied by the source voltage in order to find the desired resistor's voltage drop:

$$V_{R1} = \frac{R_1}{R_T} \times V_S$$
$$= \frac{20\ \Omega}{100\ \Omega} \times V_S$$
$$= 0.2 \times 100\ \text{V}$$
$$= 20\ \text{V}$$

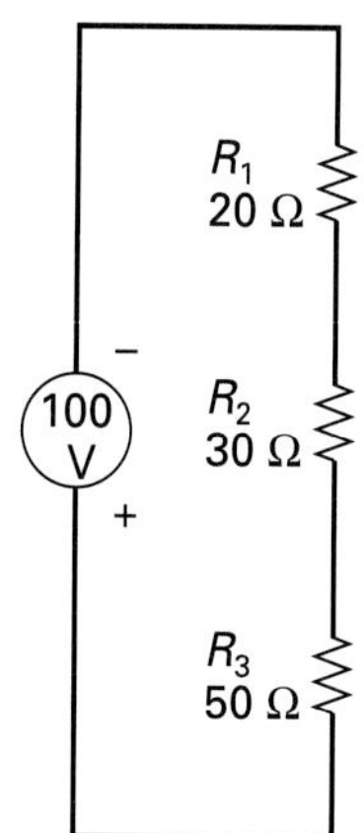

FIGURE 7-18 Series Circuit Example.

$$V_{R2} = \frac{R_2}{R_T} \times V_S$$

$$= \frac{30\ \Omega}{100\ \Omega} \times V_S$$

$$= 0.3 \times 100\ \text{V}$$

$$= 30\ \text{V}$$

$$V_{R3} = \frac{R_3}{R_T} \times V_S$$

$$= \frac{50\ \Omega}{100\ \Omega} \times V_S$$

$$= 0.5 \times 100\ \text{V}$$

$$= 50\ \text{V}$$

b. Voltage dropped across R_2 and $R_3 = 30 + 50 = 80$ V.

c. Voltage dropped across R_1, R_2, and $R_3 = 20 + 30 + 50 = 100$ V.

The voltage-divider formula can be summarized by saying that: The voltage drop across a resistor or multiple resistors in a series circuit is equal to the ratio of that resistance (R_X) to the total resistance (R_T) multiplied by the source voltage (V_S).

If three voltages of 50, 80, and 100 V were required by a piece of equipment, we could use three individual power sources, which would be very expensive, or use one 100 V voltage source connected across the resistors from Example 7-10 to divide up the 100 V and supply the three required fixed voltages, as seen in Figure 7-19.

EXAMPLE:

Figure 7-20(a) shows a simplified circuit of an oscilloscope's cathode ray tube (CRT). In this example, a three-resistor series circuit and a –600 V supply voltage are being used to produce the needed three supply voltages for the CRT's three electrodes, called the focusing anode, the control grid, and the cathode. The heated cathode emits electrons that are collected and concentrated into a

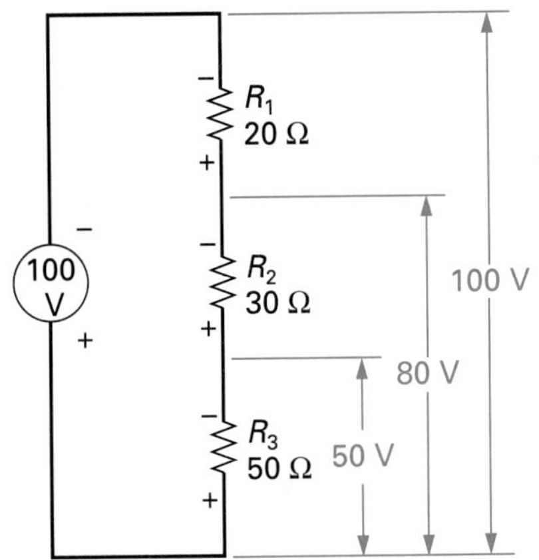

FIGURE 7-19 **Series Circuit Example.**

beam by the combined electrostatic effect of the control grid and focusing anode. This beam of electrons passes through the apertures of the control grid and focusing anode and strikes the inner surface of the CRT screen. This inner surface is coated with a phosphorescent material that emits light when it is struck by electrons. The voltage between the cathode (K) and grid (G) of the CRT (V_{KG}) determines the intensity of the electron beam and therefore the brightness of the trace seen on the screen. The voltage between the grid (G) and anode (A) of the CRT (V_{GA}) determines the sharpness of the electron beam and therefore the focus of the trace seen on the screen. For the resistance values given, calculate the voltages on each of the three CRT electrodes.

Solution:

Referring to the illustration and calculations in Figure 7-20(b) you can see how the voltage-divider formula can be used to calculate the voltage drop across R_1

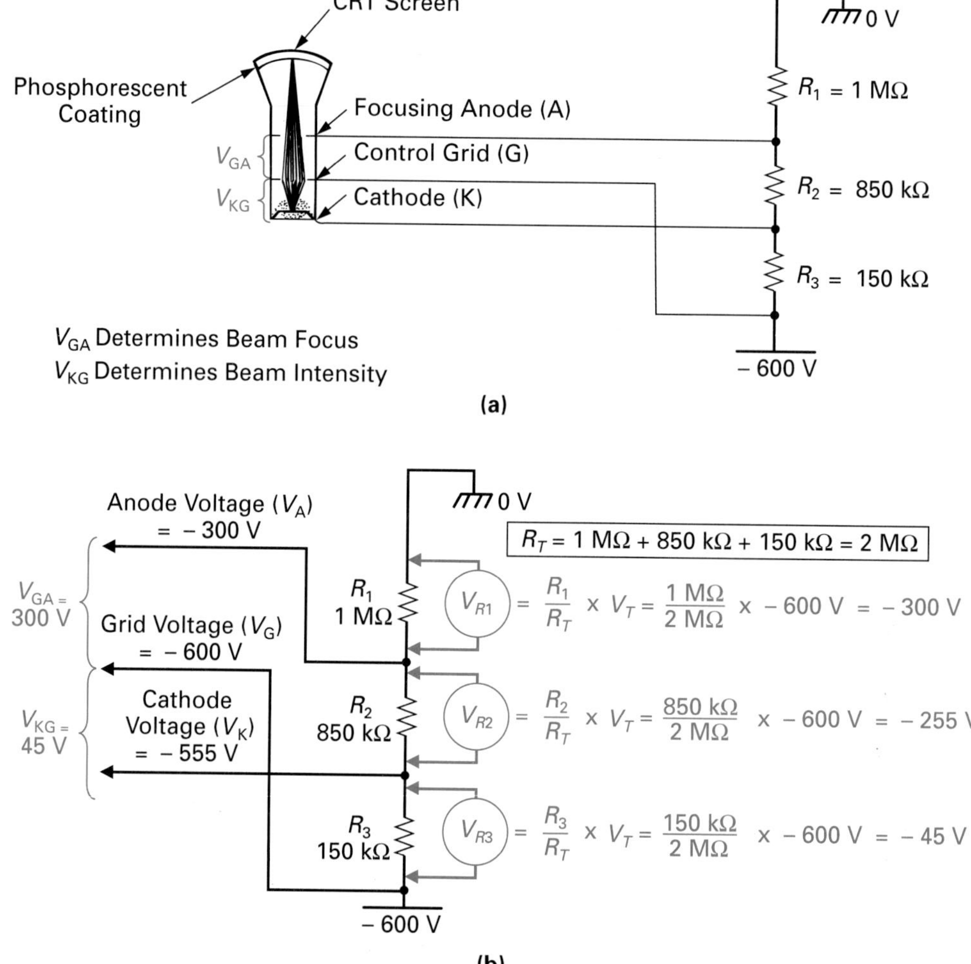

FIGURE 7-20 Fixed Voltage-Divider Circuit for Supplying Voltages to the Electrodes of a CRT.

($V_{R1} = 300$ V), R_2 ($V_{R2} = 255$ V), and R_3 ($V_{R3} = 45$ V). Since the grid of the CRT is connected directly to the –600 V supply, the grid voltage (V_G) will be

$$V_G = V_T = -600 \text{ V}$$

The cathode voltage (V_K) will be equal to the total supply voltage (V_T) minus the voltage drop across R_3 (V_{R3}), so

$$V_K = V_T - V_{R3} = (-600) - (-45) = -555 \text{ V}$$

The anode voltage (V_A) will be equal to the total supply voltage (V_T) minus the voltage drops across R_3 (V_{R3}) and R_2 (V_{R2}), and therefore

$$V_A = V_T - (V_{R3} + V_{R2}) = (-600) - (-45 + -225)$$
$$= -600 - 300 \text{ V} = -300 \text{ V}$$

The V_A, V_K, and V_G voltages are all negative voltage with respect to 0 V. The voltage V_{GA} and V_{KG}, shown on the left of Figure 7-20(b), will be the potential difference between the two electrodes. Therefore, V_{KG} is equal to the difference between V_K and V_G, and is equal to –45 V ($V_{KG} = -45 \text{ V} = V_{R3}$). The voltage between the CRT's grid and anode (V_{GA}) is equal to the difference between V_G and V_A, which will be –300 V ($V_{GA} = -300 \text{ V} = V_{R1}$).

EXAMPLE:

Figure 7-21(a) shows a 24 V voltage source driving a 10 Ω resistor that is located 1000 ft from the battery. If two 1000 ft lengths of AWG No. 13 wire are used to connect the source to the load, what will be the voltage applied across the load?

Solution:

Referring back to Table 3-2, you can see that AWG No. 13 copper cable has a resistance of 2.003 Ω for every 1000 ft. This means that our circuit should be redrawn as shown in Figure 7-21(b) to show the series resistances of wire 1 and wire 2. Using the voltage-divider formula, we can calculate the voltage drop across wire 1 and wire 2.

$$V_{W1} = \frac{R_{W1}}{R_T} \times V_T$$

$$= \frac{2\ \Omega}{14\ \Omega} \times 24 \text{ V} = 3.43 \text{ V}$$

Since the voltage drop across wire 2 will also be 3.43 V, the total voltage drop across both wires will be 6.86 V. The remainder, 17.14 V (24 V – 6.86 V = 17.14 V), will appear across the load resistor, R_L.

7-4-2 Variable Voltage Divider

When discussing variable-value resistors in Chapter 4, we talked about a potentiometer, or variable voltage divider, which consists of a fixed value of resistance between two terminals and a wiper that can be adjusted to vary resistance between its terminal and one of the other two. Figure 7-22(a) and (b) illustrate the potentiometer's schematic symbol and physical appearance.

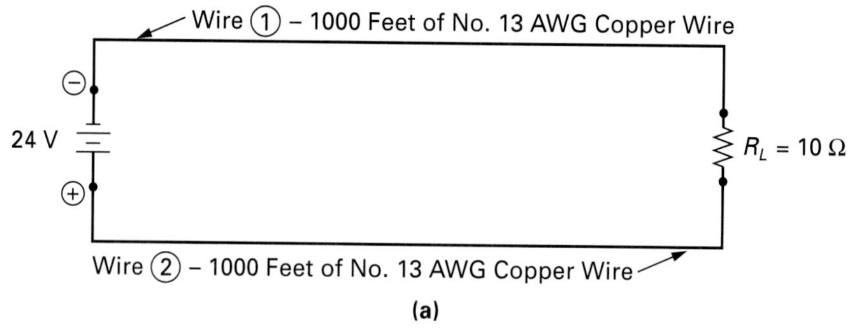

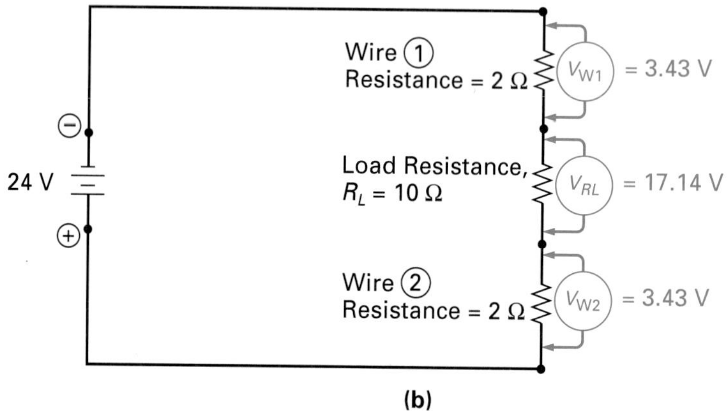

FIGURE 7-21 Series Wire Resistance.

If the wiper is moved down, as seen in Figure 7-23(a), the resistance between terminals A and B will increase, while the resistance between B and C will decrease. If the wiper is moved up, as seen in Figure 7-23(b), the resistance between terminals A and B will decrease, while the resistance between B and C will increase. The resistances between A and B and B and C are inversely proportional to one another in that if one were to increase the other would decrease, and vice versa, and can be thought of as two separate resistors, as seen in Figure 7-23(c). However, the total of the resistance between A and B (R_{AB})

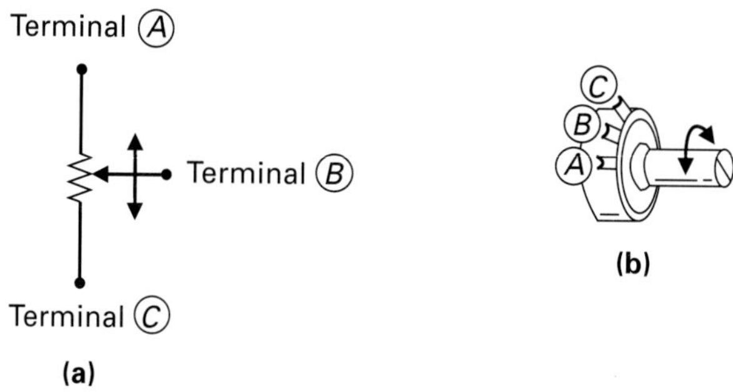

FIGURE 7-22 Potentiometer. (a) Schematic Symbol. (b) Physical Appearance.

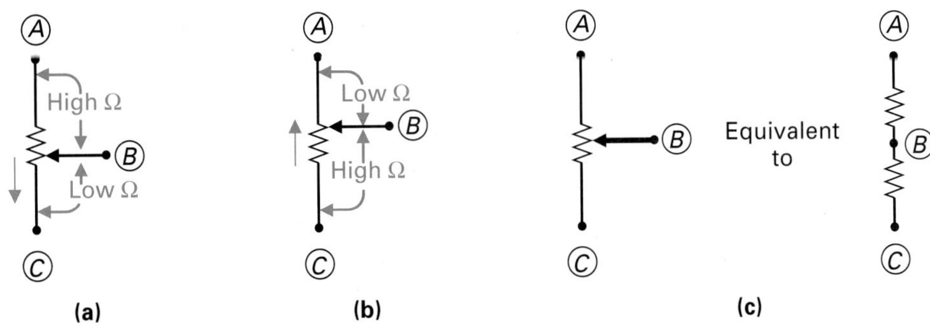

FIGURE 7-23 **Potentiometer Operation.**

and the resistance between *B* and *C* (R_{BC}) will always be equal to the rated value of the potentiometer, and equal to the resistance between *A* and *C* (R_{AC}).

$$R_{AB} + R_{BC} = R_{AC}$$

Figure 7-24(a) illustrates a 10 kΩ potentiometer that has been hooked up across a 10 V dc source with a voltmeter between terminals *B* and *C*. If the wiper terminal is positioned midway between *A* and *C*, the voltmeter should read 5 V, and the potentiometer will be equivalent to two 5 kΩ resistors in series, as shown in Figure 7-24(b). Kirchhoff's voltage law states that the entire source voltage will be dropped across the resistances in the circuit, and since the resistance values are equal, each will drop half of the source voltage, that is, 5 V.

In Figure 7-25(a), the wiper has been moved down so that the resistance between *A* and *B* is equal to 8 kΩ, and the resistance between *B* and *C* equals 2 kΩ. This will produce 2V on the voltmeter, as shown in Figure 7-25(b). The amount of voltage drop is proportional to the resistance, so a larger voltage will be dropped across the larger resistance. Using the voltage-divider formula, you can calculate that the 8 kΩ is 80% of the total resistance and therefore will drop 80% of the voltage:

$$V_{AB} = \frac{R_{AB}}{R_{\text{total}}} \times V_S = \frac{8\ \text{k}\Omega}{10\ \text{k}\Omega} \times 10\ \text{V} = 8\ \text{V}$$

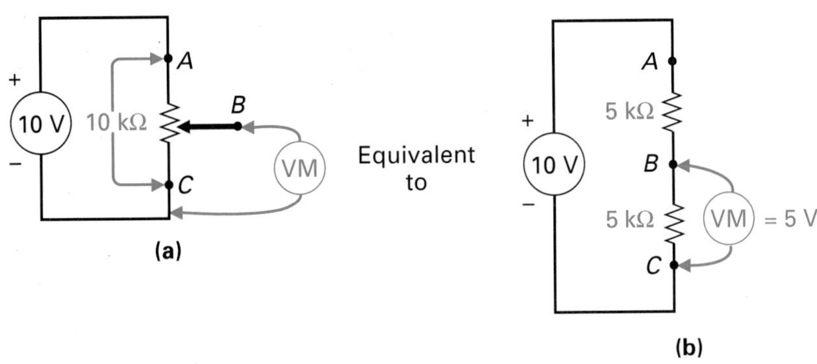

FIGURE 7-24 **Potentiometer Wiper in Mid-position.**

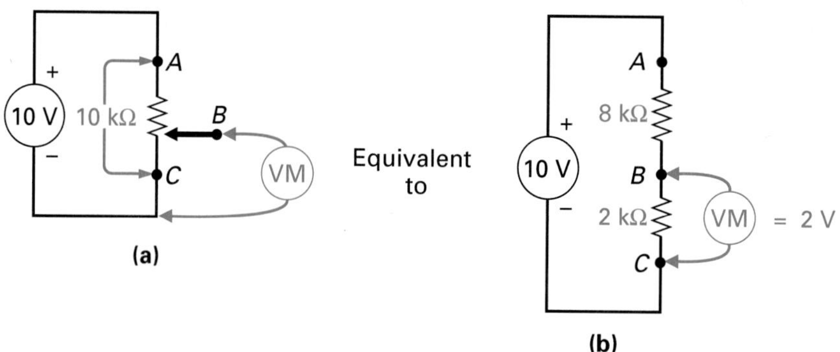

FIGURE 7-25 **Potentiometer Wiper in Lower Position.**

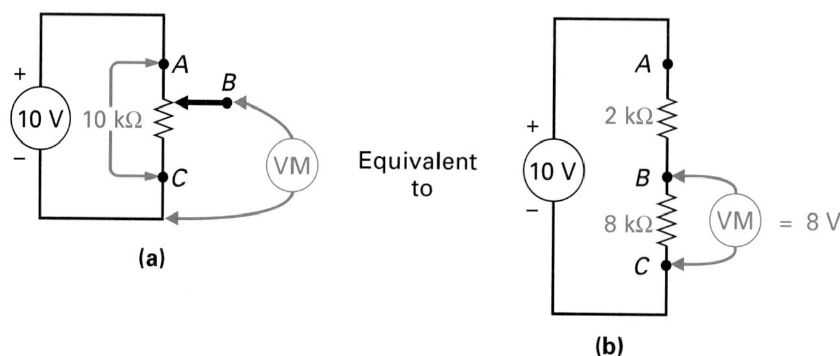

FIGURE 7-26 **Potentiometer Wiper in Upper Position.**

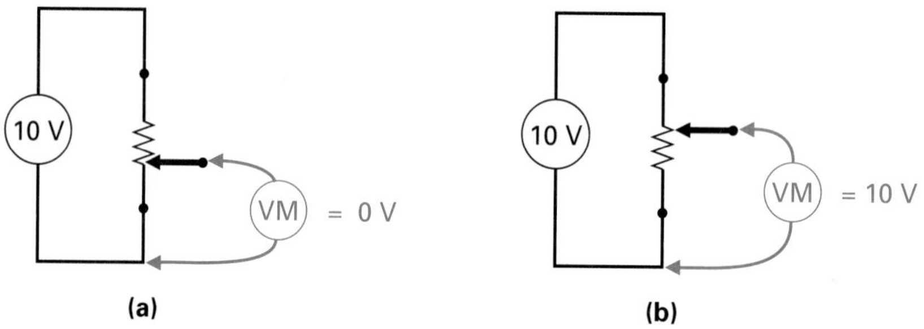

FIGURE 7-27 **Minimum and Maximum Settings of a Potentiometer.**

The 2 kΩ resistance between *B* and *C* is 20% of the total resistance and consequently will drop 20% of the total voltage:

$$V_{BC} = \frac{R_{BC}}{R_{\text{total}}} \times V_S = \frac{2\text{ k}\Omega}{10\text{ k}\Omega} \times 10\text{ V} = 2\text{ V}$$

which will be indicated on the voltmeter.

In Figure 7-26(a), the wiper has been moved up and now 2 kΩ exists between *A* and *B,* and 8 kΩ is present between *B* and *C.* In this situation, 2 V will be dropped across the 2 kΩ between *A* and *B,* and 8 V will be dropped across the 8 kΩ between *B* and *C,* as shown in Figure 7-26(b).

From this discussion, you can see that the potentiometer can be adjusted to supply different voltages on the wiper. This voltage can be decreased by moving the wiper down to supply a minimum of 0 V [Figure 7-27(a)], or the wiper can be moved up to supply a maximum of 10 V [Figure 7-27(b)], or it can be placed anywhere in between, which is why the potentiometer is known as a variable voltage divider.

EXAMPLE:

Figure 7-28 illustrates how a potentiometer can be used to control the output volume of an amplifier that is being driven by a compact disk (CD) player. The preamplifier is producing an output of 2 V, which is developed across a 50 kΩ potentiometer. If the wiper of the potentiometer is in its upper position, the full 2 V from the preamp will be applied into the input of the power amplifier. The power amplifier has a fixed voltage gain (A_V) of 12, and therefore the power amplifier's output is always 12 times larger than the input voltage. An input of 2 V (V_{in}) will therefore produce an output voltage (V_{out}) of 24 V ($V_{out} = V_{in} \times A_V = 2\text{ V} \times 12 = 24\text{ V}$). As the wiper is moved down, less of the 2 V from the preamp will be applied to the power amplifier, and therefore the output of the power amplifier and volume of the music heard will decrease. If the wiper of the potentiometer is adjusted so that a resistance of 20 kΩ exists between the wiper and the lower end of the potentiometer, what will be the input voltage to the power amplifier and output voltage to the speaker?

Solution:

By using the voltage-divider formula, we can determine the voltage developed across the potentiometer with 20 kΩ of resistance between the wiper (B) and lower end (C).

$$V_{in} = \frac{R_{BC}}{R_{AC}} \times V_{Pre}$$

$$= \frac{20\text{ k}\Omega}{50\text{ k}\Omega} \times 2\text{ V} = 0.8\text{ V (800 mV)}$$

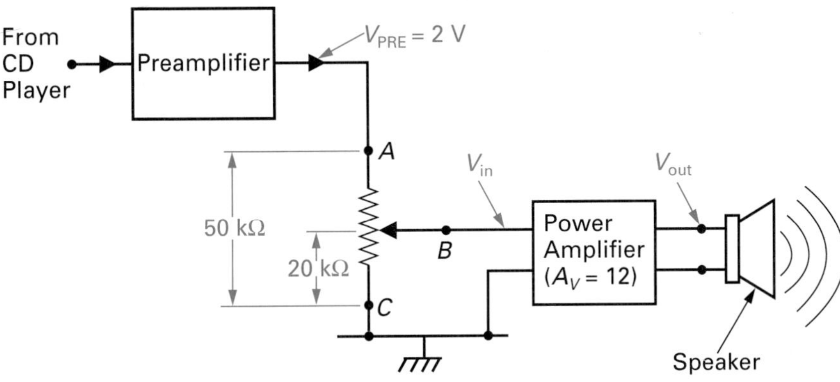

FIGURE 7-28 The Potentiometer as a Volume Control.

The voltage, V_{in} , is applied to the input of the power amplifier, which has an amplification factor or gain of 12, and therefore the output voltage will be

$$V_{out} = V_{in} \times A_V$$
$$= 0.8 \text{ V} \times 12 = 9.6 \text{ V}$$

SELF-TEST REVIEW QUESTIONS FOR SECTION 7-4

1. True or false: A series circuit is also known as a voltage-divider circuit.
2. True or false: The voltage drop across a series resistor is proportional to the value of the resistor.
3. If 6 Ω and 12 Ω resistors are connected across a dc 18 V supply, calculate I_T and the voltage drop across each.
4. State the voltage-divider formula.
5. Which component can be used as a variable voltage divider?
6. Could a rheostat be used in place of a potentiometer?

7-5 POWER IN A SERIES CIRCUIT

Power is the rate at which work is done. Work is said to have been done when energy is converted, in this case from electrical energy to heat energy. Resistors dissipate heat energy, and the rate at which they dissipate energy is called *power* and is measured in *watts* (joules per second). Resistors all have a resistive value, a tolerance, and a wattage rating. The wattage of a resistor is the amount of heat energy a resistor will dissipate per second, and this wattage is directly proportional to the resistor's size; a larger resistor will be able to dissipate more heat than a smaller resistor. Figure 7-29 reviews the size versus wattage rating of commercially available resistors. As you know, any of the power formulas can be used to calculate wattage. Resistors are manufactured in several different physical sizes, and if, for example, it is calculated that for a certain value of current and voltage a 5 W resistor is needed, and a ½ W resistor is put in its place, the ½ W resistor will burn out, because it is generating heat (5 W) faster than it can dissipate heat (½ W). A 10 W or 25 W resistor or greater could be used to replace a 5 W resistor; however, anything less than a 5 W resistor will burn out.

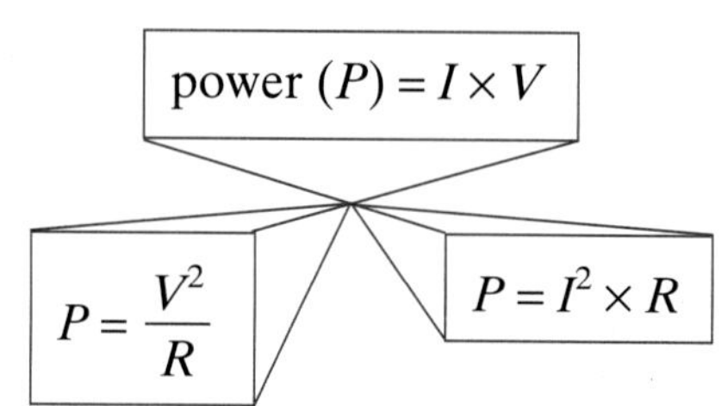

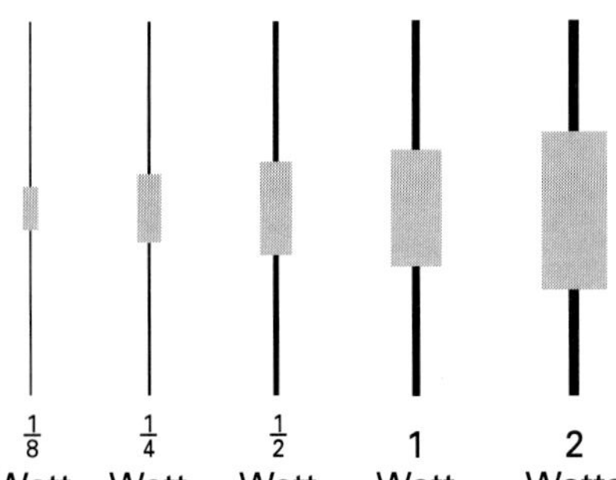

FIGURE 7-29 Resistor Wattage Ratings.

It is, therefore, necessary that we have some way of calculating which wattage rating is necessary for each specific application.

A question you may be asking is why not just use large-wattage resistors everywhere. The two disadvantages with this are:

1. The larger the wattage, the greater the cost.
2. The larger the wattage, the greater the size and area the resistor occupies within the equipment.

EXAMPLE:

Figure 7-30 illustrates a 20 V battery driving a 12 V/1 A television set. R_1 is in series with the TV set and is being used to drop 8 V of the 20 V supply, so 12 V will be applied across the set.

a. What is the wattage rating for R_1?
b. What is the series load resistance of the TV set?
c. What is the amount of power being consumed by the TV set?

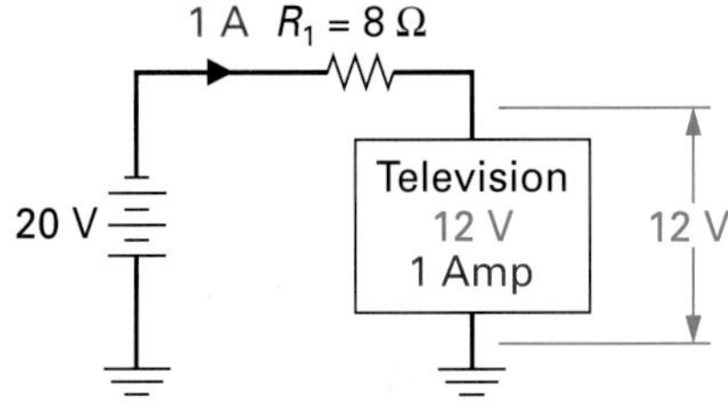

FIGURE 7-30 Series Circuit Example.

Solution:

a. Everything is known about R_1. Its resistance is 8 Ω, it has 1 A of current flowing through it, and 8 V is being dropped across it. Consequently, any one of the three power formulas can be used to calculate the wattage rating of R_1.

$$\text{power } (P) = I \times V = 1\text{ A} \times 8\text{ V} = 8\text{ W}$$

or

$$P = I^2 \times R = 1^2 \times 8 = 8\text{ W}$$

or

$$P = \frac{V^2}{R} = \frac{8^2}{8} = 8\text{ W}$$

The nearest commercially available device would be a 10 W resistor; however, if size is not a consideration, it is ideal to double the wattage needed and use a 16 W resistor.

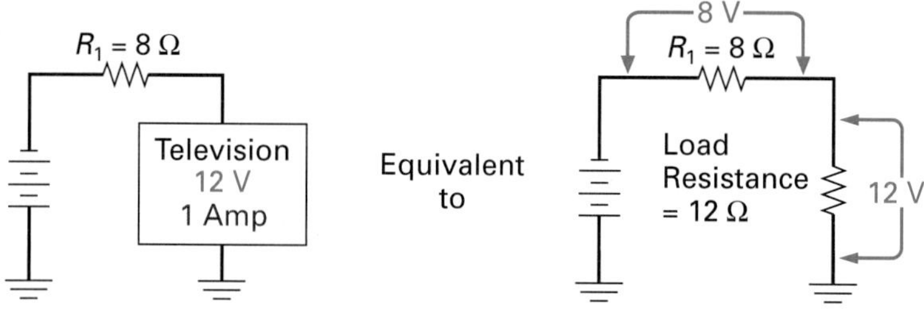

FIGURE 7-31 **Series Circuit Example with Values Inserted.**

b. You may recall that any piece of equipment is equivalent to a load resistance. The TV set has 12 V across it and is pulling 1 A of current; its load resistance can be calculated simply by using Ohm's law and deriving an equivalent circuit, as shown in Figure 7-31.

$$R_L \text{ (load resistance)} = \frac{V}{I} = \frac{12 \text{ V}}{1 \text{ A}} = 12\ \Omega$$

c. The amount of power being consumed by the TV set is

$$P = V \times I = 12 \text{ V} \times 1 \text{ A} = 12 \text{ W}$$

EXAMPLE:

Calculate the total amount of power dissipated in the series circuit in Figure 7-32.

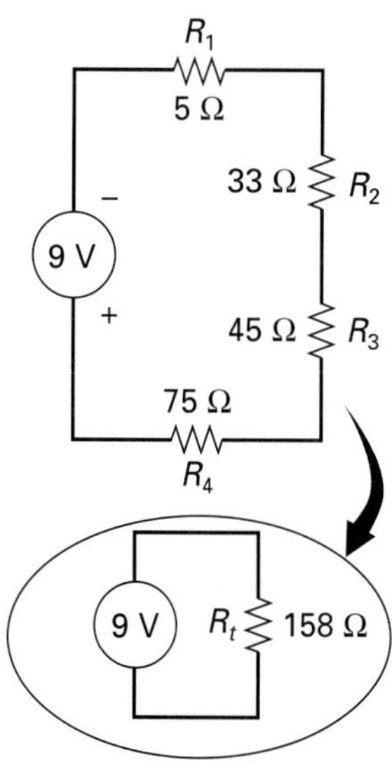

FIGURE 7-32 **Series Circuit Example.**

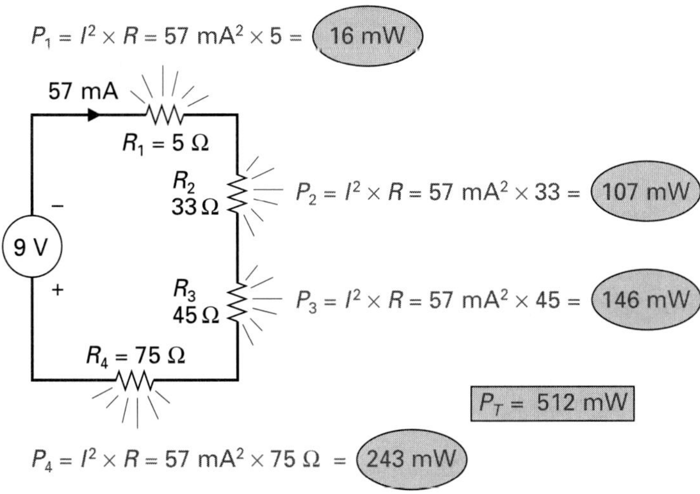

FIGURE 7-33 Series Circuit Example with Values Inserted.

Solution:

The total power dissipated in a series circuit is equal to the sum of all the power dissipated by all the resistors. The easiest way to calculate the total power is to simplify the circuit to one resistance.

$$\begin{aligned} R_T &= R_1 + R_2 + R_3 + R_4 \\ &= 5\ \Omega + 33\ \Omega + 45\ \Omega + 75\ \Omega \\ &= 158\ \Omega \end{aligned}$$

We now have total resistance and total voltage, so we can calculate the total power:

$$P_T = \frac{V_S^2}{R_T} = \frac{9\text{ V}^2}{158\ \Omega} = \frac{81}{158} = 512.7 \text{ milliwatts (mW)}$$

The longer method would have been to calculate the current through the series current:

$$I = \frac{V_S}{R_T} = \frac{9\text{ V}}{158\ \Omega} = 56.96\text{ mA} \quad \text{or} \quad 57\text{ mA}$$

We could then calculate the power dissipated by each separate resistor and add up all the individual values to gain a total power figure. This is illustrated in Figure 7-33.

total power = addition of all the individual power losses

$$P_T = P_1 + P_2 + P_3 + P_4 + \cdots$$

$$\begin{aligned} &= 16\text{ mW} + 107\text{ mW} + 146\text{ mW} + 243\text{ mW} \\ &= 512\text{ mW} \end{aligned}$$

SELF-TEST REVIEW QUESTIONS FOR SECTION 7-5

1. State the power formula.
2. Calculate the power dissipated by a 12 Ω resistor connected across a 12 V supply.
3. What fixed resistor type should probably be used for Question 2, and what would be a safe wattage rating?
4. What would be the total power dissipated if R_1 dissipates 25 W and R_2 dissipates 3800 mW?

7-6 TROUBLESHOOTING A SERIES CIRCUIT

A resistor will usually burn out and result in an open between its two leads when an excessive current flow occurs. This can normally, but not always, be noticed by a visual check of the resistor, which will appear charred due to the excessive heat. In some cases you will need to use your multimeter (combined ammeter, voltmeter, and ohmmeter) to check the circuit components to determine where a problem exists.

The two basic problems that normally exist in a series circuit are opens and shorts; however, a problem is not always as drastic as a short or an open, but may be a variation in a component's value over a long period of time, which will eventually cause a problem.

To summarize, then, we can say that one of three problems can occur to components in a series circuit:

1. A component will open (infinite resistance).
2. A component's value will change over a period of time.
3. A component will short (zero resistance).

The voltmeter is the most useful tool when checking series circuits as it can be used to measure voltage drops by connecting the meter leads across the component or resistors. Let's now analyze a problem and see if we can solve it by logically **troubleshooting** the circuit and isolating the faulty component. To begin with, let us take a look at the effects of an open component.

Troubleshooting
The process of locating and diagnosing malfunctions or breakdowns in equipment by means of systematic checking or analysis.

7-6-1 Open Component in a Series Circuit

A component is open when its resistance is the maximum possible (infinity). Let us see with a few examples what effect this will have.

EXAMPLE:

Figure 7-34(a) illustrates a TV set with a load resistance of 3 Ω. The TV set is off because R_2 has burned out and become an open circuit. How would you determine that the problem is R_2?

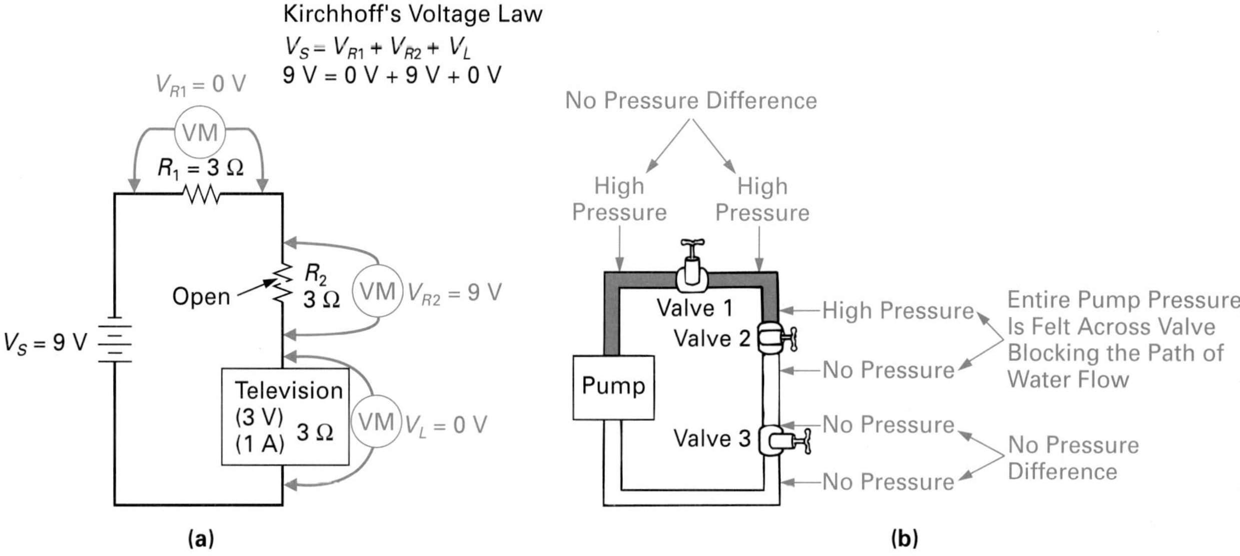

FIGURE 7-34 **Troubleshooting an Open in a Series Circuit.**

Solution:

If an open circuit ever occurs in a series circuit, due in this case to R_2 having burned out, there can be no current flow, because series circuits have only one path for current to flow and that path has been broken ($I = 0$ A). Using the voltmeter to check the amount of voltage drop across each resistor, two results will be obtained:

1. The voltage drop across a good resistor will be zero volts.
2. The voltage drop across an open resistor will be equal to the source voltage V_S.

No voltage will be dropped across a good resistor because current is zero, and if $I = 0$, the voltage drop, which is the product of I and R, must be zero ($V = I \times R = 0 \times R = 0$ V). If no voltage is being dropped across the good resistor R_1 and the TV set resistance of 3 Ω, the entire source voltage will appear across the open resistor, R_2, in order that this series circuit comply with Kirchhoff's voltage law: V_S (9 V) = V_{R1} (0 V) + V_{R2} (9 V) + V_L (0 V).

To emphasize this point further, refer to Figure 7-34(b), which illustrates the fluid analogy. Like R_2, valve 2 has completely blocked any form of flow (water flow = 0), and looking at the pressure differences across all three valves, you can see that no pressure difference occurs across valves 1 and 3, but the entire pump pressure is appearing across valve 2, the component that has opened the circuit.

EXAMPLE:

Figure 7-35 illustrates a set of three lights connected across a 9 V battery. Bulb 3 is open and therefore there is no current flow; with no current flow, all three bulbs are off, and we need to isolate which of the three is faulty.

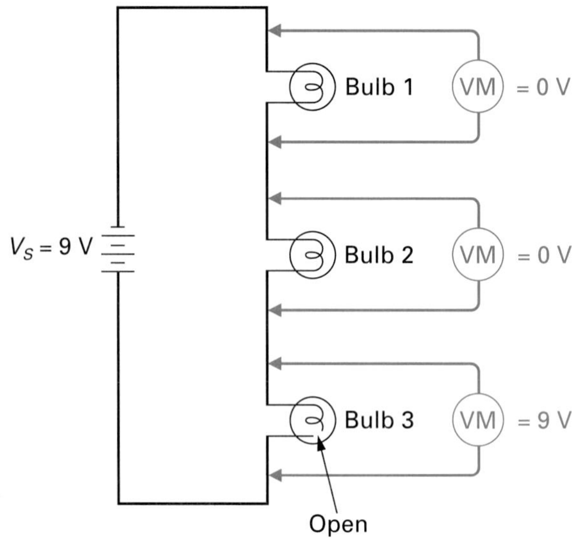

FIGURE 7-35 Troubleshooting an Open Bulb in a Series Circuit.

■ ***Solution:***

Using the voltmeter, there will be:

1. Zero volts dropped across bulb 1; 9 V appears on both sides of bulb 1, so the potential difference or voltage drop across bulb 1 is zero. Bulb 1 is OK.
2. Zero volts dropped across bulb 2; bulb 2 is OK.
3. Nine volts dropped across bulb 3; the entire source voltage is being dropped across bulb 3. Bulb 3 is open and needs to be replaced.

7-6-2 *Component Value Variation in a Series Circuit*

Resistors will rarely go completely open unless severely stressed because of excessive current flow. With age, resistors will normally change their resistance value. This occurs slowly and will generally always cause a decrease in the resistor's resistance and eventually cause a circuit problem. This lowering of resistance will cause an increase of current, which will cause an increase in the power dissipated; if the wattage of the resistor is exceeded, it can burn out. If they do not burn out but merely blow the circuit fuse due to this increase in current, the problem can be found by measuring the resistance values of each resistor or by measuring how much voltage is dropped across each resistor and comparing these to the calculated voltage, based on the parts list supplied by the manufacturer.

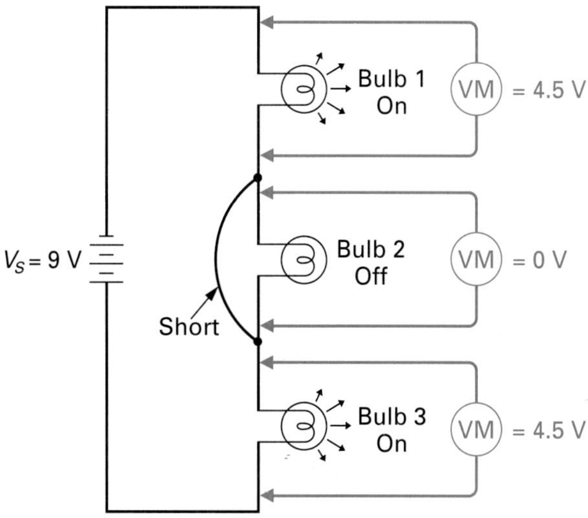

FIGURE 7-36 **Troubleshooting a Short in a Series Circuit.**

7-6-3 *Shorted Component in a Series Circuit*

A component has gone short when its resistance is 0 Ω. Let's work out a problem using the same example of the three bulbs across the 9 V battery, as shown in Figure 7-36.

EXAMPLE:

Bulbs 1 and 3 in Figure 7-36 are on and bulb 2 is off. A piece of wire exists between the two terminals of bulb 2, and the current is taking the lowest resistance path through the wire rather than through the filament resistor of the bulb. Bulb 2 therefore has no current flow through it, and light cannot be generated without current. How would you determine what the problem is?

Solution:

If bulb 2 was open (burned out), there could be no current flow in the circuit, so bulbs 1 and 2 would be off. Bulb 2 must therefore be shorted. Using the voltmeter to investigate this further, you find that bulb 2 drops 0 V across it because it has no resistance except the very small wire resistance of the bypass, which means that 4.5 V must be dropped across each working bulb (1 and 3) in order to comply with Kirchhoff's voltage law. The loss of bulb 2's resistance causes the overall circuit resistance offered by bulbs 1 and 3 to decrease, which will cause an increase in current; so bulbs 1 and 3 should glow more brightly.

To summarize opens and shorts in series circuits, we can say that:

1. The supply voltage appears across an open component.
2. Zero volts appears across a shorted component.

SELF-TEST REVIEW QUESTIONS FOR SECTION 7-6

1. List the three basic problems that can occur with components in a series circuit.
2. How can an open component be detected in a series circuit?
3. True or false: If a series-connected resistor's value were to decrease, the voltage drop across that same resistor would increase.
4. How could a shorted component be detected in a series circuit?

SUMMARY

Series Circuits (Figure 7-37)

1. A series circuit is the connecting of components end to end in a circuit to provide a single path for the current. In all cases the components are connected in succession or strung together one after another so that only one path for current exists between the negative (–) and positive (+) terminals of the supply.

Current in a Series Circuit

2. The current in a series circuit has only one path to follow and cannot divert in any other direction; consequently, the current through a series circuit is the same throughout that circuit.

Resistance in a Series Circuit

3. Resistance is the opposition to current flow, and in a series circuit every resistor in series offers opposition to the current flow.
4. The total resistance in a series-connected electronic resistive circuit is thus equal to the sum of all the individual resistances.
5. Total resistance (R_T) is the only opposition the voltage source can sense; it does not see the individual separate resistors, but one equivalent resistance. Based on its voltage and this total resistance, a value of currrent will be produced to flow through the circuit (Ohm's law, $I = V/R$).

Voltage in a Series Circuit

6. A potential difference or voltage drop will occur across each resistor in a series circuit when current is flowing. The amount of voltage drop is dependent on the value of the resistor and the amount of current flow.
7. The voltage drop across resistors can be found by utilizing Ohm's law: $V = I \times R$.
8. The series circuit divides up the applied voltage, and it appears proportionally across all the individual resistors. This characteristic was first observed by Gustav Kirchhoff in 1847; in honor of his discovery, this effect is known as Kirchhoff's voltage law, which states: The sum of the voltage drops in a series circuit is equal to the total voltage applied.

QUICK REFERENCE SUMMARY SHEET

FIGURE 7-37 Series Circuits.

Example Circuit (Resistors)

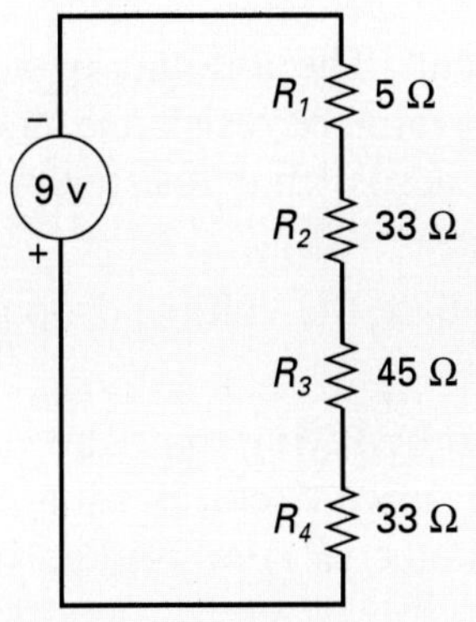

Example Circuit (Batteries)

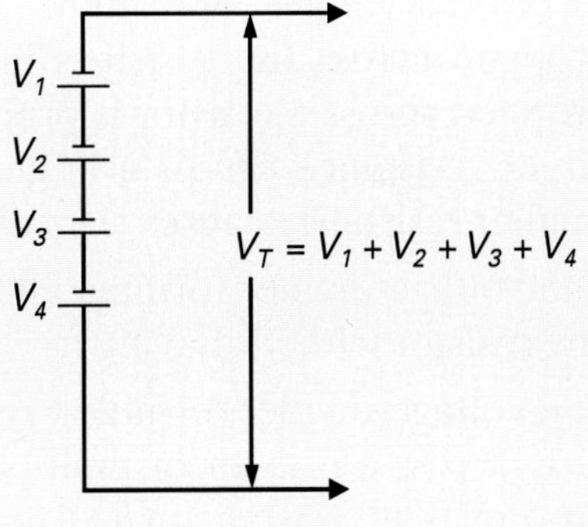

$$I = I_1 = I_2 = I_3 = \cdots$$

$$R_T = R_1 + R_2 + R_3 + \cdots$$

$$V_S = V_{R1} + V_{R2} + V_{R3} + \cdots$$

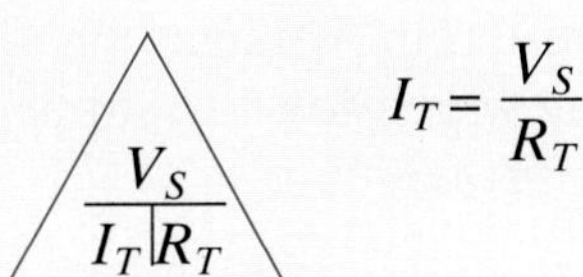

$$I_T = \frac{V_S}{R_T}$$

$V_{R1} = I_1 \times R_1$,
$V_{R2} = I_2 \times R_2$,
$V_{R3} = I_3 \times R_3$, and so on

Voltage-divider formula $$V_X = \left(\frac{R_X}{R_T}\right) \times V_S$$

$$P_T = P_1 + P_2 + P_3 + \cdots$$

$$P_T = \frac{V_S^2}{R_T}$$

$$P_T = I^2 \times R_T$$

$$P_T = V_S \times I$$

I = series current in amperes
I_1 = current through R_1, in amperes
I_2 = current through R_2, in amperes
I_3 = current through R_3, in amperes
R_T = total circuit resistance, in ohms
R_1 = resistance of R_1, in ohms
R_2 = resistance of R_2, in ohms
R_3 = resistance of R_3, in ohms
V_S = source voltage, in volts
V_{R1} = voltage drop across R_1, in volts
V_{R2} = voltage drop across R_2, in volts
V_{R3} = voltage drop across R_3, in volts
V_X = voltage drop across selected resistor(s)
R_X = selected resistor(s) ohmic value
P_T = total power dissipated, in watts
P_1 = power dissipated by R_1, in watts
P_2 = power dissipated by R_2, in watts
P_3 = power dissipated by R_3, in watts

9. If only one resistor is connected in a series circuit, the entire applied voltage appears across this resistor. The amount of current flow is determined by the value of this single resistor and remains the same throughout the circuit.
10. A series-connected circuit is often referred to as a *voltage-divider circuit.* The total voltage applied (V_T) or source voltage (V_S) is divided and dropped across the resistors in the series circuit. The amount of voltage dropped across a resistor is proportional to the value of resistance, and so a larger resistance causes a larger voltage drop across that resistor, while a smaller resistance causes a smaller voltage drop.
11. The voltage-divider formula allows you to calculate the voltage drop across any resistor without having to work out current.
12. The voltage-divider formula can be summarized by saying that: The voltage drop across a resistor or multiple resistors in a series circuit is equal to the ratio of that resistance (R_X) to the total resistance (R_T) multiplied by the source voltage (V_S).
13. A potentiometer, or variable voltage divider, consists of a fixed value of resistance between two terminals and a wiper that can be adjusted to vary resistance between its terminal and one of the other two. If the wiper is moved down the resistance between terminals *A* and *B* will increase, while the resistance between *B* and *C* will decrease. If the wiper is moved up the resistance between terminals *A* and *B* will decrease, while the resistance between *B* and *C* will increase. The resistances between *A* and *B* and *B* and *C* are inversely proportional to one another to that if one were to increase the other would decrease, and vice versa.
14. The potentiometer can be adjusted to supply different voltages on the wiper. This voltage can be decreased by moving the wiper down to supply a minimum or the wiper can be moved up to supply a maximum.

Power in a Series Circuit

15. Power is the rate at which work is done. Work is said to have been done when energy is converted, in this case from electrical energy to heat energy.
16. Resistors dissipate heat energy, and the rate at which they dissipate energy is called *power* and is measured in *watts* (joules per second).
17. Resistors all have a resistive value, a tolerance, and a wattage rating. The wattage of a resistor is the amount of heat energy a resistor will dissipate per second, and this wattage is directly proportional to the resistor's size; a larger resistor will be able to dissipate more heat than a smaller resistor.
18. The larger the wattage, the greater the cost and the greater the size and area the resistor occupies within the equipment.

Troubleshooting a Series Circuit

19. A resistor will usually burn out and result in an open between its two leads when an excessive current flow occurs. This can normally, but not always, be noticed by a visual check of the resistor, which will appear charred due

to the excessive heat. In some cases you will need to use your multimeter to check the circuit components to determine where a problem exists.

20. The two basic problems that normally exist in a series circuit are opens and shorts; however, a problem is not always as drastic as a short or an open, but may be a variation in a component's value over a long period of time, which will eventually cause a problem.
21. The voltmeter is the most useful tool when checking series circuits as it can be used to measure voltage drops by connecting the meter leads across the component or resistors.
22. A component is open when its resistance is the maximum possible (infinity).
23. Using the voltmeter to check the amount of voltage drop across each resistor, two results will be obtained.
    a. The voltage drop across a good resistor will be zero volts.
    b. The voltage drop across an open resistor will be equal to the source voltage V_S.
24. With age, resistors will normally change their resistance value. This occurs slowly and will generally always cause a decrease in the resistor's resistance and eventually cause a circuit problem. This lowering of resistance will cause an increase of current, which will cause an increase in the power dissipated; if the wattage of the resistor is exceeded, it can burn out. If they do not burn out but merely blow the circuit fuse due to this increase in current, the problem can be found by measuring the resistance values of each resistor or by measuring how much voltage is dropped across each resistor and comparing these to the calculated voltage, based on the parts list supplied by the manufacturer.
25. A component has gone short when its resistance is 0 Ω.
26. To summarize opens and shorts in series circuits, we can say that:
    a. The supply voltage appears across an open component.
    b. Zero volts appears across a shorted component.

NEW TERMS

Equivalent resistance
Fixed voltage dividers
Kirchhoff's voltage law
Series circuit
Troubleshooting series circuits
Variable voltage divider
Voltage divider

REVIEW QUESTIONS

Multiple-Choice Questions

1. A series circuit:
    a. Is the connecting of components end to end
    b. Provides a single path for current
    c. Functions as a voltage divider
    d. All of the above
2. The total current in a series circuit is equal to:
    a. $I_1 + I_2 + I_3 + \cdots$
    b. $I_1 - I_2$
    c. $I_1 = I_2 = I_3 = \cdots$
    d. All of the above

3. If R_1 and R_2 are connected in series with a total current of 2 A, what will be the current flowing through R_1 and R_2, respectively?
   a. 1 A, 1 A
   b. 2 A, 1 A
   c. 2 A, 2 A
   d. All of the above could be true on some occasions.
4. The total resistance in a series circuit is equal to:
   a. The total voltage divided by the total current
   b. The sum of all the individual resistor values
   c. $R_1 + R_2 + R_3 + \cdots$
   d. All of the above
   e. None of the above are even remotely true.
5. Which of Kirchhoff's laws applies to series circuits:
   a. His voltage law
   b. His current law
   c. His power law
   d. None of them apply to series circuits, only parallel.
6. The amount of voltage dropped across a resistor is proportional to:
   a. The value of the resistor
   b. The current flow in the circuit
   c. Both (a) and (b)
   d. None of the above
7. If three resistors of 6 kΩ, 4.7 kΩ, and 330 Ω are connected in series with one another, what total resistance will the battery sense?
   a. 11.03 MΩ **c.** 6 kΩ
   b. 11.03 Ω **d.** 11.03 kΩ
8. The voltage-divider formula states that the voltage drop across a resistor or multiple resistors in a series circuit is equal to the ratio of that ____________ to the ____________ multiplied by the ____________.
   a. Resistance, source voltage, total resistance
   b. Resistance, total resistance, source voltage
   c. Total current, resistance, total voltage
   d. Total voltage, total current, resistance
9. The ____________ can be used as a variable voltage divider.
   a. Potentiometer **c.** SPDT switch
   b. Fixed resistor **d.** None of the above
10. A resistor of larger physical size will be able to dissipate ____________ heat than a small resistor.
    a. More **c.** About the same
    b. Less **d.** None of the above
11. The ____________ is the most useful tool when checking series circuits.
    a. Ammeter **c.** Voltmeter
    b. Wattmeter **d.** Both (a) and (b)

12. When an open component occurs in a series circuit, it can be noticed because:
   a. Zero volts appears across it
   b. The supply voltage appears across it
   c. 1.3 V appears across it
   d. None of the above
13. Power can be calculated by:
   a. The addition of all the individual power figures
   b. The product of the total current and the total voltage
   c. The square of the total voltage divided by the total resistance
   d. All of the above
14. A series circuit is known as a:
   a. Current divider c. Current subtractor
   b. Voltage divider d. All of the above.
15. In a series circuit only ____________ path(s) exists for current flow, while the voltage applied is distributed across all the individual resistors.
   a. Three c. Four
   b. Several d. One

Essay Questions

16. Describe a series-connected circuit. (7-1)
17. Describe and state mathematically what happens to current flow in a series circuit. (7-2)
18. Describe how total resistance can be calculated in a series circuit. (7-3)
19. Describe why voltage is dropped around a series circuit and how each voltage drop can be calculated. (7-4)
20. Briefly describe why resistance and voltage drops are proportional to one another. (7-4)
21. Describe a fixed and a variable voltage divider. (7-4-1 and 7-4-2)
22. How can individual and total power be calculated in a series circuit? (7-5)
23. How can you recognize shorts and opens in a series circuit when troubleshooting with a voltmeter? (7-6)
24. List the three problems that can occur in a series circuit. (7-6)
25. State Kirchhoff's voltage (series circuit) law. (7-4)

Practice Problems

26. If three resistors of 1.7 kΩ, 3.3 kΩ, and 14.4 kΩ are connected in series with one another across a 24 V source as shown in Figure 7-38, calculate:
   a. Total resistance (R_T)
   b. Circuit current
   c. Individual voltage drops
   d. Individual and total power dissipated
27. If 40 Ω and 35 Ω resistors are connected across a 24 V source, what would be the current flow through the resistors, and what resistance would cause half the current to flow?

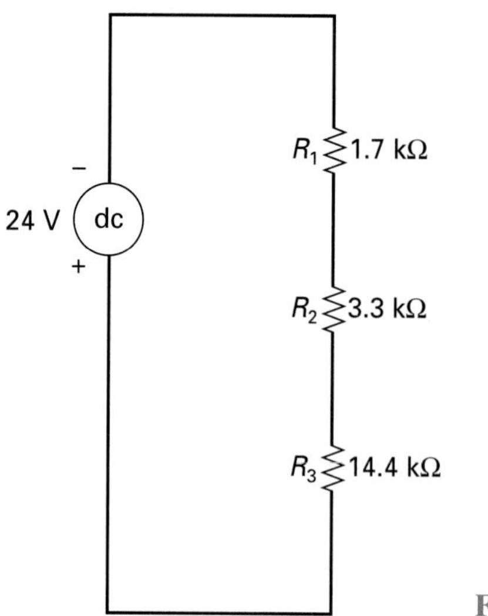

FIGURE 7-38

28. Calculate the total resistance (R_T) of the following series-connected resistors: 2.7 kΩ, 3.4 MΩ, 370 Ω, and 4.6 MΩ.
29. Calculate the value of resistors needed to divide up a 90 V source to produce 45 V and 60 V outputs, with a divider circuit current of 1 A.
30. If $R_1 = 4.7$ kΩ and $R_2 = 6.4$ kΩ and both are connected across a 9 V source, how much voltage will be dropped across R_2?
31. What current would flow through R_1 if it were one-third the ohmic value of R_2 and R_3, and all were connected in series with a total current of 6.5 mA flowing out of V_S?
32. Draw a circuit showing R_1 = 2.7 kΩ, R_2 = 3.3 kΩ, and R_3 = 0.027 MΩ in series with one another across a 20 V source. Calculate:

a. I_T	**d.** P_2	**g.** V_{R2}
b. P_T	**e.** P_3	**h.** V_{R3}
c. P_1	**f.** V_{R1}	**i.** I_{R1}

33. Calculate the current flowing through three light bulbs that are dissipating 120 W, 60 W, and 200 W when they are connected in series across a 120 V source. How is the voltage divided around the series circuit?
34. If three equal-value resistors are series connected across a dc power supply adjusted to supply 10 V, what percentage of the source voltage will appear across R_1?

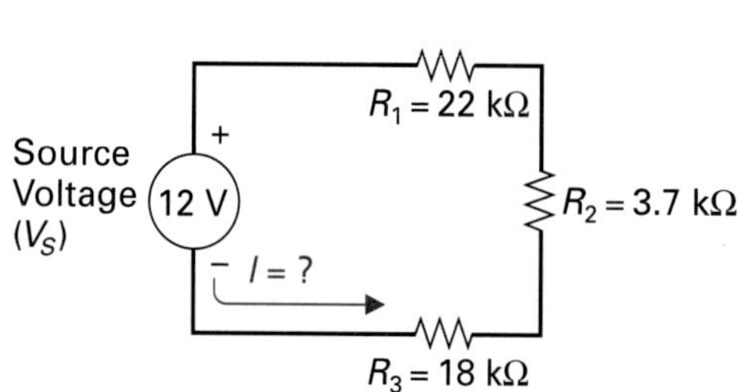

FIGURE 7-39

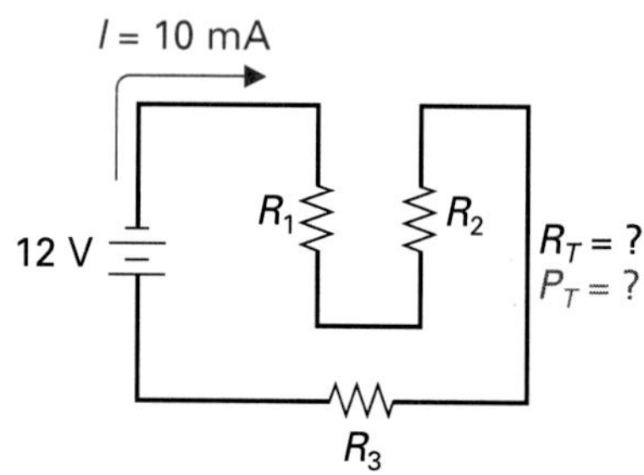

FIGURE 7-40

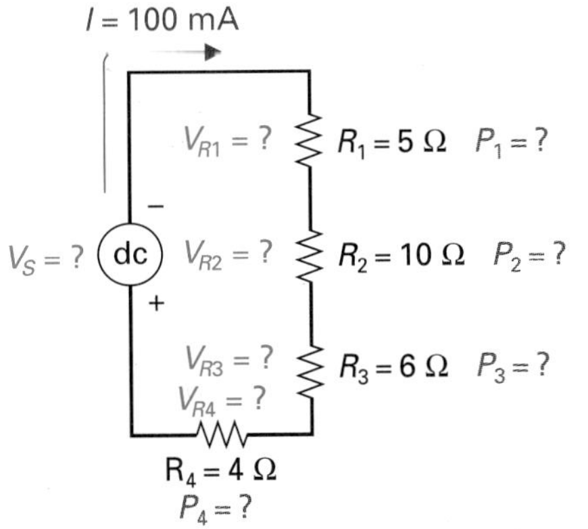

FIGURE 7-41

35. Refer to the following figures and calculate:

a. I (Figure 7-39)
b. R_T and P_T (Figure 7-40)
c. V_S, V_{R1}, V_{R2}, V_{R3}, V_{R4}, P_1, P_2, P_3, and P_4 (Figure 7-41)
d. P_T, I, R_1, R_2, R_3, and R_4 (Figure 7-42)

Troubleshooting Questions

36. If three bulbs are connected across a 9 V battery in series, and the filament in one of the bulbs burned out, causing an open in the bulb, would the other lamps be on? Explain why.

37. Using a voltmeter, how would a short be recognized in a series circuit?

38. If one of three series-connected bulbs is shorted, will the other two bulbs be on? Explain why.

39. When one resistor in a series string is open, explain what would happen to the circuit's:

a. Current
b. Resistance
c. Voltage across the open component
d. Voltage across the other components

40. When one resistor in a series string is shorted, explain what would happen to the circuit's:

a. Current
b. Resistance
c. Voltage across the shorted component
d. Voltage across the other components

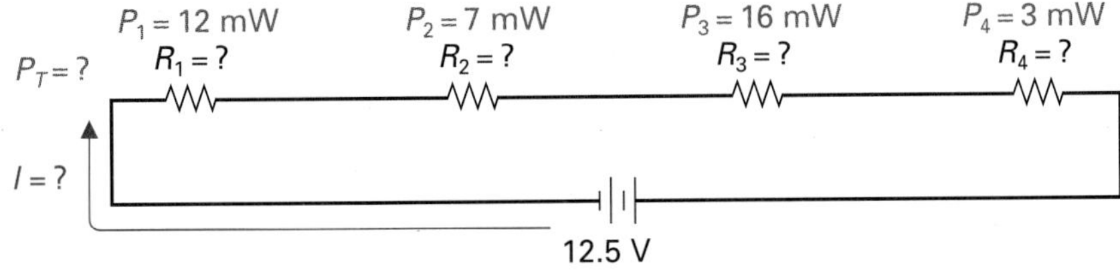

FIGURE 7-42

CHAPTER

8

OBJECTIVES

After completing this chapter, you will bc able to:

1. Describe the difference between a series and a parallel circuit.
2. Be able to recognize and determine whether circuit components are connected in series or parallel.
3. Explain why voltage measures the same across parallel-connected components.
4. State Kirchhoff's current law.
5. Describe why branch current and resistance are inversely proportional to one another.
6. Determine the total resistance of any parallel-connected resistive circuit.
7. Describe and be able to apply all formulas associated with the calculation of voltage, current, resistance, and power in a parallel circuit.
8. Describe how a short, open, or component variation will affect a parallel circuit's operation and how it can be recognized.

Parallel DC Circuits

OUTLINE

VIGNETTE

Bringing Home the Bacon

Sociable and athletic looking, Robert Noyce was the son of a small-town Iowa Congregational minister. As a college freshman in the spring of 1948, he decided to throw a Hawaiian luau for his friends. This was not a mistake, but stealing a pig and roasting it was, as the theft was uncovered. Noyce, a very promising student, was expelled and went to work for an insurance company.

After a short time, Noyce realized that he wanted to return to his studies, which proved not only profitable for him but also for the electronics industry. He graduated an honor student from MIT, became a brilliant physicist, invented the semiconductor integrated circuit, won the National Medal of Science Award, and founded Intel Corporation.

INTRODUCTION

By tracing the path of current, we can determine whether a circuit has series-connected or parallel-connected components. In a series circuit, there is only one path for current, whereas in parallel circuits the current has two or more paths. These paths are known as *branches.* A parallel circuit, by definition, is when two or more components are connected to the same voltage source so that the current can branch out over two or more paths. In a parallel resistive circuit, two or more resistors are connected to the same source, so current splits to travel through each separate branch resistance.

8-1 COMPONENTS IN PARALLEL

Parallel Circuit
Also called shunt; circuit having two or more paths for current flow.

Many components, other than resistors, can be connected in parallel, and a **parallel circuit** can easily be identified because current is split into two or more paths. Being able to identify a parallel connection requires some practice, because they can come in many different shapes and sizes. The means for recognizing series circuits is that if you can place your pencil at the negative terminal of the voltage source (battery) and follow the wire connections through components to the positive side of the battery and only have one path to follow, the circuit is connected in series. If, however, you can place your pencil at the negative terminal of the voltage source and follow the wire and at some point have a choice of two or more routes, the circuit is connected with two or more parallel

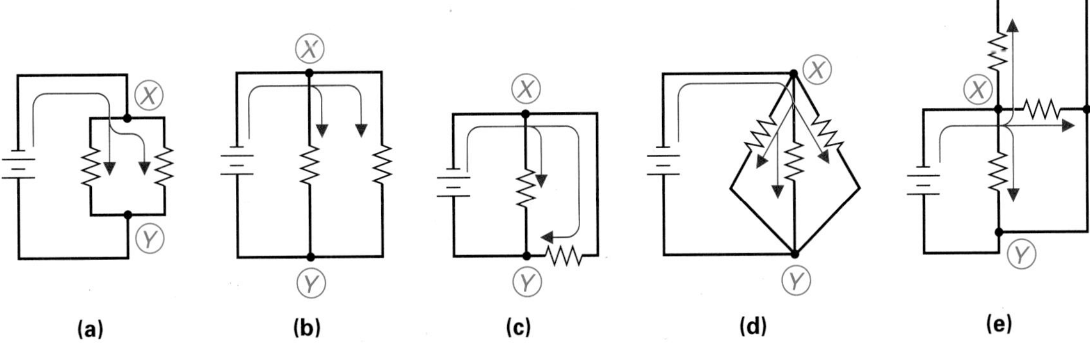

FIGURE 8-1 **Parallel Circuits.**

branches; the number of routes determines the number of parallel branches. Figure 8-1 illustrates five examples of parallel resistive circuits.

■ EXAMPLE:

Figure 8-2(a) illustrates four resistors laid out on a table top.

a. With wire leads, connect all four resistors in parallel between the negative and positive terminals of the battery.
b. Draw the schematic diagram of the parallel-connected circuit.

■ ***Solution:***

Figure 8-2(b) shows how to connect the resistors in parallel.

■ EXAMPLE:

Figure 8-3(a) shows four 1.5 V cells and three lamps. Using wires, connect all of the cells in parallel to create a 1.5 V source. Then connect all of the three lamps in parallel with one another and finally connect the 1.5 V source across the three parallel-connected lamp load.

■ ***Solution:***

In Figure 8-3(b) you can see the final circuit containing a source consisting of four parallel-connected 1.5 V cells, and a load consisting of three parallel-connected lamps. As explained in Chapter 6, when cells are connected in parallel, the total voltage remains the same as for one cell; however, the current demand can now be shared.

8-2 VOLTAGE IN A PARALLEL CIRCUIT

Figure 8-4(a) illustrates a simple circuit with four resistors connected in parallel across the voltage source of a 9 V battery. The current from the negative side of the battery will split between the four different paths or branches, yet the volt-

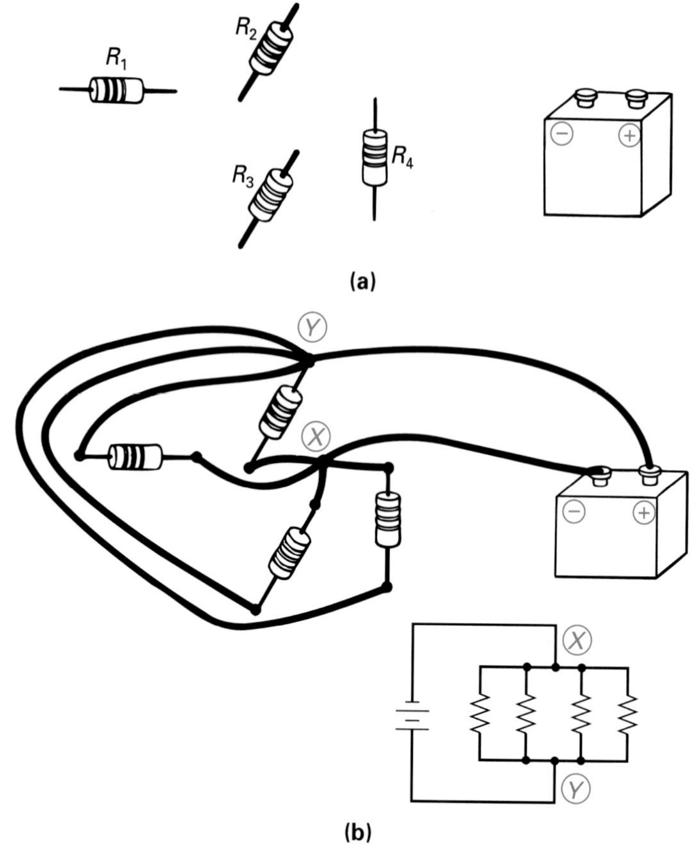

FIGURE 8-2 **Connecting Resistors in Parallel. (a) Problem. (b) Solution.**

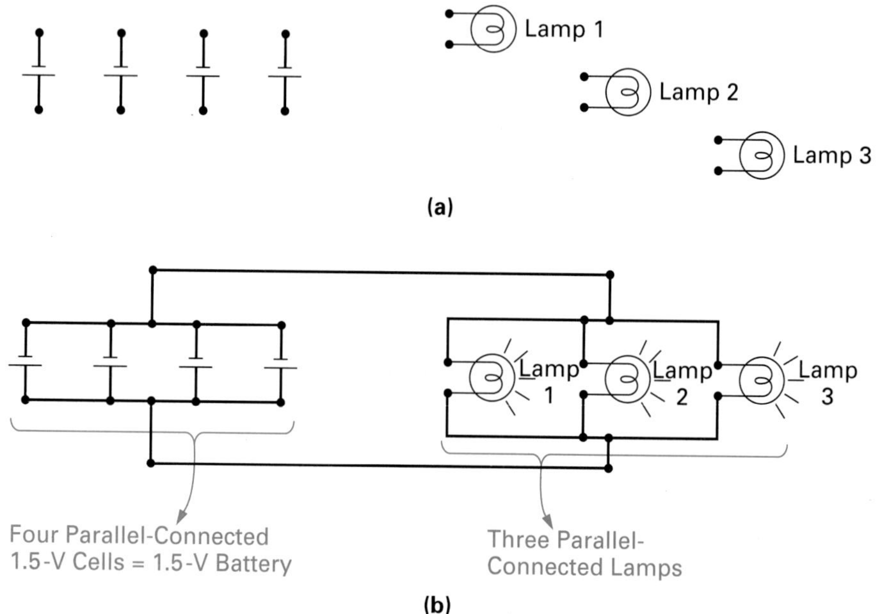

FIGURE 8-3 **Parallel-Connected Cells and Lamps.**

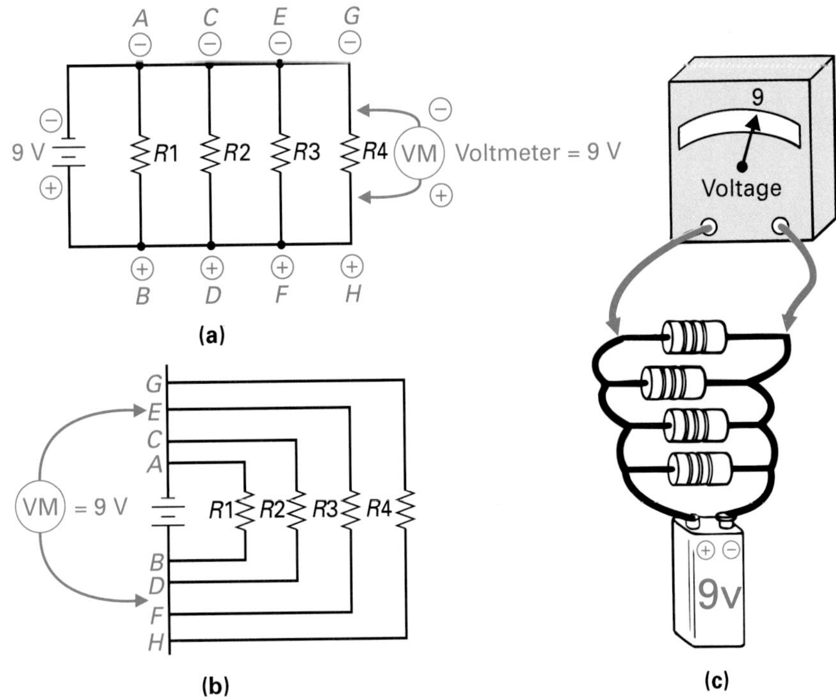

FIGURE 8-4 Voltage in a Parallel Circuit.

age drop across each branch of a parallel circuit is equal to the voltage drop across all the other branches in parallel. This means that if the voltmeter were to measure the voltage across *A* and *B* or *C* and *D* or *E* and *F* or *G* and *H*, they would all be the same or, in this example, would all drop 9 V.

It is quite easy to imagine why there will be the same voltage drop across all the resistors, seeing that points *A, C, E,* and *G* are all one connection and points *B, D, F,* and *H* are all one connection. Measuring the voltage drop with the voltmeter across any of the resistors is the same as measuring the voltage across the battery, as shown in Figure 8-4(b). As long as the voltage source remains constant, the voltage drop will always be common (9 V) across the parallel resistors, no matter what value or how many resistors are connected in parallel. The voltmeter is therefore measuring the voltage between two common points that are directly connected to the battery, and the voltage dropped across all these parallel resistors will be equal to the source voltage.

In Figure 8-5(a) and (b), the same circuit is shown in two different ways, so you can see how the same circuit can look completely different. In both examples, the voltage drop across any of the resistors will always be the same and, as long as the voltage source is not heavily loaded, equal to the source voltage. Just as you can trace the positive side of the battery to all four resistors, you can also trace the negative side to all four resistors.

Mathematically stated, we can say that in a parallel circuit

$$V_{R1} = V_{R2} = V_{R3} = V_{R4} = V_S$$

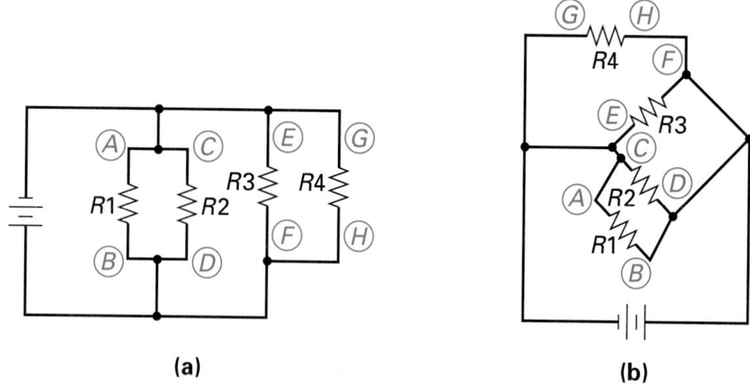

FIGURE 8-5 **Parallel-Circuit Voltage Drop.**

Voltage drop across R_1 = voltage drop across R_2 = voltage drop across R_3 (etc.) = source voltage

For the sake of discussion, we will relate this to a water analogy, as seen in Figure 8-6. The pressure across valves A and B will always be the same, even if one offers more opposition than the other. This is because the pressure measured across either valve will be the same as checking the pressure difference between piping X and Y, which run directly back to the pump, so the pressure across A and B is the same as the pressure difference across the pump.

EXAMPLE:

Refer to Figure 8-7 and calculate:

a. Voltage drop across R_1
b. Voltage drop across R_2
c. Voltage drop across R_3

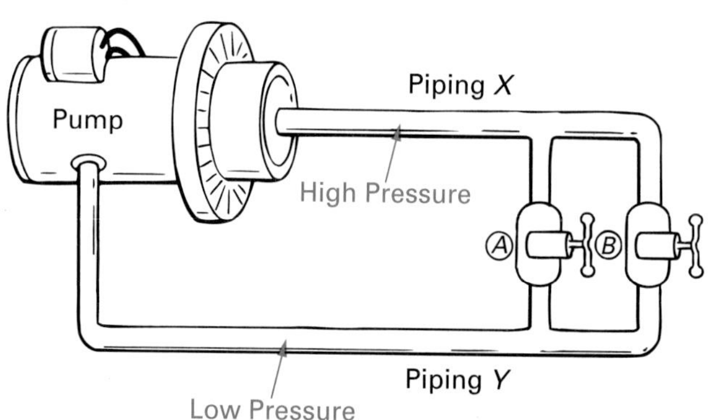

FIGURE 8-6 **Fluid Analogy of Parallel-Circuit Pressure.**

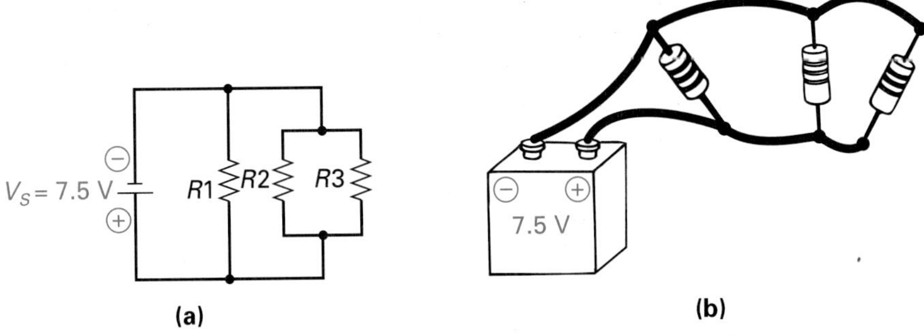

FIGURE 8-7 **(a) Schematic. (b) Pictorial.**

Solution:

Since all these resistors are connected in parallel, the voltage across every branch will be the same and equal to the source voltage applied. Therefore,

$$V_{R1} = V_{R2} = V_{R3} = V_S$$
$$7.5\text{ V} = 7.5\text{ V} = 7.5\text{ V} = 7.5\text{ V}$$

SELF-TEST REVIEW QUESTIONS FOR SECTIONS 8-1 AND 8-2

1. Describe a parallel circuit.
2. True or false: A parallel circuit is also known as a voltage-divider circuit.
3. What would be the voltage drop across R_1 if V_S = 12 V and R_1 and R_2 are both in parallel with one another and equal to 24 Ω each?
4. Can Kirchhoff's voltage law be applied to parallel circuits?

8-3 CURRENT IN A PARALLEL CIRCUIT

In addition to providing the voltage law for series circuits, Gustav Kirchhoff (in 1847) was the first to observe and prove that the sum of all the **branch currents** in a parallel circuit ($I_1 + I_2 + I_3$, etc.) was equal to the total current (I_T). In honor of his second discovery, this phenomenon is known as **Kirchhoff's current law.** It states that the sum of all the current entering a junction is equal to the sum of all the currents leaving that same junction.

Branch Current
A portion of the total current that is present is one path of a parallel circuit.

Kirchhoff's Current Law
The sum of the currents flowing into a point in a circuit is equal to the sum of the currents flowing out of that same point.

Figure 8-8(a) and (b) illustrate two examples of how this law applies. In both examples, the sum of the currents entering a junction is equal to the sum of the currents leaving that same junction. In Figure 8-8(a) the total current arrives at a junction X and splits to produce three branch currents, I_1, I_2, and I_3, which cumulatively equal the total current (I_T) that arrived at the junction X. The same three branch currents combine at junction Y, and the total current (I_T) leaving that junction is equal to the sum of the three branch currents arriving at junction Y.

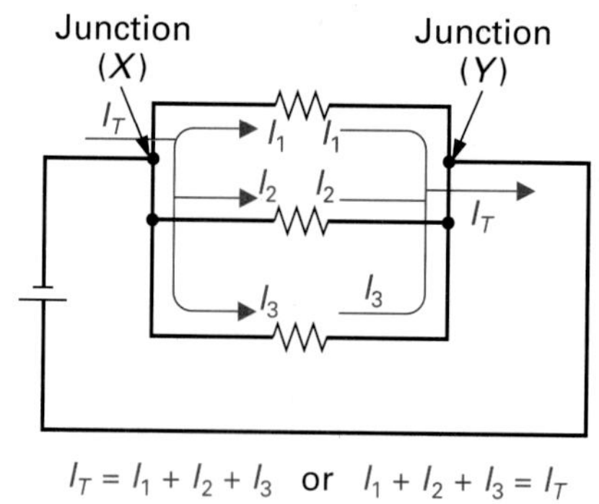

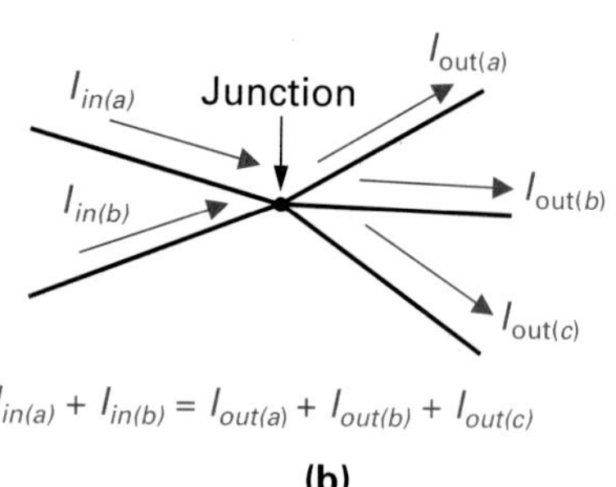

FIGURE 8-8 Kirchhoff's Current Law.

In Figure 8-8(b) you can see that there are two branch currents entering a junction [$I_{in(a)}$ and $I_{in(b)}$] and three branch currents leaving that same junction [$I_{out(a)}$, $I_{out(b)}$, and $I_{out(c)}$], and the sum of the input currents will equal the sum of the output currents: $I_{in(a)} + I_{in(b)} = I_{out(a)} + I_{out(b)} + I_{out(c)}$.

EXAMPLE:

Refer to Figure 8-9 and calculate the value of I_1.

Solution:

By Kirchhoff's current law,

$$I_T = I_1 + I_2$$
$$7\text{ A} = ? + 3\text{ A}$$
$$I_1 = 4\text{ A}$$

EXAMPLE:

Refer to Figure 8-10 and calculate the value of I_T.

Solution:

By Kirchhoff's current law,

$$I_T = I_1 + I_2 + I_3 + I_4$$
$$? = 2\text{ mA} + 17\text{ mA} + 7\text{ mA} + 37\text{ mA}$$
$$I_T = 63\text{ mA}$$

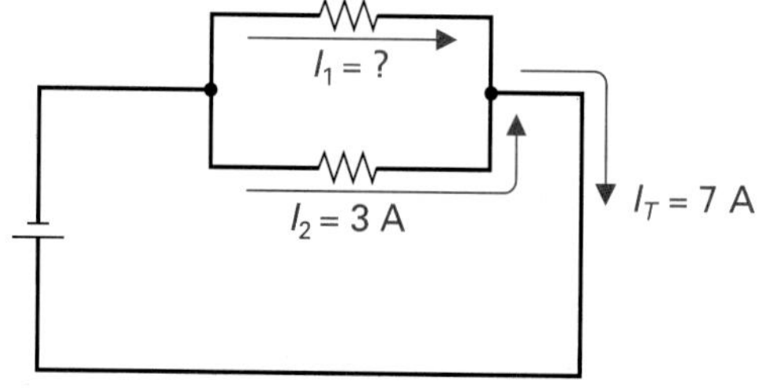

FIGURE 8-9 Calculating Branch Current.

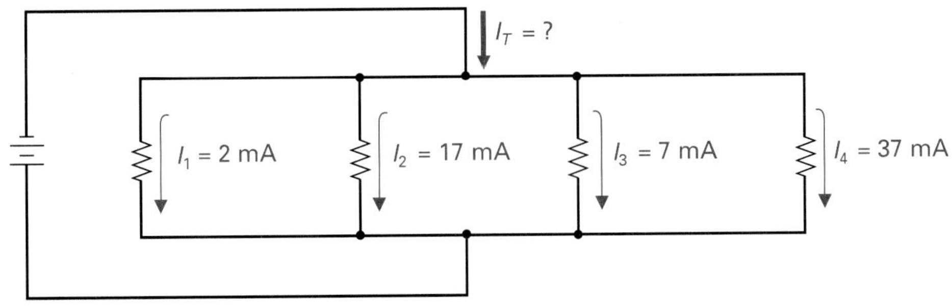

FIGURE 8-10 Calculating Total Current.

As with series circuits, to find out how much current will flow through a parallel circuit, we need to find out how much opposition or resistance is being connected across the voltage source.

$$I_T = \frac{V_S}{R_T}$$

$$\text{total current} = \frac{\text{source voltage}}{\text{total resistance}}$$

When we connect resistors in parallel, the total resistance in the circuit will actually decrease. In fact, the total resistance in a parallel circuit will always be less than the value of the smallest resistor in the circuit.

Let us now examine this point further. If two sets of identical resistors (R_1, R_2, and R_3) were used to build both a series and a parallel circuit, as seen in Figure 8-11, the total current flow in the parallel circuit would be larger than the

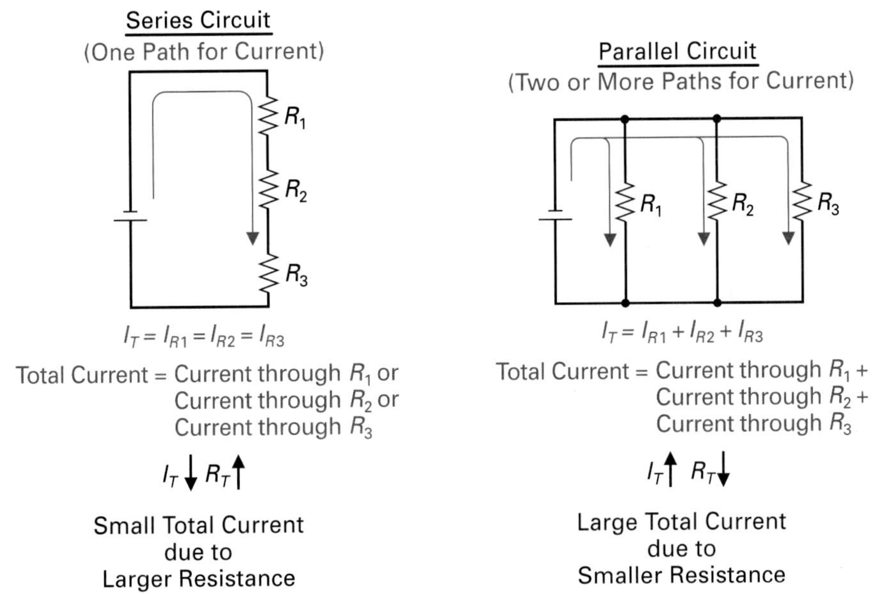

FIGURE 8-11 Series Circuit and Parallel Circuit Current Comparison.

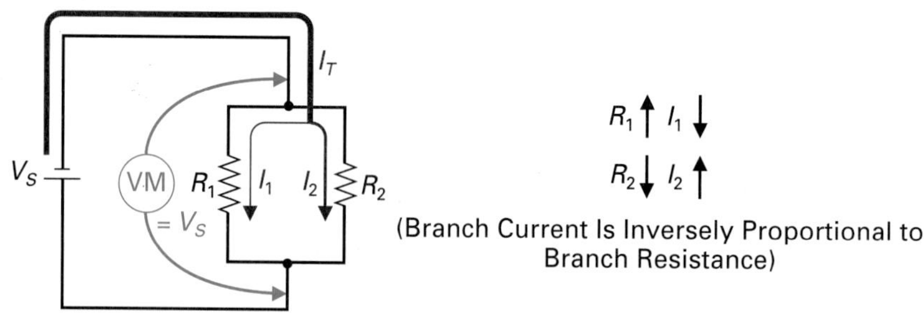

FIGURE 8-12 The Parallel Circuit Current Divider.

total current in the series circuit, because the parallel circuit has two or more paths for current to flow, while the series circuit only has one. To explain why the total current will be larger in a parallel circuit, let us take the analogy of a freeway with only one path for traffic to flow. The freeway is equivalent to a series circuit, and a certain amount of traffic is allowed to flow along this freeway. If the freeway is expanded to accommodate two lanes, a greater amount of traffic can flow along the freeway in the same amount of time; with more lanes, you have a greater total amount of traffic flow and, in parallel circuits, more branches will allow a greater total amount of current flow because there is less resistance in more paths than there is with only one path. This concept is summarized in Figure 8-11.

Current Divider A parallel network designed to proportionally divide the circuit's total current.

Just as a series circuit is often referred to as a voltage-divider circuit, a parallel circuit is often referred to as a **current-divider** circuit, because the total current arriving at a junction will divide or split into branch currents (Kirchhoff's law), as shown in Figure 8-12.

The current division is inversely proportional to the resistance in the branch seeing that the voltage across both resistors is constant and equal to the source voltage (V_S). This means that a large branch resistance will cause a small branch current ($I\downarrow = V/R\uparrow$), and a small branch resistance will cause a large branch current ($I\uparrow = V/R\downarrow$).

EXAMPLE:

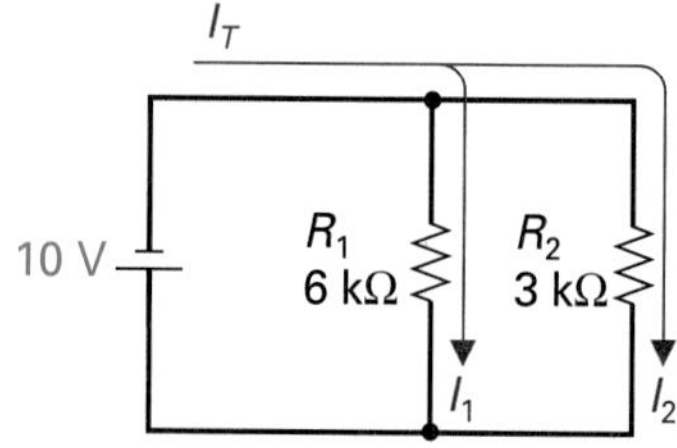

FIGURE 8-13 Parallel Circuit Example.

Calculate the following for Figure 8-13:

a. I_1
b. I_2
c. I_T

Solution:

Since R_1 and R_2 are connected in parallel across the 10 V source, the voltage across both resistors will be 10 V.

a. $I_1 = \frac{V_{R1}}{R_1} = \frac{10\text{ V}}{6\text{ k}\Omega} = 1.6\text{ mA}$ (smaller branch current through larger branch resistance).

b. $I_2 = \frac{V_{R2}}{R_2} = \frac{10\text{ V}}{3\text{ k}\Omega} = 3.3\text{ mA}$ (larger branch current through smaller branch resistance).

c. By Kirchhoff's current law,

$$\begin{aligned} I_T &= I_1 + I_2 \\ &= 1.6\text{ mA} + 3.3\text{ mA} \\ &= 4.9\text{ mA} \end{aligned}$$

By rearranging Ohm's law, we can arrive at another formula, which is called the *current-divider formula* and can be used to calculate the current through any branch of a multiple-branch parallel circuit.

$$I_x = \frac{R_T}{R_x} \times I_T$$

where I_x = branch current desired
R_T = total resistance
R_x = resistance in branch
I_T = total current

EXAMPLE:

Refer to Figure 8-14 and calculate the following if the total circuit resistance (R_T) is equal to 1 kΩ:

a. $I_1 =$
b. $I_2 =$
c. $I_3 =$

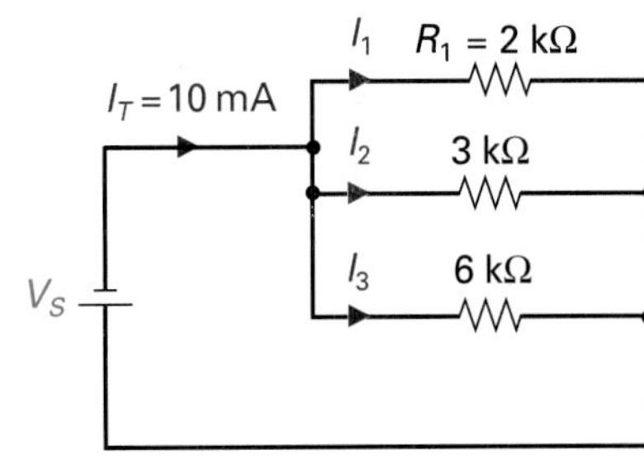

FIGURE 8-14 Parallel Example Circuit.

Solution:

Since the source and therefore the voltage across each branch resistor are not known, we will use the current-divider formula to calculate I_1, I_2, and I_3.

a. $I_1 = \frac{R_T}{R_1} \times I_T = \frac{1\text{ k}\Omega}{2\text{ k}\Omega} \times 10\text{ mA} = 5\text{ mA}$

(smallest branch resistance has largest branch current)

b. $I_2 = \frac{R_T}{R_2} \times I_T = \frac{1\text{ k}\Omega}{3\text{ k}\Omega} \times 10\text{ mA}$

$= 3.33\text{ mA}$

CALCULATOR SEQUENCE

Step	Keypad Entry	Display Response
1.	1 E 3	1.E3
2.	÷	
3.	2 E 3	2.E3
4.	×	0.5
5.	1 0 E 3 +/−	10E-3
6.	=	5.-03

c. $I_3 = \frac{R_T}{R_3} \times I_T = \frac{1\ \text{k}\Omega}{6\ \text{k}\Omega} \times 10\ \text{mA}$

$= 1.67\ \text{mA}$

(largest branch resistance has smallest branch current)

To double-check that the values for I_1, I_2, and I_3 are correct, you can apply Kirchhoff's current law, which is

$$I_T = I_1 + I_2 + I_3$$
$$10\ \text{mA} = 5\ \text{mA} + 3.33\ \text{mA} + 1.67\ \text{mA}$$
$$= 10\ \text{mA}$$

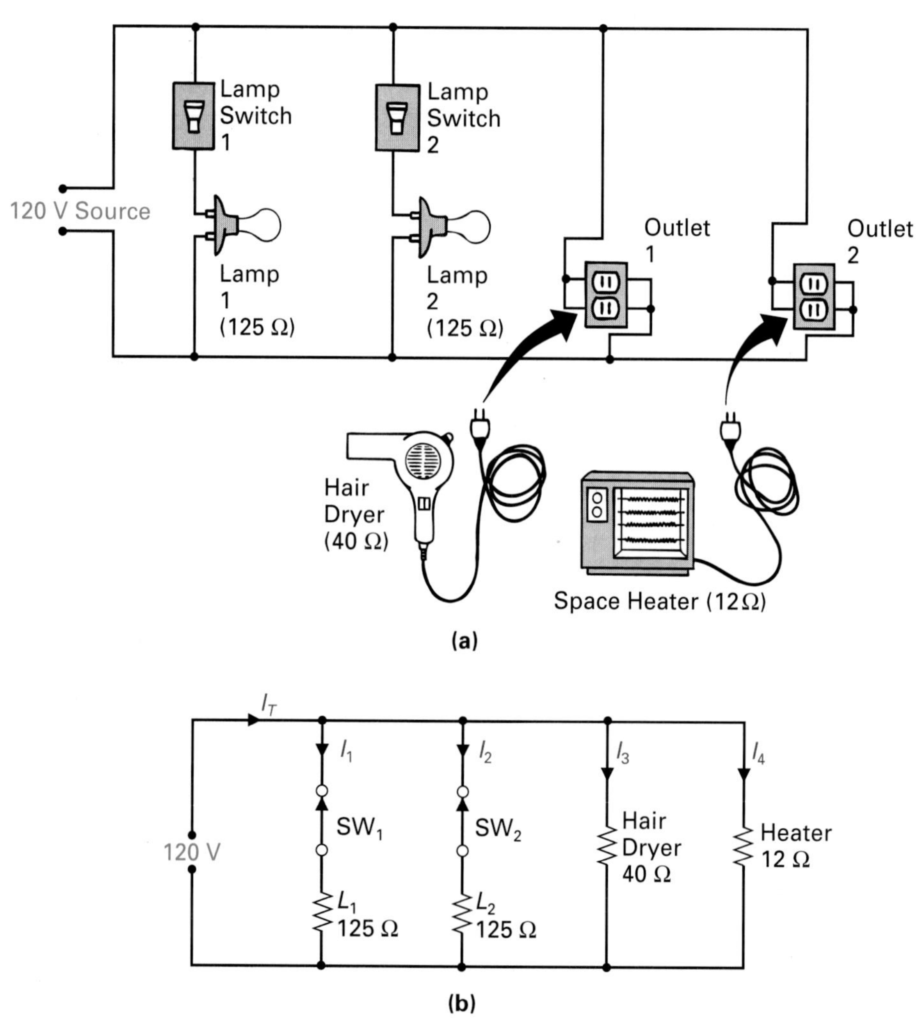

FIGURE 8-15 **Parallel Home Electrical System.**

EXAMPLE:

A common use of parallel circuits is in the residential electrical system. All of the household lights and appliances are wired in parallel, as seen in the typical room wiring circuit in Figure 8-15(a). If it is a cold winter morning, and lamps 1 and 2 are switched ON, together with the space heater and hair dryer, what will the individual branch currents be, and what will be the total current drawn from the source?

Solution:

Figure 8-15(b) shows the schematic of the pictorial in Figure 8-15(a). Since all resistances are connected in parallel across a 120 V source, the voltage across all devices will be 120 V. Using Ohm's law we can calculate the four branch currents:

$$I_1 = \frac{V_{\text{lamp1}}}{R_{\text{lamp1}}} = \frac{120\text{ V}}{125\ \Omega} = 960\text{ mA}$$

$$I_2 = \frac{V_{\text{lamp2}}}{R_{\text{lamp2}}} = \frac{120\text{ V}}{125\ \Omega} = 960\text{ mA}$$

$$I_3 = \frac{V_{\text{hairdryer}}}{R_{\text{hairdryer}}} = \frac{120\text{ V}}{40\ \Omega} = 3\text{ A}$$

$$I_4 = \frac{V_{\text{heater}}}{R_{\text{heater}}} = \frac{120\text{ V}}{12\ \Omega} = 10\text{ A}$$

By Kirchhoff's current law,

$$\begin{aligned} I_T &= I_1 + I_2 + I_3 + I_4 \\ &= 960\text{ mA} + 960\text{ mA} + 3\text{ A} + 10\text{ A} \\ &= 14.92\text{ A} \end{aligned}$$

SELF-TEST REVIEW QUESTIONS FOR SECTION 8-3

1. State Kirchhoff's current law.
2. If $I_T = 4$ A and $I_1 = 2.7$ A in a two-resistor parallel circuit, what would be the value of I_2?
3. State the current-divider formula.
4. Calculate I_1 if $R_T = 1$ kΩ, $R_1 = 2$ kΩ, and $V_T = 12$ V.

8-4 RESISTANCE IN A PARALLEL CIRCUIT

We now know that parallel circuits will have a larger current flow than a series circuit containing the same resistors due to the smaller total resistance, as already illustrated in Figure 8-11. To calculate exactly how much total current will flow, we need to calculate the total resistance that the parallel circuit develops or contains.

The ability of a circuit to conduct current is a measure of that circuit's conductance, and you will remember from Chapter 2 that conductance (G) is equal to the reciprocal of resistance and is measured in siemens.

$$G = \frac{1}{R} \quad \text{(siemens)}$$

Every resistor in a parallel circuit will have a conductance figure that is equal to the reciprocal of its resistance, and the total conductance (G_T) of the circuit will be equal to the sum of all the individual resistor conductances; therefore,

$$G_T = G_{R1} + G_{R2} + G_{R3} + \cdots$$

(Total conductance is equal to the conductance of R_1 + the conductance of R_2 + the conductance of R_3 + $\cdot\cdot\cdot$.) Once you have calculated total conductance, the reciprocal of this figure will give you total resistance. If, for example, we have two resistors in parallel, as shown in Figure 8-16, the conductance for R_1 will equal

$$G_{R1} = \frac{1}{R_1} = \frac{1}{20\ \Omega} = 0.05\ \text{S}$$

The conductance for R_2 will equal

$$G_{R2} = \frac{1}{R_2} = \frac{1}{40\ \Omega} = 0.025\ \text{S}$$

The total conductance will therefore equal

$$\begin{aligned} G_{\text{total}} &= G_{R1} + G_{R2} \\ &= 0.05 + 0.025 \\ &= 0.075\ \text{S} \end{aligned}$$

Since total resistance is equal to the reciprocal of total conductance, total resistance for the parallel circuit in Figure 8-16 will be

$$R_{\text{total}} = \frac{1}{G_{\text{total}}} = \frac{1}{0.075\ \text{S}} = 13.3\ \Omega$$

Combining these three steps (first calculate individual conductances, total conductance and then total resistance) we can arrive at the following *reciprocal formula:*

$$R_{\text{total}} = \frac{1}{(1/R_1) + (1/R_2)}$$

this formula states that the conductance of R_1 (G_{R1})
+ conductance of R_2 (G_{R2})
= total conductance (G_T),
and the reciprocal of total
conductance is equal to
total resistance

In the example for Figure 8-16, this combined general formula for total resistance can be verified by plugging in the example values.

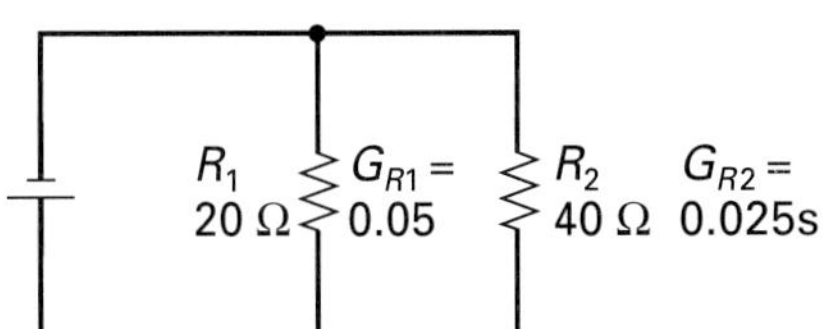

FIGURE 8-16 Parallel Circuit Conductance and Resistance.

$$R_T = \frac{1}{(1/R_1) + (1/R_2)}$$
$$= \frac{1}{(1/20) + (1/40)}$$
$$= \frac{1}{0.05 + 0.025}$$
$$= \frac{1}{0.075}$$
$$= 13.3\ \Omega$$

CALCULATOR SEQUENCE

Step	Keypad Entry	Display Response
1.	(Clear Memory)	
2.	[2] [0]	20.
3.	[1/x]	5.E-2
4.	[+]	5.E-2
5.	[4] [0]	40
6.	[1/x]	2.5E-2
7.	[=]	7.5E-2
8.	[1/x]	13.33333

The *reciprocal formula* for calculating total parallel circuit resistance for any number of resistors then is

$$R_T = \frac{1}{(1/R_1) + (1/R_2) + (1/R_3) + (1/R_4) + \cdots}$$

EXAMPLE:

Referring to Figure 8-17, calculate:

a. Total resistance
b. Voltage drop across R_2
c. Voltage drop across R_3

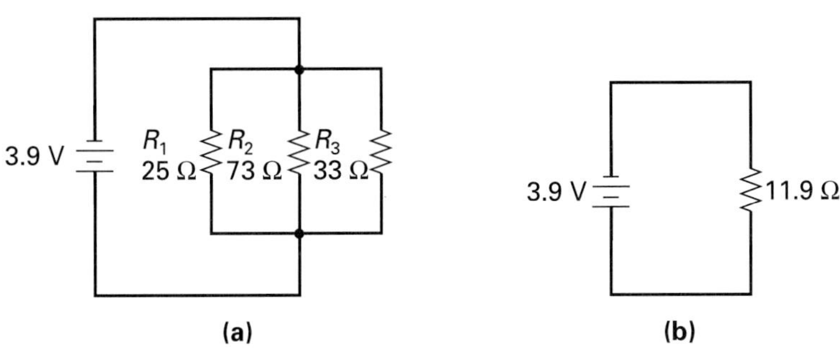

FIGURE 8-17 (a) Parallel Example Circuit. (b) Equivalent Circuit.

■ ***Solution:***

$$\text{a. } R_T = \frac{1}{(1/R_1) + (1/R_2) + (1/R_3)}$$
$$= \frac{1}{(1/25\ \Omega) + (1/73\ \Omega) + (1/33\ \Omega)}$$
$$= \frac{1}{0.04 + 0.014 + 0.03}$$
$$= 11.9\ \Omega$$

With parallel resistance circuits, the total resistance is always smaller than the smallest branch resistance. In this example the total opposition of this circuit is equivalent to 11.9 Ω, as shown in Figure 8-17(b).

With parallel resistive circuits, the voltage drop across any branch is equal to the voltage drop across each of the other branches and is equal to the source voltage, in this example 3.9 V.

8-4-1 *Two Resistors in Parallel*

If only two resistors are connected in parallel, a quick and easy formula called the *product-over-sum formula* can be used to calculate total resistance.

$$R_T = \frac{R_1 \times R_2}{R_1 + R_2}$$

$$\text{total resistance} = \frac{\text{product of both resistance values}}{\text{sum of both resistance values}}$$

Using this formula with the example in Figure 8-18, the total resistance can be calculated by the use of either parallel resistance formulas, (a) or (b), to make a comparison between the two:

$$\text{(a) } R_T = \frac{R_1 \times R_2}{R_1 + R_2}$$
$$= \frac{3.7\ \text{k}\Omega \times 2.2\ \text{k}\Omega}{3.7\ \text{k}\Omega + 2.2\ \text{k}\Omega}$$
$$= \frac{8.14\ \text{k}\Omega}{5.9\ \text{k}\Omega}$$

$$\text{(b) } R_T = \frac{1}{(1/R_1) + (1/R_2)}$$
$$= \frac{1}{(1/3.7\ \text{k}\Omega) + (1/2.2\ \text{k}\Omega)}$$
$$= \frac{1}{(270.2 \times 10^{-6}) + (454.5 \times 10^{-6})}$$

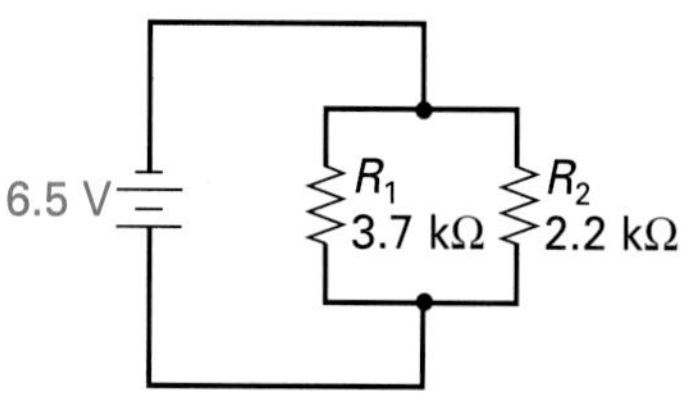

FIGURE 8-18 Two Resistors in Parallel.

$= 1.38\ k\Omega$

$$= \frac{1}{724.7 \times 10^{-6}}$$
$$= 1.38\ k\Omega$$

8-4-2 More Than Two Resistors in Parallel

You can see that the advantage of the product-over-sum parallel resistance formula (a) is its ease of use. Its disadvantage is that it can only be used for two resistors in parallel. In circuits containing more than two resistors, the previously discussed *reciprocal formula* must be used.

$$R_T = \frac{1}{(1/R_1) + (1/R_2) + (1/R_3) + (1/R_4) + \cdots}$$

8-4-3 Equal-Value Resistors in Parallel

If resistors of equal value are connected in parallel, a special case *equal-value formula* can be used to calculate the total resistance.

$$R_T = \frac{\text{value of one resistor } (R)}{\text{number of parallel resistors } (n)}$$

EXAMPLE:

Figure 8-19(a) shows how a stereo music amplifier is connected to drive two 8 Ω speakers, which are connected in parallel with one another. What is the total resistance connected across the amplifier's output terminals?

Solution:

Since both parallel-connected speakers have the same resistance [Figure 8-19(b)], the total resistance is most easily calculated by using the equal-value formula:

$$R_T = \frac{R}{n} = \frac{8\ \Omega}{2} = 4\ \Omega$$

EXAMPLE:

Refer to Figure 8-20 and calculate:

a. Total resistance in part (a)
b. Total resistance in part (b)

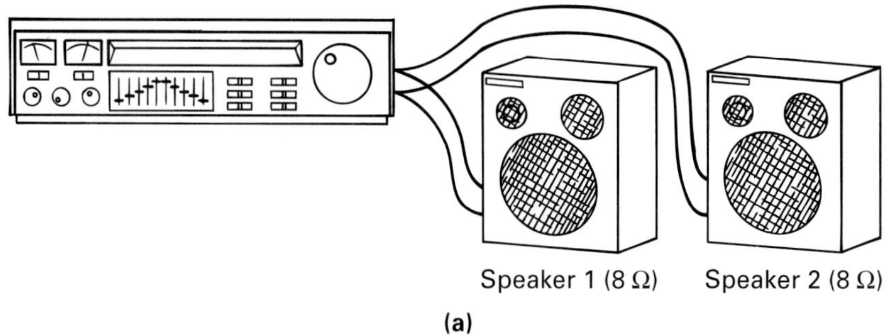

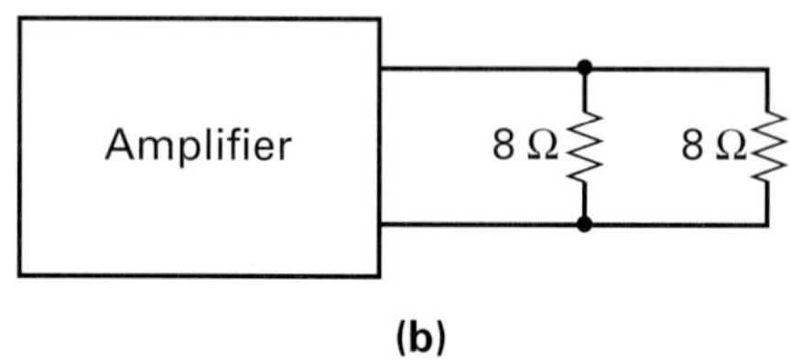

FIGURE 8-19 **Parallel-Connected Speakers.**

■ ***Solution:***

a. Figure 8-20(a) has only two resistors in parallel and therefore the two parallel resistor formula can be used; the equivalent circuit is seen in Figure 8-20(c)

$$R_T = \frac{R_1 \times R_2}{R_1 + R_2}$$

$$= \frac{4.5\ \text{M}\Omega \times 3.2\ \text{M}\Omega}{4.5\ \text{M}\Omega + 3.2\ \text{M}\Omega}$$

$$= \frac{14.4\ \text{M}\Omega}{7.7\ \text{M}\Omega}$$

$$= 1.9\ \text{M}\Omega$$

CALCULATOR SEQUENCE FOR (A)

Step	Keypad Entry	Display Response
1.	4 . 5 E 6	4.5E6
2.	×	
3.	3 . 2 E 6	3.2E6
4.	=	1.44E13
5.	STO (store in memory)	
6.	4 . 5 E 6	4.5E6
7.	+	
8.	3 . 2 E 6	3.2E6
9.	=	7.7E6
10.	C/CE	0.
11.	RM (Recall memory)	1.44E13
12.	÷	
13.	7 . 7 E 6	7.7E6
14.	=	1.87E6

b. Figure 8-20(b) has more than two resistors in parallel, and therefore the sum of conductances reciprocated formula must be used. The equivalent circuit can be seen in Figure 8-20(d).

$$R_T = \frac{1}{(1/R_1) + (1/R_2) + (1/R_3)}$$

$$= \frac{1}{(1/27\ \text{k}\Omega) + (1/10\ \text{k}\Omega) + (1/3.3\ \text{k}\Omega)}$$

$$= \frac{1}{440.0 \times 10^{-6}}$$

$$= 2.27\ \text{k}\Omega$$

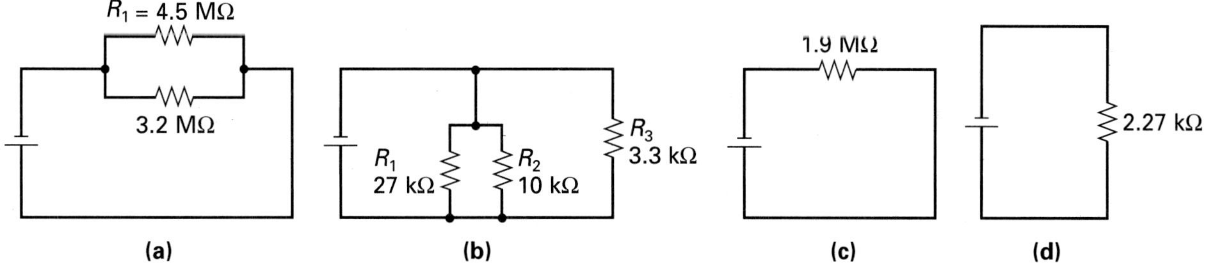

FIGURE 8-20 **Parallel Circuit Examples.**

EXAMPLE:

Find the total resistance of the parallel circuit in Figure 8-21.

Solution:

Since all four resistors are connected in parallel and are all of the same value, the equal-value resistors in parallel formula can be used.

$$R_T = \frac{R}{n} = \frac{2\text{ k}\Omega}{4} = 500\ \Omega$$

To summarize the effects of current, voltage, and resistance in parallel circuits, so far we could say:

1. Components are said to be connected in parallel when the current has to travel two or more paths between the negative and positive side of the voltage source.
2. The voltage across all the components in parallel is always the same.
3. The total current from the source is equal to the sum of all the branch currents (Kirchhoff's current law).
4. The amount of current flowing through each branch is inversely proportional to the resistance value in that branch.
5. The total resistance of a parallel circuit is always less than the value of the smallest branch resistor.

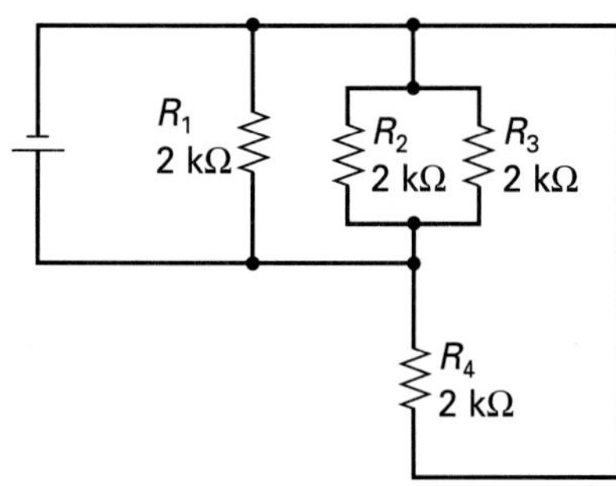

FIGURE 8-21 **Parallel Circuit Example.**

SELF-TEST REVIEW QUESTIONS FOR SECTION 8-4

State the parallel resistance formulas in questions 1 through 3 for calculating R_T when:

1. Two resistors are in parallel.
2. More than two are in parallel.
3. Equal-value resistors are in parallel.
4. Calculate total resistance when: $R_1 = 2.7\ \text{k}\Omega$, $R_2 = 24\ \text{k}\Omega$, and $R_3 = 1\ \text{M}\Omega$.

8-5 POWER IN A PARALLEL CIRCUIT

As with series circuits, the total power in a parallel resistive circuit is equal to the sum of all the power losses for each of the resistors in parallel.

$$P_T = P_1 + P_2 + P_3 + P_4 + \cdots$$

total power = addition of all the power losses

The formulas for calculating the amount of power dissipated are

$$P = I \times V$$

$$P = \frac{V^2}{R}$$

$$P = I^2 \times R$$

EXAMPLE:

Calculate the total amount of power dissipated in Figure 8-22.

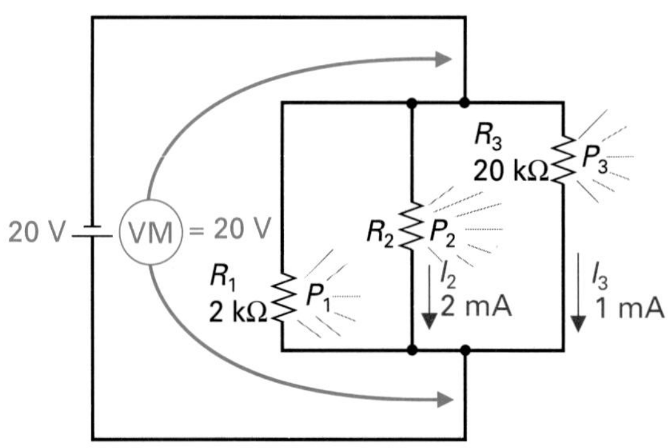

FIGURE 8-22 **Parallel Circuit Example.**

Solution:

The total power dissipated in a parallel circuit is equal to the sum of all the power dissipated by all the resistors. With P_{R1}, we only know voltage and resistance and therefore we can use the formula

$$P_{R1} = \frac{V^2}{R} = \frac{20^2}{2\text{ k}\Omega} = 0.2\text{ W} \quad \text{or} \quad 200\text{ mW}$$

With P_{R2}, we only know current and voltage, and therefore we can use the formula

$$P_{R2} = I \times V = 2\text{ mA} \times 20\text{ V} = 40\text{ mW}$$

With P_{R3}, we know V, I, and R; however, we will use the third power formula:

$$P_{R3} = I^2 \times R = 1\text{ mA}^2 \times 20\text{ k}\Omega = 20\text{ mW}$$

Total power (P_T) equals the sum of all the power or wattage losses for each resistor:

$$\begin{aligned} P_T &= P_{R1} + P_{R2} + P_{R3} \\ &= 200\text{ mW} + 40\text{ mW} + 20\text{ mW} \\ &= 260\text{ mW} \end{aligned}$$

CALCULATOR SEQUENCE

Step	Keypad Entry	Display Response
1.	2 0	20.
2.	x²	400.
3.	÷	
4.	2 E 3	2.E3
5.	=	0.2

CALCULATOR SEQUENCE

Step	Keypad Entry	Display Response
1.	2 E 3 +/−	2.E-3
2.	×	
3.	2 0	20.
4.	=	40E-3

CALCULATOR SEQUENCE

Step	Keypad Entry	Display Response
1.	1 E 3 +/− x²	1.E-6
2.	×	1.E-6
3.	2 0 E 3	20E3
4.	=	20.E-3

EXAMPLE:

In Figure 8-23, there are two ½ W (0.5 W) resistors connected in parallel. Should the wattage rating for each of these resistors be increased or decreased, or can they remain the same?

Solution:

Since current and voltage are known for both branches, the power formula used in both cases will be $P = I \times V$.

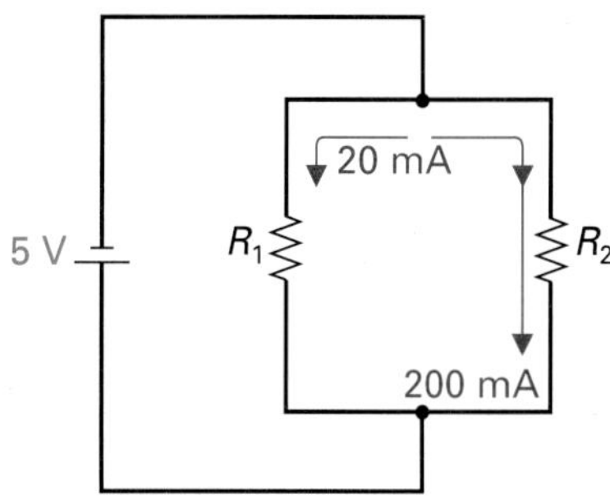

FIGURE 8-23 Parallel Circuit Example.

$$P_1 = I \times V \qquad P_2 = I \times V$$
$$= 20 \text{ mA} \times 5 \text{ V} \qquad = 200 \text{ mA} \times 5 \text{ V}$$
$$= 0.1 \text{ W} \qquad = 1 \text{ W}$$

R_1 is dissipating 0.1 W and has been designed to dissipate up to 0.5 W; it is therefore safe in this application. However, R_2 is dissipating 1 W and is only designed to dissipate 0.5 W. R_2 will consequently overheat unless it is replaced with a resistor of the same ohmic value with a 1 W or greater rating.

EXAMPLE:

Figure 8-24(a) shows a simplified diagram of an automobile external light system. A 12 V lead–acid battery is used as a source and is connected across eight parallel-connected lamps. The left and right brake lights are controlled by the brake switch, which is attached to the brake pedal. When the light switch is turned on, both the rear taillights and the low-beam headlights are brought into circuit and turned on. The high-beam set of headlights are activated only if the

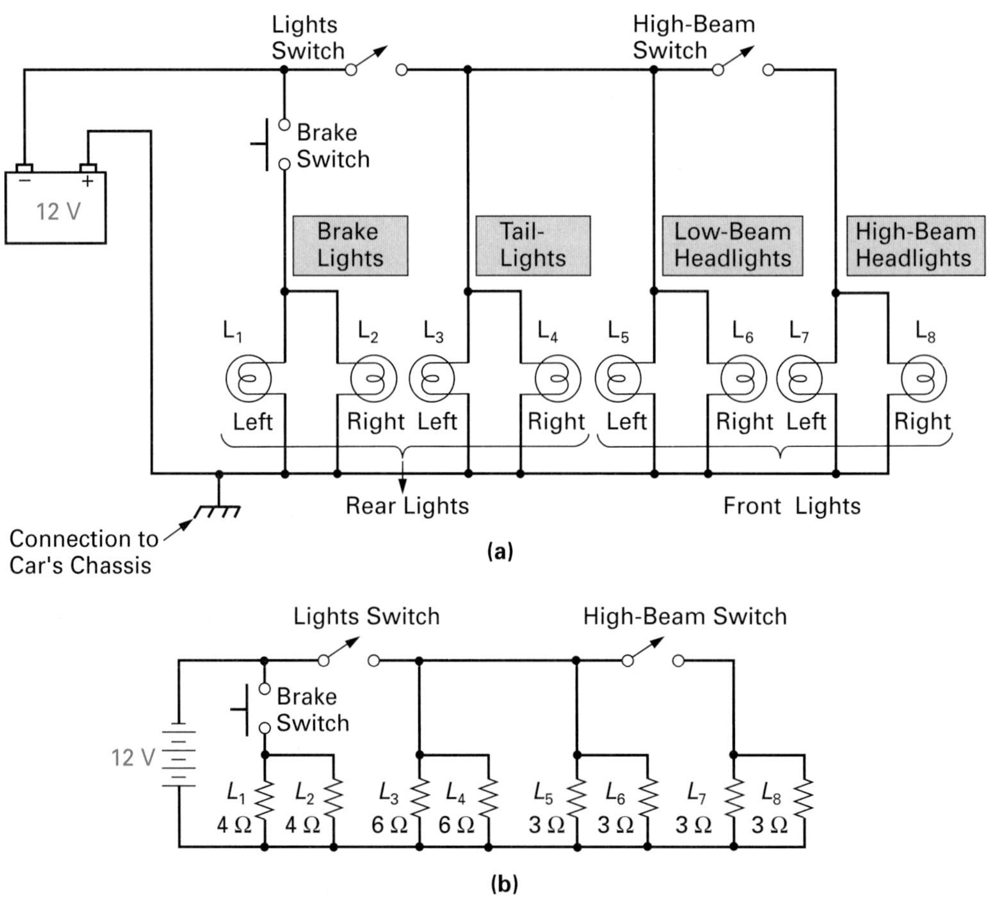

FIGURE 8-24 Parallel Automobile External Light System.

high-beam switch is closed. For the lamp resistances given, calculate the output power of each lamp, when in use.

Solution:

Figure 8-24(b) shows the schematic diagram of the pictorial in Figure 8-24(a). Since both V and R are known, we can use the V^2/R power formula.

Brake lights: Left lamp wattage $(P) = \frac{V^2}{R} = \frac{12\text{ V}^2}{4\Omega} = 36\text{ W}$

Right lamp wattage is the same as left.

Taillights: Left lamp wattage $(P) = \frac{V^2}{R} = \frac{12\text{ V}^2}{6\Omega} = 24\text{ W}$

Right lamp wattage is the same as left.

Low-beam headlights: $P = \frac{V^2}{R} = \frac{12\text{ V}^2}{3\Omega} = 48\text{ W}$

Each low-beam headlight is a 48-W lamp.

High-beam headlights: $P = \frac{V^2}{R} = \frac{12\text{ V}^2}{3\Omega} = 48\text{ W}$

Each high-beam headlight is a 48 W lamp.

SELF-TEST REVIEW QUESTIONS FOR SECTION 8-5

1. True or false: Total power in a parallel circuit can be obtained by using the same total power formula as for series circuits.
2. If $I_1 = 2$ mA and $V = 24$ V, calculate P_1.
3. If $P_1 = 22$ mW and $P_2 = 6400$ μW, $P_T = ?$
4. Is it important to observe the correct wattage ratings of resistors when they are connected in parallel?

8-6 TROUBLESHOOTING A PARALLEL CIRCUIT

In the troubleshooting discussion on series circuits, we mentioned that one of three problems can occur, and these three also apply to parallel circuits:

1. A component will open.
2. A component will short.
3. A component's value will change over a period of time.

As a technician, it is important that you know how to recognize these circuit malfunctions and know how to isolate the cause and make the repair. Since you could be encountering opens, shorts, and component value changes in the parallel circuits you construct in lab, let us step through the troubleshooting procedure you should use.

Figure 8-25(a) indicates the normal readings that should be obtained from a typical parallel circuit if it is operating correctly. Keep these normal readings in mind, since they will change as circuit problems are introduced in the following section.

8-6-1 *Open Component in a Parallel Circuit*

In Figure 8-25(b), R_1 has opened, so there can be no current flow through the R_1 branch. An open is equivalent to the maximum possible resistance (infinite ohms), and therefore the total resistance of this parallel circuit will increase, causing the total current flow to decrease. The total current flow, which was 7 mA, will actually decrease by the amount that was flowing through the now open branch (1 mA) to a total current flow of 6 mA. If you had constructed this circuit in lab, you would probably not have noticed that there was anything wrong with the circuit until you started to make some multimeter measurements. A voltmeter measurement across the parallel circuit, as seen in Figure 8-25(b), would not have revealed any problem since the voltage measure across each resistor will always be the same and equal to the source voltage. A total current measurement, on the other hand, would indicate that there is something wrong since the measured reading does not equal the expected calculated value. Using your ammeter, you could isolate the branch fault by one of the following two methods:

1. If you measure each branch current, you will isolate the problem of R_1, because there will be no current flow through R_1. However, this could take three checks.
2. If you just measure total current (one check), you will notice that total current has decreased by 1 mA, and after making a few calculations you realize that the only path that had a branch current of 1 mA was through R_1, so R_1 must have gone open.

If R_1 opens, total current decreases by 1 mA to 6 mA ($I_T = I_1 + I_2 + I_3 = 0 \text{ mA} + 4 \text{ mA} + 2 \text{ mA} = 6 \text{ mA}$).

If R_2 opens, total current decreases by 4 mA to 3 mA ($I_T = I_1 + I_2 + I_3 = 1 \text{ mA} + 0 \text{ mA} + 2 \text{ mA} = 3 \text{ mA}$).

If R_3 opens, total current decreases by 2 mA to 5 mA ($I_T = I_1 + I_2 + I_3 = 1 \text{ mA} + 4 \text{ mA} + 0 \text{ mA} = 5 \text{ mA}$).

Method 2 can be used as long as the resistors are unequal. If they are equal, method 1 may be used to locate the 0 current branch.

In most cases, the ohmmeter is ideal for locating open circuits. To make a resistance measurement, you should disconnect power from the circuit, isolate the device from the circuit, and then connect the ohmmeter across the device. Figure 8-25(c) shows why it is necessary for you to isolate the device to be tested from the circuit. With R_1 in circuit, the ohmmeter will be measuring the parallel resistance of R_2 and R_3's, even though the ohmmeter is connected across R_1. This will lead to a false reading and confuse the troubleshooting process. Disconnecting the device to be tested from the circuit, as shown in Figure 8-25(d), will give an accurate reading of R_1, R_2, and R_3's resistance and lead you to the cause of the problem.

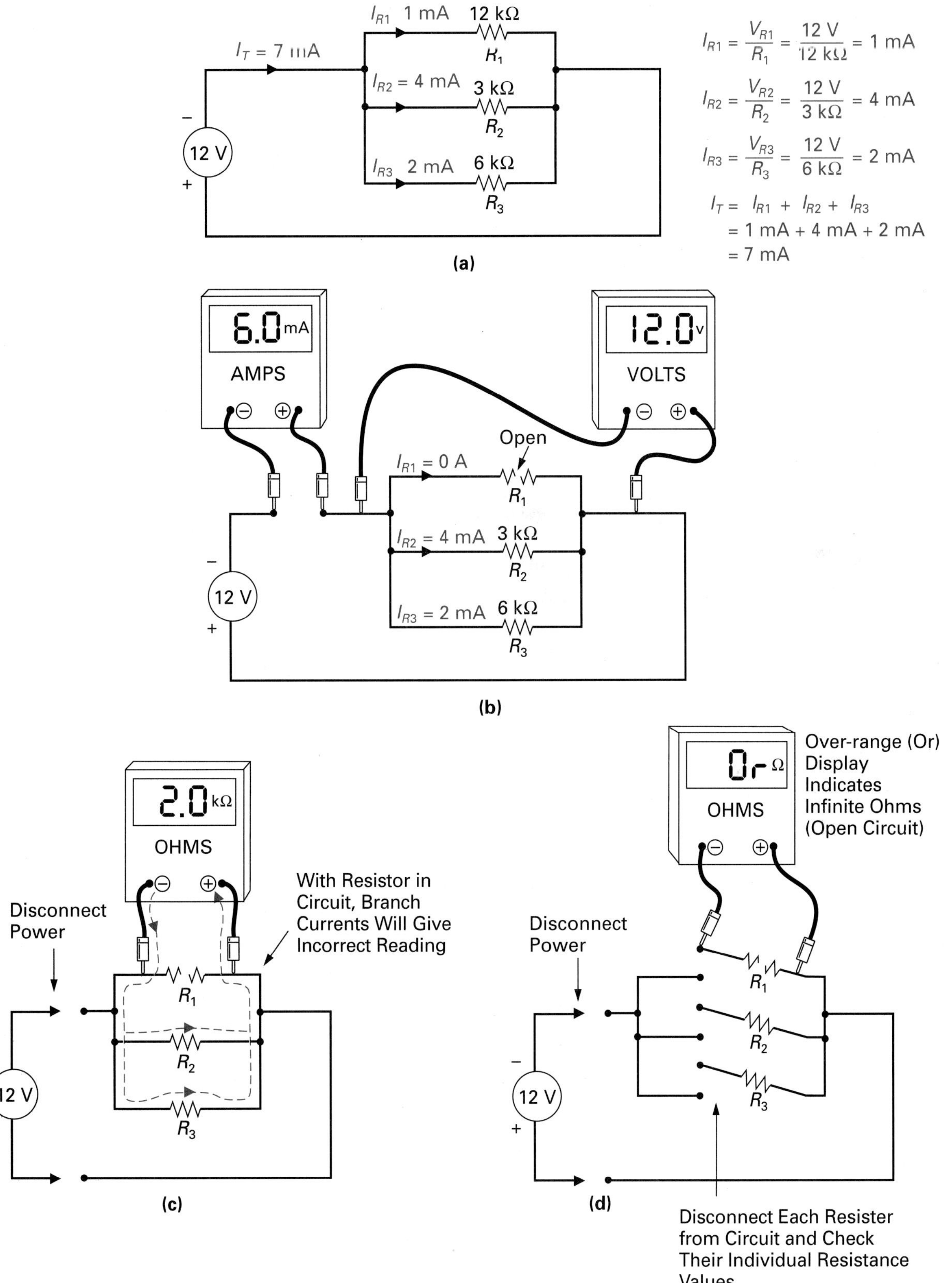

FIGURE 8-25 Troubleshooting an Open in a Parallel Circuit.

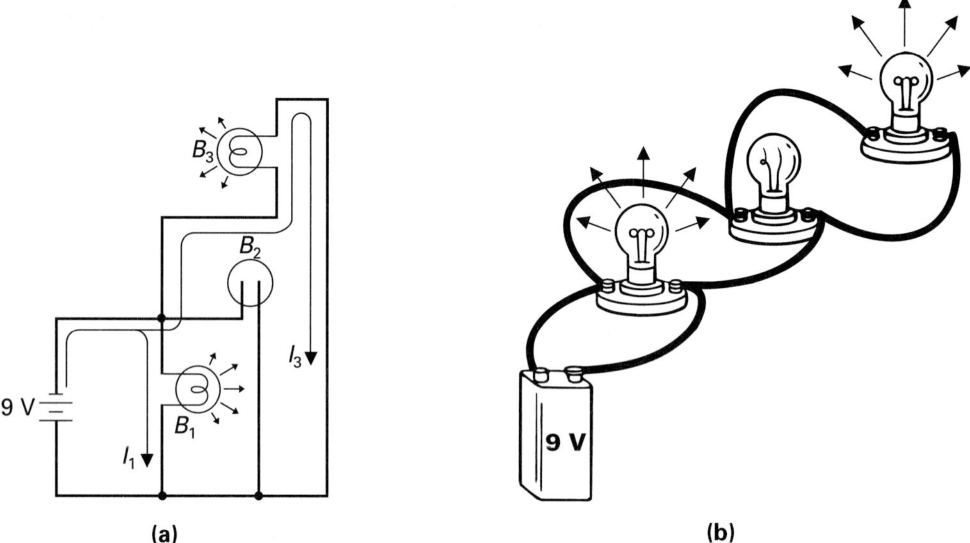

FIGURE 8-26 **Open Bulb in a Parallel Circuit. (a) Schematic. (b) Pictorial.**

EXAMPLE:

Figure 8-26 illustrates a set of three light bulbs connected in parallel across a 9 V battery. The filament in bulb 2 has burned out, causing an open, so there is no current flow through that path. However, there will be current flow through the other two, so bulbs 1 and 3 will be on. How would the faulty bulb be located?

Solution:

As always, a visual inspection would be your first approach. Since current is flowing through both B_1 and B_3, these two bulbs will be on. On the other hand, since B_2 is off, it would be easy for you to determine that bulb B_2 has blown.

Opens in parallel paths are generally quite easy to isolate since the device in the parallel path will not operate because its power source has been disconnected.

8-6-2 *Shorted Component in a Parallel Circuit*

In Figure 8-27(a), R_2, which has a resistance of 3 kΩ, has been shorted out. The only resistance in this center branch is the resistance of the wire, which would typically be a fraction of an ohm. In this example, let us assume that the resistance in the branch is 1 Ω, and therefore the current in this branch will attempt to increase to:

$$I_2 = \frac{12\text{ V}}{1\ \Omega} = 12\text{ A}$$

Before the current even gets close to 12 A, the fuse in the dc power supply will blow, as shown in the inset in Figure 8-27(a), and disconnect power from the cir-

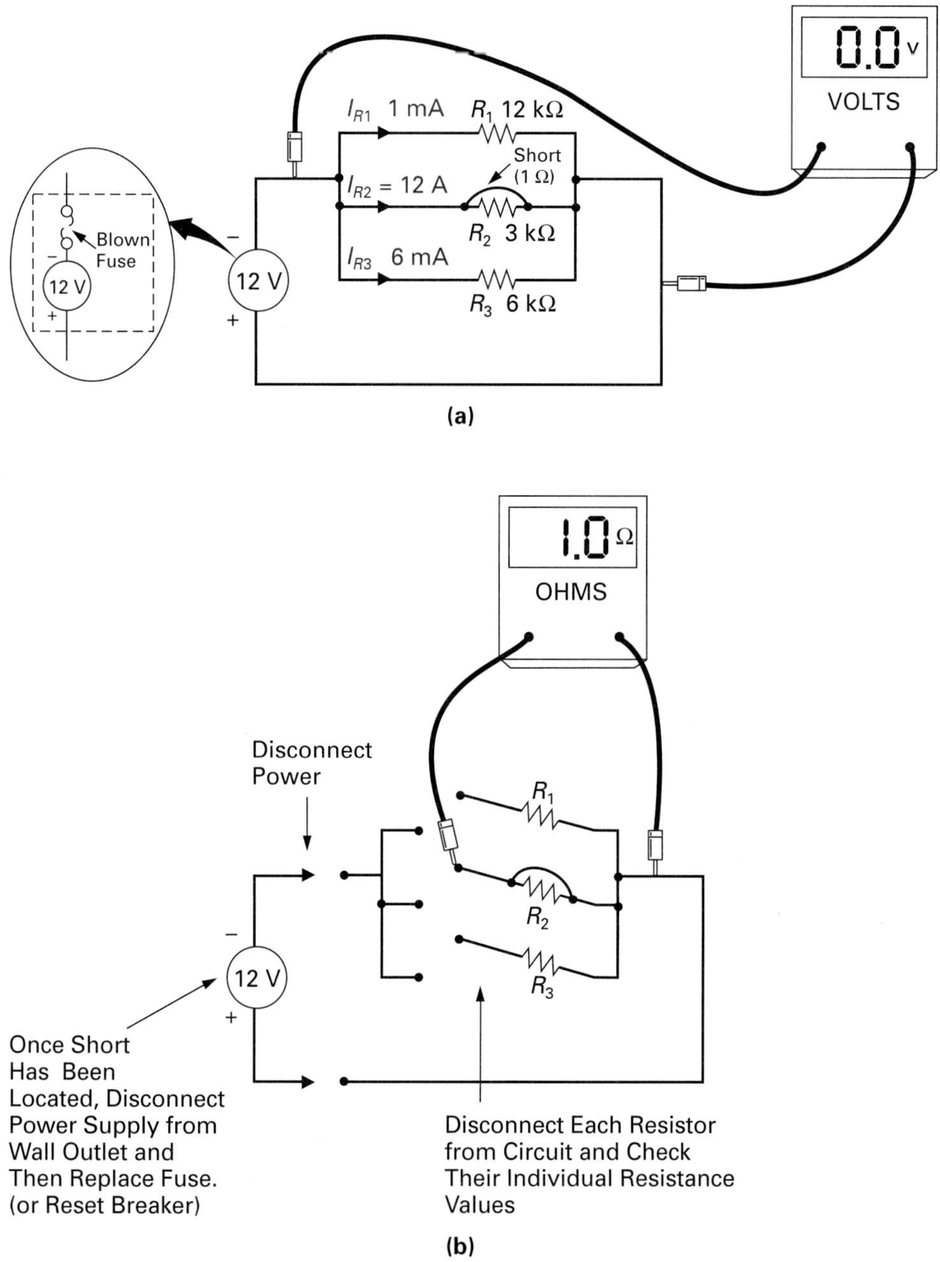

FIGURE 8-27 Troubleshooting a Short in a Parallel Circuit.

cuit. This is typically the effect you will see from a circuit short. Replacing the fuse will be a waste of time and money since the fuse will simply blow again once power is connected across the circuit. As before, the tool to use to isolate the cause is the ohmmeter, as shown in Figure 8-27(b). To check the resistance of each branch, disconnect power from the circuit, disconnect the devices to be tested from the circuit to prevent false readings, and then test each branch until the shorted branch is located. Once you have corrected the fault, you will need to change the power supply fuse or reset its circuit breaker.

8-6-3 *Component Value Variation in a Parallel Circuit*

The resistive values of resistors will change with age. This increase or decrease in resistance will cause a corresponding decrease or increase in branch current and therefore total current. This deviation from the desired value can also be checked with the ohmmeter. Be careful to take into account the resistor's tolerance because this can make you believe that the resistor's value has changed, when it is, in fact, within the tolerance rating.

8-6-4 *Summary of Parallel Circuit Troubleshooting*

1. An open component will cause no current flow within that branch. The total current will decrease, and the voltage across the component will be the same as the source voltage.
2. A shorted component will cause maximum current through that branch. The total current will increase and the voltage source fuse will normally blow.
3. A change in resistance value will cause a corresponding opposite change in branch current and total current.

SELF-TEST REVIEW QUESTIONS FOR SECTION 8-6

How would problems 1 through 3 be recognized in a parallel circuit?

1. An open component
2. A shorted component
3. A component's value variation
4. True or false: An ammeter is typically used to troubleshoot series circuits, whereas a voltmeter is typically used to troubleshoot parallel circuits.

SUMMARY

Parallel Circuits (Figure 8-28)

1. Series or parallel circuits can be determined by current.
2. In a series circuit, there is only one path for current, whereas in parallel circuits the current has two or more paths. These paths are known as *branches*.
3. A parallel circuit, by definition, is when two or more components are connected to the same voltage source so that the current can branch out over two or more paths.

QUICK REFERENCE SUMMARY SHEET

FIGURE 8-28 Parallel Circuits.

Example Circuit (Resistors)

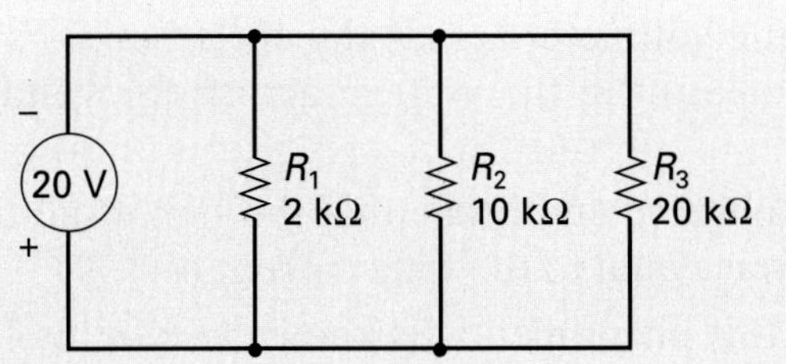

Example Circuit (Batteries)

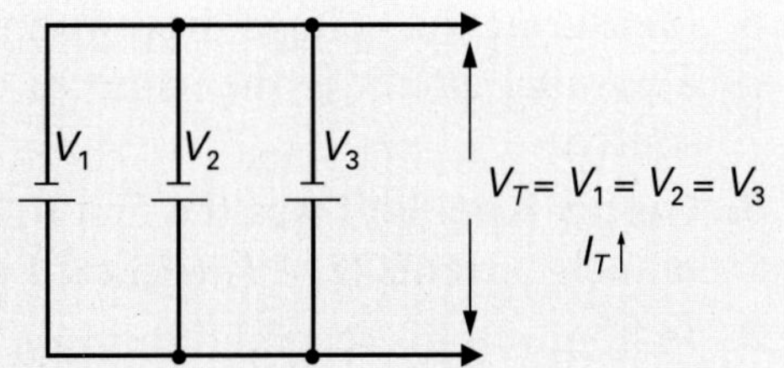

$$V_S = V_{R1} = V_{R2} = V_{R3} = \cdots$$

V_S

$I_T \mid R_T$

$$R_T = \frac{V_S}{I_T}$$

$$I_T = I_1 + I_2 + I_3 + \cdots$$

$$I_x = \frac{R_T}{R_x} \times I_T$$

$$R_T = \frac{1}{(1/R_1) + (1/R_2) + (1/R_3) + \cdots}$$

$$R_T = \frac{R_1 \times R_2}{R_1 + R_2}$$

$$R_T = \frac{R}{n}$$

$$P_T = P_1 + P_2 + P_3 + \cdots$$

$$P_T = \frac{V_S^2}{R_T} = I_T^2 \times R_T = V_S \times I_T$$

V_S = source voltage, in volts
V_{R1} = voltage across R_1, in volts
V_{R2} = voltage across R_2, in volts
V_{R3} = voltage across R_3, in volts
I_T = total current, in amperes
R_T = total resistance, in ohms
I_1 = current through R_1, in amperes
I_2 = current through R_2, in amperes
I_3 = current through R_3, in amperes
I_x = branch current desired, in amperes
R_x = resistance in branch
R_1 = resistance of R_1, in ohms
R_2 = resistance of R_2, in ohms
R_3 = resistance of R_3, in ohms
R_{EV} = resistance of one of the equivalent-value resistors
n = number of parallel resistors
P_T = total power dissipated, in watts
P_1 = power dissipated by R_1, in watts
P_2 = power dissipated by R_2, in watts
P_3 = power dissipated by R_3, in watts

4. In a parallel resistive circuit, two or more resistors are connected to the same source, so current splits to travel through each separate branch resistance.
5. Many components, other than resistors, can be connected in parallel, and a parallel circuit can easily be identified because current is split into two or more paths.
6. Measuring the voltage drop with the voltmeter across any of the resistors in a parallel circuit is the same as measuring the voltage across the voltage source.
7. Gustav Kirchhoff was the first to observe and prove that the sum of all the branch currents ($I_1 + I_2 + I_3$, etc.) was equal to the total current (I_T).
8. In honor of his second discovery, this phenomenon is known as Kirchhoff's current law. It states that the sum of all the current entering a junction is equal to the sum of all the currents leaving that same junction.
9. When we connect resistors in parallel, the total resistance in the circuit will decrease. In fact, the total resistance in a parallel circuit will always be less than the value of the smallest resistor in the circuit.
10. Just as a series circuit is often referred to as a voltage-divider circuit, a parallel circuit is often referred to as a current-divider circuit, because the total current arriving at a junction will divide or split into branch currents (Kirchhoff's law).
11. The current division is inversely proportional to the resistance in the branch seeing that the voltage across both resistors is constant and equal to the source voltage (V_S). This means that a large branch resistance will cause a small branch current ($I\downarrow = V/R\uparrow$), and a small branch resistance will cause a large branch current ($I\uparrow = V/R\downarrow$).
12. By rearranging Ohm's law, we can arrive at another formula, which is called the *current-divider formula* and can be used to calculate the current through any branch of a multiple-branch parallel circuit.
13. Parallel circuits will have a larger current flow than a series circuit containing the same resistors due to the smaller total resistance.
14. To calculate exactly how much total current will flow, we need to calculate the total resistance that the parallel circuit develops or contains.
15. The reciprocal formula allows us to calculate total parallel circuit resistance for any number of resistors.
16. With parallel resistance circuits, the total resistance is always smaller than the smallest branch resistance.
17. With parallel resistive circuits, the voltage drop across any branch is equal to the voltage drop across each of the other branches and is equal to the source voltage.
18. If only two resistors are connected in parallel, a quick and easy formula called the *product-over-sum formula* can be used to calculate total resistance.
19. If resistors of equal value are connected in parallel, a special case *equal-value formula* can be used to calculate the total resistance.

20. As with series circuits, the total power in a parallel resistive circuit is equal to the sum of all the power losses for each of the resistors in parallel.
21. An open component will cause no current flow within that branch. The total current will decrease, and the voltage across the component will be the same as the source voltage.
22. A shorted component will cause maximum current through that branch. The total current will increase and the voltage source fuse will normally blow.
23. A change in resistance value will cause a corresponding opposite change in branch current and total current.

NEW TERMS

Branch current	Fixed current divider	Parallel circuit
Current divider	Kirchhoff's current law	

REVIEW QUESTIONS

Multiple-Choice Questions

1. A parallel circuit has ____________ path(s) for current to flow.
   a. One **c.** Only three
   b. Two or more **d.** None of the above
2. If a source voltage of 12 V is applied across four resistors of equal value in parallel, the voltage drops across each resistor would be equal to:
   a. 12 V **c.** 4 V
   b. 3 V **d.** 48 V
3. What would be the voltage drop across two 25 Ω resistors in parallel if the source voltage were equal to 9 V?
   a. 50 V **c.** 12 V
   b. 25 V **d.** None of the above
4. If a four-branch parallel circuit has 15 mA flowing through each branch, the total current into the parallel circuit will be equal to:
   a. 15 mA **c.** 30 mA
   b. 60 mA **d.** 45 mA
5. If the total three-branch parallel circuit current is equal to 500 mA, and 207 mA is flowing through one branch and 153 mA through another, what would be the current flow through the third branch?
   a. 707 mA **c.** 140 mA
   b. 653 mA **d.** None of the above
6. A large branch resistance will cause a ____________ branch current.
   a. Large **c.** Medium
   b. Small **d.** None of the above are true
7. What would be the conductance of a 1 kΩ resistor?
   a. 10 mS **c.** 2 kΩ
   b. 1 mS **d.** All of the above

8. If only two resistors are connected in parallel, the total resistance equals:
   a. The sum of the resistance values
   b. Three times the value of one resistor
   c. The product over the sum
   d. All of the above
9. If resistors of equal value are connected in parallel, the total resistance can be calculated by:
   a. One resistor value divided by the number of parallel resistors
   b. The sum of the resistor values
   c. The number of parallel resistors divided by one resistor value
   d. All of the above could be true.
10. The total power in a parallel circuit is equal to the:
   a. Product of total current and total voltage
   b. Reciprocal of the individual power losses
   c. Sum of the individual power losses
   d. Both (a) and (b)
   e. Both (a) and (c)

Essay Questions

11. Describe the difference between a series and a parallel circuit. (8-1)
12. Explain and state mathematically the situation regarding voltage in a parallel circuit. (8-2)
13. State Kirchhoff's current law for parallel circuits. (8-3)
14. What is the current-divider formula? (8-3)
15. List the formulas for calculating the following total resistances: (8-4)
   a. Two resistors of different values
   b. More than two resistors of different values
   c. Equal-value resistors
16. Describe the relationship between branch current and branch resistance. (8-3)
17. Briefly describe total and individual power in a parallel resistive circuit. (8-5)
18. Discuss troubleshooting parallel circuits as applied to: (8-6)
   a. A shorted component b. An open component
19. Explain why parallel circuits have a smaller total resistance and larger total current than series circuits. (8-3)
20. Briefly describe why Kirchhoff's voltage law applies to series circuits and why Kirchhoff's current law relates to parallel circuits. (8-3)

Practice Problems

21. Calculate the total resistance of four 30 kΩ resistors in parallel.
22. Find the total resistance for each of the following parallel circuits:
   a. 330 Ω and 560 Ω
   b. 47 kΩ, 33 kΩ, and 22 kΩ
   c. 2.2 MΩ, 3 kΩ, and 220 Ω

23. If 10 V is connected across three 25 Ω resistors in parallel, what will be the total and individual branch currents?

24. If a four-branch parallel circuit has branch currents equal to 25 mA, 37 mA, 220 mA, and 0.2 A, what is the total circuit current?

25. If three resistors of equal value are connected across a 14 V supply and the total resistance is equal to 700 Ω, what is the value of each branch current?

26. If three 75 W light bulbs are connected in parallel across a 110 V supply, what is the value of each branch current? What is the branch current through the other two light bulbs if one burns out?

27. If 33 kΩ and 22 kΩ resistors are connected in parallel across a 20 V source, calculate:

    a. Total resistance
    b. Total current
    c. Branch currents
    d. Total power dissipated
    e. Individual resistor power dissipated

28. If four parallel-connected resistors are each dissipating 75 mW, what is the total power being dissipated?

29. Calculate the branch currents through the following parallel resistor circuits when they are connected across a 10 V supply:

    a. 22 kΩ and 33 kΩ **b.** 220 Ω, 330 Ω, and 470 Ω

30. If 30 Ω and 40 Ω resistors are connected in parallel, which resistor will generate the greatest amount of heat?

31. Calculate the total conductance and resistance of the following parallel circuits:

    a. Three 5 Ω resistors
    b. Two 200 Ω resistors
    c. 1 MΩ, 500 MΩ, 3.3 MΩ
    d. 5 Ω, 3 Ω, 2 Ω

32. Connect the three resistors in Figure 8-29 in parallel across a 12 V battery and then calculate the following:

    a. V_{R1}, V_{R2}, V_{R3}
    b. I_1, I_2, I_3
    c. I_T
    d. P_T
    e. P_1, P_2, P_3
    f. G_{R1}, G_{R2}, G_{R3}

33. Calculate R_T in Figure 8-30 (a), (b), (c), and (d).

34. Calculate the branch currents through four 60 W bulbs connected in parallel across 110 V. How much is the total current, and what would happen to the total current if one of the bulbs were to burn out? What change would occur in the remaining branch currents?

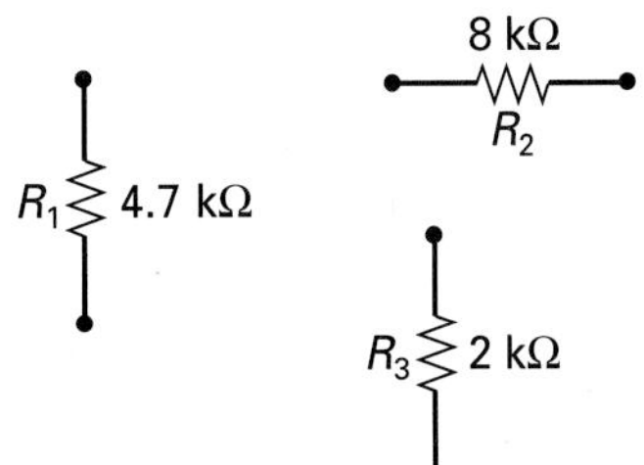

FIGURE 8-29 **Connect in Parallel Across a 12-volt Source.**

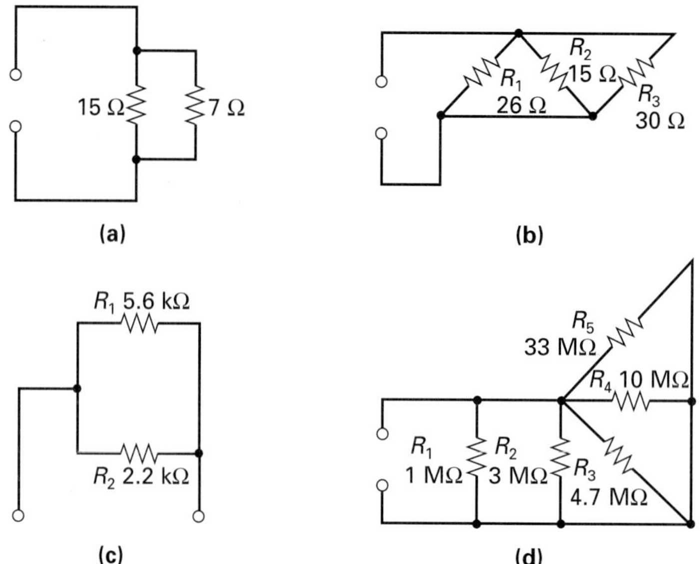

FIGURE 8-30 **Calculate Total Resistance.**

35. Calculate the following in Figure 8-31:

a. I_2 **c.** V_S, I_1, I_2
b. I_T **d.** R_2, I_1, I_2, P_T

Troubleshooting Questions

36. An open component in a parallel circuit will cause ____________ current flow within that branch, which will cause the total current to ____________.

a. Maximum, increase **c.** Maximum, decrease
b. Zero, decrease **d.** Zero, increase

37. A shorted component in a parallel circuit will cause ____________ current through a branch, and consequently the total current will ____________.

a. Maximum, increase **c.** Maximum, decrease
b. Zero, decrease **d.** Zero, increase

38. If a 10 kΩ and two 20 kΩ resistors are connected in parallel across a 20 V supply, and the total current measured is 2 mA, determine whether a problem exists in the circuit and, if it does, isolate the problem.

39. What situation would occur and how would we recognize the problem if one of the 20 kΩ resistors in Question 38 were to short?

40. With age, the resistance of a resistor will ____________, resulting in a corresponding but opposite change in ____________.

a. Increase, branch current **c.** Decrease, source resistance
b. Change, source voltage **d.** Change, branch current

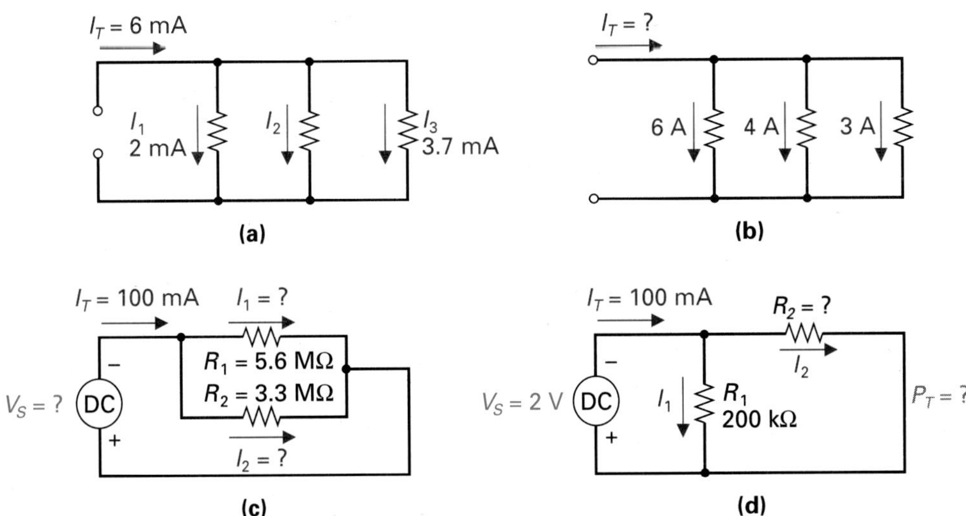

FIGURE 8-31 Calculate the Unknown.

CHAPTER

9

OBJECTIVES

After completing this chapter, you will be able to:

1. Identify the difference between a series, a parallel, and a series–parallel circuit.
2. Describe how to use a three-step procedure to determine total resistance.
3. Describe for the series–parallel circuit how to use a five-step procedure to calculate:
   a. Total resistance.
   b. Total current.
   c. Voltage division.
   d. Branch current.
   e. Total power dissipated.
4. Explain what loading effect a piece of equipment will have when connected to a voltage divider.
5. Identify and describe the Wheatstone bridge circuit in both the balanced and unbalanced condition.
6. Describe the R–$2R$ ladder circuit used for digital-to-analog conversion.
7. Explain how to identify the following problems in a series–parallel circuit:
   a. Open series resistor.
   b. Open parallel resistor.
   c. Shorted series resistor.
   d. Shorted parallel resistor.
   e. Resistor value variation.
8. Describe the differences between a voltage and a current source.
9. Analyze series–parallel networks using:
   a. The superposition theorem.
   b. Thévenin's theorem.
   c. Norton's theorem.

Series–Parallel DC Circuits

OUTLINE

V I G N E T T E

The Christie Bridge Circuit

Who invented the Wheatstone bridge circuit? It was obviously Sir Charles Wheatstone. Or was it?

The Wheatstone bridge was actually invented by S.H. Christie of the Royal Military Academy at Woolwich, England. He described the circuit in detail in the *Philosophical Transactions* paper dated February 28, 1833. Christie's name, however, was unknown and his invention was ignored.

Ten years later, Sir Charles Wheatstone called attention to Christie's circuit. Sir Charles was very well known, and from that point on, and even to this day, the circuit is known as a Wheatstone bridge. Later, Werner Siemens would modify Christie's circuit and invent the variable-resistance arm bridge circuit, which would also be called a Wheatstone bridge.

No one has given full credit to the real inventors of these bridge circuits, until now!

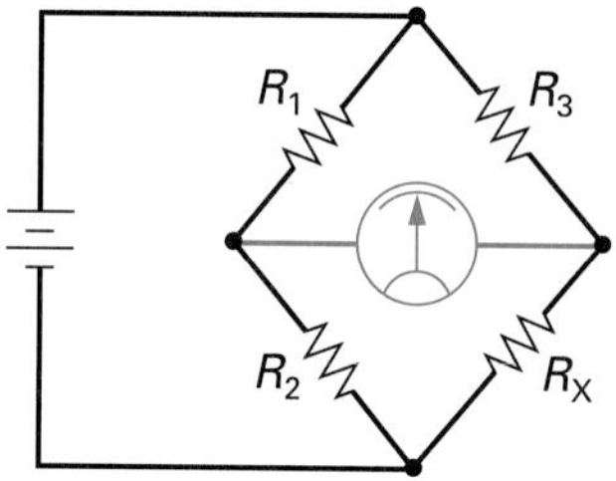

The Christie Bridge

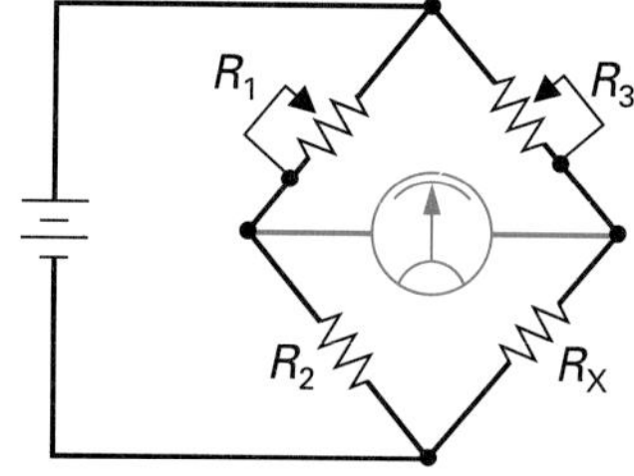

The Siemens Bridge

INTRODUCTION

Very rarely are we lucky enough to run across straightforward series or parallel circuits. In general, all electronic equipment is composed of many components that are interconnected to form a combination of series and parallel circuits. In this chapter, we will be combining our knowledge of the series and parallel circuits discussed in the previous two chapters.

9-1 SERIES- AND PARALLEL-CONNECTED COMPONENTS

Figure 9-1(a) through (f) illustrates six different examples of **series–parallel resistive circuits.** The most important point to learn is how to distinguish between the resistors that are connected in series and the resistors that are connected in parallel, which will take a little practice.

Series–Parallel Circuit
Network or circuit that contains components that are connected in both series and parallel.

One thing that you may not have noticed when examining Figure 9-1 is that:

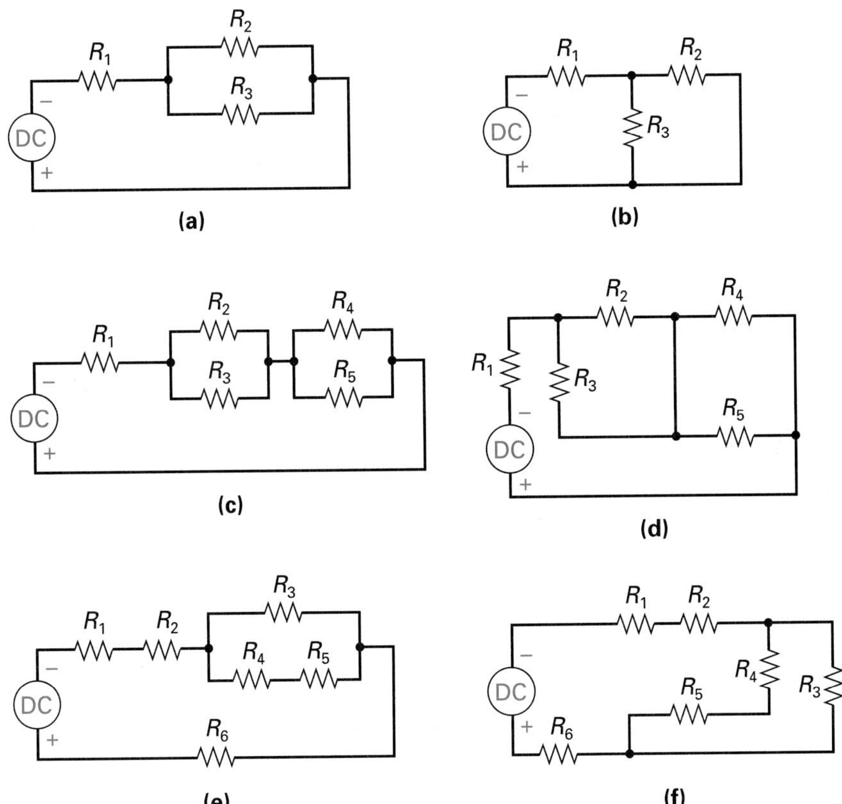

FIGURE 9-1 **Series–Parallel Resistive Circuits.**

Circuit 9-1(a) is equivalent to 9-1(b)

Circuit 9-1(c) is equivalent to 9-1(d)

Circuit 9-1(e) is equivalent to 9-1(f)

When analyzing these series–parallel circuits, always remember that current flow determines whether the resistor is connected in series or parallel. Begin at the negative side of the battery and apply these two rules:

1. If the total current has only one path to follow through a component, that component is connected in series.
2. If the total current has two or more paths to follow through two or more components, those components are connected in parallel.

Referring again to Figure 9-1, you can see that series or parallel resistor networks are easier to identify in parts (a), (c), and (e) than in parts (b), (d), and (f). Redrawing the circuit so that the components are arranged from left to right or from top to bottom is your first line of attack in your quest to identify series- and parallel-connected components.

EXAMPLE:

Refer to Figure 9-2 and identify which resistors are connected in series and which are in parallel.

Solution:

First, let's redraw the circuit so that the components are aligned either from left to right [Figure 9-3(a)] or from top to bottom [Figure 9-3(b)]. Placing your pencil at the negative terminal of the battery on whichever figure you prefer, either Figure 9-3(a) or (b), trace the current paths through the circuit toward the positive side of the battery, as illustrated in Figure 9-4.

The total current arrives first at R_1. There is only one path for current to flow, which is through R_1, and therefore R_1 is connected in series. The total current proceeds on past R_1 and arrives at a junction where current divides and travels through two branches, R_2 and R_3. Since current had to split into two paths, R_2 and

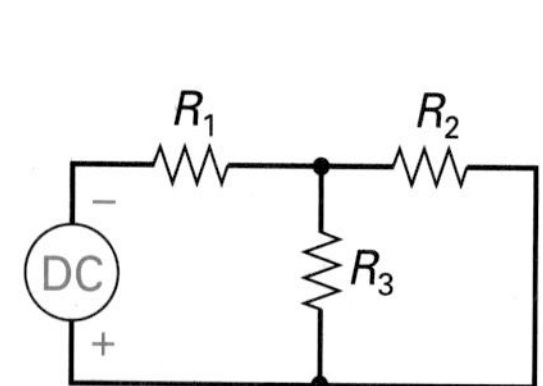

FIGURE 9-2 Series–Parallel Circuit Example.

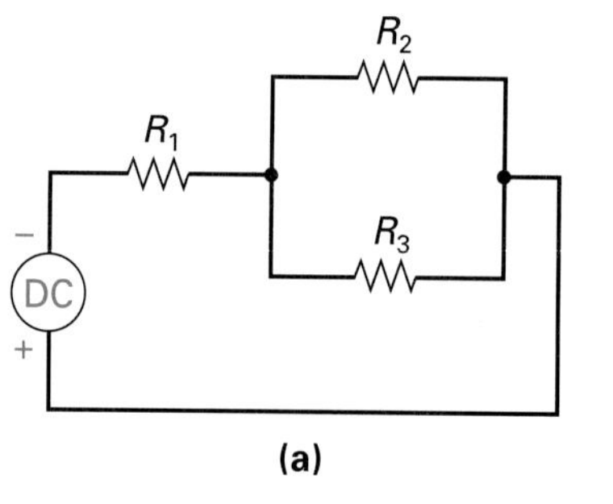

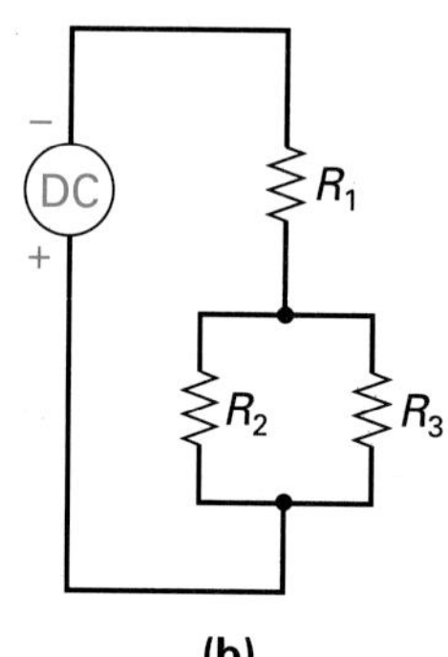

FIGURE 9-3 Redrawn Series–Parallel Circuit. (a) Left to Right. (b) Top to Bottom.

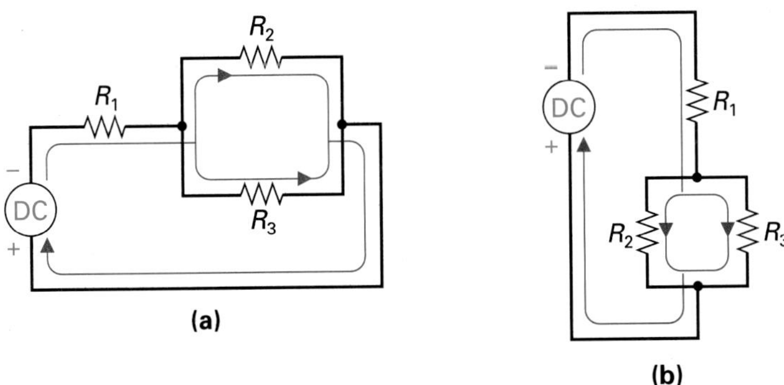

FIGURE 9-4 **Tracing Current Through a Series–Parallel Circuit.**

R_3 are therefore connected in parallel. After the parallel connection of R_2 and R_3, total current combines and travels to the positive side of the battery.

To summarize, R_1 is in series with the parallel combination of R_2 and R_3.

EXAMPLE:

Refer to Figure 9-5 and identify which resistors are connected in series and which are connected in parallel.

Solution:

Figure 9-6 illustrates the simplified, redrawn schematic of Figure 9-5. Total current leaves the negative terminal of the battery, and all of this current has to travel through R_1, which is therefore a series-connected resistor. Total current will split at junction A, and consequently, R_3 and R_4 with R_2 make up a parallel

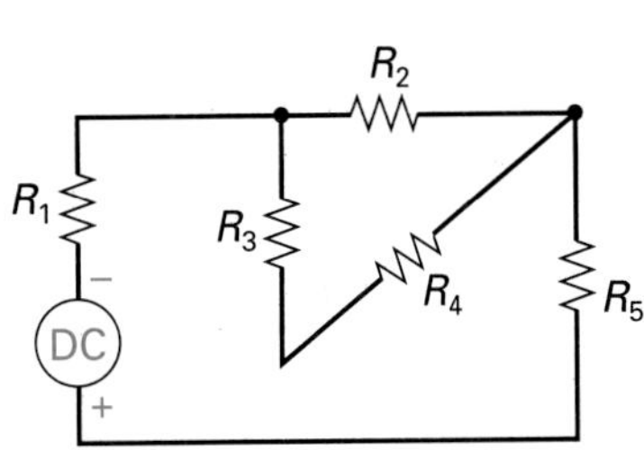

FIGURE 9-5 **Series–Parallel Circuit Example.**

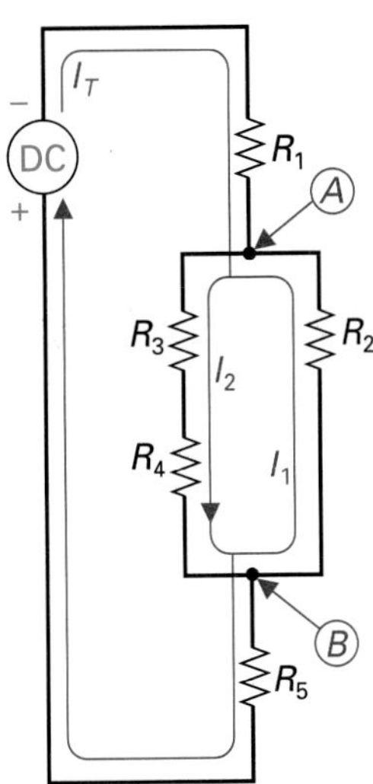

FIGURE 9-6 **Redrawn Series–Parallel Circuit Example.**

combination. The current that flows through R_3 (I_2) will also flow through R_4 and have only one path to follow; therefore, R_3 is in series with R_4. I_1 and I_2 branch currents combine at junction B to produce total current, which has only one path to follow through the series resistor R_5, and finally, to the positive side of the battery.

To summarize, R_3 and R_4 are in series with one another and both are in parallel with R_2, and this combination is in series with R_1 and R_5.

9-2 TOTAL RESISTANCE IN A SERIES–PARALLEL CIRCUIT

No matter how complex or involved the series–parallel circuit, there is a simple three-step method to simplify the circuit to a single equivalent total resistance. Figure 9-7 illustrates an example of a series–parallel circuit. Once you have analyzed and determined the series–parallel relationship, we can proceed to solve for total resistance.

The three-step method is:

Step A: Determine the equivalent resistances of all branch series-connected resistors.

Step B: Determine the equivalent resistances of all parallel-connected combinations.

Step C: Determine the equivalent resistance of the remaining series-connected resistances.

Let's now put our theory to work with the example circuit in Figure 9-7.

STEP A Solve for all branch series-connected resistors. In our examples, this applies only to R_3 and R_4, and since this is a series connection, we have to use the series resistance formula.

$$R_{3,4} = R_3 + R_4 = 8 + 2 = 10\ \Omega \qquad \text{(series resistance formula)}$$

With R_3 and R_4 solved, the circuit appears as indicated in Figure 9-8.

STEP B Solve for all parallel combinations. In this example, they are the two parallel combinations of (a) R_2 and $R_{3,4}$ and (b) R_5 and R_6 and R_7. Since these are parallel connections, use the parallel resistance formulas.

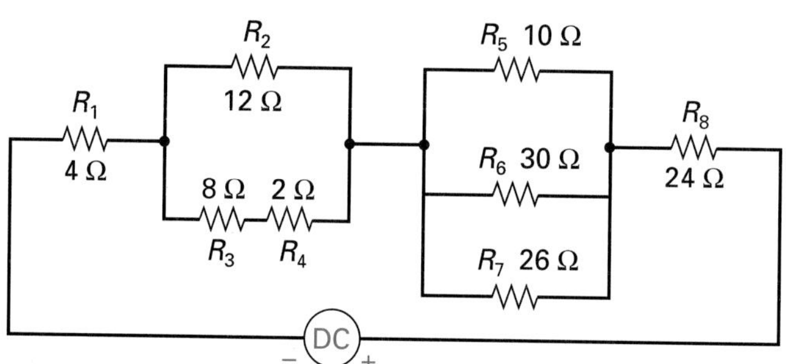

FIGURE 9-7 Total Series–Parallel Circuit Resistance.

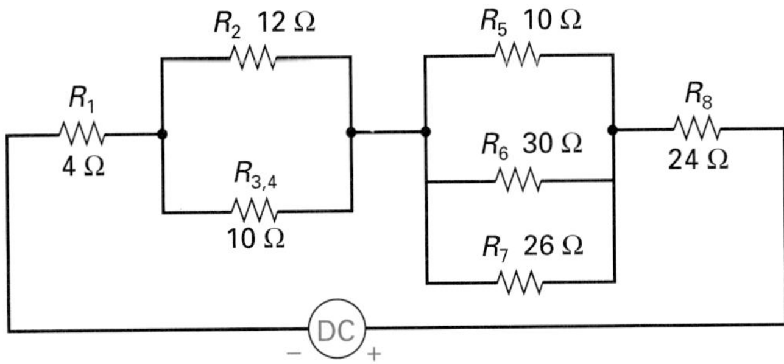

FIGURE 9-8 After Completing Step A.

$$R_{2,3,4} = \frac{R_2 \times R_{3,4}}{R_2 + R_{3,4}} = \frac{12 \times 10}{12 + 10} = 5.5\ \Omega \qquad \text{(product-over-sum formula)}$$

$$R_{5,6,7} = \frac{1}{(1/R_5) + (1/R_6) + (1/R_7)} = 5.8\ \Omega \qquad \text{(reciprocal formula)}$$

With $R_{2,3,4}$ and $R_{5,6,7}$ solved, the circuit now appears as illustrated in Figure 9-9.

STEP C Solve for the remaining series resistances. There are now four remaining series resistances, which can be reduced to one equivalent resistance (R_{eq}) or total resistance (R_T), as seen in Figure 9-10. By using the series resistance formula,

$$\begin{aligned} R_{eq} &= R_1 + R_{2,3,4} + R_{5,6,7} + R_8 \\ &= 4\ \Omega + 5.5\ \Omega + 5.8\ \Omega + 24\ \Omega \\ &= 39.3\ \Omega \end{aligned}$$

EXAMPLE:

Find the total resistance of the circuit in Figure 9-11.

Solution:

STEP A Solve for all branch series-connected resistors. This applies to R_2 and R_3 (series connection):

$$R_{2,3} = R_2 + R_3 = 1\ \text{k}\Omega + 3\ \text{k}\Omega = 4\ \text{k}\Omega$$

The resulting circuit, after completing step A, is illustrated in Figure 9-12(a).

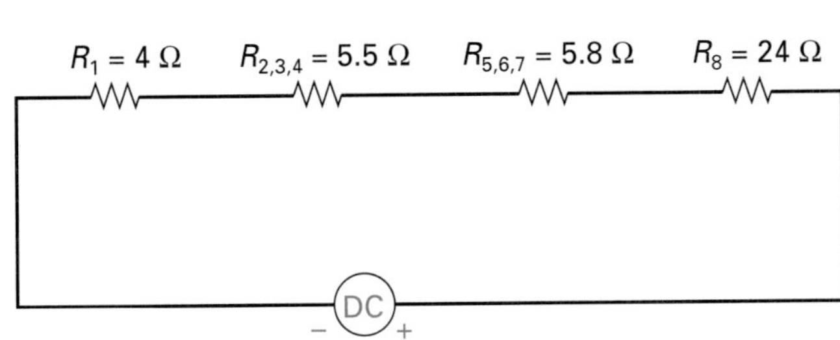

FIGURE 9-9 After Completing Step B.

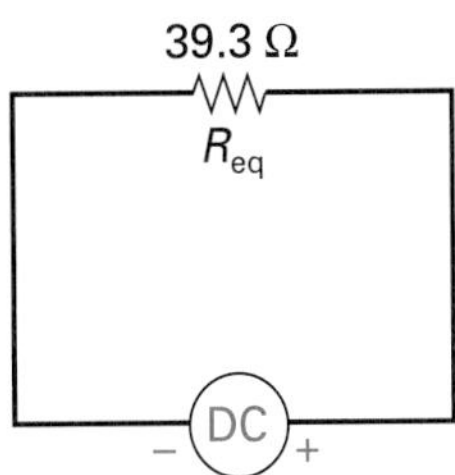

FIGURE 9-10 After Completing Step C.

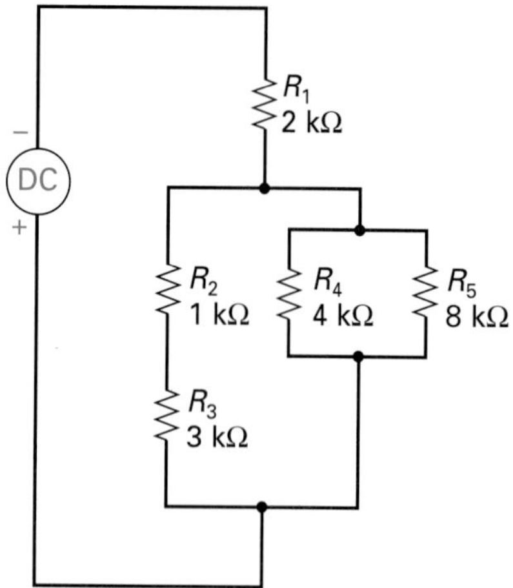

FIGURE 9-11 **Calculate Total Resistance.**

STEP B Solve for all parallel combinations. Observing the circuit resulting from step A [Figure 9-12(a)], you can see that current branches into three paths, so the parallel reciprocal formula can be used:

$$R_{2,3,4,5} = \frac{1}{(1/R_{2,3}) + (1/R_4) + (1/R_5)}$$

$$= \frac{1}{(1/4\text{ k}\Omega) + (1/4\text{ k}\Omega) + (1/8\text{ k}\Omega)} = 1.6\text{ k}\Omega$$

The resulting circuit, after completing step B, is illustrated in Figure 9-12(b).

STEP C Solve for the remaining series resistances. Observing the circuit resulting from step B [Figure 9-12(b)], you can see that there are two remaining series resistances. The equivalent resistance (R_{eq}) is equal to

$$R_{eq} = R_1 + R_{2,3,4,5} = 2\text{ k}\Omega + 1.6\text{ k}\Omega = 3.6\text{ k}\Omega$$

The total equivalent resistance, after completing all three steps, is illustrated in Figure 9-12(c).

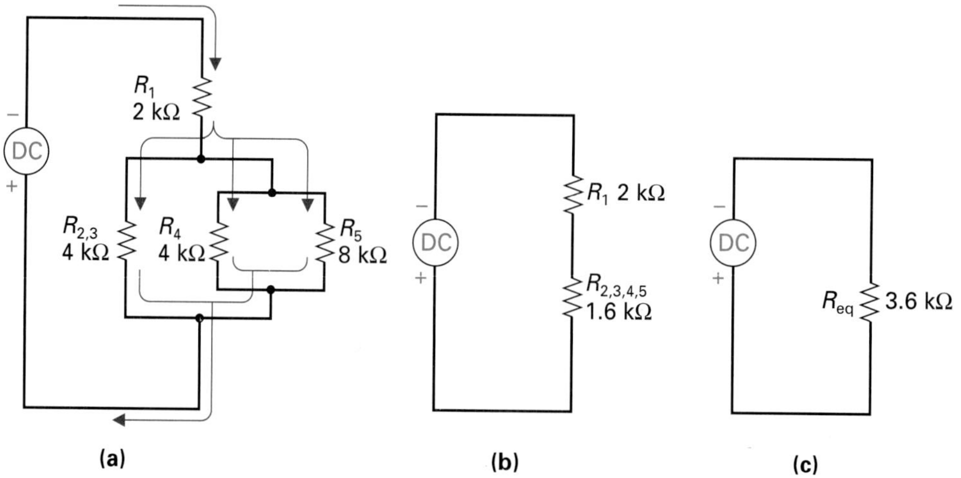

FIGURE 9-12 **Calculating Total Resistance. (a) Step A. (b) Step B. (c) Step C.**

SELF-TEST REVIEW QUESTIONS FOR SECTIONS 9-1 AND 9-2

1. Given a series–parallel circuit, how can we determine which resistors are connected in series and which are connected in parallel?
2. Calculate the total resistance if two series-connected 12 kΩ resistors are connected in parallel with a 6 kΩ resistor.
3. State the three-step procedure used to determine total resistance in a circuit made up of both series and parallel resistors.
4. Sketch the following series–parallel resistor network made up of three resistors. R_1 and R_2 are in series with each other and are connected in parallel with R_3. If $R_1 = 470\ \Omega$, $R_2 = 330\ \Omega$, and $R_3 = 270\ \Omega$, what is R_T?

9-3 VOLTAGE DIVISION IN A SERIES–PARALLEL CIRCUIT

There is a simple three-step procedure for finding the voltage drop across each part of the series–parallel circuit. Figure 9-13 illustrates an example of a series–parallel circuit to which we will apply the three-step method for determining voltage drop.

STEP 1 Determine the circuit's total resistance. This is achieved by following the three-step method used previously for calculating total resistance.

Step A: $R_{3,4} = 4 + 8 = 12\ \Omega$

Step B: $$R_{2,3,4} = \frac{1}{(1/R_2) + (1/R_{3,4})} = 6\ \Omega$$

$$R_{5,6,7} = \frac{1}{(1/R_5) + (1/R_6) + (1/R_7)} = 12\ \Omega$$

Figure 9-14 illustrates the equivalent circuit up to this point. We end up with one series resistor (R_1) and two series equivalent resistors ($R_{2,3,4}$ and $R_{5,6,7}$). R_T is therefore equal to 28 Ω.

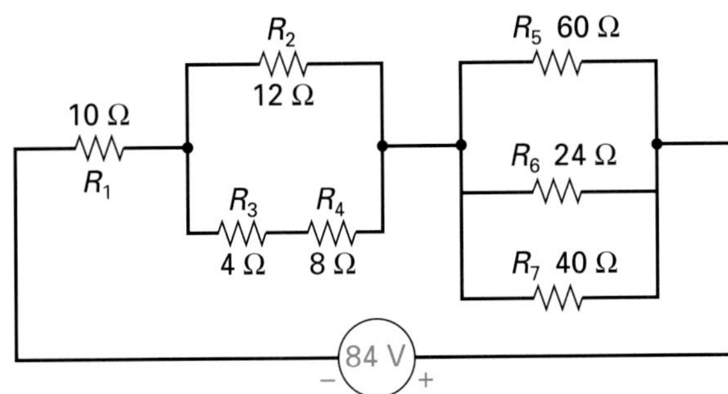

FIGURE 9-13 **Series–Parallel Circuit Example.**

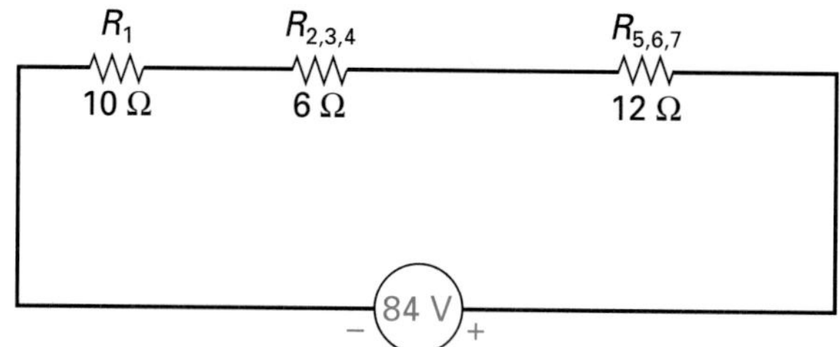

FIGURE 9-14 **After Completing Step 1.**

STEP 2 Determine the circuit's total current. This step is achieved simply by utilizing Ohm's law.

$$I_T = \frac{V_T}{R_T} = \frac{84\text{ V}}{28\ \Omega} = 3\text{ A}$$

STEP 3 Determine the voltage across each series resistor and each parallel combination (series equivalent resistor) in Figure 9-14. Since these are all in series, the same current (I_T) will flow through all three.

$$V_{R1} = I_T \times R_1 = 3\text{ A} \times 10\ \Omega = 30\text{ V}$$
$$V_{R2,3,4} = I_T \times R_{2,3,4} = 3\text{ A} \times 6\ \Omega = 18\text{ V}$$
$$V_{R5,6,7} = I_T \times R_{5,6,7} = 3\text{ A} \times 12\ \Omega = 36\text{ V}$$

The voltage drops across the series resistor (R_1) and series equivalent resistors ($R_{2,3,4}$ and $R_{5,6,7}$) are illustrated in Figure 9-15.

Kirchhoff's voltage law states that the sum of all the voltage drops is equal to the source voltage applied. This law can be used to confirm that our calculations are all correct:

$$\begin{aligned} V_T &= V_{R1} + V_{R2,3,4} + V_{R5,6,7} \\ &= 30\text{ V} + 18\text{ V} + 36\text{ V} \\ &= 84\text{ V} \end{aligned}$$

To summarize, refer to Figure 9-16, which shows these voltage drops inserted into our original circuit. As you can see from this illustration:

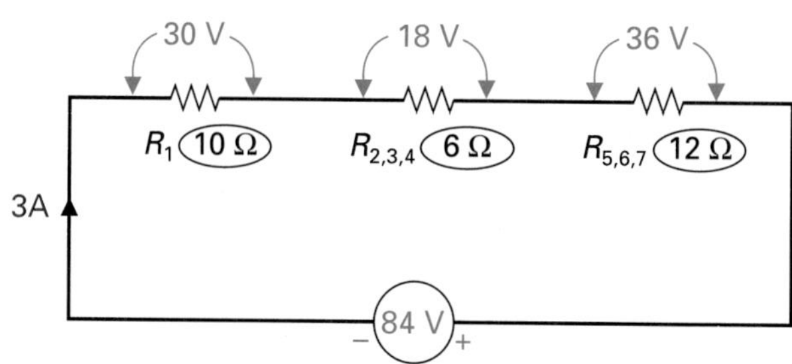

FIGURE 9-15 **After Completing Steps 2 and 3.**

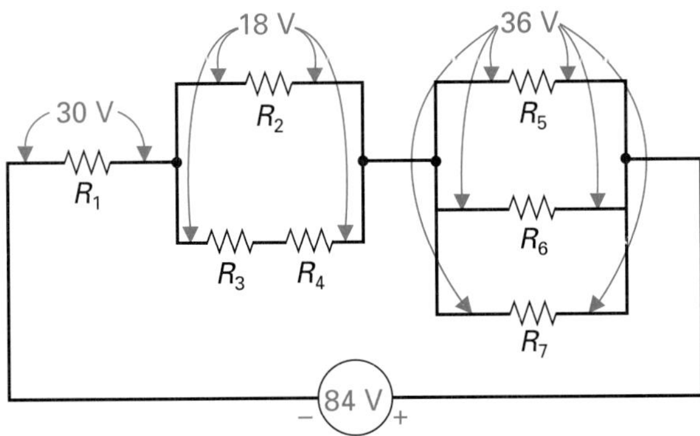

FIGURE 9-16 **Detail of Step 3.**

30 V is dropped across R_1.
18 V is dropped across R_2.
18 V is dropped across both R_3 and R_4.
36 V is dropped across R_5.
36 V is dropped across R_6.
36 V is dropped across R_7.

SELF-TEST REVIEW QUESTIONS FOR SECTION 9-3

1. State the three-step procedure used to calculate the voltage drop across each part of a series–parallel circuit.
2. Referring to Figure 9-13, double the values of all the resistors. Would the voltage drops calculated previously change, and if so, what would they be?

9-4 BRANCH CURRENTS IN A SERIES–PARALLEL CIRCUIT

In the preceding example, step 2 calculated the total current flowing in a series–parallel circuit. The next step is to find out exactly how much current is flowing through each parallel branch. This will be called step 4. Figure 9-17 shows the previously calculated data inserted in the appropriate places in our example circuit.

STEP 4 Total current (I_T) will exist at points A, B, C, and D. Between A and B, current has only one path to flow, which is through R_1. R_1 is therefore a series resistor, so $I_{R1} = I_T = 3$ A. Between points B and C, current has two paths: through R_2 (12 Ω) and through R_3 and R_4 (12 Ω).

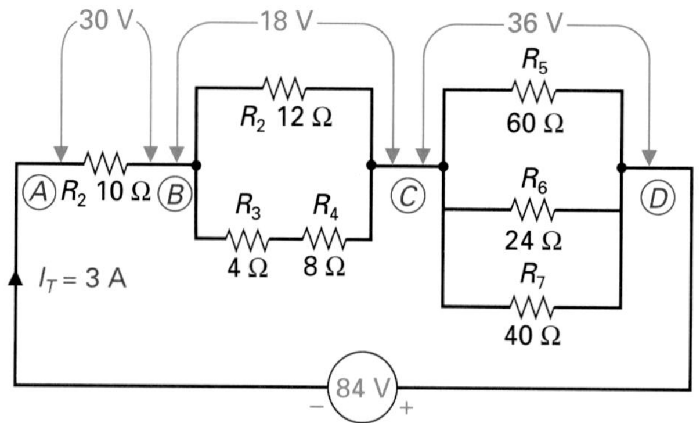

FIGURE 9-17 Series–Parallel Circuit Example with Previously Calculated Data.

$$I_{R2} = \frac{V_{R2}}{R_2} = \frac{18\text{ V}}{12\ \Omega} = 1.5\text{ A}$$

$$I_{R3,4} = \frac{V_{R3,4}}{R_{3,4}} = \frac{18\text{ V}}{12\ \Omega} = 1.5\text{ A}$$

Not surprisingly, the total current of 3 A is split equally due to both branches having equal resistance.

The two 1.5-A branch currents will combine at point C to produce once again the total current of 3 A. Between points C and D, current has three paths to flow through, R_5, R_6, and R_7.

$$I_{R5} = \frac{V_{R5}}{R_5} = \frac{36\text{ V}}{60\ \Omega} = 0.6\text{ A}$$

$$I_{R6} = \frac{V_{R6}}{R_6} = \frac{36\text{ V}}{24\ \Omega} = 1.5\text{ A}$$

$$I_{R7} = \frac{V_{R7}}{R_7} = \frac{36\text{ V}}{40\ \Omega} = 0.9\text{ A}$$

All three branch currents will combine at point D to produce the total current of 3 A ($I_T = I_{R5} + I_{R6} + I_{R7} = 0.6 + 1.5 + 0.9 = 3$ A), proving Kirchhoff's current law.

9-5 POWER IN A SERIES–PARALLEL CIRCUIT

Whether resistors are connected in series or in parallel, the total power in a series–parallel circuit is equal to the sum of all the individual power losses.

$$P_t = P_1 + P_2 + P_3 + P_4 + \cdots$$

total power = addition of all power losses

The formulas for calculating the amount of power lost by each resistor are

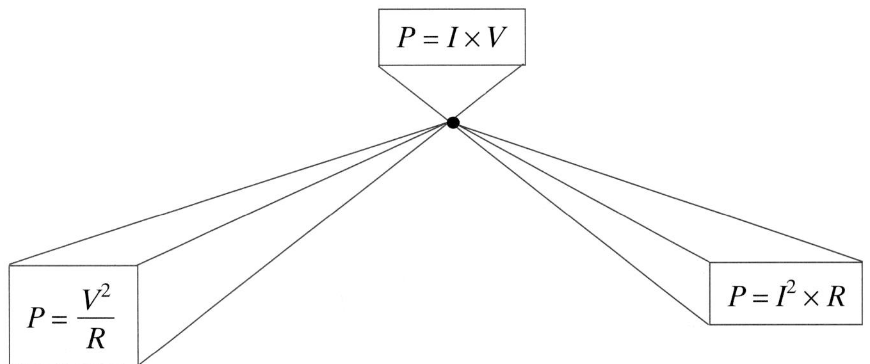

Let us calculate the power dissipated by each resistor; this final calculation will be called step 5. Since resistance, voltage, and current are known, either of the three formulas for power can be used to determine power.

$$P_{R1} = \frac{V_{R1}^{\ 2}}{R_1} = \frac{30\ \text{V}^2}{10\ \Omega} = 90\ \text{W}$$

$$P_{R2} = \frac{V_{R2}^{\ 2}}{R_2} = \frac{18\ \text{V}^2}{12\ \Omega} = 27\ \text{W}$$

$$P_{R3} = I_{R3}^{\ 2} \times R_3 = 1.5\ \text{A}^2 \times 4\ \Omega = 9\ \text{W}$$

$$P_{R4} = I_{R4}^{\ 2} \times R_4 = 1.5\ \text{A}^2 \times 8\ \Omega = 18\ \text{W}$$

$$P_{R5} = \frac{V_{R5}^{\ 2}}{R_5} = \frac{36\ \text{V}^2}{60\ \Omega} = 21.6\ \text{W}$$

$$P_{R6} = \frac{V_{R6}^{\ 2}}{R_6} = \frac{36\ \text{V}^2}{24\ \Omega} = 54\ \text{W}$$

$$P_{R7} = \frac{V_{R7}^{\ 2}}{R_7} = \frac{36\ \text{V}^2}{40\ \Omega} = 32.4\ \text{W}$$

$$\begin{aligned} P_T &= P_{R1} + P_{R2} + P_{R3} + P_{R4} + P_{R5} + P_{R6} + P_{R7} \\ &= 90 + 27 + 9 + 18 + 21.6 + 54 + 32.4 \\ &= 252\ \text{W} \end{aligned}$$

or

$$P_T = \frac{V_T^{\ 2}}{R_T} = \frac{84\ \text{V}^2}{28\ \Omega}$$

$$= 252\ \text{W}$$

The total power dissipated in this example circuit is 252 W. All the information can now be inserted in a final diagram for the example, as shown in Figure 9-18.

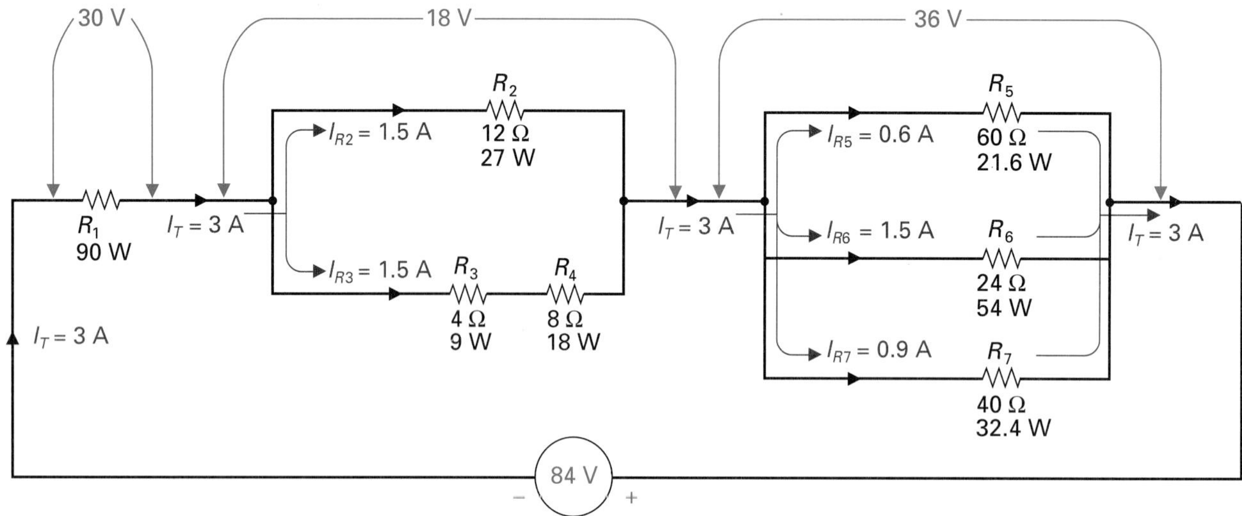

FIGURE 9-18 Series–Parallel Circuit Example with All Information Inserted.

9-6 FIVE-STEP METHOD FOR SERIES–PARALLEL CIRCUIT ANALYSIS

Let's now combine and summarize all the steps for calculating resistance, voltage, current, and power in a series–parallel circuit by solving another problem. Before we begin, however, let us review the five-step procedure.

Solving for Resistance, Voltage, Current, and Power in a Series–Parallel Circuit

STEP 1 Determine the circuit's total resistance.

Step A Solve for series-connected resistors in all parallel combinations.

Step B Solve for all parallel combinations.

Step C Solve for remaining series resistances.

STEP 2 Determine the circuit's total current.

STEP 3 Determine the voltage across each series resistor and each parallel combination (series equivalent resistor).

STEP 4 Determine the value of current through each parallel resistor in every parallel combination.

STEP 5 Determine the total and individual power dissipated by the circuit.

■ **EXAMPLE:**

Referring to Figure 9-19, calculate:

a. Total resistance

b. Voltage drop across all resistors

c. Current through each resistor

d. Total power dissipated by the circuit

Solution:

The problem has basically asked us to calculate everything about the series–parallel circuit shown in Figure 9-19.

Applying the five-step procedure to our problem in Figure 9-19, we will first proceed with step 1, calculating the circuit's total resistance.

STEP 1 Determine the circuit's total resistance.

Step A: There are no series resistors within parallel combinations.

Step B: There are two-resistor (R_2, R_3) and three-resistor (R_5, R_6, R_7) parallel combinations in this circuit.

$$R_{2,3} = \frac{1}{(1/R_2) + (1/R_3)} = 222.2\ \Omega$$

$$R_{5,6,7} = \frac{1}{(1/R_5) + (1/R_6) + (1/R_7)} = 500\ \Omega$$

Figure 9-20 illustrates the circuit resulting after step B.

Step C: Solve for the remaining four resistances to gain the circuit's total resistance (R_T) or equivalent resistance (R_{eq}).

$$\begin{aligned} R_{eq} &= R_1 + R_{2,3} + R_4 + R_{5,6,7} \\ &= 1000\ \Omega + 222.2\ \Omega + 777.8\ \Omega + 500\ \Omega \\ &= 2500\ \Omega \quad \text{or} \quad 2.5\ \text{k}\Omega \end{aligned}$$

Figure 9-21 illustrates the circuit resulting after step C.

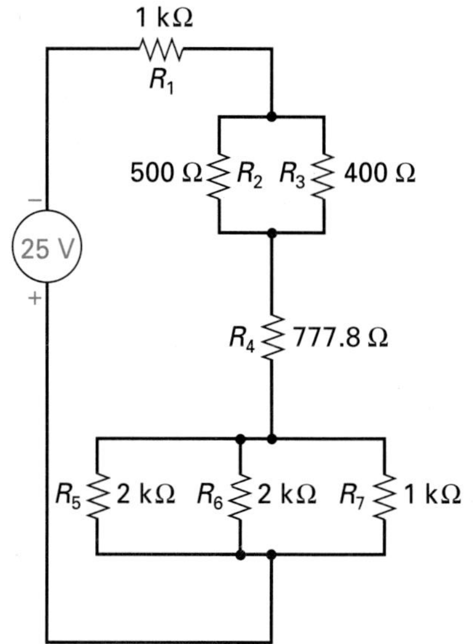

FIGURE 9-19 Apply the Five-Step Procedure to this Series–Parallel Circuit Example.

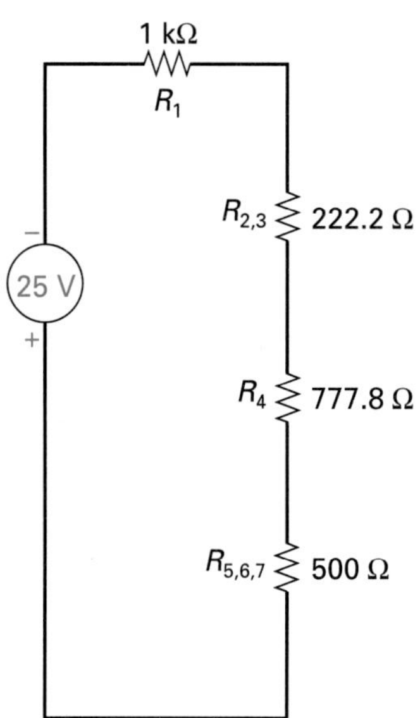

FIGURE 9-20 Circuit Resulting after Step 1B.

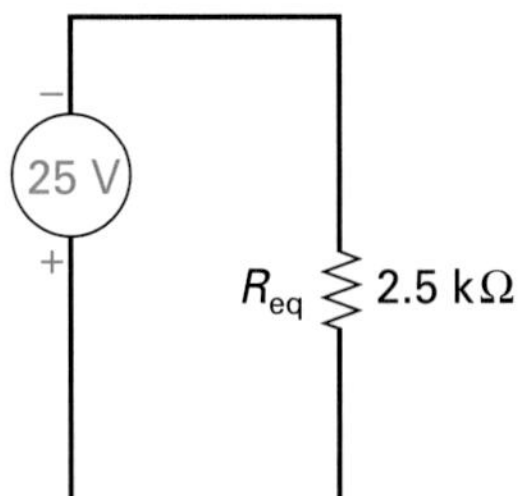

FIGURE 9-21 **Circuit Resulting after Step 1C.**

STEP 2 Determine the circuit's total current.

$$I_T = \frac{V_S}{R_T} = \frac{25\text{ V}}{2.5\text{ k}\Omega} = 10\text{ mA}$$

STEP 3 Determine the voltage across each series resistor and each series equivalent resistor. To achieve this, we utilize the diagram obtained after completing step B (Figure 9-20):

$$V_{R1} = I_T \times R_1 = 10\text{ mA} \times 1\text{ k}\Omega = 10\text{ V}$$

$$V_{R2,3} = I_T \times R_{2,3} = 10\text{ mA} \times 222.2\ \Omega = 2.222\text{ V}$$

$$V_{R4} = I_T \times R_4 = 10\text{ mA} \times 777.8\ \Omega = 7.778\text{ V}$$

$$V_{R5,6,7} = I_T \times R_{5,6,7} = 10\text{ mA} \times 500\ \Omega = 5\text{ V}$$

Figure 9-22 illustrates the results after step 3.

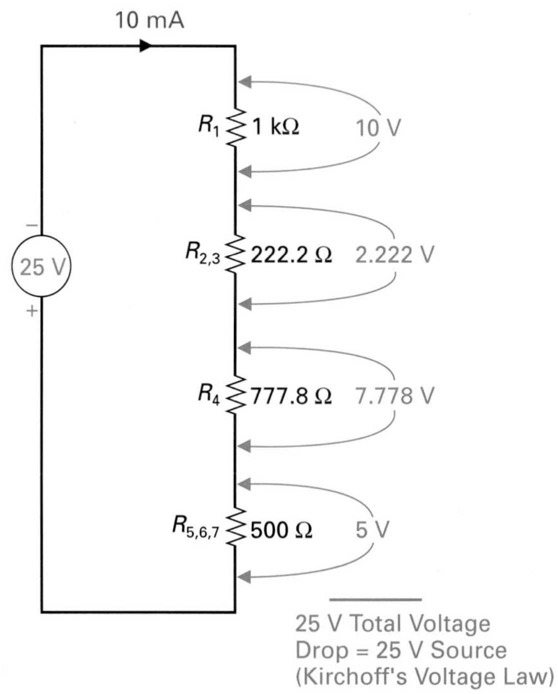

FIGURE 9-22 **Circuit Resulting after Step 3.**

STEP 4 Determine the value of current through each parallel resistor (Figure 9-23). R_1 and R_4 are series-connected resistors, and therefore their current will equal 10 mA.

$$I_{R1} = 10 \text{ mA}$$

$$I_{R4} = 10 \text{ mA}$$

The current through the parallel resistors is calculated by Ohm's law.

$$I_{R2} = \frac{V_{R2}}{R_2} = \frac{2.222 \text{ V}}{500 \text{ }\Omega} = 4.4 \text{ mA}$$

$$I_{R3} = \frac{V_{R3}}{R_3} = \frac{2.222 \text{ V}}{400 \text{ }\Omega} = 5.6 \text{ mA}$$

$$\left.\begin{aligned} I_T &= I_{R2} + I_{R3} \\ 10 \text{ mA} &= 4.4 \text{ mA} + 5.6 \text{ mA} \end{aligned}\right\} \text{ Kirchhoff's current law}$$

$$I_{R5} = \frac{V_{R5}}{R_5} = \frac{5 \text{ V}}{2 \text{ k}\Omega} = 2.5 \text{ mA}$$

$$I_{R6} = \frac{V_{R6}}{R_6} = \frac{5 \text{ V}}{2 \text{ k}\Omega} = 2.5 \text{ mA}$$

$$I_{R7} = \frac{V_{R7}}{R_7} = \frac{5 \text{ V}}{1 \text{ k}\Omega} = 5 \text{ mA}$$

$$\left.\begin{aligned} I_T &= I_{R5} + I_{R6} + I_{R7} \\ 10 \text{ mA} &= 2.5 \text{ mA} + 2.5 \text{ mA} + 5 \text{ mA} \end{aligned}\right\} \text{ Kirchhoff's current law}$$

STEP 5 Determine the total power dissipated by the circuit.

$$P_T = P_{R1} + P_{R2} + P_{R3} + P_{R4} + P_{R5} + P_{R6} + P_{R7}$$

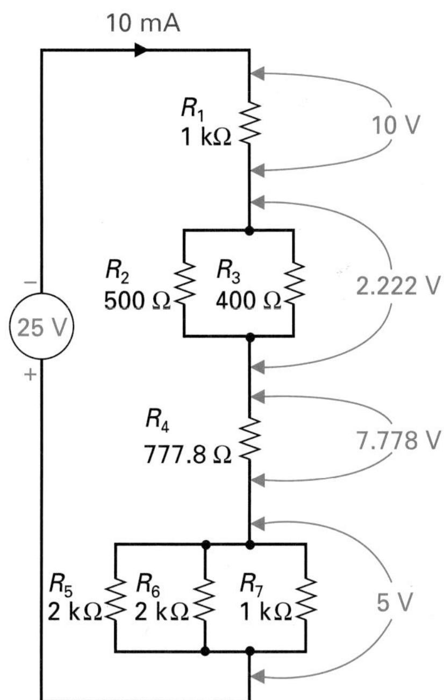

FIGURE 9-23 Series–Parallel Circuit Example with Step 1, 2, and 3 Data Inserted.

or

$$P_T = \frac{V_T^2}{R_T}$$

Each resistor's power figure can be calculated and the sum would be the total power dissipated by the circuit. Since the problem does not ask for the power dissipated by each individual resistor, but for the total power dissipated, it will be easier to use the formula:

$$P_T = \frac{V_T^2}{R_T}$$

$$= \frac{25\text{ V}^2}{2.5\text{ k}\Omega}$$

$$= 0.25\text{ W}$$

SELF-TEST REVIEW QUESTIONS FOR SECTIONS 9-4, 9-5, AND 9-6

1. State the five-step method used for series–parallel circuit analysis.
2. Design your own five-resistor series–parallel circuit, apply values, and use the five-step analysis method.

9-7 SERIES–PARALLEL CIRCUITS

9-7-1 *Loading of Voltage-Divider Circuits*

Loading
The adding of a load to a source.

The straightforward voltage divider was discussed in Chapter 7, but at that point we did not explore some changes that will occur if a load resistance is connected to the voltage divider's output. Figure 9-24 illustrates a voltage divider, and as you can see, the advantage of a voltage-divider circuit is that it can be used to produce several different voltages from one main voltage source by the use of a few chosen resistor values.

In our discussion on load resistance, we mentioned that every circuit or piece of equipment offers a certain amount of resistance, and this resistance represents how much a circuit or piece of equipment will load down the source supply.

Figure 9-25 illustrates an example voltage-divider circuit that is being used to develop a 10 V source from a 20 V battery supply. Figure 9-25(a) illustrates this circuit in the unloaded condition, and by making a few calculations you can understand everything about the circuit.

STEP 1 $R_T = R_1 + R_2 = 1\text{ k}\Omega + 1\text{ k}\Omega = 2\text{ k}\Omega$

STEP 2 $I_T = \frac{V_T}{R_T} = \frac{20\text{ V}}{2\text{ k}\Omega} = 10\text{ mA}$

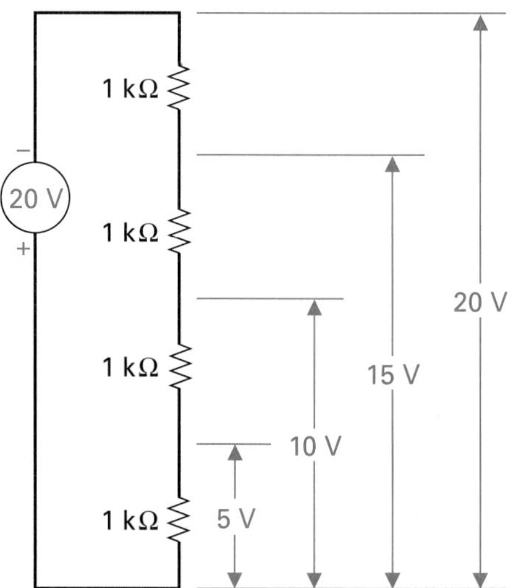

FIGURE 9-24 Voltage-Divider Circuit.

The current that flows through a voltage divider, without a load connected, is called the **bleeder current.** In this example the current is equal to 10 mA. It is called the bleeder current because it is continually drawing or bleeding this current from the voltage source.

Bleeder Current
Current drawn continuously from a voltage source. A bleeder resistor is generally added to lessen the effect of load changes or provide a voltage drop across a resistor.

STEP 3 $V_{R1} = V_{R2}$ (as resistors are the same value)

$$V_{R1} = 10 \text{ V}$$

$$V_{R2} = 10 \text{ V}$$

In Figure 9-25(b) we have connected a piece of equipment represented as a resistance (R_3) across the 10-V supply. This automatically turns the previous series circuit of R_1 and R_2 into a series–parallel circuit made up of R_1, R_2, and the 100-kΩ load resistance. By making a few more calculations, we can discover the changes that have occurred by connecting this load resistance.

STEP 1 Total resistance (R_T)

Step B:

$$R_{2,3} = \frac{R_2 \times R_3}{R_2 + R_3} = \frac{1 \text{ k}\Omega \times 100 \text{ k}\Omega}{1 \text{ k}\Omega + 100 \text{ k}\Omega} = 990.1\ \Omega$$

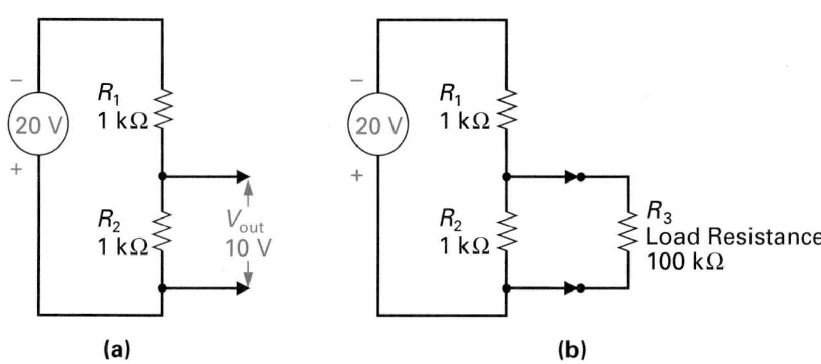

FIGURE 9-25 Voltage-Divider Circuit. (a) Unloaded Output Voltage. (b) Loaded Output Voltage.

Step C: $R_{1,2,3} = R_1 + R_{2,3}$

$$= 1\text{ k}\Omega + 990.1$$

$$= 1.99\text{ k}\Omega$$

STEP 2 Total current (I_T)

$$I_T = \frac{20\text{ V}}{1.99\text{ k}\Omega} = 10.05\text{ mA}$$

STEP 3 $V_{R1} = I_T \times R_1 = 10.05\text{ mA} \times 1\text{ k}\Omega = 10.05\text{ V}$

$$V_{R2,3} = I_T \times R_{2,3} = 10.05\text{ mA} \times 990.1\ \Omega = 9.95\text{ V}$$

STEP 4 $I_{R1} = I_T = 10.05\text{ mA}$

$$I_{R2} = \frac{V_{R2}}{R_2} = \frac{9.95\text{ V}}{1\text{ k}\Omega} = 9.95\text{ mA}$$

$$I_{R3} = \frac{V_{R3}}{R_3} = \frac{9.95\text{ V}}{100\text{ k}\Omega} = 99.5\ \mu\text{A}$$

$$\left.\begin{aligned} I_{R2} + I_{R3} &= I_T \\ 9.95\text{ mA} + 99.5\ \mu\text{A} &= 10.05\text{ mA} \end{aligned}\right\}\ \text{Kirchhoff's current law}$$

As you can see, the load resistance is pulling 99.5 μA, and this pulls the voltage down to 9.95 V from the required 10 V that was desired and is normally present in the unloaded condition.

When designing a voltage divider, design engineers need to calculate how much current a particular load will pull and then alter the voltage-divider resistor values to offset the loading effect when the load is connected.

9-7-2 The Wheatstone Bridge

Wheatstone Bridge A four-arm, generally resistive, bridge that is used to measure resistance.

In 1850, Charles Wheatstone developed a circuit to measure resistance. This circuit, which is still widely used today, is called the **Wheatstone bridge** in his honor and is illustrated in Figure 9-26(a). In Figure 9-26(b), the same circuit has been drawn differently, yet the circuit functions exactly the same.

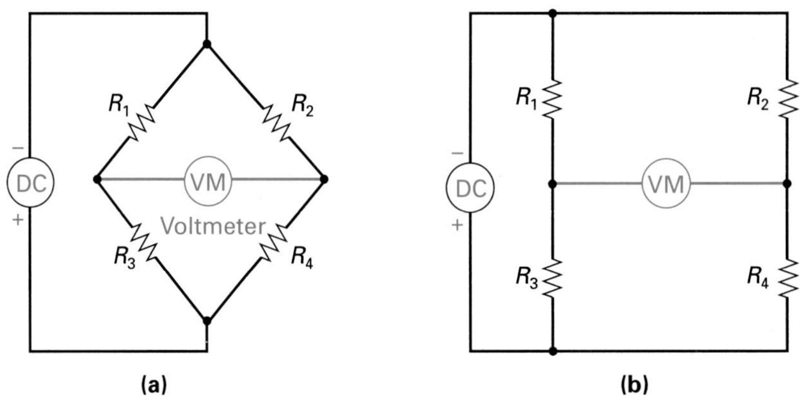

FIGURE 9-26 Wheatstone Bridge. (a) Actual Circuit. (b) Redrawn Simplified Circuit.

Balanced Bridge

Figure 9-27 illustrates an example circuit in which four resistors are connected together to form a series–parallel arrangement. Let us now use the five-step procedure to find out exactly what resistance, current, voltage, and power exist throughout the circuit.

STEP 1 Total resistance (R_T)

Step A:

$$R_{1,3} = R_1 + R_3 = 10 + 20 = 30$$

$$R_{2,4} = R_2 + R_4 = 10 + 20 = 30$$

Step B:

$$R_{Total}: (R_{1,2,3,4}) = \frac{R_{1,3} \times R_{2,4}}{R_{1,3} + R_{2,4}}$$

$$= \frac{30 \times 30}{30 + 30} = 15\ \Omega$$

Total resistance = 15 Ω

STEP 2 Total current (I_T)

$$I_T = \frac{V_T}{R_T} = \frac{30\text{ V}}{15\ \Omega} = 2\text{ A}$$

STEP 3 Since $R_{1,3}$ is in parallel with $R_{2,4}$, 30 V will appear across both $R_{1,3}$ and $R_{2,4}$.

$$V_T = V_{R1,3} = V_{R2,4} = 30\text{ V}$$

voltage-divider formula

$$V_{R1} = \frac{R_1}{R_{1,3}} \times V_T$$

$$= \frac{10}{30} \times 30 = 10\text{ V}$$

$$V_{R3} = \frac{R_3}{R_{1,3}} \times V_T = 20\text{ V}$$

$$V_{R2} = \frac{R_2}{R_{2,4}} \times V_T = 10\text{ V}$$

$$V_{R4} = \frac{R_4}{R_{2,4}} \times V_T = 20\text{ V}$$

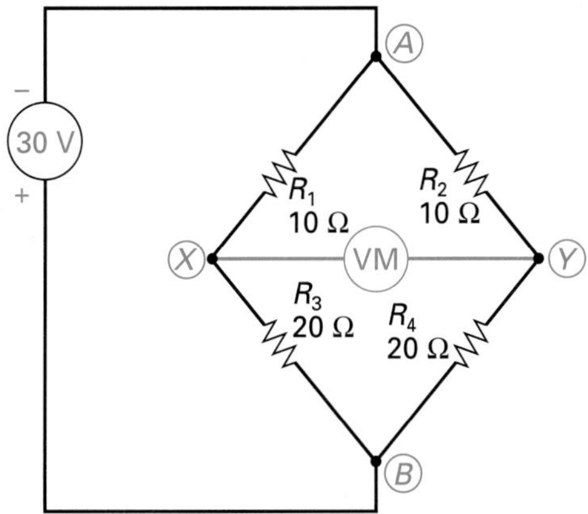

FIGURE 9-27 Wheatstone Bridge Example Circuit.

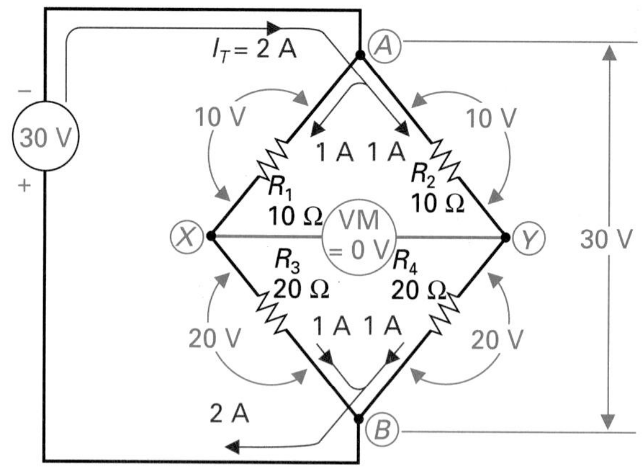

FIGURE 9-28 Balanced Wheatstone Bridge.

STEP 4 $I_{R1} = \dfrac{V_{R1,3}}{R_{1,3}} = \dfrac{30\text{ V}}{30\ \Omega} = 1\text{ A}$

$$I_{R2,4} = \frac{V_{R2,4}}{R_{2,4}} = \frac{30\text{ V}}{30\ \Omega} = 1\text{ A}$$

$$\left.\begin{aligned} I_{R1,3} + I_{R2,4} &= I_T \\ 1\text{ A} + 1\text{ A} &= 2\text{ A} \end{aligned}\right\}\ \text{Kirchhoff's current law}$$

STEP 5 Total power dissipated $(P_T) = I_T^2 \times R_T$

$$= 2\text{ A}^2 \times 15\ \Omega$$
$$= 4 \times 15\ \Omega$$
$$= 60\text{ W}$$

Figure 9-28 illustrates all of the step results inserted in the Wheatstone bridge example schematic. The Wheatstone bridge is said to be in the balanced condition since the voltage at point X will equal the voltage at point Y ($V_{R3} = V_{R4}$, 20 V = 20 V). This same voltage exists across R_3 and R_4, so the voltmeter, which is measuring the voltage difference between X and Y, will indicate 0 V potential difference.

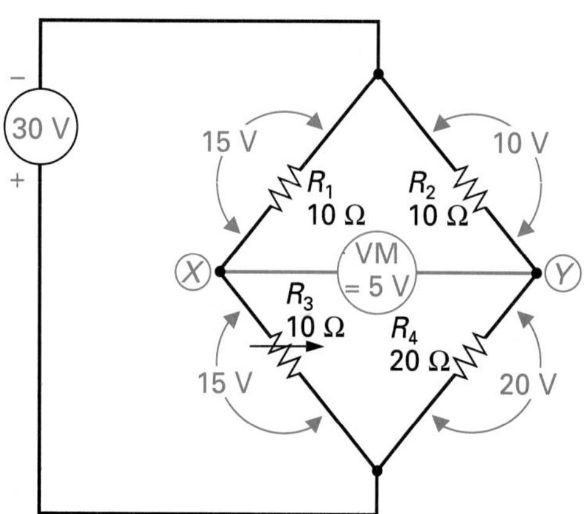

FIGURE 9-29 Unbalanced Wheatstone Bridge.

Unbalanced Bridge

In Figure 9-29 we have replaced R_3 with a variable resistor and set it to 10 Ω. The R_2 and R_4 resistor combination will not change its voltage drop; however, R_1 and R_3, which are now equal, will each split the 30 V supply, producing 15 V across R_3. The voltmeter will indicate the difference in potential (5 V) from the voltage across R_3 at point X (15 V) and across R_4 at point Y (20 V). The voltmeter is actually measuring the imbalance in the circuit, which is why this circuit in this condition is known as an *unbalanced bridge.*

Determining Unknown Resistance

Figure 9-30 illustrates a Wheatstone bridge into which an unknown resistor (R_{un}) and variable resistor (R_{va}) have been inserted. The variable value resistor is a calibrated resistor (resistance has been checked against a known, accurate resistance) in which the resistance can be adjusted and read from a calibrated dial.

The procedure to follow to find the value of the unknown resistor is as follows:

1. Adjust the variable value resistor until the voltmeter indicates that the Wheatstone bridge is balanced (voltmeter indicates 0 V).
2. Read the value of the variable value resistor. As long as $R_1 = R_2$, the variable resistance value will be the same as the unknown resistance value.

$$R_{va} = R_{un}$$

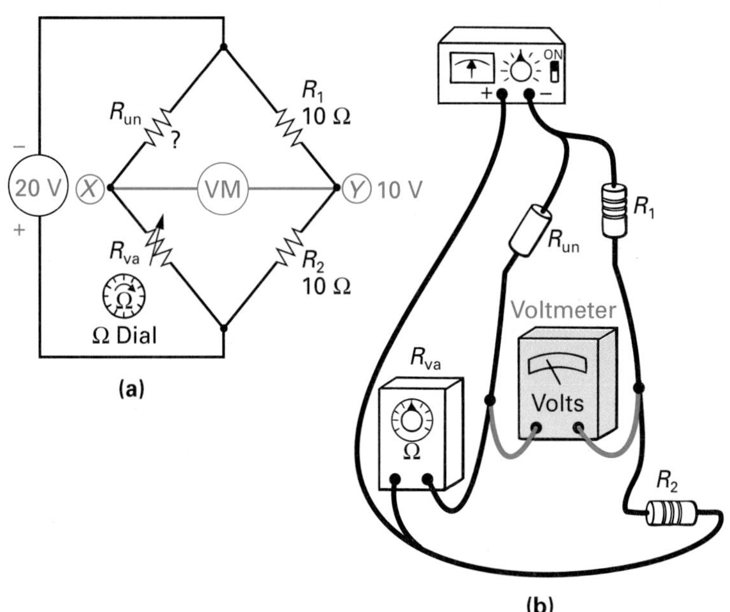

FIGURE 9-30 Using a Wheatstone Bridge to Determine an Unknown Resistance. (a) Schematic. (b) Pictorial.

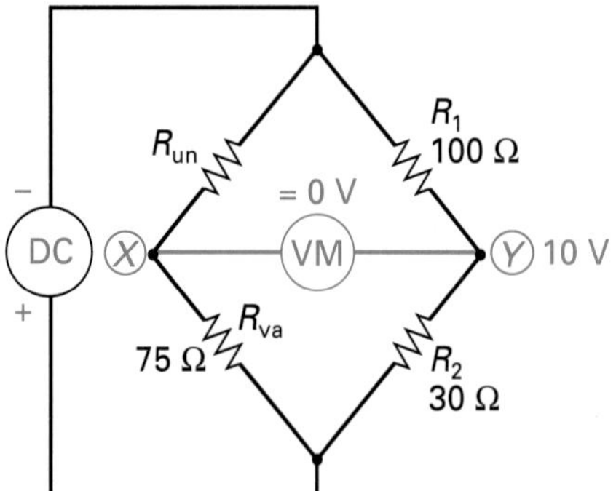

FIGURE 9-31 Wheatstone Bridge Circuit Example.

Since R_1 and R_2 are equal to one another, the voltage will be split across the two resistors, producing 10 V at point *Y*. The variable value resistor must therefore be adjusted so that it equals the unknown resistance, and therefore the same situation will occur, in that the 20 V source will be split, producing 10 V at point *X*, indicating a balanced zero-volt condition on the voltmeter. For example, if the unknown resistance is equal to 5 Ω, then only when the variable value resistor is adjusted and equal to 5 Ω would 10 V appear at point *X* and allow the circuit meter to read zero volts, indicating a balance. The variable value resistor resistance could be read (5 Ω) and the unknown resistor resistance would be known (5 Ω).

EXAMPLE:

What is the unknown resistance in Figure 9-31?

Solution:

The bridge is in a balanced condition as the voltmeter is reading a 0 V difference between points *X* and *Y*. In the previous section we discovered that if $R_1 = R_2$, then:

$$R_{va} = R_{un}$$

In this case, R_1 does not equal R_2, so a variation in the formula must be applied to take into account the ratio of R_1 and R_2.

$$R_{un} = R_{va} \times \frac{R_1}{R_2}$$

$$= 75\ \Omega \times \frac{100}{30}$$

$$= 75\ \Omega \times 3.33 = 250\ \Omega$$

Since R_1 is 3.33 times greater than R_2, then R_{un} must be 3.33 times greater than R_{va} if the Wheatstone bridge is in the balanced condition.

CALCULATOR SEQUENCE

Step	Keypad Entry	Display Response
1.	1 0 0	100
2.	÷	
3.	3 0	30
4.	=	3.333
5.	×	
6.	7 5	75
7.	=	250

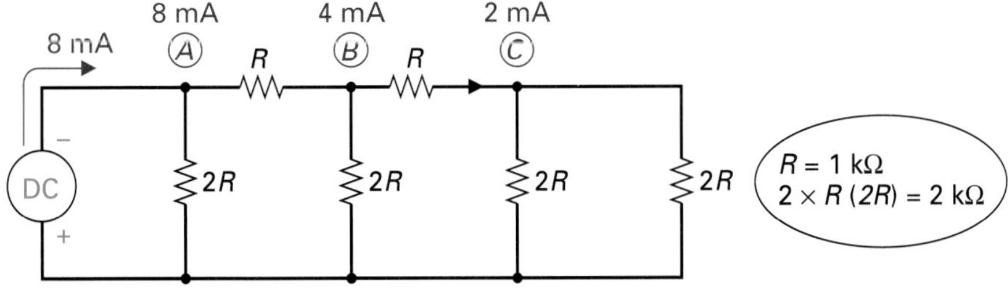

FIGURE 9-32 ***R*–2*R* Ladder Circuit.**

9-7-3 *The* R–2R *Ladder Circuit*

Figure 9-32 illustrates a ***R*–2*R* ladder circuit**, which is a series–parallel circuit used for digital-to-analog conversion. To fully understand this circuit, our first step should be to find out exactly which branches will have which values of current. This can be obtained by finding out what value of resistance the current sees when it arrives at the three junctions *A, B,* and *C.* Let us begin with point *C* first and simplify the circuit. This is illustrated in Figure 9-33(a). No specific resistance value has been chosen, but in all cases 2*R* resistors (2 × *R*) are twice the resistance of an *R* resistor.

***R*–2*R* Ladder Circuit** A network or circuit composed of a sequence of *L* networks connected in tandem. This *R*–2*R* circuit is used in digital-to-analog converters.

In Figure 9-33(a), if 2 mA of current arrives at point *C,* it sees 2*R* of resistance in parallel with a 2*R* resistance, and consequently the 2 mA of current splits and 1 mA flows through each branch. Two 2*R* resistors in parallel with one another would consequently be equivalent to one *R,* as seen in Figure 9-33(b).

In Figure 9-34(a), if 4 mA of current arrives at point *B,* it sees two series resistances to the right, which is equivalent to 2*R* [Figure 9-34(b)] and one resistance down of 2*R.* The 4 mA therefore splits, causing 2 mA down one path and 2 mA down the other. The two 2*R* resistors in parallel [Figure 9-34(b)] are equivalent to one *R,* as shown in Figure 9-34(c).

In Figure 9-35(a), if 8 mA of current arrives at point *A,* it sees two series resistors to the right, which is equivalent to 2*R* [Figure 9-35(b)], and one resis-

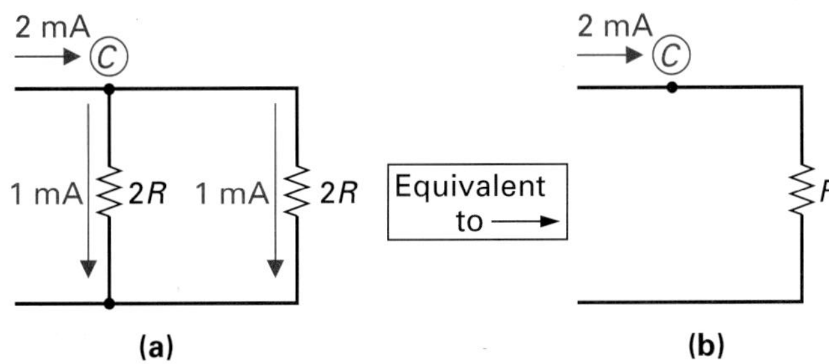

FIGURE 9-33 ***R*–2*R* Equivalent Circuit at Junction C.**

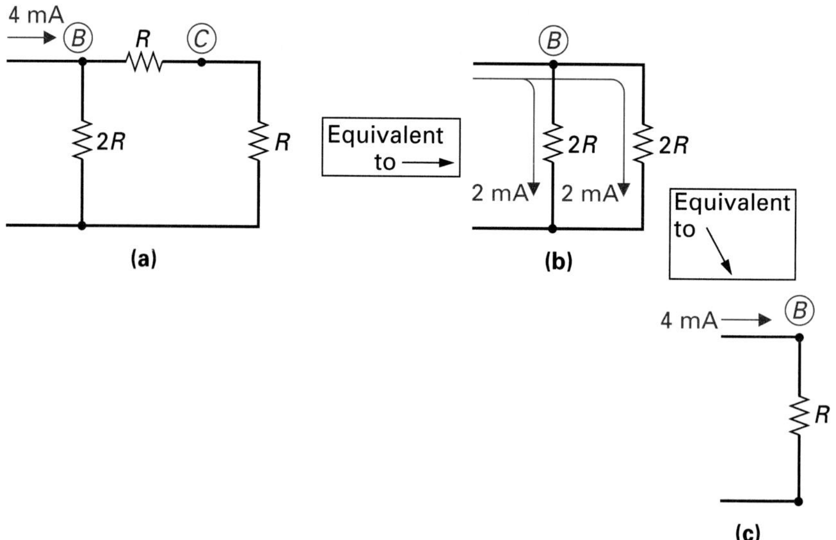

FIGURE 9-34 ***R*–2*R*** **Equivalent Circuit at Junction B.**

tance down of 2*R*. The 8 mA of current therefore splits equally, causing 4 mA down one path and 4 mA down the other. The two 2*R* resistors in parallel [Figure 9-35(b)] are equivalent to one resistance *R* [Figure 9-35(c)].

Figure 9-36(a) through (f) summarizes the step-by-step simplification of this circuit.

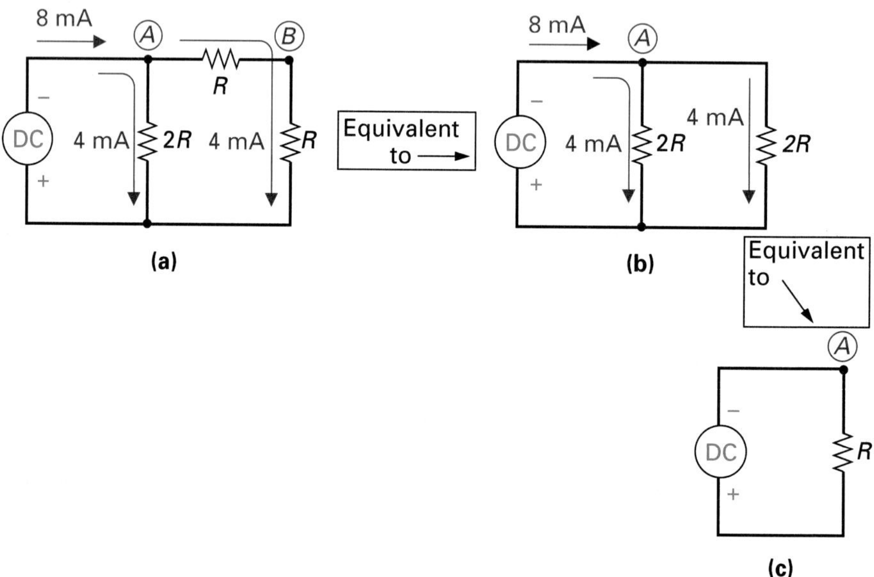

FIGURE 9-35 ***R*–2*R*** **Equivalent Circuit at Junction A.**

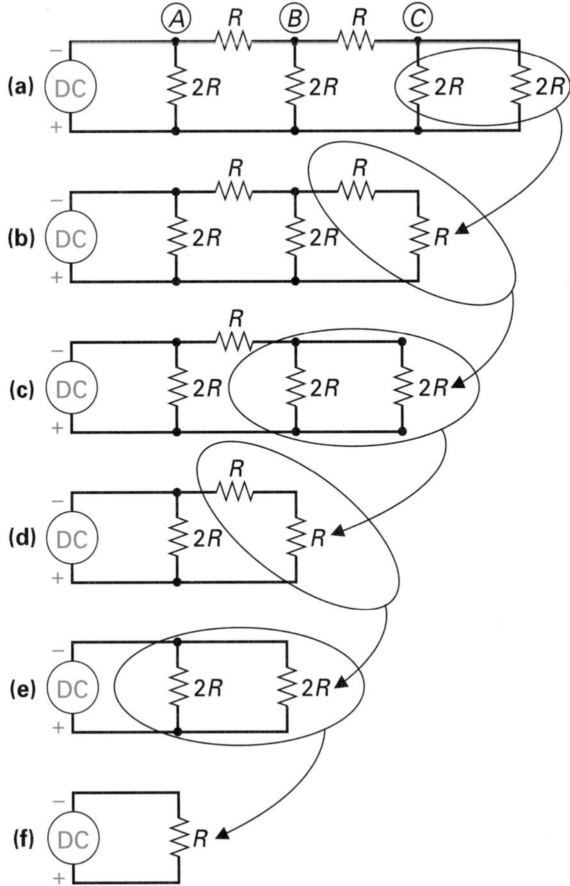

FIGURE 9-36 **Step-by-Step Simplification of an *R*–2*R* Ladder Circuit.**

The question you may have at this point is: What is the primary application of this circuit? The answer is as a current divider, as seen in Figure 9-37(a). The 8 mA of reference current is repeatedly divided by 2 as it moves from left to right, producing currents of 4 mA, 2 mA, and 1 mA. The result of this *R*–2*R* current division can be used in a circuit known as a *digital-to-analog converter* (DAC), which is illustrated in Figure 9-37(b).

Digital data or information exists within a computer, and this system expresses numbers and letters in two discrete steps; for example, on–off, high–low, open–closed, or 0–1. Only two conditions exist within the computer, and all information is represented by this two-state system.

Analog data or information exists outside a computer in our environment, and in this system, data are expressed in many different levels, as opposed to just two with digital information. Numbers are expressed in one of ten different levels (0–9), as opposed to only two (0–1) in digital.

Due to these differences, a device is needed that will interface (convert or link two different elements) the digital information within a computer to the

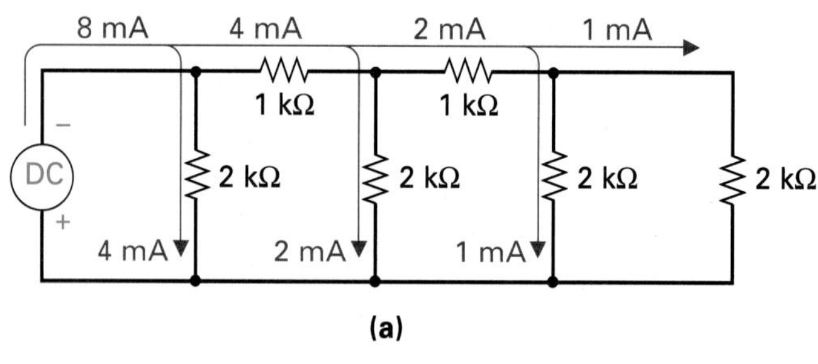

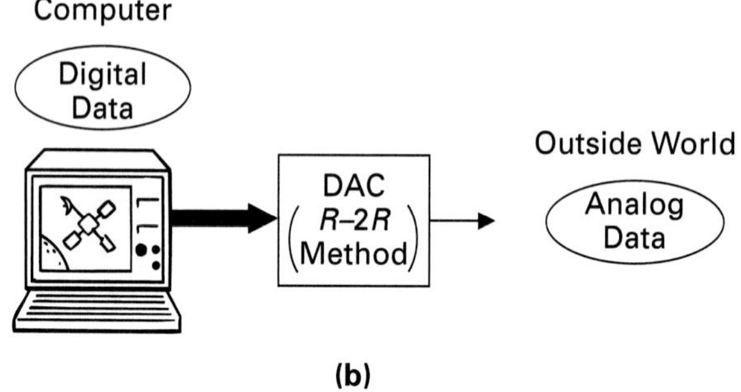

FIGURE 9-37 ***R–2R* Ladder in a Digital-to-Analog Converter.**

analog information that you and I understand. This device, called a digital-to-analog converter, uses the *R*–2*R* ladder circuit that we just discussed.

EXAMPLE:

Determine the reference current and branch currents for the circuit in Figure 9-38.

Solution:

In our simplification of the ladder circuit, we discussed previously that any *R*–2*R* ladder circuit can be simplified to one resistor equal to *R*, as seen in Figure 9-39(a). The reference or total current supplied will therefore equal

$$I_T = \frac{V_T}{R_T} = \frac{24 \text{ V}}{1 \text{ k}\Omega} = 24 \text{ mA}$$

and the current will split through each branch, as shown in Figure 9-39(b).

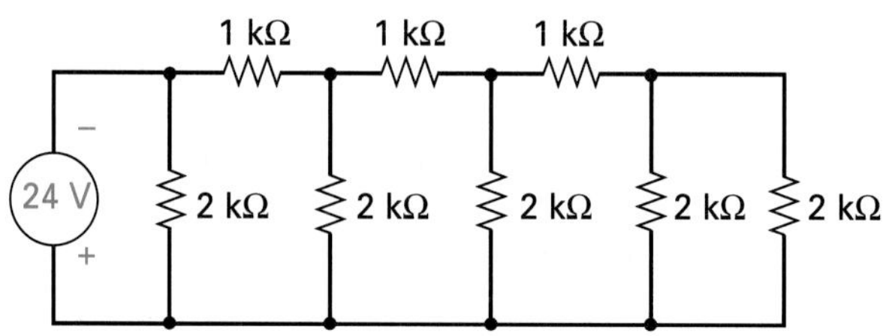

FIGURE 9-38 Calculate *R–2R* Circuit Reference Current and Branch Currents.

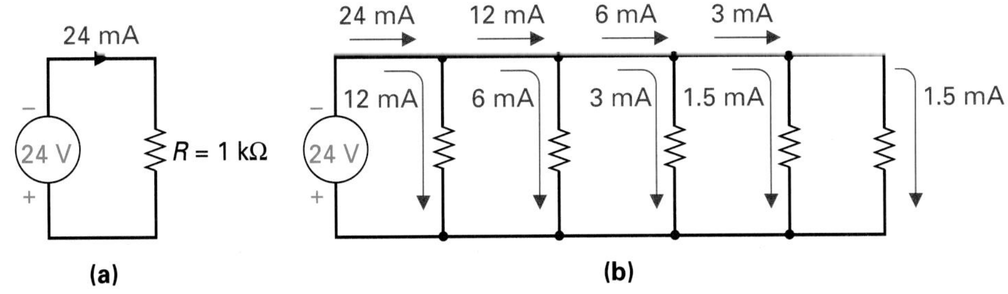

FIGURE 9-39 *R–2R* Circuit with Data Inserted.

SELF-TEST REVIEW QUESTIONS FOR SECTION 9-7

1. What is meant by loading of a voltage-divider circuit?
2. Sketch a Wheatstone bridge circuit and list an application of this circuit.
3. What value will the total resistance of an *R–2R* ladder circuit always equal?
4. For what application could the *R–2R* ladder be used?

9-8 TROUBLESHOOTING SERIES–PARALLEL CIRCUITS

Troubleshooting is defined as the process of locating and diagnosing malfunctions or breakdowns in equipment by means of systematic checking or analysis. As discussed in previous resistive-circuit troubleshooting procedures, there are basically only three problems that can occur:

1. A component will open. This usually occurs if a resistor burns out or a wire or switch contact breaks.
2. A component will short. This usually occurs if a conductor, such as solder, wire, or some other conducting material, is dropped or left in the equipment, making or connecting two points that should not be connected.
3. There is a variation in a component's value. This occurs with age in resistors over a long period of time and can eventually cause a malfunction of the equipment.

Using the example circuit in Figure 9-40, we will step through a few problems, beginning with an open component. Throughout the troubleshooting, we will use the voltmeter whenever possible, as it can measure voltage by just connecting the leads across the component, rather than the ammeter, which has to be placed in the circuit, in which case the circuit path has to be opened. In some instances, ammeter use can be difficult to accomplish.

To begin, let's calculate the voltage drops and branch current obtained when the circuit is operating normally.

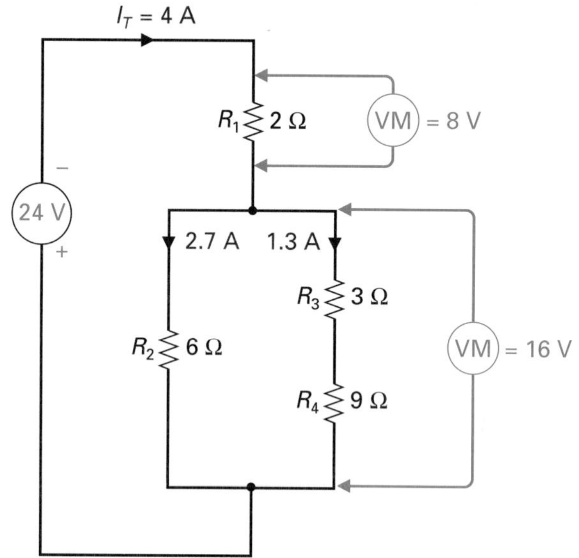

FIGURE 9-40 Series–Parallel Circuit Example.

STEP 1 (A) $R_{3,4} = R_3 + R_4 = 3\ \Omega + 9\ \Omega = 12\ \Omega$

(B) $R_{2,3,4} = \dfrac{R_2 \times R_{3,4}}{R_2 + R_{3,4}} = \dfrac{6\ \Omega \times 12\ \Omega}{6\ \Omega + 12\ \Omega} = 4\ \Omega$

(C) $R_{1,2,3,4} = R_T = R_1 + R_{2,3,4}$

$= 2\ \Omega + 4\ \Omega = 6\ \Omega$

STEP 2 $I_T = \dfrac{V_T}{R_T} = \dfrac{24\text{ V}}{6\ \Omega} = 4\text{ A}$

STEP 3 $V_{R1} = I_{R1} \times R_1 = 4\text{ A} \times 2\ \Omega = 8\text{ V}$

$V_{R2,3,4} = I_{R2,3,4} \times R_{2,3,4} = 4\text{ A} \times 4\ \Omega = 16\text{ V}$

(Kirchhoff's voltage law)

STEP 4 $I_{R1} = 4\text{ A}$ (series resistor)

$I_{R2} = \dfrac{V_{R2}}{R_2} = \dfrac{16\text{ V}}{6\ \Omega} = 2.7\text{ A}$

$I_{R3,4} = \dfrac{V_{R3,4}}{R_{3,4}} = \dfrac{16\text{ V}}{12\ \Omega} = 1.3\text{ A}$

All these results have been inserted in Figure 9-40.

9-8-1 *Open Component*

R_1 Open (Figure 9-41)

With R_1 open, there cannot be any current flow through the circuit as there is not a path from one side of the battery to the other. This fault can be recognized easily because approximately all of the applied 24 V will be measured across the open resistor (R_1), and 0 V will appear across all the other resistors.

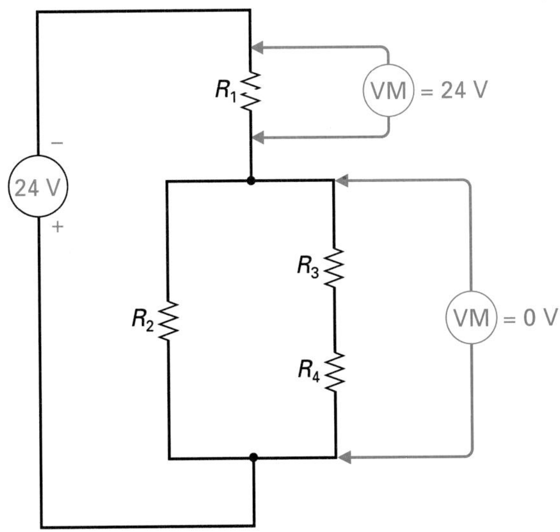

FIGURE 9-41 **Open Series-Connected Resistor in a Series–Parallel Circuit.**

R_3 Open (Figure 9-42)

With R_3 open, there will be no current through the branch made up of R_3 and R_4. The current path will be through R_1 and R_2, and therefore the total resistance will now increase ($R_T\uparrow$) from 6 Ω to

$$R_T = R_1 + R_2$$
$$= 2\ \Omega + 6\ \Omega = 8\ \Omega$$

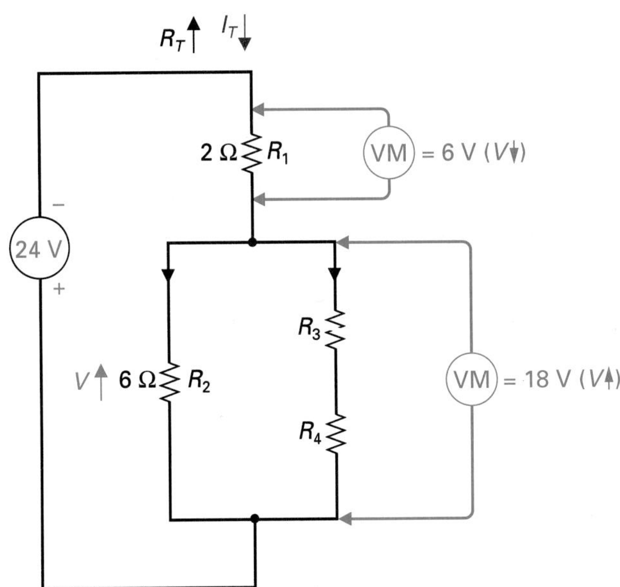

FIGURE 9-42 **Open Parallel-Connected Resistor in a Series–Parallel Circuit.**

This 8 Ω is an increase in circuit resistance from the typical resistance, which was 6 Ω, which implies that an open has occurred to increase resistance. The total current will decrease ($I_T\downarrow$) from 4 A to

$$I_T = \frac{V_T}{R_T} = \frac{24\text{ V}}{8\ \Omega} = 3\text{ A}$$

The voltage drop across the resistors will be

$$V_{R1} = I_T \times R_1 = 3\text{ A} \times 2\ \Omega = 6\text{ V}$$
$$V_{R2} = I_T \times R_2 = 3\text{ A} \times 6\ \Omega = 18\text{ V}$$

If one of the parallel branches is opened, the overall circuit resistance will always increase. This increase in the total resistance will cause an increase in the voltage dropped across the parallel branch (the greater the resistance, the greater the voltage drop), which enables the technician to localize the fault area and also to determine that the fault is an open.

The voltage measured with a voltmeter will be

$$R_1 = 6\text{ V}$$
$$R_2 = 18\text{ V}$$
$$R_3 = 18\text{ V (open)}$$
$$R_4 = 0\text{ V}$$

This identifies the problem as R_3 being open, as it drops the entire parallel circuit voltage (18 V) across itself, whereas normally the voltage would be dropped proportionally across R_3 and R_4, which are in series with one another.

9-8-2 *Shorted Component*

R_1 Shorted (Figure 9-43)

With R_1 shorted, the total circuit resistance will decrease ($R_T\downarrow$), causing an increase in circuit current ($I_T\uparrow$). This increase in current will cause an increase in the voltage dropped across the parallel branch; however, the fault can be located once you measure the voltage across R_1, which will read 0 V, indicating that this resistor has almost no resistance, as it has no voltage drop across it.

R_3 Shorted (Figure 9-44)

With R_3 shorted, there will be a decrease in the circuit's total resistance from 6 Ω to

$$R_{2,3,4} = \frac{R_2 \times R_{3,4}}{R_2 + R_{3,4}} = \frac{6 \times 9}{6 + 9} = 3.6$$

$$R_{1,2,3,4} = R_1 + R_{2,3,4}$$
$$= 2\ \Omega + 3.6\ \Omega$$
$$= 5.6\ \Omega$$

This decrease in total resistance ($R_T\downarrow$) will cause an increase in total current ($I_T\uparrow$), which implies that a short has occurred to decrease resistance. The total current will now increase from 4 A to

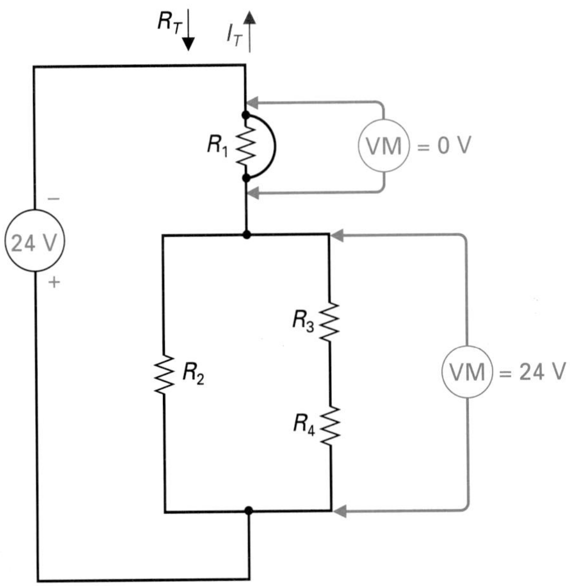

FIGURE 9-43 Shorted Series-Connected Resistor in a Series–Parallel Circuit.

$$I_T = \frac{V_T}{R_T} = \frac{24\ \text{V}}{5.6\ \Omega} = 4.3\ \text{A}$$

The voltage drops across the resistors will be

$$V_{R1} = I_T \times R_1 = 4.3\ \text{A} \times 2\ \Omega = 8.6\ \text{V}$$
$$V_{R2,3,4} = I_T \times R_{2,3,4} = 4.3\ \text{A} \times 3.6\ \Omega = 15.4\ \text{V}$$

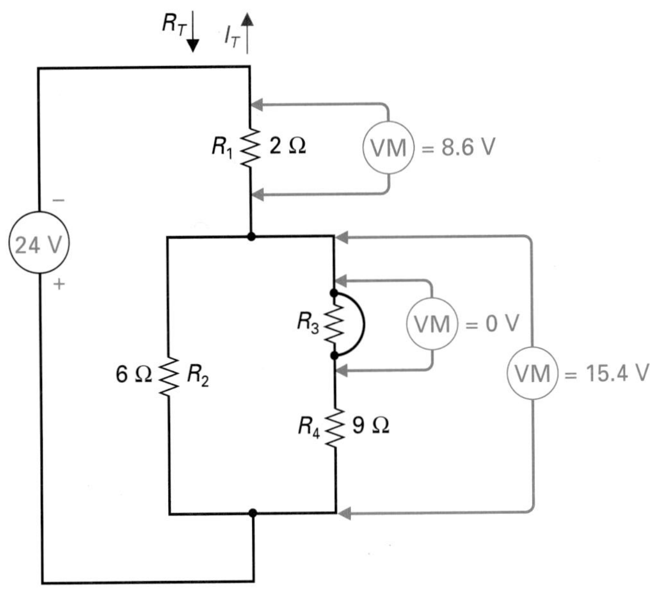

FIGURE 9-44 Shorted Parallel-Connected Resistor in a Series–Parallel Circuit.

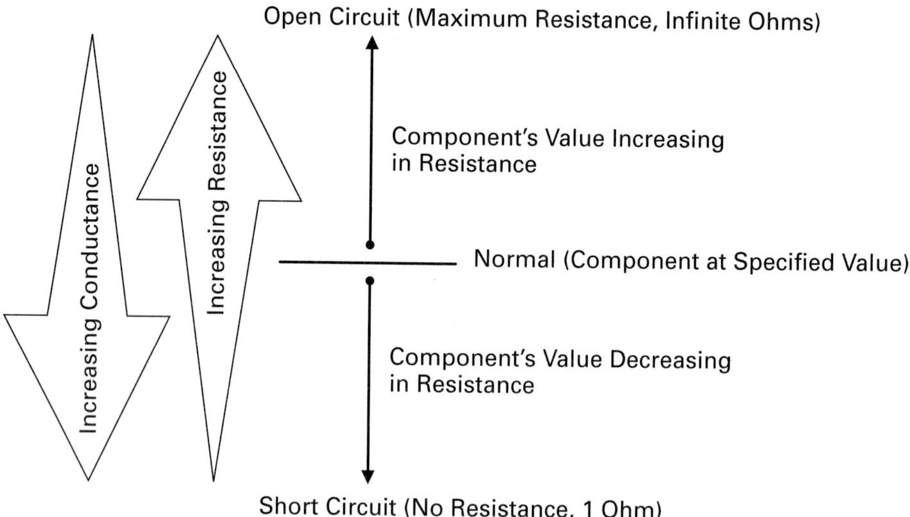

FIGURE 9-45 **Summary of Symptoms for Opens and Shorts.**

When one of the parallel branch resistors is shorted, the overall circuit resistance will always decrease. This decrease in total resistance will cause a decrease in the voltage dropped across the parallel branch (the smaller the resistance, the smaller the voltage drop), and this enables the technician to localize the faulty area and also to determine that the fault is a short.

The voltage measured with the voltmeter (VM) will be

$$R_1 = 8.6\text{ V}$$
$$R_2 = 15.4\text{ V}$$
$$R_3 = 0\text{ V (short)}$$
$$R_4 = 15.4\text{ V}$$

This identifies the problem as R_3 being a short, as 0 V is being dropped across it.

In summary, you may have noticed that (Figure 9-45):

1. An open component causes total resistance to increase ($R_T\uparrow$) and, therefore, total current to decrease ($I_T\downarrow$), and the open component, if in series, has the supply voltage across it, and if in a parallel branch, has the parallel branch voltage across it.
2. A shorted component causes total resistance to decrease ($R_T\downarrow$) and, therefore, total current to increase ($I_T\uparrow$), and the shorted component, if in series or parallel branches, will have 0 V across it.

9-8-3 Resistor Value Variation

R_2 Resistance Decreases (Figure 9-46)

If the resistance of R_2 decreases, the total circuit resistance (R_T) will decrease and the total circuit current (I_T) will increase. The result of this problem and the way in which the fault can be located is that when the voltage drop

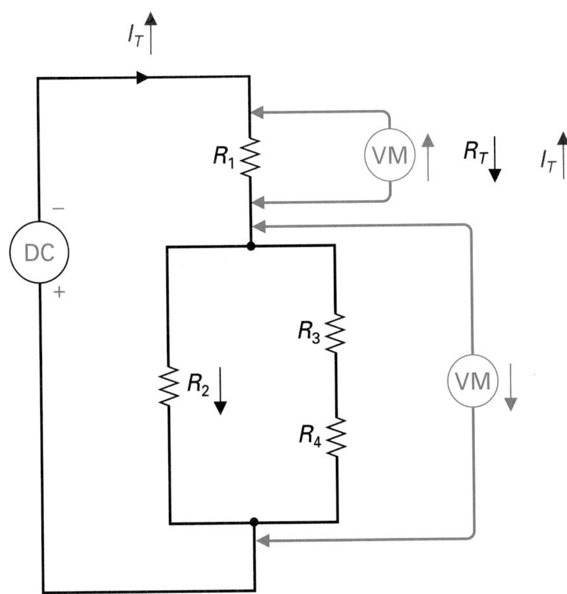

FIGURE 9-46 Parallel-Connected Resistor Value Decrease in Series–Parallel Circuit.

across R_1 and the parallel branch is tested, there will be an increased voltage drop across R_1 due to the increased current flow, and a decrease in the voltage drop across the parallel branch due to a decrease in the parallel branch resistance.

With open and shorted components, the large voltage (open) or small voltage (short) drop across a component enables the technician to identify the faulty component. A variation in a component's value, however, will vary the circuit's behavior; but with this example, the symptoms could have been caused by a combination of variations. So once the area of the problem has been localized, the next troubleshooting step is to remove each of the resistors in the suspected faulty area and verify that their resistance values are correct by measuring their resistances with an ohmmeter. In this problem we had an increase in voltage across R_1 and a decrease in voltage across the parallel branch when measuring with a voltmeter, which was caused by R_2 decreasing. The same swing in voltage readings could also have been obtained by an increase in the resistance of R_1.

R_2 Resistance Increases (Figure 9-47)

If the resistance of R_2 increases, the total circuit resistance (R_T) will increase and the total circuit current (I_T) will decrease. The voltage drop across R_1 will decrease and the voltage across the branch will increase in value. Once again, these measured voltage changes could be caused by the resistance of R_2 increasing or the resistance of R_1 decreasing.

To reinforce your understanding let's work out a few examples of troubleshooting other series–parallel circuits.

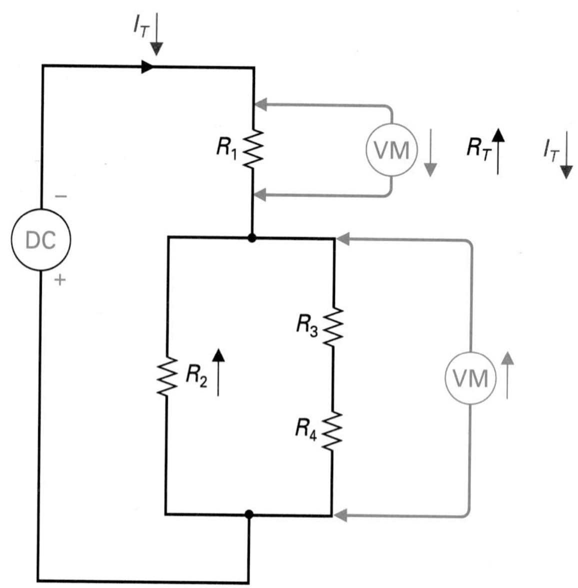

FIGURE 9-47 Series-Connected Resistor Value Decrease in Series–Parallel Circuit.

EXAMPLE:

If bulb 1 in Figure 9-48 goes open, what effect will it have, and how will the fault be recognized?

Solution:

A visual inspection of the circuit shows all bulbs off, as there is no current path from one side of the battery to the other, since bulb 1 is connected in series and has opened the only path. Since all bulbs are off, the faulty bulb cannot be visually isolated; however, you can easily localize the faulty bulb by using one of two methods:

1. Use the voltmeter and check the voltage across each bulb. Bulb 1 would have 12 V across it, while 2 and 3 would have 0 V across them. This isolates the faulty component to bulb 1, since all the supply voltage is being measured across it.
2. By analyzing the circuit diagram, you can see that only one bulb can open and cause all the bulbs to go out, and that is bulb 1. If bulb 2 opens, 1 and 3

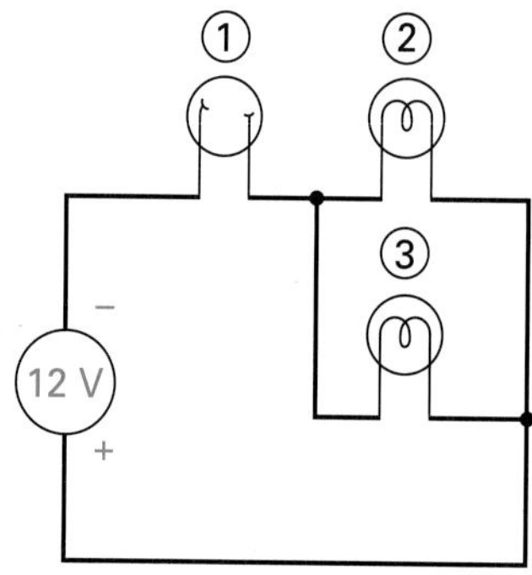

FIGURE 9-48 Series-Connected Open Bulb in Series–Parallel Circuit.

would still be on, and if bulb 3 opens, 1 and 2 would still remain on. With power off, the ohmmeter could verify this open.

EXAMPLE:

One resistor in Figure 9-49 has shorted. From the voltmeter reading shown, determine which one.

Solution:

If the supply voltage is being measured across the parallel branch of R_2 and R_3, there cannot be any other resistance in circuit, so R_1 must have shorted. The next step would be to locate the component, R_1, and determine what has caused it to short. If we were not told that a resistor had shorted, the same symptom could have been caused if R_2 and R_3 were both open, and therefore the open parallel branch would allow no current to flow and maximum supply voltage would appear across it. The individual component resistance, when checked, will isolate the problem.

EXAMPLE:

Determine if there is an open or short in Figure 9-50. If so, isolate it by the two voltage readings that are shown in the circuit diagram.

Solution:

Performing a few calculations, you should come up with a normal total circuit resistance of 14 kΩ and a total circuit current of 1 mA. This should cause 12 V across R_1 and 2 V across the parallel branch under no-fault conditions. The decrease in the voltage drop across the series resistor R_1 leads you to believe that there has been a decrease in total circuit current, which must have been caused by a total resistance increase, which points to an open component (assuming that only an open or short can occur and not a component value variation).

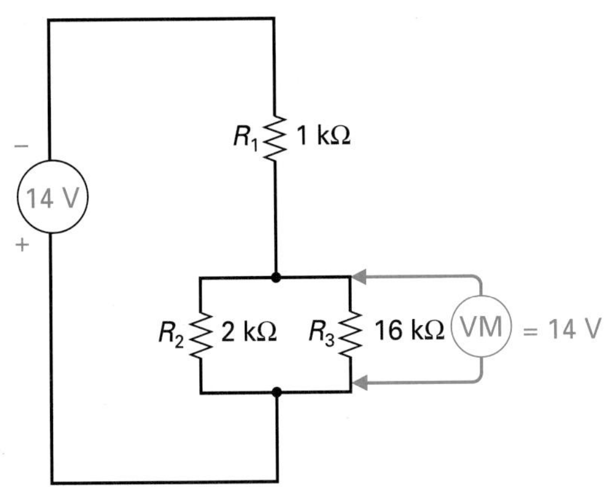

FIGURE 9-49 **Find the Shorted Resistor.**

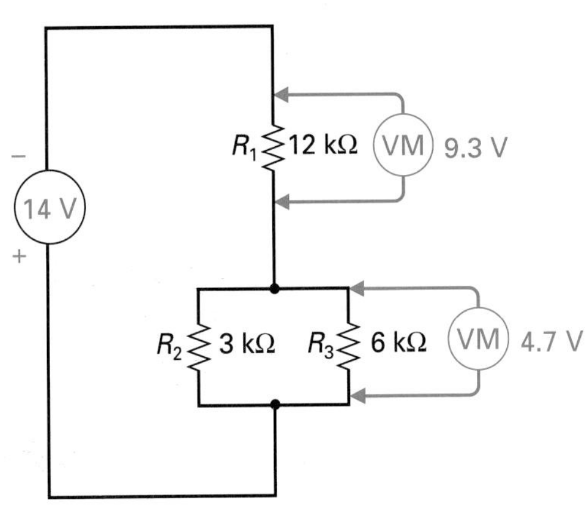

FIGURE 9-50 **Does a Problem Exist?**

If R_1 was open, all the 14 V would have been measured across R_1, which did not occur.

If R_3 opened,

$$\text{total resistance} = 15\ \text{k}\Omega$$

$$\text{total current} = 0.93\ \text{mA}\left(\frac{V_T}{R_T} = \frac{14\ \text{V}}{15\ \text{k}\Omega}\right)$$

$$V_{R1} = I_T \times R_1 = 11.16\ \text{V}$$

Since the voltage dropped across R_1 was 9.3 V, R_3 is not the open.

If R_2 opened,

$$\text{total resistance} = 18\ \text{k}\Omega$$

$$\text{total current} = 0.78\ \text{mA}\left(\frac{V}{R} = \frac{14\ \text{V}}{18\ \text{k}\Omega}\right)$$

$$V_{R1} = I_T \times R_1 = 9.36\ \text{V}$$

This circuit's problem is resistor R_2, which has burned out, causing an open circuit.

SELF-TEST REVIEW QUESTIONS FOR SECTION 9-8

Describe how to isolate the following problems in a series–parallel circuit.

1. An open component
2. A shorted component
3. A resistor value variation

9-9 THEOREMS FOR DC CIRCUITS

Series–parallel circuits can become very complex in some applications, and the more help you have in simplifying and analyzing these networks, the better. The following theorems can be used as powerful analytical tools for evaluating circuits. To begin, let's discuss the differences between voltage and current sources.

9-9-1 Voltage and Current Sources

The easiest way to understand a current source is to compare its feature to a voltage source, so let's begin by discussing voltage sources.

Voltage Source

Voltage Source
The circuit or device that supplies voltage to a load circuit.

The battery is an example of a **voltage source** that in the ideal condition will produce a fixed output voltage regardless of what load resistance is connected across its terminals. This means that even if a large load current is drawn from the battery (heavy load, due to a small load resistance) or if a small load

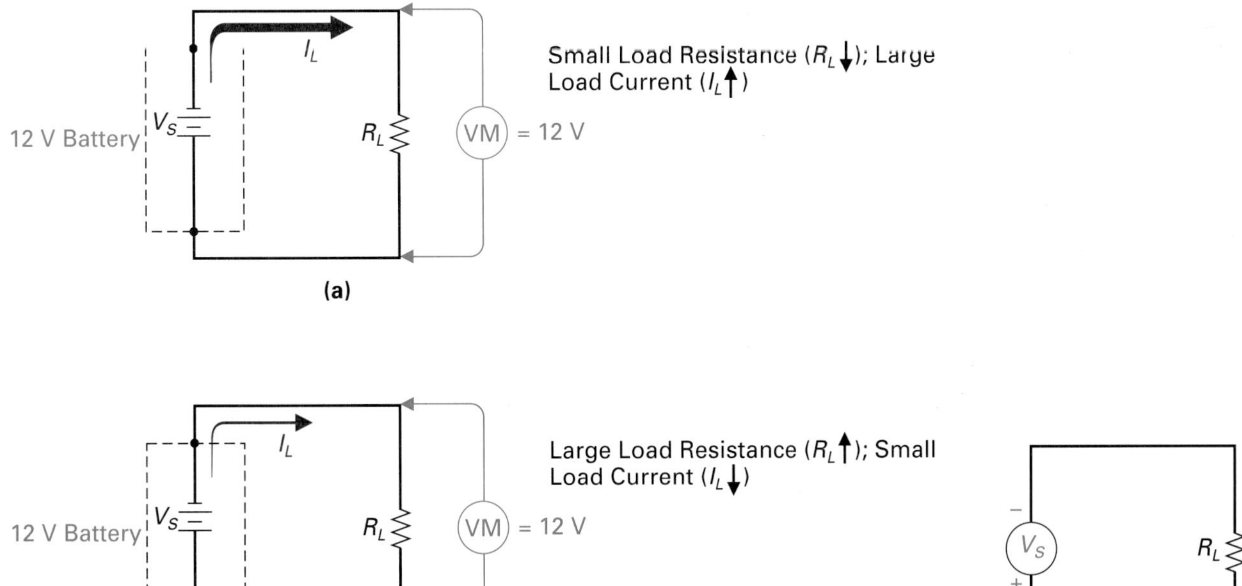

FIGURE 9-51 **Ideal Voltage Source. (a) Heavy Load. (b) Light Load. (c) Symbol.**

current results (light load, due to a large load resistance), the battery will always produce a constant output voltage, as seen in Figure 9-51.

In reality, every voltage source, whether a battery, power supply, or generator, will have some level of inefficiency and not only generate an output electrical dc voltage but also generate heat. This inefficiency is represented as an internal resistance, as seen in Figure 9-52(a), and in most cases this internal source resistance (R_{int}) is very low (several ohms) compared to the load resistance (R_L). In Figure 9-52(a), no load has been connected, so the output or open-circuit voltage will be equal to the source voltage, V_S. When a load is connected across the battery, as shown in Figure 9-52(b), R_{int} and R_L form a

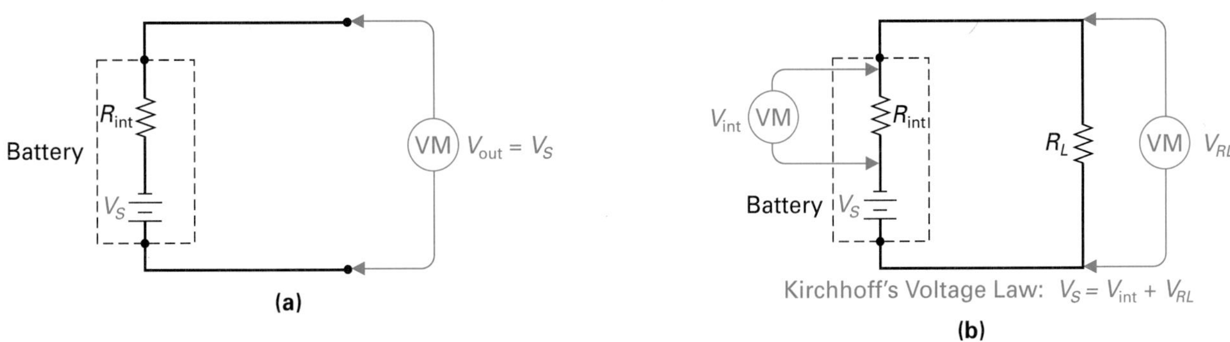

FIGURE 9-52 **Realistic Voltage Source. (a) Unloaded. (b) Loaded.**

series circuit, and some of the source voltage appears across R_{int}; so the output or load voltage is always less than V_S. Since R_{int} is normally quite small compared to R_L, the voltage source approaches ideal, as almost all the source voltage (V_S) appears across R_L.

In conclusion, a voltage source should have the smallest possible internal resistance, so that the output voltage (V_{out}) will remain constant and approximately equal to V_S independent of whether a light load (large R_L, small I_L) or heavy load (small R_L, large I_L) is connected across its output terminals.

EXAMPLE:

Calculate the output voltage in Figure 9-53 if R_L is equal to:

a. 100 Ω
b. 1 kΩ
c. 100 kΩ

Solution:

(Voltage-divider formula)

a. $R_L = 100\ \Omega$:

$$V_{out} = \frac{R_L}{R_T} \times V_S$$

$$= \frac{100\ \Omega}{110\ \Omega} \times 100\ \text{V} = 90.9\ \text{V}$$

b. $R_L = 1\ \text{k}\Omega\ (1000\ \Omega)$

$$V_{out} = \frac{1000\ \Omega}{1010\ \Omega} \times 100\ \text{V} = 99.0\ \text{V}$$

c. $R_L = 100\ \text{k}\Omega\ (100{,}000\ \Omega)$

$$V_{out} = \frac{100{,}000\ \Omega}{100{,}010\ \Omega} \times 100\ \text{V} = 99.99\ \text{V}$$

From this example you can see that the larger the load resistance, the greater the output voltage (V_{out} or V_{RL}). To explain this in a little more detail, we can say that a large R_L is considered a light load for the voltage source, as it only has to produce a small load current ($R_L\uparrow$, $I_L\downarrow$), and consequently, the heat generated by the source is small ($P_{R_{int}}\downarrow = I^2\downarrow \times R$) and the voltage source is more efficient (V_{out} almost equals V_S), approaching ideal ($V_{out} = V_S$). The voltage source in this

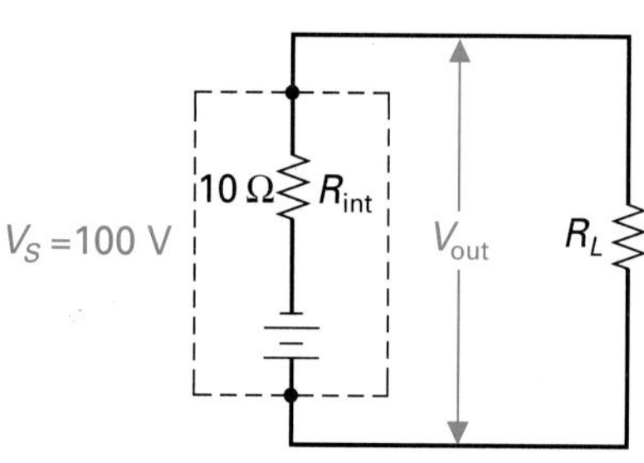

FIGURE 9-53 Voltage Source Circuit Example.

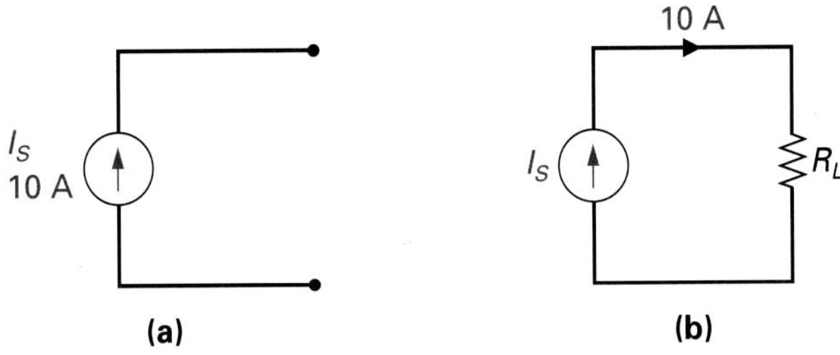

FIGURE 9-54 The Current Source Maintains a Constant Output Current Whether (a) Unloaded or (b) Loaded.

example, however, produced an almost constant output voltage (within 10% of V_S) despite the very large changes in R_L.

Current Source

Current Source
The circuit or device that supplies current to a load circuit.

Just as a voltage source has a certain voltage rating, the **current source** has a certain current rating; and just as a voltage source should deliver a constant output voltage, an ideal current source should deliver its constant rated current, regardless of what value of load resistance is connected across its output terminals, as seen in Figure 9-54.

A current source can be thought of as a voltage source with an extremely large internal resistance, as seen in Figure 9-55(a) and symbolized in Figure 9-55(b).

In conclusion, a current source should have a large internal resistance, so that, whatever the load resistance connected across the output, it will have very little effect on the total resistance and the load current will remain constant. The symbol for a constant current source has an arrow within a circle and this arrow points in the direction of current flow.

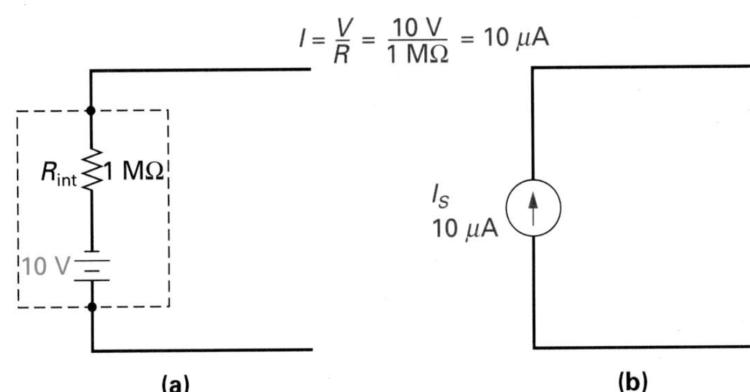

FIGURE 9-55 The Large Internal Resistance of the Current Source.

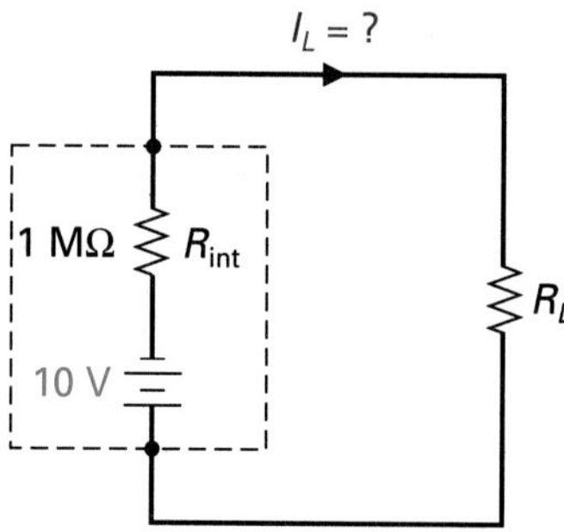

FIGURE 9-56 **Current Source Circuit Example.**

EXAMPLE:

Calculate the load current supplied by the current source in Figure 9-56 if the following values of R_L are connected across the output terminals:

a. 100 Ω

b. 1 kΩ

c. 100 kΩ

Solution:

a. $R_L = 100\ \Omega\ (R_T = R_{int} + R_L = 1{,}000{,}000\ \Omega + 100\ \Omega = 1{,}000{,}100\ \Omega)$

$$I_L = \frac{V_S}{R_T}$$

$$= \frac{10\text{ V}}{1{,}000{,}100\ \Omega} = 9.999\ \mu\text{A}$$

b. $R_L = 1\text{ k}\Omega\ (R_T = R_{int} + R_L = 1{,}001{,}000\ \Omega)$

$$I_L = \frac{10\text{ V}}{1{,}001{,}000\ \Omega} = 9.99\ \mu\text{A}$$

c. $R_L = 100\text{ k}\Omega\ (R_T = R_{int} + R_L = 1{,}100{,}000\ \Omega$

$$I_L = \frac{10\text{ V}}{1{,}100{,}000\ \Omega} = 9.09\ \mu\text{A}$$

From this example you can see that the current source delivered an almost constant output current regardless of the large load resistance change.

Superposition Theorem
In a network or circuit containing two or more voltage sources, the current at any point is equal to the algebraic sum of the individual source currents produced by each source acting separately.

9-9-2 *Superposition Theorem*

The **superposition theorem** is used not only in electronics, but also in physics and even economics. It is used to determine the net effect in a circuit that has two or more sources connected. The basic idea behind this theorem is that if two voltage sources are both producing a current through the same circuit, the net current can be determined by first finding the individual currents and then adding them together. Stated formally: *In a network containing two or more voltage sources, the current at any point is equal to the algebraic sum of the individual source currents produced by each source acting separately.*

The best way to fully understand the theorem is to apply it to a few examples. Figure 9-57(a) illustrates a simple series circuit with two resistors and two voltage sources. The 12 V source (V_1) is trying to produce a current in a clockwise direction, while the 24 V source (V_2) is trying to force current in a counterclockwise direction. What will be the resulting net current in this circuit?

STEP 1 To begin, let's consider what current would be produced in this circuit if only V_1 were connected, as shown in Figure 9-57(b).

$$I_1 = \frac{V_1}{R_T} = \frac{12\text{ V}}{9\ \Omega} = 1.33\text{ A}$$

STEP 2 The next step is to determine how much current V_2 would produce if V_1 were not connected in the circuit, as shown in Figure 9-57(c).

$$I_2 = \frac{V_2}{R_T} = \frac{24\text{ V}}{9\Omega} = 2.66\text{ A}$$

V_1 is attempting to produce 1.33 A in the clockwise direction, while V_2 is trying to produce 2.66 A in the counterclockwise direction. The net current will consequently be 1.33 A in the counterclockwise direction.

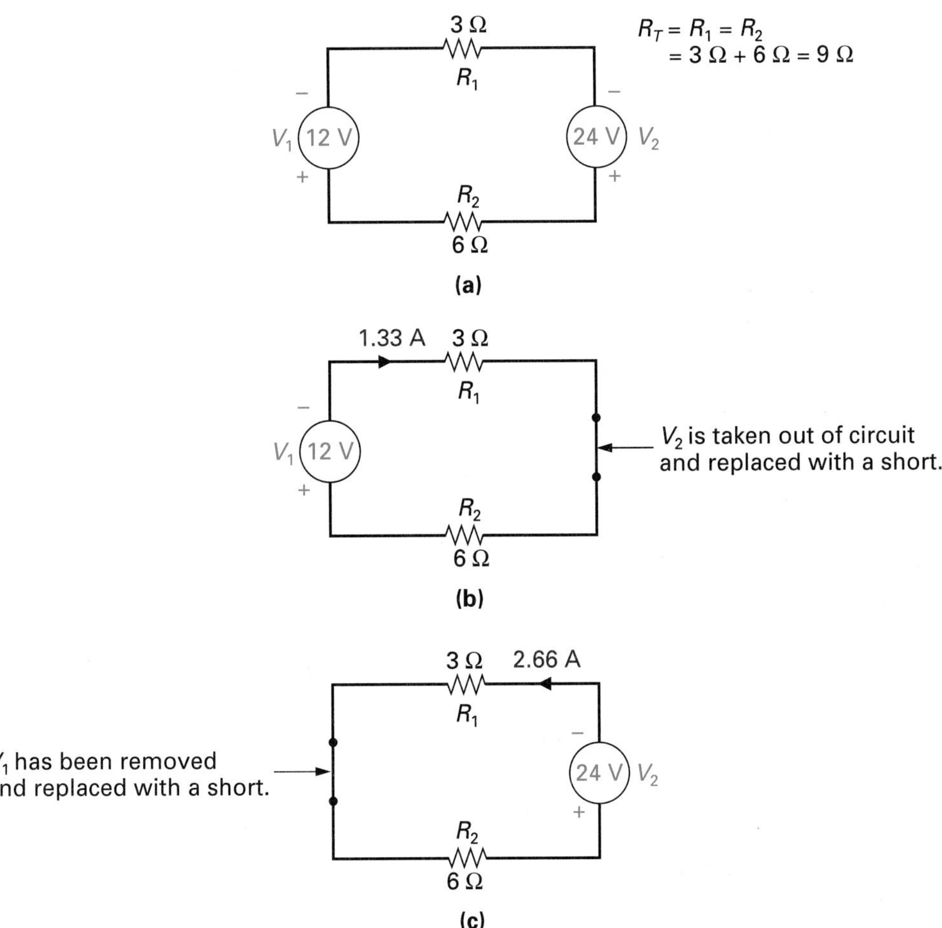

FIGURE 9-57 Superposition.

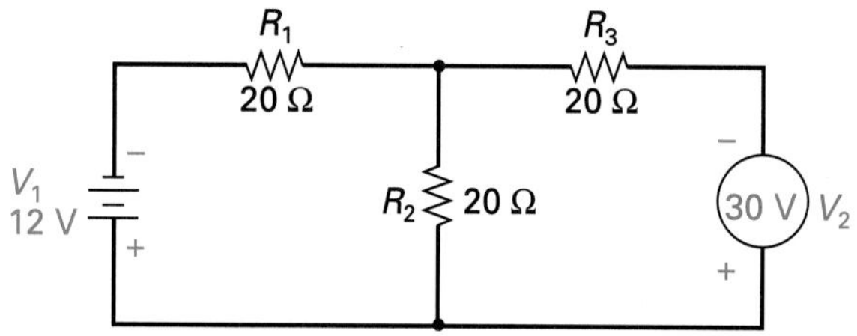

FIGURE 9-58 **Superposition Circuit Example.**

EXAMPLE:

Calculate the current through R_2 in Figure 9-58 using the superposition theorem.

Solution:

The first step is to calculate the current through R_2 due only to the voltage source V_1. This is shown in Figure 9-59(a). R_1 is a series-connected resistor, while R_2 and R_3 are connected in parallel with one another. So:

$$R_{2,3} = \frac{R_{\text{val}}}{n} = \frac{20\ \Omega}{2} = 10\ \Omega \qquad \text{(same-value parallel resistor formula)}$$

$$R_T = R_1 + R_{2,3} = 20\ \Omega + 10\ \Omega = 30\ \Omega \qquad \text{(total resistance)}$$

$$I_T = \frac{V_1}{R_T} = \frac{12\ \text{V}}{30\ \Omega} = 400\ \text{mA} \qquad \text{(total current)}$$

$$I_{R2} = \frac{R_{2,3}}{R_2} \times I_T = \frac{10\ \Omega}{20\ \Omega} \times 400\ \text{mA} = 200\ \text{mA} \qquad \text{(current-divider formula)}$$

This 200 mA of current is flowing down through R_2.

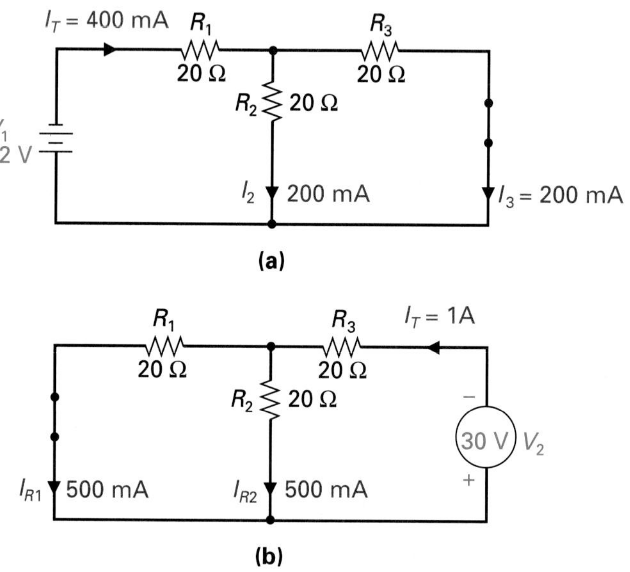

FIGURE 9-59 **Superposition Circuit Solution.**

The next step is to find the current flow through R_2 due only to the voltage source V_2. This is shown in Figure 9-59(b). In this instance, R_3 is a series-connected resistor, and R_1 and R_2 make up a parallel circuit. So:

$$R_{1,2} = 10\ \Omega$$

$$R_T = R_{1,2} + R_3 = 10\ \Omega + 20\ \Omega = 30\ \Omega$$

$$I_T = \frac{V_2}{R_T} = \frac{30\text{ V}}{30\ \Omega} = 1\text{ A} \qquad (1000\text{ mA})$$

$$I_{R2} = \frac{R_{1,2}}{R_2} \times I_T = \frac{10\ \Omega}{20\ \Omega} \times 1000\text{ mA} = 500\text{ mA}$$

This 500 mA of current will flow down through R_2.

Since both V_1 and V_2 produce a current flow down through R_2, so I_{R2} and I_{R3} have the same algebraic sign, the total current through R_2 is equal to the sum of the two currents produced by V_1 and V_2.

$$\begin{aligned} I_{R2}\text{ (total)} &= I_{R2}\text{ due to }V_1 + I_{R2}\text{ due to }V_2 \\ &= 200\text{ mA} + 500\text{ mA} \\ &= 700\text{ mA} \end{aligned}$$

EXAMPLE:

Calculate the current flow through R_1 in Figure 9-60.

Solution:

With the superposition theorem, current sources are treated differently from voltage sources in that each *current source is removed from the circuit and replaced with an open,* as illustrated in the first step of the solution shown in Figure 9-61(a). In this instance you can see that current has only one path (series circuit), so the current through R_1 is counterclockwise and equal to the I_1 source current, 100 μA.

In Figure 9-61(b), the current source I_1 has been removed and replaced with an open. R_1 and R_2 form a series circuit, so the total source current from I_2 flows through R_1 (I_{R1} =500 μA) in a clockwise direction.

Since I_1 is producing 100 μA of current through R_1 in a counterclockwise direction and I_2 is producing 500 μA of current in a clockwise direction, the resulting current through R_1 will equal

$$\begin{aligned} I_{R1} &= I_2\text{ (cw)} - I_1\text{ (ccw)} \\ &= 500\ \mu\text{A} - 100\ \mu\text{A} = 400\ \mu\text{A clockwise} \end{aligned}$$

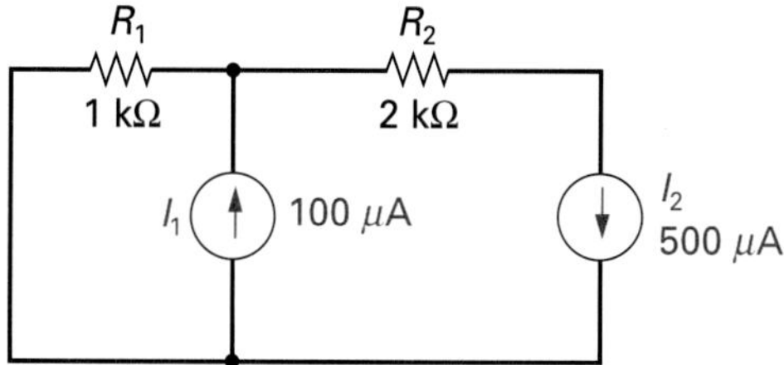

FIGURE 9-60 Superposition Circuit Example.

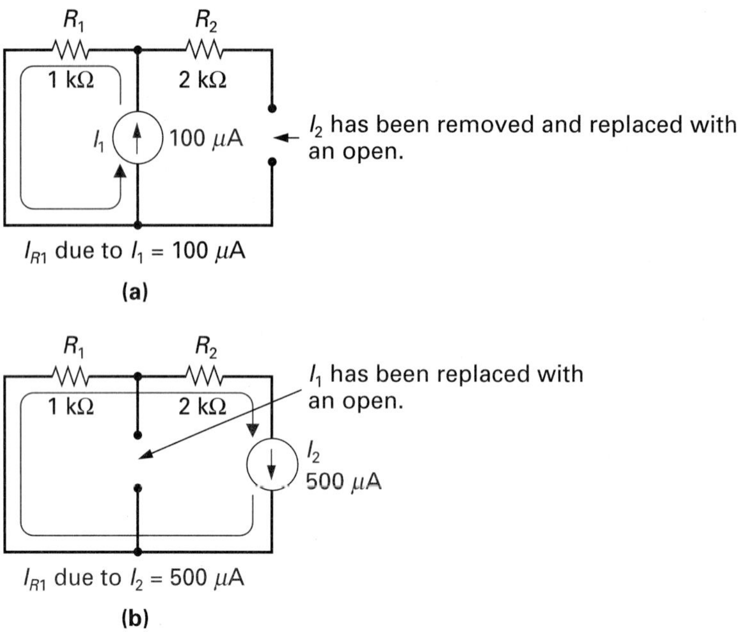

FIGURE 9-61 **Superposition Circuit Solution.**

9-9-3 *Thévenin's Theorem*

Thévenin's Theorem Any network of voltage sources and resistors can be replaced by a single equivalent voltage source (V_{TH}) in series with a single equivalent resistance (R_{TH}).

Thévenin's theorem allows us to replace the complex networks in Figure 9-62(a) with an equivalent circuit containing just one source voltage (V_{TH}) and one series-connected resistance (R_{TH}), as in Figure 9-62(b). Stated formally: *Any network of voltage sources and resistors can be replaced by a single equivalent voltage source* (V_{TH}) *in series with a single equivalent resistance* (R_{TH}).

Figure 9-63(a) illustrates an example circuit. As with any theorem, a few rules must be followed to obtain an equivalent V_{TH} and R_{TH}.

STEP 1 The first step is to disconnect the load (R_L) and calculate the voltage that will appear across points A and B, as in Figure 9-63(b). This open-circuit voltage will be the same as the voltage drop across R_2 (V_{R2}) and is called the *Thévenin equivalent voltage* (V_{TH}). First, let's calculate current:

$$I_T = \frac{V_S}{R_T} = \frac{12\text{ V}}{9\ \Omega} = 1.333\text{ A}$$

Therefore, V_{R2} or V_{TH} will equal

$$V_{R2} = I_T \times R_2 = 1.333 \times 6\ \Omega = 8\text{ V}$$

so $V_{TH} = 8$ V.

STEP 2 Now that the Thévenin equivalent voltage has been calculated, the next step is to calculate the Thévenin equivalent resistance. In this step, the source voltage is removed and replaced with a short, as seen in Figure 9-63(c), and the Thévenin equivalent resistance is equal to whatever resistance exists

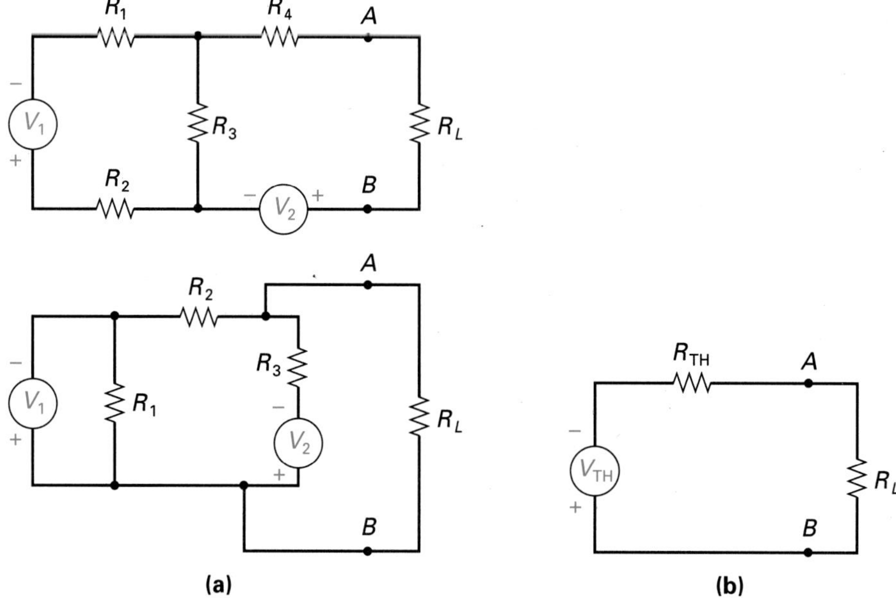

FIGURE 9-62 **Thévenin's Theorem. (a) Complex Multiple Resistors and Source Networks Are Replaced by (b) One Source Voltage (V_{TH}) and One Series-Connected Resistance (R_{TH}).**

between points A and B. In this example, R_1 and R_2 form a parallel circuit, the total resistance of which is equal to

$$R_T = \frac{R_1 \times R_2}{R_1 + R_2} = \frac{3 \times 6}{3 + 6} = \frac{18}{9} = 2\ \Omega$$

so $R_{TH} = 2\ \Omega$. The circuit to be Thévenized in Figure 9-63(a) can be represented by the Thévenin equivalent circuit shown in Figure 9-63(d).

The question you may have at this time is why we would need to simplify such a basic circuit, when Ohm's law could have been used just as easily to analyze the network. Thévenin's theorem has the following advantages:

1. If you had to calculate load current and load voltage (I_{RL} and V_{RL}) for 20 different values of R_L, it would be far easier to use the Thévenin equivalent circuit with the series-connected resistors R_{TH} and R_L, rather than applying Ohm's law to the series–parallel circuit made up of R_1, R_2 , and R_L.
2. Thévenin's theorem permits you to solve complex circuits that could not easily be analyzed using Ohm's law.

EXAMPLE:

Determine V_{TH} and R_{TH} for the circuit in Figure 9-64.

Solution:

The first step is to remove the load resistor R_L and calculate what voltage will appear between points A and B, as shown in Figure 9-65(a). Removing R_L will

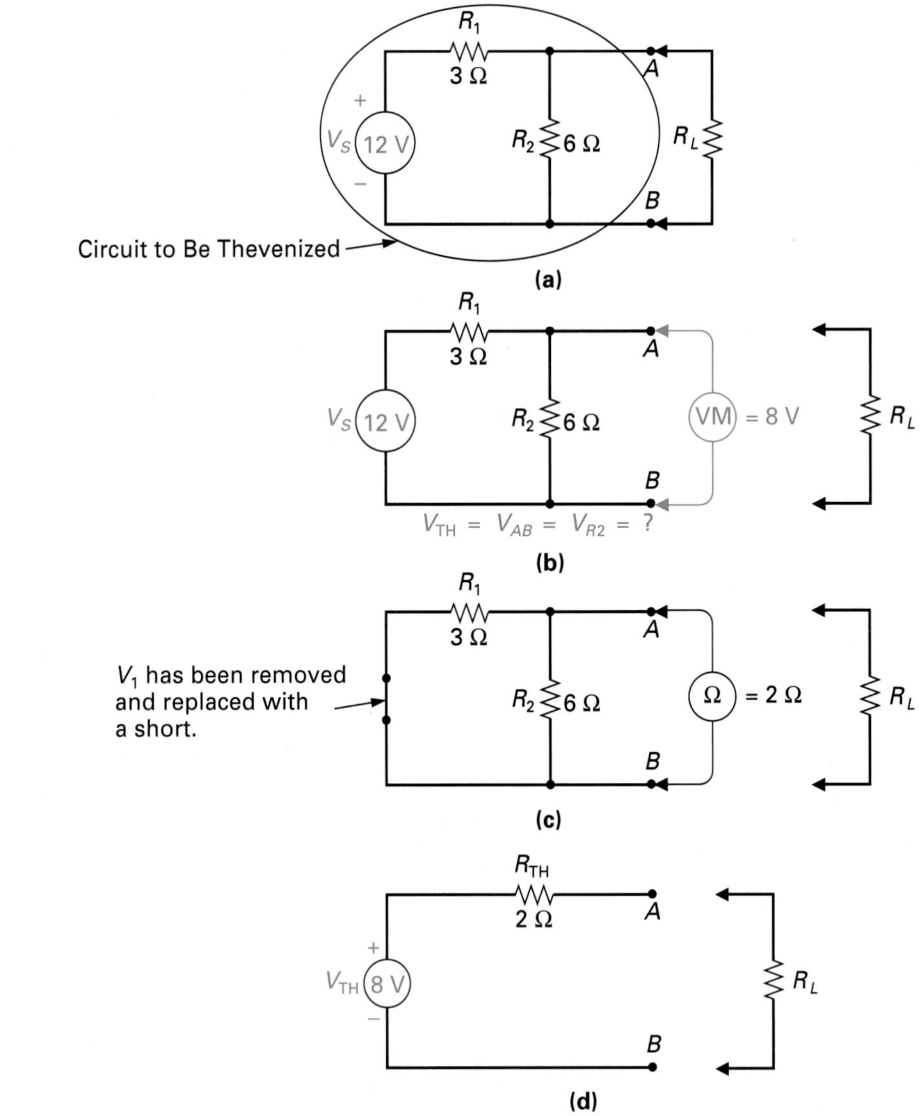

R_1
3 Ω
A
V_S 12 V
R_2 6 Ω
R_L
B
Circuit to Be Thevenized
(a)
R_1
3 Ω
A
V_S 12 V
R_2 6 Ω
VM = 8 V
R_L
B
$V_{TH} = V_{AB} = V_{R2} = ?$
(b)
R_1
3 Ω
A
V_1 has been removed and replaced with a short.
R_2 6 Ω
Ω = 2 Ω
R_L
B
(c)
R_{TH}
2 Ω
A
V_{TH} 8 V
R_L
B
(d)

FIGURE 9-63 Thévenin's Theorem. (a) Example Circuit. (b) Obtaining Thévenin Voltage (V_{TH}). (c) Obtaining Thévenin Resistance. (d) Thévenin Equivalent Circuit.

R_1
10 Ω
10 V DC
20 Ω R_2
R_4
A
10 Ω
10 Ω R_3
R_L
B

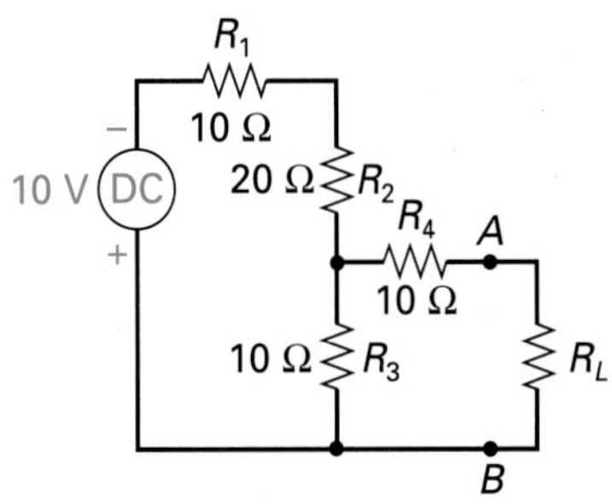

FIGURE 9-64 Thévenin Circuit Example.

open the path for current to flow through R_4, which will consequently have no voltage drop across it. The voltage between points A and B therefore will be equal to the voltage dropped across R_3, and since R_1, R_2, and R_3 form a series circuit, the voltage-divider formula can be used:

$$V_{R3} = \frac{R_3}{R_T} \times V_S$$

$$= \frac{10\ \Omega}{40\ \Omega} \times 10\ \text{V} = 2.5\ \text{V}$$

Therefore,

$$V_{AB} = V_{R3} = V_{\text{TH}} = 2.5\ \text{V}$$

The next step is to calculate the Thévenin resistance, which will equal whatever resistance appears across the terminals A and B with the voltage source having been removed and replaced with a short, as shown in Figure 9-65(b). In Figure 9-65(c), the circuit has been redrawn so that the relationship between the resistors can be seen in more detail. As you can see, R_1 and R_2 are in series with one another and both are in parallel with R_3, and this combination is in series

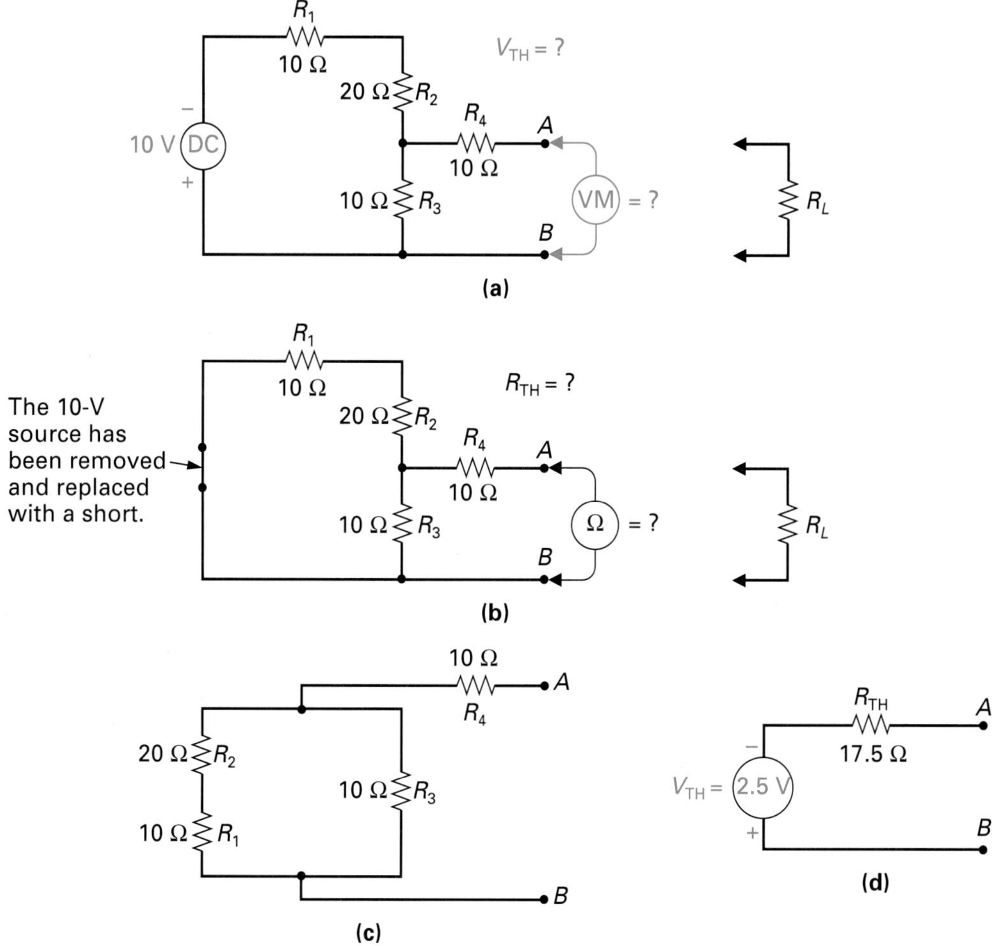

FIGURE 9-65 Thévenin Circuit Solution.

with R_4. Using our three-step procedure for calculating total resistance in a series–parallel circuit, the following results are obtained:

1. $R_{1,2} = R_1 + R_2 = 10\ \Omega + 20\ \Omega = 30\ \Omega$
2. $R_{1,2,3} = \dfrac{R_{1,2} \times R_3}{R_{1,2} + R_3} = \dfrac{30\ \Omega \times 10\ \Omega}{30\ \Omega + 10\ \Omega} = \dfrac{300\ \Omega}{40\ \Omega} = 7.5\ \Omega$
3. $R_T = R_{1,2,3} + R_4 = 7.5\ \Omega + 10\ \Omega = 17.5\ \Omega$

Figure 9-65(d) illustrates the Thévenin equivalent circuit.

9-9-4 Norton's Theorem

Norton's Theorem
Any network of voltage sources and resistors can be replaced by a single equivalent current source (T_N) in parallel with a single equivalent resistance (R_N).

Norton's theorem, like Thévenin's theorem, is a tool for simplifying a complex circuit into a more manageable one. Figure 9-66 illustrates the difference between a Thévenin equivalent and Norton equivalent circuit. Thévenin's theorem simplifies a complex network and uses an equivalent voltage source (V_{TH}) and an equivalent series resistance (R_{TH}). Norton's theorem, on the other hand, simplifies a complex circuit and represents it with an equivalent current source (I_N) in parallel with an equivalent Norton resistance (R_N), as shown in Figure 9-66.

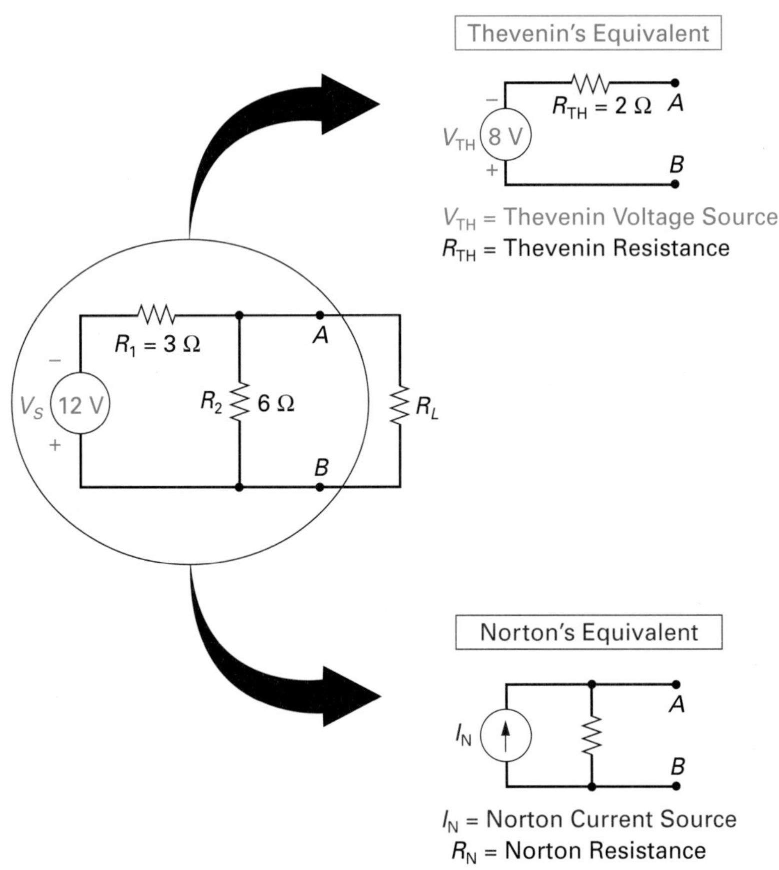

FIGURE 9-66 Comparison of Thévenin's and Norton's Circuits.

As with any theorem, a set of steps has to be carried out to arrive at an equivalent circuit. The example we will use is shown in Figure 9-67(a) and is the same example used in Figure 9-66.

STEP 1 Calculate the Norton equivalent current source, which will be equal to the current that would flow between terminals *A* and *B* if the load resistor was removed and replaced with a short, as seen in Figure 9-67(b). Placing a short between terminals *A* and *B* will short out the resistor R_2, so the only resistance in the circuit will be R_1. The Norton equivalent current source in this example will therefore be equal to

$$I_N = \frac{V_S}{R_T} = \frac{12\ V}{3\ \Omega} = 4\ A$$

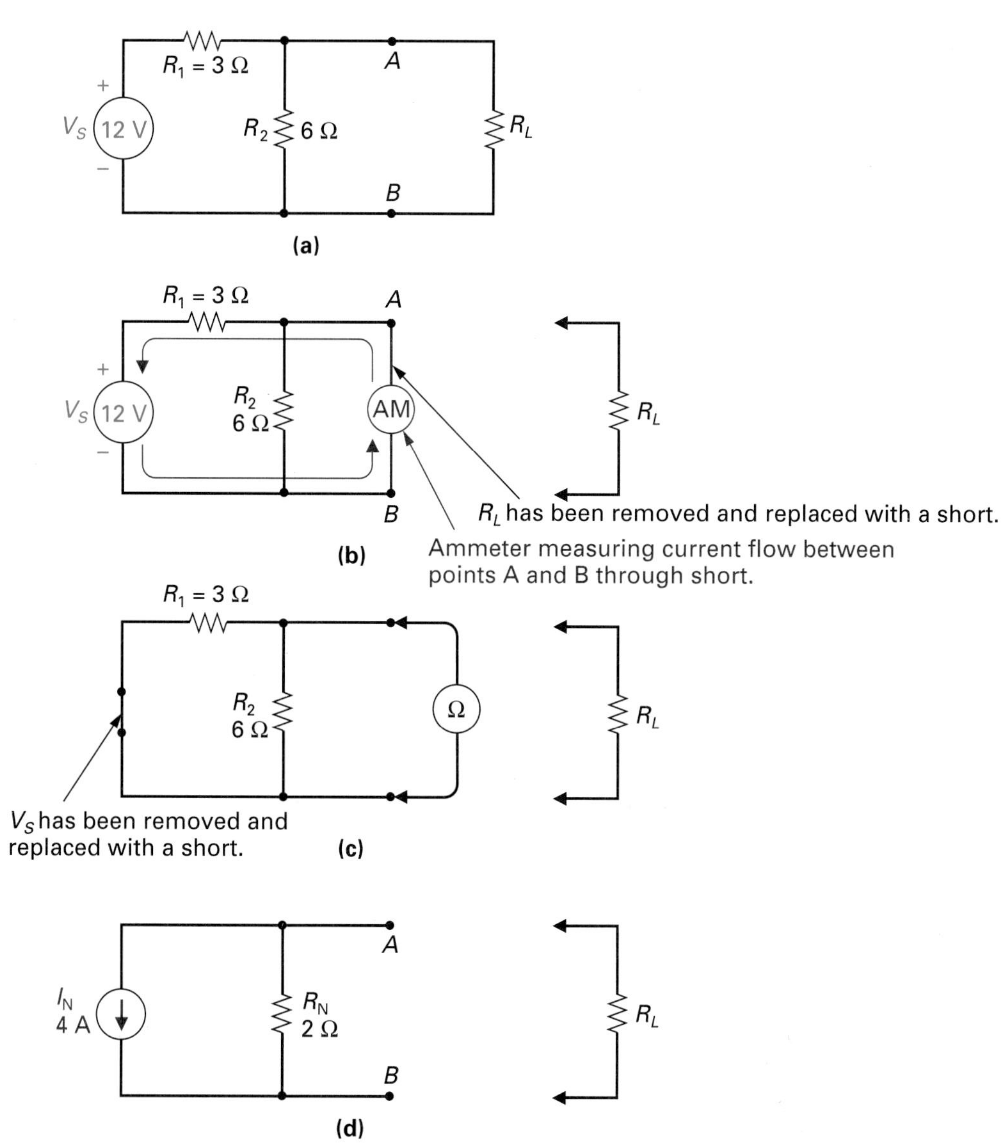

FIGURE 9-67 **Norton's Theorem. (a) Example Circuit. (b) Obtaining Norton Current. (c) Obtaining Norton Resistance (R_N). (d) Norton Equivalent Circuit.**

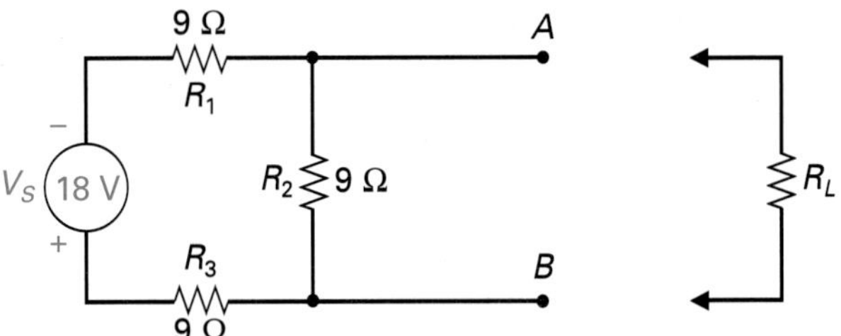

FIGURE 9-68 **Norton Circuit Example.**

STEP 2 The next step is to determine the value of the Norton equivalent resistance that will be placed in parallel with the current source, unlike Théven-in's equivalent resistance, which was placed in series. Like Thévenin's theorem, though, Norton's equivalent resistance (R_N) is equal to the resistance between terminals *A* and *B* when the voltage source is removed and replaced with a short, as shown in Figure 9-67(c). Since R_1 and R_2 form a parallel circuit, the Norton equivalent resistance will be equal to

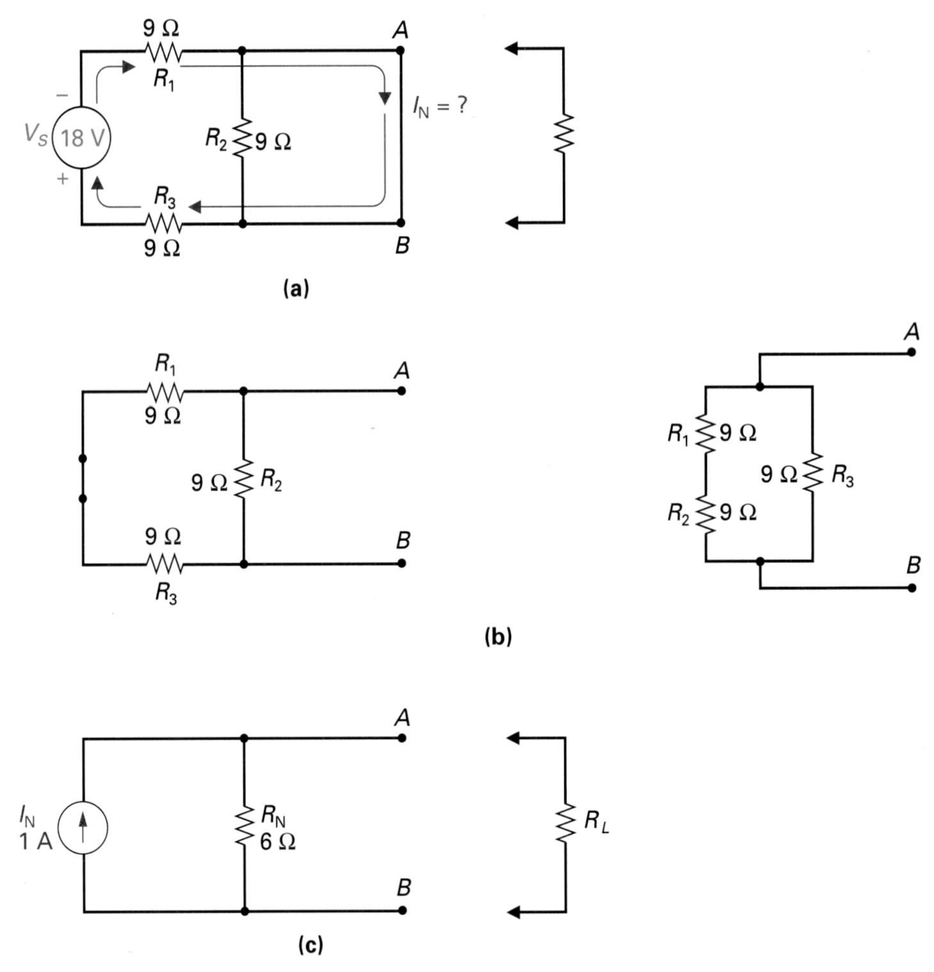

FIGURE 9-69 **Norton Circuit Solution.**

$$R_N = \frac{R_1 \times R_2}{R_1 + R_2} = \frac{3\ \Omega \times 6\ \Omega}{3\ \Omega + 6\ \Omega} = \frac{18\ \Omega}{9\ \Omega} = 2\ \Omega$$

The Norton equivalent circuit has been determined simply by carrying out these two steps and is illustrated in Figure 9-67(d).

EXAMPLE:

Determine I_N and R_N for the circuit in Figure 9-68.

Solution:

If a short is placed between terminals *A* and *B*, R_2 will be shorted out, so the current between points *A* and *B*, and therefore the Norton equivalent current, will be limited by only R_1 and R_3 and will equal [Figure 9-69(a)]

$$I_N = \frac{V_S}{R_T} = \frac{V_S}{R_1 + R_3} = \frac{18\ \text{V}}{9\ \Omega + 9\ \Omega} = 1\ \text{A}$$

Replacing V_S with a short, you can see that our Norton equivalent resistance between terminals *A* and *B* is made up of R_1 and R_3 in series with one another, and both are in parallel with R_2. R_N will therefore be equal to [Figure 9-69(b)]

$$R_{1,3} = R_1 + R_3 = 9\ \Omega + 9\ \Omega = 18\ \Omega$$

$$R_T = \frac{R_{1,3} \times R_2}{R_{1,3} + R_2} = \frac{18\ \Omega \times 9\ \Omega}{18\ \Omega + 9\ \Omega} = 6\ \Omega$$

The Norton equivalent circuit is shown in Figure 9-69(c).

SELF-TEST REVIEW QUESTIONS FOR SECTION 9-9

1. A constant voltage source will have a ____________ internal resistance, while a constant current source will have a ____________ internal resistance. (large or small)
2. The superposition theorem is a logical way of analyzing networks with more than one ____________.
3. Thévenin's theorem represents a complex two-terminal network as a single ____________ source with a series-connected single ____________.
4. Norton's theorem also allows you to analyze complex two-terminal networks as a single ____________ source in parallel with a single resistor.

SUMMARY

Series–Parallel Circuits (Figure 9-70)

1. If current has only one path to follow through a component, that component is connected in series.
2. If the total current has two or more paths to follow, these components are connected in parallel.

QUICK REFERENCE SUMMARY SHEET

FIGURE 9-70 Series–Parallel Circuits.

Example Circuit

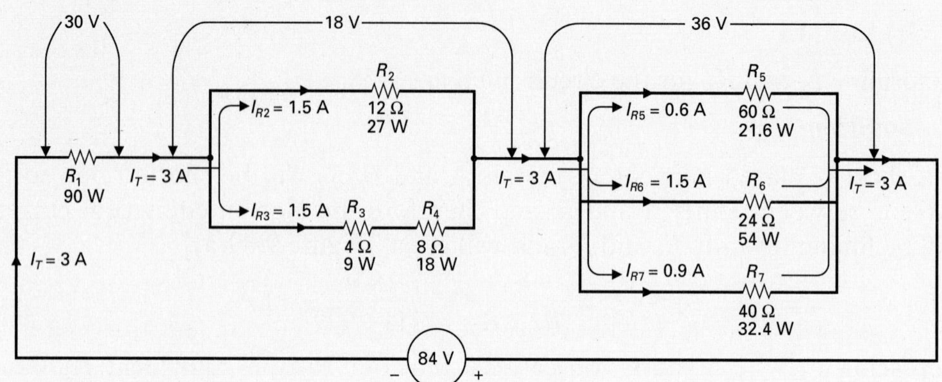

Solving for Resistance, Voltage, Current, and Power in a Series–Parallel Circuit

STEP 1 Determine the circuit's total resistance.

Step A Solve for series-connected resistors in all parallel combinations.

Step B Solve for all parallel combinations.

Step C Solve for remaining series resistances.

STEP 2 Determine the circuit's total current.

STEP 3 Determine the voltage across each series resistor and each parallel combination (series equivalent resistor).

STEP 4 Determine the value of current through each parallel resistor in every parallel combination.

STEP 5 Determine the total and individual power dissipated by the circuit.

The Wheatstone Bridge Circuit

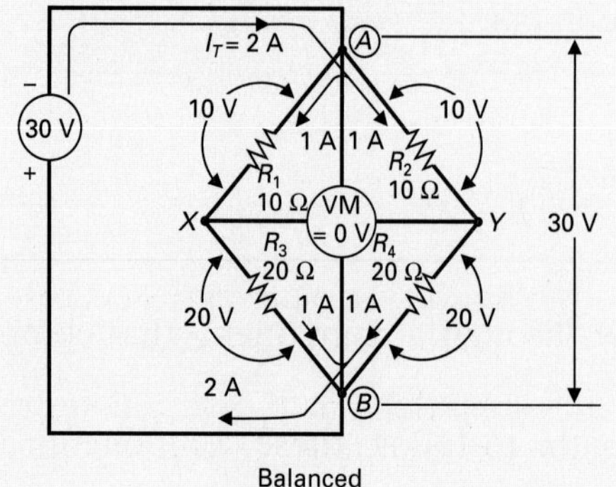

Balanced

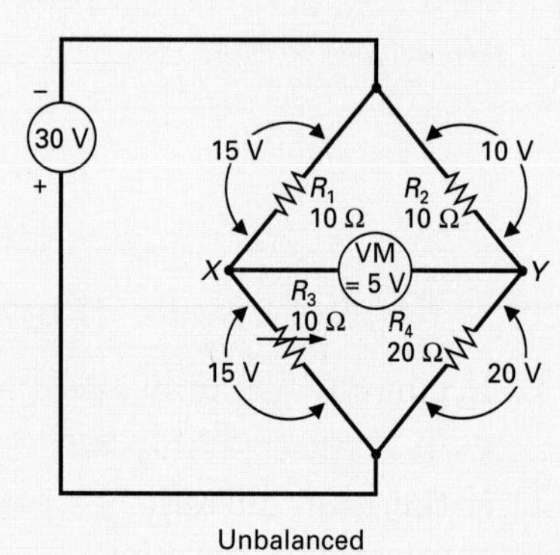

Unbalanced

QUICK REFERENCE SUMMARY SHEET

FIGURE 9-70 (continued)

Wheatstone bridge	$R_{\text{variable}} = R_{\text{unknown}}$	Branch resistances are equal.
	$R_{\text{unknown}} = R_{\text{variable}} \times \left(\frac{R_1}{R_2}\right)$	Branch resistors, R_1 and R_2, are not equal.

3. All electronic equipment is composed of many components interconnected to form a combination of series and parallel (series–parallel) circuits.
4. The total resistance in a series–parallel circuit can be calculated by:

   Step A: Determining an equivalent resistance of all series-connected resistors in parallel combinations.

   Step B: Determining the equivalent resistance of all parallel-connected combinations.

   Step C: Determining the equivalent resistance of the remaining series-connected resistances.
5. **a.** The voltage division in a series–parallel circuit can be calculated by:

   Step 1: Determining the circuit's total resistance by applying steps A, B, and C.

   Step 2: Determining the circuit's total current.

   Step 3: Determining the voltage drop across each series resistor and each parallel combination (series equivalent resistor).

   b. The fourth step is to:

   Step 4: Determine the amount of current flowing through each of the parallel combinations.

   c. The fifth and final step is to:

   Step 5: Determine the total power and individual power figures.
6. The Wheatstone bridge is a series–parallel circuit that was first developed in 1850 by Charles Wheatstone to measure resistance.
7. The *R*–2*R* ladder circuit is a series–parallel circuit used for digital-to-analog conversion.
8. Troubleshooting is the process of locating and diagnosing malfunctions or breakdowns in equipment by means of systematic checking or analysis.
9. **a.** If a series-connected resistor in a series–parallel circuit opens, there cannot be any current flow, and the source voltage will appear across the open and 0 V will appear across all the others.

b. If a parallel branch in a series–parallel circuit opens, the overall circuit resistance will increase and the total current will decrease. A greater parallel resistance will cause a greater voltage drop across that parallel circuit.

10. a. If a series-connected resistor in a series–parallel circuit shorts, the total circuit resistance will decrease, resulting in an increase in total current. The fault can be located by measuring the voltage across the shorted resistor, which will be 0 V.

    b. If a parallel branch in a series–parallel circuit shorts, the total resistance will decrease, resulting in a total current increase. This smaller parallel circuit resistance will result in a smaller voltage drop across the parallel circuit.

11. A resistor's resistance will typically change with age, resulting in a total resistance change and therefore a total current change. The faulty component can be located due to its abnormal voltage drop.

12. An ideal voltage source will have an internal resistance of 0 Ω and provide a constant output voltage regardless of what load resistance is connected across its output.

Theorems for DC Circuits

13. In reality, every practical voltage source will have some level of inefficiency, and this is represented as an internal resistance, and the output voltage will remain relatively constant despite variations in load resistance.

14. An ideal current source will have an infinite internal resistance and provide a constant output current, regardless of what load resistance is connected across its output.

15. In reality, every practical current source has a relatively high internal resistance, and the output current will remain relatively constant despite variations in load resistance.

16. The superposition theorem is a useful tool when analyzing circuits with more than one voltage source.

17. Thévenin's theorem is another handy tool that can be used to represent complex series–parallel networks as a single voltage source (V_{TH}) in series with a single resistor (R_{TH}).

18. Norton's theorem can also simplify a complex series–parallel network to an equivalent form consisting of a single current source (I_N) in parallel with a single resistor (R_N).

NEW TERMS

Analog data	Current source	Loading
Balanced bridge	Digital data	Network
Bleeder current	Digital-to-analog converter	Norton resistance
Calibrated	Interface	Norton's theorem
Computer	Ladder circuit	Norton voltage

Series–parallel circuit
Superposition theorem
Thévenin resistance
Thévenin's theorem
Thévenin voltage
Unbalanced bridge
Voltage source
Wheatstone bridge

REVIEW QUESTIONS

Multiple-Choice Questions

1. A series–parallel circuit is a combination of:
   - **a.** Components connected end to end
   - **b.** Series (one current path) circuits
   - **c.** Both series and parallel circuits
   - **d.** Parallel (two or more current path) circuits
2. Total resistance in a series–parallel circuit is calculated by applying the ____________ resistance formula to series-connected resistors and the ____________ resistance formula to resistors connected in parallel.
   - **a.** Series, parallel
   - **b.** Parallel, series
   - **c.** Series, series
   - **d.** Parallel, parallel
3. Total current in a series–parallel circuit is determined by dividing the total ____________ by the total ____________.
   - **a.** Power, current
   - **b.** Voltage, resistance
   - **c.** Current, resistance
   - **d.** Voltage, power
4. Branch current within series–parallel circuits can be calculated by:
   - **a.** Ohm's law
   - **b.** The current-divider formula
   - **c.** Kirchhoff's current law
   - **d.** All of the above
5. All voltages on a circuit diagram are with respect to ____________ unless otherwise stated.
   - **a.** The other side of the component
   - **b.** The high-voltage source
   - **c.** Ground
   - **d.** All of the above
6. A ____________ ground has the negative side of the source voltage connected to ground, while a ____________ ground has the positive side of the source voltage connected to ground.
   - **a.** Positive, negative
   - **b.** Chassis, earth
   - **c.** Positive, earth
   - **d.** Negative, positive
7. The output voltage will always ____________ when a load or voltmeter is connected across a voltage divider.
   - **a.** Decrease
   - **b.** Remain the same
   - **c.** Increase
   - **d.** All of the above could be considered true.
8. A Wheatstone bridge was originally designed to measure:
   - **a.** An unknown voltage
   - **b.** An unknown current
   - **c.** An unknown power
   - **d.** An unknown resistance
9. A balanced bridge has an output voltage:
   - **a.** Equal to the supply voltage
   - **b.** Equal to half the supply voltage
   - **c.** Of 0 V
   - **d.** Of 5 V

10. The total resistance of a ladder circuit is best found by starting at the point ____________ the source.

a. Nearest to c. Midway between
b. Farthest from d. All of the above

11. The R–$2R$ ladder circuit finds its main application as a(an):

a. Analog-to-digital converter
b. Digital-to-analog converter
c. Device to determine unknown resistance values
d. All of the above

12. The Norton equivalent resistance (R_N) is always in ____________ with the Norton equivalent current source (I_N).

a. Proportion c. Parallel
b. Series d. Either series or parallel

13. An ideal current source has ____________ internal resistance, while an ideal voltage source has ____________ internal resistance.

a. An infinite, 0 Ω of c. 0 Ω of, an infinite
b. No, a large amount d. No, an infinite

14. The Thévenin equivalent resistance (R_{TH}) is always in ____________ with the Thévenin equivalent voltage (V_{TH}).

a. Proportion c. Parallel
b. Series d. Either series or parallel

15. The superposition theorem is useful for analyzing circuits with:

a. Two or more voltage sources c. Only two voltage sources
b. A single voltage source d. A single current source

16. A resistor, when it burns out, will generally:

a. Decrease slightly in value c. Short
b. Increase slightly in value d. Open

17. In a series–parallel resistive circuit, an open series-connected resistor will cause ____________ current, whereas an open parallel-connected resistor will result in a total current ____________.

a. An increase in, decrease c. Zero, decrease
b. A decrease in, increase d. None of the above

18. In a series–parallel resistive circuit, a shorted series-connected resistor will cause ____________ current, whereas a shorted parallel-connected resistor will result in a total current ____________.

a. An increase in, increase c. An increase in, decrease
b. A decrease in, decrease d. A decrease in, increase

19. A resistor's resistance will typically ____________ with age, resulting in a total circuit current ____________.

a. Decrease, decrease c. Decrease, increase
b. Increase, increase d. All of the above

20. The maximum power transfer theorem states that maximum power is transferred from source to a load when load resistance is equal to source resistance.

a. True **b.** False

Essay Questions

21. State the five-step method for determining a series–parallel circuit's resistance, voltage, current, and power values. (9-6)

22. Illustrate the following series–parallel circuits:

a. R_1 in series with a parallel combination R_2, R_3, and R_4.
b. R_1 in series with a two-branch parallel combination consisting of R_2 and R_3 in series and R_4 in parallel.
c. R_1 in parallel with R_2, which is in series with a three-resistor parallel combination, R_3, R_4, and R_5.

23. Using the example in Question 22(c), apply values of your choice and apply the five-step procedure. (9-6)

24. Describe what is meant by "loading of a voltage-divider circuit." (9-7-1)

25. Illustrate and describe the Wheatstone bridge in the: (9-7-2)

a. Balanced condition
b. Unbalanced condition
c. Application of measuring unknown resistances

26. Describe how the ladder circuit acts as a current divider. (9-7-3)

27. Briefly describe the difference between a voltage source and a current source. (9-9-1)

28. Briefly describe the following theorems: (9-9)

a. Superposition **c.** Norton's
b. Thévenin's **d.** Maximum power transfer

29. What would be the advantages of Thévenin's and Norton's theorems to obtain an equivalent circuit? (9-9)

30. Draw the components that would exist in a: (9-9)

a. Thévenin equivalent circuit **b.** Norton equivalent circuit

31. Describe the steps involved in obtaining a: (9-9)

a. Thévenin equivalent circuit **b.** Norton equivalent circuit

32. When troubleshooting series–parallel circuits, describe what effect (9-8)

a. An open series
b. An open parallel-connected resistor

would have on total current and resistance, and how the opened resistor could be isolated.

33. When troubleshooting series–parallel circuits, describe what effect (9-8-2)

a. A shorted series
b. A shorted parallel-connected resistor

would have on total current and resistance, and how the shorted resistors could be isolated.

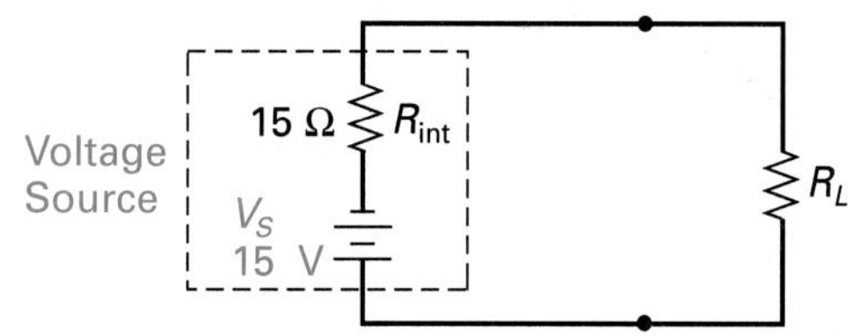

FIGURE 9-71

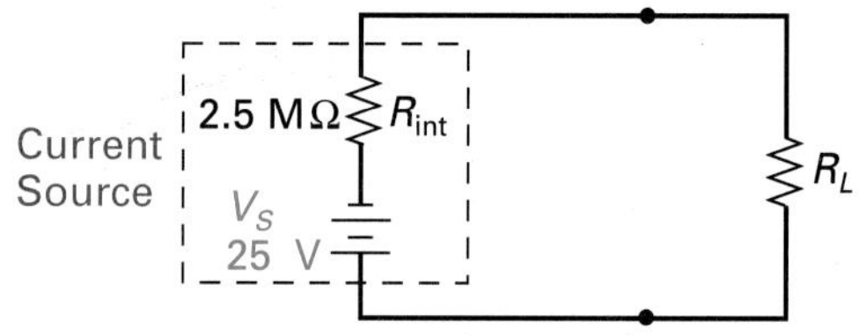

FIGURE 9-72

34. Describe what effect a resistor's value variation would have and how it could be recognized. (9-8-3)

35. Give the divider formula, Ohm's law, and Kirchhoff's laws used to determine: (9-6)

a. Branch currents
b. Voltage drops in a series–parallel circuit
(List all six.)

Practice Problems

36. R_3 and R_4 are in series with one another and are both in parallel with R_5. This parallel combination is in series with two series-connected resistors, R_1 and R_2. $R_1 = 2.5\ \text{k}\Omega$, $R_2 = 10\ \text{k}\Omega$, $R_3 = 7.5\ \text{k}\Omega$, $R_4 = 2.5\ \text{k}\Omega$, $R_5 = 2.5\ \text{M}\Omega$, and $V_S = 100$ V. For these values, calculate:

a. Total resistance
b. Total current
c. Voltage across series resistors and parallel combinations
d. Current through each resistor
e. Total and individual power figures

37. Referring to the example in Question 36, calculate the voltage at every point of the circuit with respect to ground.

38. A 10 V source is connected across a series–parallel circuit made up of R_1 in parallel with a branch made up of R_2 in series with a parallel combination of R_3 and R_4. $R_1 = 100\ \Omega$, $R_2 = 100\ \Omega$, $R_3 = 200\ \Omega$, and $R_4 = 300\ \Omega$. For these values, apply the five-step procedure, and also determine the voltage at every point of the circuit with respect to ground.

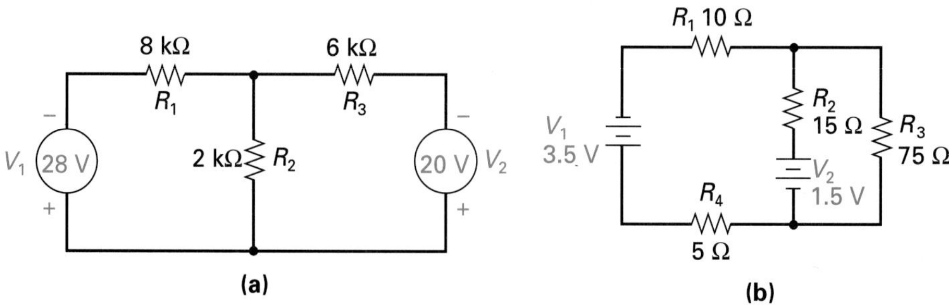

FIGURE 9-73

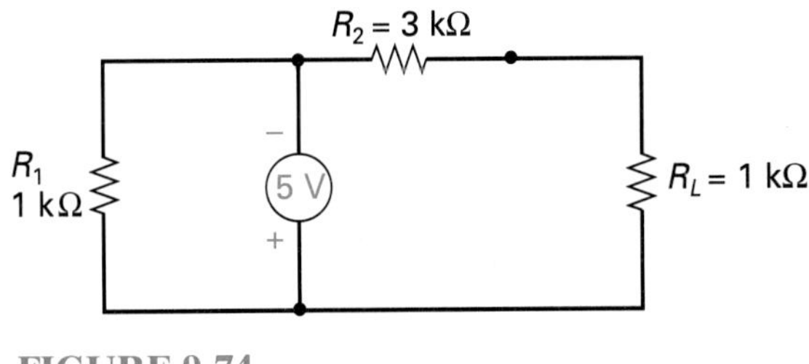

FIGURE 9-74

39. Calculate the output voltage (V_{RL}) in Figure 9-71 if R_L is equal to:

a. 25 Ω **b.** 2.5 kΩ **c.** 2.5 MΩ

40. What load current will be supplied by the current source in Figure 9-72 if R_L is equal to:

a. 25 Ω **b.** 2.5 kΩ **c.** 2.5 MΩ

41. Use the superposition theorem to calculate total current:

a. Through R_2 in Figure 9-73(a)
b. Through R_3 in Figure 9-73(b)

42. Convert the following voltage sources to equivalent current sources:

a. $V_S = 10$ V, $R_{int} = 15\ \Omega$
b. $V_S = 36$ V, $R_{int} = 18\ \Omega$
c. $V_S = 110$ V, $R_{int} = 7\ \Omega$

43. Use Thévenin's theorem to calculate the current through R_L in Figure 9-74. Sketch the Thévenin and Norton equivalent circuits for Figure 9-74.

44. Sketch the Thévenin and Norton equivalent circuits for the networks in Figure 9-75.

45. Convert the following current sources to equivalent voltage sources:

a. $I_S = 5$ mA, $R_{int} = 5$ MΩ **c.** $I_S = 0.0001$ A, $R_{int} = 2.5$ kΩ
b. $I_S = 10$ A, $R_{int} = 10$ kΩ

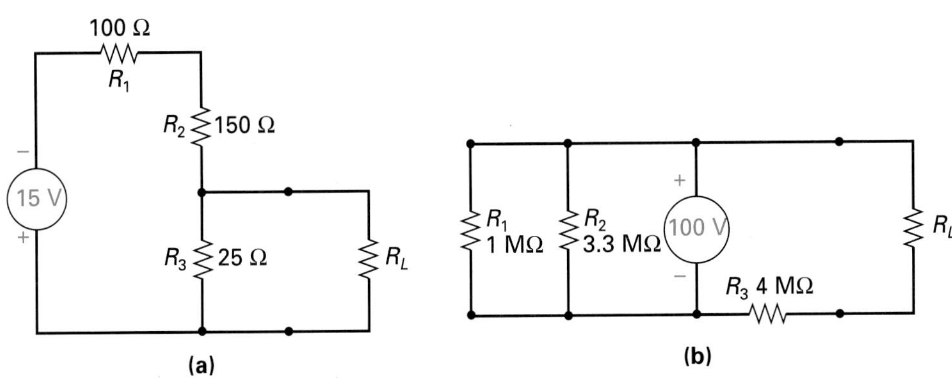

FIGURE 9-75

Troubleshooting Questions

46. Referring to the example circuit in Figure 9-49, describe the effects you would get if a resistor were to short, and if a resistor were to open.
47. Design a simple five-resistor series–parallel circuit and insert a source voltage and resistance values. Apply the five-step series–parallel circuit procedure, and then theoretically open and short all the resistors and calculate what effect would occur and how you would recognize the problem.
48. Carbon composition resistors tend to increase in resistance with age, while most other types generally decrease in resistance. What effects would resistance changes have on their respective voltage drops?
49. Use Thévenin's theorem to simplify the circuit in Figure 9-76. What effect would the following faults have on the Thévenin equivalent circuit?

    a. R_2 is shorted. **b.** R_2 is open.
50. Calculate the Norton equivalent for Figure 9-76 and describe what circuit differences will occur for the same faults listed in Question 49.

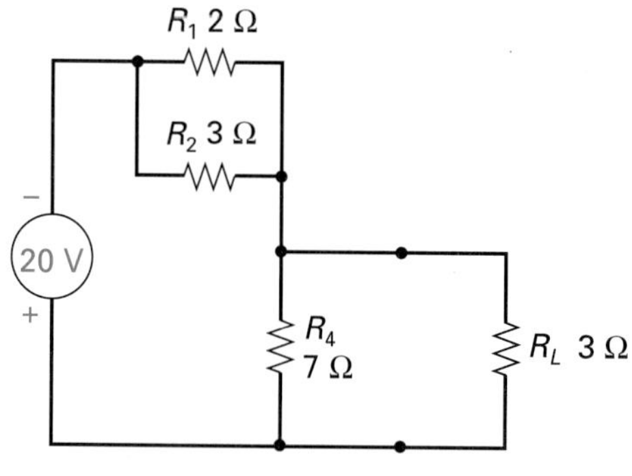

FIGURE 9-76

CHAPTER

10

OBJECTIVES

After completing this chapter, you will be able to:

1. State the difference between an analog and digital readout multimeter.
2. Explain the construction and operation of a D'Arsonval or moving coil meter movement.
3. Describe meter resistance and sensitivity.
4. Describe why, when using the ammeter to measure current:
   - **a.** Shunt resistors are used to achieve different range scales.
   - **b.** Five rules must be applied.
   - **c.** It affects the circuit in which it is connected.
5. Describe why, when using the voltmeter to measure voltage:
   - **a.** Multiplier resistors are used to achieve different range scales.
   - **b.** Voltmeter sensitivity determines meter accuracy.
   - **c.** Five rules must be applied.
   - **d.** It affects the circuit in which it is connected.
6. State the difference between an earth and chassis ground.
7. Explain why voltage measurements are usually measured with respect to ground.
8. Describe why, when using the ohmmeter to measure resistance:
   - **a.** An internal battery source is used to supply current.
   - **b.** The ohmmeter scale is nonlinear.
   - **c.** The range scales of an ohmmeter are interpreted differently from those of the voltmeter and ammeter.
   - **d.** Four rules must be applied.

The Series–Parallel Analog Multimeter

OUTLINE

VIGNETTE

The Frog Lost!

The first meter movement that indicated values of current was called a galvanometer and therefore must have been invented by the Italian physiologist Luigi Galvani. This current meter was actually invented by Frenchman Claude Pouillet in 1837; however, the inventor fell into obscurity. Later his invention was named the galvanometer after Galvani, whose name was well known. One of Galvani's students discovered that a frog's legs twitched when it was touched by metal with a static machine operating nearby. Galvani followed up this observation, and discovered that a frog's leg also twitched when two different metals were connected at one end, and the other ends touched to the dissected leg. His theory was that electricity from the nerves was discharging through the metal circuit. Although he explained the frog-leg reaction incorrectly, this experiment gained him much fame; enough, in fact, to have Pouillet's invention attached to his name.

INTRODUCTION

In the preceding three chapters we examined series circuits, parallel circuits, and finally a combination of the two: series–parallel circuits. In this chapter you will analyze the analog multimeter, which is used to measure or determine the magnitude of voltage, current, and resistance. Although the digital multimeter is more commonly used to measure these three electrical quantities, the analog multimeter is still in use and contains many good examples of practical series–parallel circuits. These applications of series–parallel circuits, therefore, will help reinforce your understanding of series–parallel networks.

Multimeter
Piece of electronic test equipment that can perform multiple tasks in that it can be used to measure voltage, current, or resistance.

10-1 THE ANALOG METER MOVEMENT

The **multimeter** is a device used for measurement and troubleshooting in electronic equipment. It is called a multimeter because it is like having three meters all rolled into one, as shown in Figure 10-1(a). The multimeter's three units are selected by the main function switch:

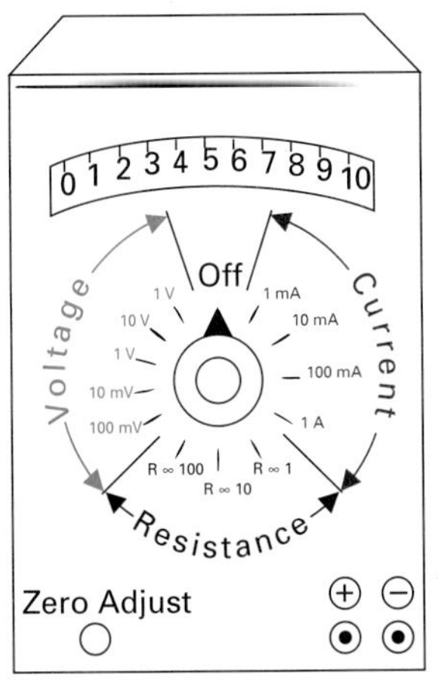

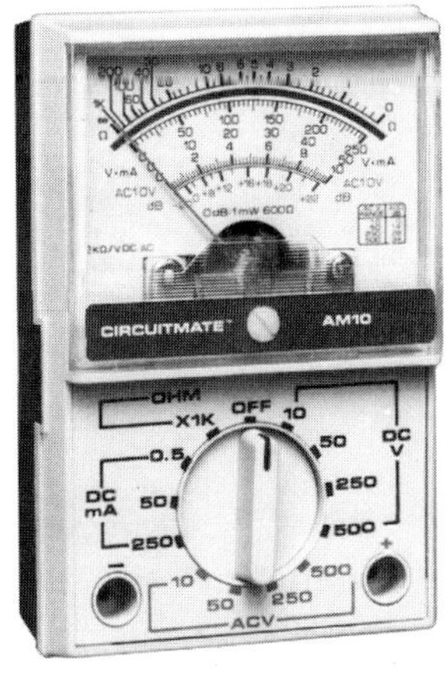

(a)

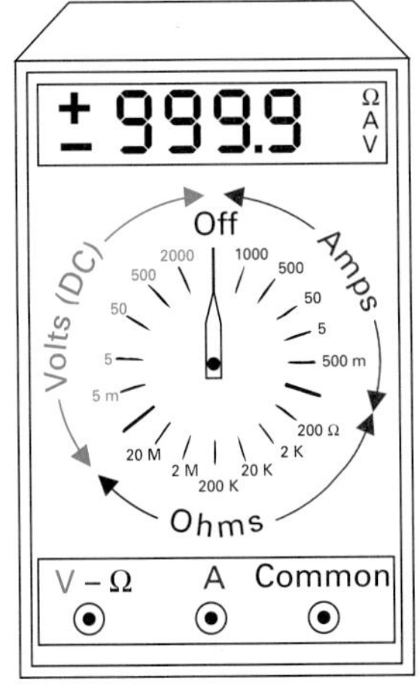

(b)

FIGURE 10-1 **Multimeters. (a) Analog. (b) Digital.**

Ammeter
Meter placed in the path of current flow to measure the amount.

Voltmeter
Instrument designed to measure the voltage or potential difference. Its scale can be graduated in kilovolts, volts, or millivolts.

Ohmmeter
Measurement device used to measure electrical resistance.

Analog Data
Information that is represented in a continuously varying form, as opposed to digital data, which have distinct or discrete values.

Analog Display
Display in which a moving pointer or some other analog device is used to display the data.

Digital Display
Display in which either light-emitting diodes (LEDs) or liquid crystal diodes (LCDs) form digits which are used to display the data.

1. The **ammeter** unit measures current.
2. The **voltmeter** unit measures voltage.
3. The **ohmmeter** unit measures resistance.

For many years, the analog readout type of multimeter shown in Figure 10-1(a) was the only type available. At present, these analog readout multimeters are still being used, but are slowly being replaced by digital readout multimeters, similar to the type shown in Figure 10-1(b).

Analog is a word that means "similar to." For example, a ball is an analog of the world. With an analog readout multimeter, the amount of pointer deflection across the scale is an analog, or similar to, the magnitude of the electrical property being measured **(analog data).** This type of indicator is therefore called an **analog display,** as opposed to a **digital display,** which makes use of deci-

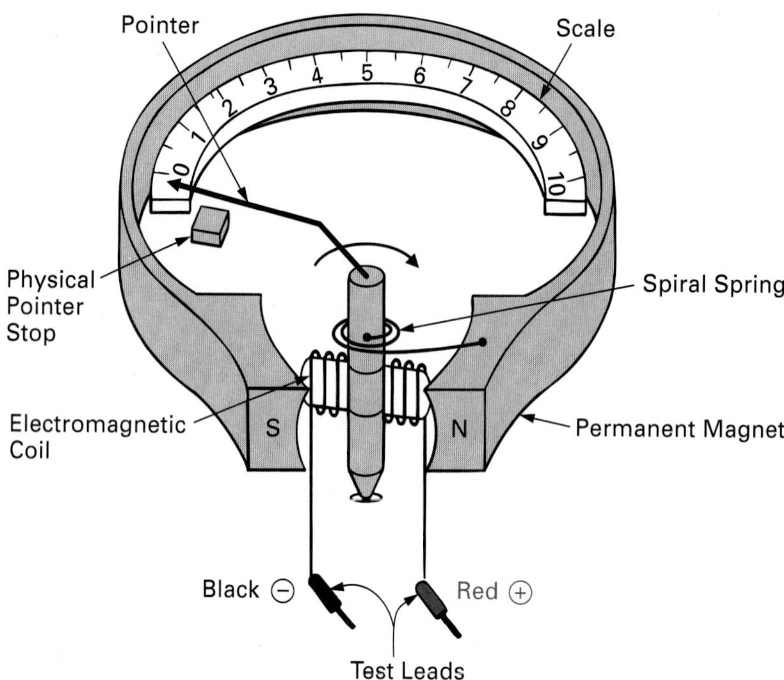

FIGURE 10-2 **Moving Coil Meter Movement.**

Digital Data
Data represented in digital form.

mal digits to display the magnitude of the electrical property **(digital data).** Let us now examine the internal workings of the analog multimeter.

10-1-1 *Construction*

The D'Arsonval or moving coil type of meter movement illustrated in Figure 10-2 was invented by Jacques Arsène D'Arsonval and can be used in instruments to measure current, voltage, or resistance. A coil of copper wire is mounted on a moving core known as an **armature,** which is free to rotate between the two poles of a stationary permanent magnet. A spiral spring is attached to the moving coil and holds the armature in a rest position so that the pointer, which is also attached to the armature, is pointing to 0 on the scale.

Armature
Rotating or moving component of a magnetic circuit.

10-1-2 *Operation*

This moving coil type of meter movement is used in the analog multimeter, and whether we are measuring current, voltage, or resistance, a current will always flow in the moving coil to represent the amount of current, voltage, or resistance being measured with our two test leads. Let's examine now what will actually happen when a current flows through the moving coil.

When a current is passed through the moving coil, a magnetic field is set up by the electromagnet, producing a north pole on the right side, and a south pole on the left side of the moving coil, as seen in Figure 10-3(a) [left-hand rule for

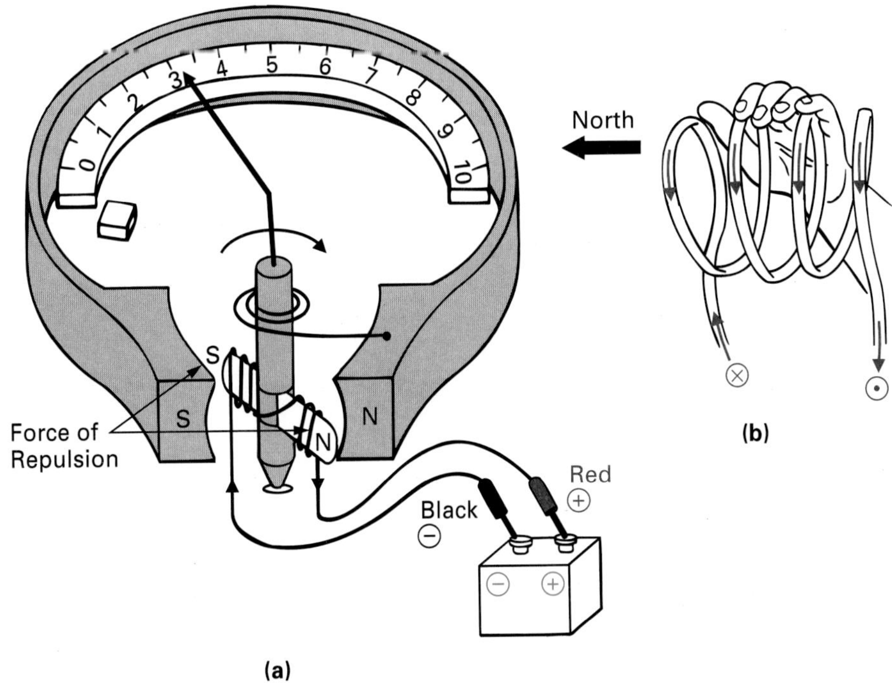

FIGURE 10-3 **Moving Coil Meter Movement Operation.**

electromagnets, Figure 10-3(b)]. The magnetic field of the electromagnet will interact with the magnetic field of the permanent magnet. This will force the armature to rotate in a clockwise direction, since like poles repel one another, as summarized in Figure 10-4.

The south pole generated in the electromagnet is repulsed by the south pole of the permanent magnet and attracted by the north pole of the permanent magnet, while the north pole generated in the electromagnet is repulsed by the north pole of the permanent magnet and attracted by the south pole of the permanent magnet. The overall effect is to overcome the tension created by the spring and move the pointer in a clockwise direction. If a greater current is passed through the electromagnet, a larger magnetic field will be generated, causing a stronger magnetic interaction between the electromagnet and permanent magnet, thus increasing the repulsion and attraction, resulting in an increased movement of the pointer across the scale in the clockwise direction. A greater current, therefore, causes a greater meter deflection.

If current flows through the electromagnet in the opposite direction, it will generate a north pole in the left side of the electromagnet and a south pole in the right side of the electromagnet. This will result in a counterclockwise movement of the pointer so that the pointer will strike the physical pointer stop, as shown in Figure 10-5. Too large a current could damage the meter, and this is why the meter leads are said to be polarity (positive and negative) sensitive. Therefore, when measuring a potential difference, the red positive test lead must always be placed on the point that is positive with respect to the other terminal, or, said another way, the negative black test lead must always probe the terminal that is negative with respect to the other terminal. Another point to remember is

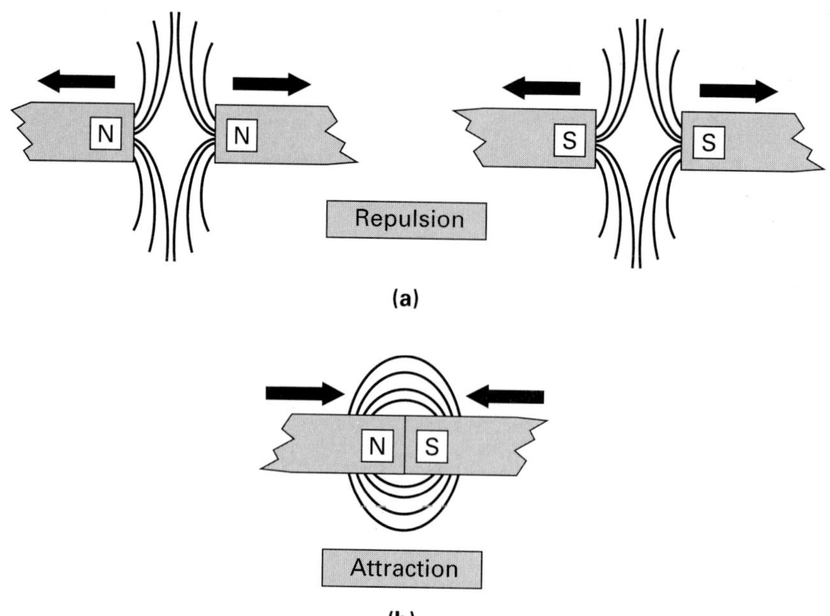

FIGURE 10-4 **Magnetic Law of Attraction and Repulsion. (a) Like Poles Repulse One Another. (b) Unlike Poles Attract One Another.**

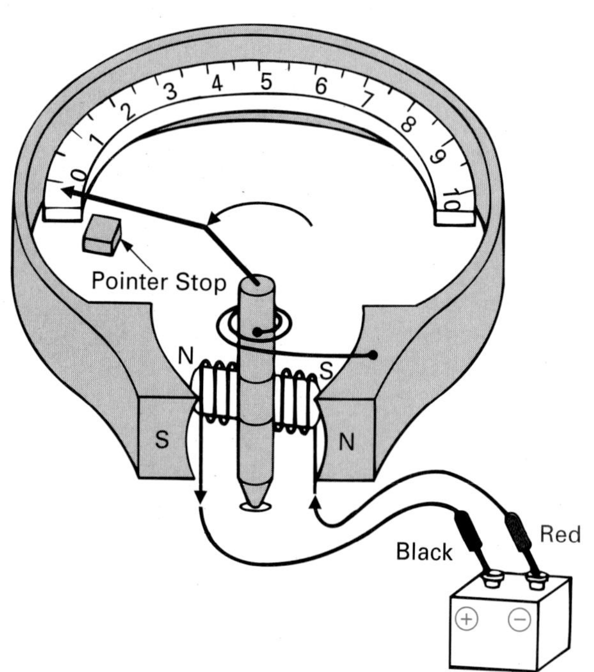

FIGURE 10-5 **Counterclockwise Meter Movement.**

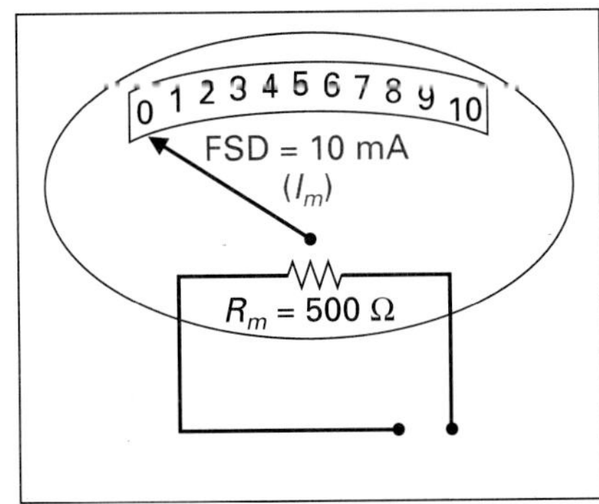

FIGURE 10-6 Simplified Diagram of Meter Movement.

that since gravity will have an effect on the pointer, the meter should be placed flat on its back so as to obtain the most accurate readings.

10-1-3 *Meter Resistance and Sensitivity*

The coil of the armature in a moving coil meter possesses some value of resistance and is called the movement's internal resistance (R_m). This internal **meter resistance** (R_m) is small and normally in the range 1 to 500 Ω.

Meter Resistance The resistance of a meter's armature coil.

The wire used to make the coil is extremely thin (about the thickness of a human hair) and can therefore not, without damage, carry very much current. The typical current rating of a meter movement is therefore small, ranging from 10 μA to 10 mA. This value specifies the **meter FSD current**, the amount of current needed to cause the meter movement to deflect the needle to its full-scale deflection (FSD) position at the far right side of the scale. The maximum current of a meter movement (I_m) will therefore tell you how sensitive a meter is; for example, if a meter movement needs only 10 μA to deflect the pointer to FSD, it is much more sensitive than a meter movement that requires 10 mA. A simplified diagram of a meter movement showing R_m and I_m is illustrated in Figure 10-6.

Meter FSD Current The value of current needed to cause the meter movement to deflect the needle to its full-scale deflection (FSD) position.

SELF-TEST REVIEW QUESTIONS FOR SECTION 10-1

1. An __________ is used to measure current, a __________ is used to measure voltage, and an __________ is used to measure __________.
2. Define:
   a. R_m
   b. I_m

10-2 THE AMMETER

Figure 10-7(a) illustrates a meter with a maximum meter movement current (I_m) of 1 mA. This meter will function perfectly for any current value in the range from 0 to 1 mA. If the current being measured exceeds 1 mA, which is the cur-

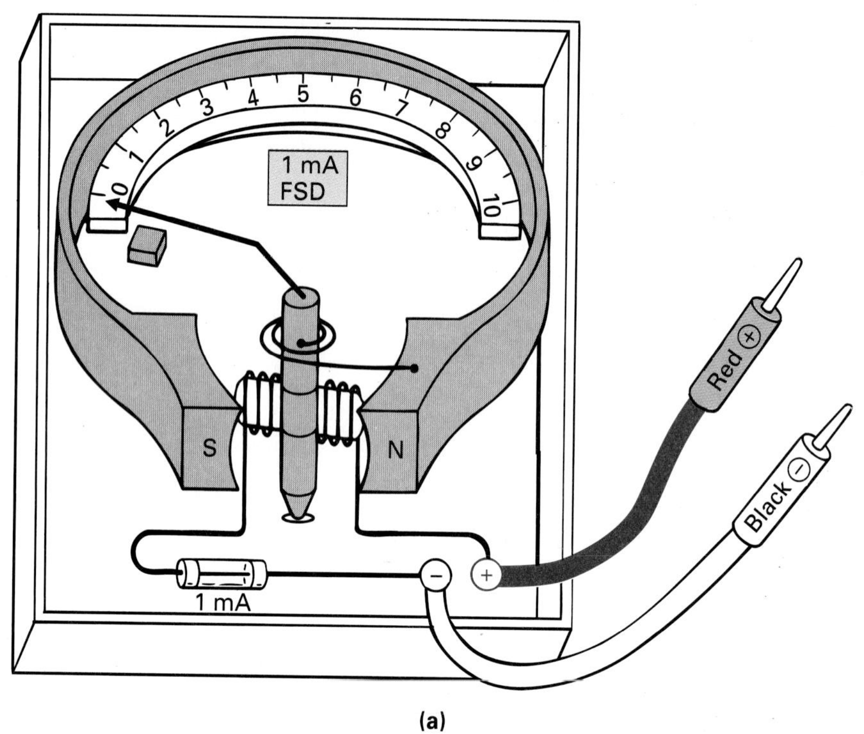

(a)

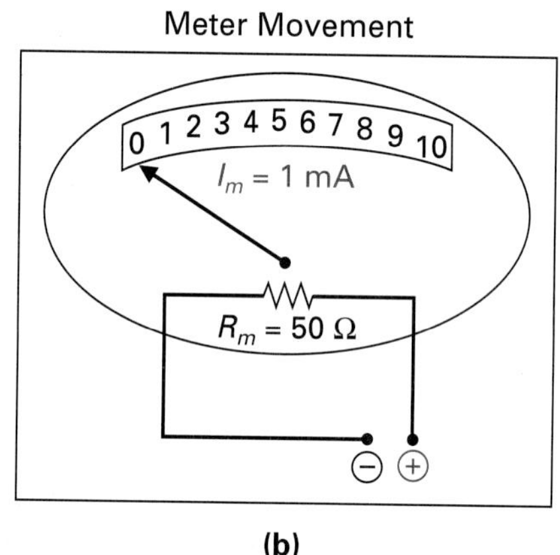

(b)

FIGURE 10-7 **1-mA Ammeter. (a) Physical View. (b) Simplified Diagram.**

rent rating of the fuse, the fuse will blow and protect the meter from damage. Some modifications are necessary if we wish to measure any current greater than the FSD current (I_m). To make this modification, we must first know the meter's internal resistance (R_m) and FSD current (I_m). In this example, we will use a 1 mA, 50 Ω meter movement shown in the simplified diagram in Figure 10-7(b).

10-2-1 Ammeter Range Scales

The maximum possible current that the meter shown in Figure 10-7(a) can measure as it is presently constructed is 1 mA; however, if we wish to measure current in the range from 0 to 10 mA, we will need some other path for the 9 mA to travel through so that only 1 mA will flow through the meter movement.

A resistor is connected in parallel or shunt with the meter movement resistance (R_m) to shunt the damaging current away from the meter movement, as shown in Figure 10-8. The next question is: What will be the value of the **shunt resistor** (R_{sh}) so that it will shunt 9 mA away from the meter movement when we are measuring a current value of 10 mA? This can be calculated by first finding the voltage across R_{sh}.

Shunt Resistor A resistor connected in parallel or shunt with the meter movement of an ammeter.

$$V_{Rm} = I_{Rm} \times R_m = 1\text{ mA} \times 50\ \Omega = 50\text{ mV}$$

Since $V_{Rm} = V_{Rsh}$ (the resistors are connected in parallel), the value of R_{sh} can therefore be calculated by applying Ohm's law:

$$R_{sh} = \frac{V_{Rsh}}{I_{Rsh}} = \frac{50\text{ mV}}{9\text{ mA}} = 5.6\ \Omega$$

If a maximum current of 10 mA is measured by the meter, the 10 mA will travel into the meter, where it will split through a parallel circuit made up of R_m and R_{sh}. The current split will be as follows:

$$I_{Rm} = \frac{V_{Rm}}{R_m} = \frac{50\text{ mV}}{50\ \Omega} = 1\text{ mA}$$

$$I_{Rsh} = \frac{V_{Rsh}}{R_{sh}} = \frac{50\text{ mV}}{5.6\ \Omega} = 9\text{ mA}$$

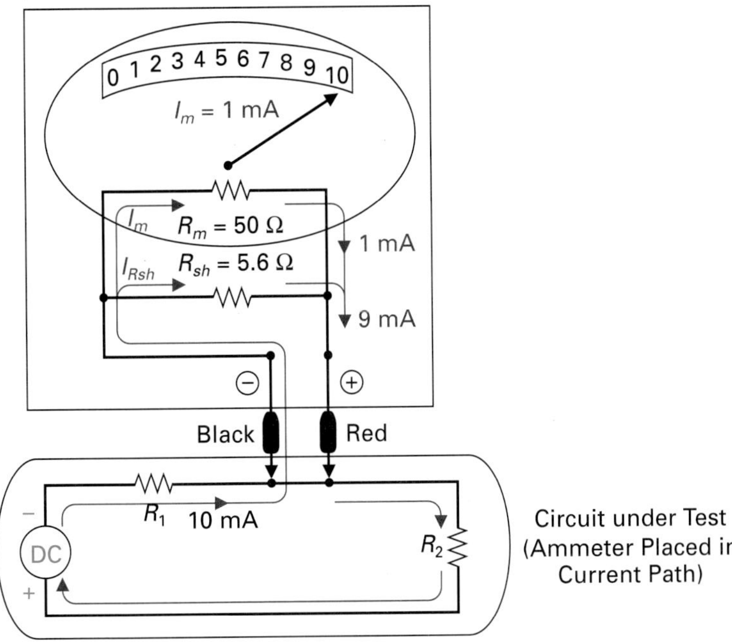

FIGURE 10-8 **10-mA Ammeter.**

For a current of 10 mA, you will have 1 mA flowing through the moving coil and this will cause FSD.

To have a multirange ammeter, there will have to be some way of switching in different values of shunt resistors for different current ranges. Figure 10-9 shows how this can be achieved. Figure 10-9(a) illustrates the different shunt resistors needed for different ranges.

If the rotary switch is placed in position *A* (0 to 1 mA range), there is no need for a shunt resistor because the FSD current for the moving coil is 1 mA. With the rotary switch in position *B* (0 to 10 mA range), we have switched in a

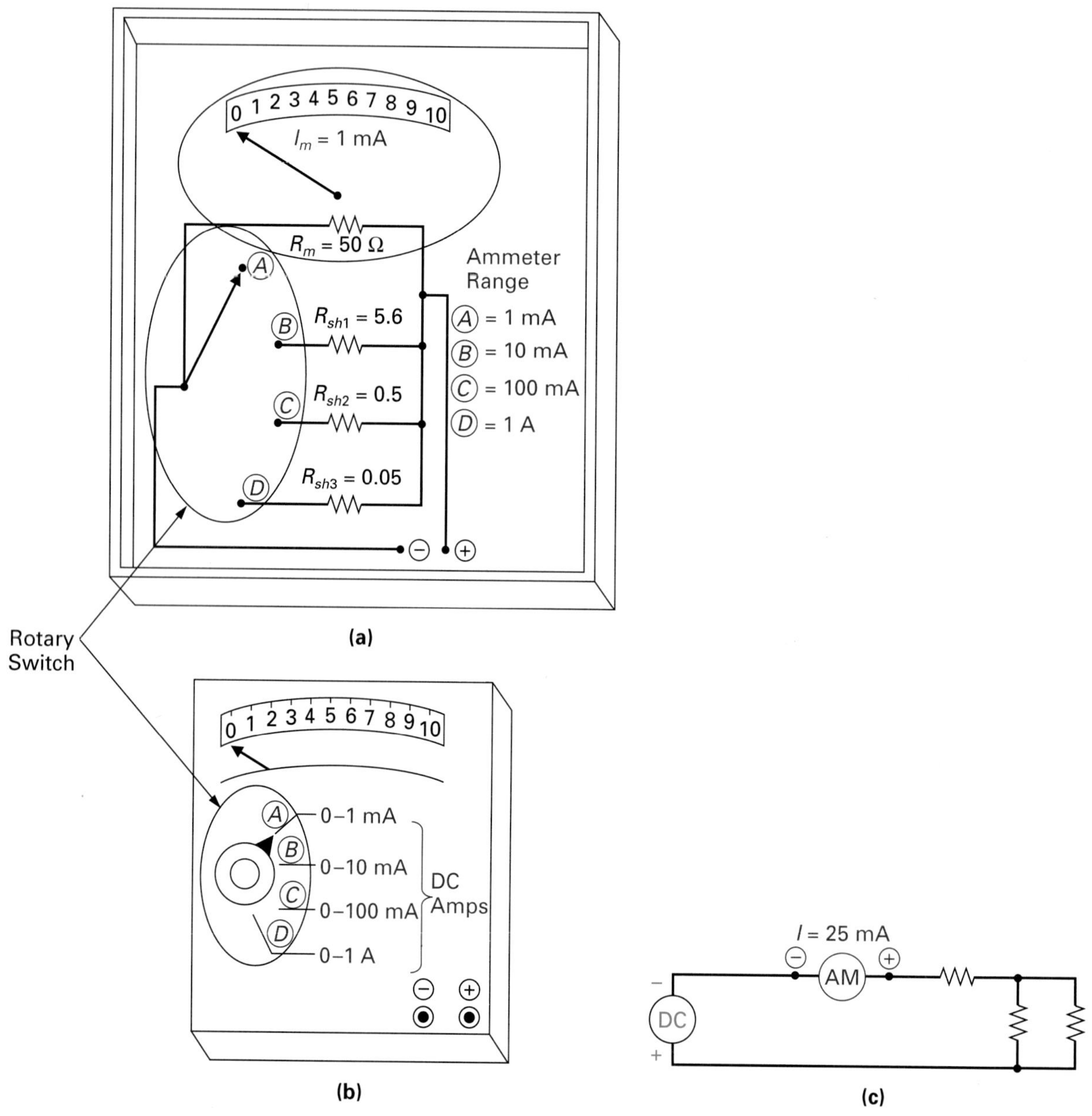

FIGURE 10-9 **Multirange Ammeter. (a) Internal Circuit. (b) External Appearance. (c) Ammeter Symbol in Circuit.**

shunt resistance of 5.6 Ω, and, as calculated previously, this 5.6 Ω resistor will shunt 9 mA and allow 1 mA through R_m when a maximum of 10 mA is being measured.

In position C (0 to 100 mA), the rotary switch has switched in the shunt resistor R_{sh2}, whose job it is to shunt 99 mA away from the moving coil if 100 mA is being measured. Since we already know that 50 mV will appear across all shunt resistors,

$$R_{sh2} = \frac{V_{Rsh2}}{I_{Rsh2}} = \frac{50 \text{ mV}}{99 \text{ mA}} = 0.5 \ \Omega$$

If the rotary switch is placed in position D (0 to 1 A or 1000 mA), R_{sh3} is switched in parallel with R_m to shunt 999 mA away from the moving coil. R_{sh3} can be calculated and is equal to

$$R_{sh3} = \frac{V_{sh3}}{I_{sh3}} = \frac{50 \text{ mV}}{999 \text{ mA}} = 0.05 \ \Omega$$

Figure 10-9(b) shows the external appearance of the rotary switch that is used by the technician to select the ammeter range scale required. Figure 10-9(c) illustrates the schematic symbol of the ammeter (AM) inserted in a circuit to measure current.

With multiple ranges, the scale has to be interpreted by the technician differently, depending on the range scale selected. For example, if the 1-mA range is selected, the 10 on the scale is now equivalent to 1 mA, the 9 on the scale is equivalent to 0.9 mA, the 8 to 0.8 mA, and so on. If the 10-mA range is selected, 10 on the scale is equal to 10 mA, 9 to 9 mA, 8 to 8 mA, and so on. If the 100-mA range is selected, a 2 on the scale is equivalent to 20 mA, all the way up to the 10 on the scale, which in this case is equal to 100 mA. On the 1-A (1000-mA) range scale, 6 on the scale is equivalent to 0.6 A or 600 mA, 7 on the scale is equal to 0.7 A or 700 mA, and so on.

Some range scales are not oriented only around 10, but can incorporate a scale of 30, as well as the 10, as seen in Figure 10-10. If the 100-mA or 1-A scales are selected, the upper scale from 0 to 10 is used, with 10 on the scale representing 100 mA or 1 A. If the 300-mA or 3-A scales are selected, the lower scale

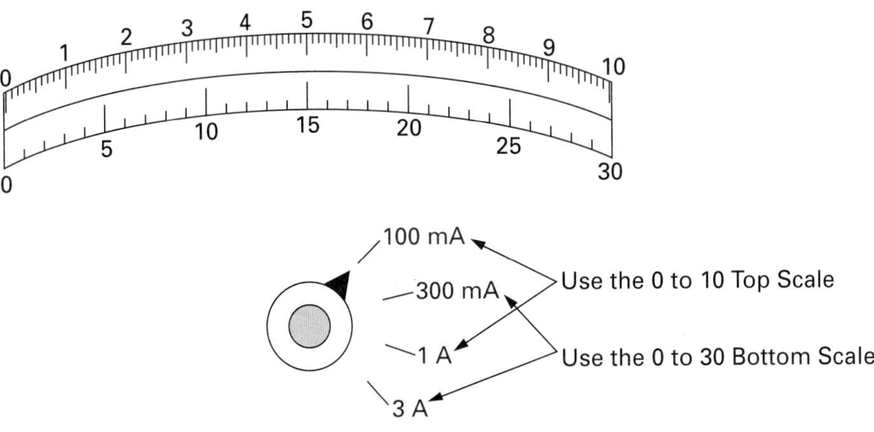

FIGURE 10-10 **Ammeter Scales.**

from 0 to 30 is used, with 30 on the scale representing 300 mA or 3 A, respectively.

10-2-2 Measuring Current

There are some important points to remember when using the ammeter to measure current in a circuit, as shown in Figure 10-11.

1. Always set the selector to the higher scale first (amperes) and then reduce as needed to the milliampere or microampere ranges, just in case a larger current than anticipated is within the circuit, in which case the ammeter could be damaged.
2. The positive lead of the ammeter (red) should be connected in the circuit so that it can be traced back to the positive side of the voltage supply, and the negative lead of the meter (black) should be connected in the circuit so that it can be traced back to the negative side of the voltage supply.
3. The ammeter must always be connected in the path of the current flow and, therefore, the circuit path must be opened and the ammeter inserted into the circuit.
4. Most analog meters have an accuracy of ±3% of full scale. For this reason, current readings should be as close to full scale as possible. For example, if 7 mA was measured on the 10-mA scale, the maximum error is ±0.3 mA. If 7 mA was measured on the 100-mA scale, however, the maximum error is ±3% of 100 or 3 mA. This means that on the 10-mA scale, a 7-mA reading could read from 6.7 to 7.3 mA, whereas on the 100-mA scale a 7-mA reading could read from 4 to 10 mA.

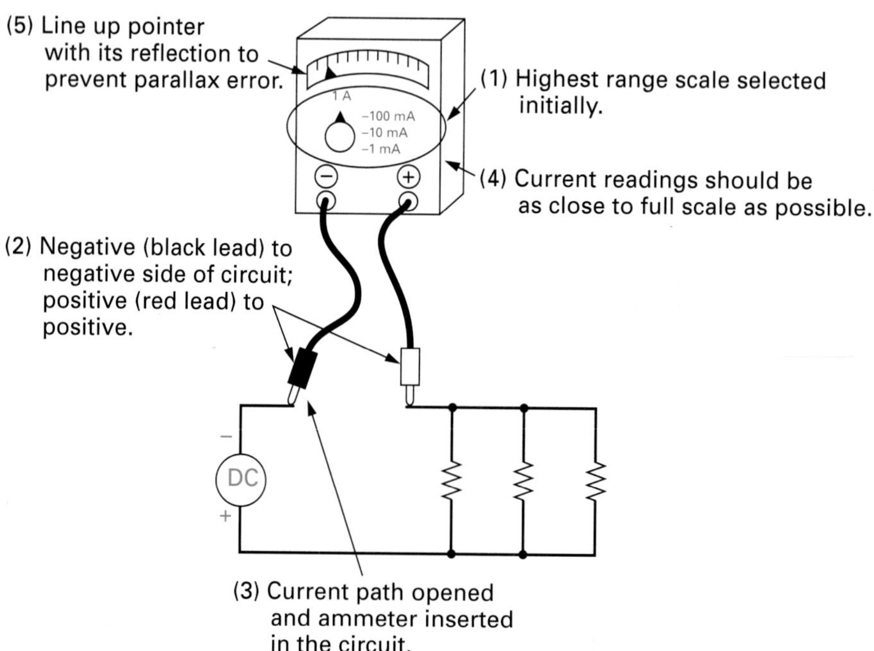

FIGURE 10-11 Connecting an Ammeter in Circuit to Measure Current.

5. Most analog meters include a mirror behind the scale and pointer. A reflection of the pointer will therefore appear in the mirror, and only when you are directly in front of the scale will the pointer line up with its reflection and enable you to read off an accurate value from the scale. The mirror is included to prevent a **parallax error.**

Parallax Error
The apparent displacement of an object's position caused by a shift in the point of observation of the object.

10-2-3 Ammeter Loading Effect

The ammeter is placed in a circuit to measure the current flowing in that circuit. The meter should ideally offer no resistance to the circuit, as any resistance in an ammeter will cause the overall circuit resistance to change, and this will change the current flow that we are trying to measure.

Figure 10-12 illustrates an example of an ammeter set up for the 10-mA range and, as you can see, the total resistance offered by the ammeter is the value of the shunt resistance in parallel with the coil resistance, a total of 5 Ω. The resistance of the circuit (10,000 Ω) is so large compared to the internal resistance of the meter (5 Ω) that this meter resistance can be disregarded, and the measured current will be relatively accurate. In this example, the total circuit resistance will equal 10,000 Ω (10 kΩ) and, therefore, the circuit current will equal 1.2 mA (12 V/10 kΩ). With the ammeter in the circuit, the total resistance will now equal 10,005 Ω and, therefore, the circuit current will equal 1.199 mA (12 V/10,005 Ω). It can be seen therefore that the ammeter has no significant effect on the circuit current it measures.

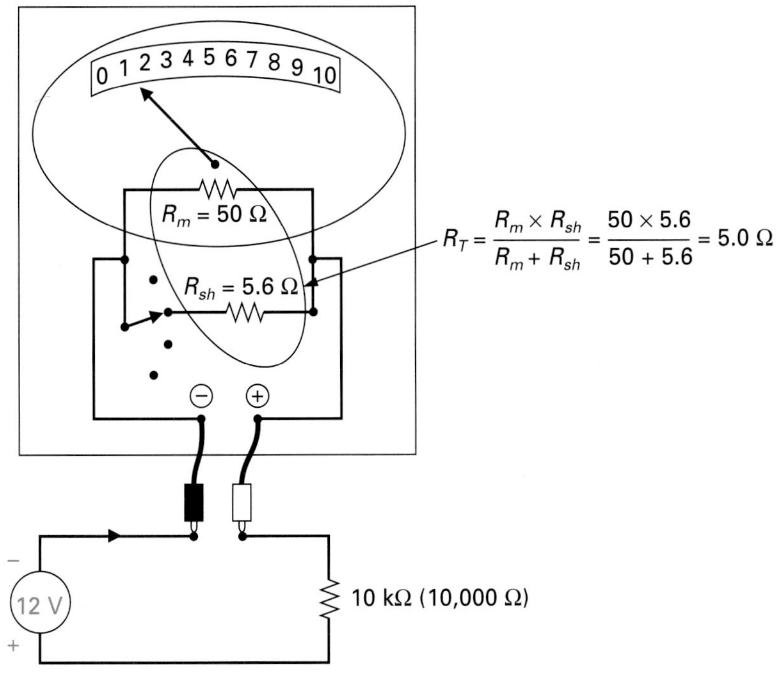

FIGURE 10-12 **Effect of Ammeter in Circuit.**

SELF-TEST REVIEW QUESTIONS FOR SECTION 10-2
1. To have a multirange ammeter, different values of ______________ resistors are connected in ______________ with the meter's coil resistance.   **2.** If $I_m = 1$ mA and $R_m = 50\ \Omega$, what value of R_{sh} would be needed to measure 30 mA?

10-3 THE VOLTMETER

A voltage increase causes a current increase and this relationship between voltage and current means that the moving coil type of meter movement discussed previously can also be used to measure voltage, as seen in Figure 10-13(a). In this explanation we will, for variety, use a different movement, where $I_m = 1$ mA and $R_m = 10\ \Omega$. If 10 mV is connected across 10 Ω of moving coil resistance, a current of 1 mA will flow:

$$I = \frac{V}{R} = \frac{10\text{ mV}}{10\ \Omega} = 1\text{ mA}$$

This will cause full-scale deflection of the meter. With this setup, the meter can be used to measure any voltage from 0 to 10 mV.

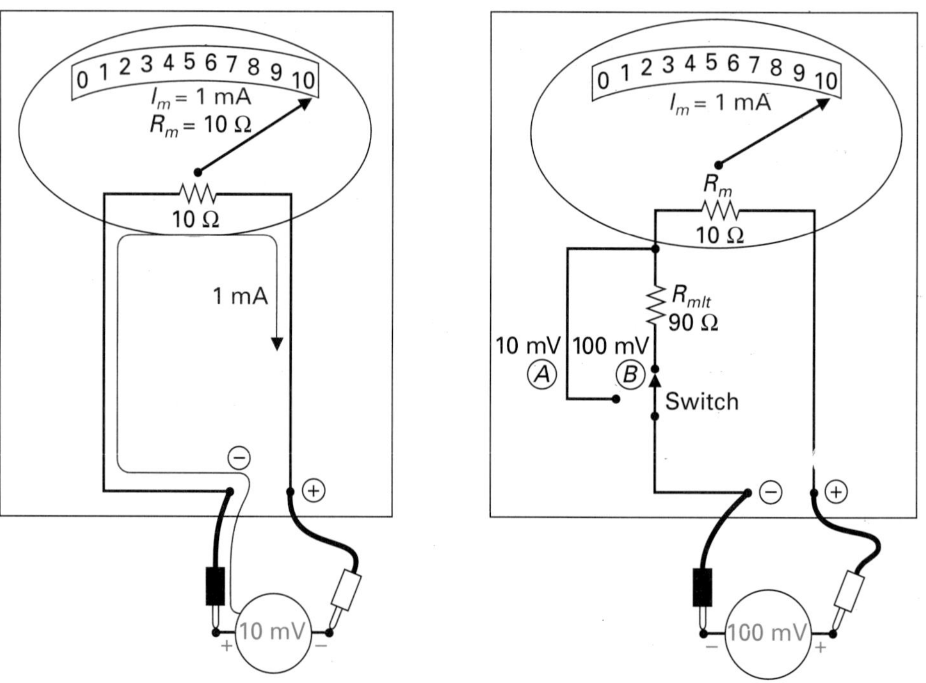

FIGURE 10-13 **Voltmeter. (a) 10 mV Voltmeter. (b) 100 mV Voltmeter.**

10-3-1 Voltmeter Range Scales

To measure a voltage greater than 10 mV, we will utilize a resistor, referred to as a **multiplier resistor,** to drop the extra voltage. For example, if we wished to measure a voltage in the range of 0 to 100 mV, we are already aware that a 10-mV drop across the moving coil resistance (R_m = 10 Ω) is all that is necessary to cause a full-scale deflection current of 1 mA (I_m = 10 mV/10 = 1 mA). To measure 100 mV, a multiplier resistor is connected in series with R_m to drop the extra 90 mV. The value of the multiplier resistor can be calculated by simply applying Ohm's law.

Multiplier Resistor
A resistor connected in series with the meter movement of a voltmeter.

$$R_{mlt} = \frac{V_{mlt}}{I_{mlt}} = \frac{90 \text{ mV}}{1 \text{ mA}} = 90\ \Omega$$

This modification with the 90 Ω multiplier resistor connected in series with the moving coil resistance will allow us to measure a voltage range from 0 to 100 mV. When 100 mV is being measured across the positive (+) and negative (–) terminals of the meter, 90 mV will be dropped across the multiplier and 10 mV across the moving coil resistor (R_m), causing full-scale deflection.

If a switch is included, as seen in Figure 10-13(b), we can either (1) switch out the multiplier (position A) to convert the meter to a 10-mV range, or (2) if position B is selected, the multiplier resistor can be brought in series with R_m to obtain a 100-mV range.

Figure 10-14(a) illustrates a multirange voltmeter in which four multiplier resistors can be switched into circuit. The voltage drop across the moving coil

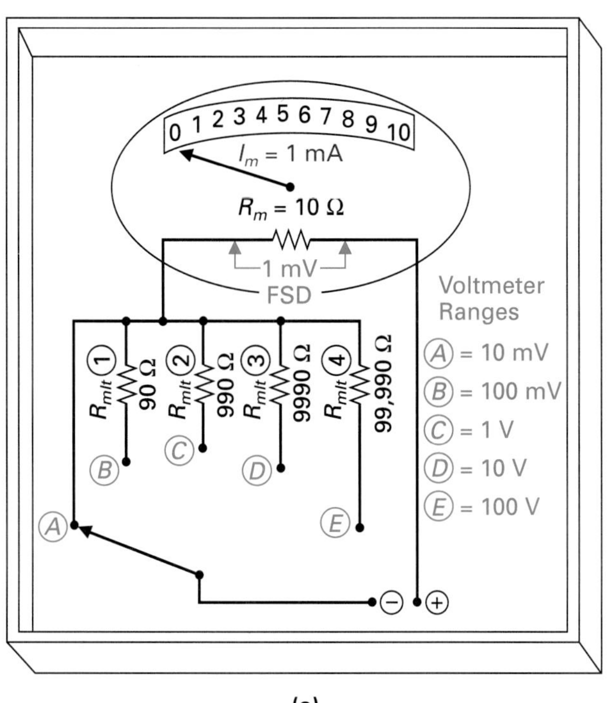

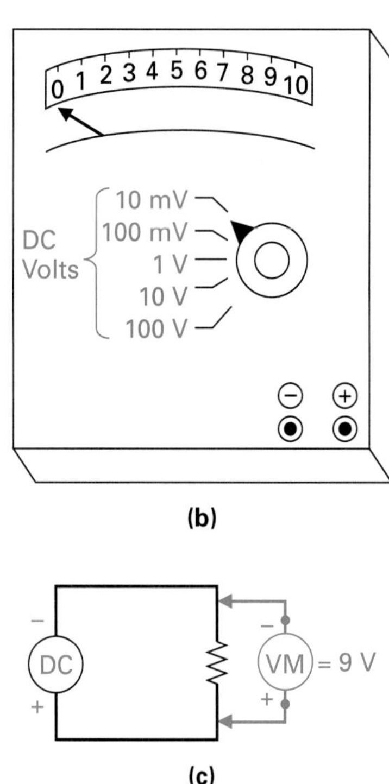

FIGURE 10-14 Multirange Voltmeter. (a) Internal Circuit. (b) External Appearance. (c) Voltmeter Symbol in Circuit.

resistance must always be 10 mV to cause FSD. The remaining voltage on whichever range has been selected will be dropped across the series-connected multiplier resistor.

Position A. If the 10-mV range is selected, there is no need for a multiplier as 5 mV across the (+) and (–) terminals will cause one-half FSD and 10 mV across the terminals will cause FSD.

Position B. If the 100-mV range is selected, the difference in voltage, which in this case is 90 mV (100 mV – 10 mV), must be dropped across a multiplier resistor (R_{mlt1}) whose resistance will equal

$$R_{mlt1} = \frac{V_{mlt}}{I_{mlt}} = \frac{90 \text{ mV}}{1 \text{ mA}} = 90 \ \Omega$$

Position C. When the 1-V (1000-mV) range is selected, the voltage difference that will have to be dropped across the multiplier resistor (R_{mlt2}) will be 990 mV (1000 mV – 10 mV). The value of R_{mlt2} will therefore be

$$R_{mlt2} = \frac{V_{mlt2}}{I_{mlt2}} = \frac{990 \text{ mV}}{1 \text{ mA}} = 990 \ \Omega$$

Position D. [10-V range (10,000 mV)] The voltage difference is equal to 10,000 mV – 10 mV = 9990 mV.

$$R_{mlt3} = \frac{V_{mlt3}}{I_{mlt3}} = \frac{9990 \text{ mV}}{1 \text{ mA}} = 9990 \ \Omega$$

Position E. [100-V range (100,000 mV)] The voltage difference is equal to 100,000 mV – 10 mV = 99,990 mV.

$$R_{mlt4} = \frac{V_{mlt4}}{I_{mlt4}} = \frac{99{,}990 \text{ mV}}{1 \text{ mA}} = 99{,}990 \ \Omega$$

Figure 10-14(b) shows the external appearance of the rotary switch used by the technician to select the voltage range scale required. Figure 10-14(c) illustrates the schematic symbol of the voltmeter being used to measure the voltage drop across a resistor.

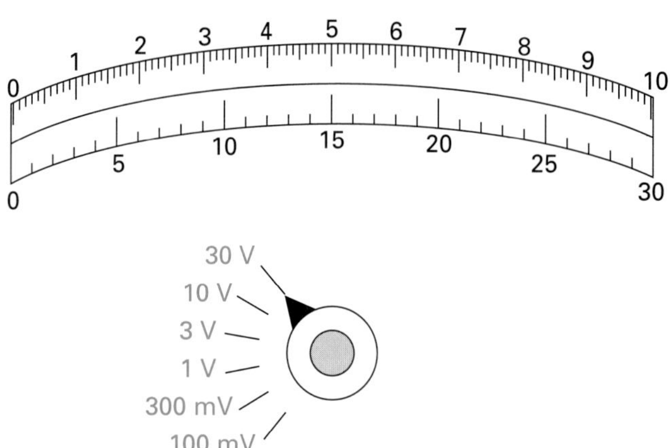

FIGURE 10-15 **Voltmeter Scales and Ranges.**

The range scales of the voltmeter are like the ammeter in that they may not only be oriented around 10, such as 10 mV, 100 mV, 1 V, 10 V, and 100 V, but also around, for example, 3, 30, or 300 V.

Figure 10-15 illustrates a multimeter with six range scales of 100 mV, 300 mV, 1 V, 3 V, 10 V, and 30 V. The top scale of 0 through 10 is used for the 100 mV, 1 V, and 10 V range scales, and the 0 through 30 scale is used for the 300 mV, 3 V, and 30 V range scales. The range scale selected refers to the FSD voltage, or maximum voltage that can be measured.

10-3-2 Voltmeter Sensitivity (Ω/V)

The sensitivity of a voltmeter is measured in a quantity known as the **ohms-per-volt rating** (Ω/V). The Ω/V rating can be calculated by taking the reciprocal of the full-scale deflection current of the moving coil (I_m). For example, a 1 mA meter will have an Ω/V rating of 1000 Ω/V.

Ohms-per-Volt Rating
Value that indicates the sensitivity of a voltmeter. The higher the ohms-per-volt rating, the more sensitive the meter.

$$\text{voltmeter sensitivity } (\Omega/\text{V}) = \frac{1\text{ V}}{\text{meter coil current } (I_m)}$$

$$= \frac{1\text{ V}}{1\text{ mA}} = 1000\ \Omega/\text{V}$$

A meter that only requires 100 μA (I_m) for FSD for the moving coil will have a sensitivity of

$$V_m \text{ sensitivity} = \frac{1\text{ V}}{I_m}$$

$$= \frac{1\text{ V}}{100\ \mu\text{A}}$$

$$= 10\text{ k}\Omega/\text{V} \quad \text{or} \quad 10{,}000\ \Omega/\text{V}$$

It can be said, therefore, that if a meter only needs a small current to cause FSD it will have a large Ω/V rating and will be classified as a more sensitive meter.

10-3-3 Measuring Voltage

There are some important points to remember and apply when measuring voltage with a voltmeter, as shown in Figure 10-16.

1. Always set the range selector to the highest range scale first (100 V) and then reduce as needed to the lower ranges (volts or millivolts).
2. Always connect the positive lead (+, red) of the voltmeter to the component end that is connected back to the positive side of the battery or source, and the negative lead (–, black) of the meter to the component end that is connected back to the negative side of the battery or source.

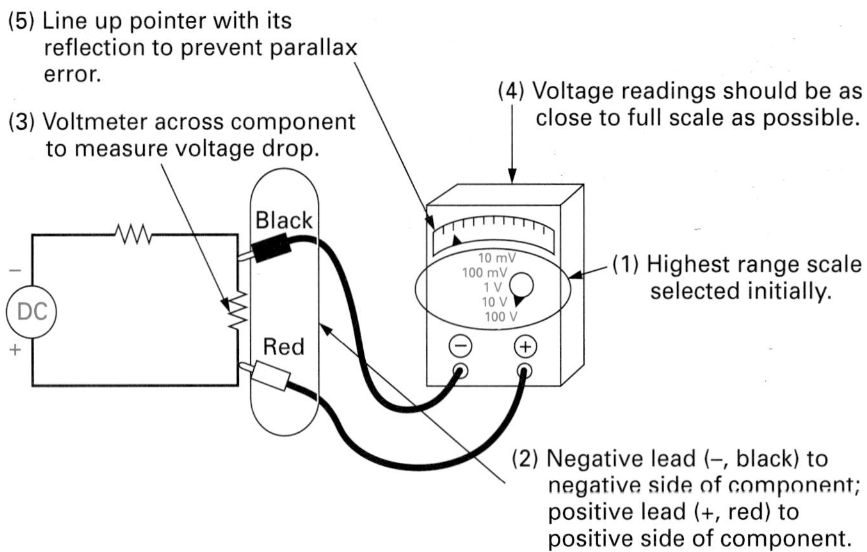

FIGURE 10-16 **Connecting a Voltmeter to Measure Voltage.**

3. The voltmeter must be connected across or in parallel with the component voltage to be measured.
4. Since most analog meters have an accuracy of ±3% of full scale, voltage readings should be as close to full scale as possible. For example, if a 7-V reading is taken on the 10-volt scale the maximum error is ±3% of 10 V, or 0.3 V. If 7 V was measured on the 100-V scale, however, a maximum error of ±3% of 100 V, or 3 V, could occur. This means that on the 10-V scale a 7 V measurement could read between 6.7 and 7.3 V, while on the 100-V scale a 7 V measurement could read between 4 and 10 V.
5. Always line the pointer up with its reflection in the mirror and then take a reading to prevent parallax error.

10-3-4 Voltmeter Loading Effect

A voltmeter is connected in parallel with the component to measure the voltage developed across that component. The voltmeter is, therefore, much easier to use for troubleshooting or testing purposes, as the circuit path does not have to be opened and the meter inserted in the current path, as with the ammeter.

When a voltmeter is connected across a component for voltage measurement, some of the circuit current flows through the voltmeter to deflect the meter movement. Figure 10-17(a) illustrates a voltmeter that has an internal resistance of 100 kΩ being used to measure the voltage drop across R_2. Without the voltmeter connected, the voltage will be dropped equally across R_1 and R_2, causing 5 V across R_2. If the voltmeter is now connected across R_2, the voltmeter resistance of 100 kΩ is now in parallel with the resistance of R_2, causing an equivalent resistance of 4.76 kΩ. The meter has loaded down the circuit and caused this change in resistance, which means that R_2 will not drop 5 V as nor-

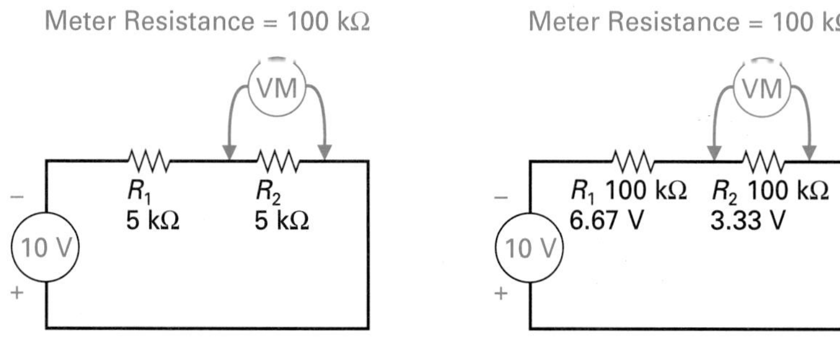

FIGURE 10-17 Voltmeter Loading Effect.

mal, but will in fact drop 4.88 V. This error of 0.12 V caused by the loading effect of the meter is very small. As long as the internal voltmeter resistance is large with respect to the resistance of the component across which the voltage is being measured, the reading will be reasonably accurate. In the first example, Figure 10-17(a), the meter's resistance is 100 kΩ, which is very large with respect to the component's resistance of 5 kΩ, so the loading effect of the meter is negligible, causing a small error of 0.12 V.

In Figure 10-17(b), the same voltmeter, with a resistance of 100 kΩ, is being used to measure the voltage drop across R_2, whose resistance is also equal to 100 kΩ. Once again, the voltage of 10 V will be divided equally across R_1 and R_2 when the meter is not connected, causing a 5 V drop across R_2.

If the voltmeter is now connected across R_2 so as to measure the voltage drop across the resistor, the voltmeter's resistance of 100 kΩ is now in parallel with the resistance of R_2, causing an equivalent resistance of 50 kΩ. The voltage drop across R_2, which is normally 5 V, is now pulled down to 3.33 V due to the loading effect of the meter, which is a 1.67 V error.

These two examples illustrate the loading effect of a voltmeter on a circuit under test. However, if a voltmeter has a large internal resistance with respect to the component's resistance (10 or more times greater), the loading effect of the meter can be disregarded and the voltmeter will be more accurate.

10-3-5 *Voltage Reference*

A voltmeter will always measure voltage with respect to some other point. If a resistor has a voltage drop of 30 V, for example, this means that one side of the resistor is 30 V more positive or negative than the other. This could occur if one side of the resistor was at +30 V and the other at +60 V (30 V difference in potential) or if one side was −100 V and the other was −70 V (30 V difference). Generally, when troubleshooting, a voltage measurement is taken with respect to 0 V or **ground,** of which there are two basic types:

Ground
An intentional or accidental conducting path between an electrical circuit or system and the earth, or some conducting body acting in place of the earth.

1. *Earth ground:* The power lines from your local electric company have a reference point referred to as earth ground; this is because one side of the voltage source is attached to a metal rod and inserted in the ground (Figure 10-18).

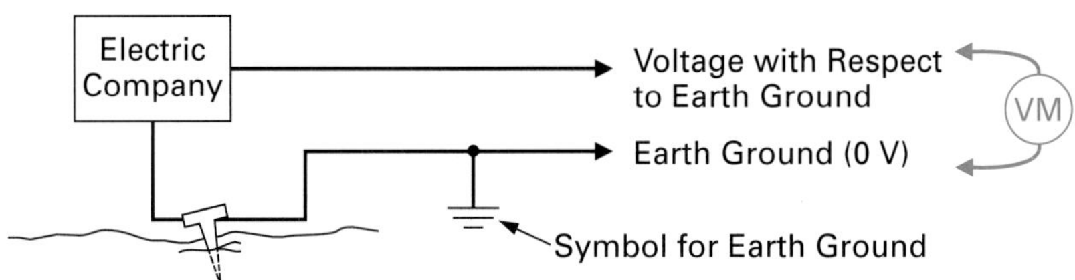

FIGURE 10-18 Earth Ground.

Positive Ground
A system whereby the positive terminal of the voltage source is connected to the system's conducting chassis or body.

Negative Ground
A system whereby the negative terminal of the voltage source is connected to the system's conducting chassis or body.

2. *Circuit or chassis ground:* In electronic equipment, the metal chassis or housing of the equipment is used as a common reference point. One side of the voltage supply is connected to the metal frame or chassis of the equipment and voltage is measured with respect to this common ground reference point, referred to as a circuit, chassis, or frame ground (Figure 10-19).

Another term that is used with all dc voltage sources is **positive ground** or **negative ground.** This refers to which polarity of the battery or source is connected to the chassis; for example, if the negative side of the battery is connected to the chassis, then positive voltages will be measured throughout the circuit with respect to the negative ground [Figure 10-20(a)]. On the other hand, if the positive side of the battery is connected to the chassis, then negative voltages

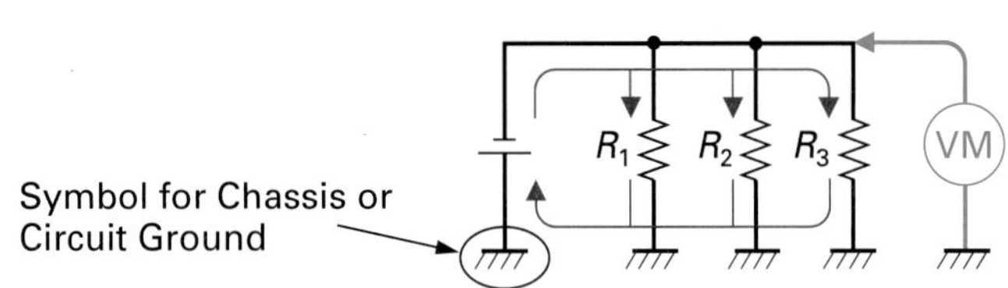

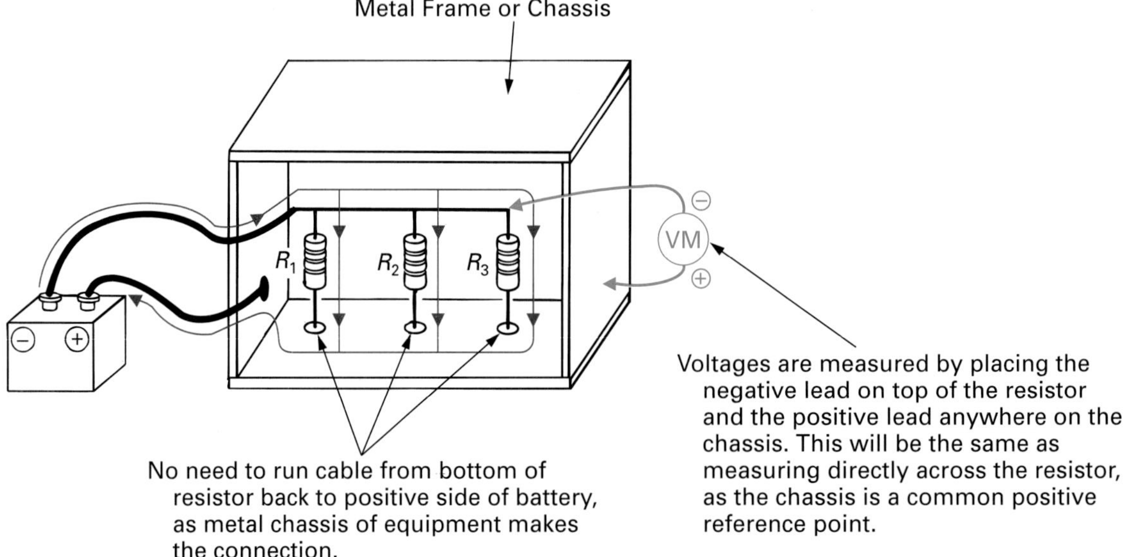

FIGURE 10-19 Chassis Ground.

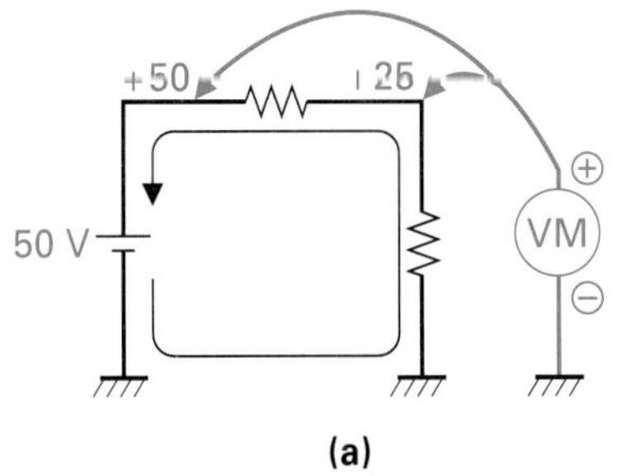

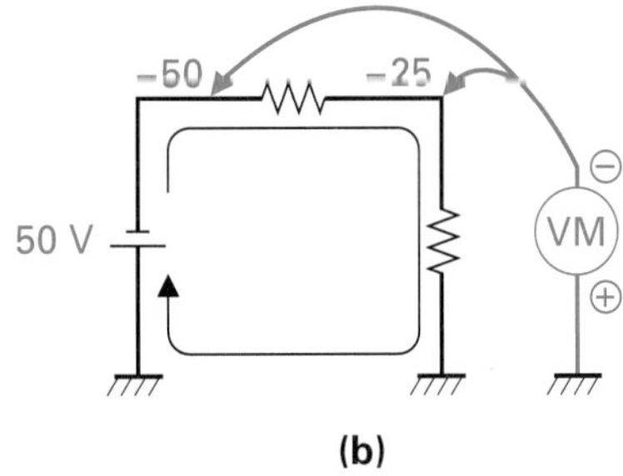

FIGURE 10-20 **Voltage Reference. (a) Negative Ground, Positive Voltage Measurements. (b) Positive Ground, Negative Voltage Measurements.**

with respect to a positive ground will be present throughout the circuit [Figure 10-20(b)]. To obtain positive readings on the voltmeter in this instance, the meter leads will have to be reversed.

If a voltmeter is normally always used to measure voltage with respect to chassis ground (0 V), how could you find the voltage dropped across R_1 in Figure 10-21? The voltmeter will measure 50 V at point A, but this is the voltage dropped across both R_1 and R_2. The voltage dropped across R_1 can be obtained by measuring the voltage at point A with respect to chassis ground (+50 V) and then at point B with respect to chassis ground (+25 V). The difference between these two readings on your voltmeter will be the potential difference between A and B; in this case, 25 V is dropped across R_1.

The next question you are probably asking is: Why would we make two voltmeter checks to find out that we have dropped 25 V across R_1 when we could have just hooked up the voltmeter across R_1 and measured 25 V [Figure 10-22(a) and (b)]. The reason why voltage measurements are almost always measured with respect to chassis ground rather than the other side of the resistor or component is convenience and fault prevention. If voltages are measured with respect to chassis ground, the negative lead of the voltmeter can be permanently attached to the chassis, and you only have to worry about moving the positive meter lead to probe the desired testing points [Figure 10-22(a)].

If you measure the voltage dropped across a resistor, as shown in Figure 10-22(b), both the positive and negative leads must be held by the technician to probe in two places, which means:

1. The technician does not have a free hand because he or she is probing with both voltmeter leads.

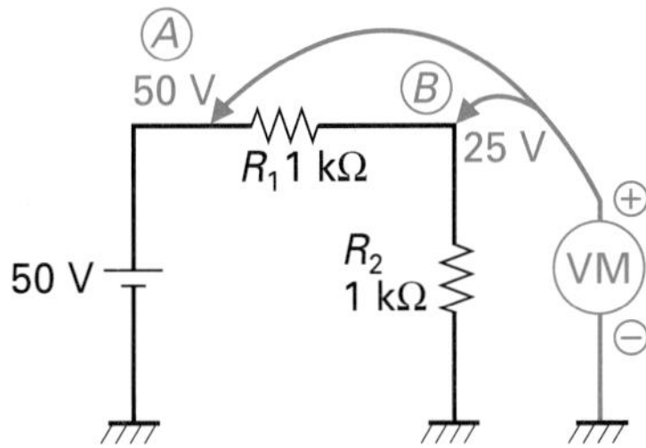

FIGURE 10-21 **Voltmeter Probing.**

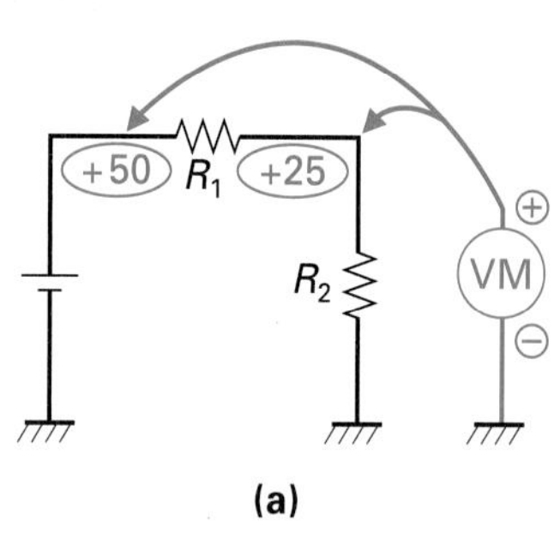

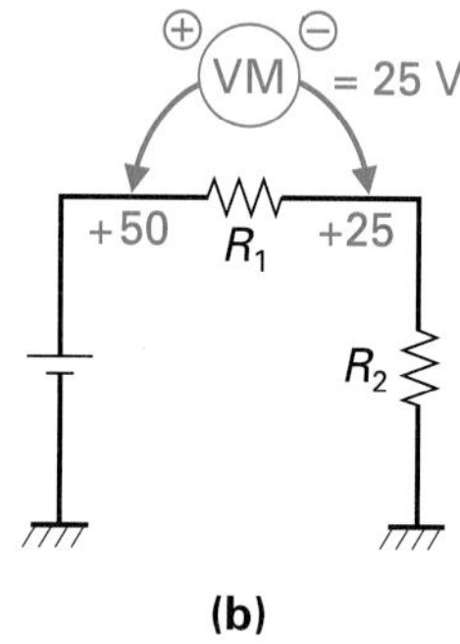

FIGURE 10-22 **Measuring Voltage Drop. (a) Two Checks to Find the Voltage Drop across R_1. (b) One Check to Find the Voltage Drop across R_1.**

2. If the technician is trying to position two probes at once in two different points, it becomes twice as difficult, and the risk of slipping and causing a short and then possibly an additional problem in the circuit under test is present.

For safety reasons, it is often recommended that the test leads be attached for voltage measurements with the equipment off; then turn it on to observe the reading. This is especially important in high-voltage systems of 300 V or more.

EXAMPLE:

Using the voltmeter with the negative lead tied to the chassis, the voltages at points *A*, *B*, and *C* are as shown in Figure 10-23. From these readings, calculate the voltage drops across R_1, R_2, R_3, and R_4.

Solution:

The difference between *A* and *B* = 40 V − 15 V = 25 V; therefore, 25 V is dropped across R_1.

The difference between *B* and *C* = 15 V − 4 V = 11 V; therefore, 11 V is dropped across R_2 and R_3 (resistors are in parallel).

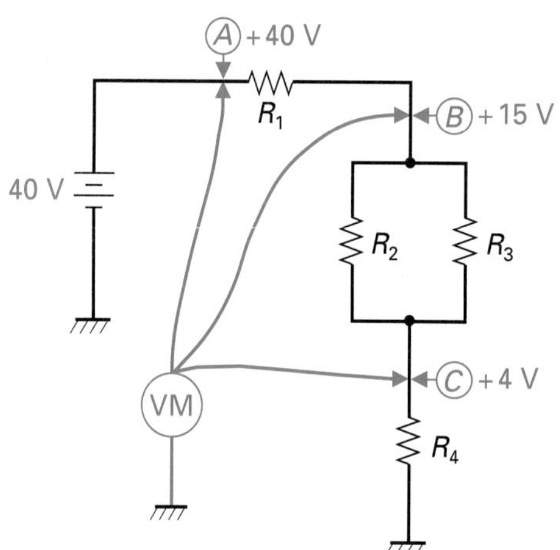

FIGURE 10-23 **Calculating Voltage Drop.**

The difference between C and chassis ground = 4 V − 0 V = 4 V; therefore, 4 V is dropped across R_4.

SELF-TEST REVIEW QUESTIONS FOR SECTION 10-3

1. To have a multirange voltmeter, different values of ____________ resistors are connected in ____________ with the meter's coil resistance.
2. Define ohms/volt and how it relates to a meter's sensitivity.

10-4 THE OHMMETER

Resistance can also be measured using the same moving coil meter movement. Figure 10-24(a) illustrates the meter configuration for measuring resistance. The only difference is that a series 1.5 V battery source is used to supply current. When the meter leads are connected together (shorted), the zero-ohms adjust resistor is adjusted to cause full-scale deflection (1 mA); this will occur when the meter's total resistance is equal to

$$R = \frac{V}{I} = \frac{1.5\ \text{V}}{1\ \text{mA}} = 1500\ \Omega$$

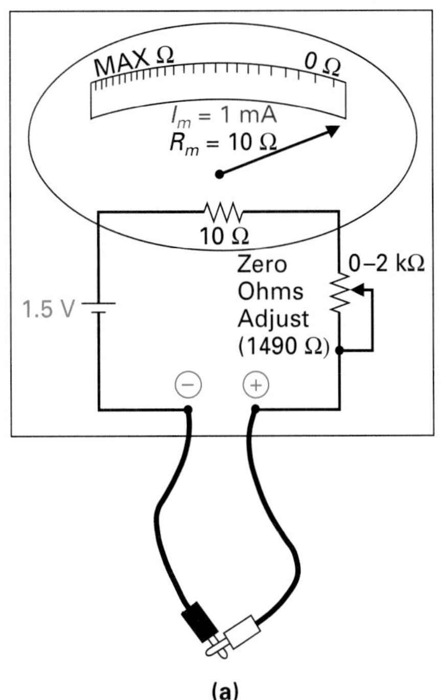

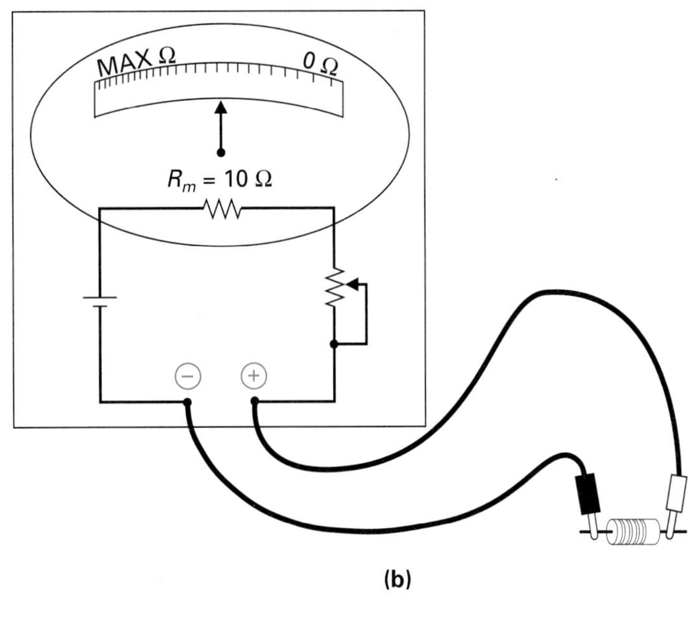

FIGURE 10-24 **Ohmmeter. (a) Zeroing the Meter. (b) Measuring Resistance.**

Since the moving coil resistance is equal to 10 Ω (R_m), the zero-ohms adjust resistance will need to be adjusted to equal 1490 Ω (1500 Ω − 10 Ω = 1490 Ω) in order to cause a total circuit resistance of 1500 Ω and, therefore, achieve full-scale deflection current (I_m) of 1 mA. Adjustment of the zero-ohms adjust resistor to obtain FSD is called **zeroing** the meter. When the meter leads are shorted together, the meter will deflect to the full-scale position (right side), indicating that there is no resistance (0 Ω) between the two meter leads. As different values of resistance are inserted between the two meter probes, different values of current will result, causing different levels of pointer deflection. A higher value of resistance will cause a smaller value of current to flow through the meter circuit, resulting in a smaller deflection of the pointer, as shown in Figure 10-24(b).

Zeroing
Calibrating a meter so that it shows a value of zero when zero is being measured.

The reason for the ohms adjust control is due to the discharge of the inter nal 1.5 V ohmmeter battery. As its voltage falls after months of use, the resistance of the zero-ohms adjust resistor is decreased to still cause an FSD current of 1 mA when the meter leads are shorted together. If, for example, the battery in the circuit of Figure 10-24(a) has discharged to 1 V, the zero-ohms adjust resistor needs to be varied to a value of 990 Ω, and therefore the total meter circuit resistance will equal

$$\text{zero-ohms resistance} + \text{moving coil resistance}$$

$$990\ \Omega + 10\ \Omega = 1000\ \Omega$$

This will cause a meter circuit current of

$$I_m = \frac{V}{R} = \frac{1\ \text{V}}{1000\ \Omega} = 1\ \text{mA}$$

The meter is now said to be zeroed, because 1 mA of meter circuit current is flowing through the moving coil, causing FSD.

When the ohmmeter's probes are shorted together, it causes FSD, which indicates to the technician that 0 Ω (a short circuit) exists between the two probes. On the other hand, if the two probes are open, the meter needle remains in its resting position, indicating that infinite (∞) resistance (maximum possible resistance or open circuit) exists between the two probes.

Figure 10-25(a) illustrates an example ohmmeter, which has a 1 mA, 10-Ω meter movement. Let's now proceed to take three examples of measured resistance and see how the ohmmeter responds to indicate these resistances.

1. If the two leads are shorted together, the total meter circuit resistance will equal

$$R_m + R_{\text{zero}} + \text{measured resistance} = 10\ \Omega + 1490\ \Omega + 0\ \Omega$$
$$= 1500\ \Omega$$

The moving coil current will consequently equal

$$I_m = \frac{V}{R} = \frac{1.5\ \text{V}}{1500\ \Omega} = 1\ \text{mA}$$

and this will cause full-scale deflection [Figure 10-25(b)].

2. If the two meter leads are connected across a 1500 Ω resistor, the total circuit resistance will equal

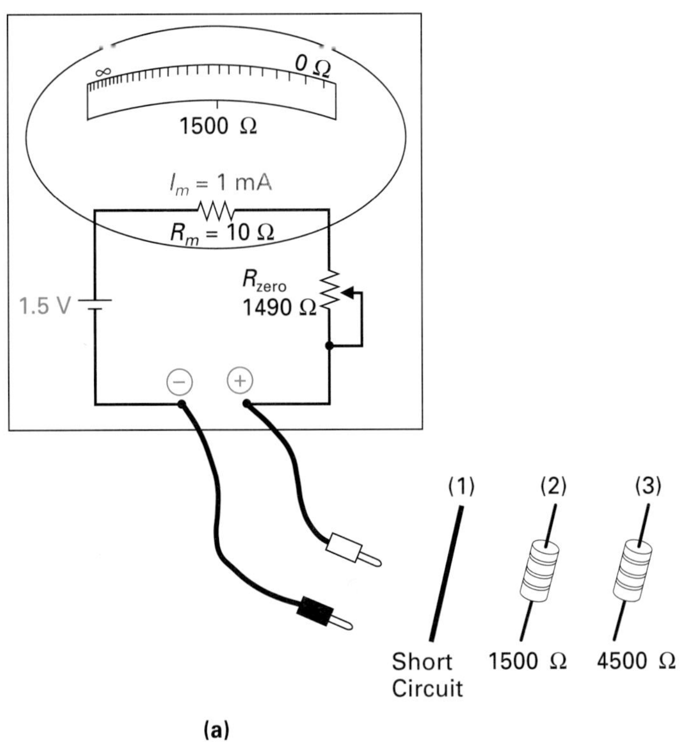

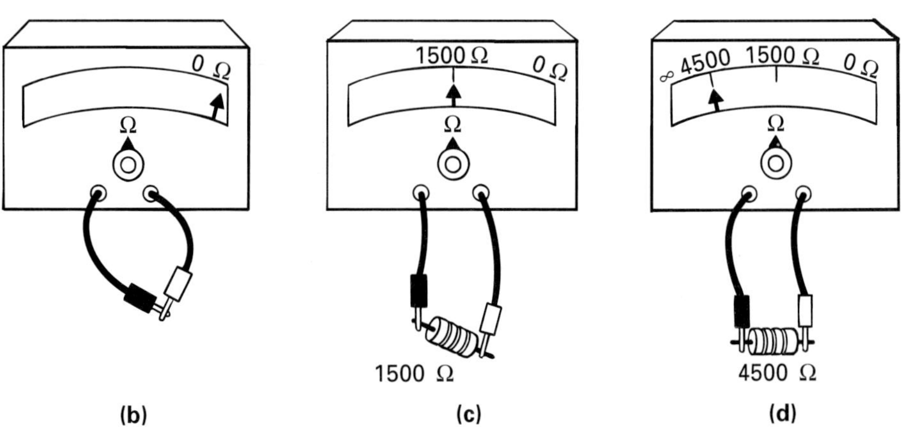

FIGURE 10-25 Ohmmeter's Response to Resistance. (a) Ohmmeter. (b) 0 Ω. (c) 1500 Ω. (d) 4500 Ω.

$$\begin{aligned} R_m + R_{\text{zero}} + \text{measured resistance} &= 10\ \Omega + 1490\ \Omega + 1500\ \Omega \\ &= 3000\ \Omega \end{aligned}$$

The moving coil current will consequently equal

$$I_m = \frac{V}{R} = \frac{1.5\text{ V}}{3000\ \Omega} = 0.5\text{ mA}$$

Since 1 mA causes full-scale deflection, 0.5 mA will cause one-half full-scale deflection [Figure 10-25(c)].

3. If we connect our ohmmeter leads across a resistor of 4500 Ω, the total meter circuit resistance will equal

$$R_m + R_{zero} + \text{measured resistance} = 10\ \Omega + 1490\ \Omega + 4500\ \Omega$$
$$= 6000\ \Omega$$

The moving coil current consequently equals

$$I_m = \frac{V}{R} = \frac{1.5\ \text{V}}{6000\ \Omega} = 0.25\ \text{mA}$$

Since 1 mA causes full-scale deflection, 0.25 mA will cause one-quarter full-scale deflection [Figure 10-25(d)].

Nonlinear Scale
A scale whose divisions are not uniformly spaced.

Linear Scale
A scale whose divisions are uniformly spaced.

The ohmmeter scale is referred to as a **nonlinear scale,** as opposed to the voltmeter and ammeter scales, which are **linear.** Figure 10-26(a) illustrates the nonlinearity of the ohmmeter, while Figure 10-26(b) shows how the voltmeter and ammeter scales follow a linear variation.

Maybe the best way to clearly see the difference between the two types of scales is first to analyze the ohmmeter's nonlinear scale. The difference with the ohmmeter, first, is that 0 begins on the right side. Between three-quarters FSD and one-half FSD on the ohmmeter's output scale is a change of 500 to 1500 on the input resistance being measured, which is a 1000 Ω change. From one-quarter FSD to one-half FSD is an input resistance change of 3000 Ω. From one-half FSD to three-quarters FSD is the same amount of distance on the output scale, yet the input resistance change is three times greater between one-quarter and one-half FSD. The output scale, therefore, does not vary in direct proportion or at the same rate as the input resistance being measured and is consequently referred to as a nonlinear scale (the divisions are not uniformly spaced).

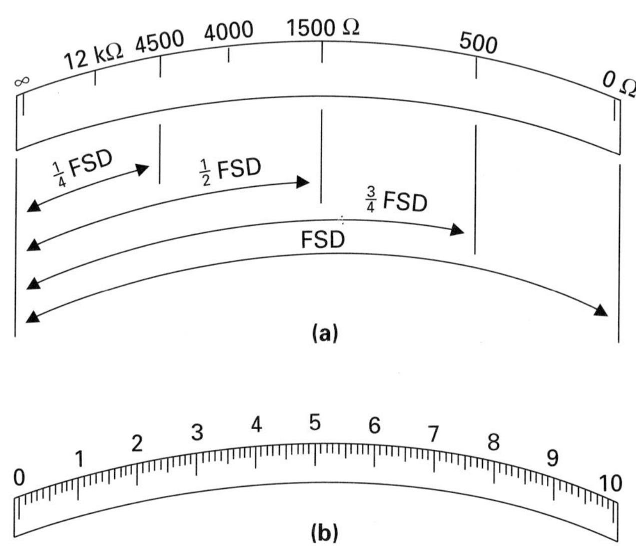

FIGURE 10-26 Nonlinear versus Linear Scale. (a) Nonlinear Ohmmeter Scale. (b) Linear Voltmeter and Ammeter Scale.

If the output scale varies in direct proportion to the input being measured, as with the voltmeter and ammeter scales, the scale is said to be linear. This can be further explained by using the same simple example.

Between one-quarter and one-half FSD is a change of 2.5 to 5 V on the output scale and is in direct proportion to the input change of 2.5 to 5 V (or amperes). Between one-half and three-quarters FSD is exactly the same amount of change as one-quarter to one-half FSD, both on the output scale and the input being measured, which is why Figure 10-26(b) is referred to as a linear scale (all divisions are evenly spaced).

10-4-1 Ohmmeter Range Scales

Figure 10-27(a) illustrates the external appearance of an ohmmeter with multiple ranges, and Figure 10-27(b) illustrates the internal series and shunt resistors that allow different range settings to be attained. The range scales of an ohmmeter are interpreted differently from those of the ammeter and voltmeter. The ohmmeter's range scales in Figure 10-27(a) of $R \times 1$, $R \times 10$, and $R \times 100$ are used as multipliers; for example, if the pointer is at 500 and the range scale selected is $R \times 10$, the resistor's resistance will be equal to

$$R \times 10 = 500 \times 10 = 5000 \text{ or } 5 \text{ k}\Omega$$

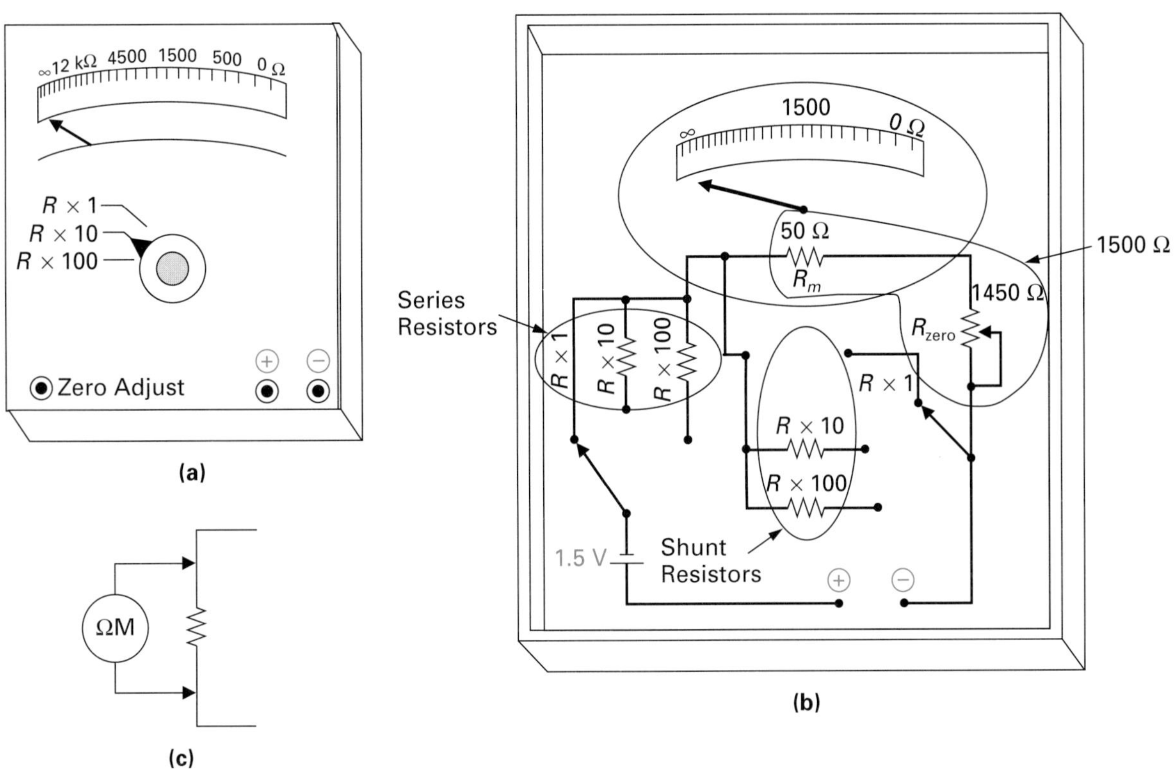

FIGURE 10-27 **Multirange Ohmmeter. (a) External Appearance. (b) Internal Circuit. (c) Ohmmeter Symbol in Circuit.**

A typical ohmmeter will usually have a small battery of 1.5 V, which is used for the $R \times 1$, $R \times 10$, and $R \times 100$ ranges, while a larger battery of typically 9 or 15 V will be used for the $R \times 1000$ range and higher.

Figure 10-27(b) shows the internal circuit of an ohmmeter. When the $R \times 1$ range is selected, no series multiplier resistor or parallel shunt resistor is connected, and the meter will operate as follows: If the meter leads are shorted together, a resistance of 0 Ω exists between them, and the meter should cause full-scale deflection of the pointer (1 mA of moving coil current through R_m) to the 0 Ω position on the far-right side of the scale.

R_m and R_{zero} are in series with one another, giving a combined resistance of 1500 Ω and causing a total circuit resistance of 1500 Ω. Meter circuit current will therefore equal

$$I = \frac{V}{R} = \frac{1.5 \text{ V}}{1500 \text{ }\Omega} = 1 \text{ mA}$$

which will flow through R_m, causing FSD of the pointer to the 0 Ω position of the scale.

If a 1500-Ω resistor is connected between the two meter leads, the meter should cause the pointer to deflect to the one-half FSD position (0.5 mA of moving coil current through R_m). With the 1500 Ω resistor between the two meter leads, a total circuit resistance of

$$\begin{aligned} R_t &= R_m + R_{zero} + R_{measured} \\ &= 50 \text{ }\Omega + 1450 \text{ }\Omega + 1500 \text{ }\Omega \\ &= 3000 \text{ }\Omega \end{aligned}$$

This will cause a circuit current of $I = V/R = 1.5 \text{ V}/3000 \text{ }\Omega = 0.5$ mA, which will cause one-half FSD of the pointer to the 1500 position on the scale.

For the $R \times 10$ and $R \times 100$ scales, an ohmmeter will switch in a different value of series and shunt resistors in order to obtain the $R \times 10$ and $R \times 100$ ohmmeter positions.

Figure 10-27(c) shows the schematic symbol for an ohmmeter.

10-4-2 *Measuring Resistance*

(See Figure 10-28.) There are a few points to remember when measuring resistance with an ohmmeter:

1. Connect (short) the positive and negative leads together and adjust the zero adjust control so that the pointer is at 0 Ω on all scales. This zeroing of the ohmmeter is also known as a confidence test since it lets you know if the meter is functioning.
2. Ensure that no power is connected to the component to be measured, as the circuit voltage and internal ohmmeter battery voltage could aid one another, causing an excessive current, which could damage the meter movement.
3. Connect the leads across the component and read off the resistance indicated by the scale, multiplying by 1, 10, 100, 1000, or more, depending on the ohms range selected.

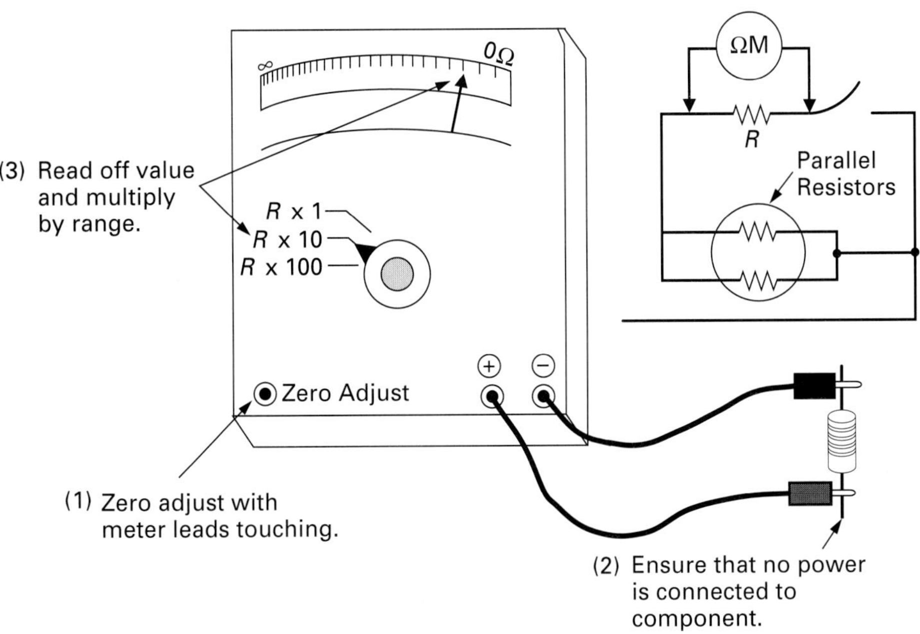

FIGURE 10-28 **Measuring Resistance.**

4. If the resistance of a component is measured while the component is connected in a circuit, an invalid reading may be obtained due to the parallel effects of other resistances in the circuit. This problem can be overcome by disconnecting one end of the component from the circuit to obtain a true reading.

SELF-TEST REVIEW QUESTIONS FOR SECTION 10-4

1. What are the major differences between an ammeter/voltmeter and an ohmmeter?
2. State the four points that should be remembered when measuring resistance.

10-5 THE ANALOG MULTIMETER

After discussing the ammeter, voltmeter, and ohmmeter separately, we will now combine all three to end up with a multipurpose meter (multimeter), or, as they are sometimes called, a **VOM meter** (volt-ohm-milliamp meter), as shown in Figure 10-29. Having all three meters in one package is both convenient and cheaper than having to buy three meters.

VOM Meter Abbreviation for volt-ohm-milliamp meter.

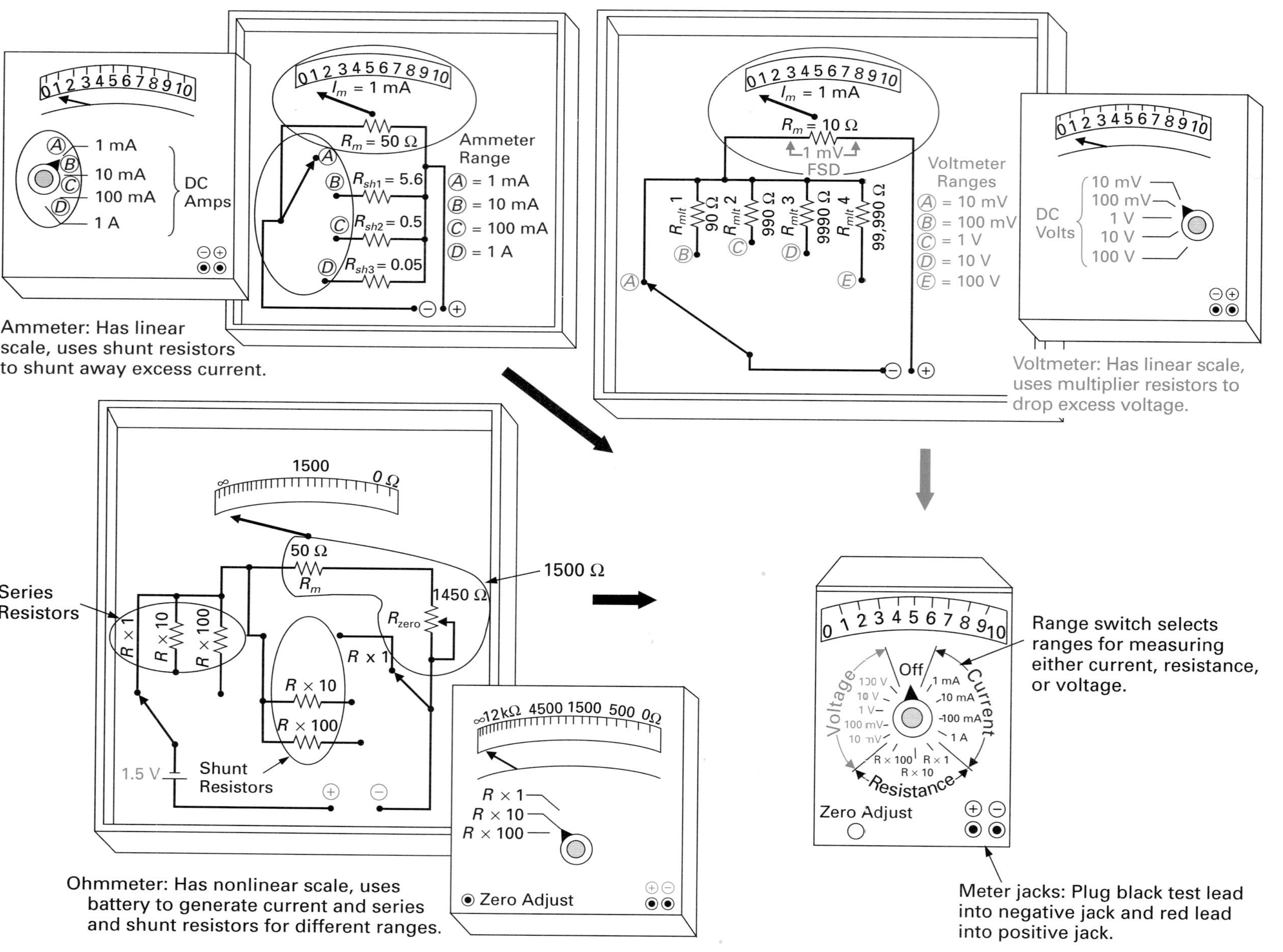

FIGURE 10-29 Multimeter.

SELF-TEST REVIEW QUESTIONS FOR SECTION 10-5

1. Which meter has:
   a. A nonlinear scale?
   b. Only shunt resistors?
   c. Only multiplier resistors?
   d. An internal voltage source?
   e. A linear scale?
2. VOM is an abbreviation for ____________.

SUMMARY

The Analog Readout Volt-Ohm-Milliamp (VOM) Multimeter (Figure 10-30)

1. The multimeter is a device used for the measurement of current, voltage, and resistance and is used for experimentation with and troubleshooting of electronic equipment.
2. The multipurpose meter has three meters all rolled into one package: the ammeter measures current, the voltmeter measures voltage, and the ohmmeter measures resistance.
3. The analog readout multimeter utilizes a moving coil, which is deflected by a current when measuring volts, ohms, or amperes.
4. The ammeter uses shunt resistors in parallel with the meter movement and has a very low internal resistance when placed in circuit to minimize the loading effect.
5. The voltmeter uses multiplier resistors in series with the meter movement and has a very large internal resistance when connected across a component to minimize the loading effect.
6. The ohmmeter has a combination of series and parallel resistors in and across the meter movement, and uses an internal current source in the form of a battery to deflect the pointer.

NEW TERMS

Analog data
Analog display
Analog readout
Armature
Chassis ground
D'Arsonval meter movement
Digital data
Digital display
Digital readout
Earth ground
Full-scale deflection (FSD)
Linear scale
Meter movement
Meter resistance
Meter sensitivity
Multimeter
Multiplier resistor
Multirange meter
Negative ground
Nonlinear scale
Ohms per volt (Ω/V)
Parallax error
Polarity sensitive
Positive ground
Sensitivity
Shunt resistor
Zeroing

QUICK REFERENCE SUMMARY SHEET

FIGURE 10-30 The Volt-Ohm-Milliamp Multimeter.

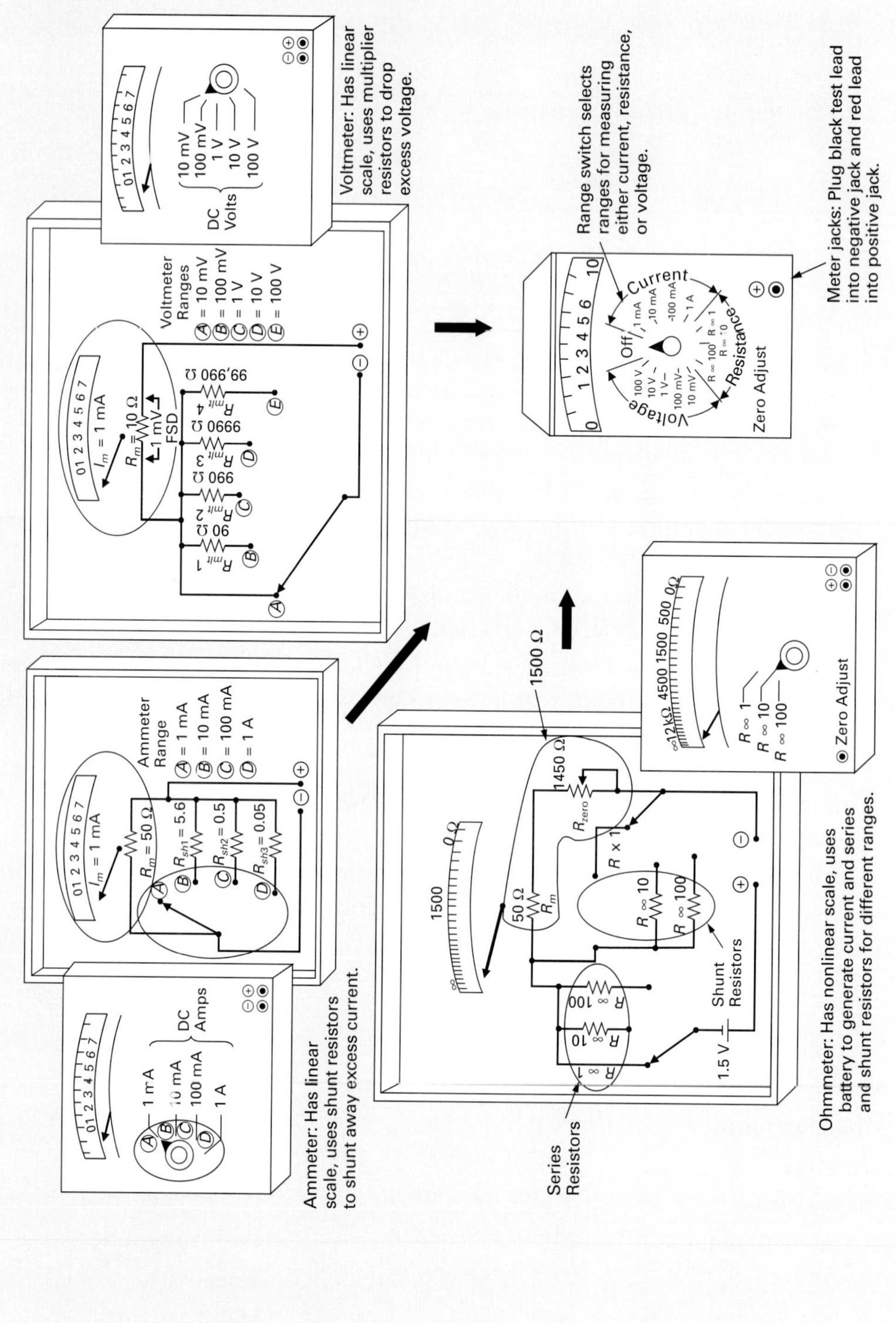

$$\text{Voltmeter sensitivity } (\Omega/\text{V}) = \frac{1\text{ V}}{\text{meter coil current } (I_m)}$$

REVIEW QUESTIONS

Multiple-Choice Questions

1. The device used to measure the number of coulombs per second passing a given point is called

   a. An ammeter c. An ohmmeter
   b. A voltmeter d. A wattmeter

2. The device used to measure the potential difference between two points is called:

   a. An ammeter c. An ohmmeter
   b. A voltmeter d. A wattmeter

3. The device used to measure the amount of opposition to current flow offered by a particular component is called:

   a. An ammeter c. An ohmmeter
   b. A voltmeter d. A wattmeter

4. The ____________ or moving coil type of meter movement was named after its inventor.

   a. Joule c. D'Arsonval
   b. Wattmeter d. Guglielmo

5. A meter movement that requires only 10 μA to deflect the pointer to FSD is ____________ than a meter that requires 10 mA.

   a. More sensitive c. Insensitive
   b. Less sensitive d. All of the above could be true.

6. Ammeter ranges are accomplished by ____________ resistors connected in ____________ with the meter movement.

   a. Multiplier, series c. Multiplier, parallel
   b. Shunt, parallel d. Shunt, series

7. Voltmeter ranges are accomplished by ____________ resistors connected in ____________ with the meter movement.

   a. Multiplier, series c. Multiplier, parallel
   b. Shunt, parallel d. Shunt, series

8. An ammeter should ideally offer ____________ resistance to the circuit in which it is connected.

   a. 100 kΩ of c. A very low
   b. A very large d. 150 Ω of

9. A voltmeter should ideally offer____________ resistance to the circuit in which it is connected.

   a. 100 kΩ of c. A very low
   b. A very large d. 150 Ω of

10. The sensitivity of a voltmeter is measured in a quantity known as the ____________ rating.

    a. Ohms/cm c. Ohms/volt
    b. Coulombs/second d. Amperes/volt

11. The only difference between the ohmmeter and the voltmeter or ammeter is that the ohmmeter uses a/an __________ to supply current.

   a. External voltage source
   b. Battery
   c. Internal voltage source
   d. Both (b) and (c)
   e. Both (a) and (b)

12. The two advantages of the digital multimeter are that it is:

   a. Easy to read and cheap in price
   b. Accurate and cheap in price
   c. Accurate and easy to interpret
   d. Inexpensive and has a quick response

13. The two advantages of the analog multimeter are that it is:

   a. Easy to read and cheap in price
   b. Accurate and cheap in price
   c. Accurate and easy to interpret
   d. Inexpensive and has a quick response

14. Which meter uses a nonlinear scale?

   a. Ammeter
   b. Voltmeter
   c. Ohmmeter
   d. Both (a) and (b)

15. A meter containing an ammeter, voltmeter, and ohmmeter is generally referred to as a:

   a. VOM
   b. Wattmeter
   c. Multimeter
   d. Both (a) and (c)

Essay Questions

16. List the four steps that should be remembered when using the ohmmeter. (10-4-2)
17. List the five steps that should be remembered when using the voltmeter. (10-3-3)
18. List the five steps that should be remembered when using the ammeter. (10-2-2)
19. Briefly describe the construction and operation of the D'Arsonval or moving coil type of meter movement. (10-1-1 and 10-1-2)
20. Briefly describe how ammeters have multiple range scales by the use of shunt resistors. (10-2-1)
21. Briefly describe how voltmeters have multiple range scales by the use of multiplier resistors (10-3-1)
22. Describe what loading effect an ammeter and voltmeter will have when measuring current and voltage in a circuit. (10-2-3 and 10-3-4)
23. How is the ohms-per-volt rating of a voltmeter obtained? (10-3-2)
24. What is the difference between an earth ground and a chassis ground? (10-3-5)
25. What is a multimeter or VOM? (10-5)

Practice Problems

26. At what position would the pointer be (1 to 10 scale) if 15 μA flows through a 60 μA meter movement?

27. If a voltmeter on the 1000-V range has a sensitivity of 60,000 Ω/V, what is its internal resistance?

28. For a 50 μA, 100-Ω ammeter meter movement, determine the value of shunt resistors for the:

a. 100 mA range **b.** 100 μA range

29. What is the value of multiplier resistors for a 50,000 Ω/V voltmeter on the:

a. 10-V range? **b.** 100-V range?

30. If an ohmmeter's pointer points to 60 on the $R \times 100$ range, the resistance of the resistor being measured is equal to how many ohms?

31. A voltmeter has a sensitivity of 20,000 Ω/V. Calculate the resistance of the voltmeter if the following voltage scales are being used:

a. 25 V **b.** 100 V **c.** 1.5 V

32. Calculate the shunt resistances needed for the ranges B, C, and D of the ammeter shown in Figure 10-9 if $I_m = 100$ μA and $R_m = 1$ kΩ.

33. Calculate the multiplier resistances needed for the voltmeter shown in Figure 10-14 if $I_m = 50$ μA and $R_m = 2000$ Ω.

34. Sketch the schematic for the following meters and indicate resistor values.

a. Ammeter $I_m = 50$ μA, $R_m = 1500$ Ω. Four range scales: 100 μA, 100 mA, 1 A, and 10 A.

b. Voltmeter $I_m = 0.5$ mA, $R_m = 1500$ Ω. Four range scales: 1 V, 3 V, 10 V, 30 V.

c. Ohmmeter $I_m = 1$ mA, $R_m = 150$ Ω, battery = 1.5 V, $R_{zero} = ?$

35. To double the current and voltage ranges of the meters in Question 34(a) and (b), what changes would have to occur?

CHAPTER 11

OBJECTIVES

After completing this chapter, you will be able to:

1. Explain the difference between alternating current and direct current.
2. Describe how ac is used to deliver power and to represent information.
3. Give the three advantages that ac has over dc from a power point of view.
4. Describe basically the ac power distribution system from the electric power plant to the home or industry.
5. Describe the three waves used to carry information between two points.
6. Explain how the three basic information carriers are used to carry many forms of information on different frequencies within the frequency spectrum.
7. Explain the different characteristics of the five basic wave shapes.
8. Describe fundamental and harmonic frequencies.
9. Define the differences between electrical equipment and electronic equipment.

Alternating Current (AC)

OUTLINE

VIGNETTE

The Laser

In 1898, H. G. Wells's famous book, *The War of the Worlds,* had Martian invaders with laserlike death rays blasting bricks, firing trees, and piercing iron as if it were paper. In 1917, Albert Einstein stated that, under certain conditions, atoms or molecules could absorb light and then be stimulated to release this borrowed energy. In 1954, Charles H. Townes, a professor at Columbia University, conceived and constructed with his students the first "maser" (acronym for "microwave amplification by stimulated emission of radiation"). In 1958, Townes and Arthur L. Shawlow wrote a paper showing how stimulated emission could be used to amplify light waves as well as microwaves, and the race was on to develop the first "laser." In 1960, Theodore H. Maiman, a scientist at Hughes Aircraft Company, directed a beam of light from a flash lamp into a rod of synthetic crystal, which responded with a burst of crimson light so bright that it outshone the sun.

An avalanche of new lasers emerged, some as large as football fields, while others were no bigger than a pinhead. They can be made to produce invisible infrared or ultraviolet light or any visible color in the rainbow, and the high-power lasers can vaporize any material a million times faster and more intensely than a nuclear blast, while the low-power lasers are safe to use in children's toys.

At present, the laser is being used by the FBI to detect fingerprints that are 40 years old, in defense programs, in compact disk players, in underground fiber optic communication to transmit hundreds of telephone conversations, to weld car bodies, to drill holes in baby-bottle nipples, to create three-dimensional images called holograms, and as a surgeon's scalpel in the operating room. Not a bad beginning for a device that when first developed was called "a solution looking for a problem."

INTRODUCTION

In this chapter you will be introduced to alternating current (ac). While direct current (dc) can be used in many instances to deliver power or represent information, there are certain instances in which ac is preferred. For example, ac is easier to generate and is more efficiently transmitted as a source of electrical power for the home or business. Audio (speech and music) and video (picture) information are generally always represented in electronic equipment as an alternating current or alternating voltage signal.

In this chapter, we will begin by describing the difference between dc and ac, and then examine where ac is used. Following this we will discuss all the characteristics of ac waveform shapes, and finally, the differences between electricity and electronics.

11-1 THE DIFFERENCE BETWEEN DC AND AC

One of the best ways to describe anything new is to begin by redescribing something known and then discuss the unknown. The known topic is **direct current.** Direct current (dc) is the flow of electrons in one DIRECTion and one direction

Direct Current (dc)
Current flow in only one direction.

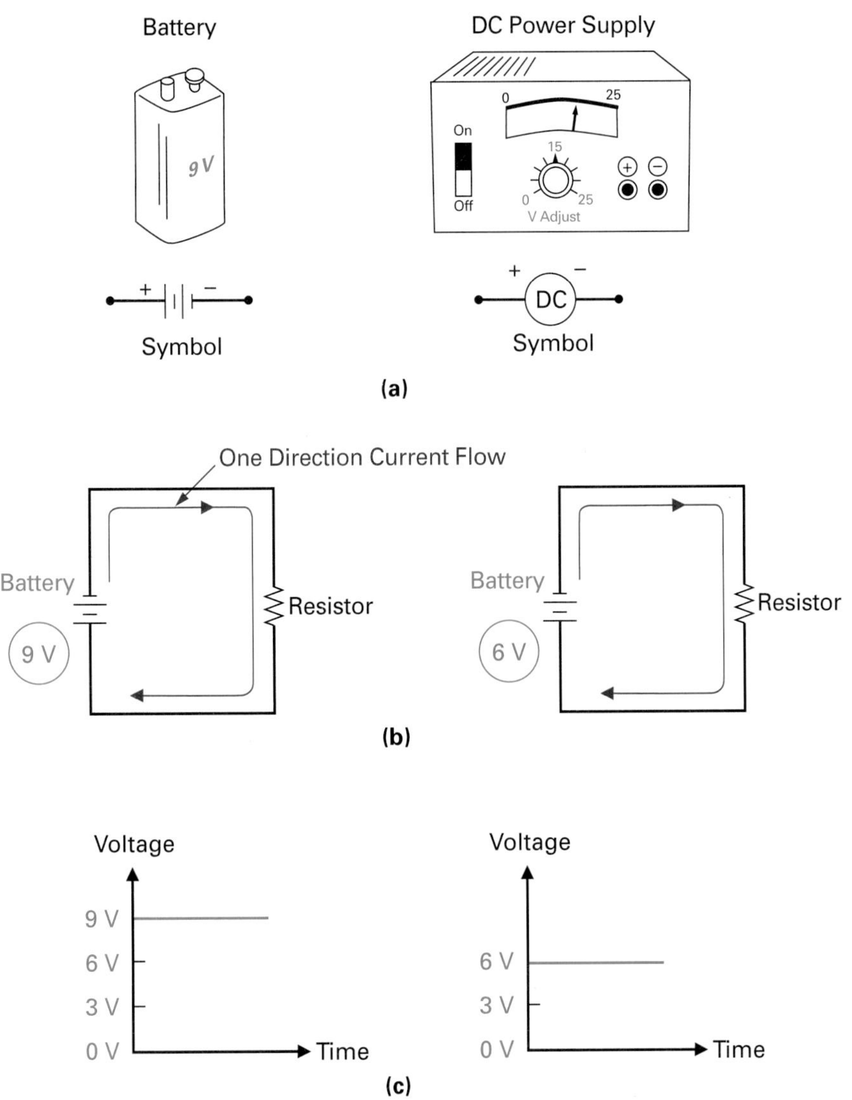

FIGURE 11-1 Direct Current. (a) DC Sources. (b) DC Flow. (c) Graphical Representation of DC.

only. Dc voltage is nonvarying and normally obtained from a battery or power supply unit, as seen in Figure 11-1(a). The only variation in voltage from a battery occurs due to the battery's discharge, but even then, the current will still flow in only one direction, as seen in Figure 11-1(b).

A dc voltage of 9 or 6 V could be illustrated graphically as shown in Figure 11-1(c). Whether 9 or 6 V, the voltage can be seen to be constant or the same at any time.

Some power supplies supply a form of dc known as pulsating dc, which varies periodically from zero to a maximum, back to zero, and then repeats. Figure 11-2(a) illustrates the physical appearance and schematic diagram of a battery charger that is connected across two series resistors. The battery charger is generating a waveform, as shown in Figure 11-2(b), known as pulsating dc. At time 1 [Figure 11-2(c)], the power supply is generating 9 V and direct current is

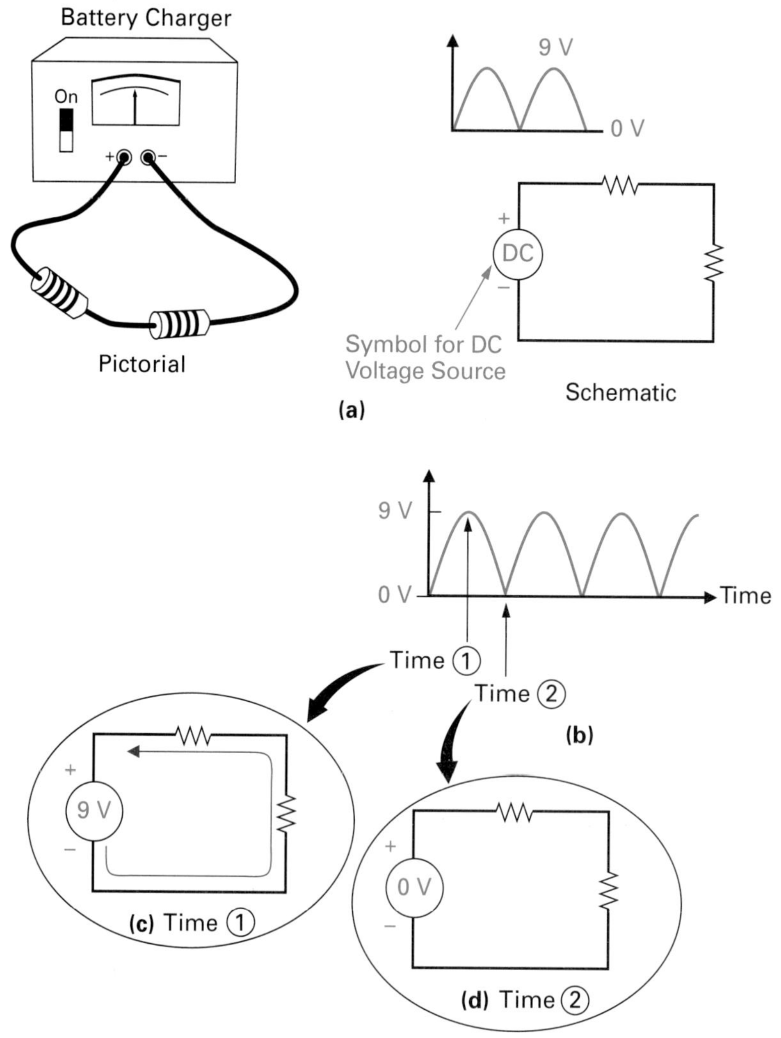

FIGURE 11-2 **Pulsating DC.**

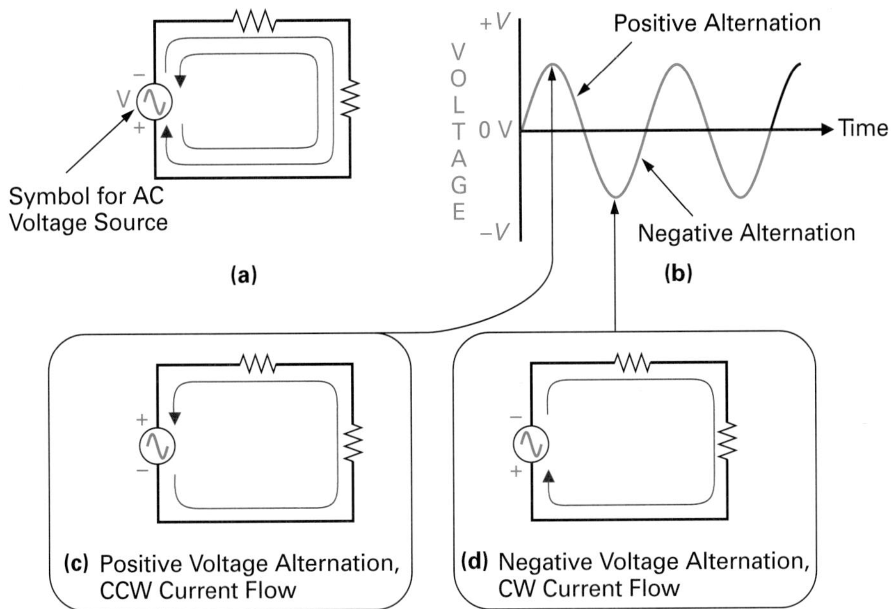

FIGURE 11-3 **Alternating Current.**

flowing from negative to positive. At time 2 [Figure 11-2(d)], the power supply is producing 0 V, and therefore no current is being produced. In between time 1 and time 2, the voltage out of the power supply will decrease from 9 V to 0 V; however, no matter what the voltage, whether 8, 7, 6, 5, 4, 3, 2, or 1 V, current will be flowing in only one direction (unidirectional) and is therefore referred to as dc.

Pulsating dc is normally supplied by a battery charger and is used to charge secondary batteries. It is also used to operate motors that convert the pulsating dc electrical energy into a mechanical rotation output. Whether steady or pulsating, direct current is current in only one DIRECTion.

Alternating current (ac) flows first in one direction and then in the opposite direction. This reversing current is produced by an alternating voltage source, as shown in Figure 11-3(a), which reaches a maximum in one direction (positive), decreases to zero, and then reverses itself and reaches a maximum in the opposite direction (negative). This is graphically illustrated in Figure 11-3(b).

Alternating Current
Electric current that rises from zero to a maximum in one direction, falls to zero, and then rises to a maximum in the opposite direction, and then repeats another cycle, the positive and negative alternations being equal.

During the time of the positive voltage alternation, the polarity of the voltage will be as shown in Figure 11-3(c), so current will flow from negative to positive in a counterclockwise direction.

During the time of the negative voltage alternation, the polarity of the voltage will reverse, as shown in Figure 11-3(d), causing current to flow once again from negative to positive, but in this case, in the opposite clockwise direction.

SELF-TEST REVIEW QUESTIONS FOR INTRODUCTION AND SECTION 11-1

1. Give the full names of the following abbreviations: (a) ac; (b) dc.
2. Is the pulsating waveform generated by a battery charger considered to be ac or dc? Why?
3. The polarity of a/an ____________ voltage source will continually reverse, and therefore so will the circuit current.
4. The polarity of a/an ____________ voltage source will remain constant, and therefore current will flow in only one direction.
5. List the two main applications of ac.
6. State briefly the difference between ac and dc.

11-2 WHY ALTERNATING CURRENT?

The question that may be troubling you at this point is: If we have been managing fine for the past chapters with dc, why do we need ac?

There are two main applications for ac:

1. *Power transfer:* to supply electrical power for lighting, heating, cooling, appliances, and machinery in both home and industry
2. *Information transfer:* to communicate or carry information, such as radio music and television pictures, between two points

To begin with, let us discuss the first of these applications, power transfer.

11-2-1 Power Transfer

There are three advantages that ac has over dc from a power point of view. Let us now discuss these in detail.

Generator
Device used to convert a mechanical energy input into an electrical energy output.

1. Flashlights, radios, and portable televisions all use batteries (dc) as a source of power. In these applications where a small current is required, batteries will last a good length of time before there is a need to recharge or replace them. Many appliances and most industrial equipment need a large supply of current, and in this situation a **generator** would have to be used to generate this large amount of current. Generators operate in the opposite way to motors, in that a generator converts a mechanical rotation input into an electrical output. Generators can be used to generate either dc or ac, but ac generators can be larger, less complex internally, and cheaper to operate, which are the first reasons why we use ac instead of dc for supplying power. Figure 11-4 illustrates a typical ac generator.
2. From a power point of view, ac is always used by electric companies when transporting power over long distances to supply both the home and industry with electrical energy. Recalling the power formula, you will remember

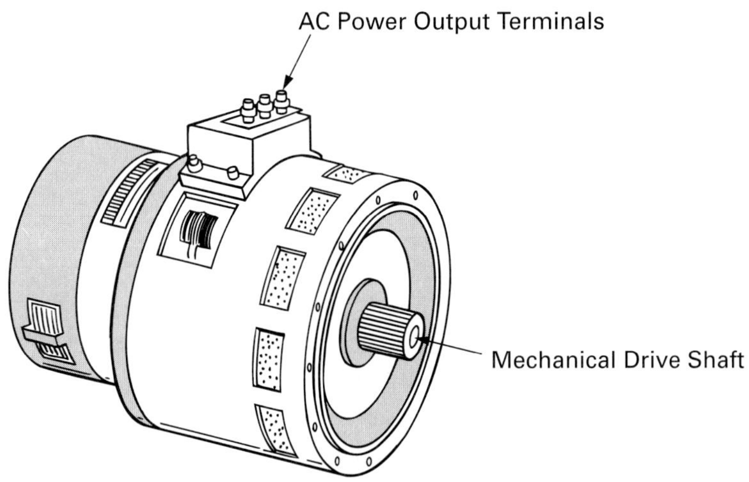

FIGURE 11-4 **AC Generator.**

that power is proportional to either current or voltage squared ($P \propto I^2$ or $P \propto V^2$), which means that to supply power to the home or industry, we would supply either a large current or voltage. As you can see in Figure 11-5, between the electric power plant and home or industry are power lines carrying the power. The amount of power lost (heat) in these power lines can be calculated by using the formula $P = I^2 \times R$, where I is the current flowing through the line and R is the resistance of the power lines. This means that the larger the current, the greater the amount of power lost in the lines in the form of heat and therefore the less the amount of power supplied to the home or industry. For this reason, power companies transport electric energy at a very high voltage between 200,000 and 600,000 V. Since the voltage is high, the current can be low and provide the same amount of power to the consumer. Yet, by keeping the current low, the amount of heat loss generated in the power lines is minimal.

Now that we have discovered why it is more efficient over a long distance to transport high voltages than high current, what does this have to do with ac? An ac voltage can easily and efficiently be transformed up or down to a higher or lower voltage by utilizing a device known as a **transformer**, and even though dc voltages can be stepped up and down, the method is inefficient and more complex.

Transformer Device consisting of two or more coils that are used to couple electric energy from one circuit to another, yet maintain electrical isolation between the two.

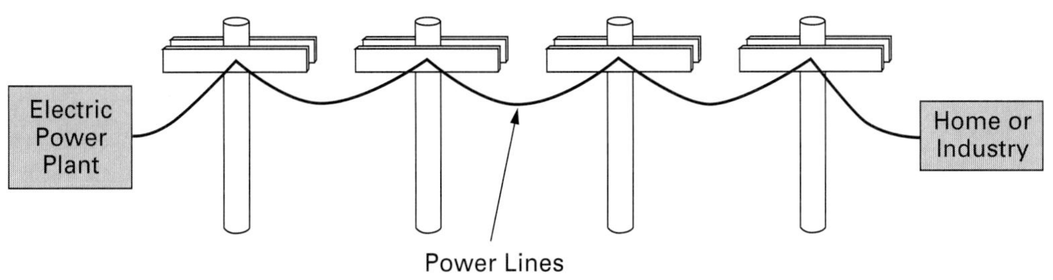

FIGURE 11-5 **Power Lines Connect Power from the Electric Company to Home and Industry.**

3. Nearly all electronic circuits and equipment are powered by dc voltages, which means that once the ac power arrives at the home or industry, in most cases it will have to be converted into dc power to operate electronic equipment. It is a relatively simple process to convert ac to dc; however, conversion from dc to ac is a complex and comparatively inefficient process.

Figure 11-6 illustrates ac power distribution from the electric power plant to the home and industry. The ac power distribution system begins at the electric power plant, which has the powerful large generators driven by turbines to generate large ac voltages. The turbines can be driven by either falling water (hydroelectric), or from steam, which is produced with intense heat by burning either coal, gas, or oil or from a nuclear reactor (thermoelectric). The turbine supplies the mechanical energy to the generator, to be transformed into ac electrical energy.

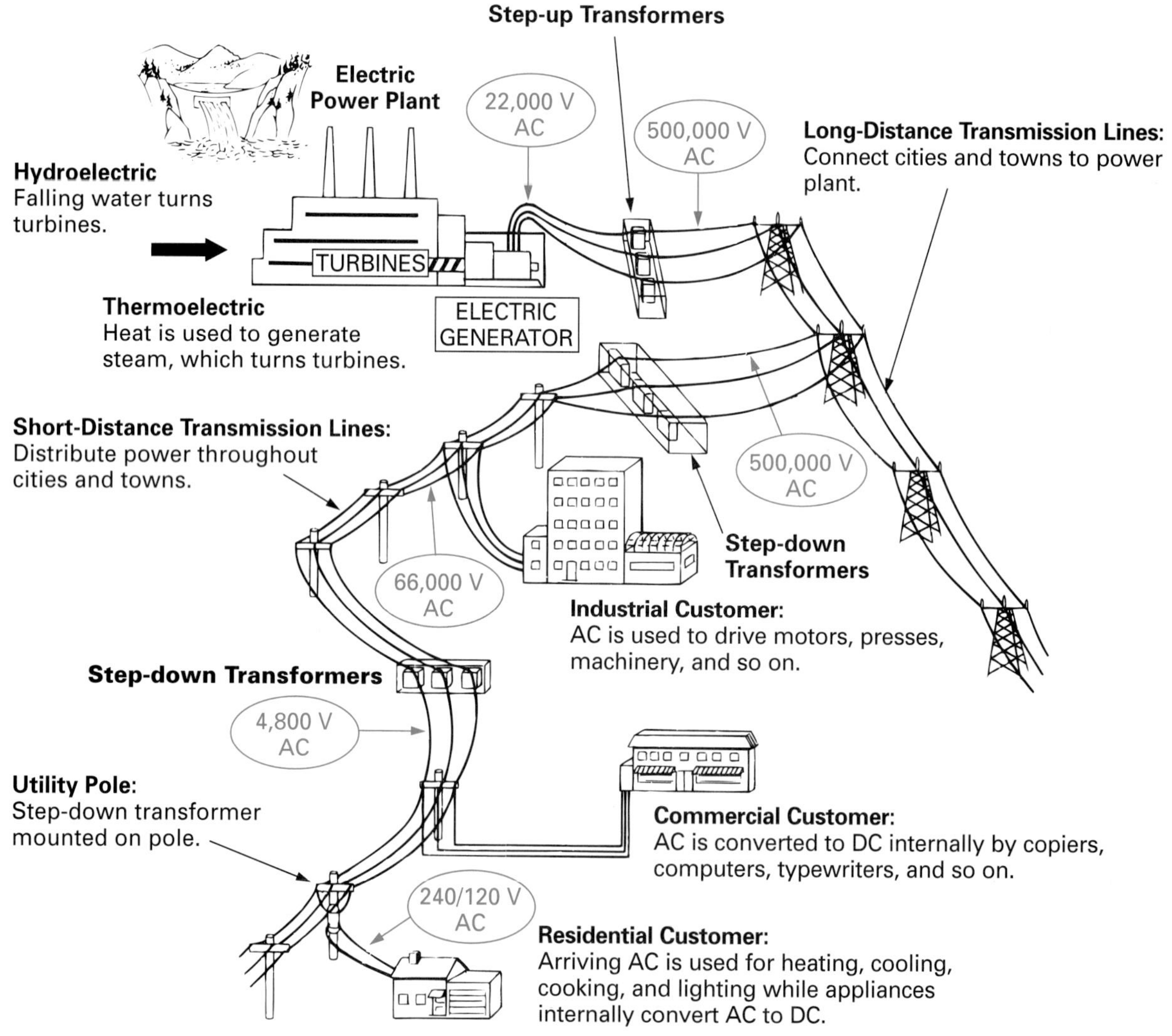

FIGURE 11-6 AC Power Distribution.

The generator generates an ac voltage of approximately 22,000 V, which is stepped up by transformers to approximately 500,000 V. This voltage is applied to the long-distance transmission lines, which connect the power plant to the city or town. At each city or town, the voltage is tapped off the long-distance transmission lines and stepped down to approximately 66,000 V and is distributed to large-scale industrial customers. The 66,000 V is stepped down again to approxi-

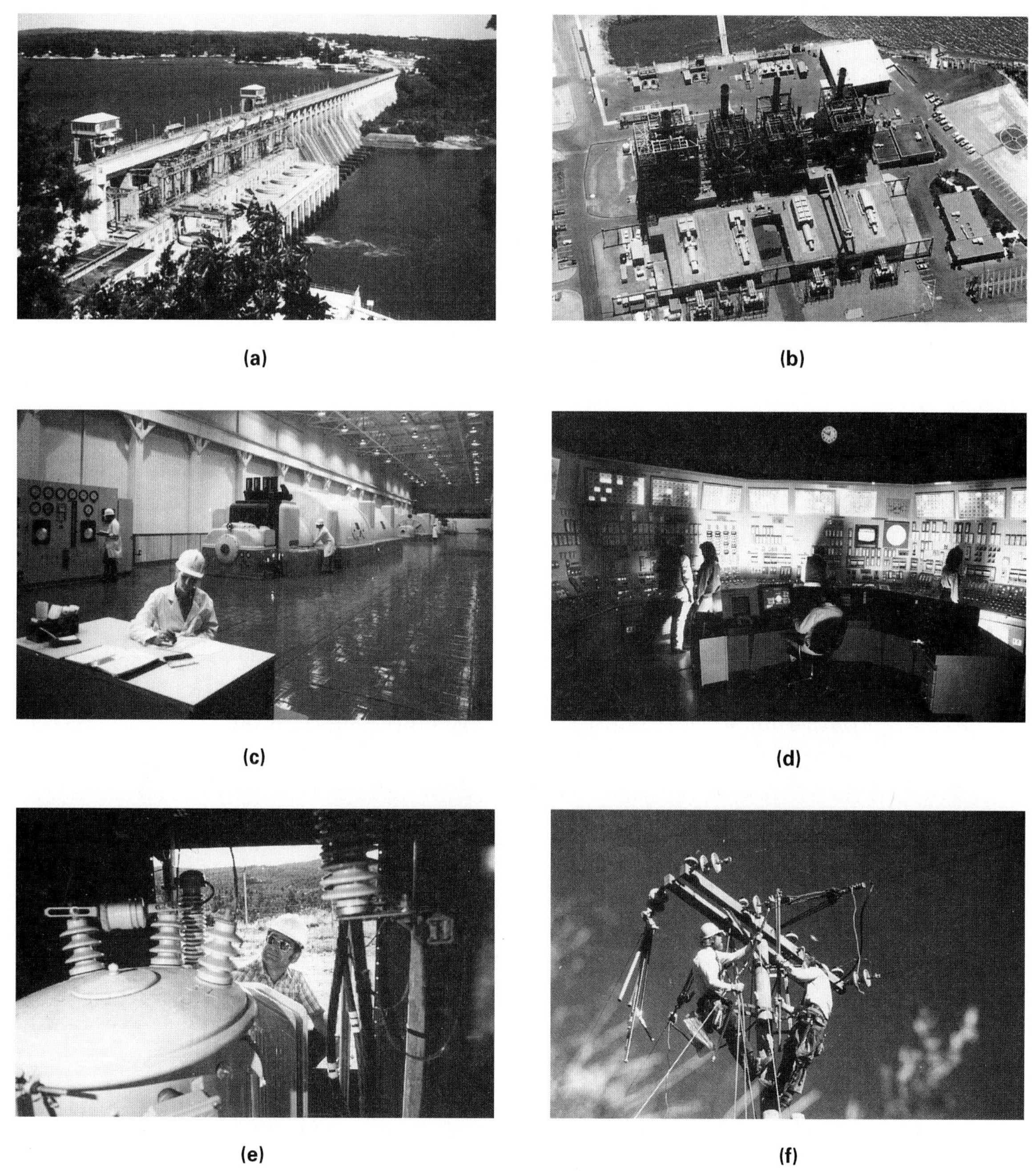

(a) (b) (c) (d) (e) (f)

FIGURE 11-7 AC Power Distribution. (a) Hydroelectric Dam. (b) Electric Power Plant. (c) Generator. (d) Control Room. (e) Step-down Transformer. (f) Utility Pole.

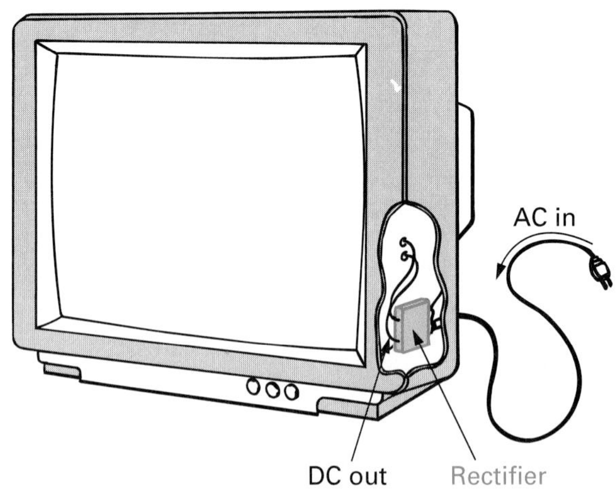

FIGURE 11-8 Rectification (Converting AC to DC) by Appliances.

mately 4800 V and distributed throughout the city or town by short-distance transmission lines. This 4800 V is used by small-scale industrial customers and residential customers who receive the ac power via step-down transformers on utility poles, which step down the 4800 V to 240 V and 120 V. The appearance of some of the devices that make up the power distribution system can be seen in Figure 11-7.

A large amount of equipment and devices within industry and the home will run directly from the ac power, such as heating, lighting, and cooling. Some equipment that runs on dc, such as televisions and computers, will accept the 120 V ac and internally convert it to the dc voltages required. Figure 11-8 illustrates a TV set that is operating from the 120 V ac from the wall outlet and internally converting it to dc so that it can power the circuits necessary to produce both audio (sound) and video (picture) information. This unit that converts ac to dc is called a **rectifier.**

Rectifier
Device that achieves rectification.

SELF-TEST REVIEW QUESTIONS FOR SECTION 11-2-1

1. In relation to power transfer, what three advantages does ac have over dc?
2. True or false: A generator converts an electrical input into a mechanical output.
3. What formula is used to calculate the amount of power lost in a transmission line?
4. What is a transformer?
5. What voltage is provided to the wall outlet in the home?
6. Most appliances internally convert the ____________ input voltage into a ____________ voltage.

11-2-2 *Information Transfer*

Information, by definition, is the property of a signal or message that conveys something meaningful to the recipient. **Communication,** which is the transfer of information between two points, began with speech and progressed to handwritten words in letters and printed words in newspapers and books. To achieve greater distances of communication, face-to-face communications evolved into telephone and radio communications.

Communication Transmission of information between two points.

A simple communication system can be seen in Figure 11-9(a). The audio or **sound waves** generated when the person on the left speaks are converted to an equivalent electrical signal or wave by the microphone. This electrical wave is fed into an amplifier, which boosts the magnitude of the electrical signal. The enlarged electrical signal emerging out of the amplifier circuit is then converted back to the original sound waves of larger amplitude by the speaker, where they are then received by the recipient.

Sound Wave Traveling wave propogated in an elastic medium that travels at a speed of approximately 1133 ft/s.

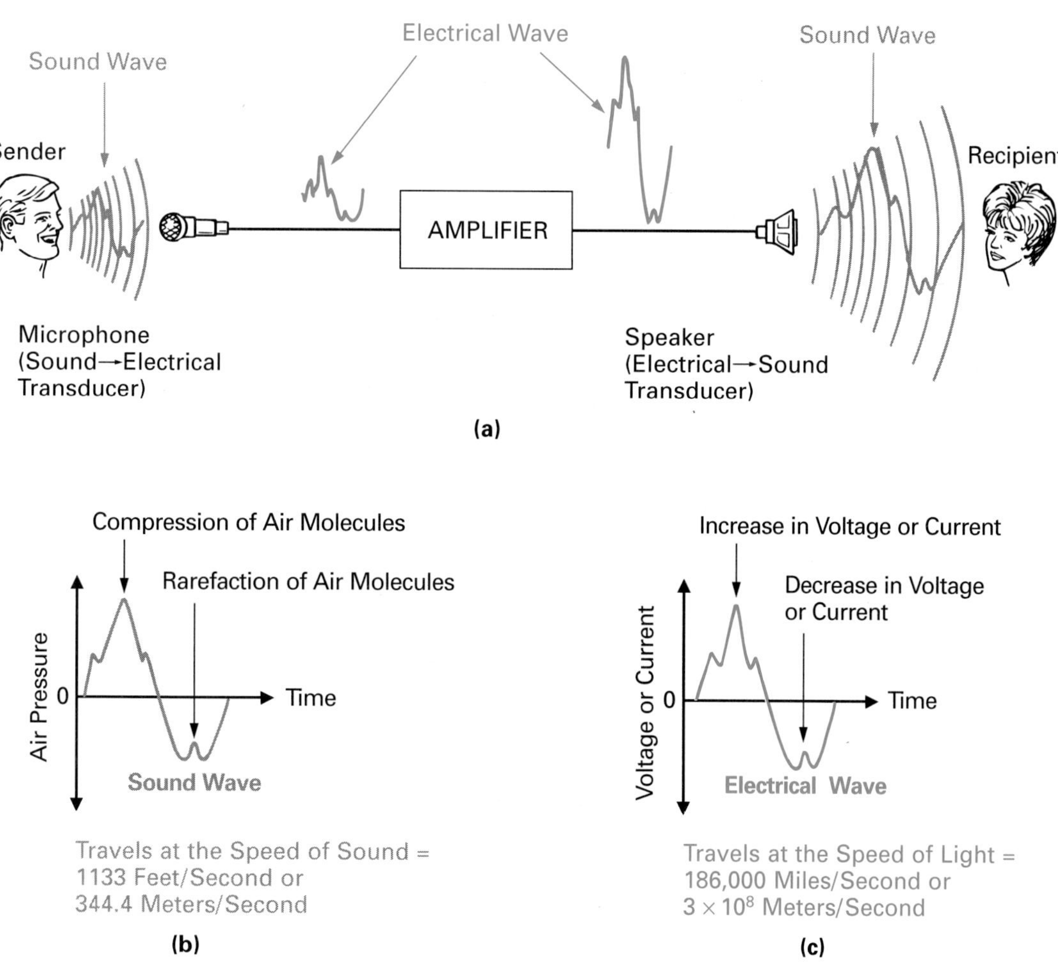

FIGURE 11-9 **Information Transfer. (a) Wire Communication System. (b) Sound Wave. (c) Electrical Wave.**

Transducer
Any device that converts energy from one form to another.

Electrical Wave
Traveling wave propagated in a conductive medium that is a variation in voltage or current and travels at slightly less than the speed of light.

The voice information or sound wave produced by the sender is a variation in air pressure, as shown in Figure 11-9(b), and travels at the speed of sound. Sound waves or sounds are normally generated by a vibrating reed or plucked string in the case of musical instruments. In this example the sender's vocal cords vibrate backward and forward, producing a rarefaction or decreased air pressure, where few air molecules exist, and a compression or increased air pressure, where many air molecules exist. Like the ripples produced by a stone falling in a pond, the sound waves produced by the sender are constantly expanding and traveling outward.

The microphone is in fact a **transducer** (energy converter), because it converts the sound wave (which is a form of mechanical energy) into electrical energy in the form of voltage and current, which varies in the same manner as the sound wave and therefore contains the sender's message or information.

The **electrical wave,** shown in Figure 11-9(c), is a variation in voltage or current and can only exist in a wire conductor or circuit. This electrical signal

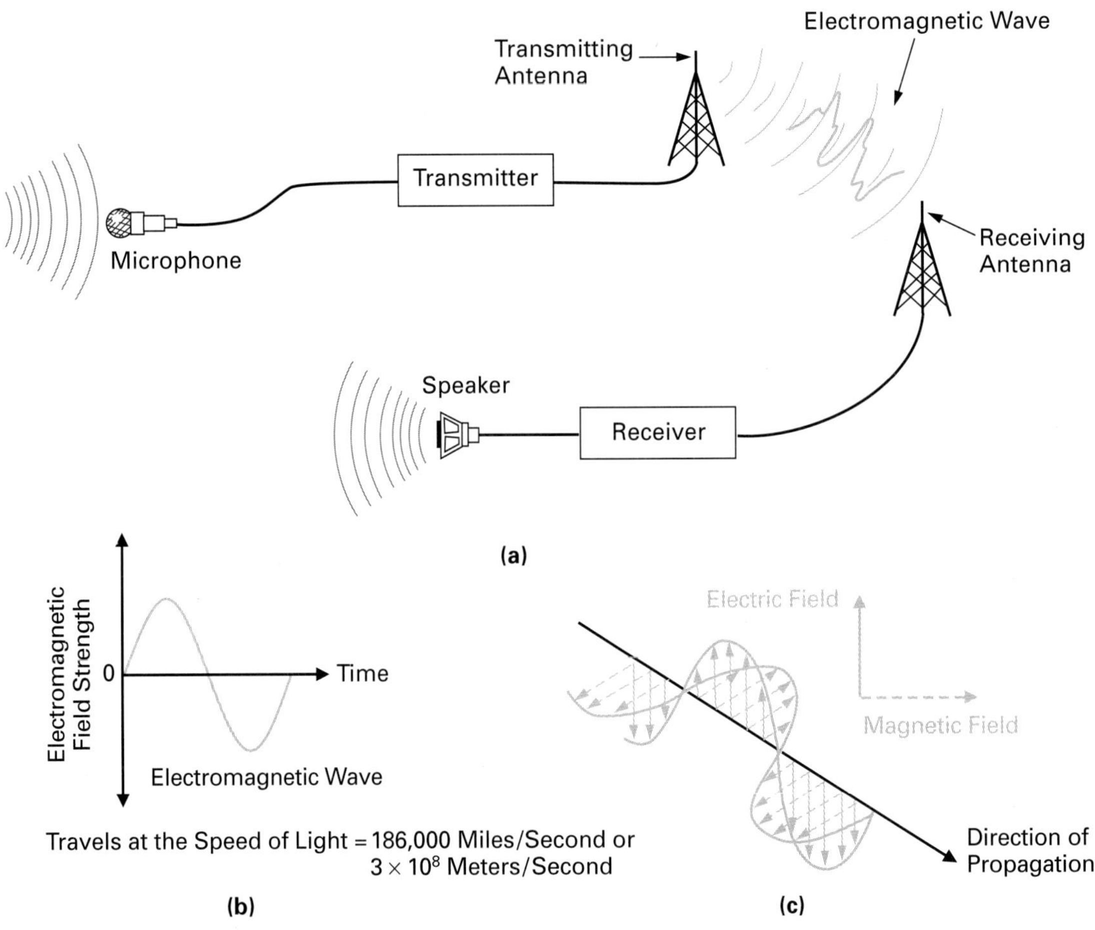

FIGURE 11-10 **Information Transfer. (a) Wireless Communication System. (b) Simplified Electromagnetic Wave. (c) Components of an Electromagnetic Wave.**

travels at the speed of light, and after being increased in magnitude by the amplifier, it is applied to a loudspeaker.

The speaker, like the microphone, is also an electroacoustical transducer that converts the electrical energy input into a mechanical soundwave output. These sound waves strike the outer eardrum, causing the ear diaphragm to vibrate, and these mechanical vibrations actuate nerve endings in the ear, which convert the mechanical vibrations into electrochemical impulses that are sent to the brain. The brain decodes this information by comparing these impulses with a library of previous sounds and so provides the sensation of hearing.

To communicate between two distant points, a wire must be connected between the microphone and speaker. However, if an electrical wave is applied to an antenna, the electrical wave is converted into a radio or electromagnetic wave, as shown in Figure 11-10(a), and communication is established without the need of a connecting wire—hence the term **wireless communication.** Antennas are designed to radiate and receive electromagnetic waves, which vary in field strength, as shown in Figure 11-10(b), and can exist in either air or space. These radio waves, as they are also known, travel at the speed of light and allow us to achieve great distances of communication.

Wireless Communication
Term describing radio communication that requires no wires between the two communicating points.

More specifically, they are composed of two basic components, as shown in Figure 11-10(c). The electrical voltage applied to the antenna is converted into an electric field and the electrical current into a magnetic field. This **electromagnetic** (electric–magnetic) **wave** is used to carry a variety of information, such as speech, radio broadcasts, television signals, and so on.

Electromagnetic (radio) wave
Wave that consists of both an electric and magnetic variation, and travels at the speed of light.

In summary, the sound wave is a variation in air pressure, the electrical wave is a variation of voltage or current, and the electromagnetic wave is a variation of electric and magnetic field strength, as seen in Figure 11-11.

EXAMPLE:

How long will it take the sound wave produced by a rifle shot to travel 9630.5 feet?

Solution:

This problem makes use of the following formula:

$$\text{Distance} = \text{velocity} \times \text{time}$$

or

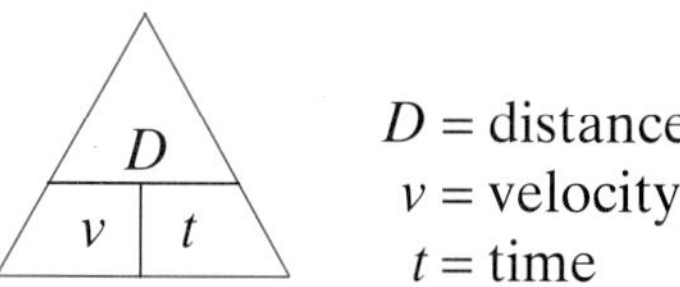

D = distance
v = velocity
t = time

If someone travels at 20 mph for 2 hours, the person will travel 40 miles ($D = v \times t = 20 \text{ mph} \times 2 \text{ hours} = 40 \text{ miles}$). In this problem, the distance (9630.5) and the

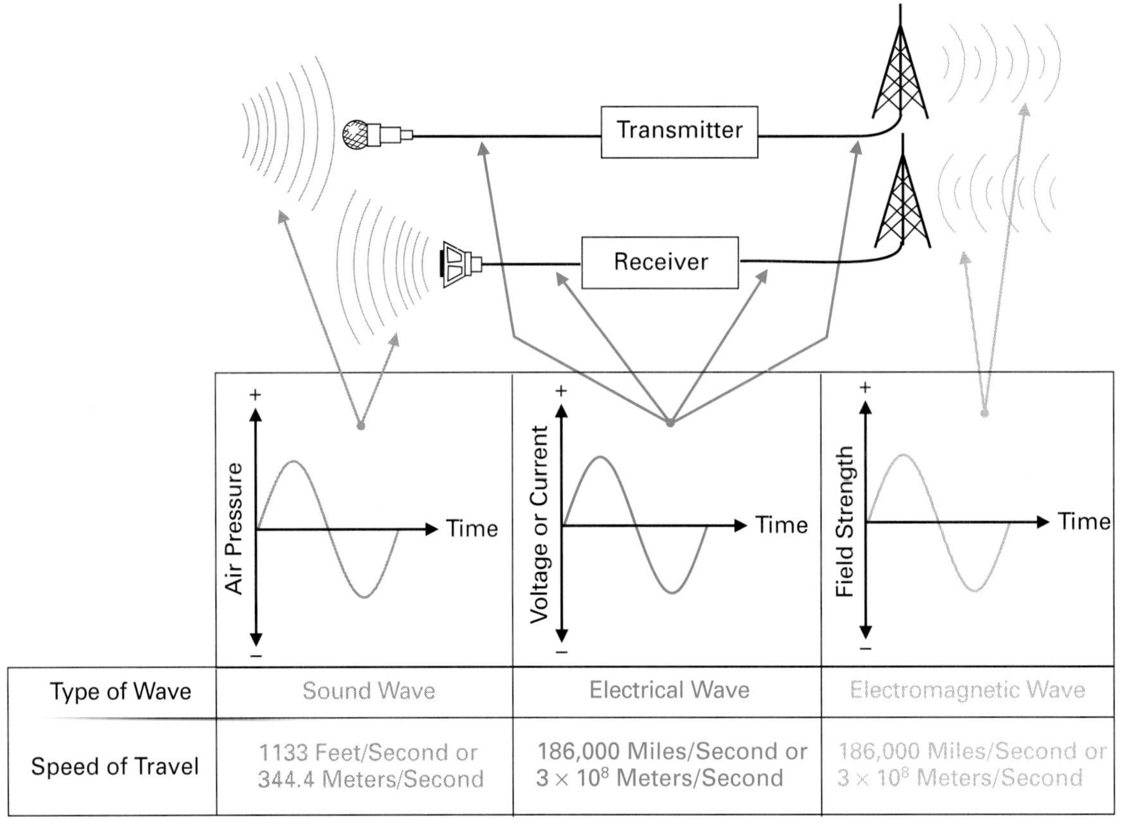

FIGURE 11-11 Summary of the Sound, Electrical, and Electromagnetic Waves.

sound wave's velocity (1130 ft/sec) are known. By rearranging the formula, we can find time:

$$\text{time} = \frac{\text{distance}}{\text{velocity}} = \frac{9630.5 \text{ ft}}{1130 \text{ ft/s}} = 8.5 \text{ s}$$

EXAMPLE:

How long will it take an electromagnetic (radio) wave to reach a receiving antenna that is 2000 miles away from the transmitting antenna?

Solution:

In this problem, both distance (2000 miles) and velocity (186,000 miles/s) are known, and time has to be calculated:

$$\text{time} = \frac{\text{distance}}{\text{velocity}} = \frac{2000 \text{ miles}}{186{,}000 \text{ miles/s}}$$

$$= 1.075 \times 10^{-2} \quad \text{or} \quad 10.8 \text{ ms}$$

SELF-TEST REVIEW QUESTIONS FOR SECTION 11-2-2

1. Define *information* and *communication.*
2. The ____________ wave is a variation in air pressure, the ____________ wave is a variation in *field* strength, and the ____________ wave is a variation of voltage or current.
3. Which of the three waves described in Question 2:
   - **a.** Can only exist in the air?
   - **b.** Can exist in either air or a vacuum?
   - **c.** Exists in a wire conductor?
4. Sound waves travel at the speed of sound, which is ____________, while electrical and electromagnetic waves travel at the speed of light, which is ____________.
5. A human ear is designed to receive ____________ waves, an antenna is designed to transmit or receive ____________ waves, and an electronic circuit is designed to pass only ____________ waves.
6. Give the names of the following energy converters or transducers:
   - **a.** Sound wave (mechanical energy) to electrical wave
   - **b.** Electrical wave to sound wave
   - **c.** Electrical wave to electromagnetic wave
   - **d.** Sound wave to electrochemical impulses

MINI-MATH REVIEW

This mini-math review is designed to remind you of the mathematical details relating to **trigonometry**, which will be used in following sections and chapters.

Trigonometry
The study of the properties of triangles and trigonometric functions and of their applications.

Introduction to Trigonometry

Electronic technicians, electronic engineers, mechanical engineers, electricians, carpenters, architects, navigators, and many other people in technical trades make use of trigonometry on a daily basis. Although volumes of theory are available about this subject, most of it is not that useful to us for everyday applications. In this mini-math review we will concentrate on the more practical part of trigonometry—the **right-angle triangle.**

Right-Angle Triangle
A triangle having a 90° or square corner.

The 3, 4, 5 Right-Angle Triangle

Like all triangles, the right-angle triangle or right triangle has three sides and three corners. Its distinguishing feature, however, is that two of the sides of this triangle are at right angles (at 90°) to one another, as shown in Figure

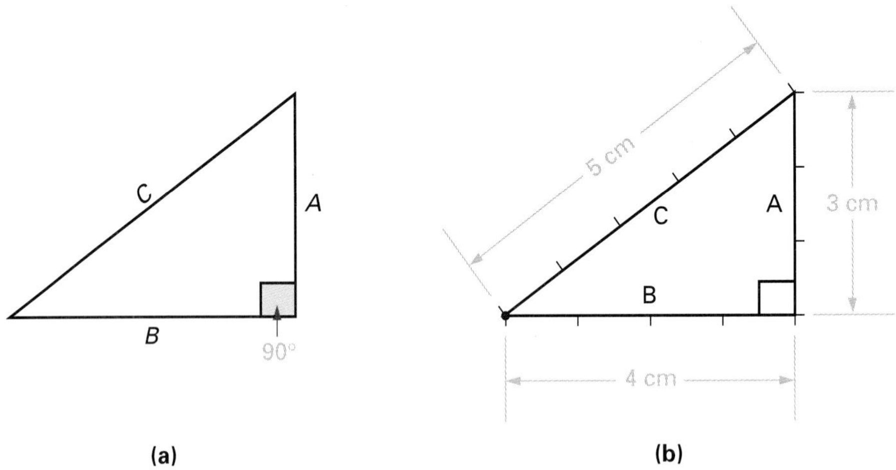

FIGURE 11-12 **Right-Angle (3, 4, 5) Triangle.**

11-12(a). The small square box within the triangle is placed in the corner to show that sides A and B are *square* or at *right angles* to one another.

If you study this right triangle you may notice two interesting facts about the relative lengths of sides A, B, and C. These observations are:

1. Side C is always longer than side A or side B.
2. The total length of side A and side B combined is always longer than side C.

The right triangle in Figure 11-12(b) has been drawn to scale to demonstrate another interesting fact about this triangle. If side A were to equal 3 cm and side B was made to equal 4 cm, side C would equal 5 cm. This demonstrates a basic relationship among the three sides and accounts for why this right triangle is sometimes referred to as a *3,4,5 triangle.* As long as the relative lengths remain the same, it makes no difference whether the sides are 3 cm, 4 cm, and 5 cm, or 30 km, 40 km, and 50 km. Unfortunately, in most applications our lengths of A, B, and C will not work out as easy as 3, 4, 5.

When a relationship exists among three quantities, we can develop a formula to calculate an unknown when two of the quantities are known. It was Pythagoras who first developed this basic formula or equation (known as the **Pythagorean Theorem**), which states:

Pythagorean Theorem
A theorem in geometry: the square of the length of the hypotenuse of a right triangle equals the sum of the squares of the lengths of the other two sides.

$$\boxed{C^2 = A^2 + B^2} \qquad \begin{pmatrix} 5^2 = 3^2 + 4^2 \\ 25 = 9 \ + 16 \end{pmatrix}$$

By using the rules of algebra, we can transpose this formula to derive formulas for C, B, and A.

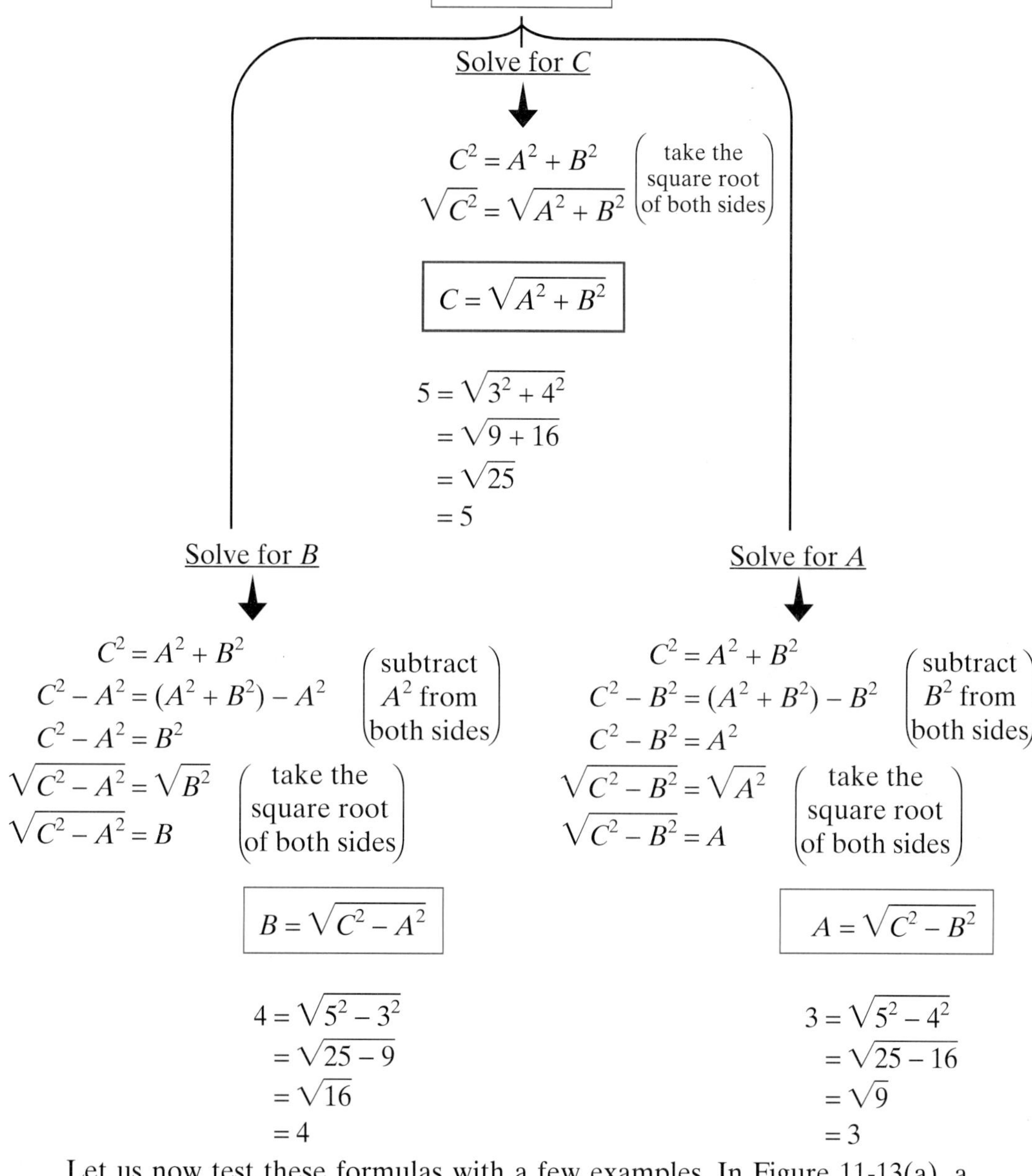

Let us now test these formulas with a few examples. In Figure 11-13(a), a 4 ft ladder has been placed in a position 2 ft from a wall. How far up the wall will the ladder reach?

In this example, A is unknown, and therefore

$$\begin{aligned} A &= \sqrt{C^2 - B^2} \\ &= \sqrt{4^2 - 2^2} \\ &= \sqrt{16 - 4} \\ &= \sqrt{12} \\ &= 3.46 \quad \text{or} \quad 3.5 \text{ ft} \end{aligned}$$

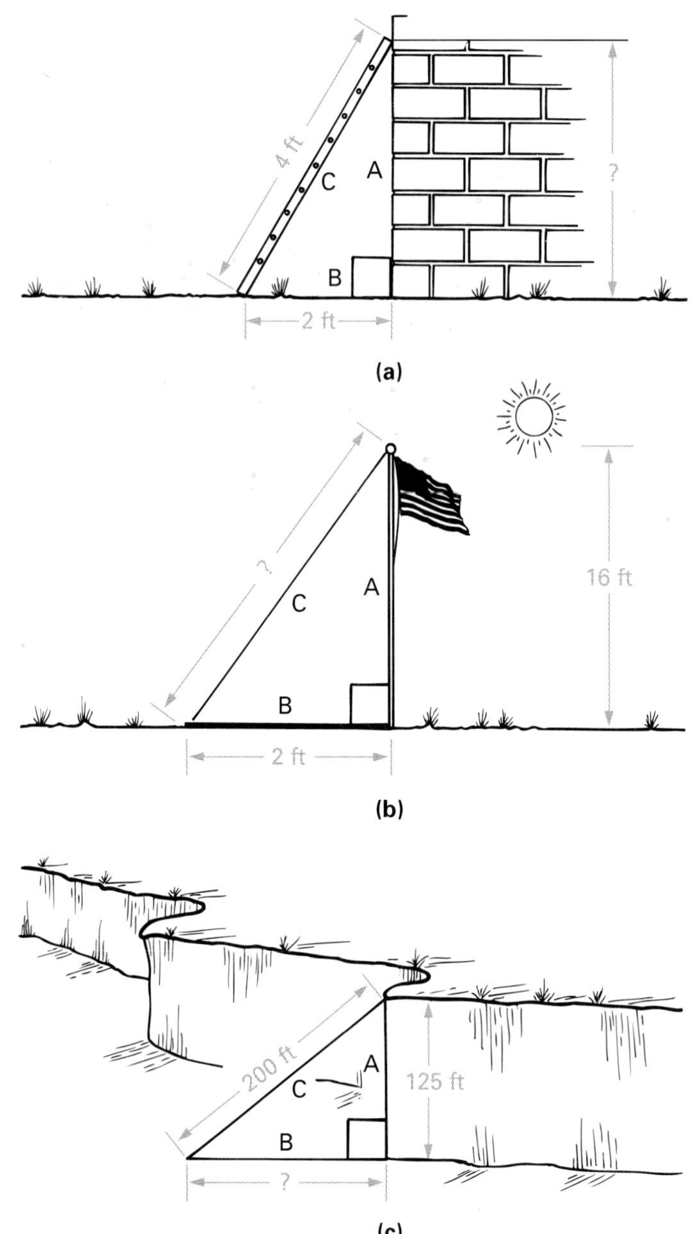

FIGURE 11-13 Right-Triangle Examples.

In Figure 11-13(b), a 16 ft flagpole is casting a 12 ft shadow. What would be the distance from the end of the shadow to the top of the flagpole?

In this example, C is unknown, and therefore

$$\begin{aligned} C &= \sqrt{A^2 + B^2} \\ &= \sqrt{16^2 + 12^2} \\ &= \sqrt{256 + 144} \\ &= \sqrt{400} \\ &= 20 \text{ ft} \end{aligned}$$

In Figure 11-13(c), a 200 ft piece of string is stretched from the top of a 125 ft cliff to a point on the beach. What would be the distance from this point to the cliff?

In this example, B is unknown, and therefore

$$\begin{aligned} B &= \sqrt{C^2 - A^2} \\ &= \sqrt{200^2 - 125^2} \\ &= \sqrt{40{,}000 - 15{,}625} \\ &= \sqrt{24{,}375} \\ &= 156 \text{ ft} \end{aligned}$$

Opposite, Adjacent, Hypotenuse, and Theta

Up until this point we have called the three sides of our right triangle *A, B,* and *C.* Figure 11-14 shows the more common names given to the three sides of a right-angle triangle. Side *C,* called the *hypotenuse side*, is always the longest of the three sides. Side *B,* called the *adjacent side,* always extends between the hypotenuse and the vertical side. An angle called *theta* (θ, a Greek letter) is formed between the hypotenuse side and the adjacent side. The side that is always opposite to the angle θ is called the *opposite side.*

Angles are always measured in degrees since all angles are part of a circle, like the one shown in Figure 11-15(a). A circle is divided into 360 small sections called degrees. In Figure 11-15(b) our right triangle has been placed within the circle. In this example the triangle occupies a 45 degree (45°) section of the circle and therefore theta equals 45 degrees ($\theta = 45°$). Moving from left to right in Figure 11-15(c) you will notice that angle θ has been increased from 5° to 85°. The length of the hypotenuse (H) in all of these examples remains the same; however, the adjacent side's length decreases ($A\downarrow$) and the opposite side's length increases ($O\uparrow$) as angle θ is increased ($\theta\uparrow$). This relationship between the relative length of a triangle's sides and theta means that we do not have to know the length of two sides to calculate a third. If we have the value of just one side and the angle theta, we can calculate the length of the other two sides.

Sine of Theta (Sin θ)

In the preceding section we discovered that a relationship exists between the relative length of a triangle's sides and the angle of theta. *Sine* is a comparison between the length of the opposite side and the length of the hypotenuse. Expressed mathematically,

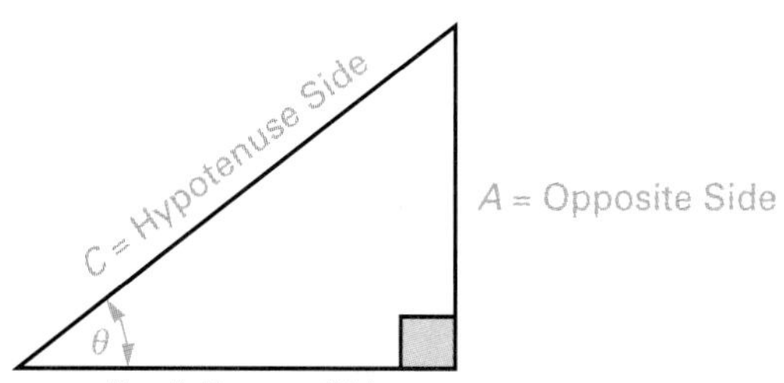

FIGURE 11-14 Names Given to the Sides of a Right Triangle.

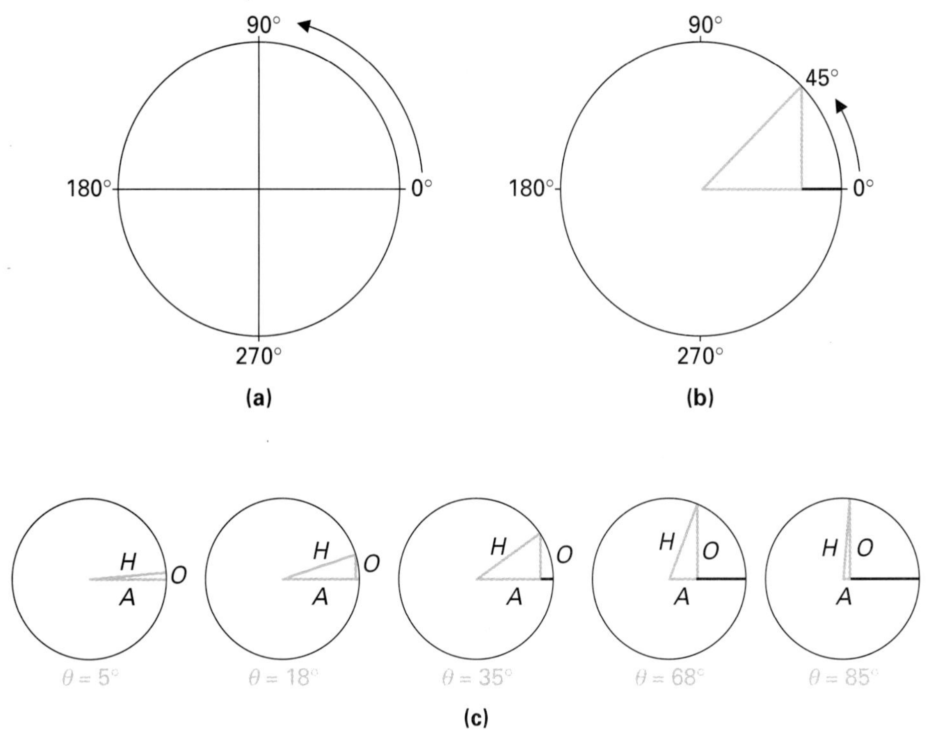

FIGURE 11-15 **Angle Theta (θ).**

$$\text{sine of theta } (\sin \theta) = \frac{\text{opposite side } (O)}{\text{hypotenuse side } (H)}$$

Since the hypotenuse is always larger than the opposite side, the result will always be less than 1 (a fraction).

Let us use this formula on a few examples to see how it works. In Figure 11-16(a), angle theta is equal to 41° while the opposite side is equal to 20 cm in length. Calculate the length of the hypotenuse.

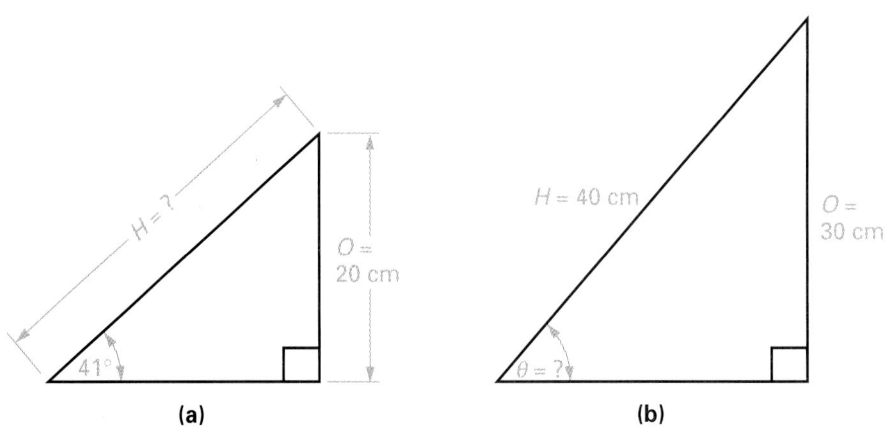

FIGURE 11-16 **Sine of Theta.**

Inserting these values in our formula, we obtain the following:

$$\text{Sin } \theta = \frac{O}{H}$$

$$\text{Sin } 41° = \frac{20 \text{ cm}}{H}$$

Looking up 41° in a sine trigonometry table or by using a scientific calculator that has all of the trigonometry tables stored permanently in its memory, you will find that the sine of 41° is 0.656. This value describes the fact that when θ = 41°, the opposite side will be 0.656, or 65.6%, as long as the hypotenuse side. By inserting this value into our formula and transposing the formula according to the rules of algebra, we can determine the length of the hypotenuse.

$$\text{Sin } 41° = \frac{20 \text{ cm}}{H} \qquad \left(\begin{array}{l}\text{calculator sequence:} \\ \quad \boxed{4}\boxed{1}\boxed{\text{SIN}}\end{array}\right)$$

$$0.656 = \frac{20 \text{ cm}}{H}$$

$$0.656 \times H = \frac{20 \text{ cm} \times \not{H}}{\not{H}} \qquad \text{(multiply both sides by } H\text{)}$$

$$\frac{\cancel{0.656} \times H}{\cancel{0.656}} = \frac{20 \text{ cm}}{0.656} \qquad \text{(divide both sides by 0.656)}$$

$$H = \frac{20 \text{ cm}}{0.656} = 30.5 \text{ cm}$$

Figure 11-16(b) illustrates another example; however, in this case the lengths of sides H and O are known but θ is not.

$$\text{Sin } \theta = \frac{O}{H}$$

$$= \frac{30 \text{ cm}}{40 \text{ cm}}$$

$$= 0.75$$

The ratio of side O to side H is 0.75, or 75%, which means that the opposite side is 75%, or 0.75, as long as the hypotenuse side. To calculate angle θ we must isolate it on one side of the equation. To achieve this we must multiply both sides of the equation by *arcsin,* or *inverse sine* (invsin), which does the opposite of sine.

$$\text{Sin } \theta = 0.75$$

$$\text{invsin (sin } \theta) = \text{invsin } 0.75$$

$$\theta = \text{invsin } 0.75 \qquad \left(\begin{array}{l}\text{calculator sequence:} \\ \boxed{.}\boxed{7}\boxed{5}\boxed{\text{INV}}\boxed{\text{SIN}}\end{array}\right)$$

$$= 48.6°$$

In summary, therefore, the sin trig functions take an angle θ and give you a number x. The inverse sin (arcsin) trig functions take a number x and give you an angle θ. In both cases, the number x is a ratio and describes what fraction the opposite side is in comparison to the hypotenuse.

Sine: angle $\theta \rightarrow$ number x

Inverse sine: number $x \rightarrow$ angle θ

Cosine of Theta (Cos θ)

As sine is a comparison between the opposite and the hypotenuse, cosine is a comparison between the adjacent and hypotenuse sides.

$$\text{cosine of theta } (\text{Cos } \theta) = \frac{\text{adjacent } (A)}{\text{hypotenuse } (H)}$$

Figure 11-17(a) illustrates an example in which the angle θ and the length of the hypotenuse are known. From this information calculate the length of the adjacent side.

$$\text{Cos } \theta = \frac{A}{H}$$

$$\text{Cos } 30° = \frac{A}{40 \text{ cm}} \quad \left(\begin{array}{c}\text{calculator sequence:} \\ \boxed{3}\boxed{0}\boxed{\text{COS}}\end{array}\right)$$

$$0.866 = \frac{A}{40 \text{ cm}}$$

Looking up the cosine of 30° you will obtain the fraction 0.866. This value states that when $\theta = 30°$, the adjacent side will always be 0.866, or 86.6%, as long as the hypotenuse. By transposing this equation we can calculate the length of the adjacent side:

$$0.866 = \frac{A}{40 \text{ cm}}$$

$$40 \times 0.866 = \frac{A}{40 \text{ cm}} \times 40 \quad \text{(multiply both sides by 40)}$$

$$A = 0.866 \times 40$$

$$= 34.64 \text{ cm}$$

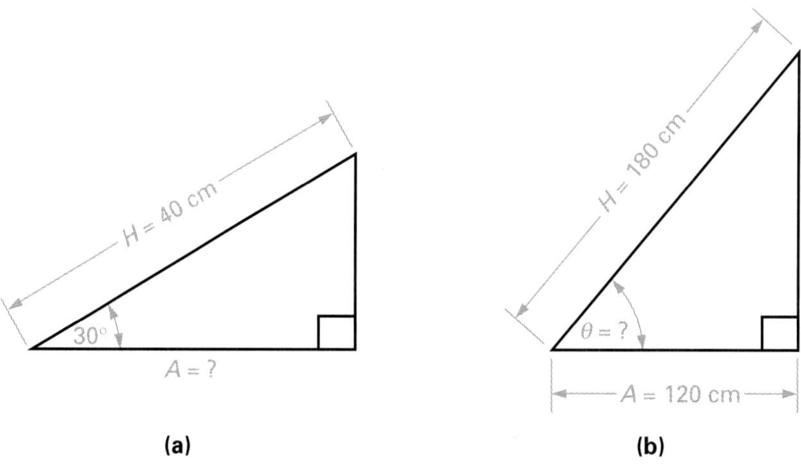

FIGURE 11-17 **Cosine of Theta.**

As another example, calculate the angle θ in Figure 11-17(b) if $A = 120$ cm and $H = 180$ cm.

$$\text{Cos } \theta = \frac{A}{H}$$

$$= \frac{120 \text{ cm}}{180 \text{ cm}}$$

$$= 0.667 \quad (A \text{ is } 66.7\% \text{ as long as } H)$$

$$\cancel{\text{invcos}}\,(\cancel{\cos}\,\theta) = \text{invcos } 0.667 \quad \text{(multiply both sides by invcos)}$$

$$\theta = \text{invcos } 0.667 \quad \left(\begin{array}{c}\text{calculator sequence:}\\ \boxed{0}\boxed{.}\boxed{6}\boxed{6}\boxed{7}\boxed{\text{INV}}\boxed{\text{COS}}\end{array}\right)$$

$$= 48.2°$$

Once again the inverse cosine trig function does the reverse operation of the cosine function, in that

Cosine: angle $\theta \rightarrow$ number x

Inverse cosine: number $x \rightarrow$ angle θ

Tangent of Theta (Tan θ)

Tangent is a comparison between the opposite side of a right triangle and the adjacent side.

$$\text{tangent of theta (Tan } \theta) = \frac{\text{opposite } (O)}{\text{adjacent } (A)}$$

Figure 11-18(a) illustrates an example in which $\theta = 65°$ and the opposite side = 43 cm. Calculate the length of the adjacent side.

$$\tan \theta = \frac{O}{A}$$

$$\tan 65° = \frac{43 \text{ cm}}{A} \quad \left(\text{calculator sequence: } \boxed{6}\boxed{5}\boxed{\text{TAN}}\right)$$

$$2.14 = \frac{43 \text{ cm}}{A} \quad \left(\begin{array}{c}\text{whenever } \theta = 65°\text{, the opposite}\\ \text{side will be 2.14 times longer}\\ \text{than the adjacent side}\end{array}\right)$$

$$2.14 \times A = \frac{43 \text{ cm}}{A} \times A \quad \text{(multiply both sides by } A)$$

$$\frac{\cancel{2.14} \times A}{\cancel{2.14}} = \frac{43 \text{ cm}}{2.14} \quad \text{(divide both sides by 2.14)}$$

$$A = \frac{43 \text{ cm}}{2.14}$$

$$= 20.1 \text{ cm}$$

As another example, calculate the angle θ in Figure 11-18(b) if $O = 30$ cm and $A = 130$ cm.

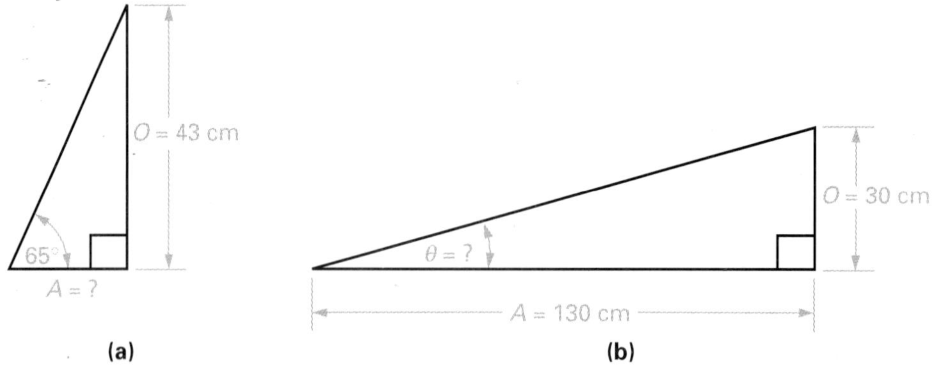

FIGURE 11-18 **Tangent of Theta.**

$$\begin{aligned}
\tan\theta &= \frac{O}{A} \\
&= \frac{30\text{ cm}}{130\text{ cm}} \\
&= 0.231 \quad (O \text{ is } 0.231 \text{ or } 23.1\% \text{ as long as } A) \\
\text{invtan}(\tan\theta) &= \text{invtan } 0.231 \quad (\text{multiply both sides by invtan}) \\
\theta &= \text{invtan } 0.231 \quad \left(\begin{array}{l}\text{calculator sequence:} \\ \boxed{0}\,\boxed{.}\,\boxed{2}\,\boxed{3}\,\boxed{1}\,\boxed{\text{INV}}\,\boxed{\text{TAN}}\end{array}\right) \\
&= 12.99 \text{ or } 13^\circ
\end{aligned}$$

Summary

As you have seen, trigonometry involves the study of the relationship between the three sides (*O, H, A*) of the right triangle, and it also relates to the relationship between the sides of the right triangle and the number of degrees contained in the angle theta (θ).

If the length of two sides of a right triangle are known, and the length of the third side is needed, remember that

$$C^2 = A^2 + B^2$$

If the angle θ is known along with the length of one side, or if angle θ is needed and the lengths of the two sides are known, one of the three formulas can be chosen based on what variables are known, and what is needed.

$$\sin\theta = \frac{O}{H} \qquad \cos\theta = \frac{A}{H} \qquad \tan\theta = \frac{O}{A}$$

When I was introduced to trigonometry, my mathematics professor spent 15 minutes having the whole class practice what he described as an old oriental war cry which went like this: "SOH CAH TOA". After he explained that it wasn't a war cry but in fact a memory aid to help us remember that SOH was in fact Sin $\theta = O/H$, CAH was Cos $\theta = A/H$, and TOA was Tan $\theta = O/A$, we understood the method in his madness.

11-3 AC WAVE SHAPES

In all fields of electronics, whether medical, industrial, consumer, or data processing, different types of information are being conveyed between two points, and electronic equipment is managing the flow of this information.

Let's now discuss the basic types of ac wave shapes. The way in which a wave varies in magnitude with respect to time describes its wave shape. All ac waves can be classified into one of six groups, and these are illustrated in Figure 11-19.

Sine Wave
Wave whose amplitude is the sine of a linear function of time. It is drawn on a graph that plots amplitude against time or radial degrees relative to the angular rotation of an alternator.

11-3-1 The Sine Wave

The **sine wave** is the most common type of waveform. It is the natural output of a generator that converts a mechanical input, in the form of a rotating shaft, into an electrical output in the form of a sine wave. In fact, for one cycle of the input shaft, the generator will produce one sinusoidal ac voltage waveform, as shown in Figure 11-20. When the input shaft of the generator is at 0°, the ac output is

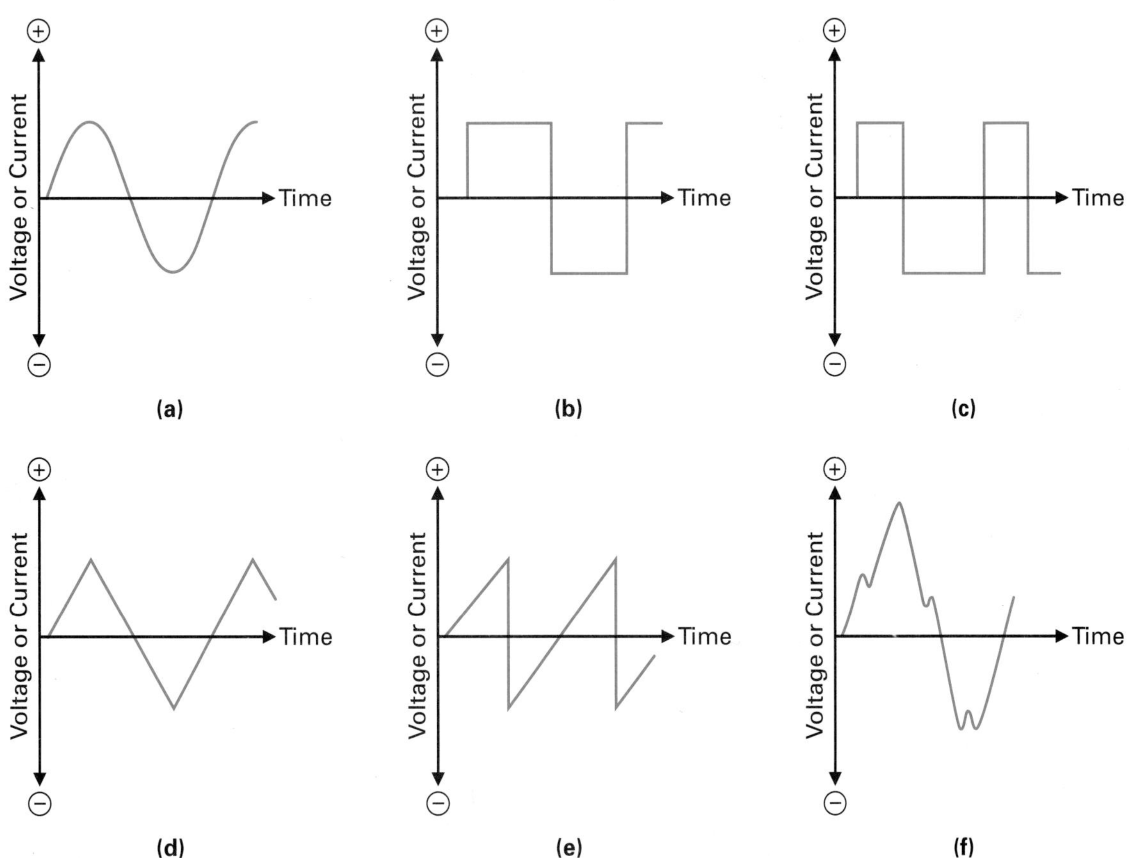

FIGURE 11-19 **AC Wave Shapes. (a) Sine Wave. (b) Square Wave. (c) Pulse Wave. (d) Triangular Wave. (e) Sawtooth Wave. (f) Irregular Wave.**

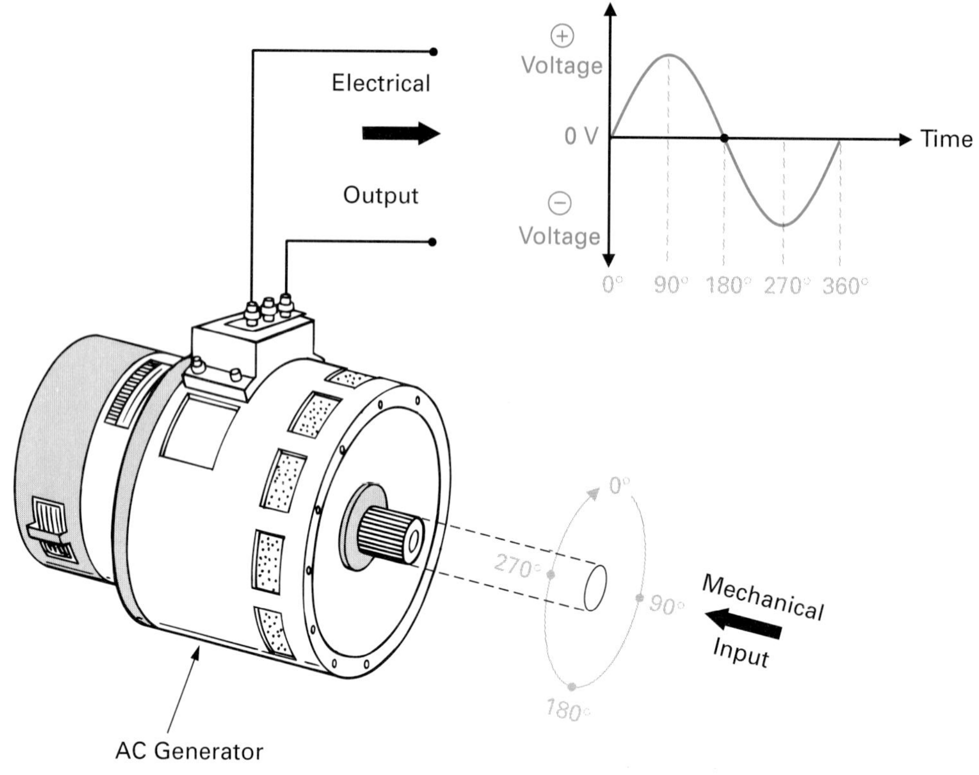

FIGURE 11-20 **Degrees of a Sine Wave.**

0 V. As the shaft is rotated through 360°, the ac output voltage will rise to a maximum positive voltage at 90°, fall back to 0 V at 180°, and then reach a maximum negative voltage at 270°, and finally return to 0 V at 360°. If this ac voltage is applied across a closed circuit, it produces a current that continually reverses or alternates in each direction.

Amplitude
Magnitude or size an alternation varies from zero.

Vector
Quantity that has both magnitude and direction. They are normally represented as a line, the length of which indicates magnitude and the orientation of which, due to the arrowhead on one end, indicates direction.

Peak Value
Maximum or highest-amplitude level.

Figure 11-21 illustrates the sine wave, with all the characteristic information inserted, which at first glance looks a bit ominous. Let's analyze and discuss each piece of information individually, beginning with the sine wave's amplitude.

Amplitude

Figure 11-22 plots direction and amplitude against time. The **amplitude** or magnitude of a wave is often represented by a **vector** arrow, also illustrated in Figure 11-22. The vector's length indicates the magnitude of the current or voltage, while the arrow's point is used to show the direction, or polarity.

Peak Value

The peak of an ac wave occurs on both the positive and negative alternation, but is only at the peak (maximum) for an instant. Figure 11-23(a) illustrates an ac current waveform rising to a positive peak of 10 A, falling to zero, and then reaching a negative peak of 10 A in the reverse direction; Figure 11-23(b) shows an ac voltage waveform reaching positive and negative peaks of 9 V.

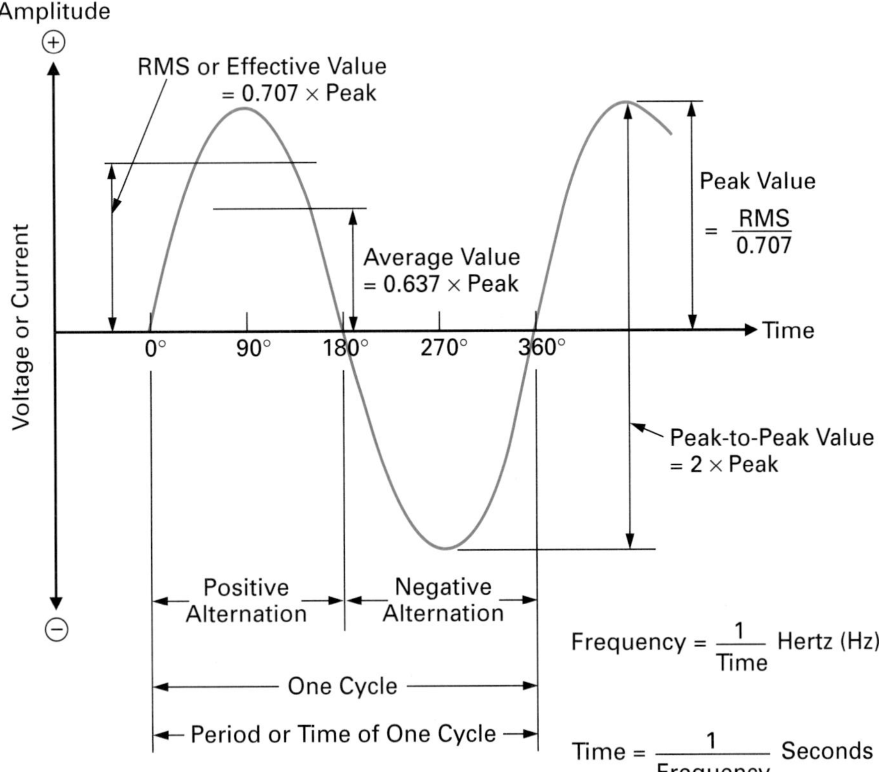

FIGURE 11-21 Sine Wave.

Peak-to-Peak Value

The **peak-to-peak value** of a sine wave is the value of voltage or current between the positive and negative maximum values of a waveform. For example, the peak-to-peak value of the current waveform in Figure 11-23(a) is equal to $I_{p-p} = 2 \times I_p = 20$ A. In Figure 11-23(b), it would be equal to $V_{p-p} = 2 \times V_p = 18$ V.

Peak-to-Peak Value Difference between the maximum positive and maximum negative values.

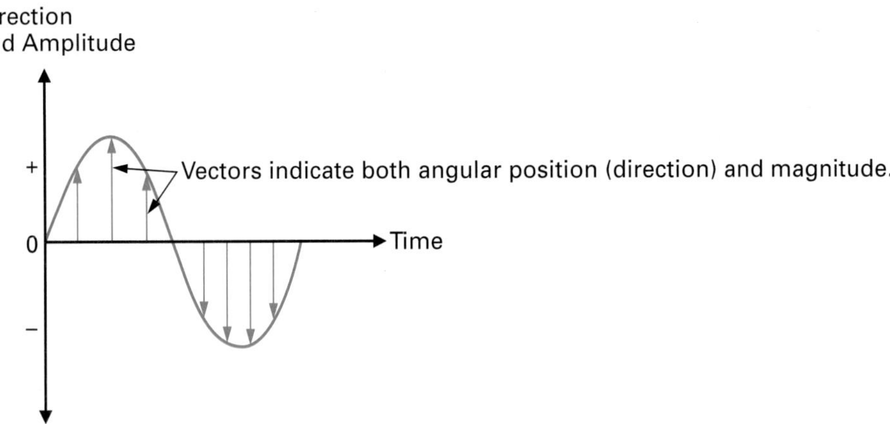

FIGURE 11-22 Sine-Wave Amplitude.

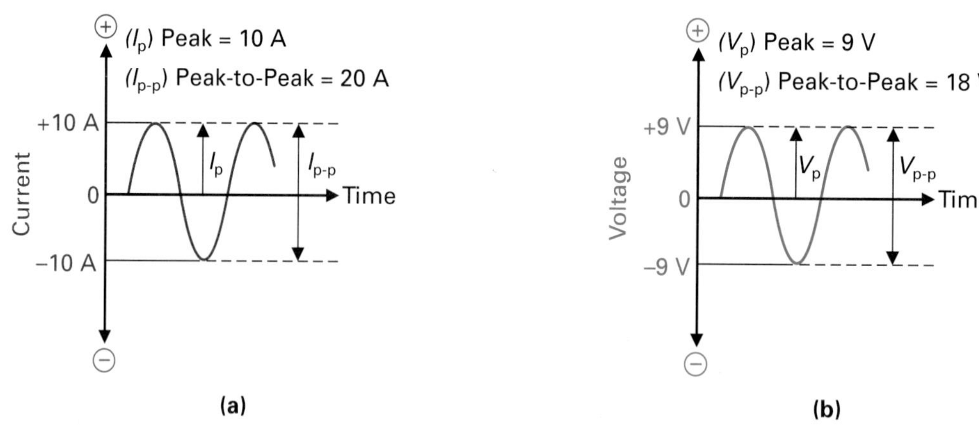

FIGURE 11-23 **Peak and Peak-to-Peak of a Sine Wave.**

$$\text{p} - \text{p} = 2 \times \text{peak}$$

RMS Value
Rms value of an ac voltage, current, or power waveform is equal to 0.707 times the peak value. The rms value is the effective or dc value equivalent of the ac wave.

RMS or Effective Value

Both the positive and negative alternation of a sine wave can accomplish the same amount of work; but the ac waveform is only at its maximum value for an instant in time, spending most of its time between peak currents. Our examples in Figure 11-23(a) and (b), therefore, cannot supply the same amount of power as a dc value of 10 A or 9 V.

The effective value of a sine wave is equal to 0.707 of the peak value. Let's now see how this value was obtained. Power is equal to either $P = I^2 \times R$ or $P = V^2/R$; said another way, power is proportional to the voltage or current squared. If every instantaneous value of either the positive or negative half-cycle of any voltage or current sinusoidal waveform is squared, as shown in Figure 11-24, and then averaged out to obtain the mean value, the square root of this mean value would be equal to 0.707 of the peak.

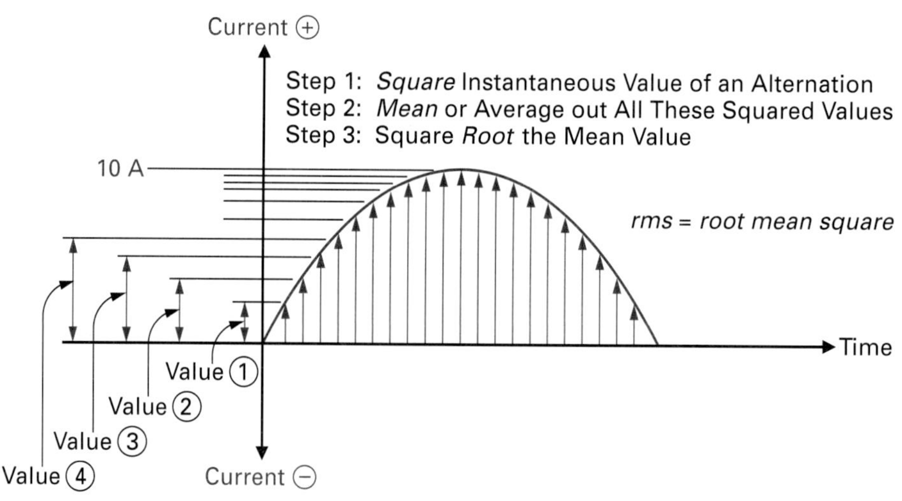

FIGURE 11-24 **Obtaining the RMS Value of 0.707 for a Sine Wave.**

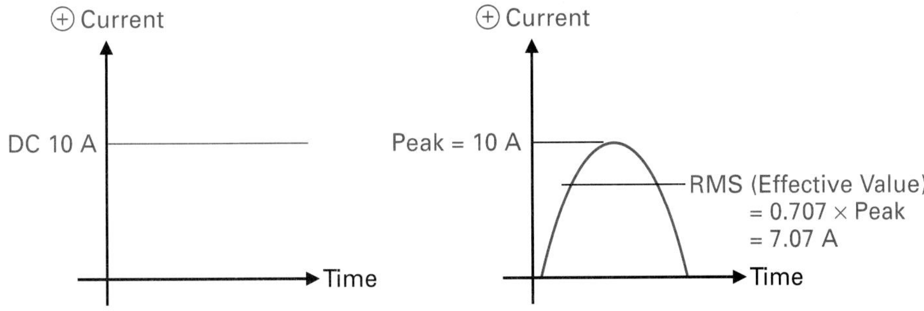

FIGURE 11-25 **Effective Equivalent.**

For example, if the process is carried out on the 10 A current waveform considered previously, the result would equal 7.07 A (0.707 × 10 A = 7.07 A), which is 0.707 of the peak value of 10 A.

$$\text{rms} = 0.707 \times \text{peak}$$

This root-mean-square (rms) result of 0.707 can always be used to tell us how effective an ac sine wave will be. For example, a 10 A dc source would be 10 A effective because it is continually present and always delivering power to the circuit to which it is connected, while a 10 A ac source would only be 7.07 A effective, as seen in Figure 11-25, because it is at 10 A for only a short period of time. As another example, a 10 V ac sine-wave alternation would be as effective or supply the same amount of power to a circuit as a 7.07 V dc source.

Unless otherwise stated, ac values of voltage or current are always given in rms, so the peak value can be calculated by transposing the original rms formula of rms = peak × 0.707, and ending up with

$$\text{peak} = \frac{\text{rms}}{0.707}$$

Since 1/0.707 = 1.414, the peak can also be calculated by

$$\text{peak} = \text{rms} \times 1.414$$

Average Value

The **average value** of the positive or negative alternation is found by taking either the positive or negative alternation, and listing the amplitude or vector length of current or voltages at 1° intervals, as shown in Figure 11-26(a). The sum of all these values is then divided by the total number of values (averaging), which for all sine waves will calculate out to be 0.637 of the peak voltage or cur-

Average Value
Mean value found when the area of a wave above a line is equal to the area of the wave below the line.

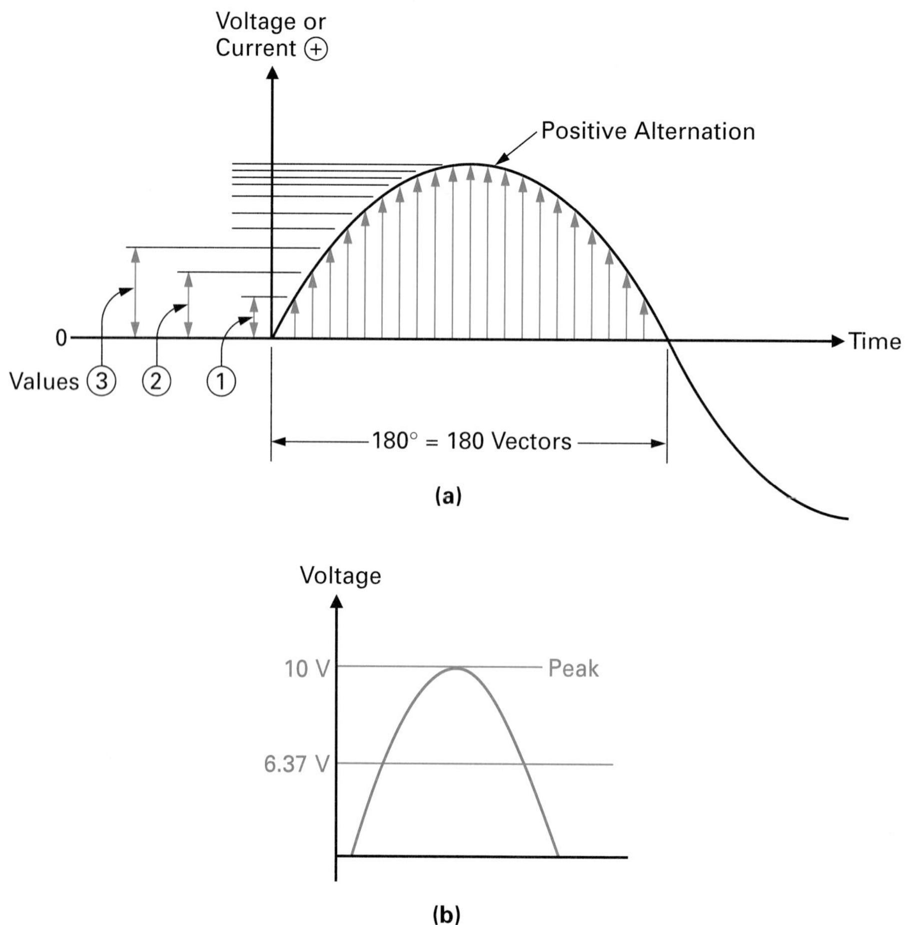

FIGURE 11-26 **Average Value of a Sine-Wave Alternation = 0.637 × Peak.**

rent. For example, the average value of a sine-wave alternation with a peak of 10 V, as seen in Figure 11-26(b), is equal to

$$\text{average} = 0.637 \times \text{peak}$$

$$= 0.637 \times 10\text{ V}$$
$$= 6.37\text{ V}$$

The average of a positive or negative alternation (half-cycle) is equal to 0.637 × peak; however, the average of the complete cycle, including both the positive and negative half-cycles, is mathematically zero as the amount of voltage or current above the zero line is equal but opposite to the amount of voltage or current below the zero line, as shown in Figure 11-27.

EXAMPLE:

Calculate V_p, V_{p-p}, V_{rms}, and V_{avg} of a 16 V peak sine wave.

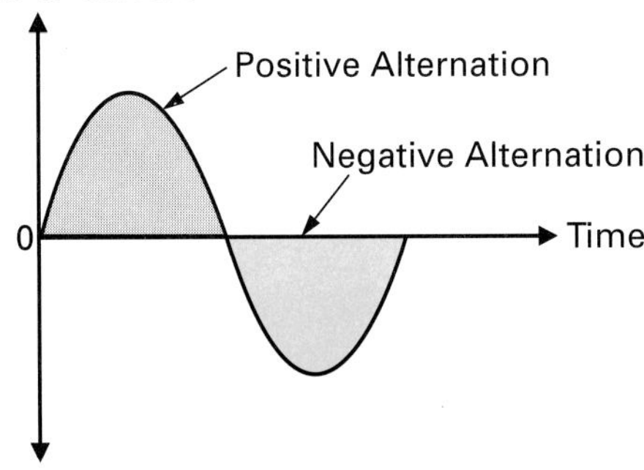

FIGURE 11-27 Average Value of a Complete Sine-Wave Cycle = 0.

Solution:

$$V_p = 16\text{ V}$$
$$V_{p-p} = 2 \times V_p = 2 \times 16\text{ V} = 32\text{ V}$$
$$V_{rms} = 0.707 \times V_p = 0.707 \times 16\text{ V} = 11.3\text{ V}$$
$$V_{avg} = 0.637 \times V_p = 0.637 \times 16\text{ V} = 10.2\text{ V}$$

EXAMPLE:

Calculate V_p, V_{p-p}, and V_{avg} of a 120 V (rms) ac main supply.

Solution:

$$V_p = \text{rms} \times 1.414 = 120\text{ V} \times 1.414 = 169.68\text{ V}$$
$$V_{p-p} = 2 \times V_p = 2 \times 169.68\text{ V} = 339.36\text{ V}$$
$$V_{avg} = 0.637 \times V_p = 0.637 \times 169.68\text{ V} = 108.09\text{ V}$$

The 120 V (rms) that is delivered to every home and business has a peak of 169.68 V; however, it will deliver the same power as 120 V dc.

Frequency and Period

The **period** (t) is the time required for one complete cycle (positive and negative alternation) of the sinusoidal current or voltage waveform. A *cycle,* by definition, is the change of an alternating wave from zero to a positive peak, to zero, then to a negative peak, and finally, back to zero (see Figure 11-28).

Frequency is the number of repetitions of a periodic wave in a unit of time. It is symbolized by f and is given the unit hertz (cycles per second), in honor of a German physicist, Heinrich Hertz.

Period
Time taken to complete one complete cycle of a periodic or repeating waveform.

Frequency
Rate or recurrences of a periodic wave normally within a unit of one second, measured in hertz (cycles/second).

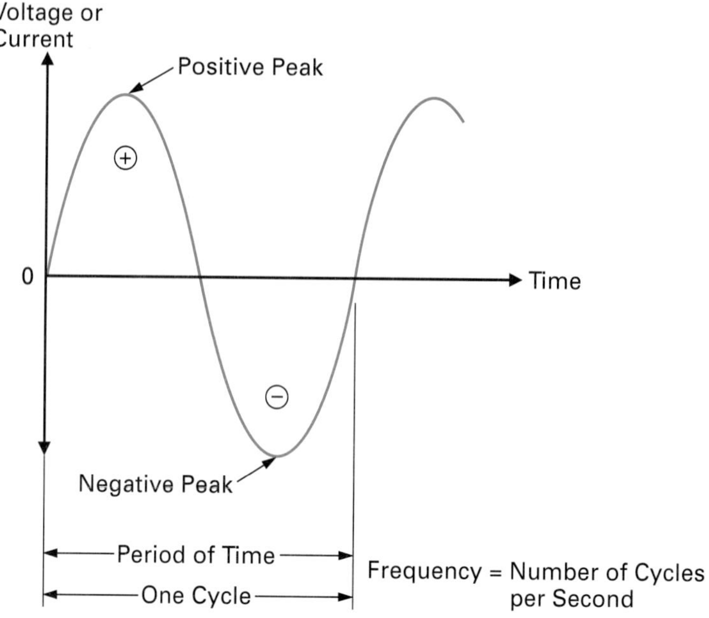

FIGURE 11-28 **Frequency and Period.**

Sinusoidal waves can take a long or a short amount of time to complete one cycle. This time is related to frequency in that period is equal to the reciprocal of frequency, and vice versa.

$$f\,(\text{hertz}) = \frac{1}{t}$$

$$t\,(\text{seconds}) = \frac{1}{f}$$

where t = period

f = frequency

For example, the ac voltage of 120 V (rms) arrives at the household electrical outlet alternating at a frequency of 60 hertz (Hz). This means that 60 cycles arrive at the household electrical outlet in 1 second. If 60 cycles occur in 1 second, as seen in Figure 11-29(a), it is actually taking 1⁄60 of a second for one of the 60 cycles to complete its cycle, which calculates out to be

$$\tfrac{1}{60} \text{ of 1 second} = \frac{1 \text{ second}}{60 \text{ cycles}} \times 1 \text{ second} = 16.67 \text{ milliseconds (ms)}$$

So the time or period of one cycle can be calculated by using the formula period $(t) = 1/f = 1/60$ Hz = 16.67 ms, as shown in Figure 11-29(b).

If the period or time of a cycle is known, the frequency can be calculated; for example,

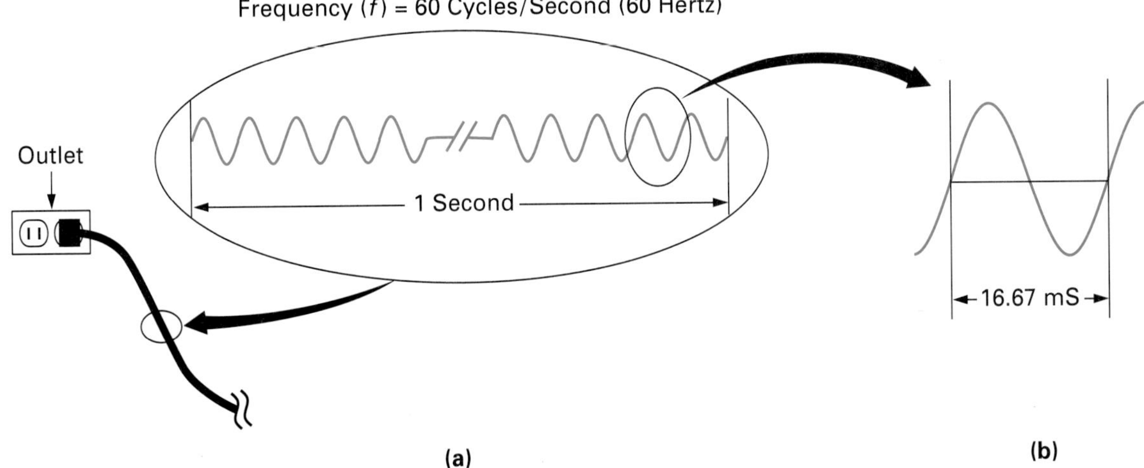

FIGURE 11-29 **120-V, 60-Hz AC Supply.**

$$\text{frequency } (f) = \frac{1}{\text{period}} = 1/16.67 \text{ ms} = 60 \text{ Hz}$$

As illustrated in Figure 11-6, all homes in the United States receive at their wall outlets an ac voltage of 120 V rms at a frequency of 60 Hz. This frequency was chosen for convenience, as a lower frequency would require larger transformers, and if the frequency were too low, the slow switching (alternating) current through the light bulb would cause it to flicker. A higher frequency than 60 Hz was found to cause an increase in the amount of heat generated in the core of all power distribution transformers due to eddy currents and hysteresis losses. Consequently, 60 Hz was chosen in the United States; however, other countries, such as England and most of Europe, use an ac power line frequency of 50 Hz (240 V).

EXAMPLE:

If a sine wave has a period of 400 μs, what is its frequency?

Solution:

$$\text{frequency } (f) = \frac{1}{\text{time } (t)} = \frac{1}{400 \text{ μs}} = 2.5 \text{ kHz}$$

EXAMPLE:

If it takes a sine wave 25 ms to complete two cycles, how many of the cycles will be received in 1 s?

Solution:

If the period of two cycles is 25 ms, one cycle period will equal 12.5 ms. The number of cycles per second or frequency will equal

$$f = \frac{1}{t} = \frac{1}{12.5 \text{ ms}} = 80 \text{ Hz} \quad \text{or} \quad 80 \text{ cycles/second}$$

EXAMPLE:

Calculate the period of the following:

a. 100 MHz
b. 40 cycles every 5 seconds
c. 4.2 kilocycles/second
d. 500 kHz

Solution:

$$f = \frac{1}{t} \qquad \text{therefore, } t = \frac{1}{f}$$

a. $t = \dfrac{1}{100\text{ MHz}} = 10$ nanoseconds (ns)

b. 40 cycles/5 s = 8 cycles/second (8 Hz)

$t = \dfrac{1}{8\text{ Hz}} = 125\text{ ms}$

c. $t = \dfrac{1}{4.2\text{ kHz}} = 238\ \mu\text{s}$

d. $t = \dfrac{1}{500\text{ kHz}} = 2\ \mu\text{s}$

Wavelength

Wavelength Distance between two points of corresponding phase and is equal to waveform velocity or speed divided by frequency.

Wavelength, as its name states, is the physical length of one complete cycle and is generally measured in meters. The wavelength (λ, lambda) of a complete cycle is dependent on the frequency and velocity of the transmission:

$$\lambda = \frac{\text{velocity}}{\text{frequency}}$$

Electromagnetic waves. Radio waves travel at the speed of light in air or a vacuum, which is 3×10^8 meters/second or 3×10^{10} cm/second.

$$\lambda\,(\text{m}) = \frac{3 \times 10^8\text{ m/s}}{f\,(\text{Hz})} \qquad \text{or} \qquad \lambda\,(\text{cm}) = \frac{3 \times 10^{10}\text{ cm/s}}{\text{frequency (Hz)}}$$

(There are 100 centimeters [cm] in 1 meter [m], therefore cm = 10^{-2}, and m = 10^0 or 1.) Subsequently, the higher the frequency, the shorter the wavelength, which is why a shortwave radio receiver is designed to receive high frequencies ($\lambda\downarrow = 3 \times 10^8/f\uparrow$).

EXAMPLE:

Calculate the wavelength of the electromagnetic waves illustrated in Figure 11-30.

Solution:

a. $\lambda = \dfrac{3 \times 10^8}{f\,(\text{Hz})}\ \text{m/s} = \dfrac{3 \times 10^8}{10\ \text{kHz}} = 30{,}000\ \text{m} \quad \text{or} \quad 30\ \text{km}$

b. $\lambda = \dfrac{3 \times 10^{10}}{f\,(\text{Hz})}\ \text{cm/s} = \dfrac{3 \times 10^{10}}{2182\ \text{kHz}} = 13{,}748.9\ \text{cm} \quad \text{or} \quad 137.489\ \text{m}$

c. $\lambda = \dfrac{3 \times 10^{10}}{f\,(\text{Hz})}\ \text{cm/s} = \dfrac{3 \times 10^{10}}{4.0\ \text{GHz}} = \dfrac{3 \times 10^{10}}{4 \times 10^9} = 7.5\ \text{cm} \quad \text{or} \quad 0.075\ \text{m}$

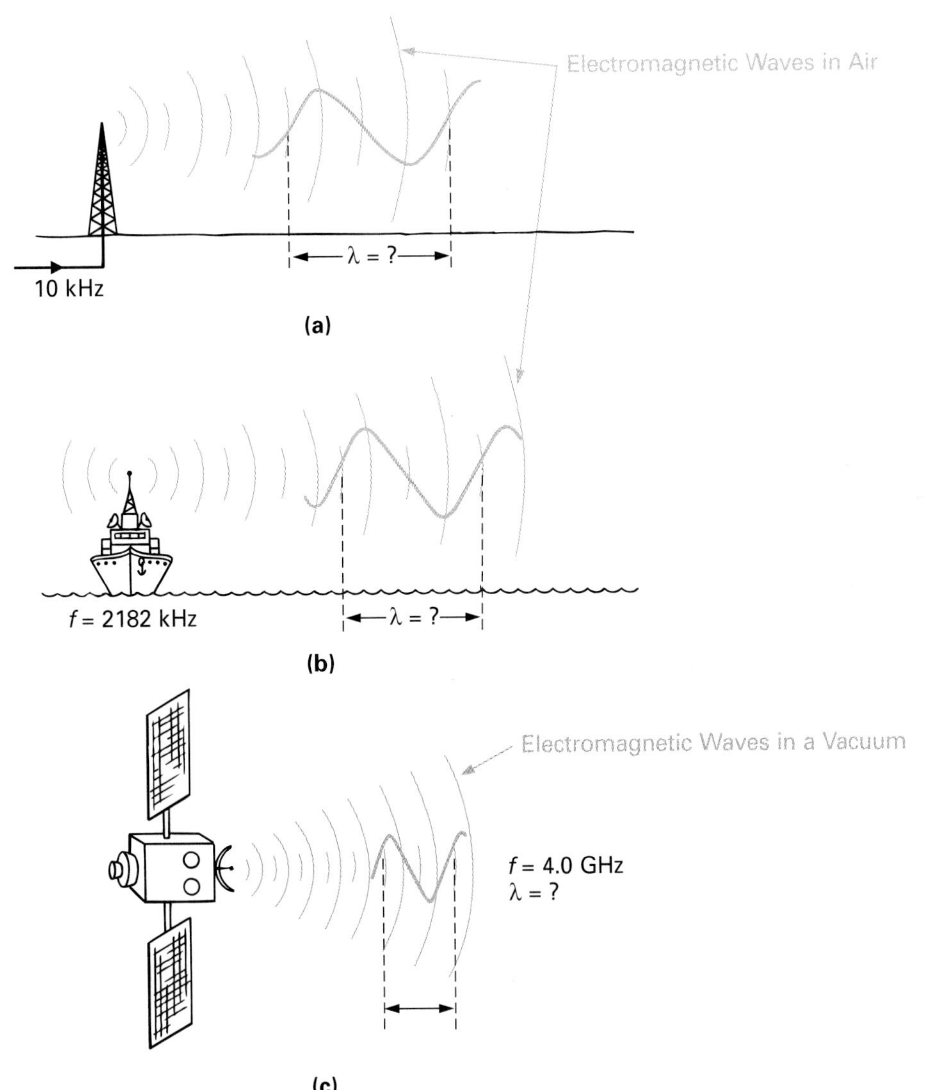

FIGURE 11-30 **Electromagnetic Wavelength Examples.**

Sound waves. Sound waves travel at a slower speed than electromagnetic waves, as their mechanical vibrations depend on air molecules, which offer resistance to the traveling wave. For sound waves, the wavelength formula will be equal to

$$\lambda\ (\text{m}) = \frac{344.4\ \text{m/s}}{f\,(\text{Hz})}$$

EXAMPLE:

Calculate the wavelength of the sound waves illustrated in Figure 11-31.

Solution:

a. $\lambda\ (\text{m}) = \frac{344.4\ \text{m/s}}{f\,(\text{Hz})} = \frac{344.4}{35\ \text{kHz}} = 9.8 \times 10^{-3}\ \text{m} = 9.8\ \text{mm}$

b. 300 Hz: $\lambda\ (\text{m}) = \frac{344.4\ \text{m/s}}{300\ \text{Hz}} = 1.15\ \text{m}$

3000 Hz: $\lambda\ (\text{m}) = \frac{344.4\ \text{m/s}}{3000\ \text{Hz}} = 0.115\ \text{m}$ or 11.5 cm

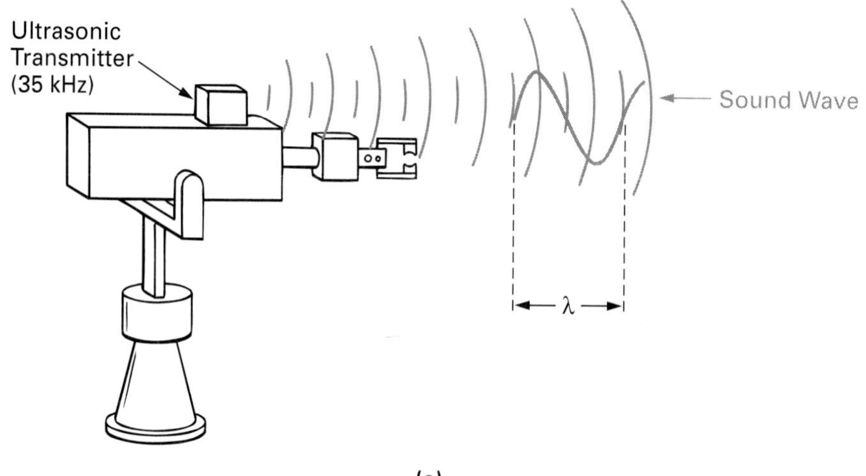

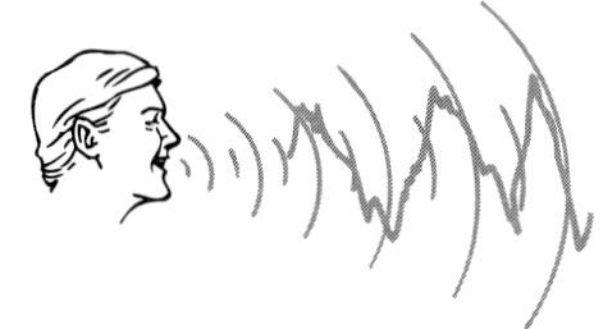

FIGURE 11-31 **Sound Wavelength Examples.**

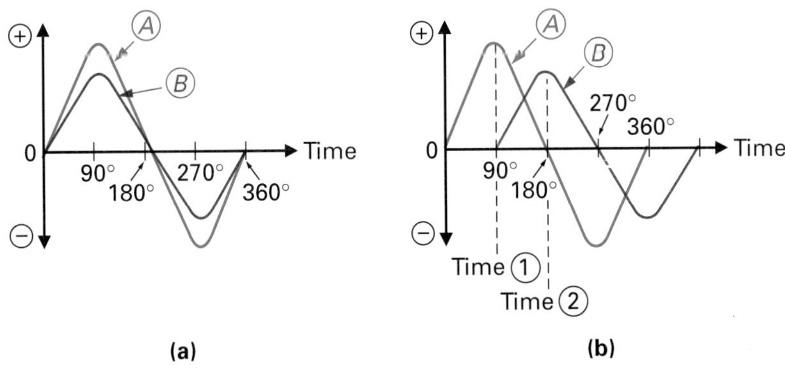

FIGURE 11-32 **Phase Relationship. (a) In Phase. (b) Out of Phase.**

Phase Relationships

The **phase** of a sine wave is always relative to another sine wave of the same frequency. Figure 11-32(a) illustrates two sine waves that are in phase with one another, while Figure 11-32(b) shows two sine waves that are out of phase with one another. Sine wave *A* is our reference, since the positive-going zero crossing is at 0°, its positive peak is at 90°, its negative-going zero crossing is at 180°, its negative peak is at 270°, and the cycle completes at 360°. In Figure 11-32(a), sine wave *B* is in phase with *A* since its peaks and zero crossings occur at the same time as sine wave *A*'s. In Figure 11-32(b), sine wave *B* has been shifted to the right by 90° with respect to the reference sine wave *A*. This **phase shift** or **phase angle** of 90° means that sine wave *A leads B* by 90°, or *B lags A* by 90°. Sine wave *A* is said to lead *B* as its positive peak, for example, occurs first at time 1, while the positive peak of *B* occurs later at time 2.

Phase
Angular relationship between two waves, normally between current and voltage in an ac circuit.

Phase Shift or Angle
Change in phase of a waveform between two points, given in degrees of lead or lag. Phase difference between two waves, normally expressed in degrees.

EXAMPLE:

What are the phase relationships between the two waveforms illustrated in Figure 11-33(a) and (b)?

Solution:

a. The phase shift or angle is 90°. Sine wave *B* leads sine wave *A* by 90°, or *A* lags *B* by 90°.

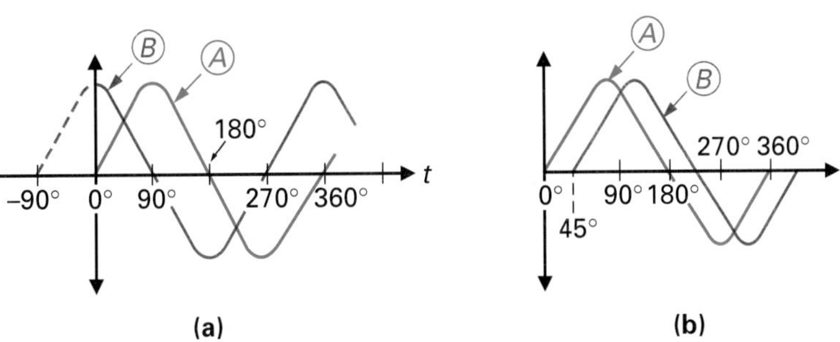

FIGURE 11-33 **Phase Relationship Examples.**

b. The phase shift or angle is 45°. Sine wave A leads sine wave B by 45°, or B lags A by 45°.

The Meaning of Sine

The square, rectangular, triangular, and sawtooth waveform shapes are given their names because of their waveform shapes. The sine wave is the most common type of waveform shape, and why the name *sine* was given to this wave needs to be explained further.

Figure 11-34(a) shows the correlation between the 360° of a circle and the 360° of a sine wave. Within the circle a triangle has been drawn to represent the right-angle triangle formed at 30°. The hypotenuse side will always remain at the same length throughout 360°, and will equal the peak of the sine wave. The opposite side of the triangle is equal to the amplitude vector of the sine wave at 30°. To calculate the amplitude of the opposite side, and therefore the amplitude of the sine wave at 30°, we use the sine of theta formula discussed previously in the mini-math review on trigonometry.

$$\sin\theta = \frac{\text{opposite}}{\text{hypotenuse}}$$

$$\sin 30° = \frac{O}{H} \quad \left(\begin{array}{c}\text{calculator sequence:}\\ \boxed{3}\boxed{0}\boxed{\text{SIN}}\end{array}\right)$$

$$0.500 = \frac{O}{H}$$

This tells us that at 30°, the opposite side is 0.500 or half the size of the hypotenuse. At 30°, therefore, the amplitude of the sine wave will be 0.5 (50%) of the peak value.

Figure 11-34(b) lists the sine values at 15° increments. Figure 11-34(c) shows an example of a 10-V-peak sine wave. At 15°, a sine wave will always be at 0.259 (sine of 15°) of the peak value, which for a 10 V sine wave will be 2.59 V (0.259 × 10 V = 2.59 V). At 30°, a sine wave will have increased to 0.500 (sine of 30°) of the peak value. At 45°, a sine wave will be at 0.707 of the peak, and so on. The sine wave is consequently called a sine wave, as it changes in amplitude at the same rate as the sine trigonometric function.

11-3-2 The Square Wave

Square Wave
Wave that alternates between two fixed values for an equal amount of time.

The **square wave** is a periodic (repeating) wave that alternates from a positive peak value to a negative peak value, and vice versa, for equal lengths of time.

In Figure 11-35 you can see an example of a square wave that is at a frequency of 1 kHz and has a peak of 10 V. If the frequency of a wave is known, its period or time of one cycle can be calculated by using the formula $T = 1/f = 1/1\text{ kHz} = 1\text{ ms}$ or $\frac{1}{1000}$ of a second. One complete cycle will take 1 ms to complete, so the positive and negative alternations will each last for 0.5 ms.

If the peak of the square wave is equal to 10 V, the peak-to-peak value of this square wave will equal $V_{p\text{-}p} = 2 \times V_p = 20$ V.

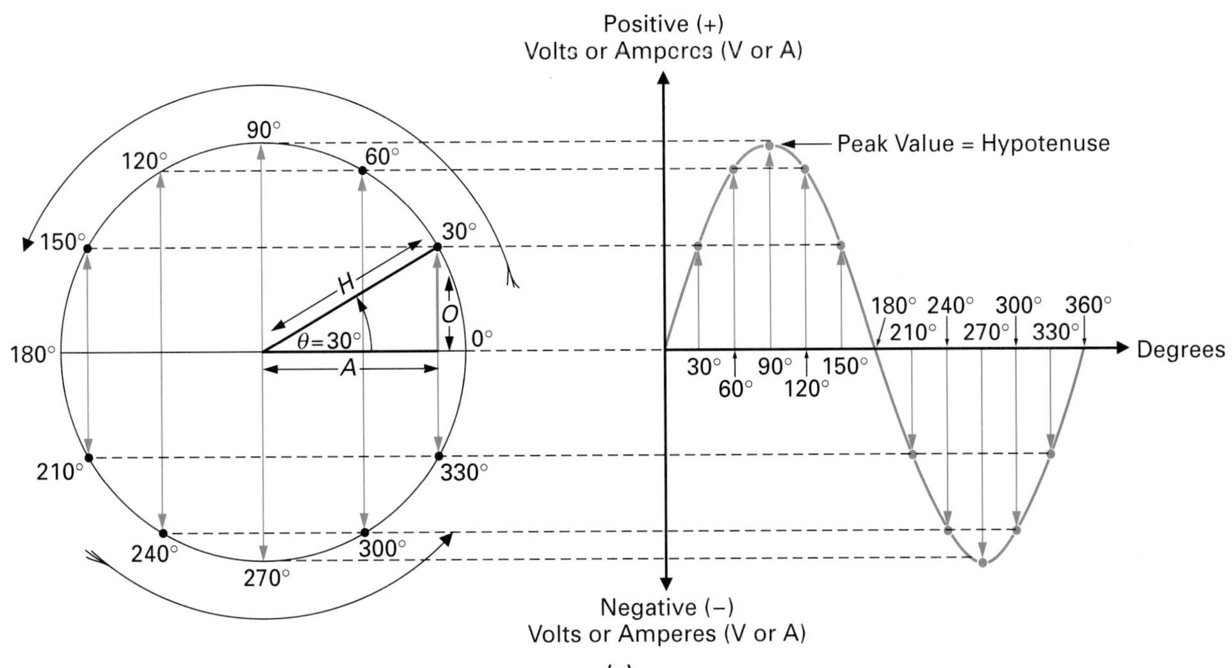

(a)

Angle	Sine
0°	0.000
15	0.259
30	0.500
45	0.707
60	0.866
75	0.966
90	1.000

(b)

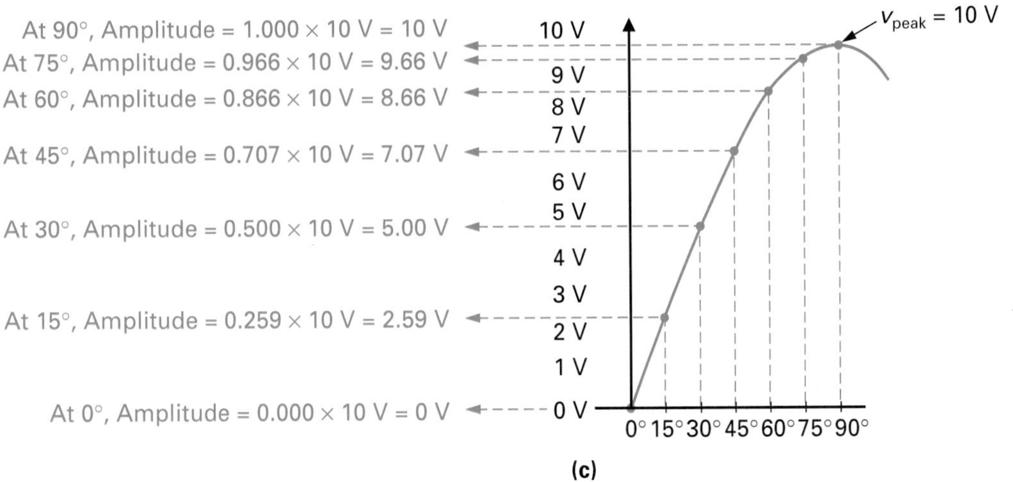

(c)

FIGURE 11-34 Meaning of Sine.

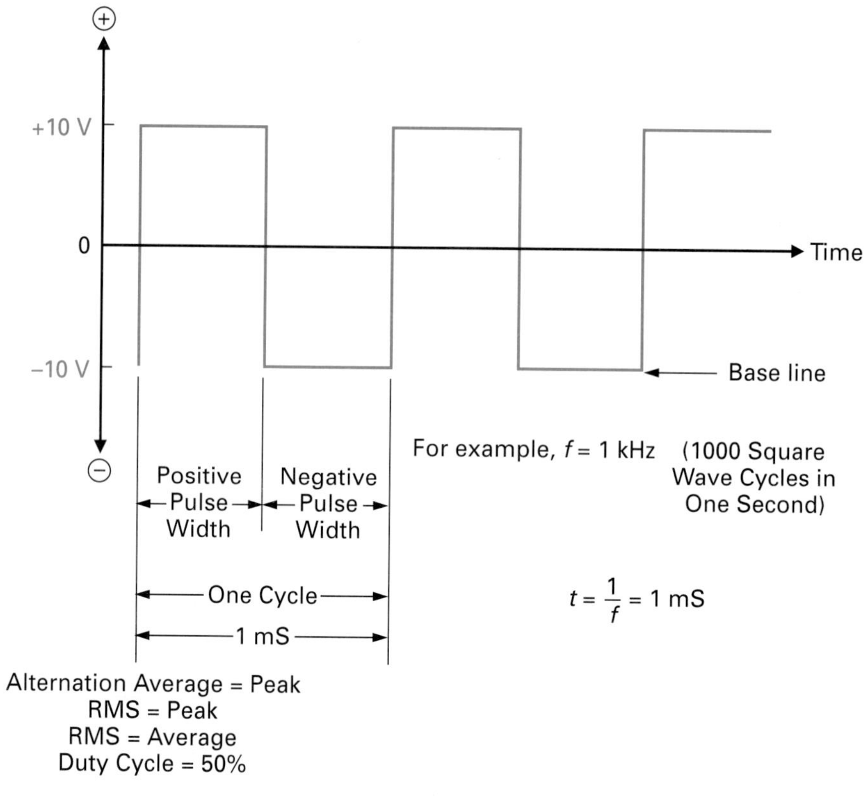

FIGURE 11-35 Square Wave.

To summarize this example, the square wave alternates from a positive peak value of +10 V to a negative peak value of −10 V for equal time lengths (half-cycles) of 0.5 ms.

Duty Cycle

Duty Cycle
A term used to describe the amount of ON time versus OFF time. ON time is usually expressed as a percentage.

Duty cycle is an important relationship, which has to be considered when discussing square waveforms. The **duty cycle** is the ratio of a pulse width (positive or negative pulse or cycle) to the overall period or time of the wave and is normally given as a percentage.

$$\text{duty cycle }(\%) = \frac{\text{pulse width }(P_w)}{\text{period }(t)} \times 100\%$$

The duty cycle of the example square wave in Figure 11-35 will equal

$$\text{duty cycle }(\%) = \frac{\text{pulse width }(P_w)}{\text{period }(p)} \times 100\%$$

$$= \frac{0.5\text{ ms}}{1\text{ ms}} \times 100\%$$

$$= 50\%$$

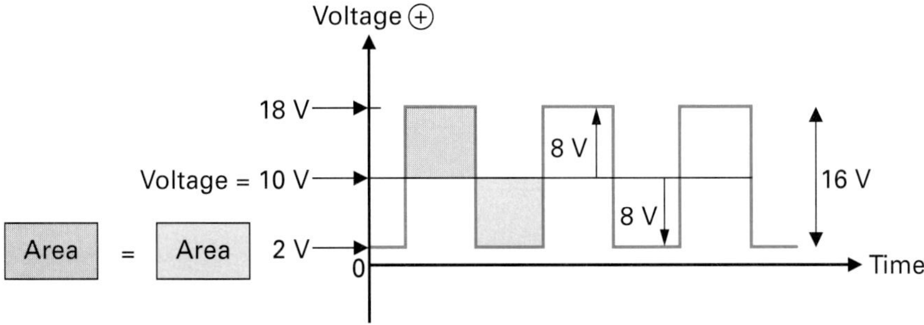

FIGURE 11-36 **2- to 18-V Square Wave.**

Since a square wave always has a positive and a negative alternation that are equal in time, the duty cycle of all square waves is equal to 50%, which actually means that the positive cycle lasts for 50% of the time of one cycle.

Average

The average or mean value of a square wave can be calculated by using the formula

$$V \text{ or } I \text{ average} = \text{baseline} + (\text{duty cycle} \times \text{peak to peak})$$

The average of the complete square wave cycle in Figure 11-35 should calculate out to be zero, as the amount above the line equals the amount below. If we apply the formula to this example, you can see that

$$\begin{aligned} V_{avg} &= \text{baseline} + (\text{duty cycle} \times \text{peak to peak}) \\ &= -10\text{ V} + (0.5 \times 20\text{ V}) \\ &= -10\text{ V} + 10\text{ V} \\ &= 0\text{ V} \end{aligned}$$

However, a square wave does not always alternate about 0; for example, Figure 11-36 illustrates a 16 V_{p-p} square wave that rests on a baseline of 2 V. The average value of this square wave is equal to

$$\begin{aligned} V_{avg} &= \text{baseline} + (\text{duty cycle} \times \text{peak to peak}) \\ &= 2\text{ V} \times (0.5 \times 16\text{ V}) \\ &= (+2) + (+8\text{ V}) \\ &= 10\text{ V} \end{aligned}$$

EXAMPLE:

Calculate the duty cycle and V_{avg} of a square wave of 0 to 5 V.

Solution:

The duty cycle of a square wave is always 0.5 or 50%. The average is:

$$V_{avg} = \text{baseline} + (\text{duty cycle} \times V_{p\text{-}p})$$
$$= 0\text{ V} + (0.5 \times 5\text{ V})$$
$$= 0\text{ V} + 2.5\text{ V} = 2.5\text{ V}$$

Up to this point, we have seen the ideal square wave, which has instantaneous transition from the negative to the positive values, and vice versa, as shown in Figure 11-37(a). In fact, the transitions from negative to positive (positive or leading edge) and from positive to negative (negative or trailing edge) are not as ideal as shown here. It takes a small amount of time for the wave to increase to its positive value (the **rise time**) and consequently an equal amount of time for a wave to decrease to its negative value (the **fall time**). Rise time (T_R), by definition, is the time it takes for an edge to rise from 10% to 90% of its full amplitude, while fall time (T_F) is the time it takes for an edge to fall from 90% to 10% of its full amplitude, as shown in Figure 11-37(b).

Rise Time
Time it takes a positive edge of a pulse to rise from 10% to 90% of its peak value.

Fall Time
Time it takes a negative edge of a pulse to fall from 90% to 10% of its peak value.

With a waveform such as that in Figure 11-37(b), it is difficult, unless a standard is used, to know exactly what points to use when measuring the width of either the positive or negative alternation. The standard width is always measured between the two 50% amplitude points, as shown in Figure 11-38.

Voltage ⊕
Positive (Rising) Edge or Step
+10 V
Negative (Falling) Edge or Step
0 V
(a)

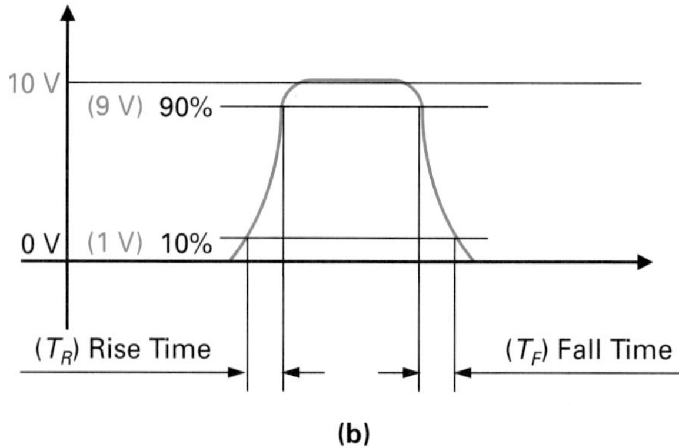

FIGURE 11-37 **Square Wave's Rise and Fall Times. (a) Ideal. (b) Actual.**

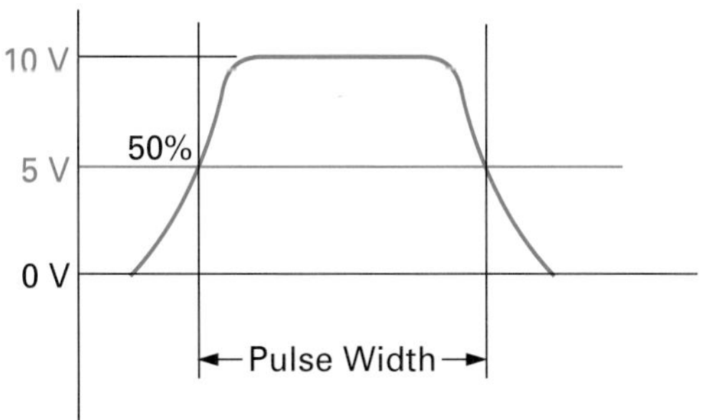

FIGURE 11-38 **Pulse Width of a Square Wave.**

Time-Domain Analysis
A method of representing a waveform by plotting its amplitude versus time.

Frequency-Domain Analysis

A *periodic wave* is a wave that repeats the same wave shape from one cycle to the next. Figure 11-39(a) is a **time-domain** representation of a periodic sine wave, which is the same way it would appear on an oscilloscope display as it plots the sine wave's amplitude against time. Figure 11-39(b) is a **frequency-domain** representation of the same periodic sine wave, and this graph, which shows the wave as it would appear on a spectrum analyzer, plots the sine wave's amplitude against frequency instead of time. This graph shows all the frequency components contained within a wave, and since, in this example, the sine wave has a period of 1 ms and therefore a frequency of 1 kHz, there is one bar at the 1 kHz point of the graph and its size represents the sine wave's amplitude. Pure sine waves have no other frequency components.

Frequency-Domain Analysis
A method of representing a waveform by plotting its amplitude versus frequency.

Fundamental Frequency
This sine wave is always the lowest frequency and largest amplitude component of any waveform shape and is used as a reference.

Other periodic wave shapes, such as square, pulse, triangular, sawtooth, or irregular, are actually made up of a number of sine waves having a particular frequency, amplitude, and phase. To produce a square wave, for instance, you would start with a sine wave, as shown in Figure 11-40(a), whose frequency is equal to the resulting square-wave frequency. This sine wave is called the **fundamental frequency,** and all the other sine waves that will be added to this fundamental are called **harmonics** and will always be lower in amplitude and higher in frequency. These harmonics or multiples are harmonically related to the funda-

Harmonic Frequency
Sine wave that is smaller in amplitude and is some multiple of the fundamental frequency.

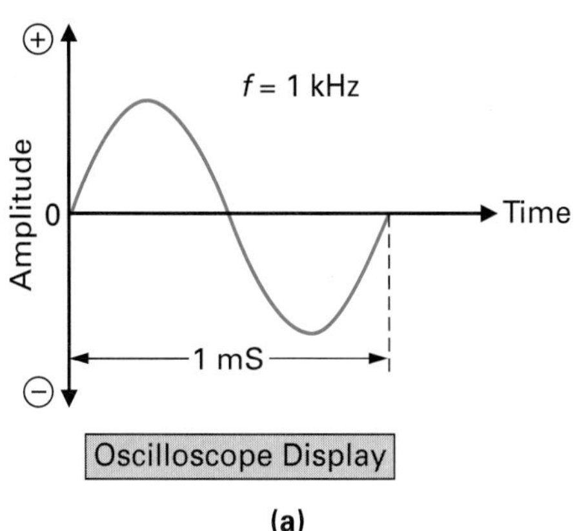

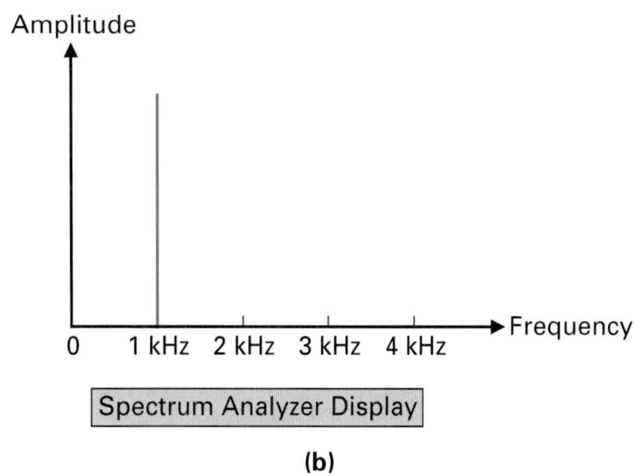

FIGURE 11-39 **Analysis of a 1-kHz Sine Wave. (a) Time Domain. (b) Frequency Domain.**

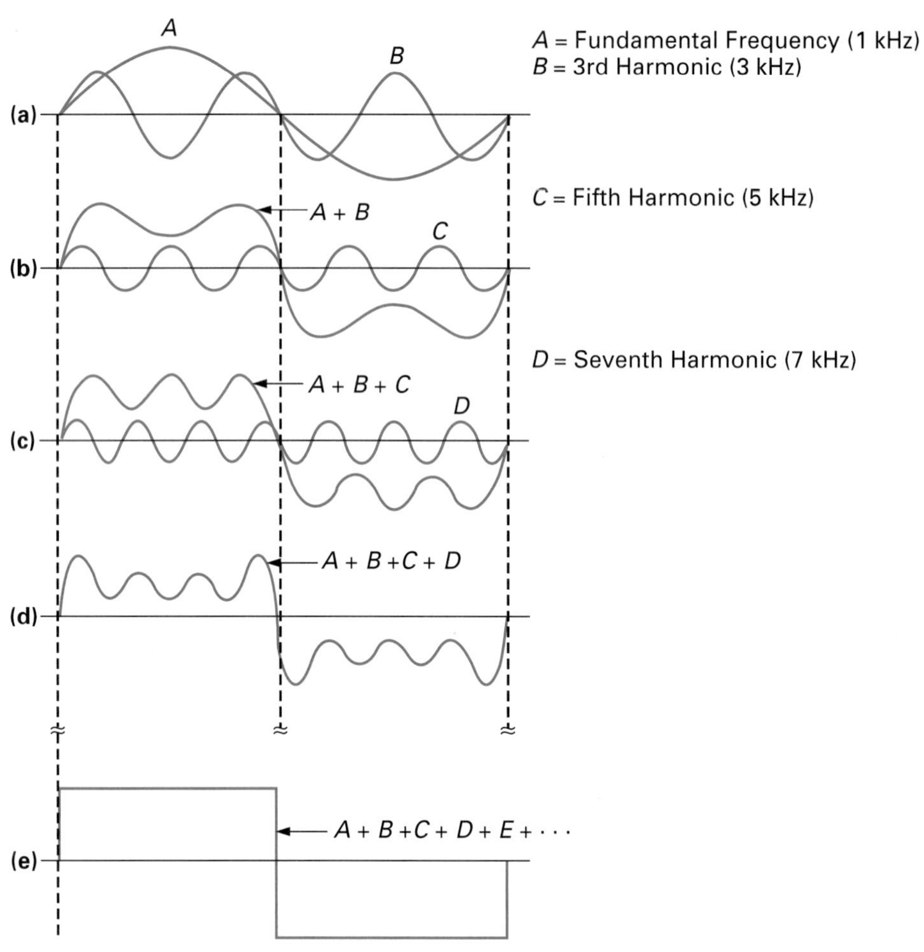

FIGURE 11-40 Square-Wave Composition.

mental, in that the second harmonic is twice the fundamental frequency, the third harmonic is three times the fundamental frequency, and so on.

Square waves are composed of a fundamental frequency and an infinite number of odd harmonics (third, fifth, seventh, and so on). If you look at the progression in Figure 11-40(a) through (d), you see that by continually adding

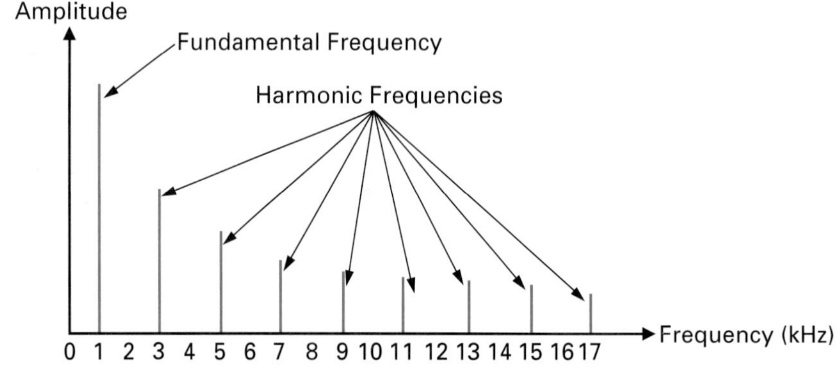

FIGURE 11-41 Frequency-Domain Analysis of a Square Wave.

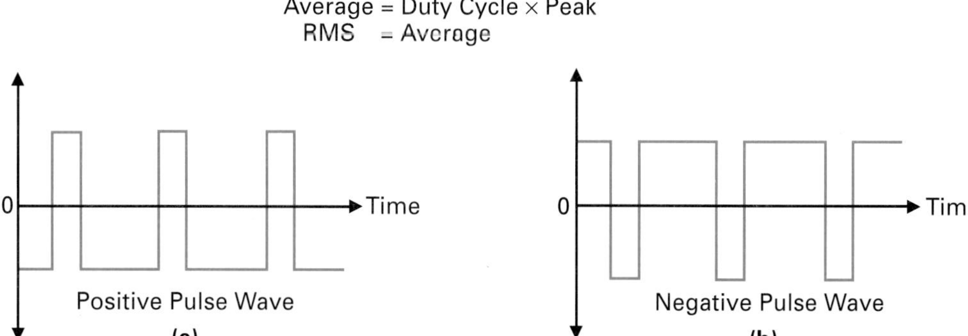

FIGURE 11-42 **Rectangular or Pulse Wave.**

these odd harmonics the waveform comes closer to a perfect square wave, as shown in Figure 11-40(e).

Figure 11-41 plots the frequency domain of a square wave, with the bars representing the odd harmonics of decreasing amplitude.

11-3-3 The Rectangular or Pulse Wave

The **rectangular wave** is similar to the square wave in many respects, in that it is a periodic wave that alternately switches between one of two fixed values. The difference in the rectangular wave is that it does not remain at the two peak values for equal lengths of time, as shown in the examples in Figure 11-42(a) and (b).

In Figure 11-42(a), the rectangular wave remains at its negative level for a longer period than its positive, while the rectangular wave in Figure 11-42(b) stays at its positive value for the longer period of time and is only momentarily at its negative value.

PRF, PRT, and Pulse Length

When discussing a rectangular wave, a few terms change. Instead of stating the cycles per second as frequency, it is called **pulse repetition frequency** (PRF), which is far more descriptive. The reciprocal of frequency is time, and with rectangular pulse waveforms the reciprocal of the PRF is **pulse repetition time** (PRT), as summarized in Table 11-1. Frequency is equivalent to PRF and time to PRT; the only difference is the name.

Let us look at the example in Figure 11-43 of a 5 V rectangular wave at a frequency of 1 kHz and a pulse width of 1 μs, and practice with these new terms. With a pulse repetition frequency of 1 kHz, the time between the leading edges of pulses (PRT) will be 1/1 kHz = 1 ms. **Pulse width** (P_w), **pulse duration** (P_d), or **pulse length** (P_l) are all terms that describe the length of time for which the pulse lasts, and in this example it is equal to 1 μs, which means that 999 μs exists between the end of one pulse and the beginning of the next.

Rectangular (Pulse) Wave
Also known as a pulse wave; it is a repeating wave that only alternates between two levels or values and remains at one of these values for a small amount of time relative to the other.

Pulse Repetition Frequency
The number of times per second that a pulse is transmitted.

Pulse Repetition Time
The time interval between the start of two consecutive pulses.

Pulse Width, Pulse Length, or Pulse Duration
The time interval between the leading edge and trailing edge of a pulse at which the amplitude reaches 50% of the peak pulse amplitude.

TABLE 11-1

SQUARE AND SINE WAVE		RECTANGULAR WAVE
Frequency (f) = $\frac{1}{\text{time } (t)}$	Equivalent to	pulse repetition frequency (PRF) = $\frac{1}{\text{pulse repetition time (PRT)}}$
Time (t) = $\frac{1}{\text{frequency } (f)}$		pulse repetition time (PRT) = $\frac{1}{\text{pulse repetition frequency (PRF)}}$

Duty Cycle

The duty cycle is calculated in exactly the same way as for the square wave and is a ratio of the pulse width to the overall time (PRT). In our example in Figure 11-43 the duty cycle will be equal to

$$\begin{aligned}\text{duty cycle } (\%) &= \frac{\text{pulse width } (P_w)}{\text{PRT}} \times 100\% \\ &= \frac{1\ \mu\text{s}}{1000\ \mu\text{s}} \\ &= \text{duty cycle figure of } 0.001 \\ &= 0.001 \times 100\% \\ &= 0.1\%\end{aligned}$$

The result tells us that the positive pulse lasts for 0.1% of the total time (PRT).

Average

The average or mean value of this waveform is calculated by using the same square-wave formula. The average of the pulse wave in Figure 11-43 will be

$$V \text{ or } I \text{ average} = \text{baseline} + (\text{duty cycle} \times \text{peak to peak})$$

$$\begin{aligned}V_{avg} &= 0\text{ V} + (0.001 \times 5\text{ V}) \\ &= 0\text{ V} + (5\text{ mV}) \\ &= 5\text{ mV}\end{aligned}$$

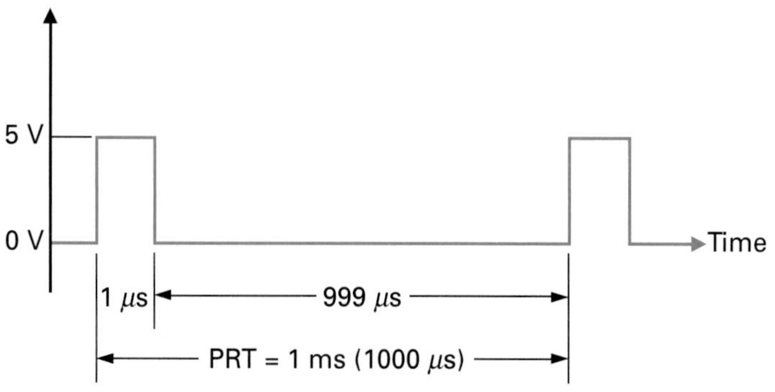

FIGURE 11-43 **PRF and PRT of a Pulse Wave.**

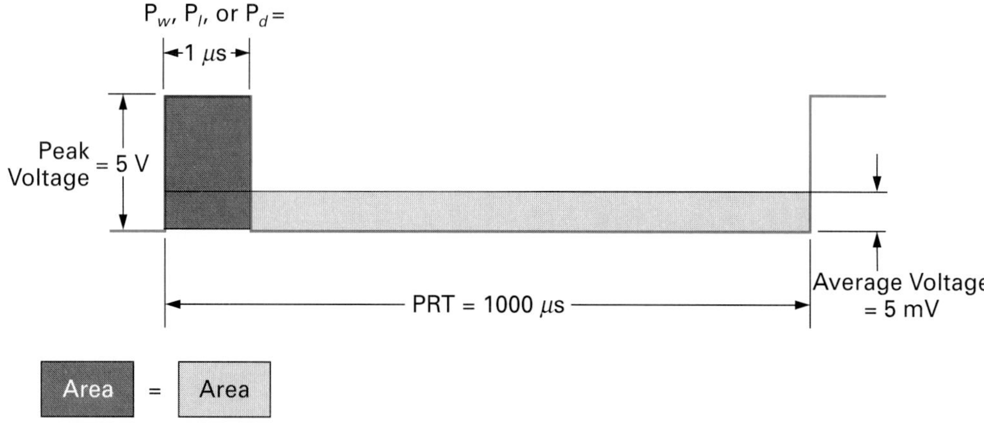

FIGURE 11-44 Average of a Pulse Wave.

Figure 11-44 illustrates the average value of this rectangular waveform. If the voltage and width of the positive pulse are taken and spread out over the entire PRT, they will have a mean level equal, in this example, to 5 mV.

EXAMPLE:

Calculate the duty cycle and average voltage of the following radar pulse waveform:

$$\text{peak voltage, } V_p = 20 \text{ kV}$$
$$\text{pulse length, } P_l = 1 \text{ μs}$$
$$\text{baseline voltage} = \text{OV}$$
$$\text{PRF} = 3300 \text{ pulses per second (pps)}$$

Solution:

$$\text{duty cycle} = \frac{\text{pulse length } (P_l)}{\text{PRT}} \times 100\%$$
$$= \frac{1 \text{ μs}}{303 \text{ μs}} \times 100\% \left(\text{PRT} = \frac{1}{\text{PRF}} = \frac{1}{3300} = 303 \text{ μs}\right)$$
$$= (3.3 \times 10^{-3}) \times 100\%$$
$$= 0.33\%$$
$$V_{avg} = \text{baseline} + (\text{duty cycle} \times V_{p-p})$$
$$= 0 \text{ V} + [(3.3 \times 10^{-3}) \times 20 \times 10^3]$$
$$= 66 \text{ V}$$

Frequency-Domain Analysis

The pulse or rectangular wave is closely related to the square wave, as shown in Figure 11-45; however, some changes occur in its harmonic content. One is that even-number harmonics are present and their amplitudes do not fall

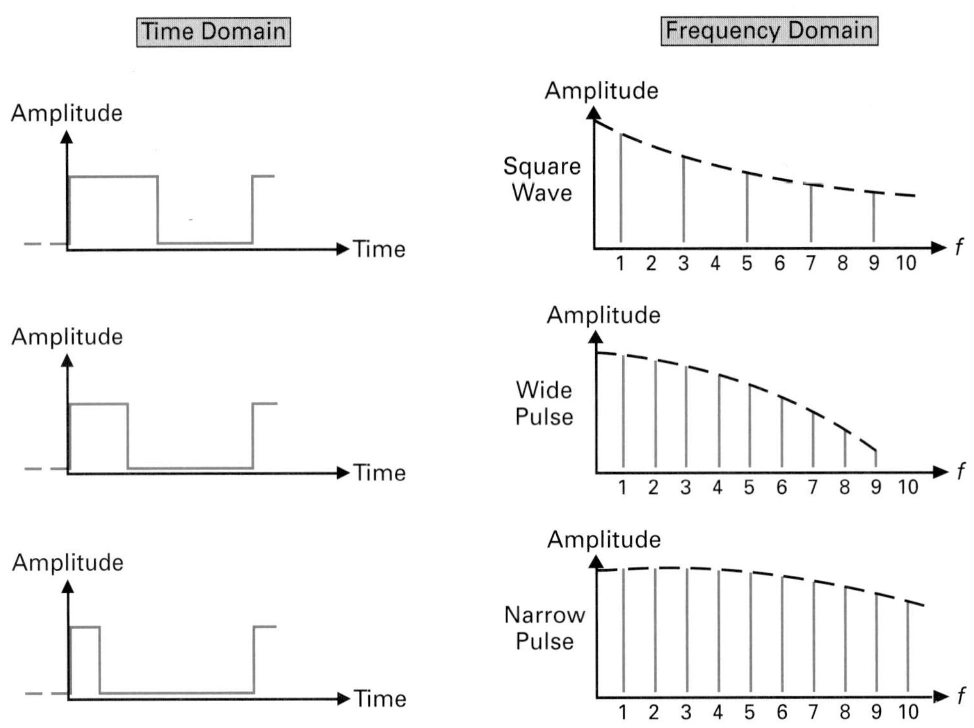

FIGURE 11-45 Time- and Frequency-Domain Analysis of a Pulse Waveform.

off as quickly as do those of the square wave. The amplitude and phase of these sine-wave harmonics are determined by the pulse width and pulse repetition frequency, and the narrower the pulse, the greater the number of harmonics present.

11-3-4 The Triangular Wave

Triangular Wave
A repeating wave that has equal positive and negative ramps that have linear rates of change with time.

Linear
Relationship between input and output in which the output varies in direct proportion to the input.

A **triangular wave** consists of a positive and negative ramp of equal values, as shown in Figure 11-46. Both the positive and negative ramps have a linear increase and decrease, respectively. Linear, by definition, is the relationship between two quantities that exists when a change in a second quantity is directly proportional to a change in the first quantity. The two quantities in this case are voltage or current and time. As shown in Figure 11-46, if the increment of change of voltage ΔV (pronounced "delta vee") is changing at the same rate as the increment of time, Δt ("delta tee"), the ramp is said to be **linear.**

With Figure 11-47(a), the voltage has risen 1 V in 1 second (time 1) and maintains that rise through to time 2 and, consequently, is known as a linearly rising slope. In Figure 11-47(b), the voltage is falling first from 6 to 5 V, which is a 1 V drop in 1 second, and in time 2 from 6 V to 2 V, which is a 4 V drop in 4 s. The rate of fall still remains the same, so the waveform is referred to as a linearly falling slope.

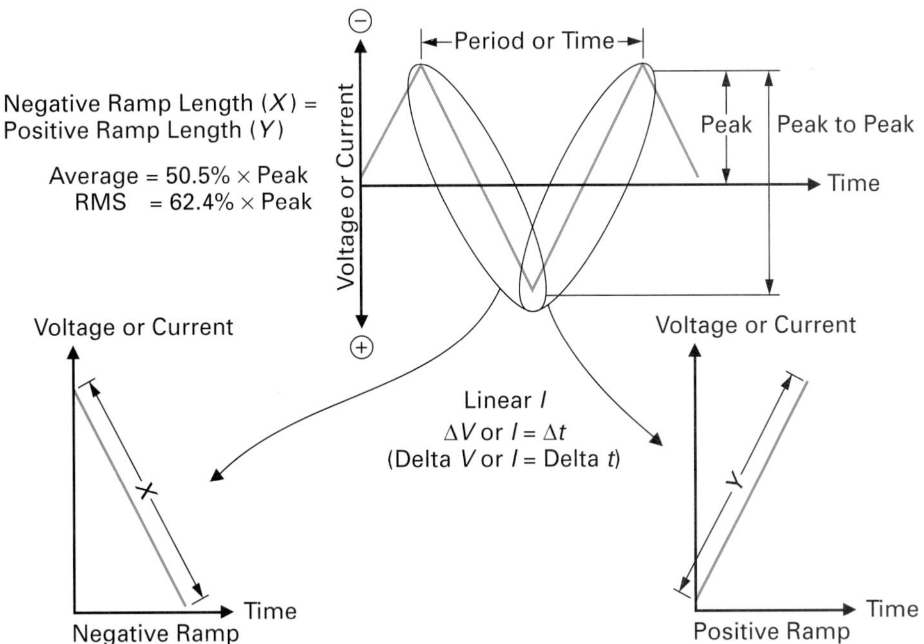

FIGURE 11-46 Triangular Wave.

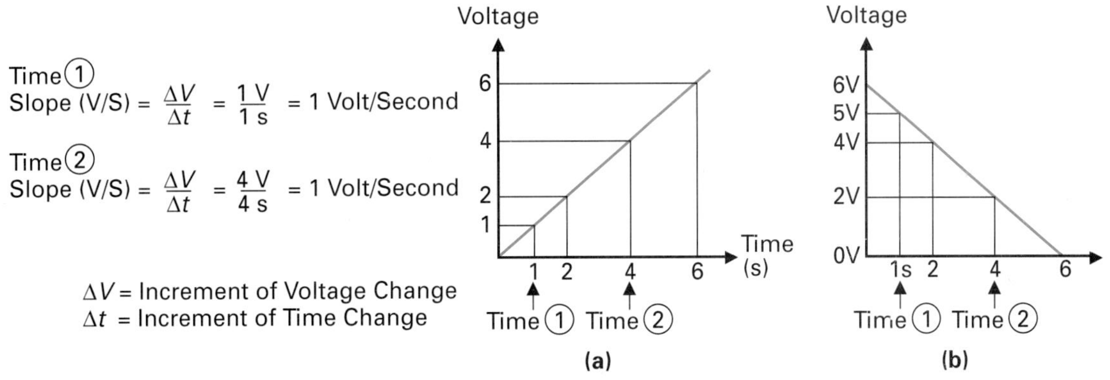

FIGURE 11-47 Linear Triangular-Wave Rise and Fall.

This formula for slope will also apply to a current waveform, and is

$$\text{slope (A/s)} = \frac{\Delta I}{\Delta t}$$

where ΔI = increment of current change

Δt = increment of time change

With triangular waves, frequency and time apply as usual, as seen in Figure 11-48, with

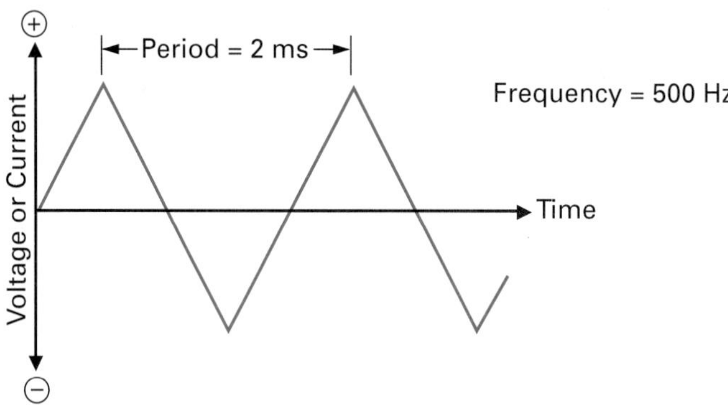

FIGURE 11-48 Triangular-Wave Period and Frequency.

$$\text{frequency} = \frac{1}{\text{time}} \qquad \text{(Hz)}$$

$$\text{time} = \frac{1}{\text{frequency}} \qquad \text{(s)}$$

Frequency-Domain Analysis

The time domain of a triangular wave is shown in Figure 11-49(a), while the frequency domain of a triangular wave is shown in Figure 11-49(b). Frequency domain analysis is often used to test electronic circuits as it tends to highlight problems that would normally not show up when a sine-wave signal is applied.

11-3-5 *The Sawtooth Wave*

Sawtooth Wave
Repeating waveform that rises from zero to a maximum value linearly and then falls to zero and repeats.

On an oscilloscope display (time-domain presentation), the **sawtooth wave** is very similar to a triangular wave, in that a sawtooth wave has a linear ramp. However, unlike the triangular wave, which reverses and has an equal but oppo-

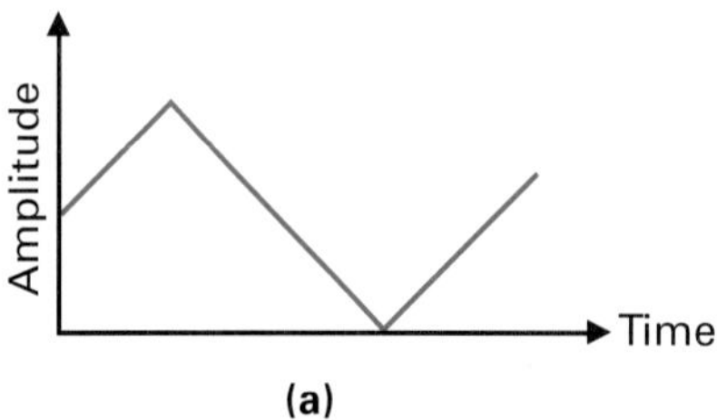

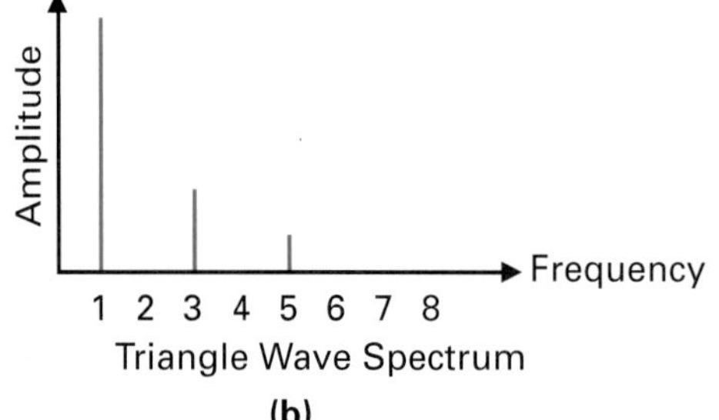

FIGURE 11-49 Analysis of a Triangular Wave. (a) Time Domain. (b) Frequency Domain.

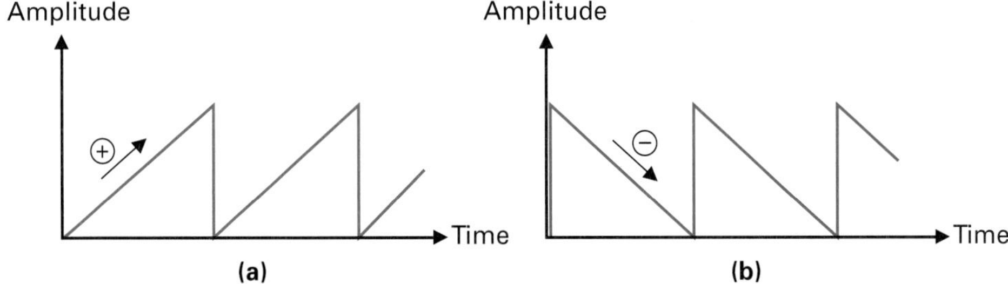

FIGURE 11-50 **Sawtooth Wave. (a) Positive Ramp. (b) Negative Ramp.**

site ramp back to its starting level, the sawtooth "flies" back to its starting point immediately and then repeats the previous ramp, as seen in Figure 11-50, which shows both a positive and a negative ramp sawtooth.

Figure 11-51 shows the odd and even harmonics contained in the frequency domain analysis of a negative-going ramp.

11-3-6 ***Other Waveforms***

The waveforms discussed so far are some of the more common types; however, since every waveform shape (except a pure sine wave) is composed of a large number of sine waves combined in an infinite number of ways, any waveform shape is possible. Figure 11-52 illustrates a variety of waveforms that can be found in all fields of electronics, and the appendix contains a detailed frequency spectrum chart detailing the applications of many electromagnetic frequencies.

SELF-TEST REVIEW QUESTIONS FOR SECTION 11-3

1. Sketch the following waveforms:
- **a.** Sine wave
- **b.** Square wave
- **c.** Rectangular wave
- **d.** Triangular wave
- **e.** Sawtooth wave

2. An oscilloscope gives a ____________-domain representation of a periodic wave, while a spectrum analyzer gives a ____________-domain representation.

3. What are the wavelength formulas for sound waves and electromagnetic waves, and why are they different?

4. A square wave is composed of an infinite number of ____________ harmonics.

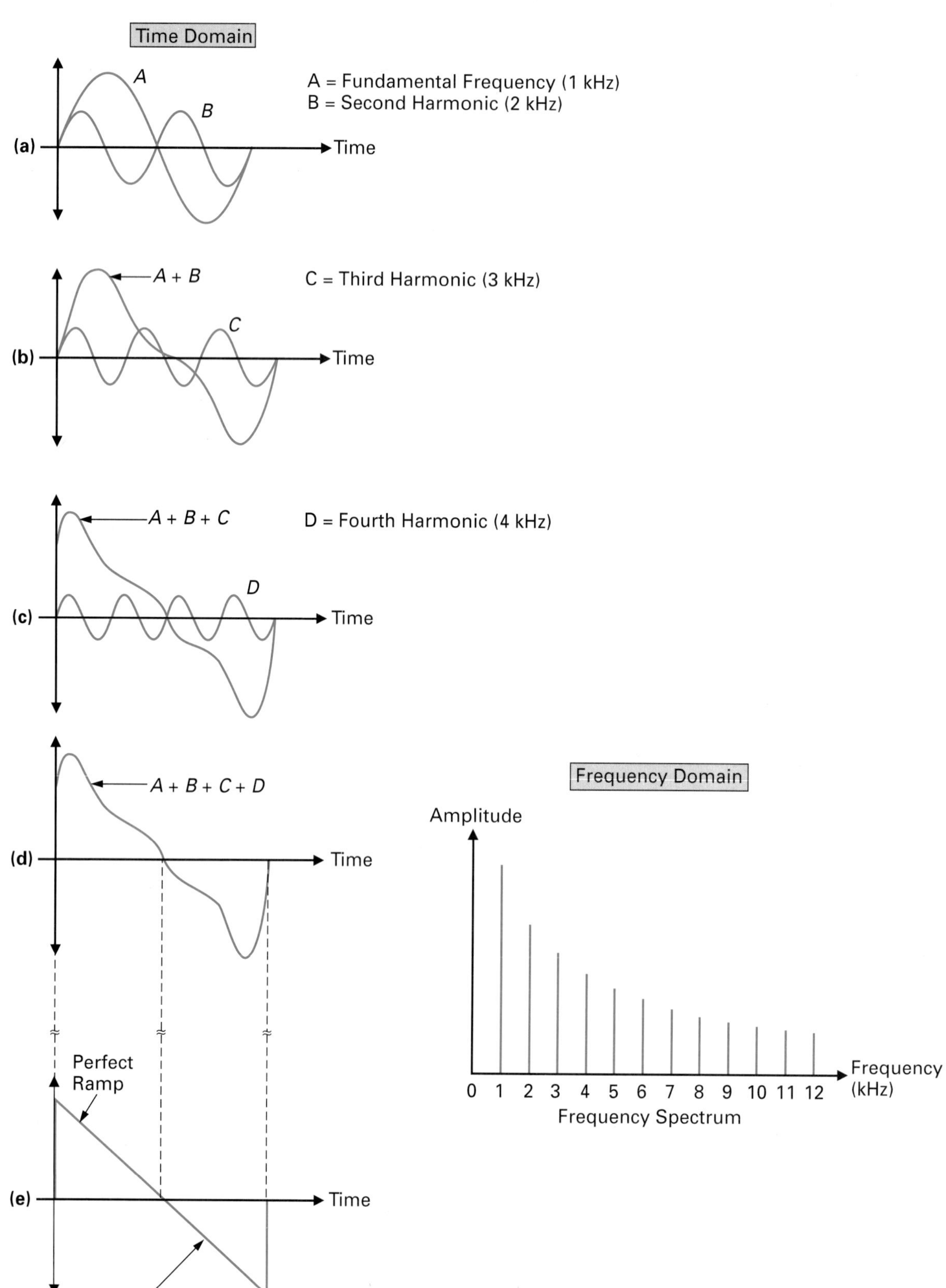

FIGURE 11-51 Analysis of a Sawtooth Wave.

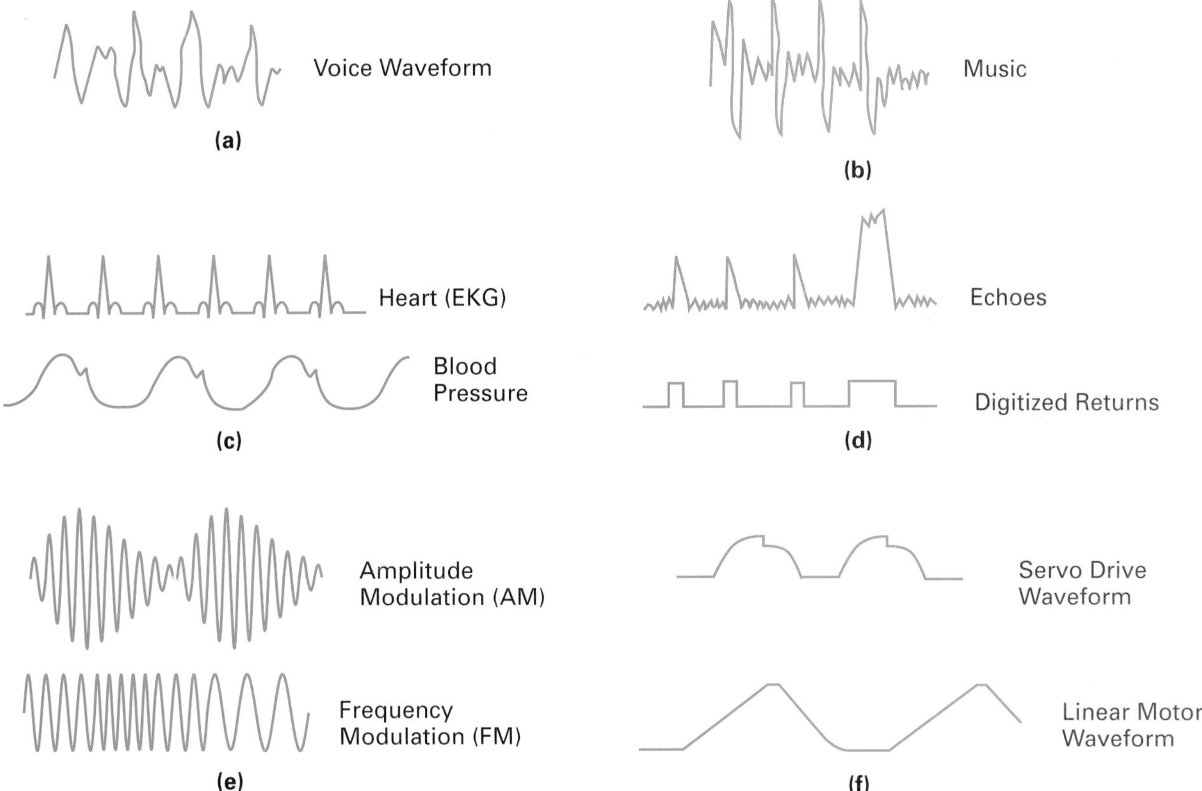

FIGURE 11-52 **Waveforms. (a) Telephone Communications. (b) Radio Broadcast. (c) Medical. (d) Radar/Sonar. (e) Communications. (f) Industrial.**

11-4 ELECTRICITY AND ELECTRONICS

In the beginning of this chapter, it was stated that ac is basically used in two applications, for (1) power transfer and (2) information transfer. These two uses for ac help define the difference between electricity and electronics. Electronic equipment manages the flow of information, while electrical equipment manages the flow of power. In summary:

EQUIPMENT	MANAGES
Electrical	Power (large values of V and I)
Electronic	Information (small values of V and I)

To use a few examples, we can say that a dc power supply is a piece of electrical equipment since it is designed to manage the flow of power. A TV set, however, is an electronic system since its electronic circuits are designed to manage the flow of audio (sound) and video (picture) information.

Since most electronic systems include a dc power supply, we can say that the electrical circuits manage the flow of power, and this power supply enables the electronic circuits to manage the flow of information.

SELF-TEST REVIEW QUESTIONS FOR SECTION 11-4

1. ___________ equipment manages the flow of information, and these ac waveforms normally have small values of current and voltage.

2. ___________ equipment manages the flow of power, and these ac waveforms normally have large values of current and voltage.

SUMMARY

Direct Current (DC) and Alternating Current (AC)

1. While direct current (dc) can be used in many instances to deliver power or represent information, there are certain instances in which ac is preferred. For example, ac is easier to generate and is more efficiently transmitted as a source of electrical power for the home or business. Audio (speech and music) and video (picture) information are generally always represented in electronic equipment as an alternating current or alternating voltage signal.
2. Direct current (dc) is the flow of electrons in one DIRECTion and one direction only. Dc voltage is nonvarying and normally obtained from a battery or power supply unit. The only variation in voltage from a battery occurs due to the battery's discharge, but even then, the current will still flow in only one direction.
3. Some power supplies supply a form of dc known as pulsating dc, which varies periodically from zero to a maximum, back to zero, and then repeats. Pulsating dc is normally supplied by a battery charger and is used to charge secondary batteries. It is also used to operate motors that convert the pulsating dc electrical energy into a mechanical rotation output. Whether steady or pulsating, direct current is current in only one DIRECTion.
4. Alternating current (ac) flows first in one direction and then in the opposite direction. This reversing current is produced by an alternating voltage source, which reaches a maximum in one direction (positive), decreases to zero, and then reverses itself and reaches a maximum in the opposite direction (negative). During the time of the positive voltage alternation, the polarity of the voltage will cause current to flow from negative to positive in one direction. During the time of the negative voltage alternation, the polarity of the voltage will reverse, causing current to flow once again from negative to positive, but in this case, in the opposite direction.
5. There are two main applications for ac:
   a. *Power transfer:* to supply electrical power for lighting, heating, cooling, appliances, and machinery in both home and industry

   b. *Information transfer:* to communicate or carry information, such as radio music and television pictures, between two points.

6. There are three advantages that ac has over dc from a power point of view.
   a. Many appliances and most industrial equipment need a large supply of current, and in this situation a generator would have to be used to generate this large amount of current. Generators can be used to generate either dc or ac, but ac generators can be larger, less complex internally, and cheaper to operate, which are the first reasons why we use ac instead of dc for supplying power.
   b. From a power point of view, ac is always used by electric companies when transporting power over long distances to supply both the home and industry with electrical energy. The amount of power lost (heat) in these power lines can be calculated by using the formula $P = I^2 \times R$, where I is the current flowing through the line and R is the resistance of the power lines. This means that the larger the current, the greater the amount of power lost in the lines in the form of heat and therefore the less the amount of power supplied to the home or industry. For this reason, power companies transport electric energy at a very high voltage between 200,000 and 600,000 V. Since the voltage is high, the current can be low and provide the same amount of power to the consumer. An ac voltage can easily and efficiently be transformed up or down to a higher or lower voltage by utilizing a device known as a transformer, and even though dc voltages can be stepped up and down, the method is inefficient and more complex.
   c. Nearly all electronic circuits and equipment are powered by dc voltages, which means that once the ac power arrives at the home or industry, in most cases it will have to be converted into dc power to operate electronic equipment. It is a relatively simple process to convert ac to dc; however, conversion from dc to ac is a complex and comparatively inefficient process.

Power Transfer

7. The ac power distribution system begins at the electric power plant, which has the powerful large generators driven by turbines to generate large ac voltages.
8. The turbines can be driven by either falling water (hydroelectric), or from steam, which is produced with intense heat by burning either coal, gas, or oil or from a nuclear reactor (thermoelectric).
9. The turbine supplies the mechanical energy to the generator, to be transformed into ac electrical energy.
10. The generator generates an ac voltage of approximately 22,000 V, which is stepped up by transformers to approximately 500,000 V.
11. At each city or town, the voltage is tapped off the long-distance transmission lines and stepped down to approximately 66,000 V and is distributed to large-scale industrial customers.

12. The 66,000 V is stepped down again to approximately 4800 V and distributed throughout the city or town by short-distance transmission lines. This 4800 V is used by small-scale industrial customers and residential customers who receive the ac power via step-down transformers on utility poles, which step down the 4800 V to 240 V and 120 V.
13. A large amount of equipment and devices within industry and the home will run directly from the ac power, such as heating, lighting, and cooling.
14. Some equipment that runs on dc, such as televisions and computers, will accept the 120 V ac and internally convert it to the dc voltages required.
15. The unit that converts ac to dc is called a rectifier.

Information Transfer

16. Information, by definition, is the property of a signal or message that conveys something meaningful to the recipient.
17. Communication, which is the transfer of information between two points, began with speech and progressed to handwritten words in letters and printed words in newspapers and books. To achieve greater distances of communication, face-to-face communications evolved into telephone and radio communications.
18. Sound waves or sounds are normally generated by a vibrating reed or plucked string or a person's vocal cords.
19. Like the ripples produced by a stone falling in a pond, the sound waves are constantly expanding and traveling outward.
20. The microphone is in fact a transducer (energy converter), because it converts the sound wave (which is a form of mechanical energy) into electrical energy in the form of voltage and current, which varies in the same manner as the sound wave and therefore contains the sender's message or information.
21. The electrical wave is a variation in voltage or current and can only exist in a wire conductor or circuit. This electrical signal travels at the speed of light.
22. The speaker, like the microphone, is also an electroacoustical transducer that converts the electrical energy input into a mechanical sound-wave output.
23. To communicate between two distant points, a wire must be connected between the microphone and speaker. However, if an electrical wave is applied to an antenna, the electrical wave is converted into a radio or electromagnetic wave.
24. Antennas are designed to radiate and receive electromagnetic waves, which vary in field strength, and can exist in either air or space. These radio waves, as they are also known, travel at the speed of light and allow us to achieve great distances of communication.
25. More specifically, radio waves are composed of two basic components. The electrical voltage applied to the antenna is converted into an electric field and the electrical current into a magnetic field. This electromagnetic (elec-

FIGURE 11-53 AC Wave Shapes.

The Sine Wave

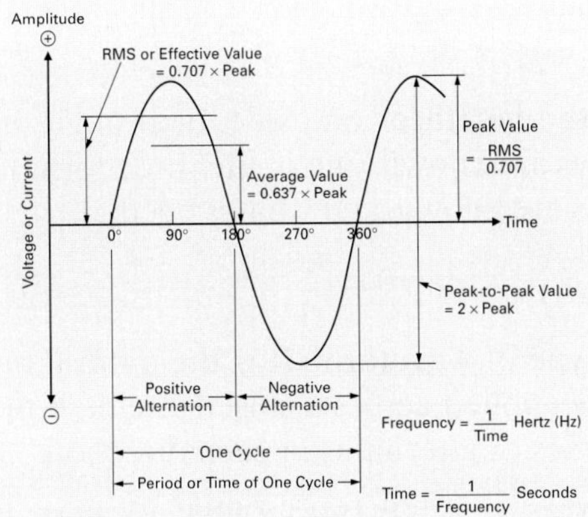

The Square Wave

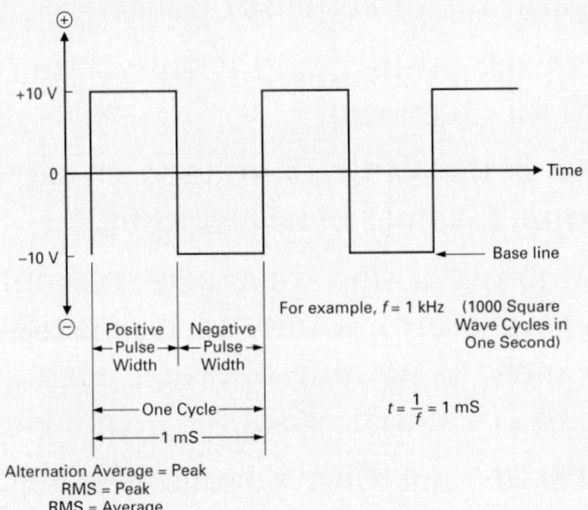

The Rectangular or Pulse Wave

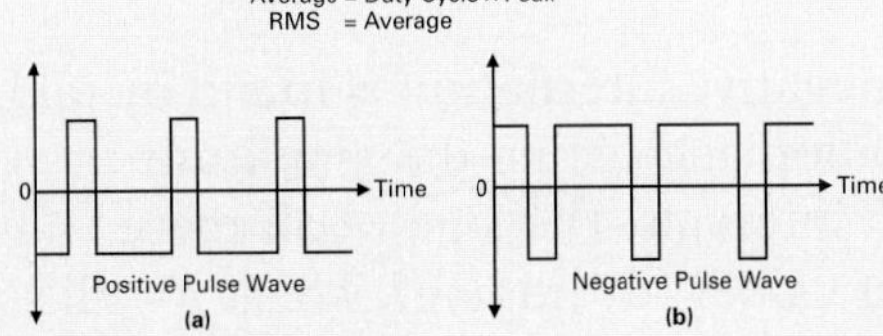

The Triangular Wave

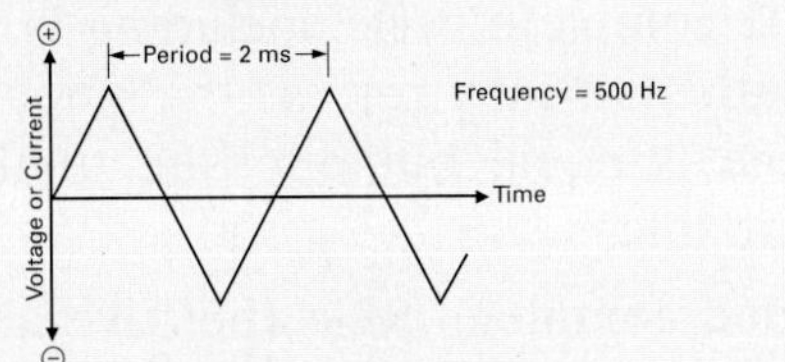

The Sawtooth Wave

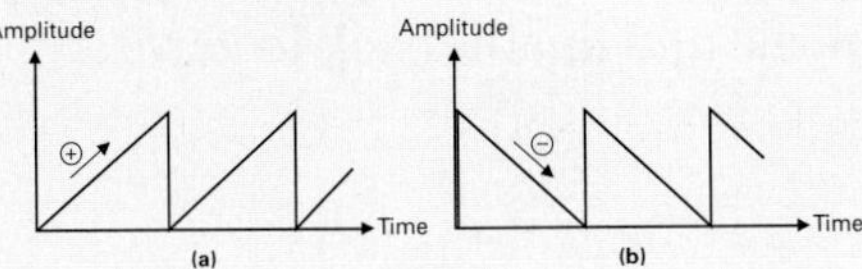

Distance = velocity × time

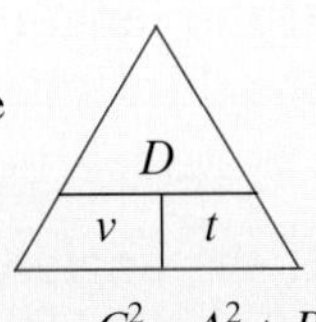

$$C^2 = A^2 + B^2$$

$$C = \sqrt{A^2 + B^2} \qquad B = \sqrt{C^2 - A^2} \qquad A = \sqrt{C^2 - B^2}$$

$$\sin\theta = \frac{O}{H} \qquad \cos\theta = \frac{A}{H} \qquad \tan\theta = \frac{O}{A}$$

$$f = \frac{1}{t} \qquad t = \frac{1}{f}$$ f = frequency, in hertz (Hz); t = period or time, in seconds (s)

Sine wave: $V_{p\text{-}p} = 2 \times V_p$ $\quad V_{p\text{-}p}$ = peak-to-peak voltage; V_p = peak voltage

$I_{p\text{-}p} = 2 \times I_p$ $\quad I_{p\text{-}p}$ = peak-to-peak current; I_p = peak current

$$\text{rms} = 0.707 \times \text{peak}$$

$$\text{Peak} = \frac{\text{rms}}{0.707} \quad \text{or} \quad 1.414 \times \text{rms}$$

$$\text{Average} = 0.637 \times \text{peak}$$

Square wave: $$\text{Duty cycle (\%)} = \frac{\text{pulse width } (P_w)}{\text{period } (t)} \times 100\%$$

$$V \text{ or } I \text{ average} = \text{baseline} + (\text{duty cycle} \times \text{peak to peak})$$

Rectangular wave: $\text{PRF} = \dfrac{1}{\text{PRT}}$ PRF = pulse repetition frequency in pulses per second (pps)

$\text{PRT} = \dfrac{1}{\text{PRF}}$ PRT = pulse repetition time, in seconds(s)

Triangular wave: Slope (A/s) = $\dfrac{\Delta I}{\Delta t}$ delta I (ΔI) = increment of current change; delta t (Δt) = increment of time

$$\text{Wavelength } (\lambda) = \frac{\text{velocity}}{\text{frequency (Hz)}}$$

Sound wave: $\lambda \text{ (m)} = \dfrac{344.4 \text{ m/s}}{f}$

Electromagnetic wave: $\lambda \text{ (cm)} = \dfrac{3 \times 10^{10} \text{ cm/s}}{f}$ or $\lambda \text{ (m)} = \dfrac{3 \times 10^{8} \text{ m/s}}{f}$

Sound, Electrical, and Electromagnetic Waves

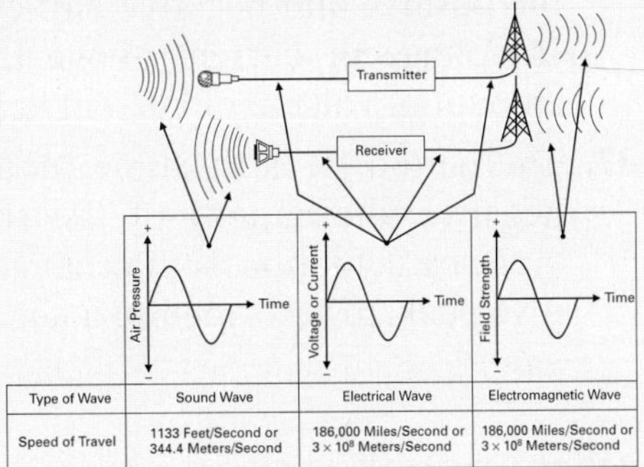

Type of Wave	Sound Wave	Electrical Wave	Electromagnetic Wave
Speed of Travel	1133 Feet/Second or 344.4 Meters/Second	186,000 Miles/Second or 3×10^8 Meters/Second	186,000 Miles/Second or 3×10^8 Meters/Second

tric–magnetic) wave is used to carry a variety of information, such as speech, radio broadcasts, television signals, and so on.

AC Wave Shapes (Figure 11-53)

26. If the length of two sides of a right triangle are known, and the length of the third side is needed, remember that

$$C^2 = A^2 + B^2$$

27. If the angle θ is known along with the length of one side, or if angle θ is needed and the lengths of the two sides are known, one of the three formulas can be chosen based on what variables are known, and what is needed.

$$\sin\theta = \frac{O}{H} \qquad \cos\theta = \frac{A}{H} \qquad \tan\theta = \frac{O}{A}$$

28. The sine wave is the most common type of waveform. It is the natural output of a generator. If this ac voltage is applied across a closed circuit, it produces a current that continually reverses or alternates in each direction.

29. The amplitude or magnitude of a wave is often represented by a vector arrow. The vector's length indicates the magnitude of the current or voltage, while the arrow's point is used to show the direction, or polarity.

30. The peak of an ac wave occurs on both the positive and negative alternation, but is only at the peak (maximum) for an instant.

31. The peak-to-peak value of a sine wave is the value of voltage or current between the positive and negative maximum values of a waveform.

32. Both the positive and negative alternation of a sine wave can accomplish the same amount of work; but the ac waveform is only at its maximum value for an instant in time, spending most of its time between peak currents. The effective value of a sine wave is equal to 0.707 of the peak value.

33. This root-mean-square (rms) result of 0.707 can always be used to tell us how effective an ac sine wave will be.

34. Unless otherwise stated, ac values of voltage or current are always given in rms.

35. The average value of the positive or negative alternation is found by taking either the positive or negative alternation and listing the amplitude or vector length of current or voltages at 1° intervals. The sum of all these values is then divided by the total number of values (averaging), which for all sine waves will calculate out to be 0.637 of the peak voltage or current.

36. The average of a positive or negative alternation (half-cycle) is equal to 0.637 × peak; however, the average of the complete cycle, including both the positive and negative half-cycles, is mathematically zero as the amount of voltage or current above the zero line is equal but opposite to the amount of voltage or current below the zero line.

37. The period (t) is the time required for one complete cycle (positive and negative alternation) of the sinusoidal current or voltage waveform. A *cycle,* by definition, is the change of an alternating wave from zero to a positive peak, to zero, then to a negative peak, and finally, back to zero.

38. Frequency is the number of repetitions of a periodic wave in a unit of time. It is symbolized by f and is given the unit hertz (cycles per second), in honor of a German physicist, Heinrich Hertz.

39. Sinusoidal waves can take a long or a short amount of time to complete one cycle. This time is related to frequency in that period is equal to the reciprocal of frequency.

40. All homes in the United States receive at their wall outlets an ac voltage of 120 V rms at a frequency of 60 Hz. This frequency was chosen for convenience, as a lower frequency would require larger transformers, and if the frequency were too low, the slow switching (alternating) current through the light bulb would cause it to flicker.

41. A higher frequency than 60 Hz was found to cause an increase in the amount of heat generated in the core of all power distribution transformers due to eddy currents and hysteresis losses.

42. Wavelength, as its name states, is the physical length of one complete cycle and is generally measured in meters. The wavelength (λ, lambda) of a complete cycle is dependent on the frequency and velocity of the transmission.

43. Electromagnetic waves or radio waves travel at the speed of light in air or a vacuum, which is 3×10^8 meters/second or 3×10^{10} cm/second.

44. Sound waves travel at a slower speed than electromagnetic waves, as their mechanical vibrations depend on air molecules, which offer resistance to the traveling wave.

45. The phase of a sine wave is always relative to another sine wave of the same frequency.

46. The sine wave is called a sine wave because it changes in amplitude at the same rate as the sine trigonometric function.

47. The square wave is a periodic (repeating) wave that alternates from a positive peak value to a negative peak value, and vice versa, for equal lengths of time.

48. The duty cycle is the ratio of a pulse width (positive or negative pulse or cycle) to the overall period or time of the wave and is normally given as a percentage.

49. Since a square wave always has a positive and a negative alternation that are equal in time, the duty cycle of all square waves is equal to 50%, which actually means that the positive cycle lasts for 50% of the time of one cycle.

50. The average of the complete square wave cycle should calculate out to be zero, as the amount above the line equals the amount below.

51. It takes a small amount of time for the wave to increase to its positive value (the rise time) and consequently an equal amount of time for a wave to decrease to its negative value (the fall time). Rise time (T_R), by definition, is the time it takes for an edge to rise from 10% to 90% of its full amplitude, while fall time (T_F) is the time it takes for an edge to fall from 90% to 10% of its full amplitude.

52. A *periodic wave* is a wave that repeats the same wave shape from one cycle to the next.
53. A time-domain representation of a periodic sine wave, which is the same way it would appear on an oscilloscope display, plots the sine wave's amplitude against time.
54. A frequency-domain representation of the same periodic sine wave, which shows the wave as it would appear on a spectrum analyzer, plots the sine wave's amplitude against frequency instead of time. Pure sine waves have no other frequency components.
55. Other periodic wave shapes, such as square, pulse, triangular, sawtooth, or irregular, are actually made up of a number of sine waves having a particular frequency, amplitude, and phase.
56. The fundamental frequency sine wave is always the lowest frequency and largest amplitude component of any waveform shape and is used as a reference.
57. Sine wave harmonic frequencies are smaller in amplitude and are some multiple of the fundamental frequency.
58. Square waves are composed of a fundamental frequency and an infinite number of odd harmonics (third, fifth, seventh, and so on).
59. The rectangular wave is similar to the square wave in many respects, in that it is a periodic wave that alternately switches between one of two fixed values. The difference in the rectangular wave is that it does not remain at the two peak values for equal lengths of time.
60. Pulse repetition frequency is the number of times per second that a pulse is transmitted.
61. Pulse repetition time is the time interval between the start of two consecutive pulses.
62. Pulse width, pulse length, or pulse duration is the time interval between the leading edge and trailing edge of a pulse at which the amplitude reaches 50% of the peak pulse amplitude.
63. The triangular wave is a repeating wave that has equal positive and negative ramps that have linear rates of change with time.
64. The term linear describes a relationship between input and output in which the output varies in direct proportion to the input.
65. The sawtooth wave is a repeating waveform that rises from zero to a maximum value linearly and then falls to zero and repeats.

Electricity and Electronics

66. AC is basically used in two applications, for (1) power transfer and (2) information transfer.
67. Electronic equipment manages the flow of information, while electrical equipment manages the flow of power.
68. A dc power supply is a piece of electrical equipment since it is designed to manage the flow of power. A TV set, however, is an electronic system since

its electronic circuits are designed to manage the flow of audio (sound) and video (picture) information.

69. Since most electronic systems include a dc power supply, we can say that the electrical circuits manage the flow of power, and this power supply enables the electronic circuits to manage the flow of information.

NEW TERMS

Adjacent
Alternating current (ac)
Amplifier
Amplitude
Antenna
Audio
Average value
Battery charger
Communication
Cosine
Delta (Δ)
Direct current
Duty cycle
Effective value
Electrical wave
Electric field
Electroacoustical transducer
Electromagnetic wave
Fall time
Frequency
Frequency domain
Fundamental frequency
Harmonics
Hypotenuse
Information
Lag
Lead
Leading edge
Microphone
Opposite
Oscilloscope
Peak-to-peak value (p–p)
Peak value (p)
Period (t)
Periodic
Phase shift
Power transfer
Pulsating dc
Pulse duration (P_d)
Pulse length (P_l)
Pulse repetition frequency (PRF)
Pulse repetition time (PRT)
Pulse waveform
Pulse width (P_w)
Pythagorean Theorem
Rectangular waveform
Rectifier
Right-angle triangle
Rise time
Root mean square (rms)
Sawtooth waveform
Sine
Sine wave
Sinusoidal
Slope
Sound wave
Speaker
Spectrum analyzer
Speed of light
Speed of sound
Square wave
Tangent
Time domain
Trailing edge
Transducer
Transformer
Transmission lines
Triangular wave
Trigonometry
Utility pole
Vector
Video
Wavelength

REVIEW QUESTIONS

Multiple-Choice Questions

1. A current that rises from zero to maximum positive, returns to zero, and then repeats is known as:

   a. Alternating current
   b. Ac
   c. Pulsating dc
   d. Steady dc

2. A current that rises from zero to maximum positive, decreases to zero, and then reverses to reach a maximum in the opposite direction (negative) is known as:

   a. Alternating current **c.** Steady direct current
   b. Pulsating direct current **d.** All of the above

3. The advantage(s) of ac over dc from a power distribution point of view is/are:

   a. Generators can supply more power than batteries
   b. Ac can be transformed to a high or low voltage easily, minimizing power loss
   c. Ac can easily be converted into dc
   d. All of the above
   e. Only (a) and (c)

4. The approximate voltage appearing on long-distance transmission lines in the ac distribution system is:

   a. 250 V **c.** 500,000 V
   b. 2500 V **d.** 250,000 V

5. The most common type of alternating wave shape is the:

   a. Square wave **c.** Rectangular wave
   b. Sine wave **d.** Triangular wave

6. ____________ equipment manages the flow of information.

   a. Electronic **c.** Discrete
   b. Electrical **d.** Integrated

7. ____________ equipment manages the flow of power.

   a. Electronic **c.** Discrete
   b. Electrical **d.** Integrated

8. The peak-to-peak value of a sine wave is equal to:

   a. Twice the rms value **c.** Twice the peak value
   b. 0.707 times the rms value **d.** 1.14 × the average value

9. The rms value of a sine wave is also known as the:

   a. Effective value **c.** Peak value
   b. Average value **d.** All of the above

10. The peak value of a 115 V (rms) sine wave is:

    a. 115 V **c.** 162.7 V
    b. 230 V **d.** Two of the above could be true.

11. The mathematical average value of a sine wave cycle is:

    a. 0.637 × peak **c.** 1.414 × rms
    b. 0.707 × peak **d.** Zero

12. The frequency of a sine wave is equal to the reciprocal of ____________.

    a. The period **d.** Both (a) and (b)
    b. One cycle time **e.** None of the above
    c. One alternation

13. What is the period of a 1-MHz sine wave?

a. 1 ms c. 10 ms
b. One millionth of a second d. 100 μs

14. The sine of 90° is:

a. 0 c. 1
b. 0.5 d. Any of the above

15. What is the frequency of a sine wave that has a cycle time of 1 ms?

a. 1 MHz c. 200 m
b. 1 kHz d. 10 kHz

16. The pulse width (P_w) is the time between the __________ points on the positive and negative edges of a pulse.

a. 10% c. 50%
b. 90% d. All of the above

17. The duty cycle is the ratio of __________ to period.

a. Peak c. Pulse length
b. Average power d. Both (a) and (c)

18. With a pulse waveform, PRF can be calculated by taking the reciprocal of:

a. The duty cycle c. P_d
b. PRT d. P_l

19. The sound wave exists in __________ and travels at approximately __________.

a. Space, 1130 ft/s c. Air, 3×10^6 m/s
b. Wires, 186,282.397 miles/s d. None of the above

20. The electrical and electromagnetic waves travel at a speed of:

a. 186,000 miles/s c. 162,000 nautical miles/s
b. 3×10^8 meters/s d. All of the above

Essay Questions

21. Describe the three advantages that ac has over dc from a power point of view. (11-2-1)

22. Describe briefly the ac power distribution system. (11-2-1)

23. What is the difference between the words electricity and electronics? (11-4)

24. What are the five basic ac information wave shapes? (11-3)

25. Describe briefly the following terms as they relate to the sine wave: (11-3-1)

a. RMS d. Average g. Period
b. Peak e. The name sine h. Wavelength
c. Peak to peak f. Frequency i. Phase

26. Describe briefly the following terms as they relate to the square wave: (11-3-2)

a. Duty cycle b. Average

27. Describe briefly the following terms as they relate to the rectangular wave: (11-3-3)

a. PRT **c.** Duty cycle
b. PRF **d.** Average

28. Briefly describe the meaning of the terms *fundamental frequency* and *harmonics.* (11-3-2)

29. List and describe all the pertinent information relating to the following information carriers: (11-3)

a. Sound wave **c.** Electromagnetic wave
b. Electrical wave

30. Describe the difference between frequency- and time-domain analysis. (11-3-2)

Practice Problems

31. Calculate the periods of the following sine-wave frequencies:

a. 27 kHZ **d.** 365 Hz
b. 3.4 MHz **e.** 60 Hz
c. 25 Hz **f.** 200 kHz

32. Calculate the frequency for each of the following values of time:

a. 16 ms **d.** 0.05 s
b. 1 s **e.** 200 μs
c. 15 μs **f.** 350 ms

33. A 22 V peak sine wave will have the following values:

a. Rms voltage = **c.** Peak-to-peak voltage =
b. Average voltage =

34. A 40 mA rms sine wave will have the following values:

a. Peak current = **c.** Average current =
b. Peak-to-peak current =

35. How long would it take an electromagnetic wave to travel 60 miles?

36. An 11 kHz rectangular pulse, with a pulse width of 10 μs, will have a duty cycle of ____________%.

37. Calculate the PRT of a 400 kHz pulse waveform.

38. Calculate the average current of the pulse waveform in Question 36 if its peak current is equal to 15 A.

39. What is the duty cycle of a 10 V peak square wave at a frequency of 1 kHz?

40. Considering a fundamental frequency of 1 kHz, calculate the frequency of its:

a. Third harmonic **c.** Seventh harmonic
b. Second harmonic

CHAPTER

12

OBJECTIVES

After completing this chapter, you will be able to:

1. Describe how the multimeter is able to measure ac as well as dc.
2. Explain some of the multimeter accessories, such as:
   a. Current clamps
   b. RF probes
   c. High-voltage probes
3. Compare the advantages and disadvantage of the analog multimeter and digital multimeter.
4. Define the function and basic operation of;
   a. A frequency counter
   b. An audio-frequency generator
   c. A radio-frequency generator
5. Describe how to troubleshoot electronic equipment by applying a six-step troubleshooting procedure.
6. Explain some frequently used troubleshooting techniques.

AC Test Equipment and Troubleshooting Techniques

OUTLINE

V I G N E T T E

Logging On

During the seventeenth century, European thinkers were obsessed with any device that could help in mathematical calculation. Scottish mathematician John Napier decided to meet this need, and in 1614 he published his new discovery of logarithms. In this book, consisting mostly of tediously computed tables, Napier stated that a logarithm is the exponent of a base number, for example, 100 is 10^2, 27 is $10^{1.43136}$, 10 is 10^1, 6 is $10^{0.77815}$, and any number no matter how large or small can be represented in this manner. He also outlined how the multiplication of two numbers could be achieved by simple addition. For example, if the logarithm of one number, 2, which is 0.30103, is added to the logarithm of another number, 4, which is 0.60206, the result is equal to the logarithm of 8, which is 0.90309. Therefore, the multiplication of two large numbers could now be achieved by looking up the logarithms of the two numbers in a log table, adding them together, and then finding the number that corresponds to that sum in an antilog (reverse log) table. In this example, the antilog of 0.9039 is 8.

Napier's table of logarithms were used by William Oughtred, who, just ten years after Napier's death in 1617, developed a hand mechanical device that could be used for rapid calculation. This device, considered the first pocket calculator, was the slide rule.

As well as being a brilliant mathematician, Napier was also interested in designing military weapons. One such unfinished project was a death ray system consisting of an arrangement of mirrors and lenses that, when aligned, would produce a concentrated lethal beam of sunlight.

I N T R O D U C T I O N

As a technician or engineer, you are going to be required to diagnose and repair a problem in the shortest amount of time possible. To aid in the efficiency of this fault finding and repair process, you can make use of certain pieces of test equipment. Humans have five kinds of sensory systems: touch, taste, sight, sound, and smell. Four can be used for electronic troubleshooting: sight, sound, touch, and smell.

Electronic test equipment can be used either to *sense* a circuit's condition or to *generate* a signal to see the response of the component or circuit to that signal. For example, the dc multimeter can be used to measure voltage, current, or resistance. When voltage or current is selected, the meter senses the power present in the circuit and gives an indication of that power as voltage or current on either an analog or digital readout display. When resistance is being measured, the multimeter utilizes an internal battery to generate a current that flows out of the meter. The amount of current flow determines the amount of meter deflection, and this current is dependent on the amount of resistance connected to the meter to be measured.

In Chapter 5 you were introduced to the four basic test instruments: the dc power supply, the multimeter, the function generator, and the oscilloscope. In this chapter we extend our studies of test instruments, and examine a six-step troubleshooting procedure that will help you to establish good troubleshooting techniques.

12-1 ADDITIONAL TEST INSTRUMENTS

12-1-1 *The AC Meter*

In Chapter 5 we discussed how a multimeter, like the one shown in Figure 12-1, could be used to measure direct current and voltage. Most multimeters can be used to measure either dc or ac. When the technician or engineer wishes to measure ac, the ac current or voltage is converted to dc internally by a circuit known as a **rectifier,** as shown in Figure 12-2, before passing on to the meter.

Rectifier
A device that converts alternating current into a unidirectional or dc current.

The dc produced by the rectification process is in fact pulsating, as shown in Figure 12-3, so the current through the meter is a series of pulses rising from zero to maximum (peak) and from maximum back to zero. Frequencies below 10 Hz (lower frequency limit) will cause the digits on a digital multimeter to continually increase in value and then decrease in value as the meter follows the

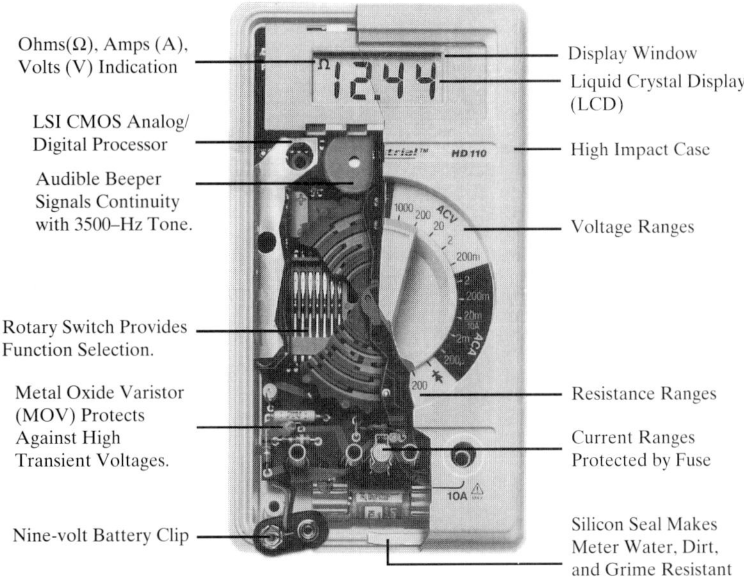

FIGURE 12-1 **Multimeter.**

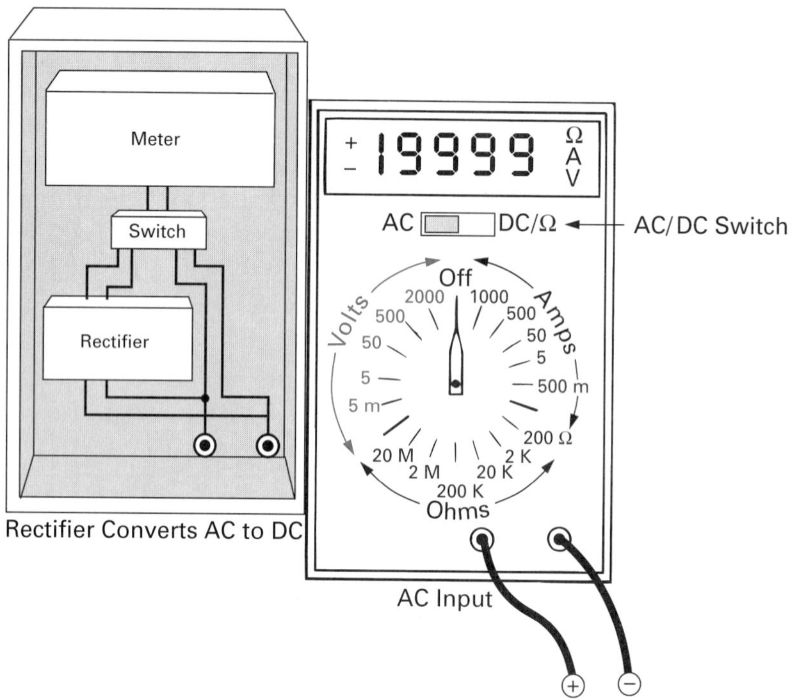

FIGURE 12-2 **AC Meter Uses a Rectifier to Produce DC.**

pulsating dc. This makes it difficult to read the meter. This effect will also occur with an analog multimeter which, from 10 Hz to approximately 2 kHz, will not be able to follow the fluctuation, causing the needle of the meter movement to remain in a position equal to the average value of the pulsating dc from the rectifier (0.637 of peak). Most meters are normally calibrated internally to indicate rms values (0.707 of peak) rather than average values, because this effective value is most commonly used when expressing ac voltage or current. The upper frequency limit of the ac meter is approximately 2 to 8 kHz. Beyond this limit the meter becomes progressively inaccurate due to the reactance of the capacitance in the rectifier of both types of meters, and the reactance of the moving coil in the analog multimeter. This reactance will result in inaccurate indications due to the change in opposition at different ac input frequencies.

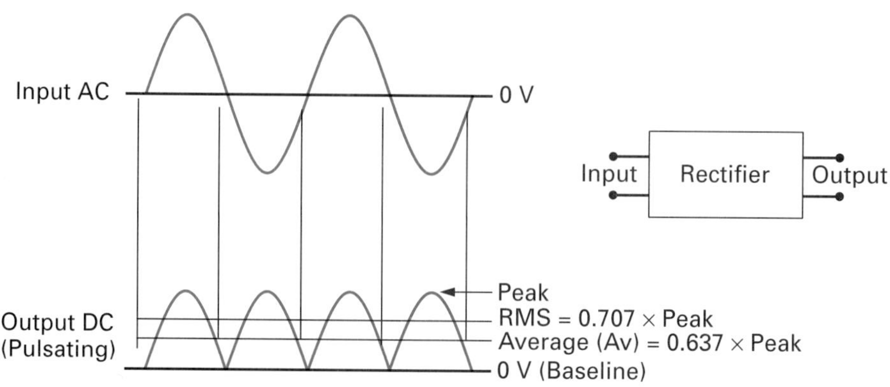

FIGURE 12-3 **Rectifier.**

Current Clamps

The voltmeter is probably the most frequently used setting on the multimeter. A meter reading can be obtained by just connecting the probes across the component or source to be measured, unlike the ammeter, which requires that the circuit current path be opened, and the ammeter inserted in the path of current flow. If a current measurement is required, a clamp can be used, as seen in Figure 12-4, which allows us to sense the amount of current flow through the conductor without opening the current path.

Current Clamp
A device used in conjunction with an ac ammeter, containing a magnetic core in the form of hinged jaws that can be snapped around the current-carrying wire.

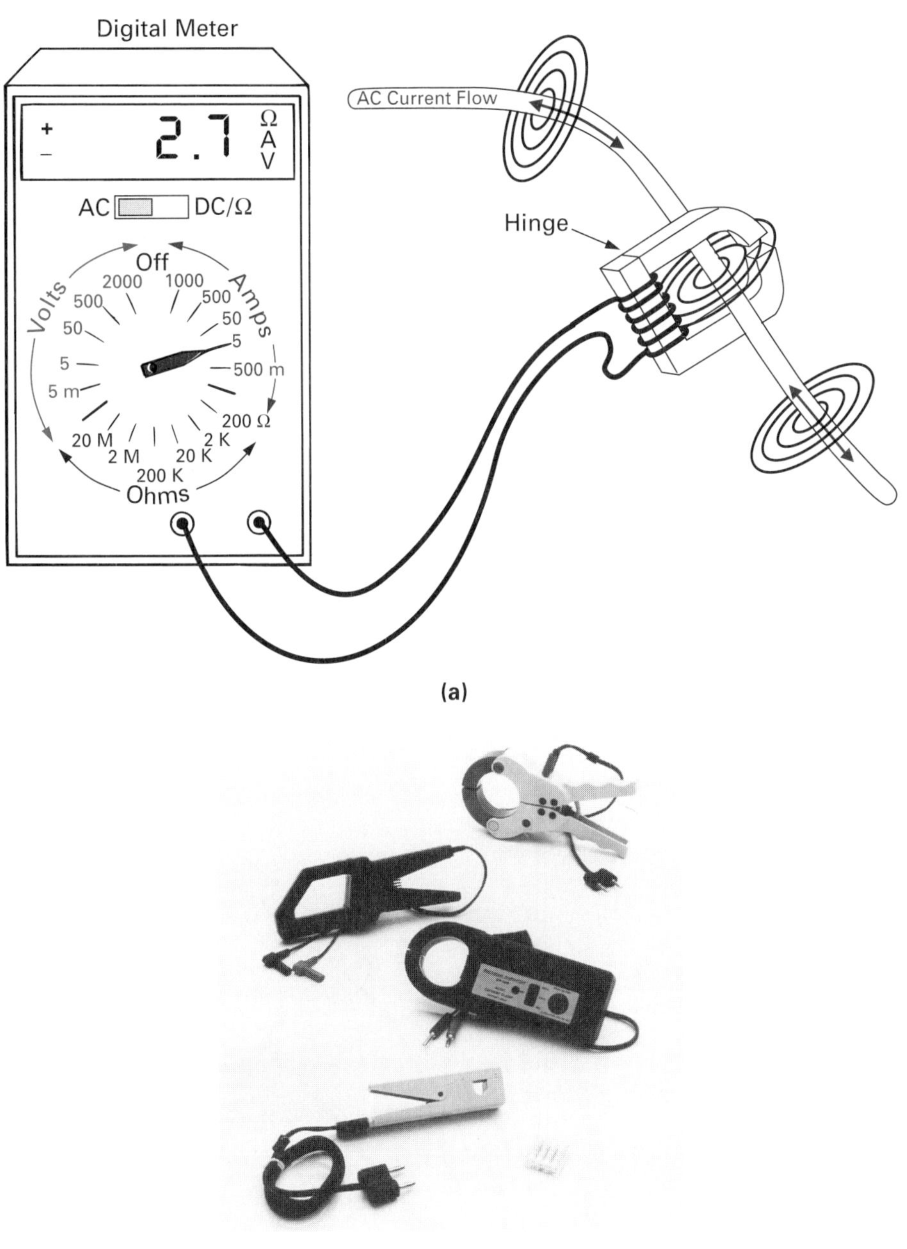

FIGURE 12-4 **Current Clamp.**

The alternating current flowing through the conductor produces an expanding and collapsing magnetic field, which cuts across the coil of wire wound around the core of the clamp, and induces an alternating voltage in the coil (1 mA induces 1 mV). The induced alternating voltage causes an alternating current to flow, which is converted to dc by the rectifier and used to operate the meter. The larger the current flowing in the conductor, the larger the magnetic field surrounding the conductor, which results in a greater induced voltage, current, and consequent meter movement. These clamps are generally ineffective at measuring smaller currents (microamperes), because the magnetic field produced by the current is too weak.

Radio-Frequency Probes

Radio-Frequency Probe
A probe used in conjunction with an ac meter to measure high-frequency RF signals.

As mentioned previously, the meter has a high range limit of approximately 2 kHz. If higher-frequency (radio frequencies) electrical waves are to be measured, a **radio-frequency** (RF) **probe,** as shown in Figure 12-5, can be used. The probe picks up the high-frequency ac voltage from a conducting point on the circuit and passes or couples it to a capacitor, which blocks any dc that could be present at the test point, as we only want to measure the high-frequency ac. The rectifier within the probe will convert this ac input into a dc output, which will be displayed as an rms value on the meter display.

High-Voltage Probe

High-Voltage Probe
Accessory to the voltmeter that has added multiplier resistors within the probe to divide up the large potential being measured by the probe.

The typical multimeter can handle voltages up to approximately 1000 V. If you wish to measure voltages higher than this, another component, known as a **high-voltage** (HV) **probe,** such as the one shown in Figure 12-6, can be used. The high-voltage probe has additional multiplier resistors to drop the extra voltage. Most high-voltage probes are designed so that 1⁄100 of voltage at the probe tip

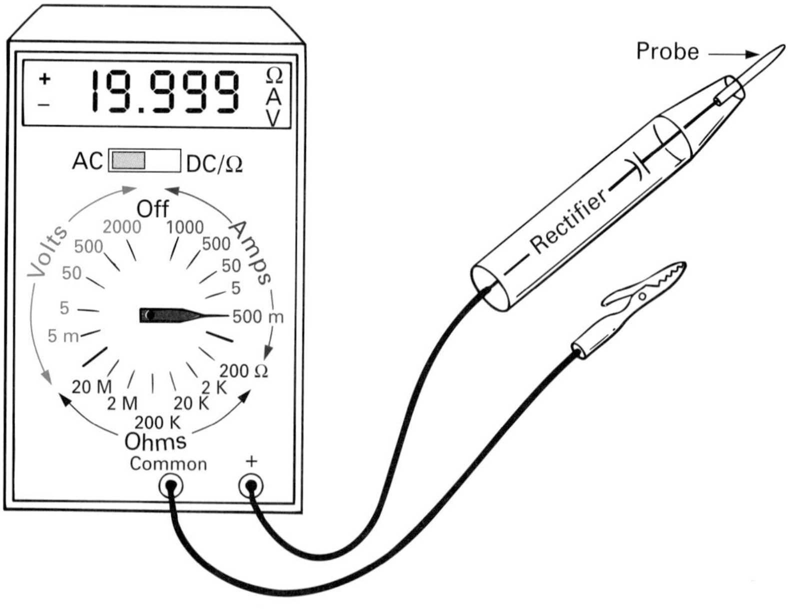

FIGURE 12-5 Radio-Frequency Probe.

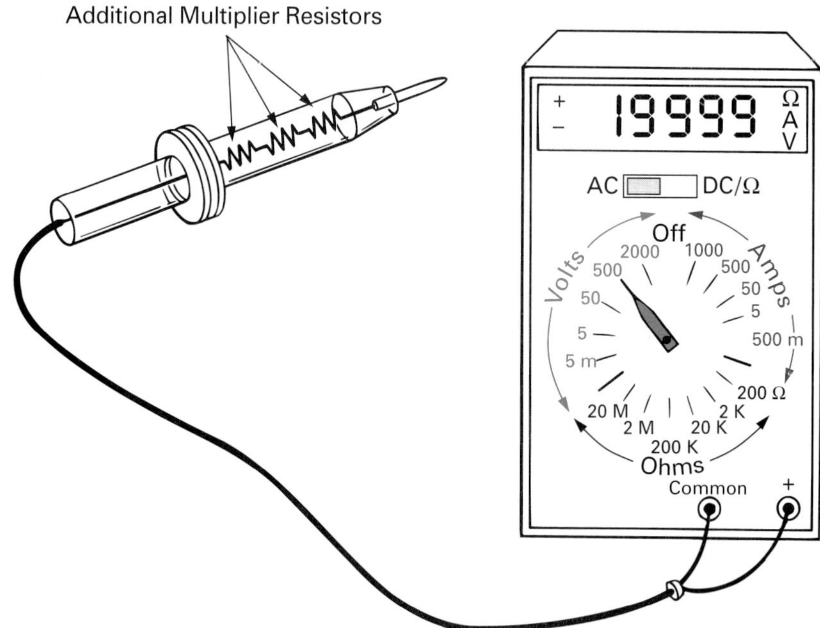

FIGURE 12-6 **High-Voltage Probe.**

from the test point will appear out of the probe and be applied to the meter. For example, if 10 kV is being measured, 100 V will appear out of the probe and be applied to the meter ($\frac{1}{100} \times 10{,}000$ V = 100 V). This probe would be called a × 100 probe because the 100 V shown on the meter display would now have to be multiplied by 100 for the operator to determine the voltage (100 V × 100 = 10 kV). The high-voltage probe is especially well insulated to protect its user, who should apply all safety precautions and exercise extreme caution.

EXAMPLE:

A DMM indicates 3.9 V on its display when a test point is probed by a × 1000 high-voltage probe. What is the voltage at this test point?

Solution:

A × 1000 probe will divide the voltage by 1000, so the displayed voltage must be multiplied by 1000 to obtain the correct value:

$$3.9 \text{ V} \times 1000 = 3.9 \text{ kV}$$

Analog and Digital Multimeters

The digital meter is superior to the analog meter in two basic ways:

1. The digital multimeter (DMM) has an easy-to-read display, with decimal points and polarity, while with the analog multimeter the value has to be interpreted by the needle position and the range selected, which may result in human errors.

2. Some of the best analog multimeters have accuracies of 1%, whereas a DMM will typically have an accuracy of 0.01%.

The analog multimeter, however, has one advantage over the DMM: When measuring low-frequency ac signals of several hertz, the analog multimeter will deflect between zero and some value due to the pulsating dc. The operator will see that the signal is pulsating at a low frequency due to the back-and-forth meter needle movement, whereas with the DMM this continual change will cause the digits on the display to change continually and therefore not allow the operator to take a reading.

When it comes to a choice between an analog or a digital multimeter, it seems that most people prefer the DMM because of its easy-to-read display and accuracy.

12-1-2 ***The Oscilloscope***

Oscilloscope
Instrument used to view signal amplitude, frequency, and shape at different points throughout a circuit.

Figure 12-7 illustrates a typical **oscilloscope** (sometimes abbreviated to *scope*), which is used primarily to display the shape and spacing of electrical signals. The oscilloscope displays the actual sine, square, rectangular, triangular, or sawtooth wave shape that is occurring at any point in a circuit. This display is made on a cathode-ray tube (CRT), which is also used in television sets and computers to display video information. From the display on the CRT, we can measure or calculate time, frequency, and amplitude characteristics such as rms, average, peak, and peak-to-peak.

The oscilloscope allows us to see what is happening at every point through a circuit. In Figure 12-8 you can see the different waveforms at different points on the circuit. There are also voltage test points that can be tested with a voltmeter if a scope is not available. A voltmeter, as you can well imagine, does not supply the technician or engineer with as much information as the oscilloscope.

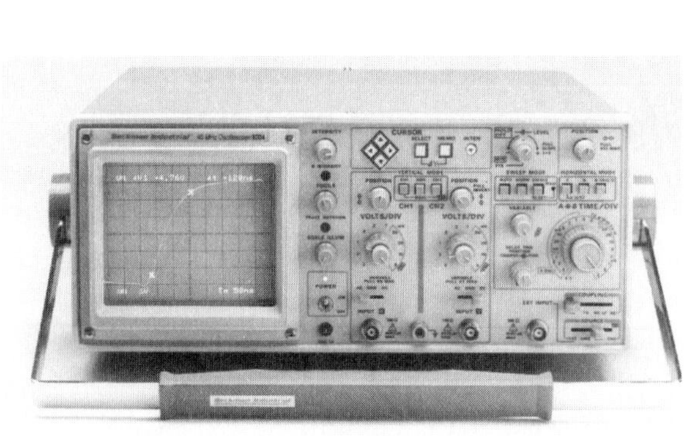

(a)

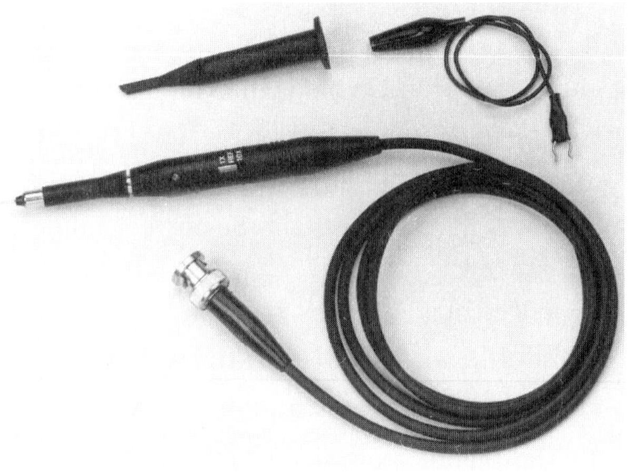

(b)

FIGURE 12-7 Typical Oscilloscope. (a) Oscilloscope. (b) Oscilloscope Probe.

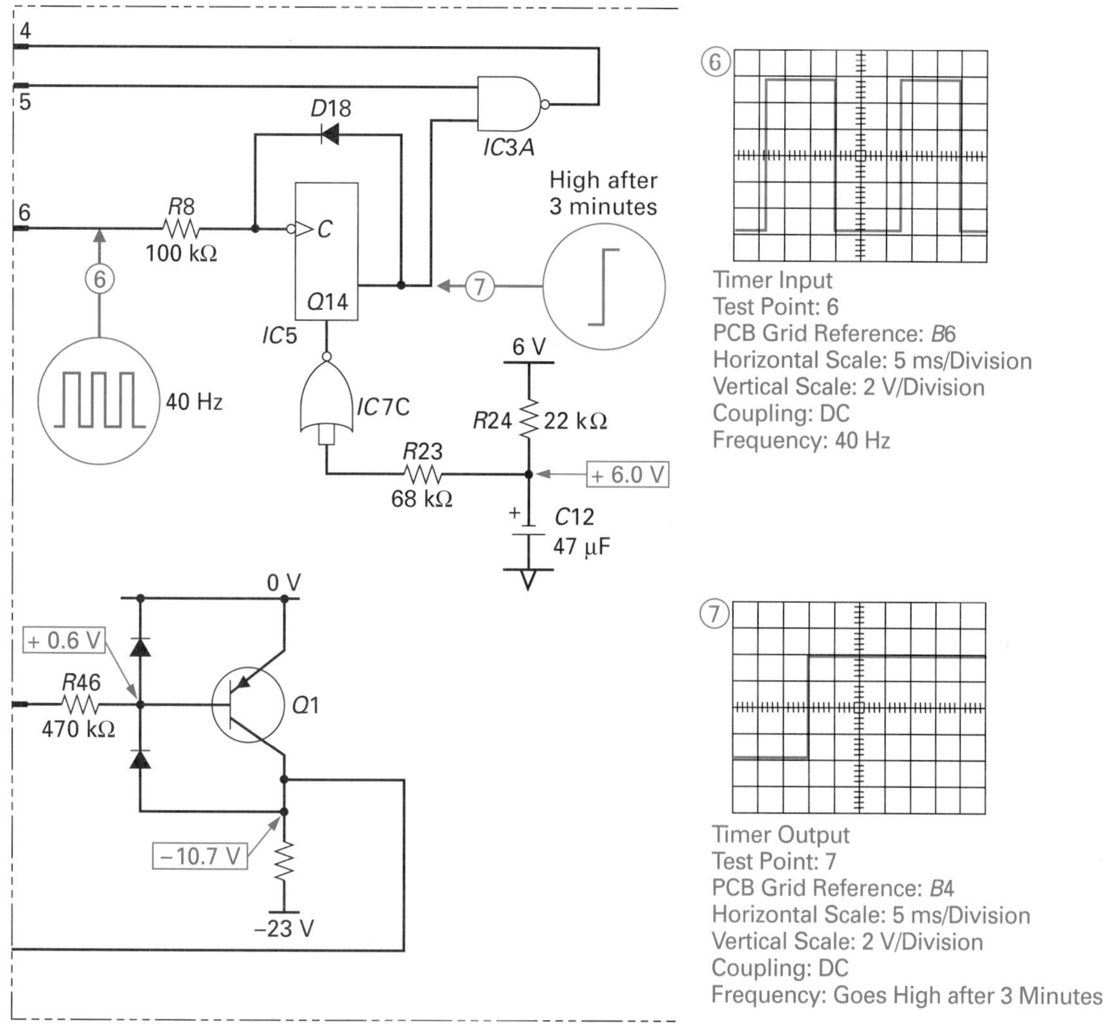

FIGURE 12-8 Schematic Diagram with Voltage and Waveform Test Points.

Controls

Oscilloscopes come with a wide variety of features and functions; however, the basic operational features are almost identical. Figure 12-9 illustrates the front panel of a typical oscilloscope; we will now discuss the various control functions. Some of these controls are difficult to understand without practice and experience, so practical experimentation is essential if you hope to gain a clear understanding of how to operate an oscilloscope.

General controls (see Figure 12-9)

Intensity control: Controls the brightness of the trace, which is the pattern produced on the screen of a CRT.

Focus control: Used to focus the trace.

Power OFF/ON: Switch will turn on oscilloscope while indicator shows when oscilloscope is turned on.

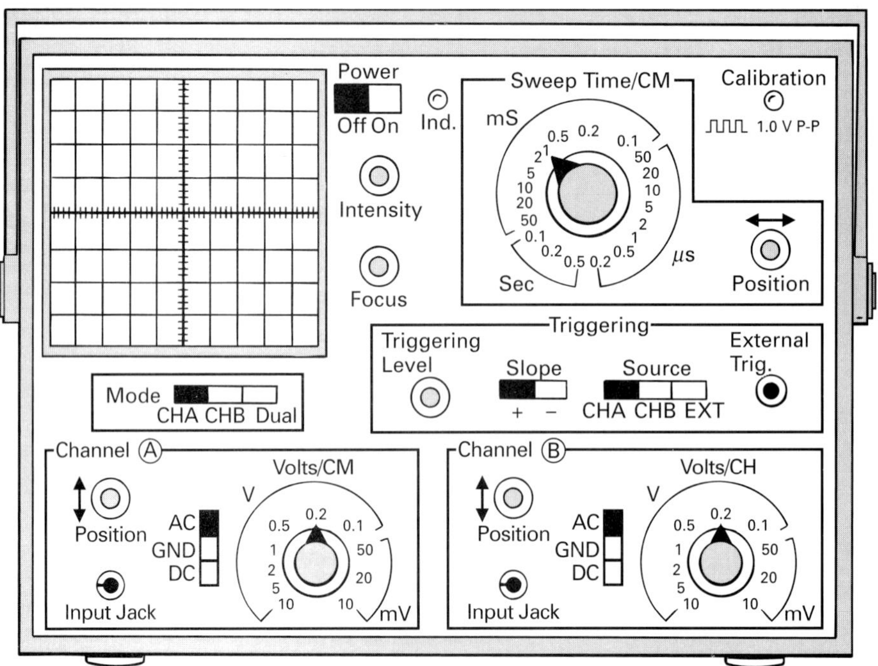

FIGURE 12-9 **Oscilloscope Controls.**

Some oscilloscopes have the ability to display more than one pattern or trace on the CRT screen, as seen by the examples in Figure 12-10. A dual-trace oscilloscope can produce two traces or patterns on the CRT screen at the same time, whereas a single-trace oscilloscope can trace out only one pattern on the screen. The dual-trace oscilloscope is very useful, as it allows us to make comparisons between the phase, amplitude, shape, and timing of two signals from two separate test points. One signal or waveform is applied to the channel *A* input of the oscilloscope, while the other waveform is applied to the channel *B* input.

Channel selection (see Figure 12-9)

Mode switch: This switch allows us to select which channel input should be displayed on the CRT screen.

CHA: The input arriving at channel *A*'s jack is displayed on the screen as a single trace.

CHB: The input arriving at channel *B*'s jack is displayed on the screen as a single trace.

Dual: The inputs arriving at jacks *A* and *B* are both displayed on the screen, as a dual trace.

Calibration
To determine, by measurement or comparison with a standard, the correct value of each scale reading.

(a) **Calibration** Output This output connection provides a point where a fixed 1-V peak-to-peak square-wave signal can be obtained at a frequency of 1 kHz. This signal is normally fed into either channel *A* or *B*'s input to test probes and the oscilloscope operation.

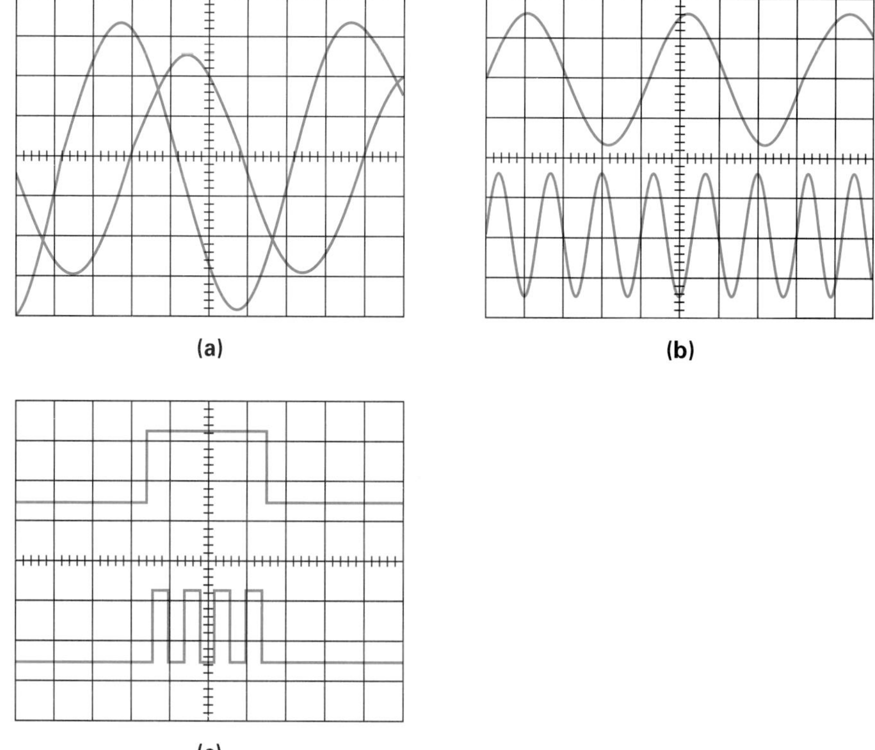

FIGURE 12-10 **Sample of Dual-Trace Oscilloscope Displays for Comparison.**

(b) Channel *A* and *B* Horizontal Controls

↔ *Position control:* This control will move the position of the one (single-trace) or two (dual-trace) waveforms horizontally (left or right) on the CRT screen.

Sweep time/cm switch: The oscilloscope contains circuits that produce a beam of light that is swept continually from the left to the right of the CRT screen. When no input signal is applied, this sweep will produce a straight horizontal line in the center of the screen. When an input signal is present, this horizontal sweep is influenced by the input signal, which moves it up and down to produce a pattern on the CRT screen the same as the input pattern (sine, square, sawtooth, and so on). This sweep time/cm switch selects the speed of the sweep from left to right, and it can be either fast (0.2 microseconds per centimeter; 0.2 μs/cm) or slow (0.5 second per centimeter; 0.5 s/cm). A low-frequency input signal (long cycle time or period) will require a long time setting (0.5 s/cm) so that the sweep can capture and display one or more cycles of the input. A number of settings are available, with lower time settings displaying fewer cycles and higher time settings showing more cycles of an input.

(c) **Triggering** Controls These provide the internal timing control between the sweep across the screen and the input waveform.

Triggering
Initiation of an action in a circuit which then functions for a predetermined time, for example, the duration of one sweep in a cathode-ray tube.

Triggering level control: This determines the point where the sweep starts.

Slope switch (+): Sweep is triggered on positive-going slope.
(–): Sweep is triggered on negative-going slope.

Source switch, CHA: The input arriving into channel *A* jack triggers the sweep.
CHB: The input arriving into channel *B* jack triggers the sweep.
EXT: The signal arriving at the external trigger jack is used to trigger the sweep.

(d) Channel *A* and *B* Vertical Controls The *A* and *B* channel controls are identical.

Volts/cm switch: This switch sets the number of volts to be displayed by each major division on the vertical scale of the screen.

↕*Position control:* Moves the trace up or down for easy measurement or viewing.

AC-DC-GND *switch:* In the AC position, a capacitor on the input will pass the ac component entering the input jack, but block any dc components.
In the GND position, the input is grounded (0 V) so that the operator can establish a reference.
In the DC position, both ac and dc components are allowed to pass on to and be displayed on the screen.

Measurements

The oscilloscope is probably the most versatile of test equipment, as it can be used to test:

Dc voltage

Ac voltage

Waveform duration

Waveform frequency

Waveform shape

(a) Voltage Measurement The screen is divided into eight vertical and ten horizontal divisions, as shown in Figure 12-11. This 8 × 10 cm grid is called the *graticule.* Every vertical division has a value depending on the setting of the volts/cm control. For example, if the volts/cm control is set to 5 V/div or 5 V/cm, the waveform shown in Figure 12-12(a), which rises up four major divisions, will have a peak positive alternation value of 20 V (4 div × 5 V/div = 20 V).

As another example, look at the positive alternation in Figure 12-12(b). The positive alternation rises up three major divisions and then extends another four subdivisions, which are each equal to 1 V because five subdivisions exist within one major division, and one major division is, in this example, equal to 5 V. The positive alternation shown in Figure 12-12(b) therefore has a peak of three major divisions (3 × 5 V/cm = 15 V), plus three subdivisions (3 × 1 V = 3 V), which equals 18 volts peak.

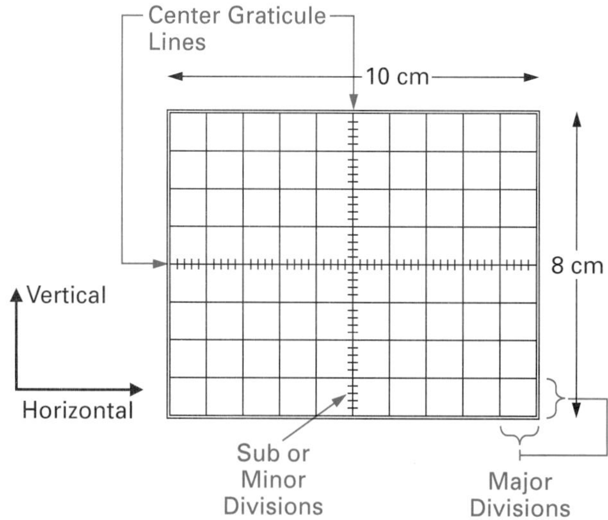

FIGURE 12-11 Oscilloscope Grid.

In Figure 12-12(c), we have selected the 10 volt/cm position, which means that each major division is equal to 10 V and each subdivision is equal to 2 V. In this example, the waveform peak will be equal to two major divisions ($2 \times 10 = 20$ V), plus four subdivisions (4×2 V $= 8$ V), which is equal to 28 V. Once the peak value of a sine wave is known, the peak to peak, average, and rms can be calculated mathematically.

When measuring a dc voltage with the oscilloscope, the volts/cm is applied in the same way, as shown in Figure 12-13. A positive dc voltage in this situation will cause deflection toward the top of the screen, whereas a negative voltage will cause deflection toward the bottom of the screen.

To determine the dc voltage, count the number of major divisions and to this add the number of minor divisions. In Figure 12-13, a major division equals 1 V/cm and, therefore, a minor division equals 0.2 V/cm, so the dc voltage being measured is interpreted as +2.6 V.

(b) Time and Frequency Measurement The frequency of an alternating wave, such as that seen in Figure 12-14(a), is inversely proportional to the amount of time it takes to complete one cycle ($f = 1/t$). Consequently, if time can be measured, frequency can be determined.

The time/cm control relates to the horizontal line on the oscilloscope graticule and is used to determine the period of a cycle so that frequency can be calculated. For example, in Figure 12-14(b), a cycle lasts five major horizontal divisions, and since the 20 μs/division setting has been selected, the period of the cycle will equal 5×20 μs/division = 100 μs. If the period is equal to 100 μs, the frequency of the waveform will be equal to $f = 1/t = 1/100$ μs = 10 kHz.

EXAMPLE:

A complete sine-wave cycle occupies four horizontal divisions and four vertical divisions from peak to peak. If the oscilloscope is set on the 20 ms/cm and 500 mV/cm, calculate:

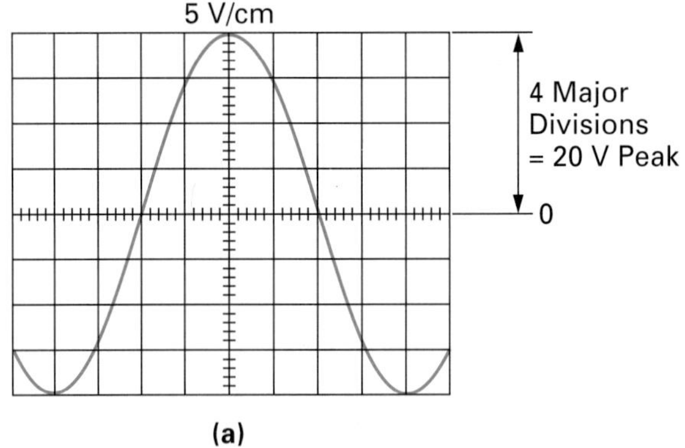

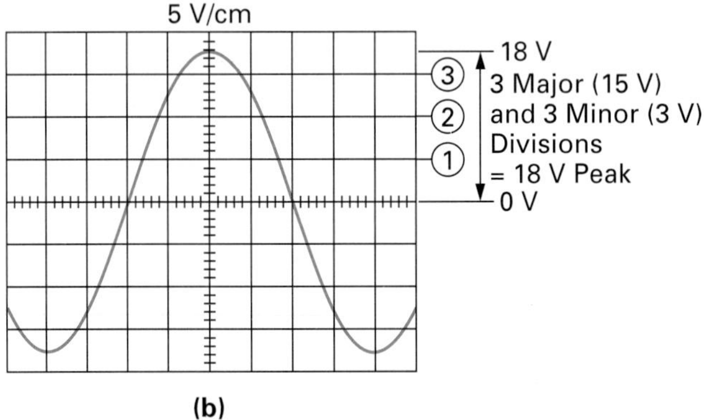

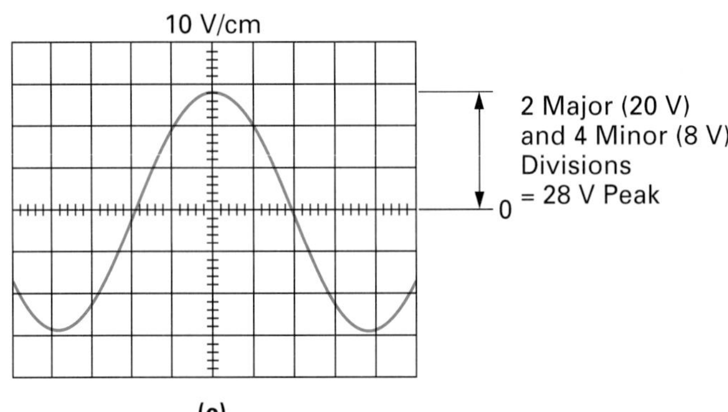

FIGURE 12-12 Measuring AC Peak Voltage.

a. $V_{\text{p-p}}$
b. t
c. f
d. V_{p}
e. V_{avg} of peak
f. V_{rms}

Solution:

$$4 \text{ horizontal divisions} \times 20 \text{ ms/div} = 80 \text{ ms}$$
$$4 \text{ vertical divisions} \times 500 \text{ mV/div} = 2 \text{ V}$$

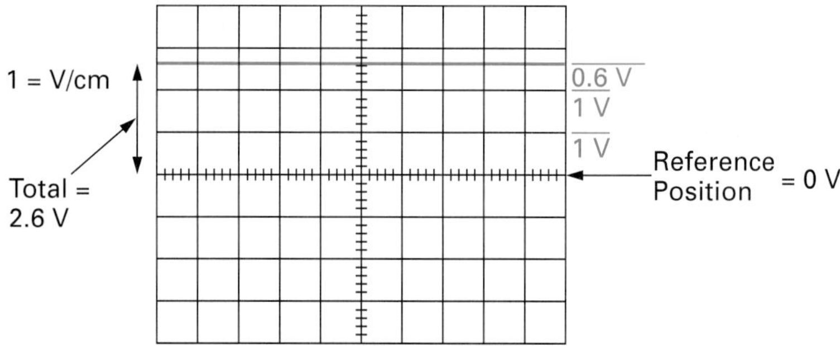

FIGURE 12-13 **Measuring DC Voltage.**

a. $V_{\text{p-p}} = 2\text{ V}$
b. $t = 80\text{ ms}$
c. $f = \dfrac{1}{t} = 12.5\text{ Hz}$
d. $V_{\text{p}} = 0.5 \times V_{\text{p-p}} = 1\text{ V}$
e. $V_{\text{avg}} = 0.637 \times V_{\text{p}} = 0.637\text{ V}$
f. $V_{\text{rms}} = 0.707 \times V_{\text{p}} = 0.707\text{ V}$

Hand-held Scopemeters

The hand-held **scopemeter**, shown in Figure 12-15, combines a multimeter, oscilloscope, frequency counter, and signal generator in one easy-to-carry battery-operated unit. For a technician, this test instrument is portable, easy to set up, and easy to use since it will automatically change its settings for the best operating mode and continue to adjust itself as the input changes. To explain this test instrument in more detail, let us examine some of its key functions.

Scopemeter A hand-held, battery-operated instrument that combines a multimeter, oscilloscope, frequency counter, and signal generator in one.

Measure menu: The measurement menu button gives you direct access to a quick pop-up menu of more than 30 measurements including: V_{rms}, V_{mean} (arithmetic average), $V_{\text{peak-to-peak}}$, frequency, time delay, rise time, phase,

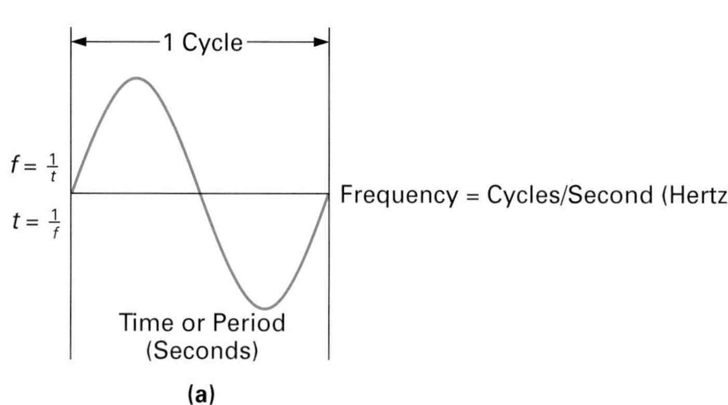

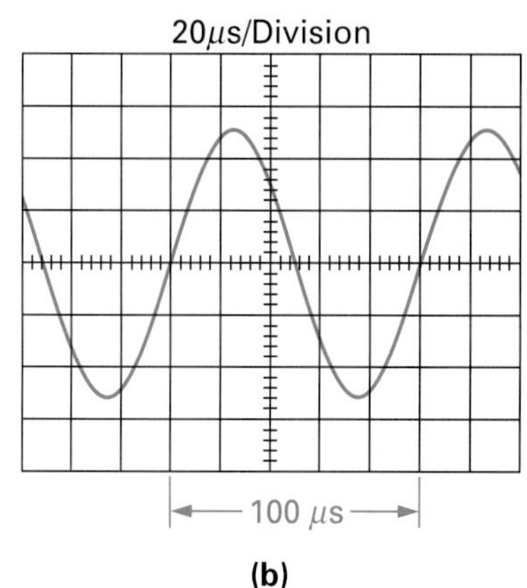

FIGURE 12-14 **Time and Frequency Measurement.**

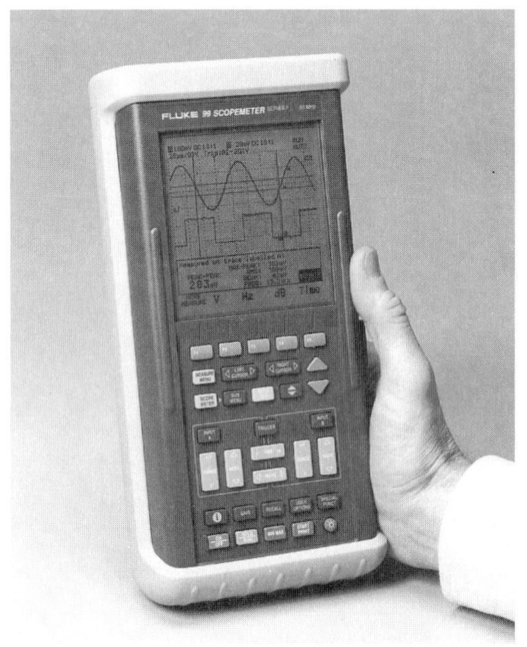

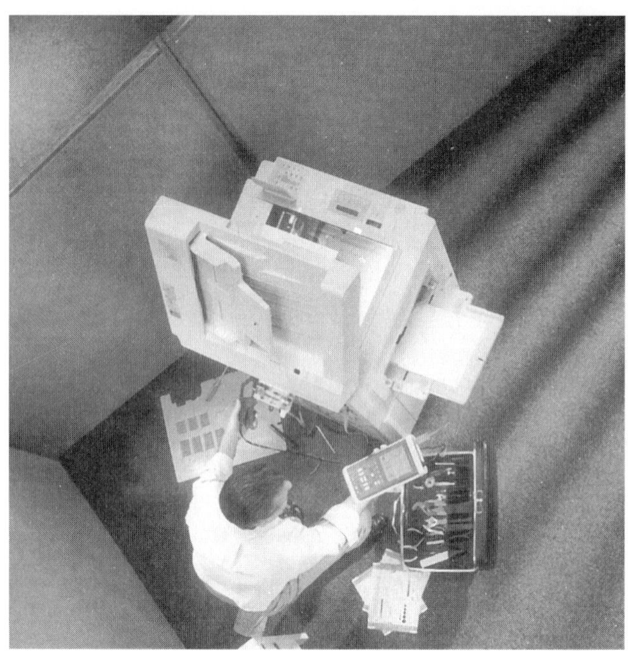

FIGURE 12-15 **The Hand-Held Scopemeter. (Courtesy of Fluke Corporation. Reproduced with permission.)**

current, and so on. Once the measurement is selected, the scopemeter automatically sets itself up and takes the measurement, as seen in the example in Figure 12-16(a).

Auto set: As you move from test point to test point, the scopemeter will handle the changing inputs automatically in this continuous autoset mode. Each time the signal input changes, the scopemeter will automatically search for the best trigger level, timebase (sweep time), and range scale to speed up measurements and reduce errors, as seen in Figure 12-16(b).

Min max trendplot: This function is used to display and record up to 40 days of signal trend which is sometimes needed in order to pinpoint intermittent problems that occur randomly. Figure 12-16(c) shows an example of how the display shows minimum, maximum, and average readings.

Save: Most scopemeters have large internal memories for saving screen images, setups, and waveforms from the field. These stored measurements can be recalled at any time, downloaded or sent to a personal computer for reports, or sent directly to a printer. Figure 12-16(d) shows how these waveforms and setups would appear on a computer screen.

Function Generator
Signal generator that can function as a sine, square, rectangular, triangular, or sawtooth waveform generator.

12-1-3 *The Function Generator*

In many cases you will wish to generate a waveform of a certain shape and frequency, and then apply this waveform to your newly constructed circuit. One of the most versatile waveform generators is the **function generator,** so called because it can function as a sine wave, square wave, rectangular wave, triangular

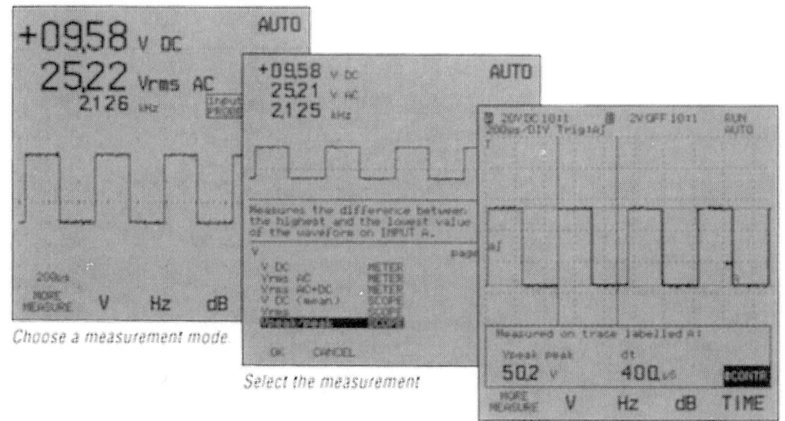

(a)

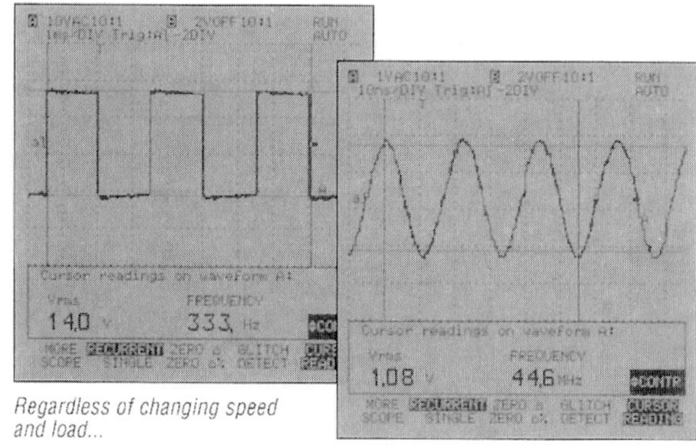

(b)

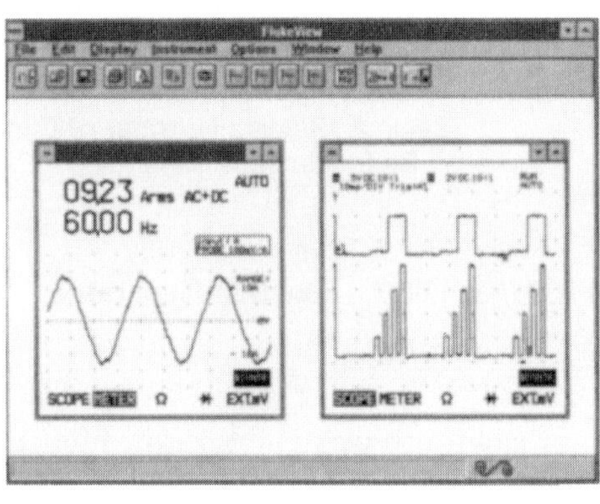

(c) (d)

FIGURE 12-16 Scopemeter Operation. (Courtesy of Fluke Corporation. Reproduced with permission.)

(a)

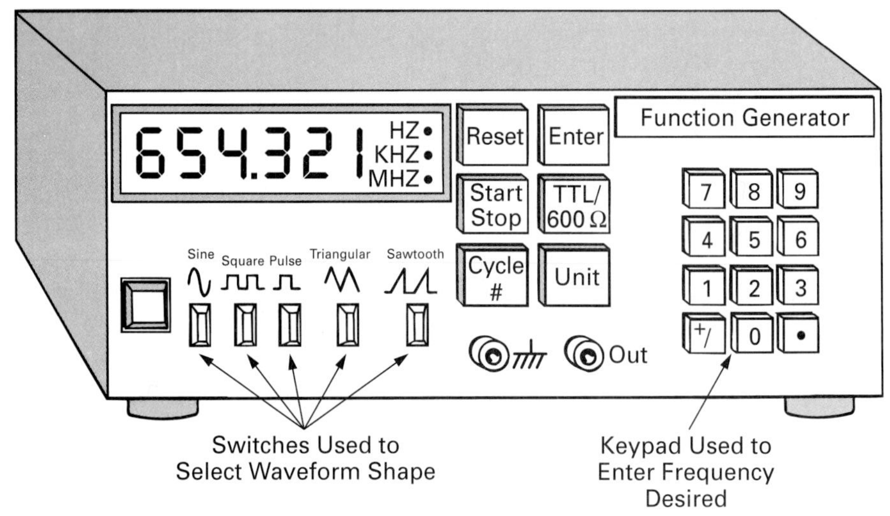

Unit Switch: Used to Select Either Hz, kHz, or MHz

TTL/600 Ω: Used to Select Either Digital (Transistor, Transistor Logic/TTL) or Analog (600 Ω-Output Impedance)

Cycle #: Selects Number of Cycles to Be Produced

(b)

FIGURE 12-17 Function Generators.

wave, or sawtooth wave generator. Figure 12-17(a) shows a photograph of a typical function generator, and Figure 12-17(b) shows a line drawing of a typical programmable function generator.

Frequency Counter Meter used to measure the frequency or cycles per second of a periodic wave.

12-1-4 The Frequency Counter

It is very important that anyone involved in the design, manufacture, and servicing of electronic equipment be able to accurately measure the frequency of a periodic wave. Without the ability to measure frequency accurately, there

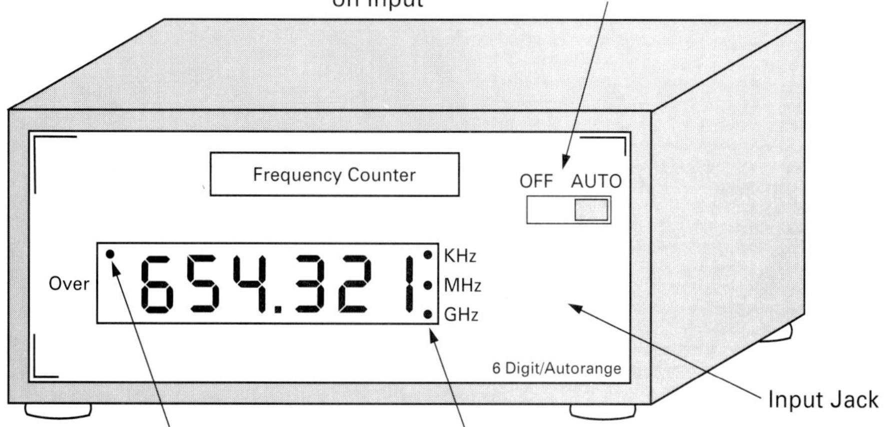

(a)

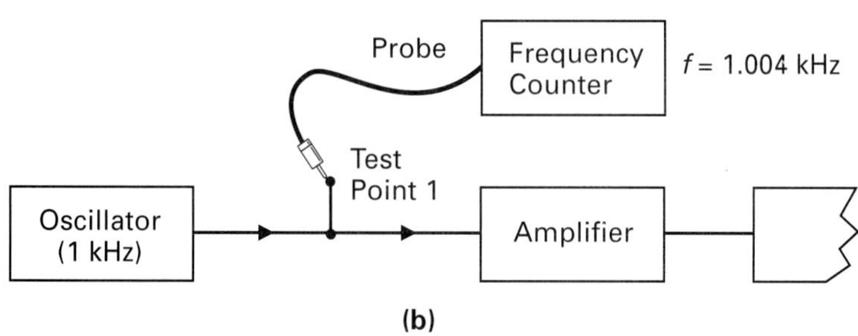

(b)

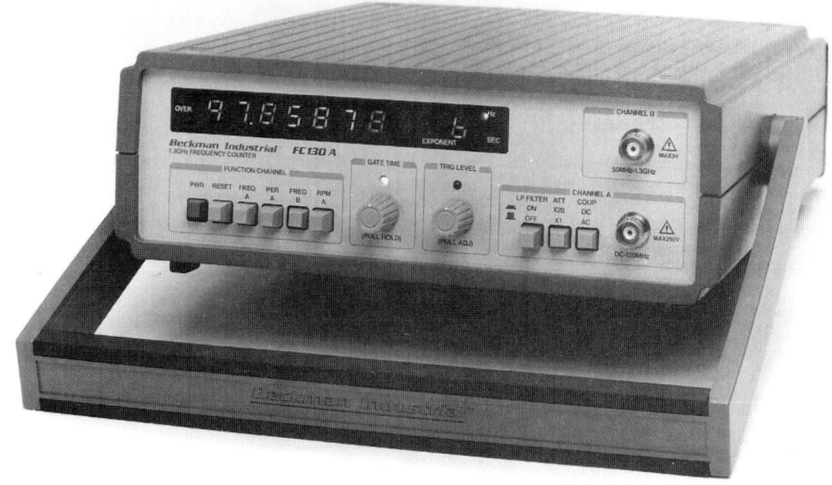

(c)

FIGURE 12-18 Frequency Counter.

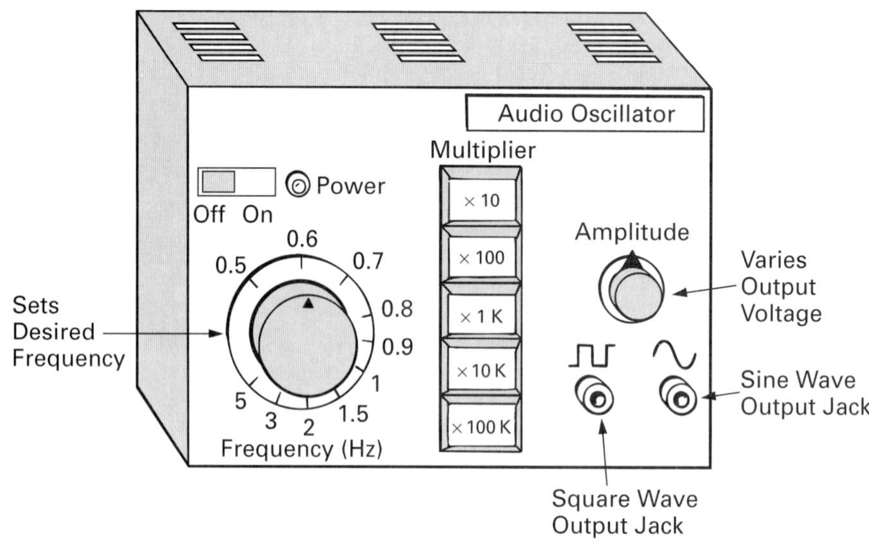

FIGURE 12-19 **Audio-Frequency Generator.**

could be no communications, home entertainment, or a great number of other systems.

The frequency meter or counter, shown in Figure 12-18(a), analyzes the frequency of the periodic wave being applied to the input jack and provides a readout on the display of its frequency. In Figure 12-18(b), the frequency counter is being used to determine the accuracy of an oscillator's output of 1 kHz. Figure 12-18(c) shows a photograph of a typical frequency counter. Most frequency counters on the market today can measure frequencies anywhere from hertz to gigahertz.

Audio-Frequency Generator
A signal generator that can be set to generate a sinusoidal AF signal voltage at any desired frequency in the audio spectrum.

12-1-5 *The Audio-Frequency Generator*

Audio is a Greek word meaning "I hear," and although we can only hear sound-wave frequencies from about 20 to 20,000 Hz, audio-frequency generators will generally produce electrical sine or square waves from 1 Hz to 1 MHz. An audio-frequency (AF) generator or oscillator can be seen in Figure 12-19.

Radio-Frequency Generator
A generator capable of supplying RF energy at any desired frequency in the radio spectrum.

12-1-6 *The Radio-Frequency Generator*

Figure 12-20(a) shows a typical RF generator that can be used to generate any frequency from 3 kHz to 3 GHz. The electrical wave produced can be either a constant-amplitude continuous wave, shown in Figure 12-20(b), or a varying-amplitude signal, known as an amplitude-modulated (AM) wave, shown in Figure 12-20(c).

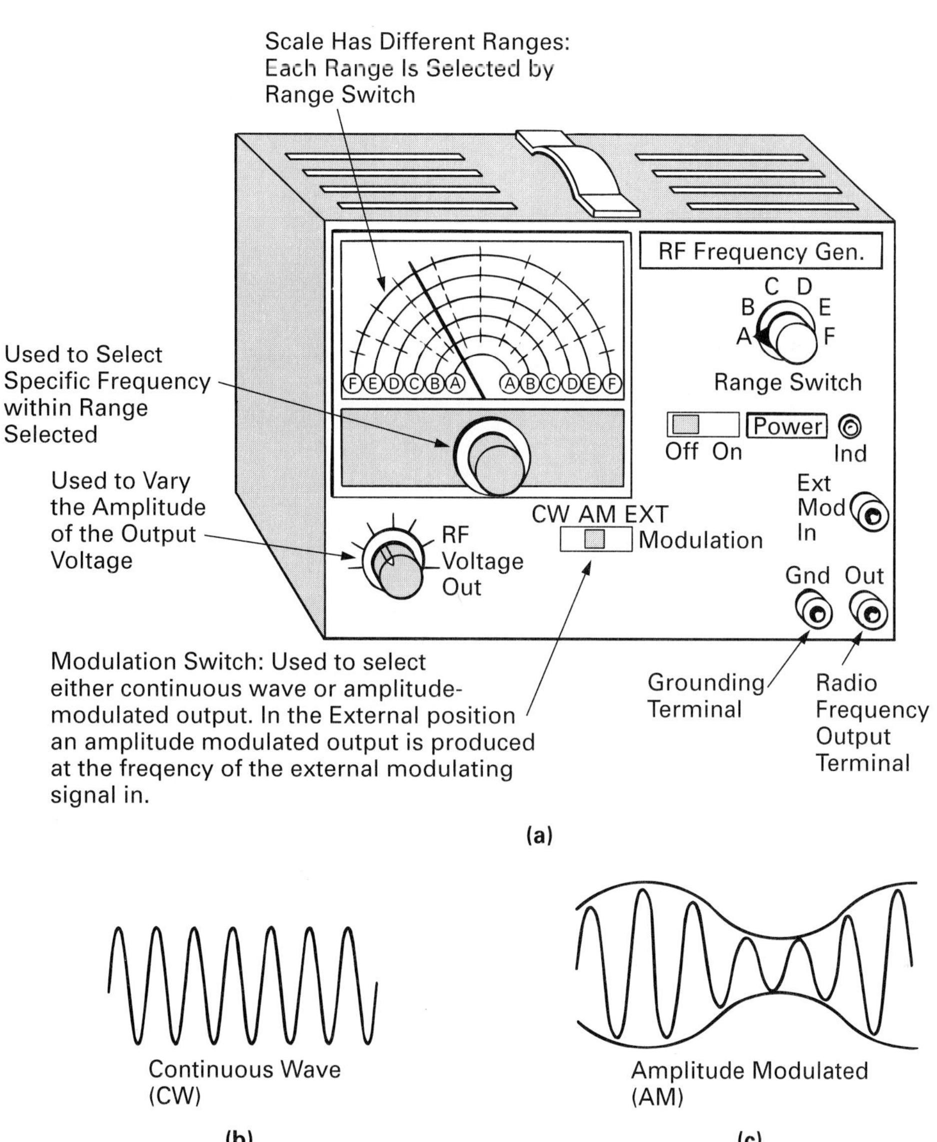

FIGURE 12-20 Radio-Frequency Generator.

SELF-TEST REVIEW QUESTIONS FOR SECTION 12-1

1. True or false: The current clamp does not require the circuit current path to be broken in order to measure current. Can it be used to measure dc amps?
2. What is the difference between an RF and an HV probe?
3. The __________ is used primarily to display the shape and spacing of electrical signals.
4. Name the device within the oscilloscope on which the waveforms can be seen.
5. On the 20-μs/div time setting, a full cycle occupies four major divisions. What is the waveform's frequency and period?
6. On the 2-V/cm voltage setting, the waveform swings up and down two major divisions (a total of 4 cm of vertical swing). Calculate the waveform's peak and peak-to-peak voltage.
7. What is the advantage(s) of the dual-trace oscilloscope?
8. List some of the waveform shapes typically generated by a function generator.
9. What is the function of a frequency counter?
10. What is the difference between an audio- and a radio-frequency generator?

12-2 TROUBLESHOOTING TECHNIQUES

Troubleshooting
The process of locating and repairing malfunctions in equipment by means of systematic analysis and testing.

Troubleshooting is the process of locating and repairing malfunctions in equipment by means of systematic analysis and testing. For you to become a good electronics technician or troubleshooter, you will need to have a thorough understanding of *electronics, test equipment, troubleshooting techniques,* and *system repair*. Your understanding of electronics and test equipment will build on the foundation you have established so far in your studies.

Your knowledge of troubleshooting techniques and system repair will improve as you continue experimenting in lab since, as you have probably already discovered, very few circuits operate properly the first time. Try to view these circuit failures as opportunities rather than disasters, and benefit from the troubleshooting experience you can obtain.

In this section we step through a basic six-step troubleshooting procedure, which is illustrated in Figure 12-21. This procedure can be used as a guide to help you quickly isolate and repair basic circuit or complex system malfunctions.

12-2-1 Step 1: Symptom Recognition

Before even touching a piece of test equipment you should familiarize yourself with the operation of the circuit or system. This information can be obtained from operator's manuals, technical manuals, and other documentation. Many technicians bypass this step and proceed directly to the detailed testing, only to

Name:	Date:	
Equipment Description:	Start Time:	Stop Time:

Step 1: Symptom Recognition

From the presentation and all other visual indications, list what functions do not operate or what these functions are doing wrong.

SYMPTOM RECOGNITION
Tools: Operator's manual, technical manual, circuit descriptions.

Step 2: List Probable Faulty Units

Referring to the overall equipment block diagram, list the probable block/board that could cause the malfunction listed in step 1.

LIST PROBABLE FAULTY UNITS
Tools: Block interconnect diagram, thought, controls, service or repair logs.

Step 3: Localize Faulty Section

In the following table, list the test point and data utilized to isolate the malfunctioning block or board.

Test Point	*Normal Reading*	*Actual Reading*

LOCALIZING FAULTY SECTION
Tools: Sight, sound, smell, test equip., knowledge, thought, block and circuit diagrams.

Step 4: Localize Faulty Circuit

In the following table, list the test points and data utilized to isolate the faulty circuit within the printed circuit board.

Test Point	*Normal Reading*	*Actual Reading*

LOCALIZING FAULTY CIRCUIT IN PCB
Tools: Sight, sound, smell, test equip., knowledge, thought, circuit diagrams, theory of operation, troubleshooting guides or trees.

Step 5: Localize Faulty Component

In the following table, list the test points and data utilized to isolate the faulty component within the circuit.

Test Point	*Normal Reading*	*Actual Reading*

LOCALIZING FAULTY COMPONENT
Tools: Sight, sound, smell, test equip., knowledge, thought, circuit diagrams, theory of operation.

FAULTY COMPONENT IS:

Step 6: Repair, Final Testing, and Documentation of Failure

In this step you should repair the circuit and then final-test the system. Also , explain why this component caused the symptoms listed in step 1 and how the problem was resolved.

FIGURE 12-21 Six-Step Troubleshooting Procedure.

Circuit Failure A malfunction or breakdown of a component in a circuit.

Operator Error An incorrect use of the controls of a circuit or system.

discover that what appeared to be a complex **circuit failure** turned out to be a simple **operator error.** Once you are completely familiar with the operation of the circuit or system, operate the controls to determine what functions do not operate or what these functions are doing wrong.

12-2-2 Step 2: List Probable Faulty Units

Referring to the circuit's schematic diagram or the system's block interconnect diagram, list the probable block or board that could cause the malfunction that was listed in step 1.

12-2-3 Step 3: Localize Faulty Section

In this step you remove equipment covers and begin testing the entire system to isolate the malfunctioning block or section, as shown in Figure 12-22.

As you proceed with this step, keep several thoughts in mind:

1. Always look for obvious errors, such as that the power is not connected, the system or circuit is not turned on, and so on.
2. Use your senses of sight, sound, and smell to check for burning components, crackling connections, and so on.
3. Use a troubleshooting technique called the **half-split method** to help quickly isolate the area of malfunction. For example, if a circuit or system has a good input and a bad output, instead of checking at each and every point between input and output, test a point midway between input and

Half-Split Method A troubleshooting technique used to isolate a faulty block in a circuit or system. In this method a mid-point is chosen and tested to determine which half is malfunctioning.

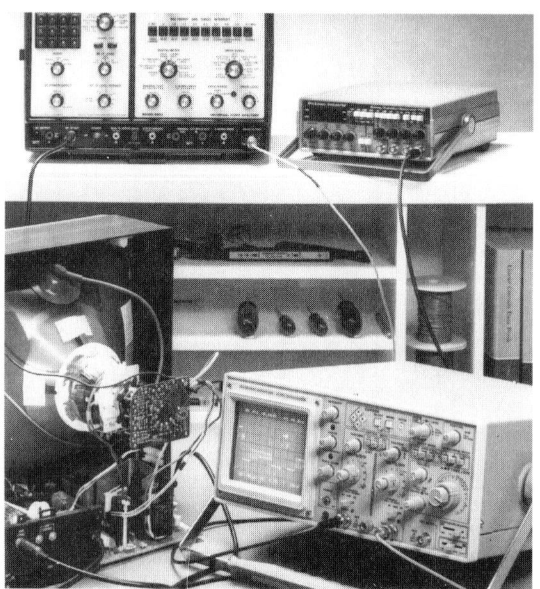

FIGURE 12-22 Localizing the Faulty Block or Board.

output. If the signal is good, the problem is in the second half of the circuit or system, whereas if the signal is bad, the problem is in the first half of the circuit or system.

4. If the entire circuit or system, or if a section of the circuit or system, is "dead," you can probably safely assume that you have a power problem. Use your voltmeter to measure the ac input to the power supply and the dc output supplies, and don't forget to check one of the most common malfunctions—a blown fuse or tripped circuit breaker.

12-2-4 Step 4: Localize Faulty Circuit

Once you have isolated the problem to a particular board, your next step is to localize the faulty circuit on the printed circuit board (PCB), as shown in Figure 12-23. Keep in mind to use your senses, check first for the obvious, and apply the half-split method. Many equipment manufacturers include PCB troubleshooting trees or guides in the system technical manual, which will help you quickly isolate the faulty section of a circuit.

12-2-5 Step 5: Localize Faulty Component

In most cases it is important to troubleshoot and repair a system quickly, since failures result in a loss of productivity, and therefore money. In some instances, therefore, it may only be necessary for you to locate the faulty PCB and install a replacement, and then isolate the faulty component at a later time. This technique is known as *board-level substitution.*

If a replacement board is not available, you will have to troubleshoot the circuit to find its faulty component. At this stage you may need to refer back to the detailed circuit description, use the troubleshooting guides, perform several tests on the circuit, and logically try to zero in on the faulty component.

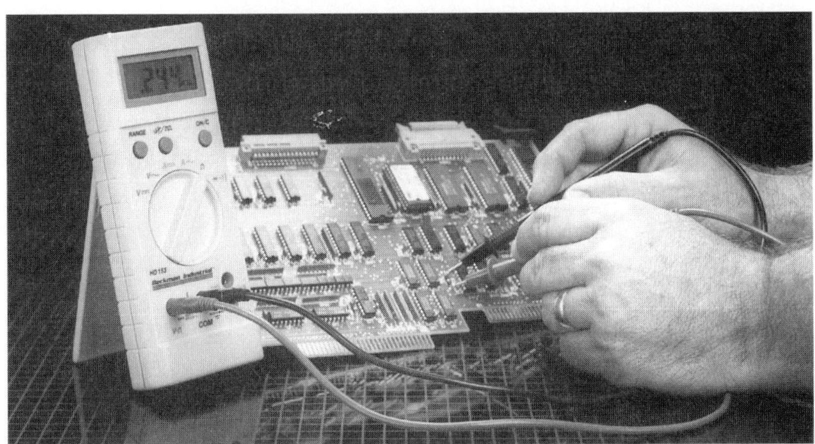

FIGURE 12-23 Localizing Faulty Circuit on Printed Circuit Board.

FIGURE 12-24 Making the Repair.

Component-level substitution can also be used to speed up the troubleshooting process. For example, once you have located the problem in a general area containing several components, replace the main components one at a time with known good components until the failure is fixed. Although this seems time consuming, it can save time with hard-to-find, elusive circuit problems.

12-2-6 *Step 6: Repair, Final Testing, and Documentation of Failure*

Making the repair may be as simple as changing a fuse, reinstalling a connector, removing a short, or repairing an open wire. Generally, though, it involves the removal and replacement of a component, as shown in Figure 12-24. If the component is mounted in a protoboard, this process will be simple; however, if the component is soldered into a PCB, you will have to follow the desoldering steps outlined in Chapter 5. When a component is removed, make sure that you make a note of its orientation (if this is important) and be sure to handle all static-sensitive components with care.

Once the repair has been made, you should final-test the circuit or system to see that the symptoms listed in step 1 have been resolved, and then document the failure.

SELF-TEST REVIEW QUESTIONS FOR SECTION 12-2

1. To become a good troubleshooter you will need a thorough knowledge of ____________, ____________, ____________, and ____________.

2. List the six-step troubleshooting procedure.

SUMMARY

1. Test equipment can be used to either sense a circuit's condition or generate a signal to see the response of the component or circuit to that signal.
2. Most multimeters can be used to measure either ac or dc.
3. When multimeters are selected to measure ac, internally a rectifier is used to convert the ac input into a dc voltage.
4. Multimeters are normally calibrated to indicate rms values of the ac being measured.
5. A current clamp allows the technician or engineer to measure ac current without opening the current path.
6. The RF probe can be used with the multimeter to more accurately measure higher frequencies above 2 kHz.
7. The high-voltage probe can be used to measure voltages in the kilovolt range by connecting additional multiplier resistors.
8. The analog multimeter unlike the DMM can be used to measure low frequency ac.
9. The digital multimeter is generally more popular because of its easy-to-read display and accuracy.
10. The oscilloscope displays the shape and spacing of electrical signals, and can therefore be used to measure dc voltage, ac voltage, waveform duration, waveform frequency, and waveform shape.
11. The oscilloscope can be used to display waveform shapes, and from this presentation we can calculate the waveform's time, frequency, and amplitude characteristics.
12. A dual-trace oscilloscope can produce two traces or waveforms on the screen, which allows the technician or engineer to make comparisons between phase, amplitude, shape, or timing.
13. The function generator can produce a sine wave, square wave, rectangular wave, triangular wave, or sawtooth wave.
14. The frequency counter measures and displays the number of cycles per second (hertz) on a digital display.
15. The audio-frequency, radio-frequency, and function generators are all used to produce a wide range of electrical waves for testing or controlling the operation of different circuits.
16. A function generator may function or perform as a sine, square, triangular, pulse, or sawtooth waveform generator.
17. To become a good technician you have to have a thorough knowledge of electronics, test equipment, troubleshooting techniques, and system repair.
18. Troubleshooting is the process of locating and repairing malfunctions in equipment by means of systematic analysis and testing.
19. The six-step troubleshooting procedure is:
    a. Learn to recognize symptoms.
    b. List probable faulty units.

c. Localize faulty section.
d. Localize faulty circuit.
e. Localize faulty component.
f. Repair, final-test, and document failure.

NEW TERMS

AC meter
Audio-frequency (AF) generator
Circuit failure
Current clamp
Frequency counter
Half-split method
High-voltage (HV) probe
Instrument
Operator error
Radio-frequency (RF) generator
Radio-frequency (RF) probe
Rectifier
Signal generator
Substitution
Troubleshooting

REVIEW QUESTIONS

Multiple-Choice Questions

1. Which of the following would be considered a sensing instrument?
   a. Ohmmeter c. Audio oscillator
   b. Voltmeter d. Two of the above
2. Which of the following generates instead of sensing voltage or current?
   a. Voltmeter c. Ohmmeter
   b. Ammeter d. Two of the above
3. When the analog multimeter is selected to measure ac:
   a. A rectifier is connected internally in circuit.
   b. A pulsating dc waveform is applied to the meter movement.
   c. The meter will indicate the rms or dc equivalent of the ac input.
   d. All of the above.
4. What is the frequency limit associated with accurately measuring ac with a multimeter?
   a. 10 Hz to 2 kHz c. 15 MHz to 10 GHz
   b. 1000 to 15,000 Hz d. All three of the above
5. When connected to a multimeter, the current clamp:
   a. Allows you to measure ac voltage with the same ease as current
   b. Indicates current flow in a conductor based on magnetic field strength
   c. Allows you to measure ac current with the same ease as voltage
   d. Both (a) and (b)
   e. Both (b) and (c)
6. Radio-frequency probes are used to detect high-frequency ____________ waves and display a rms value on the meter.
   a. Sound c. Electromagnetic
   b. Electrical d. None of the above

7. A × 1000 high-voltage probe means that ___________ of the voltage at the probe tip will appear out of the probe and be applied to the meter.

   a. One tenth c. One thousandth
   b. 1000% d. Most

8. Which meter type has the highest accuracy?

   a. The AMM (analog multimeter)
   b. The DMM (digital multimeter)
   c. The VOM (volt-ohm-milliammeter)
   d. Both (a) and (c)

9. What are two advantages of the analog multimeter?

   a. Able to read low-frequency ac and low in cost
   b. Accuracy and ability to read low-frequency ac
   c. Easy-to-read display and accuracy
   d. Low cost and accuracy

10. The oscilloscope can be used to measure:

   a. Ac and dc voltage d. Rise and fall times
   b. Frequency e. All of the above
   c. Duration

Essay Questions

11. How can a multimeter be used to measure ac current or voltage? (12-1-1)
12. Briefly describe the following multimeter accessories: (12-1-1)

    a. The current clamp c. The high-voltage probe
    b. The RF probe

13. Describe the pros and cons of analog and digital readout multimeters. (12-1-1)
14. Explain the purpose of the frequency counter. (12-1-4)
15. If a multimeter indicates the rms value of a sine wave, what does an oscilloscope indicate? (12-1-2)
16. How can the oscilloscope be used to measure: (12-1-2)

    a. Dc voltage b. Frequency

17. What advantage does a dual-trace oscilloscope have over a single-trace oscilloscope? (12-1-2)
18. Briefly describe the function of:

    a. AF generator controls (12-1-5) b. RF generator controls (12-1-6)

19. In relation to the oscilloscope, describe: (12-1-2)

    a. The sweep time/cm switch b. The volts/cm switch

20. What is troubleshooting? (12-2)
21. What understanding is necessary to be an electronics technician? (12-2)
22. List the six-step troubleshooting guide. (12-2)
23. Describe the following troubleshooting techniques: (12-2)

    a. Half-split method b. Substitution

24. What is the difference between component-level troubleshooting and board-level troubleshooting?

25. Briefly describe the six-step troubleshooting procedure. (12-2)

Practice Problems

26. If one cycle of a sine wave occupies 4 cm on the oscilloscope horizontal grid and 5 cm from peak to peak on the vertical grid, calculate frequency, period, rms, average, and peak for the following control settings:

    a. 0.5 V/cm, 20 μs/cm **c.** 50 mV/cm, 0.2 μs/cm
    b. 10 V/cm, 10 ms/cm

27. Assuming the same graticule and switch settings of the oscilloscope in Figure 12-9, what would be the lowest setting of the volts/cm and time/division switches to fully view a 6 V rms, 350 kHz sine wave?

28. If the volts/cm switch is positioned to 10 V/cm and the waveform extends 3.5 divisions from peak to peak, what is the peak-to-peak value of this wave?

29. If a square wave occupies 5.5 horizontal cm on the 1 μs/cm position, what is its frequency?

30. Which setings would you use on an autoset scopemeter to fully view an 8 V rms, 20 kHz triangular wave?

CHAPTER

13

OBJECTIVES

After completing this chapter, you will be able to:

1. Define the term *capacitance*.
2. Describe the basic capacitor construction.
3. Explain the charging and discharging process and its relationship to electrostatics.
4. State the unit of capacitance and explain how it relates to charge and voltage.
5. List and explain the factors determining capacitance.
6. Describe capacitance breakdown and capacitor leakage.
7. Calculate total capacitance in parallel and series capacitance circuits.
8. Describe the advantages and differences between the five basic types of fixed capacitors.
9. Describe the advantages and differences between the four basic types of variable capacitors.
10. Explain the characteristics and new techniques used to create the one-farad capacitor.
11. Describe the coding of capacitor values on the body by use of alphanumerics or color.
12. Explain the capacitor time constant as it relates to dc charging and discharging.
13. Explain how the capacitor charges and discharges when ac is applied.

Capacitance and Capacitors

OUTLINE

VIGNETTE

Sounds Like Electronics

Robert Moog, a self-employed engineer of Trumansburg, New York, was frequently commissioned by his close friend, composer Herbert Deutsch, to build voltage-controlled oscillators and amplifiers. Combining these devices and adding some versatility, Moog built an analog music synthesizer that allowed the operator to control many tones simultaneously and rapidly alter tones during performances.

Synthesizers, devices used to generate a number of frequencies, had been around for a number of years, but electronic music did not find wide acceptance with the popular music audience until a young musician named Walter Carlos used a Moog instrument to produce a totally synthesized hit record. The album *Switched-On Bach*, released in 1968, was an interpretation of music by eighteenth-century composer Johann Sebastian Bach, and both the album and instrument were such breakthroughs in electronic music that in no time at all composers were writing a wide variety of original music specifically for Moog synthesizers.

INTRODUCTION

Up to this point we have concentrated on circuits containing only resistance, which oppose the flow of current and then convert or dissipate power in the form of heat. Capacitance and inductance are two circuit properties that act differently from resistance in that they will charge or store the supplied energy and then return almost all the stored energy back to the circuit, rather than lose it in wasted heat. Inductance will be discussed in a following chapter.

Capacitance is the ability of a circuit or device to store electrical charge. A device or component specifically designed to have this capacity or capacitance is called a *capacitor.* A capacitor stores an electrical charge similar to a bucket holding water. Using the analogy throughout this chapter that a capacitor is similar to a bucket will help you gain a clear understanding of the capacitor's operation. For example, the capacitor holds charge in the same way that a bucket holds water. A larger capacitor will hold more charge and will take longer to charge, just as a larger bucket will hold more water and take longer to fill. A larger circuit resistance means a smaller circuit current, and therefore a larger capacitor charge time. Similarly, a smaller hose will have a greater water resistance producing a smaller water flow, and therefore the bucket will take a

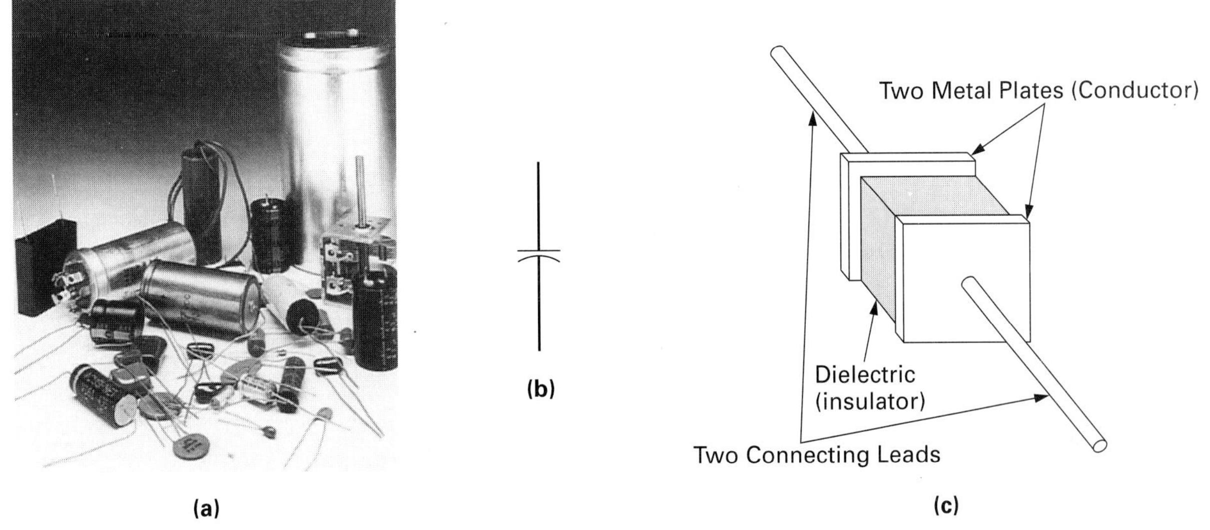

FIGURE 13-1 Capacitor. (a) Physical Appearance (Courtesy of Sencore). (b) Schematic Symbol. (c) Basic Construction.

long time to fill. Capacitors store electrons, and basically, the amount of electrons stored is a measure of the capacitor's capacitance.

13-1 CAPACITOR CONSTRUCTION

Figure 13-1 illustrates the main parts and schematic symbol of the **capacitor.** Several years ago, capacitors were referred to as condensers; however, that term is very rarely used today. Two leads are connected to two parallel metal conductive plates, which are separated by an insulating material known as a *dielectric.* The conductive plates are normally made of metal foil, while the dielectric can be paper, air, glass, ceramic, mica, or some other form of insulator.

Capacitor
Device that stores electric energy in the form of an electric field that exists within a dielectric (insulator) between two conducting plates each of which is connected to a lead. This device was originally called a condensor.

SELF-TEST REVIEW QUESTIONS FOR SECTION 13-1

1. What are the principal parts of a capacitor?
2. What is the name given to the insulating material between the two conductive plates?

13-2 CHARGING AND DISCHARGING A CAPACITOR

Like a secondary battery, a capacitor can be made to charge or discharge, and when discharging it will return all of the energy it consumed during charge.

13-2-1 *Charging a Capacitor*

Capacitance is the ability of a capacitor to store an electrical charge. Figure 13-2 illustrates how the capacitor stores an electric charge. This capacitor is shown as two plates with air acting as the dielectric.

In Figure 13-2(a), the switch is open, so no circuit current results. An equal number of electrons exist on both plates, so the voltmeter (VM) indicates zero, or no potential difference exists across the capacitor.

In Figure 13-2(b), the switch is now closed and electrons travel to the positive side of the battery, away from the right-hand plate of the capacitor. This creates a positive right-hand capacitor plate, which results in an attraction of free electrons from the negative side of the battery to the left-hand plate of the capacitor. In fact, for every electron that leaves the right-hand capacitor plate and is attracted into the positive battery terminal, another electron leaves the

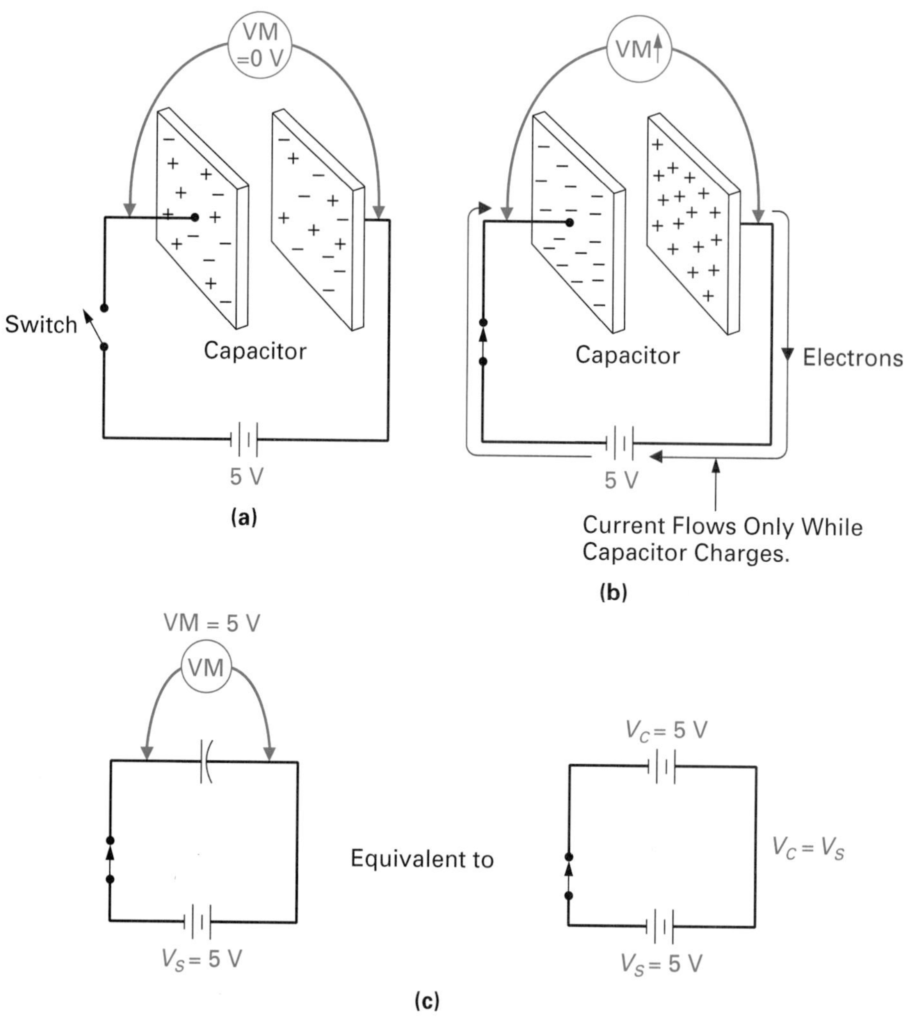

FIGURE 13-2 Charging a Capacitor. (a) Uncharged Capacitor. (b) Charging Capacitor. (c) Charged Capacitor.

negative side of the battery and travels to the left-hand plate of the capacitor. Current appears to be flowing from the negative side of the battery, around to the positive, through the capacitor. This is not really the case, because no electrons can flow through the insulator or dielectric. There appears to be one current flowing throughout the circuit, when in fact there are actually two separate currents—one from the battery to the capacitor and the other from the capacitor to the battery.

A voltmeter across the capacitor will indicate an increase in the potential difference between the plates, and the capacitor is said to be charging toward, in this example, 5 V. This potential difference builds up across the two plates until the potential difference (voltage) across the capacitor is equal to the potential difference (voltage) of the battery. In this example, when the capacitor reaches a potential difference or charge of 5 V, the capacitor will be equivalent to a 5 V battery, as seen in Figure 13-2(c). In this condition no potential difference exists between the battery and capacitor, and therefore no current flow can exist without a potential difference and the capacitor is said to be charged.

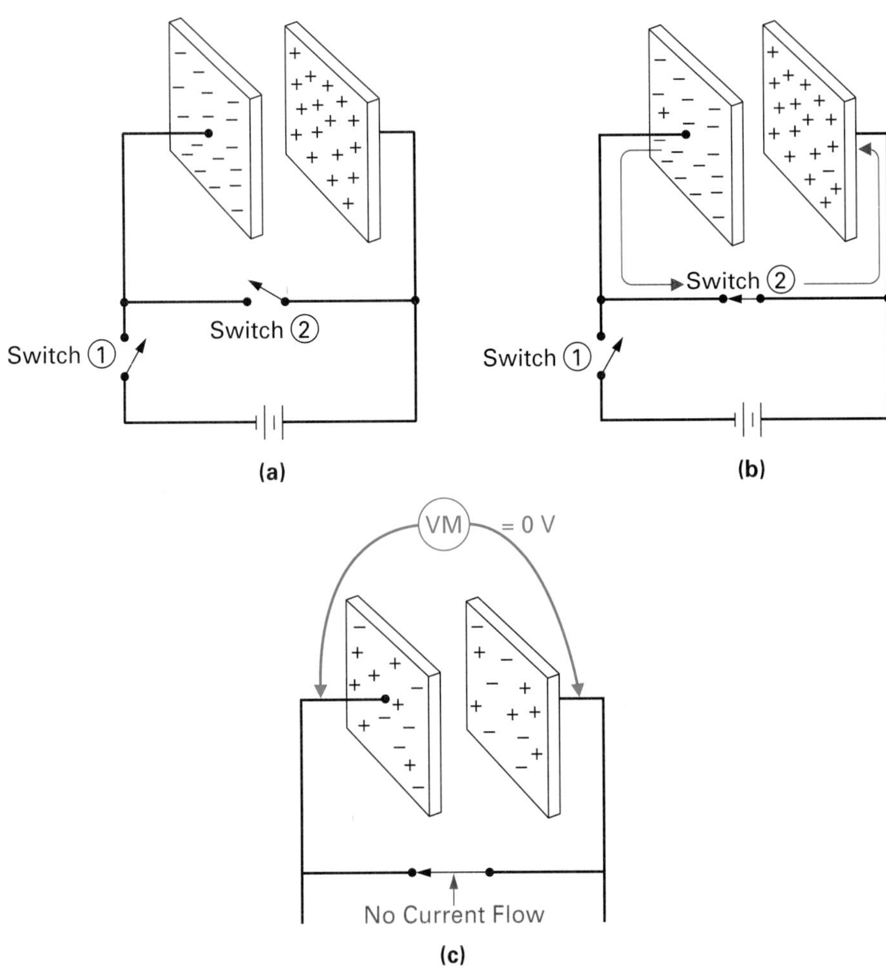

FIGURE 13-3 Discharging a Capacitor. (a) Charged Capacitor. (b) Discharging Capacitor. (c) Discharged Capacitor.

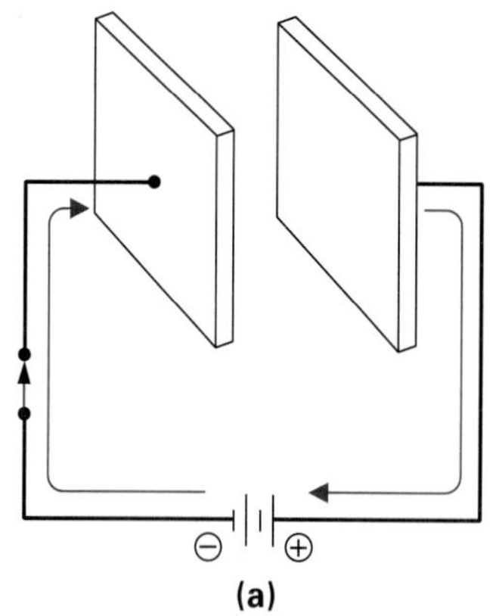

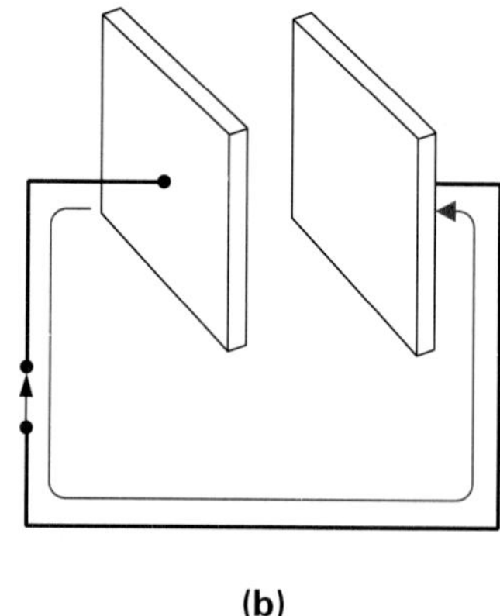

FIGURE 13-4 Capacitor Charge and Discharge Current Paths. (a) Charging. (b) Discharging.

13-2-2 *Discharging a Capacitor*

If the capacitor is now disconnected from the circuit by opening switch 1, as shown in Figure 13-3(a), it will remain in its charged condition. If switch 2 is now closed, a path exists across the charged capacitor, as shown in Figure 13-3(b), and the excess of electrons on the left plate will flow through the conducting wire to the positive plate on the right side. The capacitor is now said to be *discharging.* When equal numbers of electrons exist on both sides, the capacitor is said to be *discharged;* and since both plates have equal charge, the potential difference across the capacitor will be zero, as shown in Figure 13-3(c).

Figure 13-4(a) and (b) illustrate the capacitor's charge and discharge currents, which flow in opposite directions. In both cases the current flow is always from one plate to the other and never exists through the dielectric insulator.

SELF-TEST REVIEW QUESTIONS FOR SECTION 13-2

1. True or false: When both plates of a capacitor have an equal charge, the capacitor is said to be charged.
2. If 2 μA of current flows into one plate of a capacitor and 2 μA flows out of the other plate, how much current is flowing through the dielectric of this working capacitor?

Electrostatic or Electric Field
Force field produced by static electrical charges. Also called a voltage field, it is a field or force that exists in the space between two different potentials or voltages.

13-3 ELECTROSTATICS

Just as a magnetic field is produced by the flow of current, an **electric field** is produced by voltage.

Current generates a magnetic field

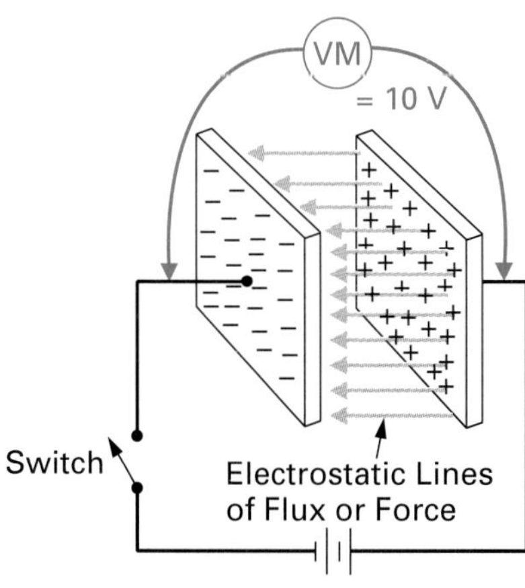

FIGURE 13-5 Electrostatic (Electric) Field between the Plates of a Charged Capacitor.

Voltage generates an electric field

Figure 13-5 illustrates an example capacitor circuit with the capacitor charged and the switch open. In this condition, the capacitor retains its charge and an invisible electric or electrostatic (voltage) field will be produced by non-moving or static electrical (electrostatic) charges of different polarities. You will remember that like charges repel and unlike charges attract. Invisible electrosta tic lines of flux or force can be illustrated to show this electrostatic force of attraction or repulsion. These lines are polarized away from a positive electro static (stationary electrical) charge and toward the negative electrostatic charge, as shown in Figure 13-6(a). If two like charges are in close proximity to one another, the electrostatic lines organize themselves into a pattern, as shown in Figure 13-6(b).

A charged capacitor has an electric or electrostatic field existing between the positively charged and negatively charged plates. The **strength of the electrostatic field** is proportional to the charge or potential difference on the plates and inversely proportional to the distance between the plates.

Field Strength
The strength of an electric, magnetic, or electromagnetic field at a given point.

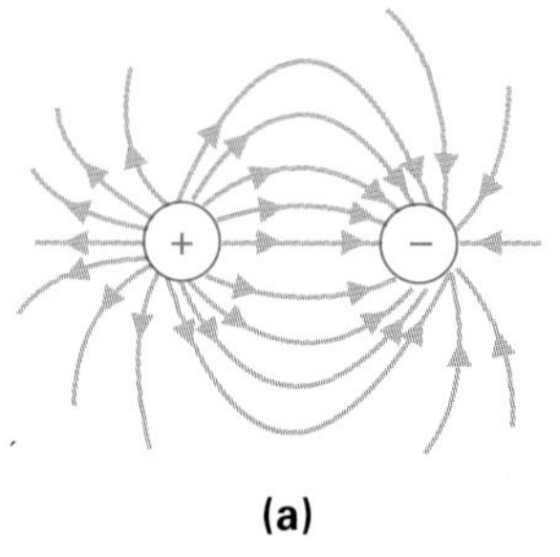

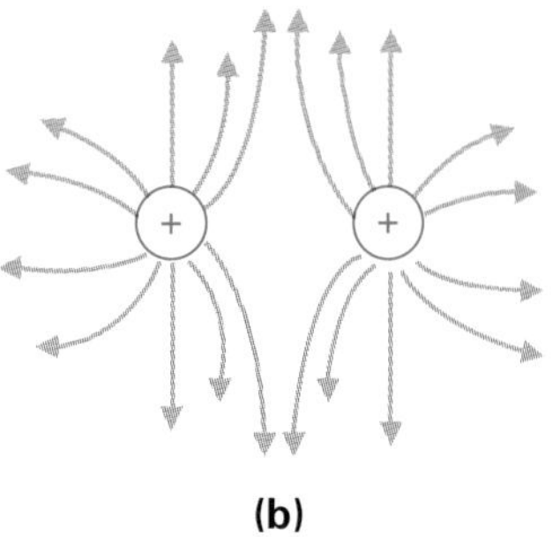

FIGURE 13-6 Electrostatic Field. (a) Electrostatic Lines of Attraction. (b) Electrostatic Lines of Repulsion.

$$\text{field strength (V/m)} = \frac{\text{charge difference } (V)\text{, volts}}{\text{distance between plates } (d)\text{, meters}}$$

The dielectric or insulator between the plates, like any other material, has its own individual atoms, and although the dielectric electrons are more tightly bound to their atoms than conductor electrons, stresses are placed on the atoms within the dielectric, as seen in Figure 13-7. The electrons in orbit around the dielectric atoms are displaced or distorted by the electric field existing between the positive and negative plate. If the charge potential across the capacitor is high enough and the distance between the plates is small enough, the attraction

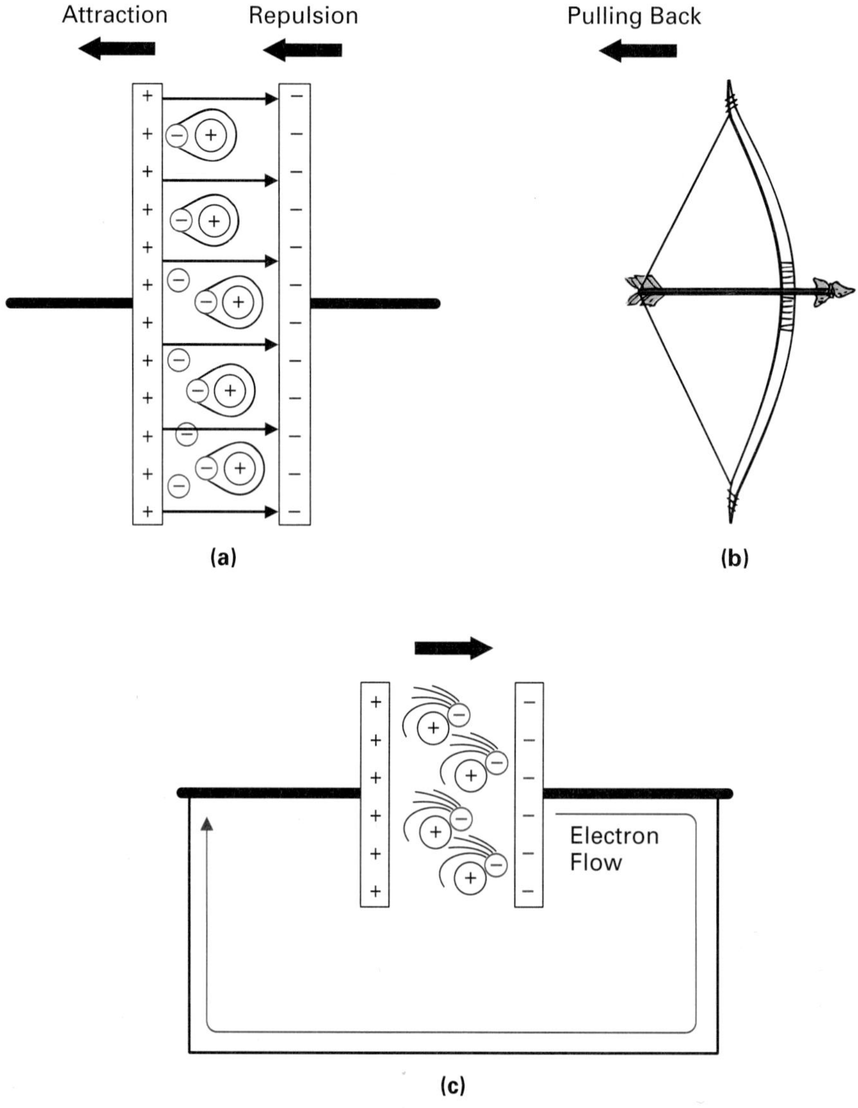

FIGURE 13-7 **Electric Polarization.**

and repulsion exerted on the dielectric atom can be large enough to free the dielectric atom's electrons. The material then becomes ionized, and a chain reaction of electrons jumping from one atom to the next in a right-to-left movement occurs. If this occurs, a large number of electrons will flow from the negative to the positive plate, and the dielectric is said to have broken down. This situation occurs if the capacitor is placed in a circuit where the voltages within the circuit exceed the voltage rating of the capacitor.

If the voltage rating of the capacitor is not exceeded, an electrostatic or electric field still exists between the two plates and causes this pulling of the atom's electrons within the dielectric toward the positive plate. This displacement, known as **electric polarization,** is similar to the pulling-back effect on a bow, as shown in Figure 13-7(b). When the capacitor is given a path for discharge, as shown in Figure 13-7(c), the electric field in the dielectric, which is causing the distortion, is the force field that drives the electrons, like the bow drives the arrow.

Electric Polarization
A displacement of bound charges in a dielectric when placed in an electric field.

To summarize electrostatics and capacitors, the charges on the plates of a capacitor produce an electric field, the electric field causes the distortion of the atoms known as *electric polarization,* and this pulling back or distortion, which is held there by the electric field, is the electron moving force (emf) that drives the electrons when a discharge path is provided. The energy in a capacitor is actually stored in the electric or electrostatic field within the dielectric.

Now that we understand these points, we can see where the word **dielectric** comes from. The dielectric is the insulating material that exists between two (di) plates and undergoes electric polarization when an electric field exists within it (dielectric).

Dielectric
Insulating material between two *(di)* plates in which the *electric* field exists.

SELF-TEST REVIEW QUESTIONS FOR SECTION 13-3

1. An ____________ field is generated by the flow of current, while an ____________ field is produced by voltage.
   a. Electric, magnetic
   b. Magnetic, electric
2. State the formula for calculating field strength.
3. What is electric polarization?
4. Describe why the term *dielectric* is used to indicate the insulating material between the plates of a capacitor.

13-4 THE UNIT OF CAPACITANCE

Capacitance is the ability of a capacitor to store an electrical charge, and the unit of capacitance is the **farad** (F), named in honor of Michael Faraday's work in 1831 in the field of capacitance. A capacitor with the capacity of 1 farad (1 F) can store 1 coulomb of electrical charge (6.24×10^{18} electrons) if 1 volt is applied across the capacitor's plates, as seen in Figure 13-8.

Farad
Unit of capacitance.

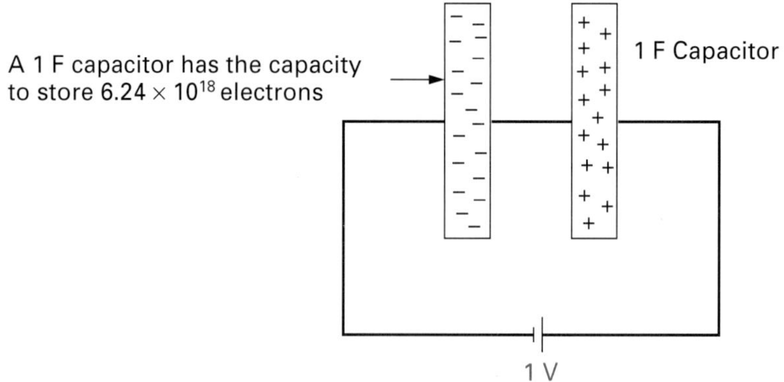

FIGURE 13-8 One Farad of Capacitance.

A 1 F capacitor is a very large value and not frequently found in electronic equipment. Most values of capacitance found in electronic equipment are in the units between the microfarad ($\mu F = 10^{-6}$) and picofarad ($pF = 10^{-12}$). A microfarad is 1 millionth of a farad (10^{-6}). So if a 1 F capacitor can store 6.24×10^{18} electrons with 1 V applied, a 1 μF capacitor, which has 1 millionth the capacity of a 1 F capacitor, can store only 1 millionth of a coulomb, or $(6.24 \times 10^{18}) \times (1 \times 10^{-6}) = 6.24 \times 10^{12}$ electrons when 1 V is applied, as shown in Figure 13-9.

EXAMPLE:

Convert the following to either microfarads or picofarads (whichever is more appropriate):

a. 0.00002 F
b. 0.00000076 F
c. 0.00047×10^{-7} F

Solution:

a. 20 μF
b. 0.76 μF
c. 47 pF

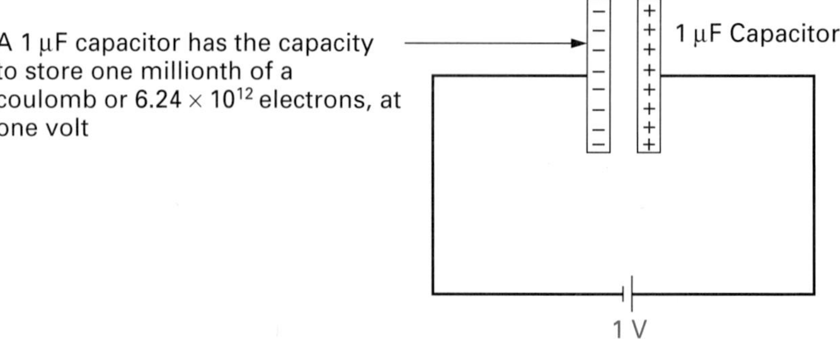

FIGURE 13-9 One-millionth of a Farad.

Since there is a direct relationship between capacitance, charge, and voltage, there must be a way of expressing this relationship in a formula.

$$\text{capacitance, } C \text{ (farads)} = \frac{\text{charge, } Q \text{ (coulombs)}}{\text{voltage, } V \text{ (volts)}}$$

where C = capacitance, in farads

Q = charge, in coulombs

V = voltage, in volts

By transposition of the formula, we arrive at the following combinations for the same formula:

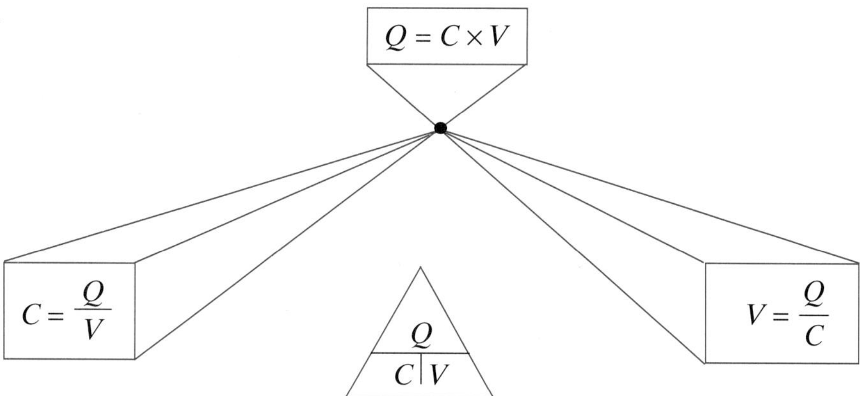

EXAMPLE:

If a capacitor has the capacity to hold 36 C ($36 \times 6.24 \times 10^{18} = 2.25 \times 10^{20}$ electrons) when 12 V is applied across its plates, what is the capacitance of the capacitor?

Solution:

$$C = \frac{Q}{V}$$
$$= \frac{36\text{ C}}{12\text{ V}}$$
$$= 3\text{ F}$$

EXAMPLE:

How many electrons could a 3 μF capacitor store when 5 V is applied across it?

Solution:

$$Q = C \times V$$
$$= 3\ \mu\text{F} \times 5\text{ V}$$
$$= 15\ \mu\text{C}$$

(15 microcoulombs is 15 millionths of a coulomb.) Since 1 C = 6.24×10^{18} electrons, 15 μC = $(15 \times 10^{-6}) \times 6.24 \times 10^{18} = 9.36 \times 10^{13}$ electrons.

EXAMPLE:

If a capacitor of 2 F has stored 42 C of charge (2.63×10^{20} electrons), what is the voltage across the capacitor?

Solution:

$$V = \frac{Q}{C} = \frac{42\text{ C}}{2\ F} = 21\text{ V}$$

SELF-TEST REVIEW QUESTIONS FOR SECTION 13-4

1. What is the unit of capacitance?
2. State the formula for capacitance in relation to charge and voltage.
3. Convert 30,000 μF to farads.
4. If a capacitor holds 17.5 C of charge when 9 V is applied, what is the capacitance of the capacitor?

13-5 FACTORS DETERMINING CAPACITANCE

The capacitance of a capacitor is determined by three factors:

1. The plate area of the capacitor
2. The distance between the plates
3. The type of dielectric used

Let's now discuss these three factors in more detail, beginning with the plate area.

13-5-1 *Plate Area (A)*

The capacitance of a capacitor is directly proportional to the plate area. This area in square inches is the area of only one plate and is calculated by multiplying length by width. This is illustrated in Figure 13-10(a) and (b). In these two examples, the (b) capacitor plate is twice as large as the (a) capacitor plate, and since capacitance is proportional to plate area ($C \propto A$), the capacitor in example (b) will have double the capacity or capacitance of the capacitor in example (a). Since the energy of a charged capacitor is in the electric field between the plates

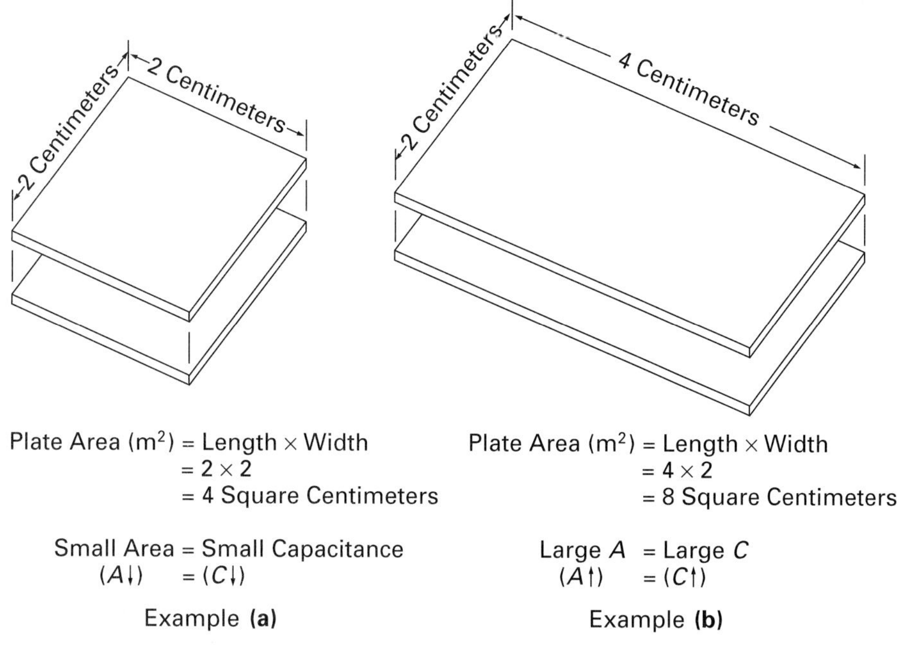

FIGURE 13-10 Capacitance Is Proportional to Plate Area.

and the plates of the capacitor (b) are double those of (a), there is twice as much area for the electric field to exist, and this doubles the capacitor's capacitance.

13-5-2 Distance between the Plates (d)

The distance or separation between the plates is dependent on the thickness of the dielectric used. The capacitance of a capacitor is inversely proportional to this distance between plates, in that an increase in the distance ($d\uparrow$) causes a

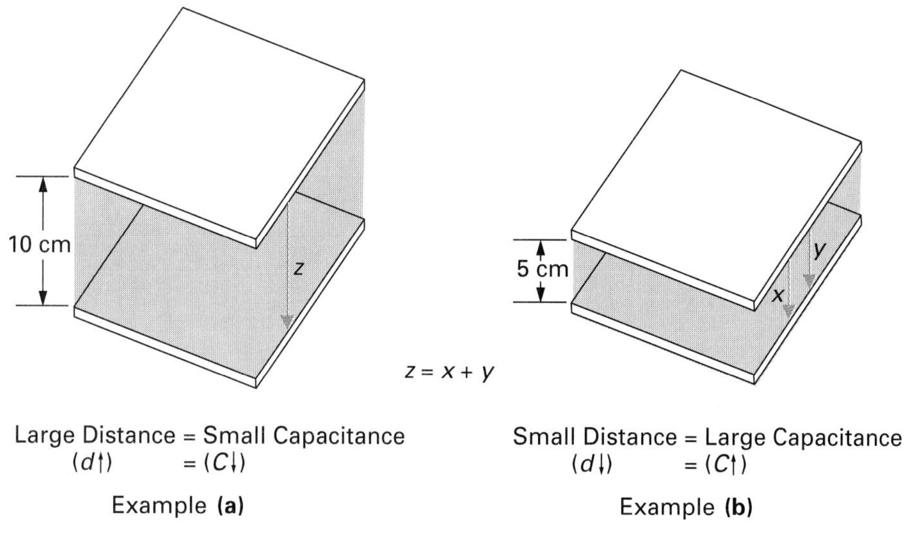

FIGURE 13-11 Capacitance Is Inversely Proportional to Plate Separation or Distance (*d*).

decrease in the capacitor's capacitance ($C\downarrow$). In Figure 13-11(a), a large distance between the capacitor plates results in a small capacitance, whereas in Figure 13-11(b) the dielectric thickness and the plate separation are half that of capacitor (a). This illustrates also how the capacitance of a capacitor can be doubled, in this case by halving the space between the plates. The gap across which the electric lines of force exist is halved in capacitor (b), and this doubles the strength of the electric field, which consequently doubles capacitance. Simply stated, an electric line of force (Z) in Figure 13-11(a) can be used to produce two electric lines of force (X and Y) in Figure 13-11(b) if the distance is half.

13-5-3 Dielectric Constant

Dielectric Constant (*K*)
The property of a material that determines how much electrostatic energy can be stored per unit volume when unit voltage is applied. Also called permittivity.

The insulating dielectric of a capacitor concentrates the electric lines of force between the two plates. Consequently, different dielectric materials can change the capacitance of a capacitor by being able to concentrate or establish an electric field with greater ease than other dielectric insulating materials. The **dielectric constant (*K*)** is the ease with which an insulating material can establish an electrostatic (electric) field. A vacuum is the least effective dielectric and has a dielectric constant of 1, as seen in Table 13-1. All the other insulators listed in this table will support electrostatic lines of force more easily than a vacuum. The vacuum is used as a reference. All the other materials have dielectric constant values that are relative to the vacuum dielectric constant of 1. For example, mica has a dielectric constant of 5.0, which means that mica can cause an electric field

TABLE 13-1 Dielectric Constants

MATERIAL	DIELECTRIC CONSTANT (*K*)[a]
Vacuum	1.0
Air	1.0006
Teflon	2.0
Wax	2.25
Paper	2.5
Amber	2.65
Rubber	3.0
Oil	4.0
Mica	5.0
Ceramic (low)	6.0
Bakelite	7.0
Glass	7.5
Water	78.0
Ceramic (high)	8000.

[a]The different material compositions can cause different values of *K*.

five times the intensity of a vacuum; and, since capacitance is proportional to the dielectric constant ($C \propto K$), the mica capacitor will have five times the capacity of the same-size vacuum dielectric capacitor. In another example, we can see that the capacitance of a capacitor can be increased by a factor of almost 8000 by merely using ceramic rather than air as a dielectric between the two plates.

13-5-4 *The Capacitance Formula*

Thus plate area, separation, and the dielectric used are the three factors that change the capacitance of a capacitor. The formula that combines these three factors is

$$C = \frac{(8.85 \times 10^{-12}) \times K \times A}{d}$$

where C = capacitance, in farads (F)

8.85×10^{-12} is a constant

K = dielectric constant

A = plate area, in square meters (m^2)

d = distance between the plates, in meters (m)

This formula summarizes what has been said in relation to capacitance. The capacitance of a capacitor is directly proportional to the dielectric constant (K) and the plates' area (A) and is inversely proportional to the dielectric thickness or distance between the plates (d).

EXAMPLE:

What is the capacitance of a ceramic capacitor with a 0.3 m^2 plate area and a dielectric thickness of 0.0003 m?

Solution:

$$C = \frac{(8.85 \times 10^{-12}) \times K \times A}{d}$$

$$= \frac{(8.85 \times 10^{-12}) \times 6 \times 0.3\ m^2}{0.0003\ m}$$

$$= 5.31 \times 10^{-8}$$

$$= 0.0531\ \mu F$$

CALCULATOR SEQUENCE

Step	Keypad Entry	Display Response
1.	8 . 8 5 E 1 2 +/−	8.85E-12
2.	×	
3.	6	6
4.	×	5.31E-11
5.	0 . 3	
6.	÷	1.59E-11
7.	0 . 0 0 0 3	
8.	=	5.31E-8

SELF-TEST REVIEW QUESTIONS FOR SECTION 13-5

1. List the three variable factors that determine the capacitance of a capacitor.
2. State the formula for capacitance.
3. If a capacitor's plate area is doubled, the capacitance will ____________?
4. If a capacitor's dielectric thickness is halved, the capacitance will ____________?

13-6 DIELECTRIC BREAKDOWN AND LEAKAGE

Capacitors store a charge just as a container or tank stores water. The amount of charge stored by a capacitor is proportional to the capacitor's capacitance and the voltage applied across the capacitor ($Q = C \times V$). The charge stored by a fixed-value capacitor (C is fixed) therefore can be increased by increasing the voltage across the plates, as shown in Figure 13-12(a). If the voltage across the capacitor is increased further, the charge held by the capacitor will increase until the dielectric between the two plates of the capacitor breaks down and a spark jumps or arcs between the plates.

Using the water analogy shown in Figure 13-12(b), if the pressure of the water being pumped in is increased, the amount of water stored in the tank will also increase until a time is reached when the tank's seams at the bottom of the tank cannot contain the large amount of pressure and break down under strain. The amount of water stored in the tank is proportional to the pressure applied,

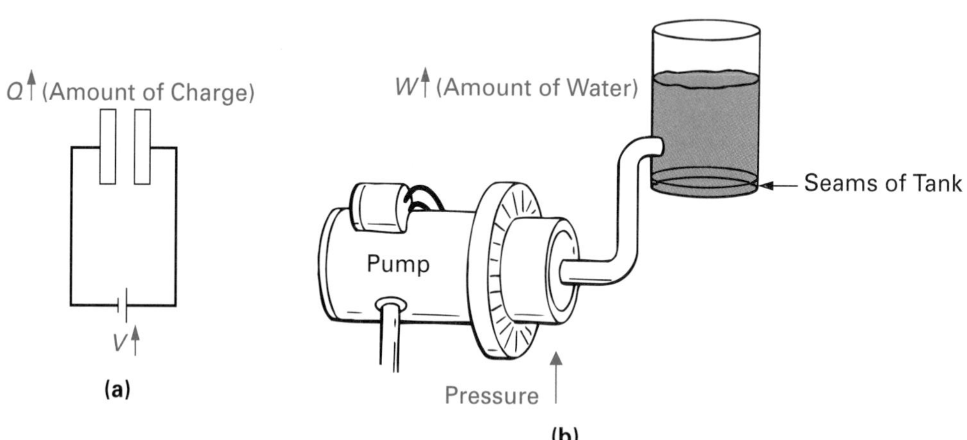

FIGURE 13-12 Breakdown. (a) Voltage Increase, Charge Increases until Dielectric Breakdown. (b) Pressure Increase, Water Increases until Seams Break Down.

TABLE 13-2 Dielectric Strengths

MATERIAL	DIELECTRIC STRENGTH (V/mm)
Air	787
Oil	12,764
Ceramic	39,370
Paper	49,213
Teflon	59,055
Mica	59,055
Glass	78,740

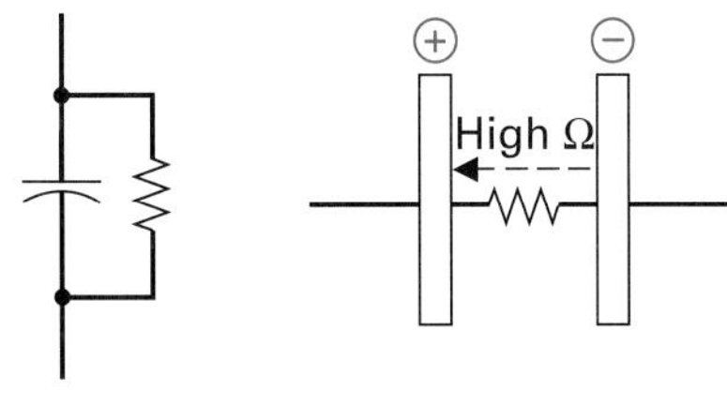

FIGURE 13-13 Capacitor Leakage.

just as the amount of charge stored in a capacitor is proportional to the amount of voltage applied.

The **breakdown voltage** of a capacitor is determined by the strength of the dielectric used. Table 13-2 illustrates some of the different strengths of many of the common dielectrics. As an example, let's consider a capacitor that uses 1 mm of air as a dielectric between its two plates. This particular capacitor can withstand any voltage up to 787 V. If the voltage is increased further, the dielectric will break down and current will flow between the plates, destroying the capacitor (air capacitors, however, can recover from ionization).

Breakdown Voltage
The voltage at which breakdown occurs in a dielectric or insulation.

The ideal or perfect insulator should have a resistance equal to infinite ohms. Insulators or the dielectric used to isolate the two plates of a capacitor are not perfect and therefore have some very high values of resistance. This means that some value of resistance exists between the two plates, as shown in Figure 13-13; although this value of resistance is very large, it will still allow a small amount of current to flow between the two plates (in most applications a few nanoamperes or picoamperes). This small current, referred to as **leakage current,** causes any charge in a capacitor to slowly, over a long period of time, discharge between the two plates. A capacitor should have a large leakage resistance to ensure the smallest possible leakage current.

Leakage Current
Small, undesirable flow of current through an insulator or dielectric.

SELF-TEST REVIEW QUESTIONS FOR SECTION 13-6

1. Which dielectric material has the best breakdown voltage figure?
2. True or false: Current flows through the dielectric at and below the breakdown voltage of the material.
3. A capacitor should have a ____________ leakage resistance to ensure a ____________ leakage current.
4. Leakage current is normally in:

   a. Milliamperes **c.** Microamperes
   b. Nanoamperes **d.** Kiloamperes

13-7 CAPACITORS IN COMBINATION

Like resistors, capacitors can be connected in either series or parallel. The rules for determining total capacitance for parallel- and series-connected capacitors are opposite to series- and parallel-connected resistors.

13-7-1 *Capacitors in Parallel*

In Figure 13-14(a), you can see a 2 μF and 4 μF capacitor connected in parallel with one another. As the top plate of capacitor *A* is connected to the top plate of capacitor *B* with a wire, and a similar situation occurs with the bottom plates, you can see that this is the same as if the top and bottom plates were touching one another, as shown in Figure 13-14(b). When drawn so that the respective plates are touching, the dielectric constant and plate separation is the same as shown in Figure 13-14(a); however, now we can easily see that the plate area is actually increased. Consequently, if capacitors are connected in parallel, the effective plate area is increased; and since capacitance is proportional to plate area [$C\uparrow = (8.85 \times 10^{-12}) \times K \times A\uparrow/d$], the capacitance will also increase. Total capacitance is actually calculated by adding the plate areas, so total capacitance is equal to the sum of all the individual capacitances in parallel.

$$C_T = C_1 + C_2 + C_3 + C_4 + \cdots$$

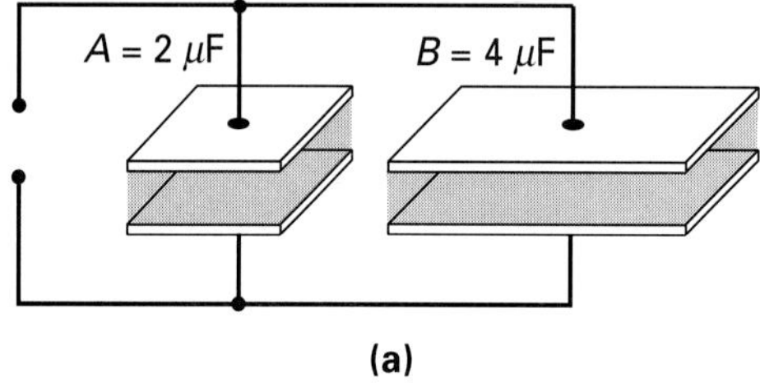

(a)

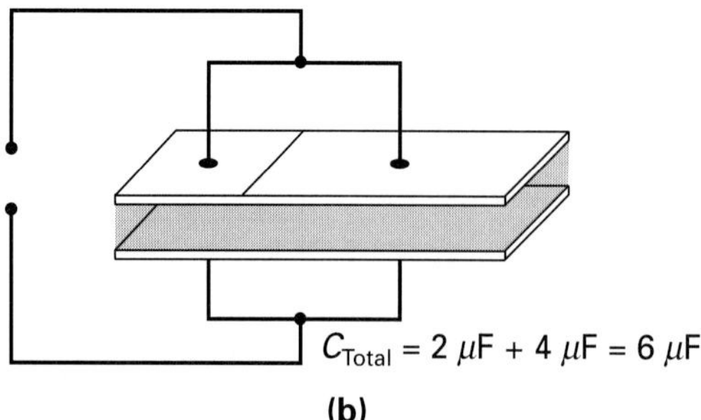

(b)

FIGURE 13-14 **Capacitors in Parallel.**

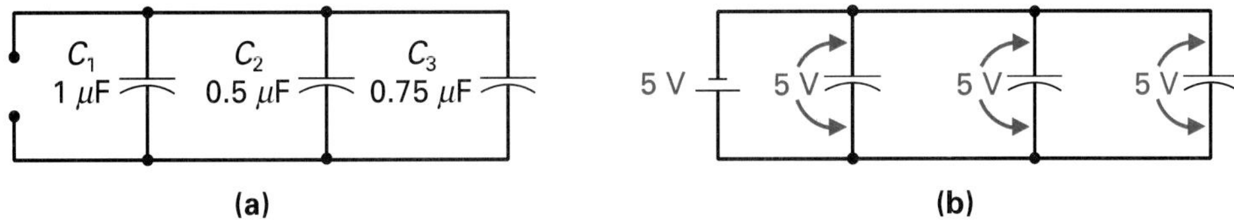

FIGURE 13-15 Example of Parallel-Connected Capacitors.

EXAMPLE:

Determine the total capacitance of the circuit in Figure 13-15(a). What will be the voltage drop across each capacitor?

Solution:

$$C_T = C_1 + C_2 + C_3$$
$$= 1\ \mu F + 0.5\ \mu F + 0.75\ \mu F$$
$$= 2.25\ \mu F$$

As with any parallel-connected circuit, the source voltage appears across all the components. If, for example, 5 V is connected to the circuit of Figure 13-15(b), all the capacitors will charge to the same voltage of 5 V because the same voltage always exists across each section of a parallel circuit.

13-7-2 Capacitors in Series

In Figure 13-16(a), we have taken the two capacitors of 2 μF and 4 μF and connected them in series. Since the bottom plate of the *A* capacitor is connected to the top plate of the *B* capacitor, they can be redrawn so that they are touching, as shown in Figure 13-16(b).

The top plate of the *A* capacitor is connected to a wire into the circuit, and the bottom plate of *B* is connected to a wire into the circuit. This connection creates two center plates that are isolated from the circuit and can therefore be disregarded, as shown in Figure 13-16(c). The first thing you will notice in this illustration is that the dielectric thickness ($d\uparrow$) has increased, causing a greater separation between the plates. The effective plate area of this capacitor has decreased, as it is just the area of the top plate only. Even though the bottom plate extends out further, the electric field can only exist between the two plates, so the surplus metal of the bottom plate has no metal plate opposite for the electric field to exist.

Consequently, when capacitors are connected in series the effective plate area is decreased ($A\downarrow$) and the dielectric thickness increased ($d\uparrow$), and both of these effects result in an overall capacitance decrease ($C\downarrow\downarrow = (8.85 \times 10^{-12}) \times K \times A\downarrow/d\uparrow$).

The plate area is actually decreased to the smallest individual capacitance connected in series, which in this example is the plate area of *A*. If the plate area were the only factor, then capacitance would always equal the smallest capacitor value. However, the dielectric thickness is always equal to the sum of all the

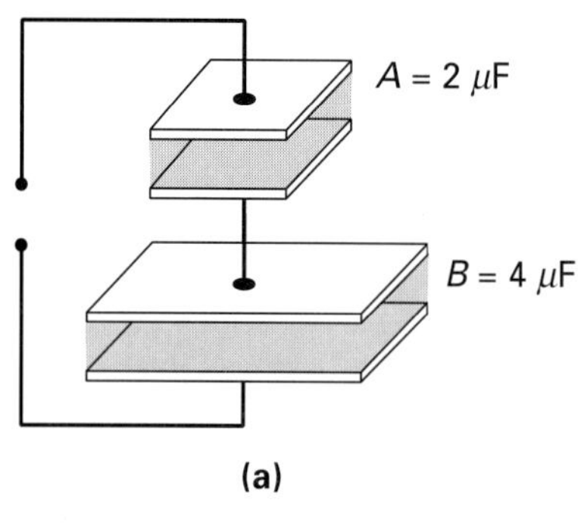

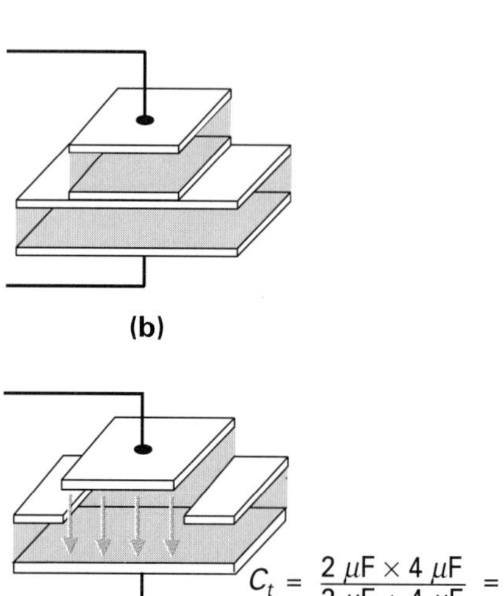

FIGURE 13-16 Capacitors in Series.

capacitor dielectrics, and this factor always causes the total capacitance (C_T) to be less than the smallest individual capacitance when capacitors are connected in series.

The total capacitance of two or more capacitors in series therefore is calculated by using the following formulas: For two capacitors in series,

$$C_T = \frac{C_1 \times C_2}{C_1 + C_2}$$

(product-over-sum formula)

For more than two capacitors in series,

$$C_T = \frac{1}{(1/C_1) + (1/C_2) + (1/C_3) + \cdots}$$

(reciprocal formula)

EXAMPLE:

Determine the total capacitance of the circuit in Figure 13-17.

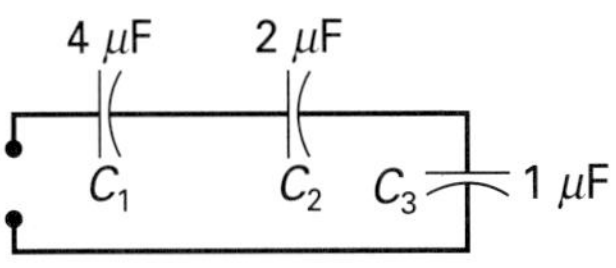

FIGURE 13-17 Example of Series-Connected Capacitors.

Solution:

$$C_T = \frac{1}{(1/C_1) + (1/C_2) + (1/C_3)}$$

$$= \frac{1}{(1/4\ \mu\text{F}) + (1/2\ \mu\text{F}) + (1/1\ \mu\text{F})}$$

$$= \frac{1}{1.75 \times 10^6} = 5.7143 \times 10^{-7}$$

$$= 0.5714\ \mu\text{F} \quad \text{or} \quad 0.6\ \mu\text{F}$$

The total capacitance for capacitors in series is calculated in the same way as total resistance when resistors are in parallel.

As with series-connected resistors, all of the voltage drops across the series-connected resistors will equal the voltage applied (Kirchhoff's voltage law). With capacitors connected in series, the charged capacitors act as a voltage divider, and therefore the voltage-divider formula can be applied to capacitors in series.

$$V_{cx} = \frac{C_T}{C_x} \times V_T$$

where V_{cx} = voltage across desired capacitor

C_T = total capacitance

C_x = desired capacitor's value

V_T = total supplied voltage

EXAMPLE:

Using the voltage-divider formula, calculate the voltage dropped across each of the capacitors in Figure 13-17 if $V_S = 24$ V.

Solution:

$$V_{C1} = \frac{C_T}{C_1} \times V_S = \frac{0.5714\ \mu\text{F}}{4\ \mu\text{F}} \times 24\ \text{V} = 3.4\ \text{V}$$

$$V_{C2} = \frac{C_T}{C_2} \times V_S = \frac{0.5714\ \mu\text{F}}{2\ \mu\text{F}} \times 24\ \text{V} = 6.9\ \text{V}$$

$$V_{C3} = \frac{C_T}{C_3} \times V_S = \frac{0.5714\ \mu\text{F}}{1\ \mu\text{F}} \times 24\ \text{V} = 13.7\ \text{V}$$

$$V_S = V_{C1} + V_{C2} + V_{C3} = 3.4 + 6.9 + 13.7 = 24\ \text{V}$$

(Kirchhoff voltage law)

If the capacitor values are the same, as seen in Figure 13-18(a), the voltage is divided equally across each capacitor, as each capacitor has an equal amount of charge and therefore has half of the applied voltage (in this example, 3 V across each capacitor).

When the capacitor values are different, the smaller value of capacitor will actually charge to a higher voltage than the larger capacitor. In the example in Figure 13-18(b), the smaller capacitor is actually half the size of the other capacitor, and it has charged to twice the voltage. Since Kirchhoff's voltage law has to apply to this and every series circuit, you can easily calculate that the voltage across C_1 will equal 4 V and is twice that of C_2, which is 2 V. To understand this fully, we must first understand that although the capacitance is different, both capacitors have an equal value of coulomb charge held within them, which in this example is 8 μC.

$$\begin{aligned} Q_1 &= C_1 \times V_1 \\ &= 2\ \mu\text{F} \times 4\ \text{V} = 8\ \mu\text{C} \\ Q_2 &= C_2 \times V_2 \\ &= 4\ \mu\text{F} \times 2\ \text{V} = 8\ \mu C \end{aligned}$$

This equal charge occurs because the same amount of current flow exists throughout a series circuit, so both capacitors are being supplied with the same number or quantity of electrons. The charge held by C_1 is large with respect to its small capacitance, whereas the same charge held by C_2 is small with respect to its larger capacitance.

If the charge remains the same (Q is constant) and the capacitance is small, the voltage drop across the capacitor will be large, because the charge is large with respect to the capacitances.

$$V\uparrow = \frac{Q}{C\downarrow}$$

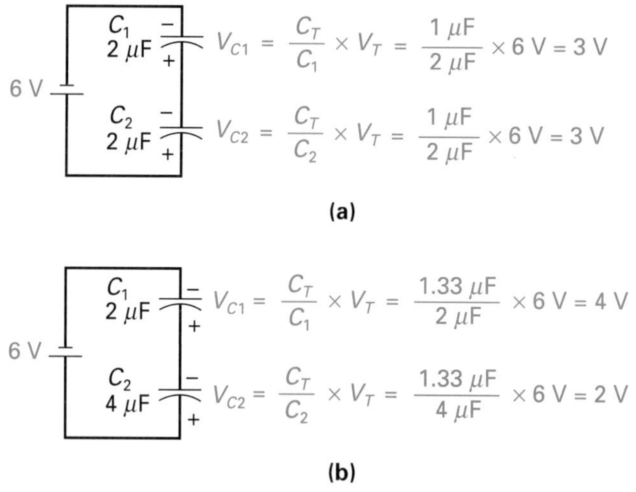

FIGURE 13-18 Voltage Drops across Series-Connected Capacitors.

On the other hand, for a constant charge, a large capacitance will have a small charge voltage because the charge is small with respect to the capacitance:

$$V\downarrow = \frac{Q}{C\uparrow}$$

We can apply the water analogy once more and imagine two series-connected buckets, one of which is twice the size of the other. Both are being supplied by the same series pipe, which has an equal flow of water throughout, and are consequently each holding an equal amount of water, for example, 1 gallon. The 1 gallon of water in the small bucket is large with respect to the size of the bucket, and a large amount of pressure exists within that bucket. The 1 gallon of water in the large bucket is small with respect to the size of the bucket, so a small amount of pressure exists within this bucket. The pressure within a bucket is similar to the voltage across a capacitor, and therefore a small bucket or capacitor will have a greater pressure or voltage associated with it, while a large bucket or capacitor will develop a small pressure or voltage.

To summarize capacitors in series, all the series-connected components will have the same charging current throughout the circuit, and because of this, two or more capacitors in series will always have equal amounts of coulomb charge. If the charge (Q) is equal, the voltage across the capacitor is determined by the value of the capacitor. A small capacitance will charge to a larger voltage ($V\uparrow = Q/C\downarrow$), whereas a large value of capacitance will charge to a smaller voltage ($V\downarrow = Q/C\uparrow$).

SELF-TEST REVIEW QUESTIONS FOR SECTION 13-7

1. If 2 μF, 3 μF, and 5 μF capacitors are connected in series, what will be the total circuit capacitance?
2. If 7 pF, 2 pF, and 14 pF capacitors are connected in parallel, what will be the total circuit capacitance?
3. State the voltage-divider formula as it applies to capacitance.
4. True or false: With resistors, the large value of resistor will drop a larger voltage, whereas with capacitors the smaller value of capacitor will actually charge to a higher voltage.

13-8 TYPES OF CAPACITORS

Capacitors come in a variety of shapes and sizes and can be either fixed or variable in their values of capacitance. Within these groups, capacitors are generally classified by the dielectric used between the plates.

A **fixed-value capacitor** is a capacitor whose capacitance value remains constant and cannot be altered. Fixed capacitors normally come in a disk or a tubular package, as seen in Figure 13-19, and consist of metal foil plates separated by

Fixed-Value Capacitor
A capacitor whose value is fixed and cannot be varied.

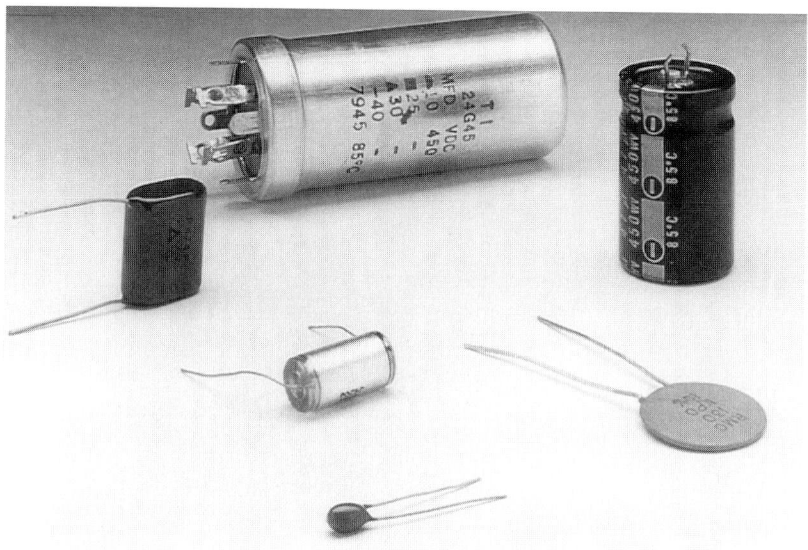

FIGURE 13-19 **Fixed-Value Capacitors (Courtesy of Sencore).**

one of the following types of insulators (dielectric), which is the means by which we classify them.

1. Mica
2. Ceramic
3. Paper
4. Plastic
5. Electrolytic

A *variable-value capacitor* is a capacitor whose capacitance value can be changed by rotating a shaft. The variable capacitor normally consists of one electrically connected movable plate and one electrically connected stationary plate. There are basically four types of variable-value capacitors, which are also classified by the dielectric used.

1. Air
2. Mica
3. Ceramic
4. Plastic

First, let's take a closer look at the different types of fixed-value capacitors.

13-8-1 *Fixed-Value Capacitors*

Mica Capacitor
Fixed capacitor that uses mica as the dielectric between its plates.

Mica

Figure 13-20 illustrates the physical appearance and construction of the **mica capacitor.** In Figure 13-20(a), which illustrates the construction of the mica capacitor, you can see that thin foil plates (normally aluminum) are alternately

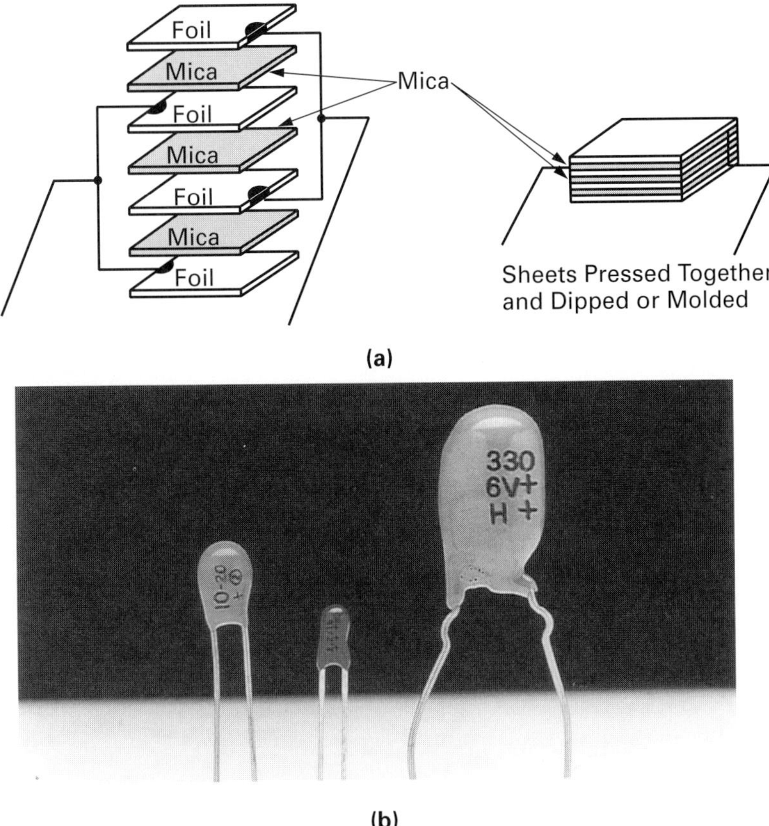

FIGURE 13-20 Mica Capacitors. (a) Construction. (b) Physical Appearance (Courtesy of Sencore).

stacked to form the plates of the capacitor, all of which are isolated from one another by a thin layer of mica dielectric. Every other plate is connected to one lead on the right, while the other metal foil plates are attached to the other connecting lead on the left. This arrangement of stacked plates provides an increase in the capacitance due to the larger overall plate area.

The complete assembly of plates and mica is then sealed inside a protective casing. Figure 13-20(b) shows some examples of dipped mica capacitors. If molding equipment is used to protect and seal the assembly the capacitor is referred to as a *molded* mica capacitor; however, if the plate and mica assembly is sealed by a dipping process, the capacitor is referred to as a *dipped* mica capacitor.

Ceramic

Figure 13-21 illustrates the construction and physical appearance of the molded and dipped types of **ceramic capacitor.** In their construction, you will notice the stacking of plates and isolation by ceramic, similar to the mica capacitor. Generally, the low-dielectric-constant ceramic ($K = 6.0$) is placed within the molded package, while the high-dielectric-constant ceramic (different composition means that $K = 8000$) is placed within the dipped package to obtain higher voltage ratings and high values of capacitance in small packages.

Ceramic Capacitor
Capacitor in which the dielectric material used between the plates is ceramic.

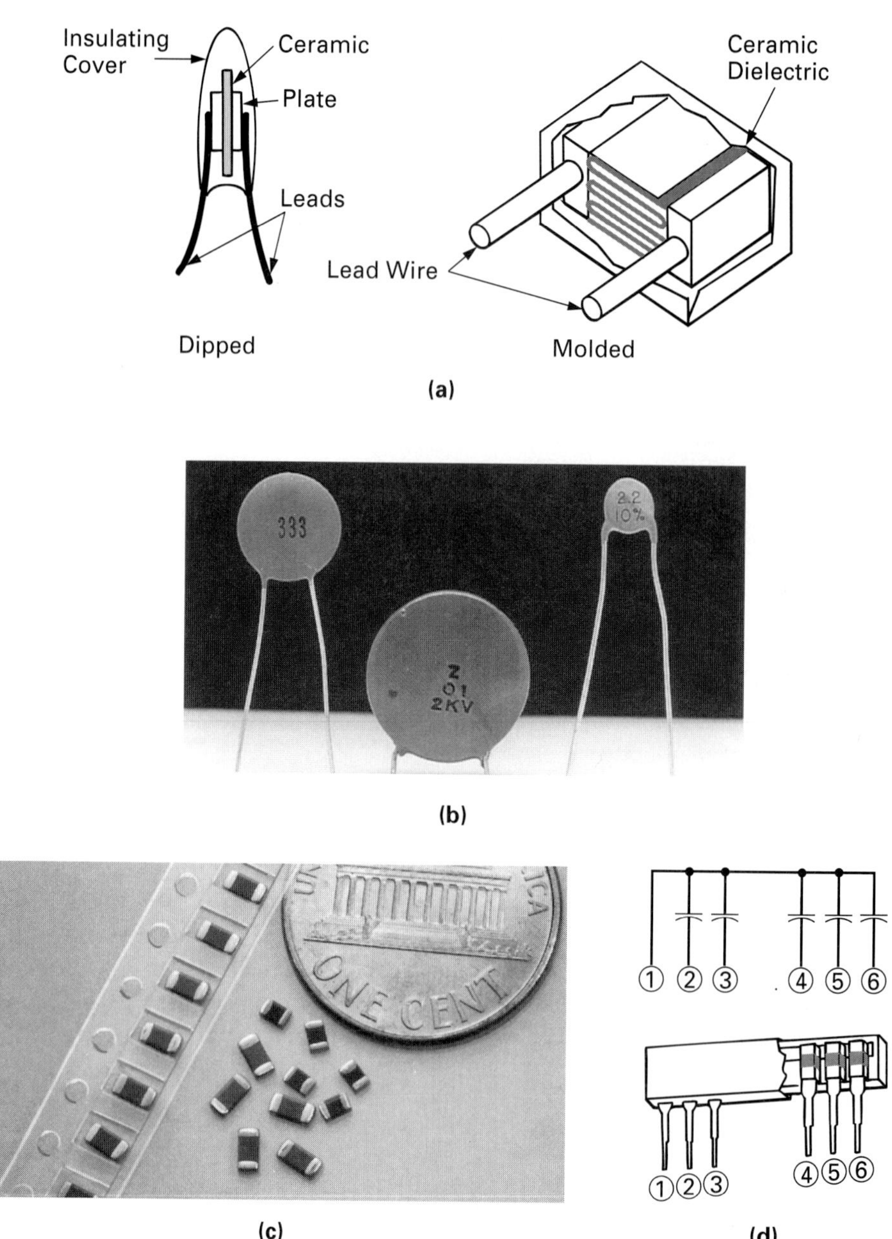

FIGURE 13-21 Ceramic Capacitors. (a) Construction. (b) Physical Appearance (Courtesy of Sencore). (c) Chip Ceramic Casing. (Product photo supplied as a courtesy of NIC Components Corp, Amityville, N.Y.) (d) SIP Casing.

Figure 13-21(c) shows the ceramic chip surface-mount capacitors that are now available on the market for both discrete and integrated types of circuits. Values typically range from 16 to 1600 pF. Different values of capacitance can be obtained by using different types of ceramic, thereby varying the dielectric constant.

Figure 13-21(d) shows how the chip ceramic capacitors can be mounted in a SIP (single in-line package) casing.

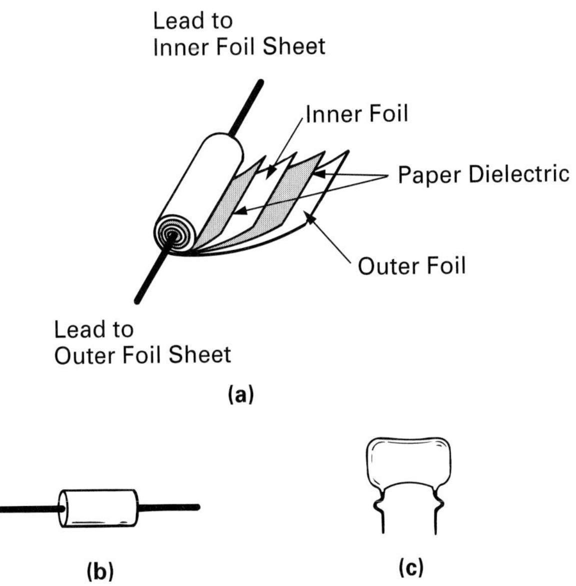

FIGURE 13-22 **Paper Capacitors. (a) Construction. (b) Axial Lead Wrapped. (c) Radial Lead Dipped.**

Paper

Figure 13-22 illustrates the construction and physical appearance of tubular **paper capacitors.** Two long strips of foil are separated by a paper dielectric that has been saturated with paraffin to ensure that the paper dielectric is a good insulator. The two foil plates separated by the paper dielectric are rolled into a tubular shape and placed in either a molded or a dipped case.

The leads of the capacitor can be either axial lead, as shown in Figure 13-22(b) (leads coming out of either end), or radial lead, as shown in Figure 13-22(c) (leads coming out of a single end).

Paper Capacitors
Fixed capacitor using oiled or waxed paper as a dielectric.

Plastic Film Capacitor
A capacitor in which alternate layers of metal aluminum foil are separated by thin films of plastic dielectric.

Electrolytic Capacitor
Capacitor having an electrolyte between the two plates; due to chemical action, a very thin layer of oxide is deposited on only the positive plate, which accounts for why this type of capacitor is polarized.

Plastic

Plastic film capacitors have almost completely replaced the older types of paper capacitors, and their method of construction is illustrated in Figure 13-23(a). The construction is identical to the paper type; however, in this instance a plastic film dielectric strip is used as the dielectric and then rolled into a cylinder. Mylar, polycarbonate, Teflon, and polypropylene are all examples of plastic that are used as a dielectric.

The final cylindrical package can be either encased in an outer plastic layer, as shown in Figure 13-23(b), or dipped, as shown in Figure 13-23(c).

Electrolytic

Figure 13-24 shows the construction and physical appearance of typical **electrolytic capacitors.** These types of capacitors are constructed in a similar manner as both paper and plastic types of capacitors, the difference being that

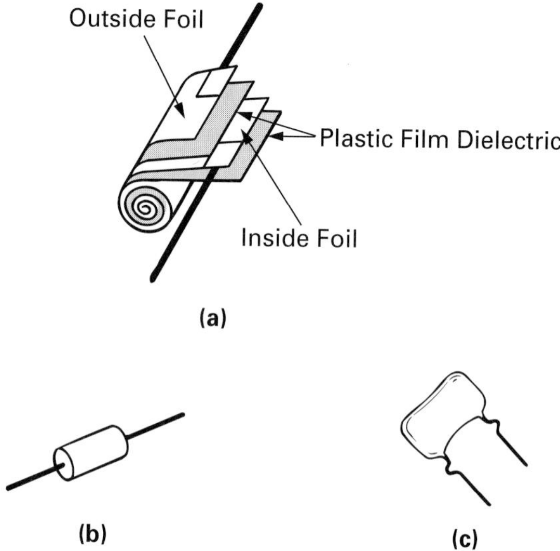

FIGURE 13-23 Plastic Capacitors. (a) Construction. (b) Axial Lead. (c) Radial Lead.

the foil plates are separated by a strip of gauze that has been saturated with a conductive fluid known as an *electrolyte.* In the manufacturing process, a dc voltage causes a current flow in one direction, which causes the electrolyte to interact chemically with the aluminum foil and create a coating of aluminum oxide on the surface of the positive aluminum foil plate, as shown in Figure 13-24(a), which causes it to change chemically. This oxide, which is formed by an electrochemical reaction, becomes the dielectric for the electrolytic capacitor, and because it is extremely thin, the electrolytic capacitor can have a large capaci-

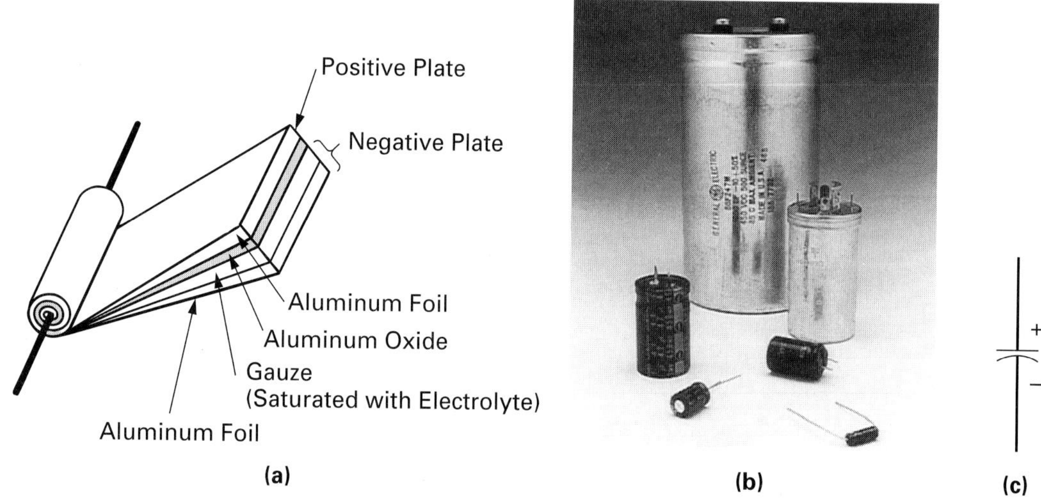

FIGURE 13-24 Electrolytic Capacitors. (a) Construction. (b) Physical Appearance (Courtesy of Sencore). (c) Schematic Symbol for Electrolytic.

Name	Construction	Approximate Range of Values and Tolerances	Characteristics	
Mica	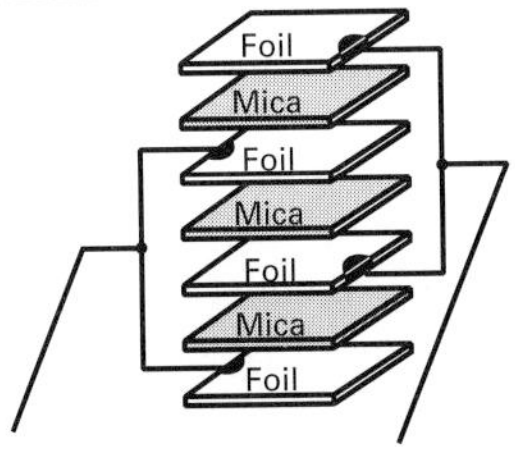 Foil, Mica, Foil, Mica, Foil, Mica, Foil	1 pF–0.1 μF ±1% to ±5%	Lower voltage rating than other capacitors of the same size	Small Capacitor Values
Ceramic	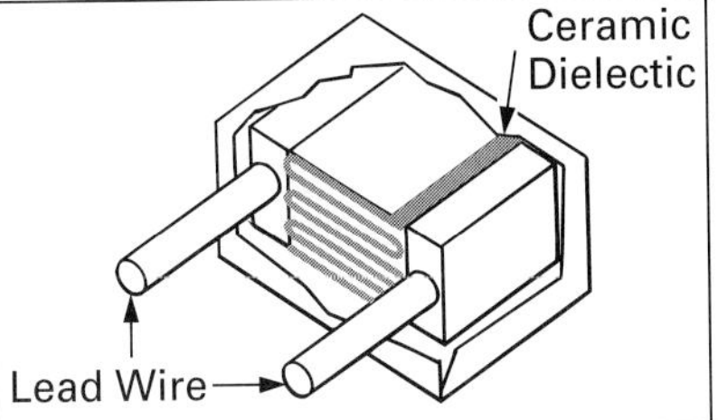 Ceramic Dielectic; Lead Wire	*Low Dielectric K*: 1 pF–0.01 μF ±0.5% to ±10% *High Dielectric K*: 1 pF–0.1 μF ±10% to ±80%	Most popular small valve capacitor due to lower cost than mica, and its ruggedness	
Paper	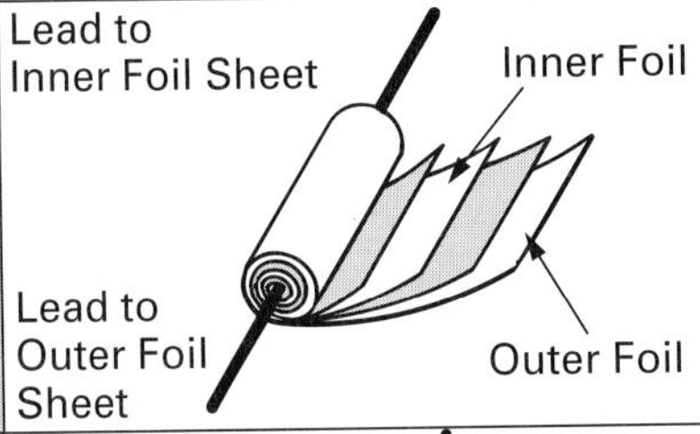 Lead to Inner Foil Sheet; Inner Foil; Lead to Outer Foil Sheet; Outer Foil	1 pF–1 μF ± 10%	Has a large plate area and therefore large capacitance for a small size	Large Capacitor Values
Plastic	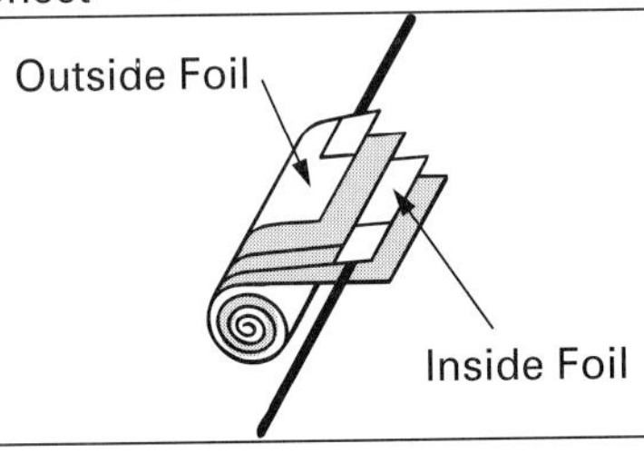 Outside Foil; Inside Foil	1 pF–10 μF ± 5% to ±10%	Has almost completely replaced paper capacitors; has large capacitance valves for small size and high voltage ratings	
Electrolytic (Aluminum and Tantalum)	(Tantalum has tantalum rather than aluminum foil plates) 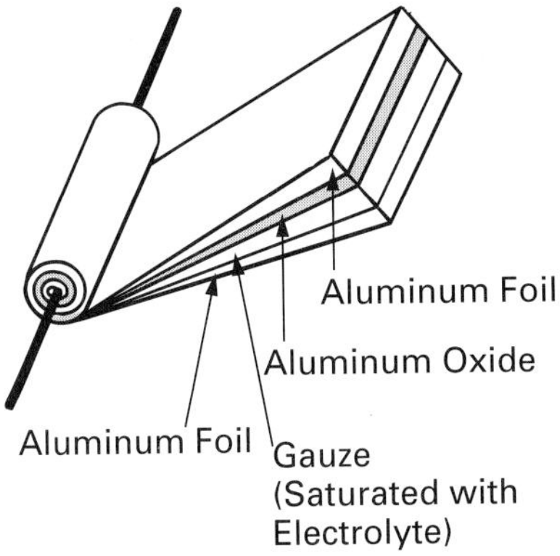 Aluminum Foil; Aluminum Oxide; Aluminum Foil; Gauze (Saturated with Electrolyte)	1 μF–1F ± 10% to ±50%	Most popular large valve capacitor: large capacitance into small area, wide range of valves. Disadvantages are: cannot be used in AC circuits as they are polarized; poor tolerances; low leakage resistance and so high leakage current. *Tantalum* advantages over aluminum include smaller size, longer life than aluminum, which has an approximate lifespan of 12 years. Disadvantages: 4 to 5 times the price.	

FIGURE 13-25 Summary of Fixed-Value Capacitors.

tance for a small size. The chemical change in the positive aluminum plate during this electrochemical process makes the electrolytic capacitor **polarized.** Thus the electrolytic capacitor must always have a positive charge applied to its positive plate and a negative charge to its negative plate. If this rule is not followed, the electrolytic capacitor becomes a safety hazard, because it can in the worst-case condition explode violently.

Polarized Electrolytic Capacitor
An electrolytic capacitor in which the dielectric is formed adjacent to one of the metal plates, creating a greater opposition to current in one direction only.

Figure 13-24(c) shows the schematic symbols for electrolytic capacitors, which are always marked with a (+) or (–) sign to indicate polarity. In Figure 13-24(b) you can see the manufacturer's method of polarity indication.

Figure 13-24(b) shows a "can" electrolytic capacitor. With this type of electrolytic capacitor, the negative outside foil plate is connected to the metal can casing, which acts as the negative capacitor lead.

Tantalum Capacitor
Electrolytic capacitor having a tantalum foil anode.

Tantalum instead of aluminum is used in some electrolytic capacitors for the plates and has advantages over the aluminum type, which are:

1. Higher capacitance per volt for a given unit volume
2. Longer life and excellent shelf life (storage)
3. Able to operate in a wider temperature range
4. Temperature stability is better
5. Construction features make them more rugged

The cost of tantulum electrolytics, however, is almost four to five times that of aluminum electrolytics, and their operating voltages are much lower.

Figure 13-25 illustrates, reviews, and includes added information on the five basic fixed capacitors discussed previously.

13-8-2 Variable-Value Capacitors

Variable-Value Capacitor
A capacitor whose value can be varied.

Variable-value capacitors are the second basic type, and within this group, four types are commercially available. Like the fixed-value type of capacitors, they are classified by dielectric.

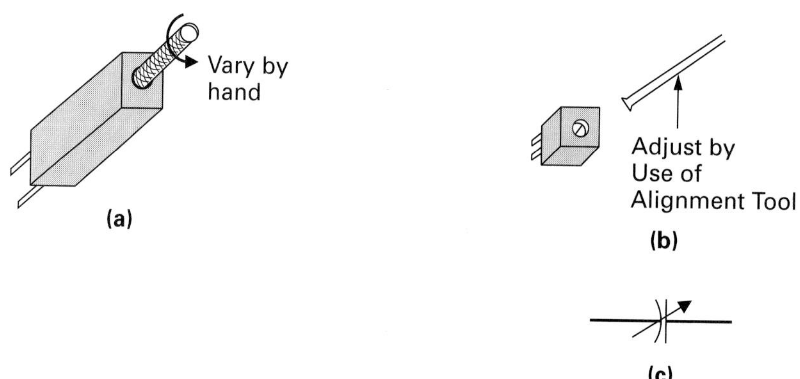

FIGURE 13-26 **(a) Variable Capacitor. (b) Adjustable Capacitor. (c) Schematic Symbol for Variable or Adjustable Capacitor.**

Variable

1. Air
2. Mica

The variable-value type uses a hand-rotated shaft to vary the effective plate area, and so capacitance, as shown in Figure 13-26(a).

Adjustable

3. Ceramic
4. Plastic

The adjustable type uses screw-in or screw-out mechanical adjustment to vary the distance between the plates, and so capacitance, as shown in Figure 13-26(b).

Capacitance is dependent on effective plate area and can therefore be varied by either changing the effective plate area or distance between the plates. The dielectric constant is fixed and dcpcnds on the particular type being used. The schematic symbol for a variable capacitor is shown in Figure 13-26(c).

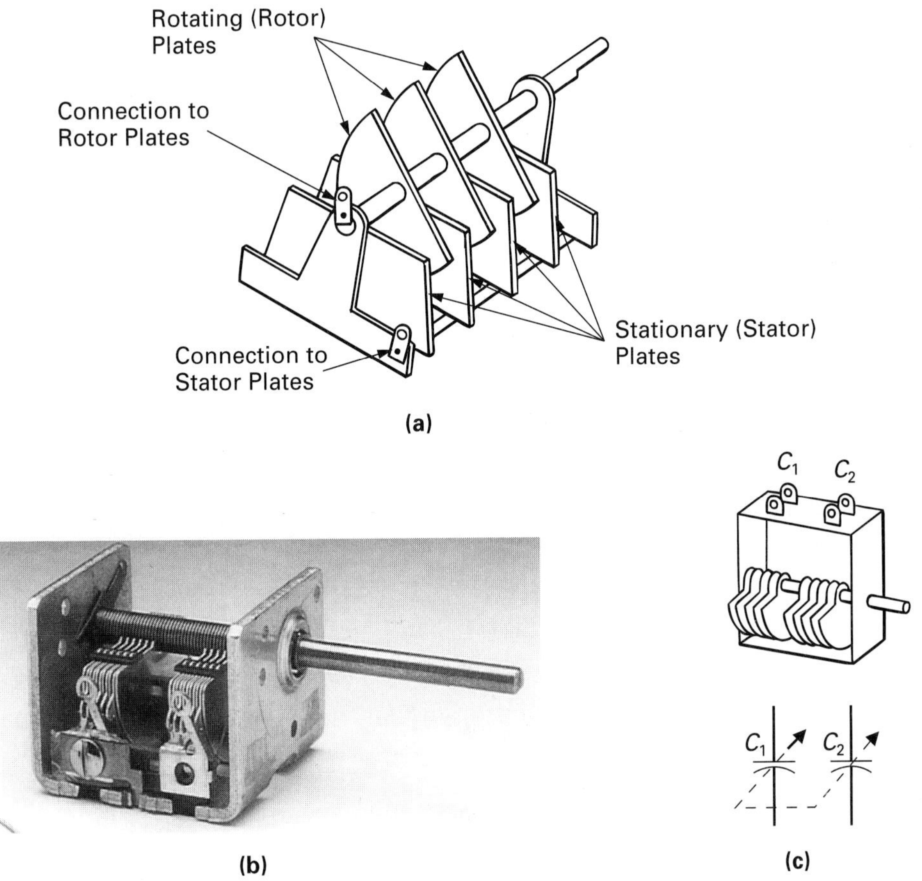

FIGURE 13-27 Variable Air-Type Capacitors. (a) Construction. (b) Physical Appearance (Courtesy of Sencore). (c) Ganged.

Air (Variable)

Figure 13-27(a) illustrates the construction of a typical air dielectric variable capacitor. With this type of capacitor, the effective plate area is adjusted to vary capacitance by causing a set of rotating plates (rotor) to mesh with a set of stationary plates (stator). When the rotor plates are fully out, the capacitance is minimum, and when the plates are fully in, the capacitance is maximum, because the maximum amount of rotor plate area is now opposite the stator plate, creating the maximum value of capacitance.

The plates are usually made of aluminum to prevent corrosion and the dielectric between the plates is air. The capacitor has to be carefully manufactured to ensure that the rotor plates do not touch the stator plates when the shaft is rotated and the plates interweave. Figure 13-27(b) illustrates a typical variable air capacitor.

In radio equipment, it is sometimes necessary to have two or more variable-value capacitors that have been constructed in such a way that a common shaft (rotor) runs through all the capacitors and varies their capacitance simultaneously, as shown in Figure 13-27(c). If you mechanically couple (gang) two or more variable capacitors so that they can all be operated from a single control, the component is known as a **ganged capacitor,** and the symbol for this arrangement is also shown in Figure 13-27(c).

Ganged Capacitor
Mechanical coupling of two or more capacitors, switches, potentiometers, or any other components so that the activation of one control will operate all sections.

Mica, Ceramic, and Plastic (Adjustable)

Figure 13-28 illustrates some of the typical packages for mica, ceramic, or plastic film types of adjustable capacitors, which are also referred to as *trimmers.* The adjustable capacitor generally has one stationary plate and one spring metal moving plate. The screw forces the spring metal plate closer or farther away from the stationary plate, varying the distance between the plates and so changing capacitance. The two plates are insulated from one another by either mica, ceramic, or plastic film, and the advantages of each are the same as for fixed-value capacitors.

These types of capacitors should only be adjusted with a plastic or nonmetallic alignment tool, because a metal screwdriver may affect the capacitance of the capacitor when nearby, making it very difficult to adjust for a specific value of capacitance.

FIGURE 13-28 **Adjustable Mica, Ceramic, or Plastic Capacitors. (a) Construction. (b) Physical Appearance (Courtesy of Sencore).**

13-8-3 *The One-Farad Capacitor*

Most capacitor values are normally measured in either microfarads or picofarads. The traditional picofarad and microfarad capacitors, which have been discussed, consist of two conductor plates separated by an insulator; when voltage is applied to the capacitor, current flows and a charge builds up within the capacitor and is stored. The amount of charge stored is dependent on the plate area, plate separation, and the dielectric constant of the insulator used. All three of these factors can be varied separately or in combination to create the capacitors that we have today.

The ceramic capacitor has become the most popular small-value capacitor due to the high dielectric constants that can be obtained from ceramic (8000 or more). The electrolytic capacitor has become the most popular large-value capacitor because the extremely thin dielectric oxide layer enables us to obtain large values for a small size.

Many techniques have been tried to extend the capacitance value up toward the farad; however, the conductive surface area is the main factor that needs to be increased in order to gain high values of capacitance, as seen in Figure 13-29(a). For example, if two pieces of aluminum the size of fingernails were held together, a fingernail thickness apart, a capacitance of approximately 1 pF would be created. This 1 pF capacitor would now have to be increased by a factor of 1,000,000,000,000 to obtain a capacitor of 1 F.

To increase surface area, a process known as *double etch* has been used, as seen in Figure 13-29(b), where, instead of having a flat plate surface, as seen in Figure 13-29(a), large pits are etched onto the surface and then smaller pits are etched onto the larger pits to create a larger surface area. This method, however, still does not get us close to 1 F.

The 1 F capacitor uses a relatively new double-layer technique, the effect of which was first noticed more than a century ago by the German scientist Hermann von Helmholtz. Helmholtz discovered that a double layer of charge will build up on the surface between a solid and a liquid.

The double-layer capacitor uses activated charcoal as the solid plates, and a liquid electrolyte between the two plates and an insulating separator allow a charge to build up within the capacitor due to the solid (charcoal) and liquid (electrolyte).

Activated charcoal is used as the plate because it has an almost infinite number of tiny particles making up its surface area, as seen in Figure 13-29(c), which yields a massive plate surface area. For example, 4 grams of activated charcoal would have an approximate surface area equivalent to a football field.

Using this technique a 1 F capacitor can be manufactured, like that seen in Figure 13-29(d). This 1 F capacitor package actually holds six 6 F capacitors connected in series, producing a total capacitance of

$$C_t \text{ (total series capacitance)} = \frac{1}{\frac{1}{6} + \frac{1}{6} + \frac{1}{6} + \frac{1}{6} + \frac{1}{6} + \frac{1}{6}}$$

$$= 1 \text{ F}$$

The reason why six 6 F capacitors are included is that the breakdown voltage of the double-layer capacitor is approximately 1 V, so by connecting six of these capacitors in series, the dielectric thickness is six times greater, increasing the

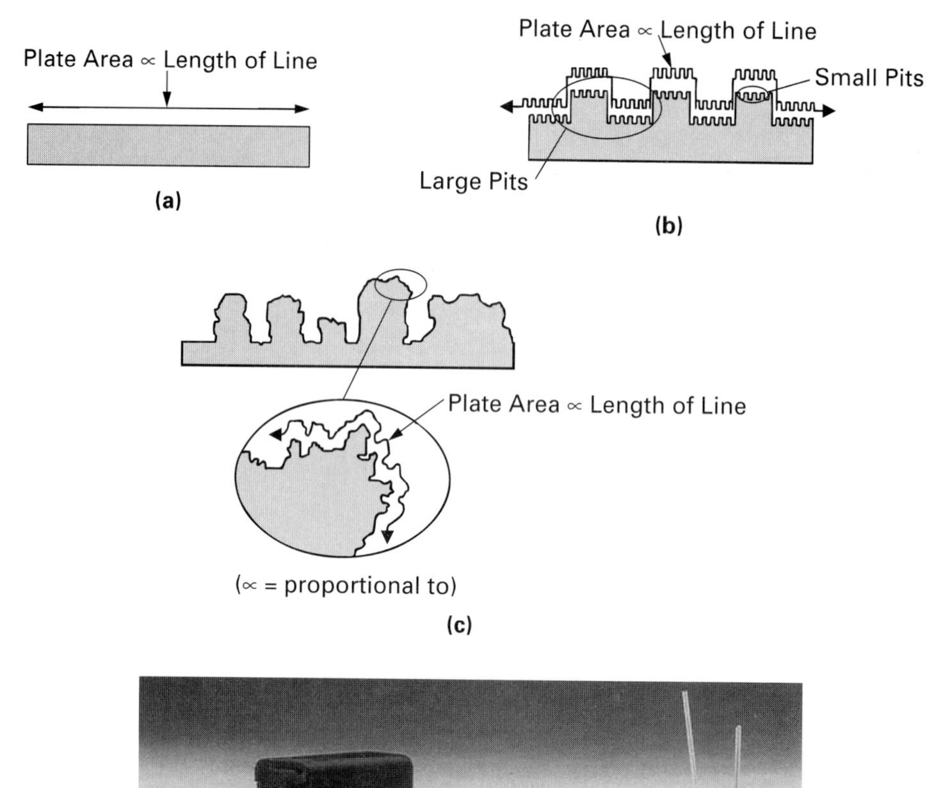

(d)

FIGURE 13-29 **The 1-F Capacitor. (a) Traditional Plate Shape. (b) Double-Etched Plate Area. (c) Activated Charcoal Surface Area. (d) Large-Farad-Value Capacitors (Courtesy of Sencore).**

breakdown voltage up to a more practical dc working voltage of 5 V. Due to the large value and the variations in activated charcoal, the tolerance of these capacitors can range from –20% to +80% of their value (0.8 to 1.8 F).

One major application for these capacitors is as a power backup for a computer's memory. When using a computer, it is important that you regularly save in a permanent memory the information you are working with in temporary memory. The reason for this is that the temporary memory will lose all of its information if there is a momentary power failure. In this situation, the 1 F capacitor could be used as a backup source of power to the computer's temporary memory during momentary power outages in order to prevent the loss of information.

SELF-TEST REVIEW QUESTIONS FOR SECTION 13-8

1. List the five types of fixed-value capacitors.
2. Which fixed-value capacitor is the most popular in applications where:
   a. Large values are required?
   b. Small values are required?
3. List the four basic types of variable capacitors.
4. Which variable capacitor types are best suited for:
   a. Large value variations?
   b. Small value variations?
5. Which capacitor type is said to be polarity conscious?

13-9 CODING OF CAPACITANCE VALUES

The capacitor value, tolerance, and voltage rating need to be shown in some way on the capacitor's exterior. Presently, two methods are used: (1) alphanumeric labels and (2) color coding.

13-9-1 *Alphanumeric Labels*

Manufacturers today most commonly use letters of the alphabet and numbers (alphanumerics) printed on either the disk or tubular body to indicate the capacitor's specifications, as illustrated in Figure 13-30.

The tubular type, which tends to be larger in size, as seen in Figure 13-30(b), is the easier of the two since the information is basically uncoded. The value of capacitance and unit, typically the microfaı ad (μF or MF), tolerance figure (preceded by ± or followed by %), and voltage rating (followed by a V for

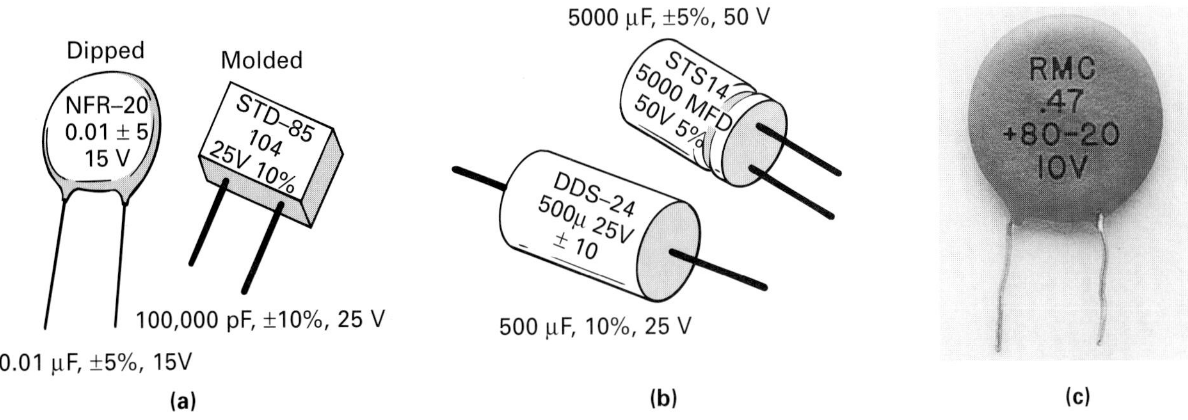

FIGURE 13-30 Alphanumeric Coding of Capacitors. (a) Disk Type. (b) Tubular. (c) 0.47 μF, +80%, –20% Tolerance, 10V (Courtesy of Sencore).

voltage) are printed on all sizes of tubular cases. The remaining letters or numbers are merely manufacturer's codes for case size, series, and the like.

With disk capacitors (dipped or molded), as seen in Figure 13-30(a), certain rules have to be applied when decoding the notations. Many capacitors of this type do not define the unit of capacitance; in this situation, try to locate a decimal point. If a decimal point exists, for example 0.01 or 0.001, the value is in microfarads (10^{-6}). If no decimal point exists, for example 50 or 220, the value is in picofarads (10^{-12}) and you must analyze the number in a little more detail. Figure 13-30(b) and (c) give additional examples.

EXAMPLE:

What is the value of the following capacitor if it is labeled 50, 50 V, ±5?

Solution:

Since no decimal point is present, the unit is in picofarads:

$$50 \text{ pF}, \quad 50 \text{ V}, \quad \pm 5\%$$

If no decimal point is present and three digits exist and the last digit is a zero, the value is as stands and in picofarads. If the third digit is a number other than 0 (1 to 9), it is a multiplier and describes the number of zeros to be added to the picofarad value.

EXAMPLE:

If two capacitors are labeled with the following coded values, how should they be interpreted?

a. 220
b. 104

Solution:

Since no decimal point is present, they are both in picofarads.

a. If the last of the three digits is a zero, the value is as it stands.

$$\underbrace{2\ 2\ 0}_{\text{three-digit value}} = 220 \text{ pF}$$

b. If the third digit is a number from 1 to 9, it is a multiplier.

$$\underbrace{1\ 0}_{\text{two-digit value}}\ \underbrace{4}_{\text{multiplier}}$$

So

$$10 \times 10^4 = 100{,}000 \text{ pF}$$

or

$$100{,}000 \times 10^{-12} = 0.1 \times 10^{-6} = 0.1\ \mu\text{F}$$

TABLE 13-13

Color	1st Digit	2nd Digit	Multiplier	Tolerance
Black		0	1	±20%
Brown	1	1	10^{-1}	±1%
Red	2	2	10^{-2}	±2%
Orange	3	3	10^{-3}	±3%
Yellow	4	4	10^{-4}	±4%
Green	5	5	10^{-5}	±5%
Blue	6	6	10^{-6}	
Violet	7	7	10^{-7}	
Gray	8	8	10^{-8}	
White	9	9	10^{-9}	±10%

The tolerance of the capacitor is sometimes clearly indicated, for example, ±5 or 10%; in other cases, a letter designation is used, such as

$$F = \pm 1\%$$
$$G = \pm 2\%$$
$$J = \pm 5\%$$
$$K = \pm 10\%$$
$$M = \pm 20\%$$
$$Z = -20\%, +80\%$$

Unfortunately, there does not seem to be a standard among capacitor manufacturers, which can cause confusion when trying to determine the value of capacitance. Therefore, if you are not completely sure, you should always consult technical data or information sheets from the manufacturer.

13-9-2 Color Coding

Table 13-3 illustrates the capacitor color code, which is almost identical to the resistor color code except for certain tolerances. Although the alphanumeric labels are now commonplace, the color code is still being used by some manufacturers.

SELF-TEST REVIEW QUESTIONS FOR SECTION 13-9

What are the following values of capacitance?

1. 470 ± 2

2. 0.47 ± 5

13-10 CAPACITIVE TIME CONSTANT

When a capacitor is connected across a dc voltage source, it will charge to a value equal to the voltage applied. If the charged capacitor was then connected across a load, the capacitor would then discharge through the load. The time it takes a capacitor to charge or discharge can be calculated if the circuit's resistance and capacitance are known. Let us now examine how we can calculate a capacitor's charge time and discharge time.

13-10-1 *DC Charging*

When a capacitor is connected across a dc voltage source, such as a battery or power supply, current will flow and the capacitor will charge up to a value equal to the dc source voltage, as shown in Figure 13-31. When the charge switch is first closed, as seen in Figure 13-31(a), there is no voltage across the capacitor at that instant and therefore a potential difference exists between the battery and capacitor. This causes current to flow and begin charging the capacitor.

Once the capacitor begins to charge, the voltage across the capacitor does not instantaneously rise to 100 V; it takes a certain amount of time before the capacitor voltage is equal to the battery voltage. When the capacitor is fully charged no potential difference exists between the voltage source and the capacitor. Consequently, no more current flows in the circuit as the capacitor has reached its full charge, as seen in Figure 13-31(b). The amount of time it takes for a capacitor to charge to the supplied voltage (in this example, 100 V) is dependent on the circuit's resistance and capacitance value. If the circuit's resistance is increased, the opposition to current flow will be increased, and it will take the capacitor a longer period of time to obtain the same amount of charge because the circuit current available to charge the capacitor is less.

If the value of capacitance is increased, it again takes a longer time to charge to 100 V because a greater amount of charge is required to build up the voltage across the capacitor to 100 V.

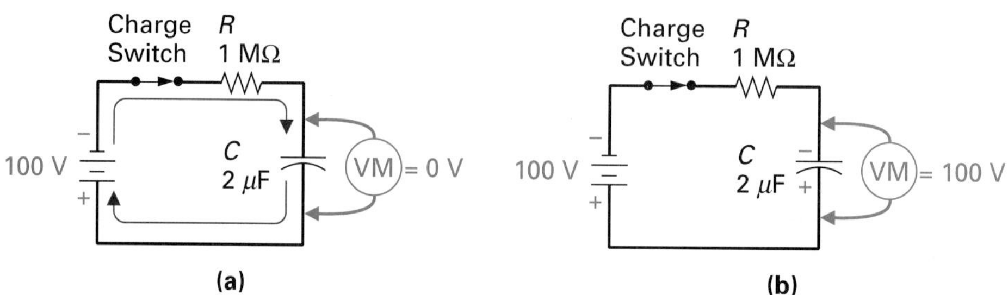

FIGURE 13-31 Capacitor Charging. (a) Switch Is Closed and Capacitor Begins to Charge. (b) Capacitor Charged.

The circuit's resistance (R) and capacitance (C) are the two factors that determine the charge time (τ). Mathematically, this can be stated as

$$\tau = R \times C$$

where τ = **time constant** (s)

R = resistance (Ω)

C = capacitance (F)

Time Constant Time needed for either a voltage or current to rise to 63.2% of the maximum or fall to 36.8% of the initial value. The time constant of an *RC* circuit is equal to the product of *R* and *C*.

In this example, we are using a resistance of 1 MΩ and a capacitance of 2 μF, which means that the time constant is equal to

$$\begin{aligned} \tau &= R \times C \\ &= 2\ \mu\text{F} \times 1\ \text{M}\Omega \\ &= (2 \times 10^{-6}) \times (1 \times 10^{6}) \\ &= 2\ \text{s} \end{aligned}$$

Two seconds is the time, so what is the constant? The constant value that should be remembered throughout this discussion is "**63.2.**"

Figure 13-32 illustrates the rise in voltage across the capacitor from 0 to a maximum of 100 V in five time constants (5 × 2 s = 10 s). So where does 63.2 come into all this?

First time constant: In 1*RC* seconds ($1 \times R \times C = 2$ s), the capacitor will charge to 63.2% of the applied voltage (63.2% × 100 V = 63.2 V).

Second time constant: In 2*RC* seconds ($2 \times R \times C = 4$ s), the capacitor will charge to 63.2% of the remaining voltage. In the example, the capacitor will be charged to 63.2 V in one time constant, and therefore the voltage remaining is equal to 100 V − 63.2 V = 36.8 V. At the end of the second time constant, therefore, the capacitor will have charged to 63.2% of the remaining voltage (63.2% × 36.8 V = 23.3 V), which means that it will have reached 86.5 V (63.2 + 23.3 = 86.5 V) or 86.5% of the applied voltage.

Third time constant: In 3*RC* seconds (6 s), the capacitor will charge to 63.2% of the remaining voltage:

$$\begin{aligned} \text{remaining voltage} &= 100\ \text{V} - 86.5\ \text{V} \\ &= 13.5\ \text{V} \end{aligned}$$

$$63.2\%\ \text{of}\ 13.5\ \text{V} = 8.532\ \text{V}$$

At the end of the third time constant, therefore, the capacitor will have charged to 86.5 V + 8.532 = 95 V, or 95% of the applied voltage.

Fourth time constant: In 4*RC* seconds (8 s), the capacitor will have charged to 63.2% of the remaining voltage (100 V − 95 V = 5 V); therefore, 63.2% of 5 V = 3.2 V. So the capacitor will have charged to 95 V + 3.2 V = 98.2 V, or 98.2% of the applied voltage.

Fifth time constant: In 5*RC* seconds (10 s), the capacitor is considered to be fully charged since the capacitor will have reached 63.2% of the remaining

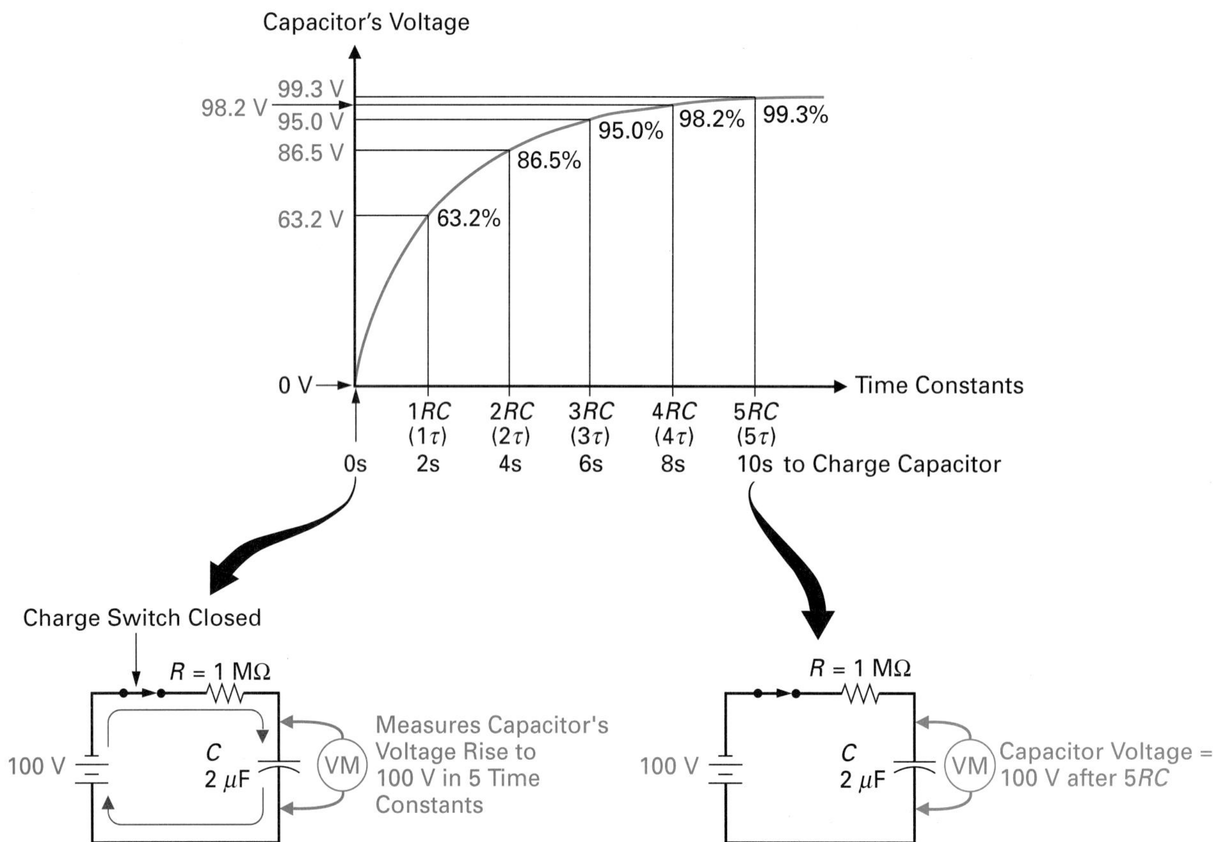

FIGURE 13-32 Charging Capacitor.

voltage (100 V − 98.2 V = 1.8 V); therefore, 63.2% of 1.8 V = 1.1 V. So the capacitor will have charged to 98.2 V + 1.1 V = 99.3 V, or 99.3% of the applied voltage.

The voltage waveform produced by the capacitor acquiring a charge is known as an *exponential* waveform, and the voltage across the capacitor is said to rise exponentially. An exponential rise is also referred to as a *natural increase.* There are many factors that exponentially rise and fall. For example, we grow exponentially, in that there is quite a dramatic change in our height in the early years and then this increase levels off and reaches a maximum.

Before the switch is closed and even at the instant the switch is closed, the capacitor is not charged, which means that there is no capacitor voltage to oppose the supply voltage and, therefore, a maximum current of *V/R,* 100 V/1 MΩ = 100 μA flows. This current begins to charge the capacitor, and a potential difference begins to build up across the plates of the capacitor, and this voltage opposes the supply voltage, causing a decrease in charging current. As the capacitor begins to charge, less of a potential difference exists between the supply voltage and capacitor voltage and so the current begins to decrease.

To calculate the current at any time, we can use the formula

$$i = \frac{V_S - V_C}{R}$$

where i = instantaneous current

V_S = source voltage

V_C = capacitor voltage

R = resistance

For example, the current flowing in the circuit after 1 time constant will equal the source voltage, 100 V, minus the capacitor's voltage, which in 1 time constant will be 63.2% of the source voltage or 63.2 V, divided by the resistance.

$$\begin{aligned} i &= \frac{V_S - V_C}{R} \\ &= \frac{100\ \text{V} - 63.2\ \text{V}}{1\ \text{M}\Omega} \\ &= 36.8\ \mu\text{A} \end{aligned}$$

As the charging continues, the potential difference across the plates exponentially rises to equal the supply voltage, as seen in Figure 13-33(a), while the current exponentially falls to zero, as shown in Figure 13-33(b). The constant of 63.2 can be applied to the exponential fall of current from 100 μA to 0 μA in 5*RC* seconds.

When the switch was closed to start charging of the capacitor, there was no charge on the capacitor; therefore, a maximum potential difference existed between the battery and capacitor, causing a maximum current flow of 100 μA ($I = V/R$).

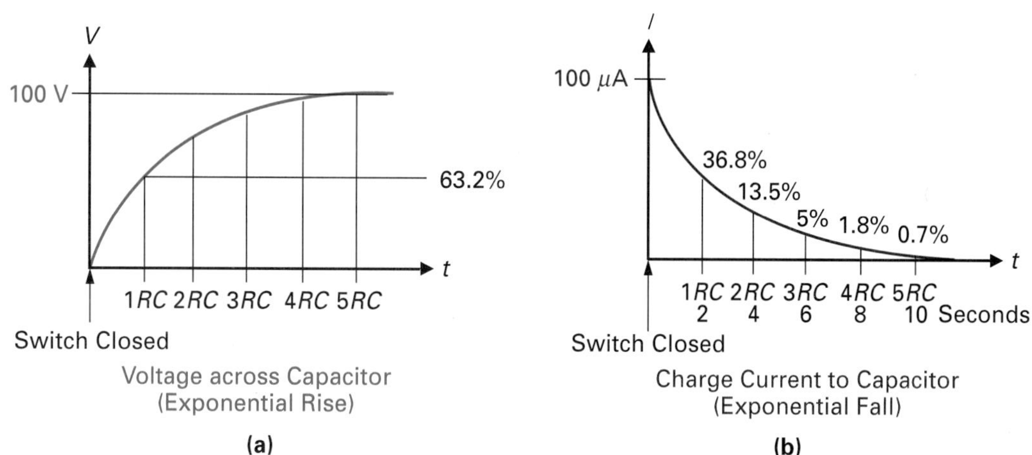

FIGURE 13-33 Exponential Rise in Voltage and Fall in Current in a Charging Capacitive Circuit.

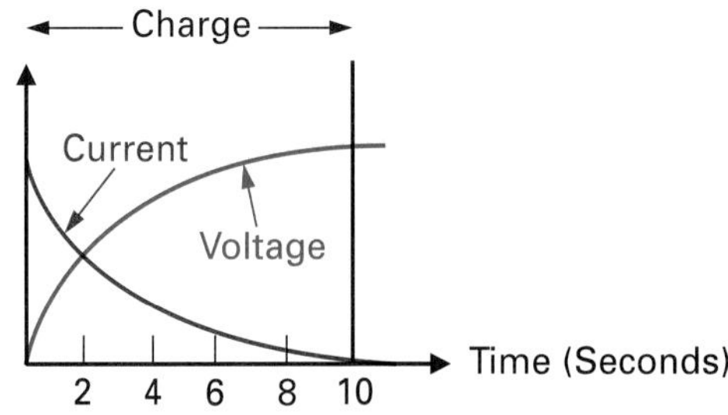

FIGURE 13-34 Phase Difference between Voltage and Current in a Charging Capacitive Circuit.

First time constant: In 1*RC* seconds, the current will have exponentially decayed to 63.2% of its maximum (63.2% of 100 μA = 63.2 μA) to a value of 36.8 μA (100 μA − 63.2 μA). In the example of 2 μF and 1 MΩ, this occurs in 2 s.

Second time constant: In 2*RC* seconds ($2 \times R \times C = 4$ s), the current will decrease 63.2% of the remaining current, which is

$$63.2\% \text{ of } 36.8\ \mu\text{A} = 23.26\ \mu\text{A}$$

The current will drop 23.26 μA from 36.8 μA and reach 13.5 μA or 13.5%.

Third time constant: In 3*RC* seconds (6 s), the capacitor's charge current will decrease 63.2% of the remaining current (13.5 μA) to 5 μA or 5%.

Fourth time constant: In 4*RC* seconds (8 s), the current will have decreased to 1.8 μA or 1.8%.

Fifth time constant: In 5*RC* seconds (10 s), the charge current is now 0.7 μA or 0.7%. At this time, the charge current is assumed to be zero and the capacitor is now charged to a voltage equal to the applied voltage.

Studying the exponential rise of the voltage and the exponential decay of current in a capacitive circuit, you will notice an interesting relationship. In a pure resistive circuit, the current flow through a resistor would be in step with the voltage across that same resistor, in that an increased current would cause a corresponding increase in voltage drop across the resistor. Voltage and current are consequently said to be *in step* or *in phase* with one another.

With the capacitive circuit, the current flow in the circuit and voltage across the capacitor are not in step or in phase with one another. When the switch is closed to charge the capacitor, the current is maximum (100 μA), while the voltage across the capacitor is zero. After five time constants (10 s), the capacitor's voltage is now maximum (100 V) and the circuit current is zero, as seen in Figure 13-34. The circuit current flow is out of phase with the capacitor voltage, and this difference is referred to as a *phase shift.* In any circuit containing capacitance, current will lead voltage.

13-10-2 **DC Discharging**

Figure 13-35 illustrates the circuit, voltage, and current waveforms that occur when a charged capacitor is discharged from 100 V to 0 V. The 2 μF capacitor, which was charged to 100 V in 10 s (5*RC*), is discharged from 100 to 0 V in the same amount of time.

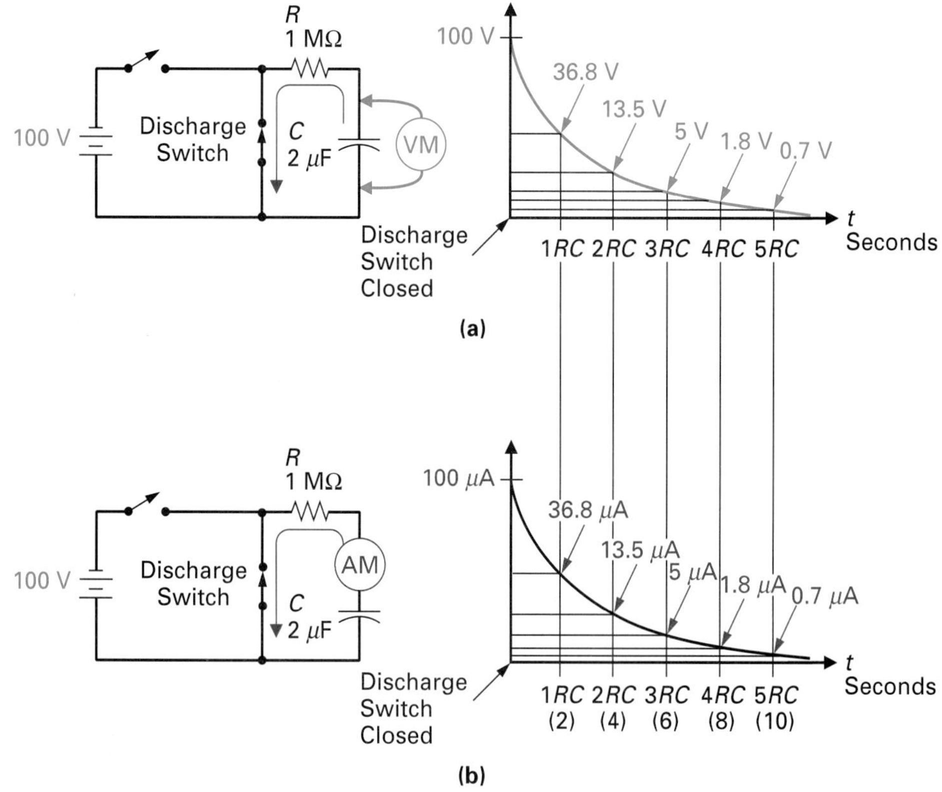

FIGURE 13-35 **Discharging Capacitor. (a) Voltage Waveform. (b) Current Waveform.**

Looking at the voltage curve, you can see that the voltage across the capacitor decreases exponentially, dropping 63.2% to 36.8 V in 1*RC* seconds, another 63.2% to 13.5 V in 2*RC* seconds, another 63.2% to 5 V in 3*RC* seconds, and so on, until zero.

The current flow within the circuit is dependent on the voltage in the circuit, which is across the 2 μF capacitor. As the voltage decreases, the current will also decrease by the same amount ($I\downarrow = V\downarrow/R$).

$$\text{discharge switch closed: } I = \frac{V}{R} = \frac{100 \text{ V}}{1 \text{ M}\Omega} = 100\ \mu\text{A} \quad \text{maximum}$$

$$1RC\ (2) \text{ seconds: } I = \frac{V}{R} = \frac{36.8 \text{ V}}{1 \text{ M}\Omega} = 36.8\ \mu\text{A}$$

$$2RC\ (4) \text{ seconds: } I = \frac{V}{R} = \frac{13.5 \text{ V}}{1 \text{ M}\Omega} = 13.5\ \mu\text{A}$$

$$3RC\ (6) \text{ seconds: } I = \frac{V}{R} = \frac{5 \text{ V}}{1 \text{ M}\Omega} = 5.0\ \mu\text{A}$$

$$4RC\ (8) \text{ seconds: } I = \frac{V}{R} = \frac{1.8 \text{ V}}{1 \text{ M}\Omega} = 1.8\ \mu\text{A}$$

$$5RC\ (10) \text{ seconds: } I = \frac{V}{R} = \frac{0.7 \text{ V}}{1 \text{ M}\Omega} = 0.7\ \mu\text{A} \quad \text{zero}$$

SELF-TEST REVIEW QUESTIONS FOR SECTION 13-10

1. What is the capacitor time constant?
2. In one time constant, a capacitor will have charged to what percentage of the applied voltage?
3. In one time constant, a capacitor will have discharged to what percentage of its full charge?
4. True or false: The charge or discharge of a capacitor follows a linear rate of change.

13-11 AC CHARGE AND DISCHARGE

Let us begin by returning to our charged capacitor that was connected across a 100-V dc source, as shown in Figure 13-36(a). When a capacitor is connected across a dc source, it charges in five time constants and then the voltage across the capacitor will oppose the source voltage, causing current to stop. Since no current flows in this circuit, the capacitor is effectively acting as an open circuit when a constant dc voltage is applied.

If the 100 V source is reversed, as shown in Figure 13-36(b), the 100 V charge on the capacitor is now aiding the path of the battery instead of pushing or reacting against it. The discharge current will flow in the opposite direction to the charge current, until the capacitor voltage is equal to 0 V, at which time it will begin to charge in the reverse direction, as shown in Figure 13-36(c).

If the battery source is once more reversed, as shown in Figure 13-36(d), the charge current will discharge the capacitor to 0 V and then begin once more to charge it in the reverse direction, as shown in Figure 13-36(e).

The result of this switching or alternating procedure is that there is current flow in the circuit at all times (except for the instant when the battery source is

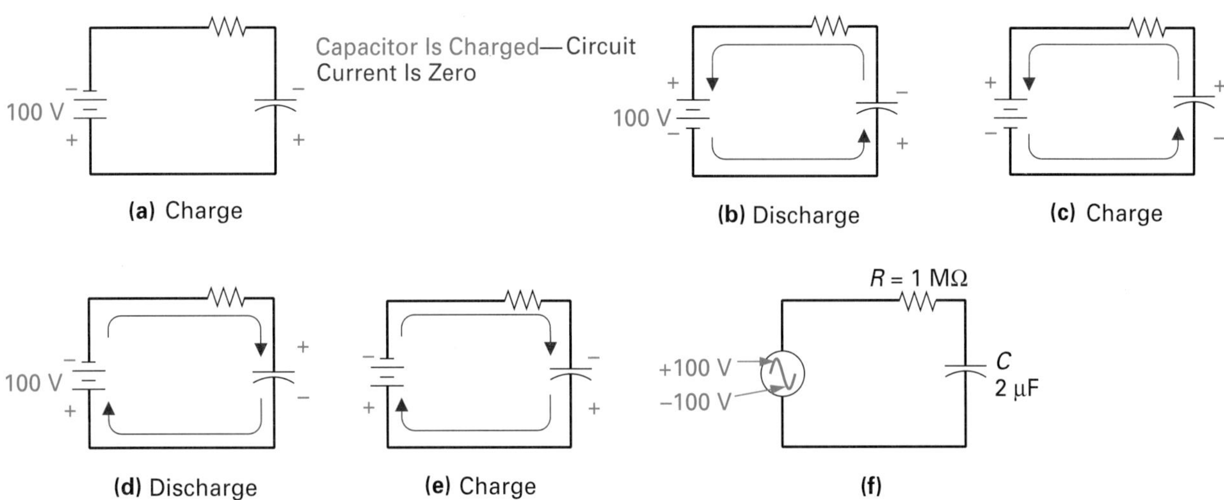

FIGURE 13-36 AC Charge and Discharge.

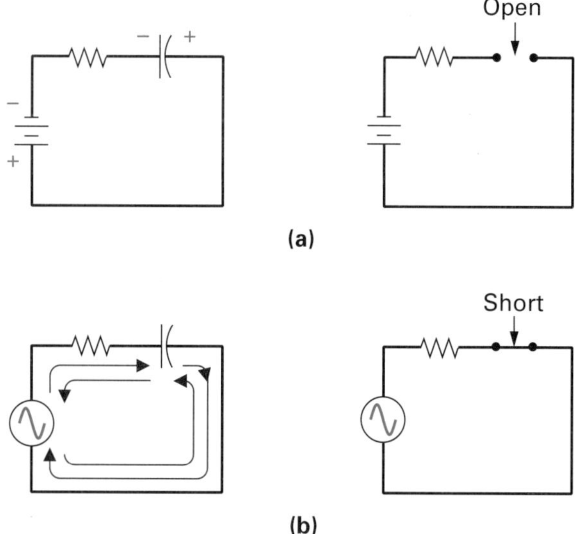

FIGURE 13-37 **Capacitors' Reaction to DC and AC. (a) Capacitor Will Block DC. (b) Capacitor Will Pass AC.**

removed and the polarity reversed). The switching of a dc voltage source in this way has almost the same effect as if we were applying an alternating voltage, as shown in Figure 13-36(f).

Figure 13-37 summarizes the capacitor's reaction to a dc source and an ac source. When a dc voltage is applied across a capacitive circuit, as shown in Figure 13-37(a), the capacitor will charge and then oppose the dc source, preventing current flow. Since no current can flow when the capacitor is charged, the capacitor acts like an open circuit, and it is this action that accounts for why capacitors are referred to as an "open to dc."

When an ac voltage is applied across a capacitive circuit, as shown in Figure 13-37(b), the continual reversal of the applied voltage will cause a continual charging and discharging of the capacitor, and therefore circuit current will always be present. Any type of ac source or in fact any fluctuating or changing dc source will cause a circuit current, and if current is present, the opposition is low. It is this action that accounts for why capacitors are referred to as a "short to ac."

This ability of a capacitor to block dc and seem to pass ac will be exploited and explained later in applications of capacitors.

SELF-TEST REVIEW QUESTIONS FOR SECTION 13-11

1. What differences occur when ac is applied to a capacitor rather than dc?
2. True or false: A capacitor's reaction to ac and dc accounts for why it is known as an ac short and dc block.

13-12 PHASE RELATIONSHIP BETWEEN CAPACITOR CURRENT AND VOLTAGE

Thinking of the applied alternating voltage as a dc source that is being continually reversed sometimes makes it easier to understand how a capacitor reacts to ac. Whether the applied voltage is dc or ac, the rule holds true in that a 90° phase shift or difference exists between circuit current and capacitor voltage. The exact relationship between ac current and voltage in a capacitive circuit is illustrated in Figure 13-38.

This phase shift that exists between the circuit current and the capacitor voltage is normally expressed in degrees. At 0°, the capacitor is fully discharged (0 V), and the source is supplying a maximum circuit charge current. From 0° to 90°, the capacitor will charge toward a maximum positive value, and this increase in capacitor voltage will oppose the source voltage, whose circuit charging current will slowly decrease to 0 A. At 90°, the capacitor is fully charged, and the circuit current is at 0 A, so the capacitor will return to the circuit the energy it consumed during the positive (+) charge cycle. This discharge circuit current is in the opposite direction to the positive charge current. At 180°, the capacitor is fully discharged (0 V), and the source is supplying a maximum circuit charge current. From 180° to 270°, the capacitor will charge toward a maximum negative value, and this negative increase in capacitor voltage will oppose the source voltage, whose circuit charging current will slowly decrease to 0 A. At 270°, the capacitor is fully charged and the circuit current is at 0 A, so the capacitor will discharge and return the energy it consumed during the negative (–) charge cycle. This discharge circuit current is in the opposite direction to the negative charge current.

Throughout this cycle, notice that the voltage across the capacitor follows the circuit current. This current, therefore, leads the voltage by 90°, and the 90° leading phase shift (current leading voltage) will occur only in a capacitive circuit.

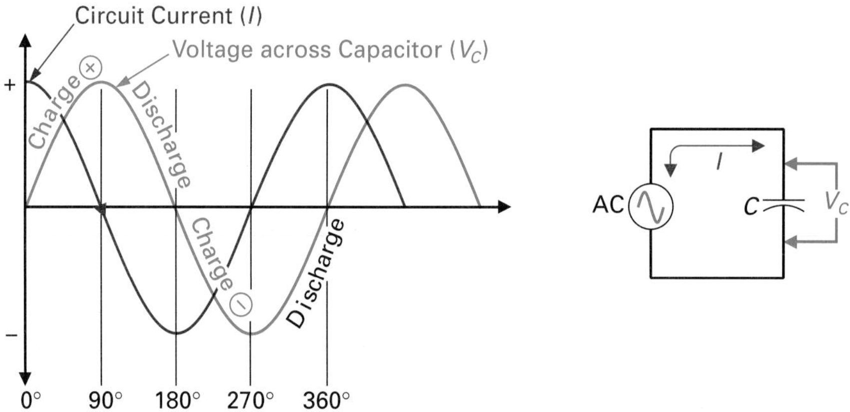

FIGURE 13-38 AC Current Leads Voltage in a Capacitive Circuit.

SELF-TEST REVIEW QUESTIONS FOR SECTION 13-12

1. What does a voltage-current phase shift mean?
2. True or false: In a resistive circuit, the current leads the voltage by exactly 90°.
3. True or false: In a purely capacitive circuit, a 90° phase difference occurs between current and voltage.
4. True or false: Phase differences between voltage and current only occur when ac is applied.

SUMMARY

Capacitance and Capacitors (Figure 13-39)

1. Resistance will oppose the flow of current and then convert or dissipate power in the form of heat.
2. Capacitance and inductance are two circuit properties that act differently from resistance in that they will charge or store the supplied energy and then return almost all the stored energy back to the circuit, rather than lose it in wasted heat.
3. Capacitance is the ability of a circuit or device to store electrical charge. A device or component specifically designed to have this capacity or capacitance is called a *capacitor*. A capacitor stores an electrical charge similar to a bucket holding water.
4. Capacitors store electrons, and basically, the amount of electrons stored is a measure of the capacitor's capacitance.
5. Capacitors, referred to in the past as condensers, have two leads connected to two parallel metal conductive plates, which are separated by an insulating material known as a *dielectric*. The conductive plates are normally made of metal foil, while the dielectric can be paper, air, glass, ceramic, mica, or some other form of insulator.
6. Like a secondary battery, a capacitor can be made to charge or discharge, and when discharging it will return all of the energy it consumed during charge.
7. Just as a magnetic field is produced by the flow of current, an electric field is produced by voltage.
8. A charged capacitor has an electric or electrostatic field existing between the positively charged and negatively charged plates. The strength of the electrostatic field is proportional to the charge or potential difference on the plates and inversely proportional to the distance between the plates.
9. The dielectric or insulator between the plates, like any other material, has its own individual atoms, and although the dielectric electrons are more tightly bound to their atoms than conductor electrons, stresses are placed on the atoms within the dielectric. The electrons in orbit around the dielectric atoms are displaced or distorted by the electric field existing between

QUICK REFERENCE SUMMARY SHEET

FIGURE 13-39 Capacitance and Capacitors.

$$\text{Field strength (V/m)} = \frac{\text{charge difference } (V) \text{ in volts}}{\text{distance between plates } (d) \text{ in meters}}$$

(V/m = volts per meter)

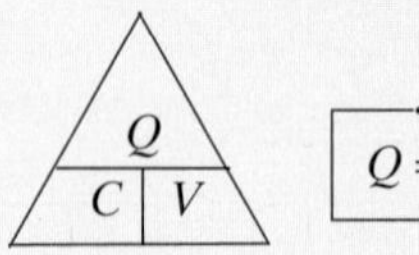

$$Q = C \times V$$

Q = change, in coulombs
C = capacitance, in farads
V = voltage, in volts

$$C = \frac{Q}{V}$$

$$V = \frac{Q}{C}$$

$$C = \frac{(8.85 \times 10^{-12}) \times K \times A}{d}$$

C = capacitance in farads
8.85×10^{-12} is a constant
K = dielectric constant
A = plate area, in square meters
d = distance between the plates, in meters

Capacitors in parallel:

$$C_T = C_1 + C_2 + C_3 + C_4 + \cdots$$

C_T = total capacitance, in farads
C_1 = capacitance of C_1, in farads
C_2 = capacitance of C_2, in farads
C_3 = and so on

Capacitors in series:

Two capacitors
$$C_T = \frac{C_1 \times C_2}{C_1 + C_2}$$

More than two capacitors
$$C_T = \frac{1}{(1/C_1) + (1/C_2) + (1/C_3) + \cdots}$$

Voltage divider formula
$$V_{cx} = \frac{C_T}{C_x} \times V_T$$

V_{cx} = desired voltage
C_T = total capacitance
C_x = selected capacitor value
V_T = supplied voltage

$$\tau = R \times C$$

τ = time constant, in seconds
R = resistance, in ohms
C = capacitance, in farads

$$i = \frac{V_S - V_C}{R}$$

i = instantaneous current, in amperes
V_S = source voltage, in volts
V_C = capacitor voltage, in volts
R = resistance, in ohms

FIGURE 13-39 (continued)

Dielectric Constants

MATERIAL	DIELECTRIC CONSTANT (K)[a]
Vacuum	1.0
Air	1.0006
Teflon	2.0
Wax	2.25
Paper	2.5
Amber	2.65
Rubber	3.0
Oil	4.0
Mica	5.0
Ceramic (low)	6.0
Bakelite	7.0
Glass	7.5
Water	78.0
Ceramic (high)	8000.

[a]The different material compositions can cause different values of K.

Dielectric Strengths

MATERIAL	DIELECTRIC STRENGTH (V/mm)[a]
Air	787
Oil	12,764
Ceramic	39,370
Paper	49,213
Teflon	59,055
Mica	59,055
Glass	78,740

Fixed-Value Capacitors

Name	Construction	Approximate Range of Values and Tolerances	Characteristics	
Mica	Foil, Mica, Foil, Mica, Foil, Mica, Foil	1 pF–0.1 μF ±1% to ±5%	Lower voltage rating than other capacitors of the same size	Small Capacitor Values
Ceramic	Ceramic Dielectic; Lead Wire	*Low Dielectric K*: 1 pF–0.01 μF ±0.5% to ±10% *High Dielectric K*: 1 pF–0.1 μF ±10% to ±80%	Most popular small valve capacitor due to lower cost than mica, and its ruggedness	Small Capacitor Values
Paper	Lead to Inner Foil Sheet; Inner Foil; Lead to Outer Foil Sheet; Outer Foil	1 pF–1 μF ± 10%	Has a large plate area and therefore large capacitance for a small size	Large Capacitor Values
Plastic	Outside Foil; Inside Foil	1 pF–10 μF ± 5% to ±10%	Has almost completely replaced paper capacitors; has large capacitance valves for small size and high voltage ratings	Large Capacitor Values
Electrolytic (Aluminum and Tantalum)	(Tantalum has tantalum rather than aluminum foil plates) Aluminum Foil; Aluminum Oxide; Aluminum Foil; Gauze (Saturated with Electrolyte)	1 μF–1F ± 10% to ±50%	Most popular large valve capacitor: large capacitance into small area, wide range of valves. Disadvantages are: cannot be used in AC circuits as they are polarized; poor tolerances; low leakage resistance and so high leakage current. *Tantalum* advantages over aluminum include smaller size, longer life than aluminum, which has an approximate lifespan of 12 years. Disadvantages: 4 to 5 times the price.	Large Capacitor Values

FIGURE 13-39 (continued)

Variable-Value Capacitors

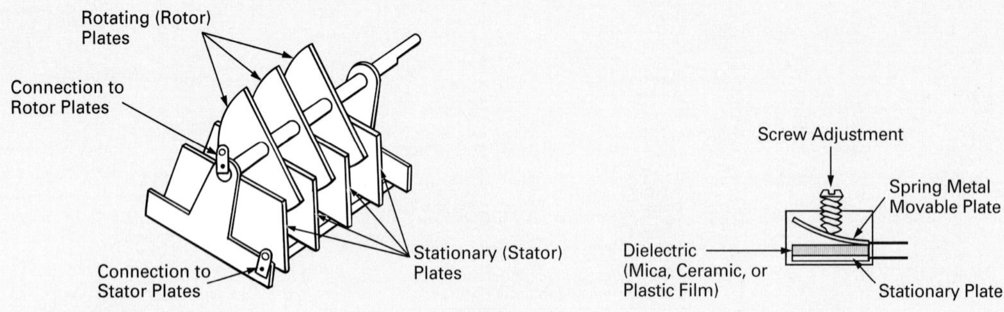

DC Charging

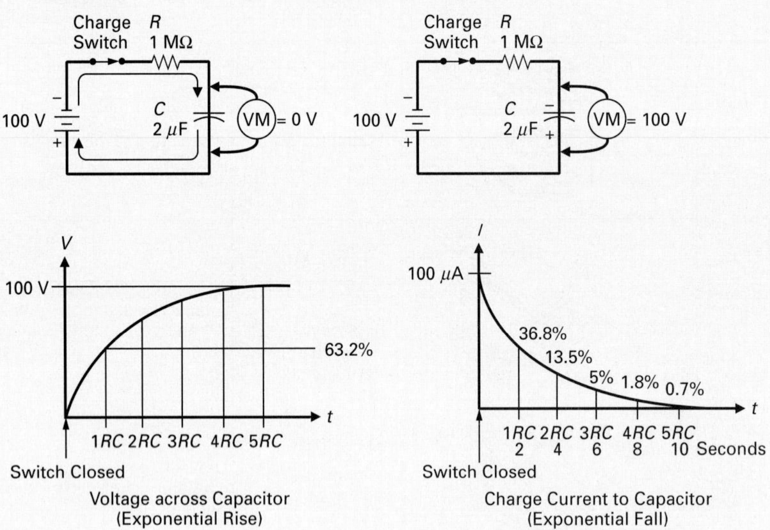

DC Discharging

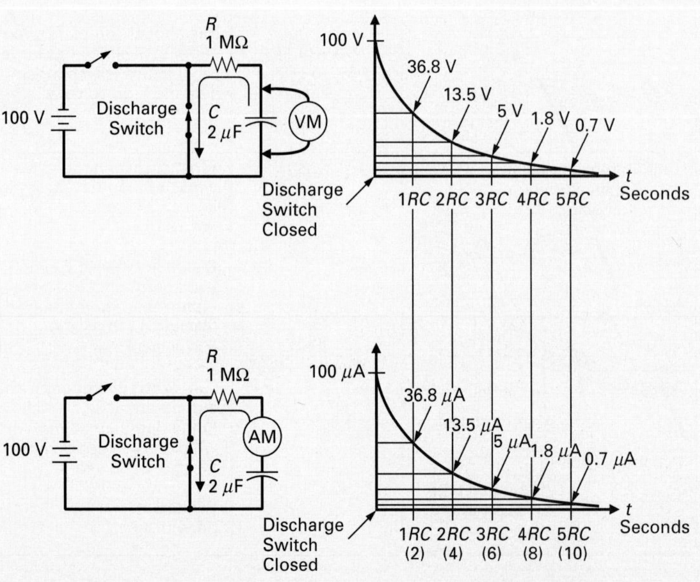

the positive and negative plate. If the charge potential across the capacitor is high enough and the distance between the plates is small enough, the attraction and repulsion exerted on the dielectric atom can be large enough to free the dielectric atom's electrons. The material then becomes ionized, and a chain reaction of electrons jumping from one atom to the next in a right-to-left movement occurs. If this occurs, a large number of electrons will flow from the negative to the positive plate, and the dielectric is said to have broken down. This situation occurs if the capacitor is placed in a circuit where the voltages within the circuit exceed the voltage rating of the capacitor.

10. If the voltage rating of the capacitor is not exceeded, an electrostatic or electric field still exists between the two plates and causes this pulling of the atom's electrons within the dielectric toward the negative plate. This displacement, known as electric polarization, is similar to the pulling-back effect on a bow. When the capacitor is given a path for discharge the electric field in the dielectric, which is causing the distortion, is the force field that drives the electrons, like the bow drives the arrow.

11. The energy in a capacitor is actually stored in the electric or electrostatic field within the dielectric.

12. *Capacitance* is the ability of a capacitor to store an electrical charge, and the unit of capacitance is the farad (F), named in honor of Michael Faraday's work in 1831 in the field of capacitance.

13. A capacitor with the capacity of 1 farad (1 F) can store 1 coulomb of electrical charge (6.24×10^{18} electrons) if 1 volt is applied across the capacitor's plates.

14. The dielectric constant (K) is the ease with which an insulating material can establish an electrostatic (electric) field. The vacuum is used as a reference. All the other materials have dielectric constant values that are relative to the vacuum dielectric constant of 1.

15. The capacitance of a capacitor is directly proportional to the dielectric constant (K) and the plates' area (A) and is inversely proportional to the dielectic thickness or distance between the plates (d).

16. Capacitors store a charge just as a container or tank stores water. The amount of charge stored by a capacitor is proportional to the capacitor's capacitance and the voltage applied across the capacitor ($Q = C \times V$).

17. If the voltage across a capacitor is increased, the charge held by the capacitor will increase until the dielectric between the two plates of the capacitor breaks down and a spark jumps or arcs between the plates.

18. The breakdown voltage of a capacitor is determined by the strength of the dielectric used.

19. The ideal or perfect insulator should have a resistance equal to infinite ohms. Insulators or the dielectric used to isolate the two plates of a capacitor are not perfect and therefore have some very high values of resistance. This means that some value of resistance exists between the two plates. Although this value of resistance is very large, it will still allow a small amount of current to flow between the two plates (in most applications a few nanoamperes or picoamperes). This small current, referred to as leakage current, causes any charge in a capacitor to slowly, over a long period of time, discharge between the two plates. A capacitor should have a large leakage resistance to ensure the smallest possible leakage current.

20. Like resistors, capacitors can be connected in either series or parallel. The rules for determining total capacitance for parallel- and series-connected capacitors are opposite to series- and parallel-connected resistors.

21. If capacitors are connected in parallel, the effective plate area is increased; and since capacitance is proportional to plate area [$C\uparrow = (8.85 \times 10^{-12}) \times K \times A\uparrow/d$], the capacitance will also increase.

22. The total capacitance for capacitors connected in parallel is actually calculated by adding the plate areas, so total capacitance is equal to the sum of all the individual capacitances in parallel.

23. When capacitors are connected in series the effective plate area is decreased ($A\downarrow$) and the dielectric thickness increased ($d\uparrow$), and both of these effects result in an overall capacitance decrease [$C\downarrow\downarrow = (8.85 \times 10^{-12}) \times K \times A\downarrow/d\uparrow$]. The plate area is actually decreased to the smallest individual capacitance connected in series. If the plate area were the only factor, then capacitance would always equal the smallest capacitor value. However, the dielectric thickness is always equal to the sum of all the capacitor dielectrics, and this factor always causes the total capacitance (C_T) to be less than the smallest individual capacitance when capacitors are connected in series.

24. The total capacitance of two or more capacitors in series therefore is calculated by using the product-over-sum formula, or the reciprocal formula.

25. As with series-connected resistors, all of the voltage drops across the series-connected resistors will equal the voltage applied (Kirchhoff's voltage law). With capacitors connected in series, the charged capacitors act as a voltage divider, and therefore the voltage-divider formula can be applied to capacitors in series.

26. Capacitors in series will have the same charging current throughout the circuit, and because of this, two or more capacitors in series will always have equal amounts of coulomb charge. If the charge (Q) is equal, the voltage across the capacitor is determined by the value of the capacitor. A small capacitance will charge to a larger voltage ($V\uparrow = Q/C\downarrow$), whereas a large value of capacitance will charge to a smaller voltage ($V\downarrow = Q/C\uparrow$).

27. A fixed-value capacitor is a capacitor whose capacitance value remains constant and cannot be altered. Fixed capacitors normally come in a disk or a tubular package and consist of metal foil plates separated by one of the following types of insulators (dielectric), which is the means by which we classify them.

a. Mica
b. Ceramic
c. Paper
d. Plastic
e. Electrolytic

28. A *variable-value capacitor* is a capacitor whose capacitance value can be changed by rotating a shaft. The variable capacitor normally consists of one electrically connected movable plate and one electrically connected stationary plate. There are basically four types of variable-value capacitors, which are also classified by the dielectric used.

    a. Air   **c.** Ceramic
    b. Mica   **d.** Plastic

29. Ceramic chip surface-mount capacitors that are now available on the market for both discrete and integrated types of circuits. Values typically range from 16 to 1600 pF.
30. The electrolytic capacitor must always have a positive charge applied to its positive plate and a negative charge to its negative plate. If this rule is not followed, the electrolytic capacitor becomes a safety hazard, because it can in the worst-case condition explode violently.
31. Tantalum instead of aluminum is used in some electrolytic capacitors for the plates and has advantages over the aluminum type, which are:

    a. Higher capacitance per volt for a given unit volume
    b. Longer life and excellent shelf life (storage)
    c. Able to operate in a wider temperature range
    d. Temperature stability is better
    e. Construction features make them more rugged

    The cost of tantulum electrolytics, however, is almost four to five times that of aluminum electrolytics, and their operating voltages are much lower.

32. The variable-value type capacitors use a hand-rotated shaft to vary the effective plate area, and so capacitance.
33. The adjustable type capacitors use screw-in or screw-out mechanical adjustment to vary the distance between the plates, and so capacitance.
34. In radio equipment, it is sometimes necessary to have two or more variable-value capacitors that have been constructed in such a way that a common shaft (rotor) runs through all the capacitors and varies their capacitance simultaneously. If you mechanically couple (gang) two or more variable capacitors so that they can all be operated from a single control, the component is known as a ganged capacitor.
35. Capacitors should only be adjusted with a plastic or nonmetallic alignment tool, because a metal screwdriver may affect the capacitance of the capacitor when nearby, making it very difficult to adjust for a specific value of capacitance.
36. The 1 F capacitor uses a relatively new double-layer technique, the effect of which was first noticed more than a century ago by the German scientist Hermann von Helmholtz. Helmholtz discovered that a double layer of charge will build up on the surface between a solid and a liquid. The double-layer capacitor uses activated charcoal as the solid plates, and a liquid electrolyte between the two plates and an insulating separator allow a charge to build up within the capacitor due to the solid (charcoal) and liquid (electrolyte).

37. The capacitor value, tolerance, and voltage rating need to be shown in some way on the capacitor's exterior. Presently, two methods are used: (a) alphanumeric labels and (b) color coding.

38. When a capacitor is connected across a dc voltage source, it will charge to a value equal to the voltage applied. If the charged capacitor was then connected across a load, the capacitor would then discharge through the load. The time it takes a capacitor to charge or discharge can be calculated if the circuit's resistance and capacitance are known.

39. Time constant is the time needed for either a voltage or current to rise to 63.2% of the maximum or fall to 36.8% of the initial value. The time constant of an RC circuit is equal to the product of R and C.

40. The voltage waveform produced by the capacitor acquiring a charge is known as an *exponential* waveform, and the voltage across the capacitor is said to rise exponentially. An exponential rise is also referred to as a *natural increase*.

41. In a pure resistive circuit, the current flow through a resistor would be in step with the voltage across that same resistor, in that an increased current would cause a corresponding increase in voltage drop across the resistor. Voltage and current are consequently said to be *in step* or *in phase* with one another.

42. With the capacitive circuit, the current flow in the circuit and voltage across the capacitor are not in step or in phase with one another. When the switch is closed to charge the capacitor, the current is maximum while the voltage across the capacitor is zero. After five time constants the capacitor's voltage is now maximum and the circuit current is zero. The circuit current flow is out of phase with the capacitor voltage, and this difference is referred to as a *phase shift*. In any circuit containing capacitance, current will lead voltage.

43. When a dc voltage is applied across a capacitive circuit the capacitor will charge and then oppose the dc source, preventing current flow. Since no current can flow when the capacitor is charged, the capacitor acts like an open circuit, and it is this action that accounts for why capacitors are referred to as an "open to dc."

44. When an ac voltage is applied across a capacitive circuit, the continual reversal of the applied voltage will cause a continual charging and discharging of the capacitor, and therefore circuit current will always be present. Any type of ac source or in fact any fluctuating or changing dc source will cause a circuit current, and if current is present, the opposition is low. It is this action that accounts for why capacitors are referred to as a "short to ac."

45. This ability of a capacitor to block dc and seem to pass ac will be exploited in applications of capacitors.

46. Thinking of the applied alternating voltage as a dc source that is being continually reversed sometimes makes it easier to understand how a capacitor reacts to ac. Whether the applied voltage is dc or ac, the rule holds true in

that a 90° phase shift or difference exists between circuit current and capacitor voltage.

47. This phase shift that exists between the circuit current and the capacitor voltage is normally expressed in degrees.

48. In a capacitive circuit, circuit current leads the capacitor voltage by 90°, and the 90° leading phase shift (current leading voltage) will occur only in a purely capacitive circuit.

NEW TERMS

Activated charcoal
Adjustable capacitor
Air capacitor
Alignment
Breakdown voltage
Capacitance
Capacitor
Capacitor breakdown
Capacitor leakage
Capacity
Ceramic capacitor
Charge difference
Charging
Condensers
Dielectric
Dielectric constant
Dielectric strength
Dielectric thickness
Discharging
Double etch
Electric field
Electric polarization
Electrolytic capacitor
Electrostatic field
Electrostatics
Exponential fall
Exponential rise
Farad
Field strength
Fixed-value capacitor
Ganged capacitor
Inductance current
Instantaneous
Leakage current
Mica capacitor
Microfarad
Paper capacitor
Picofarad
Plastic film capacitor
Plate area
Plates
Polarized
Static charge
Tantalum capacitor
Time constant
Total capacitance
Trimmer capacitor
Variable-value capacitor
Volts per meter

REVIEW QUESTIONS

Multiple-Choice Questions

1. Capacitors were originally referred to as:
   a. Vacuum tubes **b.** Condensers **c.** Inductors **d.** Suppressors

2. When a capacitor charges:
   a. The voltage across the plates rises exponentially
   b. The circuit current falls exponentially
   c. The capacitor charges to the source voltage in $5RC$ seconds
   d. All of the above

3. A ____________ field is generated by the flow of current and an ____________ field is generated by voltage.
   a. Magnetic, electrostatic **c.** Electric, magnetic
   b. Electric, electrostatic **d.** All of the above may be considered true.

4. The strength of an electric field in a capacitor is proportional to the ____________ and inversely proportional to the ____________.

   a. Plate separation, charge
   b. Plate separation, potential plate difference
   c. Plate potential difference, plate separation
   d. Both (a) and (b)

5. The plates of a capacitor are generally made out of a:

   a. Resistive material
   b. Semiconductor material
   c. Conductive material
   d. Two of the above could be true.

6. The energy of a capacitor is stored in:

   a. The magnetic field within the dielectric
   b. The magnetic field around the capacitor leads
   c. The electric field within the plates
   d. The electric field within the dielectric

7. What is the capacitance of a capacitor if it can store 24 C of charge when 6 V is applied across the plates?

   a. 2 μF b. 3 μF c. 4.7 μF d. None of the above

8. The capacitance of a capacitor is directly proportional to:

   a. The plate area
   b. The distance between the plates
   c. The constant of the dielectric used
   d. Both (a) and (c)
   e. Both (a) and (b)

9. The capacitance of a capacitor is inversely proportional to:

   a. The plate area
   b. The distance between the plates
   c. The dielectric used
   d. Both (a) and (c)
   e. Both (a) and (b)

10. The breakdown voltage of a capacitor is determined by:

    a. The type of dielectric used
    b. The size of the capacitor plates
    c. The wire gauge of the connecting leads
    d. Both (a) and (b)

11. A capacitor should have a ____________ leakage resistance to ensure a ____________ leakage current.

    a. Large, small b. Large, medium c. Small, large d. Small, small

12. Total series capacitance of two capacitors is calculated by:

    a. Using the product over sum formula
    b. Using the voltage divider formula
    c. Using the series resistance formula on the capacitors
    d. Adding all the individual values

13. Total parallel capacitance is calculated by:

    a. Using the product over sum formula
    b. Using the voltage divider formula
    c. Using the parallel resistance formula on capacitors
    d. Adding all the individual values

14. The mica and ceramic fixed capacitors have:

    a. An arrangement of stacked plates
    b. An electrolyte substance between the plates
    c. An adjustable range
    d. All of the above

15. An electrolytic capacitor:

    a. Is the most popular large-value capacitor
    b. Is polarized
    c. Can have either aluminum or tantalum plates
    d. All of the above

16. Variable capacitors normally achieve a large variation in capacitance by varying ____________, while adjustable trimmer capacitors only achieve a small capacitance range by varying ____________.

    a. Dielectric constant, plate area
    b. Plate area, plate separation
    c. Plate separation, dielectric constant
    d. Plate separation, plate area

17. In one time constant, a capacitor will charge to ____________ of the source voltage.

    a. 86.5% b. 63.2% c. 99.3% d. 98.2%

18. When ac is applied across a capacitor, a ____________ phase shift exists between circuit current and capacitor voltage.

    a. 45° b. 60° c. 90° d. 63.2°

19. A capacitor consists of:

    a. Two insulated plates separated by a conductor
    b. Two conductive plates separated by a conductor
    c. Two conductive plates separated by an insulator
    d. Two conductive plates separated by conductive spacer

20. A 47-μF capacitor charged to 6.3 V will have a stored charge of:

    a. 296.1 μC
    b. 2.96×10^{-4} C
    c. 0.296 mC
    d. All of the above
    e. Both (a) and (b)

Essay Questions

21. Describe the construction and main parts of a capacitor. (13-1)
22. What happens to a capacitor during:

    a. Charge (13-2-1) b. Discharge (13-2-2)

23. Describe how electrostatics relates to capacitance and give the formula for electric field strength. (13-3)
24. Briefly explain the relationship between capacitance, charge, and voltage. (13-4)
25. Describe the three factors affecting the capacitance of a capacitor. (13-5)

26. Briefly describe:
    a. The term *capacitor breakdown* (13-6)
    b. The term *capacitor leakage* (13-6)
    c. Why the smaller-value capacitor in a two-capacitor series circuit has the larger voltage developed across it (13-7-2)
27. List the formula(s) used to calculate total capacitance when capacitors are connected in:

    a. Parallel (13-7-1) b. Series (13-7-2)
28. Describe the following types of capacitors: (13-8)

    Fixed Value

    a. Mica b. Ceramic c. Paper d. Plastic e. Electrolytic

    Variable Value

    a. Air b. Mica, ceramic, and plastic
29. Briefly explain the technique and materials used to create the 1 F capacitor. (13-8-3)
30. Explain how the constant 63.2 is used in relation to the charge and discharge of a capacitor. (13-10)
31. How is the coding of a capacitor's value, tolerance, and voltage rating indicated on a capacitor? (13-9)
32. Describe the differences between dc and ac capacitor charging and discharging. (13-11)
33. Would a mica dielectric concentrate on electrostatic field more or less than air; if so, by how much? (13-5)
34. Calculate the total capacitance of the circuits illustrated in Figure 13-40, and describe which capacitors are in parallel and which are in series. (13-7)
35. What advantages and disadvantages do tantalum electrolytic capacitors have over the aluminum type? (13-8-1)
36. Which of the fixed-value capacitor types is polarity sensitive, and what does this mean? (13-8-2)
37. What is a ganged capacitor, and what is its application? (13-8-2)
38. Describe 1 F of capacitance. (13-4)

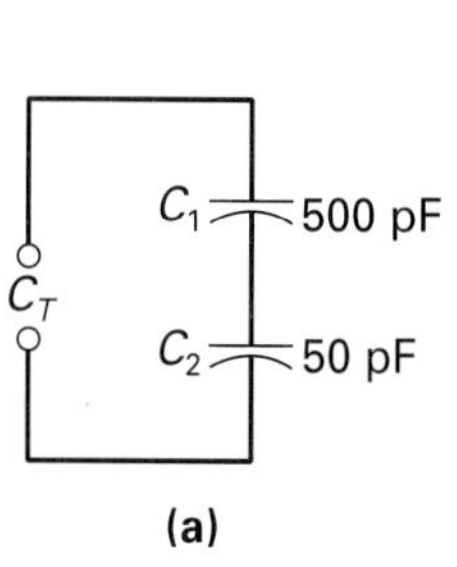

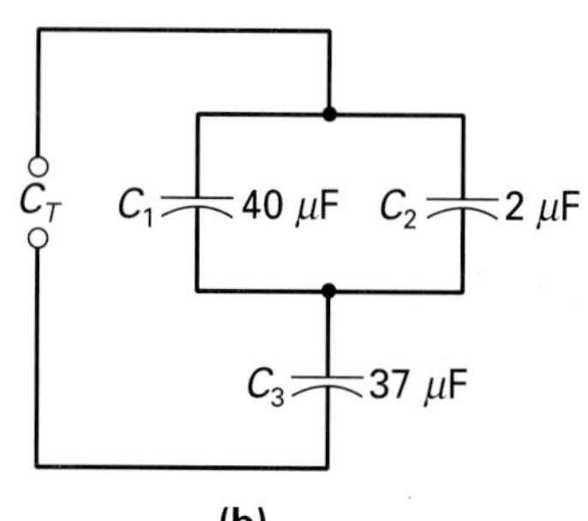

FIGURE 13-40

39. Briefly describe why series-connected capacitors are treated like parallel-connected resistors, and why parallel-connected capacitors are treated like series-connected resistors when calculating total capacitance. (13-7)

Practice Problems

40. If a 10 μF capacitor is charged to 10 V, how many coulombs of charge has it stored?

41. Calculate the electric field strength within the dielectric of a capacitor that is charged to 6 V and the dielectric thickness is 32 μm (32 millionths of a meter).

42. If a 0.006-μF capacitor has stored 125×10^{-6} C of charge, what potential difference would appear across the plates?

43. Calculate the capacitance of the capacitor that has the following parameter values: $A = 0.008\ \text{m}^2$; $d = 0.00095$ m; the dielectric used is paper.

44. Calculate the total capacitance if the following are connected in:

    a. Parallel: 1.7 μF, 2.6 μF, 0.03 μF, 1200 pF

    b. Series: 1.6 μF, 1.4 μF, 4 μF

45. If three capacitors of 0.025 μF, 0.04 μF, and 0.037 μF are connected in series across a 12-V source, as shown in Figure 13-41, what would be the voltage drop across each?

46. Give the value of the following alphanumeric capacitor value codes:

    a. 104 **b.** 125 **c.** 0.01 **d.** 220

47. Give the value of the following color-coded capacitor values:

    a. Yellow, violet, black **b.** Orange, yellow, violet

48. What would be the time constant of the following *RC* circuits?

    a. $R = 6\ \text{k}\Omega$, $C = 14\ \mu\text{F}$ **c.** $R = 170\ \Omega$, $C = 24\ \mu\text{F}$
    b. $R = 12\ \text{M}\Omega$, $C = 1400$ pF **d.** $R = 140\ \text{k}\Omega$, $C = 0.007\ \mu\text{F}$

49. If 10 V were applied across all the *RC* circuits in Question 48, what would be the voltage across each capacitor after one time constant, and how much time would it take each capacitor to fully charge?

50. In one application, a capacitor is needed to store 25 μC of charge and will always have 125 V applied across its terminals. Calculate the capacitance value needed.

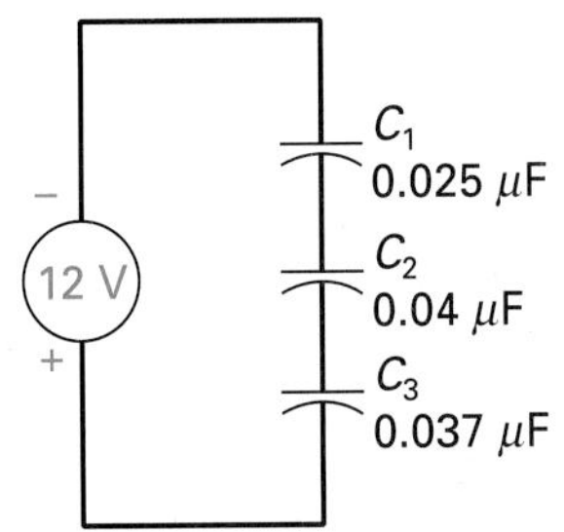

FIGURE 13-41

CHAPTER

14

OBJECTIVES

After completing this chapter, you will be able to:

1. Define and explain capacitive reactance.

2. Describe impedance, phase angle, power, and power factor as they relate to a series and parallel *RC* circuit.

3. Explain some of the more common capacitor failures and how to use an ohmmeter and capacitance analyzer to test them.

4. Describe how the capacitor and resistor in combination can be used:

a. To combine ac and dc

b. To act as a voltage divider

c. As a filter

d. As an integrator

e. As a differentiator

Capacitive Circuits, Testing, and Applications

OUTLINE

VIGNETTE

Let's Toss for It!

In 1938, Bill Hewlett and Dave Packard, close friends and engineering graduates at Stanford University, set up shop in the one-car garage behind the Packard's rented home in Palo Alto, California. In the garage, the two worked on what was to be the first product of their lifetime business together, an electronic oscillating instrument that represented a breakthrough in technology and was specifically designed to test sound equipment.

Late in the year, the oscillator (designated the 200A "because the number sounded big") was presented at a West Coast meeting of the Institute of Radio Engineers (now the Institute of Electrical and Electronics Engineers—IEEE) and the orders began to roll in. Along with the first orders for the 200A was a letter from Walt Disney Studios asking the two to build a similar oscillator covering a different frequency range. The Model 200B was born shortly thereafter, and Disney purchased eight to help develop the unique sound system for the classic film *Fantasia.* By 1940, the young company had outgrown the garage and moved into a small rented building nearby.

Over the years, the company continued a steady growth, expanding its product line to more than 10,000, including computer systems and peripheral products, test and measuring instruments, hand-held calculators, medical electronic equipment, and systems for chemical analysis. Employees have increased from 2 to almost 82,000, and the company is one of the 100 largest industrial corporations in America, with a net revenue of $8.1 billion. Whose name should go first was decided by the toss of a coin on January 1, 1939, and the outcome was Hewlett-Packard or HP.

INTRODUCTION

In Chapter 13 you discovered that when a capacitor was connected to a dc circuit, it charged up until it was at a value equal to the applied dc voltage, and then from that point on current was zero. When a capacitor was connected to an ac circuit, however, the continual reversal of the applied voltage resulted in a continuous charge and discharge of the capacitor, so circuit current was always present. This ability of the capacitor makes it an extremely useful device in ac circuits.

In this chapter we first study capacitive reactance, which is the opposition a capacitor offers to current flow. Then we study the ac circuit characteristics of a series *RC* circuit and parallel *RC* circuit. To complete our study of capacitors, in the final two sections in this chapter we discuss the testing of capacitors and then some of the more common applications of capacitors.

14-1 CAPACITIVE REACTANCE

Resistance (R), by definition, is the opposition to current flow with the dissipation of energy and is measured in ohms. Capacitors oppose current flow like a resistor; however, a resistor dissipates energy, whereas a capacitor stores energy (when it charges) and then gives back its energy into the circuit (when it discharges). Because of this difference, a new term had to be used to describe the opposition offered by a capacitor. **Capacitive reactance (X_C)**, by definition, is the opposition to current without the dissipation of energy and is also measured in ohms.

Capacitive Reactance (X_C) Measured in ohms, it is the ability of a capacitor to oppose current flow without the dissipation of energy.

If capacitive reactance is basically opposition, it is inversely proportional to the amount of current flow. If a large current is within a circuit, the opposition or reactance must be low ($I\uparrow$, $X_C\downarrow$). Conversely, a small circuit current will be the result of a large opposition or reactance ($I\downarrow$, $X_C\uparrow$).

When a dc source is connected across a capacitor, current will flow only for a short period of time ($5RC$ seconds) to charge the capacitor. After this time, there is no further current flow. Consequently, the capacitive reactance or opposition offered by a capacitor to dc is infinite (maximum).

Alternating current is continuously reversing in polarity, resulting in the capacitor continuously charging and discharging. This means that charge and discharge currents are always flowing around the circuit, and if we have a certain value of current, we must also have a certain value of reactance or opposition.

Initially, when the capacitor's plates are uncharged, they will not oppose or react against the charging current and therefore maximum current will flow ($I\uparrow$) and the reactance will be very low ($X_C\downarrow$). As the capacitor charges, it will oppose or react against the charge current, which will decrease ($I\downarrow$), so the reactance will increase ($X_C\uparrow$). The discharge current is also highest at the start of discharge ($I\uparrow$, $X_C\downarrow$) as the voltage of the charged capacitor is also high; but as the capacitor discharges, its voltage decreases and the discharge current will also decrease ($I\downarrow$, $X_C\uparrow$).

To summarize, at the start of a capacitor charge or discharge, the current is maximum, so the reactance is low. This value of current then begins to fall to zero, so the reactance increases.

If the applied alternating current is at a high frequency, as shown in Figure 14-1(a), it is switching polarity more rapidly than a lower frequency and there is very little time between the start of charge and discharge. As the charge and discharge currents are largest at the beginning of the charge and discharge of the capacitor, the reactance has very little time to build up and oppose the current, which is why the current is a high value and the capacitive reactance is small at higher frequencies. With lower frequencies, as shown in Figure 14-1(b), the applied alternating current is switching at a slower rate, and therefore the reactance, which is low at the beginning, has more time to build up and oppose the current.

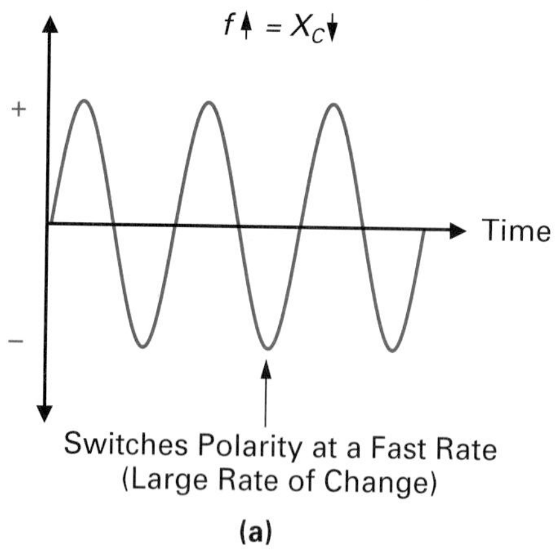

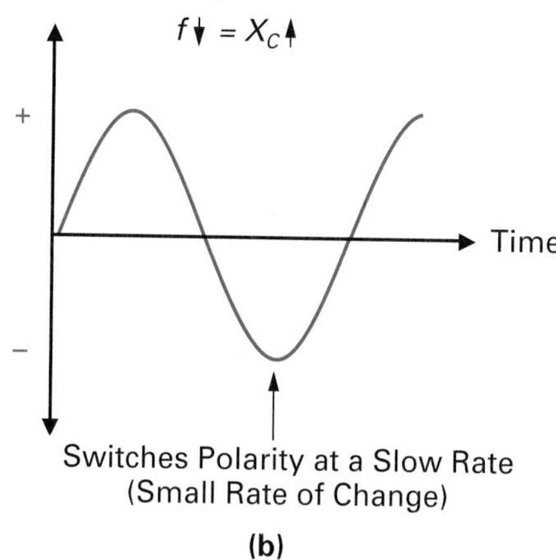

FIGURE 14-1 Capacitive Reactance Is Inversely Proportional to Frequency.

Capacitive reactance is therefore inversely proportional to frequency:

$$\text{capacitive reactance } (X_C) \propto \frac{1}{f\,(\text{frequency})}$$

Frequency, however, is not the only factor that determines capacitive reactance. Capacitive reactance is also inversely proportional to the value of capacitance. If a larger capacitor value is used a longer time is required to charge the capacitor ($t\uparrow = C\uparrow R$), which means that current will be present for a longer period of time, so the overall current will be large ($I\uparrow$); consequently, the reactance must be small ($X_C\downarrow$). On the other hand, a small capacitance value will charge in a small amount of time ($t\downarrow = C\downarrow R$) and the current is present for only a short period of time. The overall current will therefore be small ($I\downarrow$), indicating a large reactance ($X_C\uparrow$).

$$\text{capacitive reactance } (X_C) \propto \frac{1}{C\,(\text{capacitance})}$$

Capacitive reactance (X_C) is therefore inversely proportional to both frequency and capacitance and can be calculated by using the formula

$$X_C = \frac{1}{2\pi f C}$$

where X_C = capacitive reactance, in ohms
2π = constant
f = frequency, in hertz
C = capacitance, in farads

EXAMPLE:

Calculate the reactance of a 2 μF capacitor when a 10 kHz sine wave is applied.

Solution:

$$X_C = \frac{1}{2\pi f C}$$

$$= \frac{1}{2 \times \pi \times 10 \text{ kHz} \times 2\ \mu\text{F}} = 8\ \Omega$$

CALCULATOR SEQUENCE

Step	Keypad Entry	Display Response
1.	[2]	2.0
2.	[×]	
3.	[π]	3.1415927
4.	[×]	6.283185
5.	[1] [0] [EE] [3]	10E3
6.	[×]	6.2831.8
7.	[2] [EE] [6] [+/−]	2.-06
8.	[=]	0.1256637
9.	[1/x]	7.9577

SELF-TEST REVIEW QUESTIONS FOR SECTION 14-1

1. Define *capacitive reactance.*
2. State the formula for capacitive reactance.
3. Why is capacitive reactance inversely proportional to frequency and capacitance?
4. If C = 4 μF and f = 4 kHz, calculate X_C.

14-2 SERIES *RC* CIRCUIT

In a purely resistive circuit, as shown in Figure 14-2(a), the current flowing within the circuit and the voltage across the resistor are in phase with one another. In a purely capacitive circuit, as shown in Figure 14-2(b), the current flowing in the circuit leads the voltage across the capacitor by 90°.

Purely resistive: 0° phase shift (I is in phase with V)

Purely capacitive: 90° phase shift (I leads V by 90°)

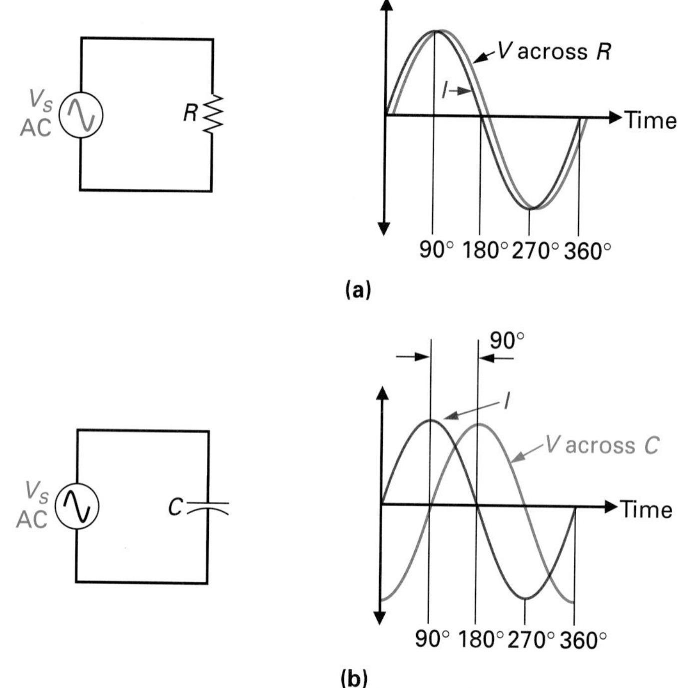

FIGURE 14-2 Phase Relationships between *V* and *I*. (a) Resistive Circuit: Current and Voltage Are in Phase. (b) Capacitive Circuit: Current Leads Voltage by 90°.

If we connect a resistor and capacitor in series, as shown in Figure 14-3(a), we have probably the most commonly used electronic circuit, which has many applications. The voltage across the resistor (V_R) is always in phase with the circuit current (I), as can be seen in Figure 14-3(b), because maximum points and zero crossover points occur at the same time. The voltage across the capacitor (V_C) lags the circuit current by 90°.

Since the capacitor and resistor are in series, the same current is supplied to both components; Kirchhoff's voltage law can be applied, which states that the

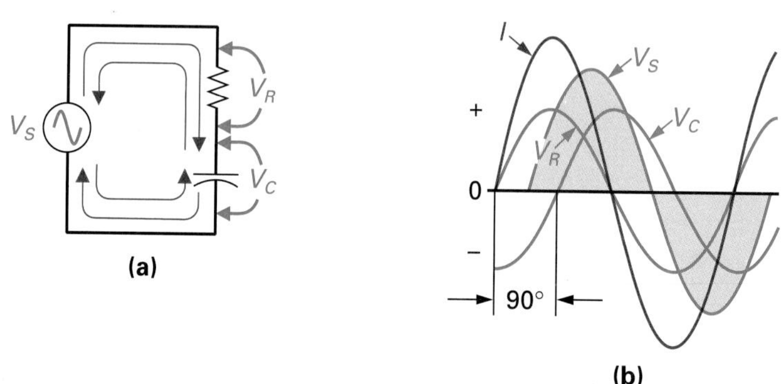

FIGURE 14-3 *RC* Series Circuit. (a) Circuit. (b) Waveforms.

sum of the voltage drops around a series circuit is equal to the voltage applied (V_S). The voltage drop across the resistor (V_R) and the voltage drop across the capacitor (V_C) are out of phase with one another, which means that their peaks occur at different times. The signal for the applied voltage (V_S) is therefore obtained by adding the values of V_C and V_R at each instant in time, plotting the results, and then connecting the points with a line; this is represented in Figure 14-3(b) by the shaded waveform.

Although the waveforms in Figure 14-3(b) indicate the phase relationship between I, V_S, V_R, and V_C, it seems difficult to understand clearly the relationship among all four because of the crisscrossing of waveforms. An easier method of representation is to return to the circle and vectors that were introduced in the trigonometry mini-math review in Chapter 11.

14-2-1 Vector Diagram

A vector (or phasor) is a quantity that has both magnitude and direction and is represented by a line terminated at the end by an arrowhead, as seen in Figure 14-4(a). Vectors are generally always used to represent a physical quantity that has two properties. For example, if you are traveling 60 miles per hour in a southeast direction, the size or magnitude of the vector would represent 60 mph, while the direction of the vector would point southeast. In an ac circuit containing a reactive component such as a capacitor, the vector is used to represent a voltage or current. The magnitude of the vector represents the value of voltage or current, while the direction of the vector represents the phase of the voltage or current.

A **vector diagram** is an arrangement of vectors to illustrate the magnitude and phase relationships between two or more quantities of the same frequency within an ac circuit. Figure 14-4(b) illustrates the basic parts of a vector diagram. As an example, the current (I) vector is at the 0° position and the size of the arrow represents the peak value of alternating current.

Vector Diagram
Arrangement of vectors showing the phase relationships between two or more ac quantities of the same frequency.

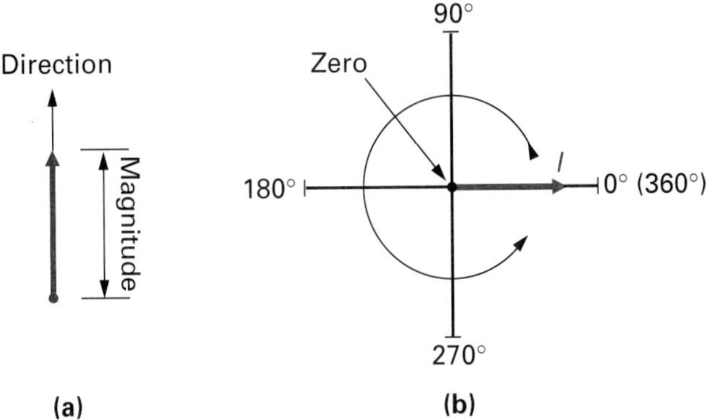

FIGURE 14-4 Vectors. (a) Vector. (b) Vector Diagram.

14-2-2 Voltage

Figure 14-5(a), (b), and (c) repeat our previous *RC* series circuit with waveforms and a vector diagram. In Figure 14-5(b), the current peak flowing in the series *RC* circuit occurs at 0° and will be used as a reference; therefore, the vector of current in Figure 14-5(c) is to the right in the 0° position. The voltage across the resistor (V_R) is in phase or coincident with the current (I), as shown in Figure 14-5(b), and the vector that represents the voltage across the resistor (V_R) over laps or coincides with the I vector, at 0°.

The voltage across the capacitor (V_C) is, as shown in Figure 14-5(b), 90° out of phase (lagging) with the circuit's current, so the V_C vector in Figure 14-5(c) is drawn at –90° (minus sign indicates lag) to the current vector, and the length of this vector represents the magnitude of the voltage across the capacitor. Since the ohmic values of the resistor (R) and the capacitor (X_C) are equal, the voltage drop across both components is the same. The V_R and V_C vectors are subsequently equal in length.

The source voltage (V_S) is, by Kirchhoff's voltage law, equal to the sum of the series voltage drops (V_C and V_R); however, since these voltages are not in phase with one another, we cannot simply add the two together. The source voltage (V_S) will be the sum of both V_C and V_R at a particular time. Studying the waveforms in Figure 14-5(b), you will notice that peak source voltage will occur at 45°. By vectorially adding the two voltages V_R and V_C in Figure 14-5(c), we obtain a resultant V_S vector that has both magnitude and phase. The angle theta (θ) formed between circuit current (I) and source voltage (V_S) will always be less than 90°, and in this example is equal to −45° because the voltage drops across R and C are equal due to R and X_C being of the same ohmic value.

If V_C and V_R are drawn to scale, the peak source voltage (V_S) can be calculated by using the same scale; however, a mathematical rather than graphical method can be used to save the drafting time.

In Figure 14-6(a), we have taken the three voltages (V_R, V_C, and V_S) and formed a right-angle triangle, as shown in Figure 14-6(b). The Pythagorean theo-

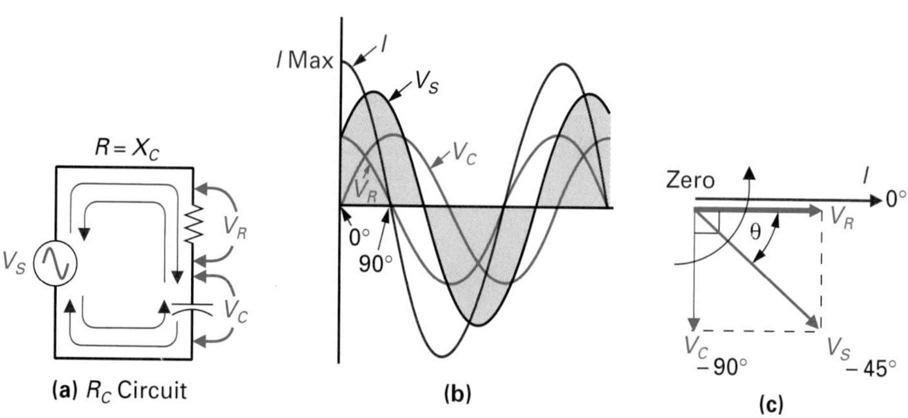

FIGURE 14-5 ***RC*** **Series Circuit Analysis.**

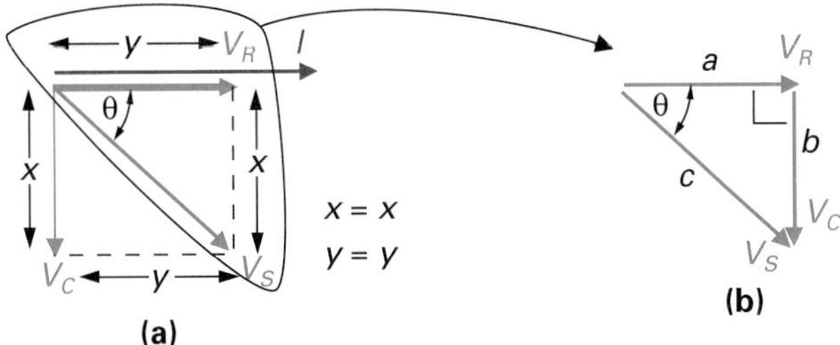

FIGURE 14-6 Voltage and Current Vector Diagram of a Series *RC* Circuit.

rem for right-angle triangles states that if you take the square of a (V_R) and add it to the square of b (V_C), the square root of the result will equal c (the source voltage, V_S).

$$V_S = \sqrt{V_R^2 + V_C^2}$$

By transposing the formula according to the rules of algebra we can calculate any unknown if two variables are known.

$$V_C = \sqrt{V_S^2 - V_R^2}$$
$$V_R = \sqrt{V_S^2 - V_C^2}$$

EXAMPLE:

Calculate the source voltage applied across an *RC* series circuit if $V_R = 12$ V and $V_C = 8$ V.

Solution:

$$V_S = \sqrt{V_R^2 + V_C^2}$$
$$= \sqrt{12\text{ V}^2 + 8\text{ V}^2}$$
$$= \sqrt{144 + 64}$$
$$= 14.42\text{ V}$$

CALCULATOR SEQUENCE

Step	Keypad Entry	Display Response
1.	1 2	12.0
2.	x^2	144.0
3.	+	
4.	8	8.0
5.	x^2	64.0
6.	=	208.0
7.	$\sqrt{x}$	14.42220

14-2-3 Impedance

Impedance (Z) Measured in ohms, it is the total opposition a circuit offers to current flow (reactive and resistive).

Since resistance is the opposition to current with the dissipation of heat, and reactance is the opposition to current without the dissipation of heat, a new term is needed to describe the total resistive and reactive opposition to current. **Impedance** (designated Z) is also measured in ohms and is the total circuit opposition to current flow. It is a combination of resistance (R) and reactance (X_C); however, in our capacitive and resistive circuit, a phase shift or difference exists, and just as V_C and V_R cannot be merely added together to obtain V_S, R and X_C cannot be simply summed to obtain Z.

If the current within a series circuit is constant (the same throughout the circuit), the resistance of a resistor (R) or reactance of a capacitor (X_C) will be directly proportional to the voltage across the resistor (V_R) or the capacitor (V_C).

$$V_R\updownarrow = I \times R\updownarrow, \qquad V_C\updownarrow = I \times X_C\updownarrow$$

A vector diagram can be drawn similarly to the voltage vector diagram to illustrate opposition, as shown in Figure 14-7(a). The current is used as a reference (0°); the resistance vector (R) is in phase with the current vector (I), since V_R is always in phase with I. The capacitive reactance (X_C) vector is at –90° to the resistance vector, due to the 90° phase shift between a resistor and capacitor. The lengths of the resistance vector (R) and capacitive reactance vector (X_C) are equal in this example. By vectorially adding R and X_C, we have a resulting impedance (Z) vector.

By using the three variables, which have again formed a right-angle triangle [Figure 14-7(b)], we can apply the Pythagorean theorem to calculate the total opposition or impedance (Z) to current flow, taking into account both R and X_C.

$$Z = \sqrt{R^2 + X_C^2}$$

$$R = \sqrt{Z^2 - X_C^2}$$

$$X_C = \sqrt{Z^2 - R^2}$$

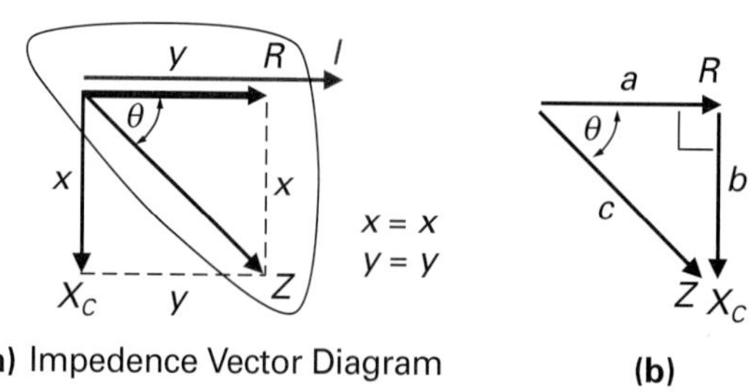

FIGURE 14-7 Resistance, Reactance, and Impedance Vector Diagram of a Series *RC* Circuit.

EXAMPLE:

Calculate the total impedance of a series RC circuit if $R = 27\ \Omega$, $C = 0.005\ \mu F$, and the source frequency = 1 kHz.

Solution:

The total opposition (Z) or impedance is equal to

$$Z = \sqrt{R^2 + X_C^2}$$

R is known, but X_C will need to be calculated.

$$X_C = \frac{1}{2\pi f C}$$

$$= \frac{1}{2 \times \pi \times 1\ \text{kHz} \times 0.005\ \mu\text{F}}$$

$$= 31.8\ \text{k}\Omega$$

Since $R = 27\ \Omega$ and $X_C = 31.8\ \text{k}\Omega$, then

$$Z = \sqrt{R^2 + X_C^2}$$

$$= \sqrt{27\ \Omega^2 + 31.8\ \text{k}\Omega^2}$$

$$= \sqrt{729 + 1 \times 10^9}$$

$$= 31.8\ \text{k}\Omega$$

As you can see in this example, the small resistance of 27 Ω has very little effect on the circuit's total opposition or impedance, due to the relatively large capacitive reactance of 31,800 Ω.

EXAMPLE:

Calculate the total impedance of a series RC circuit if $R = 45\ \text{k}\Omega$ and $X_C = 45\ \Omega$.

Solution:

$$Z = \sqrt{R^2 + X_C^2}$$

$$= \sqrt{45\ \text{k}\Omega^2 + 45^2}$$

$$= 45\ \text{k}\Omega$$

In this example, the relatively small value of X_C had very little effect on the circuit's opposition or impedance, due to the large circuit resistance.

EXAMPLE:

Calculate the total impedance of a series RC circuit if $X_C = 100\ \Omega$ and $R = 100\ \Omega$.

Solution:

$$Z = \sqrt{R^2 + X_C^2}$$

$$= \sqrt{100^2 + 100^2}$$

$$= 141.4\Omega$$

In this example R was equal to X_C.

We can define the total opposition or impedance in terms of Ohm's law, in the same way as we defined resistance.

$$I = \frac{V}{Z} \qquad Z = \frac{V}{I} \qquad V = I \times Z$$

$$\begin{array}{c} V \\ \hline I \mid Z \end{array}$$

By transposition, we can arrive at the usual combinations of Ohm's law.

14-2-4 Phase Angle or Shift (θ)

In a purely resistive circuit, the total opposition (Z) is equal to the resistance of the resistor, so the phase shift (θ) is equal to 0° [Figure 14-8(a)].

In a purely capacitive circuit, the total opposition (Z) is equal to the capacitive reactance (X_C) of the capacitor, so the phase shift (θ) is equal to –90° [Figure 14-8(b)].

When a circuit contains both resistance and capacitive reactance, the total opposition or impedance has a phase shift that is between 0 and 90°. Referring back to the impedance vector diagram in Figure 14-7 and Example 14-5, you can see that we have used a simple example where R has equaled X_C, so the phase shift (θ) has always been –45°. If the resistance and reactance are different from one another, as was the case in Examples 14-3 and 14-4, the phase shift (θ) will change, as shown in Figure 14-9. As can be seen in this illustration, the phase shift (θ) is dependent on the ratio of capacitive reactance to resistance (X_C/R). A more resistive circuit will have a phase shift between 0 and 45°, while a more reactive circuit will have a phase shift between 45 and 90°. By the use of trigonometry (the science of triangles), we can derive a formula to calculate the degree of phase shift, since two quantities X_C and R are known.

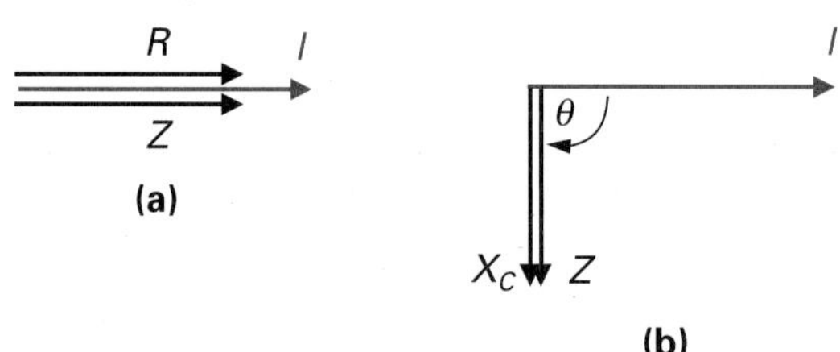

FIGURE 14-8 Phase Angles. (a) Purely Resistive Circuit. (b) Purely Capacitive Circuit.

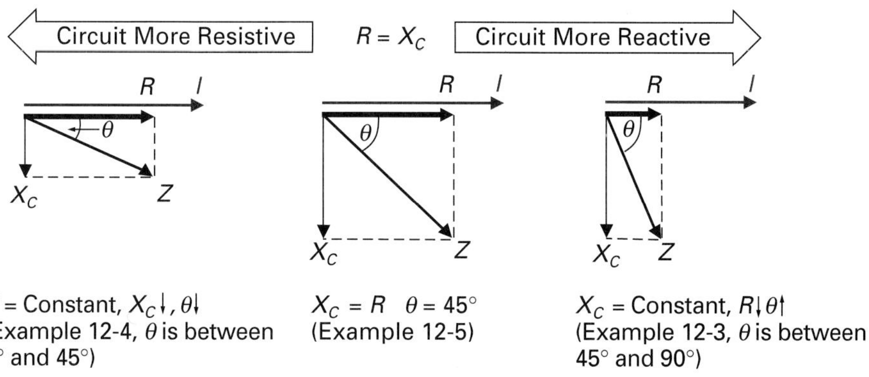

FIGURE 14-9 Phase Angle of a Series *RC* Circuit.

The phase angle, θ, is equal to

$$\theta = \text{invtan}\,\frac{X_C}{R}$$

This formula will determine by what angle Z leads R. Since X_C/R is equal to V_C/V_R, the phase angle can also be calculated if V_R and V_C are known.

$$\theta = \text{invtan}\,\frac{V_C}{V_R}$$

This formula will determine by what angle V_R leads V_S.

EXAMPLE:

Calculate the phase shift or angle in two different series *RC* circuits if:

a. $V_R = 12$ V, $V_C = 8$ V
b. $R = 27\ \Omega$, $X_C = 31.8\ \text{k}\Omega$

Solution:

a. $\theta = \text{invtan}\,\frac{V_C}{V_R}$

$= \text{invtan}\,\frac{8\ \text{V}}{12\ \text{V}}$

$= 33.7°$ (V_R leads V_S by 33.7°)

b. $\theta = \text{invtan}\,\frac{X_C}{R}$

$= \text{invtan}\,\frac{31.8\ \text{k}\Omega}{27\ \Omega}$

$= 89.95°$ (R leads Z by 89.95°)

14-2-5 *Power*

In this section we examine power in a series ac circuit. Let us begin with a simple resistive circuit and review the power formulas used previously.

Purely Resistive Circuit

In Figure 14-10 you can see the current, voltage, and power waveforms generated by applying an ac voltage across a purely resistive circuit. The applied voltage causes current to flow around the circuit, and the electrical energy is converted into heat energy. This heat or power is dissipated and lost and can be calculated by using the power formula.

$$P = V \times I$$

$$P = I^2 \times R$$

$$P = \frac{V^2}{R}$$

Voltage and current are in phase with one another in a resistive circuit, and instantaneous power is calculated by multiplying voltage by current at every instant through 360° ($P = V \times I$). The sinusoidal power waveform is totally positive, because a positive voltage multiplied by a positive current gives a positive

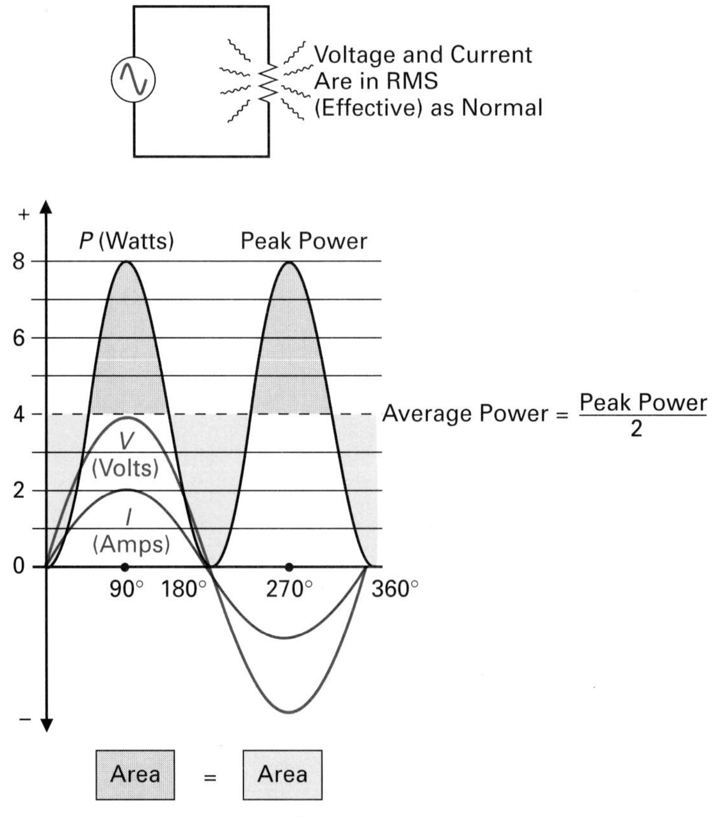

FIGURE 14-10 Power in a Purely Resistive Circuit.

value of power, and a negative voltage multiplied by a negative current will also produce a positive value of power. For these reasons, a resistor is said to generate a positive power waveform, which you may have noticed is twice the frequency of the voltage and current waveforms; two power cycles occur in the same time as one voltage and current cycle.

The power waveform has been split in half, and this line that exists between the maximum point (8 W) and zero point (0 W) is the average value of power (4 W) that is being dissipated by the resistor.

Purely Capacitive Circuit

In Figure 14-11, you can see the current, voltage, and power waveforms generated by applying an ac voltage source across a purely capacitive circuit. As expected, the current leads the voltage by 90°, and the power wave is calculated by multiplying voltage by current, as before, at every instant through 360°. The resulting power curve is both positive and negative. During the positive alternation of the power curve, the capacitor is taking power as the capacitor charges.

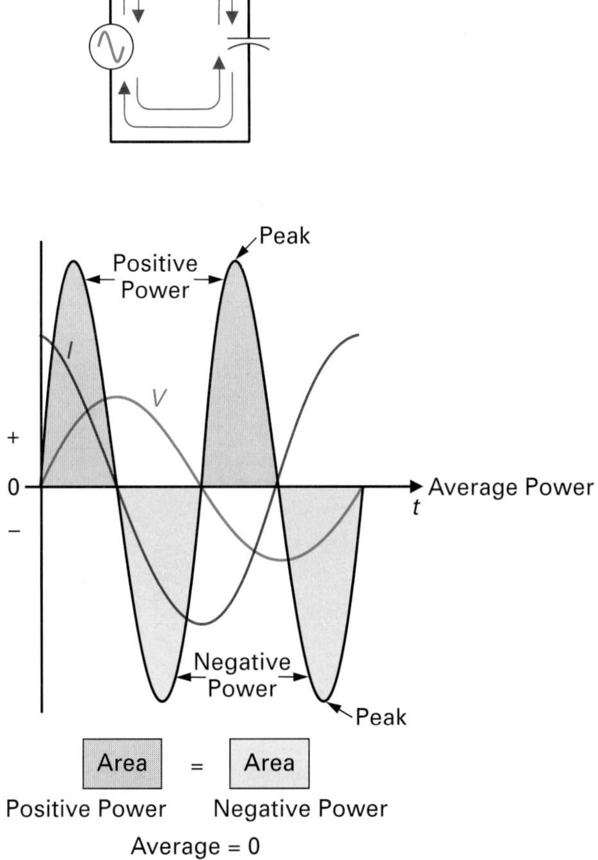

FIGURE 14-11 Power in a Purely Capacitive Circuit.

When the power alternation is negative, the capacitor is giving back the power it took as it discharges back into the circuit.

The average power dissipated is once again the value that exists between the maximum positive and maximum negative points and causes the area above this line to equal the area below. This average power level calculates out to be zero, which means that no power is dissipated in a purely capacitive circuit.

Resistive and Capacitive Circuit

In Figure 14-12 you can see the current, voltage, and power waveforms generated by applying an ac voltage source across a series-connected *RC* circuit. The current leads the voltage by some phase angle less than 90°, and the power waveform is once again determined by the product of voltage and current. The negative alternation of the power cycle indicates that the capacitor is discharging and giving back the power that it consumed during the charge.

The positive alternation of the power cycle is much larger than the negative alternation because it is the combination of both the capacitor taking power during charge and the resistor consuming and dissipating power in the form of heat. The average power being dissipated will be some positive value, due to the heat being generated by the resistor.

Power Factor

In a purely resistive circuit, all the energy supplied to the resistor from the source is dissipated in the form of heat. This form of power is referred to as **resistive power** (P_R) or **true power,** and is calculated with the formula.

Resistive Power or True Power
The average power consumed by a circuit during one complete cycle of alternating current.

$$P_R = I^2 \times R$$

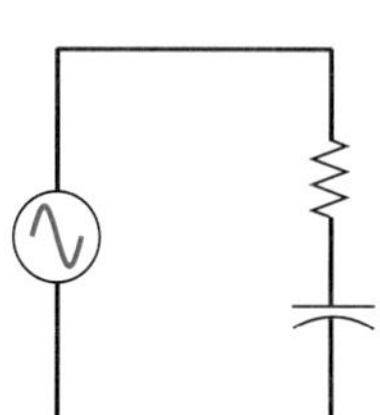

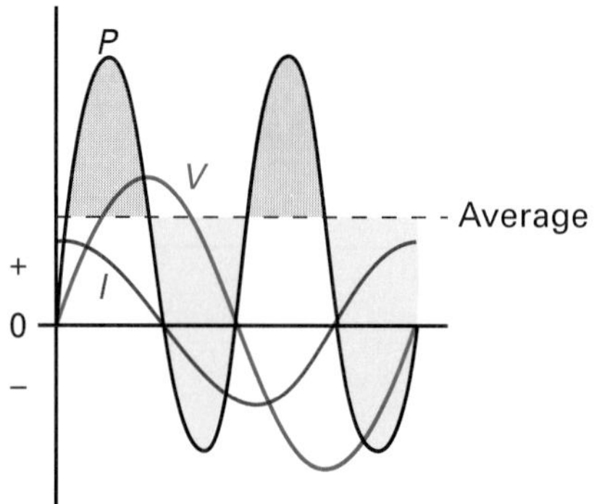

FIGURE 14-12 Power in a Resistive and Capacitive Circuit.

In a purely capacitive circuit, all the energy supplied to the capacitor is stored from the source and then returned to the source, without energy loss. This form of power is referred to as **reactive power** (P_X) or **imaginary power.**

Reactive Power or Imaginary Power Also called wattless power, it is the power value obtained by multiplying the effective value of current by the effective value of voltage and the sine of the angular phase difference between current and voltage.

$$P_X = I^2 \times X_C$$

When a circuit contains both capacitance and resistance, some of the supply is stored and returned by the capacitor and some of the energy is dissipated and lost by the resistor.

Figure 14-13(a) illustrates another vector diagram. Just as the voltage across a resistor is 90° out of phase with the voltage across a capacitor, and resistance is 90° out of phase with reactance, resistive power will be 90° out of phase with reactive power.

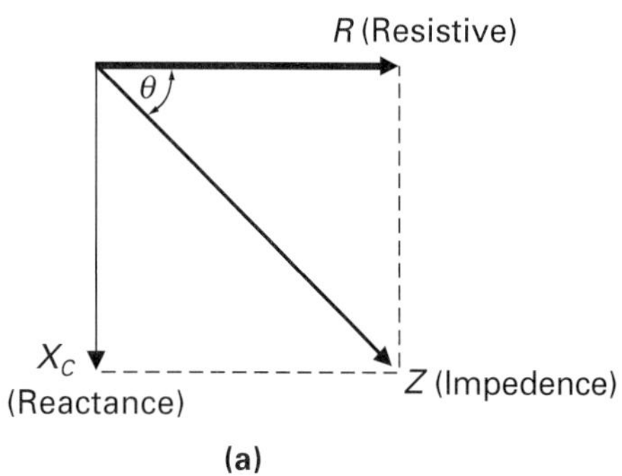

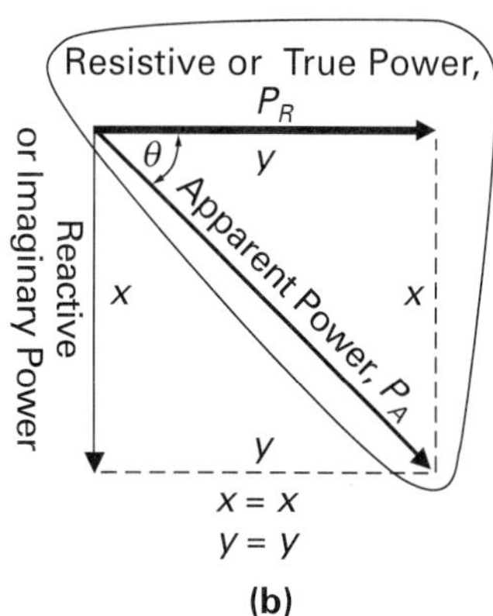

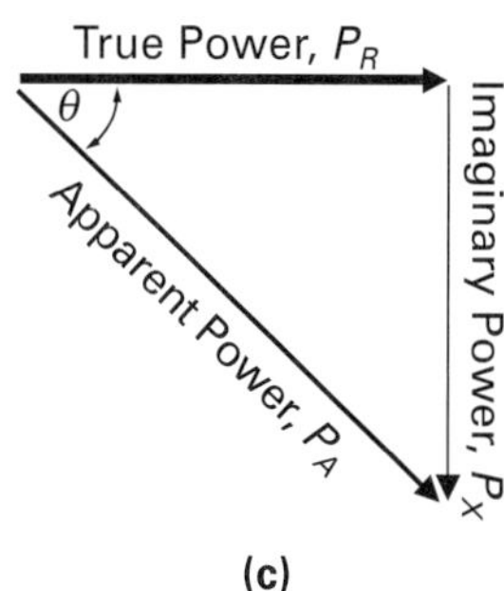

FIGURE 14-13 **Apparent Power.**

Apparent Power The power value obtained in an ac circuit by multiplying together effective values of voltage and current, which reach their peaks at different times.

If we take the three variables from Figure 14-13(b) to form a right-angle triangle as in Figure 14-13(c), we can vectorially add true power and imaginary power to produce a resultant **apparent power** vector. Apparent power is the power that appears to be supplied to the load and includes both the true power dissipated by the resistance and the imaginary power delivered to the capacitor.

Applying the Pythagorean theorem once again, we can calculate apparent power by

$$P_A = \sqrt{P_R^2 + P_X^2}$$

where P_A = apparent power, in volt-amperes (VA)
P_R = true power, in watts (W)
P_X = reactive power, in volt-amperes reactive (VAR)

Power Factor Ratio of actual power to apparent power. A pure resistor has a power factor of 1 or 100% while a capacitor has a power factor of 0 or 0%.

The **power factor** is a ratio of true power to apparent power and is a measure of the loss in a circuit. It can be calculated by using the formula

$$\text{PF} = \frac{\text{true power } (P_R)}{\text{apparent power } (P_A)}$$

Figure 14-14 helps explain what the power factor of a circuit actually indicates. In a purely resistive circuit, as shown in Figure 14-14(a), the apparent power will equal the true power ($P_A = \sqrt{P_R^2 + 0^2}$). The power factor will therefore equal 1 (PF = P_R/P_A). In a purely capacitive circuit, as shown in Figure 14-14(b), the true power will be zero, and therefore the power factor will equal 0 (PF = $0/P_A$). A power factor of 1 therefore indicates a maximum power loss (cir-

Purely Resistive

In a resistive circuit the reactive power is zero, and therefore the true power = apparent power.

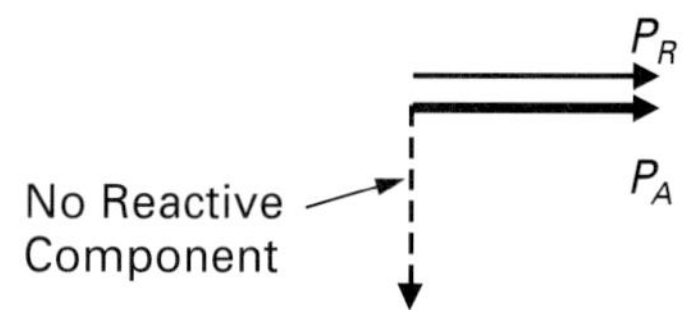

The power factor is therefore equal to:

$$\text{PF} = \frac{P_R}{P_A} = 1 \text{ (maximum value)}$$

(a)

Purely Reactive

In a reactive circuit the resistive power (P_R) is zero.

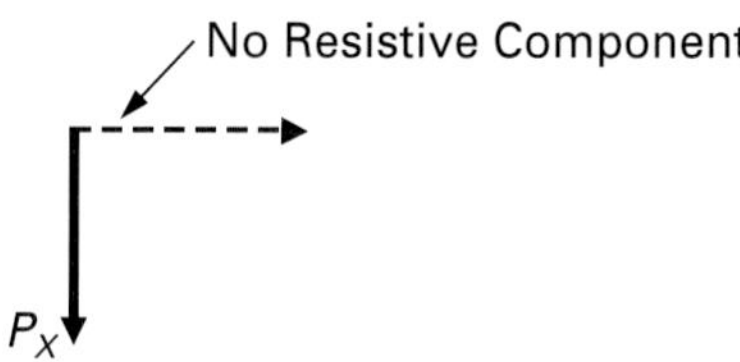

The power factor is therefore equal to:

$$\text{PF} = \frac{P_R}{P_A} = 0 \text{ (minimum value)}$$

(b)

FIGURE 14-14 Circuit's Power Factor. (a) Purely Resistive, PF = 1. (b) Purely Reactive, PF = 0.

cuit is resistive), while a power factor of 0 indicates no power loss (circuit is capacitive). With circuits that contain both resistance and reactance, the power factor will be somewhere between zero (0 = reactive) and one (1 = resistive).

Since true power is determined by resistance and apparent power is dependent on impedance, as shown in Figure 14-13(a), the power factor can also be calculated by using the formula

$$\text{PF} = \frac{R}{Z}$$

As the ratio of true power (adjacent) to apparent power (hypotenuse) determines the angle θ, the power factor can also be determined by the cosine of angle θ.

$$\text{PF} = \cos\theta$$

EXAMPLE:

Calculate the following for a series *RC* circuit if $R = 2.2\ \text{k}\Omega$, $X_C = 3.3\ \text{k}\Omega$, and $V_S = 5\ \text{V}$.

a. Z
b. I
c. θ
d. P_R
e. P_X
f. P_A
g. PF

Solution:

a. $Z = \sqrt{R^2 + X_C^{\,2}}$

$= \sqrt{2.2\ \text{k}\Omega^2 + 3.3\ \text{k}\Omega^2}$

$= 3.96\ \text{k}\Omega$

b. $I = \dfrac{V_S}{Z} = \dfrac{5\ \text{V}}{3.96\ \text{k}\Omega} = 1.26\ \text{mA}$

c. $\theta = \text{invtan}\ \dfrac{X_C}{R} = \text{invtan}\ \dfrac{3.3\ \text{k}\Omega}{2.2\ \text{k}\Omega}$

$= \text{invtan}\ 1.5 = 56.3°$

d. True power $= I^2 \times R$

$= 1.26\ \text{mA}^2 \times 2.2\ \text{k}\Omega$

$= 3.49\ \text{mW}$

e. Reactive power $= I^2 \times X_C$

$= 1.26\ \text{mA}^2 \times 3.3\ \text{k}\Omega$

$= 5.24 \times 10^{-3}$ or 5.24 mVAR

f. Apparent power $= \sqrt{P_R^2 + P_X^2}$

$$= \sqrt{3.49 \text{ mW}^2 + 5.24 \text{ mW}^2}$$

$$= 6.29 \times 10^{-3} \text{ or } 6.29 \text{ mVA}$$

CALCULATOR SEQUENCE

Step	Keypad Entry	Display Response
1.	3 . 3 E 3	3.3E3
2.	÷	
3.	2 . 2 E 3	2.2E3
4.	=	1.5
5.	inv tan	56.309932

g. Power factor $= \frac{R}{Z} = \frac{2.2 \text{ k}\Omega}{3.96 \text{ k}\Omega} = 0.55$

or

$$= \frac{P_R}{P_A} = \frac{3.49 \text{ mW}}{6.29 \text{ mW}} = 0.55$$

or

$$= \cos \theta = \cos 56.3° = 0.55$$

SELF-TEST REVIEW QUESTIONS FOR SECTION 14-2

1. What is the phase relationship between current and voltage in a series *RC* circuit?
2. What is a phasor diagram?
3. Define and state the formula for impedance.
4. What is the phase angle or shift in:
   a. A purely resistive circuit?
   b. A purely capacitive circuit?
   c. A series circuit consisting of *R* and *C*?

14-3 PARALLEL *RC* CIRCUIT

Now that we have analyzed the characteristics of a series *RC* circuit, let us connect a resistor and capacitor in parallel.

14-3-1 *Voltage*

As with any parallel circuit, the voltage across all components in parallel is equal to the source voltage; therefore,

$$V_R = V_C = V_S$$

14-3-2 *Current*

In Figure 14-15(a), you will see a parallel circuit containing a resistor and a capacitor. The current through the resistor and capacitor is simply calculated by applying Ohm's law

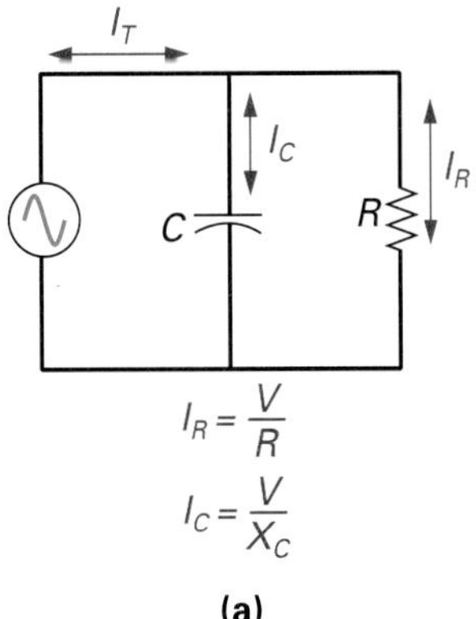

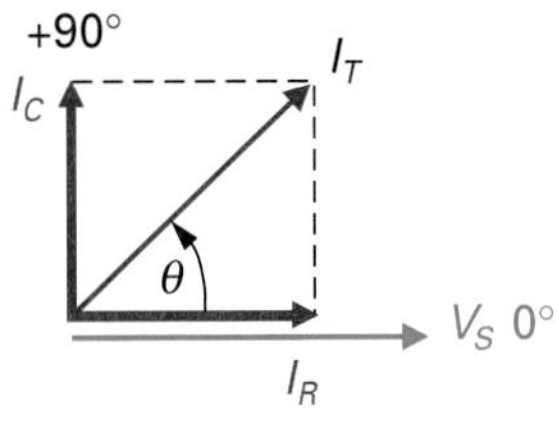

FIGURE 14-15 **Parallel *RC* Circuit.**

$$\text{(resistor current) } I_R = \frac{V_S}{R}$$

$$\text{(capacitor current) } I_C = \frac{V_S}{X_C}$$

Total current (I_T), however, is not as simply calculated. As expected, resistor current (I_R) is in phase with the applied voltage (V_S), as shown in the vector diagram in Figure 14-15(b). Capacitor current will always lead the applied voltage by 90°, and as the applied voltage is being used as our reference at 0° on the vector diagram, the capacitor current will have to be drawn at +90° in order to lead the applied voltage by 90°, since vector diagrams rotate in a counterclockwise direction.

Total current is therefore the vector sum of both the resistor and capacitor currents. Using the Pythagorean theorem, total current can be calculated by

$$I_T = \sqrt{I_R^2 + I_C^2}$$

14-3-3 Phase Angle

The angle by which the total current (I_T) leads the source voltage (V_S) can be determined with either of the following formulas:

$$\theta = \text{invtan}\ \frac{I_C\ (\text{opposite})}{I_R\ (\text{adjacent})} \qquad \theta = \text{invtan}\ \frac{R}{X_C}$$

14-3-4 Impedance

Since the circuit is both capacitive and resistive, the total opposition or impedance of the parallel *RC* circuit can be calculated by

$$Z = \frac{V_S}{I_T}$$

The impedance of a parallel *RC* circuit is equal to the total voltage divided by the total current. Using basic algebra, this basic formula can be rearranged to express impedence in terms of reactance and resistance.

$$Z = \frac{R \times X_C}{\sqrt{R^2 + X_C^2}}$$

14-3-5 Power

With respect to power, there is no difference between a series circuit and a parallel circuit. The true power or resistive power (P_R) dissipated by an *RC* circuit is calculated with the formula

$$P_R = I_R^2 \times R$$

The imaginary power or reactive power (P_X) of the circuit can be calculated with the formula

$$P_X = I_C^2 \times X_C$$

The apparent power is equal to the vector sum of the true power and the reactive power.

$$P_A = \sqrt{P_R^2 + P_X^2}$$

As with series RC circuits the power factor is calculated as

$$PF = \frac{P_R \text{ (resistive power)}}{P_A \text{ (apparent power)}}$$

A power factor of 1 indicates a purely resistive circuit, while a power factor of 0 indicates a purely reactive circuit.

EXAMPLE:

Calculate the following for a parallel RC circuit in which $R = 24\ \Omega$, $X_C = 14\ \Omega$, and $V_S = 10$ V.

a. I_R
b. I_C
c. I_T
d. Z
e. θ

Solution:

a. $I_R = \frac{V_S}{R} = \frac{10\text{ V}}{24\ \Omega} = 416.66\text{ mA}$

b. $I_C = \frac{V_S}{X_C} = \frac{10\text{ V}}{14\ \Omega} = 714.28\text{ mA}$

c. $I_T = \sqrt{I_R^2 + I_C^2}$

$= \sqrt{416.66\text{ mA}^2 + 714.28\text{ mA}^2}$

$= \sqrt{0.173 + 0.510}$

$= 826.5\text{ mA}$

d. $Z = \frac{V_S}{I_T} = \frac{10\text{ V}}{826.5\text{ mA}} = 12\ \Omega$

or

$= \frac{R \times X_C}{\sqrt{R^2 + X_C^2}} = \frac{24 \times 14}{\sqrt{24^2 + 14^2}}$

$= 12\ \Omega$

e. $\theta = \arctan \frac{I_C}{I_R} = \arctan \frac{714.28\text{ mA}}{416.66\text{ mA}}$

$= \arctan 1.714 = 59.7°$

or

$\theta = \arctan \frac{R}{X_C} = \arctan \frac{24\ \Omega}{14\ \Omega}$

$= \arctan 1.714 = 59.7°$

CALCULATOR SEQUENCE

Step	Keypad Entry	Display Response
1.	7 1 4 . 2 8 E 3 +/−	714.28E-3
2.	÷	
3.	4 1 6 . 6 6 E 3 +/−	416.66E-3
4.	=	1.7142994
5.	inv tan	59.743762

SELF-TEST REVIEW QUESTIONS FOR SECTION 14-3

1. What is the phase relationship between current and voltage in a parallel *RC* circuit?
2. Could a parallel *RC* circuit be called a voltage divider?
3. State the formula used to calculate:
   a. I_T
   b. Z
4. Will capacitor current lead or lag resistor current in a parallel *RC* circuit?

14-4 TESTING CAPACITORS

Now that you have a good understanding as to how a capacitor should function, let us investigate how to diagnose a capacitor malfunction.

14-4-1 *The Ohmmeter*

A faulty capacitor may have one of three basic problems:

1. A short, which is easy to detect and is caused by a contact from plate to plate.
2. An open, which is again quite easy to detect and is normally caused by one of the leads becoming disconnected from its respective plate.
3. A leaky dielectric or capacitor breakdown, which is quite difficult to detect, as it may only short at a certain voltage. This problem is usually caused by the deterioration of the dielectric, which starts displaying a much lower dielectric resistance than it was designed for. The capacitor with this type of problem is referred to as a *leaky capacitor.*

Capacitors of 0.5 μF and larger can be checked by using an analog ohmmeter or a digital multimeter with a bar graph display by using the procedure shown in Figure 14-16.

Step 1: Ensure that the capacitor is discharged by shorting the leads together.

Step 2: Set the ohmmeter to the highest ohms range scale. Zero the ohmmeter.

Step 3: Connect the meter to the capacitor, observing the correct polarity if an electrolytic is being tested, and observe the meter pointer. The capacitor will initially be discharged, and therefore maximum current will flow from the meter battery to the capacitor; maximum current means low resistance, which is why the meter's pointer deflects to the far right to indicate 0 Ω.

Step 4: As the capacitor charges, it will cause current flow from the meter's battery to decrease, and consequently, the meter needle will move toward the left side of the scale.

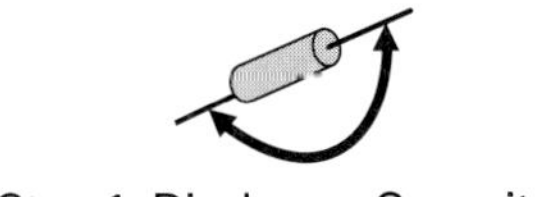
Step 1: Discharge Capacitor

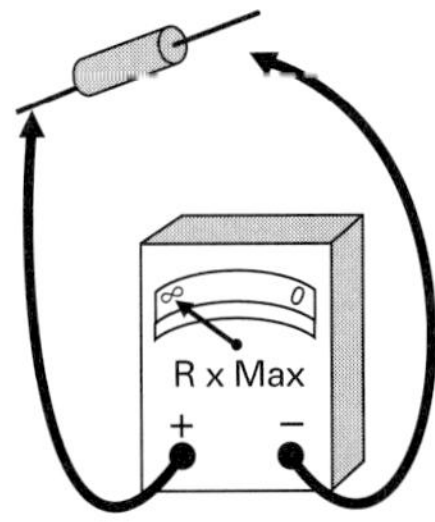

Step 2: Set Analog Ohmmeter to High Ohms Range

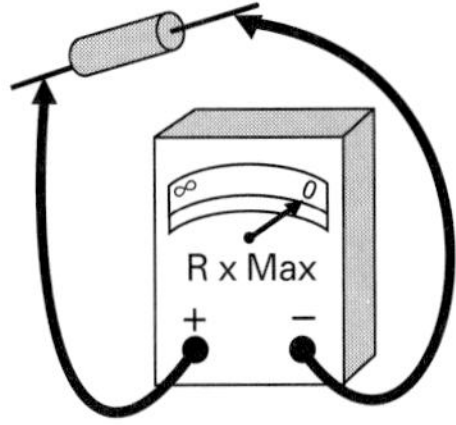

Step 3: Meter Deflects Rapidly to 0 Ohms Initially

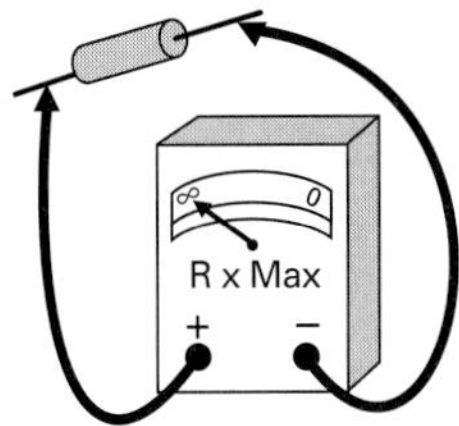

Step 4: Meter Should Then Return to Infinity as the Capacitor Charges

FIGURE 14-16 **Testing a Capacitor of More Than 0.5 μF.**

A good capacitor will cause the meter to react as just explained. A larger capacitance will cause the meter to move slowly to infinity (∞), as it will take a longer time to charge, while a smaller value of capacitance will charge at a much faster rate, causing the meter to deflect rapidly toward ∞. For this reason, the ohmmeter cannot be reliably used to check capacitors with values of less than 0.5 μF, because the capacitor charges up too quickly and the meter does not have enough time to respond.

A shorted capacitor will cause the meter to deflect to zero ohms and remain in that position. An open capacitor will cause no meter deflection (infinite resistance) because there is no path for current to flow.

A leaky capacitor will deflect to the right, as normal. If the meter pointer returns almost all the way back to ∞, only a small current is still flowing and the capacitor has a small dielectric leak. If the meter only comes back to halfway or a large distance away from infinity, a large amount of current is still flowing and the capacitor has a large dielectric leak (defect).

When using the analog ohmmeter to test capacitors, there are some other points that you should be aware of:

1. Electrolytics are noted for having a small yet noticeable amount of inherent leakage; therefore, do not expect the needle pointer to move all the way to the left (∞ ohms). Most electrolytic capacitors that are still functioning normally will show a resistance of 200 kΩ or more.
2. Some ohmmeters utilize internal batteries of up to 15 V, so be careful not to exceed the voltage rating of the capacitor.

14-4-2 *The Capacitance Meter or Analyzer*

The ohmmeter check tests the capacitor under a low-voltage condition. This may be adequate for some capacitor malfunctions; however, a problem that often occurs with capacitors is that they short or leak at a high voltage. The ohmmeter test is also adequate for capacitors of 0.5 μF or greater; however, a smaller capacitor cannot be tested because its charge time is too fast for the meter to respond. The ohmmeter cannot check for high-voltage failure, for small-value capacitance, or if the value of capacitance has changed through age or extreme thermal exposure.

Capacitance Meter
Instrument used to measure the capacitance of a capacitor or circuit.

A **capacitance meter** or analyzer, which is illustrated in Figure 14-17, can totally check all aspects of a capacitor in a range of values from approximately 1 pF to 20 F. The tests that are generally carried out include:

1. *Capacitor value change* [Figure 14-18(a)]: Capacitors will change their value over a period of time. Ceramic capacitors often change 10 to 15% within the first year as the ceramic material relaxes. Electrolytics change their value due to the electrolytic solution simply drying out. Some capacitors are simply labeled incorrectly by manufacturers or the technician cannot determine the correct value because of the labeling used. A value change accounts for approximately 25% of all defective capacitors.
2. *Capacitor leakage* [Figure 14-18(b)]: Leakage occurs due to an imperfection of the dielectric. Although the dielectric's resistance is very high, a small amount of leakage current will flow between the plates. This resistance, which is between the plates and therefore effectively in parallel with the capacitor, can become too low and cause the circuit that the capacitor is in

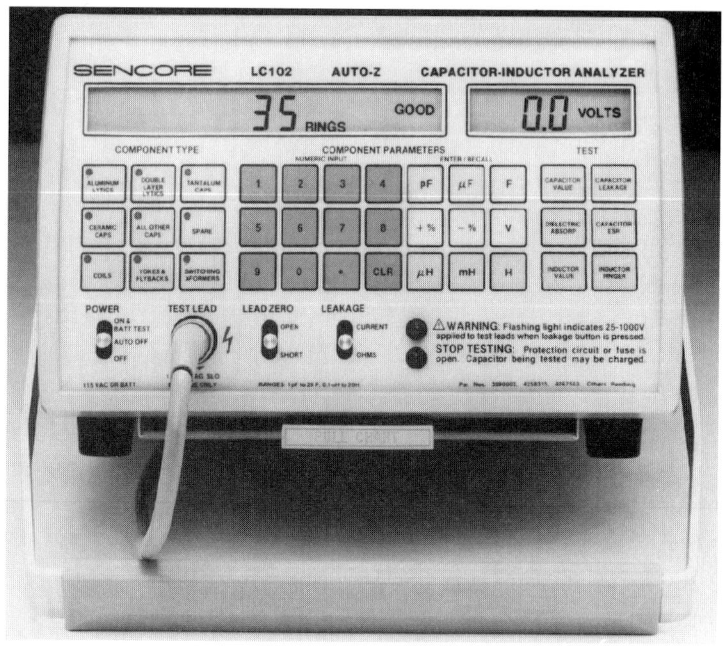

FIGURE 14-17 Capacitor Analyzer.

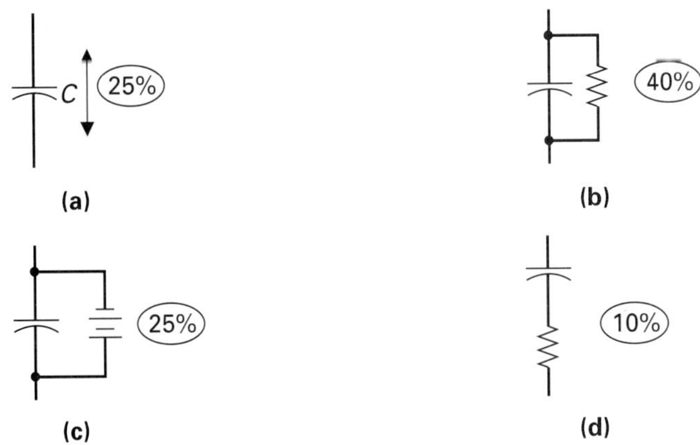

FIGURE 14-18 Problems with Capacitors. (a) Value Change. (b) Leakage. (c) Dielectric Absorption. (d) Equivalent Series Resistance.

to malfunction. Most capacitance meters will perform a leakage test with operating potentials up to 650 V. Leakage accounts for approximately 40% of all defective capacitors.

3. *Dielectric absorption* [Figure 14-18(c)]: This occurs mainly in electrolytics when they will take on a charge but will not fully discharge. This residual charge remains within the capacitor, similar to a small dc battery, and this changes the effective value of the capacitor once it is in circuit during operation. If this causes the value to change more than 15%, the capacitor should be rejected. Dielectric absorption accounts for approximately 25% of all defective capacitors.
4. *Equivalent series resistance* (ESR) [Figure 14-18(d)]: Series resistance is found in the capacitor leads, lead to plate connection, and electrolyte (almost always occurs in electrolytics) and causes the effective circuit capacitance value to change. Equivalent series resistance accounts for 10% of all defective capacitors.

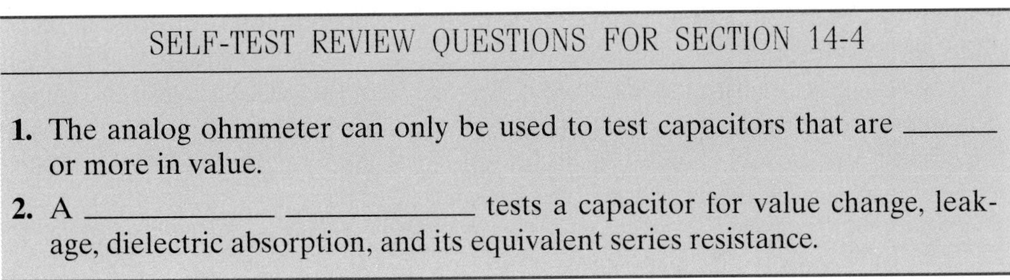

SELF-TEST REVIEW QUESTIONS FOR SECTION 14-4

1. The analog ohmmeter can only be used to test capacitors that are ______ or more in value.
2. A ____________ ____________ tests a capacitor for value change, leakage, dielectric absorption, and its equivalent series resistance.

14-5 APPLICATIONS OF CAPACITORS

There are many applications of capacitors, some of which will be discussed now; others will be presented later and in your course of electronic studies.

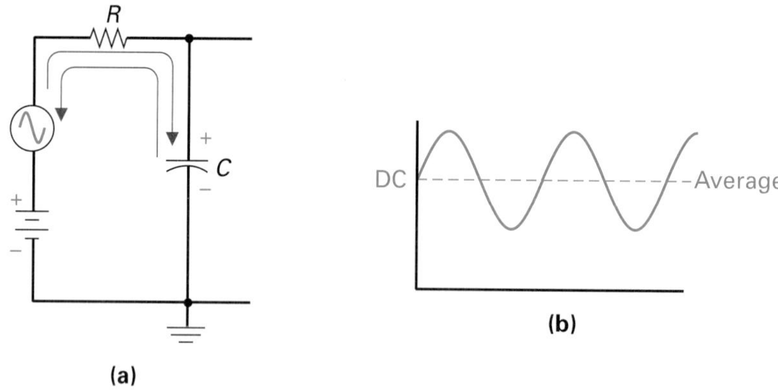

FIGURE 14-19 **Superimposing AC on a DC Level.**

14-5-1 Combining AC and DC

Figure 14-19(a) and (b) show how the capacitor can be used to combine ac and dc. The capacitor is large in value (electrolytic typically) and can be thought of as a very large bucket that will fill or charge to a dc voltage level. The ac voltage will charge and discharge the capacitor, which is similar to pouring in and pulling out (alternating) more and less water. The resulting waveform is a combination of ac and dc that varies above and below an average dc level. In this instance, the ac is said to be superimposed on a dc level.

14-5-2 The Capacitive Voltage Divider

In Chapter 13 we illustrated how capacitors could be connected in series across a dc voltage source to form a voltage divider. This circuit is repeated in Figure 14-20(a). Figure 14-20(b) shows how two capacitors can be used to divide up an ac voltage source of 30 V rms. Like a resistive voltage divider, voltage drop is proportional to current opposition, which with a capacitor is called *capacitive reactance* (X_C). The larger the capacitive reactance ($X_C\uparrow$), the larger the voltage drop ($V\uparrow$). Since capacitive reactance is inversely proportional to the capacitor value and the frequency of the ac source, a change in input frequency will cause a change in the capacitor's capacitive reactance, and this will change the voltage drop across the capacitor. Although a 90° phase shift is present between voltage and current in a purely capacitive circuit (I leads V by 90°), there is no phase difference between the input voltage and the output voltages.

Filter
Network composed of resistor, capacitor, and inductors used to pass certain frequencies yet block others through heavy attenuation.

14-5-3 RC Filters

A **filter** is a circuit that allows certain frequencies to pass but blocks other frequencies. In other words, it filters out the unwanted frequencies but passes the wanted or selected ones.

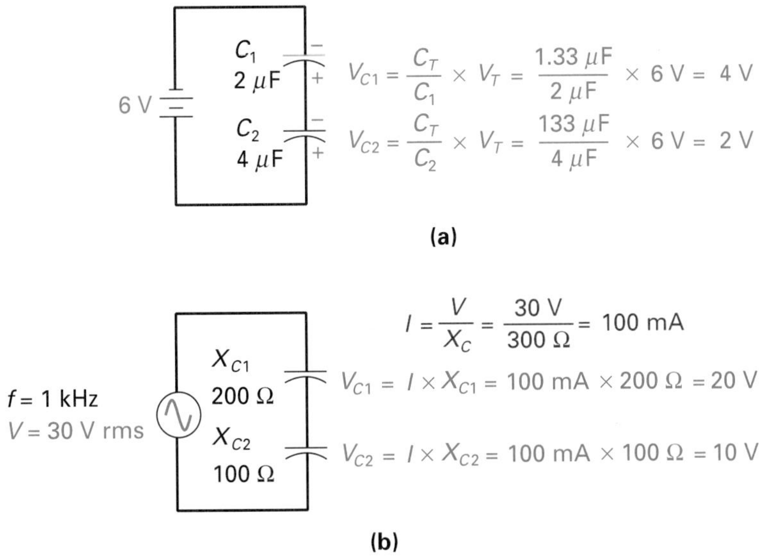

$$V_{C1} = \frac{C_T}{C_1} \times V_T = \frac{1.33\ \mu F}{2\ \mu F} \times 6\ V = 4\ V$$

$$V_{C2} = \frac{C_T}{C_2} \times V_T = \frac{133\ \mu F}{4\ \mu F} \times 6\ V = 2\ V$$

(a)

$$I = \frac{V}{X_C} = \frac{30\ V}{300\ \Omega} = 100\ mA$$

$$V_{C1} = I \times X_{C1} = 100\ mA \times 200\ \Omega = 20\ V$$

$$V_{C2} = I \times X_{C2} = 100\ mA \times 100\ \Omega = 10\ V$$

(b)

FIGURE 14-20 Capacitive Voltage Divider. (a) DC Circuit. (b) AC Circuit.

There are two basic *RC* filters:

1. The **low-pass filter,** which can be seen in Figure 14-21(a); as its name implies, it passes the low frequencies but heavily **attenuates** the higher frequencies.
2. The **high-pass filter,** which can be seen in Figure 14-21(b); as its name implies, it allows the high frequencies to pass but heavily attenuates the lower frequencies.

With either the low- or high-pass filter, it is important to remember that capacitive reactance is inversely proportional to frequency ($X_C \propto 1/f$) as stated in the center of Figure 14-21.

With the low-pass filter shown in Figure 14-21(a), the output is connected across the capacitor. As the frequency of the input increases, the amplitude of the output decreases. At dc (0 Hz) and low frequencies, the capacitive reactance is very large ($X_C\uparrow = 1/f\downarrow$) with respect to the resistor. All the input will appear across the capacitor, because the capacitor and resistor form a voltage divider, as shown in Figure 14-22(a). As with any voltage divider, the larger opposition to current flow will drop the larger voltage. Since the output voltage is determined by the voltage drop across the capacitor, almost all the input will appear across the capacitor and therefore be present at the output.

If the frequency of the input increases, the reactance of the capacitor will decrease ($X_C\downarrow = 1/f\uparrow$), and a larger amount of the signal will be dropped across the resistor. As frequency increases, the capacitor becomes more of a short circuit (lower reactance), and the output, which is across the capacitor, decreases, as shown in Figure 14-22(b).

Low-Pass Filter
Network or circuit designed to pass any frequencies below a critical or cutoff frequency and reject or heavily attenuate all frequencies above.

Attenuate
To reduce in amplitude an action or signal.

High-Pass Filter
Network or circuit designed to pass any frequencies above a critical or cutoff frequency and reject or heavily attenuate all frequencies below.

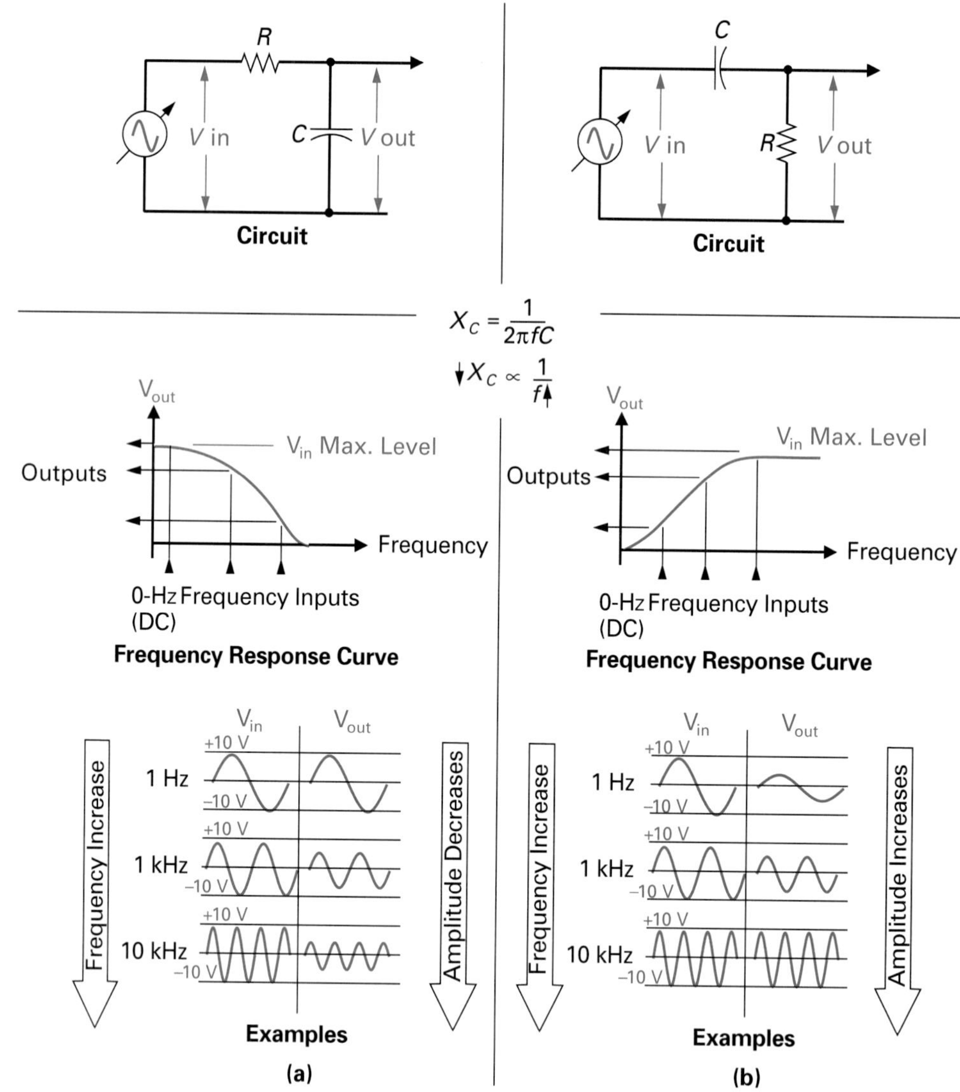

FIGURE 14-21 *RC* **Filters. (a) Low-Pass Filter. (b) High-Pass Filter.**

Frequency Response Curve
A graph indicating how effectively a circuit or device responds to the frequency spectrum.

Below the circuit of the low-pass filter in Figure 14-21(a), you will see a graph known as the **frequency response curve** for the low-pass filter. This curve illustrates that as the frequency of the input increases the voltage at the output will decrease.

With the high-pass filter, seen in Figure 14-21(b), the capacitor and resistor have traded positions to show how the opposite effect for the low-pass filter occurs.

At low frequencies, the reactance will be high and almost all of the signal in will be dropped across the capacitor. Very little signal appears across the resistor, and, consequently, the output, as shown in Figure 14-23(a). As the frequency of the input increases, the reactance of the capacitor decreases, allowing more of

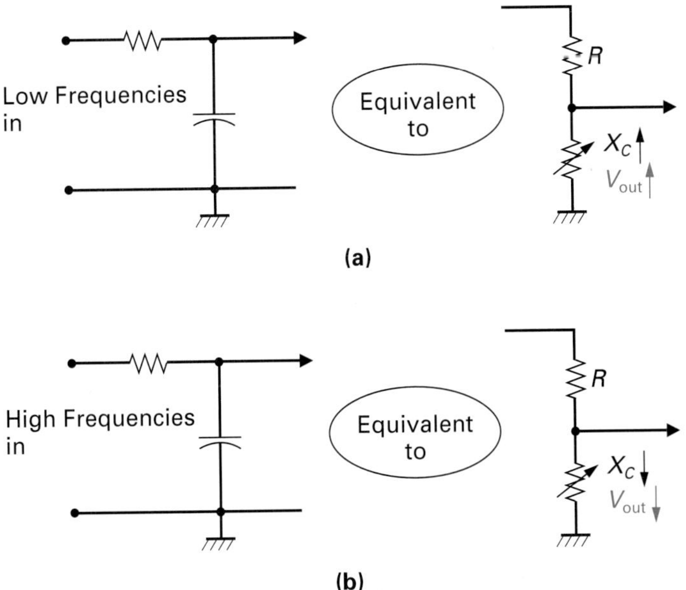

FIGURE 14-22 **Low-Pass Filter. (a) Low-Pass Filter, High X_C. (b) Low-Pass Filter, Low X_C.**

the input signal to appear across the resistor and therefore appear at the output, as shown in Figure 14-23(b).

Below the circuit of the high-pass filter in Figure 14-21(b), you will see the frequency response curve for the high-pass filter. This curve illustrates that as the frequency of the input increases the voltage at the output increases.

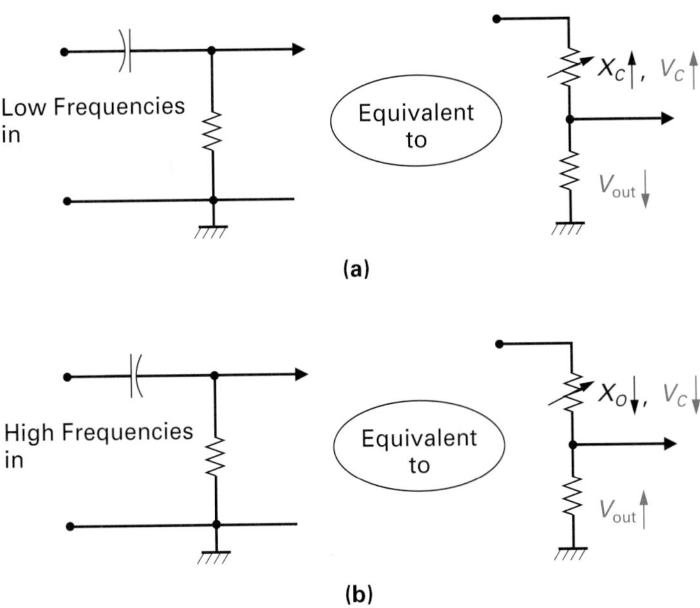

FIGURE 14-23 **High-Pass Filter. (a) High X_C. (b) Low X_C.**

14-5-4 *The RC Integrator*

Integrator
Device that approximates and whose output is proportional to an integral of the input signal.

Up until now we have analyzed the behavior of an *RC* circuit only when a sine-wave input signal was applied. In both this and the following sections we demonstrate how an *RC* circuit will react to a square-wave input, and show two other important applications of capacitors. The term **integrator** is derived from a mathematical function in calculus. This combination of *R* and *C*, in some situations, displays this mathematical function. Figure 14-24(a) illustrates an integrator circuit that can be recognized by the series connection of *R* and *C*, but mainly from the fact that the output is taken across the capacitor.

If a 10 V square wave is applied across the circuit, as shown in the waveforms in Figure 14-24(b), and the time constant of the *RC* combination calculates out to be 1 second, the capacitor will charge when the square wave input is positive toward the applied voltage (10 V) and reaches it in five time constants (5 seconds). Since the positive alternation of the square wave lasts for 6 seconds, the capacitor will be fully charged 1 second before the positive alternation ends, as shown in Figure 14-25.

When the positive half-cycle of the square-wave input ends after 6 seconds, the input falls to 0 V and the circuit is equivalent to that shown in Figure 14-26(a). The 10 V charged capacitor now has a path to discharge and in five time constants (5 seconds) is fully discharged, as shown in Figure 14-26(b).

If the same *RC* integrator circuit was connected to a square wave that has a 1-second positive half-cycle, the capacitor will not be able to charge fully toward 10 V. In fact, during the positive alternation of 1 second (one time constant),

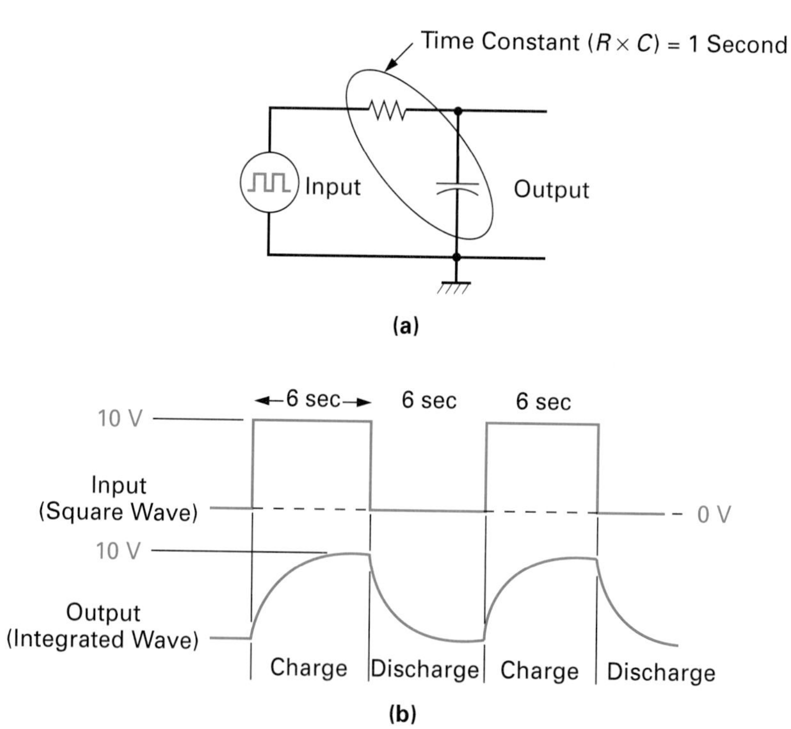

FIGURE 14-24 ***RC* Integrator. (a) Circuit. (b) Waveforms.**

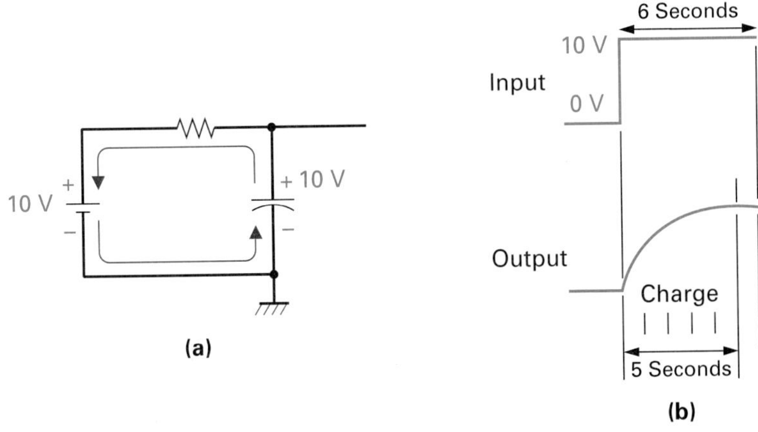

FIGURE 14-25 Integrator Response to Positive Step. (a) Equivalent Circuit. (b) Waveforms for Charge.

the capacitor will reach 63.2% of the applied voltage (6.32 V), and then during the 0 V half-cycle it will discharge to 63.2% of 6.32 V, to 2.33 V, as shown by the waveform in Figure 14-27. The voltage across the capacitor, and therefore the output voltage, will gradually build up and eventually level off to an average value of 5 V in about five time constants (5 seconds).

In summary, if the period or time of the square wave is decreased, or if the time constant is increased, the same effect results. The capacitor has less time to charge and discharge and reaches an average value of the input voltage, which for a square wave is half of its amplitude.

Figure 14-28 illustrates the circuit waveform produced when the time constant is long with respect to the period of the input waveform. If the time constant is further increased so that it is even longer with respect to the input waveform, the output will have an even smaller peak-to-peak variation.

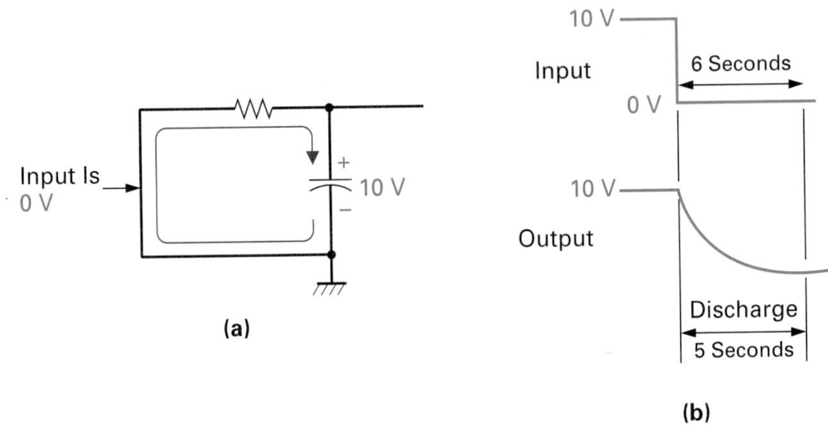

FIGURE 14-26 Integrator Response to Negative Step. (a) Equivalent Circuit. (b) Waveforms for Discharge.

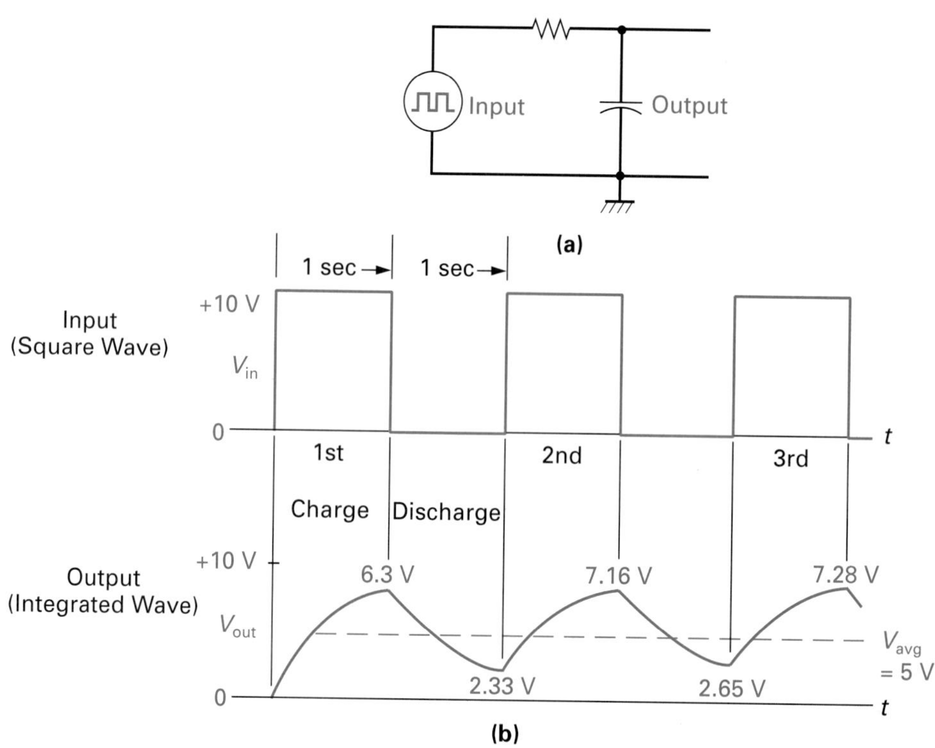

FIGURE 14-27 Integrator Response to Square Wave. (a) Circuit. (b) Waveforms.

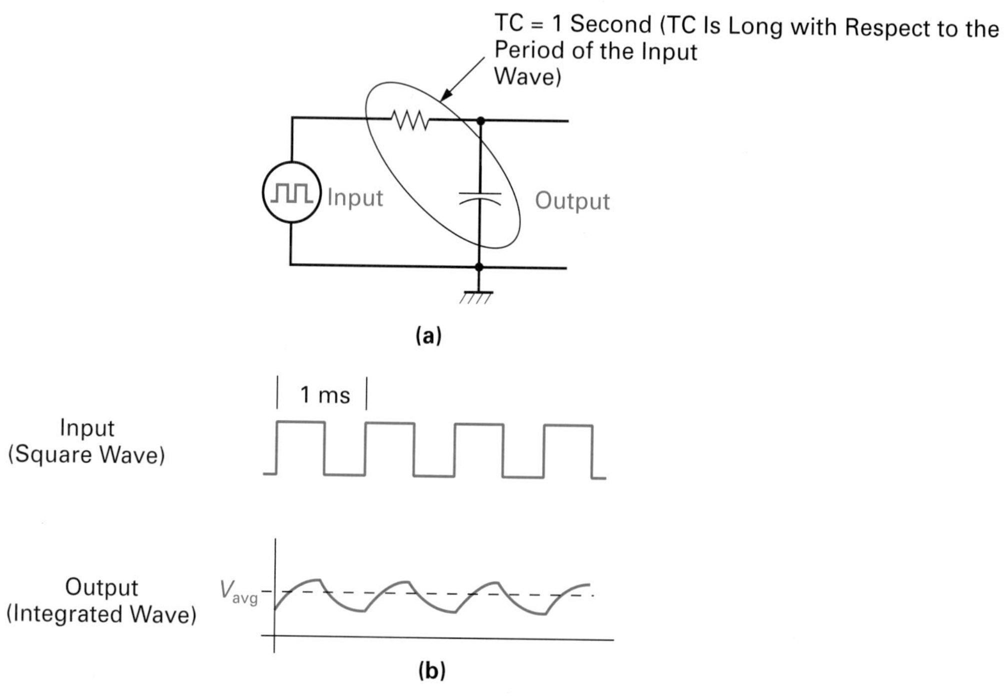

FIGURE 14-28 Integrator Response to Long Time Constant. (a) Circuit. (b) Waveforms.

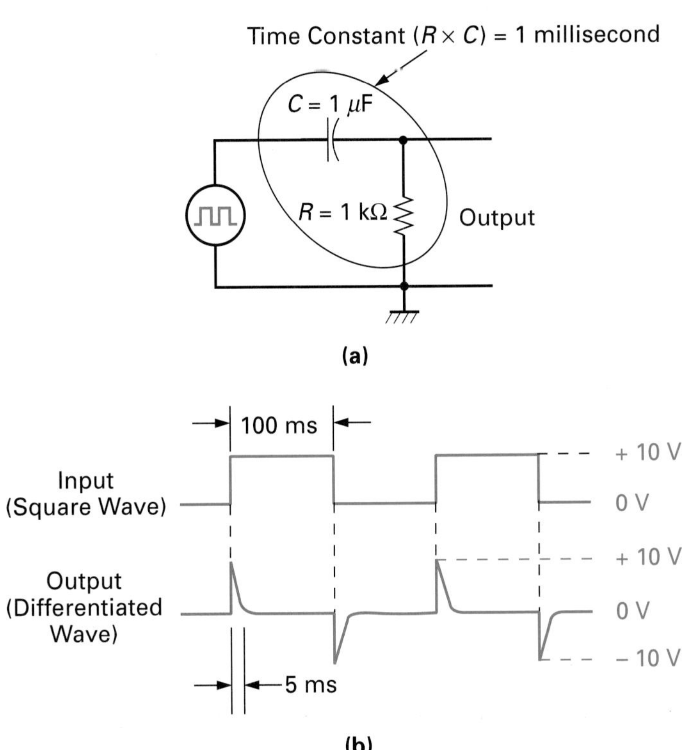

FIGURE 14-29 ***RC* Differentiator. (a) Circuit. (b) Waveforms.**

14-5-5 *The* RC *Differentiator*

Figure 14-29(a) illustrates the **differentiator** circuit, which is the integrator's opposite. In this case the output is taken across the resistor instead of the capacitor, and the time constant is always short with respect to the input square wave period.

Differentiator
A circuit whose output voltage is proportional to the rate of change of the input voltage. The output waveform is then the time derivative of the input waveform, and the phase of the output waveform leads that of the input by 90°.

The differentiator output waveform, shown in Figure 14-29(b), is taken across the resistor and is the result of the capacitor's charge and discharge.

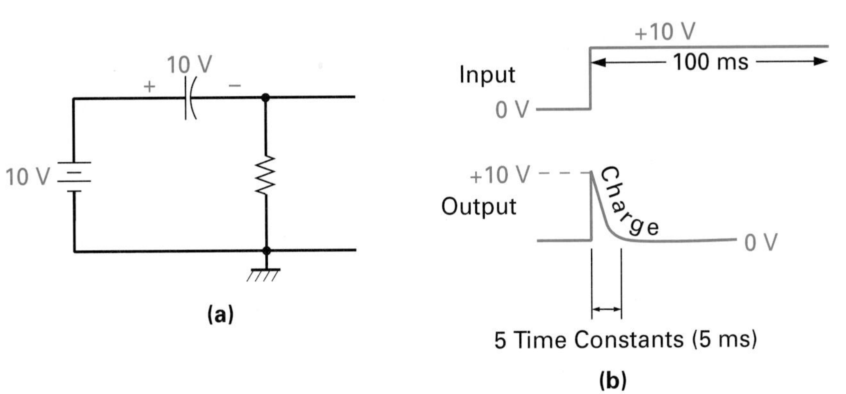

FIGURE 14-30 Differentiator's Response to Positive Step. (a) Equivalent Circuit. (b) Waveforms.

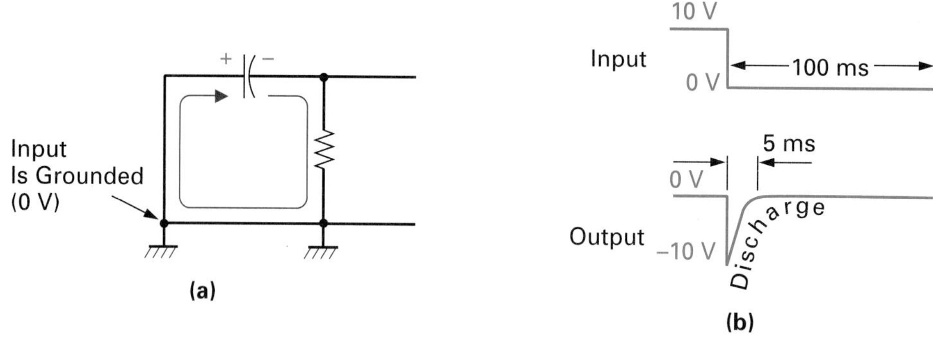

FIGURE 14-31 Differentiator's Response to a Negative Step. (a) Equivalent Circuit. (b) Waveforms.

When the square wave swings positive, the equivalent circuit is that shown in Figure 14-30(a). When the 10 V is initially applied (positive step of the square wave), all the voltage is across the resistor, and therefore at the output, as the capacitor cannot charge instantly. As the capacitor begins to charge, more of the voltage is developed across the capacitor and less across the resistor. The voltage across the capacitor exponentially increases and reaches 10 V in five time constants (5 × 1 ms = 5 ms), while the voltage across the resistor, and therefore the output, exponentially falls from its initial 10 V to 0 V in five time constants as shown in Figure 14-30(b), at which time all the voltage is across the capacitor and no voltage will be across the resistor.

When the positive half-cycle of the square wave ends and the input falls to zero, the circuit is equivalent to that shown in Figure 14-31(a). The negative

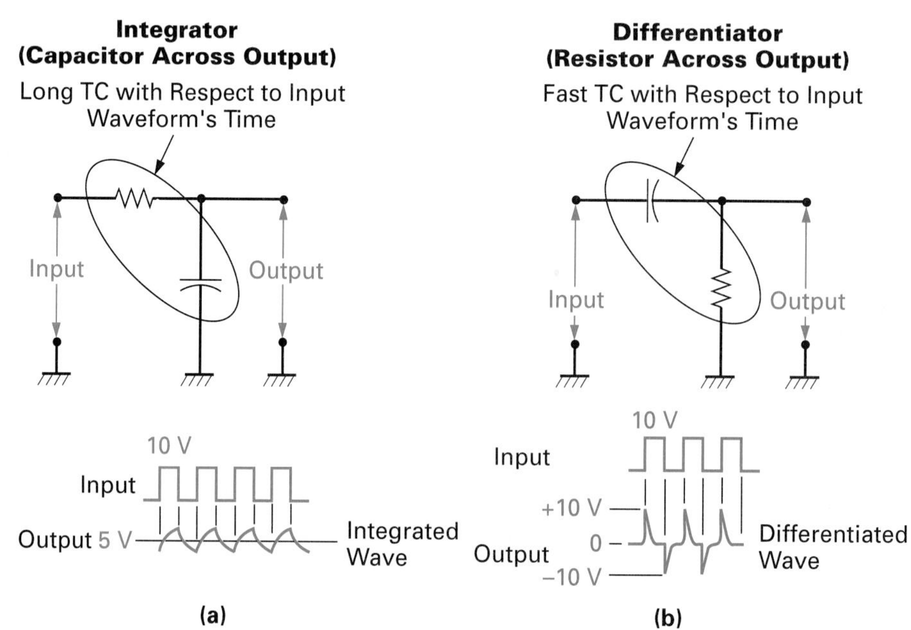

FIGURE 14-32 Summary of Integrator and Differentiator.

plate of the capacitor is now applied directly to the output. Since the capacitor cannot instantly discharge, the output drops suddenly down to –10 V as shown in Figure 14-31(b). This is the voltage across the resistor, and therefore the output. The capacitor is now in series with the resistor and therefore has a path through the resistor to discharge, which it does in five time constants to 0 V.

Figure 14-32(a) and (b) illustrate the integrator and differentiator circuits and waveforms, which are used extensively in many applications and equipment, such as computers, robots, lasers, and communications.

SELF-TEST REVIEW QUESTIONS FOR SECTION 14-5

1. Give three circuit applications for a capacitor.
2. With an RC low-pass filter, the ____________ is connected across the output. Whereas with an RC high-pass filter, the ____________ is connected across the output.
3. An integrator has a ____________ time constant compared to the period of the input square wave.
4. A ____________ circuit produces positive and negative spikes at the output when a square wave is applied at the input.

SUMMARY

Capacitive Circuits, Testing, and Applications (Figure 14-33)

1. Resistance (R), by definition, is the opposition to current flow with the dissipation of energy and is measured in ohms. Capacitors oppose current flow like a resistor; however, a resistor dissipates energy, whereas a capacitor stores energy (when it charges) and then gives back its energy into the circuit (when it discharges). Because of this difference, a new term had to be used to describe the opposition offered by a capacitor. Capacitive reactance (X_C), by definition, is the opposition to current without the dissipation of energy and is also measured in ohms.
2. If capacitive reactance is basically opposition, it is inversely proportional to the amount of current flow. If a large current is within a circuit, the opposition or reactance must be low ($I\uparrow$, $X_C\downarrow$). Conversely, a small circuit current will be the result of a large opposition or reactance ($I\downarrow$, $X_C\uparrow$).
3. When a dc source is connected across a capacitor, current will flow only for a short period of time ($5RC$ seconds) to charge the capacitor. After this time, there is no further current flow. Consequently, the capacitive reactance or opposition offered by a capacitor to dc is infinite (maximum).
4. Alternating current is continuously reversing in polarity, resulting in the capacitor continuously charging and discharging. This means that charge and discharge currents are always flowing around the circuit, and if we have a certain value of current, we must also have a certain value of reactance or opposition.

FIGURE 14-33 Capacitive Circuits, Testing, and Applications.

$$X_C = \frac{1}{2\pi f C}$$

X_C = capacitive reactance, in ohms
2π = constant
f = frequency, in hertz
C = capacitance, in farads

Series *RC* Circuits:

$$V_S = \sqrt{V_R^2 + V_C^2}$$

$$V_C = \sqrt{V_S^2 - V_R^2}$$
$$V_R = \sqrt{V_S^2 - V_C^2}$$

V_S = source voltage
V_R = voltage drop across resistor
V_C = voltage drop across capacitor

$$Z = \sqrt{R^2 + X_C^2}$$

$$R = \sqrt{Z^2 - X_C^2}$$
$$X_C = \sqrt{Z^2 - R^2}$$

Z = impedance, in ohms
R = resistance, in ohms
X_C = capacitive reactance, in ohms

$$\theta = \text{invtan}\ \frac{X_C}{R}$$

θ = phase angle or shift, in degrees

$$\theta = \text{invtan}\ \frac{V_C}{V_R}$$

$$P_A = \sqrt{P_R^2 + P_X^2}$$

P_A = apparent power, in volt-amperes (VA)
P_R = true power, in watts (W)
P_X = reactive power, in volt-amperes reactive (VAR)

$$\text{PF} = \frac{P_R}{P_A} = \frac{R}{Z} = \cos\theta$$

PF = power factor

Parallel *RC* Circuits:

$$I_R = \frac{V_S}{R}$$

$$I_C = \frac{V_S}{X_C}$$

I_R = current through resistor
I_C = current through capacitor

$$I_T = \sqrt{I_R^2 + I_C^2}$$

I_T = total circuit current

$$Z = \frac{V}{I_T}$$

$$Z = \frac{R \times X_C}{\sqrt{R^2 + X_C^2}}$$

$$\theta = \text{invtan}\ \frac{I_C}{I_R} \quad \text{or} \quad \theta = \text{invtan}\ \frac{R}{X_C}$$

FIGURE 14-33 (continued)

RC Series Circuit

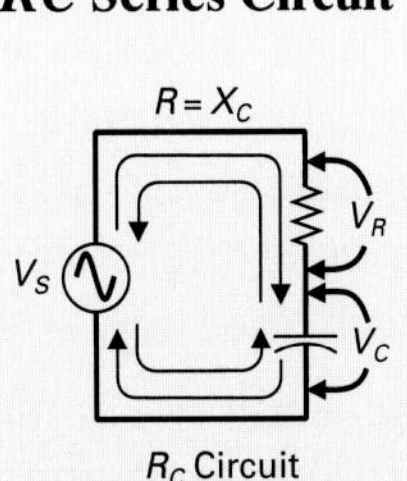

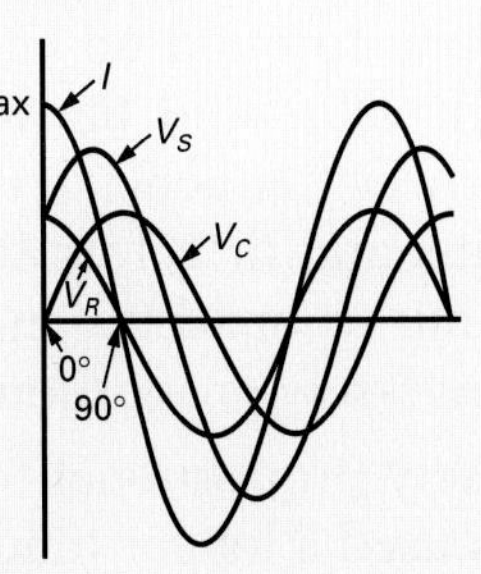

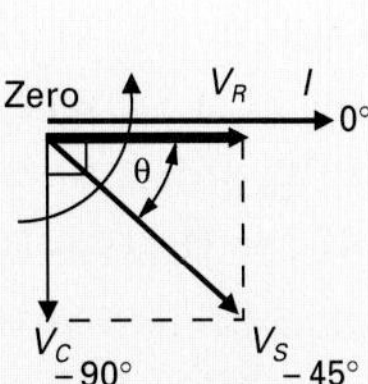

RC Parallel Circuit

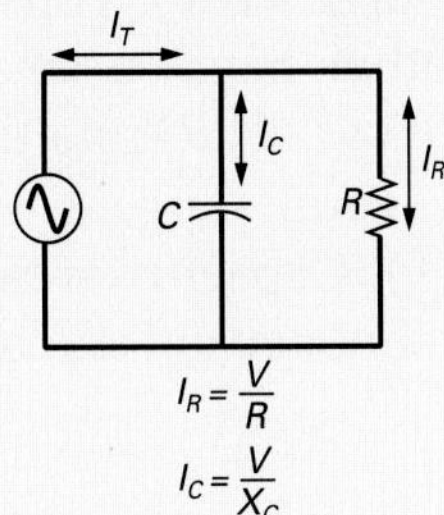

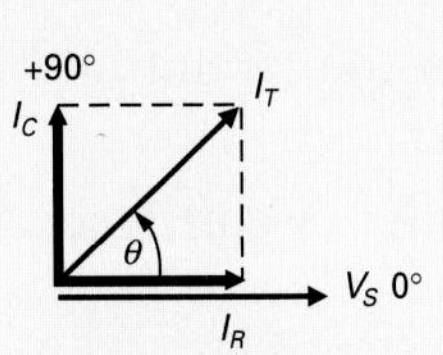

RC Filter Circuits

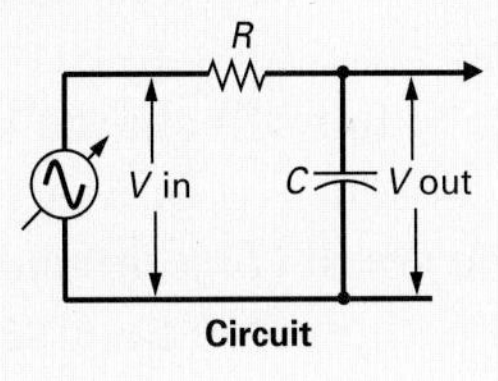

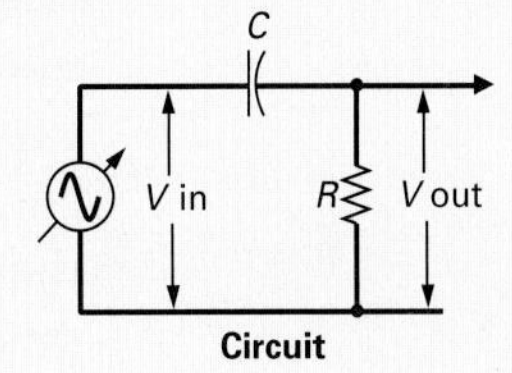

$$X_C = \frac{1}{2\pi fC}$$

$$\downarrow X_C \propto \frac{1}{f\uparrow}$$

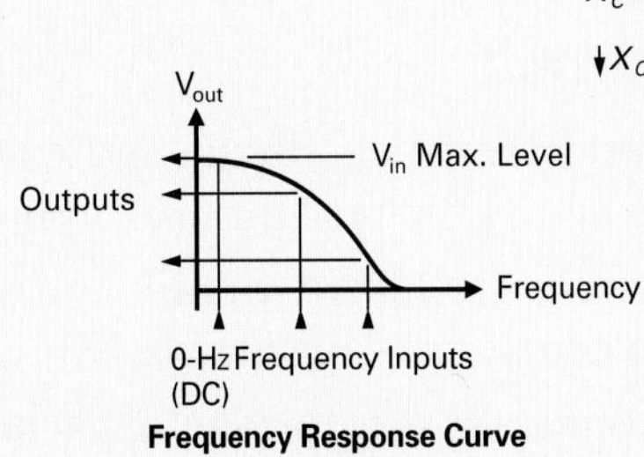

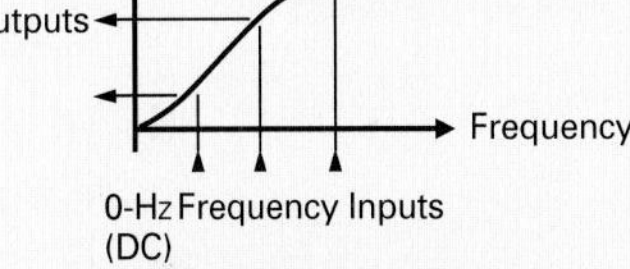

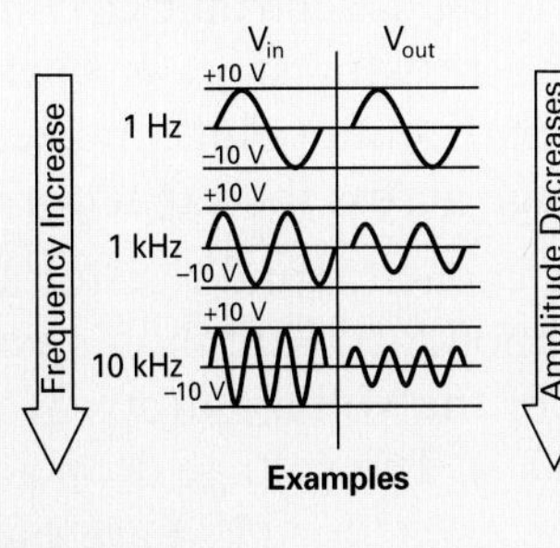

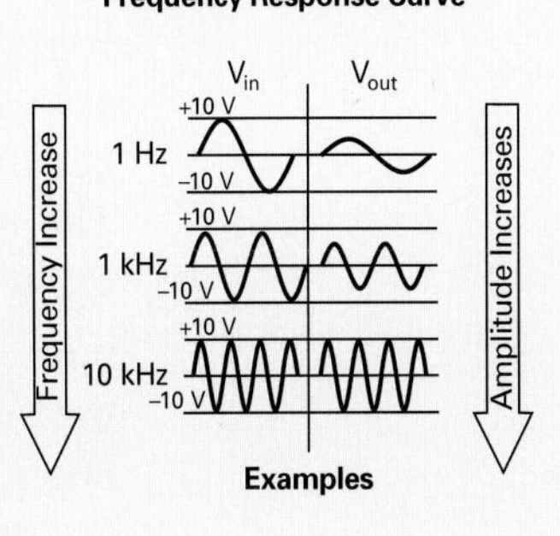

RC Integrator and Differentiator Circuit

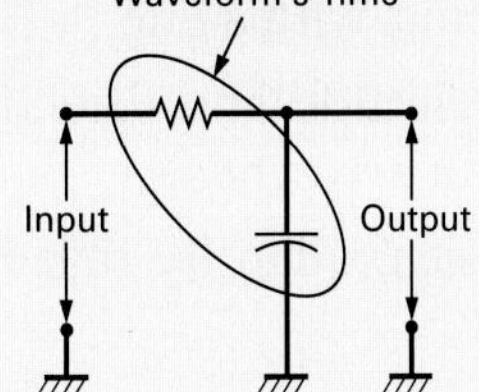

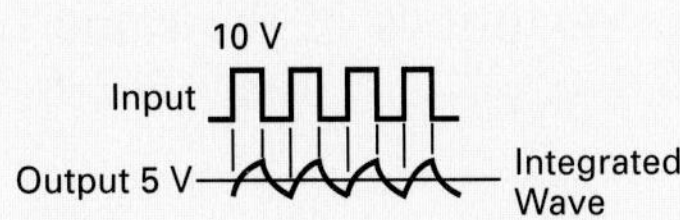

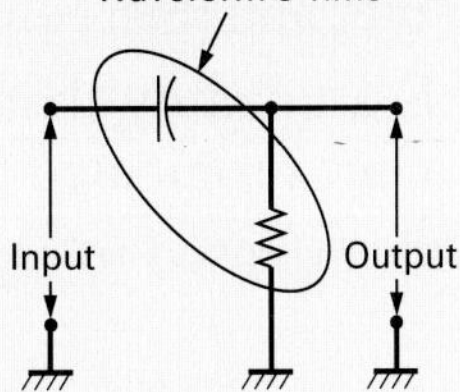

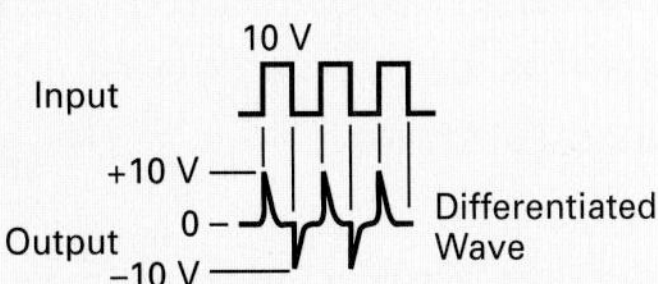

5. If the applied alternating current is at a high frequency it is switching polarity more rapidly than a lower frequency and there is very little time between the start of charge and discharge. As the charge and discharge currents are largest at the beginning of the charge and discharge of the capacitor, the reactance has very little time to build up and oppose the current, which is why the current is a high value and the capacitive reactance is small at higher frequencies. With lower frequencies, the applied alternating current is switching at a slower rate, and therefore the reactance, which is low at the beginning, has more time to build up and oppose the current. Capacitive reactance is therefore inversely proportional to frequency.
6. Capacitive reactance is also inversely proportional to the value of capacitance. If a larger capacitor value is used a longer time is required to charge the capacitor ($t\uparrow = C\uparrow R$), which means that current will be present for a longer period of time, so the overall current will be large ($I \uparrow$); consequently, the reactance must be small ($X_C\downarrow$). On the other hand, a small capacitance value will charge in a small amount of time ($t\downarrow = C\downarrow R$) and the current is present for only a short period of time. The overall current will therefore be small ($I\downarrow$), indicating a large reactance ($X_C\uparrow$). Capacitive reactance (X_C) is therefore inversely proportional to both frequency and capacitance.

Series *RC* Circuit

7. In a purely resistive circuit, the current flowing within the circuit and the voltage across the resistor are in phase with one another.
8. In a purely capacitive circuit, the current flowing in the circuit leads the voltage across the capacitor by 90°.
9. If we connect a resistor and capacitor in series, we have probably the most commonly used electronic circuit, which has many applications. The voltage across the resistor (V_R) is always in phase with the circuit current (I), because maximum points and zero crossover points occur at the same time. The voltage across the capacitor (V_C) lags the circuit current by 90°.
10. Since the capacitor and resistor are in series, the same current is supplied to both components; Kirchhoff's voltage law can be applied, which states that the sum of the voltage drops around a series circuit is equal to the voltage applied (V_S). The voltage drop across the resistor (V_R) and the voltage drop across the capacitor (V_C) are out of phase with one another, which means that their peaks occur at different times. The signal for the applied voltage (V_S) is therefore obtained by adding the values of V_C and V_R at each instant in time, plotting the results, and then connecting the points with a line.
11. A vector (or phasor) is a quantity that has both magnitude and direction and is represented by a line terminated at the end by an arrowhead.
12. Vectors are generally always used to represent a physical quantity that has two properties.
13. A vector diagram is an arrangement of vectors to illustrate the magnitude and phase relationships between two or more quantities of the same frequency within an ac circuit.

14. By vectorially adding the two voltages V_R and V_C we obtain a resultant V_S vector that has both magnitude and phase. The angle theta (θ) formed between circuit current (I) and source voltage (V_S) will always be less than 90°.
15. Since resistance is the opposition to current with the dissipation of heat, and reactance is the opposition to current without the dissipation of heat, a new term is needed to describe the total resistive and reactive opposition to current. Impedance (designated Z) is also measured in ohms and is the total circuit opposition to current flow. It is a combination of resistance (R) and reactance (X_C); however, in our capacitive and resistive circuit, a phase shift or difference exists, and just as V_C and V_R cannot be merely added together to obtain V_S, R and X_C cannot be simply summed to obtain Z.
16. By vectorially adding R and X_C, we have a resulting impedance (Z) vector.
17. In a purely resistive circuit, the total opposition (Z) is equal to the resistance of the resistor, so the phase shift (θ) is equal to 0°.
18. In a purely capacitive circuit, the total opposition (Z) is equal to the capacitive reactance (X_C) of the capacitor, so the phase shift (θ) is equal to –90°.
19. When a circuit contains both resistance and capacitive reactance, the total opposition or impedance has a phase shift that is between 0 and 90°.
20. Voltage and current are in phase with one another in a resistive circuit, and instantaneous power is calculated by multiplying voltage by current at every instant through 360° ($P = V \times I$). The sinusoidal power waveform is totally positive, because a positive voltage multiplied by a positive current gives a positive value of power, and a negative voltage multiplied by a negative current will also produce a positive value of power.
21. A resistor is said to generate a positive power waveform, which is twice the frequency of the voltage and current waveforms; two power cycles occur in the same time as one voltage and current cycle.

    The power waveform has been split in half, and the line that exists between the maximum point and zero point is the average value of power that is being dissipated by the resistor.
22. The current leads the voltage by 90° in a purely capacitive circuit, and the power wave is calculated by multiplying voltage by current, as before, at every instant through 360°. The resulting power curve is both positive and negative. During the positive alternation of the power curve, the capacitor is taking power as the capacitor charges. When the power alternation is negative, the capacitor is giving back the power it took as it discharges back into the circuit.

    The average power dissipated is once again the value that exists between the maximum positive and maximum negative points and causes the area above this line to equal the area below. This average power level calculates out to be zero, which means that no power is dissipated in a purely capacitive circuit.
23. When an ac voltage source is applied across a series-connected RC circuit, the current leads the voltage by some phase angle less than 90°, and the power waveform is once again determined by the product of voltage and current. The negative alternation of the power cycle indicates that the capac-

itor is discharging and giving back the power that it consumed during the charge.

24. The positive alternation of the power cycle is much larger than the negative alternation because it is the combination of both the capacitor taking power during charge and the resistor consuming and dissipating power in the form of heat. The average power being dissipated will be some positive value, due to the heat being generated by the resistor.
25. In a purely resistive circuit, all the energy supplied to the resistor from the source is dissipated in the form of heat. This form of power is referred to as resistive power (P_R) or true power.
26. In a purely capacitive circuit, all the energy supplied to the capacitor is stored from the source and then returned to the source, without energy loss. This form of power is referred to as reactive power (P_X) or imaginary power.
27. When a circuit contains both capacitance and resistance, some of the supply is stored and returned by the capacitor and some of the energy is dissipated and lost by the resistor.
28. Apparent power is the power that appears to be supplied to the load and includes both the true power dissipated by the resistance and the imaginary power delivered to the capacitor.
29. The power factor is a ratio of true power to apparent power and is a measure of the loss in a circuit.

Parallel *RC* Circuit

30. As with any parallel circuit, the voltage across all components in parallel is equal to the source voltage.
31. The current through the resistor and capacitor is simply calculated by applying Ohm's law.
32. Total current (I_T), however, is not as simply calculated. As expected, resistor current (I_R) is in phase with the applied voltage (V_S). Capacitor current will always lead the applied voltage by 90°, and as the applied voltage is being used as our reference at 0° on the vector diagram, the capacitor current will have to be drawn at +90° in order to lead the applied voltage by 90°, since vector diagrams rotate in a counterclockwise direction.
33. Total current is therefore the vector sum of both the resistor and capacitor currents.
34. The angle by which the total current (I_T) leads the source voltage (V_S) is call the phase angle.
35. The impedance of a parallel *RC* circuit is equal to the total voltage divided by the total current.
36. With respect to power, there is no difference between a series circuit and a parallel circuit.

Testing Capacitors

37. A faulty capacitor may have one of three basic problems:

a. A short, which is easy to detect and is caused by a contact from plate to plate.
b. An open, which is again quite easy to detect and is normally caused by one of the leads becoming disconnected from its respective plate.
c. A leaky dielectric or capacitor breakdown, which is quite difficult to detect, as it may only short at a certain voltage. This problem is usually caused by the deterioration of the dielectric, which starts displaying a much lower dielectric resistance than it was designed for. The capacitor with this type of problem is referred to as a *leaky capacitor*.

38. Capacitors of 0.5 μF and larger can be checked by using an analog ohmmeter or a digital multimeter with a bar graph display by using the following procedure:

a. Ensure that the capacitor is discharged by shorting the leads together.
b. Set the ohmmeter to the highest ohms range scale. Zero the ohmmeter.
c. Connect the meter to the capacitor, observing the correct polarity if an electrolytic is being tested, and observe the meter pointer. The capacitor will initially be discharged, and therefore maximum current will flow from the meter battery to the capacitor; maximum current means low resistance, which is why the meter's pointer deflects to the far right to indicate 0 Ω.
d. As the capacitor charges, it will cause current flow from the meter's battery to decrease, and consequently, the meter needle will move toward the left side of the scale.

39. The ohmmeter check tests the capacitor under a low-voltage condition. This may be adequate for some capacitor malfunctions; however, a problem that often occurs with capacitors is that they short or leak at a high voltage. A capacitance meter or analyzer can totally check all aspects of a capacitor in a range of values from approximately 1 pF to 20 F. The tests that are generally carried out include:

a. Capacitor value change
b. Capacitor leakage
c. Dielectric absorption
d. Equivalent series resistance

Applications of Capacitors

40. The capacitor can be used to combine ac and dc. The capacitor is large in value (electrolytic typically) and can be thought of as a very large bucket that will fill or charge to a dc voltage level. The ac voltage will charge and discharge the capacitor, which is similar to pouring in and pulling out (alternating) more and less water. The resulting waveform is a combination of ac and dc that varies above and below an average dc level. In this instance, the ac is said to be superimposed on a dc level.

41. Capacitors could be connected in series across a dc voltage source to form a voltage divider. Like a resistive voltage divider, voltage drop is proportional to current opposition, which with a capacitor is called *capacitive reactance* (X_C). The larger the capacitive reactance ($X_C\uparrow$), the larger the voltage drop ($V\uparrow$). Since capacitive reactance is inversely proportional to the capacitor value and the frequency of the ac source, a change in input frequency will cause a change in the capacitor's capacitive reactance, and this will change the voltage drop across the capacitor. Although a 90° phase shift is present between voltage and current in a purely capacitive circuit (I leads V by 90°), there is no phase difference between the input voltage and the output voltages.

42. A filter is a circuit that allows certain frequencies to pass but blocks other frequencies. In other words, it filters out the unwanted frequencies but passes the wanted or selected ones.

There are two basic *RC* filters:

a. The low-pass filter, which as its name implies, passes the low frequencies but heavily attenuates the higher frequencies.

b. The high-pass filter, which as its name implies, allows the high frequencies to pass but heavily attenuates the lower frequencies.

43. The frequency response curve for the low-pass filter illustrates that as the frequency of the input increases the voltage at the output will decrease. The frequency response curve for the high-pass filter illustrates that as the frequency of the input increases the voltage at the output increases.

44. The term integrator is derived from a mathematical function in calculus. An integrator circuit can be recognized by the series connection of R and C, but mainly from the fact that the output is taken across the capacitor. The voltage across the capacitor of an integrator, and therefore the output voltage, will gradually build up and eventually level off to an average value in about five time constants. If the period or time of the square wave input is decreased, or if the time constant is increased, the capacitor has less time to charge and discharge and reaches an average value of the input voltage, which for a square wave is half of its amplitude.

45. With the differentiator circuit, the output is taken across the resistor instead of the capacitor, and the time constant is always short with respect to the input square wave period. The differentiator output waveform is taken across the resistor and is the result of the capacitor's charge and discharge.

NEW TERMS

Apparent power
Average power
Capacitance meter
Capacitive circuit
Capacitive reactance
Dielectric absorption
Differentiator
Equivalent series resistance
Filter
Frequency response curve
High-pass filter
Imaginary power
Impedance
Integrator
Leaky capacitor
Low-pass filter
Parallel *RC* circuit
Positive power
Power factor
Pythagorean theorem
Reactance
Reactive power

Resistive circuit
Series *RC* circuit
Superimposed
Theta
True power
Vector
Vector diagram

REVIEW QUESTIONS

Multiple-Choice Questions

1. Capacitive reactance is inversely proportional to:
   a. Capacitance and resistance **c.** Capacitance and impedance
   b. Frequency and capacitance **d.** Both (a) and (c)
2. The impedance of an *RC* series circuit is equal to:
   a. The sum of R and X_C
   b. The square root of the sum of R^2 and $X_C{}^2$
   c. The square of the sum of R and X_C
   d. The sum of the square root of R and X_C
3. In a purely resistive circuit:
   a. The current flowing in the circuit leads the voltage across the capacitor by 90°.
   b. The circuit current and resistor voltage are in phase with one another.
   c. The current leads the voltage by 45°.
   d. The current leads the voltage by a phase angle between 0 and 90°.
4. In a purely capacitive circuit:
   a. The current flowing in the circuit leads the voltage across the capacitor by 90°.
   b. The circuit current and resistor voltage are in phase with one another.
   c. The current leads the voltage by 45°.
   d. The current leads the voltage by a phase angle between 0 and 90°.
5. In a series circuit containing both capacitance and resistance:
   a. The current flowing in the circuit leads the voltage across the capacitor by 90°.
   b. The circuit current and resistor voltage are in phase with one another.
   c. The current leads the voltage by 45°.
   d. Both (a) and (b).
6. In a series *RC* circuit, the source voltage is equal to:
   a. The sum of V_R and V_C **c.** The vectoral sum of V_R and V_C
   b. The difference between V_R and V_C **d.** The sum of V_R and V_C squared.
7. As the source frequency is increased, the capacitive reactance will:
   a. Increase **c.** Be unaffected
   b. Decrease **d.** Increase, depending on harmonic content
8. The phase angle of a series *RC* circuit indicates by what angle V_S ____________ V_R.
   a. Lags **b.** Leads **c.** Leads or lags **d.** None of the above

9. In a series RC circuit, the vector combination of R and X_C is the circuit's ____________.

   a. Phase angle b. Apparent power c. Source voltage d. Impedance

10. In a parallel RC circuit, the total current is equal to:

   a. The sum of I_R and I_C
   b. The difference between I_R and I_C
   c. The vectoral sum of I_R and I_C
   d. The sum of I_R and I_C squared

11. ____________ is the opposition offered by a capacitor to current flow without the dissipation of energy.

   a. Capacitive reactance
   b. Resistance
   c. Impedance
   d. Phase angle
   e. The power factor

12. ____________ is the total reactive and resistive circuit opposition to current flow.

   a. Capacitive reactance
   b. Resistance
   c. Impedance
   d. Phase angle
   e. The power factor

13. ____________ is the ratio of true (resistive) power to apparent power and is therefore a measure of the loss in a circuit.

   a. Capacitive reactance
   b. Resistance
   c. Impedance
   d. Phase angle
   e. The power factor

14. In a series RC circuit, the leading voltage will be measured across the:

   a. Resistor
   b. Capacitor
   c. Source
   d. Any of the choices are true.

15. In a series RC circuit, the lagging voltage will be measured across the:

   a. Resistor
   b. Capacitor
   c. Source
   d. Any of the choices are true.

Essay Questions

16. Give the formula and define the term *capacitive reactance.* (14-1)

17. In a series RC circuit, give the formulas for calculating: (14-2)

   a. V_S, when V_R and V_C are known
   b. Z, when R and X_C are known
   c. Z, when I and V are known
   d. θ, when X_C and R are known
   e. θ, when V_C and V_R are known
   f. Power factor, when R and Z are known
   g. Power factor, when P_R and P_A are known

18. In a parallel RC circuit, give the formulas for calculating: (14-3)

   a. I_R, when V and R are known
   b. I_C, when V and X_C are known
   c. I, when I_R and I_C are known
   d. Z, when V and I are known
   e. Z, when R and X_C are known

19. What is meant by *long* or *short time constant,* and do large or small values of RC produce a long or a short time constant? (Chapter 13)
20. Describe how the inverse relationship between frequency and capacitive reactance can be used for the application of filtering. (14-5-3)
21. Illustrate the voltage and current waveforms across a resistor and capacitor in series when a dc voltage is applied during charge and when the same resistor and capacitor are connected to discharge. (Chapter 13)
22. Describe and illustrate how the capacitor can be used in the following applications:
    - **a.** Combining ac and dc (14-5-1)
    - **b.** A voltage divider (14-5-2)
    - **c.** Filtering high and low frequencies (14-5-3)
    - **d.** Integrating a square wave (14-5-4)
    - **e.** Differentiating a square wave (14-5-5)
23. Describe the difference between reactance, resistance, and impedance. (14-2)
24. Sketch the phase relationships between: (14-2-3)
    - **a.** V_R and I in a purely resistive circuit
    - **b.** V_C and I in a purely capacitive circuit
25. What are positive power and negative power? (14-2-4)

Practice Problems

26. Calculate the capacitive reactance of the following capacitor circuits with the following parameters:
    - **a.** $f = 1$ kHz, $C = 2$ μF
    - **b.** $f = 100$ Hz, $C = 0.01$ μF
    - **c.** $f = 17.3$ MHz, $C = 47$ μF
27. In a series RC circuit, the voltage across the capacitor is 12 V and the voltage across the resistor is 6 V. Calculate the source voltage.
28. Calculate the impedance for the following series RC circuits:
    - **a.** 2.7 MΩ, 3.7 μF, 20 kHz
    - **b.** 350 Ω, 0.005 μF, 3 MHz
    - **c.** $R = 8.6$ kΩ, $X_C = 2.4$ Ω
    - **d.** $R = 4700$ Ω, $X_C = 2$ kΩ
29. In a parallel RC circuit with parameters of $V_S = 12$ V, $R = 4$ MΩ, and $X_C =$ 1.3 kΩ, calculate:

    a. I_R **b.** I_C **c.** I_T **d.** Z **e.** θ
30. Calculate the total reactance in:
    - **a.** A series circuit where $X_{C1} = 200$ Ω, $X_{C2} = 300$ Ω, $X_{C3} = 400$ Ω
    - **b.** A parallel circuit where $X_{C1} = 3.3$ kΩ, $X_{C2} = 2.7$ kΩ
31. Calculate the capacitance needed to produce 10 kΩ of reactance at 20 kHz.
32. At what frequency will a 4.7 μF capacitor have a reactance of 2000 Ω?

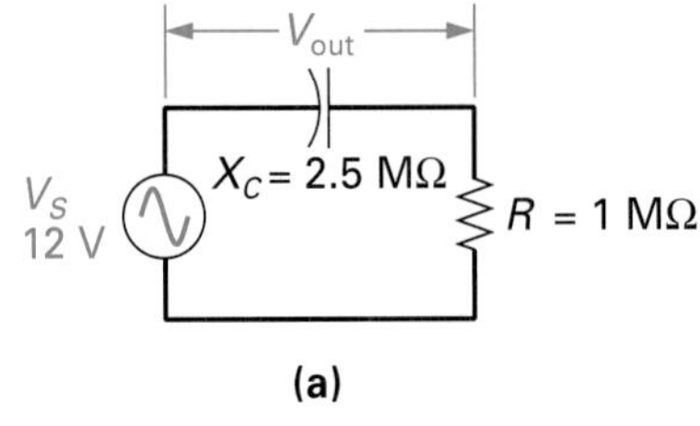

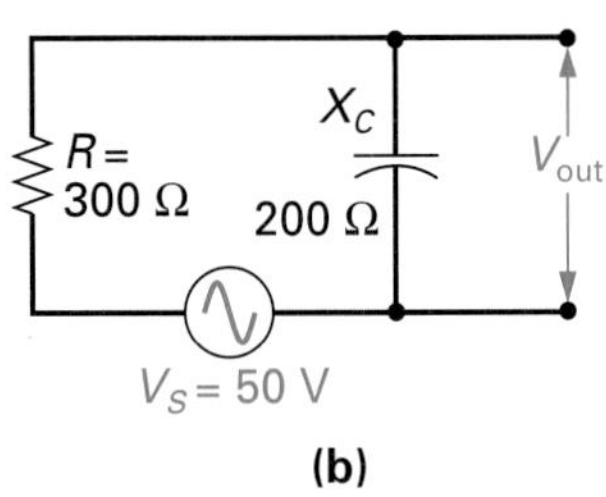

FIGURE 14-34

33. A series *RC* circuit contains a resistance of 40 Ω and a capacitive reactance of 33 Ω across a 24 V source.

    a. Sketch the schematic diagram. **b.** Calculate Z, I, V_R, V_C, I_R, I_C, and θ.

34. A parallel *RC* circuit contains a resistance of 10 kΩ and a capacitive reactance of 5 kΩ across a 100-V source.

    a. Sketch the schematic diagram. **b.** Calculate I_R, I_C, I_T, Z, V_R, V_C, and θ.

35. Calculate V_R and V_C for the circuits seen in Figure 14-34(a) and (b).

36. Calculate the impedance of the four circuits shown in Figure 14-35.

37. In Figure 14-36, the output voltage, since it is taken across the capacitor, will ____________ the voltage across the resistor by ____________ degrees.

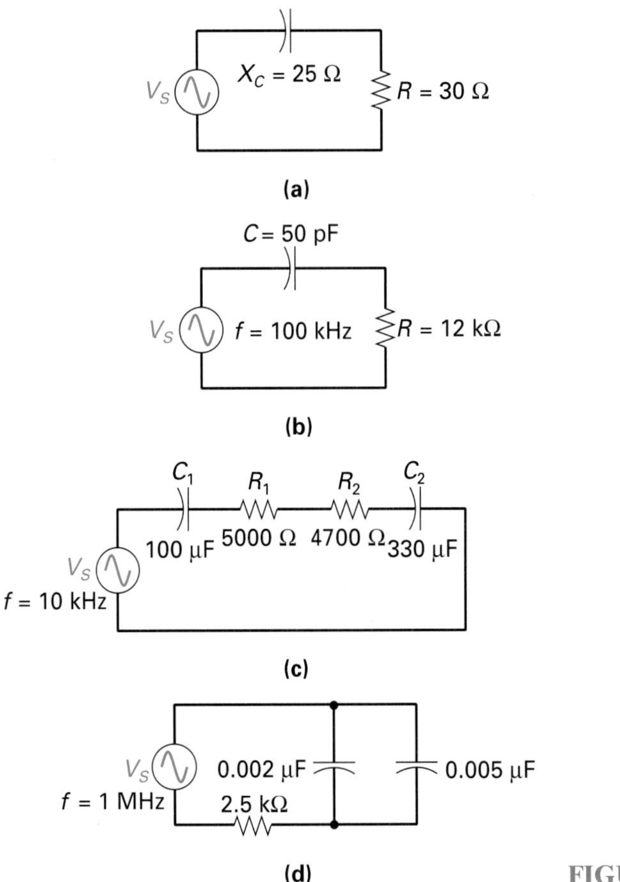

FIGURE 14-35

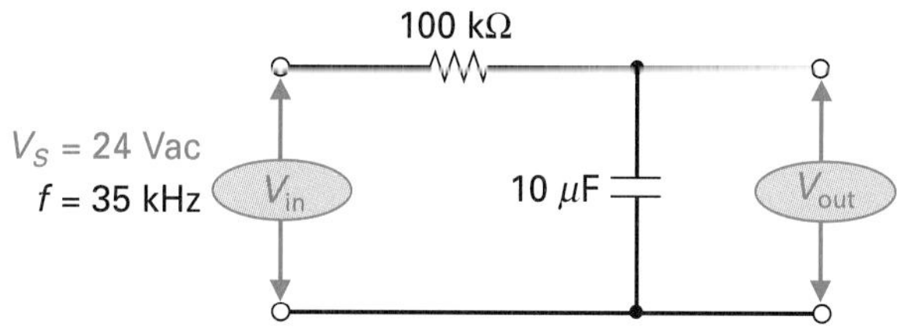

FIGURE 14-36

38. If the positions of the capacitor and resistor in Figure 14-36 are reversed, the output voltage, since it is now taken across the resistor, will ____________ the voltage across the capacitor by ____________ degrees.

39. Calculate the resistive power, reactive power, apparent power, and power factor for the circuit seen in Figure 14-36; $V_{in} = 24$ V and $f = 35$ kHz.

40. Refer to Figure 14-37 and calculate the following:

a. [Figure 14-37(a)] X_C, I, Z, I_R, θ, V_R, V_C

b. [Figure 14-37(b)] V_R, V_C, I_R, I_C, I_T, Z, θ

Troubleshooting Problems

41. Describe the three basic problems that normally occur with faulty capacitors.

42. Describe how to use the ohmmeter to check capacitors and also explain some of its limitations.

43. Describe the four basic tests performed by a capacitor meter or analyzer.

44. Which capacitor problem accounts for the largest percentage of defective capacitors? Explain exactly what this malfunction is.

45. If the needle of a meter goes to zero ohms and remains there, the capacitor is ____________ . If the capacitor is ____________, however, no charging will occur and the meter will indicate infinite ohms.

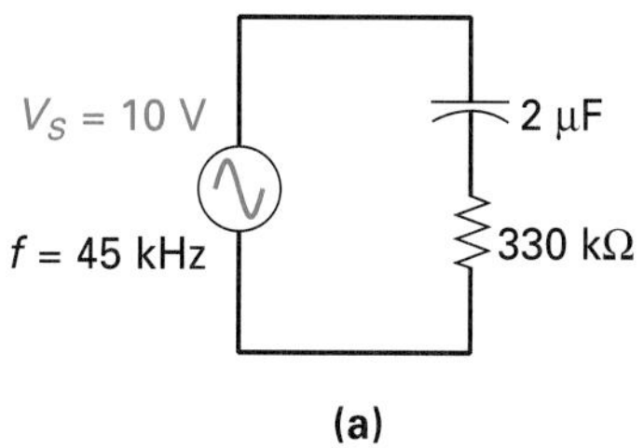

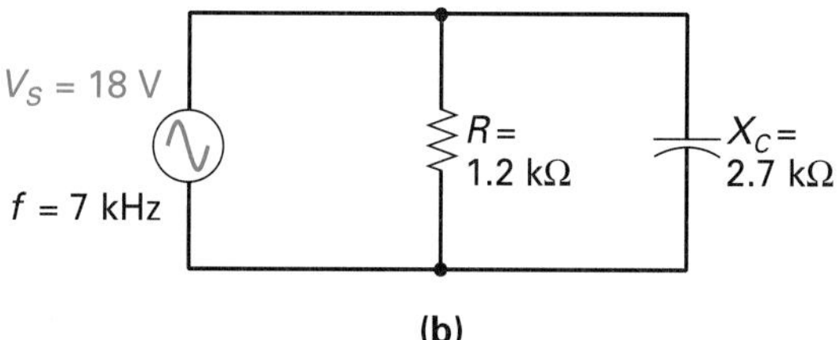

FIGURE 14-37

CHAPTER

15

OBJECTIVES

After completing this chapter, you will be able to:

1. Describe what is meant by the word *electromagnetism.*
2. Explain the atomic theory of electromagnetism.
3. Describe the left-hand rule of electromagnetism.
4. Describe how a magnetic field is generated by current flow through a:
   a. Conductor
   b. Coil
5. Describe the difference between dc and ac electromagnetism.
6. Explain the following magnetic terms:
   a. Magnetic flux
   b. Flux density
   c. Magnetizing force
   d. Magnetomotive force
   e. Reluctance
   f. Permeability (relative and absolute)
7. Explain the relationship between flux density and magnetizing force.
8. Describe the cycle known as the hysteresis loop.
9. Describe the following applications of electromagnetism:
   a. Magnetic-type circuit breaker
   b. Electric bell
   c. Relay and its applications
   d. Solenoid-type electromagnet and its application
10. Define *electromagnetic induction.*
11. State Faraday's and Lenz's laws relating to electromagnetic induction.
12. Describe the following applications of electromagnetic induction:
    a. AC generator
    b. Moving coil microphone

Electromagnetism and Electromagnetic Induction

OUTLINE

VIGNETTE

Problem-Solver

Charles Proteus Steinmetz (1865–1923) was an outstanding electrical genius who specialized in mathematics, electrical engineering, and chemistry. His three greatest electrical contributions were his investigation and discovery of the law of hysteresis, his investigations in lightning, which resulted in his theory on traveling waves, and his discovery that complex numbers could be used to solve ac circuit problems. Solving problems was in fact his specialty, and on one occasion he was commissioned to troubleshoot a failure on a large company system that no one else had been able to repair. After studying the symptoms and schematics for a short time, he chalked an X on one of the metal cabinets, saying that this was where they would find the problem, and left. He was right, and the problem was remedied to the relief of the company executives; however, they were not pleased when they received a bill for $1000. When they demanded that Steinmetz itemize the charges, he replied—$1 for making the mark and $999 for knowing where to make the mark.

The strong message this vignette conveys is that you will get $1 for physical labor and $999 for mental labor, and this is a good example as to why you should continue in your pursuit of education.

INTRODUCTION

It was during a classroom lecture in 1820 that Danish physicist Hans Christian Oersted accidentally stumbled on an interesting reaction. As he laid a compass down on a bench he noticed that the compass needle pointed to an adjacent conductor that was carrying a current, instead of pointing to the earth's north pole. It was this discovery that first proved that magnetism and electricity were very closely related to one another. This phenomenon is now called *electromagnetism* since it is now known that any conductor carrying an electro or electrical current will produce a magnetic field.

In 1831, the English physicist Michael Faraday explored further Oersted's discovery of electromagnetism and found that the process could be reversed. Faraday observed that if a conductor was passed through a magnetic field, a voltage would be induced in the conductor and cause a current to flow. This phenomenon is referred to as *electromagnetic induction.*

In this chapter we examine the terms and characteristics of electromagnetism (electricity to magnetism) and electromagnetic induction (magnetism to electricity).

15-1 ELECTROMAGNETISM

The electron plays a very important role in magnetism. However, the real key or link between electricity and a magnetic field is motion. Anytime a charged particle moves, a magnetic field is generated. Consequently, current flow, which is the movement of electrons, produces a magnetic field.

15-1-1 *Atomic Theory of Electromagnetism*

Every orbiting electron in motion around its nucleus generates a magnetic field. When electrons are forced to leave their parent atom by voltage and flow toward the positive polarity, they are all moving in the same direction and each electron's magnetic field will add to the next. The accumulation of all these electron fields will create the magnetic field around the conductor, as shown in Figure 15-1.

Electromagnetism Relates to the magnetic field generated around a conductor when current is passed through it.

A simple experiment can be performed to prove that this invisible magnetic field does in fact exist around a conductor, the setup of which is illustrated in Figure 15-2. With the switch open, as shown in Figure 15-2(a), no current will flow, and therefore no magnetic field is generated around the conductor and the iron filings on the cardboard will be disorganized.

With the switch closed, a current of 3 A ($I = V/R = 9\text{ V}/3\ \Omega = 3\text{ A}$) will flow through the circuit and a magnetic field will be set up around the conductor. The iron filings become organized in circles due to the influence of the magnetic field, as shown in Figure 15-2(b).

15-1-2 *Electromagnetism's Left-Hand Rule*

As we have seen so far, there is a relationship between the current in a conductor and the magnetic field around the conductor. Thc **left-hand rule** states that if an insulated conductor is held with your left hand so that your thumb is pointing in the direction of electron flow (to the positive potential), your fingers will be pointing in the direction of the magnetic force, as illustrated in Figure 15-3.

Left-Hand Rule If the fingers of the left hand are placed around the wire so that the thumb points in the direction of the electron flow, the fingers will be pointing in the direction of the magnetic field being produced by the conductor.

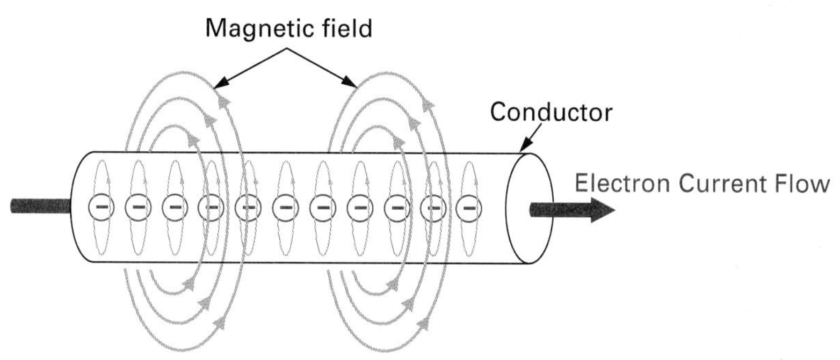

FIGURE 15-1 **Electron Magnetic Fields.**

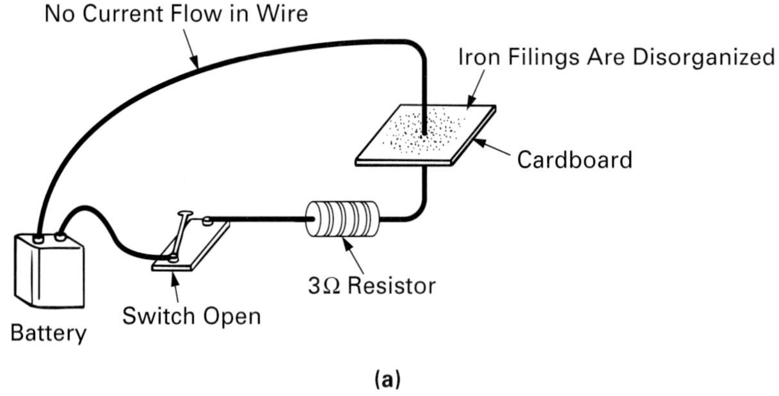

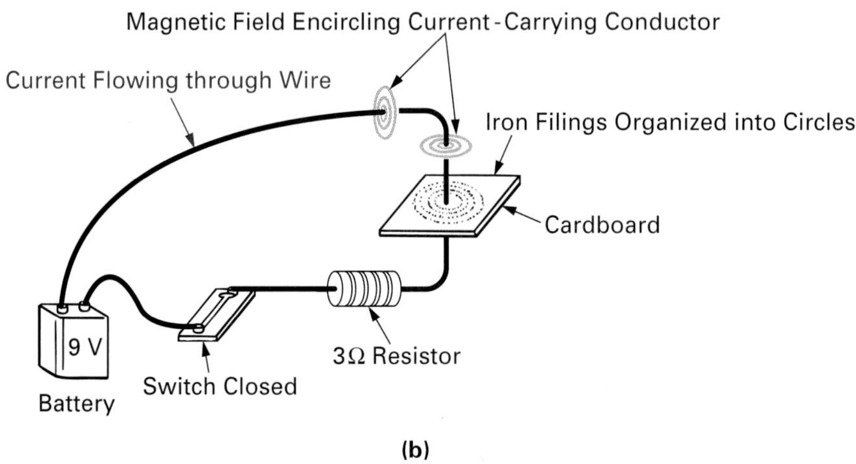

FIGURE 15-2 **Magnetic Field Experiment. (a) Switch Open, No Current Flow, No Magnetic Field. (b) Switch Closed, Current Flow, Magnetic Field.**

SELF-TEST REVIEW QUESTIONS FOR SECTIONS 15-1-1 AND 15-1-2

1. True or false: An electric field always encircles a current-carrying conductor.
2. What subatomic particle is said to have its own magnetic field?
3. For what purpose could the left-hand rule be used?
4. From which pole do the magnetic lines of force originate?

15-1-3 *The Electromagnet*

If our conductor is wound to form a coil, as illustrated in Figure 15-4(a), the conductor current will generate its own magnetic field, which will combine additively within the coil, and the net effect of all these conductor magnetic fields

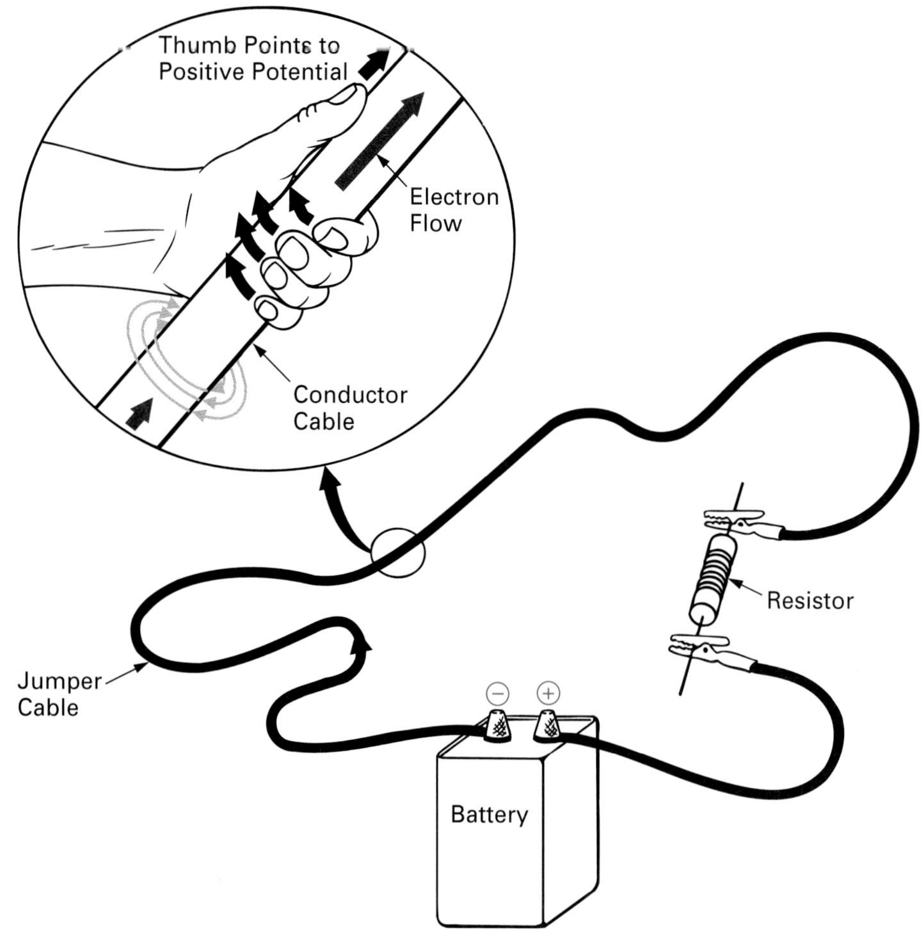

FIGURE 15-3 **Left-Hand Rule of Electromagnetism.**

can be used to generate a strong coil magnetic field. This coil of conductor which is used to generate a strong magnetic field is called an **electromagnet.**

If a greater number of loops are added, a greater or stronger magnetic field is generated, as seen in Figure 15-4(b). If the coils were wound in an opposing opposite direction, the opposing magnetic fields would cancel and so no magnetic field would be generated, as seen in Figure 15-4(c).

The left-hand rule can also be applied to electromagnets to determine which of the poles will be the north end of the electromagnet. This is illustrated in Figure 15-5, which shows that if you wrap the fingers of your left hand around the **coil** so that your fingers are pointing in the direction of current flow, your thumb will be pointing to the north end of the electromagnet.

Electromagnet
A magnet consisting of a coil wound on a soft iron or steel core. When current is passed through the coil a magnetic field is generated and the core is strongly magnetized to concentrate the magnetic field.

Coil
Number of turns of wire wound around a core to produce magnetic flux (an electromagnet) or to react to a changing magnetic flux (an inductor).

15-1-4 *DC versus AC Electromagnetism*

A magnetic field results whenever a current flows through any piece of conductor or wire, as shown in Figure 15-6(a). If a conductor is wound to form a spiral, as illustrated in Figure 15-6(b), the conductor, which is now referred to as a *coil,*

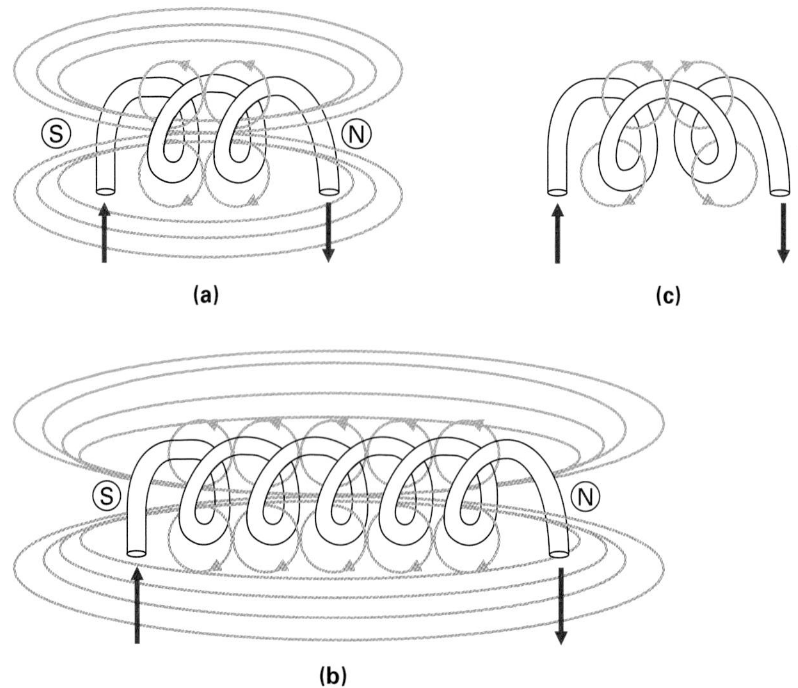

FIGURE 15-4 **Coils of an Electromagnet. (a) Small Number of Turns. (b) Large Number of Turns. (c) Opposing Turns.**

will, as a result of current flow, develop a magnetic field, which will sum or intensify within the coil. If many coils are wound in the same direction, an electromagnet is formed, as shown in Figure 15-6(c), which will produce a concentrated magnetic field whenever a current is passed through its coils.

Up until now, we have only been discussing current flow through a coil in one direction (dc). A dc voltage produces a fixed current in one direction and

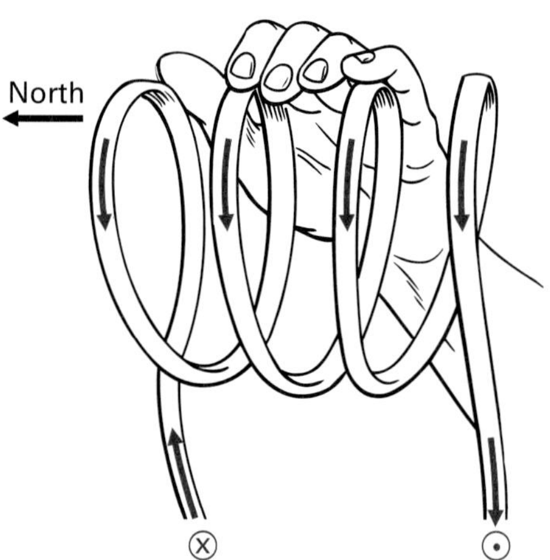

FIGURE 15-5 **Left-Hand Rule for Electromagnets.**

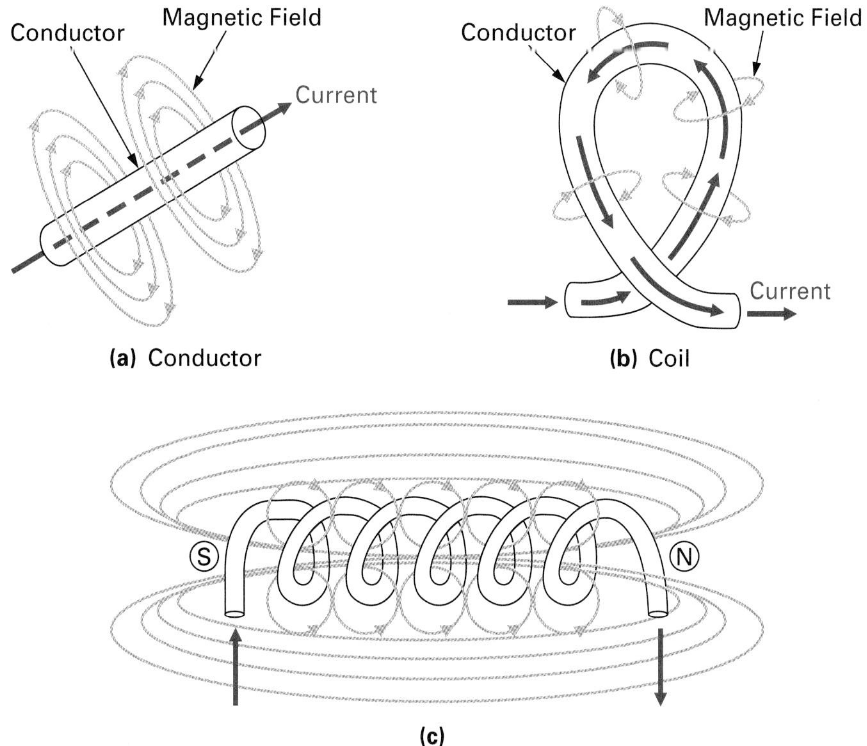

FIGURE 15-6 **Conductor and Coil Electromagnetism.**

therefore generates a magnetic field in a coil of fixed polarity, as shown in Figure 15-7(a), and as determined by the left-hand rule.

Alternating current (ac) is continually varying, and as the polarity of the magnetic field is dependent on the direction of current flow (left-hand rule), the magnetic field will also be alternating in polarity, as shown in Figure 15-7(b). Let us look at times 1 through 4 in Figure 15-7 in more detail.

Time 1: The alternating voltage has risen to a maximum positive level and causes current flow as seen in the circuit. This will cause a magnetic field with a south pole above and a north pole below.

Time 2: Between positions 1 and 2, the voltage, and therefore current, will decrease from a maximum positive value to zero. This will cause a corresponding collapse of the magnetic field from maximum (time 1) to zero (time 2).

Time 3: Voltage and consequently current increase from zero to maximum negative between positions 2 and 3. The increase in current flow causes a similar increase or buildup of magnetic flux, producing a north pole above and south pole below.

Time 4: From time 3 to time 4, the current within the circuit diminishes to zero, and the magnetic field once again collapses. The cycle then repeats.

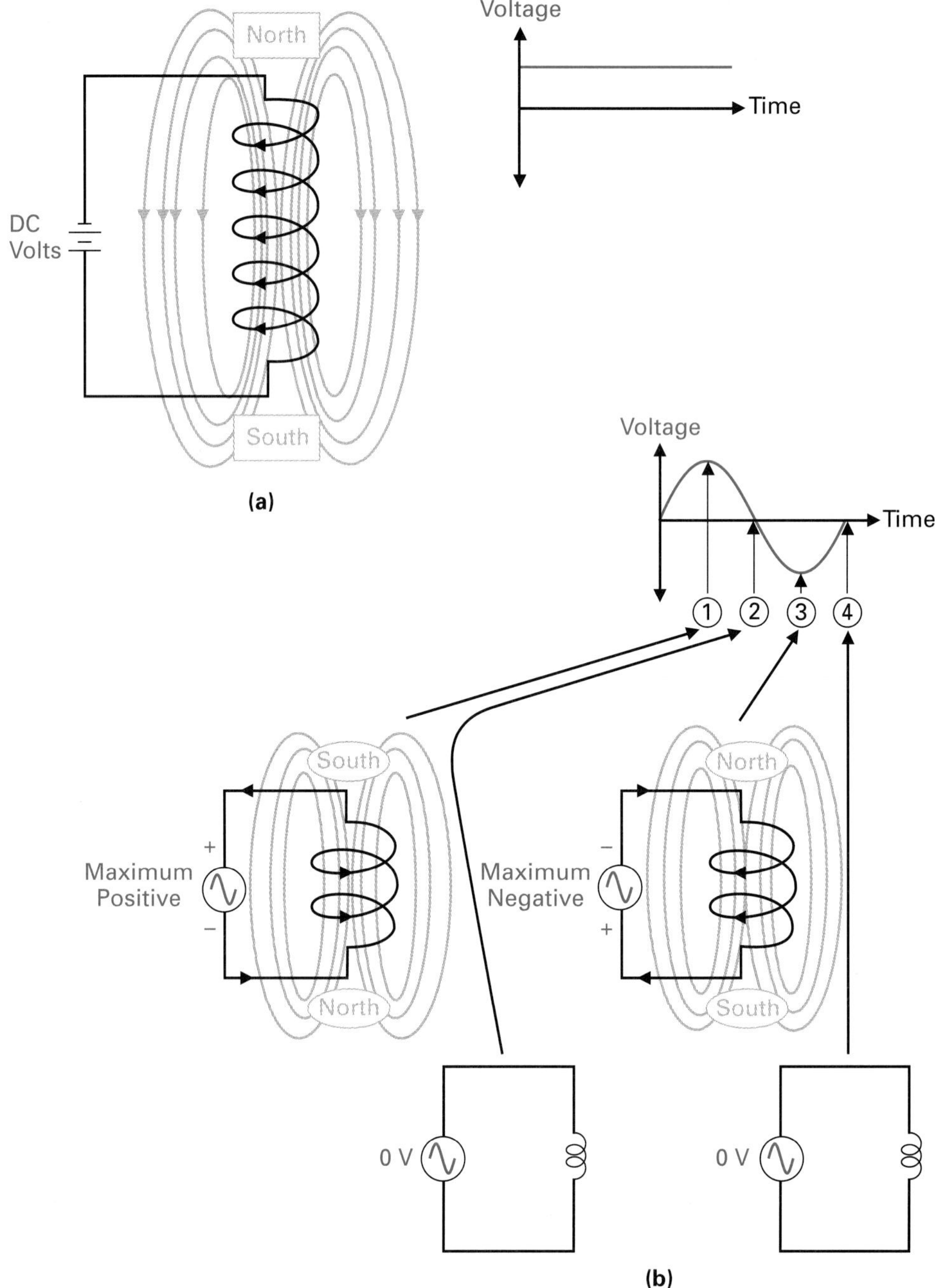

FIGURE 15-7 **Electromagnets. (a) DC Electromagnet. (b) AC Electromagnet.**

In summary:

1. Direct current (dc) produces a constant magnetic field of a fixed polarity, for example, north–south.
2. An alternating current (ac) produces an alternating magnetic field, which continuously switches polarity, for example, north–south, south–north, north–south, and so on.

SELF-TEST REVIEW QUESTIONS FOR SECTIONS 15-1-3 AND 15-1-4

1. What name could be used instead of *electromagnet?*
2. True or false: The greater the number of loops in an electromagnet, the greater or stronger the magnetic field.
3. What form of current flow produces a magnetic field that maintains a fixed magnetic polarity?
4. True or false: A magnetic field is produced only when ac is passed through a conductor.
5. Describe how the left-hand rule applies to electromagnetism.
6. What form of current flow produces a magnetic field that continuously switches in polarity?

15-1-5 Magnetic Terms

Magnetic Flux (ϕ) and Flux Density (B)

One magnetic line of flux or force is called a **maxwell,** in honor of James Maxwell's work in this field. However, this unit is too small and impractical and therefore the amount of **magnetic flux** (symbolized ϕ, phi) is measured in webers instead of maxwells. One *weber* is equal to 10^8 (100,000,000) magnetic lines of force or maxwells.

Maxwell
One magnetic line of force or flux is called a maxwell.

Magnetic Flux
The magnetic lines of force produced by a magnet.

magnetic flux (ϕ) = number of lines of force (or maxwells), in webers

If 10^8 lines of force exist in 1 square meter, and 10^8 lines of force exist in 1 square centimeter, which is the stronger magnetic field? As far as magnetic flux (ϕ) is concerned, 1 weber exists in both. So some other way is needed to specify how many lines of force exist in a given area, and this will tell us if it is a strong or weak magnetic field.

Flux density (B) is equal to the number of magnetic lines of flux (ϕ) per square meter, and it is given in the unit of tesla (T).

Flux Density
A measure of the strength of a wave.

$$\text{flux density } (B) = \frac{\text{magnetic flux } (\phi)}{\text{area } (A)}$$

where B = flux density, in teslas (T)
ϕ = Magnetic flux, in webers (Wb)
A = area, in square meters (m^2)

EXAMPLE:

If magnet *A* produces 10^8 (100,000,000) lines of flux (1 weber) in 1 square centimeter, and magnet *B* produces 10^8 lines of flux in 1 square meter, which magnet is producing the more concentrated or intense magnetic field?

Solution:

Since there are 10,000 square centimeters in 1 square meter, 1 square centimeter (1 cm^2) equals one ten-thousandth of a square meter (0.01 m^2).

$$\text{Magnet } A\text{: flux density} = \frac{1 \text{ weber}}{0.0001 \ (\text{m}^2)}$$
$$= 10{,}000 \text{ tesla}$$

$$\text{Magnet } B\text{: flux density} = \frac{1 \text{ weber}}{1 \ (\text{m}^2)}$$
$$= 1 \text{ tesla (T)}$$

Flux density determined that magnet *A* produced the more concentrated magnetic force.

Magnetomotive Force (MMF)

Magnetomotive Force
Force that produces a magnetic field.

The magnetic flux produced by an electromagnet is produced by current flowing through a coil of wire. As discussed previously, electromotive force (emf) is the pressure or voltage that forces electrons to move. **Magnetomotive force** (mmf) is the magnetic pressure that produces the magnetic field. The formula for mmf is

$$\text{mmf} = I \times N \text{ (ampere-turns)}$$

where mmf = magnetomotive force, in ampere-turns (At)
I = current, in amperes
N = number of turns in the coil

The formula basically says that if you increase the current through the coil or increase the number of turns in the coil, you will increase the magnetic pressure (magnetomotive force), and by increasing magnetic pressure, you will increase the magnetic field produced, as shown in Figure 15-8.

EXAMPLE:

What is the magnetomotive force produced when 3 A of current flows through 5 turns in a coil?

Solution:

$$\text{mmf} = I \text{ (current)} \times N \text{ (number of turns)}$$
$$= 3\text{A} \times 5 \text{ turns}$$
$$= 15 \text{ At (ampere-turns)}$$

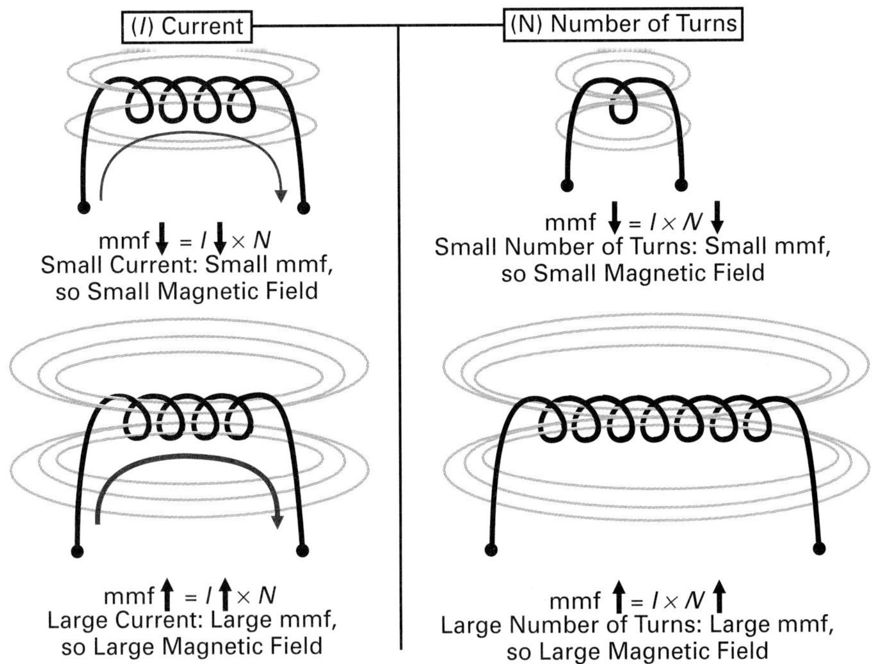

FIGURE 15-8 **Magnetomotive Force.**

Magnetizing Force (*H*)

Example 15-2 showed that magnetomotive force (mmf) is equal to the product of current and the number of turns; however, if a 20-turn coil or solenoid is stretched out to twice its length, the magnetic field or force will be half as strong, because there would be less of a reinforcing effect between the coils due to the greater distance between the coils. The length of the coil is consequently a factor that also determines the field intensity, and this new term, **magnetizing force** (*H*), is the reference we commonly use to describe the magnetic field intensity.

Magnetizing Force
Also called magnetic field strength, it is the magnetomotive force per unit length at any given point in a magnetic circuit.

$$\text{magnetizing force } (H) = \frac{I \times N}{l}$$

or

$$H = \frac{\text{mmf}}{l}$$

where H = magnetizing force in ampere-turns/meter (At/m)

I = current, in amperes (A)

N = number of coil turns

l = length of coil, in meters (m)

Reluctance ($\Re$)

Reluctance Resistance to the flow of magnetic lines of force.

Reluctance, in reference to magnetic energy, is equivalent to resistance in electrical energy. Reluctance is the opposition or resistance to the establishment of a magnetic field in an electromagnet. The formula for calculating reluctance is the magnetic energy equivalent of the electrical Ohm's law as shown:

Magnetic

$$\text{reluctance } (\Re) = \frac{\text{mmf (magnetomotive force)}}{\phi \text{ (magnetic flux)}}$$

(measured in ampere-turns/weber)

Electrical

$$\text{resistance } (R) = \frac{V \text{ (electromotive force)}}{I \text{ (current)}}$$

(measured in ohms)

Permeability (μ)

Permeability Measure of how much better a material is as a path for magnetic lines of force with respect to air, which has a permeability of 1 (symbolized by the Greek lowercase letter mu, μ).

As magnetic reluctance is equivalent to electrical resistance, magnetic permeability is equivalent to electrical conductance. **Permeability** is a measure of how easily a material will allow a magnetic field to be set up within it. Permeability is symbolized by the Greek lowercase letter mu (μ) and is measured in henrys per meter (H/m). A high permeability figure ($\mu\uparrow$) indicates that a magnetic field can easily be established within a material and therefore that this material's reluctance must be low ($\Re\downarrow$). On the other hand, a low permeability figure ($\mu\downarrow$) indicates that there will be a large reluctance ($\Re\uparrow$) to establishing a magnetic field. This is mathematically stated as:

Magnetic

$$\text{permeability } (\mu) = \frac{1}{\Re \text{ (reluctance)}}$$

(Permeability is inversely proportional to reluctance and is measured in henrys per meter.)

Electric

$$\text{conductance } (G) = \frac{1}{R \text{ (resistance)}}$$

(Conductance is inversely proportional to resistance and is measured in siemens.)

TABLE 15-1 Permeabilities of Various Materials

MATERIAL	RELATIVE PERMEABILITY (μ_r)	PERMEABILITY (μ)
Air or vacuum	1	1.26×10^{-6}
Nickel	50	6.28×10^{-5}
Cobalt	60	7.56×10^{-5}
Cast iron	90	1.1×10^{-4}
Machine steel	450	5.65×10^{-4}
Transformer iron	5,500	6.9×10^{-3}
Silicon iron	7,000	8.8×10^{-3}
Permaloy	100,000	0.126
Supermalloy	1,000,000	1.26

The permeability figures in henrys/meter for different materials are listed in Table 15-1. **Relative** (with respect to) is a word that means that a comparison has to be made. Relative permeability (μ_r) is the measure of how well another given material will conduct magnetic lines of force with respect to, or relative to, our reference material, air, which has a relative permeability value of 1. Referring to the relative permeability column in Table 15-1, you can see that magnetic lines of flux will pass through nickel 50 times easier than through air. The relative permeability (μ_r) of air, which is equal to 1, should not be confused with the absolute permeability (μ_0) of free space or air, which is equal to $4\pi \times 10^{-7}$ or 1.26×10^{-6}.

Relative
Not independent; compared or with respect to some other value of a measured quantity.

$$\text{permeability } (\mu) = \text{relative permeability } (\mu_r) \times \text{absolute permeability of air } (\mu_0)$$

Whether permeability (μ), relative permeability (μ_r), or absolute permeability (μ_0) is used as a standard makes little difference. A high permeability figure indicates a low reluctance ($\mu\uparrow$, $\Re\downarrow$), and vice versa.

EXAMPLE:

If the magnetic flux produced by a material is equal to 335 μWb and the mmf equals 15 At:

a. What is the reluctance of the material?

b. What is the material's permeability?

■ ***Solution:***

CALCULATOR SEQUENCE

Step	Keypad Entry	Display Response
1.	[1] [5]	15
2.	[÷]	
3.	[3] [3] [5] [E] [6] [+/−]	335E-6
4.	[=]	44776.1
5.	[1/x]	22.3E-6

a. $\text{Reluctance } (\Re) = \dfrac{\text{mmf (magnetomotive force)}}{\phi \text{ (magnetic flux)}}$

$= \dfrac{15 \text{ At (ampere-turns)}}{335\ \mu\text{Wb}}$

$= 44.8 \times 10^3 \text{ At/Wb}$

b. $\text{Permeability } (\mu) = \dfrac{1}{44.8 \times 10^3 \text{ (reluctance)}}$

$= 22.3 \times 10^{-6} \text{ henrys/meter (H/m)}$

Summary

The magnetic field strength of an electromagnet can be increased by increasing the number of turns in its coil, increasing the current through the electromagnet, or decreasing the length of the coil ($H = I \times N/l$). The field strength can be increased further by placing an iron core within the electromagnet, as illustrated in Figure 15-9(a). An iron core has less reluctance or opposition to the magnetic lines of force than air, so the flux density (B) is increased. Another way of saying this is that the permeability or conductance of magnetic lines of force within iron is greater than that of air, and if the permeability of iron is large, its reluctance must be small. The symbol for an iron core electromagnet is seen in Figure 15-9(b).

15-1-6 Flux Density (B) *versus Magnetizing Force* (H)

***B–H* Curve**
Curve plotted on a graph to show successive states during magnetization of a ferromagnetic material.

The ***B–H* curve** in Figure 15-10 illustrates the relationship between the two most important magnetic properties: flux density (B) and magnetizing force (H). Figure 15-10(a) illustrates the $B–H$ curve, while Figure 15-10(b) illustrates the positive rising portion of the ac current that is being applied to the iron core electromagnetic circuit in Figure 15-10(c).

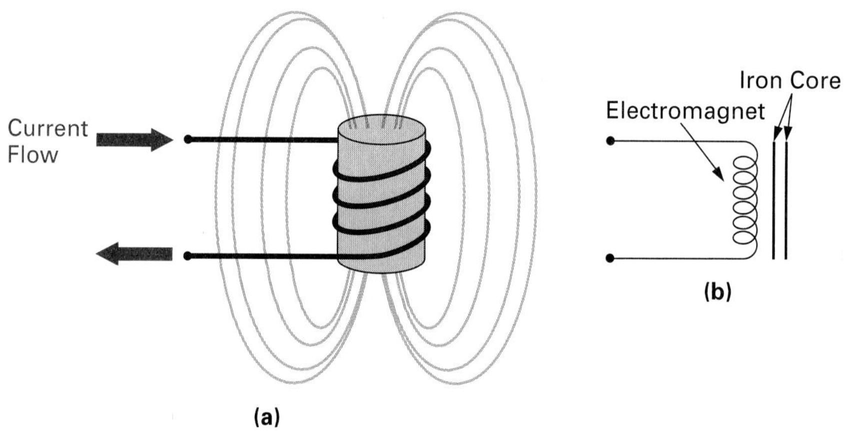

FIGURE 15-9 **Iron-Core Electromagnet. (a) Operation. (b) Symbol.**

The magnetizing force is actually equal to $H = I$ (current) $\times N$ (number of turns)/l (length of coil); but since the number of turns and length of coil are fixed for the coil being used, the magnetizing force (H) is proportional to the current (I) applied, which is shown in Figure 15-10(b). The positive rise of the current from zero to maximum positive is applied through the electromagnet and will produce a corresponding bloom or buildup in magnetic flux at a rate indicated by the B–H curve shape in Figure 15-10(a).

It is important to note that as the magnetizing force (current) is increased, there are three distinct stages in the change of flux density or magnetic flux out.

Stage 1: Up to this point, the increase in flux density is slow, as a large amount of force is required to begin alignment of the molecule magnets.

Stage 2: Increase in flux density is now rapid and almost linear as the molecule magnets are aligning easily.

Stage 3: In this state they cannot be magnetized any further because all the molecule magnets are fully aligned and no more flux density can be easily obtained. This is called the **saturation point,** and it is the state of magnetism beyond which an electromagnet is incapable of further magnetic strength, that is, the point beyond which the B–H curve is a straight, horizontal line, indicating no change.

Saturation Point The point beyond which an increase in one of two quantities produces no increase in the other.

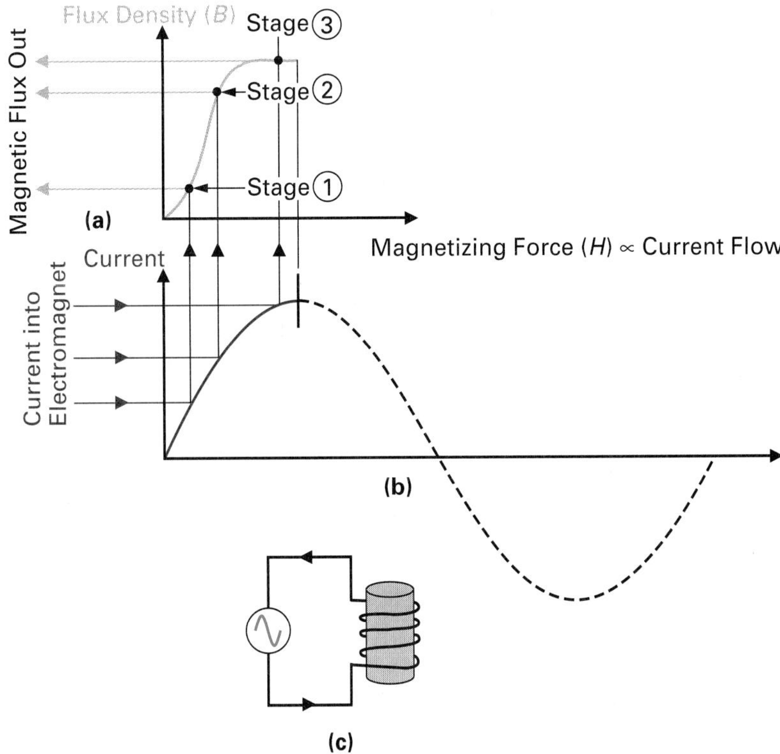

FIGURE 15-10 (a) Flux Density versus Magnetizing Force Curve (*B–H* Curve). (b) Current Applied to Circuit. (c) Circuit.

Saturation can easily be described by a simple analogy of a sponge. A dry sponge can only soak up a certain amount of water. As it continues to absorb water, a point will be reached where it will have soaked up the maximum amount of water possible. At this point, the sponge is said to be saturated with water, and no matter how much extra water you supply, it cannot hold any more.

The electromagnet is saturated at stage 3 and cannot produce any more magnetic flux, even though more magnetizing force is supplied as the sine wave continues on to its maximum positive level.

Hysteresis
A lag between cause and effect. With magnetism it is the amount that the magnetization of a material lags the magnetizing force due to molecular friction.

Looking at these three stages and the *B–H* curve that is produced, you can see that, in fact, the magnetization (setting up of the magnetic field, *B*) lags the magnetizing force (*H*) because of molecular friction. This lag or time difference between magnetizing force (*H*) and flux density (*B*) is known as **hysteresis.**

Figure 15-11(a) illustrates what is known as a *hysteresis loop,* which is formed when you plot magnetizing force (*H*) against flux density (*B*) through a complete cycle of alternating current, as seen in Figure 15-11(b). Initially, when the electric circuit switch is open, the iron core is unmagnetized; therefore, both

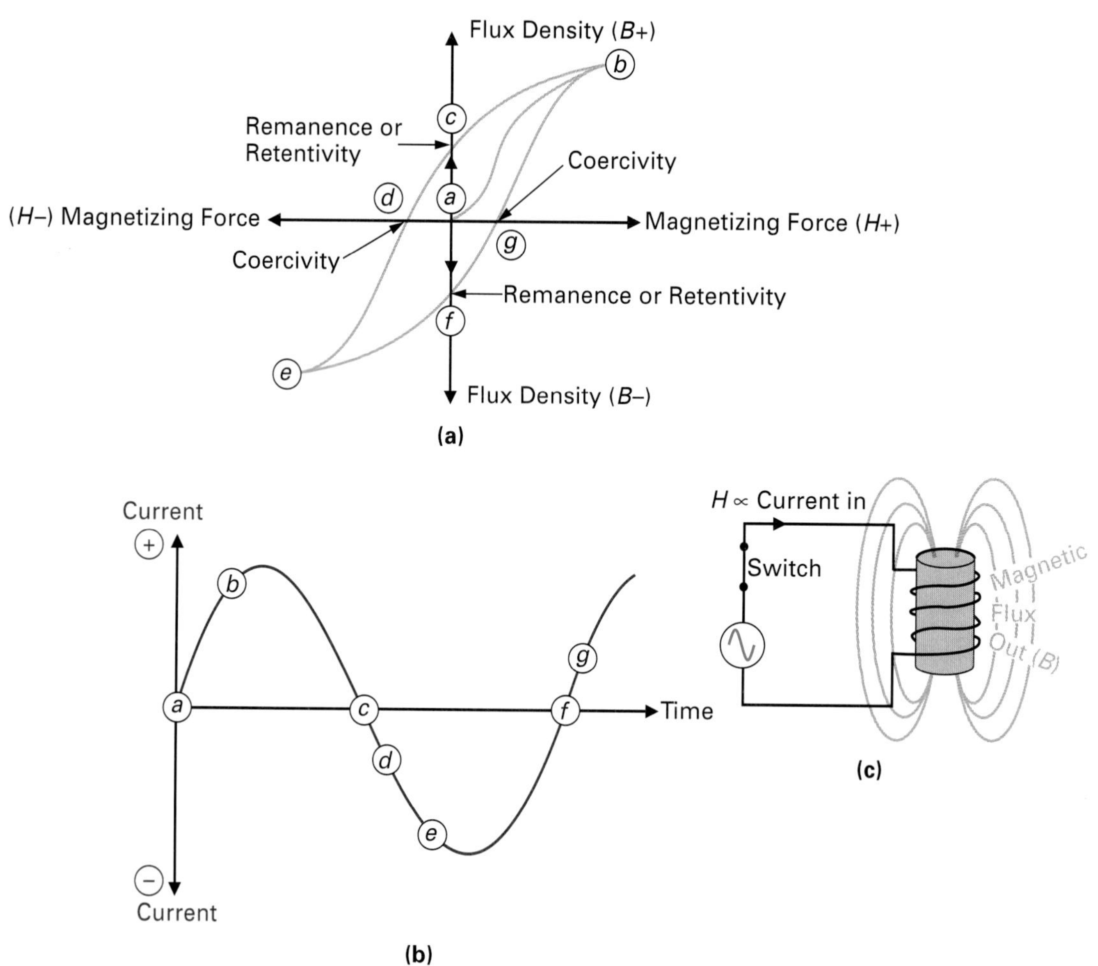

FIGURE 15-11 ***B–H* Hysteresis Loop.**

H and *B* are zero at point *a*. When the switch is closed, as seen in Figure 15-11(c), the current [Figure 15-11(b)] is increased and flux density [Figure 15-11(a)] increases until saturation point *b* is reached. This part of the waveform (from point *a* to *b*) is exactly the same as the *B–H* curve, discussed previously. The current continues on beyond saturation point *b;* however, the flux density cannot increase beyond saturation.

At point *c*, the magnetizing force (current) is zero and *B* (flux density) falls to a value *c* that is the positive magnetic flux remaining after the removal of the magnetizing force (*H*). This particular value of flux density is termed **remanence or retentivity.**

Remanence or Retentivity
Amount a material remains magnetized after the magnetizing force has been removed.

The current or magnetizing force now reverses, and the amount of current in the reverse direction that causes flux density to be brought down from *c* to zero (point *d*) after the core has been saturated is termed the **coercive force.** The current, and therefore magnetizing force, continues on toward a maximum negative until saturation in the opposite magnetic polarity occurs at point *e*.

Coercive Force
Magnetizing force needed to reduce the residual magnetism within a material to zero.

At point *f*, the magnetizing force (current) is zero and *B* falls to remanence, which is the negative magnetic flux remaining after the removal of the magnetizing force *H*. The value of current between *f* and *g* is the coercive force needed in the reverse direction to bring flux density down to zero (point *g*).

In Figure 15-12 it is easier to see how the current variation corresponds to the magnetizing force variation.

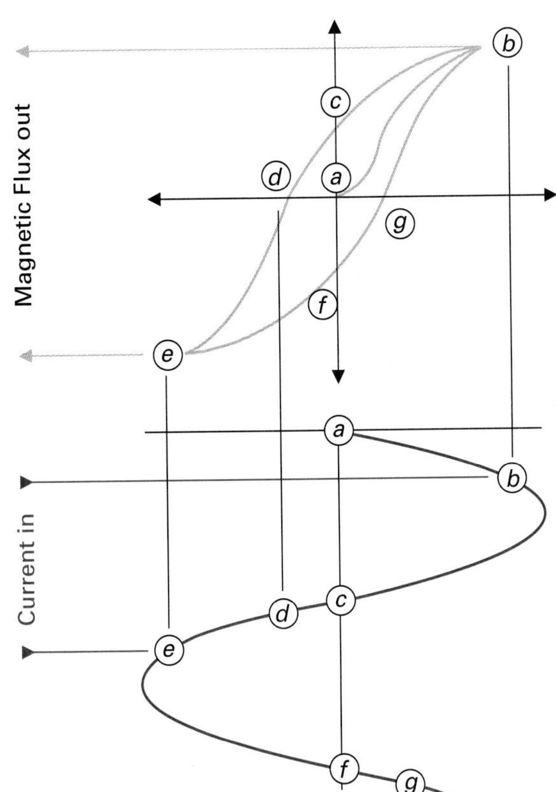

FIGURE 15-12 AC Current In, Magnetic Flux Out.

SELF-TEST REVIEW QUESTIONS FOR SECTIONS 15-1-5 AND 15-1-6

Define the following:

1. Magnetic flux
2. Flux density
3. Magnetomotive force
4. Magnetizing force
5. Reluctance
6. Permeability
7. True or false: The hysteresis loop is formed when you plot H against B through a complete cycle of ac.
8. What is magnetic saturation?

15-1-7 *Applications of Electromagnetism*

Magnetic-Type Circuit Breaker

The magnetic-type circuit breaker shown in Figure 15-13 was first discussed in Chapter 6 and is one application of an electromagnet. This and all other circuit breakers are used for current protection. If the rated current value or below is passing through the circuit breaker, the electromagnet will not generate a strong enough magnetic field to pull the iron arm to the left and release the catch so the horizontal arm will open the contacts. However, if the current rating of the circuit breaker is exceeded, the increase in current will cause a corresponding increase in magnetic flux, attracting the iron arm, opening the contacts, and protecting the equipment from the dangerous level of current.

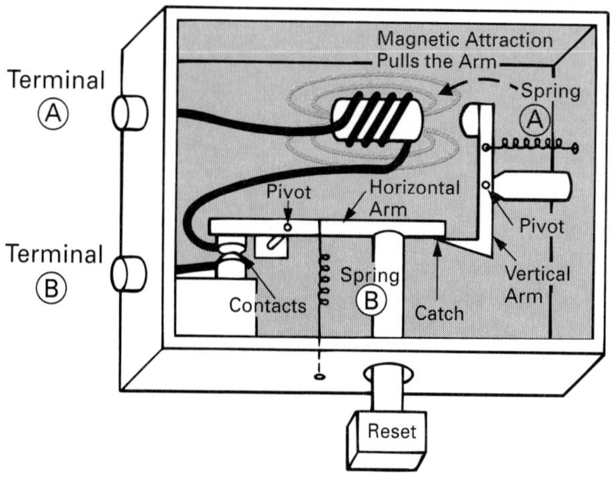

FIGURE 15-13 **Magnetic-Type Circuit Breaker.**

Electric Bell

The electric bell in Figure 15-14 utilizes a soft-iron core electromagnet, a striker, and a gong. When the bell push is pressed, a complete electrical circuit is made from the positive terminal to the electromagnet, the contact, the striker, and then back to the negative terminal. The electromagnet energizes (produces a magnetic field) and attracts the soft-iron striker, which strikes the gong and at the same time breaks the circuit, deenergizing the electromagnet. The striker is pulled back by the spring, and the circuit is reestablished, resulting in a continuous ringing of the bell.

Relay

A **relay** is an electromechanical device that either makes (closes) or breaks (opens) a circuit by moving contacts together or apart. Figure 15-15 illustrates the relay; Figure 15-15(a) shows the normally open (NO) relay and Figure 15-15(b) shows the normally closed (NC) relay.

Relay Electromechanical device that opens or closes contacts when a current is passed through a coil.

Operation. In both cases, the relay consists basically of an electromagnet connected to lines x and y, a movable iron arm known as the armature, and some contacts. When current passes through x to y, the electromagnet generates a magnetic field, or is said to be **energized,** which attracts the armature toward the electromagnet. When this occurs, it closes or makes the normally open relay contacts and opens or breaks the normally closed relay contacts.

Energized Being electrically connected to a voltage source so that the device is activated.

If the electromagnet is deenergized by stopping the current through the coil, the spring will pull back the armature to open the NO relay contacts or close the NC relay contacts between A and B.

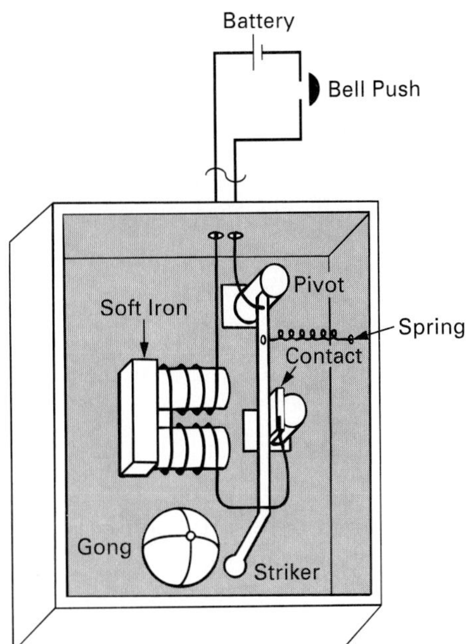

FIGURE 15-14 **Electric Bell.**

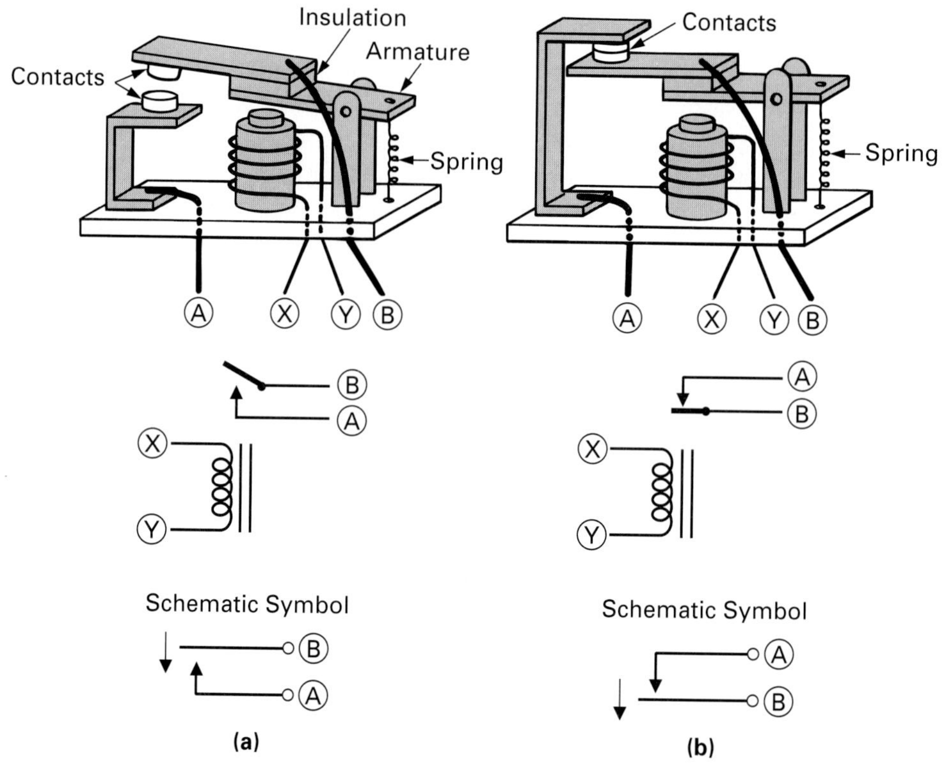

FIGURE 15-15 relays (a) Single-Pole, Single-Throw (SPST), Normally Open (NO) Relay (Contacts Are Open until Activated). (b) Single-Pole, Single-Throw (SPST), Normally Closed (NC) Relay (Contacts Are Closed until Activated).

To summarize, then, you can see that the "normal" condition for the contacts between *A* and *B* is when the electromagnet is deenergized. In the deenergized condition, the normally open relay contacts are open and the normally closed relay contacts are closed.

The two relays discussed so far are actually single-pole, single-throw relays, as they have one movable contact (single pole) and one stationary contact that the pole can be thrown to, as shown in Figure 15-15. There are actually four basic configurations for relays and all are illustrated in Table 15-2. Variations on these basic four can come in all shapes and sizes, with one relay controlling sometimes several sets of contacts. Figure 15-16 illustrates several different styles and packages of relays that are available.

Applications of a relay. The relay is generally used in two basic applications:

1. To enable one master switch to operate several remote or difficultly placed contact switches, as illustrated in Figure 15-17. When the master switch is closed, the relay is energized, closing all its contacts and turning on all the lights. The advantage of this is twofold in that, first, the master switch can turn on three lights at one time, which saves time for the operator, and second, only one set of wires need be taken from the master switch to the lights, rather than three sets for all three lights.

TABLE 15-2 Relay Types

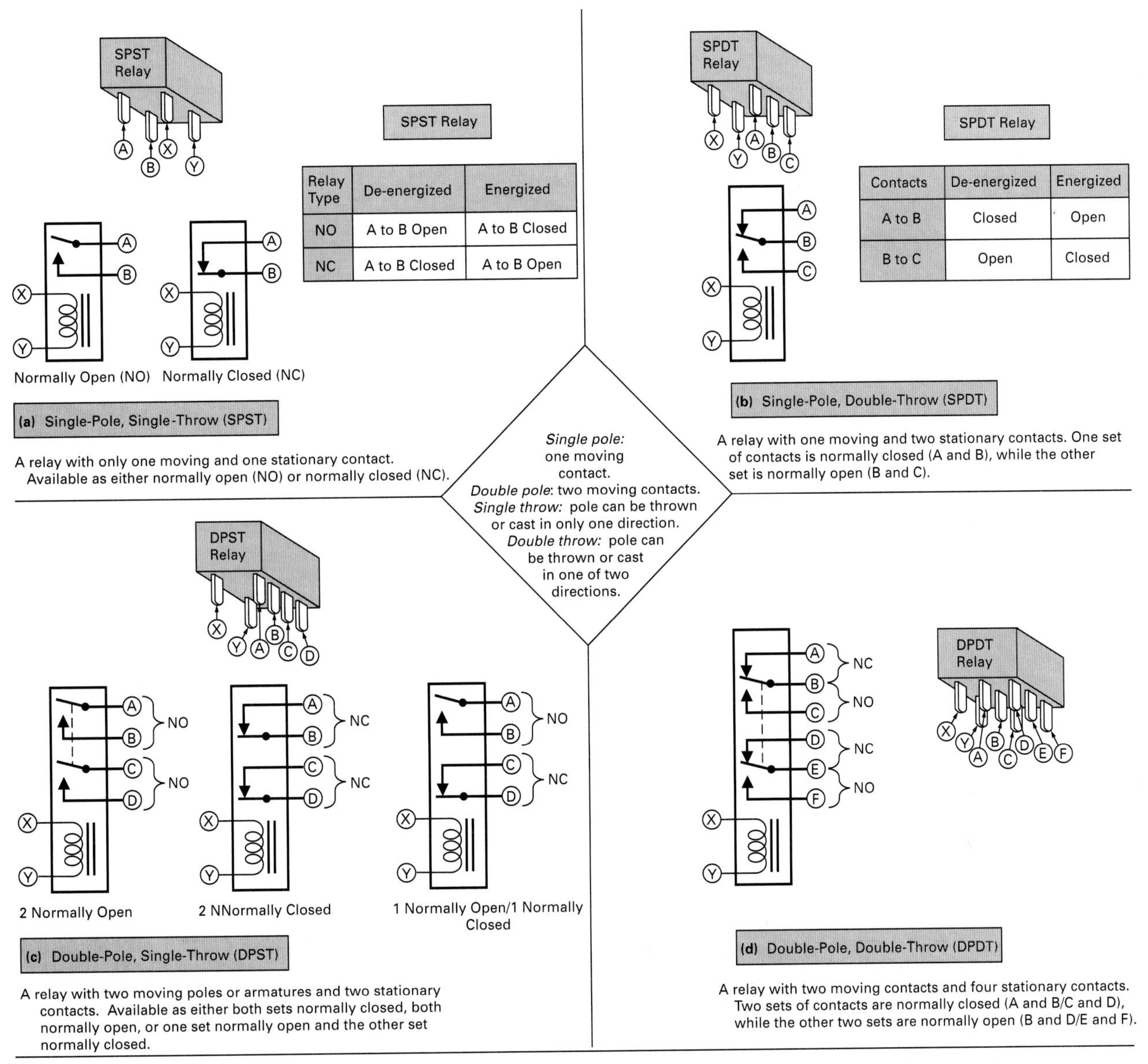

FIGURE 15-16 **Relay Styles and Packages.**

2. The second basic application of the relay is to enable a switch in a low-voltage circuit to operate relay contacts in a high-voltage circuit, as shown in Figure 15-18. The operator activates the switch in the low-voltage circuit, which will energize the relay, closing its contacts and connecting the high voltage to the motor. Figure 15-19 illustrates the circuit arrangement that would have to be used without a relay.

In a high-voltage circuit, a large current will be flowing, so a large or heavy-gauge wire will be needed to handle this current. With the relay circuit in Figure 15-18, large-gauge wire is needed only for the short distance between the relay and the motor, while in the circuit without the relay (Figure 15-19), heavy-gauge wire, which is more expensive than smaller-gauge wire, is needed for the long distance between the operator's switch and the motor.

An automobile starter circuit is illustrated in Figure 15-20. In this application of a relay, we will see how a relay can be used to supply the large dc current needed to activate a starter motor of an automobile.

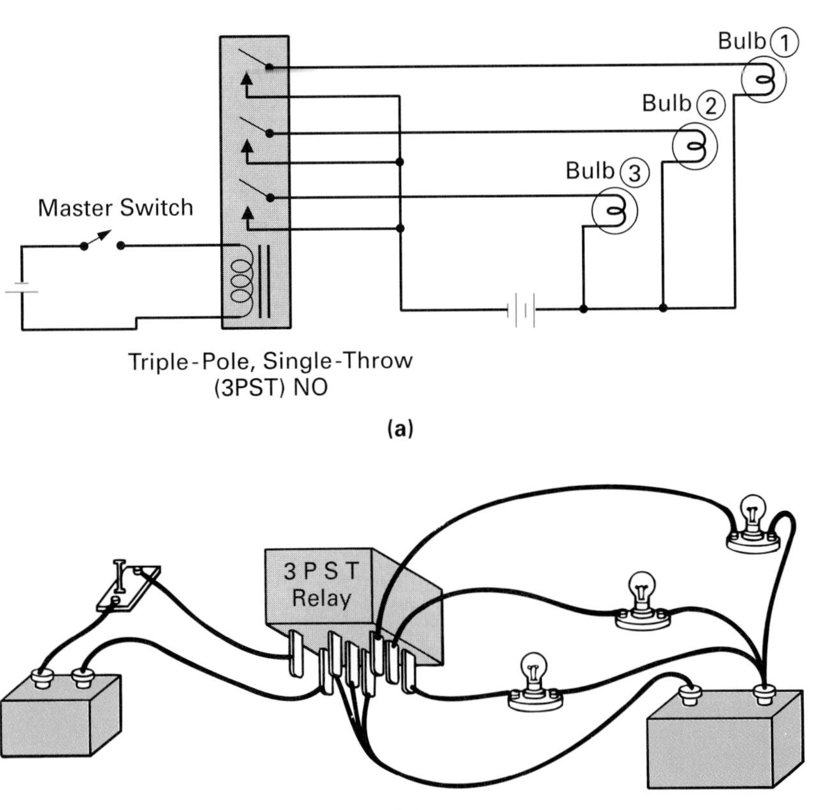

FIGURE 15-17 **One Master Switch Operating Several Remotes. (a) Schematic. (b) Pictorial.**

When the ignition switch is engaged in the passenger compartment by the driver, current flows through a light-gauge wire from the negative side of the battery, through the relay's electromagnet, through the ignition switch, and back to the positive side of the battery. This current flow through the electromagnet of the relay energizes the relay and closes the relay's contacts. Closing the relay's contacts makes a path for the current to flow through the heavy-gauge cable from the negative side of the battery, through the relay contacts and starter motor, and back to the positive side of the battery. The starter motor's output shaft spins the engine, causing it to start.

This application is a perfect example of how a relay can be used to close contacts in a heavy-current (heavy-gauge cable) circuit, while the driver has only to close contacts in a small-current (light-gauge cable) circuit.

If the relay were omitted, the driver's ignition switch would have to be used to connect the 12 V and large current to the starter motor. This would require that:

1. Heavy-gauge expensive cable be connected in a longer path between starter motor and passenger compartment.
2. The ignition switch would need to be larger to handle the heavier current.

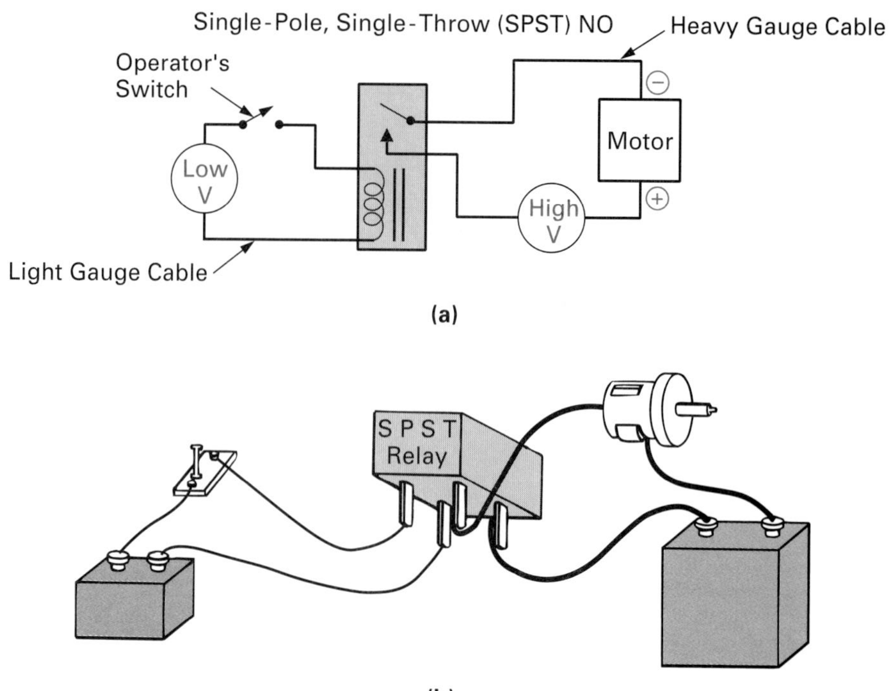

FIGURE 15-18 Low-Current Switch Enabling a High-Current Circuit. (a) Schematic. (b) Pictorial.

Reed Relay or Switch
Relay that consists of two thin magnetic strips within a glass envelope with a coil wrapped around the envelope so that when it is energized the relay's contacts or strips will snap together, making a connection between the two leads attached to each of the reed strips.

Reed relays and switches. The **reed relay or switch** consists of two flat magnetic strips mounted inside a capsule, which is normally made of glass. The reed relay and reed switch are illustrated in Figure 15-21(a) and (b). The reed relay differs from the reed switch in that a reed relay has its own energizing coil, while the reed switch needs an external magnetic force to operate.

The reed relay and reed switch are both operated by a magnetic force that is provided by a coil for the reed relay and by a separate permanent magnet for the reed switch. When the magnetic force is present, opposite magnetic polarities are induced in the overlapping, high-permeability reed blades, causing them

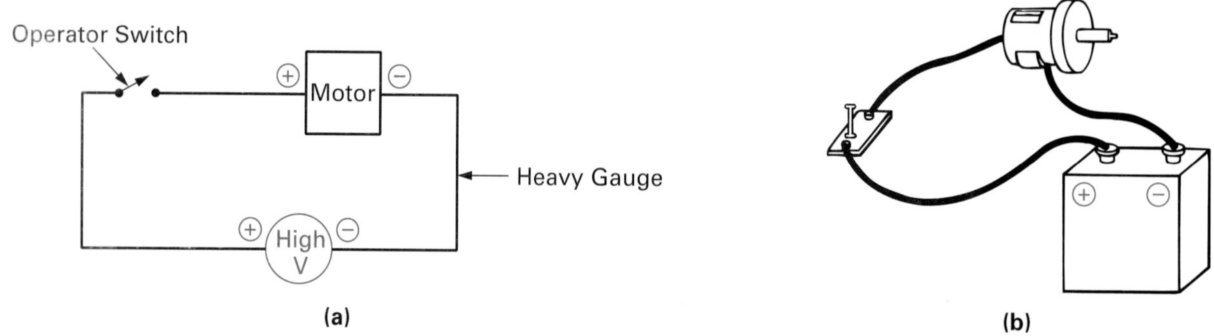

FIGURE 15-19 Without Relay (More Expensive).

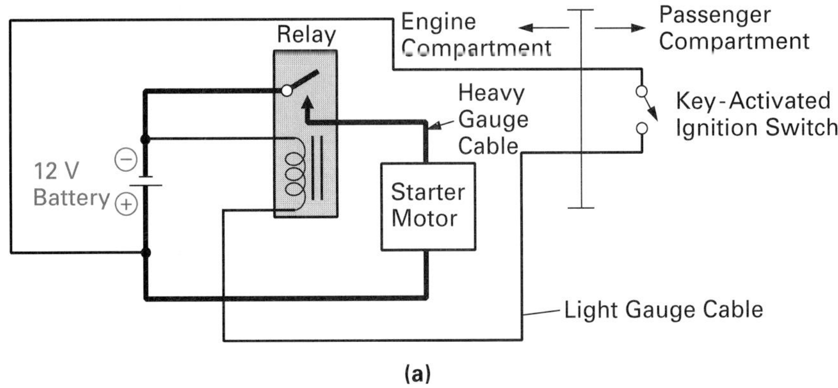

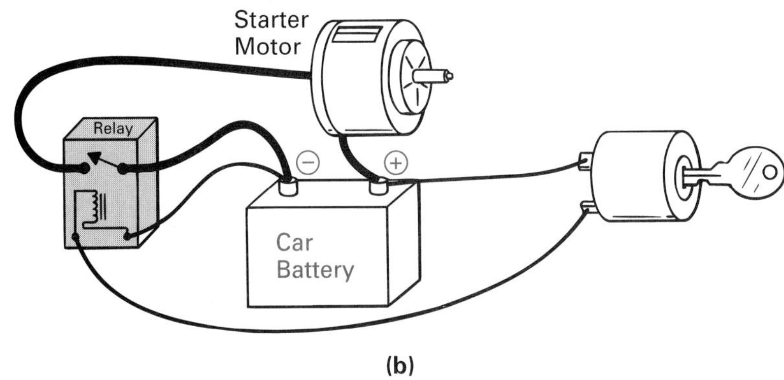

FIGURE 15-20 Automobile Starting Circuit.

Metal Contacts Are Magnetic (have high permeability)

End Terminals

End Terminals

Energizing Coil for Relay

Permanent Magnet

N

S

(a)

(b)

FIGURE 15-21 Reed Relay and Switch. (a) Reed Relay: Contacts Close When Electromagnetic Coil Is Energized. (b) Reed Switch: Contacts Close When External Magnetic Field Comes in Close Proximity, Due to Induced Magnetism in the Contacts.

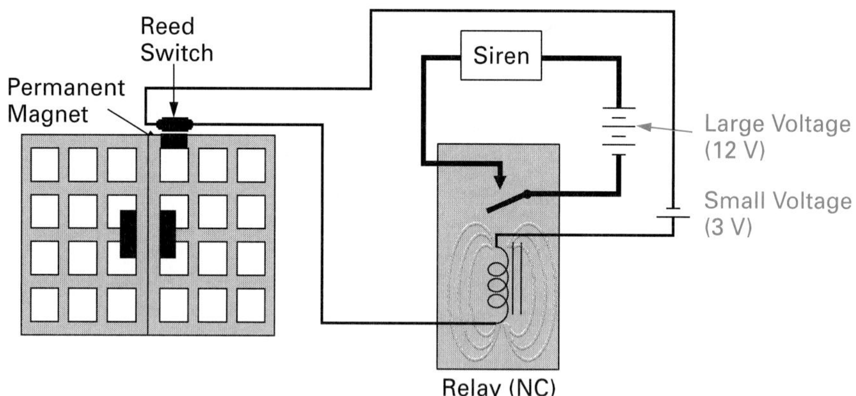

FIGURE 15-22 **Reed Switch as Part of a Home Security System.**

to attract one another by induced magnetism and snap together, thus closing the circuit. When the magnetic force is removed, the blades spring apart due to spring tension and open the circuit.

Referring to Figure 15-22 you will see a simple home security circuit. When the window is closed in the normal condition, the permanent magnet is directly adjacent to the reed switch, and therefore its contacts are closed, allowing current to flow from the negative side of the small 3-V battery, through the closed reed switch contacts, the relay's electromagnet, and back to the positive side of the 3-V battery. The relay is a normally closed (NC) SPST; but as current is flowing through the coil of the relay, the contacts are open, preventing the large positive and negative voltage from the 12-V battery from reaching the siren.

If the window is forced and opened by an intruder, the permanent magnet will no longer be in close proximity to the reed switch and the reed switch's contacts will open. When the reed switch's contacts open, the relay's coil will no longer have current flowing through it, and therefore the relay will deenergize and its contacts will return to their normal closed condition. A large current is now permitted to flow from the negative side of the large 12-V battery, through the relay's contacts, the siren, and back to the positive side of the battery. The siren will sound, as it now has 12 V applied to it, and alert the occupant of the home.

Solenoid-Type Electromagnet

Up to now, electromagnets have been used to close or open a set, or sets, of contacts to either make or break a current path. These electromagnets have used a stationary soft-iron core; however, some electromagnets are constructed with movable iron cores, as shown in Figure 15-23, which can be used to open or block the passage of a gas or liquid through a valve. These are known as **solenoid**-type electromagnets.

Solenoid
Coil and movable iron core that when energized by an alternating or direct current will pull the core into a central position.

When no current is flowing through the solenoid coil, no magnetic field is generated, so no magnetic force is exerted on the movable iron core, as shown in Figure 15-23(a), and therefore the compression spring maintains it in the up position, with the valve plug on the end of the core preventing the passage of either a liquid or gas through the valve (valve closed).

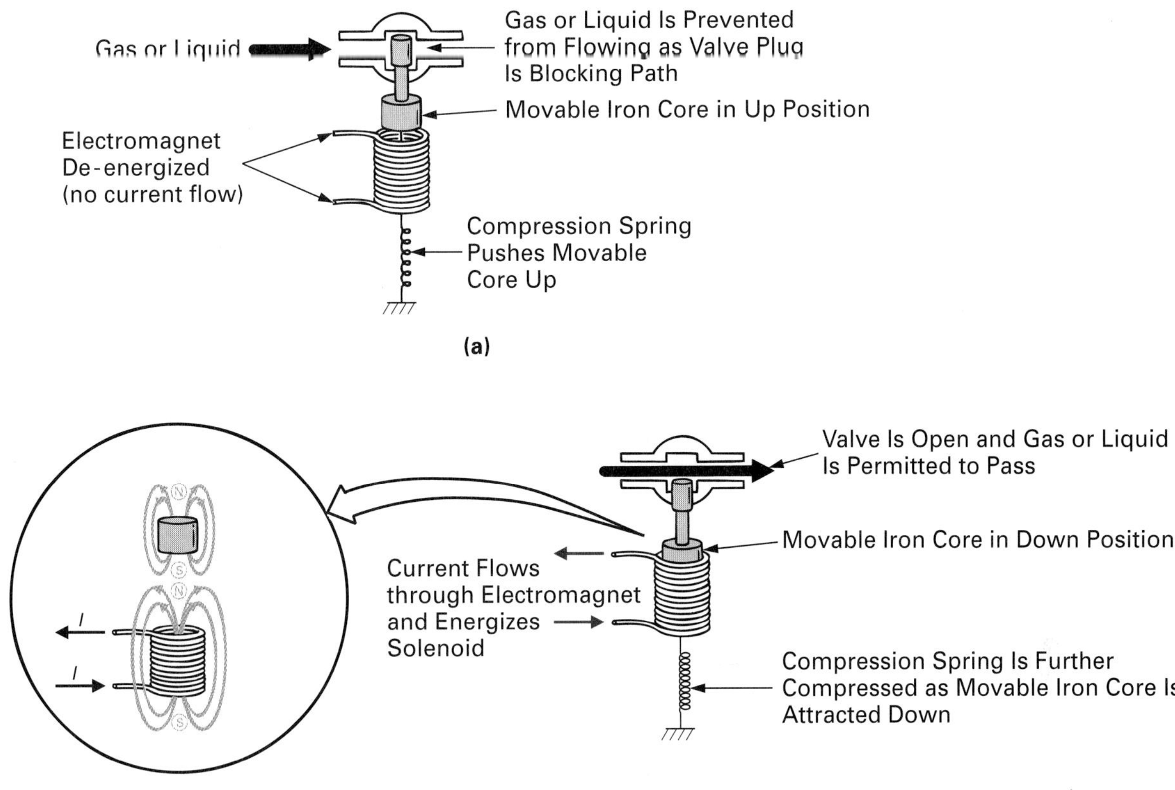

FIGURE 15-23 Solenoid-Type Electromagnet. (a) Deenergized. (b) Energized.

When a current flows through the electromagnet, the solenoid coil is energized, creating a magnetic field, as shown in Figure 15-23(b). Due to the influence of the coil's magnetic field, the movable soft-iron core will itself generate a magnetic field, as seen in the insert in Figure 15-23(b). This condition will create a north pole at the top of the solenoid coil and a south pole at the bottom of the movable core, and the resulting attraction will pull down the core (which is free to slide up and down), pulling with it the valve plug and opening the valve.

Solenoid-type electromagnets are actually constructed with the core partially in the coil and are used in washing machines to control water and in furnaces to control gas.

SELF-TEST REVIEW QUESTIONS FOR SECTION 15-1-7

1. List two applications of the electromagnet.
2. What is the difference between an NO and an NC relay?
3. What is the difference between a reed relay and a reed switch?
4. Give an application for a conventional relay and one for the reed switch.

15-2 ELECTROMAGNETIC INDUCTION

Electromagnetic Induction
The voltage produced in a coil due to relative motion between the coil and magnetic lines of force.

Electromagnetic induction is the name given to the action that causes electrons to flow within a conductor when that conductor is moved through a magnetic field. Stated another way, electromagnetic induction is the voltage or emf induced or produced in a coil as the magnetic lines of force link with the turns of a coil. Since this phenomenon was first discovered by Michael Faraday, let us begin by studying **Faraday's Law.**

Faraday's Law
1. When a magnetic field cuts a conductor, or when a conductor cuts a magnetic field, an electric current will flow in the conductor if a closed path is provided over which the current can circulate.
2. Two other laws relate to electrolytic cells.

15-2-1 *Faraday's Law*

In 1831, Michael Faraday carried out an experiment including the use of a coil, a zero center ammeter (galvanometer), and a bar permanent magnet, as shown in Figure 15-24. Faraday discovered that [Figure 15-25(a) through (f)]:

a. When the magnet is moved into a coil the magnetic lines of flux cut the turns of the coil. This action that occurs when the magnetic lines of flux link with a conductor is known as *flux linkage.* Whenever flux linkage occurs, an emf is induced in the coil known as an induced voltage, which causes current to flow within the circuit and the meter to deflect in one direction, for example, to the right. Faraday discovered, in fact, that if the magnet was moved into the coil, or if the coil was moved over the magnet, an emf or voltage was induced within the coil. What actually occurs is that the electrons within the coil are pushed to one end of the coil by the magnetic field, creating an abundance of electrons at one end of the coil (negative charge) and an absence of electrons at the other end of the coil (positive charge). This potential difference across the coil will produce a current flow if a complete path for current (closed circuit) exists.

b. When the magnet is stationary within the coil, the magnetic lines are no longer cutting the turns of the coil, and so there is no induced voltage and the meter returns to zero.

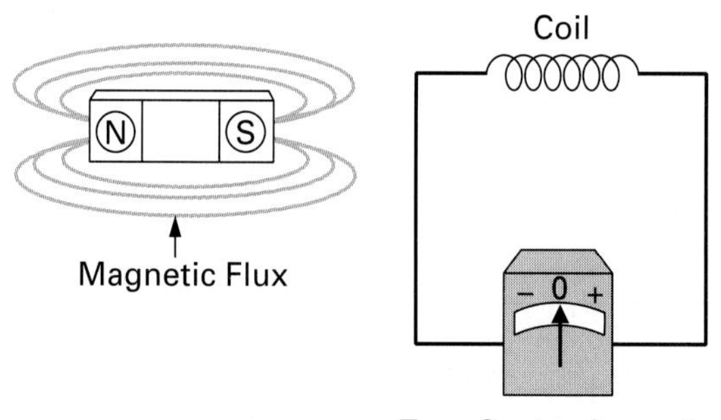

FIGURE 15-24 Faraday's Electromagnetic Induction Experiment Components.

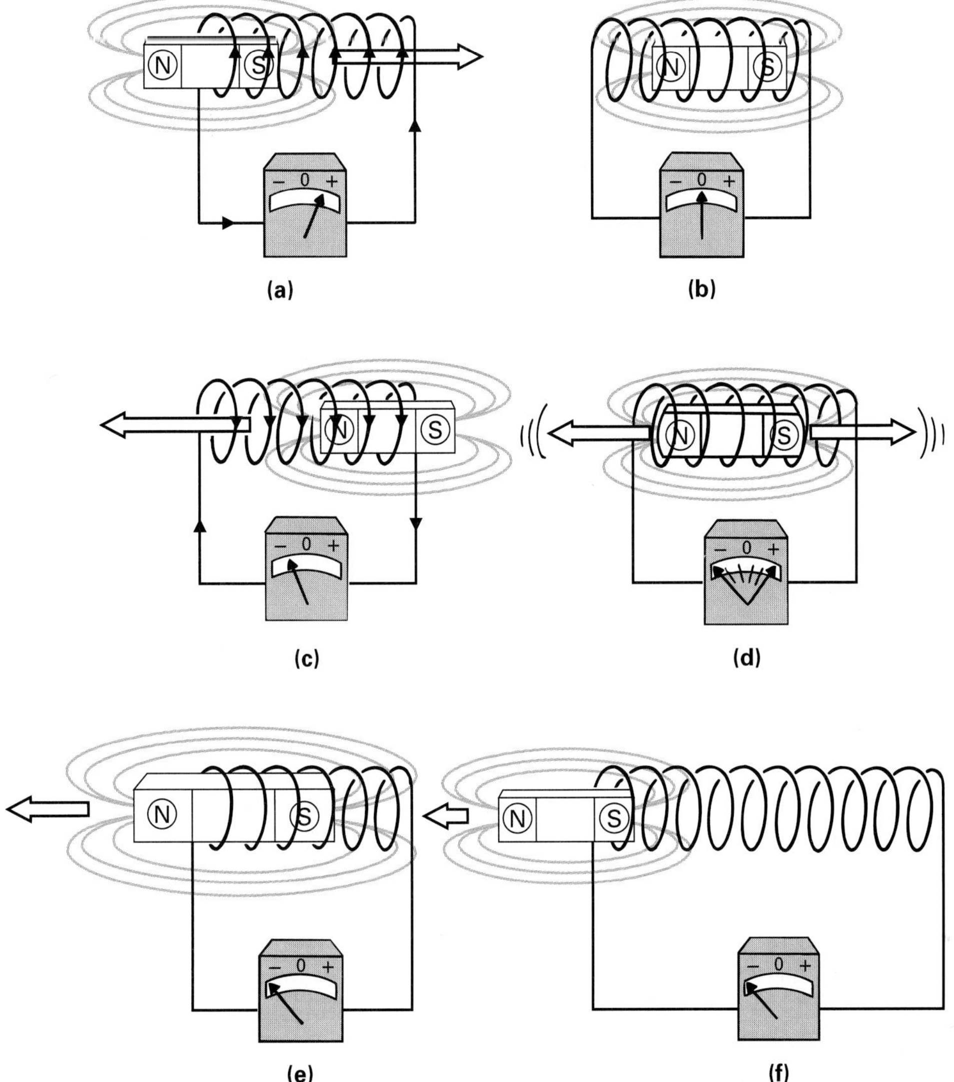

FIGURE 15-25 **Faraday's Electromagnetic Induction Discoveries.**

c. When the magnet is pulled back out of the coil, a voltage is induced that causes current to flow in the opposite direction to that of (a) and the meter deflects in the opposite direction, for example, to the left.

d. If the magnet is moved into or out of the coil at a greater speed, the voltage induced also increases, and therefore so does current.

e. If the size of the magnet and therefore the magnetic flux strength are increased, the induced voltage also increases.

f. If the number of turns in the coil is increased, the induced voltage also increases.

In summary, whenever there is relative motion or movement between the coil of a conductor and the magnetic lines of flux, a potential difference will be

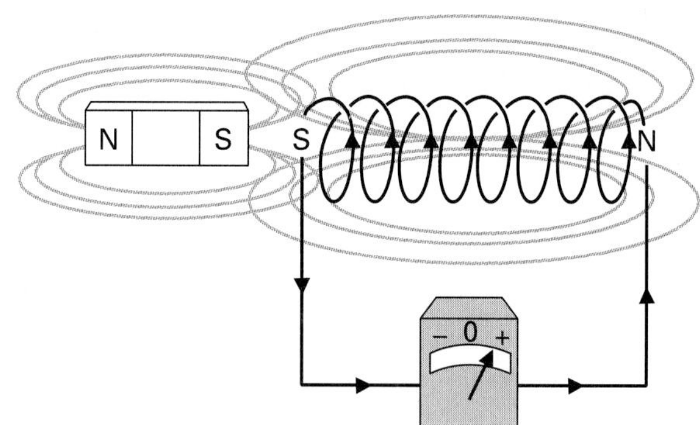

FIGURE 15-26 **Lenz's Law.**

induced and this action is called *electromagnetic induction.* The magnitude of the induced voltage depends on the number of turns in the coil, rate of change of flux linkage, and flux density.

Lenz's Law
The current induced in a circuit due to a change in the magnetic field is so directed as to oppose the change in flux, or to exert a mechanical force opposing the motion.

15-2-2 Lenz's Law

About the same time a German physicist, Heinrich Lenz, performed a similar experiment along the same lines as Faraday. His law states that the current induced in a coil due to the change in the magnetic flux is such as to oppose the cause producing it.

To explain this law further, refer to Figure 15-26. When the magnet moves into the coil, a voltage is induced such that the current flows in the coil and produces a pole at the face of the coil (left-hand rule), which opposes the entry of the magnet. In the example in Figure 15-26, we can see that as the permanent magnet moves into the coil, its magnetic lines of flux cut the turns of the coil and induce a voltage (electromagnetic induction), which causes current to flow in the coil as indicated by the meter movement. If you apply the left-hand rule to the coil, you can see that the current flow within the coil has produced a south pole on the left-hand side of the coil (electromagnetism) to oppose the entry or motion of the magnet that is producing the current.

Weber (Wb)
Unit of magnetic flux. One weber is the amount of flux that when linked with a single turn of wire for an interval of 1 second, will induce an electromotive force of 1 V.

15-2-3 The Weber

Consideration of Faraday's law enables us to take a closer look at the unit of flux (ϕ). The **weber** is equal to 10^8 magnetic lines of force, and from the electromagnetic induction point of view, if 1 weber of magnetic flux cuts a conductor for a period of 1 second, a voltage of 1 volt will be induced.

Generator
Device used to convert a mechanical energy input into an electrical energy output.

15-2-4 Applications of Electromagnetic Induction

AC Generator

The ac **generator** or alternator is an example of a device that uses electromagnetic induction to generate electricity. If you stroll around your city or town during the day or night and try to spot every piece of equipment, appliance, or device that is running from the ac electricity supplied by generators from the electric power plant, it begins to make you realize how inactive, dark, and difficult our modern society might become without ac power.

When discussing Faraday's discoveries of electromagnetic induction in Figure 15-25, the coil or conductor remained stationary and the magnet was moved. It can be operated in the opposite manner so that the magnetic field remains stationary and the coil or conductor is moved. As long as the magnetic lines of force have relative motion with respect to the conductor, an emf will be induced in the coil. Figure 15-27(a) illustrates a piece of wire wound to form a coil and attached to a galvanometer; its needle rests in the center position, indicating zero current. If the conductor remains stationary within the magnetic lines of flux being generated by the permanent magnet, there is no emf or voltage induced into the wire and so no current flow through the circuit and meter.

If the conductor is moved past the permanent magnet so that it cuts the magnetic lines of flux, as seen in Figure 15-27(b), an emf is generated within the conductor, which is known as an induced voltage, and this will cause current to flow through the wire in one direction and be indicated on the meter by the deflection of the needle to the left.

If the conductor is moved in the opposite direction past the magnetic field, it will induce a voltage of the opposite polarity and cause the meter to deflect to the right, as shown in Figure 15-27(c).

The value or amount of induced voltage is indicated by how far the meter deflects and this voltage is dependent on three factors:

1. The speed at which the conductor passes through the magnetic field
2. The strength or flux density of the magnetic field
3. The number of turns in the coil

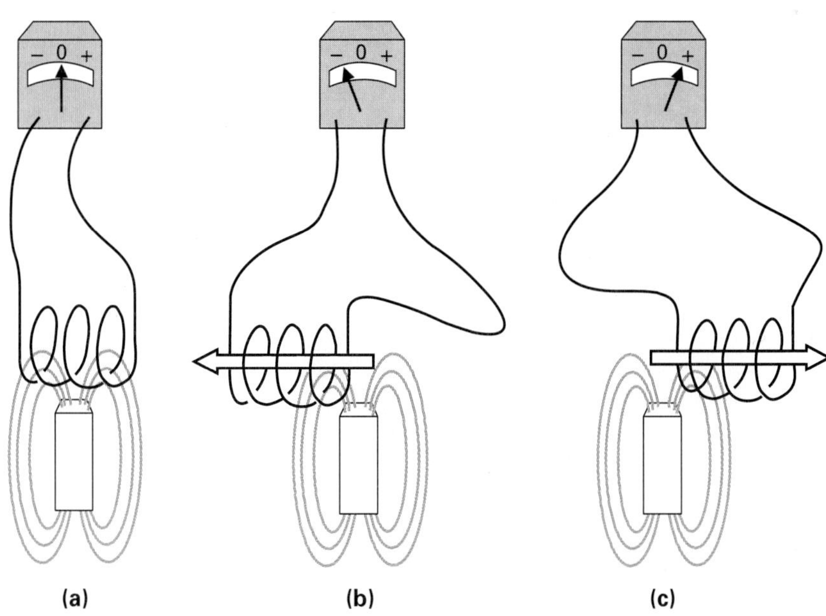

FIGURE 15-27 **(a) Stationary Conductor, No Induced Voltage. (b) Moving Conductor, Induced Voltage. (c) Moving Conductor, Induced Voltage.**

If the speed at which the conductor passes through the magnetic field or the strength of the magnetic field or the number of turns of the coil is increased, then the induced voltage will also increase. This is merely a repetition of Faraday's law, but in this case we moved the conductor instead of the magnetic field; however, the results were the same.

Basic generator. The basic generator action makes use of Faraday's discoveries of electromagnetic induction. In Chapter 11 the physical appearance of a large 700,000-kW power plant generator was shown. Figure 15-28 illustrates some smaller generators that are mobile and can therefore be used in remote locations.

Figure 15-29 illustrates the basic generator's construction. The basic principle of operation is that the mechanical drive energy input will produce ac electrical energy out by means of electromagnetic induction.

Armature
Rotating or moving component of a magnetic circuit.

A loop of conductor, known as an **armature,** is rotated continually through 360° by a mechanical drive. This armature resides within a magnetic field produced by a dc electromagnet. Voltage will be induced into the armature and will appear on slip rings, which are also being rotated. A set of stationary brushes rides on the rotating slip rings and picks off the generated voltage and applies this voltage across the load. This voltage will cause current to flow within the circuit and be indicated by the zero center ammeter.

Let's now take a closer look at the armature as it sweeps through 360°, or one complete revolution. Figure 15-30 illustrates four positions of the armature as it rotates through 360° in the clockwise direction.

Position 1: At this instant, the armature is in a position such that it does not cut any magnetic lines of force. The induced voltage in the armature conductor is equal to 0 V and there is no current flow through the circuit.

(a)

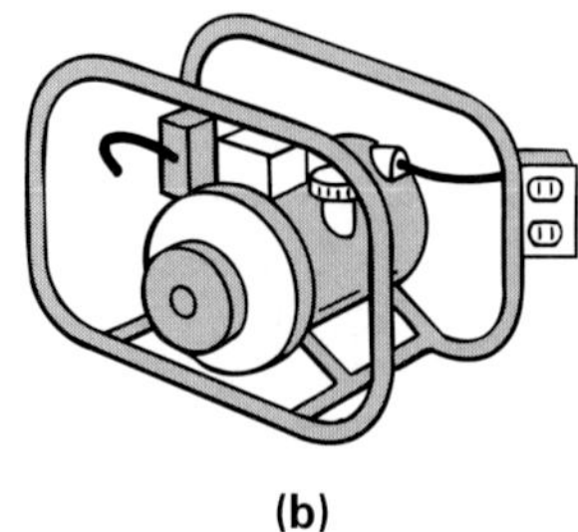

(b)

FIGURE 15-28 AC Generator. (a) 250-kW Diesel Generator. (b) Small (2 kW) Camping Generator.

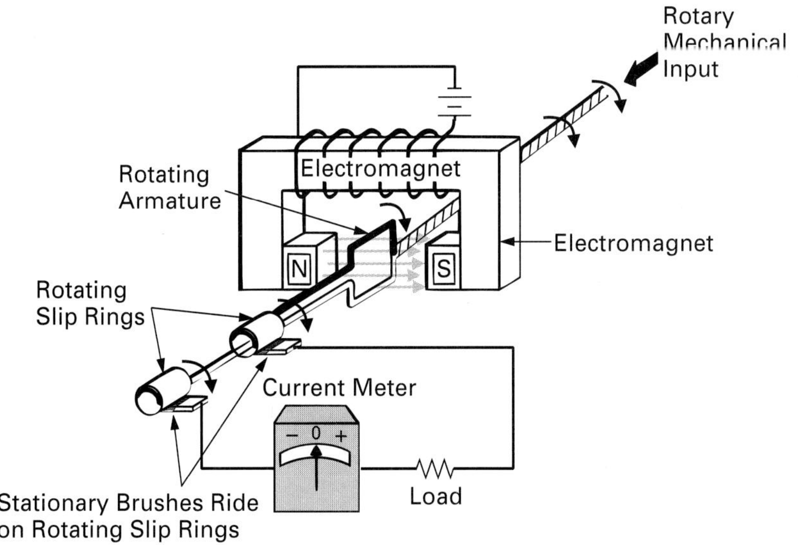

FIGURE 15-29 **Basic Generator Construction.**

Position 2: As the conducting armature moves from position 1 to position 2, you can see that more and more magnetic lines of flux will be cut, and the induced emf in the armature (being coupled off by the brushes from the slip rings) will also increase to a maximum value. The current flow throughout the circuit will rise to a maximum as the voltage increases, and this can be seen by the zero center ammeter deflection to the right. From the ac waveform in Figure 15-30, you can see the sinusoidal increase from zero to a maximum positive as the armature is rotated from 0 to 90°.

Position 3: The armature continues its rotation from 90 to 180°, cutting through a maximum quantity and then fewer magnetic lines of force. The induced voltage decreases from the maximum positive at 90° to 0 V at 180°. At the 180° position, as with the 0° position, the armature is once again perpendicular to the magnetic field, and so no lines are cut and the induced voltage is equal to 0 V.

Position 4: From 180 to 270°, the armature is still moving in a clockwise direction, and as it travels toward 270°, it cuts more and more magnetic lines of force. The direction of the cutting action between 0 and 90° causes a positive induced voltage; and since the cutting position between 180 and 270° is the reverse, a negative induced voltage will result in the armature causing current flow in the opposite direction, as indicated by the deflection of the zero center ammeter to the left. The voltage induced when the armature is at 270° will be equal but opposite in polarity to the voltage generated when the armature was at the 90° position. The current will therefore also be equal in value but opposite in its direction of flow.

From position 4 (270°), the armature turns to the 360° or 0° position, which is equivalent to position 1, and the induced voltage decreases from maximum negative to zero. The cycle then repeats.

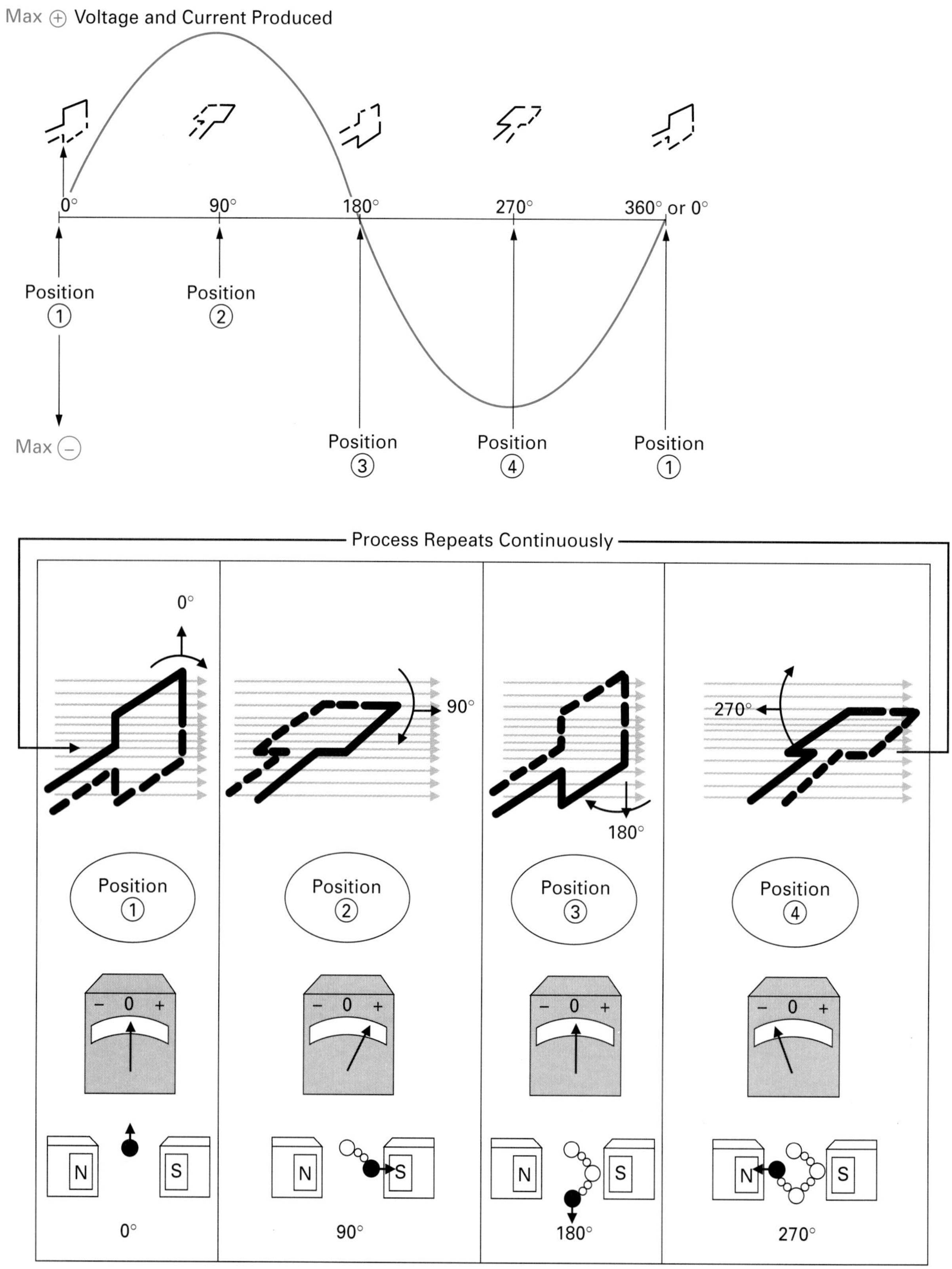

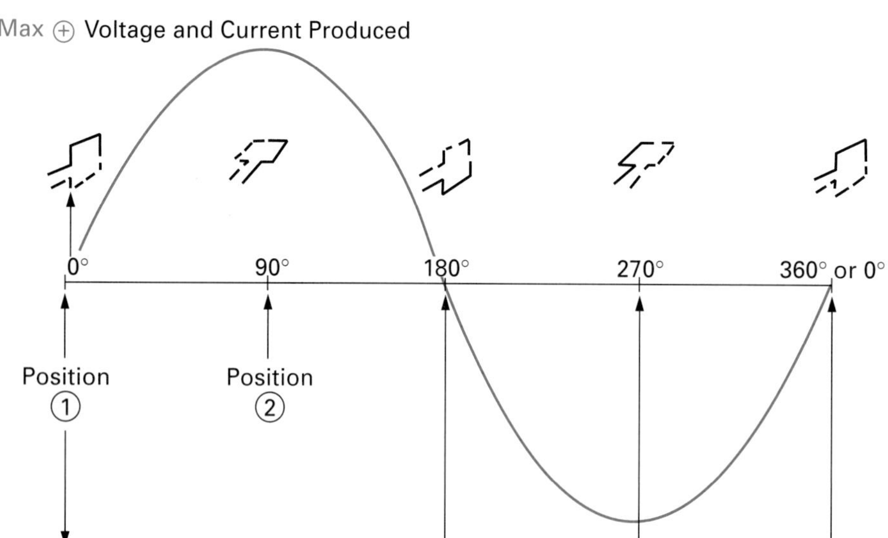

FIGURE 15-30 **360° Generator Operation.**

To summarize, in Figure 15-30, one complete revolution of the mechanical energy input causes one complete cycle of the ac electrical energy output, with a sine wave being generated as a result of circular motion.

Moving Coil Microphone

Microphone Electroacoustic transducer that responds and converts a sound wave input into an equivalent electrical wave out.

The moving coil **microphone** is an example of how electromagnetic induction is used to convert information-carrying sound waves to information-carrying electrical waves. Sound is the movement of pressure waves in the air. To create these pressure waves, a device such as a string, reed, or stretched membrane or the human vocal cords must be vibrated to compress and expand the nearby air molecules. Figure 15-31 illustrates a taut string that is vibrating back and forth and generating maximum (A) and minimum (C) pressure regions.

The frequency or pitch of the sound wave is determined by the number of complete vibrations per second (hertz or cycles/second), while the amplitude or intensity of the sound wave is determined by the amount the string shifts from left to right from its normal position (B).

A moving coil microphone converts mechanical sound waves into an electrical replica by use of electromagnetic induction. Figure 15-32 illustrates the physical appearance and construction of a moving coil type of microphone. A coil of wire is suspended in an air gap between magnetic poles and attached to a delicate diaphragm (flexible membrane). A strong magnetic field from a permanent magnet surrounds the coil, and a perforated protecting cover or shield is included to protect the delicate diaphragm.

Sound waves strike the diaphragm, causing it to vibrate back and forth. Since the coil is attached to the diaphragm, it will also be moved back and forth. This movement will cause the coil to cut the magnetic lines of force from the permanent magnet, and a resulting alternating voltage will be induced in the coil (electromagnetic induction). The electrical voltage produces an alternating current, which will have the same waveform shape (and consequently information) as the sound wave that generated it, as seen in Figure 15-32(c). This electrical

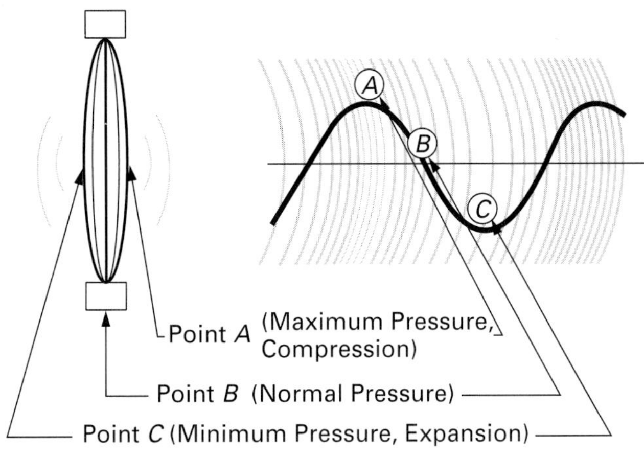

FIGURE 15-31 Sound Wave.

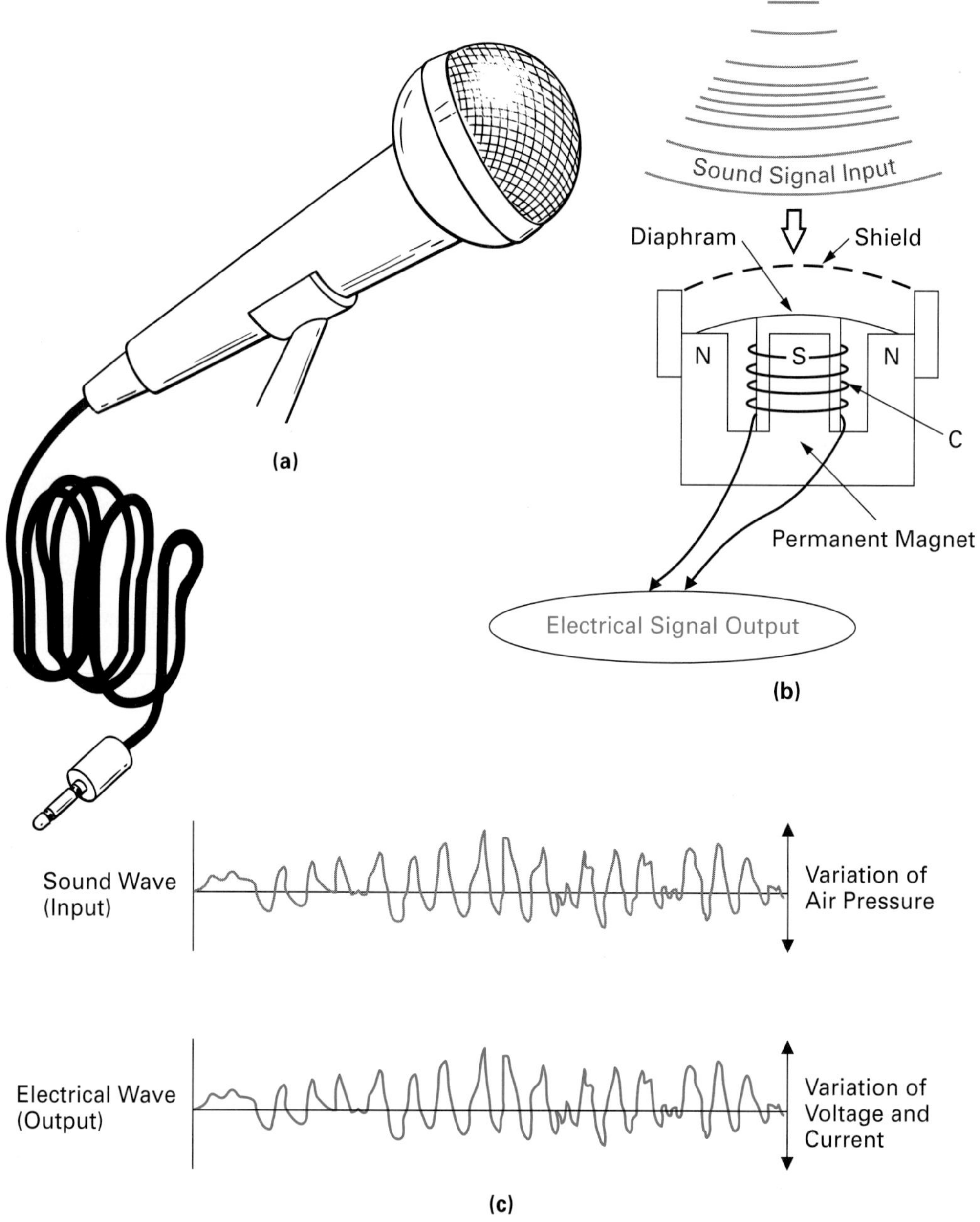

FIGURE 15-32 Moving Coil Microphone. (a) Physical Appearance. (b) Construction. (c) Input/Output Waveforms.

signal is often referred to as an *analog signal.* Analog is a term meaning "similar to," and as you can see in Figure 15-32(c), the electrical wave output of the microphone is an analog (similar to) of the sound wave input.

This microphone is sometimes called a dynamic or moving (dynamic is the opposite of static) coil microphone since it has a moving coil within it.

SELF-TEST REVIEW QUESTIONS FOR SECTION 15-2

1. Define *electromagnetic induction.*
2. Briefly describe Faraday's and Lenz's laws in relation to electromagnetic induction.
3. What waveform shape does the ac generator produce?
4. Why do you think a microphone is called an electroacoustical transducer?

SUMMARY

1. During a classroom lecture in 1820 Danish physicist Hans Christian Oersted accidentally stumbled on an interesting reaction. As he laid a compass down on a bench he noticed that the compass needle pointed to an adjacent conductor that was carrying a current, instead of pointing to the earth's north pole. It was this discovery that first proved that magnetism and electricity were very closely related to one another.
2. This phenomenon is now called *electromagnetism* since it is now known that any conductor carrying an electro or electrical current will produce a magnetic field.
3. In 1831, the English physicist Michael Faraday explored further Oersted's discovery of electromagnetism and found that the process could be reversed. Faraday observed that if a conductor was passed through a magnetic field, a voltage would be induced in the conductor and cause a current to flow. This phenomenon is referred to as *electromagnetic induction.*

Electromagnetism

4. Every orbiting electron in motion around its nucleus generates a magnetic field. When electrons are forced to leave their parent atom by voltage and flow toward the positive polarity, they are all moving in the same direction and each electron's magnetic field will add to the next. The accumulation of all these electron fields will create the magnetic field around the conductor.
5. The left-hand rule states that if an insulated conductor is held with your left hand so that your thumb is pointing in the direction of electron flow (to the positive potential), your fingers will be pointing in the direction of the magnetic force.
6. If our conductor is wound to form a coil the conductor current will generate its own magnetic field, which will combine additively within the coil, and the net effect of all these conductor magnetic fields can be used to generate a strong coil magnetic field. This coil of conductor which is used to generate a strong magnetic field is called an electromagnet.
7. If a greater number of loops are added, a greater or stronger magnetic field is generated. If the coils were wound in an opposing opposite direction, the opposing magnetic fields would cancel and so no magnetic field would be generated.

8. The left-hand rule can also be applied to electromagnets to determine which of the poles will be the north end of the electromagnet.
9. If a conductor is wound to form a spiral, the conductor, which is now referred to as a *coil,* will, as a result of current flow, develop a magnetic field, which will sum or intensify within the coil. If many coils are wound in the same direction, an electromagnet is formed, which will produce a concentrated magnetic field whenever a current is passed through its coils.
10. A dc voltage produces a fixed current in one direction and therefore generates a magnetic field in a coil of fixed polarity, and as determined by the left-hand rule.
11. Alternating current (ac) is continually varying, and as the polarity of the magnetic field is dependent on the direction of current flow (left-hand rule), the magnetic field will also be alternating in polarity.

Magnetic Terms (Figure 15-33)

12. One magnetic line of flux or force is called a maxwell, in honor of James Maxwell's work in this field. However, this unit is too small and impractical and therefore the amount of magnetic flux (symbolized ϕ, phi) is measured in webers instead of maxwells. One *weber* is equal to 10^8 (100,000,000) magnetic lines of force or maxwells.
13. Flux density (B) is equal to the number of magnetic lines of flux (ϕ) per square meter, and it is given in the unit of telsa (T).
14. The magnetic flux produced by an electromagnet is produced by current flowing through a coil of wire. Magnetomotive force (mmf) is the magnetic pressure that produces the magnetic field.
15. The length of the coil determines the field of intensity, and the new term, magnetizing force (H), is the reference we commonly use to describe the magnetic field intensity.
16. Reluctance, in reference to magnetic energy, is equivalent to resistance in electrical energy. Reluctance is the opposition or resistance to the establishment of a magnetic field in an electromagnet.
17. As magnetic reluctance is equivalent to electrical resistance, magnetic permeability is equivalent to electrical conductance. Permeability is a measure of how easily a material will allow a magnetic field to be set up within it. Permeability is symbolized by the Greek lowercase letter mu (μ) and is measured in henrys per meter (H/m).
18. Relative (with respect to) is a word that means that a comparison has to be made. Relative permeability (μ_r) is the measure of how well another given material will conduct magnetic lines of force with respect to, or relative to, our reference material, air, which has a relative permeability value of 1.
19. The relative permeability (μ_r) of air, which is equal to 1, should not be confused with the absolute permeability (μ_0) of free space or air, which is equal to $4\pi \times 10^{-7}$ or 1.26×10^{-6}.

FIGURE 15-33 Magnetic Terms.

1 magnetic line of force = 1 maxwell
10^8 lines of force or maxwells = 1 weber

$$B = \frac{\phi}{A}$$

B = flux density, in teslas (T)
ϕ = magnetic flux, in webers (Wb)
A = area, in square meters (m^2)

$$\text{mmf} = I \times N$$

mmf = magnetomotive force, in ampere turns (At)
I = current, in amperes
N = number of turns in coil

$$H = \frac{I \times N}{l}$$

H = magnetizing force, in ampere turns per meter (At/m)
l = length of coil, in meters

$$\mathfrak{R} = \frac{\text{mmf}}{\phi}$$

$\mathfrak{R}$ = reluctance, in ampere turns per weber (At/Wb)

$$\mu = \frac{1}{\mathfrak{R}}$$

μ = permeability, in henrys per meter (H/m)

$$\mu = \mu_r \times \mu_0$$

μ_r = relative permeability
μ_0 = absolute permeability = $4\pi \times 10^{-7}$

Permeabilities of Various Materials

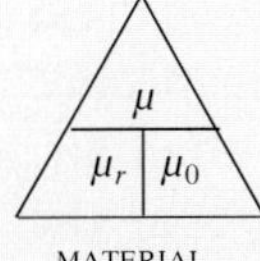

MATERIAL	RELATIVE PERMEABILITY (μ_r)	PERMEABILITY (μ)
Air or vacuum	1	1.26×10^{-6}
Nickel	50	6.28×10^{-5}
Cobalt	60	7.56×10^{-5}
Cast iron	90	1.1×10^{-4}
Transformer iron	5,500	6.9×10^{-3}
Silicon iron	7,000	8.8×10^{-3}
Permaloy	100,000	0.126
Supermalloy	1,000,000	1.26

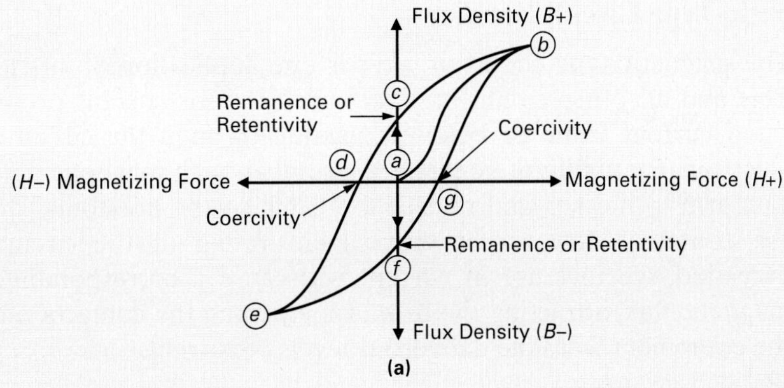

(a)

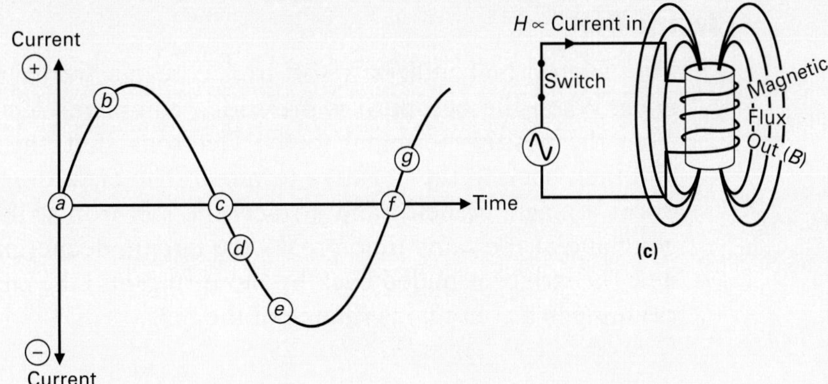

(b)

(c)

20. Whether permeability (μ), relative permeability (μ_r) or absolute permeability (μ_0) is used as a standard makes little difference. A high permeability figure indicates a low reluctance ($\mu \uparrow, \Re \downarrow$), and vice versa.

21. The magnetic field strength of an electromagnet can be increased by increasing the number of turns in its coil, increasing the current through the electromagnet, or decreasing the length of the coil ($H = I \times N/l$). The field strength can be increased further by placing an iron core within the electromagnet. An iron core has less reluctance or opposition to the magnetic lines of force than air, so the flux density (B) is increased. Another way of saying this is that the permeability or conductance of magnetic lines of force within iron is greater than that of air, and if the permeability of iron is large, its reluctance must be small.

22. The B–H curve illustrates the relationship between the two most important magnetic properties: flux density (B) and magnetizing force (H).

23. The saturation point is the state of magnetism beyond which an electromagnet is incapable of further magnetic strength, that is, the point beyond which the B–H curve is a straight, horizontal line, indicating no change.

24. Hysteresis is a lag between cause and effect. With magnetism it is the amount that the magnetization of a material lags the magnetizing force due to molecular friction.

25. Remanence or retentivity is the amount a material remains magnetized after the magnetizing force has been removed. Coercive force is the magnetizing force needed to reduce the residual magnetism within a material to zero.

Magnetic-Type Circuit Breaker

26. The magnetic-type circuit breaker is one application of an electromagnet. This and all other circuit breakers are used for current protection. If the rated current value or below is passing through the circuit breaker, the electromagnet will not generate a strong enough magnetic field to pull the iron arm to the left and release the catch so the horizontal arm will open the contacts. However, if the current rating of the circuit breaker is exceeded, the increase in current will cause a corresponding increase in magnetic flux, attracting the iron arm, opening the contacts, and protecting the equipment from the dangerous levels of current.

Electric Bell

27. The electric bell utilizes a soft-iron core electromagnet, a striker, and a gong. When the bell push is pressed, a complete electrical circuit is made from the positive terminal to the electromagnet, the contact, the striker, and then back to the negative terminal. The electromagnet energizes (produces a magnetic field) and attracts the soft-iron striker, which strikes the gong and at the same time breaks the circuit, deenergizing the electromagnet. The striker is pulled back by the spring, and the circuit is reestablished, resulting in a continuous ringing of the bell.

FIGURE 15-34 The Relay.

Relay Types

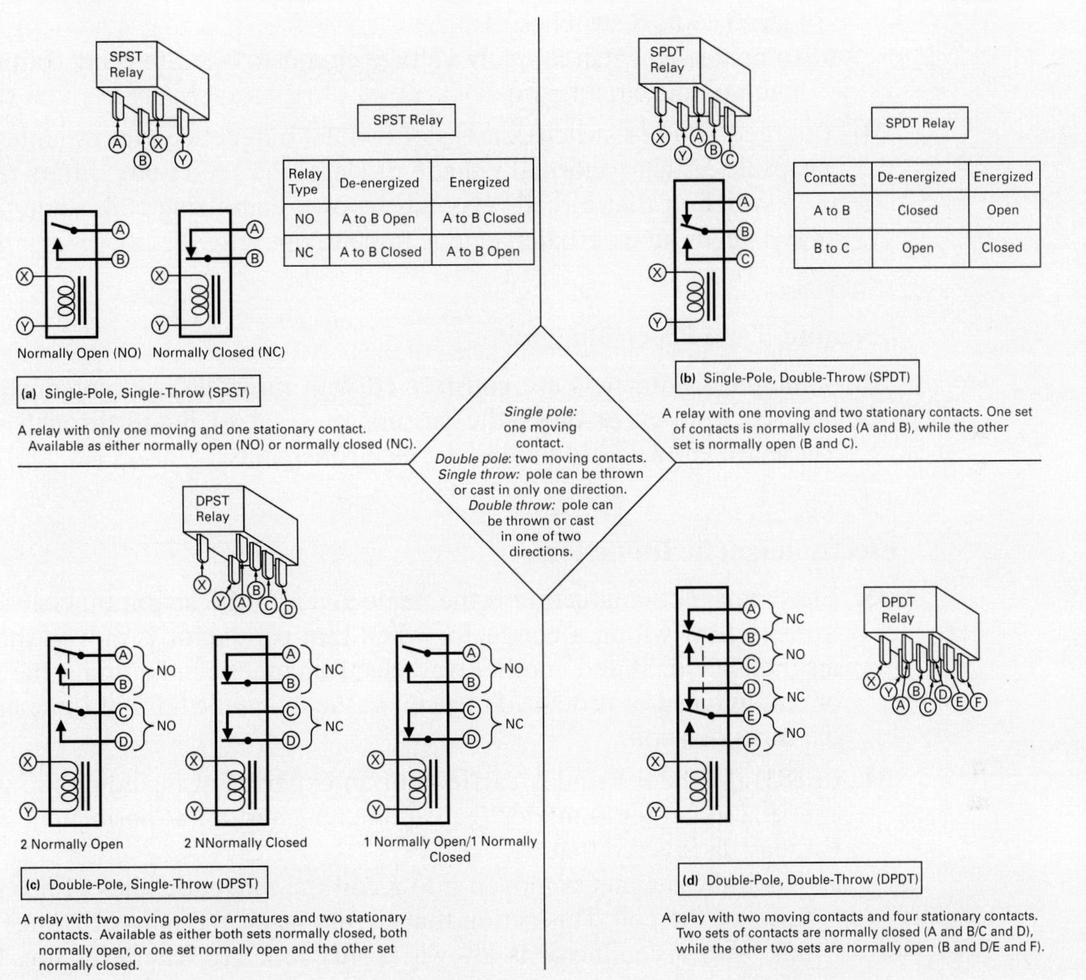

The Relay (Figure 15-34)

28. A relay is an electromechanical device that either makes (closes) or breaks (opens) a circuit by moving contacts together or apart.

29. The relay consists basically of an electromagnet connected to lines x and y, a movable iron arm known as the armature, and some contacts. When current passes through x to y, the electromagnet generates a magnetic field, or is said to be energized, which attracts the armature toward the electromagnet. When this occurs, it closes or makes the normally open relay contacts and opens or breaks the normally closed relay contacts.

If the electromagnet is deenergized by stopping the current through the coil, the spring will pull back the armature to open the NO relay contacts or close the NC relay contacts between *A* and *B*.

30. The relay is generally used in two basic applications:

a. To enable one master switch to operate several remote or difficultly placed contact switches.

b. To enable a switch in a low-voltage circuit to operate relay contacts in a high-voltage circuit.

31. The reed relay or switch consists of two flat magnetic strips mounted inside a capsule, which is normally made of glass. The reed relay differs from the reed switch in that a reed relay has its own energizing coil, while the reed switch needs an external magnetic force to operate.

Solenoid-Type Electromagnet

32. Some electromagnets are constructed with movable iron cores which can be used to open or block the passage of a gas or liquid through a valve. These are known as solenoid-type electromagnets.

Electromagnetic Induction

33. Electromagnetic induction is the name given to the action that causes electrons to flow within a conductor when that conductor is moved through a magnetic field. Stated another way, electromagnetic induction is the voltage or emf induced or produced in a coil as the magnetic lines of force link with the turns of a coil.

34. In 1831, Michael Faraday carried out an experiment including the use of a coil, a zero center ammeter (galvanometer), and a bar permanent magnet. Faraday discovered that:

a. When the magnet is moved into a coil the magnetic lines of flux cut the turns of the coil. This action that occurs when the magnetic lines of flux link with a conductor is known as *flux linkage.* Whenever flux linkage occurs, an emf is induced in the coil known as an induced voltage, which causes current to flow within the circuit and the meter to deflect in one direction.

b. When the magnet is stationary within the coil, the magnetic lines are no longer cutting the turns of the coil, and so there is no induced voltage and the meter returns to zero.

c. When the magnet is pulled back out of the coil, a voltage is induced that causes current to flow in the opposite direction and the meter deflects in the opposite direction.

d. If the magnet is moved into or out of the coil at a greater speed, the voltage induced also increases, and therefore so does current.

e. If the size of the magnet and therefore the magnetic flux strength are increased, the induced voltage also increases.

f. If the number of turns in the coil is increased, the induced voltage also increases.

35. About the same time a German physicist, Heinrich Lenz, performed a similar experiment along the same lines as Faraday. His law states that the current induced in a coil due to the change in the magnetic flux is such as to oppose the cause producing it.

AC Generator (Figure 15-35)

36. The ac generator or alternator is an example of a device that uses electromagnetic induction to generate electricity.

37. The generator's basic principle of operation is that the mechanical drive energy input will produce ac electrical energy out by means of electromagnetic induction.

38. A loop of conductor, known as an armature, is rotated continually through 360° by a mechanical drive. This armature resides within a magnetic field produced by a dc electromagnet. Voltage will be induced into the armature and will appear on slip rings, which are also being rotated. A set of stationary brushes rides on the rotating slip rings and picks off the generated voltage and applies this voltage across the load. This voltage will cause current to flow within the circuit and be indicated by the zero center ammeter.

39. One complete revolution of the mechanical energy input causes one complete cycle of the ac electrical energy output, with a sine wave being generated as a result of circular motion.

Moving Coil Microphone

40. The moving coil microphone is an example of how electromagnetic induction is used to convert information carrying sound waves to information carrying electrical waves.

41. A moving coil microphone converts mechanical sound waves into an electrical replica.

42. The operation of the microphone is as follows. Sound waves strike the diaphragm, causing it to vibrate back and forth. Since the coil is attached to the diaphragm, it will also be moved back and forth. This movement will cause the coil to cut the magnetic lines of force from the permanent magnet, and a resulting alternating voltage will be induced in the coil (electromagnetic induction). The electrical voltage produces an alternating current, which will have the same waveform shape (and consequently information) as the sound wave that generated it. This electrical signal is often referred to as an *analog signal*. Analog is a term meaning "similar to," and the electrical wave output of the microphone is an analog (similar to) of the sound wave input.

43. This microphone is sometimes called a dynamic or moving (dynamic is the opposite of static) coil microphone since it has a moving coil within it.

FIGURE 15-35 The AC Generator.

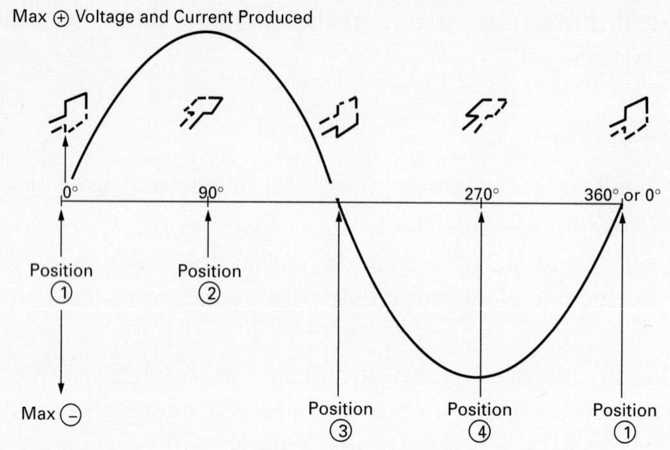

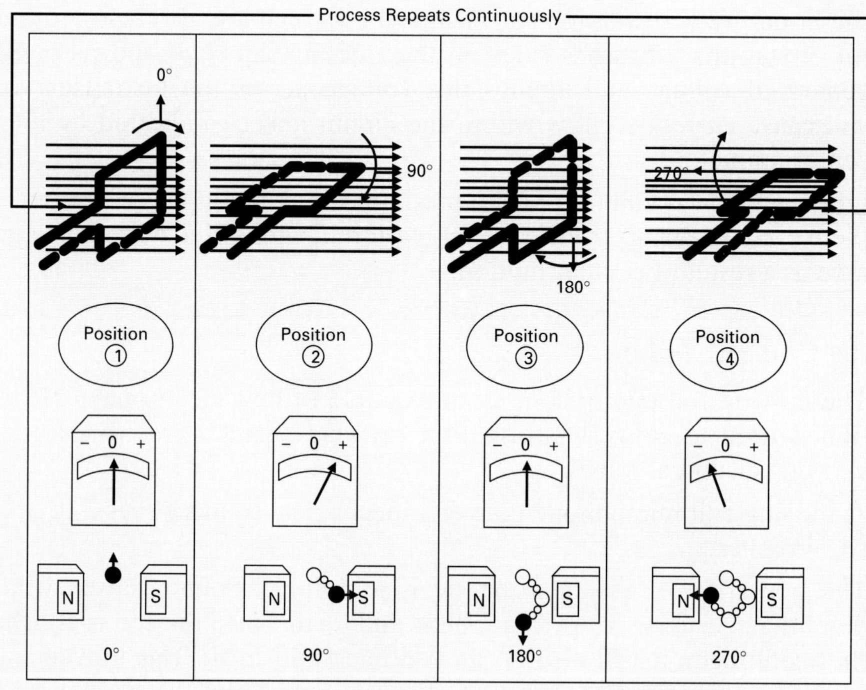

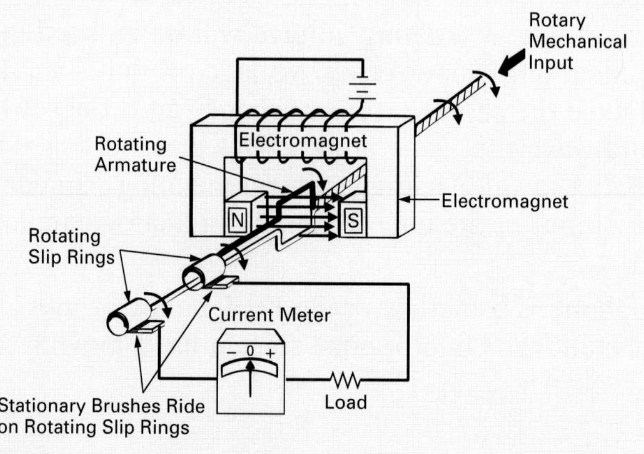

NEW TERMS

Absolute permeability (μ_0)
Alternating magnetic field
Ampere turns per meter (At/m)
Ampere turns per weber (At/Wb)
Armature
B–H Curve
Coercive force
Coercivity
Coil
Deenergized
Diaphragm
Electromagnet
Electromagnetic induction
Electromagnetism
Energized
Faraday's law
Flux density (*B*)
Flux linkage
Galvanometer
Generator
Henrys per meter (H/m)
Hysteresis loop
Left-hand rule
Lenz's law
Magnetic field
Magnetic flux (ϕ)
Magnetizing force (*H*)
Magnetomotive force (mmf)
Maxwell (Mx)
Moving coil microphone
Permeability (μ)
Reed relay
Reed switch
Relative permeability (μ_r)
Relay
Reluctance
Remanence
Retentivity
Saturation
Solenoid
Tesla
Weber

REVIEW QUESTIONS

Multiple-Choice Questions

1. Electromagnetism was first discovered by:

a. Hans Christian Oersted **c.** James Watt
b. Heinrich Hertz **d.** James Clark Maxwell

2. Every current-carrying conductor generates a(an):

a. Electric field **c.** Both (a) and (b)
b. Magnetic field **d.** None of the above

3. If a greater number of loops of a conductor in a coil are added, a _____________ magnetic field is generated.

a. Weaker **b.** Stronger

4. If a smaller current is passed through a coil, a _____________ magnetic field is generated.

a. Weaker **b.** Stronger

5. One magnetic line of flux is known as a _____________.

a. Weber **c.** Maxwell
b. Tesla **d.** Oersted

6. 10^8 magnetic lines of flux are referred to as one ____________.

a. Weber c. Maxwell
b. Tesla d. Oersted

7. Flux density is equal to the magnetic flux divided by area and is measured in ____________.

a. Webers c. Maxwells
b. Teslas d. Oersteds

8. By increasing either current or the number of turns in a coil, the mmf, which is an abbreviation for ____________, will increase.

a. Multiple magnetic formulas c. Electromotive force
b. Magnetomotive force d. None of the above

9. ____________ is a term used to describe the magnetic field intensity and is equal to the mmf divided by the length of the coil.

a. Flux density c. Magnetizing force
b. Magnetomotive force d. Reluctance

10. Reluctance is equivalent to electrical:

a. Current c. Resistance
b. Voltage d. Power

11. Permeability is equivalent to electrical:

a. Current c. Resistance
b. Conductance d. Voltage

12. The absolute permeability of air is equal to:

a. 1 b. 1.26×10^{-6} c. 4π d. None of the above

13. An electromagnet can be used in the application of:

a. A relay c. A circuit breaker
b. An electric bell d. All of the above

14. Relays can be used to:

a. Allow one master switch to enable several others
b. Allow several switches to enable one master
c. Allow a switch in a low-current circuit to close contacts in a high-current circuit
d. Both (a) and (b)
e. Both (a) and (c)

15. The starter relay in an automobile is used to:

a. Allow one master switch to enable several others
b. Allow several switches to enable one master
c. Allow a switch in a low-current circuit to close contacts in a high-current circuit
d. Both (a) and (b)

16. The ____________ uses an electromagnet around two flat magnetic strips mounted inside a glass capsule.

a. Reed switch c. Reed relay
b. Magnetic circuit breaker d. Starter relay

17. The reed switch could be used in a(an):

a. Home security system c. Magnetic-type circuit breaker
b. Automobile starter d. All of the above

18. The relative permeability of air or a vacuum is equal to:

a. 4π b. 6.26×10^{-6} c. 1 d. None of the above

19. An electromagnet is also known as a:

a. Coil d. Both (a) and (c)
b. Solenoid e. Both (a) and (b)
c. Resistor

20. A normally open relay (NO) will have:

a. Contacts closed until activated c. All contacts permanently open
b. Contacts open until activated d. All contacts permanently closed

21. Direct current produces a ____________ magnetic field of ____________ polarity.

a. Alternating, unchanging c. Constant, unchanging
b. Constant, alternating d. Both (a) and (c)

22. Magnetizing force (H) is equal to:

a. $I \times N \times l$ c. $I \times N/l$
b. $I \times N + l$ d. $N \times l/I$

23. Electromagnetism:

a. Is the magnetism resulting from electrical current flow
b. Is the electrical voltage resulting in a coil from the relative motion of a magnetic field
c. Both (a) and (b)
d. None of the above

24. Electromagnetic induction:

a. Is the magnetism resulting from electrical current flow
b. Is the electrical voltage resulting in a coil from the relative motion of a magnetic field
c. Both (a) and (b)
d. None of the above

25. The ____________ plots magnetizing force against flux density through a complete cycle of alternating current.

a. B–H curve c. Power curve
b. Coercive force d. Hysteresis loop

26. When the magnetic flux linking a conductor is changing, an emf is induced, the magnitude of which depends on the number of coil turns, rate of change of flux linkage change, and flux density. This law was discovered by:

a. Heinrich Lenz c. Michael Faraday
b. Guglielmo Marconi d. Joseph Henry

27. The current induced in a coil due to the change in the magnetic flux is such as to oppose the cause producing it. This law was discovered by:

a. Heinrich Lenz c. Michael Faraday
b. Guglielmo Marconi d. Joseph Henry

28. An ac generator uses ___________ to generate ac electricity.

a. Electromagnetism c. Magnetism
b. Electromagnetic induction d. None of the above

29. The generator converts ___________ energy into ___________ energy.

a. Electrical, electrical c. Chemical, electrical
b. Mechanical, electrical d. None of the above

30. The loop of conductor rotated through 360° in a generator is known as a(an):

a. Electromagnet c. Armature
b. Field coil d. Both (a) and (b)

31. Sound waves are a form of:

a. Electrical energy c. Magnetic energy
b. Chemical energy d. Mechanical energy

32. The moving coil microphone converts ___________ waves into ___________ waves.

a. Sound, electrical c. Electromagnetic, sound
b. Electrical, sound d. Sound, radio

33. There is a reciprocal relationship between permeability and ___________.

a. Flux density c. Remanence
b. Magnetizing force d. Reluctance

34. The amount of current in the reverse direction needed to reduce the flux density (B) to zero is termed the:

a. Coercive force c. Electromotive force
b. Magnetizing force d. All of the above

Essay Questions

35. Describe what is meant by the word *electromagnetism.* (15-1)

36. Briefly describe how the left-hand rule of electromagnetism is applied to: (15-1-2)

a. Conductors b. Coils

37. How many maxwells make up 1 weber? (15-1-5)

38. Give the formulas for the following: (15-1-5)

a. Flux density d. Reluctance
b. Magnetomotive force e. Permeability
c. Magnetizing force

39. Describe the operation of: (15-1-7)

a. The magnetic-type circuit breaker
b. The electric bell
c. The NO and NC relays

40. Describe the difference and operation of the: (15-1-7)

    a. Reed relay   b. Reed switch

41. Explain the operation and application of the solenoid-type electromagnet. (15-1-7)
42. Why is a wire looped many times to form an electromagnet more useful than a straight piece of wire? (15-1-3)
43. From which pole does the magnetic flux emerge and into which pole end does the flux return? (15-1-1)
44. What are the differences between magnetic flux and flux density? (15-1-5)
45. What effect does current have when it is passed through a coil of conductor? (15-1)
46. What effect does a magnet have when it is moved into and out of a coil? (15-2)
47. State Faraday's law. (15-2-1)
48. State Lenz's law. (15-2-2)
49. Describe the different effects when ac and dc are passed through a coil. (15-1-4)
50. Illustrate and describe all the different points on a hysteresis curve. (15-1-6)
51. Describe the meaning of the following terms: (15-1-6)

    a. Flux density   d. Coercive force
    b. Magnetizing   e. Electromagnetism
    c. Remanence   f. Electromagnetic induction

52. Briefly describe the operation of the generator. (15-2-4)
53. Briefly describe the operation of the moving coil microphone. (15-2-4)
54. Illustrate and describe the operation of the generator through 360°. (15-2-4)
55. List some other applications of ac electromagnetism. (15-1-7)

Practice Problems

56. If a magnet has a pole area of 6.4×10^{-3} m^2 and a 1200-μWb total flux, what flux density would the pole produce?
57. Calculate the magnetomotive force produced when 760 mA flow through 25 turns.
58. Calculate the magnetizing force (H) or field intensity if a 15-cm, 40-turn coil has a current of 1.2 A flowing through it.
59. Calculate the permeability (μ) of the following materials ($\mu_0 = 4\pi \times 10^{-7}$):

    a. Cast iron   b. Nickel   c. Machine steel

60. Calculate the reluctance of a magnetic circuit if the magnetic flux produced is equal to 2.3×10^{-4} Wb, and is produced by 3 A flowing through a solenoid of 36 turns.

61. If a coil of 50 turns is passing a current of 4 A, calculate the mmf.

62. Calculate the reluctance of an iron core when mmf = 150 At and ϕ = 360 μWb.

63. Calculate mmf when a 9-V battery is connected across a 50-turn, 23-Ω coil.

64. Calculate the permeability of a permalloy core.

65. Calculate the magnetizing force (H) of the coil in Question 63 if it were 0.7 m long.

CHAPTER

16

OBJECTIVES

After completing this chapter, you will be able to:

1. Describe self-induction.
2. List and explain the factors affecting inductance.
3. Give the formula for inductance.
4. Identify inductors in series and parallel and understand how to calculate total inductance when inductors are in combination.
5. List and explain the fixed and variable types of inductors.
6. Explain the inductive time constant.
7. Give the formula for inductive reactance.
8. Describe all aspects relating to a series and parallel *RL* circuit.
9. State the three typical malfunctions of inductors and explain how they can be recognized.
10. Describe how inductors can be used for the following applications:
    a. *RL* filter
    b. *RL* integrator
    c. *RL* differentiator

Inductance and Inductors

OUTLINE

VIGNETTE

What's Cooking!

As a boy, George R. Stibitz was a natural experimenter intrigued by any electrical or electronic gadget. His parents were always anxious about these tinkerings, but wishing not to discourage his imagination and keenness to learn, his father, a professor of theology, often gave him devices to occupy him. On one occasion at the age of 8, his latest experiment, in which he connected an electric motor to the home electrical outlet, caused a circuit overload and almost sent the family home in Dayton, Ohio, up in smoke.

In 1937, Stibitz, a young mathematician working at Bell Telephone Laboratories (the research division of AT&T), had another one of his brainstorms and in his spare time he began building the prototype, which consisted of telephone-system components, batteries, and other devices all interconnected by a mass of wires. The machine he was putting together would be the first system able to achieve binary arithmetic in the United States; since it was assembled on the kitchen table, he called it the Model K.

After developing a more ambitious system in 1940 with veteran Bell engineer Samuel B. Williams, the new digital calculator could handle the complex number problems that were needed for the design of long-distance telephone networks. The system was installed in Bell Labs' Manhattan, New York, headquarters and drew a large amount of attention from the American Mathematical Society, who invited Stibitz to give a presentation on the machine at Dartmouth College in Hanover, New Hampshire. Stibitz later wrote, "With my usual genius for making things more difficult for myself and others, I suggested direct telegraph operation of the complex number calculator from Hanover, and this was decided upon." Both Stibitz and Williams worked tirelessly setting up the demonstration, and on September 11, 1940, Stibitz typed in complex number problems on a keyboard in Hanover and within a minute a correct answer came racing back from New York. This was the first demonstration of long-distance computing.

Three mathematicians who would themselves greatly influence computer science sat in the audience during this presentation—John von Neumann, Norbert Weiner, and John W. Mauchly who in just a few years' time would help invent ENIAC, the world's first large-scale electronic digital computer.

INTRODUCTION

In Chapter 15 two important rules were discussed that relate to this chapter on inductance. The first was that a magnetic field will build up around any current-carrying conductor (electromagnetism), and the second was that a voltage will be induced into a conductor when it is subjected to a moving magnetic field. These two rules form the basis for a phenomenon that is covered in this chapter and is called *self-inductance* or more commonly *inductance.* Inductance, by definition, is the ability of a device to oppose a change in current flow, and the device designed specifically to achieve this function is called the *inductor.* There is, in fact, no physical difference between an inductor and an electromagnet, since they are both coils. The two devices are given different names because they are used in different applications even though their construction and principle of operation are the same. An electromagnet is used to generate a magnetic field in response to current, while an inductor is used to oppose any change in current.

In this chapter we examine all aspects of inductance and inductors, including inductive circuits, types, reactance, testing, and applications.

16-1 SELF-INDUCTION

In Figure 16-1 an external voltage source has been connected across a coil, forcing a current through the coil. This current will generate a magnetic field that will increase in field strength in a short time from zero to maximum, expanding from the center of the conductor (electromagnetism). The expanding magnetic lines of force have relative motion with respect to the stationary conductor, so an induced voltage results (electromagnetic induction). The blooming magnetic field generated by the conductor is actually causing a voltage to be induced in the conductor that is generating the magnetic field. This effect of a current-carrying coil of conductor inducing a voltage within itself is known as **self-inductance.** This phenomenon was first discovered by Heinrich Lenz, who observed that the induced voltage causes an induced (bucking) current to flow in the coil, which opposes the source current producing it.

Self-Inductance
The property that causes a counterelectromotive force to be produced in a conductor when the magnetic field expands or collapses with a change in current.

Figure 16-2(a) shows an inductor connected across a dc source. When the switch is closed, a circuit current will exist through the inductor and the resistor. As the current rises toward its maximum value, the magnetic field expands, and throughout this time of relative motion between field and conductor, an induced voltage will be present. This induced voltage will produce an induced current to oppose the change in the circuit current.

When the current reaches its maximum, the magnetic field, which is dependent on current, will also reach a maximum value and then no longer expand but remain stationary. When the current remains constant, no change will occur in the magnetic field and therefore no relative motion will exist between the conductor and magnetic field, resulting in no induced voltage or current to oppose circuit current, as shown in Figure 16-2(b). The coil has accepted electrical

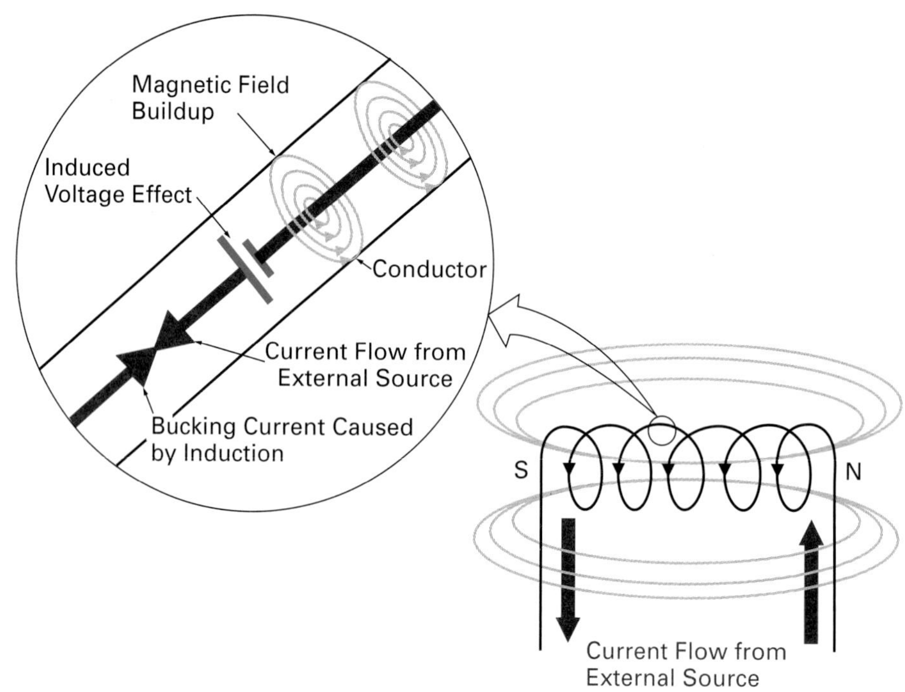

FIGURE 16-1 Self-Induction of a Coil.

Counter emf (Counter electromotive force) Abbreviated "counter emf," or "back emf," it is the voltage generated in an inductor due to an alternating or pulsating current and is always of opposite polarity to that of the applied voltage.

Inductance Property of a circuit or component to oppose any change in current as the magnetic field produced by the change in current causes an induced countercurrent to oppose the original change.

Henry Unit of inductance.

energy and is storing it in the form of a magnetic energy field, just as the capacitor stored electrical energy in the form of an electric field.

If the switch is put in position *B*, as shown in Figure 16-2(c), the current from the battery will be zero, and the magnetic field will collapse as it no longer has circuit current to support it. As the magnetic lines of force collapse, they cut the conducting coils, causing relative motion between the conductor and magnetic field. A voltage is induced in the coil, which will produce an induced current to flow in the same direction as the circuit current was flowing before the switch was opened. The coil is now converting the magnetic field energy into electrical energy and returning the original energy that it stored.

After a short period of time, the magnetic field will have totally collapsed, the induced voltage will be zero, and the induced current within the circuit will therefore also no longer be present.

This induced voltage is called a **counter emf** or *back emf.* It opposes the applied emf (or battery voltage). The ability of a coil or conductor to induce or produce a counter emf within itself as a result of a change in current is called *self-inductance,* or more commonly **inductance** (symbolized *L*). The unit of inductance is the **henry** (H), named in honor of Joseph Henry, an American physicist, for his experimentation within this area of science. The inductance of an inductor is 1 henry when a current c°hange of 1 ampere per second causes an induced voltage of 1 volt. Inductance is therefore a measure of how much counter emf (induced voltage) can be generated by an inductor for a given amount of current change through that same inductor.

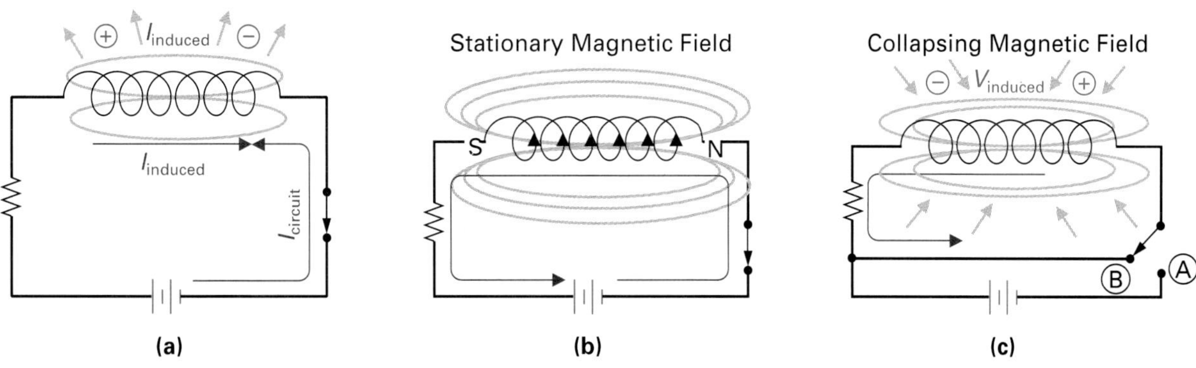

FIGURE 16-2 Self-Inductance. (a) Switch Closed. Increase in Circuit Current and, Therefore, Increase in Magnetic Field Induces Voltage in Coil. This Results in an Induced Current to Oppose Circuit Current. (b) Constant Circuit Current. Stationary Magnetic Field and, Therefore, No Induced Voltage or Current to Oppose Circuit Current. (c) Switch Opened. No Circuit Current and, Therefore, Magnetic Field Collapses and Induces Voltage in Coil. This Results in an Induced Current in Same Direction as Circuit Current.

This counter emf or induced voltage can be calculated by the formula

$$V_{\text{ind}} = L \times \frac{\Delta I}{\Delta t}$$

where L = inductance, in henrys (H)

ΔI = increment of change of current (I)

Δt = increment of change with respect to time (t)

A larger inductance ($L\uparrow$) will create a larger induced voltage ($V_{\text{ind}}\uparrow$), and if the rate of change of current with respect to time is increased ($\Delta I/\Delta t\uparrow$), the induced voltage or counter emf will also increase ($V_{\text{ind}}\uparrow$).

EXAMPLE:

What voltage is induced across an inductor of 4 H when the current is changing at a rate of:

a. 1 A/s?
b. 4 A/s?

Solution:

a. $V_{\text{ind}} = L \times \dfrac{\Delta I}{\Delta t} = 4\text{ H} \times 1\text{ A/s} = 4\text{ V}$

b. $V_{\text{ind}} = L \times \dfrac{\Delta I}{\Delta t} = 4\text{ H} \times 4\text{ A/s} = 16\text{ V}$

The faster the coil current changes, the larger the induced voltage.

SELF-TEST REVIEW QUESTIONS FOR SECTION 16-1

1. Define self-induction.
2. What is counter emf, and how can it be calculated?
3. Calculate the voltage induced in a 2mH inductor if the current is increasing at a rate of 4 kA/s.

16-2 THE INDUCTOR

Inductor
Coil of conductor used to introduce inductance into a circuit.

An **inductor** is basically an electromagnet, as its construction and principle of operation are the same. We use the two different names because they have different applications. The purpose of the electromagnet or solenoid is to generate a magnetic field, while the purpose of an inductor or coil is to oppose any change of circuit current.

In Figure 16-3 a steady value of direct current is present within the circuit and the inductor is creating a steady or stationary magnetic field. If the current in the circuit is suddenly increased (by lowering the circuit resistance), the change in the expanding magnetic field will induce a counter emf within the inductor. This induced voltage will oppose the source voltage from the battery and attempt to hold current at its previous low level.

The counter emf cannot completely oppose the current increase, for if it did, the lack of current change would reduce the counter emf to zero. Current therefore incrementally increases up to a new maximum, which is determined by the applied voltage and the circuit resistance ($I = V/R$). Once the new higher level of current has been reached and remains constant, there will no longer be a change. This lack of relative motion between field and conductor will no longer generate a counter emf, so the current will remain at its new higher constant value.

This effect also happens in the opposite respect. If current decreases (by increasing circuit resistance), the magnetic lines of force will collapse because of the reduction of current and induce a voltage in the inductor, which will produce

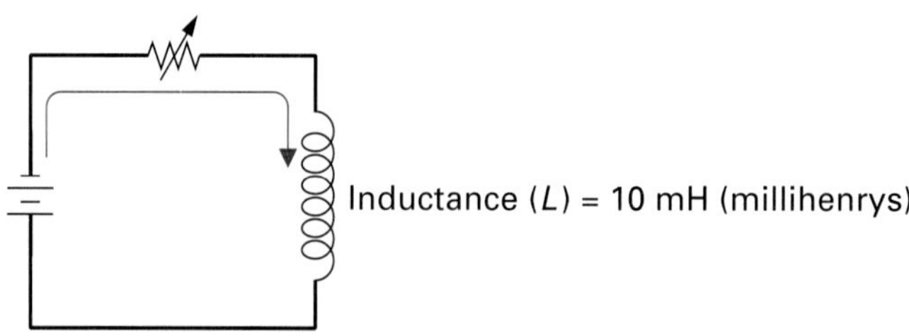

FIGURE 16-3 **Inductor's Ability to Oppose Current Change.**

an induced current in the same direction as the circuit current. These two combine and tend to maintain the current at the higher previous constant level. Circuit current will fall, however, as the induced voltage and current are only present during the change (in this case the decrease from the higher current level to the lower); and once the new lower level of current has been reached and remains constant, the lack of change will no longer induce a voltage or current. So the current will then remain at its new lower constant value.

The inductor is therefore an electronic component that will oppose any changes in circuit current, and this ability or behavior is referred to as *inductance.* Since the current change is opposed by a counter emf, inductance may also be defined as the ability of a device to induce a counter emf within itself for a change in current.

SELF-TEST REVIEW QUESTIONS FOR SECTION 16-2

1. What is the difference between an electromagnet and an inductor?
2. True or false: The inductor will oppose any changes in circuit current.

16-3 FACTORS DETERMINING INDUCTANCE

The inductance of an inductor is determined by four factors:

1. Number of turns
2. Area of the coil
3. Length of the coil
4. Core material used within the coil

Let's now discuss how these four factors can affect inductance, beginning with the number of turns.

16-3-1 *Number of Turns (N) (Figure 16-4)*

If an inductor has a greater number of turns, the magnetic field produced by passing current through the coil will have more magnetic force than an inductor with fewer turns. A greater magnetic field will cause a larger counter emf, because more magnetic lines of flux will cut more coils of the conductor, producing a larger inductance value. Inductance *(L)* is therefore proportional to the number of turns *(N):*

$$L \propto N$$

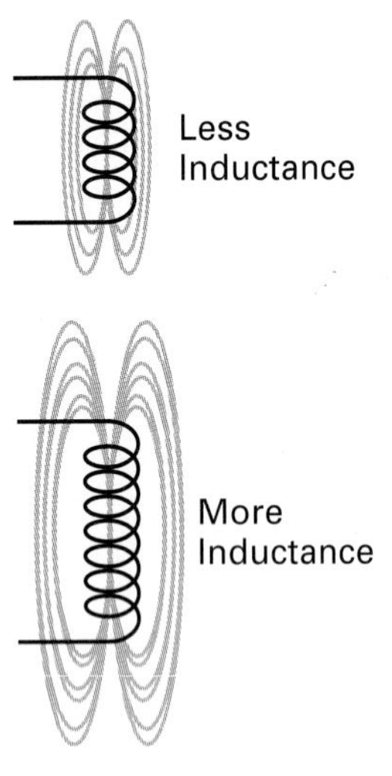

FIGURE 16-4 Inductance Is Proportional to the Number of Turns ($L \propto N$) in a Coil.

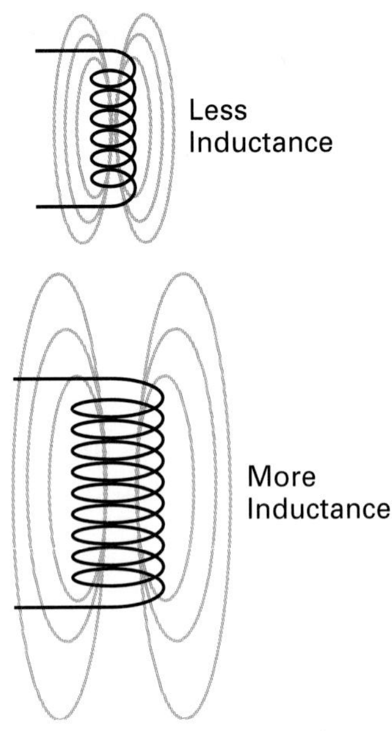

FIGURE 16-5 Inductance Is Proportional to the Area ($L \propto A$) in a Coil.

16-3-2 *Area of Coil (A) (Figure 16-5)*

If the area of the coil is increased for a given number of turns, more magnetic lines of force will be produced, and if the magnetic field is increased, the inductance will also increase. Inductance *(L)* is therefore proportional to the cross-sectional area of the coil *(A)*:

$$L \propto A$$

16-3-3 *Length of Coil (l) (Figure 16-6)*

If, for example, four turns are spaced out (long-length coil), the summation that occurs between all the individual coil magnetic fields will be small. On the other hand, if four turns are wound close to one another (short-length coil), all the

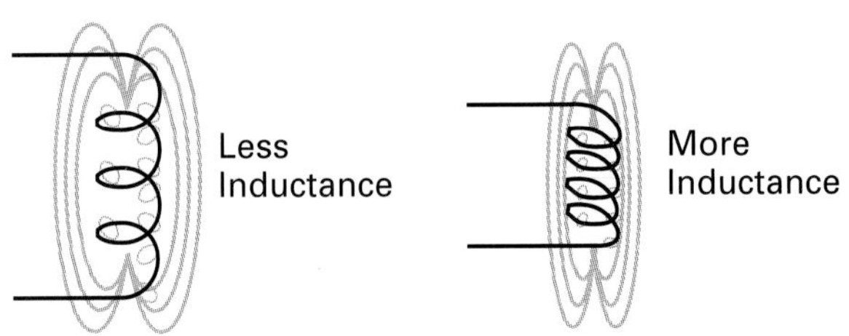

FIGURE 16-6 Inductance Is Inversely Proportional to the Length ($L \propto 1/l$) of a Coil.

individual coil magnetic fields will easily interact and add together to produce a larger magnetic field and, therefore, greater inductance. Inductance is therefore inversely proportional to the length of the coil, in that a longer coil, for a given number of turns, produces a smaller inductance, and vice versa.

$$L \propto \frac{1}{l}$$

16-3-4 *Core Material (μ) (Figure 16-7)*

Most inductors have core materials such as nickel, cobalt, iron, steel, ferrite, or an alloy. These cores have magnetic properties that concentrate or intensify the magnetic field. Permeability is another factor that is proportional to inductance and the figures for various materials are shown in Table 16-1. The greater the permeability of the core material, the greater the inductance is.

$$L \propto \mu$$

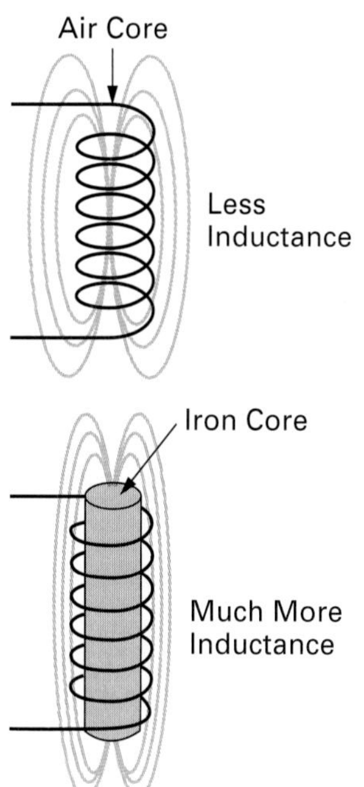

FIGURE 16-7 Inductance Is Proportional to the Permeability of the Coil's Material ($L \propto \mu$).

TABLE 16-1 Permeabilities of Various Materials

MATERIAL	PERMEABILITY (μ)
Air or vacuum	1.26×10^{-6}
Nickel	6.28×10^{-5}
Cobalt	7.56×10^{-5}
Cast iron	1.1×10^{-4}
Machine steel	5.65×10^{-4}
Transformer iron	6.9×10^{-3}
Silicon iron	8.8×10^{-3}
Permaloy	0.126
Supermalloy	1.26

16-3-5 *Formula for Inductance*

All the four factors described can be placed in a formula to calculate inductance.

$$L = \frac{N^2 \times A \times \mu}{l}$$

where L = inductance, in henrys (H)
N = number of turns
A = cross-sectional area, in square meters (m^2)
μ = permeability
l = length of core, in meters (m)

EXAMPLE:

Refer to Figure 16-8(a) and (b) and calculate the inductance of each.

Solution:

a. $L = \dfrac{5^2 \times 0.01 \times (6.28 \times 10^{-5})}{0.001} = 15.7 \text{ mH}$

b. $L = \dfrac{10^2 \times 0.1 \times (1.1 \times 10^{-4})}{0.1} = 11 \text{ mH}$

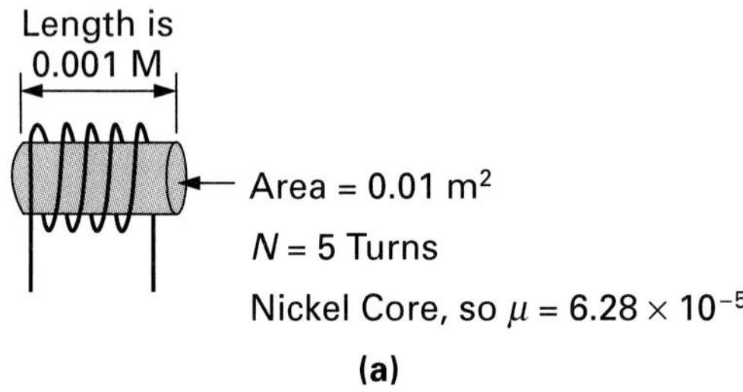

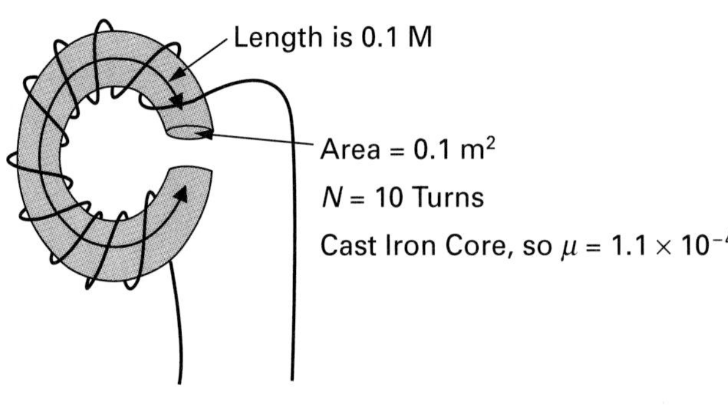

FIGURE 16-8 Inductor Examples.

SELF-TEST REVIEW QUESTIONS FOR SECTION 16-3

1. List the four factors that determine the inductance of an inductor.
2. State the formula for inductance.

16-4 INDUCTORS IN COMBINATION

Inductors oppose the change of current in a circuit and so are treated in a manner similar to resistors connected in combination. Two or more inductors in series merely extend the coil length and increase inductance. Inductors in parallel are treated in a manner similar to resistors, with the total inductance being less than that of the smallest inductor's value.

16-4-1 *Inductors in Series*

When inductors are connected in series with one another, the total inductance is calculated by summing all the individual inductances.

$$L_T = L_1 + L_2 + L_3 + \ldots$$

EXAMPLE:

Calculate the total inductance of the circuit shown in Figure 16-9.

Solution:

$$\begin{aligned} L_T &= L_1 + L_2 + L_3 \\ &= 5\text{ mH} + 7\text{ mH} + 10\text{ mH} \\ &= 22\text{ mH} \end{aligned}$$

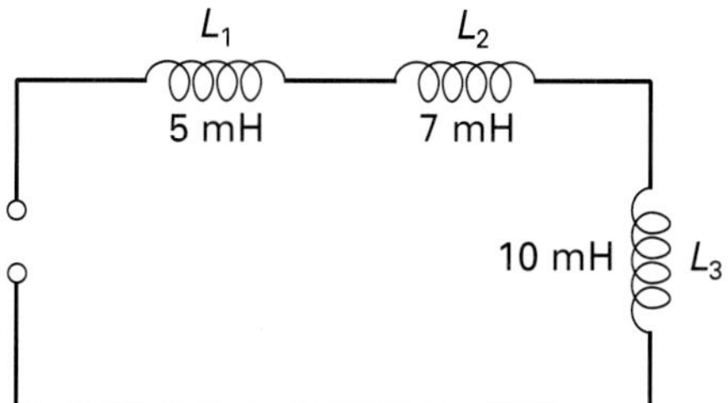

FIGURE 16-9 Inductors in Series.

16-4-2 *Inductors in Parallel*

When inductors are connected in parallel with one another, the reciprocal (two or more inductors) or product over sum (two inductors) formula can be used to find total inductance, which will always be less than the smallest inductor's value.

$$L_T = \frac{1}{(1/L_1) + (1/L_2) + (1/L_3) + \ldots}$$

$$L_T = \frac{L_1 \times L_2}{L_1 + L_2}$$

EXAMPLE:

Determine L_T for the circuits in Figure 16-10(a) and (b).

Solution:

a. General formula:

$$L_T = \frac{1}{(1/L_1) + (1/L_2) + (1/L_3)}$$

$$= \frac{1}{(1/10 \text{ mH}) + (1/5 \text{ mH}) + (1/20 \text{ mH})}$$

$$= 2.9 \text{ mH}$$

b. Product over sum:

$$L_T = \frac{L_1 \times L_2}{L_1 + L_2}$$

$$= \frac{10\ \mu\text{H} \times 2\ \mu\text{H}}{10\ \mu\text{H} + 2\ \mu\text{H}}$$

$$= \frac{20 \times 10^{-12}\text{H}}{12\ \mu\text{H}} = 1.67\ \mu\text{H}$$

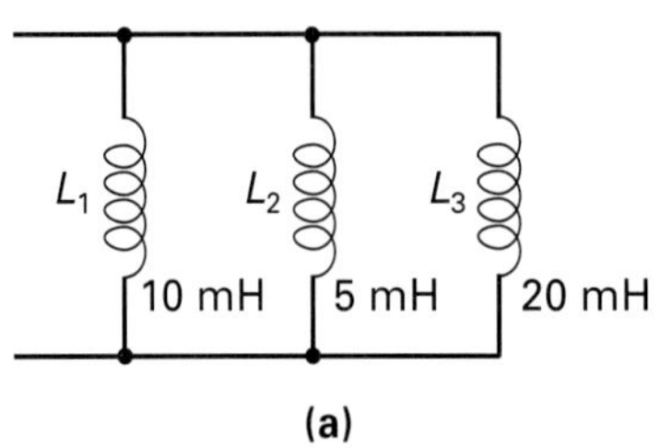

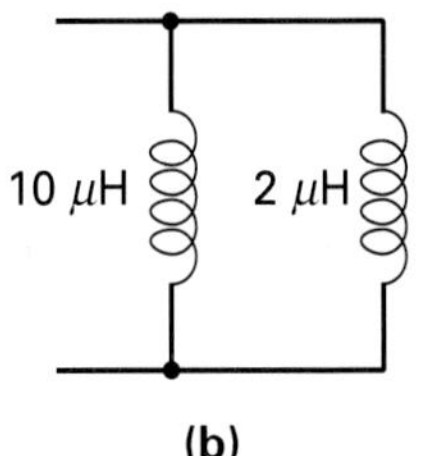

FIGURE 16-10 Inductors in Parallel.

SELF-TEST REVIEW QUESTIONS FOR SECTION 16-4

1. True or false: To calculate total inductance, inductors can be treated in the same manner as capacitors.
2. State the formula for calculating total inductance in:
   a. A series circuit
   b. A parallel circuit
3. Calculate the total circuit inductance if 4 mH and 2 mH are connected:
   a. In series
   b. In parallel

16-5 TYPES OF INDUCTORS

As with resistors and capacitors, inductors are basically divided into the two categories of fixed-value and variable-value inductors, as shown by the symbols in Figure 16-11(a) and (b). Within these two categories, inductors are generally classified by the type of core material used. Figure 4-11(c) shows a variety of different types.

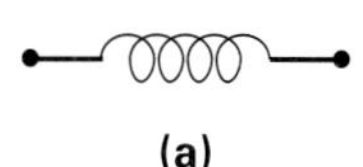

(a)

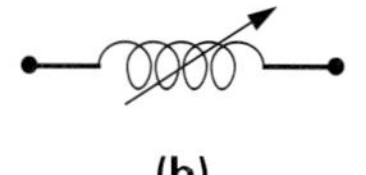

(b)

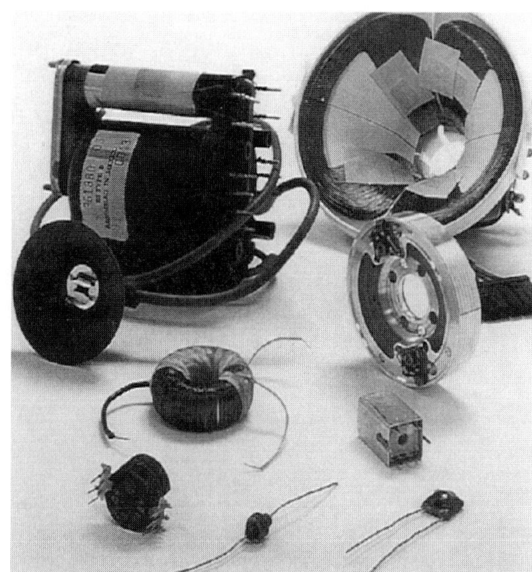

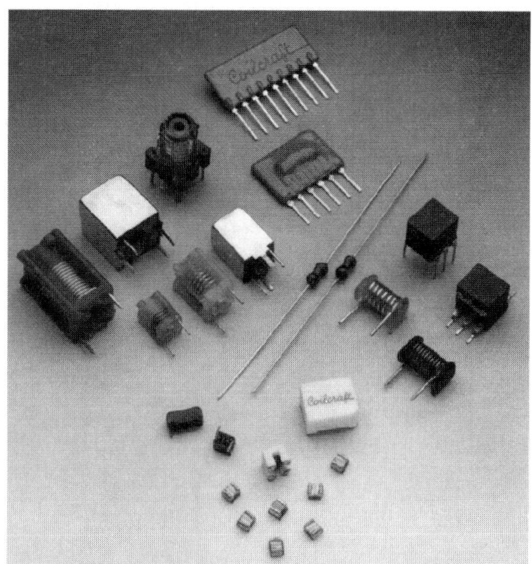

(c)

FIGURE 16-11 Inductor Types. (a) Fixed-Value Inductor Symbol. (b) Variable-Value Inductor Symbol. (c) Physical Appearance of Inductors. (Left photo courtesy of Sencore. Right photo courtesy of Coilcraft Inc., Cary, Illinois, 60013.)

16-5-1 *Fixed-Value Inductors*

Fixed-Value Inductor An inductor whose value cannot be changed.

With **fixed-value inductors,** the inductance value remains constant and cannot be altered. It is usually made of solid copper wire with an insulating enamel around the conductor. This enamel or varnish on the conductor is needed since the turns of a coil are generally wound on top of one another. The three major types of fixed-value inductors on the market today are:

1. Air core **2.** Iron core **3.** Ferrite core

Air Core

Figure 16-12 illustrates some typical air-core fixed-value inductor types. The insulated copper wire can be wound on nonmagnetic materials such as plastic, ceramic, or Bakelite, which due to their nonmagnetic properties will have no effect on the inductance value. These materials act as a support or form for the inductor, whose wire size may be too small to support itself. If a heavier or more rigid wire size is used and it can support itself, there will be no need for a nonmagnetic form.

Air-Core Inductor An inductor that has no metal core.

Inductors of this type, which make use of nonmagnetic form or do not have a form, are known as **air-core inductors.** Air-core inductors have the advantage that their inductance value does not vary with current, as do iron- and ferrite-core inductors. Their typical low values (below 10 μH) normally find application in high-frequency equipment, such as AM and FM radio, TV, and other communications transmitter and receiver circuits.

Iron Core

Iron-Core Inductor or Choke Inductor used to impede the flow of alternating or pulsating current.

Figure 16-13 illustrates a typical **iron-core inductor.** The name **choke** is used interchangeably with the word *inductor,* because choke basically describes an inductor's behavior. As we discovered previously, the inductor will oppose any change of current flow, whether it is in the form of pulsating dc or ac, because of its self-inductance, which generates a counter emf to limit or choke the flow of current.

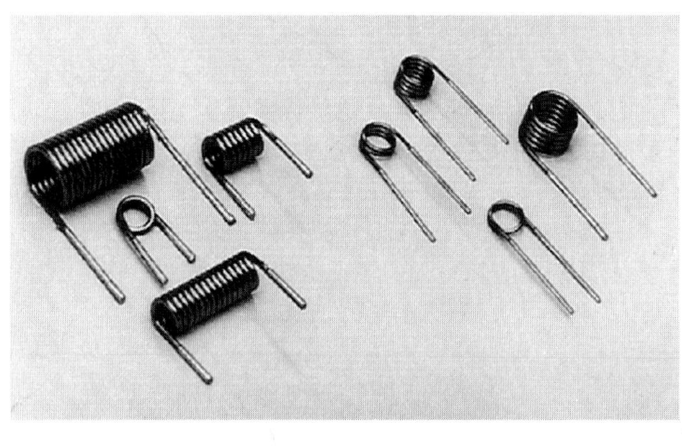

(a)

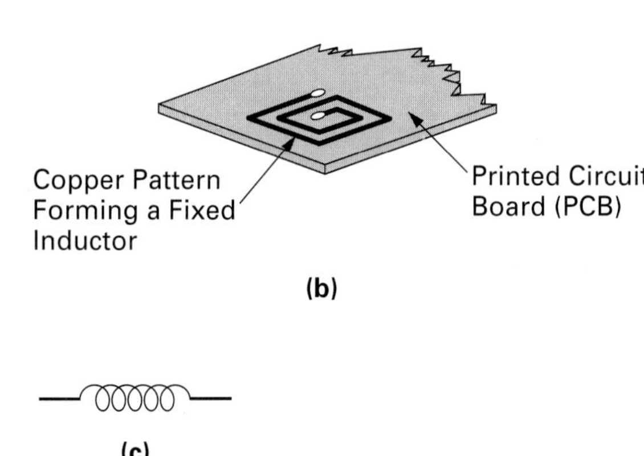

FIGURE 16-12 Air-Core Fixed-Value Inductors. (a) Air-Core Inductor (Courtesy of Dale Electronics). (b) PBC Inductor. (c) Symbol.

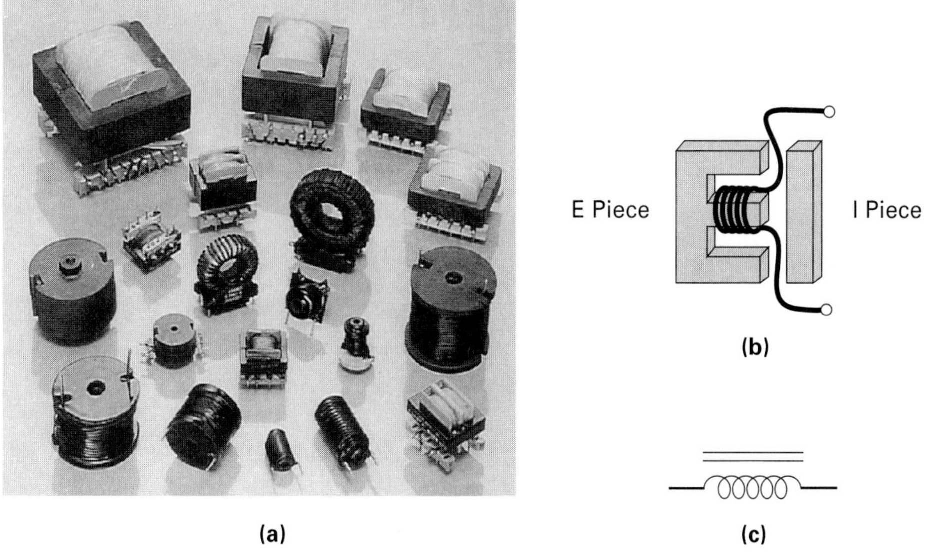

FIGURE 16-13 **Iron-Core Fixed-Value Inductors. (a) Physical Appearance (Courtesy of Coilcraft). (b) Construction. (c) Symbol.**

An iron core has a higher permeability figure than air, which means that it will concentrate the magnetic lines of force and therefore increase inductance.

$$L\uparrow = \frac{N^2 \times A \times \mu\uparrow}{l}$$

The EI **ferrous** or compound of iron core [Figure 16-13(b)] has two sections, one of which resembles the letter E and the other I. The conductor is wrapped around the center section of the E to form the coil. These inductors or chokes can be made with the highest inductance values, up to the hundreds of henrys, and are used in dc and low-frequency ac circuits. Figure 16-13(c) shows the schematic symbol for an iron-core inductor.

Ferrous
Composed of and/or containing iron. A ferrous metal exhibits magnetic characteristics, as opposed to nonferrous metals.

Ferrite Core

Figure 16-14 illustrates two typical **ferrite-core inductors. Ferrite** is a chemical compound, which is basically powdered iron oxide and ceramic. Although its permeability is less than iron, it has a higher permeability than air and can therefore obtain a higher inductance value than air for the same number of turns.

Ferrite-Core Inductor
An inductor containing a ferrite core.

Ferrite
A powdered, compressed, and sintered magnetic material having high resistivity. The high resistance makes eddy-current losses low at high frequencies.

The toroidal inductor (doughnut shaped) illustrated in Figure 16-14(a) has an advantage over the cylinder-shaped core in that the magnetic lines of force do not have to pass through air. With a cylindrical core, the magnetic lines of force merge out of one end of the cylinder, propagate or travel through air, and then enter in the other end of the cylinder. The nonmagnetic air, which has a low permeability, will reduce the strength of the field and therefore the inductance value:

$$L\downarrow = \frac{N^2 \times A \times \mu\downarrow}{l}$$

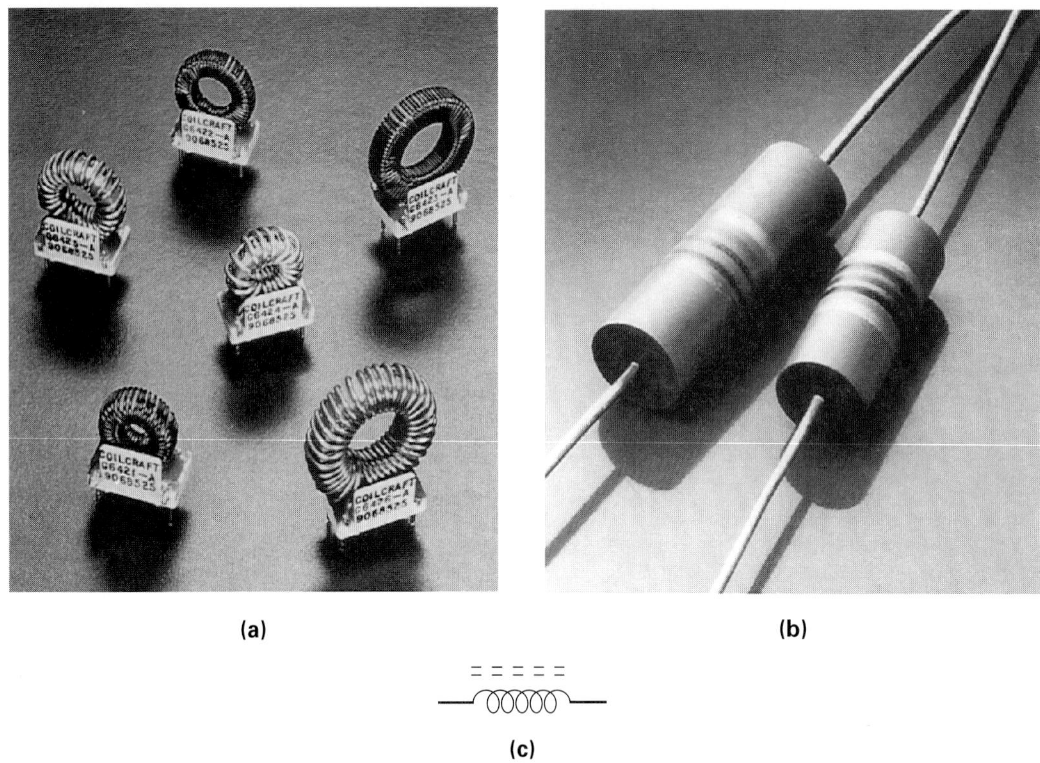

FIGURE 16-14 Ferrite-Core Fixed-Value Inductors. (a) Toroidal Inductor (Courtesy of Coilcraft). (b) Molded Inductor (Courtesy of Dale Electronics). (c) Schematic Symbol.

If one could bend around either end of the cylinder to make a toroid, the magnetic lines of force emerging from one end of the inductor's core would pass from ferrite to ferrite, which has a high permeability.

Toroidal ferrite-core inductors have a greater inductance than the cylindrical-type inductors; however, they are more expensive to manufacture. They are used in both low- and high-frequency applications.

The molded inductor, shown in Figure 16-14(b), consists of a spiral metal film deposited on a cylindrical ferrite core. The entire assembly is encapsulated, and the available values range from 1.2 μH to 10 mH, with maximum currents of about 70 mA. Due to its small size, the value is often color coded onto the body instead of using alphanumeric or typographical labels, which can be used on all of the other larger inductors. Figure 16-14(c) shows the schematic symbol for an iron-core inductor.

16-5-2 Variable-Value Inductors

Variable-Value Inductor
An inductor whose value can be varied.

With **variable-value inductors,** the inductance value can be changed by adjusting the position of the movable core with respect to the stationary coil. There is basically only one type of variable-value inductor in wide use today: the ferrite core. Figure 16-15 illustrates a typical ferrite-core variable inductor. A screw

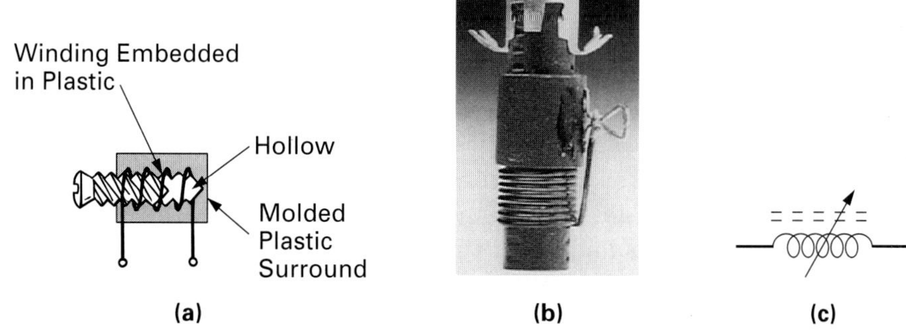

FIGURE 16-15 **Ferrite-Core Variable-Value Inductors. (a) Construction. (b) Physical Appearance (Courtesy of Sencore). (c) Symbol.**

adjustment moves a sliding ferrite core or slug farther in or out of the stationary coil. If the slug is all the way out, the permeability figure is low, because an air core is being used ($L\downarrow \propto \mu\downarrow$). If the slug is adjusted and screwed into the coil, the air is replaced by the ferrite slug, which has a higher permeability and, therefore, inductance value ($L\uparrow \propto \mu\uparrow$).

Variable inductors should only be adjusted with a plastic or nonmetallic alignment tool since any metal object in the vicinity of the inductor will interfere with the magnetic field and change the inductance of the inductor. Variable-value inductors are used extensively in radio circuits.

SELF-TEST REVIEW QUESTIONS FOR SECTION 16-5

1. List the three fixed-value inductor types.
2. Under which category would a fixed-value inductor with a nonmagnetic core come?
3. What is a ferrite compound?
4. What advantage does the toroid-shaped inductor have over a cylindrical inductor?
5. What type of variable-value inductor is the only one in wide use today?
6. What factor is varied to change the value of the variable inductor?

16-6 INDUCTIVE TIME CONSTANT

Inductors will not have any effect on a steady value of direct current (dc) from a dc voltage source. If, however, the dc is changing (pulsating), the inductor will oppose the change whether it is an increase or decrease in direct current, because a change in current causes the magnetic field to expand or contract, and in so doing it will cut the coil of the inductor and induce a voltage that will counter the applied emf.

16-6-1 DC Current Rise

Figure 16-16(a) illustrates an inductor *(L)* connected across a dc source (battery) through a switch and series-connected resistor. When the switch is closed, current will flow and the magnetic field will begin to expand around the inductor. This field cuts the coils of the inductor and induces a counter emf to oppose the rise in current. Current in an inductive circuit, therefore, cannot rise instantly to its maximum value, which is determined by Ohm's law ($I = V/R$). Current will in fact take a time to rise to maximum, as graphed in Figure 16-16(b), due to the inductor's ability to oppose change.

It will actually take five time constants (5τ) for the current in an inductive circuit to reach maximum value. This time can be calculated by using the formula

$$\tau = \frac{L}{R} \quad \text{seconds}$$

The constant to remember is the same as before: 63.2%. In one time constant ($1 \times L/R$) the current in the RL circuit will have reached 63.2% of its maximum value. In two time constants ($2 \times L/R$), the current will have increased 63.2% of the remaining current, and so on, through five time constants.

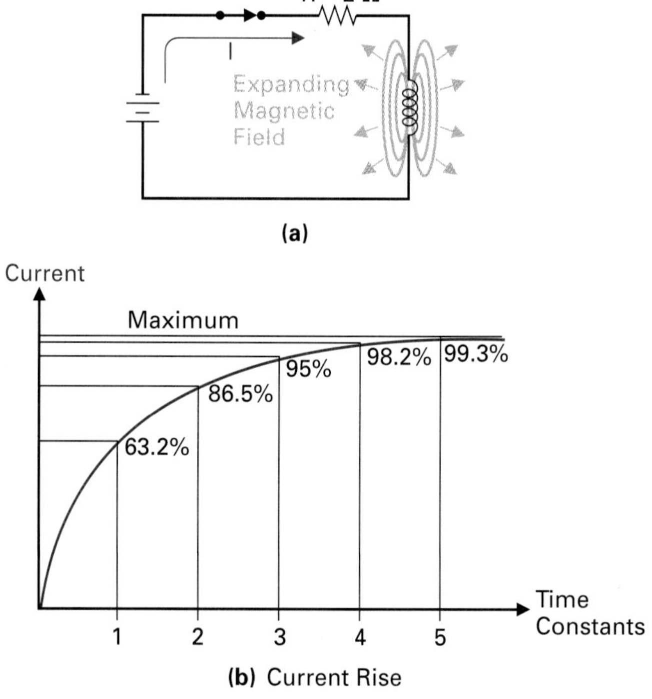

FIGURE 16-16 DC Inductor Current Rise.

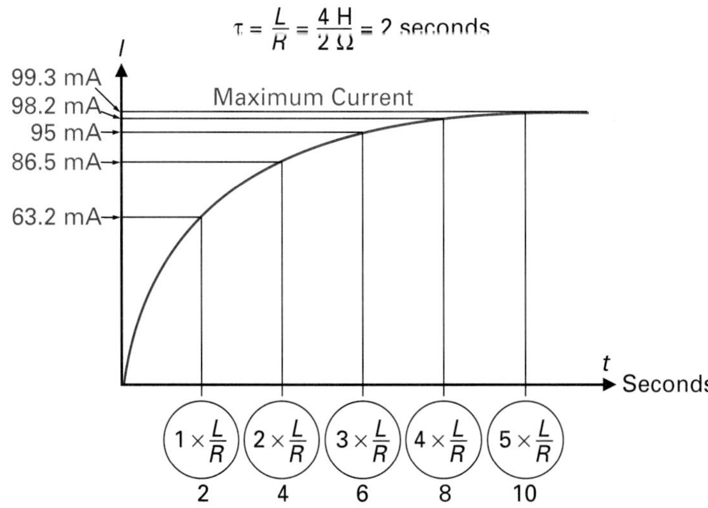

FIGURE 16-17 Exponential Current Rise.

For example, if the maximum possible circuit current is 100 mA and an inductor of 4 H is connected in series with a resistor of 2 Ω, the current will increase as shown in Figure 16-17.

Referring back to the $\tau = L/R$ formula, you will notice that how quickly an inductor will allow the current to rise to its maximum value is proportional to the inductance and inversely proportional to the resistance. A larger inductance increases the strength of the magnetic field, so the opposition or counter emf increases, and the longer it takes for current to rise to a maximum ($\tau\uparrow = L\uparrow/R$). If the circuit resistance is increased, the maximum current will be smaller, and a smaller maximum is reached more quickly than a higher ($\tau\downarrow = L/R\uparrow$).

16-6-2 *DC Current Fall*

When the inductor's dc source of current is removed, as shown in Figure 16-18(a), by placing the switch in position *B,* the magnetic field will collapse and cut the coils of the inductor, inducing a voltage and causing a current to flow in the same direction as the original source current. This current will exponentially decay, or fall from the maximum to zero level, in five time constants ($5 \times L/R = 5 \times 4/2 = 10$ seconds), as shown in Figure 16-18(b).

EXAMPLE:

Calculate the circuit current at each of the five time constants if a 12 V dc source is connected across a series *RL* circuit, and $R = 60\ \Omega$ and $L = 24$ mH. Plot the results on a graph showing current against time.

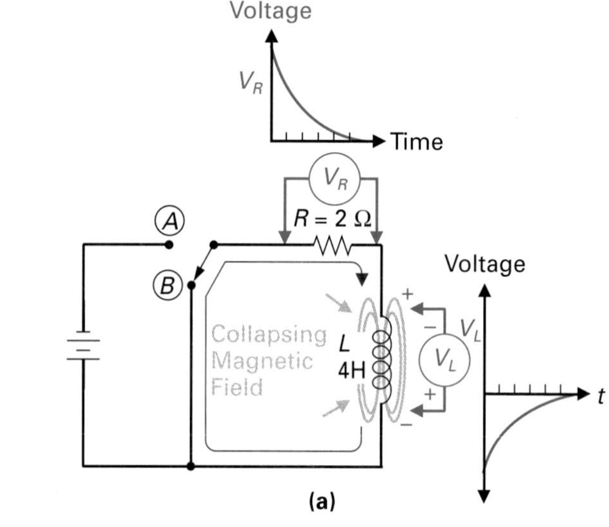

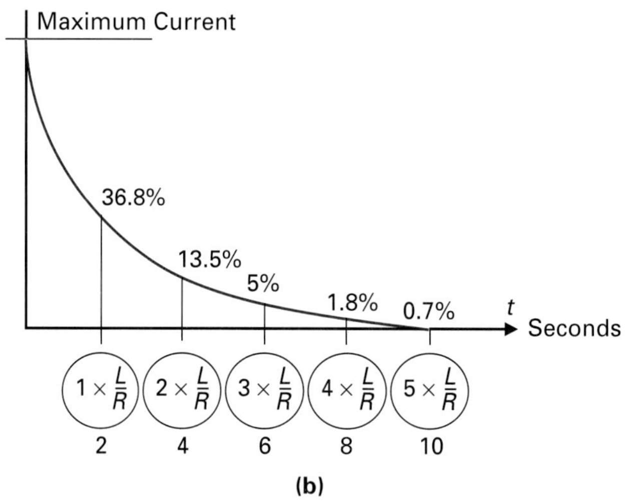

FIGURE 16-18 Exponential Current Fall.

Solution:

$$\text{maximum current, } I_{\max} = \frac{V_S}{R} = \frac{12\text{ V}}{60\ \Omega} = 200\text{ mA}$$

$$\text{time constant, } \tau = \frac{L}{R} = \frac{24\text{ mH}}{60\ \Omega} = 400\ \mu\text{s}$$

At 1 time constant (400 μs after source voltage is applied), the current will be

$$I = 63.2\% \text{ of } I_{\max}$$
$$= 0.632 \times 200\text{ mA} = 126.4\text{ mA}$$

At two time constants (800 μs after source voltage is applied):

$$I = 86.5\% \text{ of } I_{\max}$$
$$= 0.865 \times 200\text{ mA} = 173\text{ mA}$$

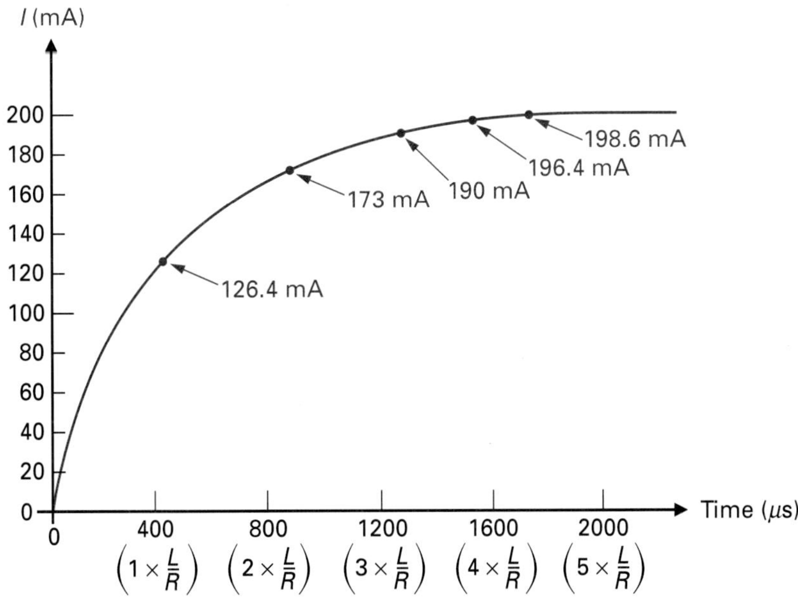

FIGURE 16-19 Exponential Current Rise for Example 16-5.

At three time constants (1200 μs or 1.2 ms):

$$I = 95\% \text{ of } I_{max}$$
$$= 0.95 \times 200 \text{ mA} = 190 \text{ mA}$$

At four time constants (1.6 ms):

$$I = 98.2\% \text{ of } I_{max}$$
$$= 0.982 \times 200 \text{ mA} = 196.4 \text{ mA}$$

At five time constants (2 ms):

$$I = 99.3\% \text{ of } I_{max}$$
$$= 0.993 \times 200 \text{ mA} = 198.6 \text{ mA, approximately maximum (200 mA)}$$

See Figure 16-19.

16-6-3 AC Rise and Fall

If an alternating (ac) voltage is applied across an inductor, as shown in Figure 16-20(a), the inductor will continuously oppose the alternating current because it is always changing. Figure 16-20(b) shows the phase relationship between the voltage across an inductor or counter emf and the circuit current.

The current in the circuit causes the magnetic field to expand and collapse and cut the conducting coils, resulting in an induced counter emf. At points *X* and *Y*, the steepness of the current waveform indicates that the current will be changing at its maximum rate, and therefore the opposition or counter emf will also be maximum. When the current is at its maximum positive or negative value, it has a very small or no rate of change (flat peaks). Therefore, the opposi-

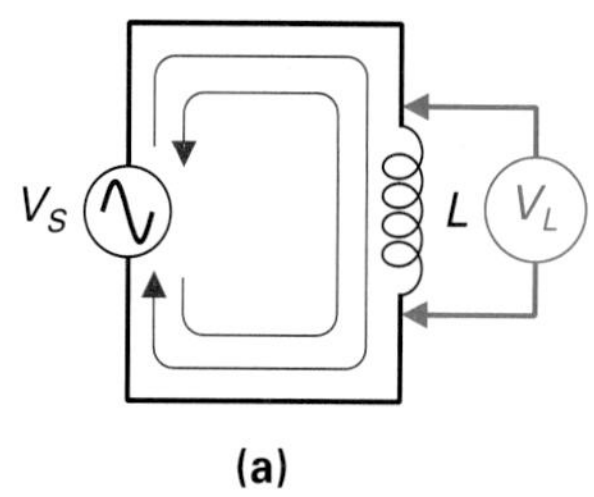

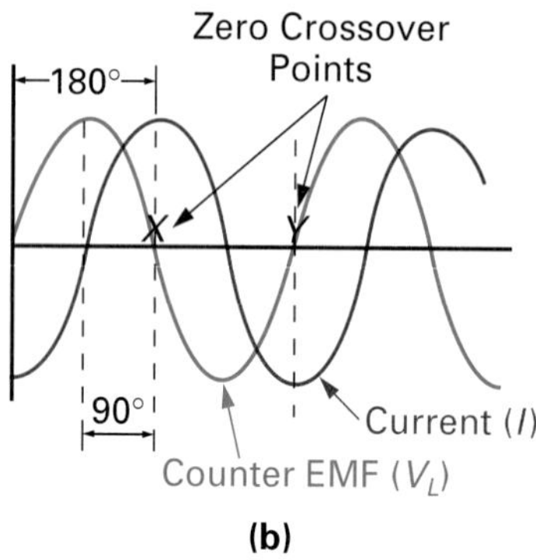

FIGURE 16-20 The Voltage across an Inductor Leads the Circuit Current by 90° in an Inductive Circuit.

tion or counter emf should be very small or zero, as can be seen by the waveforms. The counter emf is, therefore, said to be 90° out of phase with the circuit current.

To summarize, we can say that the voltage across the inductor (V_L) or counter emf leads the circuit current *(I)* by 90°.

SELF-TEST REVIEW QUESTIONS FOR SECTION 16-6

1. How does the inductive time constant relate to the capacitive time constant?
2. True or false: The greater the value of the inductor, the longer it would take for current to rise to a maximum.
3. True or false: A constant dc level is opposed continuously by an inductor.
4. What reaction does an inductor have to ac?

16-7 INDUCTIVE REACTANCE

Inductive Reactance Measured in ohms, it is the opposition to alternating or pulsating current flow without the dissipation of energy.

Reactance is the opposition to current flow without the dissipation of energy, as opposed to resistance, which is the opposition to current flow with the dissipation of energy.

Inductive reactance (X_L) is the opposition to current flow offered by an inductor without the dissipation of energy. It is measured in ohms and can be calculated by using the formula:

$$X_L = 2\pi \times f \times L$$

where X_L = inductive reactance, in ohms (Ω)

$2\pi = 2\pi$ radians, 360° or 1 cycle

f = frequency, in hertz (Hz)

L = inductance, in henrys (H)

Inductive reactance is proportional to frequency ($X_L \propto f$) because a higher frequency (fast-switching current) will cause a greater amount of current change, and a greater change will generate a larger counter emf, which is an opposition or reactance against current flow. When 0 Hz is applied to a coil (dc), there exists no change, so the inductive reactance of an inductor to dc is zero ($X_L = 2\pi \times 0 \times L = 0$).

Inductive reactance is also proportional to inductance because a larger inductance will generate a greater magnetic field and subsequent counter emf, which is the opposition to current flow.

Ohm's law can be applied to inductive circuits just as it can be applied to resistive and capacitive circuits. The current flow in an inductive circuit *(I)* is proportional to the voltage applied *(V)*, and inversely proportional to the inductive reactance (X_L). Expressed mathematically,

$$I = \frac{V}{X_L}$$

EXAMPLE:

Calculate the current flowing in the circuit illustrated in Figure 16-21.

Solution:

The current can be calculated by Ohm's law and is a function of the voltage and opposition, which in this case is inductive reactance.

$$I = \frac{V}{X_L}$$

However, we must first calculate X_L:

$$\begin{aligned} X_L &= 2\pi \times f \times L \\ &= 6.28 \times 50 \text{ kHz} \times 15 \text{ mH} \\ &= 4710\ \Omega \text{ or } 4.71\ \text{k}\Omega \end{aligned}$$

Current is therefore equal to

$$I = \frac{V}{X_L} = \frac{10 \text{ V}}{4.71\ \text{k}\Omega} = 2.12 \text{ mA}$$

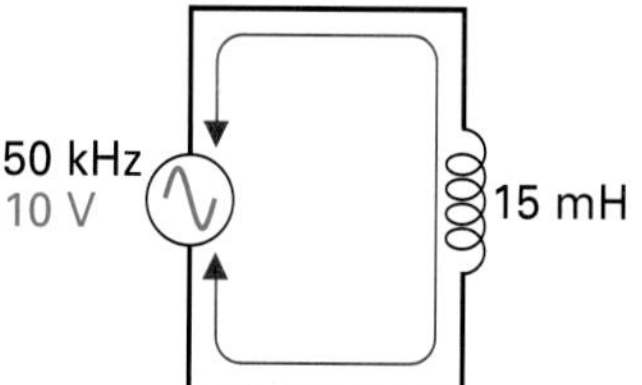

FIGURE 16-21

EXAMPLE:

What opposition or inductive reactance will a motor winding or coil offer if $V = 12$ V and $I = 4.5$ A?

Solution:

V

$I \mid X_L$

$$X_L = \frac{V}{I} = \frac{12 \text{ V}}{4.5 \text{ A}} = 2.66\ \Omega$$

SELF-TEST REVIEW QUESTIONS FOR SECTION 16-7

1. Define and state the formula for inductive reactance.
2. Why is inductive reactance proportional to frequency and the inductance value?
3. How does inductive reactance relate to Ohm's law?
4. True or false: Inductive reactance is measured in henrys.

16-8 SERIES *RL* CIRCUIT

In a purely resistive circuit, as seen in Figure 16-22, the current flowing within the circuit and the voltage across the resistor are in phase with one another. In a purely inductive circuit, as shown in Figure 16-23, the current will lag the applied voltage by 90°. If we connect a resistor and inductor in series, as shown in Figure 16-24(a), we will have the most common combination of *R* and *L* used in electronic equipment.

16-8-1 Voltage

The voltage across the resistor and inductor shown in Figure 16-24(a) can be calculated by using Ohm's law:

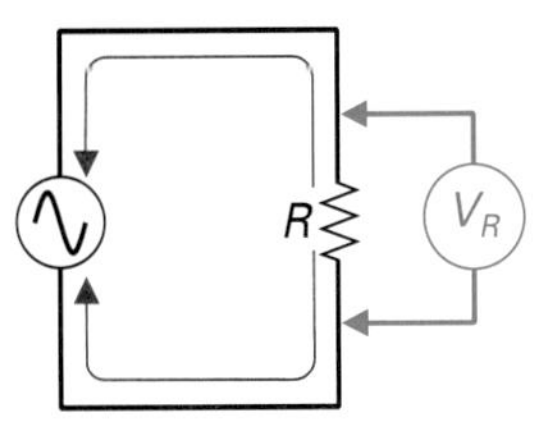

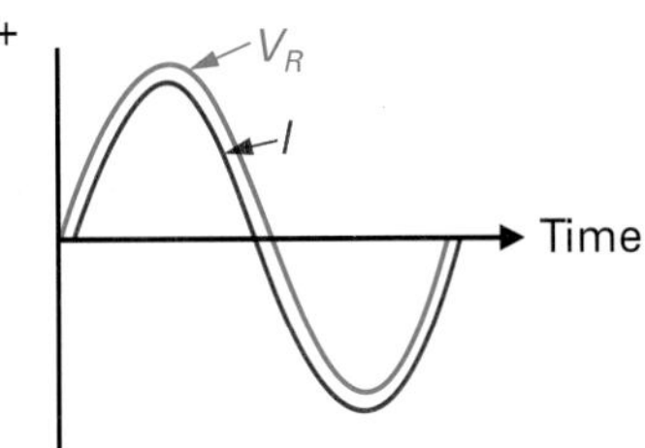

FIGURE 16-22 **Purely Resistive Circuit: Current and Voltage Are in Phase.**

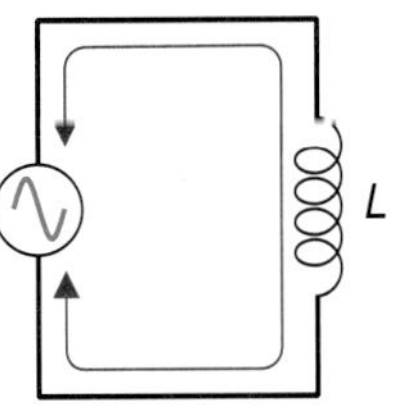

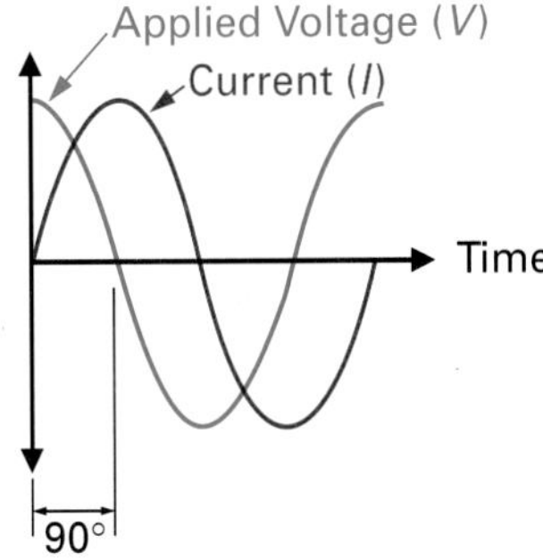

FIGURE 16-23 **Purely Inductive Circuit: Current Lags Applied Voltage by 90°.**

$$V_R = I \times R$$

$$V_L = I \times X_L$$

The vector diagram in Figure 16-24(b) illustrates current *(I)* as the reference at 0° and, as expected, the voltage across the resistor (V_R) is in phase with the circuit current. The voltage across the inductor (V_L) leads the circuit current and the voltage across the resistor (V_R) by 90°, or the circuit current vector lags the voltage across the inductor by 90°.

As with any series circuit, we have to apply Kirchhoff's voltage law when calculating the value of applied or source voltage (V_S), which, due to the phase difference between V_R and V_L, is the vector sum of all the voltage drops. By cre-

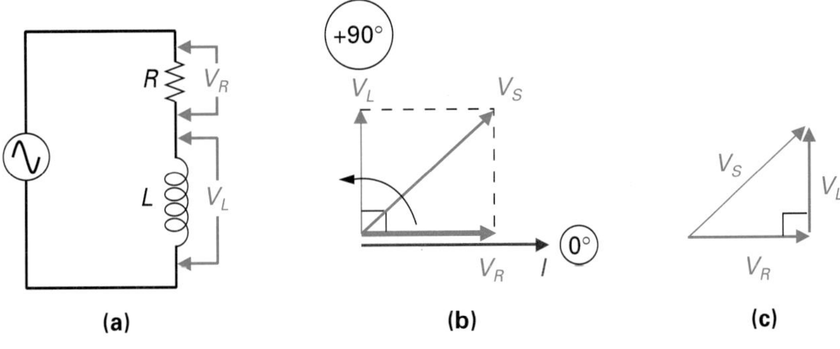

FIGURE 16-24 **Series *RL* Circuit.**

ating a right triangle from the three quantities, as shown in Figure 16-24(c), and applying the Pythagorean theorem, we arrive at a formula for source voltage.

$$V_S = \sqrt{V_R^2 + V_L^2}$$

As with any formula with three quantities, if two are known, the other can be calculated by simply rearranging the formula to

$$V_S = \sqrt{V_R^2 + V_L^2}$$

$$V_R = \sqrt{V_S^2 - V_L^2}$$

$$V_L = \sqrt{V_S^2 - V_R^2}$$

EXAMPLE:

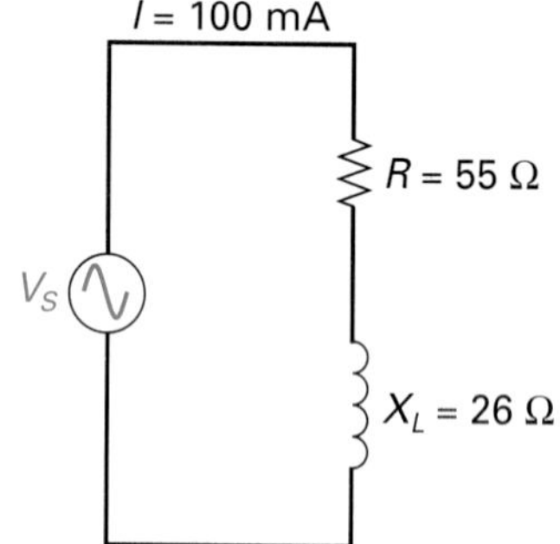

FIGURE 16-25 Voltage in a Series *RL* Circuit.

Calculate V_R, V_L, and V_S for the circuit shown in Figure 16-25.

Solution:

$$\begin{aligned} V_R &= I \times R \\ &= 100\text{ mA} \times 55\ \Omega \\ &= 5.5\text{ V} \\ V_L &= I \times X_L \\ &= 100\text{ mA} \times 26\ \Omega \\ &= 2.6\text{ V} \\ V_S &= \sqrt{V_R^2 + V_L^2} \\ &= \sqrt{5.5\text{ V}^2 + 2.6\text{ V}^2} \\ &= 6\text{ V} \end{aligned}$$

16-8-2 *Impedance (Z)*

Impedance is the total opposition to current flow offered by a circuit with both resistance and reactance. It is measured in ohms and can be calculated by using Ohm's law:

$$Z = \frac{V}{I}$$

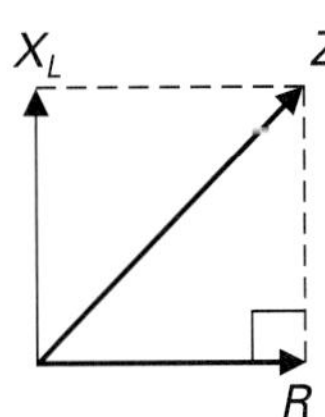

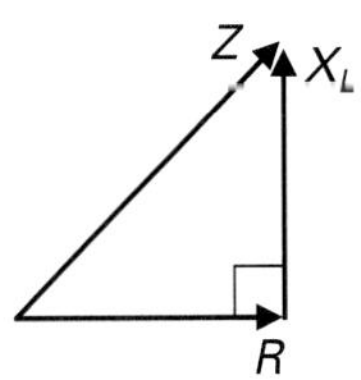

FIGURE 16-26 Impedance in a Series *RL* Circuit.

Just as a phase shift or difference exists between V_R and V_L and they cannot be added to find applied voltage, the same phase difference exists between R and X_L, so impedance or total opposition cannot be simply the sum of the two, as shown in Figure 16-26.

The impedance of a series RL circuit is equal to the square root of the sum of the squares of resistance and reactance, and by rearrangement, X_L and R can also be calculated if the other two values are known:

$$Z = \sqrt{R^2 + X_L^2}$$

$$R = \sqrt{Z^2 - X_L^2}$$

$$X_L = \sqrt{Z^2 - R^2}$$

EXAMPLE:

Referring back to Figure 16-25, calculate Z.

Solution:

$$\begin{aligned} Z &= \sqrt{R^2 + X_L^2} \\ &= \sqrt{55^2 + 26^2} \\ &= 60.8\ \Omega \end{aligned}$$

16-8-3 Phase Shift

If a circuit is purely resistive, the phase shift (θ) between the source voltage and circuit current is zero. If a circuit is purely inductive, voltage leads current by 90°; therefore, the phase shift is +90°. If the resistance and inductive reactance are equal, the phase shift will equal +45°, as shown in Figure 16-27.

The phase shift in an inductive and resistive circuit is the degrees of lead between the source voltage (V_S) and current (I), and by looking at the examples in Figure 16-27, you can see that the phase angle is proportional to

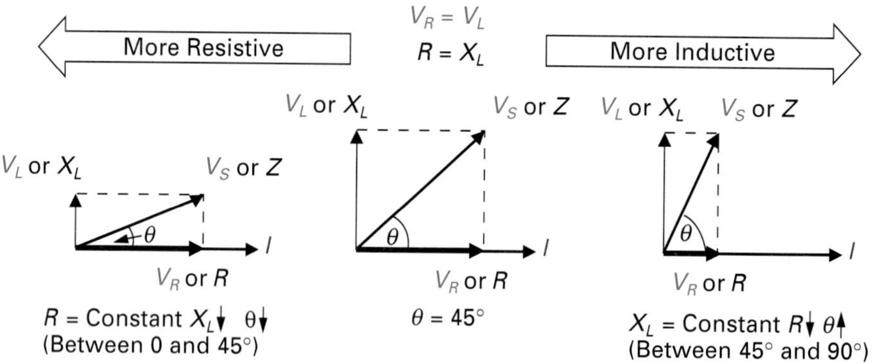

FIGURE 16-27 Phase Shift (θ) in a Series *RL* Circuit.

reactance and inversely proportional to resistance. Mathematically, it can be expressed as

$$\theta = \arctan \frac{X_L}{R}$$

As the current is the same in both the inductor and resistor in a series circuit, the voltage drops across the inductor and resistor are directly proportional to reactance and resistance:

$$V_R \updownarrow = I\,(\text{constant}) \times R\updownarrow, \quad V_L\updownarrow = I\,(\text{constant}) \times X_L\updownarrow$$

the phase shift can also be calculated by using the voltage drop across the inductor and resistor.

$$\theta = \arctan \frac{V_L}{V_R}$$

EXAMPLE:

Referring back to Figure 16-25, calculate the phase shift between source voltage and circuit current.

Solution:

Since the ratio of X_L/R and V_L/V_R are known for this example circuit, either of the phase shift formulas can be used.

$$\begin{aligned}\theta &= \arctan \frac{X_L}{R}\\ &= \arctan \frac{26\ \Omega}{55\ \Omega}\\ &= \arctan 0.4727\\ &= 25.3°\end{aligned}$$

or

$$\theta = \arctan \frac{V_L}{V_R}$$

$$= \arctan \frac{2.6 \text{ V}}{5.5 \text{ V}}$$

$$= \arctan 0.4727$$

$$= 25.3°$$

The source voltage in this example circuit leads the circuit current by 25.3°.

16-8-4 **Power**

Purely Resistive Circuit

Figure 16-28 illustrates the current, voltage, and power waveforms produced when applying an ac voltage across a purely resistive circuit. Voltage and current are in phase, and true (P_R) power in watts can be calculated by multiplying current by voltage ($P = V \times I$).

The sinusoidal power waveform is totally positive, as a positive voltage multiplied by a positive current produces a positive value of power, and a negative voltage multiplied by a negative current also produces a positive value of

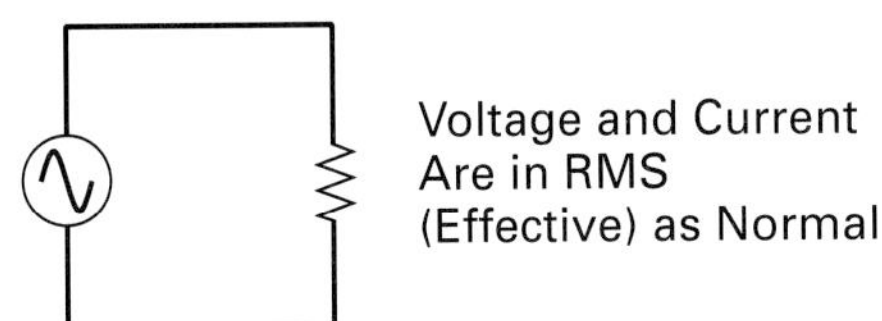

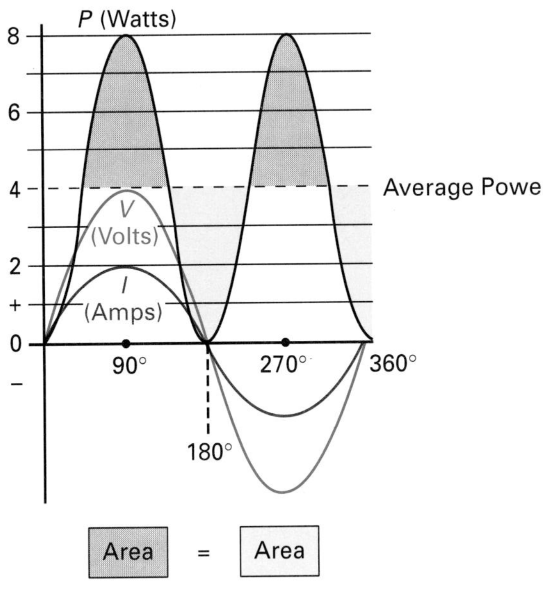

FIGURE 16-28 **Purely Resistive Circuit.**

power. For this reason, the resistor is said to develop a positive power waveform that is twice the frequency of the voltage or current waveform.

The average value of power dissipated by a purely resistive circuit is the halfway value between maximum and zero, in this example, 4 watts.

Purely Inductive Circuit

The pure inductor, like the capacitor, is a reactive component, which means that it will consume power without the dissipation of energy. The capacitor holds its energy in an electric field, while the inductor consumes and holds its energy in a magnetic field and then releases it back into the circuit.

The power curve alternates equally above and below the zero line, as seen in Figure 16-29. During the first positive power half-cycle, the circuit current is on the increase, to maximum (point A), the magnetic field is building up, and the inductor is storing electrical energy. When the circuit current is on the decline

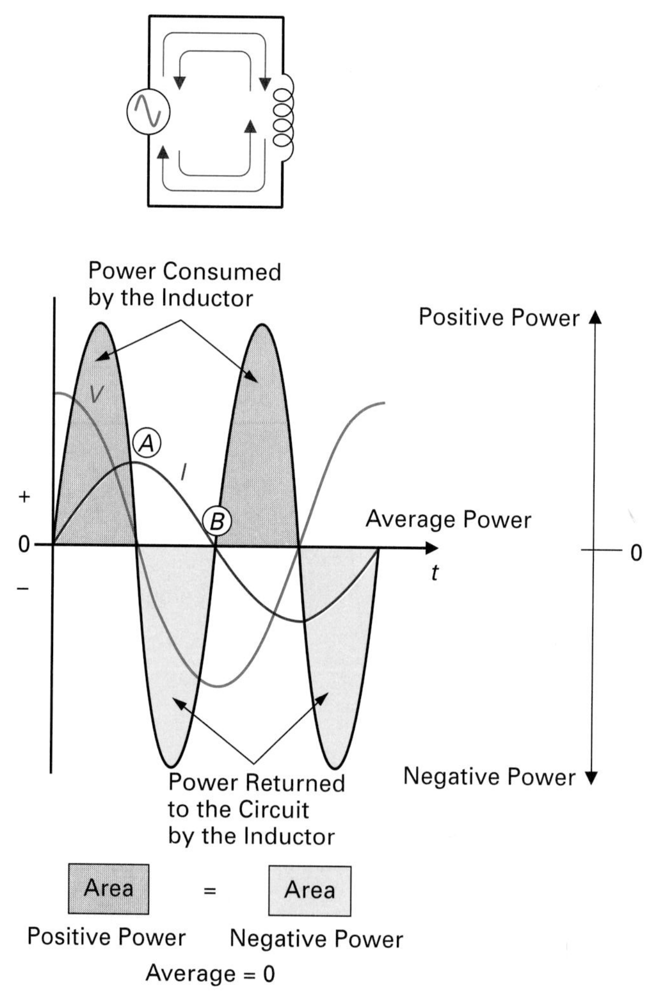

FIGURE 16-29 Purely Inductive Circuit.

between A and B, the magnetic field begins to collapse and self-induction occurs and returns electrical energy back into the circuit. The power alternation is both positive when the inductor is consuming power and negative when the inductor is returning the power back into the circuit. As the positive and negative power alternations are equal but opposite, the average power dissipated is zero.

Resistive and Inductive Circuit

An inductor is different from a capacitor in that it has a small amount of resistance no matter how pure the inductor. For this reason, inductors will never have an average power figure of zero, because even the best inductor will have some value of inductance and resistance within it, as seen in Figure 16-30. The reason an inductor has resistance is that it is simply a piece of wire, and any piece of wire has a certain value of resistance, as well as inductance. This coil resistance should be, and normally is, very small and can usually be ignored; however, in some applications even this small resistance can prevent the correct operation of a circuit, so a value or term had to be created to specify the differences in the quality of inductor available.

The **quality factor** *(Q)* of an inductor is the ratio of the energy stored in the coil by its inductance to the energy dissipated in the coil by the resistance; therefore, the higher the Q, the better the coil is at storing energy rather than dissipating it:

Quality Factor
Quality factor of an inductor or capacitor; it is the ratio of a component's reactance (energy stored) to its effective series resistance (energy dissipated).

$$\text{quality } (Q) = \frac{\text{energy stored}}{\text{energy dissipated}}$$

The energy stored is dependent on the inductive reactance (X_L) of the coil, and the energy dissipated is dependent on the resistance (R) of the coil. The quality factor of a coil or inductor can therefore also be calculated by using the formula

$$Q = \frac{X_L}{R}$$

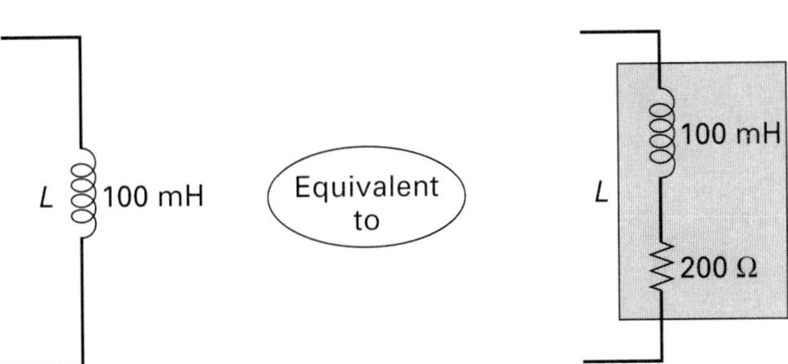

FIGURE 16-30 **Resistance within an Inductor.**

■ **EXAMPLE:**

Calculate the quality factor Q of a 22 mH coil connected across a 2 kHz, 10 V source if its internal coil resistance is 27 Ω.

■ ***Solution:***

$$Q = \frac{X_L}{R}$$

The reactance of the coil is not known but can be calculated by the formula

$$X_L = 2\pi f L$$
$$= 2\pi \times 2 \text{ kHz} \times 22 \text{ mH} = 276.5 \ \Omega$$

Therefore,

$$Q = \frac{X_L}{R} = \frac{276.5 \ \Omega}{27 \ \Omega} = 10.24$$

Inductors, therefore, will never appear as pure inductance, but rather as an inductive and resistive *(RL)* circuit, and the resistance within the inductor will dissipate *true power.*

Figure 16-31 illustrates a circuit containing R and L and the power waveforms produced when $R = X_L$; the phase shift (θ) is equal to 45°.

The positive power alternation, which is above the zero line, is the combination of the power dissipated by the resistor and the power consumed by the inductor, while circuit current is on the rise. The negative power alternation is the power that was given back to the circuit by the inductor while the inductor's magnetic field was collapsing and returning the energy that was consumed.

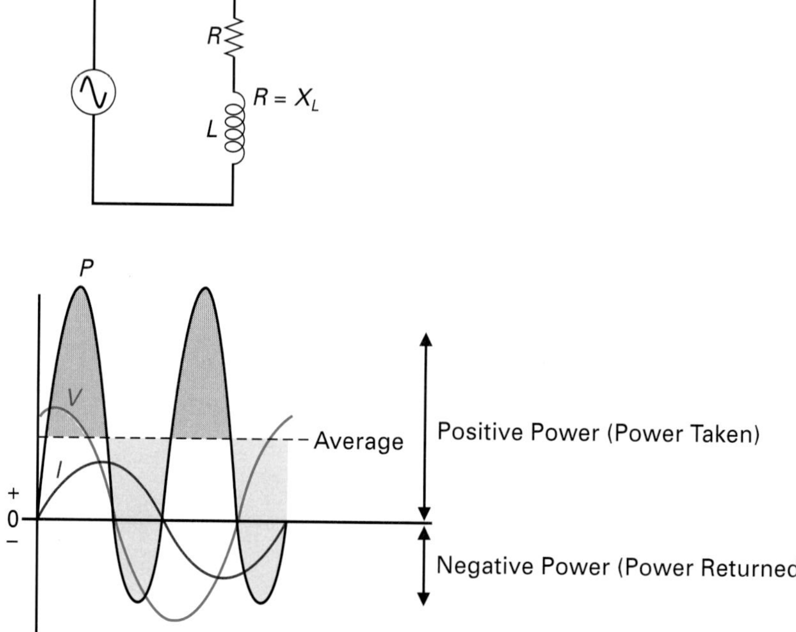

FIGURE 16-31 Power in a Resistive and Inductive Circuit.

Power Factor

When a circuit contains both inductance and resistance, some of the energy is consumed and then returned by the inductor (reactive or imaginary power), and some of the energy is dissipated and lost by the resistor (resistive or true power).

Apparent power is the power that appears to be supplied to the load and is the vector sum of both the reactive and true power; it can be calculated by using the formula

$$P_A = \sqrt{P_R^2 + P_X^2}$$

where P_A = apparent power, in volt-amperes (VA)

P_R = true power, in watts (W)

P_X = reactive power, in volt-amperes reactive (VAR)

The power factor is a ratio of the true power to the apparent power and is therefore a measure of the loss in a circuit.

$$\text{PF} = \frac{\text{true power } (P_R)}{\text{apparent power } (P_A)}$$

or

$$\text{PF} = \frac{R}{Z}$$

or

$$\text{PF} = \cos\theta$$

EXAMPLE:

Calculate the following for a series RL circuit if R = 4 … f = 20 kHz, and V_S = 6 V:

a. X_L

b. Z

c. I

d. θ

e. Apparent power

f. PF

■ ***Solution:***

a. $X_L = 2\pi fL = 2\pi \times 20\ \text{kHz} \times 450\ \text{mH}$

$= 56.5\ \text{k}\Omega$

b. $Z = \sqrt{R^2 + X_L^2}$

$= \sqrt{40\ \text{k}\Omega^2 + 56.5\ \text{k}\Omega^2} = 69.23\ \text{k}\Omega$

c. $I = \dfrac{V_S}{Z} = \dfrac{6\ \text{V}}{69.23\ \text{k}\Omega} = 86.6\ \mu\text{A}$

d. $\theta = \arctan \dfrac{X_L}{R} = \arctan \dfrac{56.5\ \text{k}\Omega}{40\ \text{k}\Omega} = 54.7°$

e. Apparent power $= \sqrt{(\text{true power})^2 + (\text{reactive power})^2}$

$P_R = I^2 \times R = 86.6\ \mu\text{A}^2 \times 40\ \text{k}\Omega = 300\ \mu\text{W}$

$P_X = I^2 \times X_L = 86.6\ \mu\text{A}^2 \times 56.5\ \text{k}\Omega = 423.7\ \mu\text{VAR}$

$P_A = \sqrt{P_R^{\ 2} + P_X^{\ 2}}$

$= \sqrt{300\ \mu\text{W}^2 + 423.7\ \mu\text{W}^2} = 519.2\ \mu\text{VA}$

f. $\text{PF} = \dfrac{P_R}{P_A} = \dfrac{300\ \mu\text{W}}{519.2\ \mu\text{W}} = 0.57$

$= \dfrac{R}{Z} = \dfrac{40\ \text{k}\Omega}{69.23\ \text{k}\Omega} = 0.57$

$= \cos\theta = \cos 54.7° = 0.57$

SELF-TEST REVIEW QUESTIONS FOR SECTION 16-8

1. True or false: In a purely inductive circuit, the current will lead the applied voltage by 90°.
2. Calculate the applied source voltage V_S in an *RL* circuit where $V_R = 4$ V and $V_L = 2$ V.
3. Define and state the formula for impedance when R and X_L are known.
4. If $R = X_L$, the phase shift will equal ______________.
5. What is positive power?
6. Define Q and state the formula when X_L and R are known.
7. What is the difference between true and reactive power?
8. State the power factor formula.

16-9 PARALLEL *RL* CIRCUIT

…we have seen the behavior of resistors and inductors in series, let us …*L* parallel circuit.

16-9-1 Current

In Figure 16-32(a) you will see a parallel combination of a resistor and inductor. The voltage across both components is equal because of the parallel connection, and the current through each branch is calculated by applying Ohm's law.

$$\text{(resistor current) } I_R = \frac{V_S}{R}$$

$$\text{(inductor current) } I_L = \frac{V_S}{X_L}$$

Total current (I_T) is equal to the vector combination of the resistor current and inductor current, as shown in Figure 16-32(b). Figure 16-32(c) illustrates the current waveforms.

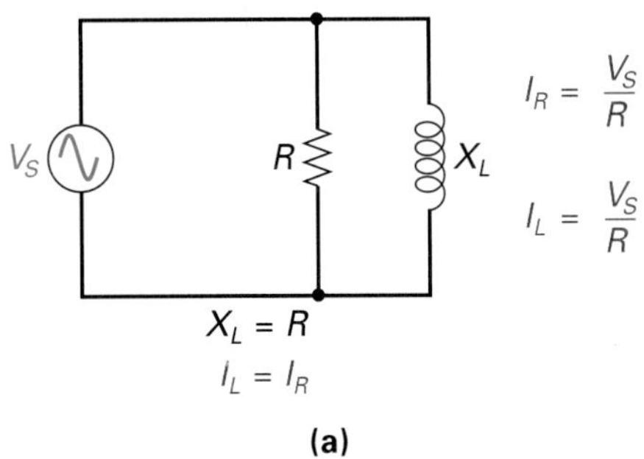

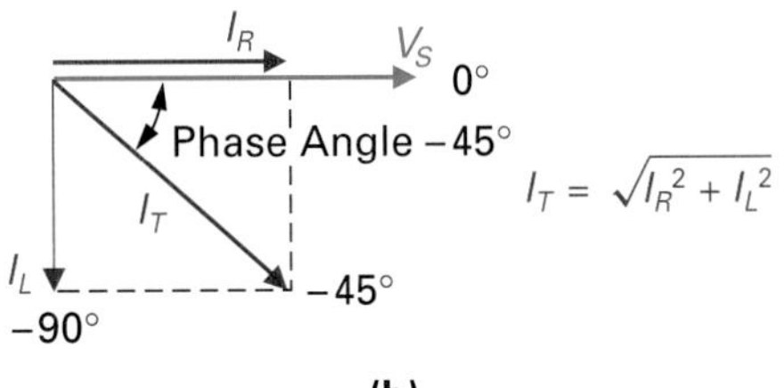

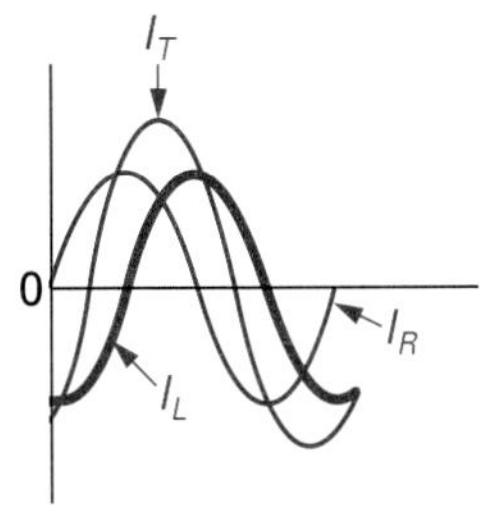

FIGURE 16-32 **Parallel *RL* Circuit.**

$$I_T = \sqrt{I_R^2 + I_L^2}$$

16-9-2 Phase Angle

The angle by which the total current (I_T) leads the source voltage (V_S) can be determined with either of the following formulas:

$$\theta = \arctan \frac{I_L}{I_R}$$

or

$$\theta = \arctan \frac{R}{X_L}$$

16-9-3 Impedance

The total opposition or impedance of a parallel *RL* circuit can be calculated by

$$Z = \frac{V_S}{I_T}$$

Using basic algebra, this formula can be rearranged to express impedance in terms of reactance and resistance.

$$Z = \frac{R \times X_L}{\sqrt{R^2 + X_L^2}}$$

16-9-4 Power

As with series *RL* circuits, resistive power and reactive power can be calculated by

$$P_R = I_R^2 \times R$$

$$P_X = I_L^2 \times X_L$$

The apparent power of the circuit is calculated by

$$P_A = \sqrt{P_R^{\,2} + P_X^{\,2}}$$

and, finally, the power factor is equal to

$$\text{PF} = \cos\theta = \frac{P_R}{P_A}$$

EXAMPLE:

Calculate the following for a parallel *RL* circuit if $R = 45\ \Omega$, $X_L = 1100\ \Omega$, and $V_S = 24$ V:

a. I_R
b. I_L
c. I_T
d. Z
e. θ

Solution:

a. $I_R = \dfrac{V_S}{R} = \dfrac{24\text{ V}}{45\ \Omega} = 533.3\text{ mA}$

b. $I_L = \dfrac{V_S}{X_L} = \dfrac{24\text{ V}}{1100\ \Omega} = 21.8\text{ mA}$

c. $I_T = \sqrt{I_R^{\,2} + I_L^{\,2}}$

$= \sqrt{533.3\text{ mA}^2 + 21.8\text{ mA}^2} = 533.7\text{ mA}$

d. $Z = \dfrac{R \times X_L}{\sqrt{R^2 + X_L^{\,2}}} = \dfrac{45\ \Omega \times 1100\ \Omega}{\sqrt{45\ \Omega^2 + 1100\ \Omega^2}}$

$= \dfrac{49.5\text{ k}\Omega}{1100.9\ \Omega} = 44.96\ \Omega$

e. $\theta = \arctan\dfrac{R}{X_L} = \arctan\dfrac{45\ \Omega}{1100\ \Omega} = 2.34°$

Therefore, I_T lags V_S by 2.34°.

SELF-TEST REVIEW QUESTIONS FOR SECTION 16-9

1. True or false: When *R* and *L* are connected in parallel with one another, the voltage across each will be different and dependent on the values of *R* and X_L.

2. State the formula for calculating total current (I_T) when I_R and I_L are known.

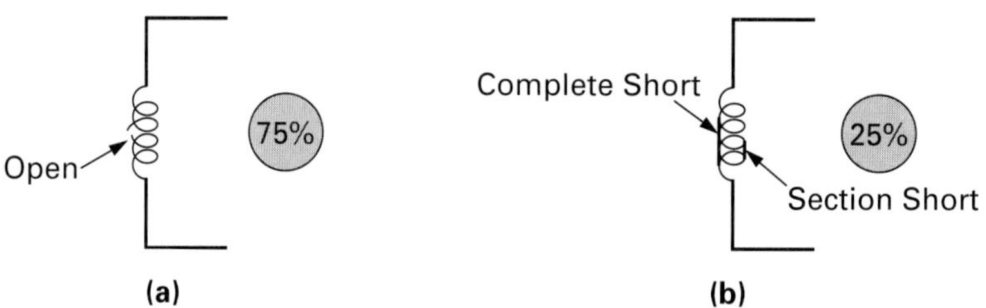

FIGURE 16-33 **Defective Inductors.**

16-10 TESTING INDUCTORS

Basically, only three problems can occur with inductors:

1. An open
2. A complete short
3. A section short (value change)

16-10-1 *Open [Figure 16-33(a)]*

This problem can be isolated with an ohmmeter. Depending on the winding's resistance, the coil should be in the range of zero to a few hundred ohms. An open accounts for 75% of all defective inductors.

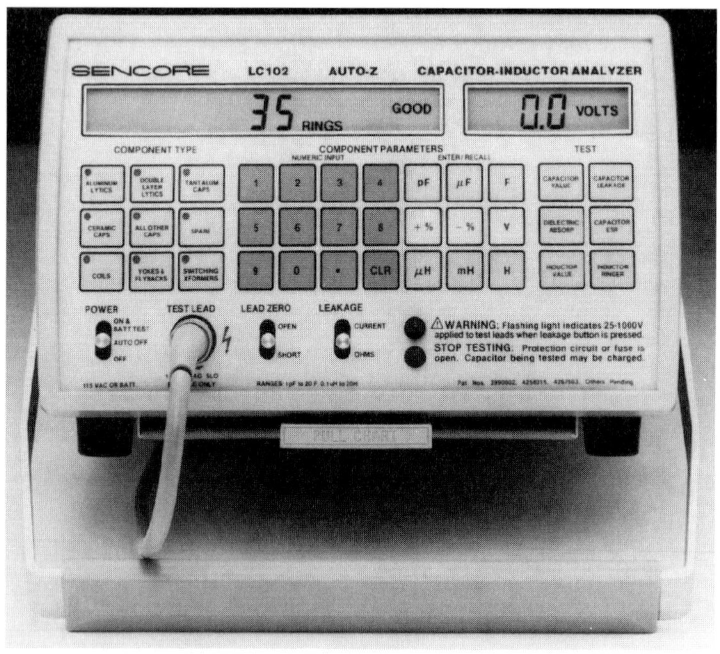

FIGURE 16-34 **Capacitor and Inductor Analyzer.**

16-10-2 Complete or Section Short [Figure 16-33(b)]

A coil with one or more shorted turns or a complete short can be checked with an ohmmeter and thought to be perfectly good because of the normally low resistance of a coil, as it is just a piece of wire. But if it is placed in a circuit with a complete or section short present, it will not function effectively as an inductor, if at all. For these checks, an **inductor analyzer** needs to be used like the one seen in Figure 16-34, which can be used to check capacitance and inductance. Complete or section shorts account for 25% of all defective inductors.

Inductor Analyzer
A test instrument designed to test inductors.

SELF-TEST REVIEW QUESTIONS FOR SECTION 16-10

1. How could the following inductor malfunctions be recognized?
   a. An open
   b. A complete or section short
2. Which inductor malfunction accounts for almost 75% of all failures?

16-11 APPLICATIONS OF INDUCTORS

16-11-1 RL Filters

The ***RL* filter** will achieve results similar to the *RC* filter in that it will pass some frequencies and block others, as seen in Figure 16-35. The inductive reactance of the coil and the resistance of the resistor form a voltage divider. Since inductive reactance is proportional to frequency ($X_L \propto f$), the inductor will drop less voltage at lower frequencies ($f\downarrow$, $X_L\downarrow$, $V_L\downarrow$) and more voltage at higher frequencies ($f\uparrow$, $X_L\uparrow$, $V_L\uparrow$).

***RL* Filter**
A selective circuit of resistors and inductors which offers little or no opposition to certain frequencies while blocking or alternating other frequencies.

With the low-pass filter shown in Figure 16-35(a) the output is developed across the resistor. If the frequency of the input is low, the inductive reactance will be low, so almost all the input will be developed across the resistor and applied to the output. If the frequency of the input increases, the inductor's reactance will increase, resulting in almost all the input being dropped across the inductor and none across the resistor and therefore the output.

With the high-pass filter, shown in Figure 16-35(b), the inductor and resistor have been placed in opposite positions. If the frequency of the input is low, the inductive reactance will be low, so almost all the input will be developed across the resistor and very little will appear across the inductor and therefore at the output. If the frequency of the input is high, the inductive reactance will be high, resulting in almost all of the input being developed across the input and therefore appearing at the output.

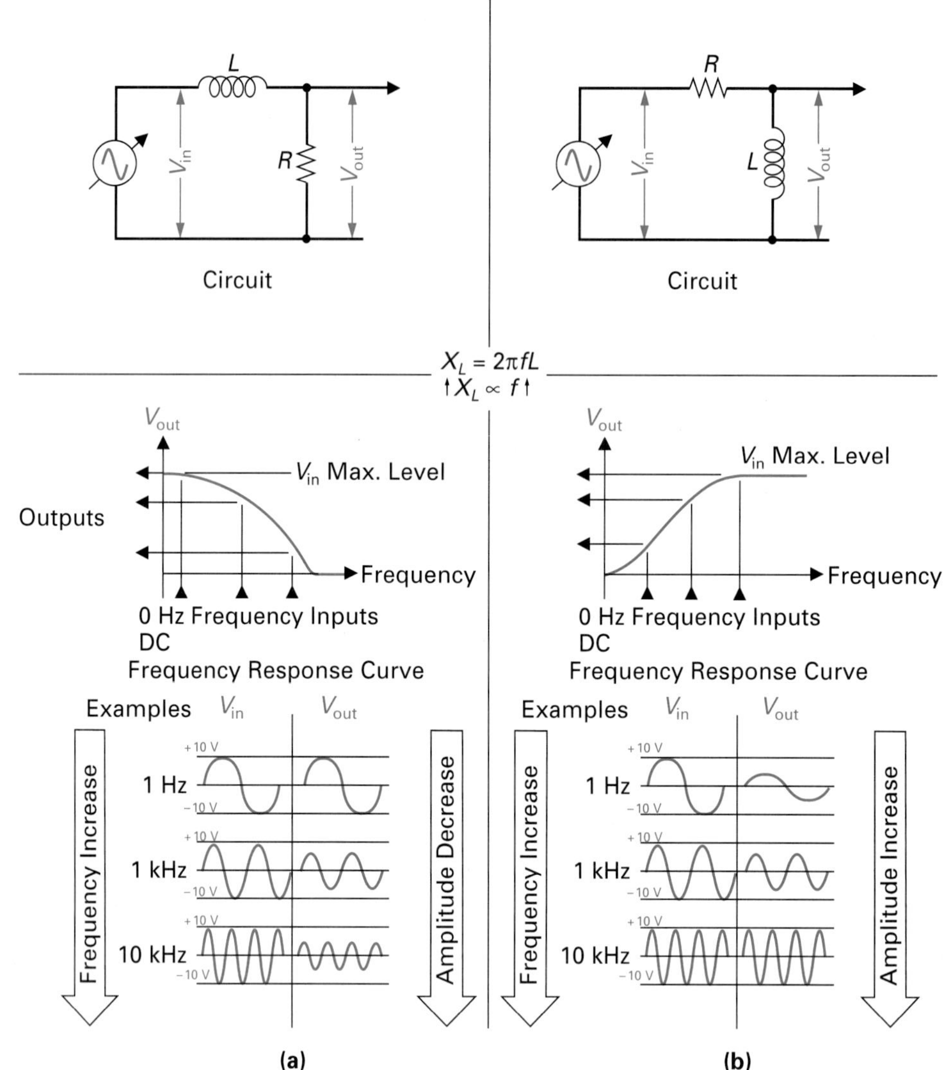

FIGURE 16-35 ***RL*** **Filter. (a) Low-Pass Filter. (b) High-Pass Filter.**

16-11-2 RL *Integrator*

***RL* Integrator**
An *RL* circuit with an output proportionate to the integral of the input signal.

In the preceding section on *RL* filters we saw how an *RL* circuit reacted to a sine-wave input of different frequencies. In this section and the following we will see how an *RL* circuit will react to a square-wave input, and show two other important applications of inductors. In the ***RL* integrator,** the output is taken across the resistor, as seen in the circuit in Figure 16-36(a). The output shown in Figure 16-36(b) is the same as the previously described *RC* integrator's output. When the input rises from 0 to 10 V at the leading positive edge of the square wave input, the situation is as seen in Figure 16-37(a).

The inductor's 10 V counter emf opposes the sudden input change from 0 to 10 V, and if 10 V is across the inductor, 0 V must be across the resistor

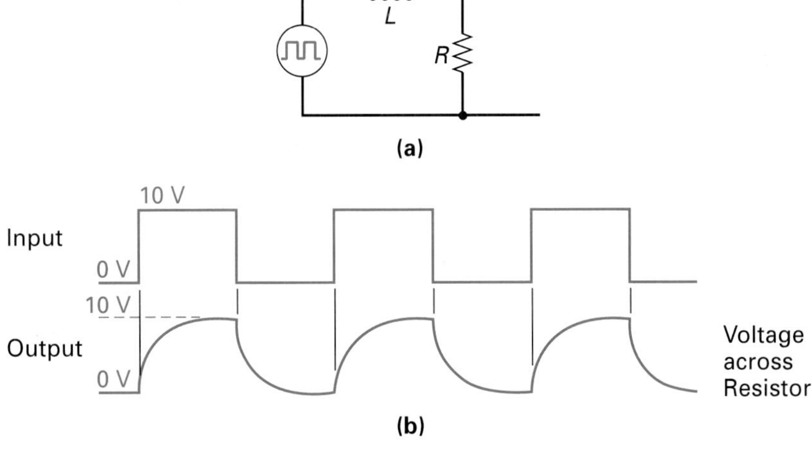

FIGURE 16-36 *RL* Integrator.

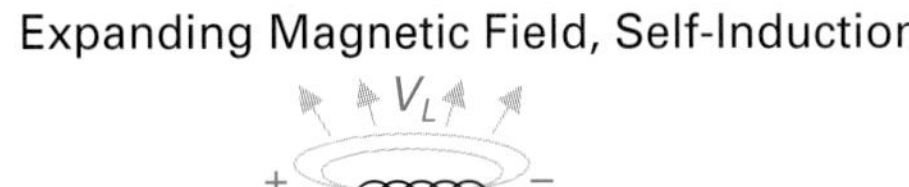

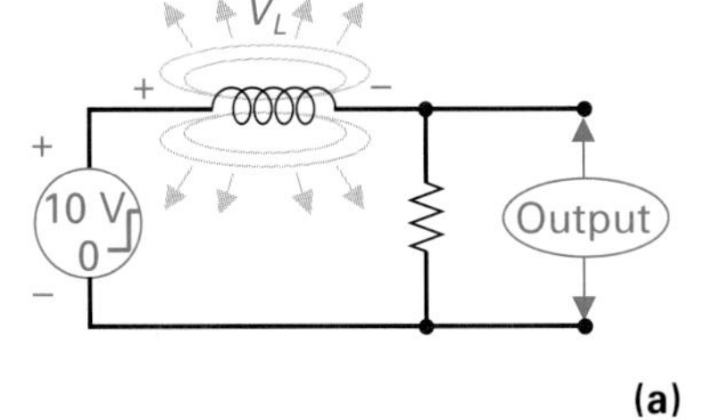

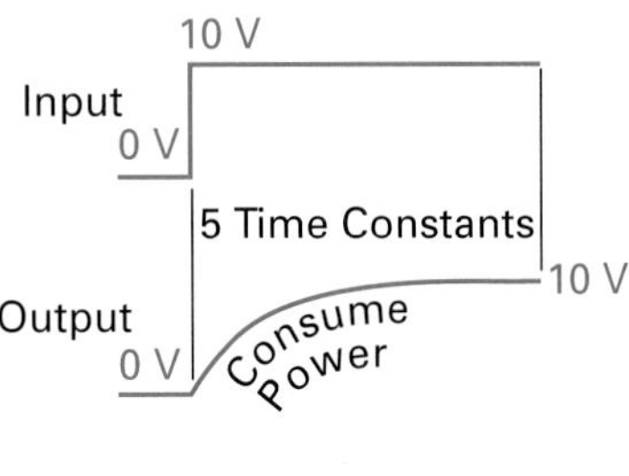

(a)

Stationary Magnetic Field, No Self-Induction

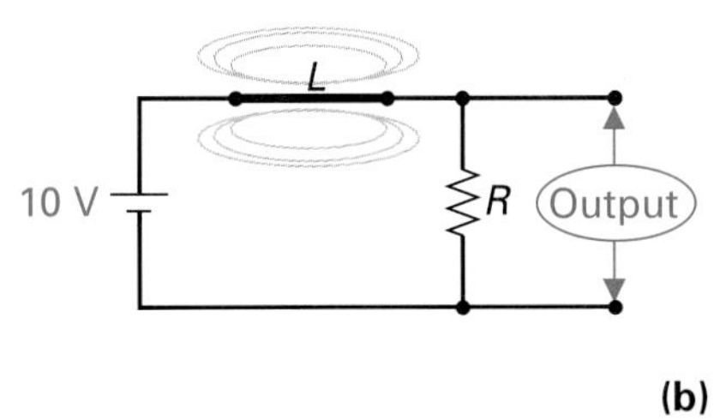

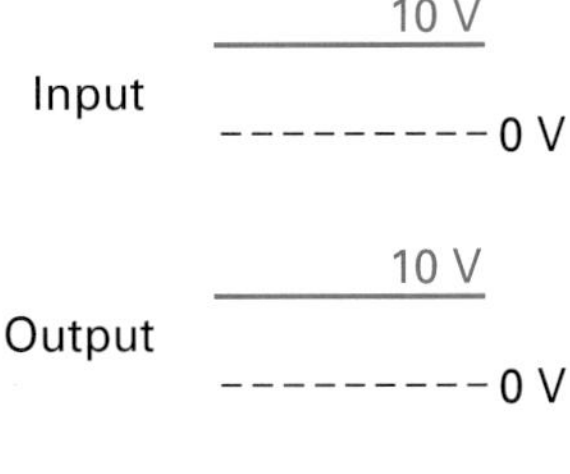

(b)

Collapsing Magnetic Field, Self-Induction

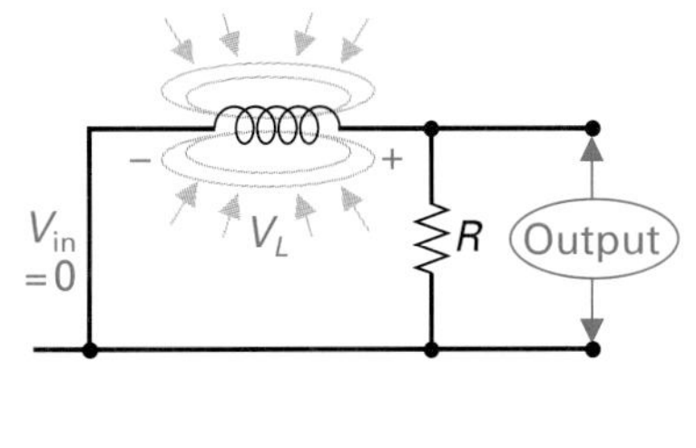

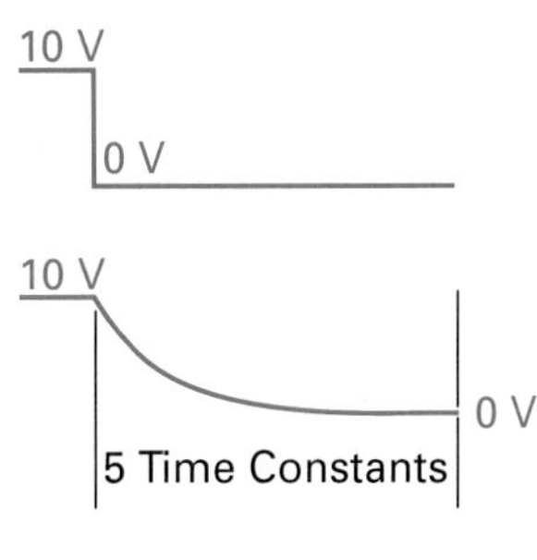

(c)

FIGURE 16-37 Input/Output Analysis of *RL* Integrator.

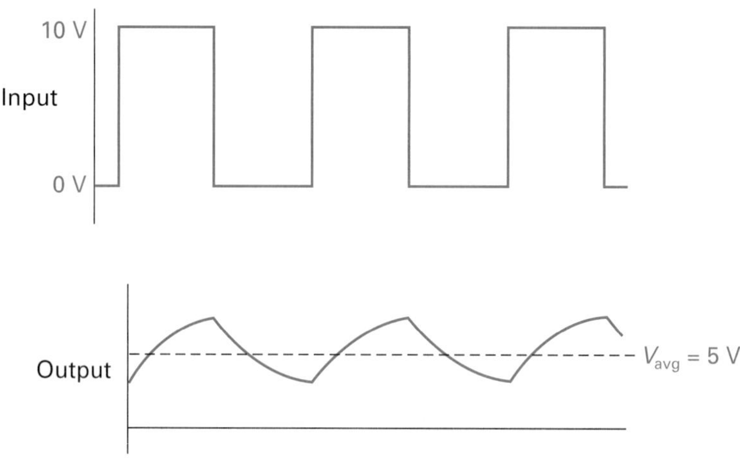

FIGURE 16-38

(Kirchhoff's voltage law), and therefore appearing at the output. After five time constants ($5 \times L/R$), the inductor's current in the circuit will have built up to maximum (V_{in}/R), and the inductor will be an equivalent short circuit, because no change and consequently no back emf exist, as shown in Figure 16-37(b). All the input voltage will now be across the resistor and therefore at the output.

When the square-wave input drops to zero, the circuit is equivalent to that seen in Figure 16-37(c), and the collapsing magnetic field will cause an induced voltage within the conductor, which will cause current to flow within the circuit for five time constants, whereupon it will reach 0.

As with the *RC* integrator, if the period of the square wave is decreased or if the time constant is increased, the output will reach an average value of half the input square wave's amplitude, as shown in Figure 16-38.

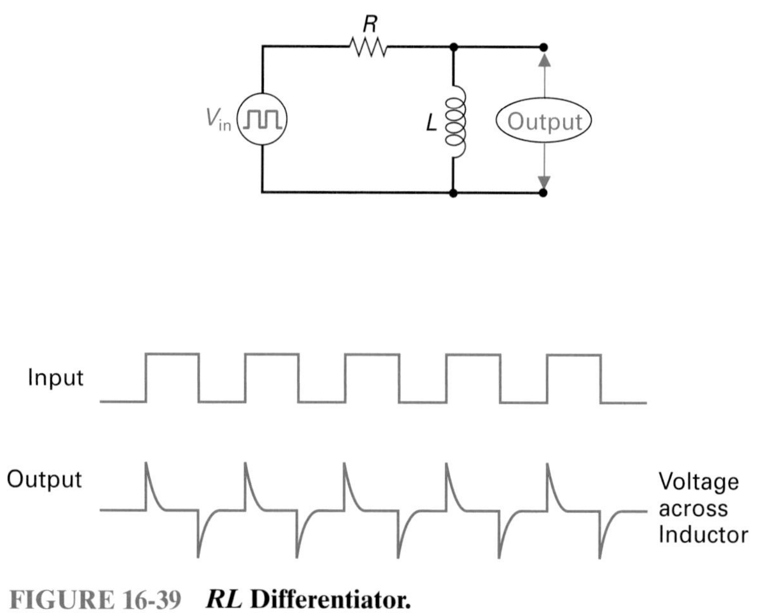

FIGURE 16-39 ***RL* Differentiator.**

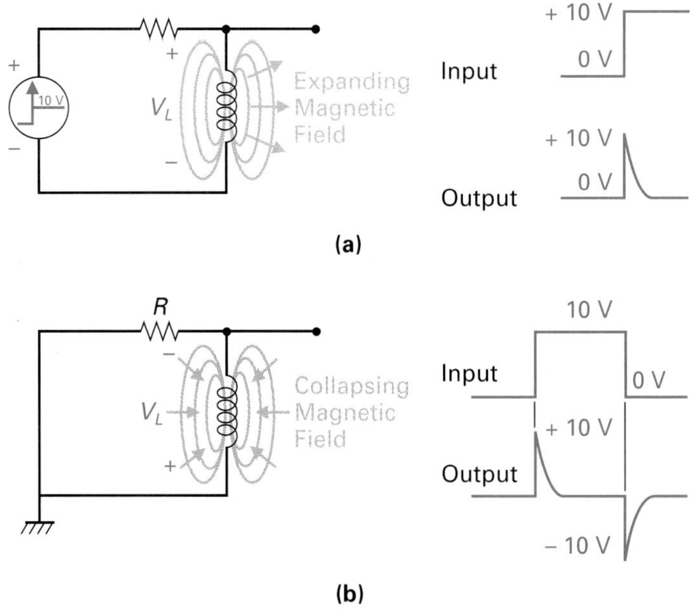

FIGURE 16-40 Input/Output Analysis of *RL* Differentiator.

16-11-3 RL *Differentiator*

With the ***RL* differentiator,** the output is taken across the inductor, as shown in Figure 16-39, and is the same as the *RC* differentiator's output. When the square-wave input rises from 0 to 10 V, the inductor will generate a 10 V counter emf across it, as shown in Figure 16-40(a), and this 10 V will all appear across the output. As the circuit current exponentially rises, the voltage across the inductor, and therefore at the output, will fall exponentially to 0 V after five time constants.

***RL* Differentiator**
An *RL* circuit whose output voltage is proportional to the rate of change of the input voltage.

When the square wave input falls from 10 to 0 V, the circuit is equivalent to that shown in Figure 16-40(b), and the collapsing magnetic field induces a counter emf. The sudden –10 V impulse at the output will decrease to 0 V as the circuit current decreases in five time constants.

SELF-TEST REVIEW QUESTIONS FOR SECTION 16-11

1. List three circuit applications of the inductor.
2. Will an *RL* filter act as a low-pass filter or high-pass filter, or can it be made to function as either?
3. What are the waveform differences between an integrator and differentiator?

SUMMARY

1. Two important rules relate to inductance. The first was that a magnetic field will build up around any current-carrying conductor (electromagnetism), and the second was that a voltage will be induced into a conductor when it is subjected to a moving magnetic field. These two rules form the basis for a phenomenon called *self-inductance* or more commonly *inductance*.
2. Inductance, by definition, is the ability of a device to oppose a change in current flow, and the device designed specifically to achieve this function is called the *inductor*.
3. There is, in fact, no physical difference between an inductor and an electromagnet, since they are both coils. The two devices are given different names because they are used in different applications even though their construction and principle of operation is the same. An electromagnet is used to generate a magnetic field in response to current, while an inductor is used to oppose any change in current.

Inductance (Figure 16-41)

4. When an external voltage source is connected across a coil, it will force a current through the coil. This current will generate a magnetic field that will increase in field strength in a short time from zero to maximum, expanding from the center of the conductor (electromagnetism). The expanding magnetic lines of force have relative motion with respect to the stationary conductor, so an induced voltage results (electromagnetic induction). The blooming magnetic field generated by the conductor is actually causing a voltage to be induced in the conductor that is generating the magnetic field. This effect of a current carrying coil of conductor inducing a voltage within itself is known as self-inductance.
5. This phenomenon was first discovered by Heinrich Lenz, who observed that the induced voltage causes an induced (bucking) current to flow in the coil, which opposes the source current producing it.
6. The induced voltage is called a counter emf or back emf. It opposes the applied emf (or battery voltage). The ability of a coil or conductor to induce or produce a counter emf within itself as a result of a change in current is called *self-inductance,* or more commonly inductance (symbolized L).
7. The unit of inductance is the henry (H), named in honor of Joseph Henry, an American physicist, for his experimentation within this area of science.
8. The inductance of an inductor is 1 henry when a current change of 1 ampere per second causes an induced voltage of 1 volt.
9. Inductance is therefore a measure of how much counter emf (induced voltage) can be generated by an inductor for a given amount of current change through that same inductor.
10. A larger inductance ($L\uparrow$) will create a larger induced voltage ($V_{ind}\uparrow$), and if the rate of change of current with respect to time is increased ($\Delta I/\Delta t\uparrow$), the induced voltage or counter emf will also increase ($V_{ind}\uparrow$).

QUICK REFERENCE SUMMARY SHEET

FIGURE 16-41 Inductance.

$$V_{\text{ind}} = L \times \frac{\Delta I}{\Delta t}$$

L = inductance, in henrys
ΔI = increment of change of current
Δt = increment of change of time

$$L = \frac{N^2 \times A \times \mu}{l}$$

N = number of turns
A = cross-sectional area, in square meters (m^2)
μ = permeability
l = length of core, in meters (m)

Series *RL* Circuit:

- $L_T = L_1 + L_2 + L_3 + \ldots$ L_T = total inductance, in henrys
- $V_R = I \times R$
- $V_L = I \times X_L$
- $V_S = \sqrt{V_R^2 + V_L^2}$ $V_R = \sqrt{V_S^2 - V_L^2}$ $V_L = \sqrt{V_S^2 - V_R^2}$
- $Z = \frac{V}{I}$ $Z = \sqrt{R^2 + X_L^2}$ $R = \sqrt{Z^2 - X_L^2}$ $X_L = \sqrt{Z^2 - R^2}$
- $\theta = \arctan \frac{X_L}{R}$ $\theta = \arctan \frac{V_L}{V_R}$

Parallel *RL* Circuit:

- $I_R = \frac{V_S}{R}$
- $I_L = \sqrt{V_S \backslash X_L}$
- $I_T = \sqrt{I_R^2 + I_L^2}$
- $Z = \frac{R \times X_L}{\sqrt{R^2 + X_L^2}}$
- $\theta = \arctan \frac{R}{X_L}$
- $L_T = \frac{1}{(1/L_1) + (1/L_2) + (1/L_3) + \ldots}$
- $L_T = \frac{L_1 \times L_2}{L_1 + L_2}$
- $P_A = \sqrt{P_R^2 + P_X^2}$ P_A = apparent power, in volt-amperes (VA)
  P_R = true power, in watts (W)
  P_X = reactive power, in volt-amperes reactive (VAR)
- $\text{PF} = \frac{\text{true power}}{\text{apparent power}}$ $\text{PF} = \frac{Z}{R}$ $\text{PF} = \cos\theta$ PF = power factor

$$\tau = \frac{L}{R}$$

τ = inductive time constant, in seconds (s)

$$X_L = 2\pi f L$$

X_L = inductive reactance, in ohms (Ω)
2π = a constant
f = frequency, in hertz (Hz)
L = inductance, in henrys (H)

$$Q = \frac{\text{energy stored}}{\text{energy dissipated}}$$

Q = quality factor of an inductor

$$Q = \frac{X_L}{R}$$

FIGURE 16-41 (continued)

Permeabilities of Various Materials

MATERIAL	PERMEABILITY (μ)
Air or vacuum	1.26×10^{-6}
Nickel	6.28×10^{-5}
Cobalt	7.56×10^{-5}
Cast iron	1.1×10^{-4}
Machine steel	5.65×10^{-4}
Transformer iron	6.9×10^{-3}
Silicon iron	8.8×10^{-3}
Permaloy	0.126
Supermalloy	1.26

DC Inductor Current Rise and Fall

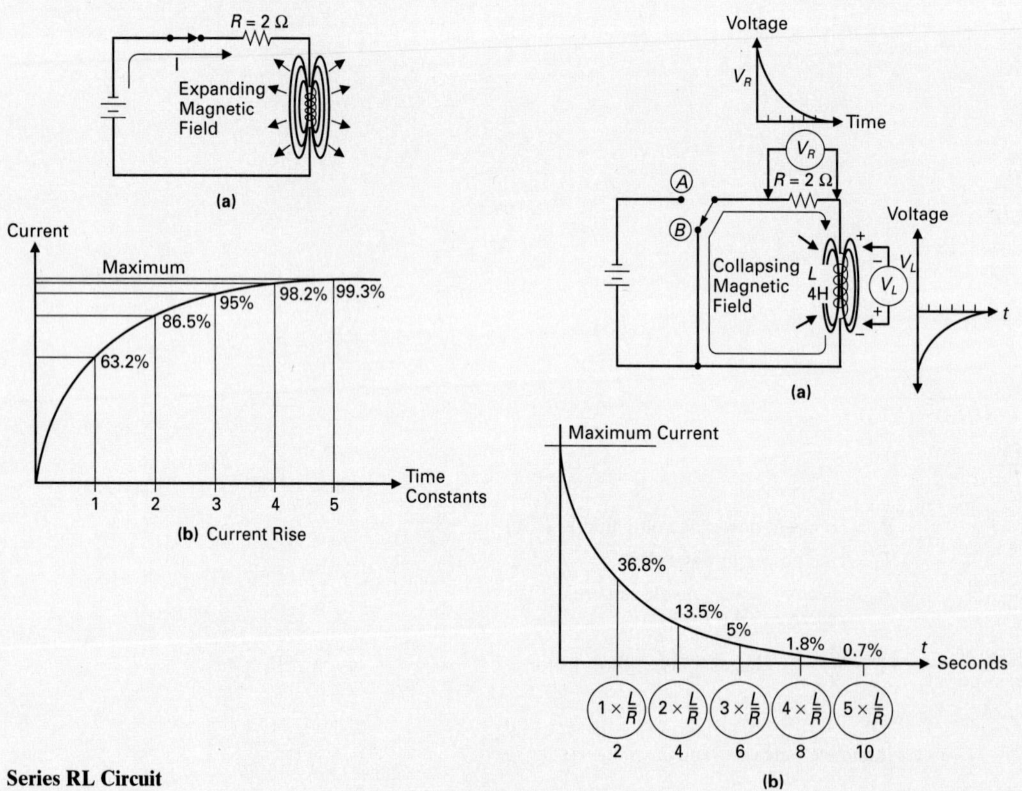

Series RL Circuit

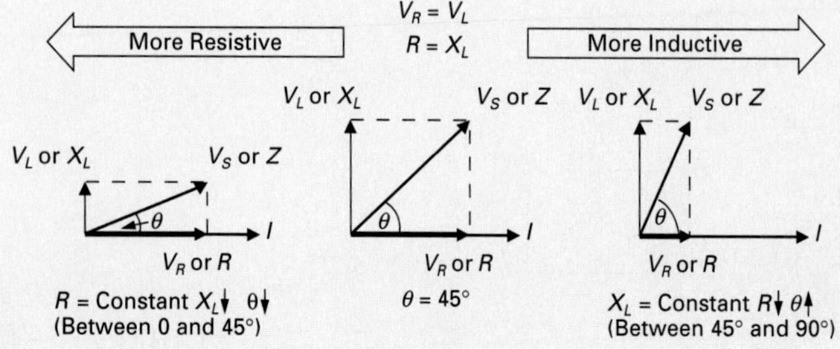

FIGURE 16-41 (continued)

Parallel RL Circuit

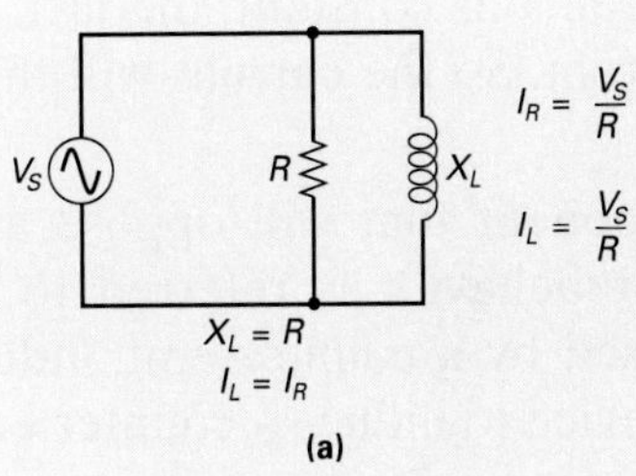

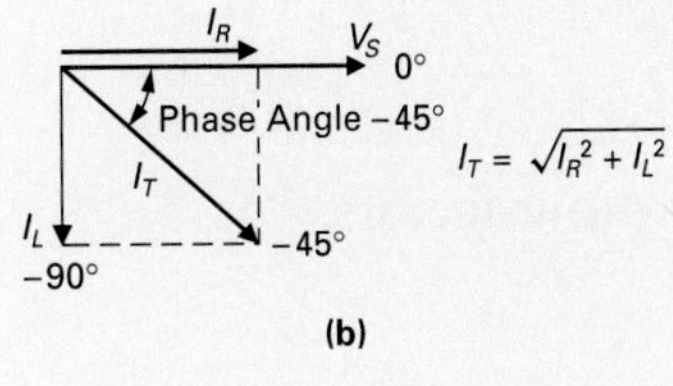

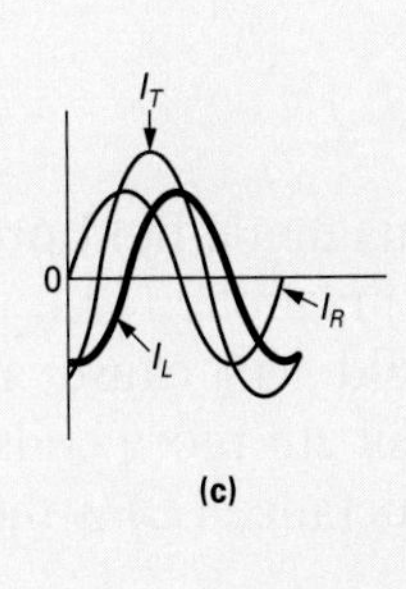

RL Filter

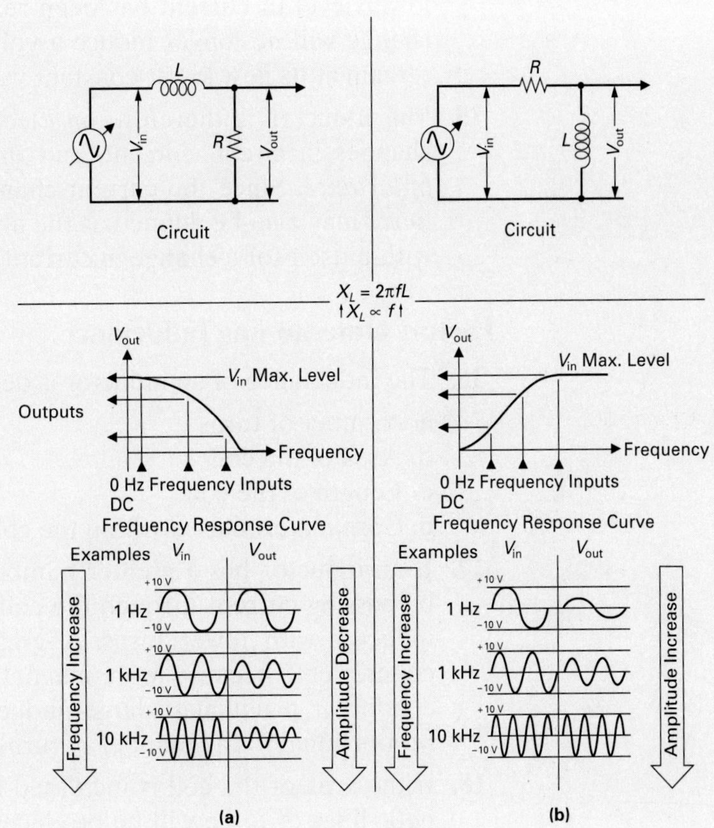

11. If the current in the circuit is suddenly increased (by lowering the circuit resistance), the change in the expanding magnetic field will induce a counter emf within the inductor. This induced voltage will oppose the source voltage from the battery and attempt to hold current at its previous low level.
12. The counter emf cannot completely oppose the current increase, for if it did, the lack of current change would reduce the counter emf to zero. Current therefore incrementally increases up to a new maximum, which is determined by the applied voltage and the circuit resistance ($I = V/R$).
13. Once the new higher level of current has been reached and remains constant, there will no longer be a change. This lack of relative motion between field and conductor will no longer generate a counter emf, so the current will remain at its new higher constant value.
14. This effect also happens in the opposite respect. If current decreases (by increasing circuit resistance), the magnetic lines of force will collapse because of the reduction of current and induce a voltage in the inductor,

which will produce an induced current in the same direction as the circuit current. These two combine and tend to maintain the current at the higher previous constant level. Circuit current will fall, however, as the induced voltage and current are only present during the change (in this case the decrease from the higher current level to the lower); and once the new lower level of current has been reached and remains constant, the lack of change will no longer induce a voltage or current. So the current will then remain at its new lower constant value.

15. The inductor is therefore an electronic component that will oppose any changes in circuit current, and this ability or behavior is referred to as *inductance.* Since the current change is opposed by a counter emf, inductance may also be defined as the ability of a device to induce a counter emf within itself for a change in current.

Factors Determining Inductance

16. The inductance of an inductor is determined by four factors:
    - **a.** Number of turns
    - **b.** Area of the coil
    - **c.** Length of the coil
    - **d.** Core material used within the coil
17. If an inductor has a greater number of turns, the magnetic field produced by passing current through the coil will have more magnetic force than an inductor with fewer turns. A greater magnetic field will cause a larger counter emf, because more magnetic lines of flux will cut more coils of the conductor, producing a larger inductance value. Inductance (L) is therefore proportional to the number of turns (N).
18. If the area of the coil is increased for a given number of turns, more magnetic lines of force will be produced, and if the magnetic field is increased, the inductance will also increase. Inductance (L) is therefore proportional to the cross-sectional area of the coil (A).
19. Inductance is inversely proportional to the length of the coil, in that a longer coil, for a given number of turns, produces a smaller inductance, and vice versa.
20. Most inductors have core materials such as nickel, cobalt, iron, steel, ferrite, or an alloy. These cores have magnetic properties that concentrate or intensify the magnetic field. Permeability is another factor that is proportional to inductance.

Inductors in Combination

21. Inductors oppose the change of current in a circuit and so are treated in a manner similar to resistors connected in combination.
22. Two or more inductors in series merely extend the coil length and increase inductance.
23. Inductors in parallel are treated in a manner similar to resistors, with the total inductance being less than that of the smallest inductor's value.

Types of Inductors

24. As with resistors and capacitors, inductors are basically divided into the two categories of fixed-value and variable-value inductors.
25. With fixed-value inductors, the inductance value remains constant and cannot be altered. It is usually made of solid copper wire with an insulating enamel around the conductor. This enamel or varnish on the conductor is needed since the turns of a coil are generally wound on top of one another. The three major types of fixed-value inductors on the market today are:

    a. Air core
    b. Iron core
    c. Ferrite core
26. The name choke is used interchangeably with the word *inductor,* because choke basically describes an inductors' behavior, which generates a counter emf to limit or choke the flow of current.
27. Ferrite is a chemical compound, which is basically powdered iron oxide and ceramic. Although its permeability is less than iron, it has a higher permeability than air and can therefore obtain a higher inductance value than air for the same number of turns.
28. The toroidal inductor (doughnut shaped) has an advantage over the cylinder-shaped core in that the magnetic lines of force do not have to pass through air. With a cylindrical core, the magnetic lines of force merge out of one end of the cyclinder, propagate or travel through air, and then enter in the other end of the cylinder. The nonmagnetic air, which has a low permeability, will reduce the strength of the field and therefore the inductance value.
29. With variable-value inductors, the inductance value can be changed by adjusting the position of the movable core with respect to the stationary coil. A screw adjustment moves a sliding ferrite core or slug farther in or out of the stationary coil. If the slug is all the way out, the permeability figure is low, because an air core is being used ($L \downarrow \propto \mu \downarrow$). If the slug is adjusted and screwed into the coil, the air is replaced by the ferrite slug, which has a higher permeability and, therefore, inductance value ($L \uparrow \propto \mu \uparrow$).
30. Variable inductors should only be adjusted with a plastic or nonmetallic alignment tool since any metal object in the vicinity of the inductor will interfere with the magnetic field and change the inductance of the inductor.

Inductive Time Constant

31. Inductors will not have any effect on a steady value of direct current (dc) from a dc voltage source. If, however, the dc is changing (pulsating), the inductor will oppose the change whether it is an increase or decrease in direct current, because a change in current causes the magnetic field to expand or contract, and in so doing it will cut the coil of the inductor and induce a voltage that will counter the applied emf.
32. Current in an inductive circuit, therefore, cannot rise instantly to its maximum value, which is determined by Ohm's law ($I = V/R$). Current will in

fact take a time to rise to maximum, due to the inductor's ability to oppose change.

33. It will actually take five time constants (5τ) for the current in an inductive circuit to reach maximum value.

34. The constant to remember is the same as before: 63.2%. In one time constant ($1 \times L/R$) the current in the LR circuit will have reached 63.2% of its maximum value. In two time constants ($2 \times L/R$), the current will have increased 63.2% of the remaining current, and so on, through five time constants.

35. When the inductor's dc source of current is removed, the magnetic field will collapse and cut the coils of the inductor, inducing a voltage and causing a current to flow in the same direction as the original source current. This current will exponentially decay, or fall from the maximum to zero level, in five time constants.

36. If an alternating (ac) voltage is applied across an inductor, the inductor will continuously oppose the alternating current because it is always changing. The current in the circuit causes the magnetic field to expand and collapse and cut the conducting coils, resulting in an induced counter emf.

37. The voltage across the inductor (V_L) or counter emf leads the circuit current (I) by 90°.

Inductive Reactance

38. Reactance is the opposition to current flow without the dissipation of energy, as opposed to resistance, which is the opposition to current flow with the dissipation of energy.

39. Inductive reactance (X_L) is the opposition to current flow offered by an inductor without the dissipation of energy.

40. Inductive reactance is proportional to frequency ($X_L \propto f$) because a higher frequency (fast switching current) will cause a greater amount of current change, and a greater change will generate a larger counter emf, which is an opposition or reactance against current flow. When 0 Hz is applied to a coil (dc), there exists no change, so the inductive reactance of an inductor to dc is zero ($X_L = 2\pi \times 0 \times L = 0$).

41. Inductive reactance is also proportional to inductance because a larger inductance will generate a greater magnetic field and subsequent counter emf, which is the opposition to current flow.

42. Ohm's law can be applied to inductive circuits just as it can be applied to resistive and capacitive circuits. The current flow in an inductive circuit (I) is proportional to the voltage applied (V), and inversely proportional to the inductive reactance (X_L).

Series *RL* Circuit

43. In a purely resistive circuit, the current flowing within the circuit and the voltage across the resistor are in phase with one another.

44. In a purely inductive circuit, the current will lag the applied voltage by 90°.

45. If we connect a resistor and inductor in series, we will have the most common combination of R and L used in electronic equipment.

46. The voltage across the resistor (V_R) is in phase with the circuit current. The voltage across the inductor (V_L) leads the circuit current and the voltage across the resistor (V_R) by 90°, or, the circuit current vector lags the voltage across the inductor by 90°.

47. As with any series circuit, we have to apply Kirchhoff's voltage law when calculating the value of applied or source voltage (V_S), which, due to the phase difference between V_R and V_L, is the vector sum of all the voltage drops.

48. Impedance is the total opposition to current flow offered by a circuit with both resistance and reactance. It is measured in ohms and can be calculated by using Ohm's law. The impedance of a series RL circuit is equal to the square root of the sum of the squares of resistance and reactance, and by rearrangement, X_L and R can also be calculated if the other two values are known.

49. If a circuit is purely resistive, the phase shift (θ) between the source voltage and circuit current is zero. If a circuit is purely inductive, voltage leads current by 90°; therefore, the phase shift is + 90°. If the resistance and inductive reactance are equal, the phase shift will equal + 45°.

50. The phase shift in an inductive and resistive circuit is the degrees of lead between the source voltage (V_S) and current (I). The phase angle is proportional to reactance and inversely proportional to resistance.

51. As the current is the same in both the inductor and resistor in a series circuit, the voltage drops across the inductor and resistor are directly proportional to reactance and resistance.

52. When applying an ac voltage across a purely resistive circuit, voltage and current are in phase, and true (P_R) power in watts can be calculated by multiplying current by voltage ($P = V \times I$).

53. The pure inductor, like the capacitor, is a reactive component, which means that it will consume power without the dissipation of energy. The capacitor holds its energy in an electric field, while the inductor consumes and holds its energy in a magnetic field and then releases it back into the circuit. The power alternation is both positive when the inductor is consuming power and negative when the inductor is returning the power back into the circuit. As the positive and negative power alternations are equal but opposite, the average power dissipated is zero.

54. An inductor is different from a capacitor in that it has a small amount of resistance no matter how pure the inductor. For this reason, inductors will never have an average power figure of zero, because even the best inductor will have some value of inductance and resistance within it.

55. The reason an inductor has resistance is that it is simply a piece of wire, and any piece of wire has a certain value of resistance, as well as inductance.

56. The quality factor (Q) of an inductor is the ratio of the energy stored in the coil by its inductance to the energy dissipated in the coil by the resistance;

therefore, the higher the Q, the better the coil at storing energy rather than dissipating it. The energy stored is dependent on the inductive reactance (X_L) of the coil, and the energy dissipated is dependent on the resistance (R) of the coil.

57. Inductors, therefore, will never appear as pure inductance, but rather as an inductive and resistive (RL) circuit, and the resistance within the inductor will dissipate *true power*.

58. When a circuit contains both inductance and resistance, some of the energy is consumed and then returned by the inductor (reactive or imaginary power), and some of the energy is dissipated and lost by the resistor (resistive or true power).

59. Apparent power is the power that appears to be supplied to the load and is the vector sum of both the reactive and true power.

Parallel *RL* Circuit

60. With a parallel combination of a resistor and inductor, the voltage across both components is equal because of the parallel connection, and the current through each branch is calculated by applying Ohm's law.

61. Total current (I_T) is equal to the vector combination of the resistor current and inductor current.

Testing Inductors

62. Basically, only three problems can occur with inductors:
    a. An open
    b. A complete short
    c. A section short (value change)

63. Depending on the winding's resistance, the coil should be in the range of zero to a few hundred ohms. An open accounts for 75% of all defective inductors.

64. A coil with one or more shorted turns or a complete short can be checked with an ohmmeter and thought to be perfectly good because of the normally low resistance of a coil, as it is just a piece of wire.

65. An inductor analyzer can be used to check capacitance and inductance. Complete or section shorts account for 25% of all defective inductors.

RL Filters

66. The RL filter will achieve results similar to the RC filter in that it will pass some frequencies and block others. The inductive reactance of the coil and the resistance of the resistor form a voltage divider. Since inductive reactance is proportional to frequency ($X_L \propto f$), the inductor will drop less voltage at lower frequencies ($f\downarrow$, $X_L\downarrow$, $V_L\downarrow$) and more voltage at higher frequencies ($f\uparrow, X_L\uparrow, V_L\uparrow$).

RL Integrator

67. In the *RL* integrator, the output is taken across the resistor. The output is the same as the previously described *RC* integrator's output. As with the *RC* integrator, if the period of the square wave is decreased or if the time constant is increased, the output will reach an average value of half the input square wave's amplitude.
68. With the *RL* differentiator, the output is taken across the inductor.

NEW TERMS

Air-core inductor	Henry	*RL* differentiator
Choke	Inductance	*RL* filter
Counter emf	Inductive reactance	*RL* integrator
Delta	Inductive time constant	Self-inductance
Ferrite-core inductor	Inductor	Slug
Ferrous	Iron-core inductor	Toroidal inductor
Form	*Q* or quality factor	

REVIEW QUESTIONS

Multiple-Choice Questions

1. Self-inductance is a process by which a coil will induce a voltage within ______________.
   a. Another inductor **c.** Itself
   b. Two or more close proximity inductors **d.** Both (a) and (b)
2. Mutual inductance is a process by which a coil will induce a voltage within ______________.
   a. Another inductor **c.** Itself
   b. Two or more close proximity inductors **d.** Both (a) and (b)
3. The inductor stores electrical energy in the form of a ______________ field, just as a capacitor stores electrical energy in the form of a ______________ field.
   a. Electric, magnetic **b.** Magnetic, electric
4. The inductor is basically:
   a. An electromagnet
   b. A coil of wire
   c. A coil of conductor formed around a core material
   d. All of the above
   e. None of the above
5. The inductance of an inductor is proportional to ______________, and inversely proportional to ______________.
   a. $N, A, \mu; l$ **c.** $\mu, l, N; A$
   b. $A, \mu, l; N$ **d.** $N, A, l; \mu$

6. The total inductance of a series circuit is:

a. Less than the value of the smallest inductor
b. Equal to the sum of all the inductance values
c. Equal to the product over sum
d. All of the above

7. The total inductance of a parallel circuit can be calculated by:

a. Using the product-over-sum formula
b. Using L divided by N for equal-value inductors
c. Using the reciprocal resistance formula
d. All of the above

8. Air-core fixed-value inductors can use air or nonmagnetic forms, such as:

a. Iron, cardboard **c.** Ceramic, cardboard
b. Ceramic, copper **d.** Silicon, germanium

9. Ferrite is a chemical compound, which is basically powdered:

a. Iron oxide and ceramic **c.** Mylar and iron
b. Iron and steel **d.** Gauze and electrolyte

10. The ferrite-core variable inductor varies inductance by changing:

a. μ **c.** N
b. l **d.** A

11. It will actually take ____________ time constants for the current in an inductive circuit to reach a maximum value.

a. 63.2 **c.** 1.414
b. 1 **d.** 5

12. The time constant for a series inductive/resistive circuit is equal to:

a. $L \times R$ **c.** V/R
b. L/R **d.** $2\pi \times f \times L$

13. Inductive reactance (X_L) is proportional to:

a. Time or period of the ac applied
b. Frequency of the ac applied
c. The stray capacitance that occurs due to the air acting as a dielectric between two turns of a coil
d. The value of inductance
e. Two of the above are true.

14. In a series RL circuit, the source voltage (V_S) is equal to:

a. The square root of the sum of V_R^2 and V_L^2
b. The vector sum of V_R and V_L
c. $I \times Z$
d. Two of the above are partially true.
e. Answers (a), (b), and (c) are correct.

15. In a purely resistive circuit, the phase shift is equal to ____________, whereas in a purely inductive or capacitive circuit, the phase shift is ____________ degrees.

a. 45, 0 **c.** 45, 90
b. 90, 0 **d.** None of the above

16. With an RL integrator, the output is taken across the:

a. Inductor c. Resistor
b. Capacitor d. Transformer's secondary

17. With an RL differentiator, the output is taken across the:

a. Inductor c. Resistor
b. Capacitor d. Transformer's secondary.

18. An inductor or choke between the input and output forms a:

a. High-pass filter b. Low-pass filter

19. An inductor or choke connected to ground or in shunt forms a:

a. Low-pass filter b. High-pass filter

20. Lenz's law states that when current is passed through a conductor a self-induced voltage in a coil will:

a. Aid the applied source voltage
b. Aid the increasing current from the source
c. Produce an opposing current
d. Both (a) and (c)

21. When tested with an ohmmeter, an open coil would show:

a. Zero resistance c. A 100 Ω to 200 Ω resistance
b. An infinite resistance d. Both (b) and (c)

22. Inductive reactance:

a. Increases with frequency
b. Is proportional to inductance
c. Reduces the amplitude of alternating current
d. All of the above

23. The current through an inductor ____________ the voltage across the same inductor by ____________.

a. Lags, 90° c. Leads, 90°
b. Lags, 45° d. Leads, 45°

24. The phasor combination of X_L and R is the circuit's:

a. Reactance c. Power factor
b. Total resistance d. Impedance

25. In a series RL circuit, where $V_R = 200$ mV and $V_L = 0.2$ V, $\theta =$:

a. 45° c. 0°
b. 90° d. 1°

Essay Questions

26. Briefly describe the terms: (Chapter 15, introduction)

a. Electromagnetism b. Electromagnetic induction

27. What is self-induction, and how does it relate to an inductor? (16-1)

28. Give the formula for inductance, and explain the four factors that determine inductance. (16-3)

29. List all the formulas for calculating total inductance when inductors are connected in: (16-4)

    a. Series **b.** Parallel

30. What are a fixed- and a variable-value inductor? (16-5)

31. List the important factors of the: (16-5)

    a. Air-, iron-, and ferrite-core fixed-value inductors
    b. Ferrite-core variable-value inductor

32. Describe the current rise and fall through an inductor when: (16-6)

    a. Dc is applied **b.** Ac is applied

33. Define the following terms:

    a. Inductive reactance (16-7)
    b. Impedance (16-8-2)
    c. Phase shift (16-8-3)
    d. Q factor (16-8-4)
    e. Power factor (16-8-4)

34. With illustrations, describe how an inductor and resistor could be used for the following applications: (16-11)

    a. Filtering high and low frequencies
    b. Integration
    c. Differentiation

35. Explain briefly why an inductor acts as an open to an instantaneous change. Why does an inductor act like a short to dc? (16-7)

Practice Problems

36. Convert the following:

    a. 0.037 H to mH
    b. 1760 μH to mH
    c. 862 mH to H
    d. 0.256 mH to μH

37. Calculate the impedance (Z) of the following series RL combinations:

    a. 22 MΩ, 25 μH, $f = 1$ MHz
    b. 4 kΩ, 125 mH, $f = 100$ kHz
    c. 60 Ω, 0.05 H, $f = 1$ MHz

38. Calculate the voltage across a coil if: ($d = \Delta$ or delta)

    a. $d_i/d_t = 120$ mA/ms and $L = 2$ μH
    b. $d_i/d_t = 62$ μA/μs and $L = 463$ mH
    c. $d_i/d_t = 4$ A/s and $L = 25$ mH

39. Calculate the total inductance of the following series circuits:

    a. 75 μH, 61 μH, 50 mH **b.** 8 mH, 4 mH, 22 mH

40. Calculate the total inductance of the following parallel circuits:

    a. 12 mH, 8 mH **b.** 75 μH, 34 μH, 27 μH

41. Calculate the total inductance of the following series–parallel circuits:

    a. 12 mH in series with 4 mH, and both in parallel with 6 mH
    b. A two-branch parallel arrangement made up of 6 μH and 2 μH in series with one another, and 8 μH and 4 μH in series with one another

c. Two parallel arrangements in series with one another, made up of 1 μH and 2 μH in parallel and 4 μH and 15 μH in parallel

42. Determine the time constant of all the examples in Question 41, and state how long it will take in each example for current to build up to maximum.

43. In a series *RL* circuit, if $V_L = 12$ V and $V_R = 6$ V, calculate:

a. V_S

b. I if $Z = 14$ kΩ

c. Phase angle

d. Q

e. Power factor

44. What value of inductance is needed to produce 3.3 kΩ of reactance at 15 kHz?

45. At what frequency will a 330 μH inductor have a reactance of 27 kΩ?

46. Calculate the impedance of the circuits seen in Figure 16-42.

47. Referring to Figure 16-43, calculate the voltage across the inductor for all five time constants after the switch has been closed.

48. Referring to Figure 16-44, calculate:

a. L_T

b. X_L

c. Z

d. I_{R1}, I_{L1}, and I_{L2}

e. θ

f. True power, reactive power, and apparent power

g. Power factor

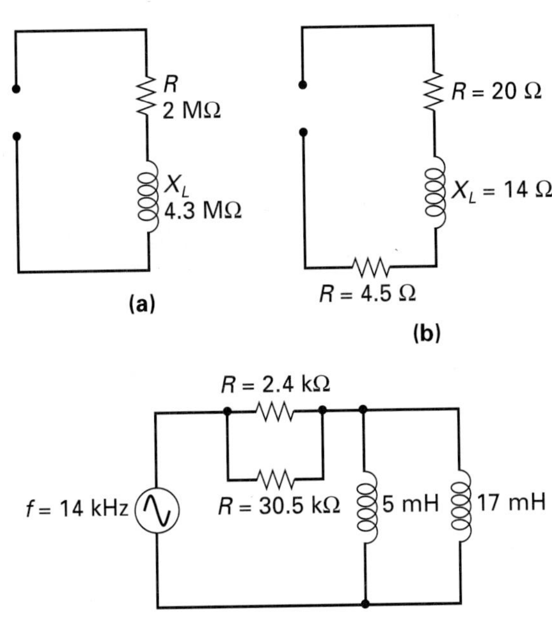

FIGURE 16-42 **Calculating Impedance.**

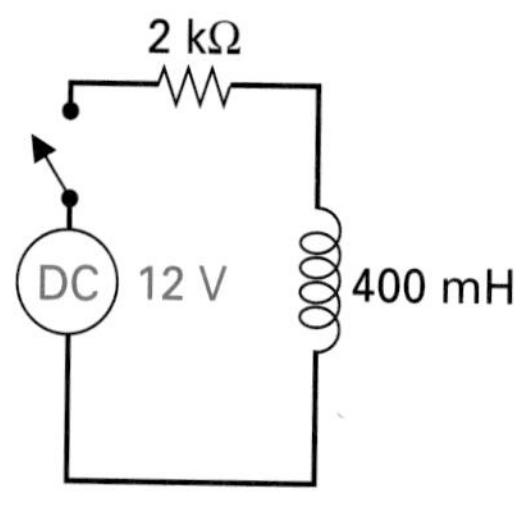

FIGURE 16-43 Inductive Time Constant Example.

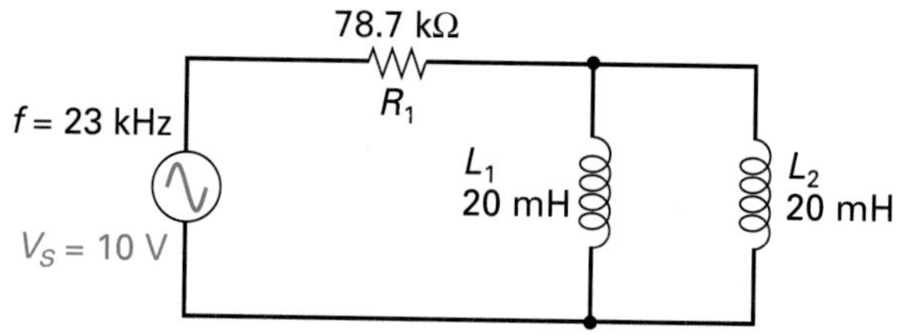

FIGURE 16-44 Series *RL* Circuit.

49. Referring to Figure 16-45, calculate:

a. R_T
b. L_T
c. X_L
d. Z
e. V_{RT}, V_{LT}
f. I_{RT}, I_{LT}
g. I_T
h. θ
i. Apparent power
j. PF

Troubleshooting Problem

50. What problems can occur with inductors, and how can the ohmmeter be used to determine these problems?

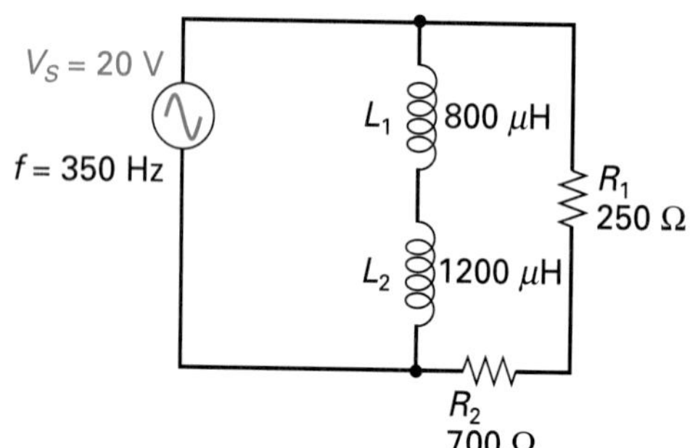

FIGURE 16-45 Parallel *RL* Circuit.

CHAPTER

17

OBJECTIVES

After completing this chapter, you will be able to:

1. Define mutual inductance and how it relates to transformers.
2. Describe the basic operation of a transformer.
3. Explain the differences between a loaded and unloaded transformer.
4. State the meaning of the coefficient of coupling.
5. List the three basic applications of transformers.
6. Describe how a transformer's turns ratio can be used to step up or step down voltage or current, or match impedances.
7. Describe the transformer dot convention as it relates to windings and phase.
8. List and explain the fixed and variable transformer types.
9. Explain how to test the windings of a transformer for opens, partial shorts, or complete shorts.
10. Describe the three basic transformer power losses.

Transformers

OUTLINE

VIGNETTE

Woolen Mill Makes Minis

At the age of 24 in 1950, Kenneth Olsen went to work at MIT's Digital Computer Laboratory as a research assistant. For the next seven years, Olsen worked on the SAGE (Semi Automatic Ground Environment) project, which was a new computer system designed to store the constantly changing data obtained from radars tracking long-range enemy bombers capable of penetrating American air space. IBM won the contract to build the SAGE network and Olsen traveled to the plant in New York to supervise production. After completing the SAGE project, Olsen was ready to move on to bigger and better things; however, he didn't want to go back to MIT and was not happy with the rigid environment at a large corporation like IBM.

In September 1957, Olsen and fellow MIT colleague Harlan Anderson leased an old brick woolen mill in Maynard, Massachusetts, and after some cleaning and painting they began work in their new company, which they named Digital Equipment Corporation or DEC. After three years, the company came out with their first computer, the PDP 1 (programmed data processor), and its ease of use, low cost, and small size made it an almost instant success. In 1965, the desktop PDP 8 was launched, and since in that era miniskirts were popular, it was probably inevitable that the small machine was nicknamed a mini-computer. In 1969, the PDP 11 was unveiled, and like its predecessors, this number-crunching data processor went on to be accepted in many applications, such as tracking the millions of telephone calls received on the 911 emergency line, directing welding machines in automobile plants, recording experimental results in laboratories, and processing all types of data in offices, banks, and department stores. Just 20 years after introducing low-cost computing in the 1960s, DEC was second only to IBM as a manufacturer of computers in the United States.

INTRODUCTION

Transformer Device consisting of two or more coils that are used to couple electric energy from one circuit to another, yet maintain electrical isolation between the two.

When alternating current was introduced in Chapter 11, it was mentioned that an ac voltage could be stepped up to a larger voltage or stepped down to a smaller voltage by a device called a **transformer.** The transformer is an electrical device that makes use of electromagnetic induction to transfer alternating current from one circuit to another. The transformer consists of two inductors that are placed in very close proximity to one another. When an alternating current flows through the first coil or **primary winding** the inductor sets up a magnetic field. The expanding and contracting magnetic field produced by the primary cuts across the windings of the second inductor or **secondary winding** and induces a voltage in this coil.

Primary Winding First winding of a transformer that is connected to the source.

Secondary Winding Output winding of a transformer that is connected across the load.

By changing the ratio between the number of turns in the secondary winding to the number of turns in the primary winding, some characteristics of the ac signal can be changed or transformed as it passes from primary to secondary. For example, a low ac voltage could be stepped up to a higher ac voltage, or a high ac voltage could be stepped down to a lower ac voltage.

In this chapter we examine the operation, characteristics, types, testing, and applications of transformers.

17-1 MUTUAL INDUCTANCE

As was discussed in the previous chapter, self-inductance is the process by which a coil induces a voltage within itself. The principle on which a transformer is based is an inductive effect known as **mutual inductance,** which is the process by which an inductor induces a voltage in another inductor.

Mutual Inductance Ability of one inductor's magnetic lines of force to link with another inductor.

Figure 17-1 illustrates two inductors that are magnetically linked, yet electrically isolated from one another. As the alternating current continually rises, falls, and then rises in the opposite direction, a magnetic field will build up, collapse, and then build up in the opposite direction.

If a second inductor, or secondary coil (L_2), is in close proximity with the first inductor or primary coil (L_1), which is producing the alternating magnetic field, a voltage will be induced into the nearby inductor, which causes current to flow in the secondary circuit through the load resistor. This phenomenon is known as mutual inductance or transformer action.

As with self-inductance, mutual inductance is dependent on change. Direct current (dc) is a constant current and produces a constant or stationary magnetic field that does not change, as reviewed in Figure 17-2(a). Alternating current, however, is continually varying, and as the polarity of the magnetic field is dependent on the direction of current flow (left-hand rule), the magnetic field will also be alternating in polarity, as seen in Figure 17-2(b); it is this continual

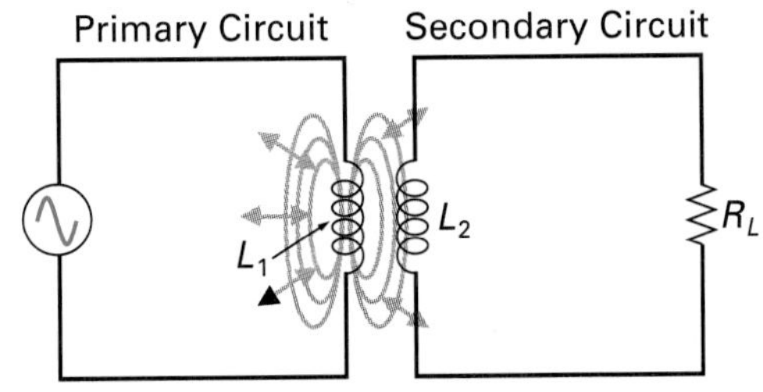

FIGURE 17-1 **Mutual Inductance.**

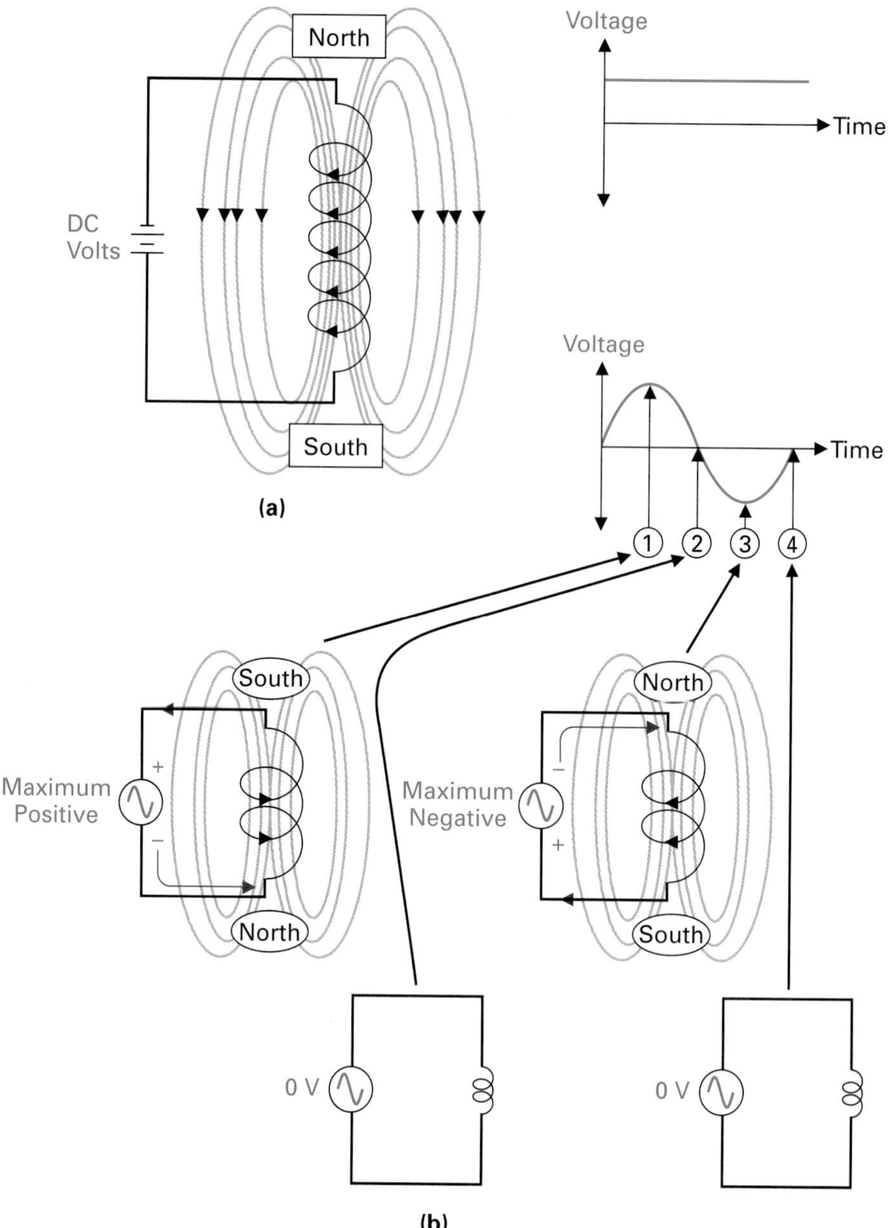

FIGURE 17-2 **Magnetic Field Produced by (a) DC across an Inductor and (b) AC across an Inductor.**

building up and collapsing of the magnetic field that cuts the adjacent inductor's conducting coils and induces a voltage in the secondary circuit. Mutual induction is possible only with ac and cannot be achieved with dc due to the lack of change.

Self-induction is a measure of how much voltage an inductor can induce within itself. Mutual inductance is a measure of how much voltage is induced in the secondary coil due to the change in current in the primary coil.

SELF-TEST REVIEW QUESTIONS FOR SECTION 17-1

1. Define mutual inductance and how it differs from self-inductance.
2. True or false: Mutual induction is possible only with direct current flow.

17-2 BASIC TRANSFORMER

Figure 17-3(a) illustrates a basic transformer, which consists of two coils within close proximity to one another, to ensure that the second coil will be cut by the magnetic flux lines produced by the first coil, and thereby ensure mutual inductance. The ac voltage source is electrically connected (through wires) to the primary coil or winding, and the load (R_L) is electrically connected to the secondary coil or winding.

In Figure 17-3(b), the ac voltage source has produced current flow in the primary circuit, as illustrated. This current flow produces a north pole at the top of the primary winding and, as the ac voltage input swings more negative, the current increase causes the magnetic field being developed by the primary winding to increase. This expanding magnetic field cuts the coils of the secondary winding and induces a voltage, and a subsequent current flows in the secondary circuit, which travels up through the load resistor. The ac voltage follows a sinu-

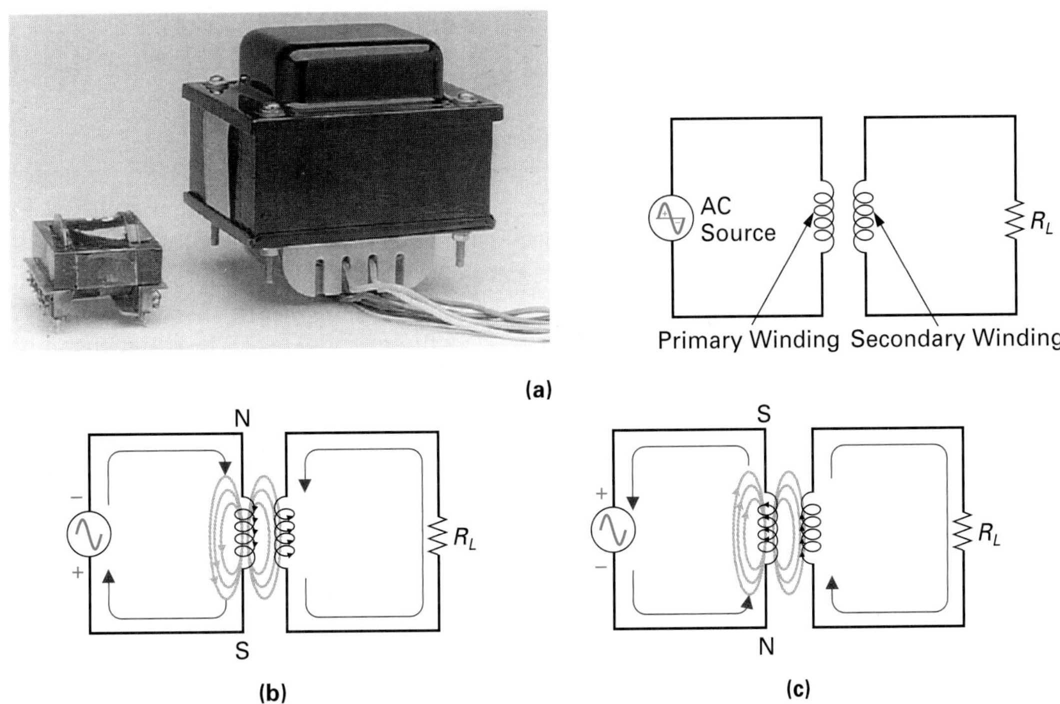

FIGURE 17-3 Transformer Action.

soidal pattern and moves from a maximum negative to zero and then begins to build up toward a maximum positive.

In Figure 17-3(c), the current flow in the primary circuit is in the opposite direction due to the ac voltage increase in the positive direction. As voltage increases, current increases, and the magnetic field expands and cuts the secondary winding, inducing a voltage and causing current to flow in the reverse direction down through the load resistor.

You may have noticed a few interesting points about the basic transformer just discussed.

1. As primary current increases, secondary current increases, and as primary current decreases, secondary current also decreases. It can therefore be said that the frequency of the alternating current in the secondary is the same as the frequency of the alternating current in the primary.
2. Although the two coils are electrically isolated from one another, energy can be transferred from primary to secondary, because the primary converts electrical energy into magnetic energy, and the secondary converts magnetic energy back into electrical energy.

SELF-TEST REVIEW QUESTIONS FOR SECTION 17-2

1. True or false: A transformer achieves electrical isolation.
2. True or false: The transformer converts electrical energy to magnetic, and then from magnetic back to electrical.
3. What are the names given to the windings of a transformer?

17-3 TRANSFORMER LOADING

Let's now carry our discussion of the basic transformer a little further and see what occurs when the transformer is not connected to a load, as shown in Figure 17-4. Primary circuit current is determined by $I = V/Z$, where Z is the impedance of the primary coil (both its inductive reactance and resistance) and V is the applied voltage. Since no current can flow in the secondary, because an open in the circuit exists, the primary acts as a simple inductor, and the primary current

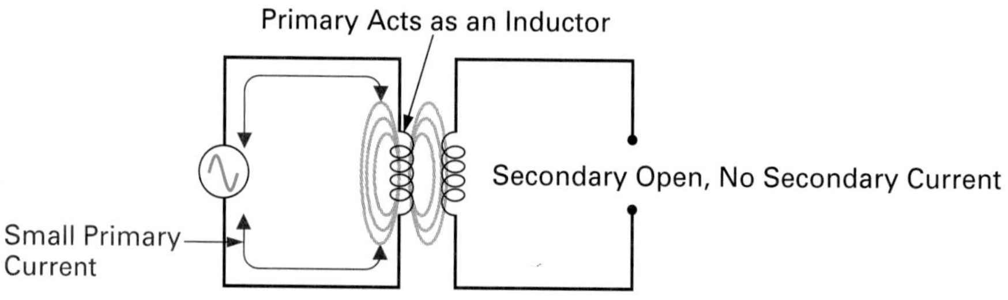

FIGURE 17-4 Unloaded Transformer.

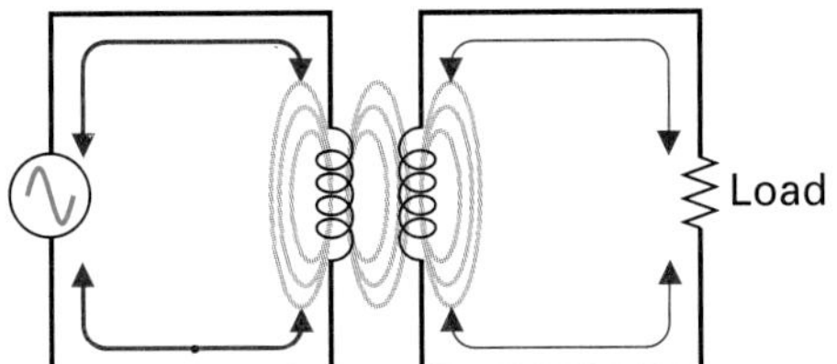

FIGURE 17-5 **Loaded Transformer.**

is small due to the inductance of the primary winding. This small primary current lags the applied voltage due to the counter emf by approximately 90° because the coil is mainly inductive and has very little resistance.

When a load is connected across the secondary, as shown in Figure 17-5, a change in conditions occurs and the transformer acts differently. The important point that will be observed is that as we go from a no-load to a load condition the primary current will increase due to mutual inductance. Let's follow the steps one by one.

1. The ac applied voltage sets up an alternating magnetic field in the primary winding.
2. The continually changing flux of this primary field induces and produces a counter emf into the primary to oppose the applied voltage.
3. The primary's magnetic field also induces a voltage in the secondary winding, which causes current to flow in the secondary circuit through the load.
4. The current in the secondary winding produces another magnetic field that is opposite to the field being produced by the primary.
5. This secondary magnetic field feeds back to the primary and induces a voltage that tends to cancel or weaken the counter emf that was set up in the primary by the primary current.
6. The primary's counter emf is therefore reduced, so primary current can now increase.
7. This increase in primary current is caused by the secondary's magnetic field; consequently, the greater the secondary current, the stronger the secondary magnetic field, which causes a reduction in the primary's counter emf, and therefore a primary current increase.

In summary, an increase in secondary current ($I_s\uparrow$) causes an increase in primary current ($I_p\uparrow$), and this effect in which the primary induces a voltage in the secondary (V_s) and the secondary induces a voltage into the primary (V_p) is known as *mutual inductance.*

SELF-TEST REVIEW QUESTIONS FOR SECTION 17-3

1. True or false: An increase in secondary current causes an increase in primary current.
2. True or false: The greater the secondary current, the greater the primary's counter emf.

17-4 COEFFICIENT OF COUPLING (k)

The voltage induced into the secondary winding is dependent on the mutual inductance between the primary and secondary, which is determined by how much of the magnetic flux produced by the primary actually cuts the secondary winding.

Coefficient of Coupling The degree of coupling that exists between two circuits.

The **coefficient of coupling** (k) is a ratio of the number of magnetic lines of force that cut the secondary compared to the total number of magnetic flux lines being produced by the primary and is a figure between 0 and 1.

$$k = \frac{\text{flux linking secondary coil}}{\text{total flux produced by primary}}$$

If, for example, all the primary flux lines cut the secondary winding, the coefficient of coupling will equal 1. If only half of the total flux lines being produced cut the secondary winding, k will equal 0.5.

EXAMPLE:

A primary coil is producing 65 μWb (microwebers) of magnetic flux. Calculate the coefficient of coupling if 52 μWb links with the secondary coil.

Solution:

$$k = \frac{52\ \mu\text{Wb}}{65\ \mu\text{Wb}} = 0.8$$

This means that 80% of the magnetic flux lines being generated by the primary coil are linking with the secondary coil.

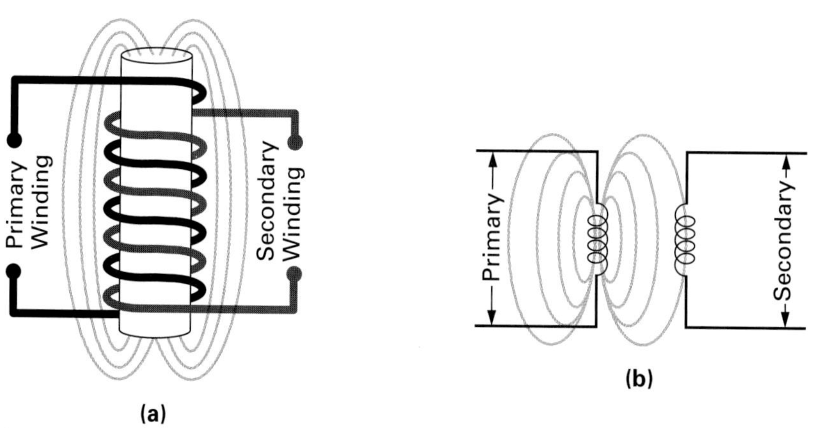

FIGURE 17-6 Coefficient of Coupling. (a) Wound on Same Core, Small Distance Between Primary and Secondary, High k. (b) Winding Far Apart, Small k.

The coefficient of coupling depends on:

1. How close together the primary and secondary are to one another
2. The type of core material used

Figure 17-6 illustrates why the primary and secondary have to be in close proximity to one another to achieve a high coefficient of coupling.

SELF-TEST REVIEW QUESTIONS FOR SECTION 17-4

1. Define and state the formula for the coefficient of coupling.
2. True or false: If $k = 0.75$, then 75% of the total flux lines produced by the primary are cutting the coils of the secondary.
3. List the two factors that determine k.

17-5 TRANSFORMER RATIOS AND APPLICATIONS

Basically, transformers are used for one of three applications:

1. To step up (increase) or step down (decrease) voltage
2. To step up (increase) or step down (decrease) current
3. To match impedances

In all three cases, any of the applications can be achieved by changing the ratio of the number of turns in the primary winding compared to the number of turns in the secondary winding. This ratio is appropriately called the turns ratio.

17-5-1 *Turns Ratio*

The **turns ratio** is the ratio between the number of turns in the secondary winding (N_s) and the number of turns in the primary winding (N_p).

Turns Ratio
Ratio of the number of turns in the secondary winding to the number of turns in the primary winding of a transformer.

$$\text{turns ratio} = \frac{N_s}{N_p}$$

Let us use a few examples to see how the turns ratio can be calculated.

EXAMPLE:

If the primary has 200 turns and the secondary has 600, what is the turns ratio (Figure 17-7)?

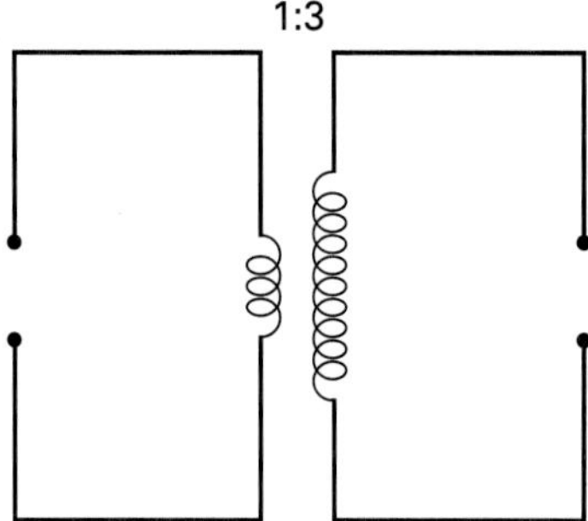

FIGURE 17-7 **Step-Up Transformer Example.**

■ ***Solution:***

$$\text{turns ratio} = \frac{N_s}{N_p} = \frac{600}{200} = \frac{3\text{ (secondary)}}{1\text{ (primary)}} = 3$$

This simply means that there are three windings in the secondary to every one winding in the primary. Moving from a small number (1) to a larger number (3) means that we *stepped up* in value. Stepping up always results in a turns ratio figure greater than 1, in this case, 3.

■ **EXAMPLE:**

If the primary has 120 turns and the secondary has 30 turns, what is the turns ratio (Figure 17-8)?

■ ***Solution:***

$$\text{turns ratio} = \frac{N_s}{N_p} = \frac{30}{120} = \frac{1\text{ (secondary)}}{4\text{ (primary)}} = 0.25$$

Said simply, there are four primary windings to every one secondary winding. Moving from a larger number (4) to a smaller number (1) means that we *stepped down* in value. Stepping down always results in a turns ratio figure of less than 1, in this case 0.25.

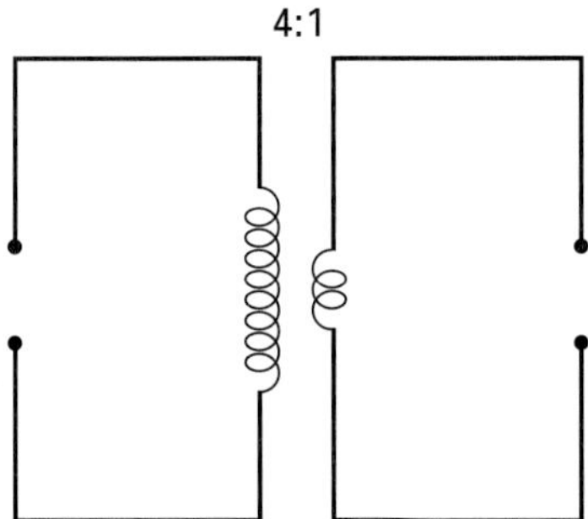

FIGURE 17-8 **Step-Down Transformer Example.**

17-5-2 *Voltage Ratio*

Transformers are used within the power supply unit of almost every piece of electronic equipment to step up or step down the 115 V ac from the outlet. Some electronic circuits require lower-power supply voltages, while other devices may require higher-power supply voltages. The transformer is used in both instances to convert the 115 V ac to the required value of voltage.

Step Up

If the secondary voltage (V_s) is greater than the primary voltage (V_p), the transformer is called a **step-up transformer** ($V_s > V_p$), as shown in Figure 17-9. The voltage is stepped up or increased in much the same way as a generator voltage can be increased by increasing the number of turns.

Step-Up Transformer Transformer in which the ac voltage induced in the secondary is greater (due to more secondary windings) than the ac voltage applied to the primary.

If the ac primary voltage is 100 V and the turns ratio is a 1:5 step up, the secondary voltage will be five times that of the primary voltage, or 500 V, because the magnetic flux established by the primary cuts more turns in the secondary and therefore induces a larger voltage.

In this example, you can see that the ratio of the secondary voltage to the primary voltage is equal to the turns ratio; in other words,

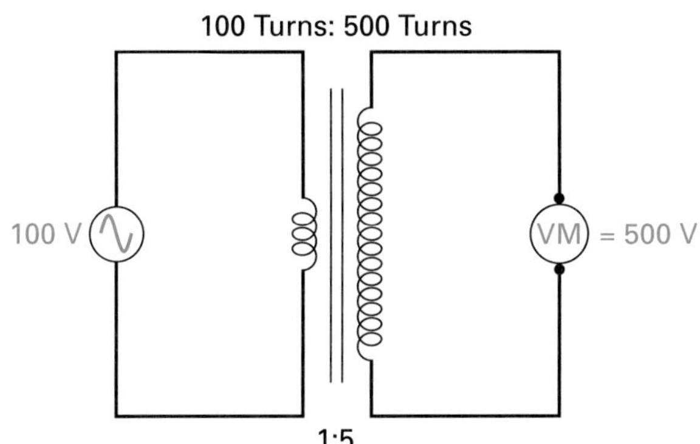

FIGURE 17-9 **Step-Up Transformer.**

$$\frac{V_s}{V_p} = \frac{N_s}{N_p}$$

or

$$\frac{500}{100} = \frac{500}{100}$$

To calculate V_s, therefore, we can rearrange the formula and arrive at

$$V_s = \frac{N_s}{N_p} \times V_p$$

In our example, this is

$$V_s = \frac{500}{100} \times 100 \text{ V}$$
$$= 500 \text{ V}$$

Step Down

Step-Down Transformer Transformer in which the ac voltage induced in the secondary is less (due to fewer secondary windings) than the ac voltage applied to the primary.

If the secondary voltage (V_s) is smaller than the primary voltage (V_p), the transformer is called a **step-down transformer** ($V_s < V_p$), as shown in Figure 17-10. The secondary voltage will be equal to

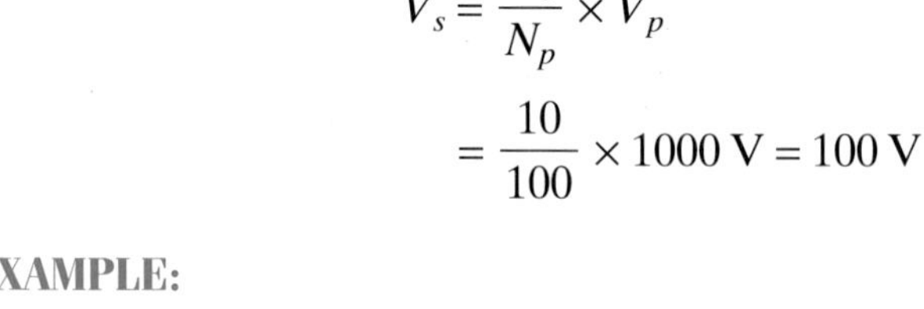

$$V_s = \frac{N_s}{N_p} \times V_p$$
$$= \frac{10}{100} \times 1000 \text{ V} = 100 \text{ V}$$

■ **EXAMPLE:**

Calculate the secondary voltage (V_s) if a 1 : 6 step-up transformer has 24 V ac applied to the primary.

100 Turns: 10 Turns

1000 V

VM = 100 V

10:1

FIGURE 17-10 **Step-Down Transformer.**

■ *Solution:*

$$V_s = \frac{N_s}{N_p} \times V_p$$

$$= \frac{6}{1} \times 24\text{ V} = 144\text{ V}$$

The coupling coefficient (k) in this formula is always assumed to be 1, which for most iron-core transformers is almost always the case. This means that all the primary magnetic flux is linking the secondary, and the secondary voltage is dependent on the number of secondary turns that are being cut by the primary magnetic flux.

The transformer can be used to transform the primary ac voltage into any other voltage, either up or down, merely by changing the transformer's turns ratio.

17-5-3 Power and Current Ratio

The power in the secondary of the transformer is equal to the power in the primary ($P_p = P_s$). Power, as we know, is equal to $P = V \times I$, and if voltage is stepped up or down, the current automatically is stepped down or up, respectively, in the opposite direction to voltage to maintain the power constant.

For example, if the secondary voltage is stepped up ($V_s\uparrow$), the secondary current is stepped down ($I_s\downarrow$), so the output power is the same as the input power.

$$P_s = V_s\uparrow \times I_s\downarrow$$

This is an equal but opposite change. Therefore, $P_s = P_p$; and you cannot get more power out than you put in. The current ratio is therefore inversely proportional to the voltage ratio:

$$\boxed{\frac{V_s}{V_p} = \frac{I_p}{I_s}}$$

If the secondary voltage is stepped up, the secondary current goes down:

$$\frac{V_s\uparrow}{V_p} = \frac{I_p}{I_s\downarrow}$$

If the secondary voltage is stepped down, the secondary current goes up:

$$\frac{V_s\downarrow}{V_p} = \frac{I_p}{I_s\uparrow}$$

If the current ratio is inversely proportional to the voltage ratio, it is also inversely proportional to the turns ratio:

$$\frac{I_p}{I_s} = \frac{V_s}{V_p} = \frac{N_s}{N_p}$$

By rearranging the current and turns ratio, we can arrive at a formula for secondary current, which is

$$I_s = \frac{N_p}{N_s} \times I_p$$

EXAMPLE:

The step-up transformer in Figure 17-11 has a turns ratio of 5 to 1. Calculate:

a. Secondary voltage (V_s)
b. Secondary current (I_s)
c. Primary power (P_p)
d. Secondary power (P_s)

Solution:

The secondary has five times as many windings as the primary, and, consequently, the voltage will be stepped up by 5 between primary and secondary. If the secondary voltage is going to be five times that of the primary, the secondary current is going to decrease to one-fifth of the primary current.

a. $V_s = \frac{N_s}{N_p} \times V_p$

$= \frac{5}{1} \times 100 \text{ V}$

$= 500 \text{ V}$

b. $I_s = \frac{N_p}{N_s} \times I_p$

$= \frac{1}{5} \times 10 \text{ A}$

$= 2 \text{ A}$

c. $P_p = V_p \times I_p = 100 \text{ V} \times 10 \text{ A} = 1000 \text{ VA}$

d. $P_s = V_s \times I_s = 500 \text{ V} \times 2 \text{ A} = 1000 \text{ VA}$

Therefore, $P_p = P_s$.

Maximum Power Transfer
The maximum power will be absorbed by the load from the source, when the impedance of the load is equal to the impedance of the source.

17-5-4 *Impedance Ratio*

The **maximum power transfer** theorem, which has been discussed previously and is summarized in Figure 17-12, states that maximum power is transferred from source (ac generator) to load (equipment) when the impedance of the load is

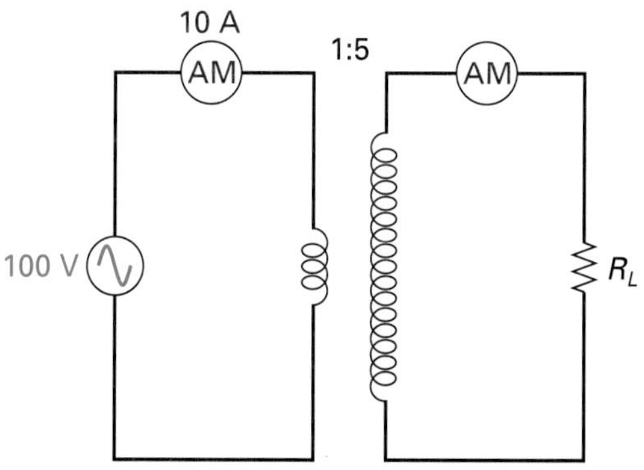

FIGURE 17-11 Step-Up Transformer Example.

equal to the internal impedance of the source. If these impedances are different, a large amount of power could be wasted.

In most cases it is required to transfer maximum power from a source that has an internal impedance (Z_s) that is not equal to the load impedance (Z_L). In this situation, a transformer can be inserted between the source and the load to make the load impedance appear to equal the source's internal impedance.

For example, let's imagine that your car stereo system (source) has an internal impedance of 100 Ω and is driving a speaker (load) of 4 Ω impedance, as seen in Figure 17-13.

By choosing the correct turns ratio, the 4 Ω speaker can be made to appear as a 100 Ω load impedance, which will match the 100 Ω internal source impedance of the stereo system, resulting in maximum power transfer.

The turns ratio can be calculated by using the formula

$$\text{turns ratio} = \sqrt{\frac{Z_{Load}}{Z_{Source}}}$$

Z_{Load} = Load impedance in ohms.

Z_{Source} = Source impedance in ohms.

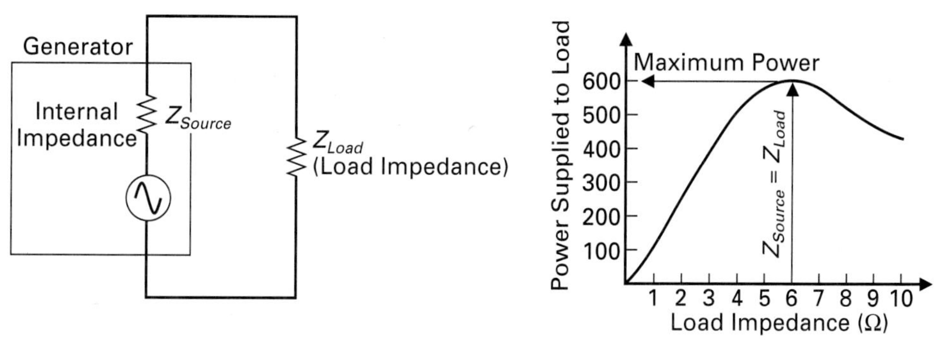

FIGURE 17-12 Maximum Power Transfer Theorem.

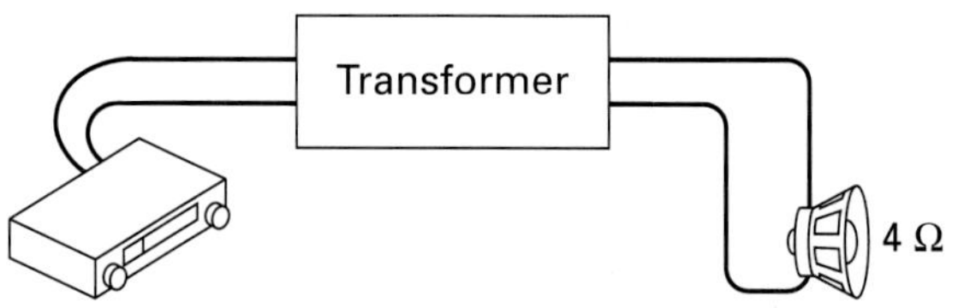

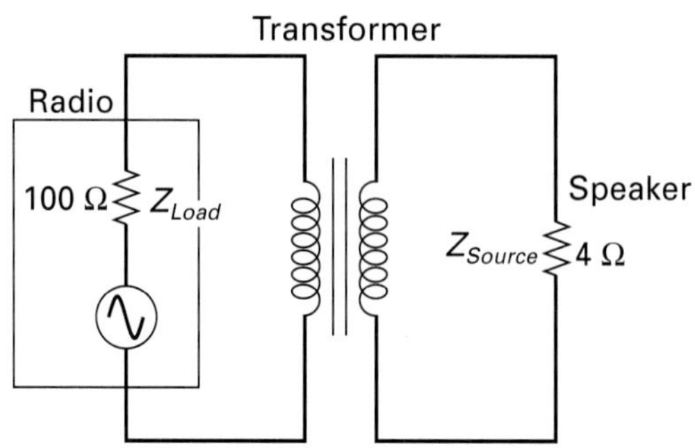

FIGURE 17-13 Impedance Matching.

In our example, this will calculate out to be

$$\text{turns ratio} = \sqrt{\frac{Z_L}{Z_s}}$$

$$= \sqrt{\frac{4}{100}} = \frac{\sqrt{4}}{\sqrt{100}}$$

$$= \frac{2}{10} = \frac{1}{5}$$

$$= 0.2$$

If the turns ratio is less than 1, a step-down transformer is required. A turns ratio of 0.2 means a step-down transformer is needed with a turns ratio of 5:1 ($\frac{1}{5}$ = 0.2).

EXAMPLE:

Calculate the turns ratio needed to match the 22.2 Ω output impedance of an amplifier to two 16 Ω speakers connected in parallel.

Solution:

The total load impedance of two 16 Ω speakers in parallel will be

$$Z_{Load} = \frac{\text{product}}{\text{sum}} = \frac{16 \times 16}{16 + 16} = 8\ \Omega$$

The turns ratio will be

$$\text{turns ratio} = \sqrt{\frac{Z_{Load}}{Z_{Source}}}$$

$$= \sqrt{\frac{8\ \Omega}{22.2\ \Omega}}$$

$$= \sqrt{0.36} = 0.6$$

Therefore, a step-down transformer is needed with a turns ratio of 1.67 : 1.

SELF-TEST REVIEW QUESTIONS FOR SECTION 17-5

1. What is the turns ratio of a 402-turn primary and 1608-turn secondary, and is this transformer step up or step down?
2. State the formula for calculating the secondary voltage (V_s).
3. True or false: A transformer can, by adjusting the turns ratio, be made to step up both current and voltage between primary and secondary.
4. State the formula for calculating the secondary current (I_s).
5. What turns ratio is needed to match a 25 Ω source to a 75 Ω load?
6. Calculate V_s if $N_s = 200$, $N_p = 112$, and $V_p = 115$ V.

17-6 WINDINGS AND PHASE

The way in which the primary and secondary coils are wound around the core determines the polarity of the voltage induced into the secondary relative to the polarity of the primary. In Figure 17-14, you can see that if the primary and secondary windings are both wound in a clockwise direction around the core, the voltage induced in the primary will be in phase with the voltage induced in the secondary. Both the input and output will also be in phase with one another if the primary and secondary are both wound in a counterclockwise direction.

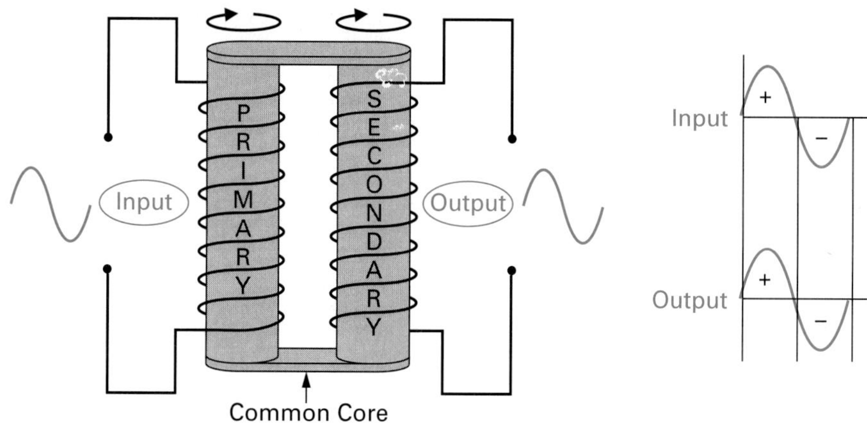

FIGURE 17-14 **Primary and Secondary in Phase.**

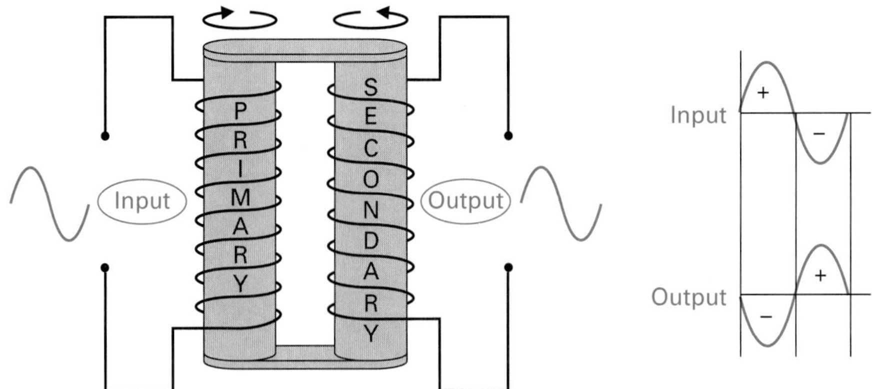

FIGURE 17-15 **Primary and Secondary out of Phase.**

In Figure 17-15, you will notice that in this case we have wound the primary and secondary in opposite directions, the primary being wound in a clockwise direction and the secondary in a counterclockwise direction. In this situation, the output ac sine-wave voltage is 180° out of phase with respect to the input ac voltage.

In a schematic diagram, there has to be a way of indicating to the technician that the secondary voltage will be in phase or 180° out of phase with the input. The **dot convention** is a standard used with transformers and is illustrated in Figure 17-16(a) and (b).

Dot Convention
A standard used with transformer symbols to indicate whether the secondary voltage will be in phase or out of phase with the primary voltage.

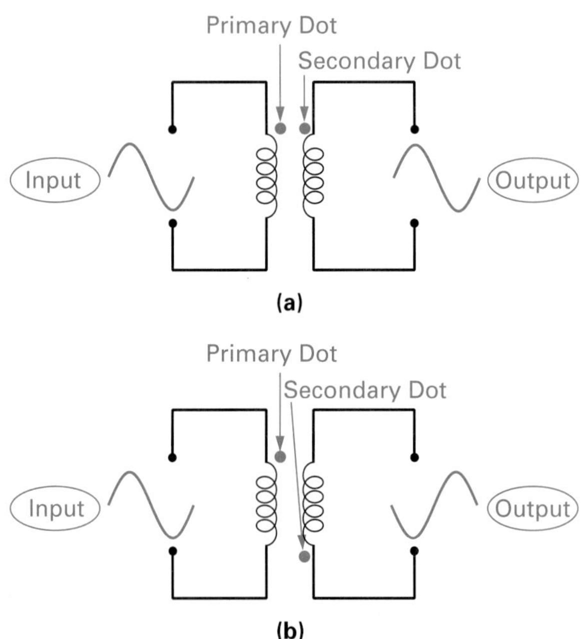

FIGURE 17-16 **Dot Convention. (a) In-Phase Dots. (b) Out-of-Phase Dots.**

A positive on the primary dot causes a positive on the secondary dot, and similarly, since ac is applied, a negative on a primary dot will cause a negative on the secondary dot. In Figure 17-16(a), you can see that when the top of the primary swings positive or negative, the top of the secondary will also follow suit and swing positive or negative, respectively, and you will now be able to determine that the transformer secondary voltage is in phase with the primary voltage.

In Figure 17-16(b), however, the dots are on top and bottom, which means that as the top of the primary winding swings positive, the bottom of the secondary will go positive, which means the top of the secondary will actually go negative, resulting in the secondary ac voltage being out of phase with the primary ac voltage.

SELF-TEST REVIEW QUESTIONS FOR SECTION 17-6

1. How does the winding of the primary and secondary affect the phase of the output with respect to the input?
2. Describe the transformer dot convention used on schematics.

17-7 TRANSFORMER TYPES

We will treat transformers the same as every other electronic component and classify them as either fixed or variable turns ratio.

17-7-1 *Fixed Turns Ratio Transformers*

Fixed turns ratio transformers have a turns ratio that cannot be varied. They are generally wound on a common core to ensure a high k and can be classified by the type of core material used. The two types of fixed transformers are:

1. Air core [Figure 17-17(a)]
2. Iron or ferrite core [Figure 17-17(b)]

The air-core transformers typically have a nonmagnetic core, such as ceramic or a cardboard hollow shell, and are used in high-frequency applications. The electronic circuit symbol just shows the primary and secondary coils. The more common iron- or ferrite-core transformers concentrate the magnetic lines of force, resulting in improved transformer performance; they are symbolized by two lines running between the primary and secondary. The iron-core transformer's lines are solid, while the ferrite-core transformer's lines are dashed.

Figure 17-18 illustrates the physical appearance of a few types of fixed transformers.

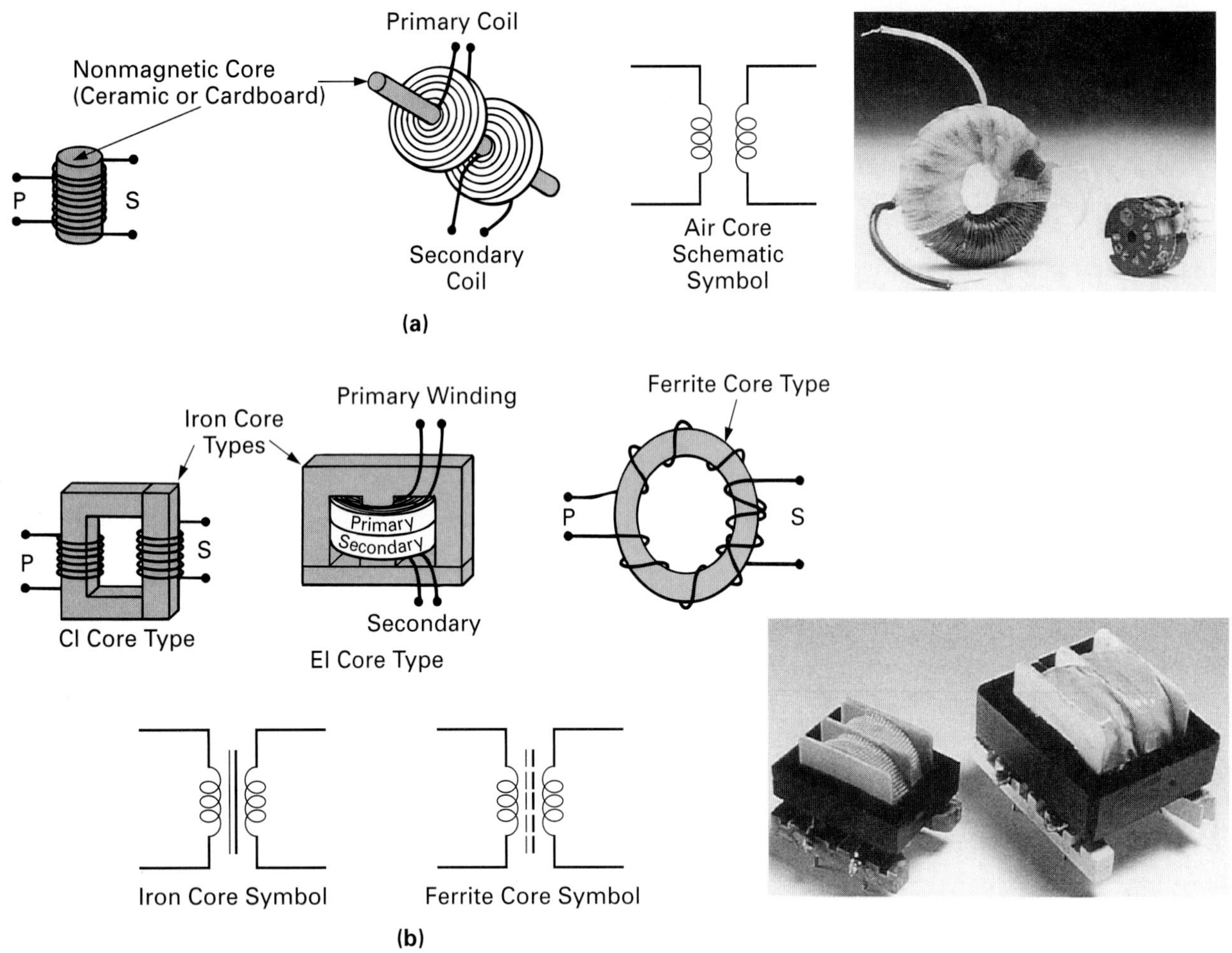

FIGURE 17-17 Fixed-Value Transformers. (a) Air-Core Transformers (Courtesy of Sencore). (b) Iron- or Ferrite-Core Transformers (Courtesy of Coilcraft).

17-7-2 Variable Turns Ratio Transformers

Variable turns ratio transformers have a turns ratio that can be varied. They can be classified as follows:

1. Center-tapped secondary
2. Multiple-tapped secondary
3. Multiple winding
4. Single winding

Figure 17-19 illustrates the first two of these, the center-tapped and the multiple-tapped secondary types. Transformers can often have tapped secondaries. A tapped winding will have a lead connected to one of the loops other than the two end connections.

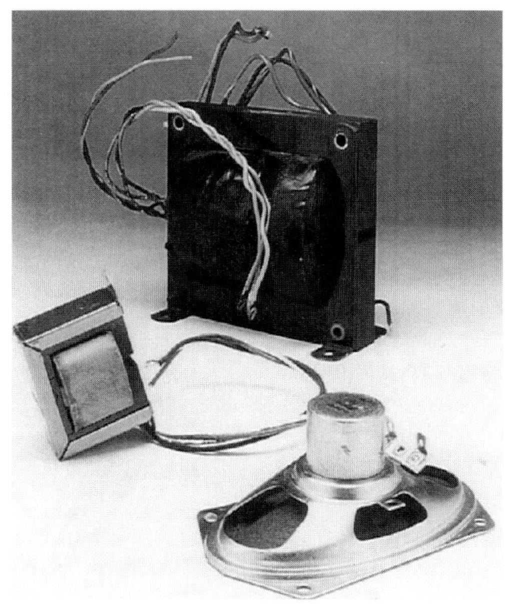

FIGURE 17-18 **Physical Appearance of Some Transformer Types (Courtesy of Sencore).**

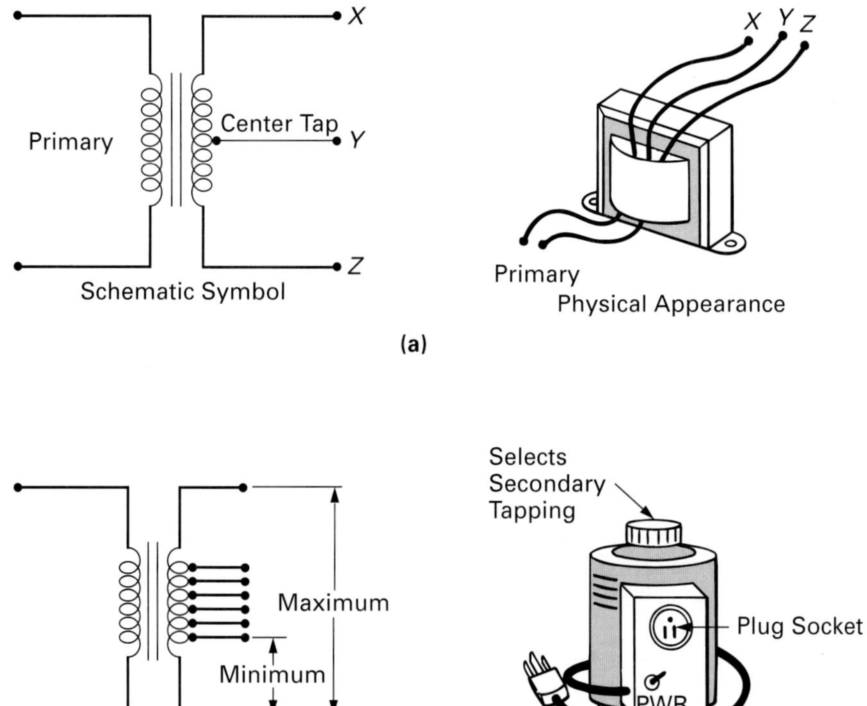

FIGURE 17-19 **Secondary Tapped Transformers. (a) Center Tapped (Courtesy of Sencore). (b) Multiple Tapped.**

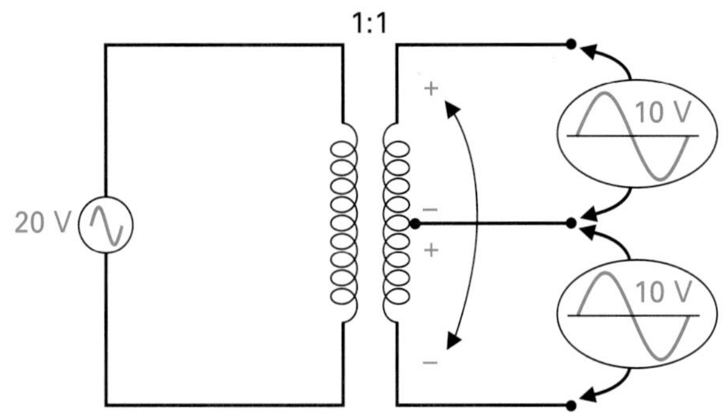

FIGURE 17-20 **Center-Tapped Secondary Transformer.**

Center-Tapped Secondary

Center-Tapped Transformer
A transformer that has a connection at the electrical center of the secondary winding.

If the tapped lead is in the exact center of the secondary, the transformer is said to have a center-tapped secondary, as shown in Figure 17-20. With the **center-tapped transformer,** the two secondary voltages are each half of the total secondary voltage. If we assume a 1 : 1 turns ratio (primary turns = secondary turns) and a 20 V ac primary voltage and therefore secondary voltage, each of the output voltages between either end of the secondary and the center tap will be 10-V waveforms, as shown in Figure 17-20. The two secondary outputs will be 180° out of phase with one another from the center tap. When the top of the secondary swings positive (+), the bottom of the secondary will be negative, and vice versa.

Figure 17-21 illustrates a typical power company utility pole, which as you can see makes use of a center-tapped secondary transformer. These transformers are designed to step down the high 4.8 kV from the power line to 120 V/240 V for commercial and residential customers. Within the United States, the actual secondary voltage may be anywhere between 225 and 245 V, depending on the demand. The demand is actually higher during the day and in the winter months (source voltage is pulled down by smaller customer load resistance). The multiple taps on the primary are used to change the turns ratio and compensate for power line voltage differences.

Neutral Wire
The conductor of a polyphase circuit or of a single-phase three-wire circuit that is intended to have a ground potential. The potential differences between the neutral and each of the other conductors are approximately equal in magnitude and are also equally spaced in phase.

The center tap of the secondary winding is connected to a copper earth grounding rod. This center tap wire is called the **neutral wire,** and within the building it is color coded with a white insulation. The two outside wires are color coded black and red. Residential or commercial appliances or loads that are designed to operate at 240 V are connected between the red wire and the black wire, while loads that are designed to operate at 120 V are connected between the red wire and the white wire (neutral) or the black wire and the white wire (neutral).

Multiple-Tapped Secondary

If several leads are attached to the secondary, the transformer is said to have a multiple-tapped secondary. This multitapped transformer will tap or pick off different values of ac voltage, as seen in Figure 17-22.

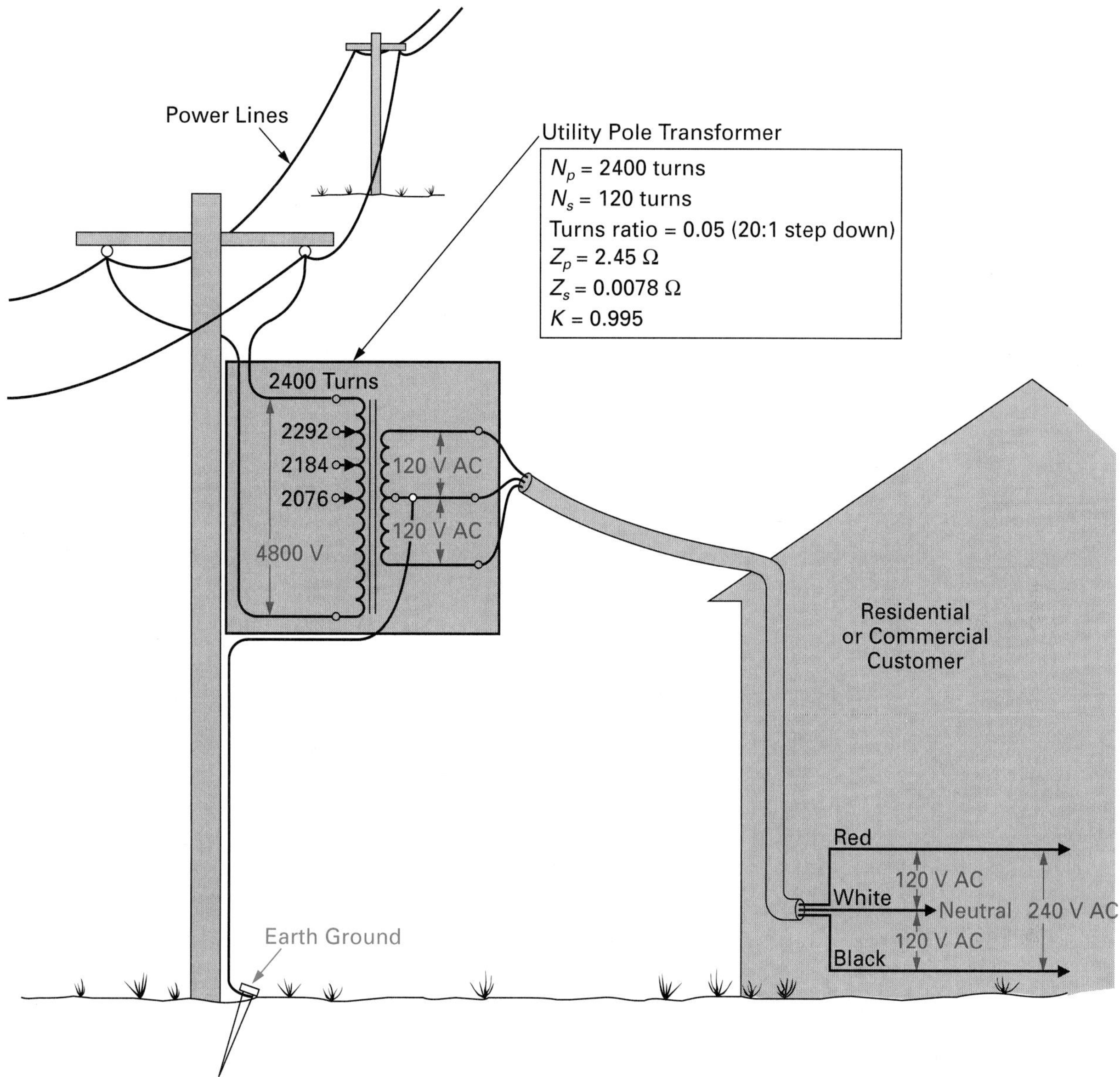

FIGURE 17-21 **Center-Tapped Utility Pole Transformer.**

Multiple Windings

The simple transformer has only one primary and one secondary. In some applications, transformers can have multiple primaries and even multiple secondaries.

Multiple primaries. The major application of this arrangement is to switch between two primary voltages and obtain the same secondary voltage. In Figure 17-23(a), the two primaries are connected in parallel, so the primary only has an

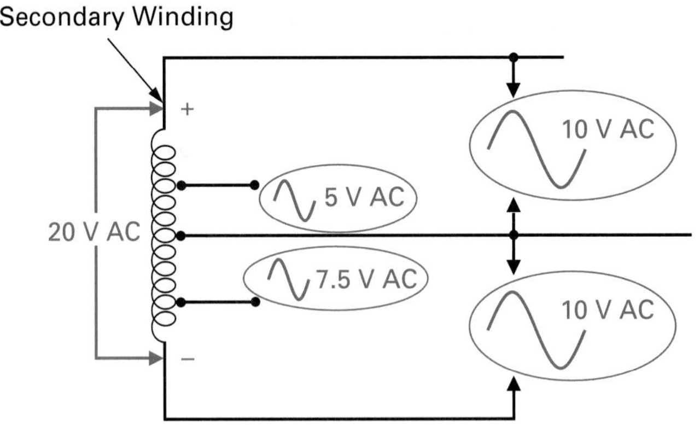

FIGURE 17-22 Multiple-Tapped Secondary Transformer.

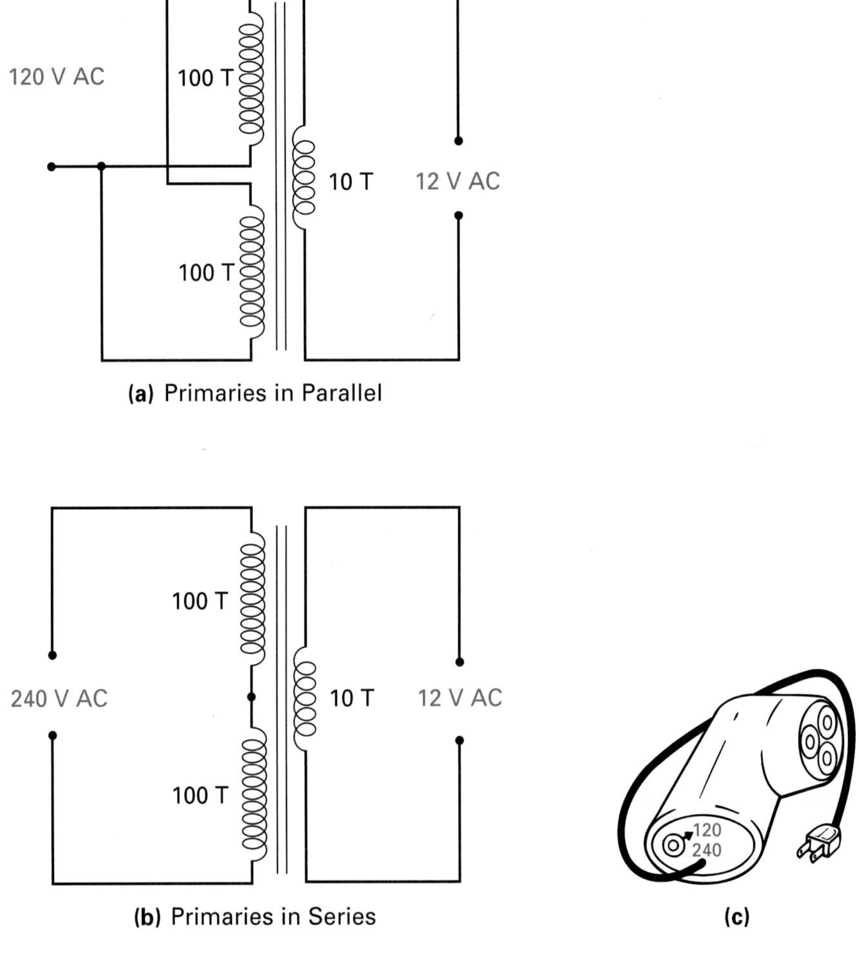

FIGURE 17-23 Multiple Primary Windings Transformer.

equivalent 100 turns; and since the secondary has 10 turns, this 10:1 step-down ratio will result in a secondary voltage of

$$V_s = \frac{N_s}{N_p} \times V_p$$

$$= \frac{10}{100} \times 120 \text{ V}$$

$$= 12 \text{ V}$$

If the two primaries are connected in series, as shown in Figure 17-23(b), the now 200-turn primary will result in a secondary voltage of

$$V_s = \frac{N_s}{N_p} \times V_p$$

$$= \frac{10}{200} \times 240 \text{ V}$$

$$= 12 \text{ V}$$

Some portable electrical or electronic equipment, such as radios or shavers, have a switch that allows you to switch between 120 V ac (U.S. wall outlet voltage) and 240 V ac (European wall outlet voltage), as shown in Figure 17-23(c). By activating the switch, you can always obtain the correct voltage to operate the equipment, whether the wall socket is supplying 240 or 120 V.

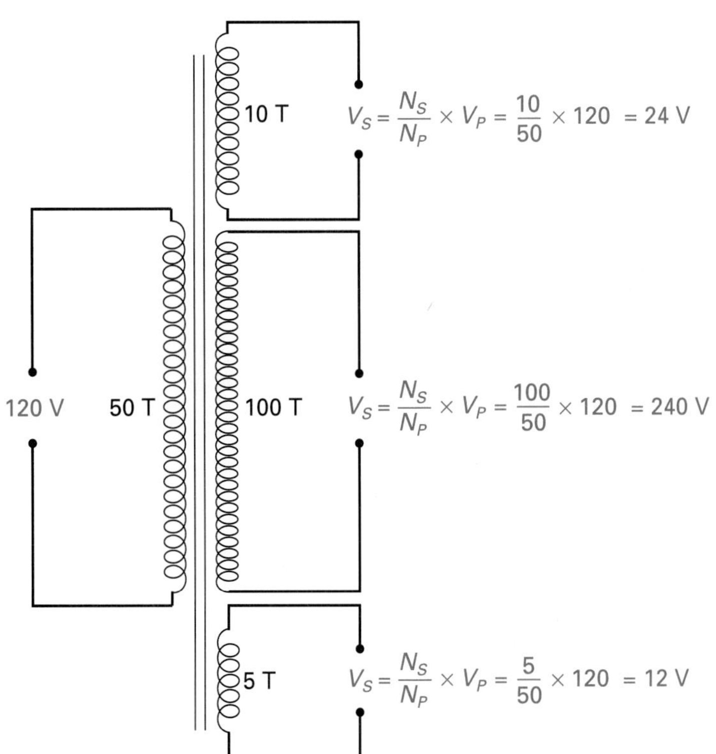

FIGURE 17-24 Multiple Secondary Windings Transformer.

Multiple secondaries. In some applications, more than one secondary is wound onto a common primary and core. The advantage of this arrangement can be seen in Figure 17-24, where many larger and smaller voltages can be acquired from one primary voltage.

Single Winding (Autotransformer)

The single winding transformer or autotransformer is used in the automobile ignition system (switched dc pulse system) to raise the low 12 V battery voltage up to about 20 kV for the spark plugs, as shown in Figure 17-25(a). Autotransformers are constructed by winding one continuous coil onto a core, which acts as both the primary and secondary. This yields three advantages for the single winding transformer:

1. They are smaller.
2. They are cheaper.
3. They are lighter than the normal separated primary/secondary transformer types.

Its disadvantage, however, is that no electrical isolation exists between primary and secondary. It is normally used as a step-up transformer for the automobile, or in televisions to obtain 20 kV. This transformer type can also be arranged to step down voltage, as shown in Figure 17-25(b).

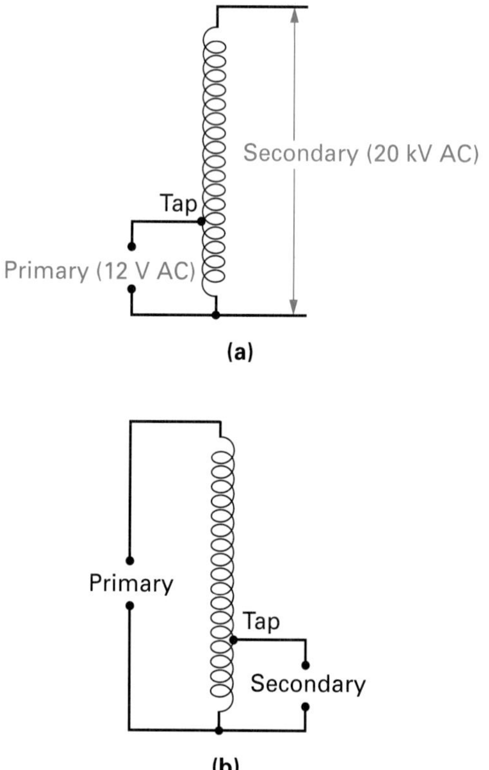

FIGURE 17-25 Single-Winding Transformer. (a) Step-Up. (b) Step-Down.

SELF-TEST REVIEW QUESTIONS FOR SECTION 17-7

1. List the two types of fixed transformers.
2. List the four basic classifications for variable transformers.
3. True or false: The two outputs of a center-tapped secondary transformer will always be 90° out of phase with one another.
4. In what application would a multiple primary transformer be used?
5. In what application would a single-winding transformer be used?
6. What are the advantages of a single-winding transformer?

17-8 TRANSFORMER RATINGS

A typical transformer rating could read 1 kVA, 500/100, 60 Hz. The 500 normally specifies the maximum primary voltage, the 100 normally specifies the maximum secondary voltage, and the 1 kVA is the apparent power rating. In this example, the maximum load current will equal

$$I_s = \frac{\text{apparent power } (P_A)}{\text{secondary voltage } (V_s)} \qquad \left(P = V \times I; \text{ therefore, } I = \frac{P}{V}\right)$$

$$= \frac{1 \text{ kVA}}{100 \text{ V}} = 10 \text{ A}$$

With this secondary voltage at 100 V and a maximum current of 10 A, the smallest load resistor that can be connected across the output of the secondary is

$$R_L = \frac{V_s}{I_s}$$

$$= \frac{100 \text{ V}}{10 \text{ A}}$$

$$= 10 \, \Omega$$

Exceeding the rating of the transformer will cause overheating and even burning out of the windings.

EXAMPLE:

Calculate the smallest value of load resistance that can be connected across a 3 kVA, 600/200, 60 Hz step-down transformer.

■ ***Solution:***

$$I = \frac{\text{apparent power } (P_A)}{\text{secondary voltage } (V_s)}$$

$$\frac{3 \text{ kVA}}{200 \text{ V}} = 15 \text{ A}$$

$$R_L = \frac{V_s}{I_s} = \frac{200 \text{ V}}{15 \text{ A}} = 13.3 \ \Omega$$

SELF-TEST REVIEW QUESTIONS FOR SECTION 17-8

1. What do each of the values mean when a transformer is rated as a 10 kVA, 200/100, 60 Hz?
2. If a 100 Ω resistor is connected across the secondary of a transformer that is to supply 1 kV and is rated at a maximum current of 8 A, will the transformer overheat and possibly burn out?

17-9 TESTING TRANSFORMERS

Like inductors, transformers, which are basically two or more inductors, can develop one of three problems:

1. An open winding
2. A complete short in a winding
3. A short in a section of a winding

17-9-1 *Open Primary or Secondary Winding*

An open in the primary winding will prevent any primary current, and therefore there will be no induced voltage in the secondary and therefore no voltage will be present across the load. An open secondary winding will prevent the flow of secondary current, and once again, no voltage will be present across the load.

An open in the primary or secondary winding is easily detected by disconnecting the transformer from the circuit and testing the resistance of the windings with an ohmmeter. Like an inductor, a transformer winding should have a low resistance, in the tens to hundreds range. An open will easily be recognized because of its infinite (maximum) resistance.

17-9-2 *Complete Short or Section Short in Primary or Secondary Winding*

A partial or complete short in the primary winding of a transformer will result in an excessive source current that will probably blow the circuit's fuse or trip the circuit's breaker. A partial or complete short in the secondary winding will cause an excessive secondary current, which in turn will result in an excessive primary current.

In both instances, a short in the primary or secondary will generally burn out the primary winding unless the fuse or circuit breaker opens the excessive current path. An ohmmeter can be used to test for partial or complete shorts in transformer windings; however, all coils have a naturally low resistance which can be mistaken for a short. An inductive or reactive analyzer can accurately test transformer windings, checking the inductance value of each coil.

It is virtually impossible to repair an open, partially shorted, or completely shorted transformer winding, and therefore defective transformers are always replaced with a transformer that has an identical rating.

SELF-TEST REVIEW QUESTIONS FOR SECTION 17-9

1. Which test instrument could you use to test for a suspected open primary winding?
2. If a secondary winding had a resistance of 50 Ω, would this value indicate a problem?

17-10 TRANSFORMER LOSSES

Internal losses cause the power delivered from the secondary winding of a transformer in reality to be less than the power fed into the primary winding. $P_s < P_p$ in reality, due to transformer losses. There are basically three types of internal transformer loss: **copper loss,** core loss, and magnetic leakage.

Copper Loss
Also called I^2R loss, it is the power lost in transformers, generators, connecting wires, and other parts of a circuit because of the current flow (I) through the resistance (R) of the conductors.

17-10-1 *Copper Losses*

Due to ohmic resistance of the windings, heat ($I^2 \times R$) is dissipated in both the primary and secondary windings.

17-10-2 *Core Losses*

Hysteresis

Hysteresis
Amount that the magnetization of a material lags the magnetizing force due to molecular friction.

During each ac cycle, the core is taken through a cycle of magnetization; hence energy is lost due to **hysteresis** and appears as heat in the core. This loss is proportional to frequency and is minimized by using a core of soft iron or an alloy such as stalloy or permalloy.

Eddy-Current Loss

Eddy Currents
Small currents induced in a conducting core due to the variations in alternating magnetic flux.

The continuously changing flux induces voltages in the conducting core, which results in local **eddy currents** in the core that combine to produce a large circulating current, as shown in Figure 17-26(a), that opposes the main flux and also generates heat. This loss, which is proportional to frequency, is minimized by using a laminated core, as shown in Figure 17-26(b).

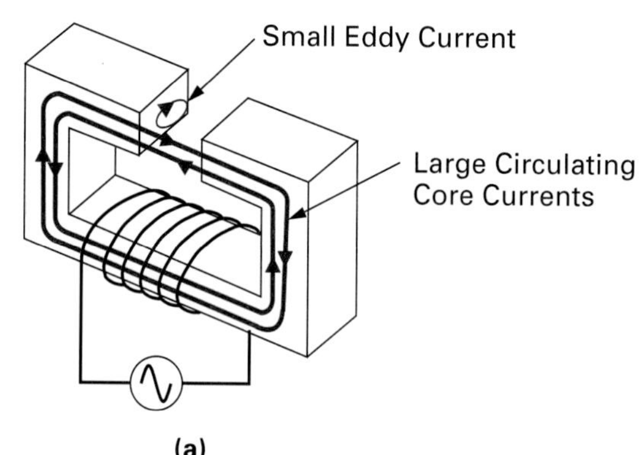

(a)

Each metal lamination has a higher resistance and so eddy currents are small and power loss is lower

Insulating Varnish between Metal Lamination

(b)

FIGURE 17-26 Eddy Currents. (a) Solid Core, Large Eddy Currents. (b) Laminated Core, Small Eddy Currents.

17-10-3 *Magnetic Leakage*

Not all the flux lines produced will cut the secondary winding. The result is that for a given turns ratio, the secondary terminal voltage will be slightly lower than expected due to **magnetic leakage.**

Magnetic Leakage
The passage of magnetic flux outside the path along which it can do useful work.

SELF-TEST REVIEW QUESTIONS FOR SECTION 17-10

1. List the four types of transformer losses.
2. Why will a laminated transformer core reduce eddy current loss?

SUMMARY

Transformers (Figure 17-27)

1. The transformer is an electrical device that makes use of electromagnetic induction to transfer alternating current from one circuit to another.
2. The transformer consists of two inductors that are placed in very close proximity to one another. When an alternating current flows through the first coil or primary winding the inductor sets up a magnetic field. The expanding and contracting magnetic field produced by the primary cuts across the windings of the second inductor or secondary winding and induces a voltage in this coil.
3. By changing the ratio between the number of turns in the secondary winding to the number of turns in the primary winding, some characteristics of the ac signal can be changed or transformed as it passes from primary to secondary.

Mutual Inductance

4. The principle on which a transformer is based is an inductive effect known as mutual inductance, which is the process by which an inductor induces a voltage in another inductor.
5. As with self-inductance, mutual inductance is dependent on change. Direct current (dc) is a constant current and produces a constant or stationary magnetic field that does not change. Alternating current, however, is continually varying, and as the polarity of the magnetic field is dependent on the direction of current flow (left-hand rule), the magnetic field will also be alternating in polarity.
6. Self-induction is a measure of how much voltage an inductor can induce within itself. Mutual inductance is a measure of how much voltage is induced in the secondary coil due to the change in current in the primary coil.
7. The basic transformer consists of two coils within close proximity to one another, to ensure that the second coil will be cut by the magnetic flux lines

QUICK REFERENCE SUMMARY SHEET

FIGURE 17-27 Transformers.

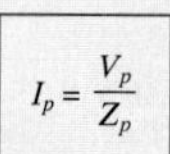

$$I_p = \frac{V_p}{Z_p}$$

$$\text{Coupling coefficient } (k) = \frac{\text{flux linking secondary coil}}{\text{total flux produced by primary}}$$

$$\text{Turns ratio} = \frac{N_s}{N_p}$$

N_s = number of turns in the secondary
N_p = number of turns in the primary

$$V_s = \frac{N_s}{N_p} \times V_p$$

V_s = secondary voltage
V_p = primary voltage
Z_p = impedance of primary coil

$$I_s = \frac{N_p}{N_s} \times I_p$$

I_s = secondary current
I_p = primary current

$$\text{Turns ratio} = \sqrt{\frac{Z_L}{Z_s}}$$

Z_L = load impedance
Z_s = source impedance

$$I_s = \frac{P_A}{V_s}$$

P_A = apparent power, in volt-amperes (VA)
R_L = load resistance, in ohms (Ω)

$$R_L = \frac{V_s}{I_s}$$

Transformer Action

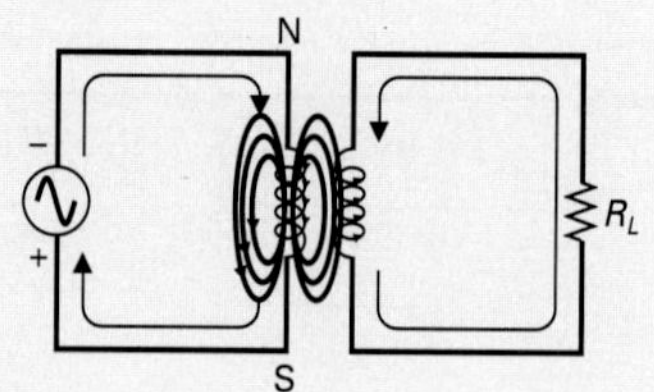

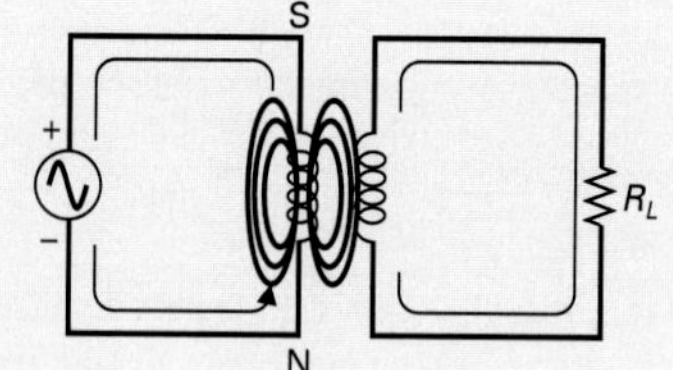

Step-Up Transformer

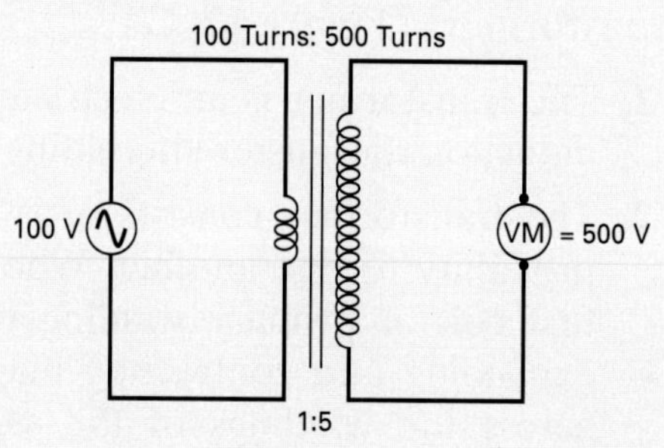

Maximum Power Transfer

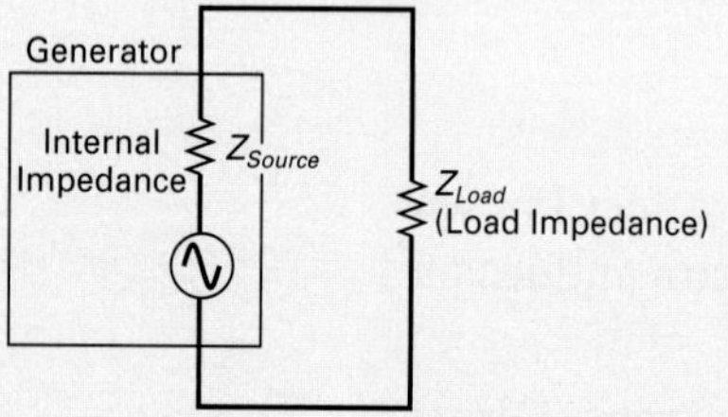

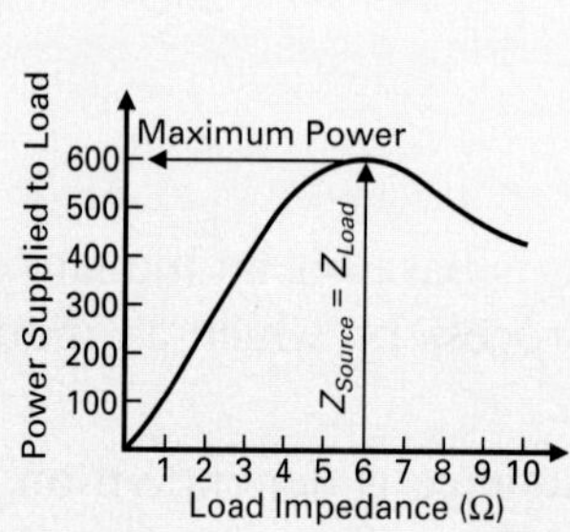

Step-Down Transformer

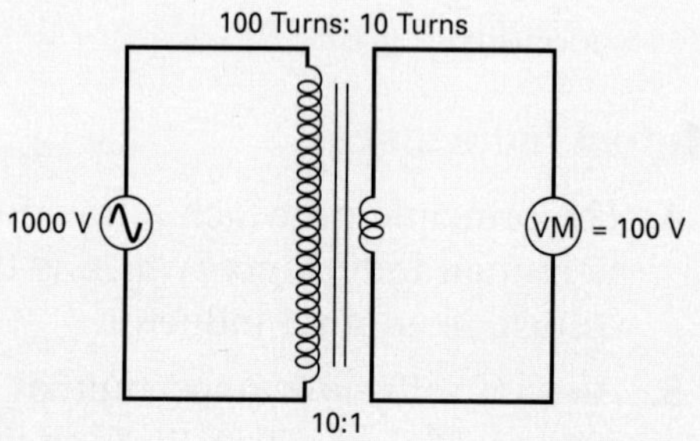

Center-Tapped Secondary

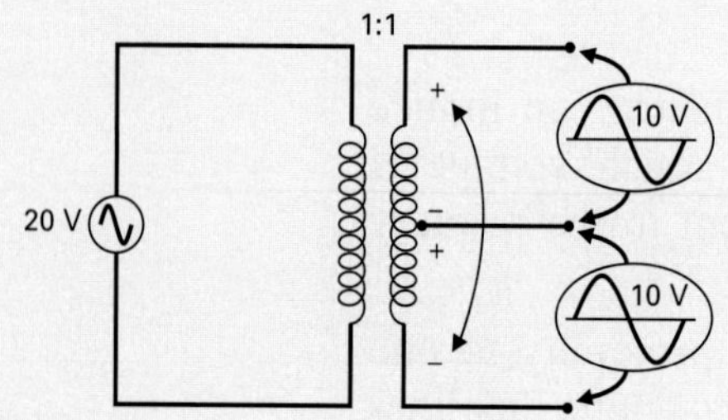

Multiple-Tapped Secondary

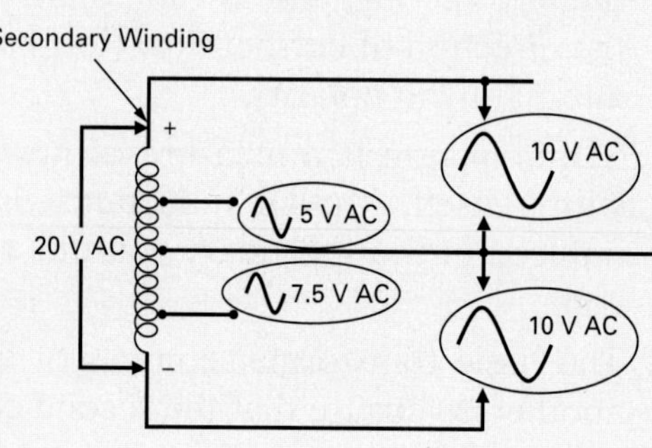

FIGURE 17-27 (*continued*)

Multiple Primary

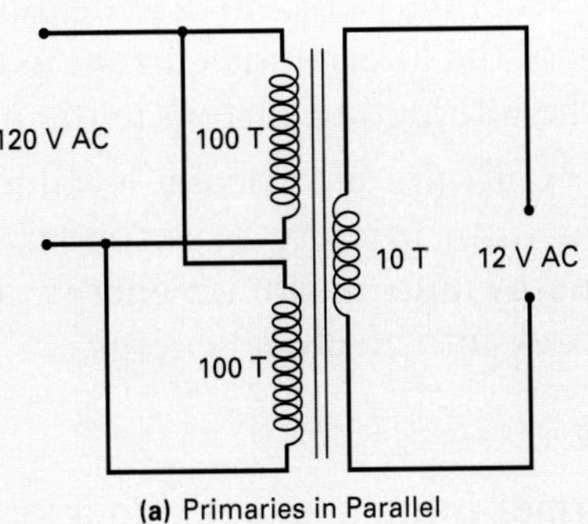

(a) Primaries in Parallel

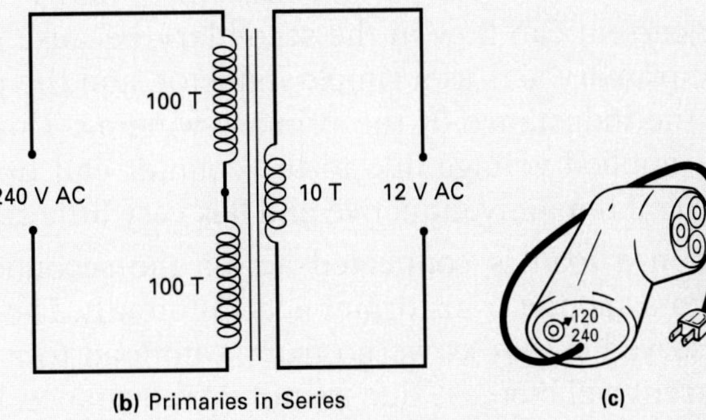

(b) Primaries in Series

(c)

Multiple Secondary

120 V 50 T

10 T $V_S = \frac{N_S}{N_P} \times V_P = \frac{10}{50} \times 120 = 24 \text{ V}$

100 T $V_S = \frac{N_S}{N_P} \times V_P = \frac{100}{50} \times 120 = 240 \text{ V}$

5 T $V_S = \frac{N_S}{N_P} \times V_P = \frac{5}{50} \times 120 = 12 \text{ V}$

Single Winding

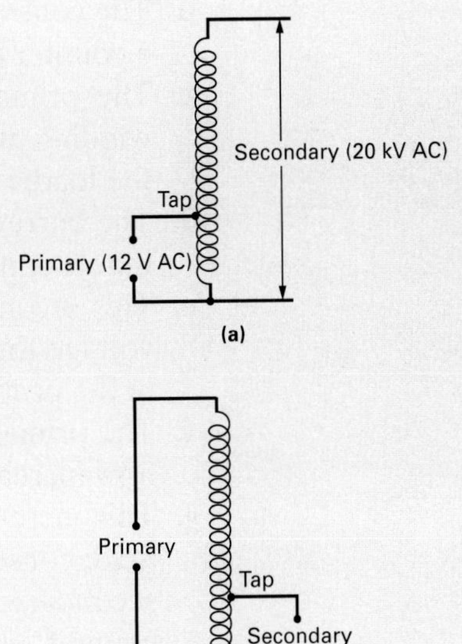

produced by the first coil, and thereby ensure mutual inductance. The ac voltage source is electrically connected (through wires) to the primary coil or winding, and the load (R_L) is electrically connected to the secondary coil or winding.

8. As primary current increases, secondary current increases, and as primary current decreases, secondary current also decreases. It can therefore be said that the frequency of the alternating current in the secondary is the same as the frequency of the alternating current in the primary.
9. Although the two coils are electrically isolated from one another, energy can be transferred from primary to secondary, because the primary converts electrical energy into magnetic energy, and the secondary converts magnetic energy back into electrical energy.

Transformer Loading

10. When the transformer is not connected to a load, primary circuit current is determined by $I = V/Z$, where Z is the impedance of the primary coil (both its inductive reactance and resistance) and V is the applied voltage. Since no current can flow in the secondary, because an open in the circuit exists, the primary acts as a simple inductor, and the primary current is small due to the inductance of the primary winding. This small primary current lags the applied voltage due to the counter emf by approximately 90° because the coil is mainly inductive and has very little resistance.
11. When a load is connected across the secondary, a change in conditions occurs and the transformer acts differently. The important point that will be observed is that as we go from a no-load to a load condition the primary current will increase due to mutual inductance. Let's follow the steps one by one.
    - **a.** The ac applied voltage sets up an alternating magnetic field in the primary winding.
    - **b.** The continually changing flux of this primary field induces and produces a counter emf into the primary to oppose the applied voltage.
    - **c.** The primary's magnetic field also induces a voltage in the secondary winding, which causes current to flow in the secondary circuit through the load.
    - **d.** The current in the secondary winding produces another magnetic field that is opposite to the field being produced by the primary.
    - **e.** This secondary magnetic field feeds back to the primary and induces a voltage that tends to cancel or weaken the counter emf that was set up in the primary by the primary current.
    - **f.** The primary's counter emf is therefore reduced, so primary current can now increase.
    - **g.** This increase in primary current is caused by the secondary's magnetic field; consequently, the greater the secondary current, the stronger the secondary magnetic field, which causes a reduction in the primary's counter emf, and therefore a primary current increase.
12. In summary, an increase in secondary current ($I_s\uparrow$) causes an increase in primary current ($I_p\uparrow$), and this effect in which the primary induces a voltage in

the secondary (V_s) and the secondary induces a voltage into the primary (V_p) is known as *mutual inductance.*

Coefficient of Coupling (k)

13. The voltage induced into the secondary winding is dependent on the mutual inductance between the primary and secondary, which is determined by how much of the magnetic flux produced by the primary actually cuts the secondary winding.
14. The coefficient of coupling (k) is a ratio of the number of magnetic lines of force that cut the secondary compared to the total number of magnetic flux lines being produced by the primary and is a figure between 0 and 1.
15. The coefficient of coupling depends on:
    - **a.** How close together the primary and secondary are to one another
    - **b.** The type of core material used

Transformer Ratios and Applications

16. Basically, transformers are used for one of three applications:
    - **a.** To step up (increase) or step down (decrease) voltage
    - **b.** To step up (increase) or step down (decrease) current
    - **c.** To match impedances

    In all three cases, any of the applications can be achieved by changing the ratio of the number of turns in the primary winding compared to the number of turns in the secondary winding. This ratio is appropriately called the turns ratio.
17. The turns ratio is the ratio between the number of turns in the secondary winding (N_s) and the number of turns in the primary winding (N_p).
18. Transformers are used within the power supply unit of almost every piece of electronic equipment to step up or step down the 115 V ac from the outlet.
19. Some electronic circuits require lower-power supply voltages, while other devices may require higher-power supply voltages. The transformer is used in both instances to convert the 115 V ac to the required value of voltage.
20. If the secondary voltage (V_s) is greater than the primary voltage (V_p), the transformer is called a step-up transformer ($V_s > V_p$).
21. If the secondary voltage (V_s) is smaller than the primary voltage (V_p), the transformer is called a step-down transformer ($V_s < V_p$).
22. The coupling coefficient (k) in this formula is always assumed to be 1, which for most iron-core transformers is almost always the case. This means that all the primary magnetic flux is linking the secondary, and the secondary voltage is dependent on the number of secondary turns that are being cut by the primary magnetic flux.
23. The power in the secondary of the transformer is equal to the power in the primary ($P_p = P_s$). Power, as we know, is equal to $P = V \times I$, and if voltage is stepped up or down, the current automatically is stepped down or up,

respectively, in the opposite direction to voltage to maintain the power constant.

24. The maximum power transfer theorem states that maximum power is transferred from source (ac generator) to load (equipment) when the impedance of the load is equal to the internal impedance of the source. If these impedances are different, a large amount of power could be wasted.
25. In most cases it is required to transfer maximum power from a source that has an internal impedance (Z_s) that is not equal to the load impedance (Z_L). In this situation, a transformer can be inserted between the source and the load to make the load impedance appear to equal the source's internal impedance.
26. The way in which the primary and secondary coils are wound around the core determines the polarity of the voltage induced into the secondary relative to the polarity of the primary.
27. If the primary and secondary windings are both wound in a clockwise direction around the core, the voltage induced in the primary will be in phase with the voltage induced in the secondary. Both the input and output will also be in phase with one another if the primary and secondary are both wound in a counterclockwise direction.
28. If both the primary and secondary are wound in opposite directions, the primary being wound in a clockwise direction and the secondary in a counterclockwise direction, the output ac sine-wave voltage is 180° out of phase with respect to the input ac voltage.
29. The dot convention is a standard used with transformer symbols to indicate whether the secondary voltage will be in phase or out of phase with the primary voltage.

Transformer Types

30. Fixed turns ratio transformers have a turns ratio that cannot be varied. They are generally wound on a common core to ensure a high k and can be classified by the type of core material used. The two types of fixed transformers are:

    a. Air core **b.** Iron or ferrite core

31. The air-core transformers typically have a nonmagnetic core, such as ceramic or a cardboard hollow shell, and are used in high-frequency applications.
32. The more common iron- or ferrite-core transformers concentrate the magnetic lines of force, resulting in improved transformer performance; they are symbolized by two lines running between the primary and secondary. The iron-core transformer's lines are solid, while the ferrite-core transformer's lines are dashed.
33. Variable turns ratio transformers have a turns ratio that can be varied. They can be classified as follows:

    a. Center-tapped secondary **c.** Multiple winding
    b. Multiple-tapped secondary **d.** Single winding

34. If the tapped lead is in the exact center of the secondary, the transformer is said to have a center-tapped secondary. With the center-tapped transformer, the two secondary voltages are each half of the total secondary voltage.
35. If several leads are attached to the secondary, the transformer is said to have a multiple-tapped secondary. This multitapped transformer will tap or pick off different values of ac voltage.
36. The simple transformer has only one primary and one secondary. In some applications, transformers can have multiple primaries and even multiple secondaries.
37. The single winding transformer or autotransformer is used in the automobile ignition system (switched dc pulse system) to raise the low 12 V battery voltage up to about 20 kV for the spark plugs.

Transformer Ratings

38. A typical transformer rating could read 1 kVA, 500/100, 60 Hz. The 500 normally specifies the maximum primary voltage, the 100 normally specifies the maximum secondary voltage, and the 1 kVA is the apparent power rating.

Testing Transformers

39. Like inductors, transformers, which are basically two or more inductors, can develop one of three problems.
    a. An open winding
    b. A complete short in a winding
    c. A short in a section of a winding
40. An open in the primary winding will prevent any primary current, and therefore there will be no induced voltage in the secondary and therefore no voltage will be present across the load.
41. An open secondary winding will prevent the flow of secondary current, and once again, no voltage will be present across the load.
42. An open in the primary or secondary winding is easily detected by disconnecting the transformer from the circuit and testing the resistance of the windings with an ohmmeter. Like an inductor, a transformer winding should have a low resistance, in the tens to hundreds range. An open will easily be recognized because of its infinite (maximum) resistance.
43. A partial or complete short in the primary winding of a transformer will result in an excessive source current that will probably blow the circuit's fuse or trip the circuit's breaker.
44. A partial or complete short in the secondary winding will cause an excessive secondary current, which in turn will result in an excessive primary current.
45. In both instances, a short in the primary or secondary will generally burn out the primary winding unless the fuse or circuit breaker opens the excessive current path.

46. An ohmmeter can be used to test for partial or complete shorts in transformer windings; however, all coils have a naturally low resistance which can be mistaken for a short.
47. An inductive or reactive analyzer can accurately test transformer windings, checking the inductance value of each coil.
48. It is virtually impossible to repair an open, partially shorted, or completely shorted transformer winding, and therefore defective transformers are always replaced with a transformer that has an identical rating.

Transformer Losses

49. Internal losses cause the power delivered from the secondary winding of a transformer in reality to be less than the power fed into the primary winding.
50. $P_s < P_p$ in reality, due to transformer losses. There are basically three types of internal transformer loss: copper loss, core loss, and magnetic leakage.
51. Due to ohmic resistance of the windings, heat ($I^2 \times R$) is dissipated in both the primary and secondary windings.
52. During each ac cycle, the core is taken through a cycle of magnetization; hence energy is lost due to hysteresis and appears as heat in the core.
53. The continuously changing flux induces voltages in the conducting core, which results in local eddy currents in the core that combine to produce a large circulating current that opposes the main flux and also generates heat.
54. Not all the flux lines produced will cut the secondary winding. The result is that for a given turns ratio, the secondary terminal voltage will be slightly lower than expected due to magnetic leakage.

NEW TERMS

Autotransformer
Center-tapped secondary
Coefficient of coupling
Copper losses
Core losses
Current ratio
Dot convention
Eddy-current losses
Fixed transformer
Impedance ratio
Magnetic leakage
Multiple-tapped secondary
Multiple winding
Mutual inductance
Neutral wire
Power ratio
Primary winding
Ratings
Secondary winding
Single-winding transformer
Step-down transformer
Step-up transformer
Transformer
Transformer action
Turns ratio
Variable transformer
Voltage ratio

REVIEW QUESTIONS

Multiple-Choice Questions

1. Transformer action is based on:
   - **a.** Self-inductance
   - **b.** Air between the coils
   - **c.** Mutual capacitance
   - **d.** Mutual inductance
2. An increase in transformer secondary current will cause a(an) ________ in primary current.
   - **a.** Decrease **b.** Increase
3. If 50% of the magnetic lines of force produced by the primary were to cut the secondary coil, the coefficient of coupling would be:
   - **a.** 75 **b.** 50 **c.** 0.5 **d.** 0.005
4. A step-up transformer will always have a turns ratio ________, while a step-down transformer has a turns ratio ________.
   - **a.** $< 1, > 1$ **b.** $> 1, > 1$ **c.** $> 1, < 1$ **d.** $< 1, < 1$
5. With an 80-V ac secondary voltage center-tapped transformer, what would be the voltage at each output, and what would be the phase relationship between the two secondary voltages?
   - **a.** 20 V, in phase with one another
   - **b.** 30 V, 180° out of phase
   - **c.** 40 V, in phase
   - **d.** 40 V, 180° out of phase
6. One application of the autotransformer would be:
   - **a.** To obtain the final anode high-voltage supply for the cathode-ray tube in a television
   - **b.** To obtain two outputs, 180° out of phase with one another
   - **c.** To tap several different voltages from the secondary
   - **d.** To obtain the same secondary voltage for different voltages
7. Alternating current can be used only with transformers because:
   - **a.** It produces an alternating magnetic field.
   - **b.** It produces a fixed magnetic field.
   - **c.** Its magnetic field is greater than that of dc.
   - **d.** Its rms is 0.707 of the peak.
8. Eddy-current losses are reduced with laminated iron cores because:
   - **a.** The air gap is always kept to a minimum.
   - **b.** The resistance of iron is always low.
   - **c.** The laminations are all insulated from one another.
   - **d.** Current cannot flow in iron.
9. Assuming 100% efficiency, the output power, P_s, is always equal to:
   - **a.** P_p
   - **b.** $V_s \times I_s$
   - **c.** $0.5 \times P_p$
   - **d.** Both (a) and (c)
   - **e.** Both (a) and (b)
10. If the primary winding of a transformer were open, the result would be:
    - **a.** No flux linkage between primary and secondary
    - **b.** No primary or secondary current
    - **c.** No induced voltage in the secondary
    - **d.** All of the above

Essay Questions

11. What is a transformer? (Introduction)
12. Describe mutual inductance and how it relates to transformers. (17-1)
13. Why does loading the transformer's secondary circuit affect primary transformer current? (17-3)
14. Describe what a coefficient of coupling figure of 0.9 means. (17-4)
15. List the three basic applications of transformers, and then describe how each of these applications can be achieved by merely changing the transformer's turns ratio. (17-5)
16. Briefly describe how the dot convention is used in schematics to describe phase. (17-6)
17. Illustrate the schematic symbol and main points relating to the following transformers:
    a. Fixed air core (17-7-1)
    b. Fixed iron core (17-7-1)
    c. Fixed ferrite core (17-7-1)
    d. Center-tapped secondary (17-7-2)
    e. Four-output tapped secondary (17-7-2)
    f. Multiple primary and multiple secondary (17-7-2)
    g. Autotransformer (17-7-2)
18. Illustrate and explain the following transformer applications: (17-7)
    a. 120/240-V primary voltage to constant secondary voltage
    b. 20-kV secondary voltage using the autotransformer
19. Briefly describe what is meant by copper losses with transformers. (17-10)
20. Briefly describe the following iron and core losses and how they can be reduced: (17-10)
    a. Hysteresis **b.** Eddy current
21. Briefly describe the transformer loss known as magnetic leakage. (17-10)
22. What meter could be used to check a suspected open primary coil? (17-9)
23. Could a transformer be considered a dc block? (17-1)
24. Would a step-up voltage transformer step up or step down current? What would happen to secondary power? (17-5)
25. Briefly describe how transformers are used to reduce I^2R losses in ac power distribution. (Chapter 11)

Practice Problems

26. Calculate the turns ratio of the following transformers and state whether they are step up or step down:
    a. P = 12T, S = 24T **c.** P = 24T, S = 5T
    b. P = 3T, S = 250T **d.** P = 240T, S = 120T
27. Calculate the secondary ac voltage for all the examples in Question 26 if the primary voltage equals 100 V.

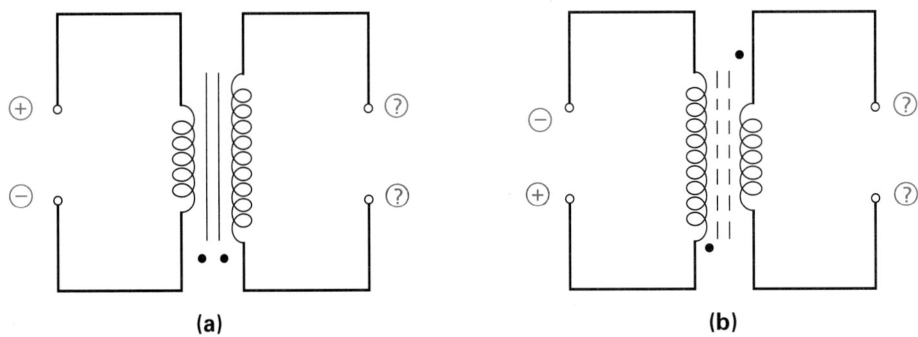

FIGURE 17-28 **Dot Convention Examples.**

28. Calculate the secondary ac current for all the examples in Question 26 if the primary current equals 100 mA.

29. What turns ratio would be needed to match a source impedance of 24 Ω to a load impedance of 8 Ω?

30. What turns ratio would be needed to step:

a. 120 V to 240 V **b.** 240 V to 720 V **c.** 30 V to 14 V **d.** 24 V to 6 V

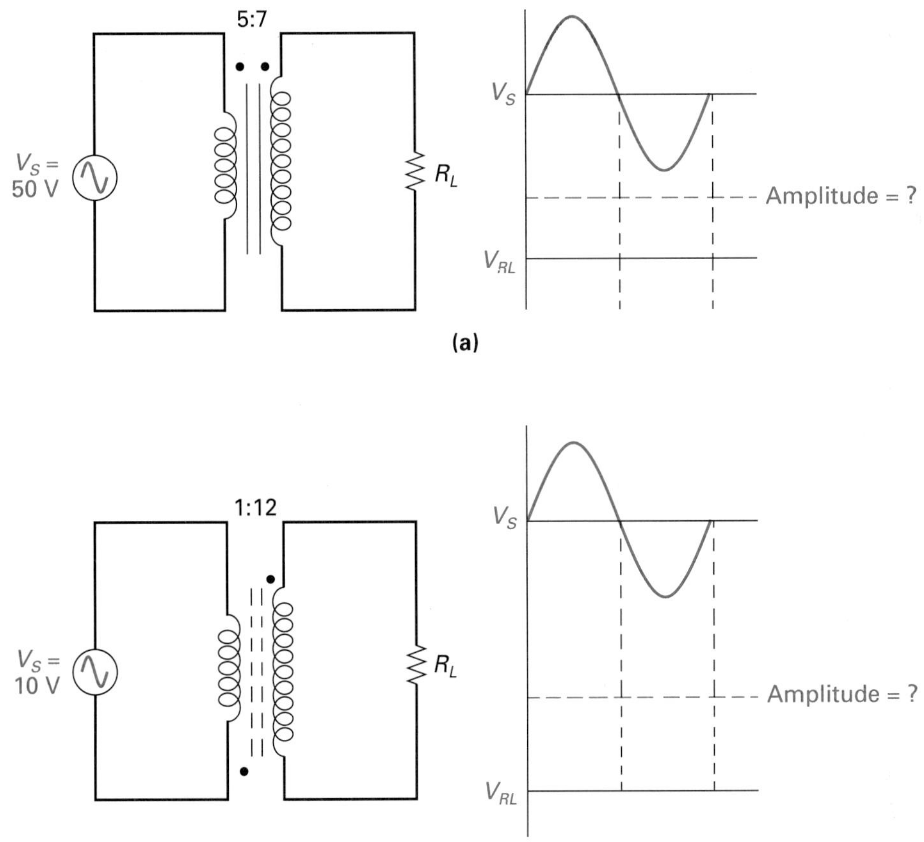

FIGURE 17-29 **Input/Output Polarity Examples.**

31. For a 24 V, 12-turn primary, and 16-, 2-, 1-, and 4-turn multiple secondary transformer, calculate each of the secondary voltages.

32. If a 2 : 1 step-down transformer has a primary input voltage of 120 V, 60 Hz, and a 2 kVA rating, calculate the maximum secondary current and smallest load resistor that can be connected across the output.

33. Indicate the polarity of the secondary voltages in Figure 17-28(a) and (b).

34. Referring to Figure 17-29(a) and (b), sketch the outputs, showing polarity and amplitude with respect to the inputs.

35. If a transformer is rated at 500 VA, 60 Hz, the primary voltage is 240 V ac, and the secondary voltage is 600 V ac, calculate:

a. Maximum load current **b.** Smallest value of R_L

CHAPTER

18

OBJECTIVES

After completing this chapter, you will be able to:

1. Identify the difference between a series and parallel *RLC* circuit.
2. Explain the following as they relate to series *RLC* circuits:
   - **a.** Impedance
   - **b.** Current
   - **c.** Voltage
   - **d.** Phase angle
   - **e.** Power
3. Explain the following as they relate to parallel *RLC* circuits:
   - **a.** Voltage
   - **b.** Current
   - **c.** Impedance
   - **d.** Phase angle
   - **e.** Power
4. Define resonance, and explain the characteristics of:
   - **a.** Series resonance
   - **b.** Parallel resonance
5. Identify and explain the following *RLC* circuit applications:
   - **a.** Low-pass filter
   - **b.** High-pass filter
   - **c.** Bandpass filter
   - **d.** Band-stop filter
6. Describe complex numbers in both rectangular and polar form.
7. Perform complex number arithmetic.
8. Describe how complex numbers apply to ac circuits containing series–parallel *RLC* components.

Resistive, Inductive, and Capacitive (*RLC*) Circuits

OUTLINE

VIGNETTE

The Fairchildren

On December 23, 1947, John Bardeen, Walter Brattain, and William Shockley first demonstrated how a semiconductor device, named the *transistor,* could be made to amplify. However, the device had mysterious problems and was very unpredictable. Shockley continued his investigations, and in 1951 he presented the world with the first reliable junction transistor. In 1956, the three shared the Nobel Prize in physics for their discovery, and much later, in 1972, Bardeen would win a rare second Nobel Prize for his research at the University of Illinois in the field of superconductivity.

Shockley left Bell Labs in 1955 to start his own semiconductor company near his home in Palo Alto and began recruiting personnel. He was, however, very selective, only hiring those who were bright, young, and talented. The company was a success, although many of the employees could not tolerate Shockley's eccentricities, such as posting everyone's salary and requiring that the employees rate one another. Two years later, eight of Shockley's most talented defected. The "traitorous eight," as Shockley called them, started their own company only a dozen blocks away, named Fairchild Semiconductor.

More than 50 companies would be founded by former Fairchild employees. One of the largest was started by Robert Noyce and two other colleagues from the group of eight Shockley defectors; they named their company Intel, which was short for "intelligence."

INTRODUCTION

In this chapter we combine resistors (R), inductors (L), and capacitors (C) into series and parallel ac circuits. Resistors, as we have discovered, operate and react to voltage and current in a very straightforward way; the voltage across a resistor is in phase with the resistor current.

Inductors and capacitors operate in essentially the same way, in that they both store energy and then return it back to the circuit. However, they have completely opposite reactions to voltage and current. To help you remember the phase relationships between voltage and current for capacitors and inductors, you may wish to use the following memory phrase:

ELI the ICE man

This phrase states that voltage (symbolized E) leads current (I) in an inductive (L) circuit (abbreviated by the word "ELI"), while current (I) leads voltage (E) in a capacitive (C) circuit (abbreviated by the word "ICE").

In this chapter we study the relationships between voltage, current, impedance, and power in both series and parallel *RLC* circuits. We also examine the important *RLC* circuit characteristic called resonance, and see how *RLC* circuits can be made to operate as filters. In the final section we discuss how complex numbers can be used to analyze series and parallel ac circuits containing resistors, inductors, and capacitors.

18-1 SERIES *RLC* CIRCUIT

Figure 18-1 begins our analysis of series *RLC* circuits by illustrating the current and voltage relationships. The circuit current is always the same throughout a series circuit and can therefore be used as a reference. Studying the waveforms and vector diagrams shown alongside the components, you can see that the voltage across a resistor is always in phase with the current, while the voltage across the inductor leads the current by 90° and the voltage across the capacitor lags the current by 90°.

Now let's analyze the impedance, current, voltage, and power distribution of this circuit in a little more detail.

18-1-1 Impedance

Impedance is the total opposition to current flow and is a combination of both reactance (X_L, X_C) and resistance (R). An example circuit is illustrated in Figure 18-2(a).

Capacitive reactance can be calculated by using the formula

$$X_C = \frac{1}{2\pi fC}$$

In the example,

$$X_C = \frac{1}{2\pi \times 60 \times 10\ \mu\text{F}} = 265.3\ \Omega$$

Inductive reactance is calculated by using the formula

$$X_L = 2\pi fL$$

In the example,

$$X_L = 2\pi \times 60 \times 20\ \text{mH} = 7.5\ \Omega$$

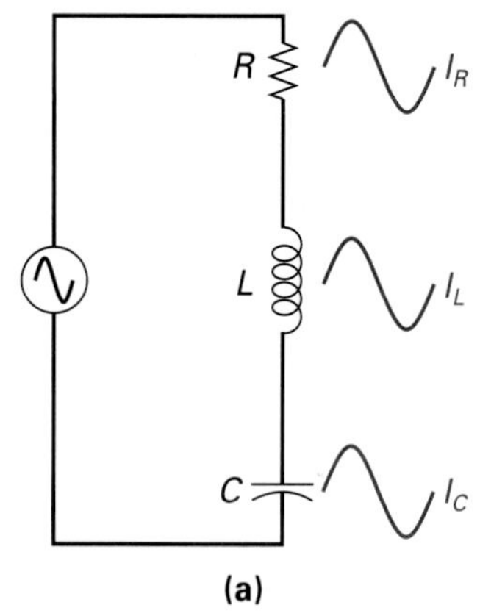

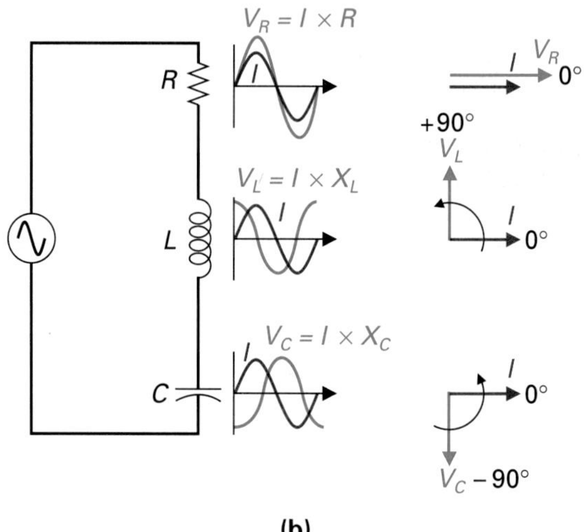

FIGURE 18-1 **Series *RLC* Circuit. (a) *RLC* Series Circuit Current: Current Flow Is Always the Same in All Parts of a Series Circuit. (b) *RLC* Series Circuit Voltages: *I* Is in Phase with V_R, *I* Lags V_L by 90°, and *I* Leads V_C by 90°.**

Resistance in the example is equal to $R = 33\ \Omega$. Figure 18-2(b) illustrates these values of resistance and reactance in a vector diagram. In this vector diagram you can see that X_L is drawn 90° ahead of *R*, and X_C is drawn 90° behind *R;* the capacitive and inductive reactances are 180° out of phase with one another and counteract to produce a vector diagram, as shown in Figure 18-2(c). The difference between X_L and X_C is equal to 257.8, and since X_C is greater than X_L, the resultant reactive vector is capacitive. Reactance, however, is not in phase with resistance, and impedance is the vector sum of the reactive (*X*) and resistance (*R*) vectors. The formula, based on the Pythagorean theorem, illustrated in Figure 18-2(d), is

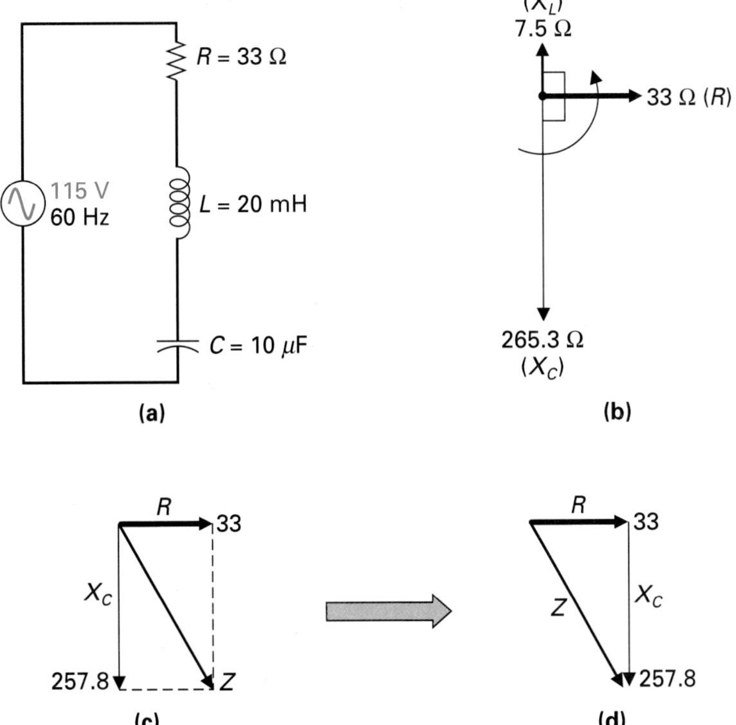

FIGURE 18-2 Series Circuit Impedance.

$$Z = \sqrt{R^2 + X^2}$$

In this example, therefore, the circuit impedance will be equal to

$$Z = \sqrt{R^2 + X^2} = \sqrt{33^2 + 257.8^2} = 260\ \Omega$$

Since reactance (X) is equal to the difference between (symbolized $\sim$) X_L and X_C ($X_L \sim X_C$), the impedance formula can be modified slightly to incorporate the calculation to determine the difference between X_L and X_C.

$$Z = \sqrt{R^2 + (X_L \sim X_C)^2}$$

Using our example with this new formula, we arrive at the same value of impedance, and since the difference between X_L and X_C resulted in a capacitive vector, the circuit is said to act capacitively.

$$\begin{aligned} Z &= \sqrt{R^2 + (X_L \sim X_C)^2} \\ &= \sqrt{33^2 + (7.5 \sim 265.3)^2} \\ &= \sqrt{33^2 + 257.8^2} \\ &= 260\ \Omega \end{aligned}$$

If, on the other hand, the component values were such that the difference was an inductive vector, then the circuit would be said to act inductively.

18-1-2 *Current*

The current in a series circuit is the same at all points throughout the circuit, and therefore

$$I = I_R = I_L = I_C$$

Once the total impedance of the circuit is known, Ohm's law can be applied to calculate the circuit current:

$$I = \frac{V_S}{Z}$$

In the example, circuit current is equal to

$$\begin{aligned} I &= \frac{V_S}{Z} \\ &= \frac{115 \text{ V}}{260 \ \Omega} \\ &= 0.44 \text{ A} \quad \text{or} \quad 440 \text{ mA} \end{aligned}$$

18-1-3 *Voltage*

Now that you know the value of current flowing in the series circuit, you can calculate the voltage drops across each component, as shown in Figure 18-3(a).

$$V_R = I \times R$$

$$V_L = I \times X_L$$

$$V_C = I \times X_C$$

Since none of these voltages are in phase with one another as shown in Figure 18-3(b), they must be added vectorially to obtain the applied voltage.

The formula, based on the Pythagorean theorem and illustrated in Figure 18-3(c) and (d), can be used to calculate V_S.

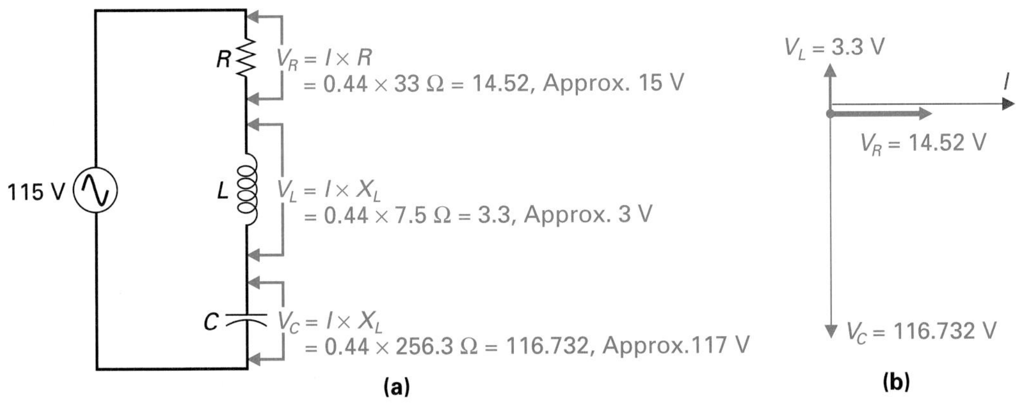

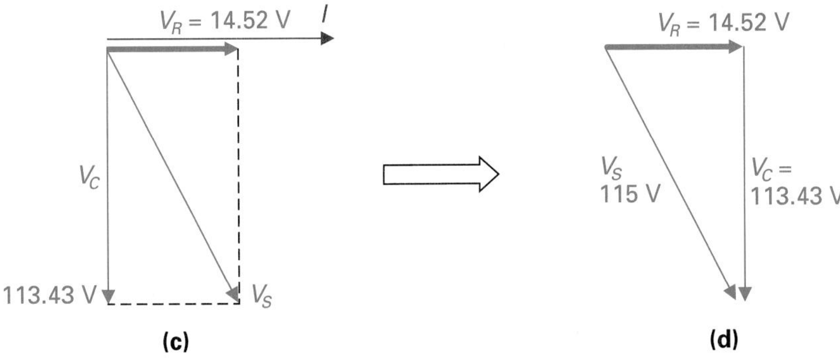

FIGURE 18-3 Series Voltage Drops.

$$V_S = \sqrt{V_R{}^2 + (V_L \sim V_C)^2}$$

In the example circuit, the applied voltage is, as we already know, 115 V.

$$\begin{aligned} V_S &= \sqrt{15^2 + (3 \sim 117)^2} \\ &= \sqrt{225 + 12{,}996} \\ &= \sqrt{13{,}221} \\ &= 115 \text{ V} \end{aligned}$$

18-1-4 Phase Angle

As can be seen in Figure 18-3(c), there is a phase difference between V_S and I. This phase difference can be calculated with either of the following formulas.

$$\theta = \arctan \frac{V_L \sim V_C}{V_R}$$

$$\theta = \arctan \frac{X_L \sim X_C}{R}$$

In the example circuit, θ is

$$\theta = \arctan \frac{3.3\text{V} \sim 116.732\text{V}}{14.52\text{V}}$$
$$= \arctan 7.812$$
$$= 82.7°$$

Since the example circuit is capacitive (ICE), the phase angle will be –82.7° since V_S lags I in a circuit that acts capacitively.

18-1-5 Power

The true power or resistive power (P_R) dissipated by a circuit can be calculated using the formula

$$P_R = I^2 \times R$$

which in our example will be

$$P_R = 0.44^2 \times 33\ \Omega = 6.4\ \text{W}$$

The apparent power (P_A) consumed by the circuit is calculated by

$$P_A = V_S \times I$$

which in our example will be

$$P_A = 115\ \text{V} \times 0.44 = 50.6\ \text{volt-amperes (VA)}$$

The true or actual power dissipated by the resistor is, as expected, smaller than the apparent power that seems to be being used.

The power factor can be calculated, as usual, by

$$\text{PF} = \cos\theta = \frac{R}{Z} = \frac{P_R}{P_A}$$

PF of 0 = reactive circuit
PF of 1 = resistive circuit

In the example circuit, PF = 0.126, indicating that the circuit is mainly reactive.

EXAMPLE:

For a series circuit where $R = 10\ \Omega$, $L = 5$ mH, $C = 0.05\ \mu$F, and $V_S = 100$ V/2 kHz, calculate:

a. X_C
b. X_L
c. Z
d. I
e. V_R, V_C, and V_L
f. Apparent power
g. True power
h. Power factor
i. Phase angle

Solution:

a. $X_C = \dfrac{1}{2\pi fC} = 1.6\ \text{k}\Omega$

b. $X_L = 2\pi fL = 62.8\ \Omega$

c. $Z = \sqrt{R^2 + (X_L \sim X_C)^2}$

$= \sqrt{(10\ \Omega)^2 + (1.6\ \text{k}\Omega \sim 62.8\ \Omega)^2}$

$= \sqrt{10\ \Omega^2 + 1.54\ \text{k}\Omega^2}$

$= 1.54\ \text{k}\Omega$ (capacitive circuit due to high X_C)

d. $I = \dfrac{V_S}{Z} = \dfrac{100\ \text{V}}{1.54\ \text{k}\Omega} = 64.9$ mA

e. $V_R = I \times R = 64.9\ \text{mA} \times 10\ \Omega = 0.65$ V

$V_C = I \times X_C = 64.9\ \text{mA} \times 1.6\ \text{k}\Omega = 103.9$ V

$V_L = I \times X_L = 64.9\ \text{mA} \times 62.8\ \Omega = 4.1$ V

f. Apparent power $= V_s \times I = 100\ \text{V} \times 64.9\ \text{mA} = 6.49$ VA

g. True power $= I^2 \times R = 64.9^2 \times 10\ \Omega = 42.17$ mW

h. $\text{PF} = \dfrac{R}{Z} = \dfrac{10\ \Omega}{1.5\ \text{k}\Omega} = 0.006$ (reactive circuit)

i. $\theta = \arctan\ \dfrac{V_L \sim V_C}{V_R}$

$= \arctan\ \dfrac{4.1\ \text{V} \sim 103.9\ \text{V}}{0.65\ \text{V}}$

$= \arctan 153.54$

$= 89.63°$

Capacitive circuit (ICE); therefore, V_S lags I by $-89.63°$.

SELF-TEST REVIEW QUESTIONS FOR SECTION 18-1

1. List in order the procedure that should be followed to fully analyze a series *RLC* circuit.
2. State the formulas for calculating the following in relation to a series *RLC* circuit.
   - **a.** Impedance
   - **b.** Current
   - **c.** Apparent power
   - **d.** V_S
   - **e.** True power
   - **f.** V_R
   - **g.** V_L
   - **h.** V_C
   - **i.** Phase angle (θ)
   - **j.** Power factor (PF)

18-2 PARALLEL *RLC* CIRCUIT

Now that the characteristics of a series circuit are understood, let us connect a resistor, inductor, and capacitor in parallel with one another. Figure 18-4(a) and (b) show the current and voltage relationships of a parallel *RLC* circuit.

18-2-1 Voltage

As can be seen in Figure 18-4(a), the voltage across any parallel circuit will be equal and in phase. Therefore,

$$V_R = V_L = V_C = V_S$$

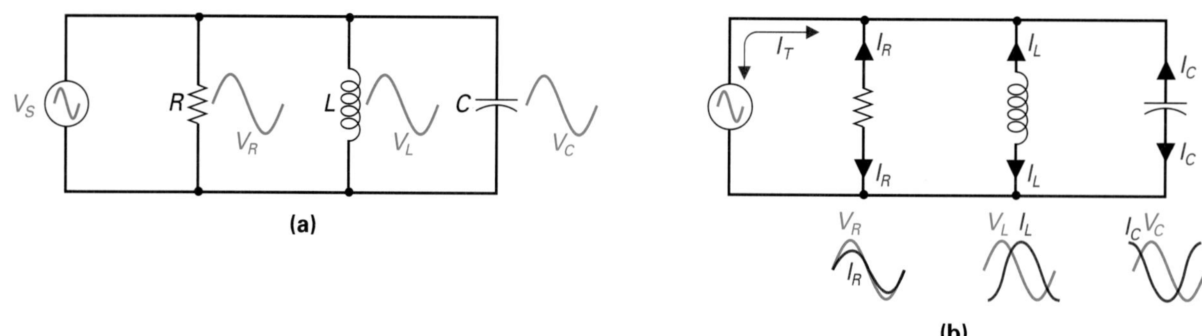

FIGURE 18-4 Parallel *RLC* Circuit. (a) *RLC* Parallel Circuit Voltage: Voltages across Each Component Are All Equal and in Phase with One Another in a Parallel Circuit. (b) *RLC* Parallel Circuit Currents: I_R Is in Phase with V_R, I_L Lags V_L by 90°, and I_C Leads V_C by 90°.

18-2-2 *Current*

With current, we must first calculate the individual branch currents (I_R, I_L, and I_C) and then calculate the total circuit current (I_T). An example circuit is illustrated in Figure 18-5(a), and the branch currents can be calculated by using the formulas

$$I_R = \frac{V}{R}$$

$$I_L = \frac{V}{X_L}$$

$$I_C = \frac{V}{X_C}$$

Figure 18-5(b) illustrates these branch currents vectorially, with I_R in phase with V_S, I_L lagging by 90°, and I_C leading I_R by 90°. Thc 180° phase difference between I_C and I_L results in a cancellation, as shown in Figure 18-5(c).

The total current (I_T) can be calculated by using the Pythagorean theorem on the right triangle, as illustrated in Figure 18-5(d).

$$I_T = \sqrt{I_R^2 + I_X^2}$$

$$(I_X = I_L \sim I_C)$$

$$= \sqrt{3.5^2 + 14.9^2}$$
$$= 15.3 \text{ A}$$

18-2-3 *Phase Angle*

As shown in Figure 18-5(b), there is a phase difference between the source voltage (V_S) and the circuit current (I_T). This phase difference can be calculated using the formula

$$\theta = \arctan \frac{I_L \sim I_C}{I_R}$$

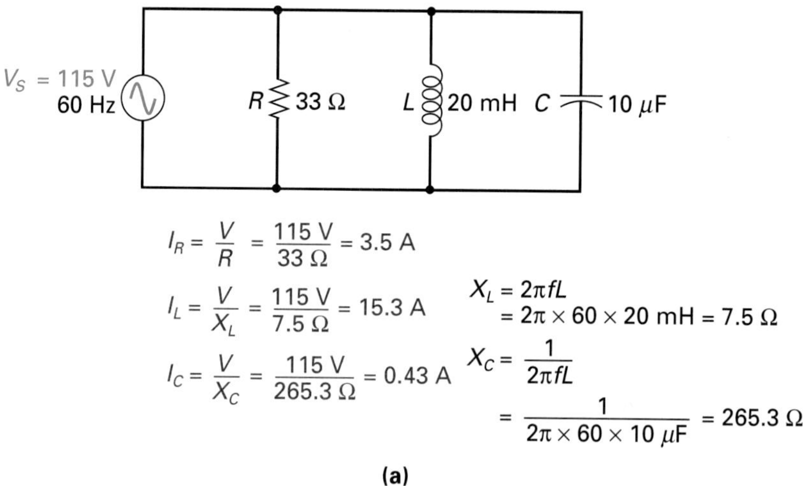

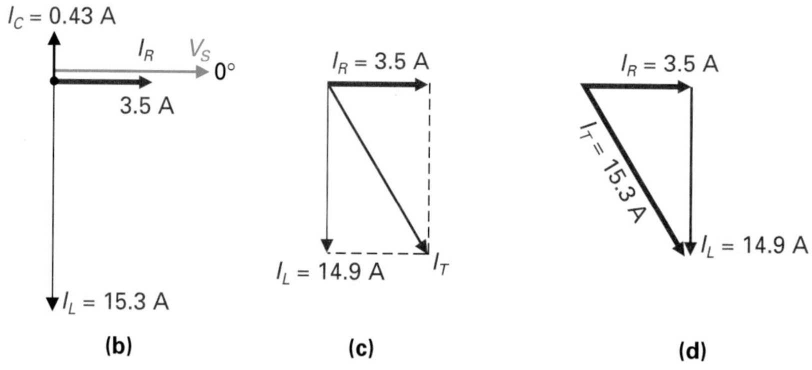

FIGURE 18-5 Example *RLC* Parallel Circuit.

$$= \arctan \frac{15.3 \text{ A} - 0.43 \text{ A}}{3.5 \text{ A}}$$
$$= \arctan 4.25$$
$$= 76.7°$$

Since this is an inductive circuit (ELI), the total current (I_T) will lag the source voltage (V_S) by –76.7°.

18-2-4 *Impedance*

Since the total current (I_T) is known, the impedance of all three components in parallel can be calculated by the formula

$$Z = \frac{V}{I_T}$$

$$= \frac{115 \text{ V}}{15.3 \text{ A}}$$

$$= 7.5\ \Omega$$

18-2-5 Power

The true power dissipated can be calculated using

$$P_R = I_R^{\,2} \times R$$

$$= 3.5^2 \times 33\ \Omega$$

$$= 404.3 \text{ W}$$

the apparent power consumed by the circuit is calculated by

$$P_A = V_S \times I_T$$

$$= 115 \text{ V} \times 15.3 \text{ A}$$

$$= 1759.5 \text{ volt-amperes (VA)}$$

and the power factor can be calculated, as usual, with

$$\text{PF} = \cos\theta = \frac{P_R}{P_A}$$

In the example circuit, PF = 0.23.

SELF-TEST REVIEW QUESTIONS FOR SECTION 18-2

1. State the formulas for calculating the following in relation to a parallel *RLC* circuit:

a. I_R **d.** I_L
b. I_T **e.** θ
c. I_C **f.** Z

2. State the formulas for:

a. P_R **c.** P_A
b. P_X **d.** PF

18-3 RESONANCE

Resonance
Circuit condition that occurs when the inductive reactance (X_L) is equal to the capacitive reactance (X_C).

Resonance is a circuit condition that occurs when the inductive reactance (X_L) and the capacitive reactance (X_C) have been balanced. Figure 18-6 illustrates a parallel- and a series-connected LC circuit. If a dc voltage is applied to the input of either circuit, the capacitor will act as an open (X_C = infinite Ω) and the inductor will act as a short ($X_L = 0\ \Omega$).

If a low-frequency ac is now applied to the input, X_C will decrease from maximum, and X_L will increase from zero. As the ac frequency is increased further, the capacitive reactance will continue to fall ($X_C \downarrow \propto 1/f\uparrow$) and the inductive reactance to rise ($X_L\uparrow \propto f\uparrow$), as shown in Figure 18-7.

As the input ac frequency is increased further, a point will be reached where X_L will equal X_C, and this condition is known as *resonance.* The frequency at which $X_L = X_C$ in either a parallel or a series LC circuit is known as the *resonant frequency* (f_0) and can be calculated by the following formula, which has been derived from the capacitive and inductive reactance formulas:

$$f_0 = \frac{1}{2\pi\sqrt{LC}}$$

where f_0 = resonant frequency, in hertz (Hz)
L = inductance, in henrys (H)
C = capacitance, in farads (F)

EXAMPLE:

Calculate the resonant frequency (f_0) of a series LC circuit if L = 750 mH and C = 47 μF.

Solution:

$$f_0 = \frac{1}{2\pi\sqrt{L \times C}}$$

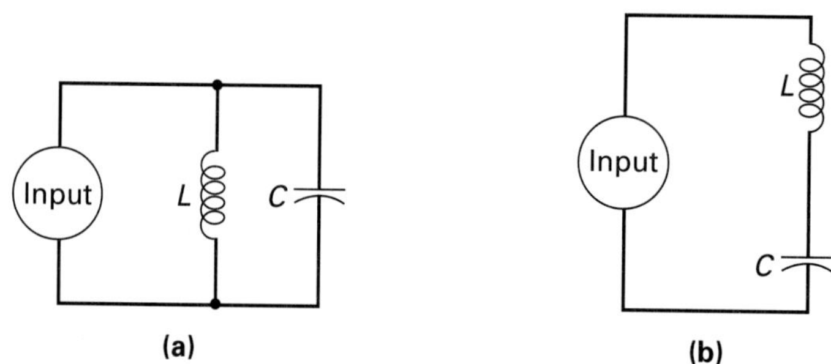

FIGURE 18-6 **Resonance. (a) Parallel *LC* Circuit. (b) Series *LC* Circuit.**

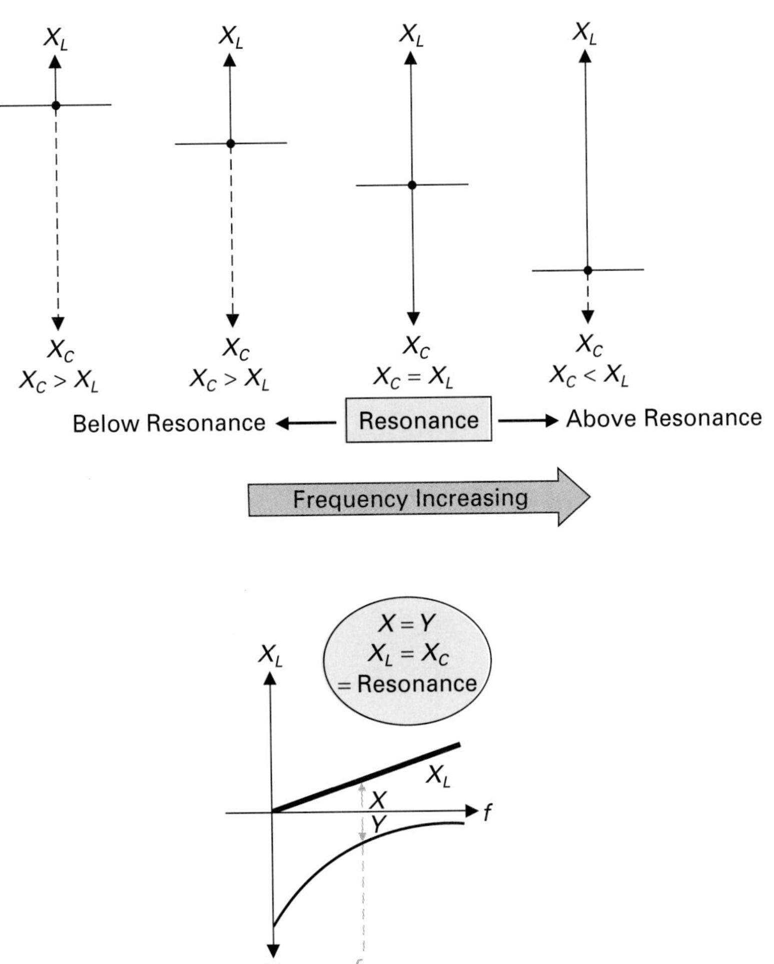

FIGURE 18-7 **Frequency versus Reactance.**

$$= \frac{1}{2\pi\sqrt{(750 \times 10^{-3}) \times (47 \times 10^{-6})}}$$

$$= 26.8 \text{ Hz}$$

18-3-1 *Series Resonance*

Figure 18-8(a) illustrates a series *RLC* circuit at resonance ($X_L = X_C$), or **series resonant circuit.** The ac input voltage causes current to flow around the circuit, and since all the components are connected in series, the same value of current (I_S) will flow through all the components. Since *R*, X_L, and X_C are all equal to 100 Ω and the current flow is the same throughout, the voltage dropped across each component will be equal, as illustrated vectorially in Figure 18-8(b).

Series Resonant Circuit
A resonant circuit in which the capacitor and coil are in series with the applied ac voltage.

The voltage across the resistor is in phase with the series circuit current (I_S); however, since the voltage across the inductor (V_L) is 180° out of phase with the

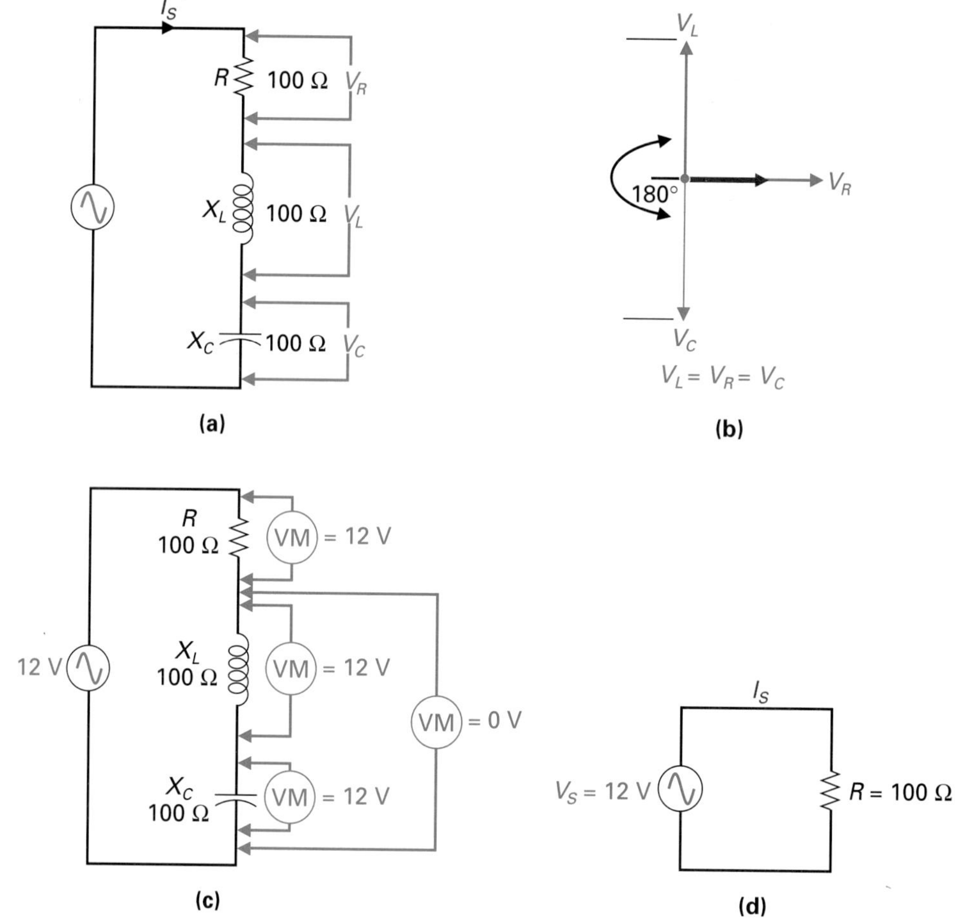

FIGURE 18-8 Series Resonant Circuit.

voltage across the capacitor (V_C), and both are equal to one another, V_L cancels V_C, when both are measured in series.

Three unusual characteristics occur when a circuit is at resonance, which do not occur at any other frequency.

(1). The first is that if V_L and V_C cancel, the voltage across L and C will measure 0 V on a voltmeter. Since there is effectively no voltage being dropped across these two components, all the voltage must be across the resistor ($V_R =$ 12 V). This is true; however, since the same current flows throughout the series circuit, a voltmeter will measure 12 V across C, 12 V across L, and 12 V across R, as shown in Figure 18-8(c). It now appears that the voltage drops around the series circuit (36 V) do not equal the voltage applied (12 V). This is not true, as V_L and V_C cancel, because they are out of phase with one another, so Kirchhoff's voltage law is still valid.

(2). The second unusual characteristic of resonance is that because the total opposition or impedance (Z) is equal to

$$Z = \sqrt{R^2 + (X_L \sim X_C)^2}$$

and the difference between X_L and X_C is 0 ($Z = \sqrt{R^2 + 0}$), the impedance of a series circuit at resonance is equal to the resistance value R ($Z = \sqrt{R^2} = R$). Consequently, the applied ac voltage of 12 V is forcing current to flow through this series RLC circuit. Since current is equal to $I_s = V/Z$ and $Z = R$, the circuit current at resonance is dependent only on the value of resistance. The capacitor and inductor are invisible and are seen by the source as simply a piece of conducting wire with no resistance, as illustrated in Figure 18-8(d). Since only resistance exists in the circuit, current (I_S) and voltage (V_S) are in phase with one another, and as expected for a purely resistive circuit, the power factor will be equal to 1.

(3). To emphasize the third strange characteristic of series resonance, we will take another example, shown in Figure 18-9. The circuit current in this example is equal to $I = V/R = 12\text{ V}/10 = 1.2\text{ A}$, since $Z = R$ at resonance. Since the same current flows throughout a series circuit, the voltage across each component can be calculated.

$$V_R = I \times R = 1.2\text{ A} \times 10 = 12\text{ V}$$
$$V_L = I \times X_L = 1.2\text{ A} \times 100 = 120\text{ V}$$
$$V_C = I \times X_C = 1.2\text{ A} \times 100 = 120\text{ V}$$

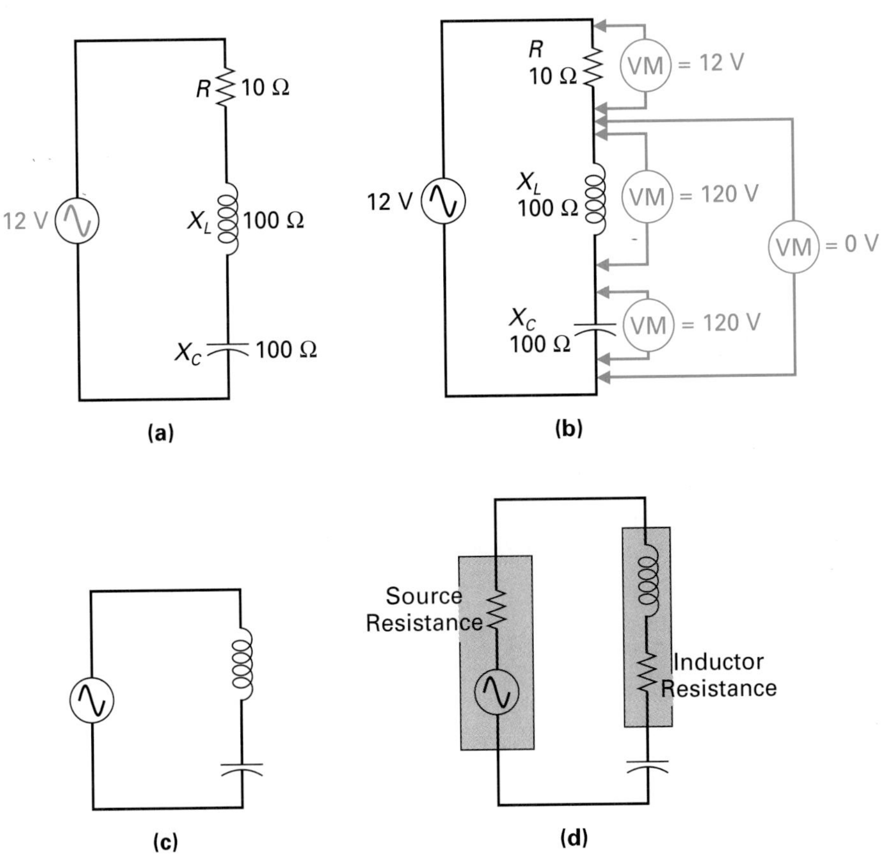

FIGURE 18-9 Circuit Effects at Resonance.

As V_L is 180° out of phase with V_C, the 120 V across the capacitor cancels with the 120 V across the inductor, resulting in 0 V across L and C combined, as shown in Figure 18-9(b). Since L and C have the ability to store energy, the voltage across them individually will appear larger than the applied voltage.

If the resistance in the circuit is removed completely, as shown in Figure 18-9(c), the circuit current, which is determined by the resistance only, will increase to a maximum ($I\uparrow = V/R\downarrow$) and, consequently, cause an infinitely high voltage across the inductor and capacitor ($V\uparrow = I\uparrow \times R$). In reality, the ac source will have some value of internal resistance, and the inductor, which is a long length of thin wire ($R\uparrow$), will have some value of resistance, as shown in Figure 18-9(d), which limits the series resonant circuit current.

In summary, we can say that in a series resonant circuit:

1. The inductor and capacitor electrically disappear due to their equal but opposite effect, resulting in 0-V drops across the series combination, and the circuit consequently seems purely resistive.
2. The current flow is large because the impedance of the circuit is low and equal to the series resistance (R), which has the source voltage developed across it.
3. The individual voltage drops across the inductor or capacitor can be larger than the source voltage if R is smaller than X_L and X_C.

Quality Factor

As discussed previously regarding inductance, the Q factor is a ratio of inductive reactance to resistance and is used to express how efficiently an inductor will store rather than dissipate energy. In a series resonant circuit, the Q factor indicates the quality of the series resonant circuit, or is the ratio of the reactance to the resistance.

$$Q = \frac{X_L}{R}$$

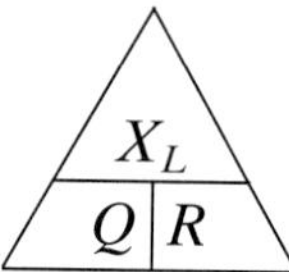

or, since $X_L = X_C$,

$$Q = \frac{X_C}{R}$$

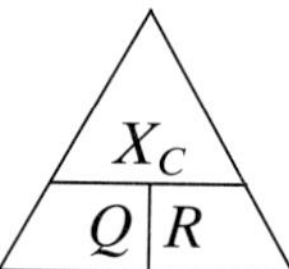

Another way to calculate the Q of a series resonant circuit is by using the formula

$$Q = \frac{V_L}{V_R} = \frac{V_C}{V_R}$$

(at resonance only)

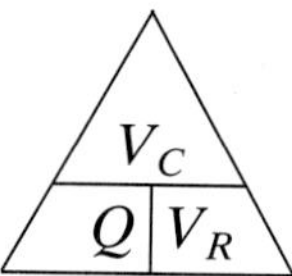

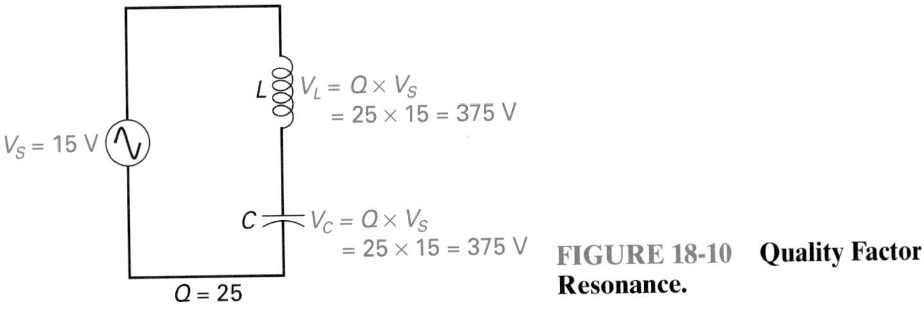

FIGURE 18-10 Quality Factor at Resonance.

or, since $V_R = V_S$,

$$Q = \frac{V_L}{V_S} = \frac{V_C}{V_S} \quad \text{(at resonance only)}$$

V_L / Q | V_S V_C / Q | V_S

If the Q and source voltage are known, the voltage across the inductor or capacitor can be found by transposition of the formula, as can be seen in the example in Figure 18-10.

The Q of a resonant circuit is almost entirely dependent on the inductor's coil resistance, because capacitors tend to have almost no resistance figure at all, only reactive, which makes them very efficient.

The inductor has a Q value of its own, and if only L and C are connected in series with one another, the Q of the series resonant circuit will be equal to the Q of the inductor, as shown in Figure 18-11. If the resistance is added in with L and C, the Q of the series resonant circuit will be less than that of the inductor's Q.

EXAMPLE:

Calculate the resistance of the series resonant circuit illustrated in Figure 18-12.

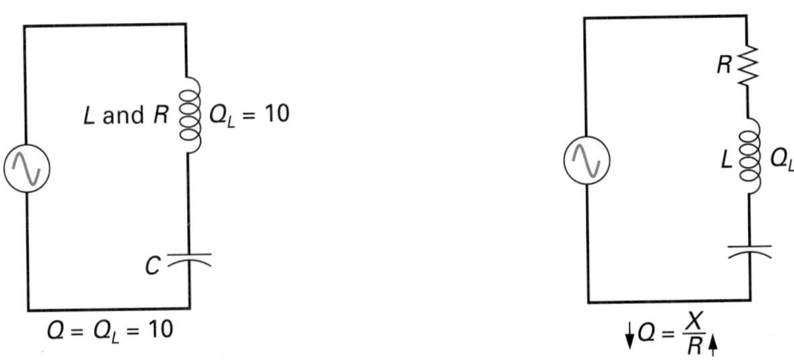

FIGURE 18-11 Resistance within Inductor.

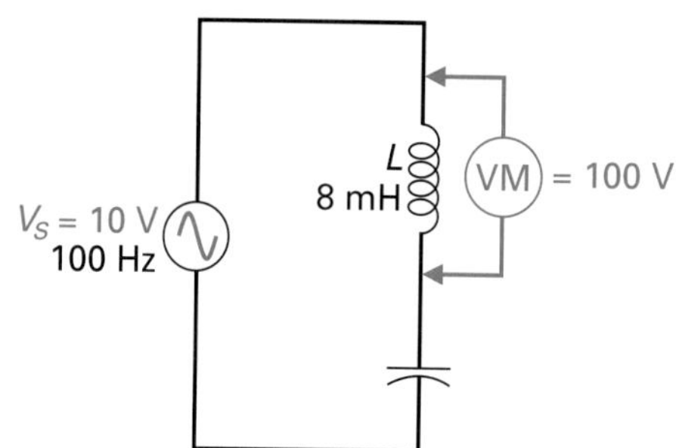

FIGURE 18-12 **Series Resonant Circuit Example.**

■ *Solution:*

$$Q = \frac{V_L}{V_S} = \frac{100\text{ V}}{10\text{ V}} = 10$$

Since $Q = X_L/R$, $R = X_L/Q$; so if the inductive reactance can be found, the R can be determined.

$$\begin{aligned} X_L &= 2\pi \times f \times L \\ &= 2\,\pi \times 100 \times 8\text{ mH} \\ &= 5\ \Omega \end{aligned}$$

R will consequently equal

$$\begin{aligned} R &= \frac{X_L}{Q} \\ &= \frac{5}{10} \\ &= 0.5\ \Omega \end{aligned}$$

Bandwidth

Bandwidth
Width of the group or band of frequencies between the half-power points.

Frequency Response Curve
A graph indicating a circuit's response to different frequencies.

Cutoff Frequency
Frequency at which the gain of the circuit falls below 0.707 of the maximum current or half-power (–3 dB).

A series resonant circuit is selective in that frequencies at resonance or slightly above or below will cause a larger current than frequencies well above or below the circuit's resonant frequency. The group or band of frequencies that causes the larger current is called the circuit's **bandwidth.**

Figure 18-13 illustrates a series resonant circuit and its bandwidth. The × marks on the curve illustrate where different frequencies were applied to the circuit and the resulting value of current measured in the circuit. The resulting curve produced is called a **frequency response curve,** as it illustrates the circuit's response to different frequencies. At resonance, $X_L = X_C$ and the two cancel, which is why maximum current was present in the circuit (100 mA) when the resonant frequency (100 Hz) was applied.

The bandwidth includes the group or band of frequencies that cause 70.7% or more of the maximum current to flow within the series resonant circuit; in this example, frequencies from 90 to 110 Hz cause 70.7 mA or more, which is 70.7% of maximum (100 mA), to flow. The bandwidth in this example is equal to

$$\text{BW} = 110 - 90 = 20\text{ Hz}$$

(110 Hz and 90 Hz are known as **cutoff frequencies**).

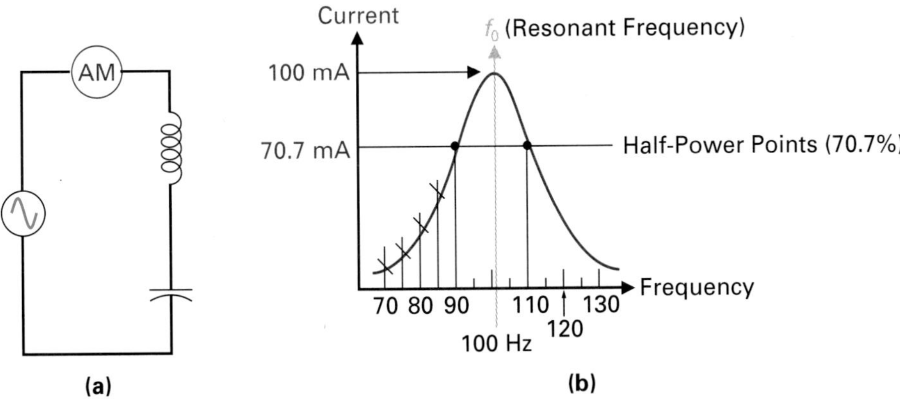

FIGURE 18-13 Series Resonant Circuit Bandwidth. (a) Circuit. (b) Frequency Response Curve.

Referring to the bandwidth curve in Figure 18-13(b), you may notice that 70.7% is also called the **half-power points,** although it does not exist halfway between 0 and maximum. This value of 70.7% is not the half-current point but the half-power point, as we can prove with a simple example.

Half-Power Point
A point at which power is 50%. This half-power point corresponds to 70.7% of the total current.

EXAMPLE:

$R = 2\ \text{k}\Omega$ and $I = 100\ \text{mA}$; therefore, power $= I^2 \times R = 100\ \text{mA}^2 \times 2\ \text{k}\Omega = 20\ \text{W}$. If the current is now reduced so that it is 70.7% of its original value, calculate the power dissipated.

Solution:

$$P = I^2 \times R = 70.7\ \text{mA}^2 \times 2\ \text{k}\Omega = 10\ \text{W}$$

In summary, the 70.7% current points are equal to the 50% or half-power points. A circuit's bandwidth is the band of frequencies that exists between the 70.7% current points or half-power points.

The bandwidth of a series resonant circuit can also be calculated by use of the formula

$$\text{BW} = \frac{f_0}{Q_{f0}}$$

f_0 = Resonant Frequency
Q_{f0} = Quality factor at resonance

This formula states that the BW is proportional to the resonant frequency of the circuit and inversely proportional to the Q of the circuit.

Figure 18-14 illustrates three example response curves. In these three examples, the value of R is changed from 100 Ω to 200 to 400 Ω. This does not vary the resonant frequency, but simply alters the Q and therefore the BW. The resistance value will determine the Q of the circuit, and since Q is inversely pro-

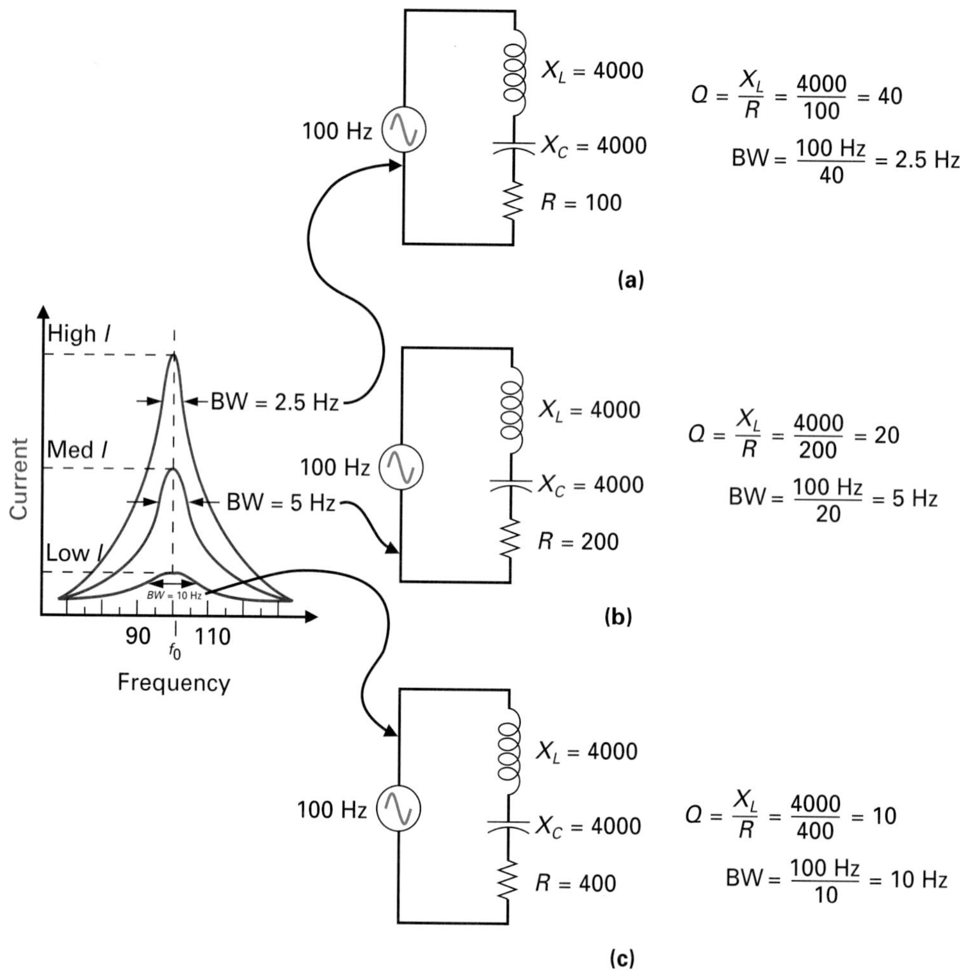

FIGURE 18-14 Bandwidth of a Series Resonant Circuit.

portional to resistance, Q is proportional to current; consequently, a high value of Q will cause a high value of current.

In summary, the bandwidth of a series resonant circuit will increase as the Q of the circuit decreases ($\text{BW}\uparrow = f_0/Q\downarrow$), and vice versa.

SELF-TEST REVIEW QUESTIONS FOR SECTION 18-3-1

1. Define *resonance.*
2. What is series resonance?
3. In a series resonant circuit, what are the three rather unusual circuit phenomena that take place?
4. How does Q relate to series resonance?
5. Define *bandwidth.*
6. Calculate BW if $f_0 = 12$ kHz and $Q = 1000$.

18-3-2 *Parallel Resonance*

The **parallel resonant circuit** acts differently from the series resonant circuit, and these different characteristics need to be analyzed and discussed. Figure 18-15 illustrates a parallel resonant circuit. The inductive current could be calculated by using the formula

Parallel Resonant Circuit
Circuit having an inductor and capacitor in parallel with one another, offering a high impedance at the frequency of resonance.

$$I_L = \frac{V_L}{X_L} = \frac{10\text{ V}}{1\text{ k}\Omega} = 10\text{ mA}$$

The capacitive current could be calculated by using the formula

$$I_C = \frac{V_C}{X_C} = \frac{10\text{ V}}{1\text{ k}\Omega} = 10\text{ mA}$$

Looking at the vector diagram in Figure 18-15(b), you can see that I_C leads the source voltage by 90° (ICE) and I_L lags the source voltage by 90° (ELI), creating a 180° phase difference between I_C and I_L. This means that when 10 mA of current flows up through the inductor, 10 mA of current will flow in the opposite direction down through the capacitor, as shown in Figure 18-16(a). During the opposite alternation, 10 mA will flow down through the inductor and 10 mA will travel up through the capacitor, as shown in Figure 18-16(b).

If 10 mA arrives into point *X* and 10 mA of current leaves point *X*, no current can be flowing from the source (V_S) to the parallel *LC* circuit; the current is simply swinging or oscillating back and forth between the capacitor and inductor.

The source voltage (V_S) is needed initially to supply power to the *LC* circuit and start the oscillations; but once the oscillating process is in progress (assuming the ideal case), current is only flowing back and forth between induc-

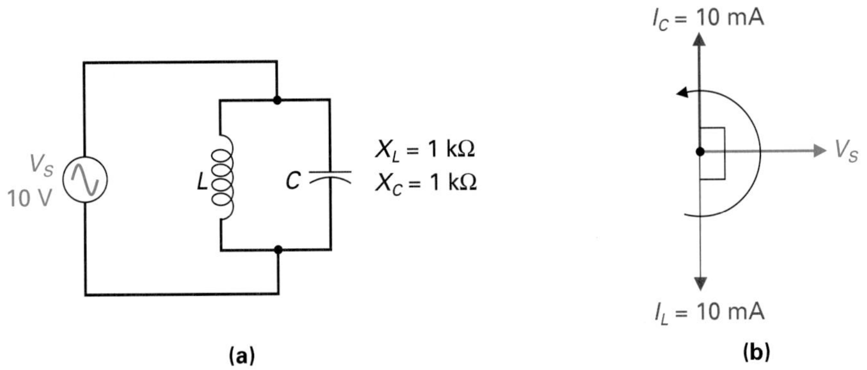

FIGURE 18-15 Parallel Resonant Circuit. (a) Circuit. (b) Vector Diagram.

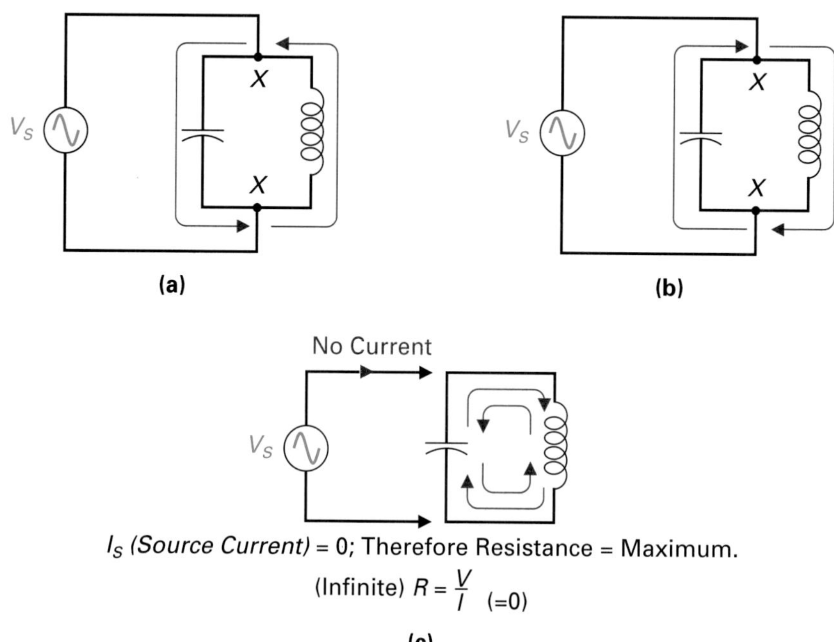

FIGURE 18-16 **Current in a Parallel Resonant Circuit.**

tor and capacitor, and no current is flowing from the source. So the *LC* circuit appears as an infinite impedance and the source can be disconnected, as shown in Figure 18-16(c).

Flywheel Action

Flywheel Action
Sustaining effect of oscillation in an *LC* circuit due to the charging and discharging of the capacitor and the expansion and contraction of the magnetic field around the inductor.

Let's discuss this oscillating effect, called **flywheel action,** in a little more detail. The name is derived from the fact that it resembles a mechanical flywheel, which, once started, will keep going continually until friction reduces the magnitude of the rotations to zero.

The electronic equivalent of the mechanical flywheel is a resonant parallel-connected *LC* circuit. Figure 18-17(a) through (h) illustrate the continual energy transfer between capacitor and inductor, and vice versa. The direction of the circulating current reverses each half-cycle at the frequency of resonance. Energy is stored in the capacitor in the form of an electric field between the plates on one half-cycle, and then the capacitor discharges, supplying current to build up a magnetic field on the other half-cycle. The inductor stores its energy in the form of a magnetic field, which will collapse, supplying a current to charge the capacitor, which will then discharge, supplying a current back to the inductor, and so on. Due to the "storing action" of this circuit, it is sometimes related to the fluid analogy and referred to as a **tank circuit.**

Tank Circuit
Circuit made up of a coil and capacitor that is capable of storing electric energy.

The Reality of Tanks

Under ideal conditions a tank circuit should oscillate indefinitely if no losses occur within the circuit. In reality, the resistance of the coil reduces that 100% efficiency, as does friction with the mechanical flywheel. This coil resis-

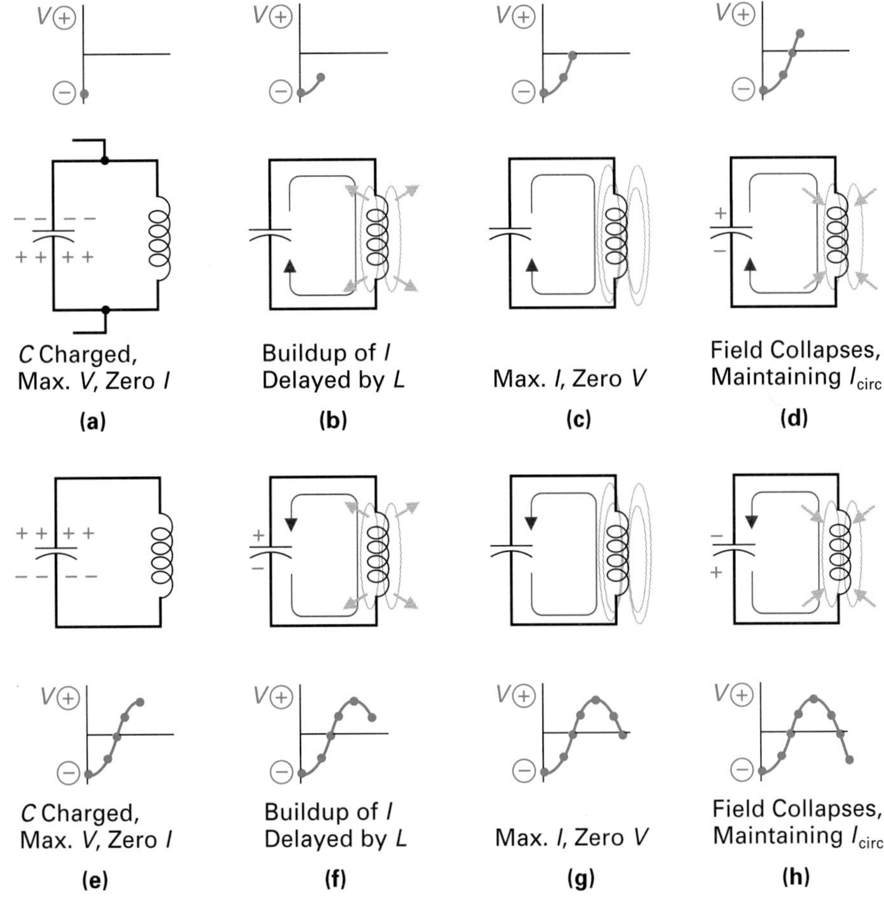

FIGURE 18-17 Energy and Current in an *LC* Parallel Circuit at Resonance.

tance is illustrated in Figure 18-18(a), and, unlike reactance, resistance is the opposition to current flow, with the dissipation of energy in the form of heat. As a small part of the energy is dissipated with each cycle, the oscillations will be reduced in size and eventually fall to zero, as shown in Figure 18-18(b).

If the ac source is reconnected to the tank, as shown in Figure 18-18(c), a small amount of current will flow from the source to the tank to top up the tank or replace the dissipated power. The higher the coil resistance is, the higher the loss and the larger the current flow from source to tank to replace the loss.

Quality Factor

In the series resonant circuit, we were concerned with voltage drops since current remains the same throughout a series circuit, so

$$Q = \frac{V_C \text{ or } V_L}{V_S} \quad \text{(at resonance only)}$$

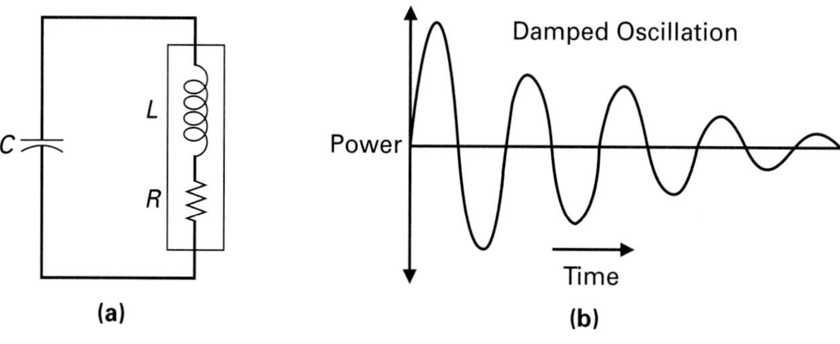

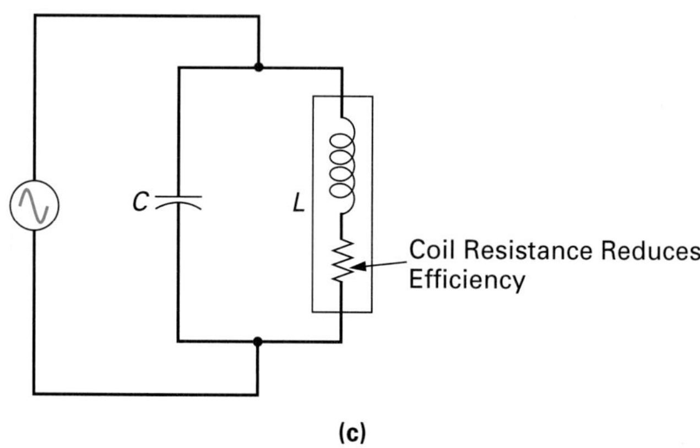

FIGURE 18-18 Losses in Tanks.

In a parallel resonant circuit, we are concerned with circuit currents rather than voltage, so

$$Q = \frac{I_{\text{tank}}}{I_S}$$

(at resonance only)

The quality factor, Q, can also be expressed as the ratio between reactance and resistance:

$$Q = \frac{X_L}{R} \quad \text{(at any frequency)}$$

Another formula, which is the most frequently used when discussing and using parallel resonant circuits, is

$$Q = \frac{Z_{\text{tank}}}{X_L}$$

(at resonance only)

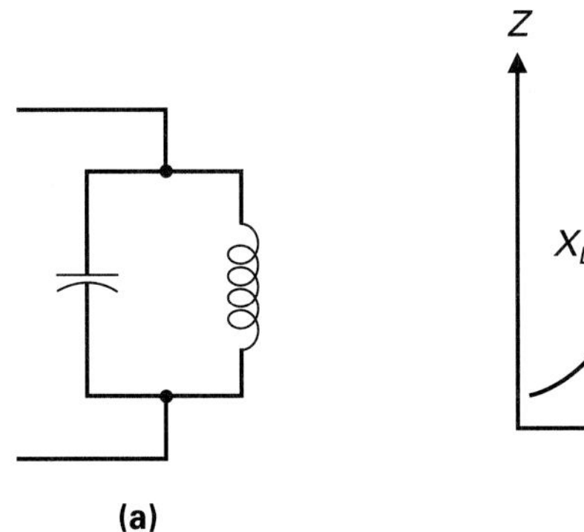

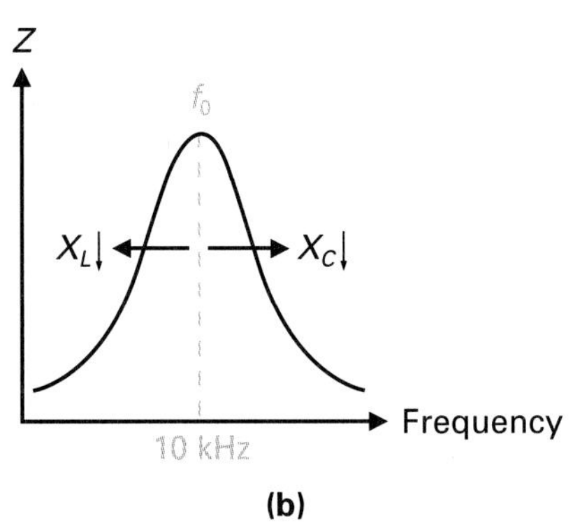

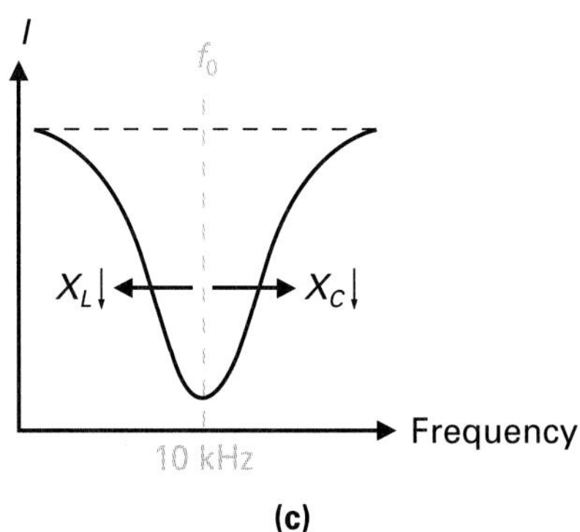

FIGURE 18-19 **Parallel Resonant Circuit Bandwidth.**

This formula states that the Q of the tank is proportional to the tank impedance. A higher tank impedance results in a smaller current flow from source to tank. This assures less power is dissipated, and that means a higher-quality tank.

Of all the three Q formulas for parallel resonant circuits, $Q = I_{\text{tank}}/I_S$, $Q = X_L/R$, or $Q = Z_{\text{tank}}/X_L$, the latter is the easiest to use as both X_L and the tank impedance can easily be determined in most cases where C, L, and R internal for the inductor are known.

Bandwidth

Figure 18-19 illustrates a parallel resonant circuit and two typical response curves. These response curves summarize what we have described previously, in that a parallel resonant circuit has maximum impedance [Figure 18-19(b)] and minimum current [Figure 18-19(c)] at resonance. The current versus frequency response curve shown in Figure 18-19(c) is the complete opposite to the series resonant response curve. At frequencies below resonance (< 10 kHz), X_L is low and X_C is high, and the inductor offers a low reactance, producing a high current path and low impedance. On the other hand, at frequencies above resonance (> 10 kHz), the capacitor displays a low reactance, producing a high current path and low impedance. The parallel resonant circuit is like the series resonant circuit in that it responds to a band of frequencies close to its resonant frequency.

The bandwidth (BW) can be calculated by use of the formula

$$\text{BW} = \frac{f_0}{Q_{f_0}}$$

f_0 = Resonant frequency

Q_{f_0} = Quality factor at resonance

EXAMPLE:

Calculate the bandwidth of the circuit illustrated in Figure 18-20.

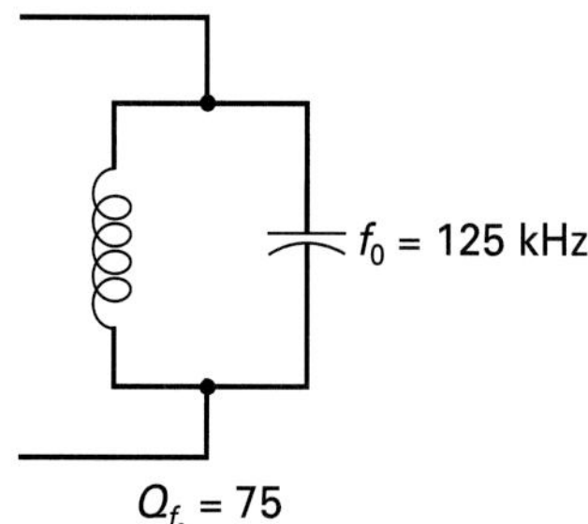

FIGURE 18-20 Bandwidth Example.

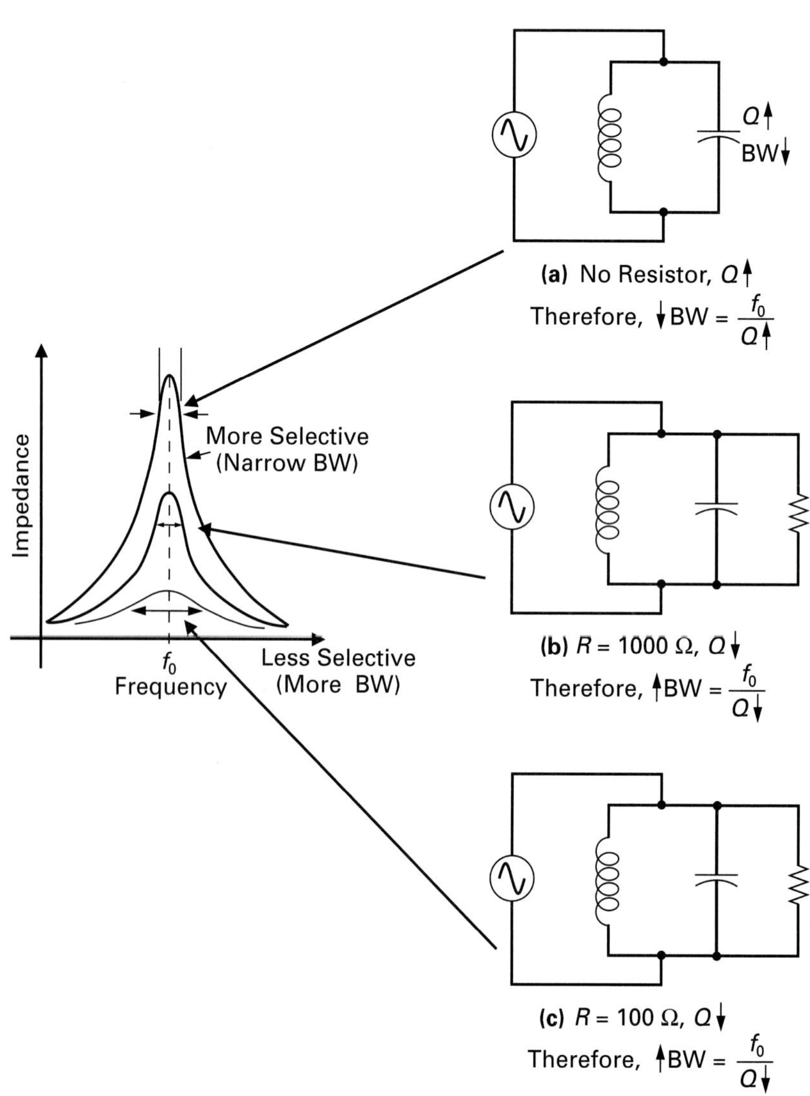

FIGURE 18-21 Varying Bandwidth by Loading the Tank Circuit.

■ *Solution:*

$$BW = \frac{f_0}{Q_{f0}}$$

$$= \frac{125 \text{ kHz}}{75}$$

$$= 1.7 \text{ kHz}$$

$$\frac{1.7 \text{ kHz}}{2} = 0.85 \text{ kHz}$$

therefore the bandwidth extends from

$$f_0 + 0.85 \text{ kHz} = 125.85 \text{ kHz}$$

$$f_0 - 0.85 \text{ kHz} = 124.15 \text{ kHz}$$

$$BW = 124.15 \text{ kHz to } 125.85 \text{ kHz}$$

Selectivity

Circuits containing inductance and capacitance are often referred to as **tuned circuits** since they can be adjusted to make the circuit responsive to a particular frequency (the resonant frequency). **Selectivity,** by definition, is the ability of a tuned circuit to respond to a desired frequency and ignore all others. Parallel resonant *LC* circuits are sometimes too selective, as the *Q* is too large, producing too narrow a bandwidth, as shown in Figure 18-21(a) ($BW\downarrow = f_0/Q\uparrow$).

In this situation, because of the very narrow response curve, a high resistance value can be placed in parallel with the *LC* circuit to provide an alternative path for line current. This process is known as *loading* or *damping* the tank and will cause an increase in line current and decrease in *Q* ($Q\downarrow = I_{tank}/I_{line}\uparrow$). The decrease in *Q* will cause a corresponding increase in BW ($BW\uparrow = f_0/Q\downarrow$), as shown by the examples in Figure 18-21, which illustrates a 1000-Ω loading resistor [Figure 18-21(b)] and a 100-Ω loading resistor [Figure 18-21(c)].

In summary, a parallel resonant circuit can be made less selective with a broader bandwidth if a resistor is added in parallel, providing an increase in current and a decrease in impedance, which widens the bandwidth.

Tuned Circuit
Circuit that can have its components' values varied so that the circuit responds to one selected frequency yet heavily attenuates all other frequencies.

Selectivity
Characteristic of a circuit to discriminate between the wanted signal and the unwanted signal.

SELF-TEST REVIEW QUESTIONS FOR SECTION 18-3-2

1. What are the differences between a series and a parallel resonant circuit?
2. Describe flywheel action.
3. Calculate the value of *Q* of a tank if $X_L = 50\ \Omega$ and $R = 25\ \Omega$.
4. When calculating bandwidth for a parallel resonant circuit, can the series resonant bandwidth formula be used?
5. What is selectivity?

18-4 APPLICATIONS OF *RLC* CIRCUITS

In the previous chapters, you saw how *RC* and *RL* filter circuits are used as low- or high-pass filters to pass some frequencies and block others. There are basically four types of filters:

1. Low-pass filter, which passes frequencies below a cutoff frequency
2. High-pass filter, which passes frequencies above a cutoff frequency
3. Bandpass filter, which passes a band of frequencies
4. Band-stop filter, which stops a band of frequencies

18-4-1 Low-Pass Filter

Figure 18-22(a) illustrates how an inductor and capacitor can be connected to act as a low-pass filter. At low frequencies, X_L has a small value compared to the load resistor (R_L), so nearly all the low-frequency input is developed and appears at the output across R_L. Since X_C is high at low frequencies, nearly all the current passes through R_L rather than *C*.

At high frequencies, X_L increases and drops more of the applied input across the inductor rather than the load. The capacitive reactance, X_C, aids this low output at high-frequency effect by decreasing its reactance and providing an alternative path for current to flow.

Since the inductor basically blocks alternating current and the capacitor shunts alternating current, the net result is to prevent high-frequency signals from reaching the load. The way in which this low-pass filter responds to frequencies is graphically illustrated in Figure 18-22(b).

18-4-2 High-Pass Filter

Figure 18-23(a) illustrates how an inductor and capacitor can be connected to act as a high-pass filter. At high frequencies, the reactance of the capacitor (X_C) is low while the reactance of the inductor (X_L) is high, so all the high frequen-

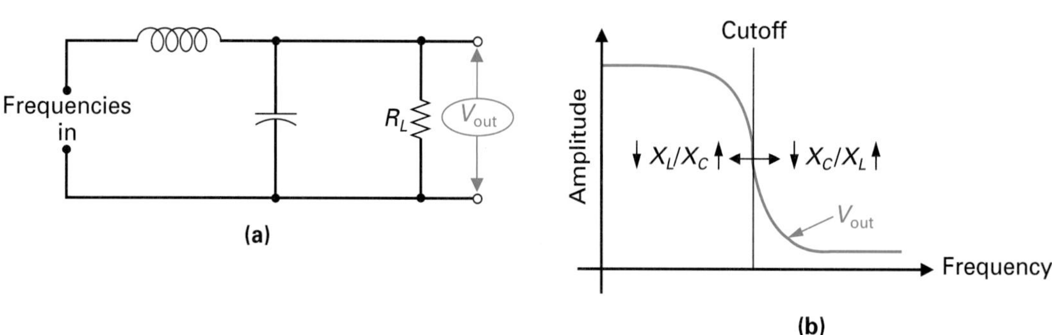

FIGURE 18-22 Low-Pass Filter. (a) Circuit. (b) Frequency Response.

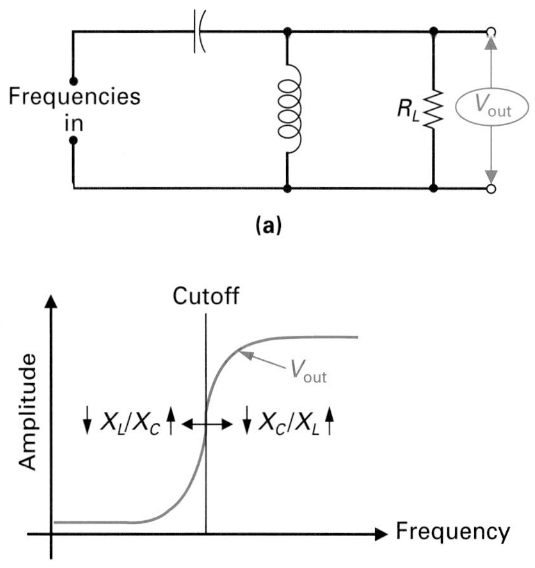

FIGURE 18-23 High-Pass Filter. (a) Circuit. (b) Frequency Response.

cies are easily passed by the capacitor and blocked by the inductor, so they all are routed through to the output and load.

At low frequencies, the reverse condition exists, resulting in a low X_L and a high X_C. The capacitor drops nearly all the input, and the inductor shunts the signal current away from the output load.

18-4-3 Bandpass Filter

Figure 18-24(a) illustrates a series resonant **bandpass filter,** and Figure 18-24(b) shows a parallel resonant bandpass filter. Figure 18-24(c) shows the frequency response curve produced by the bandpass filter. At resonance, the series resonant *LC* circuit has a very low impedance and will consequently pass the resonant frequency to the load with very little drop across the *L* and *C* components.

Bandpass Filter
Filter circuit that passes a group or band of frequencies between a lower and an upper cutoff frequency, while heavily attenuating any other frequency outside this band.

Below resonance, X_C is high, and the capacitor drops a large amount of the input signal; above resonance, X_L is high and the inductor drops most of the input frequency voltage. This circuit will therefore pass a band of frequencies centered around the resonant frequency of the series *LC* circuit and block all other frequencies above and below this resonant frequency.

Figure 18-24(b) illustrates how a parallel resonant *LC* circuit can be used to provide a bandpass response. The series resonant circuit was placed in series with the output, whereas the parallel resonant circuit will have to be placed in parallel with the output to provide the same results. At resonance, the parallel resonant circuit or tank has a high impedance, so very little current will be shunted away from the output; it will be passed on to the output, and almost all the input will appear at the output across the load.

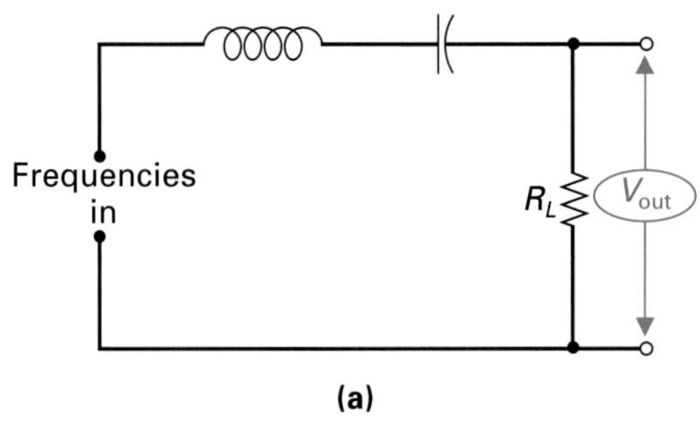

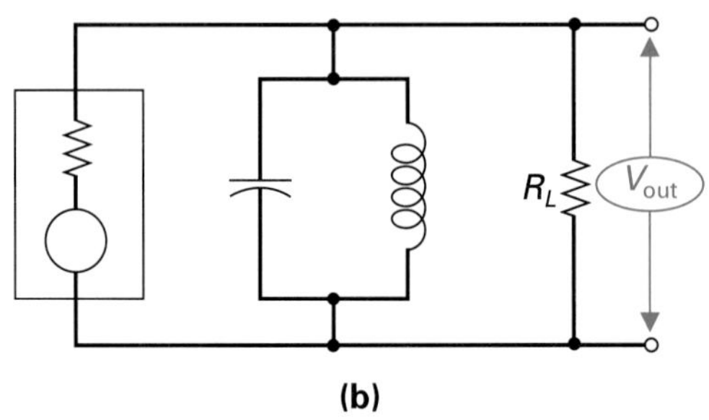

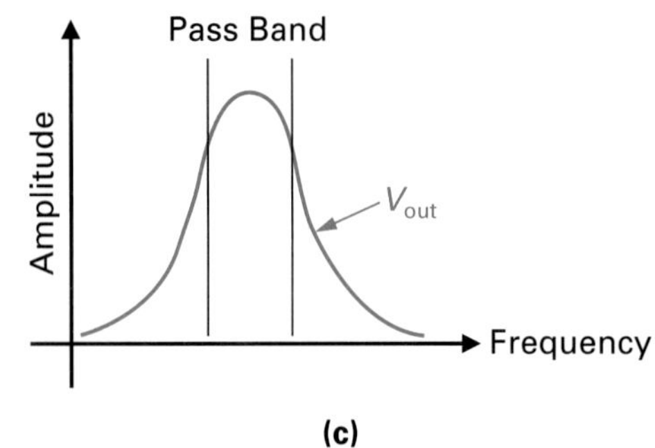

FIGURE 18-24 Bandpass Filter. (a) Series Resonant Bandpass Filter. (b) Parallel Resonant Bandpass Filter. (c) Frequency Response.

Above resonance, X_C is small, so most of the input is shunted away from the output by the capacitor; below resonance, X_L is small, and the shunting action occurs again, but this time through the inductor.

Figure 18-25 illustrates how a transformer can be used to replace the inductor to produce a bandpass filter. At resonance, maximum flywheel current flows within the parallel circuit made up of the capacitor and the primary of the transformer (L), which is known as a *tuned transformer.* With maximum flywheel current, there will be a maximum magnetic field, which means that there will be maximum power transfer between primary and secondary. So nearly all the input will be coupled to the output (coupling coefficient $k = 1$) and appear across the load at and around a small band of frequencies centered on resonance.

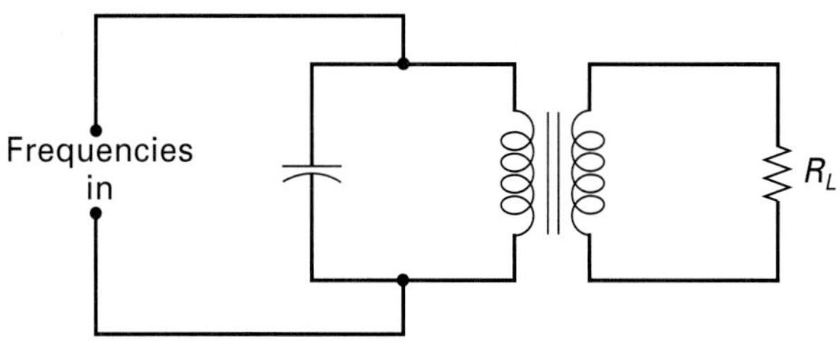

FIGURE 18-25 **Parallel Resonant Bandpass Circuit Using a Transformer.**

Above and below resonance, current within the parallel resonant circuit will be smaller. So the power transfer ability will be less, effectively keeping the frequencies outside the bandpass from appearing at the output.

Band-Stop Filter
A filter that attenuates alternating currents whose frequencies are between given upper and lower cutoff values while passing frequencies above and below this band.

18-4-4 Band-Stop Filter

Figure 18-26(a) illustrates a series resonant and Figure 18-26(b) a parallel resonant **band-stop filter.** Figure 18-26(c) shows the frequency response curve produced by a band-stop filter. The band-stop filter operates exactly the opposite to

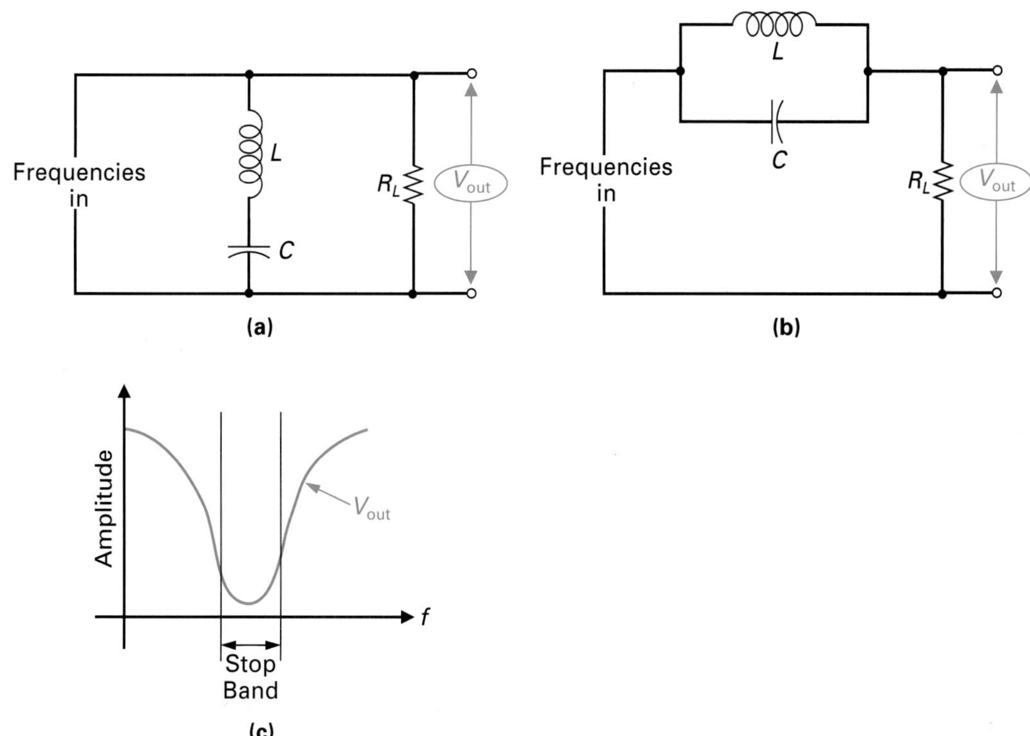

FIGURE 18-26 **Band-Stop Filter. (a) Series Resonant Band-Stop Filter. (b) Parallel Resonant Band-Stop Filter. (c) Frequency Response.**

FIGURE 18-27 Tuning in of Station by Use of a Bandpass Filter.

a bandpass filter in that it blocks or attenuates a band of frequencies centered on the resonant frequency of the *LC* circuit.

In the series resonant circuit in Figure 18-26(a), the *LC* impedance is very low at and around resonance, so these frequencies are rejected or shunted away from the output. Above and below resonance, the series circuit has a very high impedance, which results in almost no shunting of the signal away from the output.

In the parallel resonant circuit, in Figure 18-26(b), the *LC* circuit is in series with the load and output. At resonance, the impedance of a parallel resonant circuit will be very high, and the band of frequencies centered around resonance will be blocked. Above and below resonance, the impedance of the tank is very low, so nearly all the input is developed across the output.

Filters are necessary in applications such as television or radio, where we need to tune in (select or pass) one frequency that contains the information we desire, yet block all the millions of other frequencies that are also carrying information, as shown in Figure 18-27.

SELF-TEST REVIEW QUESTIONS FOR SECTION 18-4

1. Of the four types of filters, which:
   - **a.** Would utilize the inductor as a shunt?
   - **b.** Would utilize the capacitor as a shunt?
   - **c.** Would use a series resonant circuit as a shunt?
   - **d.** Would use a parallel resonant circuit as a shunt?
2. In what applications can filters be found?

18-5 COMPLEX NUMBERS

After reading this section you will realize that there is really nothing complex about **complex numbers.** The complex number system allows us to determine the *magnitude* and *phase angle* of electrical quantities by adding, subtracting, multiplying, and dividing phasor quantities, and is an invaluable tool in ac circuit analysis.

Complex Numbers
Numbers composed of a real number part and an imaginary number part.

18-5-1 The Real Number Line

Real numbers can be represented on a horizontal line, known as the real number line, as in Figure 18-28. Referring to this line, you can see that positive numbers exist to the right of the center point corresponding to zero, while negative numbers exist to the left. This representation satisfied most mathematicians for a short time, as they could indicate numbers such as 2 or 5 as a point on the line. Numbers corresponding to the $\sqrt{9}$ could also be represented, as three points to the right of zero ($\sqrt{9} = +3$). However, a problem was reached if they wished to indicate a point corresponding to $\sqrt{-9}$. The −9 is not + 3 [since $(+3) \times (+3) = +9$], and it is not −3 [since $(-3) \times (-3) = +9$]. So it was eventually realized that the square root of a negative number could not be indicated on the real number line, as it is not a real number.

Real Number
Numbers that have no imaginary parts.

18-5-2 The Imaginary Number Line

Mathematicians decided to call the square root of a negative number, such as $\sqrt{-4}$ or $\sqrt{-9}$, **imaginary numbers,** which are not fictitious or imaginary, but simply a particular type of number.

Imaginary Number
A complex number whose imaginary part is not zero.

Just as real numbers can be represented on a real number line, imaginary numbers can be represented on an imaginary number line, as shown in Figure 18-29. The imaginary number line is vertical, to distinguish it from the real number line, and when working with electrical quantities a $\pm j$ prefix, known as the ***j* operator**, is used for values that appear on the imaginary number line.

***j* Operator**
A prefix used to indicate an imaginary number.

18-5-3 The Complex Plane

A complex number is the combination of a real and imaginary number and is represented on a two-dimensional plane called the **complex plane,** shown in Figure 18-30. Generally, the real number appears first, followed by the imaginary number. Here are some examples of complex numbers.

Complex Plane
A plane whose points are identified by means of complex numbers.

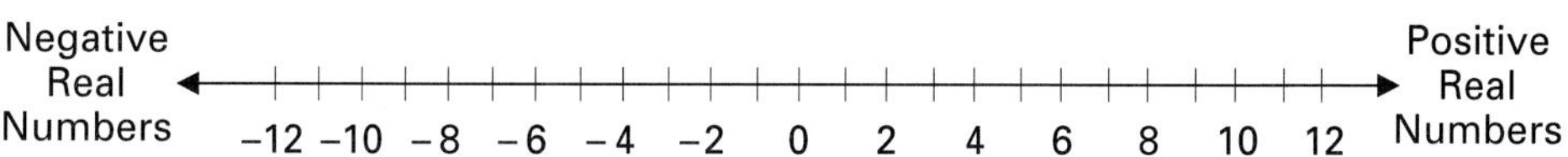

FIGURE 18-28 Real Number Line.

FIGURE 18-29 Imaginary Number Line.

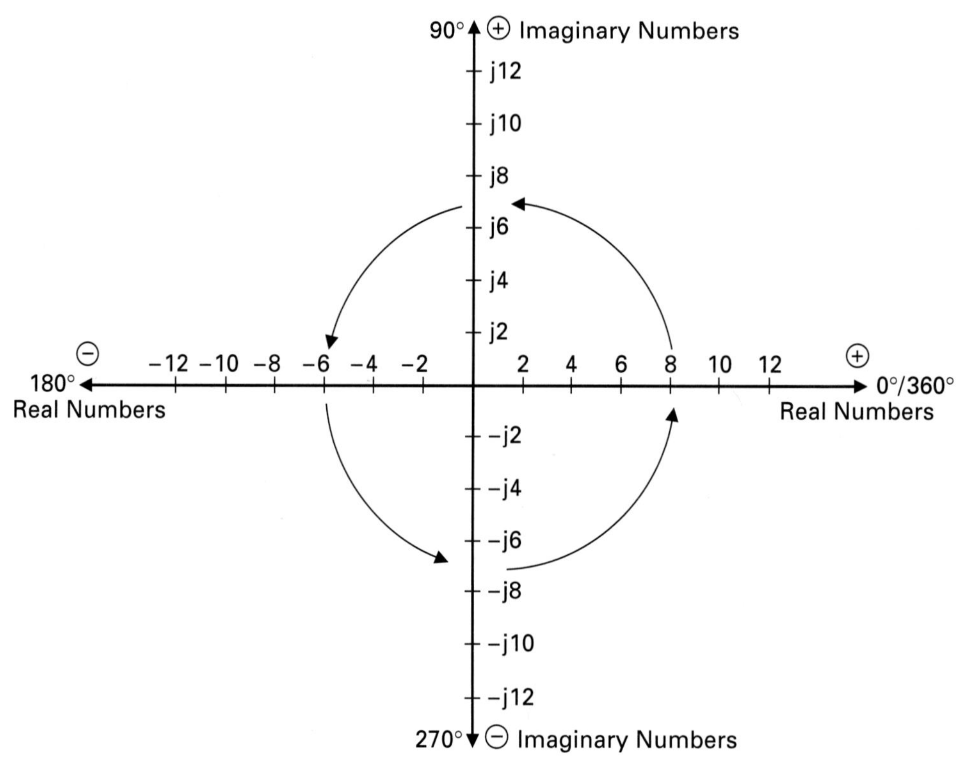

FIGURE 18-30 Complex Plane.

REAL NUMBERS	IMAGINARY NUMBERS
3	$+j4$
−2	$+j4$
−3	$-j2$

Complex numbers, therefore, are merely terms that need to be added as phasors, and all you have to do basically is draw a vector representing the real number and then draw another vector representing the imaginary number.

EXAMPLE:

Find the points in the complex plane in Figure 18-31 that correspond to the following complex numbers.

$$W = 3 + j4$$
$$X = 5 - j7$$
$$Y = -4 + j6$$
$$Z = -3 - j5$$

Solution:

By first locating the point corresponding to the real number on the horizontal line and then plotting it against the imaginary number on the vertical line, the points can be determined as shown in Figure 18-31.

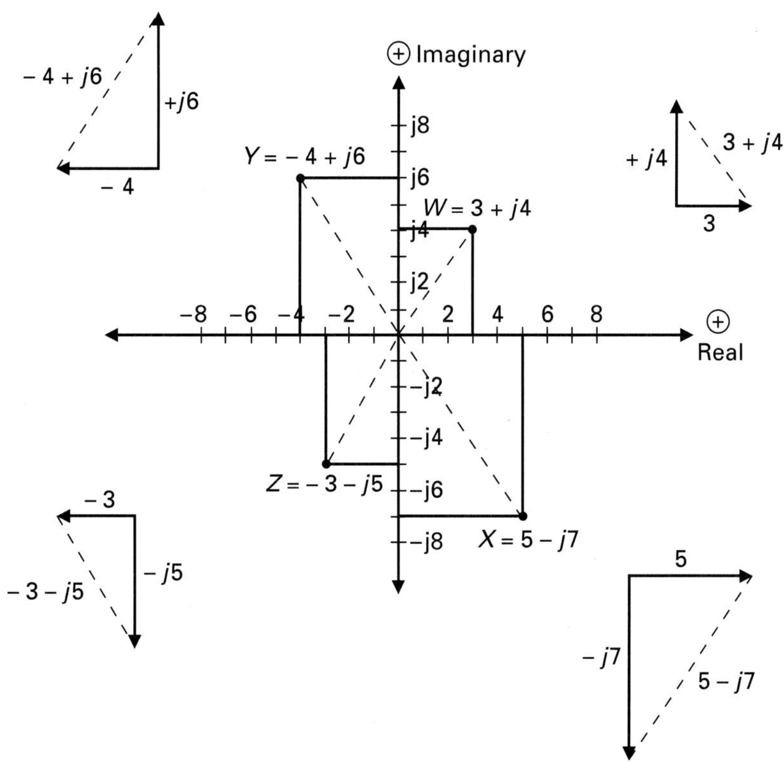

FIGURE 18-31 **Complex Numbers Examples.**

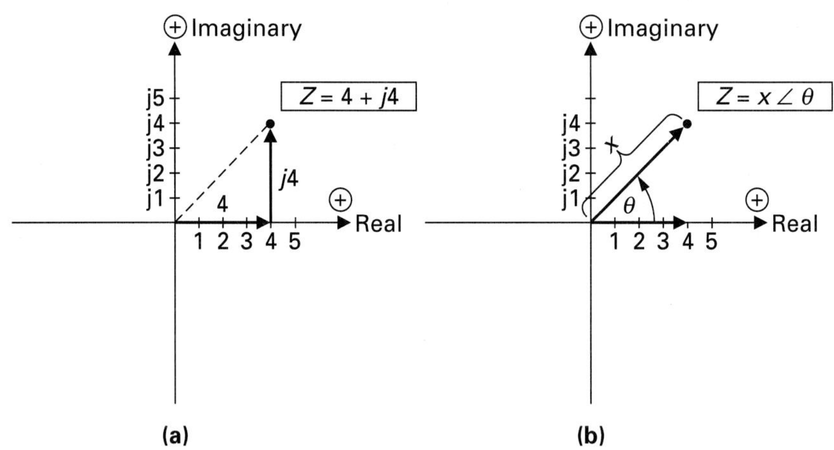

FIGURE 18-32 **Representing Phasors. (a) Rectangular Notation. (b) Polar Notation.**

Rectangular Coordinates A Cartesian coordinate of a Cartesian coordinate system whose straight-line axes or coordinate planes are perpendicular.

Polar Coordinates Either of two numbers that locate a point in a plane by its distance from a fixed point on a line and the angle this line makes with a fixed line.

A number like 3 + *j*4 specifies two phasors in **rectangular coordinates,** so this system is the *rectangular representation of a complex number.* There are several other ways to describe a complex number, one of which is the polar representation of a complex number, using **polar coordinates,** which will be discussed next.

18-5-4 Polar Complex Numbers

Phasors can also be expressed in polar form, as shown in Figure 18-32, which compares rectangular and polar notation. With the rectangular notation in Figure 18-32(a), the horizontal coordinate is the real part and the vertical coordinate is the imaginary part of the complex number. With the polar notation shown in Figure 18-32(b), the magnitude of the phasor (*x*, or size) and the angle ($\angle \theta$, meaning "angle theta") relative to the positive real axis (measured in a counterclockwise direction) are stated.

EXAMPLE:

Sketch the following polar numbers:

a. $5 \angle 60°$
b. $3 \angle 220°$

Solution:

As you can see in Figure 18-33, an equivalent negative angle, which is calculated by subtracting the given positive angle from 360°, can also be used.

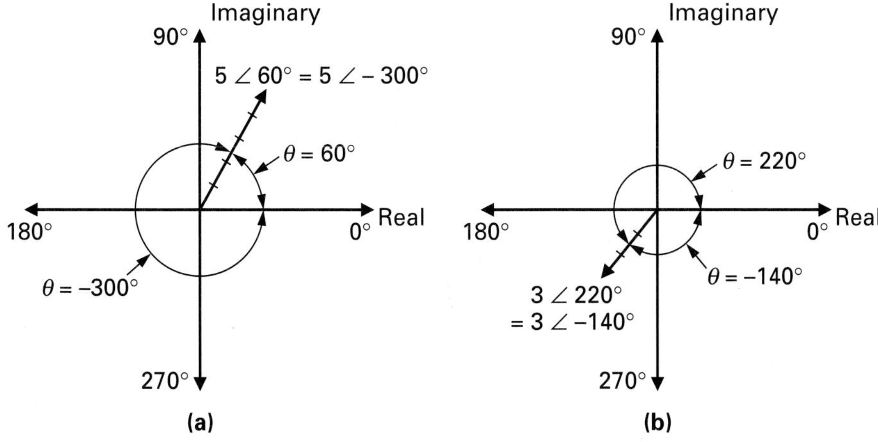

FIGURE 18-33 **Polar Number Examples.**

18-5-5 Rectangular/Polar Conversions

Many scientific calculators have a feature that allows you to convert rectangular numbers to polar numbers, and vice versa. These conversions are based on the Pythagorean theorem and trigonometric functions, discussed previously in a mini-math review.

Polar-to-Rectangular Conversion

The polar notation states the magnitude and angle, as shown in Figure 18-34(a). The following examples show how this conversion can be achieved.

EXAMPLE:

Convert the following polar numbers to rectangular form:

a. $5 \angle 30°$ b. $18 \angle -35°$ c. $44 \angle 220°$

Solution:

a. Real number = $5 \cos 30° = 4.33$
Imaginary number = $5 \sin 30° = j2.5$
Polar number, $5 \angle 30°$ = rectangular number, $4.33 + j2.5$

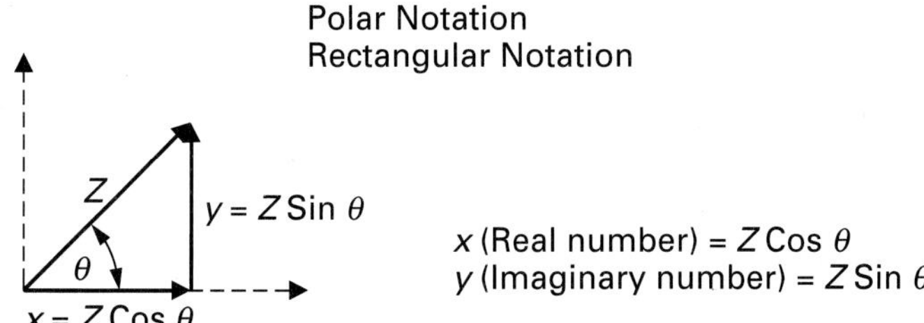

FIGURE 18-34 **Polar-to-Rectangular Conversion.**

b. Real number = 18 cos (–35°) = 14.74
Imaginary number = 18 sin (–35°) = $-j10.32$
Polar number, 18 $\angle -35°$ = rectangular number, $14.74 - j10.32$

c. Real number = 44 cos 220° = –33.7
Imaginary number = 44 sin 220° = $-j28.3$
Polar number, 44 $\angle$ 220° = rectangular number, $-33.7 - j28.3$

Rectangular-to-Polar Conversion

The rectangular notation states the horizontal (real) and vertical (imaginary) sides of a triangle, as shown in Figure 18-35. The following examples show how the conversion can be achieved.

EXAMPLE:

Convert the following rectangular numbers to polar form:

a. $4 + j3$ b. $16 - j14$

Solution:

a. Magnitude = $\sqrt{4^2 + 3^2} = 5$
Angle = arctan (3/4) = 36.9°
Rectangular number, $4 + j3$ = polar number, 5 $\angle$ 36.9°

b. Magnitude = $\sqrt{16^2 + (-14)^2} = 21.3$
Angle = arctan (–14/16) = – 41.2°
Rectangular number, $16 - j14$ = polar number, 21.3 $\angle -41.2°$

18-5-6 Complex Number Arithmetic

Since a phase difference exists between real and imaginary (j) numbers, certain rules should be applied when adding, subtracting, multiplying, or dividing complex numbers.

Addition

The sum of two complex numbers is equal to the sum of their real and imaginary parts.

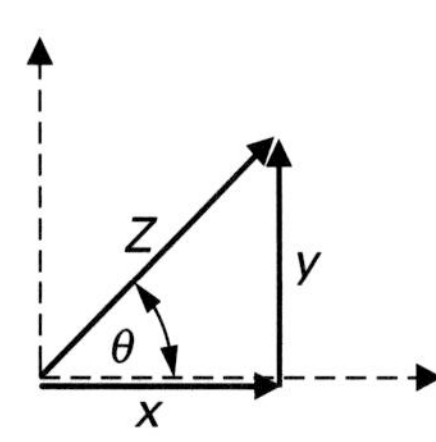

FIGURE 18-35 Rectangular-to-Polar Conversion.

EXAMPLE:

Add the following complex numbers:

a. $(3 + j4) + (2 + j5)$
b. $(4 + j5) + (2 - j3)$

Solution:

a. $(3 + j4) + (2 + j5) = (3 + 2) + (j4 + j5) = 5 + j9$
b. $(4 + j5) + (2 - j3) = (4 + 2) + (j5 - j3) = 6 + j2$

Subtraction

The difference of two complex numbers is equal to the difference between the separate real and imaginary parts.

EXAMPLE:

Subtract the following complex numbers:

a. $(4 + j3) - (2 + j2)$
b. $(12 + j6) - (6 - j3)$

Solution:

a. $(4 + j3) - (2 + j2) = (4 - 2) + j(3 - 2) = 2 + j1$
b. $(12 + j6) - (6 - j3) = (12 - 6) + j[6 - (-3)] = 6 + j9$

Multiplication

Multiplication of two complex numbers is achieved more easily if they are in polar form. The simple rule to remember is to multiply the magnitudes and then add the angles algebraically.

EXAMPLE:

Multiply the following complex numbers:

a. $5 \angle 35° \times 7 \angle 70°$
b. $4 \angle 53° \times 12 \angle -44°$

Solution:

a. Multiply the magnitudes: $5 \times 7 = 35$.
Algebraically add the angles: $\angle (35° + 70°) = \angle 105° = 35 \angle 105°$.
b. Multiply the magnitudes: $4 \times 12 = 48$.
Algebraically add the angles: $\angle [53° + (-44°)] = \angle 9° = 48 \angle 9°$.

Division

Division is also more easily carried out in polar form. The rule to remember is to divide the magnitudes, and then subtract the denominator angle from the numerator angle.

EXAMPLE:

Divide the following complex numbers:

a. $60 \angle 30°$ by $30 \angle 15°$ b. $100 \angle 20°$ by $5 \angle -7°$

Solution:

a. Divide the magnitudes: $60/30 = 2$.
Subtract the denominator angle from the numerator angle: $\angle (30° - 15°) = \angle 15° = 2 \angle 15°$.

b. Divide the magnitudes: $100/5 = 20$.
Subtract the angles: $\angle [20° - (-7°)] = \angle 27° = 20 \angle 27°$.

18-5-7 *How Complex Numbers Apply to AC Circuits*

Complex numbers find an excellent application in ac circuits due to all the phase differences that occur between different electrical quantities, such as X_L, X_C, R, and Z as shown in Figure 18-36. The positive real number line, at an angle of 0°, is used for resistance, which in this example is 3 Ω, as shown in Figure 18-36(a) and (b).

On the positive imaginary number line, at an angle of 90° ($+j$), inductive reactance (X_L) is represented, which in this example is $j4$ Ω ($X_L = 4$ Ω). The voltage drop across an inductor (V_L) is proportional to its inductive reactance (X_L), and both are represented on the $+j$ imaginary number line, since the voltage drop across an inductor will always lead the current (which in a series circuit is always in phase with the resistance) by 90°.

On the negative imaginary number line, at an angle of −90° or 270° ($-j$), capacitive reactance (X_C) is represented, which in this example is $-j2$ ($X_C = 2$ Ω). The voltage drop across a capacitor (V_C) is proportional to its capacitive reactance (X_C), and both are represented on the $-j$ imaginary number line since the voltage drop across a capacitor always lags current (charge and discharge) by −90°.

Series AC Circuits

Referring to Figure 18-36(a) and (b) once again, we can calculate total impedance simply by adding the phasors.

Z_T (Rectangular). Total series impedance is equal to the sum of all the resistances and reactances:

$$\begin{aligned} Z &= R + (jX_L \sim jX_C) \\ &= 3 + (+j4 \sim -j2) \\ &= 3 + j2 \end{aligned}$$

Z_T (Polar). The total series impedance can be converted from rectangular to polar form:

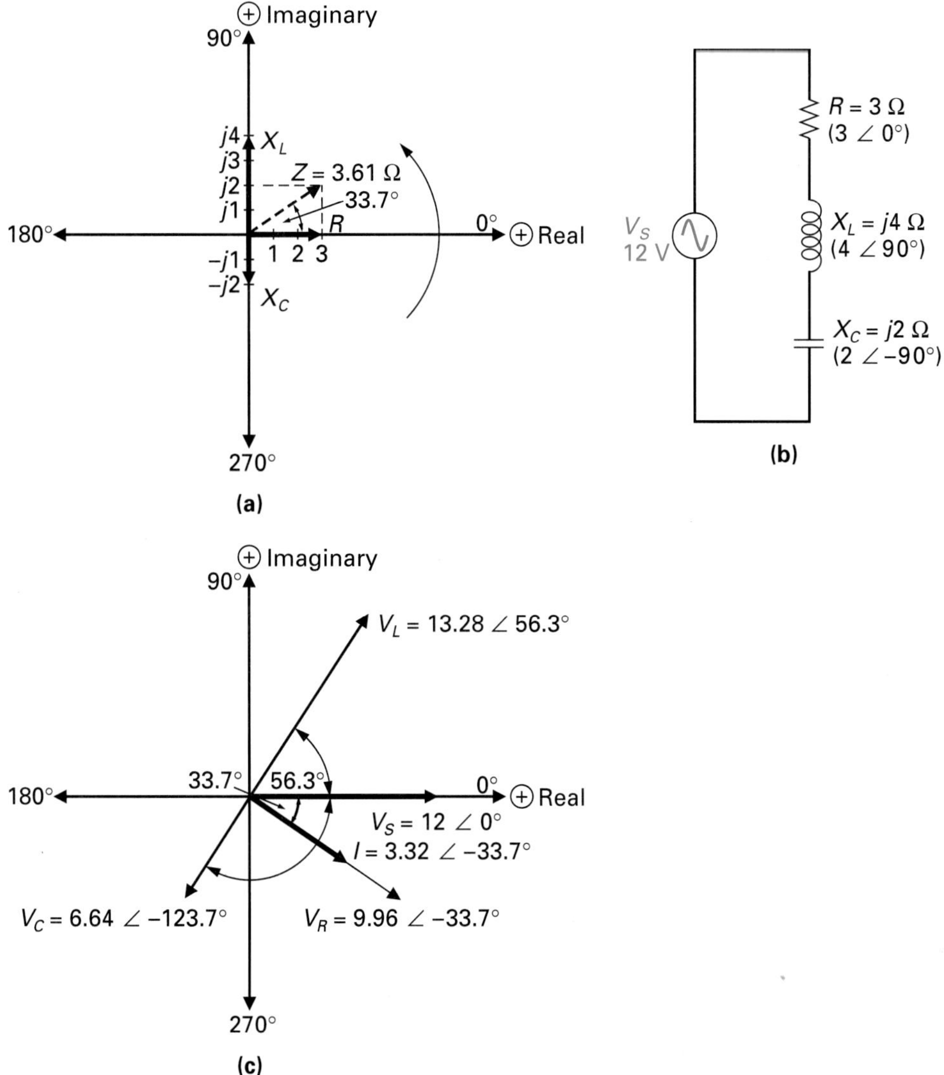

FIGURE 18-36 **Applying Complex Numbers to Series AC Circuits. (a) Impedance Phasors. (b) Series Circuit. (c) Voltage and Current Phasors.**

$$\text{magnitude} = \sqrt{3^2 + 2^2} = 3.61\ \Omega$$

$$\text{angle} = \arctan\frac{2}{3} = 33.7^\circ$$

$$= 3.61 \angle 33.7^\circ$$

Current. Once the magnitude of Z_T is known (3.61 Ω), I can be calculated. The source voltage of 12 V is a real positive number (0°) and is therefore represented as 12 ∠ 0°. Current is equal to

$$I = \frac{V_s}{Z_T} = \frac{12 \angle 0^\circ}{3.61 \angle 33.7^\circ}$$

Polar division Divide the magnitudes: $\frac{12}{3.61} = 3.32$ A

Subtract the angles: $0° - 33.7° = -33.7°$

$I = 3.32 \angle -33.7°$

Phase angle. The circuit current has an angle of –33.7°, which means that it lags V_T (inductive circuit, therefore ELI). This negative phase angle is expected since this series circuit is inductive ($X_L > X_C$), in which case current should lag voltage by some phase angle. This phase angle is less than 45° because the net reactance is less than the circuit resistance, and is shown in Figure 18-36(c).

Voltage drops. The component voltage drops are calculated with the following formulas and shown in Figure 18-36(c).

$V_R = I \times R = (3.32 \angle -33.7°) \times (3 \angle 0°)$

Multiply the magnitudes: $3.32 \times 3 = 9.96$ V

Algebraically add the angles: $\angle (-33.7° + 0°) = -33.7°$

$V_R = 9.96 \angle -33.7°$

$V_L = I \times X_L = (3.32 \angle -33.7°) \times (4 \angle 90°)$

Multiply the magnitudes: $3.32 \times 4 = 13.28$ V

Algebraically add the angles: $\angle (-33.7 + 90°) = 56.3°$

$V_L = 13.28 \angle 56.3°$

$V_C = I \times X_C = (3.32 \angle -33.7°) \times (2 \angle -90°)$

Multiply the magnitudes: $3.32 \times 2 = 6.64$ V

Algebraically add the angles: $\angle (-33.7 + -90°) = -123.7°$

$V_C = 6.64 \angle -123.7°$

Phase relationships. As shown in Figure 18-36(c), the voltage across an inductor leads the circuit current by +90°, while the voltage across a capacitor lags the circuit current by –90°. The source voltage acts as the zero reference phase and leads the circuit current and the voltage across the resistor (V_R) by 33.7°.

Source voltage. Although the source voltage is known, it can be checked to verify all the previous calculations, since the sum of all the individual voltage drops should equal the source voltage.

POLAR	⟶ RECTANGULAR
$V_L = 13.28\angle 56.3$	$= 7.37 + j11.05$
$V_R = 9.96\angle -33.7°$	$= 8.29 - j5.53$
$V_C = 6.64\angle -123.7°$	$= -3.68 - j5.52$
	$11.98 + j0$

Series–Parallel AC Circuits

Figure 18-37(a) illustrates a series–parallel ac circuit containing an R_L branch, an R_C branch, and an RLC branch.

Impedance of each branch. The three branches will each have a value of impedance that will be equal to:

RECTANGULAR ⟶ POLAR
$Z_1 = 10 + j5 \ = 11.2 \angle 26.6° \ \Omega$
$Z_2 = 25 - j15 = 29.2 \angle -31.0° \ \Omega$

The third branch is capacitive since the difference between $-j30$ (X_{C2}) and $+j10$ (X_{L2}) is $-j20$.

$$Z_3 = 20 - j20 = 28.3 \angle -45° \ \Omega$$

Branch currents. The three branch currents, I_1, I_2, and I_3, are calculated by dividing the source voltage (V_s) by the individual branch impedances.

$$I_1 = \frac{V_S}{Z_1} = \frac{30 \angle 0°}{11.2 \angle 26.6°}$$

Divide magnitudes: $30 \div 11.2 = 2.68$

Subtract angles: $\angle (0° - 26.6°) = -26.6°$

$I_1 = 2.68 \angle -26.6° = 2.4 - j1.2\text{A}$

$$I_2 = \frac{V_S}{Z_2} = \frac{30 \angle 0°}{29.2 \angle -31°} = 1.03 \angle +31° = 0.88 + j0.5 \text{ A}$$

$$I_3 = \frac{V_S}{Z_3} = \frac{30 \angle 0°}{28.3 \angle -45°} = 1.06 \angle +45° = 0.75 + j0.7 \text{ A}$$

Total current.

$$\begin{aligned} I_T &= I_1 + I_2 + I_3 \\ &= (2.4 - j1.2) + (0.88 + j0.5) + (0.75 + j0.7) \\ &= (2.4 + 0.88 + 0.75) + [-j1.2 + (+j0.5) + (+j0.7)] \\ &= 4.03 \text{ A} \end{aligned}$$

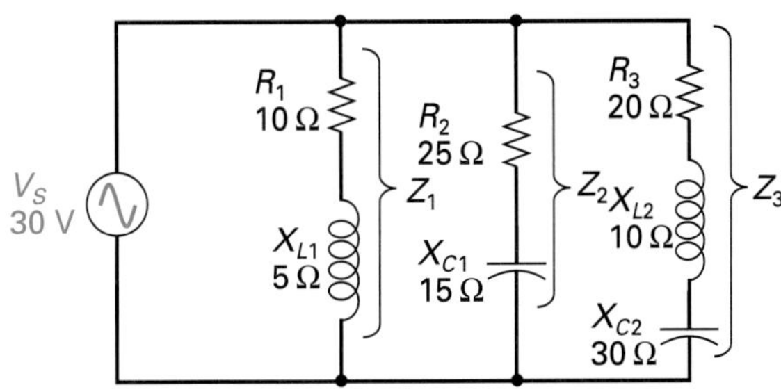

FIGURE 18-37 **Applying Complex Numbers to Series–Parallel AC.**

In polar form, this will equal 4.03 ∠ 0° A.

Total impedance.

$$Z_T = \frac{V_S}{I_T} = \frac{30 \angle 0°}{4.03 \angle 0°} = 7.44 \angle 0° \ \Omega$$

POLAR ⟶ RECTANGULAR
$7.44 \angle 0° = 7.44 + j0$

The complex ac circuit seen in Figure 18-37 is therefore equivalent to a 7.44-Ω resistor in series with no reactance.

SELF-TEST REVIEW QUESTIONS FOR SECTION 18-5

1. In complex numbers, resistance is a/an ____________ term and reactance is a/an ____________ term. (imaginary/real)
2. Convert the following rectangular number to polar form: $5 + j6$.
3. Convert the following polar number to rectangular form: 33 ∠ 25°.
4. What is a complex number?

SUMMARY

Resistive, Inductive, and Capacitive (*RLC*) Circuits (Figure 18-38)

1. Resistors, as we have discovered, operate and react to voltage and current in a very straightforward way; the voltage across a resistor is in phase with the resistor current.
2. Inductors and capacitors operate in essentially the same way, in that they both store energy and then return it back to the circuit. However, they have completely opposite reactions to voltage and current.
3. The phrase "ELI the ICE man" states that voltage (symbolized E) leads current (I) in an inductive (L) circuit (abbreviated by the word "ELI"), while current (I) leads voltage (E) in a capacitive (C) circuit (abbreviated by the word "ICE").

Series *RLC* Circuit

4. The circuit current is always the same throughout a series circuit and can therefore be used as a reference.
5. The voltage across a resistor is always in phase with the current, while the voltage across the inductor leads the current by 90° and the voltage across the capacitor lags the current by 90°.
6. Impedance is the total opposition to current flow and is a combination of both reactance (X_L, X_C) and resistance (R).
7. The capacitive and inductive reactances are 180° out of phase with one another and counteract.

FIGURE 18-38 *RLC* Circuits.

- **Series *RLC* Circuits**

$$X_C = \frac{1}{2\pi fC}$$

$$X_L = 2\pi fL$$

$$Z = \sqrt{R^2 + (X_L \sim X_C)^2}$$

$$I = \frac{V_s}{Z}$$

$$V_R = I \times R \quad V_L = I \times X_L \quad V_C = I \times X_C$$

$$V_s = \sqrt{V_R^2 + (V_L \sim V_C)^2}$$

- **Parallel *RLC* Circuits**

$$I_R = \frac{V}{R} \quad I_L = \frac{V}{X_L} \quad I_C = \frac{V}{X_C}$$

$$I_T = \sqrt{I_R^2 + I_X^2}$$

$$Z = \frac{V}{I_T}$$

- **Resonance ($X_L = X_C$)**

$$f_0 = \frac{1}{2\pi\sqrt{LC}}, \ f_0 = \text{resonant frequency}$$

$$Z = R$$

$$\text{BW} = \frac{f_0}{Q}, \ \text{BW} = \text{bandwidth}$$

- **Series Resonance**

$$Q = \frac{X_L}{R} = \frac{X_C}{R} = \frac{V_L}{V_R} = \frac{V_C}{V_R} = \frac{V_L}{V_s} = \frac{V_C}{V_s}$$

- **Parallel Resonance**

$$Q = \frac{I_{\text{tank}}}{I_s} = \frac{Z_{\text{tank}}}{X_L}$$

- **Power**

Apparent power = $V_s \times I$ (volt-amperes)

True power = $I^2 \times R$ (watts)

$$\text{PF} = \cos\theta = \frac{R}{Z} = \frac{P_R}{P_A}$$

- **Complex Numbers**

Polar-to-rectangular: Real number = magnitude × cos θ
Imaginary number = magnitude × sin θ

Rectangular-to-polar:

Magnitude = $\sqrt{(\text{real number})^2 + (\text{imaginary number})^2}$

Angle = $\arctan\left(\frac{\text{imaginary number}}{\text{real number}}\right)$

Addition = sum of the real and sum of the imaginary parts (rectangular form)

Subtraction = difference between the real and the difference between imaginary parts (rectangular form)

Multiplication = multiply the magnitudes and then add the angles algebraically (polar form)

Division = divide the magnitudes and then subtract the denominator angle from the numerator angle (polar form)

FIGURE 18-38 (*continued*)

Series *RLC* Circuits

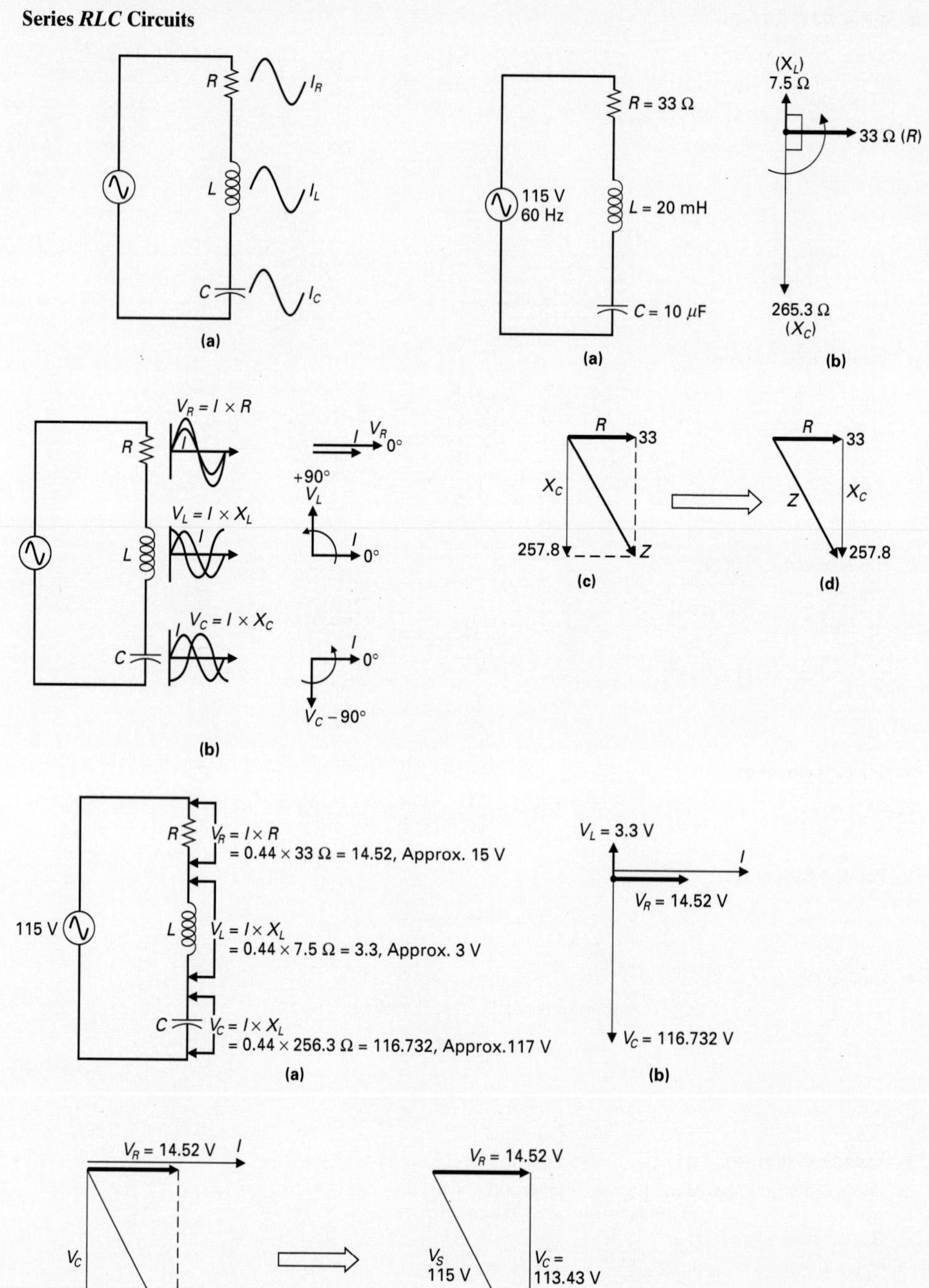

FIGURE 18-38 (*continued*)

Parallel *RLC* Circuits

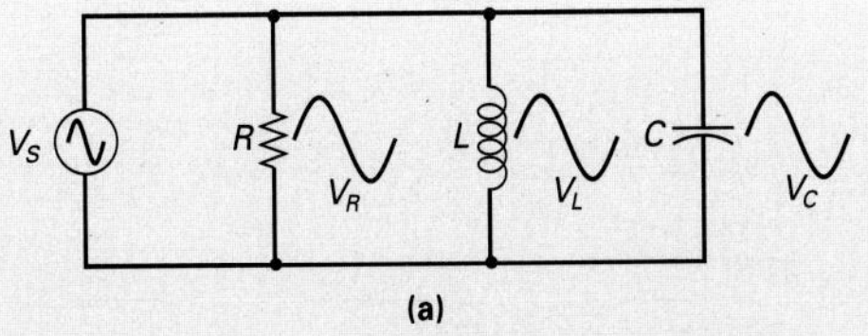

(a)

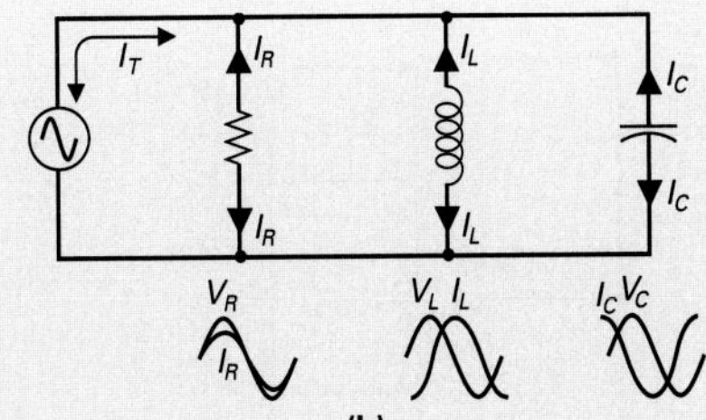

(b)

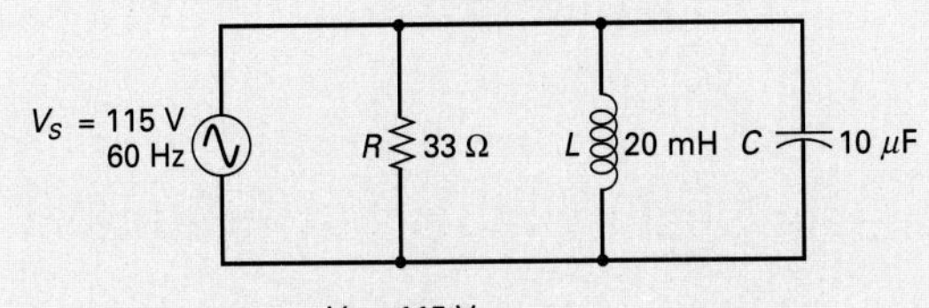

$$I_R = \frac{V}{R} = \frac{115\text{ V}}{33\ \Omega} = 3.5\text{ A}$$

$$I_L = \frac{V}{X_L} = \frac{115\text{ V}}{7.5\ \Omega} = 15.3\text{ A}$$

$$I_C = \frac{V}{X_C} = \frac{115\text{ V}}{265.3\ \Omega} = 0.43\text{ A}$$

$$X_L = 2\pi fL = 2\pi \times 60 \times 20\text{ mH} = 7.5\ \Omega$$

$$X_C = \frac{1}{2\pi fL} = \frac{1}{2\pi \times 60 \times 10\ \mu\text{F}} = 265.3\ \Omega$$

(a)

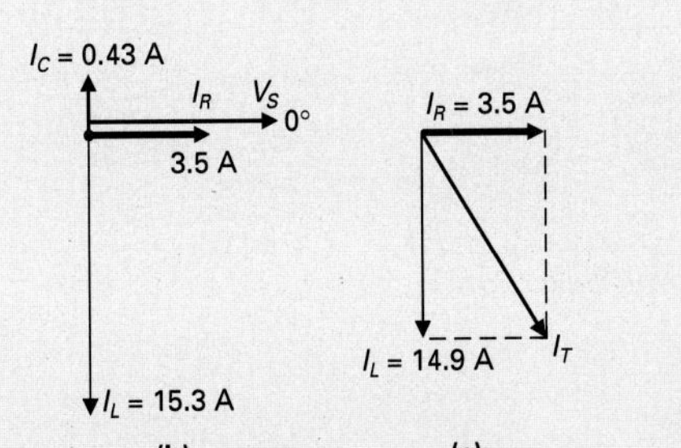

(b)

I_R = 3.5 A
I_L = 14.9 A
I_T

(c)

I_R = 3.5 A
I_T = 15.3 A
I_L = 14.9 A

(d)

8. Impedance is the vector sum of the reactive (X) and resistance (R) vectors.
9. Since none of these voltages are in phase with one another in a series *RLC* circuit, they must be added vectorially to obtain the applied voltage.

Parallel *RLC* Circuit

10. The voltage across any parallel circuit will be equal and in phase.
11. With current, we must first calculate the individual branch currents (I_R, I_L, and I_C) and then calculate the total circuit current (I_T).
12. There is a phase difference between the source voltage (V_S) and the circuit current (I_T).

FIGURE 18-38 ***(continued)***

Series Resonance

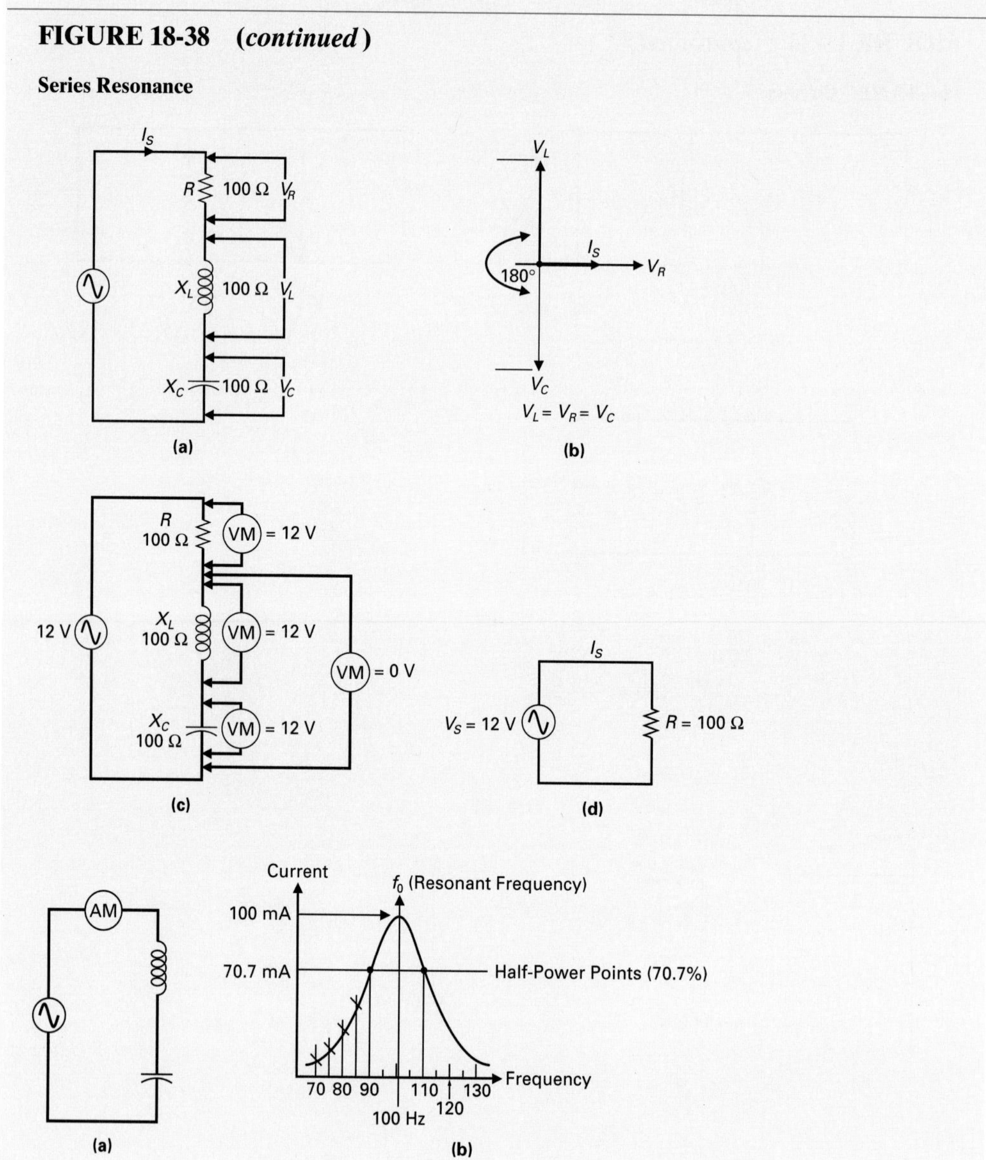

13. Once the total current (I_T) is known, the impedance of all three components in parallel can be calculated.

Resonance

14. Resonance is a circuit condition that occurs when the inductive reactance (X_L) and the capacitive reactance (X_C) have been balanced.
15. As the input ac frequency is increased, a point will be reached where X_L will equal X_C, and this condition is known as *resonance.* The frequency at which $X_L = X_C$ in either a parallel or a series LC circuit is known as the *resonant frequency* (f_0).

QUICK REFERENCE SUMMARY SHEET

FIGURE 18-38 ***(continued)***

Parallel Resonance

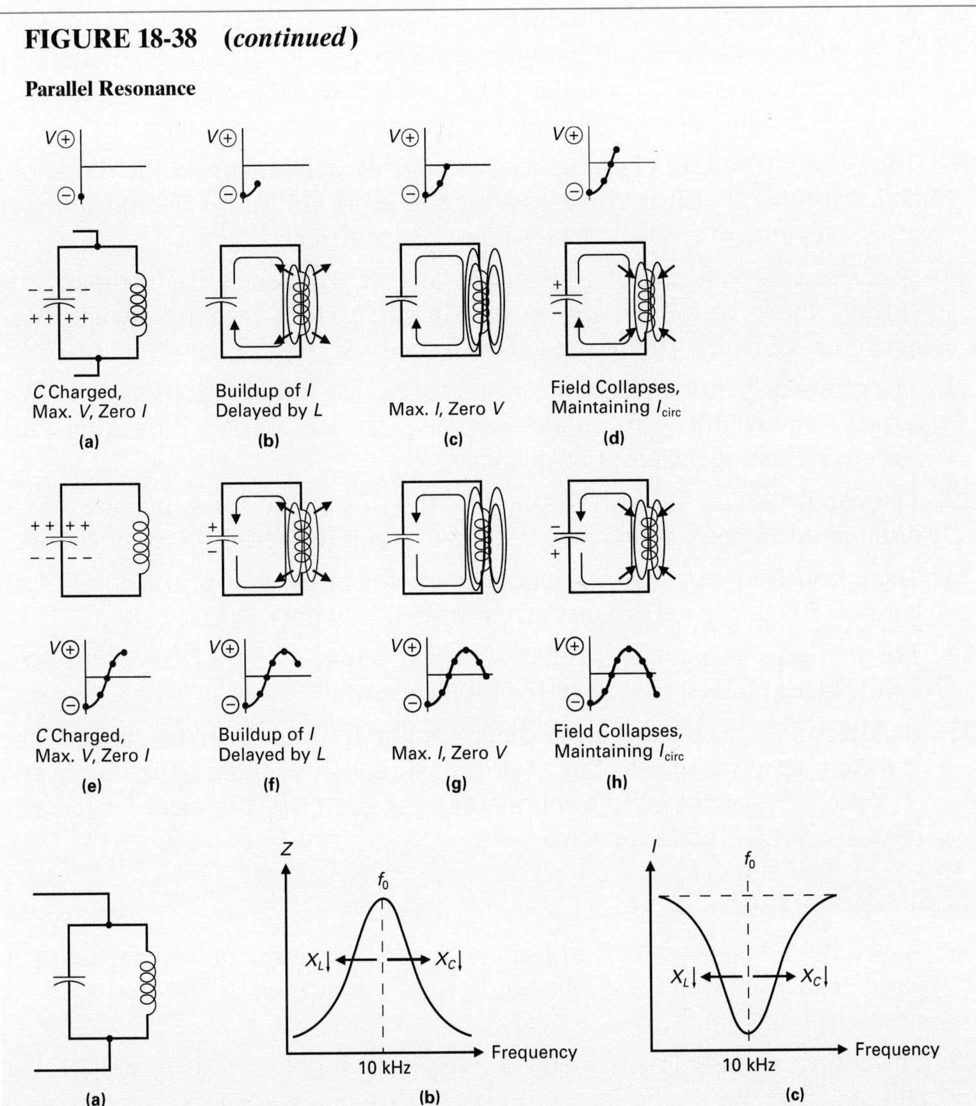

Series Resonance

16. A series resonant circuit is a resonant circuit in which the capacitor and coil are in series with the applied ac voltage.

17. In summary, we can say that in a series resonant circuit:

a. The inductor and capacitor electrically disappear due to their equal but opposite effect, resulting in 0 V drops across the series combination, and the circuit consequently seems purely resistive.

b. The current flow is large because the impedance of the circuit is low and equal to the series resistance (R), which has the source voltage developed across it.

c. The individual voltage drops across the inductor or capacitor can be larger than the source voltage if R is smaller than X_L and X_C.

18. The Q factor is a ratio of inductive reactance to resistance and is used to express how efficiently an inductor will store rather than dissipate energy. In a series resonant circuit, the Q factor indicates the quality of the series resonant circuit, or is the ratio of the reactance to the resistance.

19. The Q of a resonant circuit is almost entirely dependent on the inductor's coil resistance, because capacitors tend to have almost no resistance figure at all, only reactive, which makes them very efficient.

20. A series resonant circuit is selective in that frequencies at resonance or slightly above or below will cause a larger current than frequencies well above or below the circuit's resonant frequency.

21. The group or band of frequencies that causes the larger current is called the circuit's bandwidth. A frequency response curve is a graph indicating a circuit's response to different frequencies.

22. The bandwidth includes the group or band of frequencies that cause 70.7% or more of the maximum current to flow within the series resonant circuit.

23. The cutoff frequency is the frequency at which the gain of the circuit falls below 0.707 of the maximum current or half-power (– 3 dB).

24. The half-power point is a point at which power is 50%. This half-power point corresponds to 70.7% of the total current.

25. Bandwidth is proportional to the resonant frequency of the circuit and inversely proportional to the Q of the circuit. In summary, the bandwidth of a series resonant circuit will increase as the Q of the circuit decreases ($\text{BW}\uparrow = f_0/Q\downarrow$), and vice versa.

Parallel Resonance

26. A parallel resonant circuit is a circuit having an inductor and capacitor in parallel with one another, offering a high impedance at the frequency of resonance.

27. The source voltage (V_S) is needed initially to supply power to the LC circuit and start the oscillations; but once the oscillating process is in progress (assuming the ideal case), current is only flowing back and forth between inductor and capacitor, and no current is flowing from the source.

28. Flywheel action is a sustaining effect of oscillation in an LC circuit due to the charging and discharging of the capacitor and the expansion and contraction of the magnetic field around the inductor. The name is derived from the fact that it resembles a mechanical flywheel, which, once started, will keep going continually until friction reduces the magnitude of the rotations to zero.

29. The electronic equivalent of the mechanical flywheel is a resonant parallel-connected LC circuit.

30. The direction of the circulating current reverses each half-cycle at the frequency of resonance. Energy is stored in the capacitor in the form of an

electric field between the plates on one half-cycle, and then the capacitor discharges, supplying current to build up a magnetic field on the other half-cycle. The inductor stores its energy in the form of a magnetic field, which will collapse, supplying a current to charge the capacitor, which will then discharge, supplying a current back to the inductor, and so on. Due to the "storing action" of this circuit, it is sometimes related to the fluid analogy and referred to as a tank circuit.

31. Under ideal conditions a tank circuit should oscillate indefinitely if no losses occur within the circuit. In reality, the resistance of the coil reduces that 100% efficiency, as does friction with the mechanical flywheel.

32. As a small part of the energy is dissipated with each cycle, the oscillations will be reduced in size and eventually fall to zero.

33. If the ac source is reconnected to the tank, a small amount of current will flow from the source to the tank to top up the tank or replace the dissipated power. The higher the coil resistance is, the higher the loss and the larger the current flow from source to tank to replace the loss.

34. In the series resonant circuit, we were concerned with voltage drops since current remains the same throughout a series circuit. In a parallel resonant circuit, we are concerned with circuit currents rather than voltage.

35. Of all the three Q formulas for parallel resonant circuits, $Q = I_{\text{tank}}/I_S$, $Q = X_L/R$, or $Q = Z_{\text{tank}}/X_L$, the latter is the easiest to use as both X_L and the tank impedance can easily be determined in most cases where C, L, and R internal for the inductor are known.

36. The current versus frequency response curve is the complete opposite to the series resonant response curve. At frequencies below resonance X_L is low and X_C is high, and the inductor offers a low reactance, producing a high current path and low impedance. On the other hand, at frequencies above resonance the capacitor displays a low reactance, producing a high current path and low impedance.

37. The parallel resonant circuit is like the series resonant circuit in that it responds to a band of frequencies close to its resonant frequency.

38. Circuits containing inductance and capacitance are often referred to as tuned circuits since they can be adjusted to make the circuit responsive to a particular frequency (the resonant frequency).

39. Selectivity, by definition, is the ability of a tuned circuit to respond to a desired frequency and ignore all others.

40. Parallel resonant LC circuits are sometimes too selective, as the Q is too large, producing too narrow a bandwidth. In this situation, because of the very narrow response curve, a high resistance value can be placed in parallel with the LC circuit to provide an alternative path for line current. This process is known as *loading* or *damping* the tank and will cause an increase in line current and decrease in Q ($Q\downarrow = I_{\text{tank}}/I_{\text{line}}\uparrow$). The decrease in Q will cause a corresponding increase in BW ($\text{BW}\uparrow = f_0/Q\downarrow$).

RLC Filter Circuits (Figure 18-39)

41. There are basically four types of filters:

a. Low-pass filter, which passes frequencies below a cutoff frequency
b. High-pass filter, which passes frequencies above a cutoff frequency
c. Bandpass filter, which passes a band of frequencies
d. Band-stop filter, which stops a band of frequencies

42. An inductor and capacitor can be connected to act as a low-pass filter. At low frequencies, X_L has a small value compared to the load resistor (R_L), so nearly all the low-frequency input is developed and appears at the output across R_L. Since X_C is high at low frequencies, nearly all the current passes through R_L rather than C.

At high frequencies, X_L increases and drops more of the applied input across the inductor rather than the load. The capacitive reactance, X_C, aids this low output at high-frequency effect by decreasing its reactance and providing an alternative path for current to flow.

Since the inductor basically blocks alternating current and the capacitor shunts alternating current, the net result is to prevent high-frequency signals from reaching the load.

43. An inductor and capacitor can be connected to act as a high-pass filter. At high frequencies, the reactance of the capacitor (X_C) is low while the reactance of the inductor (X_L) is high, so all the high frequencies are easily passed by the capacitor and blocked by the inductor, so they all are routed through to the output and load.

At low frequencies, the reverse condition exists, resulting in a low X_L and a high X_C. The capacitor drops nearly all the input, and the inductor shunts the signal current away from the output load.

44. A bandpass filter circuit passes a group or band of frequencies between a lower and an upper cutoff frequency, while heavily attenuating any other frequency outside this band. At resonance, the series resonant *LC* circuit has a very low impedance and will consequently pass the resonant frequency to the load with very little drop across the *L* and *C* components.

45. Below resonance, X_C is high, and the capacitor drops a large amount of the input signal; above resonance, X_L is high and the inductor drops most of the input frequency voltage. This circuit will therefore pass a band of frequencies centered around the resonant frequency of the series *LC* circuit and block all other frequencies above and below this resonant frequency.

46. A parallel resonant *LC* circuit can be used to provide a bandpass response. The series resonant circuit was placed in series with the output, whereas the parallel resonant circuit will have to be placed in parallel with the output to provide the same results. At resonance, the parallel resonant circuit or tank has a high impedance, so very little current will be shunted away from the output; it will be passed on to the output, and almost all the input will appear at the output across the load.

Above resonance, X_C is small, so most of the input is shunted away from the output by the capacitor; below resonance, X_L is small, and the shunting action occurs again, but this time through the inductor.

FIGURE 18-39 *RLC* **Filter Circuits.**

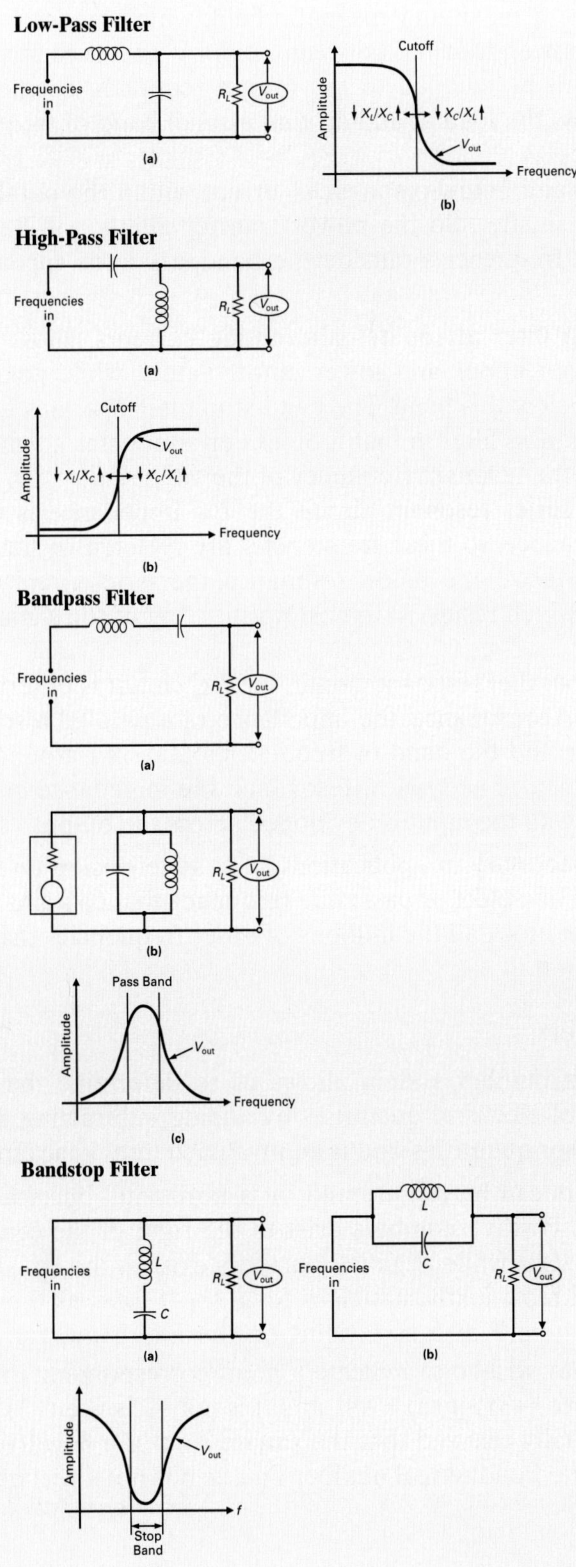

47. A transformer can be used to replace the inductor to produce a bandpass filter. At resonance, maximum flywheel current flows within the parallel circuit made up of the capacitor and the primary of the transformer (L), which is known as a *tuned transformer.* With maximum flywheel current, there will be a maximum magnetic field, which means that there will be maximum power transfer between primary and secondary. So nearly all the input will be coupled to the output (coupling coefficient $k = 1$) and appear across the load at and around a small band of frequencies centered on resonance.

    Above and below resonance, current within the parallel resonant circuit will be smaller. So the power transfer ability will be less, effectively keeping the frequencies outside the bandpass from appearing at the output.

48. A band-stop filter attenuates alternating currents whose frequencies are between given upper and lower cutoff values while passing frequencies above and below this band. The band-stop filter operates exactly the opposite to a bandpass filter in that it blocks or attenuates a band of frequencies centered on the resonant frequency of the LC circuit.

    In the series resonant circuit the LC impedance is very low at and around resonance, so these frequencies are rejected or shunted away from the output. Above and below resonance, the series circuit has a very high impedance, which results in almost no shunting of the signal away from the output.

    In the parallel resonant circuit, the LC circuit is in series with the load and output. At resonance, the impedance of a parallel resonant circuit will be very high, and the band of frequencies centered around resonance will be blocked. Above and below resonance, the impedance of the tank is very low, so nearly all the input is developed across the output.

49. Filters are necessary in applications such as television or radio, where we need to tune in (select or pass) one frequency that contains the information we desire, yet block all the millions of other frequencies that are also carrying information.

Complex Numbers

50. The complex number system allows us to determine the *magnitude* and *phase angle* of electrical quantities by adding, subtracting, multiplying, and dividing phasor quantities, and is an invaluable tool in ac circuit analysis.

51. Real numbers can be represented on a horizontal line, known as the real number line. Positive numbers exist to the right of the center point corresponding to zero, while negative numbers exist to the left. This representation satisfied most mathematicians for a short time, as they could indicate numbers such as 2 or 5 as a point on the line. However, a problem was reached if they wished to indicate a point corresponding to $\sqrt{-9}$. The $\sqrt{-9}$ is not +3 [since $(+3) \times (+3) = +9$], and it is not –3 [since $(-3) \times (-3) = +9$]. So it was eventually realized that the square root of a negative number could not be indicated on the real number line, as it is not a real number.

52. Mathematicians decided to call the square root of a negative number, such as $\sqrt{-4}$ or $\sqrt{-9}$, imaginary numbers, which are not fictitious or imaginary, but simply a particular type of number. Just as real numbers can be represented on a real number line, imaginary numbers can be represented on an imaginary number line. The imaginary number line is vertical, to distinguish it from the real number line, and when working with electrical quantities a $\pm j$ prefix, known as the *j operator*, is used for values that appear on the imaginary number line.

53. A complex number is the combination of a real and imaginary number and is represented on a two-dimensional plane called the complex plane. Generally, the real number appears first, followed by the imaginary number.

54. Complex numbers are merely terms that need to be added as phasors, and all you have to do basically is draw a vector representing the real number and then draw another vector representing the imaginary number.

55. A number like $3 + j4$ specifies two phasors in rectangular coordinates, so this system is the *rectangular representation of a complex number.*

56. There are several other ways to describe a complex number, one of which is the polar representation of a complex number, using polar coordinates.

57. With the rectangular notation the horizontal coordinate is the real part and the vertical coordinate is the imaginary part of the complex number. With the polar notation the magnitude of the phasor (x, or size) and the angle ($\angle\ \theta$, meaning "angle theta") relative to the positive real axis (measured in a counterclockwise direction) are stated.

58. Since a phase difference exists between real and imaginary (j) numbers, certain rules should be applied when adding, subtracting, multiplying, or dividing complex numbers.

59. The sum of two complex numbers is equal to the sum of their real and imaginary parts.

60. The difference of two complex numbers is equal to the difference between the separate real and imaginary parts.

61. Multiplication of two complex numbers is achieved more easily if they are in polar form. The simple rule to remember is to multiply the magnitudes and then add the angles algebraically.

62. Division is also more easily carried out in polar form. The rule to remember is to divide the magnitudes, and then subtract the denominator angle from the numerator angle.

63. Complex numbers find an excellent application in ac circuits due to all the phase differences that occur between different electrical quantities, such as X_L, X_C, R, and Z. The positive real number line, at an angle of 0°, is used for resistance.

64. On the positive imaginary number line, at an angle of 90° ($+j$), inductive reactance (X_L) is represented.

65. The voltage drop across an inductor (V_L) is proportional to its inductive reactance (X_L), and both are represented on the $+j$ imaginary number line,

since the voltage drop across an inductor will always lead the current (which in a series circuit is always in phase with the resistance) by 90°.

66. On the negative imaginary number line, at an angle of –90° or 270° (–*j*), capacitive reactance (X_C) is represented.

67. The voltage drop across a capacitor (V_C) is proportional to its capacitive reactance (X_C), and both are represented on the –*j* imaginary number line since the voltage drop across a capacitor always lags current (charge and discharge) by –90°.

NEW TERMS

Bandpass filter	Imaginary number	Resonant frequency
Band-stop filter	*j* operator	Selectivity
Bandwidth	Parallel resonance	Series resonance
Complex numbers	Polar coordinate	Tank
Complex plane	Real number	Tank circuit
Flywheel action	Rectangular coordinate	Tuned transformer
Frequency response curve	Resonance	

REVIEW QUESTIONS

Multiple-Choice Questions

1. Capacitive reactance is __________ to frequency and capacitance, while inductive reactance is __________ to frequency and inductance.
   a. Proportional, inversely proportional
   b. Inversely proportional, proportional
   c. Proportional, proportional
   d. Inversely proportional, inversely proportional

2. Resonance is a circuit condition that occurs when:
   a. V_L equals V_C **d.** Both (a) and (c)
   b. X_L equals X_C **e.** Both (a) and (b)
   c. L equals C

3. As frequency is increased, X_L will __________, while X_C will __________.
   a. Decrease, increase **c.** Remain the same, decrease
   b. Increase, decrease **d.** Increase, remain the same

4. In an *RLC* series resonant circuit, with $R = 500\ \Omega$ and $X_L = 250\ \Omega$, what would be the value of X_C?
   a. 2 Ω **c.** 250 Ω
   b. 125 Ω **d.** 500 Ω

5. At resonance, the voltage drop across both a series-connected inductor and a capacitor will equal:
   a. 70.7 V **c.** 10 V
   b. 50% of the source **d.** Zero

6. In a series resonant circuit the current flow is ____________, as the impedance is ____________ and equal to ____________.

   a. Large, small, R c. Large, small, X
   b. Small, large, X d. Small, large, R

7. A circuit's bandwidth includes a group or band of frequencies that cause ____________ or more of the maximum current, or more than ____________ of the maximum power to appear at the output.

   a. 110, 90 c. 70.7%, 50%
   b. 50%, 70.7% d. Both (a) and (c)

8. The bandwidth of a circuit is proportional to the:

   a. Frequency of resonance c. Tank current
   b. Q of the tank d. Two of the above

9. Series or parallel resonant circuits can be used to create:

   a. Low-pass filters c. Bandpass and band-stop filters
   b. Low-pass and high-pass filters d. All of the above

10. Flywheel action occurs in:

    a. A tank circuit d. Two of the above
    b. A parallel LC circuit e. None of the above
    c. A series LC circuit

11. $25 \angle 39°$ is an example of a complex number in:

    a. Polar form c. Algebraic form
    b. Rectangular form d. None of the above

12. $3 + j10$ is an example of a complex number in:

    a. Polar form c. Algebraic form
    b. Rectangular form d. None of the above

13. Which complex number form is usually more convenient for addition and subtraction?

    a. Rectangular b. Polar

14. Which complex number form is usually more convenient for multiplication and division?

    a. Rectangular b. Polar

15. In complex numbers, *resistance* is a real term, while *reactance* is a/an ____________.

    a. j term c. Value appearing on the vertical axis
    b. Imaginary term d. All of the above

Essay Questions

16. Illustrate with phasors and describe the current and voltage relationships in a series RLC circuit. (18-1)
17. Describe the procedure for the analysis of a series RLC circuit. (18-1)
18. Define resonance and give the formula for calculating the frequency of resonance. (18-3)

19. Describe the three unusual characteristics of a circuit that is at resonance. (18-3)

20. Define the following: (18-3)

    a. Flywheel action
    b. Quality figure
    c. Bandwidth
    d. Selectivity

21. Describe a frequency response curve. (18-4)

22. Illustrate with phasors and describe the current and voltage relationships in a parallel *RLC* circuit. (18-2)

23. Describe the differences between a series and a parallel resonant circuit. (18-3)

24. Explain how loading a tank affects bandwidth and selectivity. (18-3-2)

25. Illustrate the circuit and explain the operation of the following, with their corresponding response curves. (18-4)

    a. Low-pass filter
    b. High-pass filter
    c. Bandpass filter
    d. Band-stop filter

26. Describe why capacitive reactance is written as $-jX_C$ and inductive reactance is written as jX_L. (18-5)

27. How are capacitive and inductive reactances written in polar form? (18-5-4)

28. List the rules used to perform complex numbers: (18-5-6)

    a. Addition (rectangular)
    b. Subtraction (rectangular)
    c. Multiplication (polar)
    d. Division (polar)

29. Describe briefly how the real number and imaginary number lines are used for ac circuit analysis and what electrical phasors are represented at 0°, 90°, and –90°. (18-5-7)

30. Referring to Figure 18-36, describe why the series circuit current (I) is not in phase with the source voltage (V_S). (18-5-7)

Practice Problems

31. Calculate the values of capacitive and inductive reactance for the following when connected across a 60 Hz source:

    a. 0.02 μF
    b. 18 μF
    c. 360 pF
    d. 2700 nF
    e. 4 mH
    f. 8.18 H
    g. 150 mH
    h. 2H

32. If a 1.2 kΩ resistor, a 4 mH inductor, and an 8 μF capacitor are connected in series across a 120 V/60 Hz source, calculate:

    a. X_C
    b. X_L
    c. Z
    d. I
    e. V_R
    f. V_L
    g. V_C
    h. Apparent power
    i. True power
    j. Resonant frequency
    k. Circuit quality factor
    l. Bandwidth

33. If a 270 Ω resistor, a 150 mH inductor, and a 20 μF capacitor are all connected in parallel with one another across a 120 V/60 Hz source, calculate:

a. X_L f. I_T
b. X_C g. Z
c. I_R h. Resonant frequency
d. I_L i. Q factor
e. I_C j. Bandwidth

34. Calculate the impedance of a series circuit if $R = 750\ \Omega$, $X_L = 25\ \Omega$, and $X_C = 160\ \Omega$.

35. Calculate the impedance of a parallel circuit with the same values as those of Question 34 when a 1 V source voltage is applied.

36. State the following series circuit impedances in rectangular and polar form:

a. $R = 33\ \Omega, X_C = 24\ \Omega$ b. $R = 47\ \Omega, X_L = 17\ \Omega$

37. Convert the following impedances to rectangular form:

a. $25 \angle 37°$ c. $114 \angle -114°$
b. $19 \angle -20°$ d. $59 \angle 99°$

38. Convert the following impedances to polar form:

a. $-14 + j14$ c. $-33 - j18$
b. $27 + j17$ d. $7 + j4$

39. Add the following complex numbers:

a. $(4 + j3) + (3 + j2)$ b. $(100 - j50) + (12 + j9)$

40. Perform the following mathematical operations:

a. $(35 \angle -24°) \times (13 \angle 50°)$ c. $(98 \angle 80°) \div (40 \angle 17°)$
b. $(100 - j25) - (25 + j5)$

41. State the impedances of the circuits seen in Figure 18-40 in rectangular and polar form. What is Z_T in ohms and its phase angle?

42. Calculate in polar form the impedances of both circuits shown in Figure 18-41. Then combine the two impedances as if the circuits were parallel connected, using the product-over-sum method. Express the combined impedance in polar form.

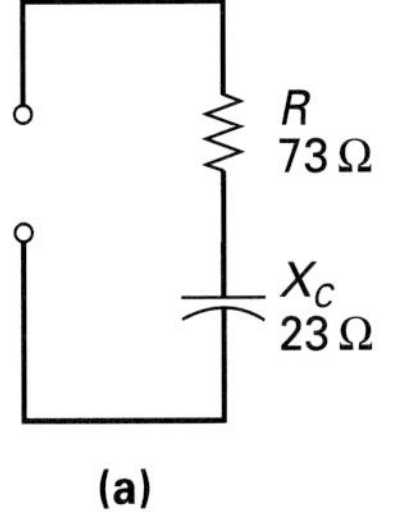

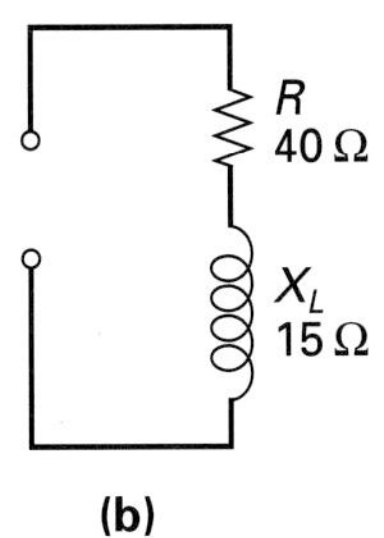

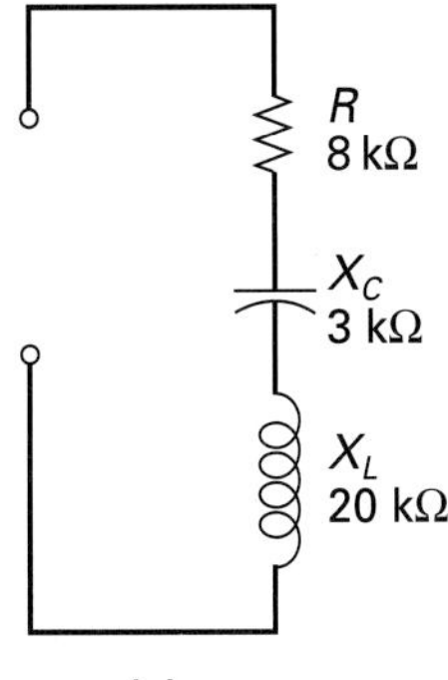

FIGURE 18-40

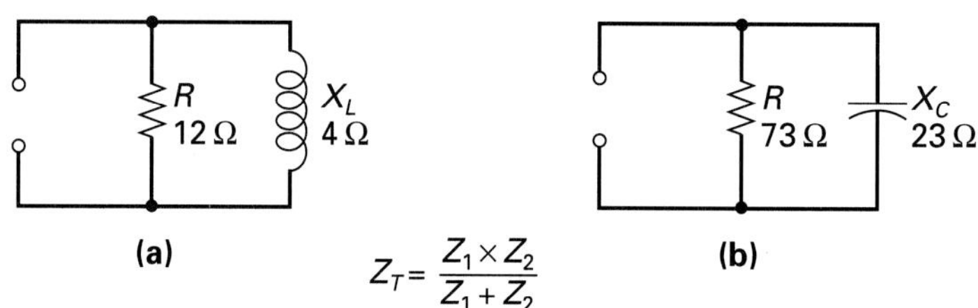

FIGURE 18-41

43. Referring to Figure 18-42, calculate:

a. Z_T (rectangular and polar)
b. Circuit current and phase angle
c. Voltage drops
d. V_C, V_L, and V_R phase relationships

44. Sketch an impedance and voltage phasor diagram for the circuit in Figure 18-40.

45. Referring to Figure 18-43, calculate:

a. Impedance of the two branches
b. Branch currents
c. Total current
d. Total impedance

46. Sketch an impedance and current phasor diagram for the circuit shown in Figure 18-41.

47. Referring to Figure 18-40, verify that the sum of all the individual voltage drops is equal to the total voltage.

48. Referring to Figure 18-41, verify that the sum of all the branch currents is equal to the total current.

49. Determine whether Questions 43 and 45 would be easier to answer with or without the use of complex numbers.

50. Is the circuit in Figure 18-41 more inductive or capacitive?

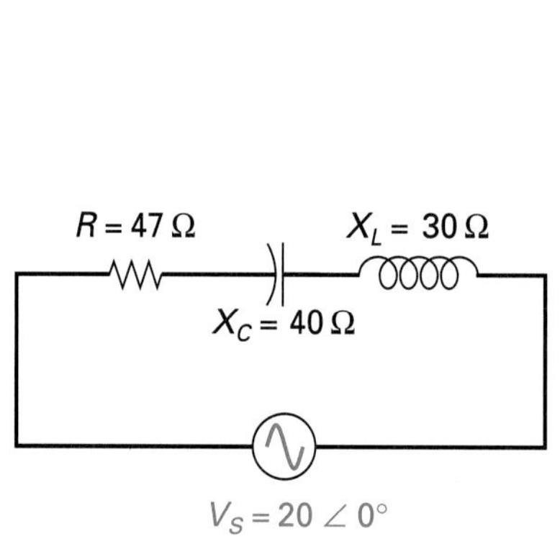

FIGURE 18-42

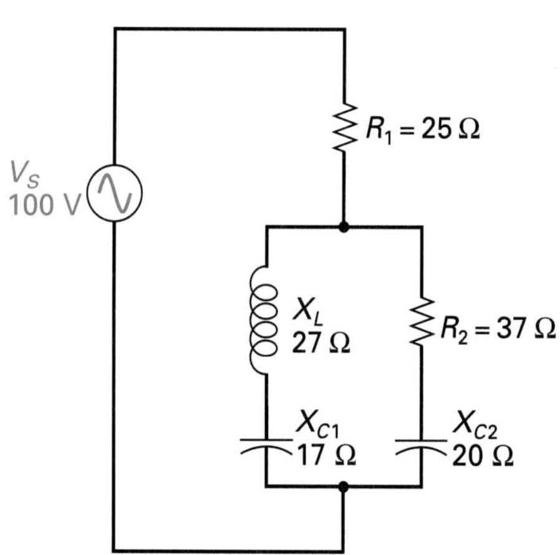

FIGURE 18-43

CHAPTER

19

OBJECTIVES

After completing this chapter, you will be able to:

1. Compare semiconductor devices to their predecessor the vacuum tube.
2. Describe why semiconductor devices have almost completely replaced vacuum tubes in most applications.
3. Explain the atom's subatomic particles.
4. Describe the terms:
   - **a.** Neutral atom
   - **b.** Negative ion
   - **c.** Positive ion
   - **d.** Valence shell
5. State the laws of attraction and repulsion.
6. Explain the difference between an atom, an element, a molecule, and a compound.
7. State the relationship between the number of valence electrons in an atom and its conductivity.
8. List the three semiconductor elements that are used to construct components for electrical and electronic circuit applications.
9. Describe why semiconductor atoms will form a crystal lattice structure due to covalent bonding, and define the term "intrinsic."
10. Explain an atom's energy gaps and energy levels along with the terms:
    - **a.** Energy gap
    - **b.** Conduction band
    - **c.** Excited state
    - **d.** Hole
    - **e.** Electron-hole pair
    - **f.** Recombination
    - **g.** Lifetime
11. Explain why semiconductor materials, and therefore semiconductor devices, have a negative temperature coefficient of resistance.
12. Define the term "hole flow," and describe why the total current in a semiconductor is equal to the sum of the electron-flow and hole-flow currents.
13. Explain why pure semiconductor materials are doped and why they remain electrically neutral.
14. Explain the similarities and differences between *n*-type and *p*-type semiconductor materials, and define the terms:
    - **a.** Extrinsic semiconductor
    - **b.** Majority carriers
    - **c.** Minority carriers
15. Describe the following in relation to the P-N junction:
    - **a.** The junction
    - **b.** The depletion region
    - **c.** The barrier voltage
16. Explain how P-N junctions can be:
    - **a.** Forward biased
    - **b.** Reverse biased
17. Calculate and define the terms:
    - **a.** Forward voltage (V_F) drop
    - **b.** Forward current (I_F)
    - **c.** Reverse voltage (V_R) drop
18. Define the following terms:
    - **a.** Diffusion current
    - **b.** Leakage or reverse current (I_R)

Semiconductor Principles

OUTLINE

VIGNETTE

The Turing Enigma

During the Second World War, the Germans developed a cipher generating apparatus called "Enigma." This electromechanical teleprinter would scramble messages with several randomly spinning rotors that could be set to a predetermined pattern by the sender. This key and plug pattern was changed three times a day by the Germans and cracking the secrets of Enigma became of the utmost importance to British Intelligence. With this objective in mind, every brilliant professor and eccentric researcher was gathered at a Victorian estate near London called Bletchley Park. They specialized in everything from engineering to literature and were collectively called the Backroom Boys.

By far the strangest and definitely most gifted of the group was an unconventional theoretician from Cambridge University named Alan Turing. He wore rumpled clothes and had a shrill stammer and crowing laugh that aggravated even his closest friends. He had other legendary idiosyncrasies that included setting his watch by sighting on a certain star from a specific spot and then mentally calculating the time of day. He also insisted on wearing his gas mask whenever he was out, not for fear of a gas attack, but simply because it helped his hay fever.

Turing's eccentricities may have been strange but his genius was indisputable. At the age of twenty-six he wrote a paper outlining his "universal machine" that could solve any mathematical or logical problem. The data or, in this case, the intercepted enemy messages could be entered into the machine on paper tape and then compared with known Enigma codes until a match was found.

In 1943 Turing's ideas took shape as the Backroom Boys began developing a machine that used 2,000 vacuum tubes and incorporated five photoelectric readers that could process 25,000 characters per second. It was named "Colossus," and it incorporated the stored program and other ideas from Turing's paper written seven years earlier.

Turing could have gone on to accomplish much more. However, his idiosyncrasies kept getting in his way. He became totally preoccupied with abstract questions concerning machine intelligence. His unconventional personal lifestyle led to his arrest in 1952 and, after a sentence of psychoanalysis, his suicide two years later.

Before joining the Backroom Boys at Bletchley Park, Turing's genius was clearly apparent at Cambridge. How much of a role he played in the development of Colossus is still unknown and remains a secret guarded by the British Official Secrets Act. Turing was never fully recognized for his important role in the development of this innovative machine, except by one of his Bletchley Park colleagues at his funeral who said, "I won't say what Turing did made us win the war, but I daresay we might have lost it without him."

INTRODUCTION

Materials can be divided into three main types according to the way they react to current when a voltage is applied across them. **Insulators** (nonconductors), for example, are materials that have a very high resistance and therefore oppose current, whereas **conductors** are materials that have a very low resistance and therefore pass current easily. The third type of material is the **semiconductor** which, as its name suggests, has properties that lie between the insulator and the conductor. Semiconductor materials are not good conductors or insulators and so the next question is: What characteristic do they possess that makes them so useful in electronics? The answer is that they can be controlled to either increase their resistance and behave more like an insulator or decrease their resistance and behave more like a conductor. *It is this ability of a semiconductor material to vary its resistive properties that makes it so useful in electrical and electronic applications.*

Insulators
Materials that have a very high resistance and oppose current.

Conductors
Materials that have a very low resistance and pass current easily.

Semiconductors
Materials that have properties that lie between insulators and conductors.

In this chapter we will examine the characteristics of semiconductor materials so that we can better understand the operation and characteristics of semiconductor devices. Before we begin semiconductors, however, let us first investigate why semiconductor devices have almost completely replaced their predecessor, the vacuum tube.

19-1 SEMICONDUCTOR DEVICES VERSUS VACUUM TUBE DEVICES

Semiconductor materials such as *germanium* and *silicon* are used to construct semiconductor devices like the *diodes, transistors,* and *integrated circuits* (*ICs*) shown in Figure 19-1. These devices are used in electrical and electronic circuits

(a)

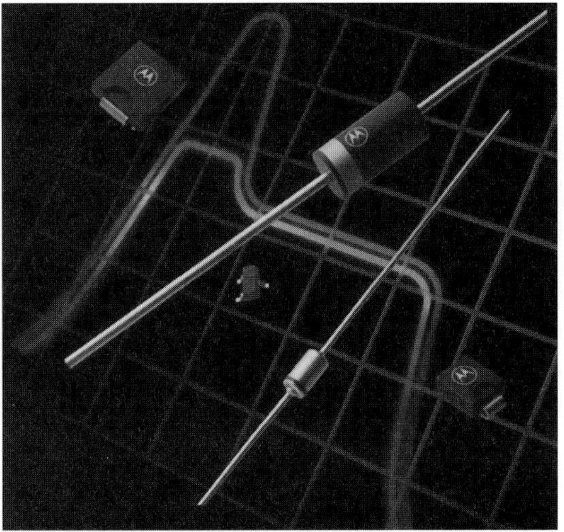

(b)

FIGURE 19-1 **Semiconductor Devices. (Copyright of Motorola, Inc. Used by permission.)**

to control current and voltage, so as to produce a desired result. For example, a diode could be used as the controlling element in a rectifier circuit that would convert ac to pulsating dc. A transistor, on the other hand, could be made to act like a variable resistance so it could amplify a radio signal. Conversely, an integrated circuit could be used to generate an oscillating signal or be made to perform arithmetic operations.

These semiconductor devices first became available in 1960. From about 1920 to 1960 the controlling element in all electrical and electronic circuits was the vacuum tube. Historians say that the only way to fully understand the present and the near future is to be familiar with the past. Let us now consider the specific reasons for this transition from vacuum tubes to semiconductor devices.

19-1-1 Vacuum Tube Devices

Vacuum Tube or Thermionic Valve An electron tube evacuated to such a degree that its electrical characteristics are essentially unaffected by the presence of residual gas or vapor. Eventually replaced by the transistor for amplification and rectification.

Triode A three-electrode vacuum tube that has an anode, cathode, and control grid.

The term **vacuum tube** or **thermionic valve** is used to describe a variety of special devices that made possible radio, television, telecommunications, radar, sonar, computers, and many more systems between 1920 and 1960. These vacuum tubes were all enclosed in a sealed glass container that had been pumped free of air (hence the name vacuum) and had electric connections to its internal parts through the base of the container. Two of the most frequently used vacuum tubes were the *diode* and **triode** which are shown in Figure 19-2.

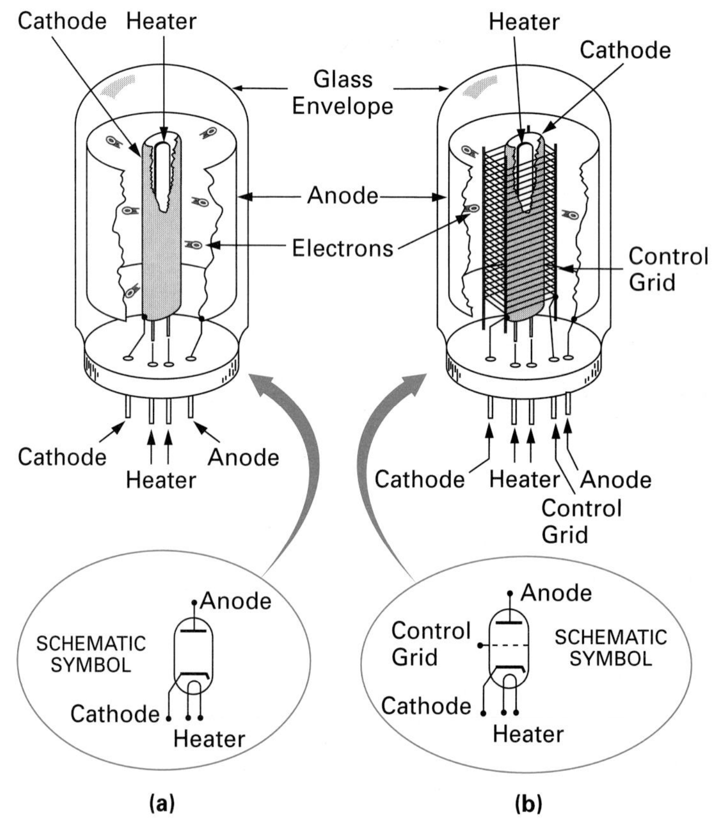

FIGURE 19-2 Vacuum Tube Devices. (a) Diode Vacuum Tube. (b) Triode Vacuum Tube.

All vacuum tubes contain a *thermionic electron emitter,* which is a specially treated metal electrode that will emit electrons when heated by a heater. This effect was first demonstrated in 1877 by British physicist Joseph Thomson. It was not until 1904 that the first vacuum tube was invented by British engineer John Fleming. He called it the diode since it contained two (*di*) electrodes (*ode*). These two electrodes were the thermionic electron emitter that released the electrons, the **cathode,** and a plate electrode that collected the electrons, the **anode.** The main elements of the diode valve, or *Fleming diode,* as it was named at that time, are shown in Figure 19-2(a). This diode was used as a rectifier in power supply circuits (where it converted ac to dc), and as a *detector* in receiver circuits (where it extracted the radio signal from the carrier wave). The diode's main disadvantage was that it could not amplify or increase the amplitude of small signals. This disadvantage was overcome by U.S. scientist Lee De Forest in 1907 when he introduced a third electrode called the *control grid* between the cathode and anode. This control grid was used to increase or decrease the number of electrons reaching the anode. By controlling the grid, the triode valve, named because it contained three (*tri*) electrodes (*ode*), could be made to amplify (increase in amplitude) small radio signals. The triode valve was used in the first commercial radio receivers in the 1920s, and would later make possible television and other communication systems. The main elements of the triode valve are shown in Figure 19-2(b).

Cathode
A negative electrode or terminal.

Anode
A positive electrode or terminal.

In the late 1920s other vacuum tubes emerged such as the **tetrode** and **pentode,** and improvements were introduced to the existing types. These electronic devices dominated the electronics industry until 1960 when the transistor took over.

Tetrode
A four-electrode electron tube that has an anode, cathode, control grid, and an additional electrode.

Pentode
A five-electrode electron tube that has an anode, cathode, control grid, and two additional electrodes.

Vacuum Tube Limitations

The vacuum tube was replaced by the transistor in so many applications because it suffered from a number of physical drawbacks. The first is that vacuum tubes are fragile and can be easily damaged by shock or vibration. The second is that the heater required a great deal of power, and the third is their relatively large size. The second and third disadvantages made the vacuum tube unsuitable for battery operated portable equipment, and, in addition, the tube heater had only a certain life span which meant that the tubes needed to be replaced regularly. To illustrate these disadvantages, Figure 19-3 compares the *electronic numeric integrator and computer (ENIAC),* which was unveiled in 1946, to today's programmable pocket calculator. The ENIAC weighed 38 tons, measured 18 feet wide and 88 feet long and used 17,486 vacuum tubes. These vacuum tubes produced a great deal of heat and developed frequent faults requiring constant maintenance. Today's programmable pocket calculator on the other hand, makes use of semiconductor integrated circuits and is far more powerful, portable, and reliable than the ENIAC. Another advantage is cost: in 1946 the ENIAC cost $400,000 to produce whereas a scientific calculator can be purchased today for less than $40.

Present Day Vacuum Tube Applications

Despite all of the vacuum tube's limitations, it is still used today in some areas of science and technology. In industry, 100 kW triode vacuum tubes are made mechanically and electrically very rugged and are used to generate Radio

FIGURE 19-3 Solid State Systems versus Vacuum Tube Systems. (a) The Electronic Numerical Integrator and Computer (ENIAC). (b) Today's Pocket Calculator.

Frequency (RF) power at frequencies from 100 kHz to around 30 MHz. In communications, vacuum tubes are used to generate the high-frequency and high-power outputs needed for radio and television transmitters. In science, vacuum tubes are used in fusion research and linear accelerators where experimental tubes are being operated at 100 kV. **Cathode ray tubes (CRTs)** are still used extensively in televisions and computer monitors; however, the semiconductor color-active matrix display is beginning to take over.

Cathode Ray Tube
A vacuum tube in which electrons emitted by a hot cathode are focused into a narrow beam by an electron gun and then applied to a fluorescent screen. The beam can be varied in intensity and position to produce a pattern or picture on the screen.

Solid State Device
Uses a solid semiconductor material, such as silicon, between the input and output whereas a vacuum tube has vacuum between input and output.

19-1-2 Solid State Devices

The most significant development in electronics since World War II has been a small semiconductor device called the transistor. It was first introduced in 1948 by its inventors William Schockley, Walter Bratten, and John Bardeen in the Bell Telephone Laboratories and was described as a **solid state device.** This term was used because the transistor contained a solid semiconductor material between its input and output pins, unlike its predecessor the vacuum tube which had a vacuum between its input and output pins.

The first *point-contact transistor* unveiled in 1948 was extremely unreliable, and it took its inventors another twelve years to develop the superior *bipolar junction transistor* (*BJT*) and make it available in commercial quantities.

In 1960, many electronic system manufacturers began to use the bipolar junction transistor instead of the vacuum tube in low-power and low-frequency applications. Research and development into semiconductor or solid state devices mushroomed and a variety of semiconductor devices began to appear. A different type of transistor emerged called the *field effect transistor* (*FET*), which had characteristics similar to those of the vacuum tube. Once it was discovered that semiconductor materials could also generate and sense light, a new line of *optoelectronic devices* became available. Later it was discovered that semiconductor materials could sense magnetism, temperature, and pressure and, as a result, a variety of sensor devices or transducers (energy converters) appeared on the market. Along with all these different types of semiconductor devices, a wide variety of semiconductor diodes emerged that could rectify, regulate, and oscillate at high frequencies. Even to this day it is clear that we have not yet seen all the potential value of semiconductors. Figure 19-4 illustrates many of these semiconductor or solid state devices.

Although semiconductor diodes and transistors are still widely used as individual or **discrete components,** in 1959 Robert Noyce discovered that more than one transistor could be constructed on a single piece of semiconductor material. Soon other components such as resistors, capacitors, and diodes were added with transistors and then interconnected to form a complete circuit on a single chip or piece of semiconductor material. This integrating of various components on a single chip of semiconductor was called an *integrated circuit* (*IC*) or *IC chip.* Today the IC is used extensively in every branch of electronics with hundreds of thousands of transistors and other components being placed on a chip of semiconductor no bigger than this ■. Figure 19-5 illustrates some of the different types of integrated circuits.

Discrete Components
Seperate active and passive devices that were manufactured before being used in a circuit.

Like an evolving species, semiconductors have come to dominate the products of which they used to be only a part. For example, there used to be 400 components in a typical cellular telephone. Now there are 40, and soon only 3 or 4 IC chips will make up the entire phone circuitry. Today the semiconductor business—once regarded as a technical sideshow—occupies center stage and is key to the development of new products for all industries.

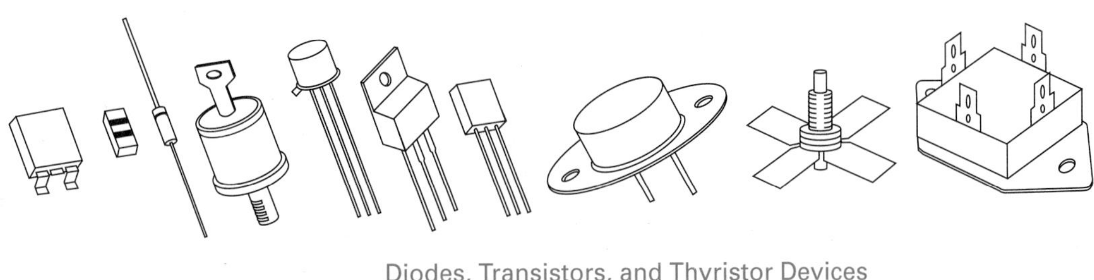

Diodes, Transistors, and Thyristor Devices

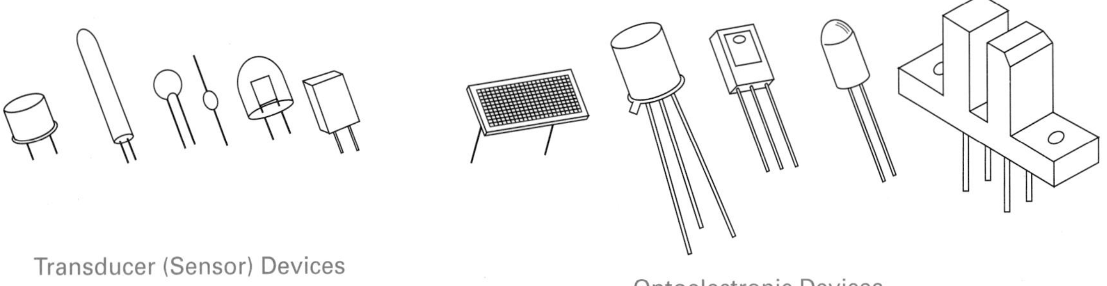

Transducer (Sensor) Devices

Optoelectronic Devices

FIGURE 19-4 Discrete Semiconductor (Solid State) Devices.

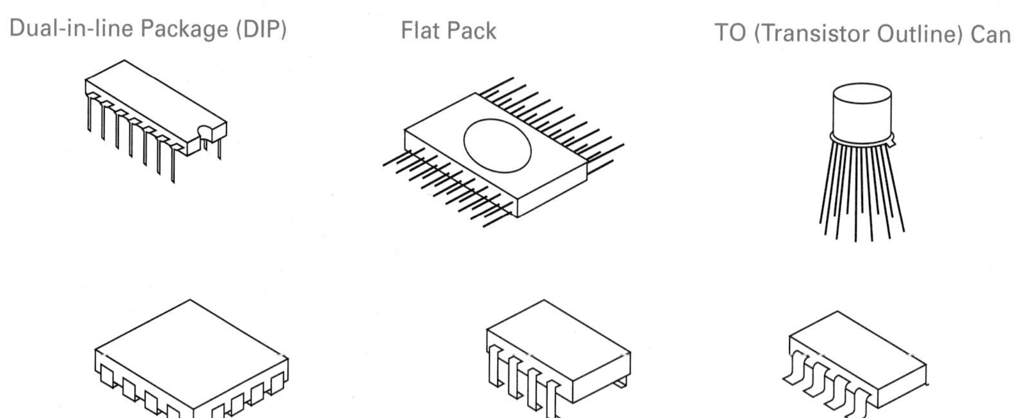

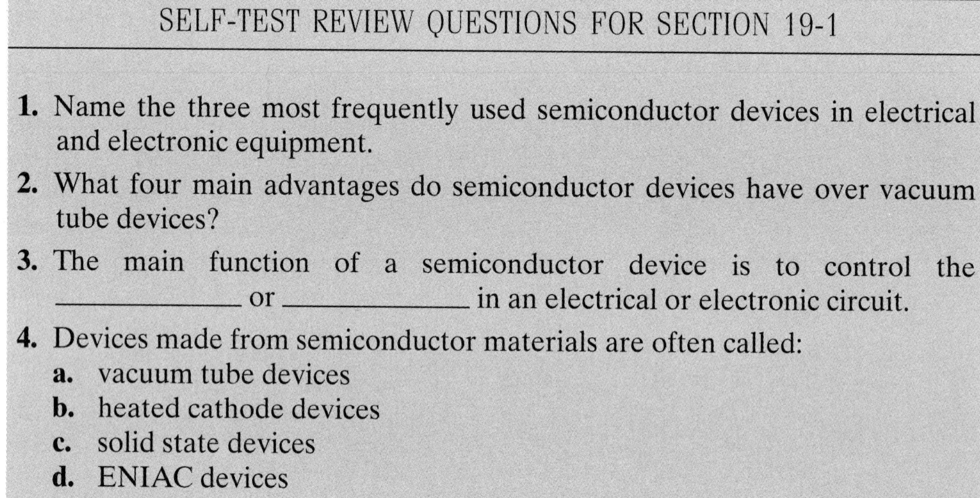

FIGURE 19-5 Semiconductor Integrated Circuits (ICs).

SELF-TEST REVIEW QUESTIONS FOR SECTION 19-1

1. Name the three most frequently used semiconductor devices in electrical and electronic equipment.
2. What four main advantages do semiconductor devices have over vacuum tube devices?
3. The main function of a semiconductor device is to control the ____________ or ____________ in an electrical or electronic circuit.
4. Devices made from semiconductor materials are often called:
   a. vacuum tube devices
   b. heated cathode devices
   c. solid state devices
   d. ENIAC devices

19-2 THE STRUCTURE OF MATTER

Element
There are 107 different chemical substances, or elements, that exist on earth. These can be categorized as gas, solid, or liquid.

Atom
The smallest particle of an element.

All of the matter on the earth and in the air surrounding the earth can be classified as being either a solid, liquid, or gas. A total of approximately 107 different elements are known to exist in, on, and around the earth. An **element,** by definition, is a substance consisting of only one type of atom; in other words, every element has its own distinctive atom, which makes it different from all the other elements. This **atom** is the smallest particle into which an element can be divided without losing its identity, and a group of identical atoms is called an element, shown in Figure 19-6.

For the sake of discussion, let us take a small amount of either a solid, a liquid, or a gas and divide it into two pieces. Then we divide a resulting piece into two pieces, and keep repeating the process until we finally end up with a

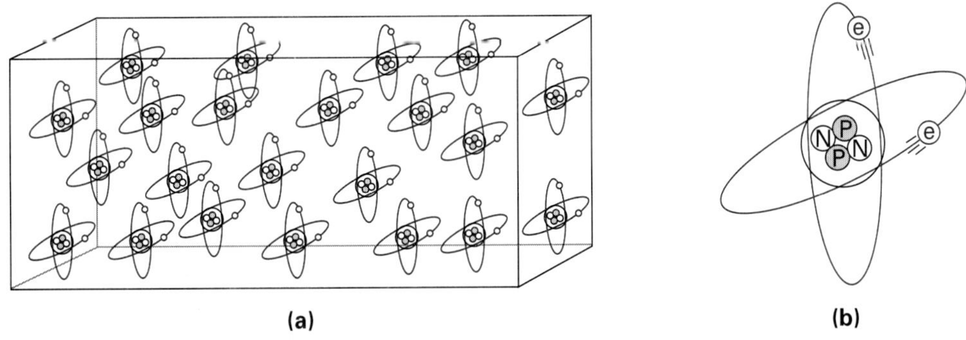

FIGURE 19-6 Elements and Atoms. (a) Element: Many Similar Atoms. (b) Atom: Smallest Unit.

tiny remaining part. Viewing the part under the microscope, as shown in Figure 19-7, the substance can still be identified as the original element as it is still made up of many of the original solid, liquid, or gas atoms. A small amount of gold, for example, the size of a pinpoint, will still contain several billion atoms. If the element subdivision is continued, however, a point will be reached at which a single atom will remain. Let us now analyze the atom in more detail.

19-2-1 *The Atom*

The word atom is a Greek word meaning a particle that is too small to be subdivided. At present, we cannot clearly see the atom; however, physicists and researchers do have the ability to record a picture as small as 12 billionths of an

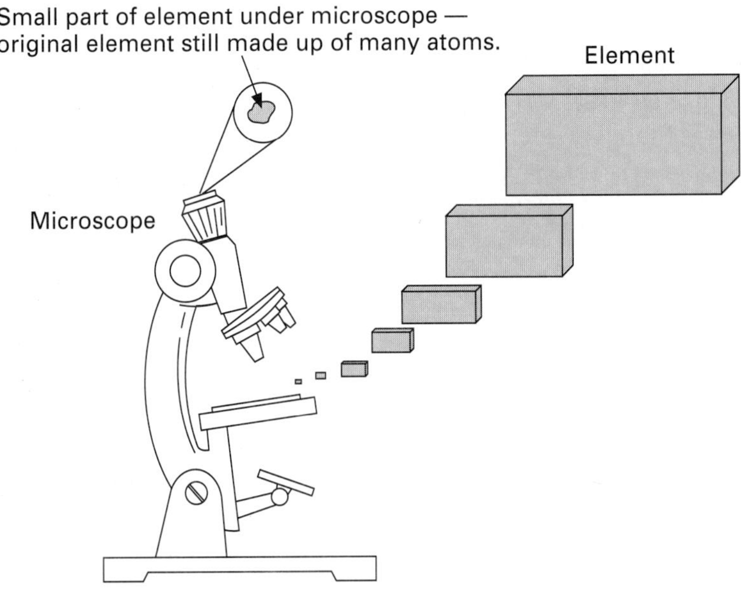

FIGURE 19-7 An Element under the Microscope.

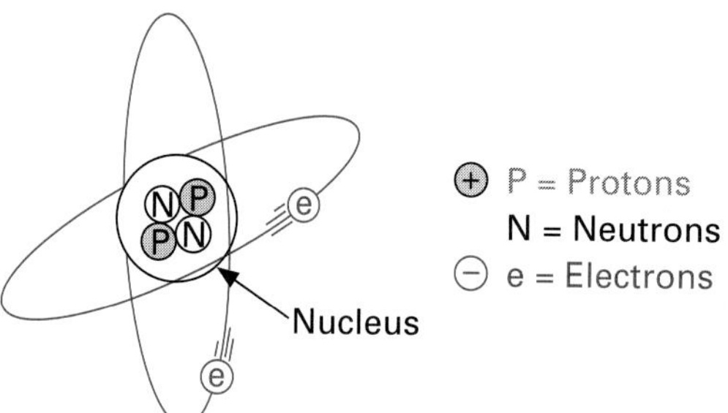

FIGURE 19-8 The Atom.

inch (about the diameter of one atom), and this image displays the atom as a white fuzzy ball.

Subatomic
Particles such as electrons, protons, and neutrons that are smaller than atoms.

In 1913, a Danish physicist, Neils Bohr, put forward a theory about the atom, and his basic model outlining the **subatomic** particles that make up the atom is still in use today and is illustrated in Figure 19-8. Bohr actually combined the ideas of Lord Rutherford's (1871–1937) nuclear atom with Max Planck's (1858–1947) and Albert Einstein's (1879–1955) quantum theory of radiation.

The three important particles of the atom are the *proton,* which has a positive charge, the *neutron,* which is neutral or has no charge, and the *electron,* which has a negative charge. Referring to Figure 19-8, you can see that the atom consists of a positively charged central mass called the *nucleus,* which is made up of protons and neutrons surrounded by a quantity of negatively charged orbiting electrons.

Atomic Number
Number of positive charges, or protons, in the nucleus of an atom.

Table 19-1 lists the periodic table of the elements, in order of their atomic number. The **atomic number** of an atom describes the number of protons that exist within the nucleus.

The proton and the neutron are almost 2000 times heavier than the very small electron, so if we ignore the weight of the electron, we can use the fourth column in Table 19-1 (weight of an atom) to give us a clearer picture of the protons and neutrons within the atom's nucleus. For example, a hydrogen atom, shown in Figure 19-9(a), is the smallest of all atoms and has an atomic number of 1, which means that hydrogen has a one-proton nucleus. Helium, however [Figure 19-9(b)], is second on the table and has an atomic number of 2, indicating that two protons are within the nucleus. The **atomic weight** of helium, however, is 4, meaning that two protons and two neutrons make up the atom's nucleus.

Atomic Weight
The relative weight of a neutral atom of an element, based on a neutral carbon atom having an

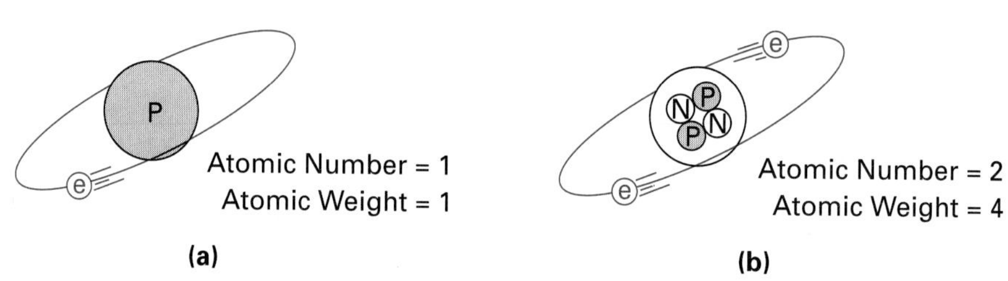

FIGURE 19-9 (a) Hydrogen Atom. (b) Helium Atom.

TABLE 19-1 Periodic Table of the Elements

ATOMIC NUMBER	ELEMENT NAME	SYMBOL	ATOMIC WEIGHT	ELECTRONS/SHELL K	L	M	N	O	P	Q	DISCOVERED	COMMENT
1	Hydrogen	H	1.007	1							1766	Active gas
2	Helium	He	4.002	2							1895	Inert gas
3	Lithium	Li	6.941	2	1						1817	Solid
4	Beryllium	Be	9.01218	2	2						1798	Solid
5	Boron	B	10.81	2	3						1808	Solid
6	Carbon	C	12.011	2	4						Ancient	Semiconductor
7	Nitrogen	N	14.0067	2	5						1772	Gas
8	Oxygen	O	15.9994	2	6						1774	Gas
9	Fluorine	F	18.998403	2	7						1771	Active gas
10	Neon	Ne	20.179	2	8						1898	Inert gas
11	Sodium	Na	22.98977	2	8	1					1807	Solid
12	Magnesium	Mg	24.305	2	8	2					1755	Solid
13	Aluminum	Al	26.98154	2	8	3					1825	Metal conductor
14	Silicon	Si	28.0855	2	8	4					1823	Semiconductor
15	Phosphorus	P	30.97376	2	8	5					1669	Solid
16	Sulfur	S	32.06	2	8	6					Ancient	Solid
17	Chlorine	Cl	35.453	2	8	7					1774	Active gas
18	Argon	Ar	39.948	2	8	8					1894	Inert gas
19	Potassium	K	39.0983	2	8	8	1				1807	Solid
20	Calcium	Ca	40.08	2	8	8	2				1808	Solid
21	Scandium	Sc	44.9559	2	8	9	2				1879	Solid
22	Titanium	Ti	47.90	2	8	10	2				1791	Solid
23	Vanadium	V	50.9415	2	8	11	2				1831	Solid
24	Chromium	Cr	51.996	2	8	13	1				1798	Solid
25	Manganese	Mn	54.9380	2	8	13	2				1774	Solid
26	Iron	Fe	55.847	2	8	14	2				Ancient	Solid (magnetic)
27	Cobalt	Co	58.9332	2	8	16	2				1735	Solid
28	Nickel	Ni	58.70	2	8	16	2				1751	Solid
29	Copper	Cu	63.546	2	8	18	1				Ancient	Metal conductor
30	Zinc	Zn	65.38	2	8	18	3				1746	Solid
31	Gallium	Ga	69.72	2	8	18	4				1875	Liquid
32	Germanium	Ge	72.59	2	8	18	4				1886	Semiconductor
33	Arsenic	As	74.9216	2	8	18	5				1649	Solid
34	Selenium	Se	78.96	2	8	18	6				1818	Photosensitive
35	Bromine	Br	79.904	2	8	18	8				1898	Liquid
36	Krypton	Kr	83.80	2	8	18	8				1898	Inert gas
37	Rubidium	Rb	85.4678	2	8	18	8	1			1861	Solid
38	Strontium	Sr	87.62	2	8	18	8	2			1790	Solid
39	Yttrium	Y	88.9059	2	8	18	9	2			1843	Solid
40	Zirconium	Zr	91.22	2	8	18	10	2			1789	Solid
41	Niobium	Nb	92.9064	2	8	18	12	1			1801	Solid
42	Molybdenum	Mo	95.94	2	8	18	13	1			1781	Solid
43	Technetium	Tc	98.0	2	8	18	14	1			1937	Solid
44	Ruthenium	Ru	101.07	2	8	18	15	1			1844	Solid
45	Rhodium	Rh	102.9055	2	8	18	16	1			1803	Solid
46	Palladium	Pd	106.4	2	8	18	18	0			1803	Solid
47	Silver	Ag	107.868	2	8	18	18	1			Ancient	Metal conductor
48	Cadmium	Cd	112.41	2	8	18	18	2			1803	Solid

TABLE 19-1 *(continued)*

ATOMIC NUMBER[a]	ELEMENT NAME	SYMBOL	ATOMIC WEIGHT	ELECTRONS/SHELL K L M N O P Q	DISCOVERED	COMMENT
49	Indium	In	114.82	2 8 18 18 3	1863	Solid
50	Tin	Sn	118.69	2 8 18 18 4	Ancient	Solid
51	Antimony	Sb	121.75	2 8 18 18 5	Ancient	Solid
52	Tellurium	Te	127.60	2 8 18 18 6	1783	Solid
53	Iodine	I	126.9045	2 8 18 18 7	1811	Solid
54	Xenon	Xe	131.30	2 8 18 18 8	1898	Inert gas
55	Cesium	Cs	132.9054	2 8 18 18 8 1	1803	Liquid
56	Barium	Ba	137.33	2 8 18 18 8 2	1808	Solid
57	Lanthanum	La	138.9055	2 8 18 18 9 2	1839	Solid
72	Hafnium	Hf	178.49	2 8 18 32 10 2	1923	Solid
73	Tantalum	Ta	180.9479	2 8 18 32 11 2	1802	Solid
74	Tungsten	W	183.85	2 8 18 32 12 2	1783	Solid
75	Rhenium	Re	186.207	2 8 18 32 13 2	1925	Solid
76	Osmium	Os	190.2	2 8 18 32 14 2	1804	Solid
77	Iridium	Ir	192.22	2 8 18 32 15 2	1804	Solid
78	Platinum	Pt	195.09	2 8 18 32 16 2	1735	Solid
79	Gold	Au	196.9665	2 8 18 32 18 1	Ancient	Solid
80	Mercury	Hg	200.59	2 8 18 32 18 2	Ancient	Liquid
81	Thallium	Tl	204.37	2 8 18 32 18 3	1861	Solid
82	Lead	Pb	207.2	2 8 18 32 18 4	Ancient	Solid
83	Bismuth	Bi	208.9804	2 8 18 32 18 5	1753	Solid
84	Polonium	Po	209.0	2 8 18 32 18 6	1898	Solid
85	Astatine	At	210.0	2 8 18 32 18 7	1945	Solid
86	Radon	Rn	222.0	2 8 18 32 18 8	1900	Inert gas
87	Francium	Fr	223.0	2 8 18 32 18 8 1	1945	Liquid
88	Radium	Ra	226.0254	2 8 18 32 18 8 2	1898	Solid
89	Actinium	Ac	227.0278	2 8 18 32 18 9 2	1899	Solid

[a]Rare earth series 58–71 and 90–107 have been omitted

The number of neutrons within an atom's nucleus can subsequently be calculated by subtracting the atomic number (protons) from the atomic weight (protons and neutrons). For example, Figure 19-10 illustrates a beryllium atom:

Beryllium
Atomic number: 4 (protons)
Atomic weight: 9 (protons and neutrons)

Neutral Atom
An atom in which the number of positive charges in the nucleus (protons) is equal to the number of negative charges (electrons) that surround the nucleus.

If the number of protons is 4, the number of neutrons is 5 (9 – 4 = 5).

A **neutral atom** or *balanced atom* is one that has an equal number of protons and orbiting electrons, so the net positive proton charge is equal but opposite to the net negative electron charge, resulting in a balanced or neutral state. For example, Figure 19-11 illustrates a copper atom, which is the most commonly used metal in the field of electronics. It has an atomic number of 29, meaning that 29 protons and 29 electrons exist within the atom when it is in its neutral state.

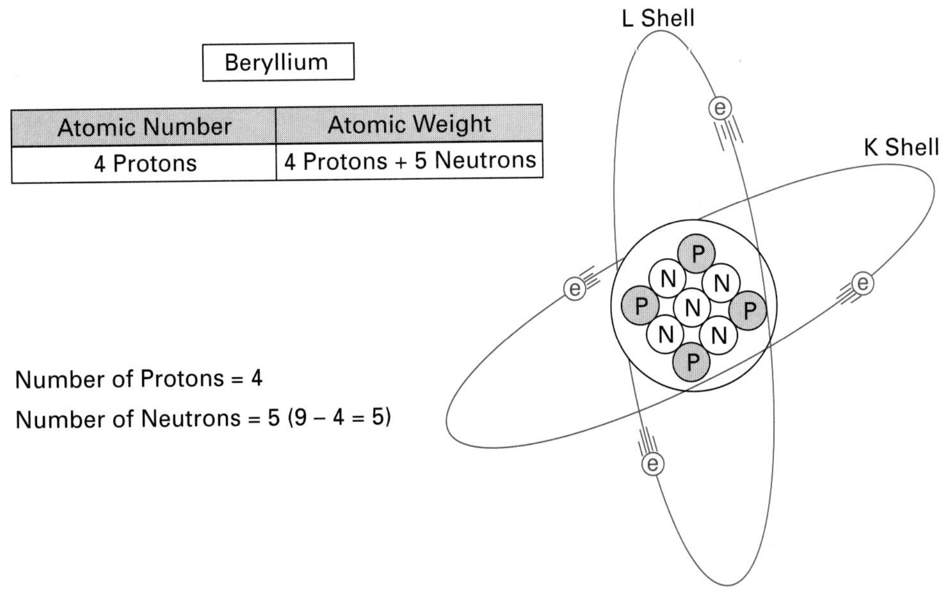

FIGURE 19-10 **Beryllium Atom.**

Orbiting electrons travel around the nucleus at varying distances from the nucleus, and these orbital paths are known as **shells or bands.** The orbital shell nearest the nucleus is referred to as the first or K shell. The second is known as the L, the third is M, the fourth is N, the fifth is O, the sixth is P, and the seventh is referred to as the Q shell. There are seven shells available for electrons (K, L, M, N, O, P, and Q) around the nucleus, and each of these seven shells can only hold a certain number of electrons, as shown in Figure 19-12. The outermost electron-occupied shell is referred to as the **valence shell or ring,** and these electrons are termed *valence electrons.* In the case of the copper atom in Figure 19-10, a single valence electron exists in the valence N shell.

Shells or Bands
An orbital path containing a group of electrons that have a common energy level.

Valence Shell or Ring
The outermost shell formed by electrons.

All matter exists in one of three basic states: solids, liquids, and gases. The atoms of a solid are fixed in relation to one another but vibrate in a back-and-forth motion, unlike liquid atoms, which can flow over each other. The atoms of a gas move rapidly in all directions and collide with one another. The far-right column of Table 19-1 indicates whether the element is a gas, a solid, or a liquid at room temperature and normal pressure.

19-2-2 *Laws of Attraction and Repulsion*

For the sake of discussion and understanding, let us theoretically imagine that we are able to separate some positive and negative subatomic particles. Using these separated protons and electrons, let us carry out a few experiments. Studying Figure 19-13, you will notice that:

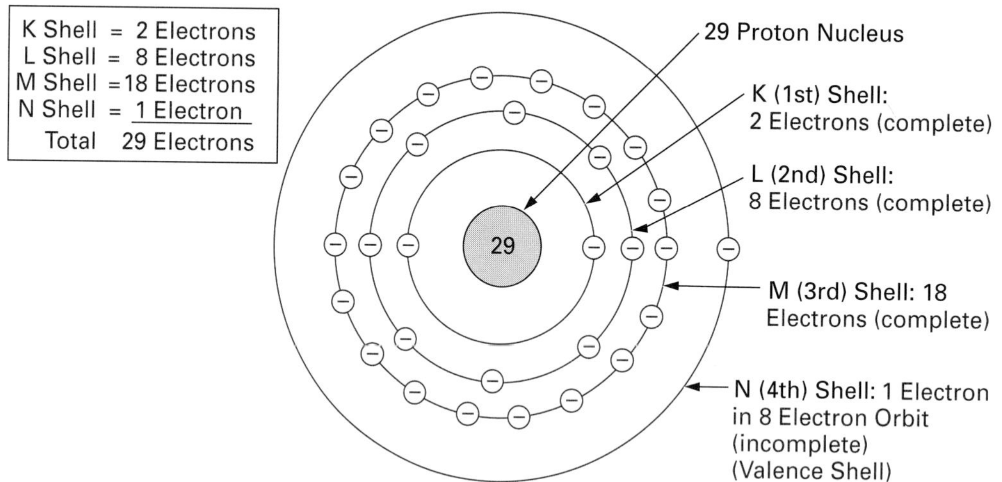

FIGURE 19-11 **Copper Atom.**

1. *Like charges* (positive and positive or negative and negative) repel one another. (Figures 19-13(a) and (b))
2. *Unlike charges* (positive and negative or negative and positive) attract one another. (Figure 19-13(c))

Orbiting negative electrons are therefore attracted toward the positive nucleus, which leads us to the question of why the electrons do not fly into the atom's nucleus. The answer is that the orbiting electrons remain in their stable orbit due to two equal but opposite forces. The centrifugal outward force exerted on the electrons due to the orbit counteracts the attractive inward force (centripetal) trying to pull the electrons toward the nucleus due to the unlike charges.

Due to their distance from the nucleus, valence electrons are described as being loosely bound to the atom. The electrons can, therefore, easily be dislodged from their outer orbital shell by any external force to become a free electron.

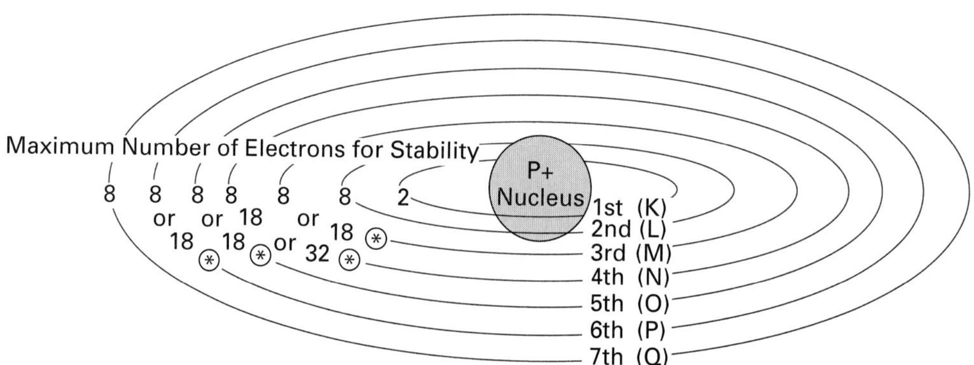

*The maximum number of electrons in these shells is dependent on the element's place in the periodic table.

FIGURE 19-12 **Electrons and Shells.**

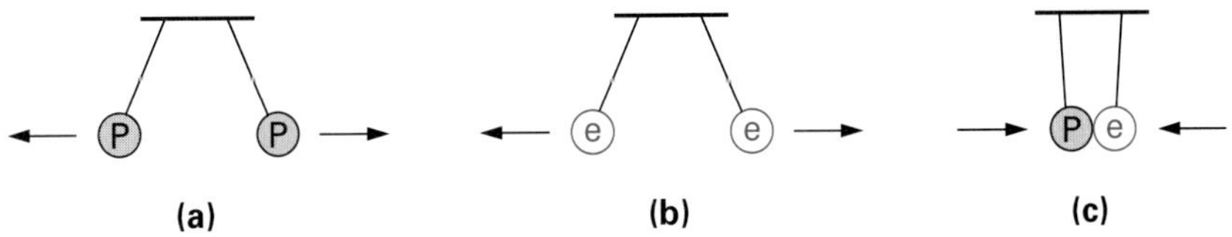

P = Proton (positive)
e = Electron (negative)

FIGURE 19-13 **Attraction and Repulsion. (a) Positive Repels Positive. (b) Negative Repels Negative. (c) Unlike Charges Attract.**

19-2-3 *The Molecule*

An atom is the smallest unit of a natural element, and an element is a substance consisting of a large number of the same atom. Combinations of elements are known as **compounds,** and the smallest unit of a compound is called a **molecule,** just as the smallest unit of an element is an atom. Figure 19-14 summarizes how elements are made up of atoms, and compounds are made up of molecules.

Water is an example of a liquid compound in which the molecule (H_2O) is a combination of an explosive gas (hydrogen) and a very vital gas (oxygen). Table salt is another example of a compound; here the molecule is made up of a highly poisonous gas atom (chlorine) and a potentially explosive solid atom

Compound
A material composed of united combinations of elements.

Molecule
The smallest particle of a compound that still retains its chemical characteristics.

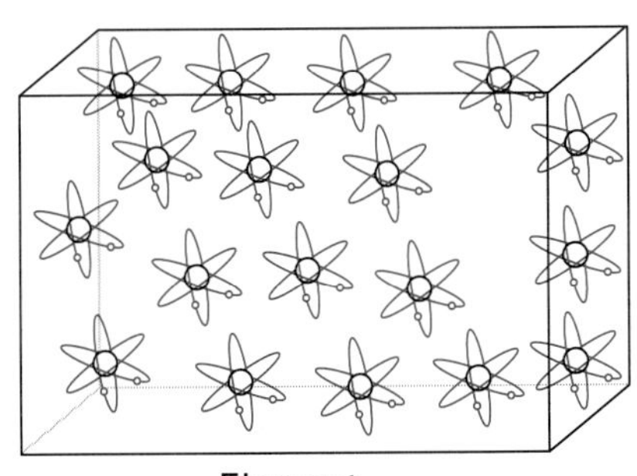

Element

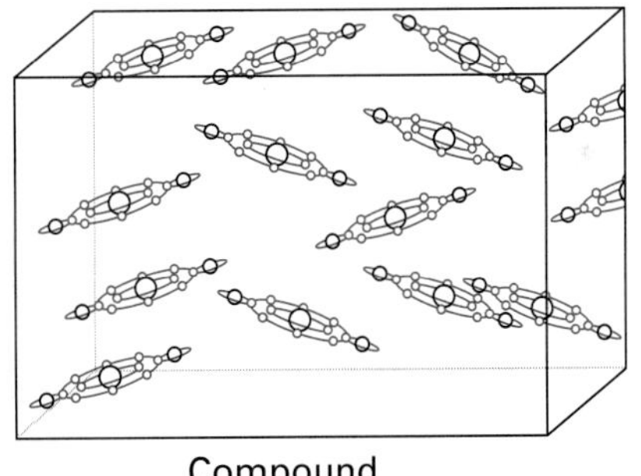

Compound

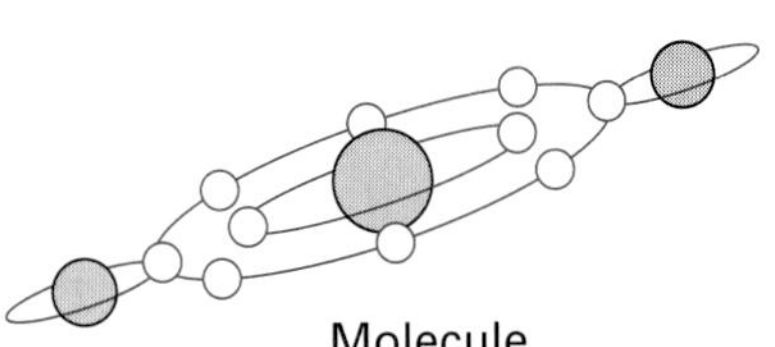

(a)

(b)

FIGURE 19-14 **(a) An Element Is Made Up of Many Atoms. (b) A Compound Is Made Up of Many Molecules.**

(sodium). These examples of compounds each contain atoms that, when alone, are both poisonous and explosive, yet when combined the resulting substance is as ordinary and basic as water and salt.

SELF-TEST REVIEW QUESTIONS FOR SECTION 19-2

1. Define the difference between an element and a compound.
2. Name the three subatomic particles that make up an atom.
3. What is the most commonly used metal in the field of electronics?
4. State the laws of attraction and repulsion.

19-3 SEMICONDUCTOR MATERIALS

A semiconductor material is one that is neither a conductor nor a nonconductor (insulator). This means simply that it will not conduct current as well as a conductor or block current as well as an insulator. Some semiconductor materials are pure or natural elements such as carbon (C), germanium (Ge), and silicon (Si), while other semiconductor materials are compounds.

Silicon and germanium are used most frequently in the construction of semiconductor devices for electrical and electronic applications. Germanium is a brittle grayish-white element that may be recovered from the ash of certain types of coals. Silicon, the most popular semiconductor material due to its superior temperature stability, is a white element normally derived from sand. Let us now examine the silicon, germanium, and carbon semiconductor atoms in more detail.

19-3-1 Semiconductor Atoms

Figure 19-15 illustrates the silicon, germanium, and carbon atoms. The silicon atom has 14 protons in its nucleus and 14 electrons in three orbital paths distributed as 2, 8, and then 4 electrons in its valence shell. The germanium atom has 32 protons within its nucleus and 32 electrons in four orbital paths distributed as 2, 8, 18, and finally 4 electrons in the valence band or shell. The carbon atom has 6 protons in its nucleus and 6 orbiting electrons in two orbital paths distributed as 2 and 4 electrons in the valence shell. The question is: What do all these atoms have in common? The answer is that all semiconductor atoms have *four valence electrons.*

The valence shell of an atom can contain up to 8 electrons, and it is the number of electrons in this valence shell that determines the conductivity of the atom. For example, an atom with only 1 valence electron would be classed as a good conductor whereas an atom having 8 valence electrons, and therefore a complete valence shell, would be classed as an insulator.

To summarize, Figure 19-15 shows three semiconductor atoms, all of which contain four valence electrons. Since the number of valence electrons determines

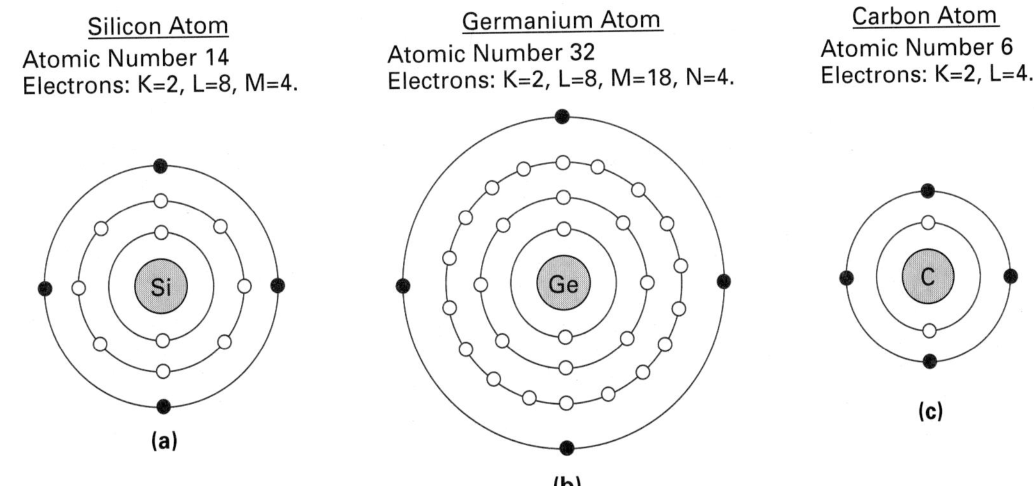

FIGURE 19-15 Semiconductor Atoms with Their 4 Valence Shell Electrons.

the conductivity of the element, semiconductor atoms are midway between conductors (which have 1 valence electron) and insulators (which have 8 valence electrons). Silicon and germanium are used to manufacture semiconductor devices, whereas carbon is combined with other elements to construct resistors.

19-3-2 Crystals and Covalent Bonding

So far we have discussed only isolated atoms. When two or more similar semiconductor atoms are combined to form a solid element, they automatically arrange themselves into an orderly lattice-like structure or pattern known as a **crystal,** as shown in Figure 19-16(a). This pattern is formed because each atom shares its four valence electrons with its four neighboring atoms. Since each atom shares one electron with a neighboring atom, two atoms will share two, or a pair, of electrons between the two cores. These two atom cores are pulling the two electrons with equal but opposite force and it is this pulling action that holds the atoms together in this solid crystal-lattice structure. The joining together of two semiconductor atoms is called an **electron-pair bond** or **covalent bond.** When many atoms combine, or bond, in this way the result is a crystal (smooth, glassy, solid) lattice structure. To illustrate this bonding process, each atom in Figure 19-16(a) has been drawn as a square and each valence shell has been drawn as an octagon (eight-sided figure) so that we can easily see which electrons belong to which atom. As you can see, the atom in the center of the diagram has 4 valence electrons (shown at the corners of the Si square), and shares one electron from each of its four neighbors.

Crystal
A solid element with an orderly lattice-like structure.

Electron-Pair Bond or Covalent Bond
A pair of electrons shared by two neighboring atoms.

Figure 19-16(b) shows a larger view of a silicon crystal structure. All of the atoms in this structure are electrically stable because all of their valence shells are complete (they all contain eight electrons). These completed valence shells cause the pure semiconductor crystal structure to act as an insulator since it will not easily give up or accept electrons. Pure semiconductor materials, which are often called **intrinsic** materials, are therefore very poor conductors. Once this

Intrinsic Semiconductor Materials
Pure semiconductor materials.

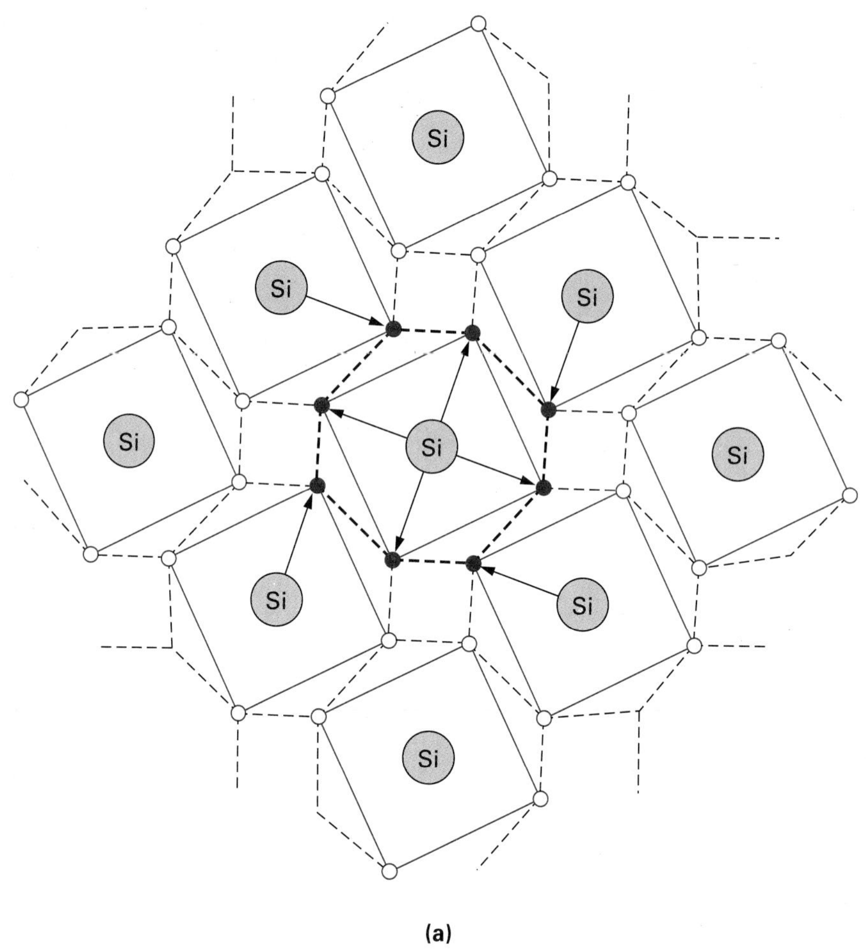

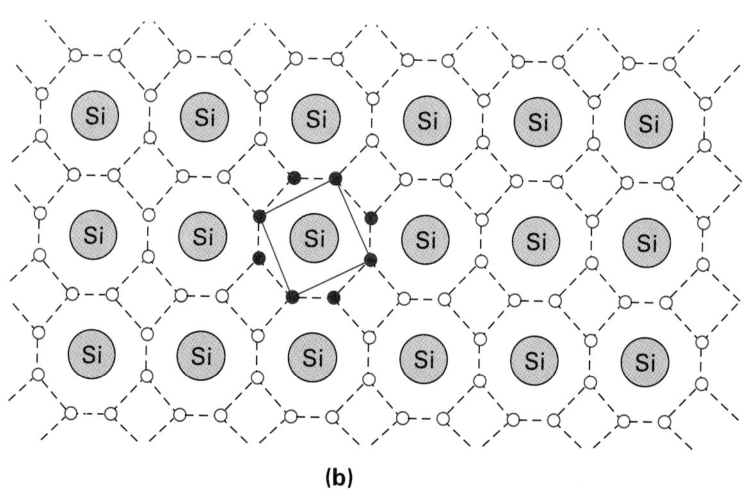

FIGURE 19-16 **Covalent Bonding. (a) Silicon Atoms Sharing Valence Electrons. (b) Silicon Crystal Lattice Structure.**

pure material is available, it must then be modified by a *doping* process to give it the qualities necessary to construct semiconductor devices. Silicon is most frequently used to construct solid state or semiconductor devices such as diodes and transistors because germanium has poor temperature stability and carbon crystals (diamonds) are too expensive to use.

19-3-3 Energy Gaps and Energy Levels

Let us now examine the relationships between electrons and orbital shells in a little more detail so that we can better understand charge and conduction within a semiconductor material.

As mentioned previously, there are seven shells available for electrons (K, L, M, N, O, P, and Q) around the nucleus. Electrons must travel or orbit in one of these orbital paths because they cannot exist in any of the spaces between orbital shells. Each orbital shell has its own specific energy level. Therefore, electrons traveling in a specific orbital shell will contain the shell's energy level. Figure 19-17 shows an example of an atom's orbital shell energy levels. The energy levels for each shell increase as you move away from the nucleus of the atom. The valence shell and the valence electrons will always have the highest energy level for a given atom. The space between any two orbital shells is called the **energy gap.** Electrons can jump from one shell to another if they absorb enough energy to make up the difference between their initial energy level and the energy level of the shell that they are jumping to. For example, in Figure 19-17 the valence shell has an energy level of 1.0 **electron-volts (eV).** Because this atom has three orbital shells, the valence shell will be energy level 3 (e3). The second energy level or orbital shell (e2) has an energy level of 0.6 eV. Therefore, for an electron to jump from energy level 2 (shell 2) to energy level 3 (e3 or valence shell), it will have to absorb a value of energy equal to the difference between e2 and e3. This will equal:

$$1.0 \text{ eV} - 0.6 \text{ eV} = 0.4 \text{ eV}$$

In this example, when either heat, light, or electrical energy was applied, one of the electrons in shell 2 (e2) absorbed 0.4 electron volts of energy and jumped to valence shell (e3).

If a valence (e3) electron absorbs enough energy it can jump from the valence shell into the **conduction band.** The conduction band is an energy band in which electrons can move freely or wander within a solid. When an electron jumps from the valence shell into the conduction band, it is released from the atom and no longer travels in one of its orbital paths. The electron is now free to move within the semiconductor material and is said to be in the **excited state.** An excited electron in the conduction band will eventually give up the energy it absorbed in the form of light or heat and return to its original energy level in the atom's valence shell.

When an electron jumps from the valence shell or band to the conduction band, it leaves a gap in the covalent bond called a **hole.** This action is shown in Figure 19-18(a). A hole is created every time an electron enters the conduction band. This action creates an **electron-hole pair.**

It only takes a few microseconds before a free electron in the conduction

Energy Gap
The space between two orbital shells.

Electron-Volt (eV)
A unit of energy equal to the energy acquired by an electron when it passes through a potential difference of 1 V in a vacuum.

Conduction Band
An energy band in which electrons can move freely within a solid.

Excited State
An energy level in which a nucleus may exist if given sufficient energy to reach this state from a lower state.

Hole
The gap in the covalent bond left when an electron jumps from the valence shell or band to the conduction band.

Electron-Hole Pair
When an electron jumps from the valence shell or band to the conduction band, it leaves a gap in the covalent bond called a hole. This action creates an electron-hole pair.

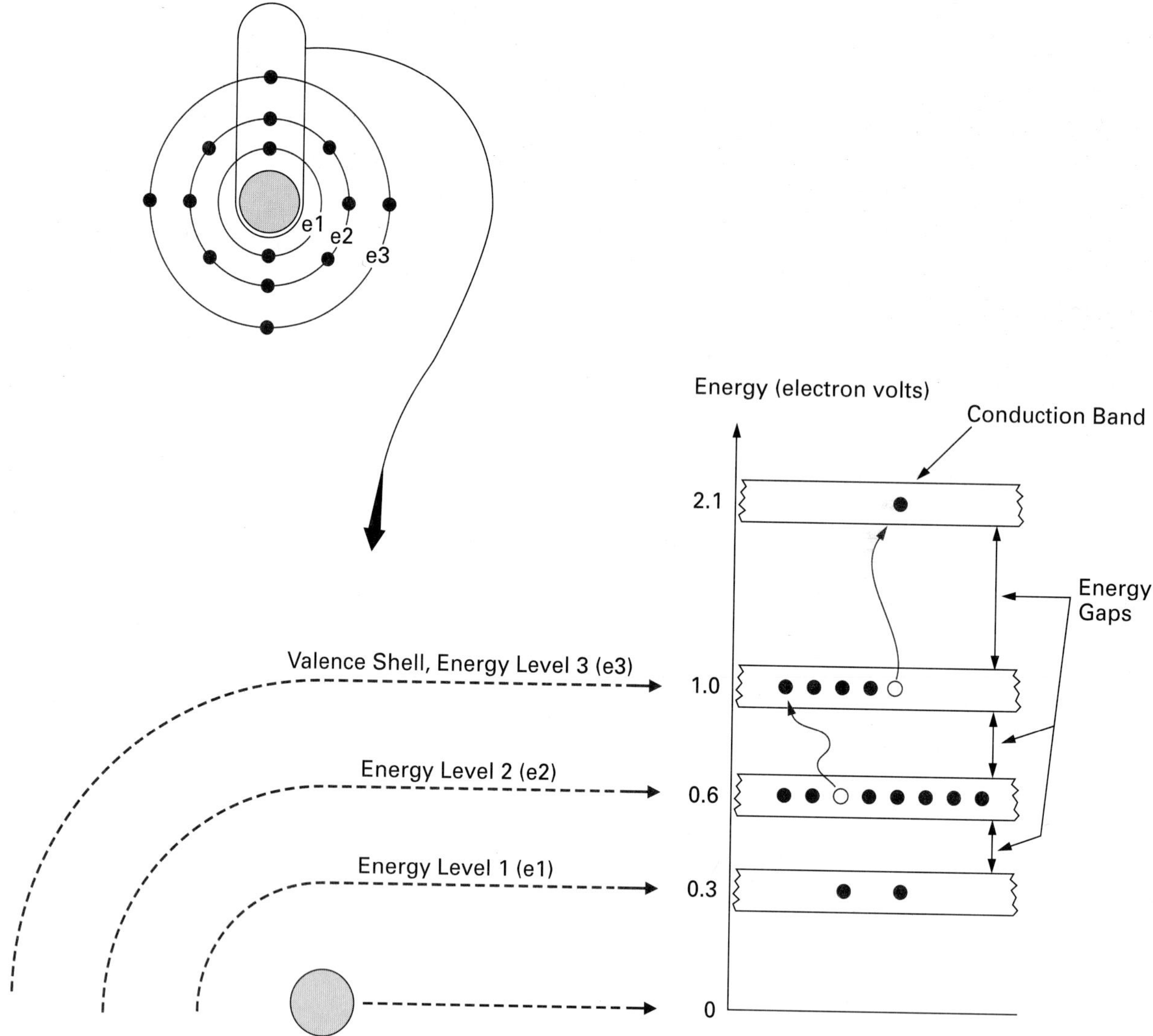

FIGURE 19-17 **An Atom's Orbital Shell Energy Levels.**

Recombination
Action occurring in microseconds as an electron in the conduction band gives up its energy and falls into one of the valence shell holes in the covalent bond.

Lifetime
The time difference between an electron jumping into the conduction band and then falling back into a hole.

band will give up its energy and fall into one of the valence shell holes in the covalent bond. This action is called **recombination** and is shown in Figure 19-18(b). The time difference between an electron jumping into the conduction band (becoming a free electron) and then falling back into a hole (recombination) is called the **lifetime** of the electron-hole pair.

19-3-4 Temperature Effects on Semiconductor Materials

At extremely low temperatures the valence electrons are tightly bound to their parent atoms, preventing valence electrons from drifting between atoms. Therefore, pure or intrinsic semiconductor materials function as insulators at temperatures close to absolute zero (−273.16°C or −459.69°F).

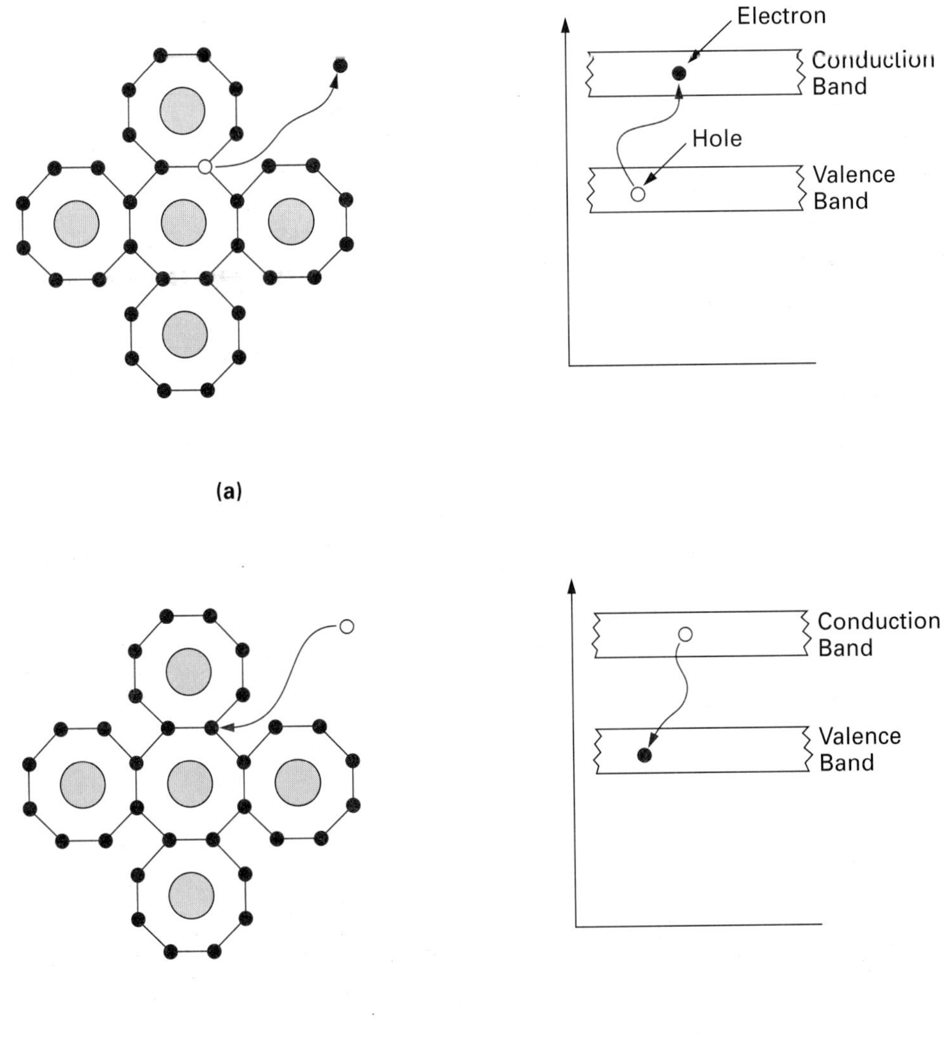

FIGURE 19-18 Valence Band and Conduction Band Actions. (a) Generating an Electron-Hole Pair. (b) Recombination.

At room temperature, however, the valence electrons absorb enough heat energy to break free of their covalent bonds creating electron-hole pairs, as shown in Figure 19-19. Therefore, *the conductivity of a semiconductor material is directly proportional to temperature, in that an increase in temperature will cause an increase in the semiconductor material's conductance.* This means: *an increase in temperature (T↑) will cause an increase in a semiconductor's conductivity (C↑) and current (I↑).* This is why all circuits containing a semiconductor device tend to consume more current once they have warmed up.

Stated another way, *semiconductor materials, and therefore semiconductor devices, have a negative temperature coefficient of resistance which means as temperature increases (T ↑), their resistance decreases (R ↓).*

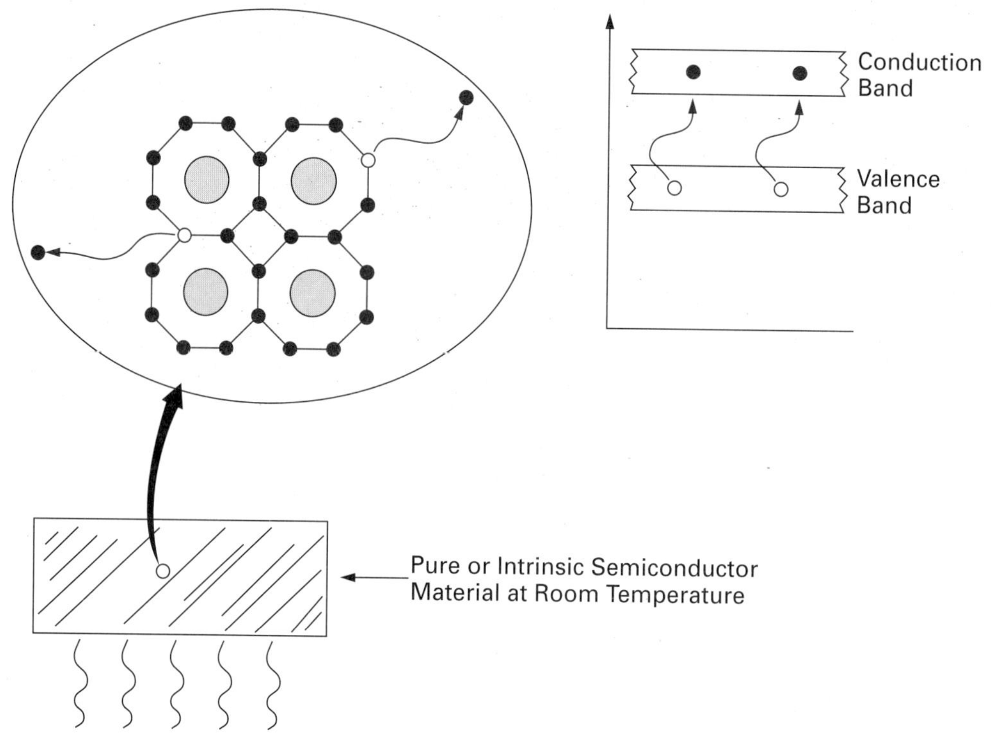

FIGURE 19-19 Temperature Effects on Semiconductor Materials.

19-3-5 *Applying a Voltage Across a Semiconductor*

Free Electrons
An electron that is able to move freely when an external force is applied.

If a voltage was applied across a room-temperature section of intrinsic semiconductor material, **free electrons** in the conduction band would make up a small electrical current as shown in Figure 19-20. In this illustration you can see how the negatively charged free electrons are attracted to the positive terminal of the voltage source. For every free electron that leaves the semiconductor material on the right side and travels to the positive terminal of the source, another electron is generated at the negative terminal of the voltage source and is injected into the left side of the semiconductor material. These injected electrons are captured by holes in the semiconductor material (recombination). As you can see from this illustration, current in a semiconductor material is made up of both electrons and holes. The holes act like positively charged particles while the electrons act like negatively charged particles. As electrons jump between atoms in a migration to the positive terminal of the source voltage, they leave behind them holes which are then filled by other advancing electrons. These advancing electrons leave behind them other holes, making it appear as though these holes are traveling towards the negative terminal of the source voltage. This **hole flow** is a new phenomenon to us, and it is one of the key differences between a semiconductor and a conductor. With conductors we were only interested in free-electron flow, but with semiconductors we must consider the movement of free electrons (negative charge carriers) and the apparent movement of holes (positive charge carriers).

Hole Flow
Conduction in a semiconductor when electrons move into holes when a voltage is applied.

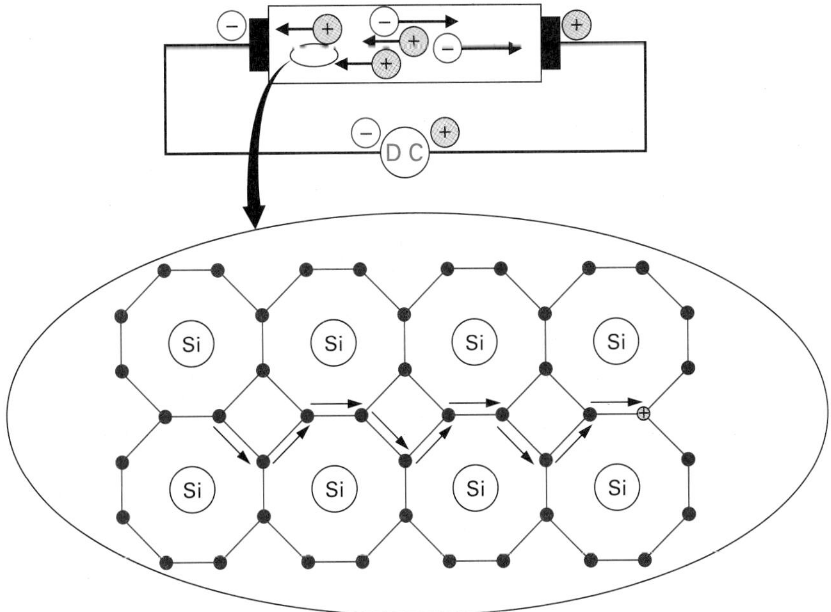

FIGURE 19-20 Electron Flow and Hole Flow in an Intrinsic Semiconductor.

In summary, therefore, *when a potential difference is applied across a semiconductor, the electrons move towards the positive potential and the holes travel towards the negative potential. The total current flow is equal to the sum of the electron flow and the hole flow currents.*

SELF-TEST REVIEW QUESTIONS FOR SECTION 19-3

1. What do all semiconductor atoms have in common?
2. The electrical conductivity of an element is determined by the number of electrons in the valence shell. Semiconductor atoms are midway between conductors, which have ______________ valence electron(s), and insulators, which have ______________ valence electron(s).
3. Each of the semiconductor atoms in a crystal lattice shares its electrons with four neighboring atoms. This joining of atoms is called a ______________ ______________.
4. Semiconductor materials have a ______________ temperature coefficient of resistance, which means as temperature increases, resistance ______________.
5. The number of electron-hole pairs within a semiconductor will increase as temperature ______________.
6. When a pure or ______________ semiconductor is connected across a voltage, free electrons travel towards the ______________ terminal of the applied voltage, whereas holes appear to travel towards the ______________ terminal of the applied voltage.

19-4 DOPING SEMICONDUCTOR MATERIALS

Doping
The process wherein impurities are added to the intrinsic semiconductor material either to increase the number of free electrons or to increase the number of holes.

Extrinsic Semiconductor
A semiconductor whose electrical properties are dependent on impurities added to the semiconductor crystal.

At room temperature pure or intrinsic semiconductors will not permit a large enough value of current. Therefore, some modification has to be applied in order to increase the semiconductor's current carrying capability or conductivity. **Doping** is a process wherein impurities are added to the intrinsic semiconductor material either to increase the number of free electrons (negative doping) or to increase the number of holes (positive doping).

Basically, there are two types of impurities that can be added to semiconductor crystals. One type of impurity is called a *pentavalent material* because its atom has five (*penta*) valence electrons. The second type of impurity is called a *trivalent material* because its atoms have three (*tri*) valence electrons. A doped semiconductor material is referred to as an **extrinsic semiconductor** material because it is no longer pure.

19-4-1 n-Type Semiconductor

Figure 19-21(a) shows how a semiconductor material's atoms will appear after pentavalent atom impurities have been added. The pentavalent atoms, which are listed in Figure 19-21(b), can be added to molten silicon to create, when cooled, a crystalline structure that has an extra electron due to the pentavalent (5 valence-electron impurity) atoms. The fifth pentavalent electron is not part of the covalent bonding and requires little energy to break free and enter the conduction band, as shown in Figure 19-21(c). Because millions of pentavalent atoms are added to the pure semiconductor, there will be millions of free electrons available for flow through the material.

***n*-Type Semiconductor**
A material that has more valence-band electrons than valence-band holes.

Majority Carriers
The type of carrier that makes up more than half the total number of carriers in a semiconductor device.

Minority Carriers
The type of carrier that makes up less than half of the total number of carriers in a semiconductor device.

Even though the doped semiconductor material has millions of free electrons, the material is still electrically neutral. This is because each arsenic atom has the same number of protons as electrons and so do the silicon atoms. Therefore, the overall number of protons and electrons in the semiconductor is still equal and the result is a net charge of zero. However, because we now have more electrons than valence-band holes, the material is called an ***n*-type semiconductor.** *n*-Type semiconductors have more conduction-band electrons than valence-band holes. The electrons are therefore called the **majority carriers** and the valence-band holes are called the **minority carriers.** In Figure 19-21(c) you can see the abundance of conduction-band electrons. The holes in the valence band are few and are generated by thermal energy because the semiconductor is at room temperature.

When a voltage is applied across an *n*-type semiconductor, as shown in Figure 19-21(d), the additional free conduction-band electrons travel toward the positive terminal of the dc source. The applied voltage will cause extra electrons to break away from their covalent bonds to create holes, resulting in an increase in current and conductivity. Although the total current flow in this *n*-type semiconductor is the sum of the electron and hole currents, the conduction band electrons make up the majority of the flow.

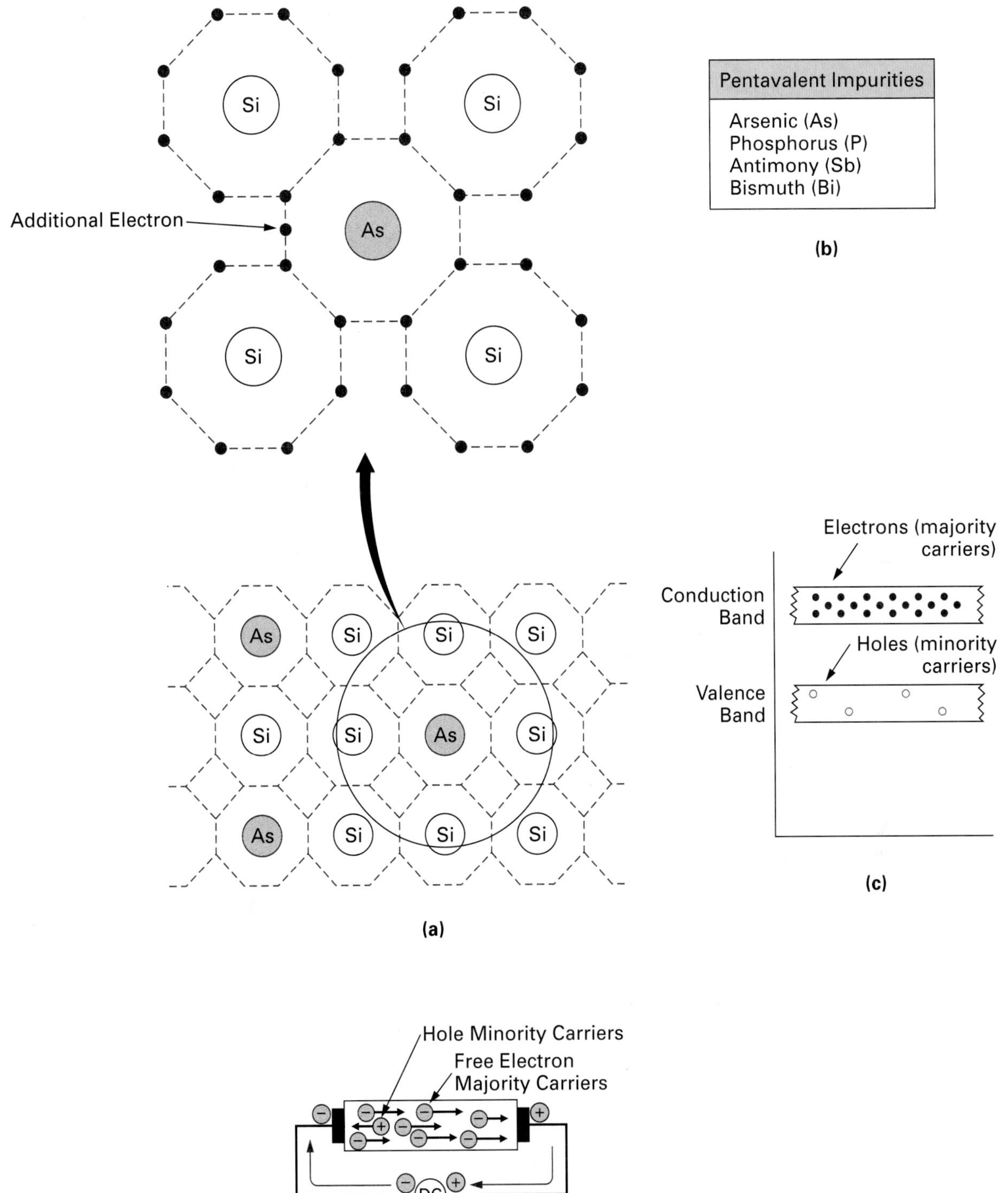

FIGURE 19-21 Adding Pentavalent Impurities to Create an *n*-Type Semiconductor Material.

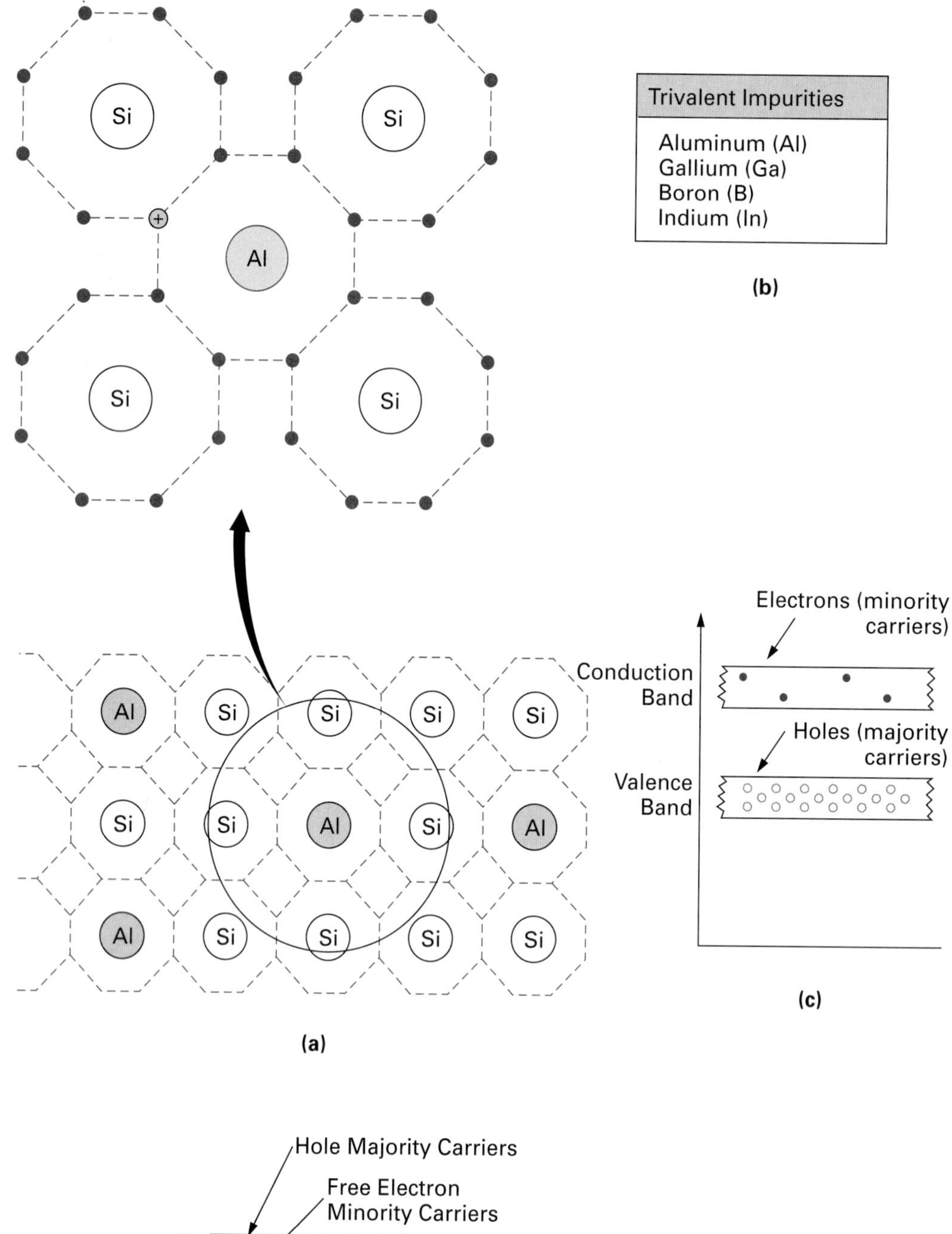

FIGURE 19-22 Adding Trivalent Impurities to Create a *p*-Type Semiconductor Material.

19-4-2 p-Type Semiconductor

Figure 19-22(a) shows how a semiconductor material's atoms will appear after trivalent atom impurities have been added. The trivalent atoms, which are listed in Figure 19-22(b), can be added to molten silicon to create, when cooled, a crystalline structure that has a hole in the valence band of every trivalent (3 valence-electron impurity) atom. Instead of an excess of electrons, we now have an excess of holes. Because millions of trivalent atoms are added to the pure semiconductor, there will be millions of holes available for flow through the material.

Even though the doped semiconductor material has millions of holes, the material is still electrically neutral. This is because each aluminum atom has the same number of protons as electrons and so do the silicon atoms. Therefore the overall number of protons and electrons in the semiconductor is still equal and the result is a net charge of zero. However, because we now have more valence band holes than electrons the material is called a ***p*-type semiconductor.** *p*-Type semiconductors have more valence-band holes than conduction-band electrons. The holes are called the *majority carriers* and the electrons are called the *minority carriers*. In Figure 19-22(c) you can see the abundance of valence-band holes. The few electrons in the conduction band are generated by thermal energy because the semiconductor is at room temperature.

***p*-Type Semiconductor**
A material that has more valence-band holes than valence-band electrons.

When a voltage is applied across a *p*-type semiconductor, as illustrated in Figure 19-22(d), the large number of holes within the material will attract electrons from the negative terminal of the dc source into the *p*-type semiconductor. These holes appear to move because each time an electron moves into a hole it creates a hole behind it, and the holes appear to move in the opposite direction to the electrons (towards the negative terminal of the dc source). The applied voltage will cause some electrons to break away from the covalent bond resulting in an increased current and conductivity. Although the total current flow in this *p*-type semiconductor is the sum of the hole and electron currents, the valence-band holes make up the majority of the flow.

SELF-TEST REVIEW QUESTIONS FOR SECTION 19-4

1. Why are impurities added to pure semiconductor materials?
2. Pentavalent atoms add __________ to semiconductor crystals, to create __________ type semiconductors.
3. Trivalent atoms add __________ to semiconductor crystals, to create __________ type semiconductors.
4. In an *n*-type semiconductor the majority carriers are __________, whereas in a *p*-type semiconductor the majority carriers are __________.

19-5 THE P-N JUNCTION

P-N Junction
The point at which two opposite doped materials come in contact with one another.

On their own, *n*-type semiconductor materials and *p*-type semiconductor materials are of little use. Together, however, these two form a **P-N semiconductor junction.** Semiconductor devices such as diodes and transistors are constructed using these P-N junctions, which give specific current flow characteristics. In this section we will examine the characteristics of the P-N junction in detail.

19-5-1 *The Depletion Region*

Figure 19-23(a) shows the individual *n*-type and *p*-type materials. The *n*-type material is represented as a block containing an excess of electrons (solid circles), while the *p*-type material is represented as a block containing an excess of holes (open circles). The energy diagrams below the two semiconductor sections show the differences between the two materials. Because different impurity atoms were added to the pure semiconductor material, the atomic make-up of the *n*-type and *p*-type materials is slightly different, which is why the valence bands and conduction bands are at slightly different energy levels.

Figure 19-23(b) shows the two *n*-type and *p*-type semiconductor sections joined together. A manufacturer of semiconductor devices would not join two individual pieces in this way to create a P-N junction. Instead, a single piece of pure semiconductor material would have each of its halves doped to create a *p*-type and *n*-type section.

The point at which the two oppositely doped materials come in contact with one another is called the *junction*. This junction of the two materials now permits the free electrons in the *n*-type material to combine with the holes in the *p*-type material as shown in Figure 19-23(c). As free electrons in the *n* material cross the junction and combine with holes in the *p* material, they create negative ions (atoms with more electrons than protons) in the *p* material, and leave behind positive ions (atoms with less electrons than protons) in the *n* material, as shown in Figure 19-23(d). An area or region on either side of the junction becomes emptied or depleted of free electrons and holes. This small layer containing positive and negative ions is called the **depletion region.**

Depletion Region
A small layer on either side of the junction that becomes empty, or depleted, of free electrons or holes.

As the ion layer on either side of the junction builds up, it has the effect of diminishing and eventually preventing any further recombination of free electrons and holes across the junction. In other words, the negative ions in the *p* region near the junction repel and prevent free electrons in the *n* region from recombination. This action prevents the depletion region from becoming larger and larger.

Barrier Potential or Barrier Voltage
The potential difference, or voltage, that exists across the junction.

These positive ions or charges and negative ions or charges accumulate a certain potential. Since these charges are opposite in polarity, a potential difference or voltage called the **barrier potential** or **barrier voltage** exists across the junction as shown in Figure 19-23(e). At room temperature, the barrier voltage of a silicon P-N junction is approximately 0.7 V, and a germanium P-N junction is approximately 0.3 V.

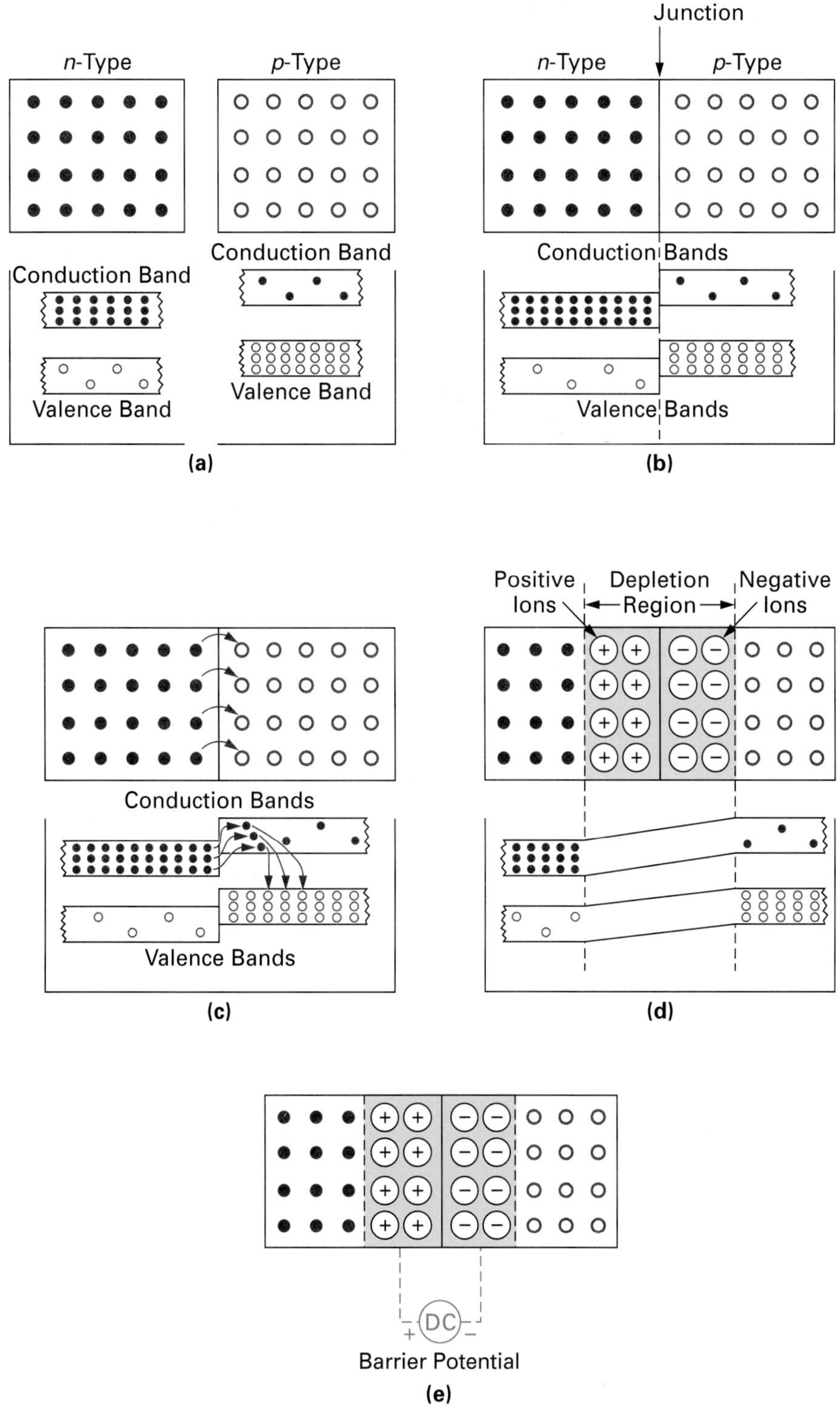

FIGURE 19-23 The Depletion Region.

19-5-2 *Biasing a P-N Junction*

Bias Voltages
The dc voltages applied to control a device's operation.

Semiconductor devices are constructed using P-N junctions. These P-N junctions need voltages of a certain amplitude and polarity to control their operation. These voltages, which incline or cause the device to operate in a certain manner, are known as **bias voltages.** Bias voltages control the width of the depletion region, which in turn controls the resistance of the P-N junction and, therefore, the amount of current that can pass through the P-N junction or semiconductor device.

Forward Biased
A small depletion region at the junction will offer a small resistance and permit a large current. Such a junction is forward biased.

To be specific, a small depletion region ($dr\downarrow$) will offer a small P-N junction resistance ($R\downarrow$) and therefore permit a large P-N junction current ($I\uparrow$). In this instance, the P-N semiconductor junction is said to be **forward biased** and acts like a conductor.

Reverse Biased
A large depletion region at the junction will offer a large resistance and permit only a small current. Such a junction is reverse biased.

On the other hand, a large depletion region ($dr\uparrow$) will offer a large P-N junction resistance ($R\uparrow$) and therefore only permit a small P-N junction current ($I\downarrow$). In this instance the P-N semiconductor junction is said to be **reverse biased** and acts like an insulator.

Forward Biasing a P-N Junction

Figure 19-24 shows in detail why a forward biased P-N junction will pass current with almost no opposition (act like a conductor). To begin, Figure 19-24(a) shows a P-N junction with wires attached. A resistor has been included to limit the amount of current passing through the P-N junction to a safe level. Energy diagrams have also been included on the right side of each part of Figure 19-24 to show the relationship between the conduction band and valence band of the *p* and *n* regions.

Let us now connect a dc voltage across the P-N junction to see how it reacts. This is shown in Figure 19-24(b). The negative potential of the dc source has been applied to the *n* region and the positive potential of the dc source has been applied to the *p* region. Referring to the energy diagram in Figure 19-24(b), you can see that the conduction band electrons in the *n* region are repelled by the negative voltage source towards the junction. On the opposite side, valence band holes in the *p* region are repelled by the positive voltage source towards the junction. A forward-conducting current will begin to flow if the external source voltage is large enough to overcome the internal barrier voltage of the P-N junction. In this example we will assume that the dc source voltage is 10 volts and therefore this will be more than enough to overcome the silicon P-N junction's barrier potential of 0.7 volts.

Forward Voltage Drop (V_F)
The forward voltage drop is equal to the junction's barrier voltage.

Conduction through the P-N junction is shown in Figure 19-24(c). When forward biased, a P-N junction will act as a conductor and have a low but finite resistance value that will cause a corresponding voltage drop across its terminals. This **forward voltage (V_F) drop** is approximately equal to the P-N junction's barrier voltage:

$$\text{Forward Voltage } (V_F) \text{ for Silicon} = 0.7\text{ V}$$

$$\text{Forward Voltage } (V_F) \text{ for Germanium} = 0.3\text{ V}$$

Figure 19-24(c) shows how a voltmeter can be used to measure the forward voltage drop of 0.7 V across a silicon P-N junction when it is forward biased.

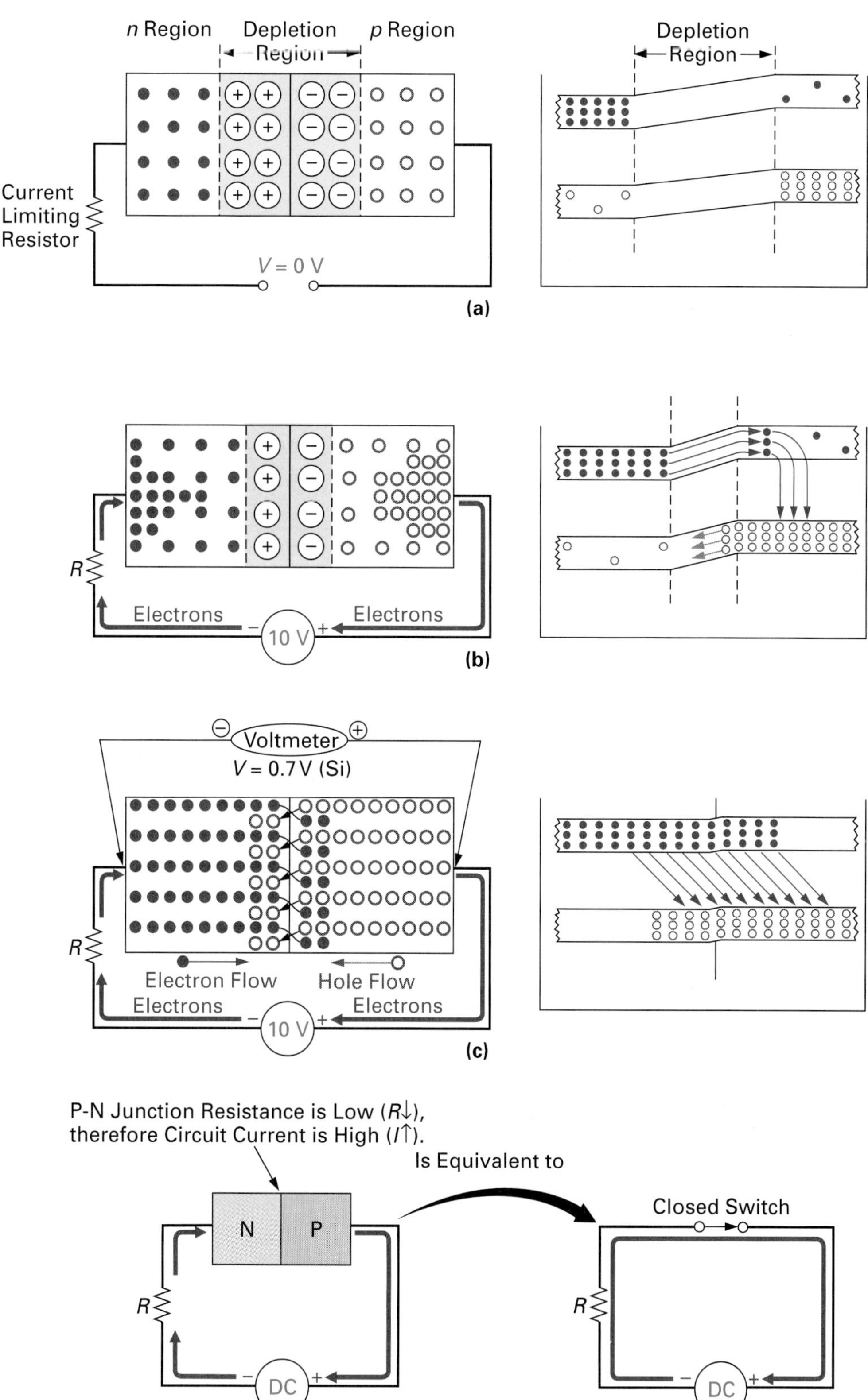

FIGURE 19-24 Forward Biasing a P-N Semiconductor Junction.

To summarize, Figure 19-24(d) shows that when a P-N junction is forward biased ($+V \rightarrow p$ region, $-V \rightarrow n$ region) the P-N junction resistance is low ($R\downarrow$), and therefore the circuit current is high ($I \uparrow$). When forward biased therefore, the P-N junction acts like a conductor and is equivalent to a closed switch.

EXAMPLE:

Calculate the current for the circuit in Figure 19-25.

Solution:

The silicon P-N junction is forward biased ($+V \rightarrow p$ region, $-V \rightarrow n$ region). The applied voltage of 10 V will be more than enough to overcome the silicon P-N junction forward voltage drop of 0.7 volts ($V_F = 0.7$ V for silicon). Since 10 volts is applied, and the P-N junction is dropping 0.7 V, the remaining voltage of 9.3 V is being dropped across the 1 kΩ resistor. Consequently, the forward-biased current (I_F) will equal:

$$I_F = \frac{V_S - V_{P\text{-}N}}{R}$$

$$= \frac{10\text{ V} - 0.7\text{ V}}{1\text{ k}\Omega}$$

$$= 9.3\text{ mA}$$

Reverse Biasing a P-N Junction

Figure 19-26 shows in detail why a reverse biased P-N junction will reduce current to almost zero (act like an insulator). To begin, Figure 19-26(a) shows a P-N junction with wires attached and no voltage being applied. Energy diagrams have again been included to show the relationship between the conduction band and valence band of the *p* and *n* regions.

Let us now connect a dc voltage across the P-N junction to see how it reacts. This is shown in Figure 19-26(b). The positive potential of the dc source is now being applied to the *n* region, and the negative potential of the dc source has been applied to the *p* region.

A forward biased P-N junction is able to conduct current because the external bias voltage forces the majority carriers in the *n* and *p* regions to combine at the junction. In this instance, however, the dc bias voltage polarity has been reversed, causing free electrons in the *n* region to travel to the positive ter-

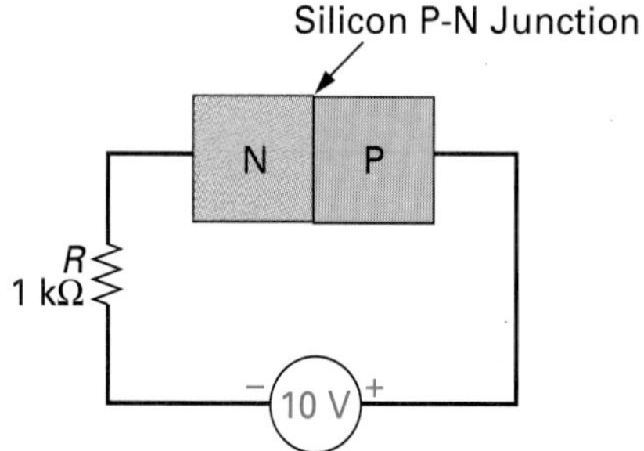

FIGURE 19-25 A P-N Junction Circuit.

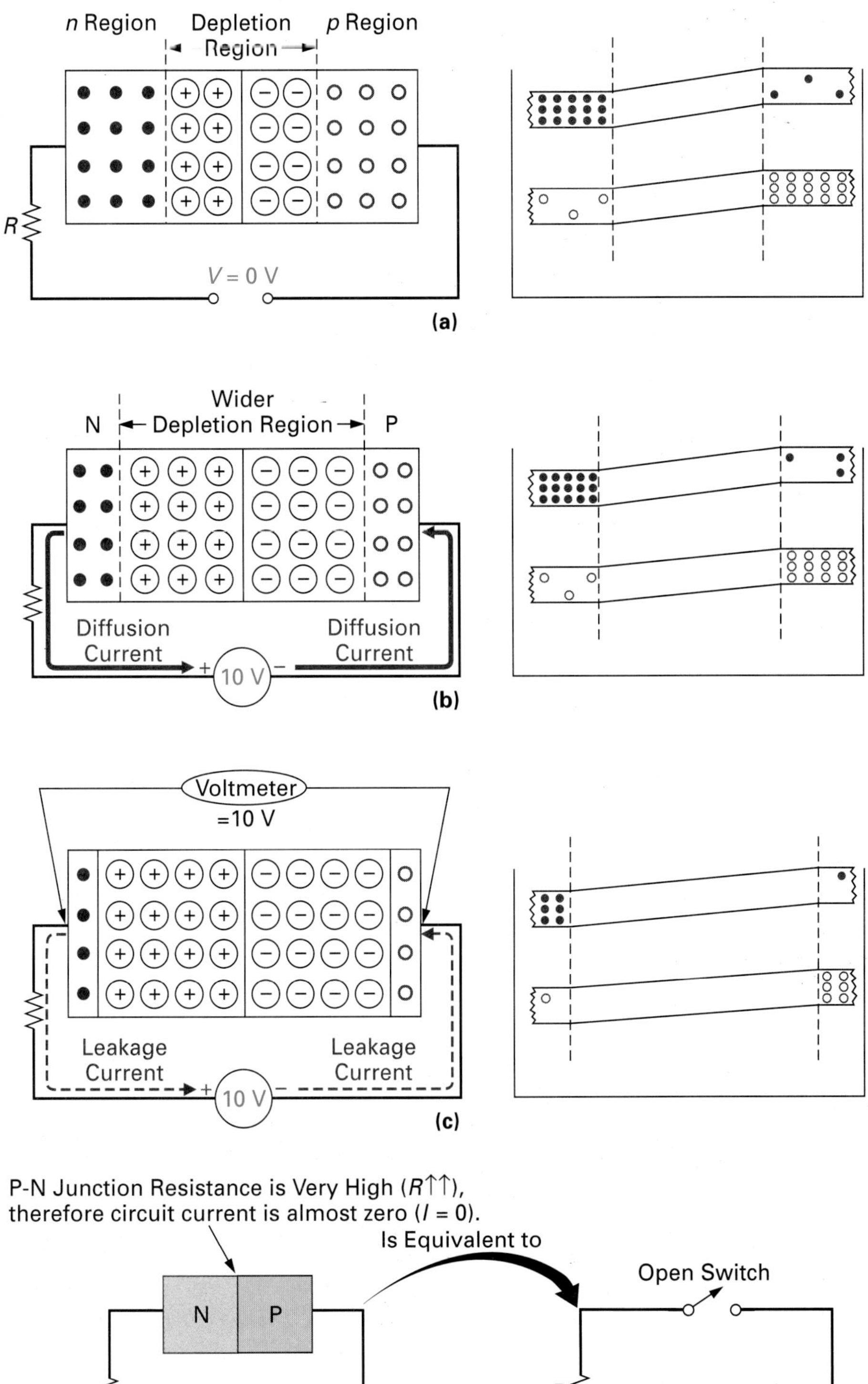

FIGURE 19-26 Reverse Biasing a P-N Semiconductor Junction.

minal of the voltage source leaving behind a large number of positive ions at the junction. This increases the width of the depletion region. At the same time, electrons from the negative terminal of the source are attracted to the holes in the *p* region of the P-N junction. These electrons fill the holes in the *p* region near the junction creating a large number of negative ions. This further increases the width of the depletion region. The current that is present at the time the depletion layer is expanding is called the **diffusion current.** Referring to Figure 19-26(b), you can see that the depletion region is now wider than the unbiased P-N junction shown in Figure 19-26(a).

Diffusion Current
The current that is present when the depletion layer is expanding.

The ions on either side of the junction build up until the P-N junction's internal-barrier voltage is equal to the external-source voltage, as shown by the voltmeter in Figure 19-26(c). When reverse biased, therefore, the **reverse voltage (V_R) drop** across a P-N junction is equal to the source or applied voltage. At this time the resistance of the junction has been increased to a point that current drops to zero.

Reverse Voltage Drop (V_R)
The reverse voltage drop is equal to the source voltage (applied voltage).

Actually, an extremely small current called the **leakage current** or **reverse current (I_R)** will pass through the P-N junction, as shown in Figure 19-26(c). It is present because the minority carriers (holes in the *n* region, electrons in the *p* region) are forced towards the junction where they combine, producing a constant small current. The current in the P-N junction is still considered to be at zero because the leakage or reverse current is so small (nanoamps in silicon diodes).

Leakage Current or Reverse Current (I_R)
The extremely small current present at the junction.

To summarize, Figure 19-26(d) shows that when a P-N junction is reverse biased ($+V \rightarrow n$ region, $-V \rightarrow p$ region), the P-N junction resistance is extremely high ($R\uparrow\uparrow$), and the circuit current is effectively zero ($I = 0$ amps). When reverse biased, therefore, the P-N junction acts like an insulator and is equivalent to an open switch.

EXAMPLE:

Referring to Figure 19-27(a) and (b), calculate each circuit's:

a. current value
b. P-N junction voltage drop

Solution:

The P-N junction in Figure 19-27(a) is reverse biased ($+V \rightarrow n$ region, $-V \rightarrow p$ region); therefore, the P-N junction resistance is extremely high ($R\uparrow\uparrow$), and the circuit current is effectively zero ($I = 0$). When reverse biased, the P-N junction acts like an insulator and is equivalent to an open switch, and the voltage developed across the open P-N junction will equal the source voltage applied. For Figure 19-27(a):

$$\text{Circuit Current} = 0$$
$$\text{P-N Junction Voltage Drop} = V_S = 6\text{ V}$$

The P-N junction in Figure 19-27(b) is forward biased ($+V \rightarrow p$ region, $-V \rightarrow n$ region). Therefore, the P-N junction resistance is low ($R\downarrow$), and the circuit current is high ($I\uparrow$). When forward biased, the P-N junction acts like a conductor and is equivalent to a closed switch, and the P-N junction's voltage drop

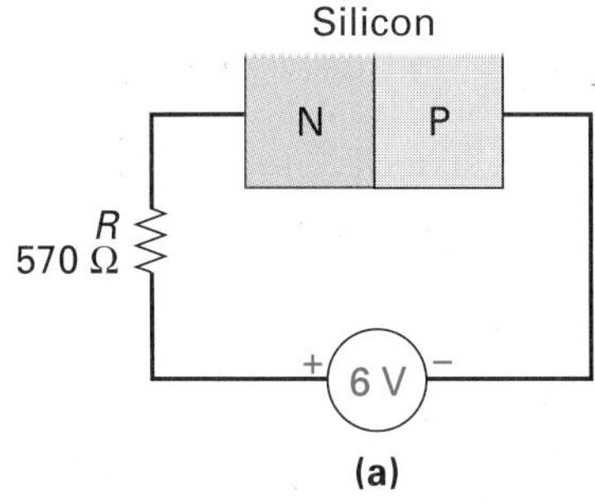

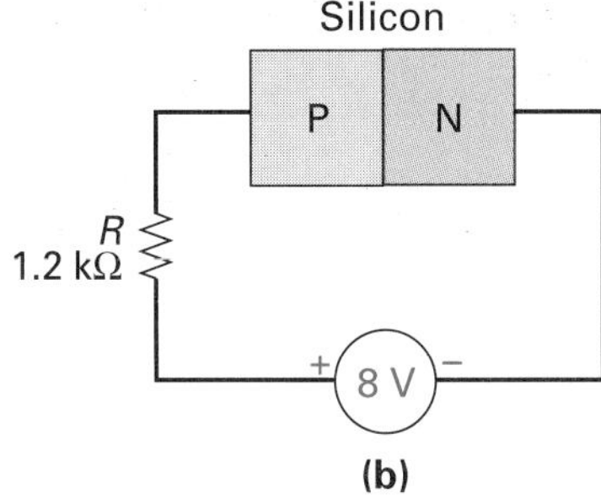

FIGURE 19-27 P-N Junction Circuit Examples.

will equal the forward voltage (V_F) for a silicon P-N junction. For Figure 19-27(b):

$$\text{Circuit Current} = I_F = \frac{V_S - V_{P\text{-}N}}{R}$$

$$= \frac{8\text{ V} - 0.7\text{ V}}{1.2\text{ k}\Omega}$$

$$= 6.08\text{ mA}$$

$$\text{P-N Junction Voltage Drop} = V_F = 0.7\text{ V}$$

SELF-TEST REVIEW QUESTIONS FOR SECTION 19-5

1. When a P-N junction is formed, a ____________ region is created on either side of the junction.
2. The barrier voltage within a silicon diode is:
   a. 700 mV **b.** 7.0 V **c.** 0.3 V **d.** None of the above
3. A P-N junction is forward biased when its P terminal is made positive relative to its N terminal. (True/False)
4. A reverse biased P-N junction acts like a/an ____________ switch, whereas a forward biased P-N junction acts like a/an ____________ switch.

SUMMARY

1. Materials can be divided into three main types according to the way they react to current when a voltage is applied across them. Insulators (nonconductors) are materials that have a very high resistance and therefore oppose current, whereas conductors are materials that have a very low resistance and therefore pass current easily. The third type of material is the semiconductor which, as its name suggests, has properties that lie between the insulator and conductor.
2. Semiconductor materials are neither a good conductor nor insulator. Their advantage is that they can be controlled to either increase their resistance and behave more like an insulator, or decrease their resistance and behave more like a conductor. It is this ability of a semiconductor material to vary its resistive properties that makes it so useful in electrical and electronic applications.
3. Semiconductor materials such as germanium and silicon are used to construct semiconductor devices such as diodes, transistors, and integrated circuits (ICs). These devices are used in electrical and electronic circuits to control current and voltage, so as to produce a desired result.
4. Semiconductor devices first became available in 1960. From about 1920 to 1960, the controlling element in all electrical and electronic circuits was the vacuum tube.
5. The term vacuum tube or thermionic valve is used to describe a variety of special devices that made possible radio, television, telecommunications, radar, sonar, computer, and many more systems between 1920 to 1960. These vacuum tubes were all enclosed in a sealed glass container, which had been pumped free of air (hence the name vacuum), and had electric connections to its internal parts through the base of the container. Two of the most frequently used vacuum tubes were the diode and triode.
6. The vacuum tube was replaced by the transistor in so many applications because it suffered from a number of physical drawbacks. The first is that vacuum tubes are fragile and can be easily damaged by shock or vibration. The second is that the heater required a great deal of power, and the third is their relatively large size. The second and third disadvantages made the vacuum tube unsuitable for battery-operated portable equipment, and, in addition, the tube heater had only a certain life span, which meant that the tubes needed to be replaced regularly.
7. Despite all of the vacuum tube limitations, it is still used today in many areas of science and technology. In industry, 100 kW triode vacuum tubes are made mechanically and electrically very rugged and are used to generate radio frequency (RF) power at frequencies from 100 kHZ to around 30 MHZ. In communications, vacuum tubes are used to generate the high-frequency and high-power outputs needed for radio and television transmitters. In science, vacuum tubes are used in fusion research and linear accelerators where experimental tubes are being operated at 100 kV. Cathode Ray Tubes (CRTs) are still used extensively in televisions and computer monitors.

8. The transistor was first introduced in 1948 by its inventors William Schockley, Walter Bratten, and John Bardeen and was described as a solid state device. This term was used because the transistor contained a solid semiconductor material between its input and output pins, unlike its predecessor the vacuum tube, which had a vacuum between its input and output pins.
9. The first point-contact transistor unveiled in 1948 was extremely unreliable, and it took its inventors another twelve years to develop the superior Bipolar Junction Transistor (BJT) and make it available in commercial quantities.
10. Research and development into semiconductor or solid state devices has resulted in a different type of transistor called the Field Effect Transistor (FET) which has characteristics similar to those of the vacuum tube. Once it was discovered that semiconductor materials could also generate and sense light, a new line of optoelectronic devices became available. Later it was discovered that semiconductor materials could sense magnetism, temperature, and pressure and, as a result, a variety of sensor devices or transducers (energy converters) appeared on the market. Along with all these different types of semiconductor devices, a wide variety of semiconductor diodes emerged that could rectify, regulate, and oscillate at high frequencies.
11. Although semiconductor diodes and transistors are still widely used as individual or discrete components, in 1959 Robert Noyce discovered that more than one transistor could be constructed on a single piece of semiconductor material. Soon other components such as resistors, capacitors, and diodes were added with transistors and then interconnected to form a complete circuit on a single chip or piece of semiconductor material. This integrating of various components on a single chip of semiconductor was called an Integrated Circuit (IC) or IC chip. The IC is used extensively in every branch of electronics with hundreds of thousands of transistors and other components being placed on a chip of semiconductor.
12. All matter on, in, and around the earth can be classified as being either a solid, a liquid, or a gas.
13. An element is a material consisting of only one type of atom.
14. Protons, neutrons, and electrons are subatomic particles that make up the atom.
15. The atomic number of an atom describes the number of protons within the nucleus, whereas the atomic weight of an atom can be used to describe the number of protons and neutrons within the atom's nucleus.
16. Elliptically orbiting electrons travel in paths or shells that are labeled K, L, M, N, O, P, and Q and extend out from the nucleus.
17. Like charges repel one another, while unlike charges attract.
18. An atom is the smallest particle of an element, whereas a molecule (which is the combination of two or more atoms) is the smallest part of a compound.

QUICK REFERENCE SUMMARY SHEET

FIGURE 19-28 Semiconductor Atoms.

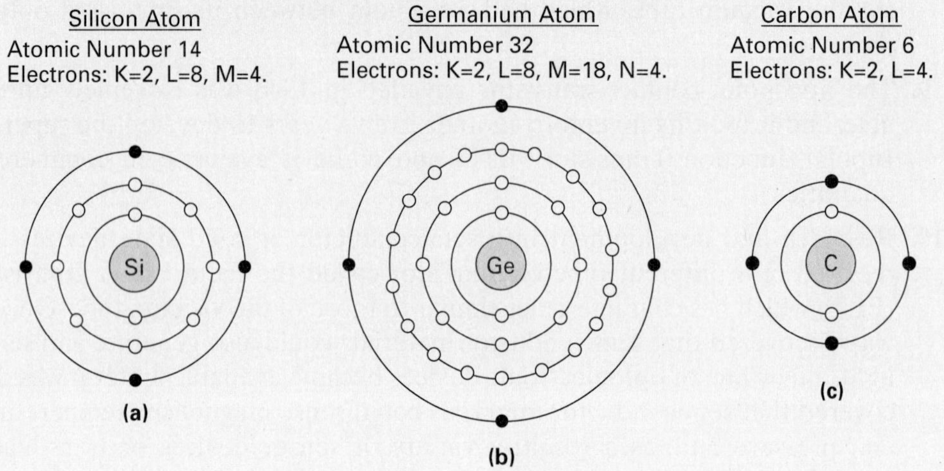

Silicon Atom	Germanium Atom	Carbon Atom
Atomic Number 14	Atomic Number 32	Atomic Number 6
Electrons: K=2, L=8, M=4.	Electrons: K=2, L=8, M=18, N=4.	Electrons: K=2, L=4.

Si

(a)

Ge

(b)

C

(c)

19. A semiconductor material is one that is neither a conductor nor a nonconductor (insulator). This means that it will not conduct current as well as a conductor or block current as well as an insulator.
20. Some semiconductor materials are pure or natural elements such as carbon (C), germanium (Ge), and silicon (Si), while other semiconductor materials are compounds.

Semiconductor Atoms (Figure 19-28)

21. All semiconductor atoms have four valence electrons. The valence shell of an atom can contain up to 8 electrons, and it is the number of electrons in this valence shell that determines the conductivity of the atom. An atom with only 1 valence electron would be classed as a good conductor whereas an atom having 8 valence electrons, and therefore a complete valence shell, would be classed as an insulator.
22. When two or more similar semiconductor atoms are combined to form a solid element, they automatically arrange themselves into an orderly lattice-like structure or pattern known as a crystal. Since each atom shares one electron with a neighboring atom, two atoms will share two, or a pair, of electrons between the two cores. These two-atom cores are pulling the two electrons with equal but opposite force and it is this pulling action that holds the atoms together in this solid crystal lattice structure. The joining together of two semiconductor atoms is called an electron-pair bond or covalent bond.
23. All of the atoms in pure semiconductors are electrically stable because all of their valence shells are complete (they all contain eight electrons). Pure semiconductor materials, called intrinsic materials, are therefore very poor conductors.

24. A pure semiconductor must be modified by a doping process to give it the qualities necessary to construct semiconductor devices.

25. Silicon is most frequently used to construct solid state or semiconductor devices such as diodes and transistors because germanium has poor temperature stability and carbon crystals (diamonds) are too expensive to use.

26. Electrons must travel, or orbit, in one of the orbital paths since they cannot exist in any of the spaces between orbital shells. Each orbital shell has its own specific energy level. The energy levels for each shell increase as you move away from the nucleus of the atom. The valence shell, and therefore the valence electrons, will always have the highest energy level for a given atom. The space between any two orbital shells is called the energy gap.

27. Electrons can jump from one shell to another if they absorb enough energy to make up the difference between their initial energy level and the energy level of the shell that they are jumping to.

28. If a valence electron absorbs enough energy it can jump from the valence shell into the conduction band. The conduction band is an energy band in which electrons can move freely or wander within a solid. When an electron jumps from the valence shell into the conduction band, it is released from the atom and no longer travels in one of its orbital paths. The electron is now free to move within the semiconductor material and is said to be in the excited state.

29. When an electron jumps from the valence shell, or band, to the conduction band, it leaves a gap in the covalent bond called a hole.

30. The conductivity of a semiconductor material is directly proportional to temperature, in that an increase in temperature will cause an increase in the semiconductor material's conductance. Stated another way, semiconductor materials, and therefore devices, have a negative temperature coefficient of resistance that means as temperature increases ($T \uparrow$), their resistance decreases ($R \downarrow$).

31. When a potential difference is applied across a semiconductor, the electrons move towards the positive potential and the holes travel towards the negative potential, and the total current flow is equal to the sum of the electron flow and the hole flow currents. The holes act like positively charged particles while the electrons act like negatively charged particles.

32. At room temperature, pure or intrinsic semiconductors will not permit a large enough value of current. Doping is a process whereby impurities are added to the intrinsic semiconductor material either to increase the number of free electrons (negative doping) or to increase the number of holes (positive doping).

Creating an *n*-Type and *p*-Type Semiconductor Material (Figures 19-29 and 19-30)

33. There are two types of impurities that can be added to semiconductor crystals. One type of impurity is called a pentavalent material because its atom has five (*penta*) valence electrons. The second type of impurity is called a

QUICK REFERENCE SUMMARY SHEET

FIGURE 19-29 Creating an *n*-Type Semiconductor Material.

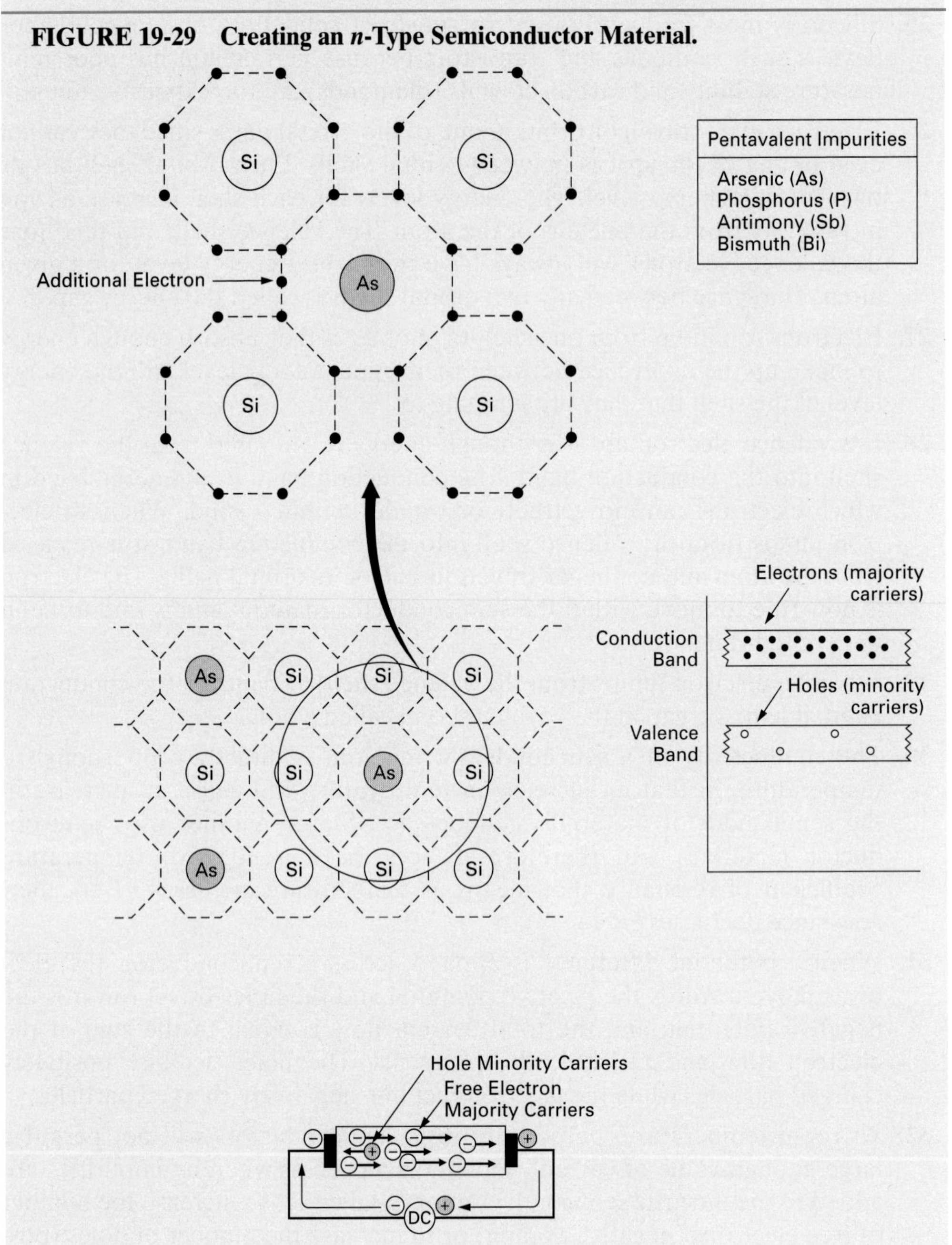

trivalent material because its atoms have three (*tri*) valence electrons. Because a doped semiconductor material is no longer pure it is referred to as an extrinsic semiconductor material.

34. When a voltage is applied across an *n*-type semiconductor, the additional free conduction-band electrons travel toward the positive terminal of the dc source. The applied voltage will cause extra electrons to break away

QUICK REFERENCE SUMMARY SHEET

FIGURE 19-30 Creating a *p*-Type Semiconductor Material.

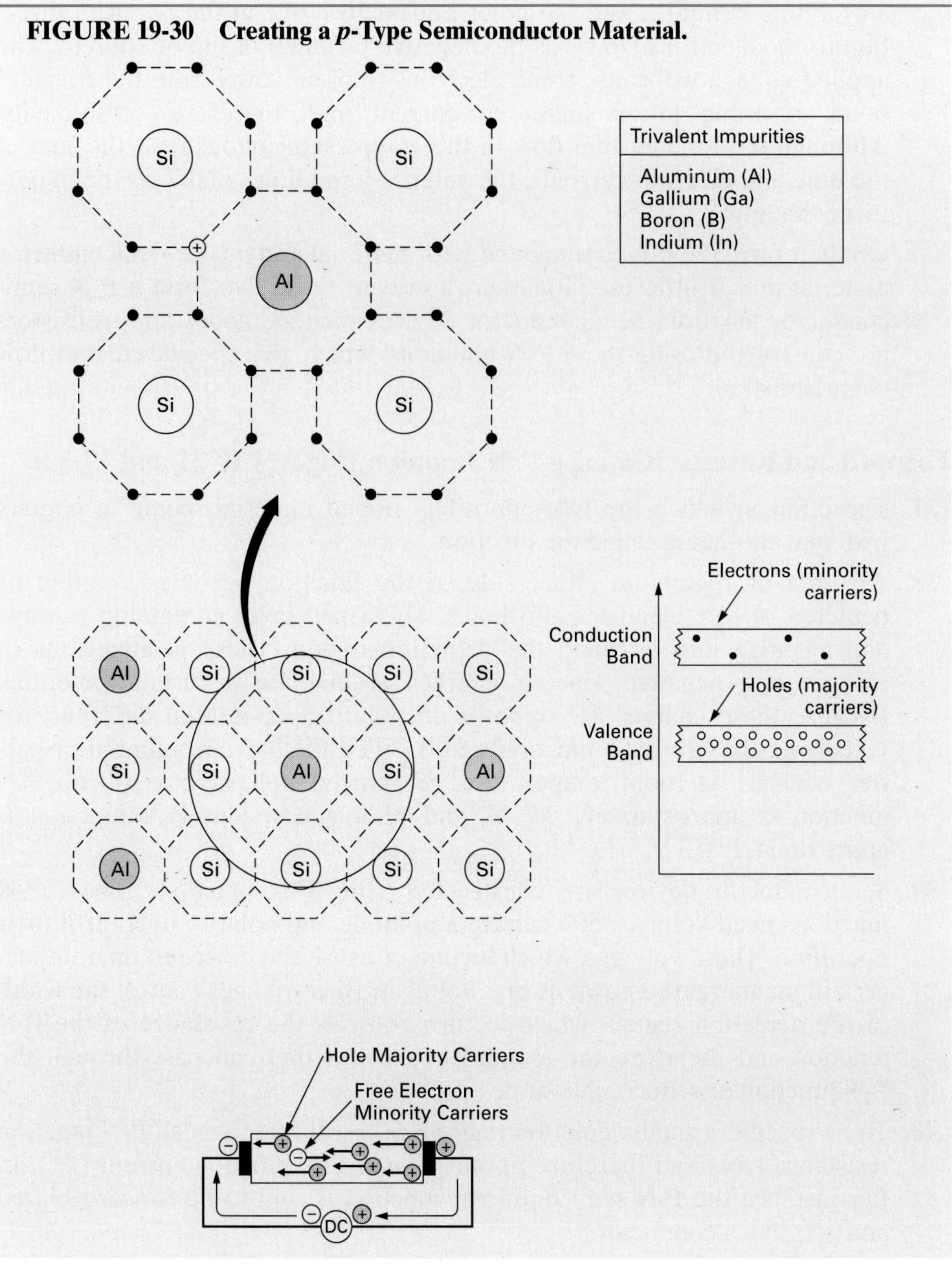

from their covalent bonds to create holes, resulting in an increase in current and therefore conductivity. Although the total current flow in this *n*-type semiconductor is the sum of the electron and hole currents, the conduction-band electrons make up the majority of the flow.

35. When a voltage is applied across a *p*-type semiconductor, the large number of holes within the material will attract electrons from the negative terminal of the dc source into the p-type semiconductor. These holes appear to move because each time an electron moves into a hole, it creates a hole behind it and the holes appear to move in the opposite direction to the electrons (towards the negative terminal of the dc source). The applied voltage will cause some electrons to break away from the covalent bond, resulting in an increased current and, therefore, conductivity. Although the total current flow in this p-type semiconductor is the sum of the hole and electron currents, the valence-band holes make up the majority of the flow.

36. On their own, the n-type semiconductor material and p-type semiconductor material are of little use. Together, however, these two form a P-N semiconductor junction. Semiconductor devices such as diodes and transistors are constructed using these P-N junctions, which give specific current flow characteristics.

Forward and Reverse Biasing a P-N Junction (Figures 19-31 and 19-32)

37. The point at which the two oppositely doped materials come in contact with one another is called the junction.

38. An area or region on either side of the junction becomes emptied or depleted of free electrons and holes. This small layer containing positive and negative ions is called the depletion region. These positive ions or charges and negative ions or charges accumulate a certain potential. Because these charges are opposite in polarity, a potential difference or voltage exists across the junction and is called the barrier potential or barrier voltage. At room temperature, the barrier voltage of a silicon P-N junction is approximately 0.7 V, and of a germanium P-N junction is approximately 0.3 V.

39. Semiconductor devices are constructed using P-N junctions. These P-N junctions need voltages of a certain amplitude and polarity to control their operation. These voltages, which incline or cause the device to operate in a certain manner, are known as bias voltages. Bias voltages control the width of the depletion region, which in turn controls the resistance of the P-N junction and therefore the amount of current that can pass through the P-N junction or semiconductor device.

40. To be specific, a small depletion region ($dr\downarrow$) will offer a small P-N junction resistance ($R\downarrow$) and therefore permit a large P-N junction current ($I\uparrow$). In this instance the P-N semiconductor junction is said to be forward biased and acts like a conductor.

41. On the other hand, a large depletion region ($dr\uparrow$) will offer a large P-N junction resistance ($R\uparrow$) and therefore only permit a small P-N junction current ($I\downarrow$). In this instance the P-N semiconductor junction is said to be reverse biased and acts like an insulator.

42. When a P-N junction is forward biased ($+V \rightarrow p$ region, $-V \rightarrow n$ region), the P-N junction resistance is low ($R\downarrow$); therefore, the circuit current is high ($I\uparrow$). When forward biased, the P-N junction acts like a conductor and is equivalent to a closed switch.

FIGURE 19-31 Forward Biasing a P-N Junction.

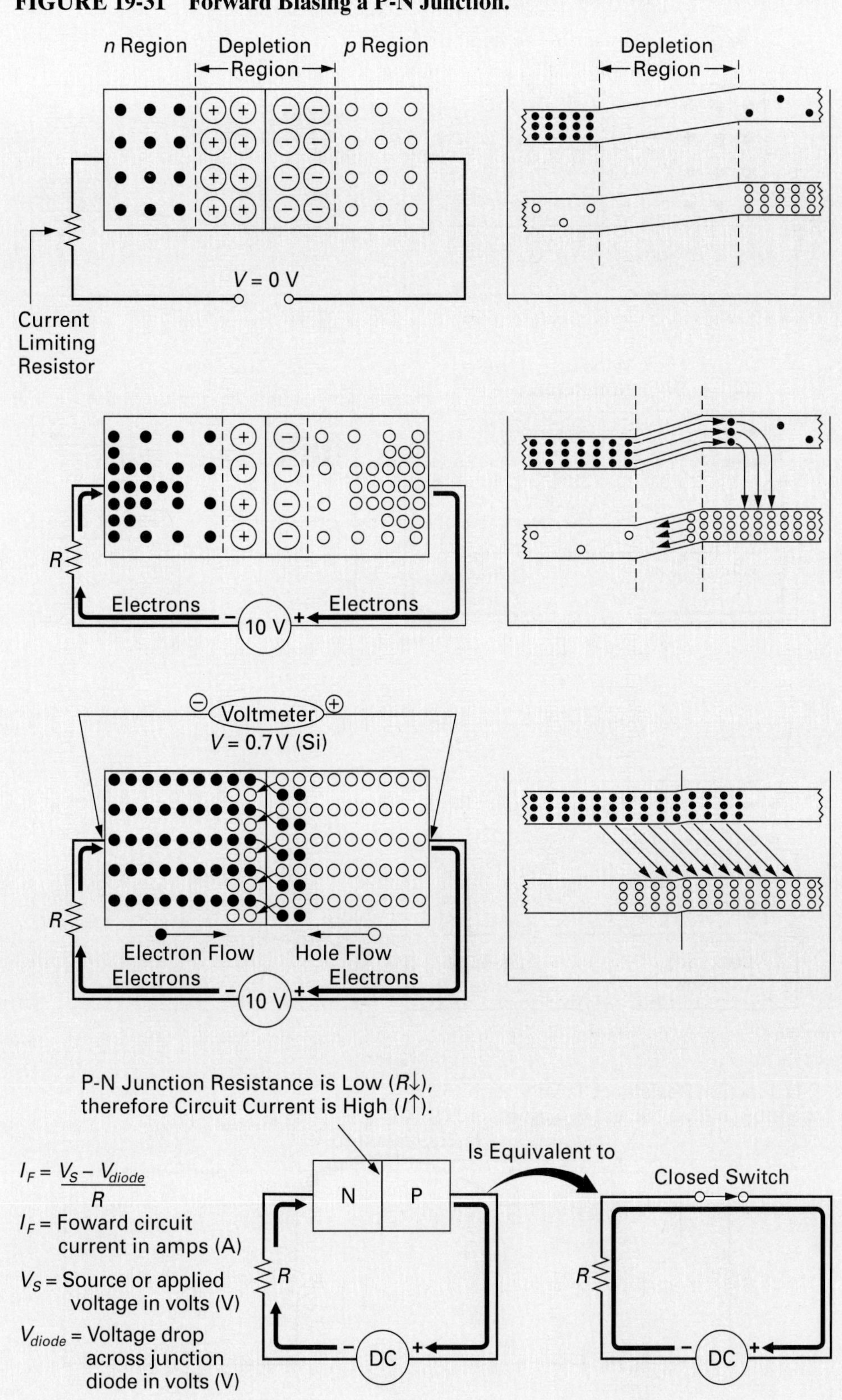

$I_F = \frac{V_S - V_{diode}}{R}$

I_F = Foward circuit current in amps (A)

V_S = Source or applied voltage in volts (V)

V_{diode} = Voltage drop across junction diode in volts (V)

R = Value of resistor in ohms (Ω)

FIGURE 19-32 Reverse Biasing a P-N Junction.

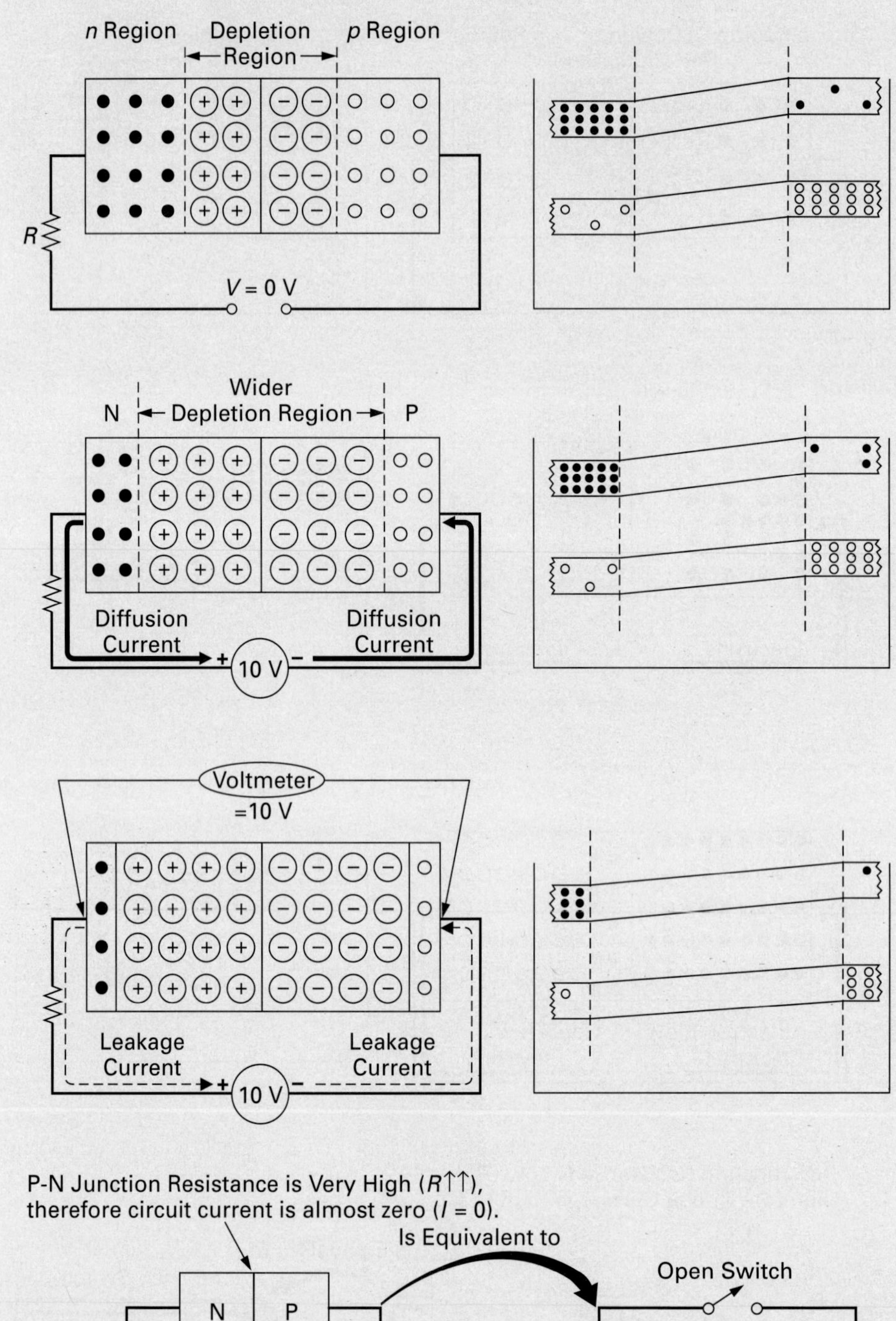

43. When a P-N junction is reverse biased ($+V \rightarrow n$ region, $-V \rightarrow p$ region), the P-N junction resistance is extremely high ($R\uparrow\uparrow$); therefore, the circuit current is effectively zero ($I = 0$ A). When reverse biased, the P-N junction acts like an insulator and is equivalent to an open switch.

NEW TERMS

Anode
Atom
Atomic Number
Atomic Weight
Bands
Barrier Potential
Barrier Voltage
Bias Voltages
Biasing
Bipolar Junction Transistor (BJT)
Carbon
Cathode
Compound
Conduction Band
Conductors
Control Grid
Covalent Bond
Crystal
Depletion Region
Detector
Diffusion Current
Diodes
Discrete Components
Doping
Electron
Electron Flow
Electron-Hole Pair
Electron-Pair Bond
Electronic Numeric Integrator and Computer (ENIAC)
Element
Energy Gap
Energy Level
Excited State
Field Effect Transistor (FET)
Fleming Diode
Forward Biased
Forward Voltage Drop
Free Electron
Germanium
Hole
Hole Flow
IC Chip
Impurity Atom
Insulators
Integrated Circuit (IC)
Intrinsic
Junction
Leakage Current
Lifetime
Like Charges
Majority Carrier
Minority Carrier
Molecule
Negative Ion
Negative Temperature Coefficient of Resistance
Neutral Atom
Neutron
n-Type Semiconductor
Optoelectronic Devices
Pentavalent Material
Pentode
P-N Semiconductor Junction
Point-Contact Transistor
Positive Ion
Proton
p-Type Semiconductor
Recombination
Reverse Biased
Reverse Current
Reverse Voltage Drop
Semiconductor
Shells
Silicon
Solid State Device
Subatomic
Tetrode
Thermionic Electron Emitter
Thermionic Valve
Transistors
Triode
Trivalent Material
Unlike Charges
Vacuum Tube
Valence Electrons
Valence Ring
Valence Shell

REVIEW QUESTIONS

Multiple-Choice Questions

1. What is the atomic number of silicon?

   a. 14 **b.** 16 **c.** 10 **d.** 32

2. How many valence electrons are normally present in the valence shell of a semiconductor material?

   a. 2 **b.** 4 **c.** 6 **d.** 8

3. Adding trivalent impurities to an intrinsic semiconductor will produce a/an ____________ material.

   a. Extrinsic **b.** *n*-type **c.** *p*-type **d.** Both (a) and (b) are true

4. What is the majority carrier in an *n*-type material?

   a. Holes **b.** Electrons **c.** Neutrons **d.** Protons

5. Adding pentavalent impurities to an intrinsic semiconductor will produce a/an ____________ material.

   a. Extrinsic **b.** *n*-type **c.** *p*-type **d.** Both (a) and (b) are true

6. What are the majority carriers in a *p*-type semiconductor?

   a. Holes **b.** Electrons **c.** Neutrons **d.** Protons

7. A semiconductor material has a ____________ temperature coefficient of resistance, which means that as temperature increases its resistance ____________.

   a. Positive, increases **c.** Negative, increases
   b. Positive, decreases **d.** Negative, decreases

8. A hole is considered to be ____________.

   a. Negative **b.** Positive **c.** Neutral **d.** Both (b) and (c) are true

9. Intrinsic semiconductors are doped to increase their ____________.

   a. Resistance **b.** Conductance **c.** Inductance **d.** Reactance

10. As temperature increases, a semiconductor acts more like a/an ____________.

    a. Conductor **b.** Insulator

11. A negative ion has more:

    a. Protons than electrons **c.** Neutrons than protons
    b. Electrons than protons **d.** Neutrons than electrons

12. A positive ion has:

    a. Lost some of its electrons **c.** Lost neutrons
    b. Gained extra protons **d.** Gained more electrons

13. The resistance of a semiconductor material is more than the resistance of:

    a. Glass **b.** Copper **c.** Ceramic **d.** Both (a) and (c) are true

14. The basic function of a semiconductor device in an electrical or electronic circuit is to:

    a. Control current **c.** Increase the price of the equipment
    b. Control voltage **d.** Both (a) and (b) are true

15. For a silicon P-N junction, $V_F = ?$

 a. The value of the applied voltage **b.** 300 mV **c.** 0.7 V **d.** 10 V

Essay Questions

16. What key advantages do solid state devices have over vacuum tube devices? (19-1)
17. What two semiconductor materials are most frequently used in the manufacture of semiconductor devices? (19-3)
18. What do all semiconductor atoms have in common? (19-3-1)
19. Define the following terms:
    - **a.** Covalent Bond (19-3-2)
    - **b.** Intrinsic Semiconductor (19-3-2)
    - **c.** Conduction Band (19-3-3)
    - **d.** Hole Flow (19-3-5)
    - **e.** Doping (19-4)
    - **f.** P-N Semiconductor Junction (19-5)
    - **g.** Depletion Region (19-5-1)
    - **h.** P-N Junction (19-5-1)
    - **i.** Majority Carriers (19-4)
    - **j.** Minority Carriers (19-4)
20. Why do semiconductor materials, and therefore devices, have a negative temperature coefficient of resistance? (19-3-4)
21. Semiconductor devices are made from intrinsic or pure semiconductor materials. True or False? (19-4)
22. What is the relationship between the number of valence electrons in an atom and the conductivity of an element? (19-3-1)
23. Describe the differences between:
    - **a.** An *n*-type semiconductor material (19-4-1)
    - **b.** A *p*-type semiconductor material (19-4-2)
24. Why are doped semiconductor materials still electrically neutral? (19-4-1)
25. How is a P-N junction formed? (19-5)
26. What is a depletion region, and how is it formed? (19-5-1)
27. What is the barrier voltage of:

    a. A silicon P-N junction (19-5-1) **b.** A germanium P-N junction (19-5-1)
28. Describe what occurs when a P-N junction is:

    a. Forward biased (19-5-2) **b.** Reverse biased (19-5-2)
29. Define the following:

    a. Forward Voltage Drop (19-5-2) **b.** Leakage Current (19-5-2)
30. Describe why a P-N junction acts like a switch. (19-5-2)

Practice Problems

31. Which of the silicon P-N junctions in Figure 19-33 are forward biased, and which are reverse biased?

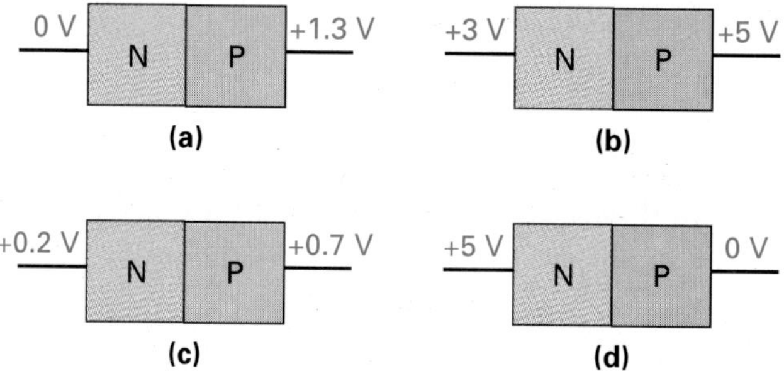

FIGURE 19-33 **Biased P-N Junctions.**

32. Determine the current for the circuits shown in Figure 19-34.
33. What would be the voltage drop (V_F) across each of the P-N junctions shown in Figure 19-34?
34. What would be the voltage drop across each of the resistors in Figure 19-34?
35. Which of the P-N junctions in Figure 19-34 are equivalent to open switches, and which are equivalent to closed switches?

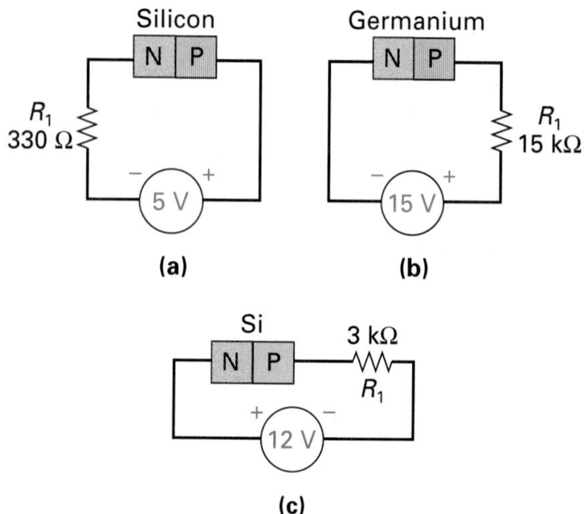

FIGURE 19-34 **P-N Junction Circuit.**

CHAPTER

20

OBJECTIVES

After completing this chapter, you will be able to:

1. Name and identify the terminals of a junction diode package and its schematic symbol.
2. Determine whether a diode is forward or reverse biased by observing the applied bias voltage's polarity.
3. Show how the diode can be used in a basic circuit application.
4. Explain how a junction diode contains a P-N junction.
5. Describe how the *p*-type and *n*-type material of a diode react to forward and reverse bias voltages.
6. Explain the forward and reverse characteristics of junction diodes.
7. Interpret the graphically plotted voltage-current *(V-I)* and temperature characteristic curves for a typical junction diode.
8. Describe how the junction diode is used in the following circuit applications:
   a. A half-wave rectifier circuit
   b. A logic OR gate
9. Determine whether a junction diode is functioning normally by testing it with an ohmmeter.

Junction Diodes

OUTLINE

VIGNETTE

A Problem with Early Mornings

René Descartes was born in Brittany, France, in 1596. At the age of eight he had surpassed most of his teachers at school and was sent on to the Jesuit College in La Flèche, one of the best in Europe. It was here that his genius in mathematics became apparent; however, due to his extremely delicate health, his professors allowed him to study in bed until midday.

In 1616 he had an urge to see the world and so he joined the army which made use of Descartes' mathematical genius in military engineering. While traveling, Descartes met Dutch philosopher Isaac Beekman who convinced him to leave the army and, in his words, "turn his mind back to science and more worthier occupations."

After leaving the army Descartes traveled looking for some purpose, and then on November 10th, 1619, he found it. Descartes was in Neuberg, Germany, where he had shut himself in a well-heated room for the winter. It was on the eve of St. Martin's that a freezing blizzard forced Descartes to retire early. That night he described having an extremely vivid dream that clarified his purpose and showed him that physics and all sciences could be reduced to geometry and were therefore all interconnected like a chain.

In his time, and to this day, he is heralded as an analytical genius. In fact, Descartes' procedure can still be used as a guide to solving any problem.

Descartes' four-step procedure for solving a problem:

1. Never accept anything as true unless it is clear and distinct enough to exclude doubt from your mind.
2. Divide the problem in as many parts as necessary to reach a solution.
3. Start with the simplest things and proceed step by step towards the complex.
4. Review the solution so completely and generally that you are sure nothing was omitted.

For me, this four-step procedure has been especially helpful as a troubleshooting guide for system and circuit malfunctions.

Descartes' fame was so renowned that he was asked in 1649 to tutor Queen Christina of Sweden. The Queen demanded that her lessons begin at 5 o'clock in the morning, which conflicted with Descartes' lifetime practice of remaining in bed until midday. After several unsuccessful attempts to change her majesty's mind, and with pressure being applied by the French ambassador, Descartes agreed to the early morning lessons. A short time later on his way to the palace one cold winter morning, Descartes caught a severe chill and died within two weeks.

INTRODUCTION

It was philosopher René Descartes who first stated that to solve any problem you should start with the simple and then proceed to the complex. For Descartes, there were three approximations: *The first approximation was the simplest, the second approximation contained more detail, and the third approximation was the complex.* I have applied this method to the problem of learning any new topic.

In this chapter you will be introduced to your first semiconductor component: the *diode*. To help you gain a clear understanding of this device, we will begin with a *first approximation description of the diode,* in which we will discuss the diode's schematic symbol, physical appearance, basic operation, and basic application. Following this basic complete picture description, the *second approximation description of the diode* will cover the diode in more detail addressing its characteristics, analog and digital circuit applications, data sheet specifications, and testing procedure.

As a technician, you will not need to examine the *third approximation description of a diode,* since these topics include more detail on specific device specifications and how to implement diodes into new circuit designs. This area of understanding is only needed for engineering students specializing in design. Throughout this text we will be concentrating on semiconductor-device operation, characteristics, applications, and testing, along with analog and digital circuit applications and troubleshooting—the necessary knowledge for a good *electronics technician*.

20-1 FIRST APPROXIMATION DESCRIPTION OF A DIODE

The first diode was accidentally created by Edison in 1883 when he was experimenting with his light bulb. At this time he did not place any importance on the device and its effect, as he could not see any practical application for it. The word *diode* is derived from the fact that the device has two (*di*) electrodes (*ode*).

Once the importance of diodes was realized, construction of the device began. The first diodes were vacuum-tube devices having a hot-filament negative cathode, which released free electrons that were collected by a positive plate called the anode. Today's diode is made of a P-N semiconductor junction but still operates on the same principle. The *n*-type region (cathode) is used to supply free electrons, which are then collected by the *p*-type region (anode). The operation of both the vacuum tube and semiconductor diode is identical in that the device will only pass current in one direction. That is, it will act as a conductor and pass current easily in one direction when the bias voltage across it is of one polarity, yet it will block current and imitate an insulator when the bias voltage applied is of the opposite polarity.

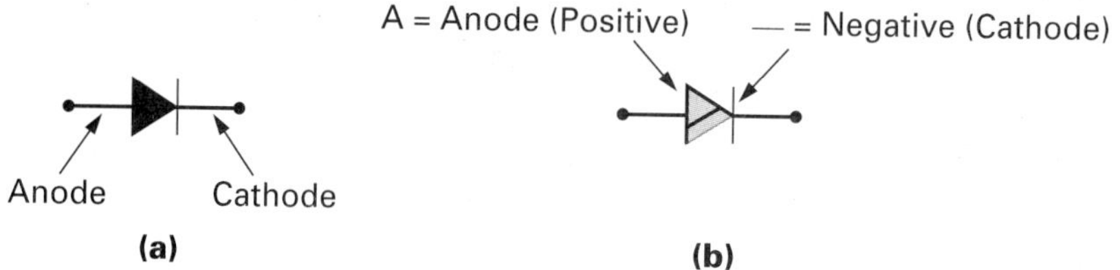

FIGURE 20-1 **Schematic Symbol of a Diode.**

20-1-1 *Diode Schematics Symbol and Packaging*

The two electrodes or terminals of the diode are called the *anode* and *cathode*, as seen in Figure 20-1(a) which shows the schematic symbol of a diode. To help you remember which terminal is the anode and which is the cathode, and which terminal is positive and which is negative, Figure 20-1(b) shows how a line drawn through the triangle section of the symbol will make the letter "A" and indicate the "anode" terminal. Similarly, if the vertical flat side of the diode symbol is aligned horizontally "—", as in Figure 20-1(b), it becomes the "negative" symbol. This memory system helps us to remember that the anode terminal of the diode is next to the triangle part of the symbol and is positive, while the cathode terminal of a diode is next to the vertical line of the symbol and is negative.

The diode is generally mounted in one of the three basic packages shown in Figure 20-2. These packages are designed to protect the diode from mechanical stresses and the environment. The difference in the size of the packages is due to the different current rating of the diode. A black band or stripe is generally placed on the package closest to the cathode terminal for identification purposes, as seen in Figure 20-2(a) and (b). Larger diode packages, like the one seen in Figure 20-2(c), usually have the diode symbol stamped on the package to indicate anode/cathode terminals.

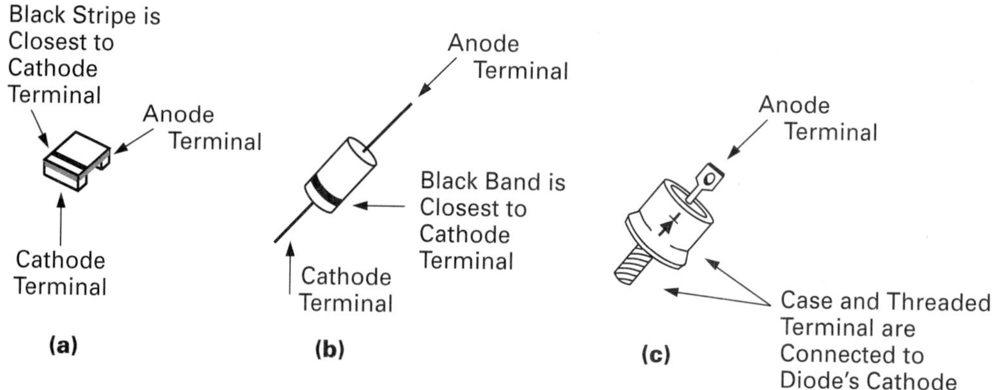

FIGURE 20-2 **Diode Packaging. (a) Chip Package—1/4 A. (b) Small Current Package—Less than 3 A. (c) Large Current Package—Greater than 3 A.**

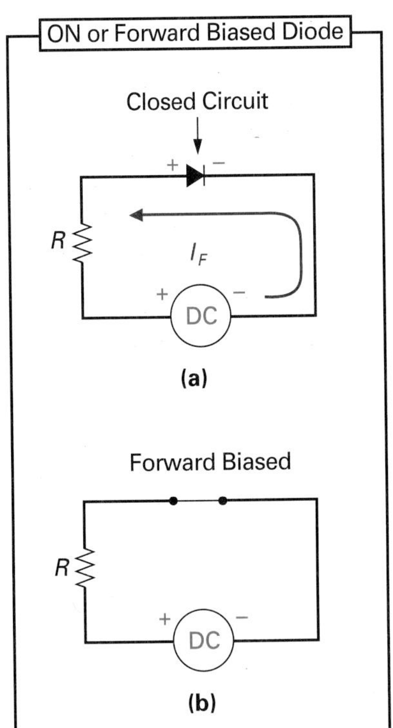

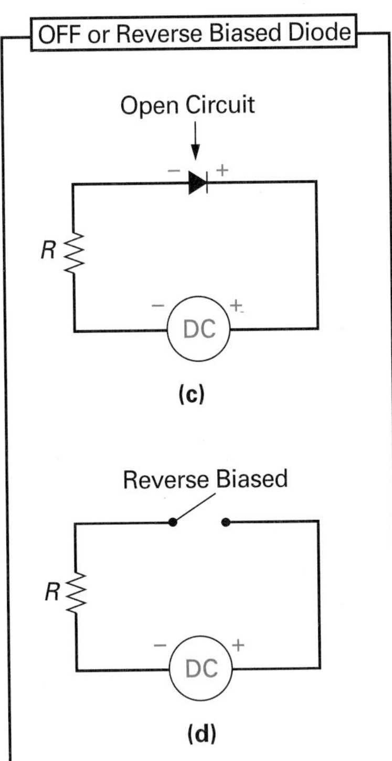

FIGURE 20-3 Diode Operation. (a)(b) Forward Biased (ON) Diode. (c)(d) Reverse Biased (OFF) Diode.

20-1-2 Diode Operation

As far as operation is concerned, the diode operates like a switch. If you give the diode what it wants, that is make the anode terminal positive with respect to the cathode terminal as seen in Figure 20-3(a), the device is equivalent to a closed switch as seen in Figure 20-3(b). In this condition, the diode is said to be ON or *forward biased.*

On the other hand, if you do not give the diode what it wants, that is make the anode terminal negative with respect to the cathode as seen in Figure 20-3(c), the device is equivalent to an open switch as seen in Figure 20-3(d). In this condition the diode is said to be OFF or *reverse biased.*

EXAMPLE:

Determine whether the diodes in Figure 20-4 are ON or OFF.

Solution:

a. Diode is ON since anode is positive relative to cathode.
b. Diode is OFF since anode is negative relative to cathode.
c. Diode is OFF since anode is less positive than cathode.
d. Diode is ON since anode is more positive than cathode.

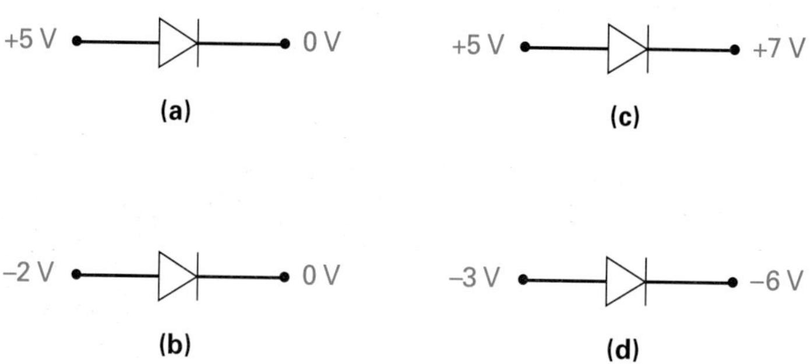

FIGURE 20-4 **ON or OFF Diodes?**

20-1-3 Diode Application

Encoder Circuit
A circuit that produces different output voltage codes, depending on the position of a rotary switch.

As an application, Figure 20-5 shows how the diode can be used as a switch within an **encoder circuit.** The pull-up resistors R_1, R_2, and R_3 ensure that lines A, B, and C are normally all at +5 V. This is the output voltage on each line when the rotary switch is in position 2, as seen in the table in Figure 20-5.

When the rotary switch is turned to position 1, D_1 is connected in-circuit and, because its anode is made positive via R_2 and its cathode is at 0 V, the diode D_1 will turn ON and be equivalent to a closed switch. The 0 V on the cathode of D_1 will be switched through to line B (all of the five volts will be dropped across R_2) producing an output voltage code of A = +5 V, B = 0 V, C = +5 V as seen in the table in Figure 20-5.

When the rotary switch is turned to position 3, D_2 and D_3 are connected in circuit and because both anodes are made positive via R_1 and R_3, and both diode cathodes are at 0 V, D_2 and D_3 will turn ON. These forward biased diodes

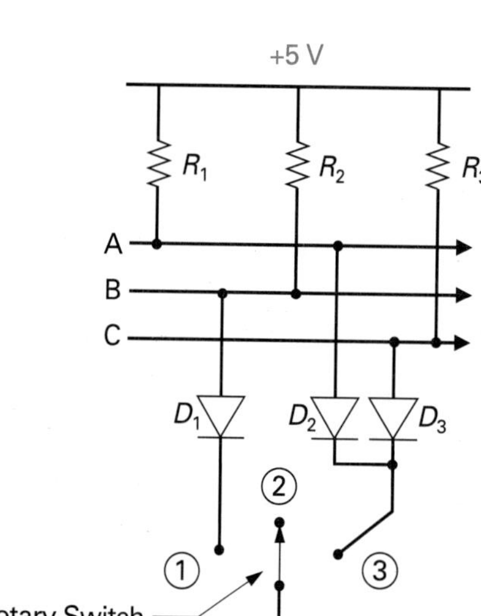

Input	Output		
Switch Position	A	B	C
①	+5 V	0 V	+5 V
②	+5 V	+5 V	+5 V
③	0 V	+5 V	0 V

FIGURE 20-5 **Diode Application: A Switch Encoder Circuit.**

will switch 0 V through to lines A and C, producing an output voltage code of $A = 0$ V, $B = +5$ V, $C = 0$ V, as seen in the table in Figure 20-5.

This *code generator* or *encoder* circuit will produce three different output voltage codes for each of the three positions of the rotary switch. These codes could then be used to initiate one of three different operations based on the operator setting of the rotary control switch.

SELF-TEST REVIEW QUESTIONS FOR SECTION 20-1

1. Name the two terminals of a diode.
2. The diode operates like a ____________.
3. An ON diode is said to be ____________ biased, while an OFF diode is said to be ____________ biased.
4. Would a diode be ON or OFF if its anode had +6 V applied and its cathode had +9 V applied?

20-2 SECOND APPROXIMATION DESCRIPTION OF A JUNCTION DIODE

Now that we understand the basic operation and application of the diode, let us examine its characteristics in more detail.

20-2-1 *The P-N Junction Within a Junction Diode*

In the previous chapter, we discussed how a pure or intrinsic semiconductor material could be doped with a pentavalent element or trivalent element to obtain the two basic semiconductor types. The first type is called an *n*-type semiconductor because its majority carriers are electrons, while the second type is called *p*-type semiconductor because its majority carriers are holes. On their own, the *n*-type semiconductor and *p*-type semiconductor are of little use. Together, however, these two form a P-N semiconductor junction. A manufacturer of semiconductor devices would not join two individual pieces to create a P-N junction. Instead, a single piece of pure semiconductor material would have each of its halves doped to create a *p*-type and *n*-type section or region.

Semiconductor devices such as diodes and transistors are constructed using these P-N junctions. A diode, for example, has only one P-N junction and is created by doping a single piece of pure semiconductor to produce an *n*-type and *p*-type region. A bipolar junction transistor, on the other hand, has two P-N junctions and is created by doping a single piece of pure semiconductor with three alternate regions (NPN or PNP). As mentioned in Chapter 19, the point at which these two opposite-doped materials come in contact with each other is

Junction Diode
A semiconductor diode whose ON/OFF characteristics occur at a junction between the *n*-type and *p*-type semiconductor materials.

called a junction, which is why these devices are called **junction diodes** and *bipolar junction transistors.*

Figure 20-6 illustrates the schematic symbol for the *junction diode,* and the inset shows how it contains one P-N junction. The *n*-type region is called the *cathode,* while the *p*-type region is called the *anode.*

Since a junction diode is basically a P-N semiconductor junction, the diode will operate in exactly the same way as the P-N junction described in Chapter 19.

20-2-2 *Biasing a Junction Diode*

Bias Voltage
Voltage that inclines or causes the diode to operate in a certain manner.

Semiconductor diodes are constructed using P-N junctions. These P-N junctions need voltages of a certain amplitude and polarity to control their operation. These voltages, which incline or cause the diode to operate in a certain manner, are known as **bias voltages.** Bias voltages control the width of the depletion region, which in turn controls the resistance of the junction and, therefore, the amount of current that can pass through the P-N junction diode.

Forward Biasing a Diode

Figure 20-7(a) shows how a junction diode can be forward biased. Like the P-N junction, the junction diode's operation is determined by the polarity of the applied voltage. In this figure the negative terminal of the applied voltage is connected to the *n* region of the diode, and the positive terminal of the applied voltage is connected to the *p* region of the diode ($+V \rightarrow p$ region, $-V \rightarrow n$ region). Free electrons are repelled from the *n* region by the negative source and attracted to the positive terminal of the voltage source. This forward-conducting electron flow will only occur if the external source voltage is large enough to overcome the internal barrier voltage of the junction diode. For a silicon diode, the external source voltage must be equal to or greater than 0.7 V, whereas for a germanium diode, the applied voltage must be equal to or greater than 0.3 V. The resistor is added in series with the diode to limit the forward current to a safe level because an excessive current will generate more heat than the diode can dissipate, causing the diode to burn out.

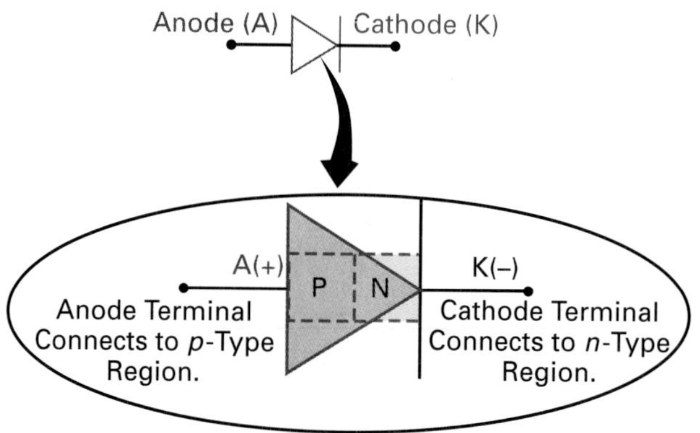

FIGURE 20-6 The P-N Junction Within a Junction Diode.

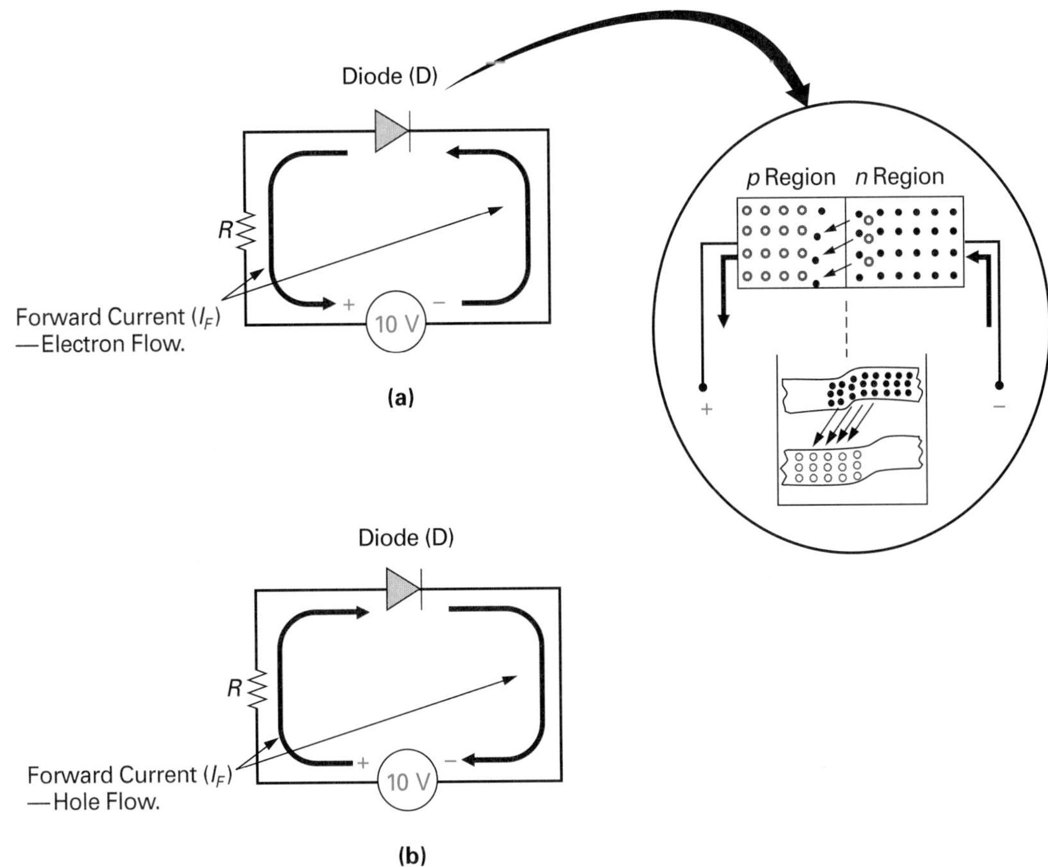

FIGURE 20-7 A Forward Biased P-N Junction Diode. (a) Electron Flow. (b) Hole (Conventional) Flow.

A forward biased diode will conduct current as long as the external bias voltage is of the correct polarity and amplitude. Figure 20-7(a) shows the direction of forward current *(I_F)*. This electron-flow current is from the negative terminal of the applied voltage to the *n* region of the diode, through the diode to the *p* region, and then to the positive terminal of the applied voltage source. This means that forward (electron flow) current passes through the diode symbol from the bar to the triangle. In other words, forward electron flow is actually traveling against the arrow formed by the diode's symbol. To further explain this, let us compare Figure 20-7(a), which shows forward electron flow, to Figure 20-7(b), which shows forward hole flow. As previously discussed in Chapter 19, apparent hole flow is in the opposite direction to electron flow. That is, electron flow is from negative to positive, while hole flow, or conventional current flow, is from positive to negative. The diode symbol actually points in the direction of forward hole flow, and therefore the symbol reflects conventional flow. Remember, when a P-N junction or diode is forward biased, the electrons move towards the positive terminal of the applied voltage, and the holes travel towards the negative terminal of the applied voltage. The total current flow is equal to the sum of the electron flow and the hole flow currents.

Like the P-N junction, a diode, when forward biased, has a low but finite resistance value that will cause a corresponding voltage drop across its terminals. This voltage drop is approximately equal to the barrier voltage of the diode (Si = 0.7 V, Ge = 0.3 V). Knowing the diode's forward voltage drop, the value of applied voltage, and the value of circuit resistance, we can calculate the value of forward current. You will recognize this formula because it is identical to the one used in Chapter 19 for the P-N junction.

$$I_F = \frac{V_S - V_{diode}}{r}$$

I_F = Forward circuit current in amps (A)
V_S = Source or applied voltage in volts (V)
V_{diode} = Voltage drop across junction diode in volts (V)
R = Value of resistor in ohms (Ω)

EXAMPLE:

Calculate the value of current for the circuit shown in Figure 20-8.

Solution:

The diode is forward biased because the applied voltage is connected so that its positive terminal is applied to the *p* region (anode) and the negative terminal is applied to the *n* region (cathode). Because a silicon diode is being used, the forward voltage drop will be 0.7 V. With an applied voltage of 8.5 V and a circuit resistance of 1.2 kΩ, the circuit current will equal

$$I_F = \frac{V_S - V_{diode}}{R}$$

$$I_F = \frac{8.5\text{ V} - 0.7\text{ V}}{1.2\text{ k}\Omega}$$

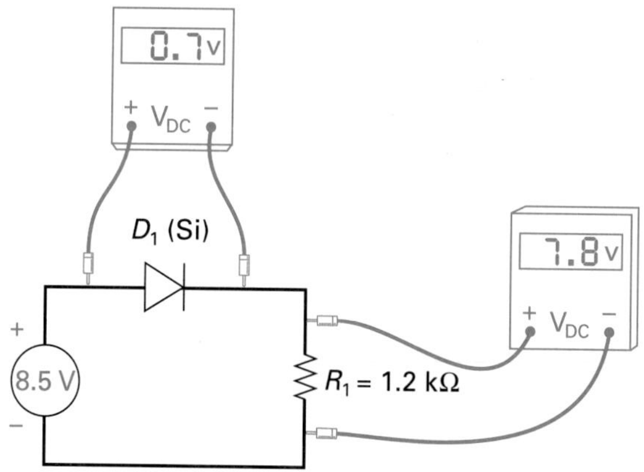

FIGURE 20-8 A P-N Junction Diode Circuit.

$$I_F = \frac{7.8\text{ V}}{1.2\text{ k}\Omega}$$

$$I_F = 6.5\text{ mA}$$

Reverse Biasing a Diode

Figure 20-9 shows how a junction diode can be reverse biased. Once again, the junction diode's operation is determined by the polarity of the applied voltage. In this figure the positive terminal of the applied voltage is connected to the *n* region of the diode, and the negative terminal of the applied voltage is connected to the *p* region of the diode ($+V \rightarrow n$ region, $-V \rightarrow p$ region). This applied voltage polarity will reverse bias the junction diode since free electrons in the *n* region are attracted and therefore travel to the positive terminal of the applied source voltage. At the same time, holes feel the attraction of the negative terminal of the applied voltage. As a result, the depletion region will increase in width until its internal voltage is equal, but opposite, to the external applied voltage. At this time the diode will stop conducting, cutting off all current flow.

In actual fact, a very small current leaks through the diode when it is reverse biased. This current is extremely small (microamps to nanoamps in value) and in most cases can be ignored. It is caused by minority carriers, which are holes in the *n* region and electrons in the *p* region, that are forced towards the junction by the applied source voltage.

EXAMPLE:

Calculate the voltage drop across the diodes and the resistors in Figure 20-10.

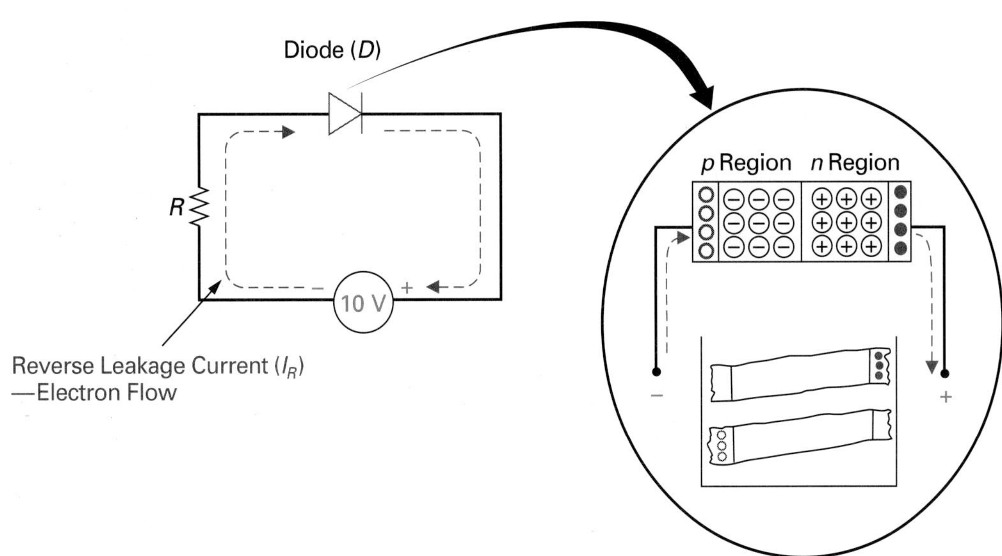

FIGURE 20-9 A Reverse Biased P-N Junction Diode.

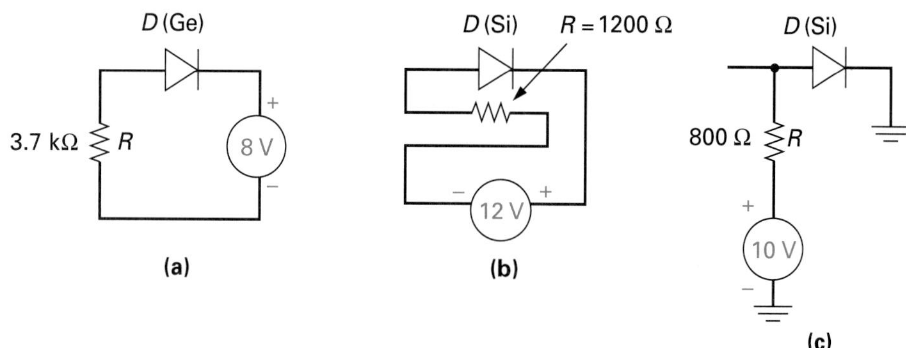

FIGURE 20-10 **P-N Junction Diode Biasing Examples.**

Solution:

a. The germanium diode in Figure 20-10(a) is reverse biased ($+V \rightarrow n$ region, $-V \rightarrow p$ region), and is therefore equivalent to an open switch. Since all of the applied voltage will always appear across an open in a series circuit

$$V_{diode} = V_S = 8\text{ V}$$
$$V_R = 0\text{ V}$$

b. The silicon diode in Figure 20-10(b) is also reverse biased ($+V \rightarrow n$ region, $-V \rightarrow p$ region), and is therefore equivalent to an open switch. Since all of the applied voltage will always appear across an open in a series circuit:

$$V_{diode} = V_S = 12\text{ V}$$
$$V_R = 0\text{ V}$$

c. The silicon diode in Figure 20-10(c) is forward biased ($+V \rightarrow p$ region, $-V$ or ground $\rightarrow n$ region), and is therefore equivalent to a closed switch. The voltage drop across a forward biased silicon diode is approximately equal to the barrier voltage of the diode, which is 0.7 V. The voltage drop across the resistor will therefore be equal to the difference between the applied source voltage (V_S) and the voltage drop across the diode (V_{diode}).

$$V_{diode} = 0.7\text{ V}$$
$$V_R = V_S - V_{diode} = 10\text{ V} - 0.7\text{ V} = 9.3\text{ V}$$

20-2-3 *The Junction Diode's Characteristic Curve*

Now that we know how the junction diode operates, it is time to examine the diode's characteristics in a little more detail. To help us analyze the P-N junction diode's voltage, current, and temperature characteristics at various values, we will plot the diode's characteristics on a graph, since a picture is generally worth a thousand words.

Graph Origin
Center of the graph where the horizontal axis and vertical axis cross.

The graph in Figure 20-11 shows how much current will pass through a typical junction diode when it is forward biased or reverse biased. The center of the graph, where the horizontal axis and vertical axis cross, is called the **graph ori-**

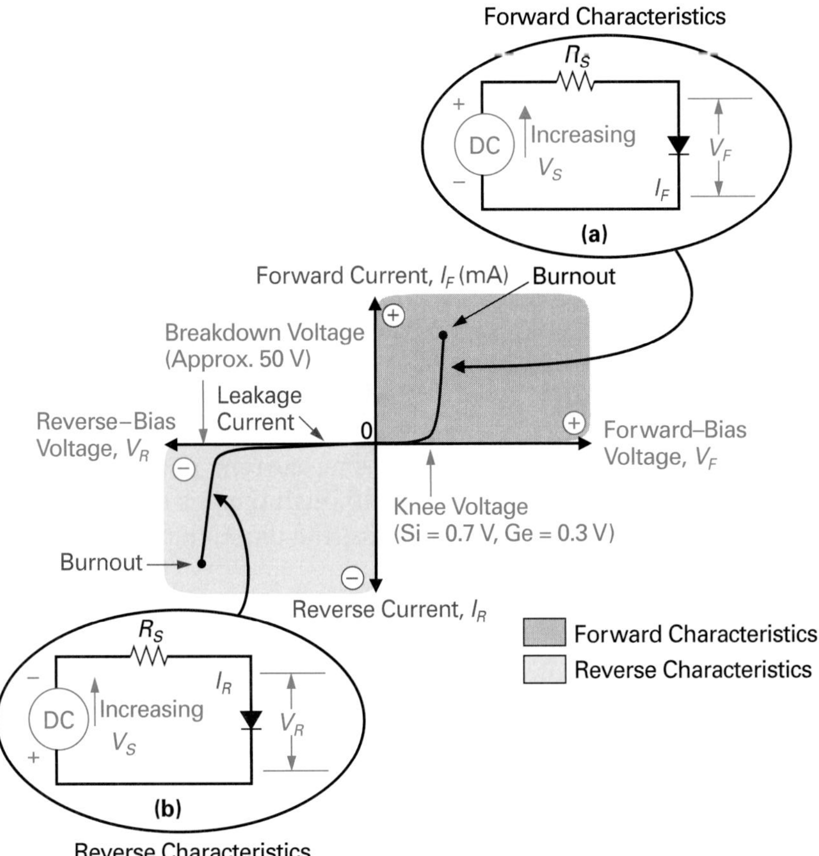

FIGURE 20-11 **The Junction Diode Voltage/Current Characteristic Curve.**

gin. This origin is a zero point for both voltage and current. For instance, voltage is plotted on the horizontal axis, with forward bias voltages *(V_F)* increasing positively to the right of the origin and reverse bias voltages *(V_R)* increasing negatively to the left of the origin or zero voltage point. Conversely, current is plotted on the vertical axis of this graph, with forward current *(I_F)* increasing positively above the origin and reverse current *(I_R)* increasing negatively below the origin or zero current point. Manufacturers of diodes create a graph like the one seen in Figure 20-11 by applying various values of forward and reverse voltages. The result is a continuous curve called a voltage-current or *V-I* characteristic curve. Let us now examine the forward diode characteristics (upper right quadrant) and the reverse diode characteristics (lower left quadrant) in more detail.

Forward Characteristics

The upper right quadrant of the four sections in Figure 20-11 shows what forward current will pass through the diode when a forward bias voltage is applied. As you can see from the inset, the diode is forward biased by applying a positive potential to its anode and a negative potential to its cathode. To review, when the forward bias voltage exceeds the diode's internal barrier voltage, its resistance drops to almost zero, resulting in a rapid increase in forward current. In this instance the diode is said to be ON, and is equivalent to a closed switch.

These characteristics can be seen in the forward curve in Figure 20-11. Beginning at the graph origin and following the curve into the forward quadrant, you can see that the forward current through a diode is extremely small until the forward bias voltage exceeds the diode's internal barrier voltage, which for silicon is 0.7 V and for germanium is 0.3 V. Once the forward bias voltage exceeds the diode's internal barrier voltage, the forward current through the diode increases rapidly at a linear rate. The point on the forward voltage scale at which the curve suddenly rises resembles the shape of a human knee, which is why this point is called the **knee voltage.** This knee voltage is just another name for the diode's internal barrier voltage, which for a silicon diode is about 0.7 V.

Knee Voltage
The point in the forward voltage scale of the *V-I* characteristic curve at which the curve suddenly rises and resembles the shape of a human knee.

Referring back to the linearly increasing current portion of the forward curve in Figure 20-11, you will notice that although there is a large change in forward current, the forward voltage drop across the diode remains almost constant between 0.7 V to 0.75 V.

The amount of heat produced in the diode is proportional to the value of current through the diode ($P\uparrow = I^2\uparrow \times R$). For example, an IN4001 diode, which is a commonly used low-power silicon diode, has a manufacturer's maximum forward (I_F max.) rating of 1 A. If this value of current is exceeded, the diode will begin generating more heat than it can dissipate and burn out. A series current limiting resistor (R_S) is generally always included to limit the forward current, as shown in the inset in Figure 20-11. Although the series resistor will limit forward current, it cannot prevent a damaging forward current if enough pressure or forward voltage is applied ($V\uparrow = I\uparrow \times R$).

Reverse Characteristics

The lower left quadrant of the four sections in Figure 20-11 shows what reverse current will pass through the diode when a reverse bias voltage is applied. As you can see from the inset, a diode is reverse biased by applying a negative potential to its anode and a positive potential to its cathode. To review, when reverse biased, the internal barrier voltage of a diode will increase until it is equal to the external voltage. In this instance, current is effectively reduced to zero and the diode is said to be OFF and equivalent to an open switch.

These characteristics can be seen in the reverse curve in Figure 20-11. Beginning at the graph origin and following the curve into the reverse quadrant, you can see that the reverse current through the diode increases only slightly (approximately 100 μA). Throughout this part of the curve the diode is said to be blocking current because the leakage current is generally so small and is ignored for most practical applications. If the reverse voltage (V_R) is further increased, a point will be reached where the diode will break down, resulting in a sudden increase in current. This excessive current is due to the large external reverse bias voltage that is now strong enough to pull valence electrons away from their parent atoms, resulting in a large increase in minority carriers. The point on the reverse voltage scale at which the diode breaks down and there is a sudden increase in reverse current is called the **breakdown voltage.** Referring to

Breakdown Voltage or Peak Inverse Voltage (PIV)
The point on the reverse voltage scale at which the diode breaks down and there is a sudden increase in the reverse current.

the reverse curve in Figure 20-11, you can see that most silicon diodes break down as the reverse bias voltage approaches 50 V. For example, the IN4001 low power silicon diode has a reverse breakdown voltage (which is sometimes referred to as the **Peak Inverse Voltage** or **PIV**) of 50 V listed on its manufacturer's data sheet. If this reverse bias voltage is exceeded, an avalanche of continuously rising current will eventually generate more heat than can be dissipated, resulting in the destruction of the diode.

Temperature Characteristics

As discussed previously in Chapter 19, semiconductor materials, and therefore diodes, have a negative temperature coefficient of resistance. This means as temperature increases ($T\uparrow$), their resistance decreases ($R\downarrow$). Let us now consider these temperature effects on diodes since they may be critical in some applications.

Figure 20-12 shows how the forward and reverse currents through a diode can be affected by changes in temperature. The forward voltage drop across a conducting diode is inversely proportional to temperature, which means that the voltage drop across a diode will be less at higher temperatures. This change in forward voltage drop, however, is very small and does not greatly affect the diode's operation in most applications.

On the other hand, the diode's reverse current is greatly affected by changes in temperature. To review, the leakage current in a diode is caused by the minority carriers in the *n* and *p* regions. At low temperatures, the reverse leakage current is almost zero. At room temperature, the reverse leakage current has increased; however, it is still too small to have any adverse effects on circuit applications. At higher temperatures, however, the reverse leakage current increases to a point that the reverse biased diode is no longer equivalent to an open switch. This means that if a circuit malfunction generated excessive

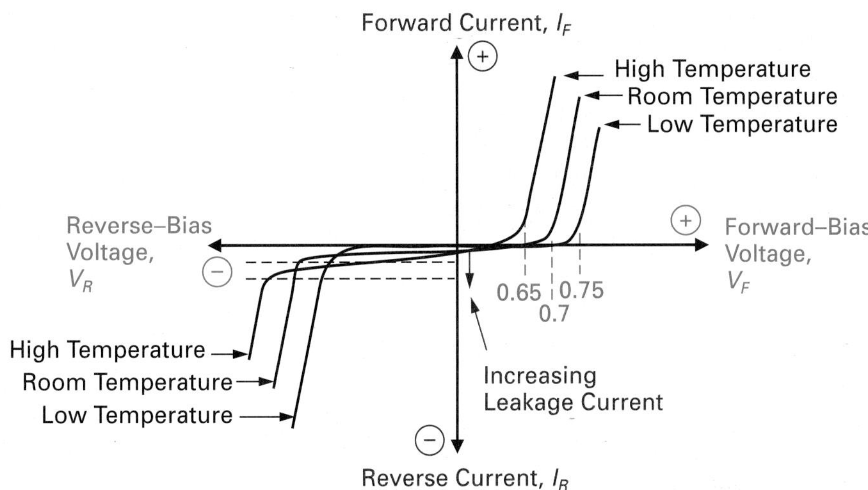

FIGURE 20-12 Temperature Effects on a Junction Diode's *V-I* Characteristic Curve.

heat, or if a system's cooling fan failed, the system's diodes might not switch OFF when reverse biased, causing additional system problems. As a general rule, the reverse leakage current tends to double for every temperature increase of 10° C.

20-2-4 A Junction Diode's Specification Sheet

Specification Sheet or Data Sheet Details the characteristics and maximum and minimum values of operation of a device.

A manufacturer's **specification sheet** or **data sheet** details the characteristics and maximum and minimum values of operation for a given device. Generally, engineers will study these details to determine whether a specific device can be incorporated into a circuit design. As a technician, you should be familiar with some of the basic operating limits of a device so that you can isolate component malfunctions within a circuit by determining whether the device is operating to specifications. For example, if a specific diode is placed in a circuit in which its maximum rating is exceeded in some instances, the fault is not with the component but with the circuit design. In this situation, simply replacing the component will not solve the problem. The device will have to be replaced with a diode which has a greater maximum rating.

Figure 20-13 shows the specific specification sheet for the IN4001 through IN4007 series of silicon junction diodes. This data sheet serves as a good example for showing the amount of details that are normally supplied to design engineers by device manufacturers. As a technician, you would mainly be interested in the diode's maximum reverse voltage, maximum forward current, and average voltage drop. For the IN4001, these values are:

Maximum reverse voltage (V_R) = 50 V

Maximum forward current (I_O) = 1 A

Average voltage drop ($V_{F(av)}$) = 0.8 V

EXAMPLE:

Would any of the maximum ratings of the IN4001 diode in Figure 20-14 be exceeded?

Solution:

The maximum reverse voltage for an IN4001 is 50 V, and since this reverse voltage is not being exceeded by the applied voltage, the diode is being operated within this specification.

The maximum forward current will be:

$$I = \frac{V_S - V_{diode}}{R}$$

$$I = \frac{12\text{ V} - 0.8\text{ V}}{10\ \Omega} = 1.12\text{ A}$$

Since this is in excess of the 1 A maximum listed in the specification sheet, the diode will more than likely burn out due to excessive current and, therefore, heat.

DEVICE: IN4001 Through IN4007— Silicon Junction Diodes

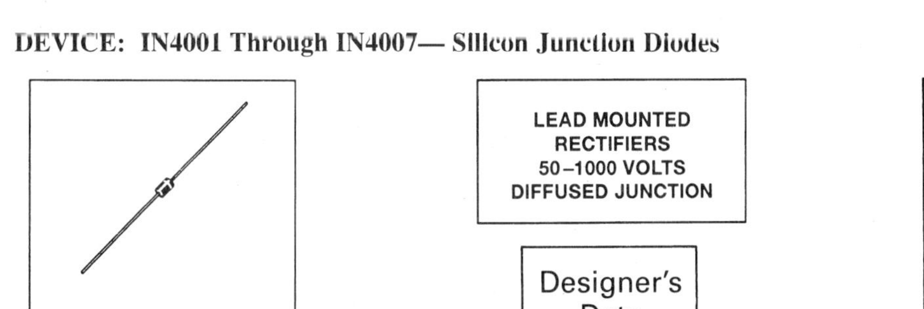

Designer's Data Sheet

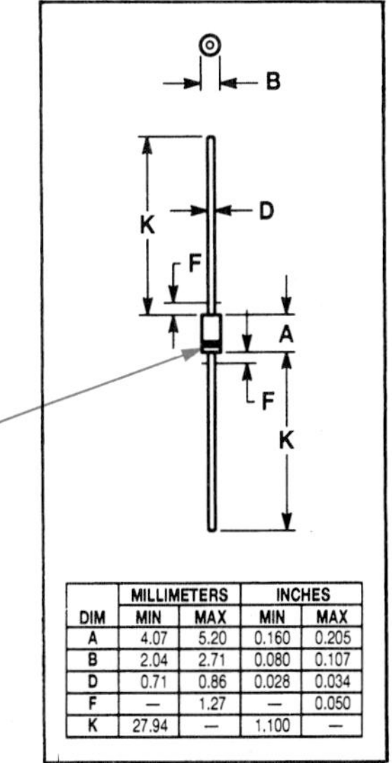

DIM	MILLIMETERS MIN	MILLIMETERS MAX	INCHES MIN	INCHES MAX
A	4.07	5.20	0.160	0.205
B	2.04	2.71	0.080	0.107
D	0.71	0.86	0.028	0.034
F	—	1.27	—	0.050
K	27.94	—	1.100	—

Mechanical Characteristics

- Case: Epoxy, Molded
- Weight: 0.4 gram (approximately)
- Finish: All External Surfaces Corrosion Resistant and Terminal Leads are Readily Solderable
- Lead and Mounting Surface Temperature for Soldering Purposes: 220°C Max. for 10 Seconds, 1/16″ from case
- Shipped in plastic bags, 1000 per bag.
- Available Tape and Reeled, 5000 per reel, by adding a "RL" suffix to the part number
- Polarity: Cathode Indicated by Polarity Band
- Marking: 1N4001, 1N4002, 1N4003, 1N4004, 1N4005, 1N4006, 1N4007

The **Peak Repetitive Reverse Voltage** is the maximum allowable reverse voltage; i.e., IN4001 will probably break down if the reverse voltage exceeds 50 V. This rating is also called **Working Peak Reverse Voltage** and **DC Blocking Voltage.**

MAXIMUM RATINGS

Rating	Symbol	1N4001	1N4002	1N4003	1N4004	1N4005	1N4006	1N4007	Unit
*Peak Repetitive Reverse Voltage Working Peak Reverse Voltage DC Blocking Voltage	V_{RRM} V_{RWM} V_R	50	100	200	400	600	800	1000	Volts
*Non–Repetitive Peak Reverse Voltage (halfwave, single phase, 60 Hz)	V_{RSM}	60	120	240	480	720	1000	1200	Volts
*RMS Reverse Voltage	$V_{R(RMS)}$	35	70	140	280	420	560	700	Volts
*Average Rectified Forward Current (single phase, resistive load, 60 Hz, see Figure 8, T_A = 75°C)	I_O	1.0							Amp
*Non–Repetitive Peak Surge Current (surge applied at rated load conditions, see Figure 2)	I_{FSM}	30 (for 1 cycle)							Amp
Operating and Storage Junction Temperature Range	T_J T_{stg}	– 65 to +175							°C

Maximum allowable nonrepeating reverse voltage.

The RMS of the Peak Repetitive Reverse Voltage. $V_{rms} = 0.707 \times V_{peak}$.

The maximum average forward current value.

The maximum surge current value.

The diode can be operated and stored at any temperature between –65° to +175° C.

ELECTRICAL CHARACTERISTICS*

Rating	Symbol	Typ	Max	Unit
Maximum Instantaneous Forward Voltage Drop (i_F = 1.0 Amp, T_J = 25°C) Figure 1	v_F	0.93	1.1	Volts
Maximum Full–Cycle Average Forward Voltage Drop (I_O = 1.0 Amp, T_L = 75°C, 1 inch leads)	$V_{F(AV)}$	—	0.8	Volts
Maximum Reverse Current (rated dc voltage) (T_J = 25°C) (T_J = 100°C)	I_R	 0.05 1.0	 10 50	μA
Maximum Full–Cycle Average Reverse Current (I_O = 1.0 Amp, T_L = 75°C, 1 inch leads)	$I_{R(AV)}$	—	30	μA

The maximum voltage drop that will appear across the diode when forward biased.

The maximum average forward voltage drop.

Maximum reverse current at different temps.

This is the maximum average value of reverse current.

FIGURE 20-13 Specific Specification Sheet for the IN4001 through IN4007 Junction Diodes.

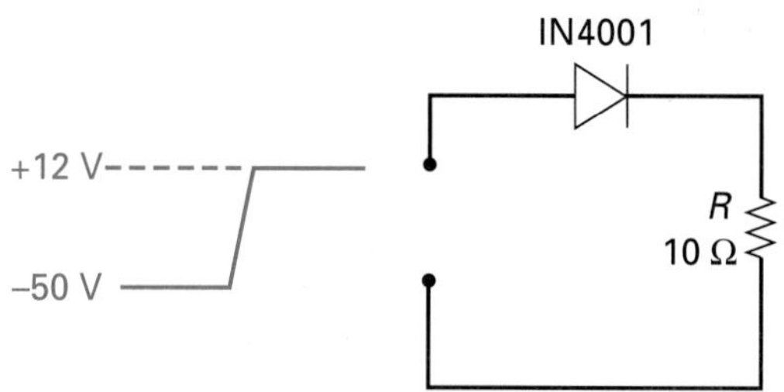

FIGURE 20-14 An IN4001 Diode Example Circuit.

20-2-5 *Junction Diode Applications*

It would be safe to say that the junction diode is used in almost every electronic and electrical system. Like resistors and capacitors, the list of diode applications is endless. However, certain uses of the diode predominate. As you proceed through this text you will see the junction diode used in a wide variety of analog and digital circuit applications. For now, let us see how the junction diode is most used in a basic analog circuit and a basic digital circuit.

Analog Circuit Application—Rectifier Circuit

Rectification
Diode's ability to pass current in only one direction and block current in the reverse direction.

Rectifier Circuit
Achieves rectification.

Rectifier Diodes or Rectifiers
Junction diodes that achieve rectification.

The diode's ability to pass current in only one direction, and block current from passing in the reverse direction, is utilized to achieve **rectification.** Rectification is achieved by a **rectifier circuit,** which is a circuit that converts alternating current (ac) into direct current (dc). Rectifier circuits are probably the biggest application for junction diodes, which accounts for why junction diodes are sometimes referred to as **rectifier diodes** or **rectifiers** as in Figure 20-13.

A basic rectifier circuit is constructed by connecting a junction diode between an ac input and a load, as shown in Figure 20-15(a). In this circuit, the 120 V_{rms} (169.7 V_{peak}) 60 Hz ac voltage from the wall outlet will be converted to a pulsating dc voltage.

When the ac voltage swings positive, as shown in Figure 20-15(b), the anode of the diode is made positive, causing the diode to turn ON and connect the positive half-cycle of the ac input voltage across the load (R_L).

When the ac voltage swings negative, as shown in Figure 20-15(c), the anode of the diode is made negative, causing the diode to turn OFF and prevent any circuit current, and therefore any voltage, from being developed across the load (R_L).

Half-Wave Rectifier
Circuit in which half the input wave appears at the output.

Referring to the input and output waveforms shown in Figure 20-15(c), you can see why the circuit is called a **half-wave rectifier.** Comparing the voltage input (V_{in}) to the voltage output (V_{out}), you can see that only half of the input wave appears at the output. Since the diode only connects the positive half-cycle of the ac input across the load (R_L), the output voltage (V_{RL}) is a positive pulsating dc waveform. The peak voltage developed across the load will, of course, be 0.7 V less than the peak input voltage due to the 0.7 V drop across the silicon junction diode, as shown in Figure 20-15(d).

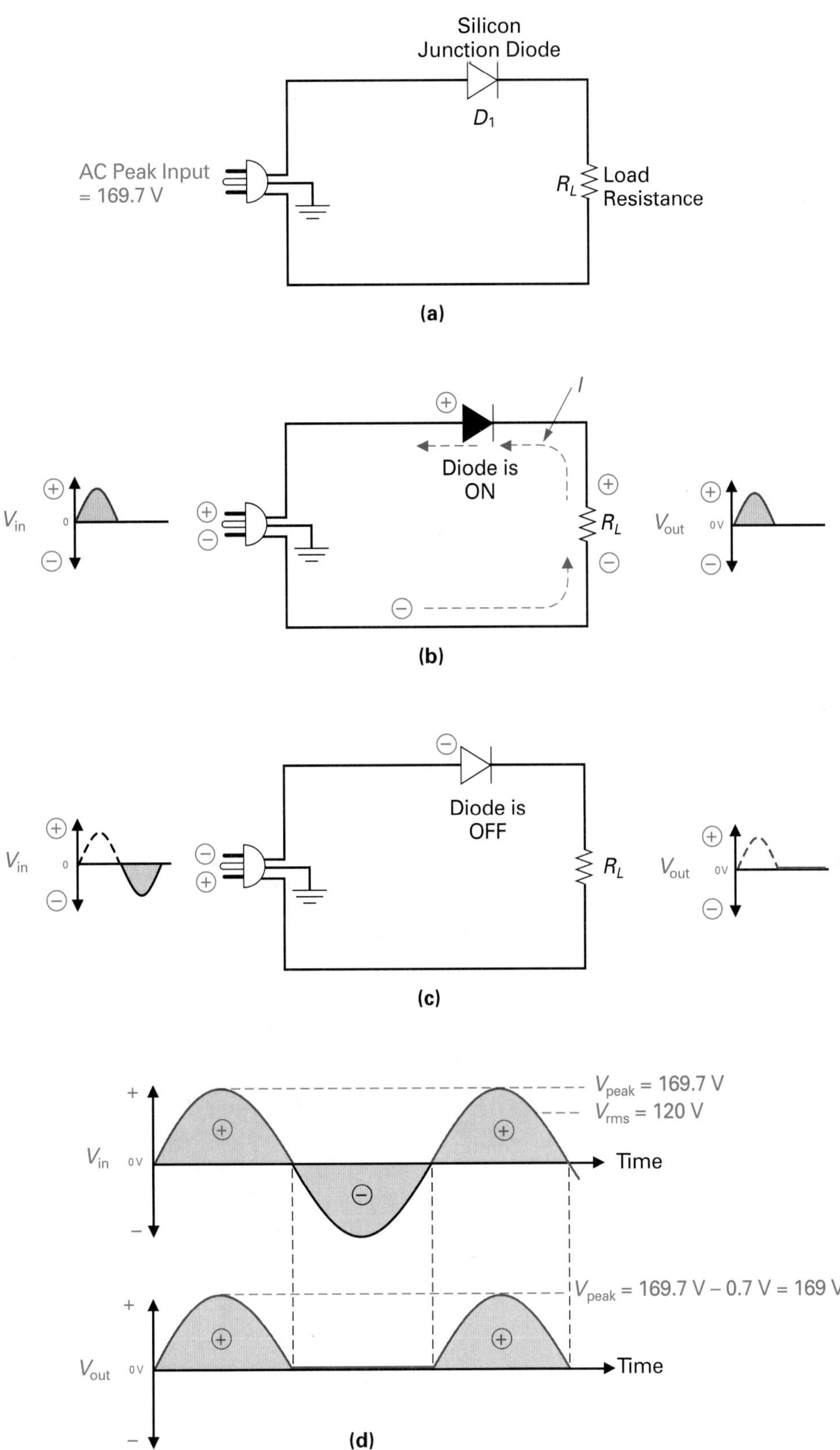

FIGURE 20-15 The Diode Being Used in an Analog Electrical Circuit. (a) Basic Rectifier Circuit. (b) Positive Input Half Cycle Operation. (c) Negative Input Half Cycle Operation. (d) Input/Output Waveforms.

EXAMPLE:

Which of the IN4000X silicon rectifier diodes should be used in the half-wave rectifier circuit shown in Figure 20-15, considering the value of reverse voltage?

Solution:

The diode is reverse biased when the ac input swings negative. At this time, the entire negative supply voltage will appear across the open or OFF diode. The maximum reverse breakdown voltage of the diode must therefore be larger than the peak of the ac input voltage. Referring to Figure 20-13, you can see that to withstand a reverse voltage of 169.7 V (peak of the ac input), we will have to use an IN4003 because it has a V_R maximum rating of 200 V.

Further details on half-wave and full-wave rectifiers will be given in Chapter 22 which discusses all types of rectifier circuits and how they are used in dc power supplies.

Digital Circuit Application—Logic Gate Circuit

Logic Gate Circuits Two-state (ON/OFF) circuits used for descision-making functions in digital logic circuits.

Within the computer, diodes are used to construct **logic gate circuits,** logic gates are used to construct flip-flop circuits, flip-flops are used to construct register and counter circuits, and register and counter circuits are the building blocks of our digital electronic computer. The logic gate is therefore our basic building block for all digital circuits. Let us examine one of these logic gates to see how it operates and in what applications it can be used.

A basic logic gate is constructed using two junction diodes and a resistor, as shown in Figure 20-16(a). This circuit is called a logic gate because it will always produce a *logical* or *predictable* output, and this output will depend on the condition of its inputs. For example, Figure 20-16(b) has a table which shows how this logic gate will react to all different input possibilities. The two binary states 0 and 1 are represented in the circuit as two voltages

$$\text{Binary } 0 = 0 \text{ volts (LOW voltage)}$$

$$\text{Binary } 1 = 5 \text{ volts (HIGH voltage)}$$

If you study the table in Figure 20-16(b), you can see that when both inputs are LOW, or at 0 V (A = 0, B = 0), both diodes will be OFF because the anodes are at 0 volts (due to the inputs), and the cathodes of the diodes are at 0 volts (due to the pull-down resistor, *R*). This input combination of A = 0 and B = 0 will therefore turn both diodes OFF and always produce an output at Y of 0.

In all other combinations in the table in Figure 20-16(b), a HIGH or +5 V (logical 1) input is applied to either or both of the inputs A and B. Any HIGH input will always turn on its associated diode since the anode will be at +5 V and the cathode will be at 0 V via *R*. When ON, a diode is equivalent to a closed switch, and the +5 V input is switched through to the output making Y equal to +5 V or logical 1. Therefore, if *A input OR B input is HIGH, the output Y will be HIGH.* This behavior accounts for why this circuit is called an **OR gate.**

OR Gate When either input A OR B is HIGH, the output will be HIGH.

The schematic symbol for the OR gate is shown in Figure 20-16(c). To show how the OR gate circuit could be used, consider the simple security system shown in Figure 20-16(d). In this application, if either the window *OR* the door

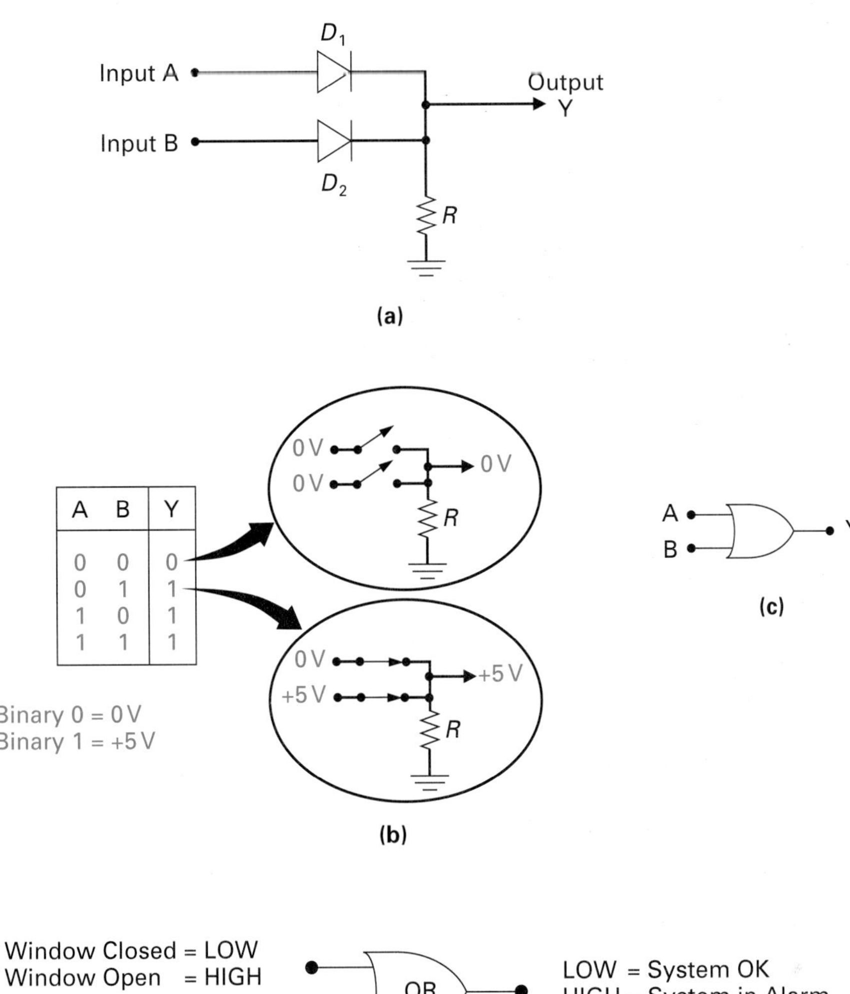

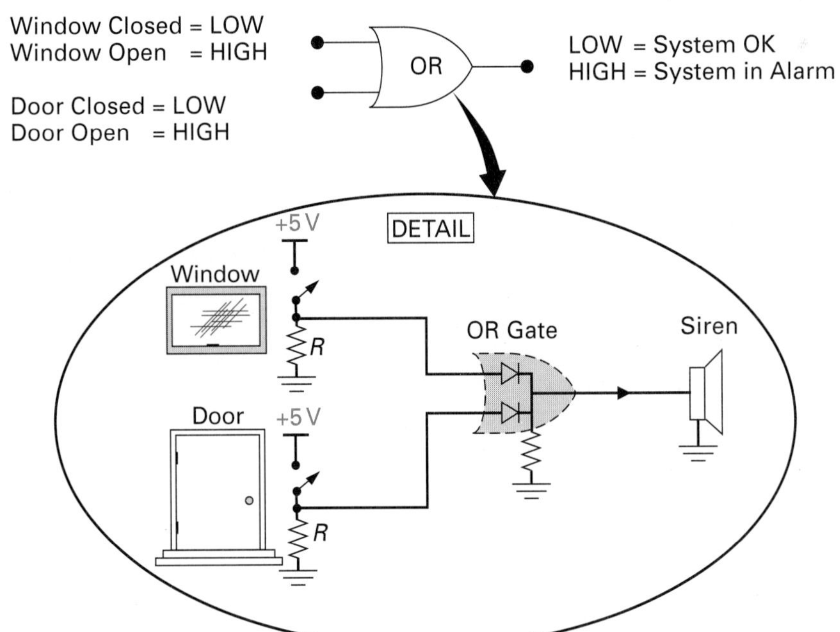

FIGURE 20-16 The Diode Being Used in a Digital Electronic Circuit. (a) Basic OR Gate Circuit. (b) OR Gate Function Table. (c) OR Gate Schematic Symbol. (d) OR Gate Security System Application.

was to open, a switch contact would close and deliver a HIGH input to the OR gate. As we know, any HIGH input to an OR gate will always generate a HIGH output and in this circuit the +5 V out will activate the siren.

In summary, therefore, a logic gate accepts inputs in the form of HIGH or LOW voltages, judges these input combinations based on a predetermined set of rules, and then produces a single output in the form of a HIGH or LOW voltage. The term logic is used because the output is predictable or logical, and the term gate is used because only certain input combinations will unlock the gate. For example, any HIGH input to an OR gate will unlock the gate and allow the HIGH at the input to pass through to the output.

The OR gate and all of the other different types of logic gates will be discussed later in the digital circuits chapter.

20-2-6 Testing Junction Diodes

Diodes that are operated beyond their maximum forward and reverse ratings will, more than likely, malfunction. These malfunctions can result in one of two types of failure. The diode may burn out and then act as a permanently open switch (less common), or effectively melt the semiconductor material and act as a permanently closed switch (more common). A defective diode will not be able to be switched ON or OFF. Instead, it will either remain permanently OFF or permanently ON.

Figure 20-17 shows how an ohmmeter can be used to check whether a diode has malfunctioned or is operating correctly. A good diode should display a very low resistance when it is biased ON, and a very high resistance when it is biased OFF. Figure 20-17(a) shows how a diode can be forward biased by an ohmmeter's internal battery (+ lead to anode, – lead to cathode), and if good, should display a low value of resistance (typically less than 10 Ω). Figure 20-17(b) shows how the diode is then flipped over and reverse biased by the ohmmeter's internal battery (+ lead to cathode, – lead to anode). If the diode is good, the ohmmeter should display a very high resistance (typically greater than 1000 MΩ). Since this value is generally off the ohmmeter's scale, you will probably have the display showing OL or OR. This is what you should expect since it means that the reverse biased diode's resistance is so high that it is over the range, or off the scale, selected.

A diode that has been damaged and is permanently shorted between its anode and cathode will display a LOW resistance on the ohmmeter when it is both forward and reverse biased by the ohmmeter. Finding the shorted diode in a circuit may not be the end of your troubles, since it may be one of the neighboring components that initially caused the diode's ratings to be exceeded. Furthermore, the shorted diode may have allowed a damaging current to pass through to another part of the circuit causing additional damage.

A diode that has been damaged and is permanently opened between its anode and cathode will display a HIGH resistance on the ohmmeter when it is both forward and reverse biased by the ohmmeter. Once again, the destruction of the diode generally occurs when the maximum voltage or current ratings of the diode have been exceeded. Remember that the diode may not have just

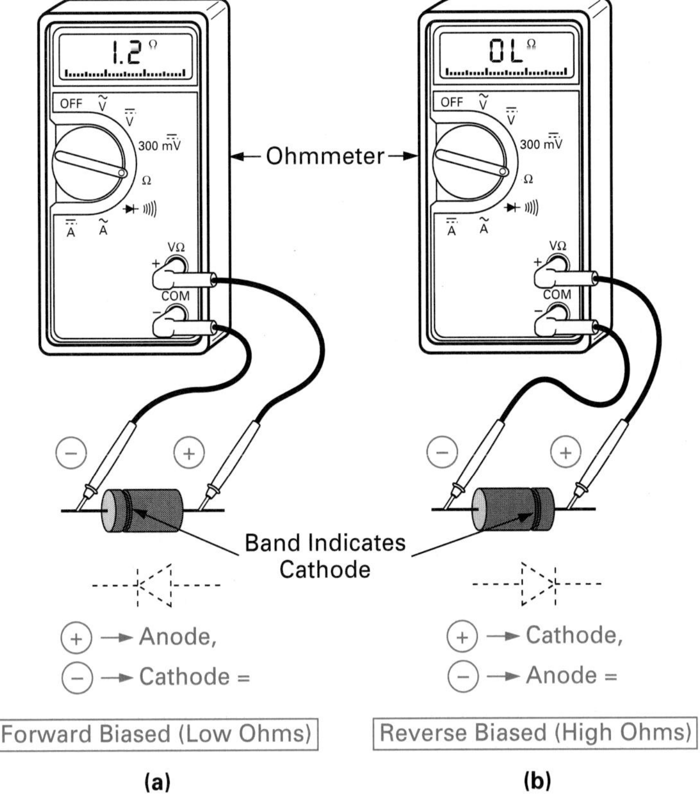

FIGURE 20-17 **Testing Diodes with an Ohmmeter.**

malfunctioned on its own; the problem may have been caused by one of the neighboring components. In this instance, however, an open diode rarely damages other associated components because current is blocked by the fuse-like action of the diode's destruction.

As you proceed through this text, you will see diodes used in a variety of circuit applications, and you will see what symptoms a malfunctioning diode will display in these circuits.

EXAMPLE:

Figure 20-18 shows the results from testing four diodes with an ohmmeter. Which of the diodes tested good, which tested bad, and for what reason?

Solution:

a. The diode in Figure 20-18(a) has a low forward resistance and a high reverse resistance and is switching ON and OFF correctly.
b. The diode in Figure 20-18(b) has a low forward resistance and a high reverse resistance and is switching ON and OFF correctly.
c. The diode in Figure 20-18(c) has a high forward resistance and a high reverse resistance and is not switching ON and OFF. It seems to be remaining permanently OFF, indicating an open between its anode and cathode terminals.

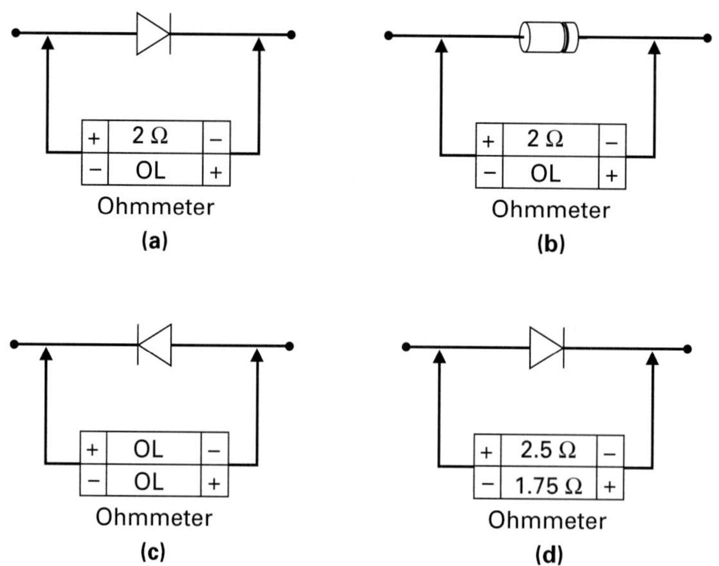

FIGURE 20-18 Testing Diode Examples.

d. The diode in Figure 20-18(d) has a low forward resistance and a low reverse resistance and is not switching ON and OFF. It seems to be remaining permanently ON, indicating a short between its anode and cathode terminals.

SELF-TEST REVIEW QUESTIONS FOR SECTION 20-2

1. What is the typical forward voltage drop across a silicon diode?
2. What barrier voltage has to be overcome in order to forward bias a silicon diode?
3. How many P-N junctions are within a junction diode?
4. What is PIV, and what would it be for a typical silicon diode?
5. As the temperature of a diode increases, the forward voltage drop ____________. (increases or decreases)

SUMMARY

1. The first diode was accidentally created by Edison in 1883 when he was experimenting with his light bulb.
2. The word *diode* is derived from the fact that the device has two (*di*) electrodes (*ode*).
3. A diode will act as a conductor and pass current easily in one direction when the bias voltage across it is of one polarity, yet it will block current and imitate an insulator when the bias voltage applied is of the opposite polarity.

The Junction Diode (Figure 20-19)

4. The two electrodes or terminals of the diode are called the anode and cathode.
5. A black band or stripe is generally placed on the package closest to the cathode terminal for identification purposes.
6. If you make the anode terminal of a diode positive with respect to the cathode terminal, the device is equivalent to a closed switch and is said to be ON, or forward biased.
7. If you make the anode terminal of a diode negative with respect to the cathode, the device is equivalent to an open switch and is said to be OFF, or reverse biased.
8. Semiconductor devices such as diodes and transistors are constructed using P-N junctions. A diode for example, has only one P-N junction and is created by doping a single piece of pure semiconductor to produce an *n*-type and *p*-type region. A bipolar junction transistor, on the other hand, has two P-N junctions and is created by doping a single piece of pure semiconductor with three alternate regions (NPN or PNP).
9. The *n*-type region is called the cathode, while the *p*-type region is called the anode.

Junction Diode Characteristics (Figure 20-20)

10. The voltages, which incline or cause the diode to operate in a certain manner, are known as bias voltages. Bias voltages control the width of the depletion region, which in turn controls the resistance of the junction and therefore the amount of current that can pass through the P-N junction diode.
11. Like the P-N junction, the junction diode's operation is determined by the polarity of the applied voltage.
12. A diode is forward biased when the negative terminal of the applied voltage is connected to the *n* region of the diode, and the positive terminal of the applied voltage is connected to the *p* region of the diode ($+V \rightarrow p$ region, $-V \rightarrow n$ region). When forward biased, free electrons are repelled from the *n* region by the negative source and attracted to the positive terminal of the voltage source. This forward-conducting electron flow will only occur if the external source voltage is large enough to overcome the internal barrier voltage of the junction diode. For a silicon diode, the external source voltage must be equal to or greater than 0.7 V, whereas for a germanium diode the applied voltage must be equal to or greater than 0.3 V.
13. The diode symbol actually points in the direction of forward hole flow and reflects conventional flow.
14. A diode is reverse biased when the positive terminal of the applied voltage is connected to the *n* region of the diode, and the negative terminal of the applied voltage is connected to the *p* region of the diode ($+V \rightarrow n$ region, $-V \rightarrow p$ region). This applied voltage polarity will reverse bias the junction

QUICK REFERENCE SUMMARY SHEET

FIGURE 20-19 The Junction Diode.

Schematic Symbol

Memory Aid for Terminal Identification

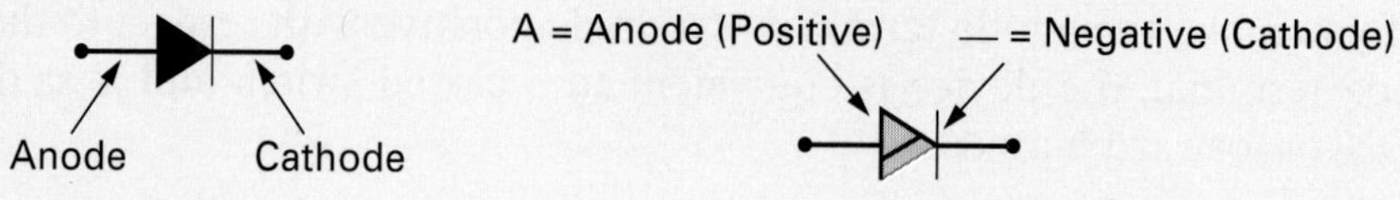

Package Types

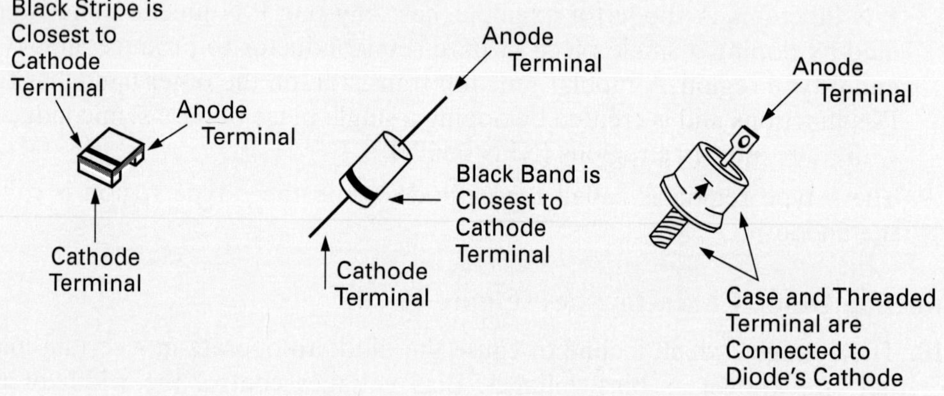

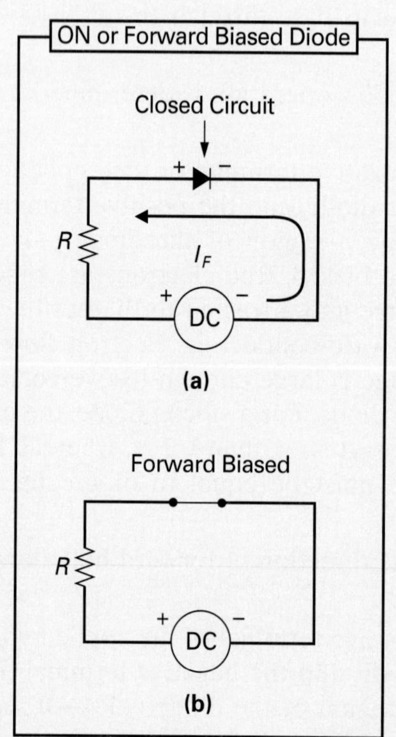

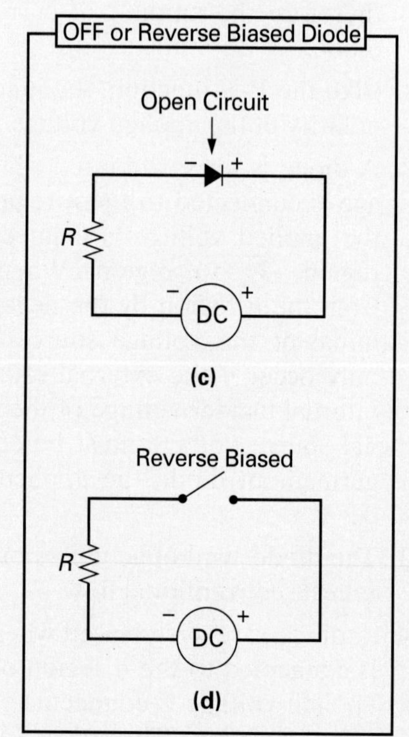

FIGURE 20-20 Junction Diode Characteristics.

Voltage/Current Characteristics

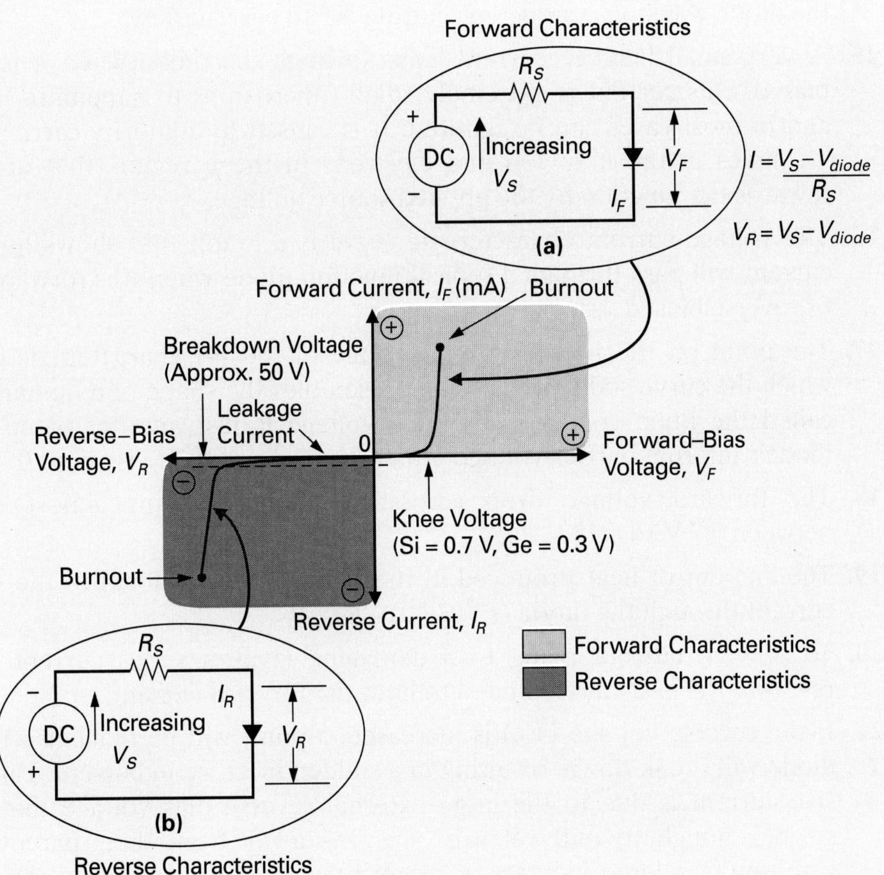

Temperature Characteristics

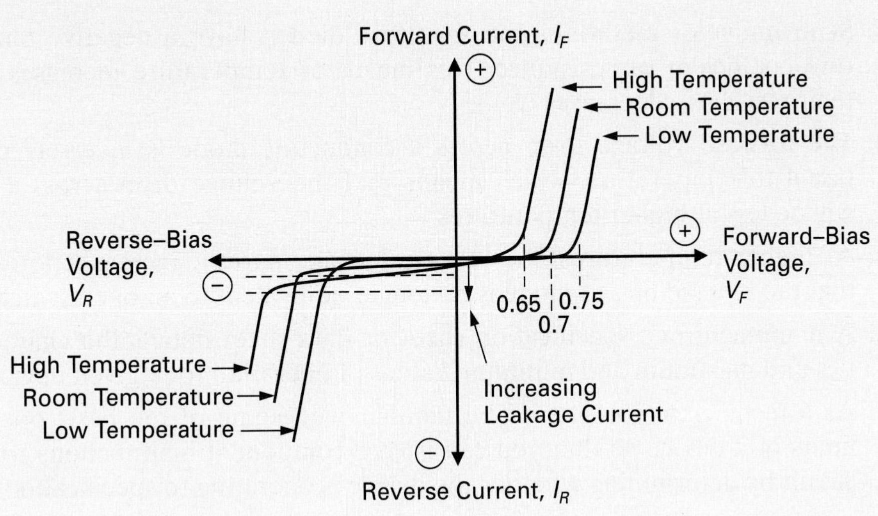

diode because free electrons in the n region are attracted and therefore travel to the positive terminal of the applied source voltage. At the same time, holes feel the attraction of the negative terminal of the applied voltage. As a result, the depletion region will increase in width until its internal voltage is equal but opposite to the external applied voltage. At this time, the diode will stop conducting, cutting off all current flow.

15. A very small leakage current leaks through the diode when it is reverse biased. This current is extremely small (microamps to nanoamps in value) and in most cases can be ignored. It is caused by minority carriers, which are holes in the n region and electrons in the p region, that are forced towards the junction by the applied source voltage.

16. The voltage-current characteristic curve is a graph that shows how much current will pass through a typical junction diode when it is forward biased or reverse biased.

17. The point on the forward voltage scale of the *V-I* characteristic curve at which the curve suddenly rises and resembles the shape of a human knee is called the knee voltage. This knee voltage is just another name for the diode's internal barrier voltage, which for a silicon diode is about 0.7 V.

18. The forward voltage drop across the diode remains almost constant between 0.7 V to 0.75 V.

19. The amount of heat produced in the diode is proportional to the value of current through the diode ($P\uparrow = I^2\uparrow \times R$).

20. To prevent current rising to a damaging level, a series current limiting resistor (R_S) is always included to limit the forward current.

21. If the reverse voltage (V_R) is increased, a point will be reached where the diode will break down, resulting in a sudden increase in current. This excessive current is due to the large external reverse bias voltage that is now strong enough to pull valence electrons away from their parent atoms, resulting in a large increase in minority carriers. The point on the reverse voltage *V-I* characteristic scale at which the diode breaks down and there is a sudden increase in reverse current is called the breakdown voltage, or peak inverse voltage (PIV).

22. Semiconductor materials, and therefore diodes, have a negative temperature coefficient of resistance. This means as temperature increases ($T\uparrow$), their resistance decreases ($R\downarrow$).

23. The forward voltage drop across a conducting diode is inversely propor tional to temperature, which means that the voltage drop across a diode will be less at higher temperatures.

24. At higher temperatures, the reverse leakage current increases to a point that the reverse biased diode is no longer equivalent to an open switch.

25. A manufacturer's specification sheet or data sheet details the characteristics and maximum and minimum values of operation for a given device.

26. As a technician, you should be familiar with some of the basic operating limits of a device so that you can isolate component malfunctions within a circuit by determining whether the device is operating to specifications.

Junction Diode Analog Circuit Application—A Rectifier Circuit (Figure 20-21)

27. The diode's ability to pass current in only one direction and block current from passing in the reverse direction is utilized to achieve rectification. Rectification is achieved by a rectifier circuit, which is a circuit that converts alternating current (ac) into direct current (dc). Rectifier circuits are probably the biggest application for junction diodes, which accounts for why junction diodes are sometimes referred to as rectifier diodes or rectifiers.

Junction Diode Digital Circuit Application—A Logic Gate Circuit (Figure 20-22)

28. Within the computer, diodes are used to construct logic gate circuits, logic gates are used to construct flip-flop circuits, flip-flops are used to construct register and counter circuits, and register and counter circuits are the building blocks of our digital electronic computer. The logic gate is, therefore, our basic building block for all digital circuits.

29. A logic gate accepts inputs in the form of HIGH or LOW voltages, judges these input combinations based on a predetermined set of rules, and then produces a single output in the form of a HIGH or LOW voltage. The term logic is used because the output is predictable or logical, and the term gate is used because only certain input combinations will unlock the gate.

Junction Diode Testing (Figure 20-23)

30. Diodes that are operated beyond their maximum forward and reverse ratings will, more than likely, malfunction. These malfunctions can result in one of two types of failure. The diode may burn out and then act as a permanently open switch (less common), or effectively melt the semiconductor material and act as a permanently closed switch (more common). A defective diode will no longer be able to be switched ON or OFF. Instead, it will either remain permanently OFF or permanently ON.

31. A good diode should display a very low resistance when it is biased ON and a very high resistance when it is biased OFF. If good, when a diode is forward biased by an ohmmeter it should display a low value of resistance (typically less than 10 Ω). When a diode is then flipped over and reverse biased by the ohmmeter's internal battery (+ lead to cathode, – lead to anode), the ohmmeter should display a very high resistance (typically greater than 1000 MΩ). Since this value is generally off the ohmmeter's scale, you will probably have the display showing OL or OR.

32. A diode that has been damaged and is permanently shorted between its anode and cathode will display a LOW resistance on the ohmmeter when it is both forward and reverse biased by the ohmmeter.

33. A diode that has been damaged and is permanently opened between its anode and cathode will display a HIGH resistance on the ohmmeter when it is both forward and reverse biased by the ohmmeter.

FIGURE 20-21 Junction Diode Analog Circuit Application—A Rectifier Circuit.

Basic Circuit

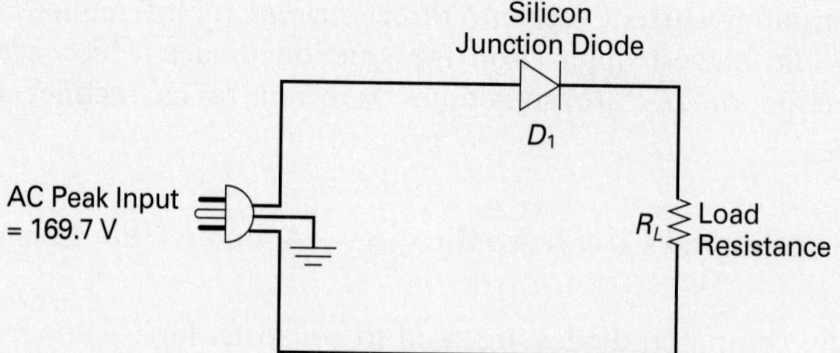

Operation of Circuit During Positive Alternation

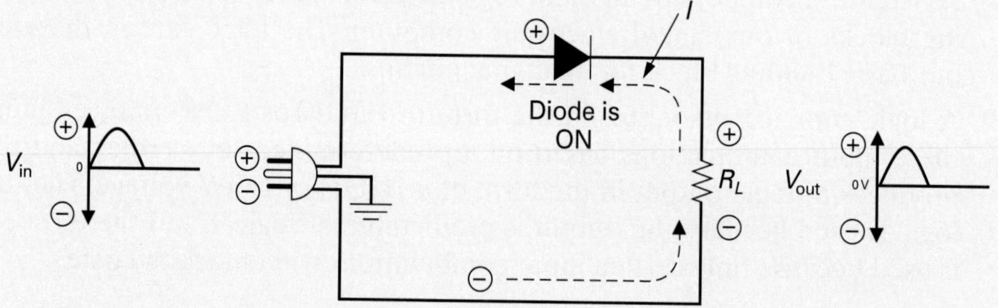

Operation of Circuit During Negative Alternation

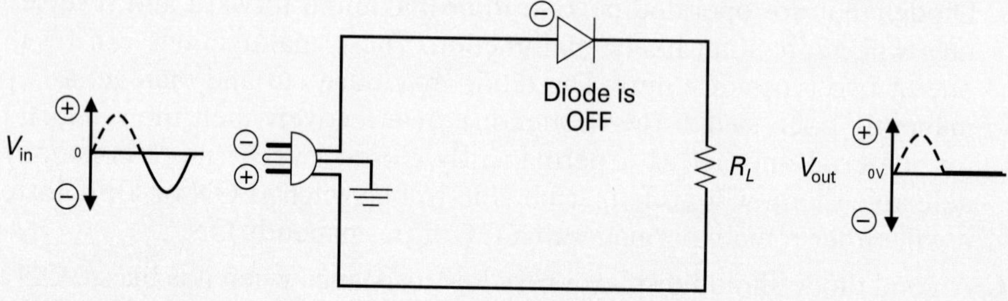

Input/Output Waveforms

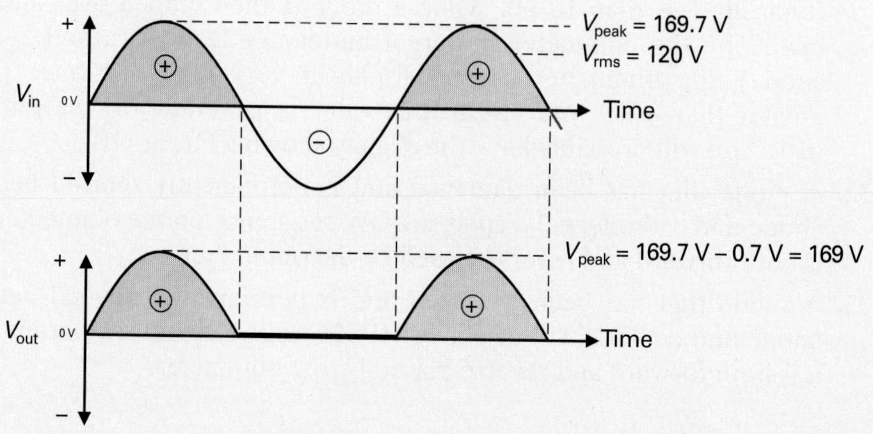

FIGURE 20-22 Junction Diode Digital Circuit Application—A Logic Gate Circuit.

Basic OR Gate Circuit

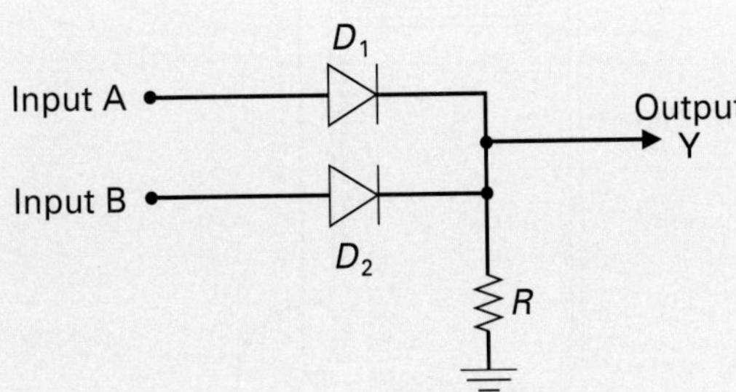

OR Gate Function Table

OR Gate Schematic Symbol

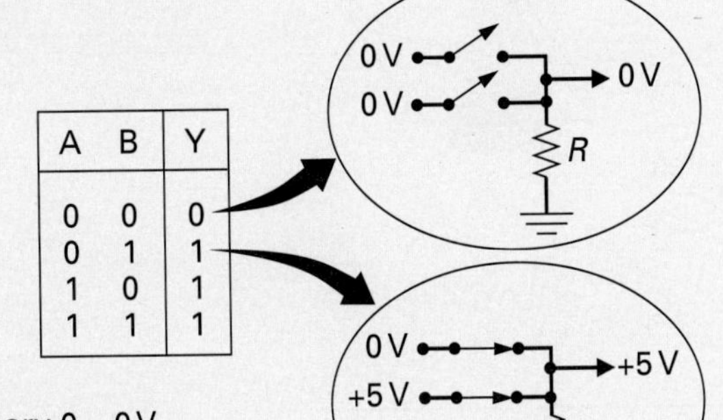

A	B	Y
0	0	0
0	1	1
1	0	1
1	1	1

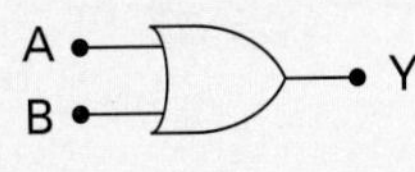

OR Gate Application — Basic Security System

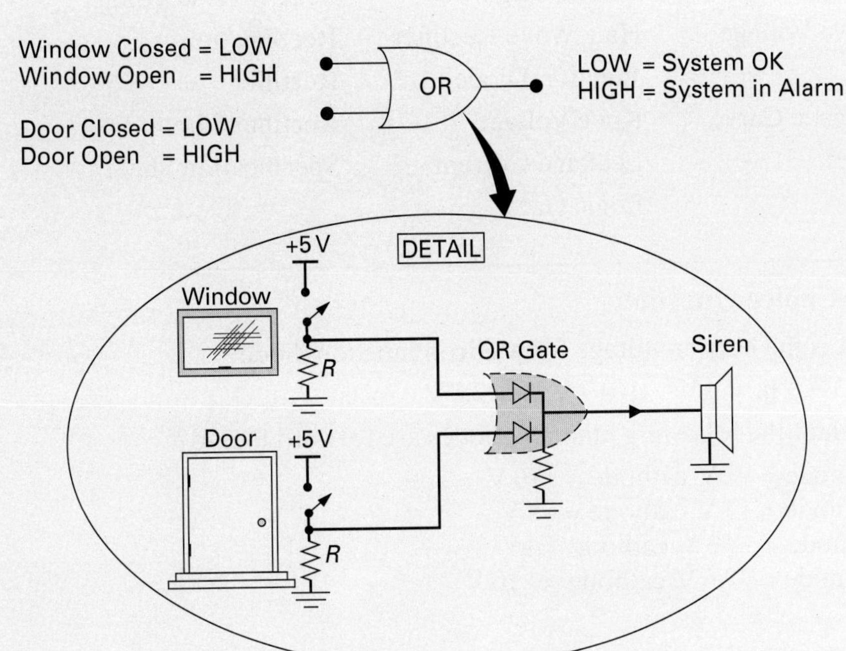

QUICK REFERENCE SUMMARY SHEET

FIGURE 20-23 Junction Diode Testing.

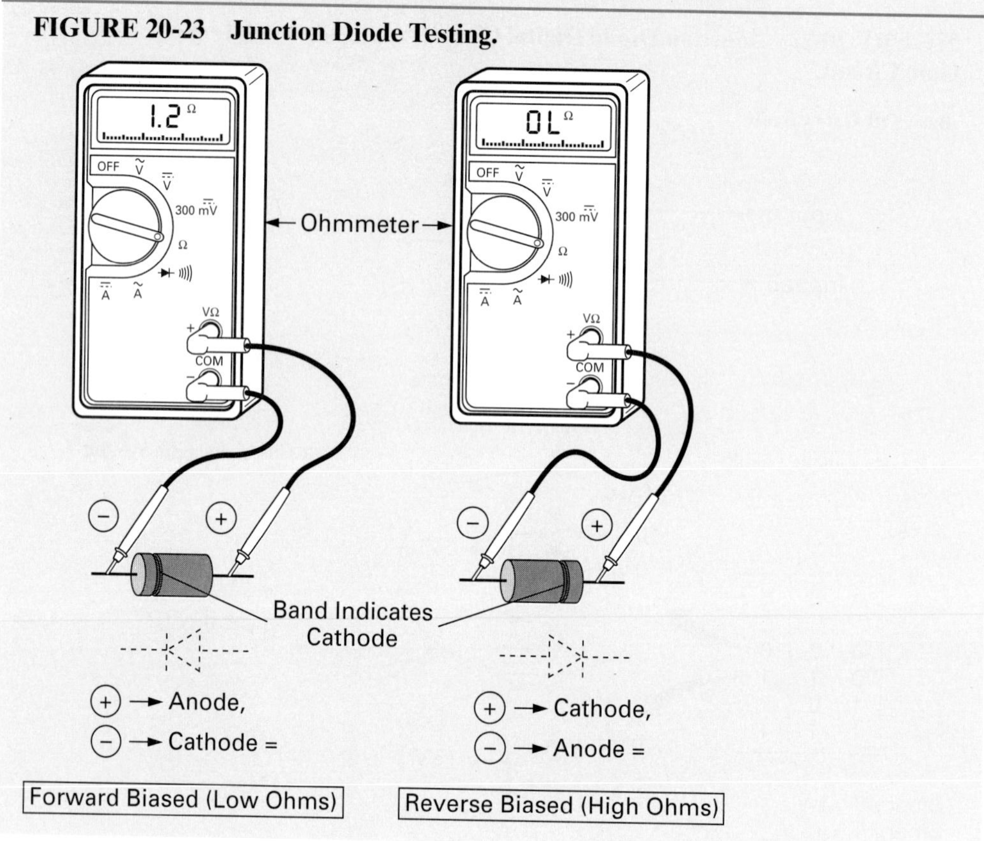

NEW TERMS

Anode
Bias Voltage
Breakdown Voltage
Cathode
Characteristic Curve
Data Sheet
Diode
Encoder Circuit
Graph Origin
Half-Wave Rectifier
Junction Diode
Knee Voltage
Leakage Current
Logic Gate
OR Gate
Peak Inverse Voltage
Rectification
Rectifier
Rectifier Circuit
Specification Sheet

REVIEW QUESTIONS

Multiple-Choice Questions

1. What is the barrier voltage for a silicon junction diode?
   a. 0.3 V **b.** 0.4 V **c.** 0.7 V **d.** 2.0 V
2. Which of the following junction diodes are forward biased?
   a. Anode = +7 V, cathode = +10 V
   b. Anode = +5 V, cathode = +3 V
   c. Anode = +0.3 V, cathode = +5 V
   d. Anode = –9.6 V, cathode = –10 V

3. The junction diode ____________ current when it is forward biased, and ____________ current when it is reverse biased.

   a. Blocks, conducts c. Blocks, prevents
   b. Conducts, passes d. Conducts, blocks

4. The n-type region of a junction diode is connected to the ____________ terminal and the p-type region is connected to the ____________.

   a. Cathode, anode b. Anode, cathode

5. Semiconductor devices need voltages of a certain amplitude and polarity to control their operation. These voltages are called:

   a. Barrier potentials c. Knee voltages
   b. Depletion voltages d. Bias voltages

6. When reverse biased, a junction diode has a leakage current passing through it which is typically measured in:

   a. Amps b. Milliamps c. Microamps d. Kiloamps

7. What happens to the forward voltage drop across the diode (V_F) if temperature increases? V_F will:

   a. Decrease c. Remain the same
   b. Increase d. Be unpredictable

8. When forward biased, a junction diode is equivalent to a/an ____________ switch, whereas when it is reverse biased it is equivalent to a/an ____________ switch.

   a. Open, closed b. Closed, closed c. Open, open d. Closed, open

9. The black band on a diode's package is always closest to the ____________.

   a. Anode b. Cathode c. p-type material d. Both (a) and (c) are true

10. When current dramatically increases, the voltage point on the diode's forward V-I characteristic curve is called the:

    a. Breakdown voltage c. Barrier voltage
    b. Knee voltage d. Both (b) and (c) are true

11. When current dramatically increases, the voltage point on the diode's reverse V-I characteristic curve is called the:

    a. Breakdown voltage c. Barrier voltage
    b. Knee voltage d. Both (b) and (c) are true

12. A logic gate is:

    a. A circuit that converts ac to dc c. A two-state decision making circuit
    b. An analog circuit d. A circuit that converts dc to ac

13. A rectifier is:

    a. A circuit that converts ac to dc c. A two-state decision making circuit
    b. An analog circuit d. A circuit that converts dc to ac

14. What resistance should a good diode have when it is reverse biased?

    a. Less than 10 Ω c. Between 120 Ω and 1.2 kΩ
    b. More than 1000 MΩ d. Both (b) and (c) are true

15. What resistance should a good diode have when it is forward biased?

a. Less than 10 Ω **c.** Between 120 Ω and 1.2 kΩ
b. More than 1000 MΩ **d.** Both (b) and (c) are true

Essay Questions

16. What is a junction diode? (20-1)
17. Sketch the schematic symbol for a diode and label its terminals. (20-1-1)
18. Describe the basic operation of a diode. (20-1-2)
19. What is the relationship between a P-N junction and a diode? (20-2-1)
20. Explain in detail the differences between a forward biased diode and a reverse biased diode. (20-2-2)
21. Referring to the junction diode's *V-I* characteristic curve, describe the: (20-2-3)

    a. Forward bias curve **b.** Reverse bias curve

22. Describe the temperature characteristics of a junction diode. (20-2-3)
23. List the following specifications for the IN4004 junction diode: (20-2-4)

    a. V_R **b.** $I_{F(av)}$ **c.** $I_{R(av)}$

24. What is a rectifier circuit? (20-2-5)
25. Sketch a half-wave rectifier circuit and its input and output wave shapes. (20-2-5)
26. A diode is a ______________ within a protective package. (20-2-1)
27. What is a logic gate? (20-2-5)
28. Sketch an OR gate circuit and a table showing what outputs will be produced for all combinations of digital inputs. (20-2-5)
29. What type of diode malfunction is most common? (20-2-6)
30. How can an ohmmeter be used to test for an open or short within a diode? (20-2-6)

Practice Problems

31. Which of the silicon diodes in Figure 20-24 are forward biased and which are reverse biased?
32. Calculate I_F for the circuits in Figure 20-25.
33. What would be the voltage drop across each of the diodes in Figure 20-25?
34. What would be the voltage drop across each of the resistors in Figure 20-25?

Troubleshooting Questions

35. Referring to the switch encoder circuit in Figure 20-26, first determine what digital codes will be generated at A, B, and C for each of the switch positions 1, 2 and 3. Next, determine what problems would occur for each of the following circuit malfunctions or conditions:

    a. What would happen if D_1 were to open permanently?
    b. Would a 200 mA fuse protect the +5 V supply voltage?

0 V +1.3 V (a)
+3 V +5 V (b)
+0.2 V +0.7 V (c)
+5 V 0 V (d)
−12 V −15 V (e)
+6 V +8.3 V (f)

FIGURE 20-24 **Biased Junction Diodes.**

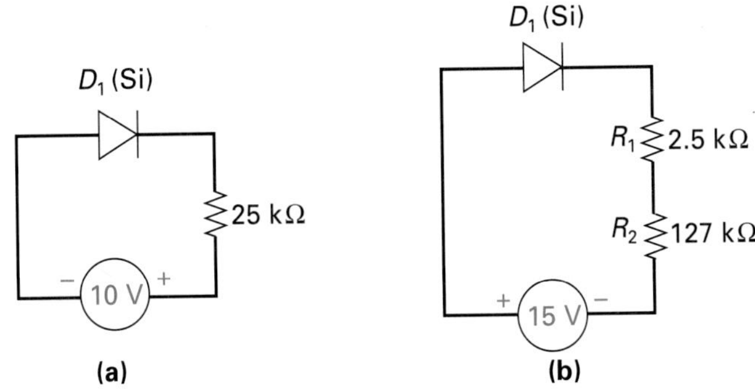

FIGURE 20-25 **Forward Current Examples.**

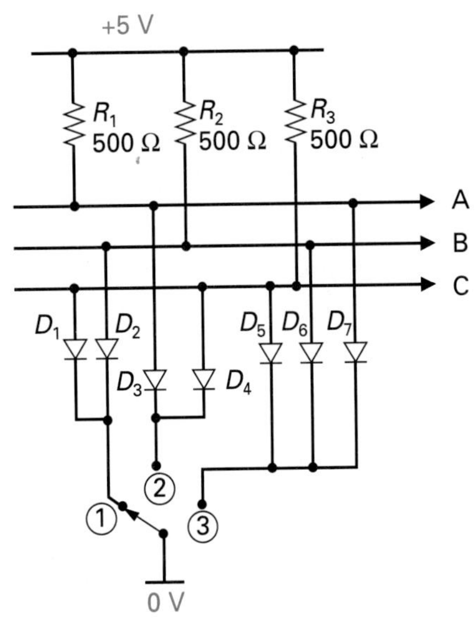

Input	Output		
Switch Position	A	B	C
1			
2			
3			

FIGURE 20-26 **A Switch Encoder Circuit.**

CHAPTER

21

OBJECTIVES

After completing this chapter, you will be able to:

1. Name the three terminals of the bipolar junction transistor.
2. Describe the difference between the construction and schematic symbol of an NPN and PNP bipolar transistor.
3. Identify the base, collector, and emitter terminals of typical low-power and high-power transistor packages.
4. Describe the two basic actions of a bipolar transistor:
   - **a.** ON/OFF switching action
   - **b.** Variable-Resistor action
5. Define the terms transistor and transistance.
6. Explain how the transistor can be used in the following basic applications:
   - **a.** A digital (two-state) circuit, such as a logic gate
   - **b.** An analog (linear) circuit, such as an amplifier
7. Describe in more detail the bipolar junction transistor action.
8. Explain in more detail how the transistor can be used as an amplifier.
9. Identify and list the characteristics of the three transistor circuit configurations, namely:
   - **a.** Common base
   - **b.** Common emitter
   - **c.** Common collector
10. Explain the meaning of the following:
    - **a.** Transistor voltage and current abbreviations
    - **b.** DC alpha
    - **c.** DC beta
    - **d.** Collector characteristic curve
    - **e.** AC beta
    - **f.** Input resistance or impedance
    - **g.** Output resistance or impedance
11. Calculate a transistor circuit's
    - **a.** DC current gain
    - **b.** AC current gain
    - **c.** Voltage gain
    - **d.** Power gain
12. Interpret the specifications given in a manufacturer's data sheet for the following:
    - **a.** A general purpose transistor
    - **b.** A high-voltage transistor
    - **c.** A high-current transistor
    - **d.** A high-power transistor
13. Describe the function of a transistor tester.
14. Explain how an ohmmeter can be used to check the P-N junctions of a transistor for opens and shorts.
15. Calculate the different values of circuit voltage and current for the following transistor biasing methods:
    - **a.** Base biasing
    - **b.** Voltage-Divider biasing
16. Define the following terms:
    - **a.** DC load line
    - **b.** Cutoff point
    - **c.** Saturation point
    - **d.** Quiescent point

Bipolar Junction Transistors (BJTs)

OUTLINE

VIGNETTE

There's No Sleeping When He's Around!

Carl Friedrich Gauss was born April 30, 1777, to poor, uneducated parents in Brunswick, Germany. He was a child of precocious abilities, particularly in mental computation. In elementary school he soon impressed his teachers, who said that mathematical ability came easier to Gauss than speech.

In secondary school he rapidly distinguished himself in ancient languages and mathematics. At 14, Gauss was presented to the court of the duke of Brunswick, where he displayed his computing skill. Until his death in 1806, the duke generously supported Gauss and his family, encouraging the boy with textbooks and a laboratory.

In the early years of the nineteenth century, Gauss's interest was in astronomy, and his accumulated work on celestial mechanics was published in 1809. In 1828, at a conference in Berlin, Gauss met physicist Wilhelm Weber, who would eventually become famous for his work on electricity. They worked together for many years and became close friends, investigating electromagnetism and the use of a magnetic needle for current measurement. In 1833 they constructed an electric telegraph system that could communicate across Göttingen from Gauss's observatory to Weber's physics laboratory. (This telegraph system of communication was later developed independently by U.S. inventor Samuel Morse.)

Gauss conceived almost all of his fundamental mathematical discoveries between the ages of 14 and 17. There are many stories of his genius in his early years, one of which involved a sarcastic teacher who liked giving his students long-winded problems and then resting, or on some occasions sleeping, in class. On his first day with Gauss, who was 8 years old, the teacher began, as usual, by telling the students to find the sum of all the numbers from 1 to 100. The teacher barely had a chance to sit down before Gauss raised his hand and said "5050." The dumbfounded teacher, who believed Gauss must have heard the problem before and memorized the answer, asked Gauss to explain how he had solved the problem. He replied: "The numbers 1, 2, 3, 4, 5, and so on to 100 can be paired as 1 and 100, 2 and 99, 3 and 98, and so on. Since each pair has a sum of 101, and there are 50 pairs, the total is 5050."

INTRODUCTION

In 1948, a component known as a transistor sparked a whole new era in electronics, the effects of which have not been fully realized even to this day. A transistor is a three-element device made of semiconductor materials used to control electron flow, the amount of which can be controlled by varying the voltages applied to its three elements. Having the ability to control the amount of current through the transistor allows us to achieve two very important applications: switching and amplification.

Like the diode, transistors are formed by *p* and *n* regions and, as we are already aware, the point at which a *p* and an *n* region join is known as a junction. Transistors in general are classified as being either the *bipolar* or *unipolar* type. The bipolar type has two P-N junctions, while unipolar transistors have only one P-N junction. In this chapter we will study all of the details relating to the *bipolar* transistor, or as it is also known, the *bipolar junction transistor* or *BJT*.

21-1 FIRST APPROXIMATION DESCRIPTION OF A BIPOLAR TRANSISTOR

In most cases it is easier to build a jigsaw puzzle when you can refer to the completed picture on the box. The same is true whenever anyone is trying to learn anything new, especially a science that contains many small pieces. These first approximation descriptions are a means for you to quickly see the complete picture without having to wait until you connect all of the pieces. Like the diode's first approximation description, this general overview will cover the transistor's basic construction, schematic symbol, physical appearance, basic operation, and main applications.

NPN Transistor
A thin, lightly doped *p*-type region (base) is sandwiched between two *n*-type regions (emitter and collector).

21-1-1 Transistor Types (NPN and PNP)

Base
The region that lies between an emitter and a collector of a transistor and into which minority carriers are injected.

Like the diode, a bipolar transistor is constructed from a semiconductor material. However, unlike the diode, which has two oppositely doped regions and one P-N junction, the transistor has three alternately doped semiconductor regions and two P-N junctions. These three alternately doped regions are arranged in one of two different ways, as shown in Figure 21-1.

Emitter
A transistor region from which charge carriers are injected into the base.

With the **NPN transistor** shown in Figure 21-1(a), a thin lightly doped *p*-type region known as the **base** (symbolized *B*) is sandwiched between two *n*-type regions called the **emitter** (symbolized *E*) and the **collector** (symbolized *C*). Looking at the NPN transistor's schematic symbol in Figure 21-1(b), you can see that an arrow is used to indicate the emitter lead. As a memory aid for the NPN transistor's schematic symbol, you may want to remember that when the emitter arrow is "**N**ot **P**ointing i**N**" to the base, the transistor is an **"NPN."** An easier method is to think of the arrow as a diode, with the tip of the arrow or cathode pointing to an *n* terminal and the back of the arrow or anode pointing to a *p* terminal, as seen in the inset in Figure 21-1(b).

Collector
A semiconductor region through which a flow of charge carriers leaves the base of the transistor.

The **PNP transistor** can be seen in Figure 21-1(c). With this transistor type, a thin, lightly doped *n*-type region (base) is placed between two *p*-type regions (emitter and collector). Figure 21-1(d) illustrates the PNP transistor's schematic symbol. Once again, if you think of the emitter arrow as a diode, as shown in the inset in Figure 21-1(d), the tip of the arrow or cathode is pointing to an *n* terminal and the back of the arrow or anode is pointing to a *p* terminal.

PNP Transistor
A thin, lightly doped *n*-type region (base) is placed between two *p*-type regions (emitter and collector).

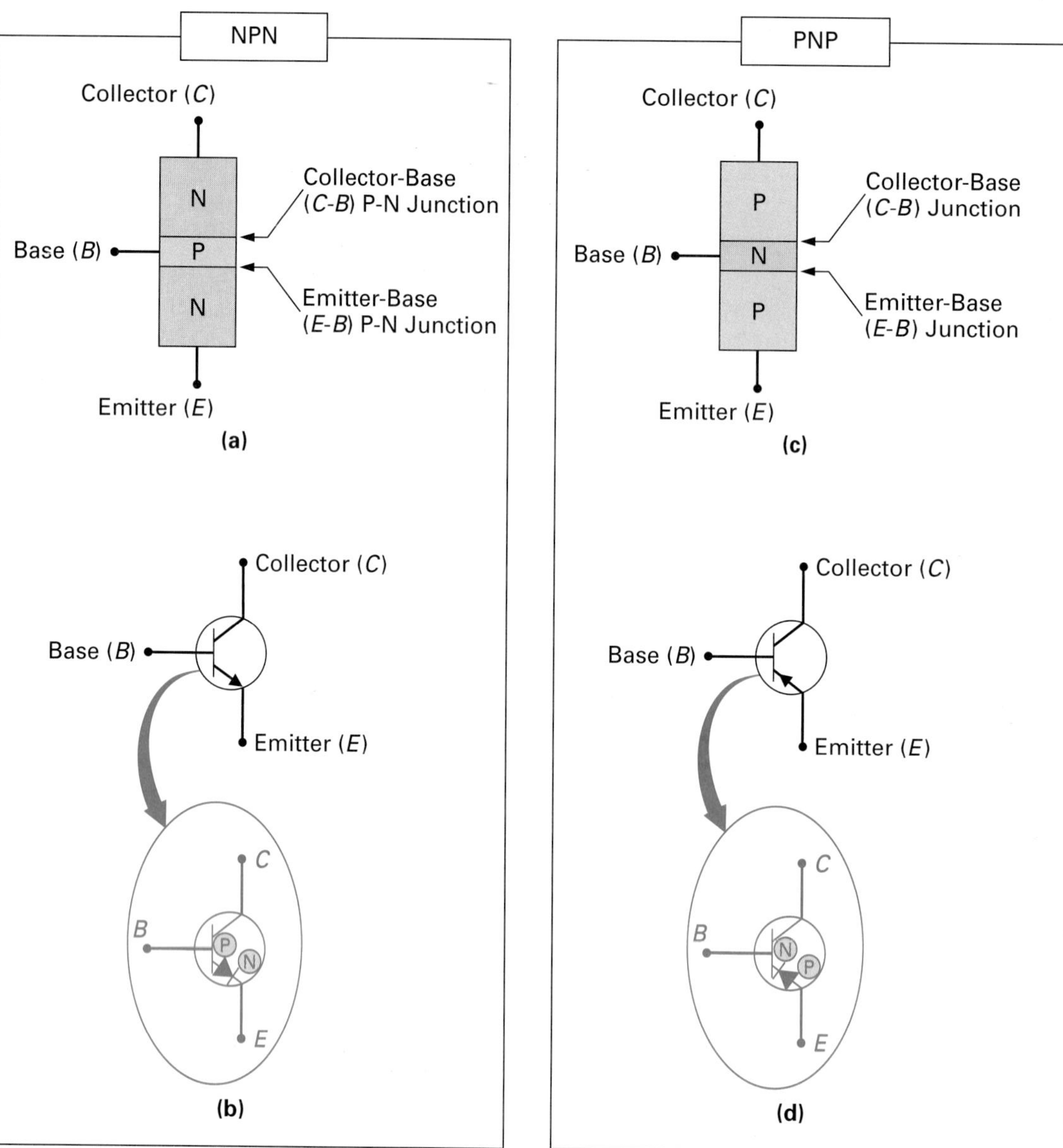

FIGURE 21-1 **Bipolar Junction Transistor (BJT) Types.**

21-1-2 Transistor Construction and Packaging

Like the diode, the three layers of an NPN or PNP transistor are not formed by joining three alternately doped regions. These three layers are formed by a "diffusion process," which first melts the base region into the collector region, and then melts the emitter region into the base region. For example, with the NPN transistor shown in Figure 21-2(a), the construction process would begin by diffusing or melting a *p*-type base region into the *n*-type collector region. Once this *p*-type base region is formed, an *n*-type emitter region is diffused or melted into the newly diffused *p*-type base region to form an NPN transistor. Keep in mind

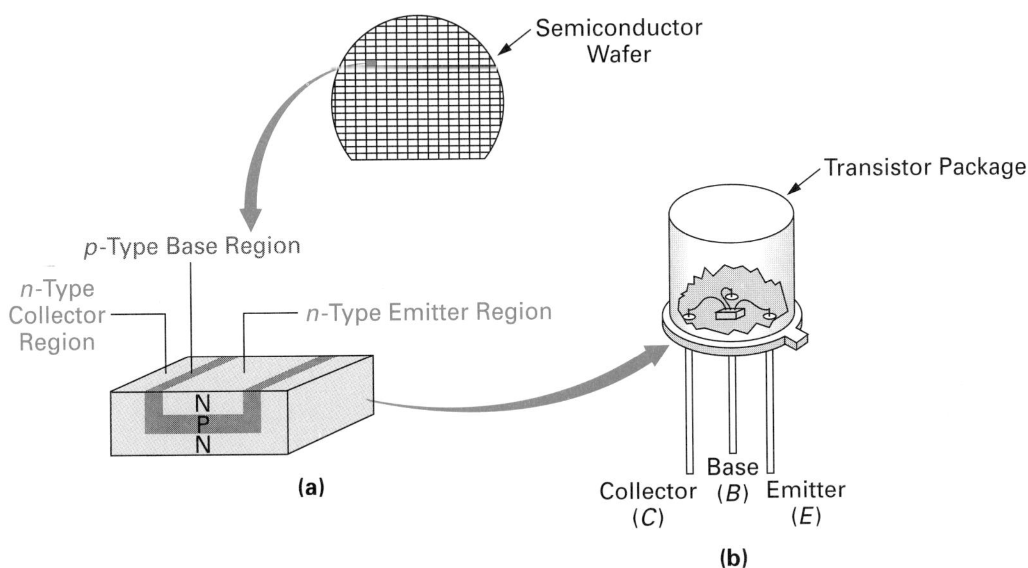

FIGURE 21-2 **Bipolar Junction Transistor Construction and Packaging.**

that manufacturers will generally construct thousands of these transistors simultaneously on a thin semiconductor wafer or disc, as shown in Figure 21-2(a). Once tested, these discs, which are about 3 inches in diameter, are cut to separate the individual transistors. Each transistor is placed in a package, as shown in Figure 21-2(b). The package will protect the transistor from humidity and dust, provide a means for electrical connection between the three semiconductor regions and the three transistor terminals, and serve as a heat sink to conduct away any heat generated by the transistor.

Figure 21-3 illustrates some of the typical low-power and high-power transistor packages. Most low-power, small-signal transistors are hermetically sealed in a metal, plastic, or epoxy package. Four of the low-power packages shown in Figure 21-3(a) have their three leads protruding from the bottom of the package because these package types are usually inserted and soldered into holes in printed circuit boards (PCBs). The surface mount technology (SMT) low-power transistor package, on the other hand, has flat metal legs that mount directly onto the surface of the PCB. These transistor packages are generally used in high component density PCBs because they use less space than a "through-hole" package. To explain this in more detail, a through-hole transistor package needs a hole through the PCB and a connecting pad around the hole to make a connection to the circuit. With an SMT package, however, no holes are needed, only a small connecting pad. Without the need for holes, pads on printed circuit boards can be smaller and placed closer together, resulting in considerable space saving.

The high-power packages, shown in Figure 21-3(b), are designed to be mounted onto the equipment's metal frame or chassis so that the additional metal will act as a heat sink and conduct the heat away from the transistor. With these high-power transistor packages, two or three leads may protrude from the package. If only two leads are present, the metal case will serve as a collector connection, and the two pins will be the base and emitter.

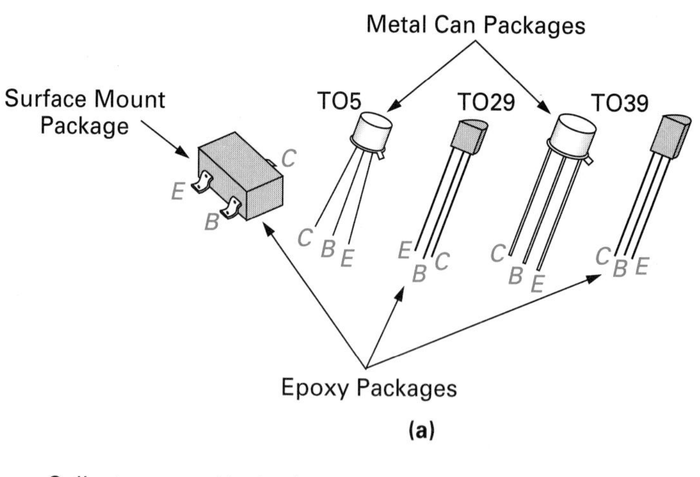

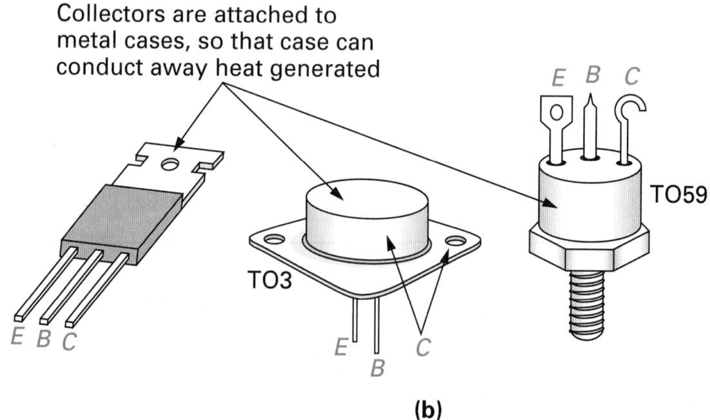

FIGURE 21-3 **Bipolar Junction Transistor Package Types. (a) Low-Power. (b) High-Power.**

Transistor package types are normally given a reference number. These designations begin with the letters "TO," which stands for transistor outline and are followed by a number. Figure 21-3 includes some examples of TO reference designators.

21-1-3 *Transistor Operation*

Figure 21-4 shows an NPN bipolar transistor, and the inset shows how a transistor can be thought of as containing two diodes: a *base-to-collector diode* and a *base-to-emitter diode.* With an NPN transistor, both diodes will be back-to-back and "**N**ot be **P**ointing i**N**" (NPN) to the base, as shown in the inset in Figure 21-4. For a PNP transistor, the base-collector and base-emitter diode will both be pointing into the base.

Transistors are basically controlled to operate as a switch, or they are controlled to operate as a variable resistor. Let us now examine each of these operating modes.

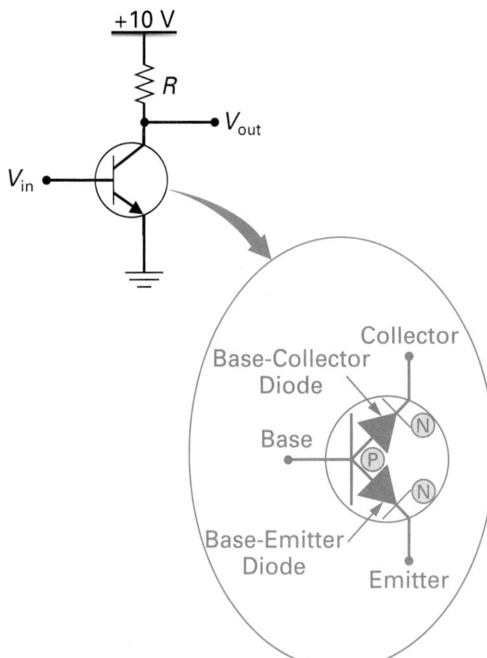

FIGURE 21-4 The Base-Collector and Base-Emitter Diodes Within a Bipolar Transistor.

The Transistor's ON/OFF Switching Action

Figure 21-5 illustrates how the transistor can be made to operate as a switch. This ON/OFF switching action of the transistor is controlled by the transistor's base-to-emitter (*B-E*) diode. If the *B-E* diode of the transistor is forward biased, the transistor will turn ON; if the *B-E* diode of the transistor is reverse biased, the transistor will turn OFF.

To begin with, let us see how the transistor can be switched ON. In Figure 21-5(a), the *B-E* diode of the transistor is forward biased (anode at base is +5 V, cathode at emitter is 0 V), and the transistor will turn ON. Its collector and emitter output terminals will be equivalent to a closed switch, as shown in Figure 21-5(b). This low resistance between the transistor's collector and emitter will cause a current (*I*), as shown in Figure 21-5(b). The output voltage in this condition will be zero volts because all of the +10 V supply voltage will be dropped across *R*. Another way to describe this would be to say that the low resistance path between the transistor's emitter and collector connects the zero volt emitter potential through to the output.

Now let us see how the transistor can be switched OFF. In Figure 21-5(c), the transistor has 0 V being applied to its base input. In this condition, the *B-E* diode of the transistor is reverse biased (anode at base is 0 V, cathode at emitter is 0 V), and so the transistor will turn OFF and its collector and emitter output terminals will be equivalent to an open switch, as shown in Figure 21-5(d). This high resistance between the transistor's collector and emitter will prevent any current and any voltage drop, resulting in the full +10 V supply voltage being applied to the output, as shown in Figure 21-5(d).

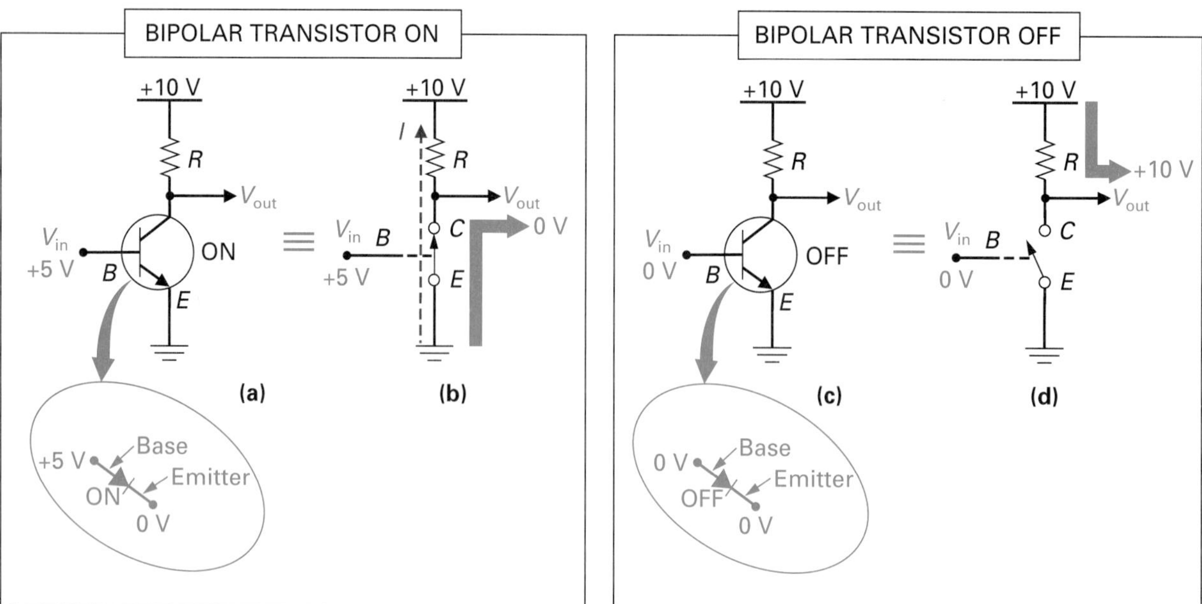

FIGURE 21-5 The Bipolar Transistor's ON/OFF Switching Action.

The Transistor's Variable-Resistor Action

In the previous section we saw how the transistor can be biased to operate in one of two states: ON or OFF. When operated in this two-state way, the transistor is being switched ON and OFF in almost the same way as a junction diode. The transistor, however, has another ability that the diode does not have—it can also function as a variable resistor, as shown in the equivalent circuit in Figure 21-6(a). In Figure 21-5 we saw how +5 V base input bias voltage would result in a low resistance between emitter and collector (closed switch) and how a 0 V base input bias would result in a high resistance between emitter and collector (open switch). The table in Figure 21-6(b) shows an example of the relationship between base input bias voltage (V_B) and emitter-to-collector resistance ($R_{C\text{-}E}$). In this table, you can see that the transistor is not only going to be driven between the two extremes of fully ON and fully OFF. When the base input voltage is at some voltage level between +5 V and 0 V, the transistor is partially ON; therefore, the transistor's emitter-to-collector resistance is somewhere between 0 Ω and maximum Ω. For example, when V_B = +4 V, the transistor is not fully ON, and its emitter-to-collector resistance will be slightly higher, at 100 Ω. If the base input bias voltage is further reduced to +3 V, for example, you can see in the table that the emitter-to-collector resistance will further increase to 10 kΩ. Further decreases in base input voltage ($V_B\downarrow$) will cause further increases in emitter-to-collector resistance ($R_{C\text{-}E}\uparrow$) until $V_B = 0$ V and $R_{C\text{-}E}$ = maximum Ω.

As a matter of interest, the name transistor was derived from the fact that through base control we can "transfer" different values of "resistance" between the emitter and collector. This effect of "transferring resistance" is known as **transistance** and the component that functions in this manner is called the transistor.

Transistance The effect of transferring resistance.

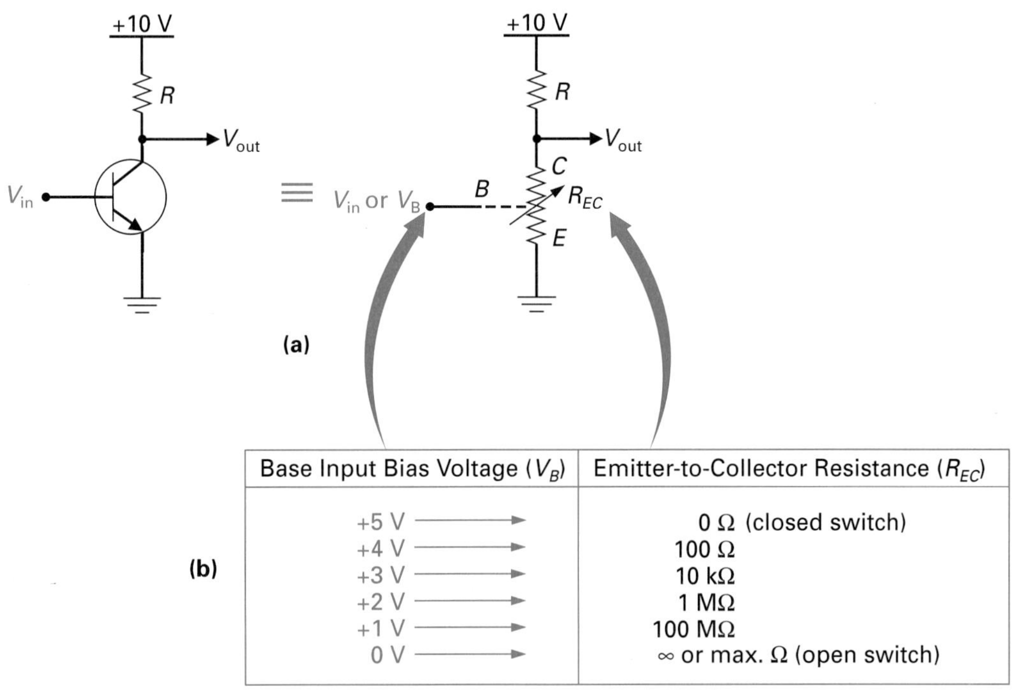

FIGURE 21-6 The Bipolar Transistor's Variable-Resistor Action.

Now that we have seen how the transistor can be made to operate as either a switch or as a variable resistor, let us see how these characteristics can be made use of in circuit applications.

21-1-4 *Transistor Applications*

The transistor's impact on electronics has been phenomenal. It initiated the multibillion dollar semiconductor industry and was the key element behind many other inventions, such as integrated circuits (*IC*s), optoelectronic devices, and digital computer electronics. In all of these applications, however, the transistor is basically made to operate in one of two ways: as a switch or as a variable resistor. Let us now briefly examine an example of each.

Digital Logic Gate Circuit

A digital logic gate circuit makes use of the transistor's ON/OFF switching action. Digital circuits are often referred to as "switching" or "two-state" circuits because their main control device (the transistor) is switched between the two states of ON and OFF. The transistor is at the very heart of all digital electronic circuits. For example, transistors are used to construct logic gate circuits, gates are used to construct flip-flop circuits, flip-flops are used to construct register and counter circuits, and these circuits are used to construct microprocessor, memory, and input/output circuits—the three basic blocks of a digital computer.

In Chapter 20 you were introduced to the OR gate, which would produce a HIGH (logic 1) output when either input *A* OR *B* was HIGH. Figure 21-7(a) shows how the transistor can be used to construct another type of digital logic gate. This gate is called the NOT gate, or more commonly, the INVERTER gate. The basic NOT gate circuit is constructed using one NPN transistor and two resistors. This logic gate has only one input (*A*) and one output (*Y*), and its schematic symbol is shown in Figure 21-7(b). Figure 21-7(c) shows how this logic gate will react to the two different input possibilities. When the input is 0 V (logic 0), the transistor's base-emitter P-N diode will be reverse biased and so the transistor will turn OFF. Referring to the inset for this circuit condition in Figure 21-7(c), you can see that the OFF transistor is equivalent to an open switch between emitter and collector, and therefore the +5 V supply voltage will be connected to the output. In summary, a logic 0 input (0 V) will be converted to a logic 1 output (+5 V). On the other hand, when the input is +5 V (logic 1), the transistor's base-emitter P-N diode will be forward biased and so the transis-

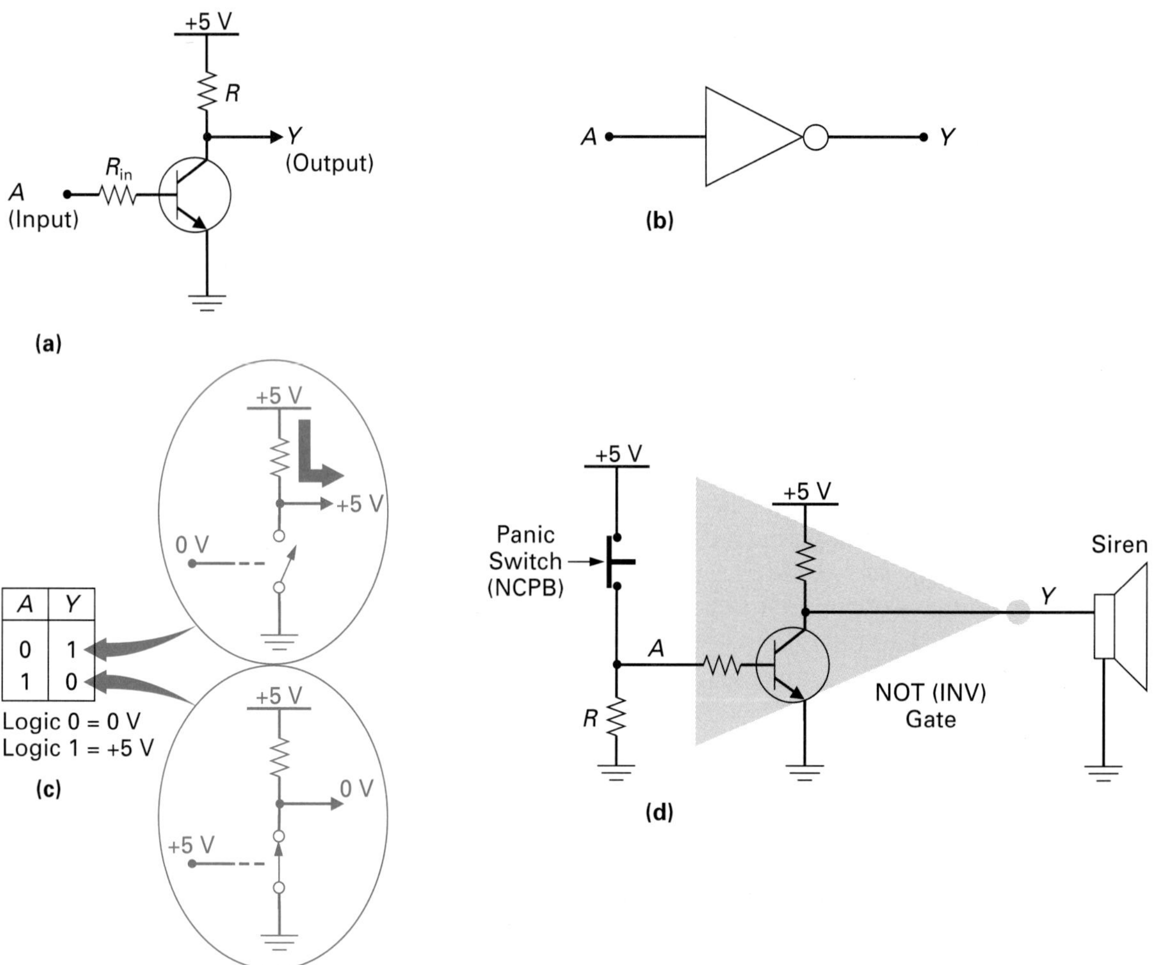

FIGURE 21-7 The Transistor Being Used in a Digital Electronic Circuit. (a) Basic NOT or INVERTER Gate Circuit. (b) NOT Gate Schematic Symbol. (c) NOT Gate Function Table. (d) NOT Gate Security System Application.

tor will turn ON. Referring to the inset for this circuit condition in Figure 21-7(c), you can see that the ON transistor is equivalent to a closed switch between emitter and collector, and therefore 0 V will be connected to the output. In summary, a logic 1 input (+5 V) will be converted to a logic 0 output (0 V).

Referring to the function table in Figure 21-7(c), you can see that the output logic level is "NOT" the same as the input logic level—hence the name NOT gate.

As an application, Figure 21-7(d) shows how a NOT or INV gate can be used to invert an input control signal. In this circuit, you can see that a normally closed push button (NCPB) switch is used as a panic switch to activate a siren in a security system. Because the push button is normally closed, it will produce +5 V at *A* when it is not in alarm. If this voltage were connected directly to the siren, the siren would be activated incorrectly. By including the NOT gate between the switch circuit and the siren, the normally HIGH output of the NCPB will be inverted to a LOW, and not activate the siren when we are not in alarm. When the panic switch is pressed however, the NCPB contacts will open producing a LOW input voltage to the NOT gate. This LOW input will be inverted to a HIGH output and activate the siren.

Analog Amplifier Circuit

When used as a variable resistor, the transistor is the controlling element in many analog or linear circuits applications such as amplifiers, oscillators, modulators, detectors, regulators, and so on. The most important of these applications is **amplification,** which is the boosting in strength or increasing in amplitude of electronic signals.

Amplification Boosting in strength, or increasing amplitude, of electronic signals.

Figure 21-8(a) shows a simplified transistor amplifier circuit, while Figure 21-8(b) shows the voltage waveforms present at different points in the circuit. As you can see, the transistor is labeled Q_1 because the letter "Q" is the standard letter designation used for transistors.

Before applying an ac sine wave input signal, let us determine the dc voltage levels at the transistor's base and collector. The 6.3 kΩ/3.7 kΩ resistance ratio of the voltage divider R_1 and R_2 causes the +10 V supply voltage to be proportionally divided, producing +3.7 V dc across R_2. This +3.7 V dc will be applied to the base of the transistor, causing the base-emitter junction of Q_1 to be forward biased and Q_1 to turn ON. With transistor Q_1 ON, a certain value of resistance will exist between the transistor's collector and emitter (R_{CE} or R_{EC}), and this resistance will form a voltage divider with R_E and R_C, as seen in the inset in Figure 21-8(a). The dc voltage at the collector of Q_1 relative to ground (V_C) will be equal to the voltage developed across Q_1's collector-emitter resistance (R_{CE}) and R_E. In this example circuit, with no ac signal applied, a V_b of +3.7 V dc will cause R_{CE} and R_E to cumulatively develop +8 V at the collector of Q_1. The transistor has a base bias voltage (V_b) that is +3.7 V dc relative to ground and a collector voltage (V_C) that is +8 V dc relative to ground. Capacitors C_1 and C_2 are included to act as dc blocks, with C_1 preventing the +3.7 V dc base bias voltage (V_b) from being applied back to the input (V_{in}) and C_2 preventing the +8 V dc collector reference voltage (V_C) from being applied across the output (V_{R_L} or V_{out}).

Let us now apply an input signal and see how it is amplified by the amplifier circuit in Figure 21-8(a). The alternating input sine wave signal (V_{in}) is

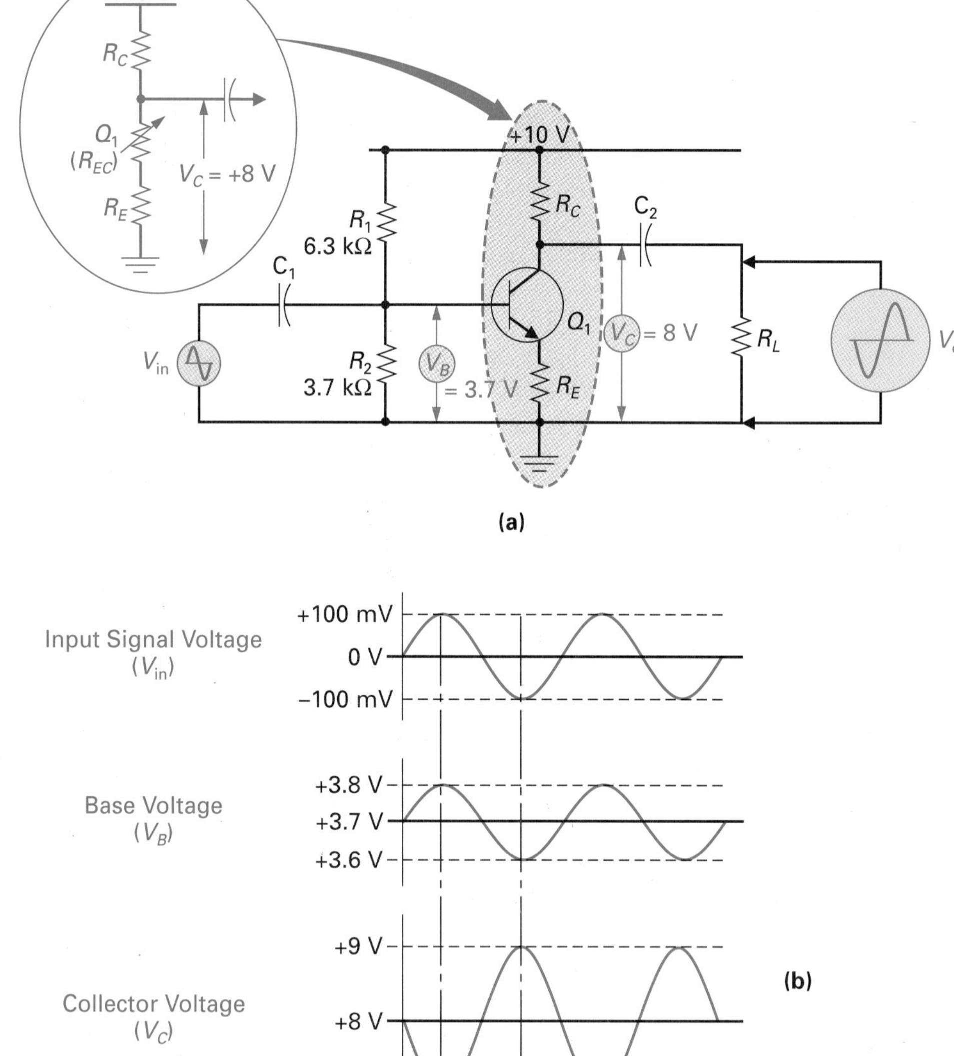

FIGURE 21-8 The Transistor Being Used in an Analog Electronic Circuit. (a) Basic Amplifier Circuit. (b) Input/Output Voltage Waveforms.

applied to the base of Q_1 via C_1, which, like most capacitors, offers no opposition to this ac signal. This input signal, which has a peak-to-peak voltage change of 200 mV, is shown in the first waveform in Figure 21-8(b). The alternating input signal will be superimposed on the +3.7 V dc base bias voltage and cause the +3.7 V dc at the base of Q_1 to increase by 100 mV (3.7 V + 100 mV = 3.8 V), and decrease by 100 mV (3.7 V – 100 mV = 3.6 V), as seen in the second waveform in Figure 21-8(b). An increase in the input signal ($V_{in}\uparrow$), and therefore the base voltage ($V_B\uparrow$), will cause an increase in the emitter diode's forward bias, causing Q_1 to turn more ON and the emitter-to-collector resistance of Q_1 to decrease ($R_{EC}\downarrow$). Because voltage drop is always proportional to resistance, a decrease in $R_{EC}\downarrow$ will cause a decrease in the voltage drop across R_{CE} and R_E ($V_C\downarrow$), and this decrease in V_C will be coupled to the output via C_2, causing a decrease in the output voltage developed across the load ($V_{out}\downarrow$).

Now let us examine what will happen when the sine wave input signal decreases. A decrease in the input signal ($V_{in}\downarrow$), and therefore the base voltage ($V_B\downarrow$), will cause a decrease in the emitter diode's forward bias, causing Q_1 to turn less ON and the emitter-to-collector resistance of Q_1 to increase ($R_{EC}\uparrow$). Because voltage drop is always proportional to resistance, an increase in $R_{EC}\uparrow$ will cause an increase in the voltage drop across R_{EC} and R_E ($V_C\uparrow$). This increase in V_C will be coupled to the output via C_2, causing an increase in the output voltage developed across the load ($V_{out}\uparrow$).

Comparing the input signal voltage (V_{in}) to the output signal voltage (V_{out}) in Figure 21-8(b), you can see that a change in the input signal voltage produces a corresponding greater change in the output signal voltage. The ratio (comparison) of output signal voltage change to input signal voltage change is a measure of this circuit's **voltage gain (A_V)**. In this example, the **output signal voltage change (ΔV_{out})** is between +1 V and –1 V, and the **input signal voltage change (ΔV_{in})** is between +100 mV and –100 mV. The circuit's voltage gain between input and output will therefore be:

Voltage Gain (A_V)
The ratio of the output signal voltage change to input signal voltage change.

Output Signal Voltage Change
Change in output signal voltage in response to a change in the input signal voltage.

Input Signal Voltage Change
The input voltage change that causes a corresponding change in the output voltage.

$$\text{Voltage Gain } (A_V) = \frac{\text{Output Voltage Change } (\Delta V_{out})}{\text{Input Voltage Change } (\Delta V_{in})}$$

$$A_V = \frac{+1 \text{ to } -1 \text{ V}}{+100 \text{ mV to } -100 \text{ mV}} = \frac{2 \text{ V}}{200 \text{ mV}} = 10$$

A voltage gain of 10 means that the output voltage is ten times larger than the input voltage. The transistor does not produce this gain magically within its NPN semiconductor structure. The gain or amplification is achieved by the input signal controlling the conduction of the transistor, which takes energy from the collector supply voltage and develops this energy across the load resistor. Amplification is achieved by having a small input voltage control a transistor and its large collector supply voltage, so that a small input voltage change results in a similar but larger output voltage change.

Comparing the input signal to the output signal at time 1 and time 2 in Figure 21-8(b), you can see that this circuit will invert the input signal voltage in the same way that the NOT gate inverts its input voltage (positive input voltage swing produces a negative output voltage swing, and vice versa). This inversion always occurs with this particular transistor circuit arrangement; however, it is not a problem since the shape of the input signal is still preserved at the output (both input and output signals are sinusoidal).

SELF-TEST REVIEW QUESTIONS FOR SECTION 21-1

1. What are the two basic types of bipolar transistor?
2. Name the three terminals of a bipolar transistor.
3. What are the two basic ways in which a transistor is made to operate?
4. Which of the modes of operation mentioned in question 3 is made use of in digital circuits and which is made use of in analog circuits?

21-2 SECOND APPROXIMATION DESCRIPTION OF A BIPOLAR TRANSISTOR

Now that we have a good understanding of the bipolar junction transistor's (BJT's) general characteristics, operation and applications, let us examine all of these aspects in a little more detail.

21-2-1 *Basic Bipolar Transistor Action*

When describing diodes previously, we saw how the P-N junction of a diode could be either forward or reverse biased to either permit or block the flow of current through the device. The transistor must also be biased correctly; however, in this case, two P-N junctions rather than one must have the correct external supply voltages applied.

A Correctly Biased NPN Transistor Circuit

Figure 21-9(a) shows how an NPN transistor should be biased for normal operation. In this circuit, a +10 V supply voltage is connected to the transistor's collector (C) via a 1 kΩ collector resistor (R_C). The emitter (E) of the transistor is connected to ground via a 1.5 kΩ emitter resistor (R_E), and, as an example, an input voltage of +3.7 V is being applied to the base (B). The output voltage (V_{out}) is taken from the collector, and this collector voltage (V_C) will be equal to the voltage developed across the transistor's collector-to-emitter and the emitter resistor R_E.

As previously mentioned in the first approximation description of the transistor, the transistor can be thought of as containing two diodes, as shown in Figure 21-9(b). In normal operation, *the transistor's emitter diode or junction is for-*

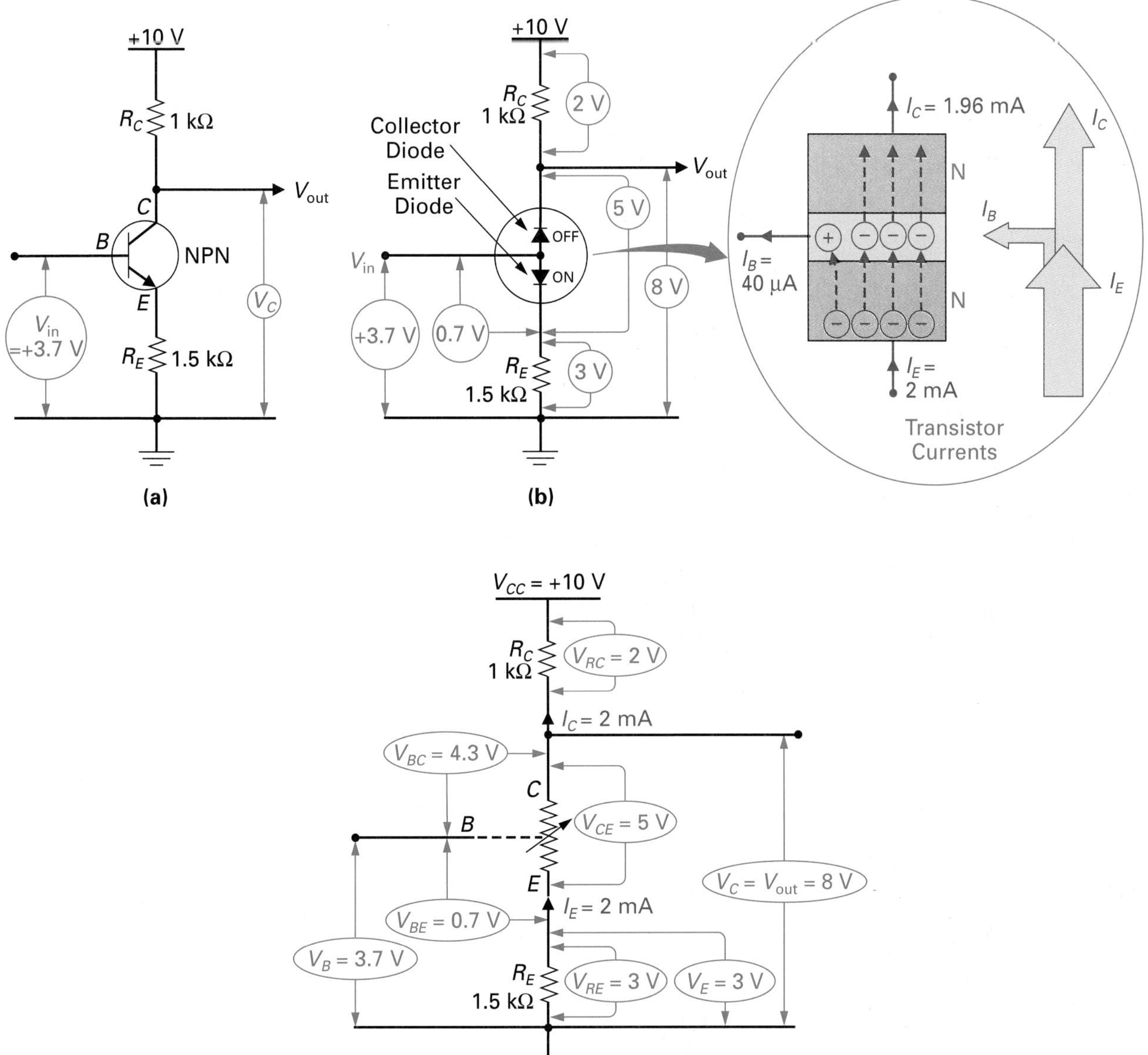

FIGURE 21-9 A Correctly Biased NPN Transistor Circuit.

ward biased, while the transistor's collector diode or junction is reverse biased. To explain how these junctions are biased ON and OFF simultaneously, let us see how the input voltage of +3.7 V will affect this transistor circuit. An input voltage of +3.7 V is large enough to overcome the barrier voltage of the emitter diode (base-emitter junction), and so it will turn ON (base or anode is +, emitter or cathode is connected to ground or 0 V). Like any forward biased silicon diode, the emitter diode will drop 0.7 V between base and emitter, and so the +3.7 V at the base will produce +3.0 V at the emitter. Knowing the voltage drop across the emitter resistor (V_{R_E} = 3 V) and the resistance of the emitter resistor (R_E = 1.5 kΩ), we can calculate the value of current through the emitter resistor.

$$I_{R_E} = \frac{V_{R_E}}{R_E} = \frac{3\text{ V}}{1.5\text{ k}\Omega} = 2\text{ mA}$$

This emitter resistor current of 2 mA will leave ground, travel through R_E, and then enter the transistor's *n*-type emitter region. This current at the transistor's emitter terminal is called the **emitter current (I_E).** The forward biased emitter diode will cause the steady stream of electrons entering the emitter to head toward the base region, as shown in the inset in Figure 21-9(b). The base is a very thin, lightly doped region with very few holes in relation to the number of electrons entering the transistor from the emitter. Consequently, only a few electrons combine with the holes in the base region and flow out of the base region. This relatively small current at the transistor's base terminal is called the **base current (I_B).** Because only a few electrons combine with holes in the base region, there is an accumulation of electrons in the base's *p* layer. These free electrons, feeling the attraction of the large positive collector supply voltage (+10 V), will travel through the *n*-type collector junction and out of the transistor to the positive external collector supply voltage. The current emerging out of the transistor's collector is called the **collector current (I_C).** Because both the collector current and base current are derived from the emitter current, we can state that:

Emitter Current (I_E)
The current at the transistor's emitter terminal.

Base Current (I_B)
The relatively small current at the transistor's base terminal.

Collector Current (I_C)
The current emerging out of the transistor's collector.

$$I_E = I_B + I_C$$

In the example in the inset in Figure 21-9(b), you can see that this is true because

$$I_E = I_B + I_C$$
$$I_E = 40\ \mu\text{A} + 1.96\text{ mA} = 2\text{ mA}\ \ (40\ \mu\text{A} = 0.04\text{ mA})$$

Stated another way, we can say that the collector current is equal to the emitter current minus the current that is lost out of the base.

$$I_C = I_E - I_B$$
$$I_C = 2\text{ mA} - 40\ \mu\text{A} = 1.96\text{ mA}$$

Approximately 98% of the electrons entering the emitter of a transistor will arrive at the collector. Because of the very small percentage of current flowing out of the base (I_B equals about 2% of I_E), we can approximate and assume that I_C is equal to I_E.

$$I_C \cong I_E$$

(I_C approximately equals I_E)

The Current-Controlled Transistor

In the previous section, we discovered that because the collector and base currents (I_C and I_B) are derived from the emitter current (I_E), an increase in the emitter current ($I_E\uparrow$), for example, will cause a corresponding increase in collector and base current ($I_C\uparrow$, $I_B\uparrow$). Looking at this from a different angle, an increase in the applied base voltage (base input increases to +3.8 V) will increase the forward bias applied to the emitter diode of the transistor, which will draw more electrons up from the emitter and cause an increase in I_E, I_B, and I_C. Similarly, a decrease in the applied base voltage (base input decreases to +3.6 V) will decrease the forward bias applied to the emitter diode of the transistor, which will decrease the number of electrons being drawn up from the emitter and cause a decrease in I_E, I_B, and I_C. The applied input base voltage will control the amount of base current, which will in turn control the amount of emitter and collector current, and therefore the conduction of the transistor. This is why *the bipolar transistor is known as a current-controlled device.*

Continuing our calculations for the example circuit in Figure 21-9(b), let us apply this current relationship and assume that I_C is equal to I_E, which, as we previously calculated, is equal to 2 mA. Knowing the value of current for the collector resistor (I_{R_C} = 2 mA) and the resistance of the collector resistor (R_C = 1 kΩ), we can calculate the voltage drop across the collector resistor.

$$V_{R_C} = I_{R_C} \times R_C = 2\text{ mA} \times 1\text{ k}\Omega = 2\text{ V}$$

With 2 V being dropped across R_C, the voltage at the transistor's collector (V_C) will be:

$$V_C = +10\text{ V} - V_{R_C} = 10\text{ V} - 2\text{ V} = 8\text{ V}$$

Because the voltage at the transistor's collector relative to ground is applied to the output, the output voltage will also be equal to 8 V.

$$V_C = V_{out} = 8\text{ V}$$

At this stage, we can determine a very important point about any correctly biased NPN transistor circuit. *A properly biased transistor will have a forward biased base-emitter junction (emitter diode is ON), and a reverse biased base-collector junction (collector diode is OFF).* We can confirm this with our example circuit in Figure 21-9(b), because we now know the voltages at each of the transistor's terminals.

Emitter diode (base-emitter junction) is forward biased (ON) because
Anode (base) is connected to +3.7 V (V_{in})
Cathode (emitter) is connected to 0 V via R_E.

Cathode diode (base-collector junction) is reverse biased (OFF) because
Anode (base) is connected to +3.7 V (V_{in})
Cathode (collector) is at +8 V (due to 2 V drop across R_C)

Keep in mind that even though the collector diode (base-collector junction) is reverse biased, current will still flow through the collector region. This is because most of the electrons traveling from emitter-to-base (through the forward biased emitter diode) do not find many holes in the thin, lightly doped base region, and therefore the base current is always very small. Almost 98% of the electrons accumulating in the base region feel the strong attraction of the positive collector supply voltage and flow up into the collector region and then out of the collector as collector current.

With the example circuit in Figure 21-9(a) and (b), the emitter diode is ON and the collector diode is OFF, and the transistor is said to be operating in its normal, or *active region.*

Operating a Transistor in the Active Region

Active Operation or in the Active Region When the base-emitter junction is forward biased and the base-collector junction is reverse biased. In this mode, the transistor is equivalent to a variable-resistor between collector and emitter.

A transistor is said to be in **active operation,** or in the **active region,** when its base-emitter junction is forward biased (emitter diode is ON), and the base-collector junction is reverse biased (collector diode is OFF). In this mode, the transistor is equivalent to a variable-resistor between collector and emitter.

In Figure 21-9(c), our transistor circuit example has been redrawn with the transistor this time being shown as a variable-resistor between collector and emitter and with all of our calculated voltage and current values inserted. Before we go any further with this circuit, let us discuss some of the letter abbreviations used in transistor circuits. To begin with, the term V_{CC} is used to denote the "stable collector voltage" and this dc supply voltage will typically be positive for an NPN transistor. Two *C*s are used in this abbreviation ($+V_{CC}$) because V_C (V sub single C) is used to describe the voltage at the transistor's collector relative to ground. The doubling up of letters such as V_{CC}, V_{EE}, or V_{BB} is used to denote a constant dc bias voltage for the collector (V_{CC}), emitter (V_{EE}), and base (V_{BB}). A single sub letter abbreviation such as V_C, V_E, or V_B is used to denote a transistor terminal voltage relative to ground. The other voltage abbreviations, V_{CE}, V_{BE}, and V_{CB}, are used for the voltage difference between two terminals of the transistor. For example, V_{CE} is used to denote the potential difference between the transistor's collector and emitter terminals. Finally, I_E, I_B, and I_C are, as previously stated, used to denote the transistor's emitter current (I_E), base current (I_B), and collector current (I_C).

Because the transistor's resistance between emitter and collector in Figure 21-9(c) is in series with R_C and R_E, we can calculate the voltage drop between collector and emitter (V_{CE}) because V_{RC} and V_{RE} are known.

$$V_{CE} = V_{CC} - (V_{R_E} + V_{R_C})$$

$$V_{CE} = 10\text{ V} - (3\text{ V} + 2\text{ V}) = 10\text{ V} - 5\text{ V} = 5\text{ V}$$

Now that we know the voltage drop between the transistor's collector and emitter (V_{CE}), we can calculate the transistor's equivalent resistance between collector and emitter (R_{C_E}) because we know that the current through the transistor is 2 mA.

$$R_{C_E} = \frac{V_{CE}}{I_C} = \frac{5\text{ V}}{2\text{ mA}} = 2.5\text{ k}\Omega$$

Operating the Transistor in Cutoff and Saturation

Figure 21-10 shows the three basic ways in which a transistor can be operated. As we have already discovered, the bias voltages applied to a transistor control the transistor's operation by controlling the two P-N junctions (or diodes) in a bipolar transistor. For example, the center column reviews how a transistor will operate in the active region. As you can see, our previous circuit example with all of its values has been used. To summarize: *when a transistor is operated in the active region, its emitter diode is biased ON, its collector diode is biased OFF, and the transistor is equivalent to a variable-resistor between the collector and the emitter.*

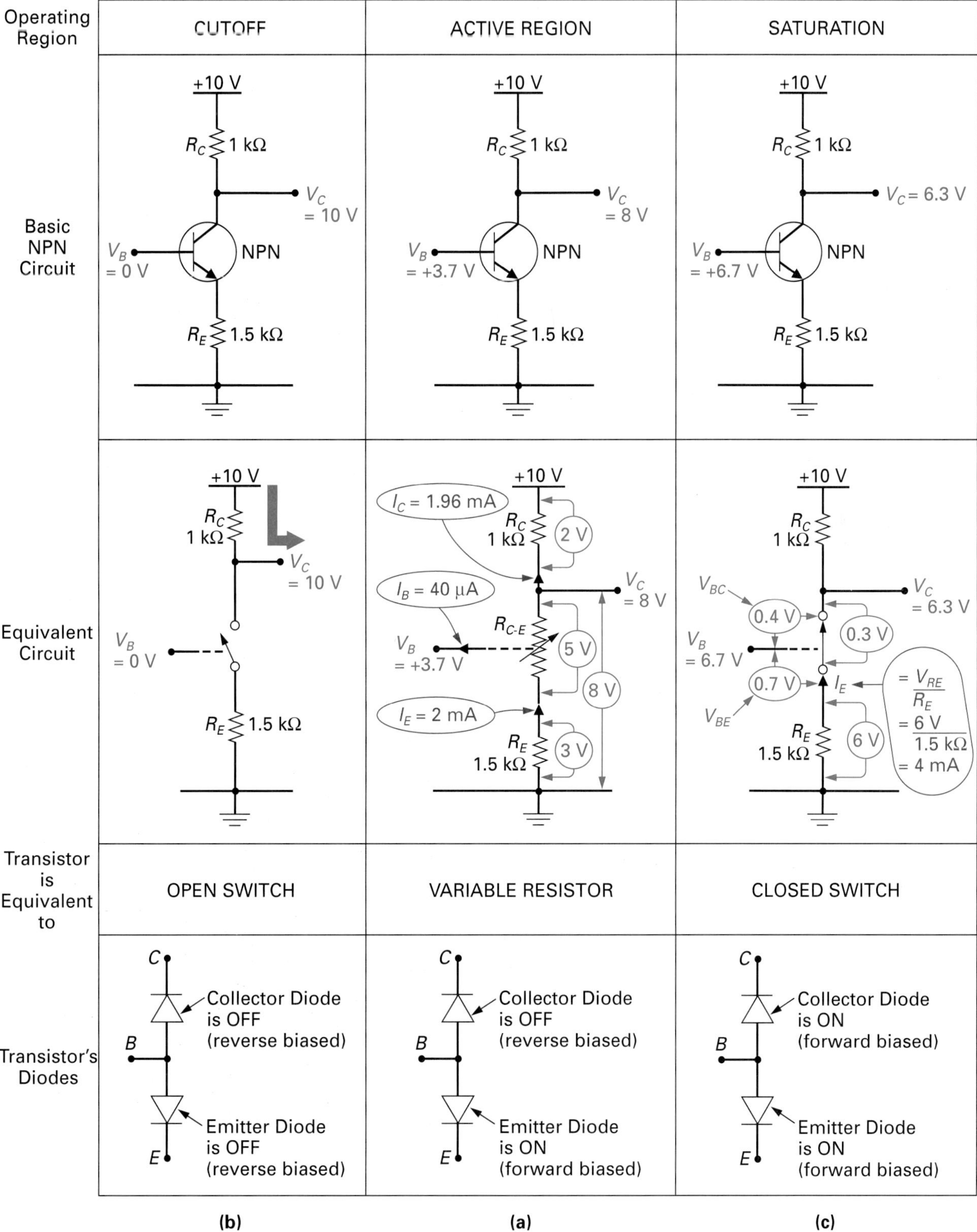

FIGURE 21-10 The Three Bipolar Transistor Operating Regions.

Cutoff
A transistor is in cutoff when the bias voltage is reduced to a point that it stops current in the transistor.

Saturation
A transistor is in saturation when the bias voltage is increased to such a point that further increase will not cause any increase in current through the transistor.

The left column in Figure 21-10 shows how the same transistor circuit can be driven into **cutoff.** A transistor is in cutoff when the bias voltage is reduced to a point that it stops current in the transistor. In this example circuit, you can see that when the base input bias voltage (V_B) is reduced to 0 V, the transistor is cut off. *In cutoff, both the emitter and the collector diode of the transistor will be biased OFF, the transistor is equivalent to an open switch between the collector and the emitter, and the transistor current is zero.*

The right column in Figure 21-10(c) shows how the same transistor circuit can be driven into **saturation.** A transistor is in saturation when the bias voltage is increased to such a point that any further increase in bias voltage will not cause any further increase in current through the transistor. In the equivalent circuit in Figure 21-10(c), you can see that when the base input bias voltage (V_B) is increased to +6.7 V, the emitter diode of the transistor will be heavily forward biased and the emitter current will be large.

$$I_E = \frac{V_B - V_{BE}}{R_E} = \frac{6.7\text{ V} - 0.7\text{ V}}{1.5\text{ k}\Omega} = 4\text{ mA}$$

Because I_B and I_C are both derived from I_E, an increase in I_E will cause a corresponding increase in both I_B and I_C. These high values of current through the transistor account for why a transistor operating in saturation is said to be equivalent to a closed switch (high conductance, low resistance). Although the transistor's resistance between the collector and emitter (R_{CE}) is assumed to be 0 Ω, there is still some small value of R_{CE}. Typically, a saturated transistor will have a 0.3 V drop between the collector and emitter (V_{CE} = 0.3 V), as shown in the equivalent circuit in Figure 21-10(c). If V_{BE} = 0.7 V and V_{CE} = 0.3 V, then the voltage drop across the collector diode of a saturated transistor (V_{BC}) will be 0.4 V. This means that the base of the transistor (anode) is now +0.4 V relative to the collector (cathode), and there is not enough reverse bias voltage to turn OFF the collector diode. *In saturation both the emitter and collector diodes are said to be forward biased, the transistor is equivalent to a closed switch, and any further increase in bias voltage will not cause any further increase in current through the transistor.*

Biasing PNP Bipolar Transistors

Generally the PNP transistor is not employed as much as the NPN transistor in most circuit applications. The only difference that occurs with PNP transistor circuits is that the polarity of V_{CC} and the base bias voltage (V_B) need to be reversed to a negative voltage, as shown in Figure 21-11. The PNP transistor has the same basic operating characteristics as the NPN transistor, and all of the previously discussed equations still apply. Referring to the inset in Figure 21-11, you will see that the –3.7 V base bias voltage will forward bias the emitter diode, and the –10 V V_{CC} will reverse bias the collector diode, so that the transistor is operating in the active region. Also, the electron transistor currents are in the opposite direction. This, however, makes no difference because the fact that the sum of the collector current entering the collector and base current entering the base is equal to the value of emitter current leaving the emitter, so $I_E = I_B + I_C$ still applies.

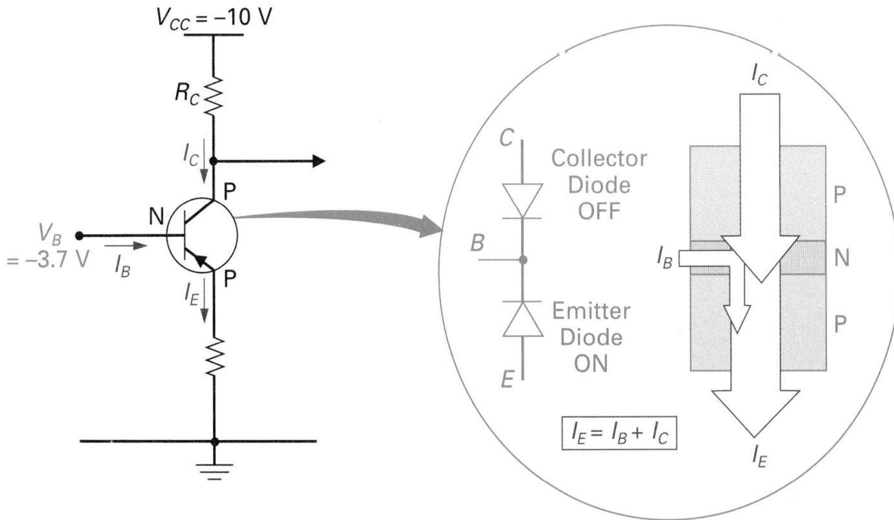

FIGURE 21-11 A Correctly Biased PNP Transistor Circuit.

21-2-2 *Bipolar Transistor Circuit Configurations and Characteristics*

In the previous sections, we have seen how the bipolar junction transistor can be used in digital two-state switching circuits and analog or linear circuits such as the amplifier. In all of these different circuit interconnections or **configurations,** the bipolar transistor was used as the main controlling element, with one of its three leads being used as a common reference and the other two leads being used as an input and an output. Although there are many thousands of different bipolar transistor circuit applications, all of these circuits can be classified in one of three groups based on which of the transistor's leads is used as the **common** reference. These three different circuit configurations are shown in Figure 21-12. With the *common-emitter (C-E)* bipolar transistor circuit configuration, shown in Figure 21-12(a), the input signal is applied between the base and emitter, while the output signal appears between the transistor's collector and emitter. With this circuit arrangement, the input signal controls the transistor's base current, which in turn controls the transistor's output collector current, and the emitter lead is common to both the input and output. Similarly, with the *common-base (C-B)* circuit configuration shown in Figure 21-12(b), the input signal is applied between the transistor's emitter and base, the output signal is developed across the transistor's collector and base, and the base is common to both input and output. Finally, with the *common-collector (C-C)* circuit configuration shown in Figure 21-12(c), the input is applied between the base and collector, the output is developed across the emitter and collector, and the collector is common to both the input and output.

Configurations
Different circuit interconnections.

Common
Shared by two or more services, circuits, or devices. Although the term "common ground" is frequently used to describe two or more connections sharing a common ground, the term "common" alone does not indicate a ground connection, only a shared connection.

To begin with, we will discuss the common-emitter circuit configuration characteristics because it has been this circuit arrangement that we have been using in all of the circuit examples in this chapter.

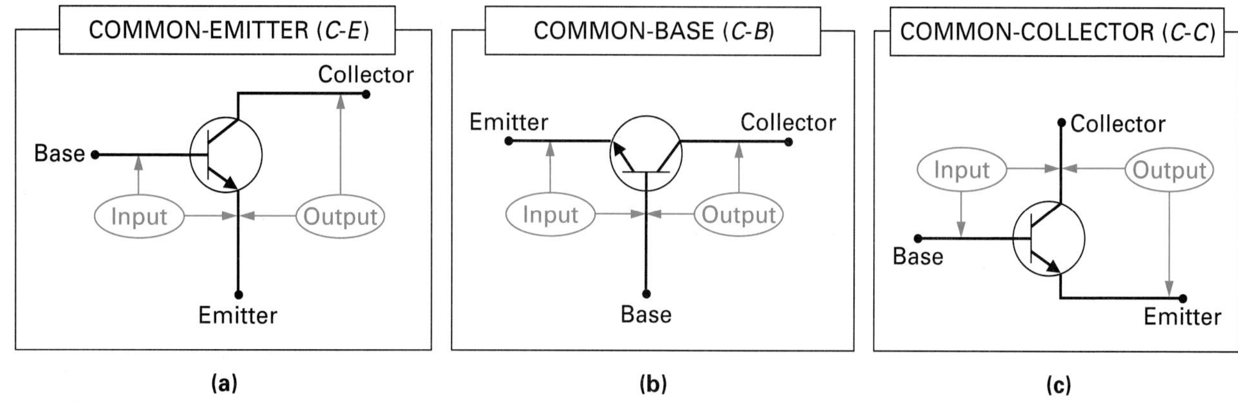

FIGURE 21-12 Bipolar Transistor Circuit Configurations.

Common-Emitter Circuits

Common-Emitter (*C-E*) Circuit Configuration in which the input signal is applied between the base and the emitter, while the output signal appears between the transistor's collector and emitter.

With the **common-emitter circuit,** the transistor's emitter lead is common to both the input and output signal. In this circuit configuration, the base serves as the input lead, and the collector serves as the output lead. Figure 21-13 contains a basic common-emitter (*C-E*) circuit, its associated input/output voltage and current waveforms, characteristic curves, and table of typical characteristics. Using this illustration, we will examine the operation and characteristics of the *C-E* circuit configuration.

DC current gain. Referring to the C-E circuit in Figure 21-13(a), and its associated waveforms in Figure 21-13(b), let us now examine this circuit's basic operation.

Before applying the ac sine wave input signal (V_{in}), let us assume that V_{in} = 0 V and examine the transistor's dc operating characteristics, or the "no input signal" condition. The voltage divider R_1 and R_2 will divide the V_{CC} supply voltage, producing a positive dc base bias voltage across R_2. This base bias voltage will be applied to the base of Q_1, and because it is generally greater than 0.7 V, it will forward bias the transistor's base-emitter junction, turning Q_1 ON (in the example in Figure 21-13, V_B = 3 V dc). Capacitor C_1 is included to act as a dc block, preventing the base bias voltage (V_B) from being applied back to the input (V_{in}). The value of dc base bias voltage will determine the value of base current (I_B) flowing out of the transistor's base, and this value of I_B will in turn determine the value of collector current (I_C) flowing out of the transistor's collector and through R_C. Because the transistor's output current (I_C) is so much larger than the transistor's very small input current (I_B), the circuit produces an increase in current, or a **current gain.** The current gain in a common-emitter configuration is called the transistor's **beta** (symbolized β). A transistor's **dc beta** (β_{DC}) indicates a common-emitter transistor's "dc current gain," and it is the ratio of its output current (I_C) to its input current (I_B). This ratio can be expressed mathematically as:

Current Gain The increase in current produced by the transistor circuit.

Beta (β) The transistor's current gain in a common-emitter configuration.

DC Beta (β_{DC}) The ratio of a transistor's dc output current to its input current.

$$\beta_{DC} = \frac{I_C}{I_B}$$

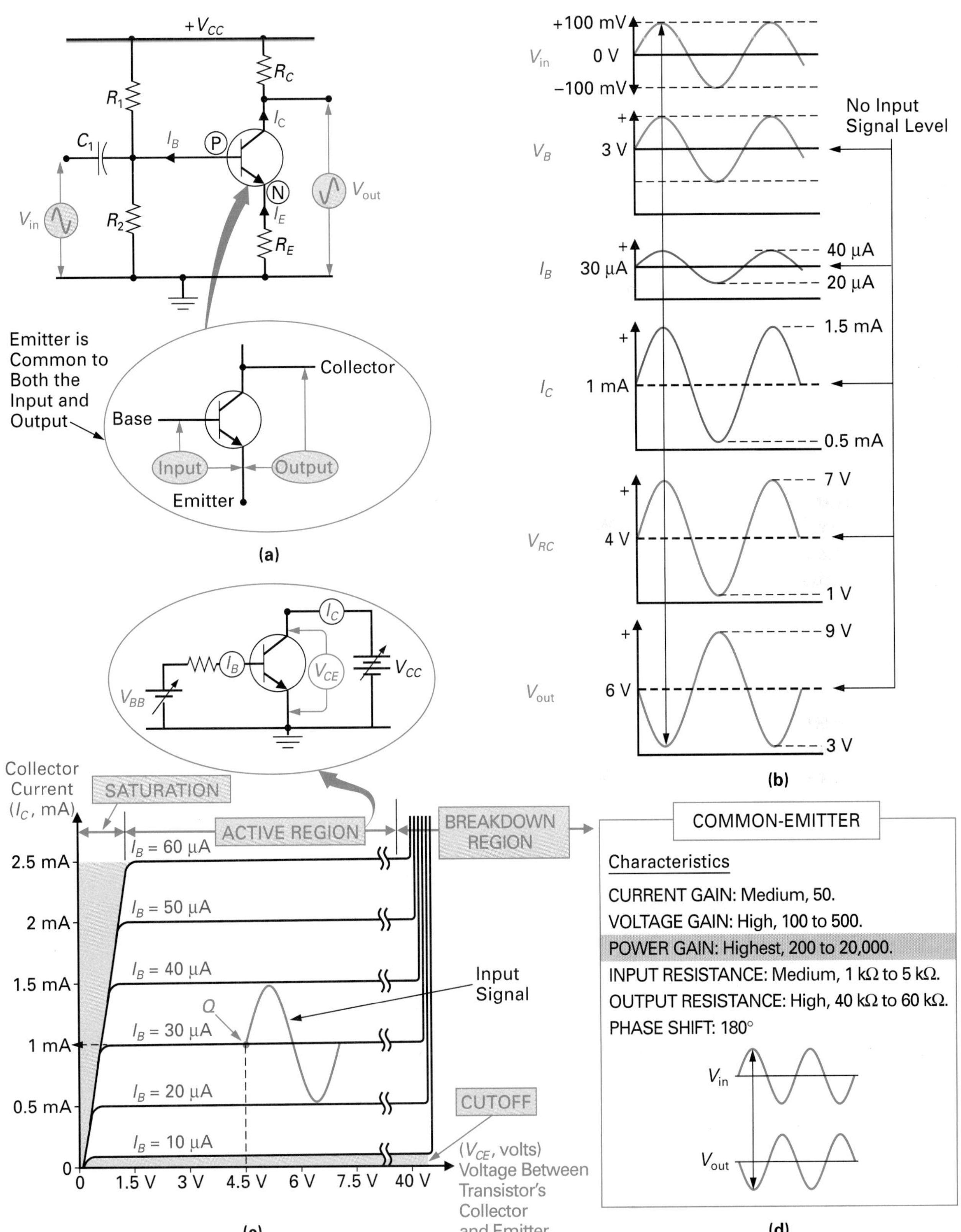

FIGURE 21-13 **Common-Emitter (C-E) Circuit Configuration Characteristics.**

EXAMPLE:

As can be seen in Figure 21-13(b), the "no input signal" level of $I_C = 1$ mA, and $I_B = 30$ μA. What is the transistor's dc beta?

Solution:

$$\beta_{DC} = \frac{I_C}{I_B}$$

$$= \frac{1 \text{ mA}}{30 \text{ μA}}$$

$$= 33.3$$

This value indicates that I_C is 33.3 times greater than I_B, therefore the dc current gain between input and output is 33.3.

Another form of system analysis, known as "hybrid parameters," uses the term *hfe* instead of β_{DC} to indicate a transistor's dc current gain.

AC current gain. When amplifying an ac waveform, a transistor has applied to it both dc voltages to make it operational and the ac signal voltage to be amplified that varies the base bias voltage and the base current. Referring to the first four waveforms in Figure 21-13(b), let us now examine what will happen when we apply a sine wave input signal. As V_{in} increases above 0 V to a peak positive voltage ($V_{in}\uparrow$), it will cause the forward base bias voltage applied to the transistor to increase ($V_B\uparrow$) above its dc reference or "no input signal level." This increase in V_B will increase the forward conduction of the transistor's emitter diode, resulting in an increase in the input base current ($I_B\uparrow$) and a corresponding larger increase in the output collector current ($I_C\uparrow$). The ratio of output collector current change (ΔI_C) to input base current change (ΔI_B) is the transistor's ac current gain or **AC beta (β_{AC})**. The formula for calculating a transistor's ac current gain is:

AC Beta (β_{AC}) The ratio of a transistor's ac output current to input current.

$$\beta_{AC} = \frac{\Delta I_C}{\Delta I_B}$$

EXAMPLE:

Calculate the ac current gain of the example in Figure 21-13(b).

Solution:

$$\beta_{AC} = \frac{\Delta I_C}{\Delta I_B}$$

$$= \frac{1.5 \text{ to } 0.5 \text{ mA}}{40 \text{ to } 20 \text{ μA}}$$

$$= \frac{1 \text{ mA}}{20 \text{ μA}}$$

$$= 50$$

This means that the alternating collector current at the output is 50 times greater than the alternating base current at the input.

Common-emitter transistors will typically have beta, or current gain values, of 50.

Voltage gain. The common-emitter circuit is not only used to increase the level of current between input and output. It can also be used to increase the amplitude of the input signal voltage, or produce a **voltage gain.** This action can be seen by examining the C-E circuit in Figure 21-13(a) and by following the changes in the associated waveforms in Figure 21-13(b). An input signal voltage increase from 0 V to a positive peak ($V_{in}\uparrow$) causes an increase in the dc base bias voltage ($V_B\uparrow$), causing the emitter diode of Q_1 to turn more ON and result in an increase in both I_B and I_C. Because the I_C flows through R_C, and because voltage drop is proportional to current, an increase in I_C will cause an increase in the voltage drop across R_C ($V_{R_C}\uparrow$). The output voltage is equal to the voltage developed across Q_1's collector-to-emitter resistance (R_{CE}) and R_E. Because R_C is in series with Q_1's collector-to-emitter resistance (R_{CE}) and R_E, an increase in $V_{R_C}\uparrow$ will cause a decrease in the voltage developed across R_{CE} and R_E, which is $V_{out}\downarrow$. This action is summarized with Kirchhoff's voltage law which states that the sum of the voltages in a series circuit is equal to the voltage applied ($V_{R_C} + V_{out} = V_{CC}$). Using the example in Figure 21-13(b), you can see that when:

Voltage Gain (A_V) The ratio of the output signal voltage change to input signal voltage change.

$V_{R_C}\uparrow$ to 7 V, $V_{out}\downarrow$ to 3 V. $\quad (V_{R_C} + V_{out} = V_{CC}, 7\text{ V} + 3\text{ V} = 10\text{ V})$

$V_{R_C}\downarrow$ to 1 V, $V_{out}\uparrow$ to 9 V. $\quad (V_{R_C} + V_{out} = V_{CC}, 1\text{ V} + 9\text{ V} = 10\text{ V})$

Although the input voltage (V_{in}) and output voltage (V_{out}) are out of phase with one another, you can see from the example values in Figure 21-13(b) that there is an increase in the signal voltage between input and output. This voltage gain between input and output is possible because the output current (I_C) is so much larger than the input current (I_B). The amount of **voltage gain** (which is symbolized A_V) can be calculated by comparing the output voltage change (ΔV_{out}) to the input voltage change (ΔV_{in}).

$$A_V = \frac{\Delta V_{out}}{\Delta V_{in}}$$

EXAMPLE:

Calculate the voltage gain of the circuit and its associated waveforms in Figure 21-13(a) and (b).

Solution:

$$A_V = \frac{\Delta V_{out}}{\Delta V_{in}} = \frac{+9\text{ V } to +3\text{ V}}{+100\text{ mV } to -100\text{ mV}} = \frac{6\text{ V}}{200\text{ mV}} = 30$$

This value indicates that the output ac signal voltage is 30 times larger than the ac input signal voltage.

Most common-emitter transistor circuits have high voltage gains between 100 to 500.

Power gain. As we have seen so far, the common-emitter circuit provides both current gain and voltage gain. Because power is equal to the product of current and voltage ($P = V \times I$), it is not surprising that the C-E circuit configuration also provides **power gain (A_P)**. The power gain of a circuit can be calculated by dividing the output signal power (P_{out}) by the input signal power (P_{in}).

Power Gain (A_P) The ratio of the output signal power to the input signal power.

$$A_P = \frac{P_{out}}{P_{in}}$$

To calculate the amount of input power (P_{in}) applied to the C-E circuit, we will have to multiply the change in input signal voltage (ΔV_{in}) by the accompanying change in input signal current (ΔI_{in} or ΔI_B).

$$P_{in} = \Delta V_{in} \times \Delta I_{in}$$

To calculate the amount of output power (P_{out}) delivered by the C-E circuit, we will have to multiply the change in output signal voltage (ΔV_{out}) produced by the change in output signal current (ΔI_{out} or ΔI_C).

$$P_{out} = \Delta V_{out} \times \Delta I_{out}$$

The power gain of a common-emitter circuit is therefore calculated with the formula:

$$A_P = \frac{P_{out}}{P_{in}} = \frac{\Delta V_{out} \times \Delta I_{out}}{\Delta V_{in} \times \Delta I_{in}}$$

EXAMPLE:

Calculate the power gain of the example circuit in Figure 21-13.

Solution:

$$A_P = \frac{P_{out}}{P_{in}} = \frac{\Delta V_{out} \times \Delta I_{out}}{\Delta V_{in} \times \Delta I_{in}}$$

$$= \frac{(9\text{ V to }3\text{ V}) \times (1.5\text{ mA to }0.5\text{ mA})}{(+100\text{ mV to }-100\text{ mV}) \times (40\ \mu\text{A to }20\ \mu\text{A})}$$

$$= \frac{6\text{ V} \times 1\text{ mA}}{200\text{ mV} \times 20\ \mu\text{A}} = \frac{6\text{ mW}}{4\ \mu\text{W}} = 1500$$

In this example, the common-emitter circuit has increased the input signal power from 4 μW to 6 mW—a power gain of 1500.

The power gain of the circuit in Figure 21-13 can also be calculated by multiplying the previously calculated C-E circuit voltage gain (A_V) by the previously calculated C-E circuit current gain (β_{AC}).

Since $A_V = \dfrac{\Delta V_{out}}{\Delta V_{in}}$, and $\beta_{AC} = \dfrac{\Delta I_{out}\ (or\ \Delta I_C)}{\Delta I_{in}\ (or\ \Delta I_B)}$

$$A_P = V \times I = \frac{\Delta V_{out}}{\Delta V_{in}} \times \frac{\Delta I_{out}\ (or\ \Delta I_C)}{\Delta I_{in}\ (or\ \Delta I_B)} \text{ or } A_P = A_V \times \beta_{AC}$$

$$A_P = A_V \times \beta_{AC}$$

For the example circuit in Figure 21-13, this will be:

$$A_P = A_V \times \beta_{AC} = 30 \times 50 = 1500$$

indicating that the common-emitter circuit's output power in Figure 21-13 is 1500 times larger than the input power.

The power gain of common-emitter transistor circuits can be as high as 20,000, making this characteristic the circuit's key advantage.

Collector characteristic curves. One of the easiest ways to compare several variables is to combine all of the values in a graph. Figure 21-13(c) shows a special graph for a typical common-emitter transistor circuit called the **collector characteristic curves.** The data for this graph are obtained by using the transistor test circuit, shown in the inset in Figure 21-13(c), which will apply different values of base bias voltage (V_{BB}) and collector bias voltage (V_{CC}) to an NPN transistor. The two ammeters and one voltmeter in this test circuit are used to measure the circuit's I_B, I_C, and V_{CE} response to each different circuit condition. The values obtained from this test circuit are then used to plot the transistor's collector current (I_C in mA) in the vertical axis against the transistor's collector-emitter voltage (V_{CE}) drop in the horizontal axis for various values of base current (I_B in μA). This graph shows the relationship between a transistor's input base current, output collector current, and collector-to-emitter voltage drop.

Collector Characteristic Curves
Graph for a typical common-emitter transistor circuit.

Let us now examine the typical set of collector characteristic curves shown in Figure 21-13(c). When V_{CE} is increased from zero, by increasing V_{CC}, the collector current rises very rapidly, as indicated by the rapid vertical rise in any of the curves. When the collector diode of the transistor is reverse biased by the voltage V_{CE}, the collector current levels off. At this point, any one of the curves can be followed based on the amount of base current, which is determined by the value of base bias voltage (V_{BB}) applied. This flat part of the curve is known as the transistor's **active region.** The transistor is normally operated in this region, where it is equivalent to a variable-resistor between the collector and emitter.

Active Region
Flat part of the collector characteristic curve. A transistor is normally operated in this region, where it is equivalent to a variable-resistor between the collector and emitter.

As an example, let us use these curves in Figure 21-13(c) to calculate the value of output current (I_C) for a given value of input current (I_B) and collector-to-emitter voltage (V_{CE}). If V_{BB} is adjusted to produce a base current of 30 μA, and V_{CC} is adjusted until the voltage between the transistor's collector and emitter (V_{CE}) is 4.5 V, the output collector current (I_C) will be equal to approximately 1 mA. This is determined by first locating 4.5 V on the horizontal axis ($V_{CE} = 4.5$ V), following this point directly up to the $I_B = 30$ μA curve, and then moving directly to the left to determine the value of output current on the vertical axis ($I_C = 1$ mA). When these values of V_{CE} and I_B are present, the transistor is said to be operating at point "Q." This dc operating point is often referred to as a **quiescent operating point (Q point),** which means a dc steady-state or no-input signal operating point. The Q point of a transistor is set by the circuit's dc bias components and supply voltages. For instance, in the circuit in Figure 21-13(a), R_1 and R_2 were used to set the dc base bias voltage (V_B, and therefore I_B), and the values of V_{CC}, R_C, and R_E were chosen to set the transistor's dc collector-emitter voltage (V_{CE}). At this dc operating point, we can calculate the transistor's dc current gain (β_{DC}), since both I_B and I_C are known:

Quiescent Operating Point (Q point)
The voltage or current value that sets up the no-signal input or operating point bias voltage.

$$\beta_{DC} = \frac{I_C}{I_B} = \frac{1 \text{ mA}}{30 \text{ µA}} = 33.3$$

If a sine wave signal was applied to the circuit, as shown in the waveforms in Figure 21-13(b) and the characteristic curves in Figure 21-13(c), it would cause the transistor's input base current, and therefore output collector current, to alternate above and below the transistor's Q point (dc operating point). This ac input signal voltage will cause the input base current (I_B) to increase between 20 µA and 40 µA, and this input base current change will generate an output collector current change of 0.5 mA to 1.5 mA. The transistor's ac current gain (β_{AC}) will be:

$$\beta_{AC} = \frac{\Delta I_C}{\Delta I_B}$$

$$= \frac{1.5 \text{ to } 0.5 \text{ mA}}{40 \text{ to } 20 \text{ µA}}$$

$$= \frac{1 \text{ mA}}{20 \text{ µA}}$$

$$= 50$$

Returning to the collector characteristic curve in Figure 21-13(c), you can see that if the collector supply voltage (V_{CC}) is increased to an extreme, a point will be reached where the V_{CE} voltage across the transistor will cause the transistor to break down, as indicated by the rapid rise in I_C. This section of the curve is called the **breakdown region** of the graph, and the damaging value of current through the transistor will generally burn out and destroy the device. As an example, for the 2N3904 bipolar transistor, breakdown will occur at a V_{CE} voltage of 40 V.

Breakdown Region
The point at which the collector supply voltage will cause a damaging value of current through the transistor.

There are two shaded sections shown in the set of collector characteristic curves in Figure 21-13(c). These two shaded sections represent the other two operating regions of the transistor. To begin with, let us examine the vertically shaded **saturation region.** If the base bias voltage (V_{BB}) is increased to a large positive value, the emitter diode of the transistor will turn ON heavily, I_B will be a large value, and the transistor will be operating in saturation. In this operating region, the transistor is equivalent to a closed switch between its collector and emitter (both the emitter and collector diode are forward biased), and therefore the voltage drop between collector and emitter will be almost zero (V_{CE} = typically 0.3 V, when transistor is saturated), and I_C will be a large value that is limited only by the externally connected components. The horizontally shaded section represents the **cutoff region** of the transistor. If the base bias voltage (V_{BB}) is decreased to zero, the emitter diode of the transistor will turn OFF, I_B will be

Saturation Region
The point at which the collector supply voltage has the transistor operating at saturation.

Cutoff Region
The point at which the collector supply voltage has the transistor operating in cutoff.

zero, and the transistor will be operating in cutoff. In this operating region, the transistor is equivalent to an open switch between its collector and emitter (both the emitter and collector diode are reverse biased), and the voltage drop between collector and emitter will be equal to V_{CC} and I_C will be zero.

A set of collector characteristic curves are therefore generally included in a manufacturer's device data sheet, and can be used to determine the values of I_B, I_C and V_{CE} at any operating point.

Input resistance. The **input resistance (R_{in})** of a common-emitter transistor is the amount of opposition offered to an input signal by the input base-emitter junction (emitter diode). Because the base-emitter junction is normally forward biased when the transistor is operating in the active region, the opposition to input current is relatively small. However, the extremely small base region will only support a very small input base current. On average, if no additional components are connected in series with the transistor's base-emitter junction, the input resistance of a C-E transistor circuit is typically a medium value between 1 kΩ to 5 kΩ. This typical value is an average because the transistor's input resistance is a "dynamic or changing quantity" that will vary slightly as the input signal changes the conduction of the C-E transistor's emitter diode, and this changes I_B ($R\updownarrow = V/I\updownarrow$).

Input Resistance (R_{in})
The amount of opposition offered to an input signal by the input base-emitter junction (emitter diode).

Because the transistor has a small value of input P-N junction capacitance and input terminal inductance, the opposition to the input signal is not only resistive but, to a small extent, reactive. For this reason, the total opposition offered by the transistor to an input signal is often referred to as the **input impedance (Z_{in})** because impedance is the total combined resistive and reactive input opposition.

Input Impedance (Z_{in})
The total opposition offered by the transistor to an input signal.

Output resistance. The **output resistance (R_{out})** of a common-emitter transistor is the amount of opposition offered to an output signal by the output base-collector junction (collector diode). This junction is normally reverse biased when the transistor is operating in the active region, and therefore the C-E transistor's output resistance is relatively high. However, because a unique action occurs within the transistor and allows current to flow through this reverse biased junction (electron accumulation at the base and then conduction through collector diode due to attraction of $+V_{CC}$), the output current (I_C) is normally large, and so the output resistance is not an extremely large value. On average, if no load resistor is connected in series with the transistor's collector diode, the output resistance of a C-E transistor circuit is typically a high value between 40 kΩ to 60 kΩ.

Output Resistance (R_{out})
The amount of opposition offered to an output signal by the output base-collector junction (collector diode).

Output Impedance (Z_{out})
The total opposition offered by the transistor to the output signal.

Because the transistor has a small value of output P-N junction capacitance and output terminal inductance, the opposition to the output signal is not only resistive but also reactive. For this reason, the total opposition offered by the transistor to the output signal is often referred to as the **output impedance (Z_{out}).**

Common-Base (C-B) Circuit
Configuration in which the input signal is applied between the transistor's emitter and base, while the output is developed across the transistor's collector and base.

Common-Base Circuits

With the **common-base circuit,** the transistor's base lead is common to both the input and output signal. In this circuit configuration, the emitter serves as the input lead, and the collector serves as the output lead. Figure 21-14 contains

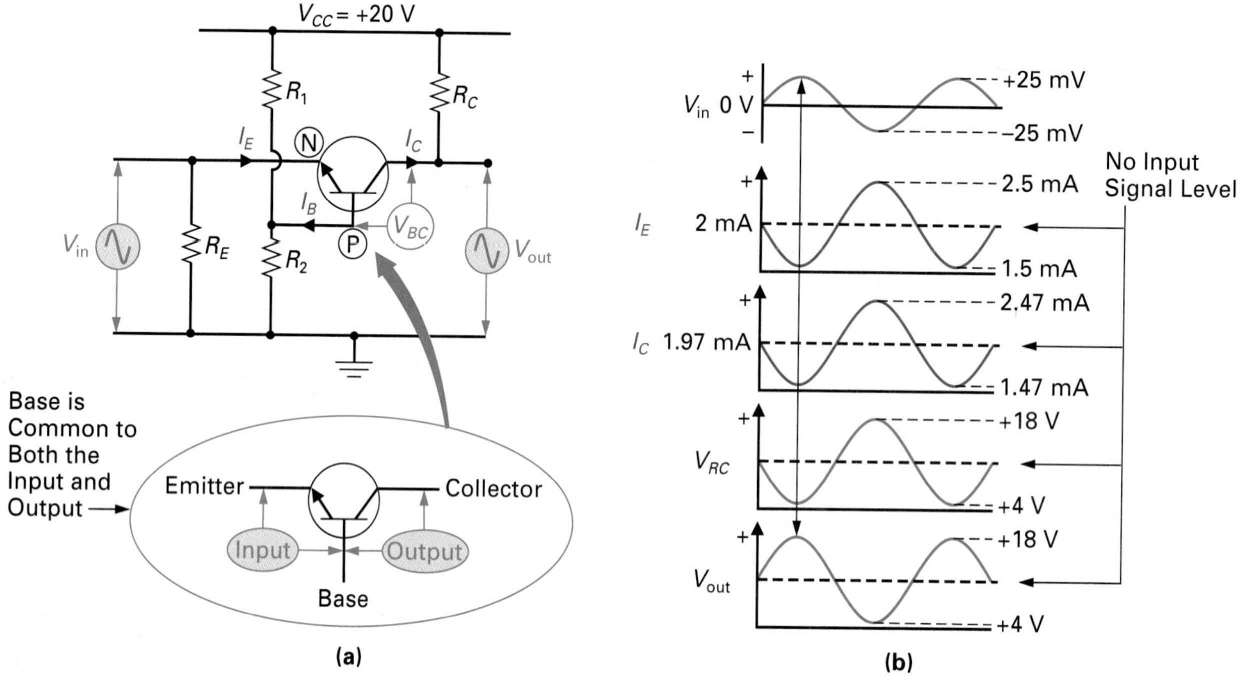

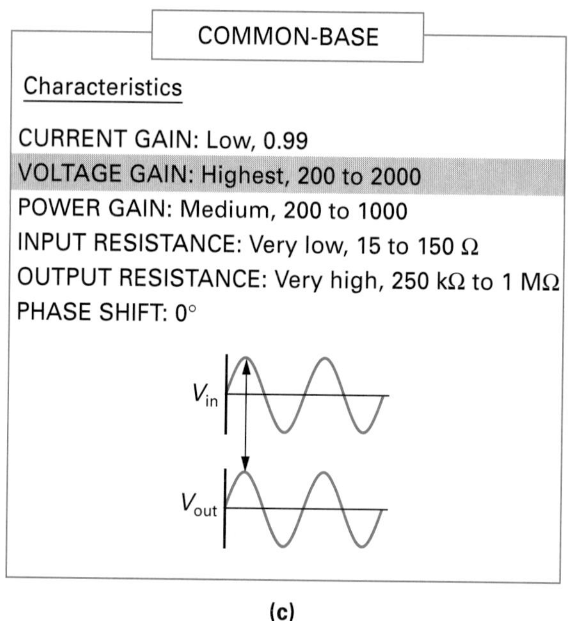

(c)

FIGURE 21-14 Common-Base (*C-B*) Circuit Configuration Characteristics.

a basic common-base (C-B) circuit, its associated input/output voltage and current waveforms, and table of typical characteristics. Using this illustration, we will examine the operation and characteristics of the C-B circuit configuration.

DC current gain. Referring to the C-B circuit in Figure 21-14(a), and its associated waveforms in Figure 21-14(b), let us now examine this circuit's basic operation.

Before applying the ac sine wave input signal (V_{in}), let us assume that V_{in} = 0 V and examine the transistor's dc operating characteristics, or the "no input signal" condition. The voltage divider R_1 and R_2 will divide the V_{CC} supply voltage, producing a positive dc base bias voltage across R_2. This base bias voltage will be applied to the base of Q_1. Because it is generally greater than 0.7 V, it will forward bias the transistor's base-emitter junction, turning Q_1 ON. Because the common-base circuit's input current is I_E and its output current is I_C, the current gain between input and output will be determined by the ratio of I_C to I_E. This ratio for calculating a C-B transistor's dc current gain is called the transistor's **DC alpha (α_{DC})** and is equal to

DC Alpha (α_{DC})
The ratio for calculating a C-B transistor's dc current gain.

$$\alpha_{DC} = \frac{I_C}{I_E}$$

The "no input signal" or "steady state" dc levels of I_E and I_C are determined by the value of voltage developed across R_2, which is controlling the conduction of the transistor's forward biased base-emitter junction (emitter diode). Because the output current I_C is always slightly lower than the input current I_E (due to the small I_B current flow out of the base), the C-B transistor circuit does not increase current between input and output. In fact, there is a slight loss in current between input and output, which is why the C-B circuit configuration is said to have a current gain that is less than 1.

EXAMPLE:

Calculate the dc alpha of the circuit in Figure 21-14(a), if I_C = 1.97mA and I_E = 2 mA.

Solution:

$$\alpha_{DC} = \frac{I_C}{I_E} = \frac{1.97\text{ mA}}{2\text{ mA}} = 0.985$$

This value of 0.985 indicates that I_C is 98.5% of I_E ($0.985 \times 100 = 98.5$).

As you can see from this example, the difference between I_C and I_E is generally so small that we always assume that the dc alpha is 1, which means that $I_C = I_E$.

AC current gain. When amplifying an ac waveform, a transistor has applied to it both dc voltages to make it operational and the ac signal voltage to be amplified that varies the base-emitter bias and the input emitter current. Referring to the waveforms in Figure 21-14(b), let us now examine what will happen when we apply a sine wave input signal. As mentioned previously, the positive voltage developed across R_2 will make the NPN transistor's base positive with respect to the emitter and forward bias the P-N base-emitter junction.

As the input voltage swings positive ($V_{in}\uparrow$), it will reduce the forward bias across the transistor's P-N base-emitter junction. For example, if $V_B = +5$ V and the transistor's *n*-type emitter is made positive, the P-N base-emitter diode will be turned more OFF. Turning the transistor's emitter diode less ON will cause a decrease in emitter current ($I_E\downarrow$), a decrease in collector current ($I_C\downarrow$), and a decrease in the voltage developed across R_C ($V_{R_C}\downarrow$). Because V_{R_C} and V_{out} are connected in series across V_{CC}, a decrease in $V_{R_C}\downarrow$ must be accompanied by an increase in $V_{out}\uparrow$. To explain this another way, the decrease in I_E and the subsequent decrease in both I_C and I_B means that the conduction of the transistor has decreased. This decrease in conduction means that the normally forward biased base-emitter junction has turned less ON, and the normally reverse biased base-collector junction has turned more OFF. Because the transistor's base-collector junction is in series with R_C and R_2 across V_{CC}, an increase in the transistor's base-collector resistance ($R_{BC}\uparrow$) will cause an increase in the voltage developed across the transistor's base-collector junction ($V_{BC}\uparrow$), which will cause an increase in $V_{out}\uparrow$.

Similarly, as the input voltage swings negative ($V_{in}\downarrow$), it will increase the forward bias across the transistor's P-N base-emitter junction. For example, if $V_B = -5$ V and the transistor's *n*-type emitter is made negative, the P-N base-emitter diode will be turned more ON. Turning the transistor's emitter diode more ON will cause an increase in emitter current ($I_E\uparrow$), an increase in collector current ($I_C\uparrow$), and an increase in the voltage developed across R_C ($V_{R_C}\uparrow$). Because V_{R_C} and V_{out} are connected in series across V_{CC}, an increase in $V_{R_C}\uparrow$ must be accompanied by a decrease in $V_{out}\downarrow$.

Now that we have seen how the input voltage causes a change in input current (I_E) and output current (I_C), let us examine the *C-B* circuit's ac current gain. The ratio of input emitter current change (ΔI_E) to output collector current change (ΔI_C) is the *C-B* transistor's ac current gain or **AC alpha (α_{AC})**. The formula for calculating a *C-B* transistor's ac current gain is:

AC Alpha (α_{AC}) The ratio of input emitter current change to output collector current change.

$$\alpha_{AC} = \frac{\Delta I_C}{\Delta I_E}$$

EXAMPLE:

Calculate the ac current gain of the example in Figure 21-14(b).

Solution:

$$\alpha_{AC} = \frac{\Delta I_C}{\Delta I_E}$$

$$= \frac{2.47 \text{ to } 1.47 \text{ mA}}{2.5 \text{ to } 1.5 \text{ mA}}$$

$$= \frac{1 \text{ mA}}{1 \text{ mA}}$$
$$= 1$$

This means that the change in output collector current is equal to the change in input emitter current, and therefore the ac current gain is 1. (The output is 1 times larger than the input, 1 mA × 1 = 1 mA.)

Common-base transistors will typically have an ac alpha, or ac current gain, of 0.99.

Voltage gain. Although the common-base circuit does not achieve any current gain, it does make up for this disadvantage by achieving a very large voltage gain between input and output. Returning to the *C-B* circuit in Figure 21-14(a) and its waveforms in Figure 21-14(b), let us see how this very high voltage gain is obtained. Only a small input voltage (V_{in}) is needed to control the conduction of the transistor's emitter diode, and therefore the input emitter current (I_E) and output collector current (I_C). Even though I_C is slightly lower than I_E, it is still a relatively large value of current and will develop a large voltage change across R_C for a very small change in V_{in}. Because V_{R_C} and V_{out} are in series and connected across V_{CC}, a large change in voltage across R_C will cause a large change in the voltage developed across the transistor output (V_{BC}) and V_{out}. As before, the amount of **voltage gain** (which is symbolized $\boldsymbol{A_V}$) can be calculated by comparing the output voltage change (ΔV_{out}) to the input voltage change (ΔV_{in}).

$$A_V = \frac{\Delta V_{out}}{\Delta V_{in}}$$

EXAMPLE:

Calculate the voltage gain of the circuit and its associated waveforms in Figure 21-14(a) and (b).

Solution:

$$A_V = \frac{\Delta V_{out}}{\Delta V_{in}} = \frac{+18 \text{ V } to +4 \text{ V}}{+25 \text{ mV } to -25 \text{ mV}} = \frac{14 \text{ V}}{50 \text{ mV}} = 280$$

This value indicates that the output ac signal voltage is 280 times larger than the ac input signal voltage.

Most common-base transistor circuits have very high voltage gains between 200 to 2000. While on the topic of comparing the input voltage to output voltage, you can see by looking at Figure 21-14(b) that, unlike the *C-E* circuit, the common-base circuit has no phase shift between input and output (V_{out} is in phase with V_{in}).

Power gain. Although the common-base circuit achieves no current gain, it does have a very high voltage gain and therefore can provide a medium amount of power gain ($P\uparrow = V\uparrow \times I$). The power gain of a circuit can be calculated by dividing the output signal power (P_{out}) by the input signal power (P_{in}).

$$A_P = \frac{P_{out}}{P_{in}}$$

To calculate the amount of input power (P_{in}) applied to the *C-B* circuit, we will have to multiply the change in input signal voltage (ΔV_{in}) by the accompanying change in input signal current (ΔI_{in} or ΔI_E).

$$P_{in} = \Delta V_{in} \times \Delta I_{in}$$

$$P_{in} = 50 \text{ mV} \times 1 \text{ mA} = 50 \text{ µW}$$

To calculate the amount of output power (P_{out}) delivered by the *C-B* circuit, we will have to multiply the change in output signal voltage (ΔV_{out}) produced by the change in output signal current (ΔI_{out} or ΔI_C).

$$P_{out} = \Delta V_{out} \times \Delta I_{out}$$

$$P_{out} = 14 \text{ V} \times 1 \text{ mA} = 14 \text{ mW}$$

The power gain of a common-base circuit is calculated with the formula:

$$A_P = \frac{P_{out}}{P_{in}} = \frac{\Delta V_{out} \times \Delta I_{out}}{\Delta V_{in} \times \Delta I_{in}}$$

EXAMPLE:

Calculate the power gain of the example circuit in Figure 21-14.

Solution:

$$A_P = \frac{P_{out}}{P_{in}} = \frac{\Delta V_{out} \times \Delta I_{out}}{\Delta V_{in} \times \Delta I_{in}}$$

$$= \frac{14 \text{ V} \times 1 \text{ mA}}{50 \text{ mV} \times 1 \text{ mA}} = \frac{14 \text{ mW}}{50 \text{ µW}} = 280$$

In this example, the common-base circuit has increased the input signal power from 50 µW to 14 mW—a power gain of 280.

The power gain of the circuit in Figure 21-14 can also be calculated by multiplying the previously calculated *C-B* circuit voltage gain (A_V) by the previously calculated *C-B* circuit current gain (α_{AC}).

$$A_P = A_V \times \alpha_{AC}$$

For the example circuit in Figure 21-14, this will be

$$A_P = A_V \times \alpha_{AC} = 280 \times 1 = 280$$

indicating that the common-base circuit's output power in Figure 21-14 is 280 times larger than the input power.

Typical common-base circuits will have power gains from 200 to 1000.

Input resistance. The input resistance (R_{in}) of a common-base transistor is the amount of opposition offered to an input signal by the input base-emitter junction (emitter diode). Because the base-emitter junction is normally forward biased and the input emitter current (I_E) is relatively large, the input signal sees a very low input resistance. On average, if no additional components are connected in series with the transistor's base-emitter junction, the input resistance of a *C-B* transistor circuit is typically a low value between 15Ω to 150Ω. This typical value is an average because the transistor's input resistance is a dynamic or changing quantity that will vary slightly as the input signal changes the conduction of the *C-B* transistor's emitter diode, and this changes I_E ($R\updownarrow = V/I\updownarrow$).

Output resistance. The output resistance (R_{out}) of a common-base transistor is the amount of opposition offered to an output signal by the output base-collector junction (collector diode). This junction is normally reverse biased when the transistor is operating in the active region, and therefore the *C-B* transistor's output resistance is relatively high. On average, if no load resistor is connected in series with the transistor's collector diode, the output resistance of a *C-B* transistor circuit is typically a very high value between 250 kΩ to 1 MΩ.

Common-Collector Circuits

Common-Collector (C-C) Circuit
Configuration in which the input signal is applied between the transistor's base and collector, while the output is developed across the transistor's collector and emitter.

With the **common-collector circuit,** the transistor's collector lead is common to both the input and output signal. In this circuit configuration therefore, the base serves as the input lead, and the emitter serves as the output lead. Figure 21-15 contains a basic common-collector (*C-C*) circuit, its associated input/output voltage and current waveforms, and table of typical characteristics. Using this illustration, we will examine the operation and characteristics of the *C-C* circuit configuration.

DC current gain. Referring to the *C-C* circuit in Figure 21-15(a) and its associated waveforms in Figure 21-15(b), let us now examine this circuit's basic operation.

Before applying the ac sine wave input signal (V_{in}), let us assume that V_{in} = 0 V and examine the transistor's dc operating characteristics, or the "no input signal" condition. The voltage divider R_1 and R_2 will divide the V_{CC} supply voltage, producing a positive dc base bias voltage across R_2. This base bias voltage will be applied to the base of Q_1 and, because it is generally greater than 0.7 V, it will forward bias the transistor's base-emitter junction, turning Q_1 ON. Because the common-collector (*C-C*) circuit's input current I_B is much smaller than the output current I_E, the circuit provides a high current gain. In fact, the common-collector circuit provides a slightly higher gain than the *C-E* circuit because the common-collector's output current (I_E) is slightly higher than the *C-E*'s output current (I_C).

Like any circuit configuration, the dc current gain is equal to the ratio of output current to input current. For the *C-C* circuit, this is equal to the ratio of I_E to I_B:

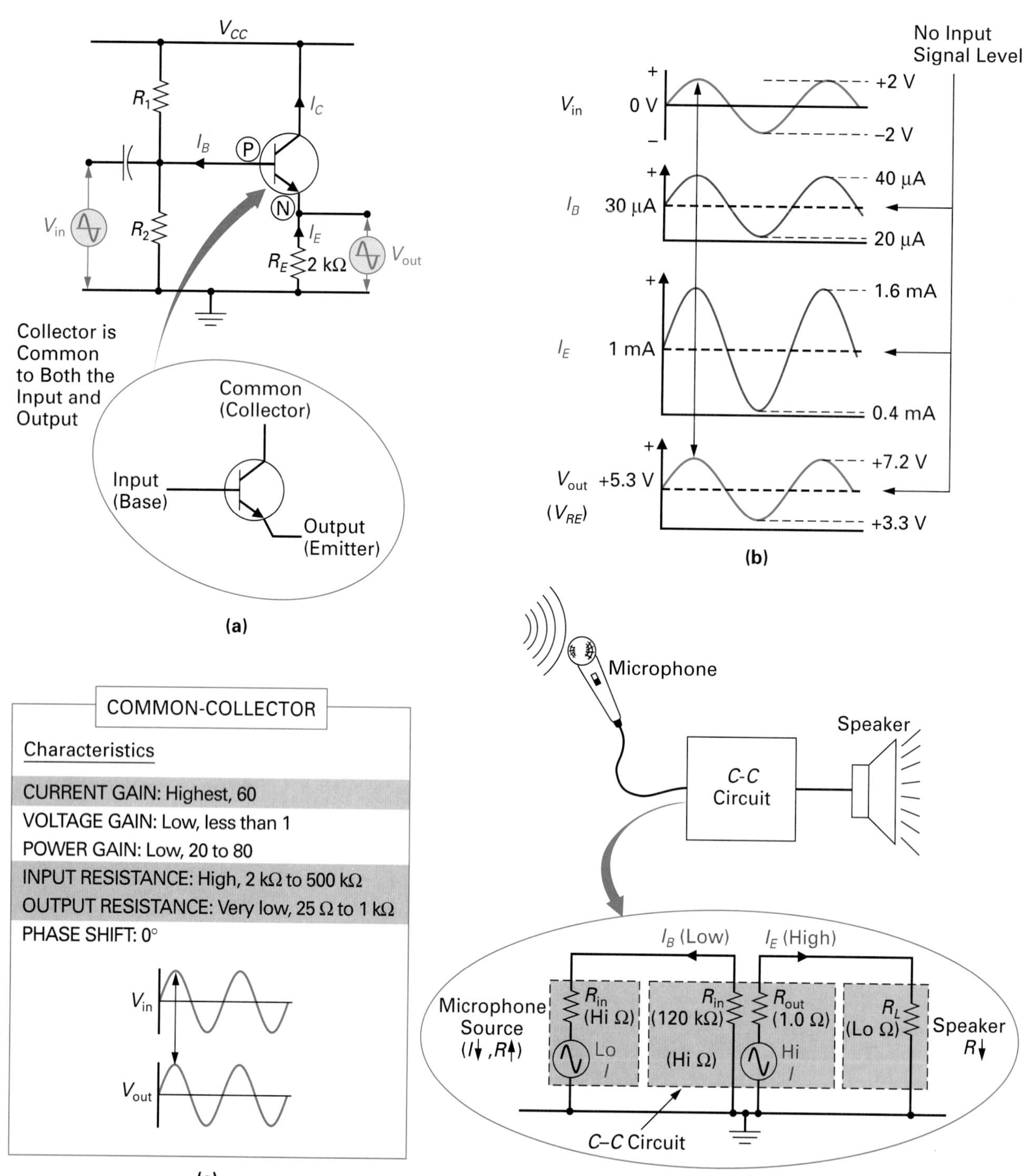

FIGURE 21-15 Common-Collector *(C-C)* Circuit Configuration Characteristics.

$$DC\ Current\ Gain = \frac{I_E}{I_B}$$

Transistor manufacturers will generally not provide specifications for all three circuit configurations. In most cases, because the common-emitter (*C-E*) circuit configuration is most frequently used, manufacturers will give the transistor's characteristics for only the *C-E* circuit configuration. In these instances, we will have to convert this *C-E* circuit data to equivalent specifications for other configurations. For example, in most data sheets the transistor's dc current gain will be listed as β_{DC}. As we know, dc beta is the measure of a *C-E* circuit's current gain because it compares input current I_B to output current I_C. How then, can we convert this value so that it indicates the dc current gain of a common-collector circuit? The answer is as follows:

$$\text{Common-Collector DC Current Gain} = \frac{\text{Output Current}}{\text{Input Current}} = \frac{I_E}{I_B}$$

Since $I_E = I_B + I_C$,

$$\text{DC Current Gain} = \frac{I_E}{I_B} = \frac{(I_B + I_C)}{I_B}$$

Since $I_B \div I_B = 1$,

$$\text{DC Current Gain} = 1 + \frac{I_C}{I_B}$$

Since $\frac{I_C}{I_B} = \beta_{DC}$,

$$\text{DC Current Gain} = 1 + \frac{I_C}{I_B} = 1 + \beta_{DC}$$

$$\text{DC Current Gain} = 1 + \beta_{DC}$$

EXAMPLE:

Calculate the dc current gain of the *C-C* circuit in Figure 21-15, if the transistor's $\beta_{DC} = 32.33$.

Solution:

$$\text{DC Current Gain} = 1 + \frac{I_C}{I_B} = 1 + \frac{(I_E - I_B)}{I_B} = 1 + \frac{1\text{ mA} - 30\text{ µA}}{30\text{ µA}}$$

$$= 1 + \frac{970\text{ µA}}{30\text{ µA}} = 1 + 32.33 = 33.33$$

or,

$$\text{DC Current Gain} = 1 + \beta_{DC} = 1 + 32.33 = 33.33$$

As you can see in this example, the dc current gain of a common-collector circuit ($\beta_{DC} + 1$) is slightly higher than the dc current gain of a *C-E* circuit (β_{DC}). In most instances, the extra 1 makes so little difference when the transistor's dc current gain is a large value of about 30, as in this example, that we assume that the current gain of a common-collector circuit is equal to the current gain of a *C-E* circuit.

$$C\text{-}C \text{ DC Current Gain} \cong C\text{-}E \text{ DC Current Gain } (\beta_{DC})$$

AC current gain. When amplifying an ac waveform, a transistor has applied to it both dc voltages to make it operational and the ac signal voltage that varies the base-emitter bias and the input base current. Referring to the waveforms in Figure 21-15(b), let us now examine what will happen when we apply a sine wave input signal. As mentioned previously, the positive voltage developed across R_2 will make the NPN transistor's base positive with respect to the emitter, and therefore forward bias the P-N base-emitter junction.

As the input voltage swings positive ($V_{in}\uparrow$), it will add to the forward bias applied across the transistor's P-N base-emitter junction. This means that the transistor's emitter diode will turn more ON, cause an increase in the $I_B\uparrow$, and therefore a proportional but much larger increase in the output current $I_E\uparrow$.

Similarly, as the input voltage swings negative ($V_{in}\downarrow$), it will subtract from the forward bias applied across the transistor's P-N base-emitter junction. This means that the transistor's emitter diode will turn less ON, cause a decrease in the $I_B\downarrow$, and therefore a proportional but larger decrease in the output current $I_E\downarrow$.

The ac current gain of a common-collector transistor is calculated using the same formula as dc current gain. However, with an ac current, we will compare the output current change (ΔI_E) to the input current change (ΔI_B).

$$\text{AC Current Gain} = \frac{\Delta I_E}{\Delta I_B}$$

EXAMPLE:

Calculate the ac current gain of the circuit in Figure 21-15(a), using the values in Figure 21-15(b).

Solution:

$$\text{AC Current Gain} = \frac{\Delta I_E}{\Delta I_B} = \frac{1.6 \text{ mA} - 0.4 \text{ mA}}{40\ \mu\text{A} - 20\ \mu\text{A}} = 60$$

Like the common-collector's dc current gain, because there is so little difference between I_E and I_C, we can assume that the ac current gain is equivalent to β_{AC}.

$$C\text{-}C \text{ AC Current Gain} \cong C\text{-}E \text{ AC Current Gain } (\beta_{AC})$$

Common-collector transistor circuit configurations can have current gains as high as 60, indicating that I_E is sixty times larger than I_B.

Voltage gain. Although the common-collector circuit has a very high current gain rating, it cannot increase voltage between input and output. Returning to the *C-C* circuit in Figure 21-15(a) and its waveforms in Figure 21-15(b), let us see why this circuit has a very low voltage gain.

As the input voltage swings positive ($V_{in}\uparrow$), it will add to the forward bias applied across the transistor's P-N base-emitter junction ($V_{BE}\uparrow$). As the transistor's emitter diode turns more ON, it will cause an increase in $I_B\uparrow$, a proportional but larger increase in $I_E\uparrow$, and therefore an increase in the voltage developed across R_E (V_{R_E}, V_{out}, or $V_E\uparrow$). This increase in the voltage developed across R_E has a **degenerative effect** because an increase in the emitter voltage ($V_E\uparrow$) will counter the initial increase in base voltage ($V_B\uparrow$), and therefore the voltage difference between the transistor's base and emitter will remain almost constant (V_{BE} is almost constant). In other words, if the base goes positive and then the emitter goes positive, there is almost no increase in the potential difference between the base and the emitter and so the change in forward bias is almost zero. There is, in fact, a very small change in forward bias between base and emitter, and this will cause a small change in I_B and I_E, and therefore a small output voltage will be developed across R_E. Comparing the input and output voltage signals in Figure 21-15(b), you can see that both are about 4 V pk-pk, and both are in phase with one another. The common-collector circuit is often referred to as an **emitter-follower** or **voltage-follower** because the emitter output voltage seems to track or follow the phase and amplitude of the input voltage.

Degenerative Effect
An effect that causes a reduction in amplification due to negative feedback.

Emitter-Follower or Voltage-Follower
The common-collector circuit in which the emitter output voltage seems to track or follow the phase and amplitude of the input voltage.

As with all circuit configurations, the amount of **voltage gain (A_V)** can be calculated by comparing the output voltage change (ΔV_{out}) to the input voltage change (ΔV_{in}).

$$A_V = \frac{\Delta V_{out}}{\Delta V_{in}}$$

EXAMPLE:

Calculate the voltage gain of the circuit and its associated waveforms in Figure 21-15(a) and (b).

Solution:

$$A_V = \frac{\Delta V_{out}}{\Delta V_{in}} = \frac{+7.2\text{ V to }+3.3\text{ V}}{+2\text{ V to }-2\text{ V}} = \frac{3.9\text{ V}}{4\text{ V}} = 0.975$$

This value indicates that the output ac signal voltage is 0.975 or 97.5% of the ac input signal voltage ($0.975 \times 4\text{ V} = 3.9\text{ V}$).

Most common-collector transistor circuits have a voltage gain that is less than 1. However, in most circuit examples it is assumed that output voltage change equals input voltage change.

Power gain. Although the common-collector circuit achieves no voltage gain, it does have a very high current gain, and therefore can provide a medium amount of power gain ($P\uparrow = V \times I\uparrow$). As before, the power gain of a circuit can be calculated by dividing the output signal power (P_{out}) by the input signal power (P_{in}).

$$A_P = \frac{P_{out}}{P_{in}}$$

To calculate the amount of input power (P_{in}) applied to the *C-C* circuit, we will have to multiply the change in input signal voltage (ΔV_{in}) by the accompanying change in input signal current (ΔI_{in} or ΔI_B).

$$P_{in} = \Delta V_{in} \times \Delta I_{in}$$

$$P_{in} = 4\ \text{V} \times 20\ \mu\text{A} = 80\ \mu\text{W}$$

To calculate the amount of output power (P_{out}) delivered by the *C-C* circuit, we will have to multiply the change in output signal voltage (ΔV_{out}) produced by the change in output signal current (ΔI_{out} or ΔI_E).

$$P_{out} = \Delta V_{out} \times \Delta I_{out}$$

$$P_{out} = 3.9\ \text{V} \times 1.2\ \text{mA} = 4.68\ \text{mW}$$

The power gain of a common-collector circuit is therefore calculated with the formula:

$$A_P = \frac{P_{out}}{P_{in}} = \frac{\Delta V_{out} \times \Delta I_{out}}{\Delta V_{in} \times \Delta I_{in}}$$

EXAMPLE:

Calculate the power gain of the example circuit in Figure 21-15.

Solution:

$$A_P = \frac{P_{out}}{P_{in}} = \frac{\Delta V_{out} \times \Delta I_{out}}{\Delta V_{in} \times \Delta I_{in}}$$

$$= \frac{3.9\ \text{V} \times 1.2\ \text{mA}}{4\ \text{V} \times 20\ \mu\text{A}} = \frac{4.68\ \text{mW}}{80\ \mu\text{W}} = 58.5$$

In this example, the common-collector circuit has increased the input signal power from 80 μW to 4.68 mW—a power gain of 58.5.

The power gain of the circuit in Figure 21-15 can also be calculated by multiplying the previously calculated *C-C* circuit voltage gain (A_V) by the previously calculated *C-C* circuit current gain.

$$A_P = A_V \times \text{AC Current Gain}$$

For the example circuit in Figure 21-15, this will be

$$A_P = A_V \times \text{AC Current Gain} = 0.975 \times 60 = 58.5$$

indicating that the common-collector circuit's output power in Figure 21-15 is 58.5 times larger than the input power.

Typical common-collector circuits will have power gains from 20 to 80.

Input resistance. An input signal voltage will see a very large input resistance when it is applied to a common-collector circuit. This is because the input signal sees the very large emitter connected resistor ($R_E\uparrow\uparrow$) and, to a smaller extent, the resistance of the forward biased base-emitter junction ($R_{in}\uparrow = V_{in}/I_B\downarrow$: R_{in} is large because I_{in} or I_B is small). Using these two elements, we can derive a formula for calculating the input resistance of a *C-C* transistor circuit.

$$R_{in} = R_E \times \text{AC Current Gain}$$

Since

$$C\text{-}C \text{ AC Current Gain} \cong C\text{-}E \text{ AC Current Gain } (\beta_{AC})$$

the input resistance can also be calculated with the formula:

$$R_{in} = R_E \times \beta_{AC}$$

EXAMPLE:

Calculate the input resistance of the circuit in Figure 21-15, assuming $\beta_{AC} = 60$.

Solution:

$$R_{in} = R_E \times \beta_{AC} = 2\text{ k}\Omega \times 60 = 120\text{ k}\Omega$$

This means that an input voltage signal will see this *C-C* circuit as a resistance of 120 kΩ.

The input resistance of a *C-C* transistor circuit is typically a very large value between 2 kΩ to 500 kΩ.

Output resistance. The output signal from a common-collector circuit sees a very low output resistance, as proved by this circuit's very high output current gain. Like this circuit's input resistance, the output resistance is largely dependent on the value of the emitter resistor R_E.

The output resistance of a typical *C-C* transistor circuit is a very low value between 25 Ω to 1 kΩ.

Impedance or resistance matching. Do not be misled into thinking that the very high input resistance and low output resistance of the common-collector transistor circuit are disadvantages. On the contrary, the very high input resistance and low output resistance of this configuration are made use of in many circuit applications, along with the *C-C* circuit's other advantage of high current gain.

To explain why a high input resistance and a low output resistance are good circuit characteristics, refer to the application circuit in Figure 21-15(d). In this example, a microphone is connected to the input of a *C-C* amplifier, and the output of this circuit is applied to a speaker. As we know, the sound wave input to the microphone will physically move a magnet within the microphone, which will in turn interact to induce a signal voltage into a stationary coil. This voice signal voltage from the microphone, which is our source, is then applied across the input resistance of our example *C-C* circuit, which is our load.

In the inset in Figure 21-15(d), you can see that the microphone has been represented as a low-current ac source with a high internal resistance, and the input resistance of the *C-C* circuit is shown as a high value (in the previous example, 120 kΩ) resistor. Remembering our previous discussion on sources and loads, we know that a small load resistance will cause a large current to be drawn from the source, and this large current will drain or pull down the source voltage. Many signal sources, such as microphones, can only generate a small signal source voltage because they have a high internal resistance. If this small signal source voltage is applied across an amplifier with a small input resistance, a large current will be drawn from the source. This heavy load will pull the signal voltage down to such a small value that it will not be large enough to control the amplifier. A large amplifier input resistance ($R_{in}\uparrow$), on the other hand, will not load the source. Therefore the input voltage applied to the amplifier will be large enough to control the amplifier circuit, to vary its transistor currents, and achieve the gain between amplifier input and output. In summary, the high input resistance of the *C-C* circuit can be connected to a high resistance source because it will not draw an excessive current and pull down the source voltage.

Referring again to the inset in Figure 21-15(d), you can see that at the output end, the *C-C*'s output circuit has been represented as a high current source with a low value internal output resistor, and the speaker has been represented as a low resistance load. The low output resistance of the *C-C* circuit means that this circuit can deliver the high current output that is needed to drive the low resistance load.

Impedance Matching Circuit
A circuit that can match, or isolate, a high resistance (low current) source.

Buffer Current Amplifier
The *C-C* circuit that can ensure that power is efficiently transferred from source to load.

As you will see later in application circuits, most *C-C* circuits are used as a resistance or **impedance matching circuit** that can match, or isolate, a high resistance (low current) source, such as a microphone, to a low resistance (high current) load, such as the speaker. By acting as a **buffer current amplifier,** the *C-C* circuit can ensure that power is efficiently transferred from source to load.

21-2-3 Bipolar Junction Transistor Data Sheets

Like the diode's data sheets, manufacturer's bipolar transistor data sheets list the typical dc and ac operating characteristics of the device. To illustrate some of the many different types of transistors available,

a. Figure 21-16 shows the data sheet of a typical *general purpose switching or amplifying bipolar transistor.*
b. Figure 21-17 shows the data sheet of a typical *high-voltage bipolar transistor.*
c. Figure 21-18 shows the data sheet of a typical *high-current bipolar transistor.*
d. Figure 21-19 (p. 1197) shows the data sheet of a typical *high-power bipolar transistor.*

DEVICE: 2N3903 and 2N3904—NPN Silicon Switching and Amplifier Transistors

Maximum continuous collector current (I_c) = 200 mA.

MAXIMUM RATINGS

Rating	Symbol	Value	Unit
Collector-Emitter Voltage	V_{CEO}	40	Vdc
Collector-Base Voltge	V_{CBO}	60	Vdc
Emitter-Base Voltage	V_{EBO}	6.0	Vdc
Collector Current — Continuous	I_C	200	mAdc
Total Device Dissipation @ T_A = 25°C Derate above 25°C	P_D	625 5.0	mW mW/°C
*Total Device Dissipation @ T_C = 25°C Derate above 25°C	P_D	1.5 12	Watts mW/°C
Operating and Storage Junction Temperature Range	T_J, T_{stg}	−55 to +150	°C

***THERMAL CHARACTERISTICS**

Characteristic	Symbol	Max	Unit
Thermal Resistance, Junction to Ambient	$R_{\theta JA}$	200	°C/W
Thermal Resistance, Junction to Case	$R_{\theta JC}$	83.3	°C/W

2N3903
2N3904★

CASE 29-04, STYLE 1
TO-92 (TO-226AA)

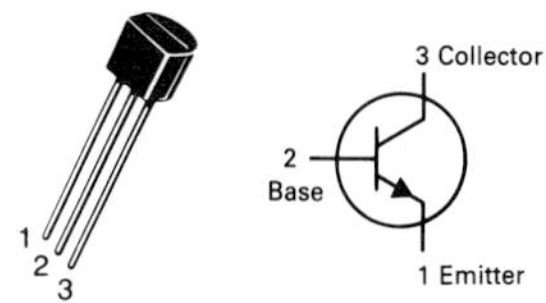

GENERAL PURPOSE TRANSISTORS

NPN SILICON

★This is a Motorola designated preferred device.

ELECTRICAL CHARACTERISTICS (T_A = 25°C unless otherwise noted.)

OFF Characteristic (operated in cutoff)		Symbol	Min	Max	Unit
Collector-Emitter Breakdown Voltage(1) (I_C = 1.0 mAdc, I_B = 0)		$V_{(BR)CEO}$	40	—	Vdc
Collector-Base Breakdown Voltage (I_C = 10 μAdc, I_E = 0)		$V_{(BR)CBO}$	60	—	Vdc
Emitter-Base Breakdown Voltage (I_E = 10 μAdc, I_C = 0)		$V_{(BR)EBO}$	6.0	—	Vdc
Base Cutoff Current (V_{CE} = 30 Vdc, V_{EB} = 3.0 Vdc)		I_{BL}	—	50	nAdc
Collector Cutoff Current (V_{CE} = 30 Vdc, V_{EB} = 3.0 Vdc)		I_{CEX}	—	50	nAdc
ON Characteristic (operated in active and saturation region)					
DC Current Gain(1) (I_C = 0.1 mAdc, V_{CE} = 1.0 Vdc)	2N3903 2N3904	h_{FE}	20 40	— —	—
(I_C = 1.0 mAdc, V_{CE} = 1.0 Vdc)	2N3903 2N3904		35 70	— —	
(I_C = 10 mAdc, V_{CE} = 1.0 Vdc)	2N3903 2N3904		50 100	150 300	
(I_C = 50 mAdc, V_{CE} = 1.0 Vdc)	2N3903 2N3904		30 60	— —	
(I_C = 100 mAdc, V_{CE} = 1.0 Vdc)	2N3903 2N3904		15 30	— —	
Collector-Emitter Saturation Voltage(1) (I_C = 10 mAdc, I_B = 1.0 mAdc) (I_C = 50 mAdc, I_B = 5.0 mAdc)		$V_{CE(sat)}$	— —	0.2 0.3	Vdc
Base-Emitter Saturation Voltage(1) (I_C = 10 mAdc, I_B = 1.0 mAdc) (I_C = 50 mAdc, I_B = 5.0 mAdc)		$V_{BE(sat)}$	0.65 —	0.85 0.95	Vdc

NOTE: The "O" following CBO, CEO, EBO indicates the third terminal is "open." For example, $V_{(BR)\,CEO}$ *means the breakdown voltage between collector and emitter with the base open.*

$h_{FE} = \beta_{DC}$, dc current gain is measured at different values of I_c.

Maximum base-emitter voltage (V_{BE}) when transistor is saturated

Maximum value of voltage between collector and emitter (V_{CE}) when transistor is in saturation.

FIGURE 21-16 **A General Purpose NPN Silicon Transistor. (Copyright of Motorola. Used by permission.)**

DEVICE: BFW43—PNP Silicon High Reverse Voltage Transistor

MAXIMUM RATINGS

Rating	Symbol	Value	Unit
Collector-Emitter Voltage (High Reverse Voltage Ratings)	V_{CEO}	150	Vdc
Collector-Base Voltage	V_{CBO}	150	Vdc
Emitter-Base Voltage	V_{EBO}	6.0	Vdc
Collector Current — Continuous	I_C	0.1	Adc
Total Device Dissipation @ $T_A = 25°C$ Derate above 25°C	P_D	0.4 2.28	Watt mW/°C
Total Device Dissipation @ $T_C = 25°C$ Derate above 25°C	P_D	1.4 8.0	Watt mW/°C
Operating and Storage Junction Temperature Range	T_J, T_{stg}	−65 to +200	°C

APPLICATIONS: High voltage circuits found in televisions and computer monitors.

BFW43

CASE 22-03, STYLE 1
TO-18 (TO-206AA)

3 Collector
2 Base
1 Emitter

HIGH VOLTAGE TRANSISTOR

PNP SILICON

FIGURE 21-17 A High-Voltage Transistor. (Copyright of Motorola. Used by permission.)

DEVICE: MPS6714—NPN Silicon High Current (I_C) Transistor

MAXIMUM RATINGS

Rating	Symbol	Value	Unit
Collector-Emitter Voltage MPS6714 MPS6715	V_{CEO}	 30 40	Vdc
Collector-Base Voltage MPS6714 MPS6715	V_{CBO}	 40 50	Vdc
Emitter-Base Voltage	V_{EBO}	5.0	Vdc
Collector Current — Continuous (High I_C Rating)	I_C	1.0	Adc
Total Device Dissipation @ $T_A = 25°C$ Derate above 25°C	P_D	1.0 8.0	Watt mW/°C
Total Device Dissipation @ $T_C = 25°C$ Derate above 25°C	P_D	2.5 20	Watts mW/°C
Operating and Storage Junction Temperature Range	T_J, T_{stg}	−55 to +150	°C

Applications: Current Regulator Circuits

MPS6714
MPS6715

CASE 29-05, STYLE 1
TO-92 (TO-226AE)

3 Collector
2 Base
1 Emitter

ONE WATT AMPLIFIER TRANSISTORS

NPN SILICON

FIGURE 21-18 A High-Current Transistor. (Copyright of Motorola. Used by permission.)

DEVICE: BUX48—NPN Silicon High Power Dissipation Transistor

SWITCHMODE II SERIES
NPN SILICON POWER TRANSISTORS

The BUX 48/BUX 48A transistors are designed for high-voltage, high-speed, power switching in inductive circuits where fall time is critical. They are particularly suited for line-operated switchmode applications such as:

- Switching Regulators
- Inverters
- Solenoid and Relay Drivers ← Applications
- Motor Controls
- Deflection Circuits

Fast Turn-Off Times
60 ns Inductive Fall Time — 25°C (Typ)
120 ns Inductive Crossover Time — 25°C (Typ)

Operating Temperature Range -65 to +200°C
100°C Performance Specified for:
Reverse-Biased SOA with Inductive Loads
Switching Times with Inductive Loads
Saturation Voltage
Leakage Currents (125°C)

BUX48
BUX48A

15 AMPERES
NPN SILICON
POWER TRANSISTORS
400 AND 450 VOLTS
V(BR)CEO
850 – 1000 VOLTS
V(BR)CEX
175 WATTS

CASE 1-07
TO-204AA
(TO-3)

MAXIMUM RATINGS

Rating	Symbol	BUX48	BUX48A	Unit
Collector-Emitter Voltage	$V_{CEO(sus)}$	400	450	Vdc
Collector-Emitter Voltage (V_{BE} = -1.5V)	V_{CEX}	850	1000	Vdc
Emitter Base Voltage	V_{EB}	7		Vdc
Collector Current — Continuous — Peak (1) — Overload	I_C I_{CM} I_{OI}	15 30 60		Adc
Base Current — Continuous — Peak (1)	I_B I_{BM}	5 20		Adc
Total Power Dissipation — T_C = 25°C — T_C = 100°C Derate above 25°C (High power dissipation rating.)	P_D	175 100 1		Watts W/°C
Operating and Storage Junction Temperature Range	T_J, T_{stg}	-65 to +200		°C

THERMAL CHARACTERISTICS

Characteristic	Symbol	Max	Unit
Thermal Resistance, Junction to Case	$R_{\theta JC}$	1	°C/W
Maximum Lead Temperature for Soldering Purposes: 1/8″ from Case for 5 Seconds	T_L	275	°C

FIGURE 21-19 A High-Power Transistor. (Copyright of Motorola. Used by permission.)

As before, notes are included in these data sheets to call out important characteristics and to explain some of the terms that have not been previously used.

21-2-4 *Testing Bipolar Junction Transistors*

Although transistors are exceptionally more reliable than their counterpart, the vacuum tube, they still will malfunction. These failures are normally the result of excessive temperature, current, or mechanical abuse and generally result in one of three problems:

1. An open between two or three of the transistor's leads
2. A short between two or three of the transistor's leads
3. A change in the transistor's characteristics

Transistor Tester

Transistor Tester Special test instrument that can be used to test both NPN and PNP bipolar transistors.

The **transistor tester** shown in Figure 21-20 is a special test instrument that can be used to test both NPN and PNP bipolar transistors. This special meter can be used to determine whether an open or short exists between any of the transistor's three terminals, the transistor's dc current gain (β_{DC}), and whether an undesirable value of leakage current is present through one of the transistor's junctions.

FIGURE 21-20 Transistor Tester. (Courtesy of Sencore, Inc.—Test Equipment for the Professional Servicer. 1-800-SENCORE.)

Ohmmeter Transistor Test

If the transistor tester is not available, the ohmmeter can be used to detect open and shorted junctions, which are the most common transistor failure. Figure 21-21 illustrates how the ohmmeter can be used to check the emitter and collector diode of an NPN or PNP transistor. Referring to this diagram, notice that reverse biasing either the collector or emitter diode of any good transistor

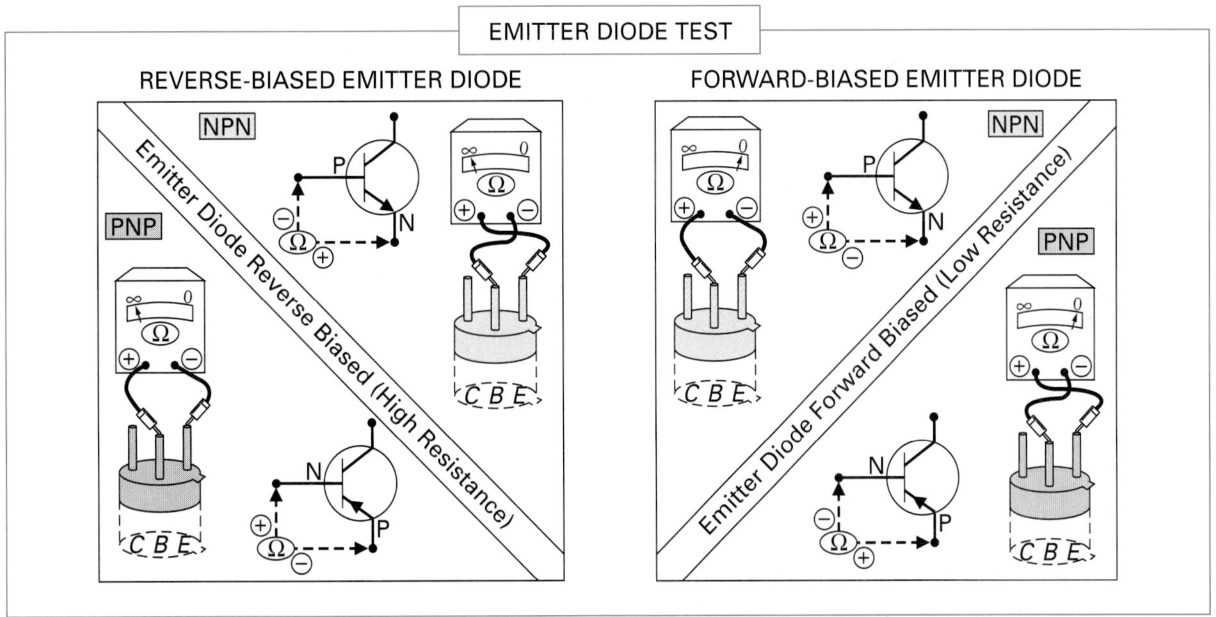

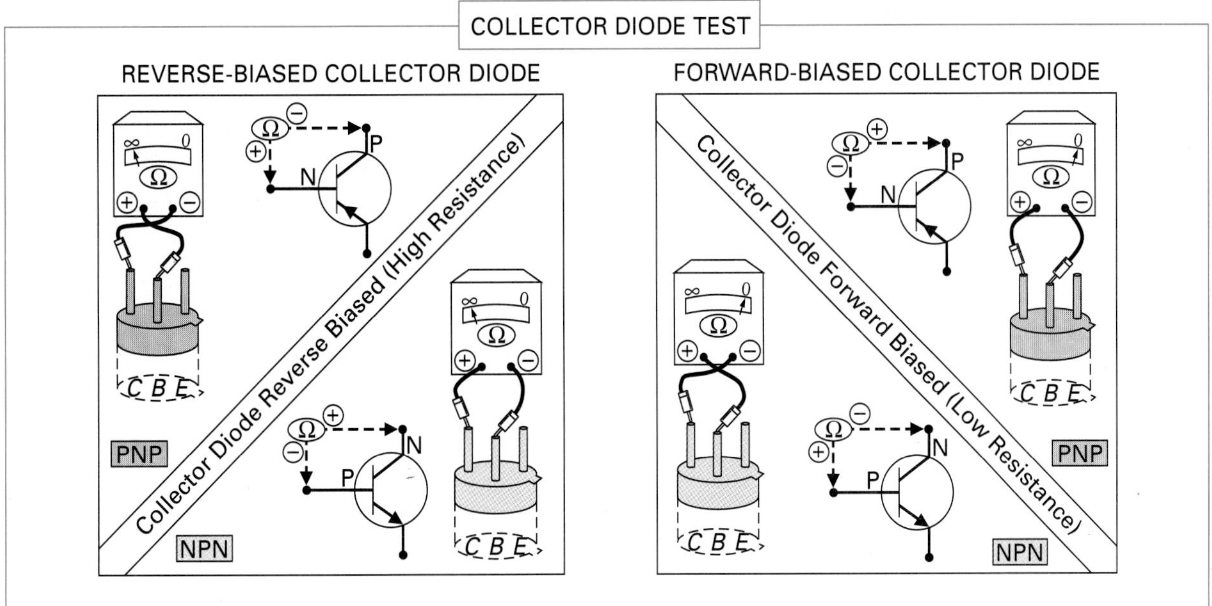

FIGURE 21-21 Using the Ohmmeter to Test a Bipolar Junction Transistor.

should cause the ohmmeter to display a relatively high resistance (several hundred thousand ohms or more). If a reverse biased emitter or collector diode has a low ohms reading, the respective transistor junction can be assumed to be shorted. Conversely, forward biasing either the collector or emitter diode of any good transistor should cause the ohmmeter to exhibit a low resistance (several hundred ohms or less). If a forward biased emitter or collector diode has a high ohms reading, the respective transistor junction can be assumed to be an open.

Transistor Testing Procedure

Figure 21-22 shows the step-by-step procedure for testing an NPN transistor. Following through this test procedure, we begin by reverse biasing the collector diode and then the emitter diode, and then forward biasing the collector diode and then the emitter diode. The table in Figure 21-22 shows the order and action to be performed for each step and the reading that should result if the NPN transistor junction is operating correctly.

Transistor Lead Identification

In most instances, to identify the bipolar transistor's leads, we would turn the transistor upside down, and then locate the emitter lead by determining which lead is closest to the side tag, as seen in Figure 21-22. Once the emitter is located, we would then assume that the lead at the other end of the case is the

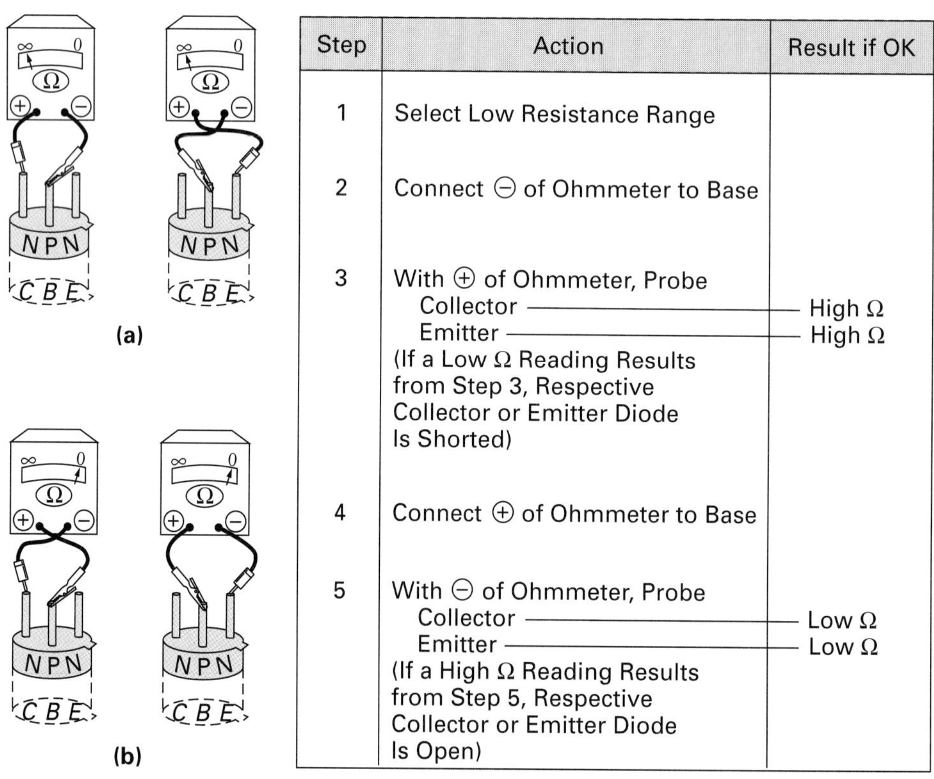

Step	Action	Result if OK
1	Select Low Resistance Range	
2	Connect ⊖ of Ohmmeter to Base	
3	With ⊕ of Ohmmeter, Probe Collector Emitter (If a Low Ω Reading Results from Step 3, Respective Collector or Emitter Diode Is Shorted)	 High Ω High Ω
4	Connect ⊕ of Ohmmeter to Base	
5	With ⊖ of Ohmmeter, Probe Collector Emitter (If a High Ω Reading Results from Step 5, Respective Collector or Emitter Diode Is Open)	 Low Ω Low Ω

FIGURE 21-22 **NPN Ohmmeter Test Procedure. (a) Reverse-Biasing Collector, Then Emitter Diode. (b) Forward-Biasing Collector, Then Emitter Diode.**

collector and that the center lead is the base. Although this procedure will most often be true, in some instances the base lead is not always between the emitter and the collector. For example, the transistor leads may be organized as *ECB* (emitter-collector-base) instead of *EBC* (emitter-base-collector). To be sure which lead is which, you should look up the device in the manufacturer's data book. If this is not available, the ohmmeter can be used to determine the transistor type (NPN or PNP) and the transistor's lead orientation *(EBC, ECB)*. The procedure, which will only work if the transistor is fully operational, is as follows.

Choosing the center lead, and assuming that it is the base, connect the negative lead of the ohmmeter and touch the two other transistor leads with the positive meter lead. If a low-resistance reading is observed in both cases, the center lead is definitely the base and the transistor is of the PNP type. The emitter lead is nearest to the tag and the collector is the remaining lead. If a high-resistance reading is observed in both cases, place the positive lead of the meter on the center lead of the transistor and probe the other two. If a low reading results in both instances, the center lead is the base and the transistor is an NPN type.

If a low resistance does not occur in both cases, the center lead is not the base and another lead should be selected to act as the assumed base. Repeating this procedure and observing the results will eventually reveal the lead configuration and transistor type.

21-2-5 Bipolar Transistor Biasing Circuits

As we discovered in the previous discussion on transistor circuit configurations, the ac operation of a transistor is determined by the "dc bias level," or "no-input signal level." This steady-state or dc operating level is set by the value of the circuit's dc supply voltage (V_{CC}) and the value of the circuit's biasing resistors. This single supply voltage and the one or more biasing resistors set up the initial dc values of transistor current (I_B, I_E and I_C) and transistor voltage (V_{BE}, V_{CE} and V_{BC}).

In this section, we will examine some of the more commonly used methods for setting the "initial dc operating point" of a bipolar transistor circuit. As you encounter different circuit applications in later chapters, you will see that many of these circuits include combinations of these basic biasing techniques and additional special purpose components for specific functions. Because the common-emitter (*C-E*) circuit configuration is used more extensively than the *C-B* and the *C-C,* we will use this configuration in all of the following basic biasing circuit examples.

Base Biasing

Figure 21-23(a) shows how a common-emitter transistor circuit could be base biased. With **base biasing,** the emitter diode of the transistor is forward biased by applying a positive base-bias voltage ($+V_{BB}$) via a current-limiting resistor (R_B) to the base of Q_1. In Figure 21-23(b), the transistor circuit from Figure 21-23(a) has been redrawn so as to simplify the analysis of the circuit. The transistor is now represented as a diode between base and emitter (emitter

Base Biasing
A transistor biasing method in which the dc supply voltage is applied to the base of the transistor via a base bias resistor.

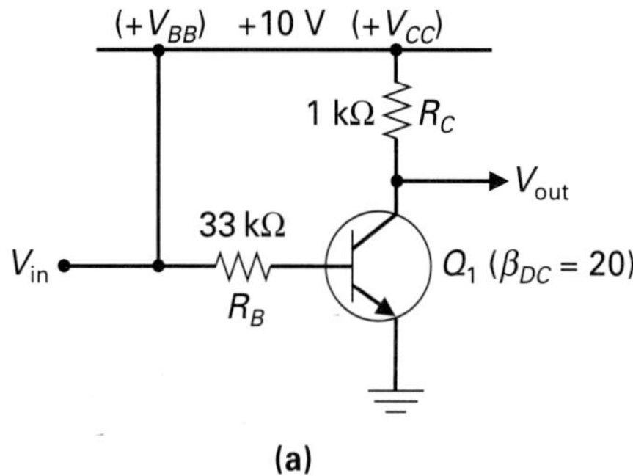

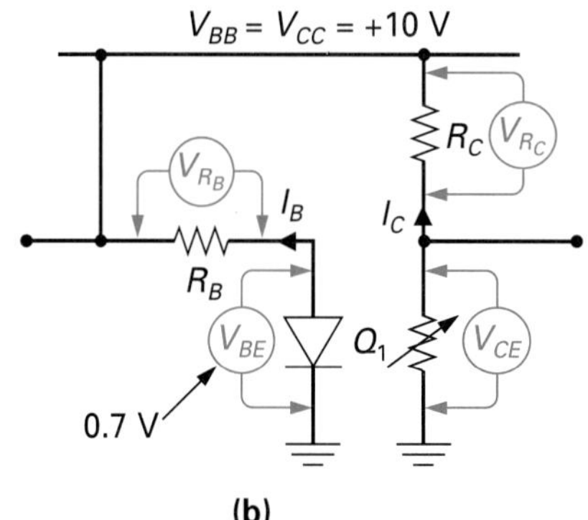

FIGURE 21-23 A Base Biased Common Emitter Circuit. (a) Basic Circuit. (b) Simplified Equivalent Circuit.

diode), and the transistor's emitter to collector has been represented as a variable resistor. Assuming Q_1 is a silicon bipolar transistor, the forward biased emitter diode will have a standard base-emitter voltage drop of 0.7 V (emitter diode drop = 0.7 V).

$$V_{BE} = 0.7 \text{ V}$$

The base bias resistor (R_B) and the transistor's emitter diode form a series circuit across V_{BB}, as seen in Figure 21-23(b). Therefore, the voltage drop across R_B (V_{R_B}) will be equal to the difference between V_{BB} and V_{BE}.

$$\begin{aligned} V_{R_B} &= V_{BB} - V_{BE} \\ &= V_{BB} - 0.7 \text{ V} \end{aligned}$$

$$V_{R_B} = V_{BB} - V_{BE} = 10 \text{ V} - 0.7 \text{ V} = 9.3 \text{ V}$$

Now that the resistance and voltage drop across R_B are known, we can calculate the current through R_B (I_{R_B}). Because a series circuit is involved, the current through R_B (I_{R_B}) will also be equal to the transistor base current I_B.

$$I_B = \frac{V_{RB}}{R_B}$$

$$I_B = \frac{V_{RB}}{R_B} = \frac{9.3\text{ V}}{33\text{ k}\Omega} = 282\ \mu\text{A}$$

Because the transistor's dc current gain (β_{DC}) is given in Figure 21-23(a), we can calculate I_C because β_{DC} tells us how much greater the output current I_C is compared to the input current I_B.

$$I_C = I_B \times \beta_{DC}$$

$$I_C = I_B \times \beta_{DC} = 282\ \mu\text{A} \times 20 = 5.6\text{ mA}$$

Because the current through R_C is I_C, we can now calculate the voltage drop across R_C (V_{R_C}).

$$V_{R_C} = I_C \times R_C$$

$$V_{R_C} = I_C \times R_C = 5.6\text{ mA} \times 1\text{ k}\Omega = 5.6\text{ V}$$

Now that V_{R_C} is known, we can calculate the voltage drop across the transistor's collector-to-emitter because V_{CE} and V_{R_C} are in series and will be equal to the applied voltage V_{CC}.

$$V_{CE} = V_{CC} - V_{R_C}$$

$$V_{CE} = V_{CC} - V_{R_C} = 10\text{ V} - 5.6\text{ V} = 4.4\text{ V}$$

Combining the previous two equations, we can obtain the following V_{CE} formula:

$$V_{CE} = V_{CC} - V_{RC}$$

Since

$$V_{R_C} = I_C \times R_C$$

$$V_{CE} = V_{CC} - (I_C \times R_C)$$

$$V_{CE} = V_{CC} - (I_C \times R_C) = 10\text{ V} - (5.6\text{ mA} \times 1\text{ k}\Omega) = 4.4\text{ V}$$

Using the above formulas, which are all basically Ohms' law, you can calculate the current and voltage values in a base biased circuit.

DC load line. In a transistor circuit, such as the example in Figure 21-23, V_{CC} and V_{R_C} are constants. On the other hand, the input current I_B and the output current I_C are variables. Using the example circuit in Figure 21-23, let us calculate what collector-to-emitter voltage drops (V_{CE}) will result for different values of I_C.

a. When Q_1 is OFF, $I_C = 0$ mA, and therefore V_{CE} equals:

$$V_{CE} = V_{CC} - (I_C \times R_C) = 10 \text{ V} - (0 \text{ mA} \times 1 \text{ k}\Omega) = 10 \text{ V} - 0 \text{ V} = 10 \text{ V}$$

This would make sense because Q_1 would be equivalent to an open switch between collector and emitter when it is OFF, and therefore all of the 10 V V_{CC} supply voltage would appear across the open. Figure 21-24 shows how this point would be plotted on a graph (point A).

b. When $I_C = 1$ mA,

$$V_{CE} = 10 \text{ V} - (1 \text{ mA} \times 1 \text{ k}\Omega) = 10 \text{ V} - 1 \text{ V} = 9 \text{ V (point } B\text{)}$$

c. When $I_C = 2$ mA,

$$V_{CE} = 10 \text{ V} - (2 \text{ mA} \times 1 \text{ k}\Omega) = 10 \text{ V} - 2 \text{ V} = 8 \text{ V (point } C\text{)}$$

d. When $I_C = 3$ mA,

$$V_{CE} = 10 \text{ V} - (3 \text{ mA} \times 1 \text{ k}\Omega) = 10 \text{ V} - 3 \text{ V} = 7 \text{ V (point } D\text{)}$$

e. When $I_C = 4$ mA, $V_{CE} = 6$ V (point E)

f. When $I_C = 5$ mA, $V_{CE} = 5$ V (point F)

g. When $I_C = 6$ mA, $V_{CE} = 4$ V (point G)

h. When $I_C = 7$ mA, $V_{CE} = 3$ V (point H)

i. When $I_C = 8$ mA, $V_{CE} = 2$ V (point I)

j. When $I_C = 9$ mA, $V_{CE} = 1$ V (point J)

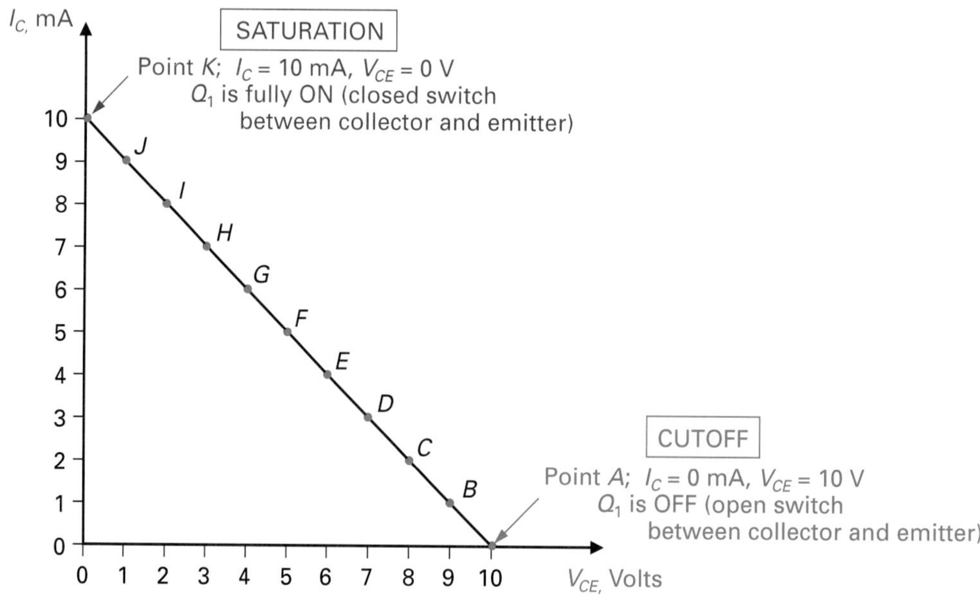

FIGURE 21-24 **A Transistor DC Load Line with Cutoff and Saturation Points.**

k. When $I_C = 10$ mA, the only resistance is that of R_C because Q_1 is fully ON and is equivalent to a closed switch between collector and emitter. It is not a surprise that the voltage drop across Q_1's collector-to-emitter is almost 0 V.

$$V_{CE} = 10\text{ V} - (10\text{ mA} \times 1\text{ k}\Omega) = 10\text{ V} - 10\text{ V} = 0\text{ V (point } K)$$

The line drawn in the graph in Figure 21-24 is called the **DC load line** because it is a line representing all the dc operating points of the transistor for a given load resistance. In this example, the transistor's load was the 1 kΩ collector connected resistor R_C in Figure 21-23.

DC Load Line
A line representing all the dc operating points of the transistor for a given load resistance.

Cutoff and saturation points. Let us now examine the two extreme points in a transistor's dc load line, which in the example in Figure 21-24 were points A and K. If a transistor's base input bias voltage is reduced to zero, its input current I_B will be zero, Q_1 will turn OFF and be equivalent to an open switch between the collector and emitter, the output current I_C will be 0 mA, and a V_{CE} will be 10 V. This point in the transistor dc load line is called *cutoff* (point A in Figure 21-24) because the output collector current is reduced to zero, or cut off. In summary, at cutoff:

$$I_{C(Cutoff)} = 0\text{ mA}$$
$$V_{CE(Cutoff)} = V_{CC}$$

In the example circuit in Figure 21-23 and its dc load line in Figure 21-24, with Q_1 cut OFF:

$$I_{C(Cutoff)} = 0\text{ mA},\ V_{CE(Cutoff)} = V_{CC} = 10\text{ V}$$

If the base input bias voltage is increased to a large positive value, the transistor's collector diode (which is normally reverse biased) will be forward biased. In this condition, I_B will be at its maximum, Q_1 will be fully ON and equivalent to a closed switch between the collector and emitter, I_C will be at its maximum of 10 mA, and V_{CE} will be 0 V. This point in the transistor's dc load line is called *saturation* (point K in Figure 21-24) because, just as a point is reached where a wet sponge is saturated and cannot hold any more water, the transistor at saturation cannot increase I_C beyond this point. In summary, at saturation:

$$I_{C(Sat.)} = \frac{V_{CC}}{R_C}$$
$$V_{CE(Sat.)} = 0\text{ V}$$

In the example circuit in Figure 21-23 and its dc load line in Figure 21-24, with Q_1 saturated:

$$I_{C(Sat.)} = \frac{V_{CC}}{R_C} = \frac{10\text{ V}}{1\text{ k}\Omega} = 10\text{ mA}$$
$$V_{CE(Sat.)} = 0\text{ V}$$

Rearranging the formula $\beta_{DC} = I_C/I_B$, we can calculate the value of input base current that causes the output saturation current:

$$\beta_{DC} = \frac{I_C}{I_B}, \text{ therefore,}$$

$$I_{B(Sat.)} = \frac{I_{C(Sat.)}}{\beta_{DC}}$$

In the example circuit in Figure 21-23 and its dc load line in Figure 21-24, the input current that will cause saturation will be

$$I_{B(Sat.)} = \frac{I_{C(Sat.)}}{\beta_{DC}} = \frac{10 \text{ mA}}{20} = 500 \text{ }\mu\text{A}$$

Figure 21-25 summarizes all of our base bias circuit calculations so far by including the dc load line from Figure 21-24 in a set of collector characteristic curves for the transistor circuit example in Figure 21-23. As you can see in the graph in Figure 21-24, at cutoff: $I_B = 0$ μA, $I_C = 0$ mA, and $V_{CE} = V_{CC}$ which is 10 V. On the other hand, at saturation: $I_B = 500$ μA, $I_C = 10$ mA, and $V_{CE} = 0$ V.

Quiescent point. Generally, the value of the base biased resistor (R_B) is chosen so that the value of base current (I_B) is near the middle of the dc load line. For example, if a base bias resistance of 37.2 kΩ was used in the example circuit in Figure 21-23 ($R_B = 37.2$ kΩ), it would produce a base current of 250 μA ($I_B = 9.3$ V/37.2 kΩ = 250 μA). Referring to the dc load line in Figure 21-25, you can see that this value of base current is half way between cutoff at 0 μA, and saturation at 500 μA. This point is called the *quiescent* (at rest) or *Q point* and is

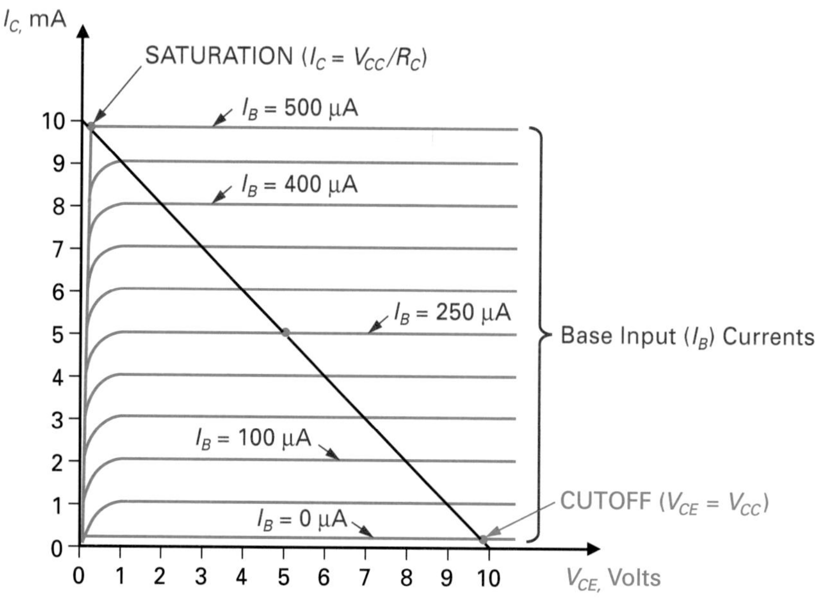

FIGURE 21-25 Transistor Input/Output Characteristic Graph.

defined as *the dc bias point at which the circuit rests when no ac input signal is applied.* An ac input signal voltage will vary I_B above and below this Q point, resulting in a corresponding but larger change in I_C.

EXAMPLE:

Complete the following for the circuit shown in Figure 21-26.

a. Calculate I_B
b. Calculate I_C
c. Calculate V_{CE}
d. Sketch the circuit's dc load line with saturation and cutoff points
e. Indicate where the Q point is on the circuit's dc load line

Solution:

a. Since $V_{BE} = 0.7$ V and $V_{BB} = 12$ V,

$$V_{R_B} = 12\text{ V} - 0.7\text{ V} = 11.3\text{ V}$$

$$I_B = \frac{V_{R_B}}{R_B} = \frac{11.3\text{ V}}{220\text{ k}\Omega} = 51.4\ \mu\text{A}$$

b.

$$I_C = I_B \times \beta_{DC} = 51.4\ \mu\text{A} \times 80 = 4.1\text{ mA}$$

c.

$$V_{R_C} = I_C \times R_C = 4.1\text{ mA} \times 1.2\text{ k}\Omega = 4.92\text{ V}$$

$$V_{CE} = V_{CC} - V_{R_C} = 12\text{ V} - 4.92\text{ V} = 7.08\text{ V}$$

d. At cutoff, the transistor is OFF and therefore equivalent to an open switch between collector and emitter. All of the V_{CC} supply voltage will therefore be across Q_1.

At cutoff, $V_{CE} = V_{CC} = 12$ V (see cutoff in the dc load line in Figure 21-27).

At saturation, the transistor is fully ON and therefore equivalent to a closed switch between the collector and emitter. The only resistance is that of R_C, and so:

At saturation,

$$I_{C(Sat.)} = \frac{V_{CC}}{R_C} = \frac{12\text{ V}}{1.2\text{ k}\Omega} = 10\text{ mA}$$

(see saturation in the dc load line in Figure 21-27).

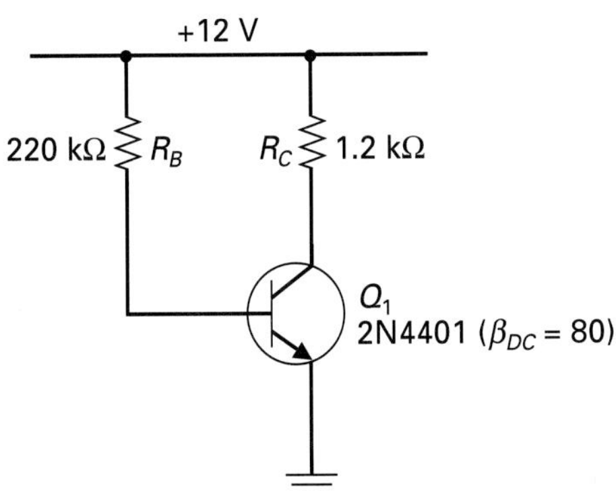

FIGURE 21-26 Bipolar Transistor Example.

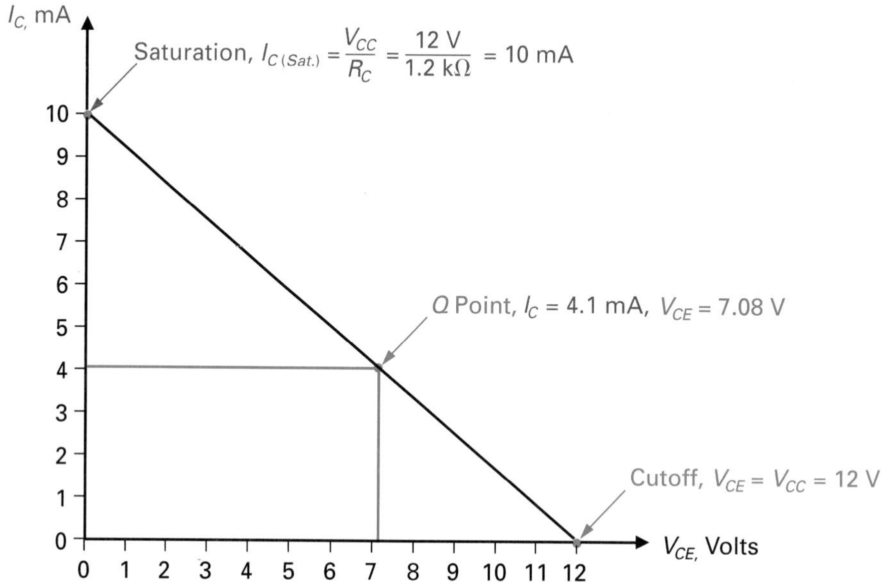

FIGURE 21-27 **The DC Load Line for the Circuit in Figure 21-26.**

e. The operating point or Q point of this circuit is set by the base bias resistor R_B. This Q point will be at

$$I_C = 4.1\text{ mA}$$

which produces a

$$V_{CE} = 7.08\text{ V}$$

This quiescent *(Q)* point is also shown on Figure 21-27.

EXAMPLE:

Calculate the current through the lamp in Figure 21-28.

Solution:

$$V_{BE} = 0.7\text{ V}, V_{in} = +5\text{ V, therefore,}$$

$$V_{R_B} = V_{in} - 0.7\text{ V}$$

$$= 5\text{ V} - 0.7\text{ V} = 4.3\text{ V}$$

$$I_B = \frac{V_{R_B}}{R_B} = \frac{4.3\text{ V}}{10\text{ k}\Omega} = 430\ \mu\text{A}$$

$$I_C = I_B \times \beta_{DC}$$

$$= 433\ \mu\text{A} \times 125 = 53.75\text{ mA}$$

An input of zero volts ($V_{in} = 0$ V) will turn OFF Q_1, and therefore lamp L_1. On the other hand, an input of +5 V will turn ON Q_1 and permit a collector current, and therefore lamp current, of 53.75 mA.

Base-biasing applications. Base-bias circuits are used in switching circuit applications like the two-state ON/OFF lamp circuit discussed in the previous example. In these circuits, the bipolar transistor is equivalent to a switch and is

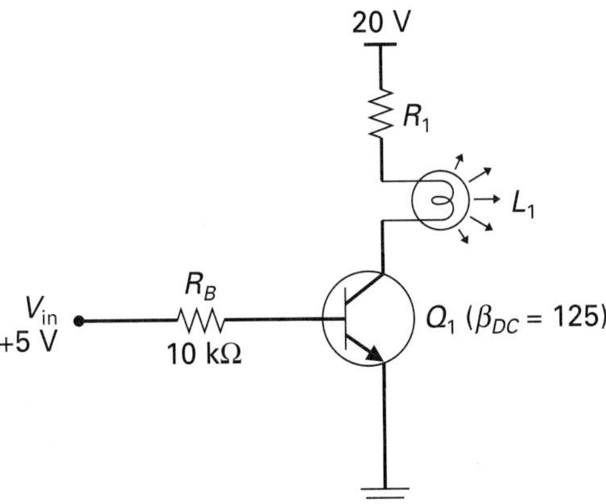

FIGURE 21-28 Two-State Lamp Circuit.

controlled by a HIGH/LOW input voltage that drives the transistor between the two extremes of cutoff and saturation.

The advantage of this biasing technique is circuit simplicity because only one resistor is needed to set the base bias voltage. The disadvantage of the base-biased circuit is that it cannot compensate for changes in its dc bias current due to changes in temperature. To explain this in more detail, a change in temperature will result in a change in the internal resistance of the transistor (all semiconductor devices have a negative temperature coefficient of resistance—temperature ↑ causes internal resistance ↓). This change in the transistor's internal resistance will change the transistor's dc bias currents (I_B and I_C), which will change or shift the transistor's dc operating point or Q point away from the desired midpoint.

Base-bias troubleshooting. Figure 21-29 repeats our original base-bias circuit example, and the previously calculated values of circuit voltage and current under normal operating conditions. As in any troubleshooting exercise, remember to apply the three step troubleshooting procedure of "diagnose, isolate, and repair."

Before beginning with any circuit troubleshooting, let us first list some of the obvious errors that should not be overlooked:

a. Is the power supplied to the circuit (+10 V) present and of the correct value?

b. Is ground connected to the circuit?

c. If this is a newly constructed circuit, are all of the components connected correctly, especially the three terminals of the transistor?

Once you have determined that the problem is, in fact, within the transistor circuit, perform a visual check of the circuit to look for

a. Shorts caused by badly connected components, loose clippings of wire, solder bridges, and so on.

b. Incorrect component values, such as resistors.

c. Sign of excessive heat on wire insulation or circuit components.

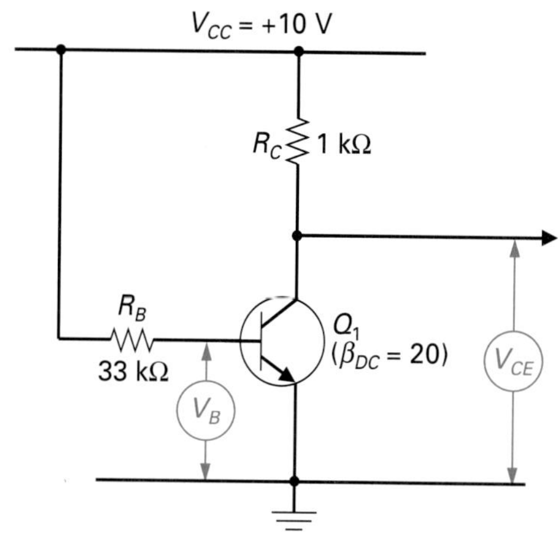

V_{BE} = 0.7 V
V_{R_B} = 9.3 V
V_{R_C} = 5.6 V
V_{CE} = 4.4 V

I_B = 282 μA
I_C = 5.6 mA
$I_{C(Sat.)}$ = 10 mA
$I_{B(Sat.)}$ = 500 μA

FIGURE 21-29 Troubleshooting a Base-Biased Circuit.

The next step would be to use test equipment to determine what "effect" the problem is causing, and what could be the "cause." To give you some examples, some of the more common problems that can occur within a base-biased common-emitter bipolar transistor circuit are listed below.

CAUSE	EFFECT
R_C Open	Collector resistor opens—there is no path for current in collector circuit. Transistor is cut off. V_{CC} will appear across open. V_{R_C} = 10 V, V_{CE} = 0 V.
R_C Short	Collector resistor short—value of R_C is chosen so that I_C is kept within safe limits: An R_C short will cause excessive collector current and burn out Q_1.
R_B Open	When base resistor opens, no biasing current can flow in base. Transistor will cut off. V_B = 0 V, V_{CE} = 10 V.
R_B Short	Base resistor short—base current increases to a maximum, transistor goes into heavy saturation (fully ON). Base will be 0.7 V above emitter voltage (V_B = +0.7 V). V_{CE} = 0 V. Transistor could burn out due to heavy internal transistor currents.
$Q_{1\,(B\text{-}E)}$ Open	Transistor cuts OFF—no circuit current. $V_B = V_{CC}$, $V_C = V_{CC}$. To verify, disconnect power, perform ohmmeter check on transistor and R_B to isolate problem.
Q_1 Leaky	Transistor is leaky, which means partially shorted. Transistor appears to be in saturation, with symptoms the same as R_B short. Disconnect power, perform ohmmeter check on transistor and R_B to isolate problem.

Voltage-Divider Biasing

Figure 21-30(a) shows how a common-emitter transistor circuit could be **voltage-divider biased.** The name of this biasing method comes from the two resistor series-voltage divider (R_1 and R_2) connected to the transistor's base. In this most widely used biasing method, the emitter diode of Q_1 is forward biased by the voltage developed across R_2 (V_{R_2}), as seen in the simplified equivalent circuit in Figure 21-30(b). To calculate the voltage developed across R_2, and therefore the voltage applied to Q_1's base, we can use the voltage divider formula.

Voltage-Divider Biasing A biasing method used with amplifiers in which a series arrangement of two fixed-value resistors is connected across the voltage source. The result is that a desired fraction of the total voltage is obtained at the center of the two resistors and is used to bias the amplifier.

$$V_{R_2} \text{ or } V_B = \frac{R_2}{R_1 + R_2} \times V_{CC}$$

$$V_{R_2} \text{ or } V_B = \frac{R_2}{R_1 + R_2} \times V_{CC} = \frac{10\text{ k}\Omega}{20\text{ k}\Omega + 10\text{ k}\Omega} \times 20\text{ V} = 0.333 \times 20\text{ V} = 6.7\text{ V}$$

Because the current through R_1 and R_2 (from ground to $+V_{CC}$) is generally more than 10 times greater than the base current of Q_1 (I_B), it is normally assumed that I_B will have no effect on the voltage divider current through R_1 and R_2. The R_1 and R_2 voltage divider can be assumed to be independent of Q_1, and the previous voltage divider formula can be used to calculate V_{R_2} or V_B.

Because $V_B = 6.7$ V, the emitter diode of Q_1 will be forward biased. Assuming a 0.7 V drop across the transistor's base-emitter junction ($V_{BE} = 0.7$ V), the voltage at the emitter terminal of Q_1 (V_E) will be:

$$V_{R_E} \text{ or } V_E = V_B - 0.7\text{ V}$$

$$V_{R_E} \text{ or } V_E = V_B - 0.7\text{ V} = 6.7\text{ V} - 0.7\text{ V} = 6\text{ V}$$

Now that the voltage drop across R_E (V_{R_E}) is known, along with its resistance, we can calculate the current through R_E and the value of current being injected into the transistor's emitter.

$$I_{R_E} = I_E = \frac{V_{R_E}}{R_E}$$

$$I_{R_E} = I_E = \frac{V_{R_E}}{R_E} = \frac{6\text{ V}}{5\text{ k}\Omega} = 1.2\text{ mA}$$

Because we know that a transistor collector current (I_C) is approximately equal to the emitter current (I_E), we can state that

$$I_E \cong I_C$$

$$I_E \cong I_C = 1.2\text{ mA}$$

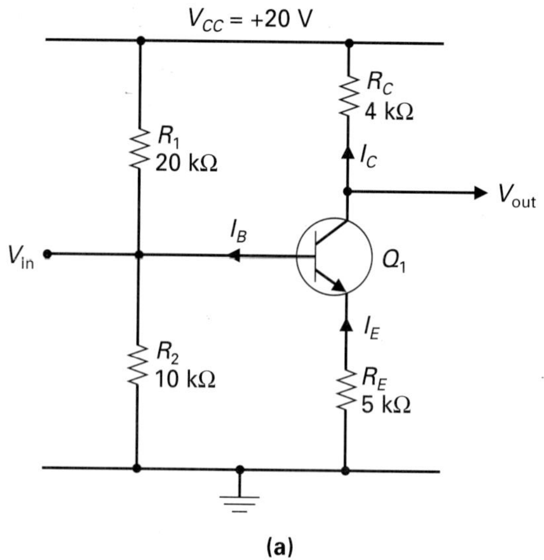

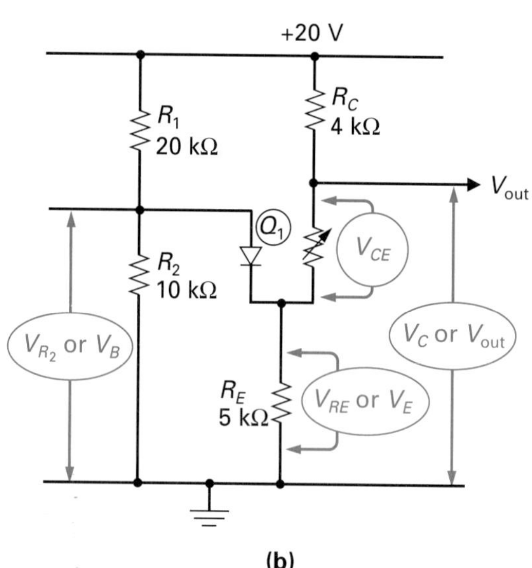

FIGURE 21-30 A Voltage Divider Biased Common Emitter Circuit. (a) Basic Circuit. (b) Simplified Equivalent Circuit.

Now that I_C is known, we can calculate the voltage drop across R_C (V_{R_C}) because both its resistance and current are known.

$$V_{R_C} = I_C \times R_C$$

$$V_{R_C} = I_C \times R_C = 1.2 \text{ mA} \times 4 \text{ k}\Omega = 4.8 \text{ V}$$

The dc quiescent voltage at the collector of Q_1 with respect to ground (V_C), which is also V_{out}, will be equal to the dc supply voltage (V_{CC}) minus the voltage drop across R_C.

$$V_C \text{ or } V_{out} = V_{CC} - V_{R_C}$$

$$V_C \text{ or } V_{out} = V_{CC} - V_{R_C} = 20 \text{ V} - 4.8 \text{ V} = 15.2 \text{ V}$$

Because V_{CC} is connected across the series voltage divider formed by R_C, Q_1's collector-to-emitter resistance (R_{CE}), and R_E, we can calculate V_{CE} if both V_{R_C} and V_E are known:

$$V_{CE} = V_{CC} - (V_{R_C} + V_E)$$

$$V_{CE} = V_{CC} - (V_{R_C} + V_E) = 20 \text{ V} - (4.8 \text{ V} + 6 \text{ V}) = 20 \text{ V} - 10.8 \text{ V} = 9.2 \text{ V}$$

DC load line. Figure 21-31 shows the dc load line for the example circuit in Figure 21-30. Referring to the dc load line's two end points, let us examine this circuit's saturation and cutoff points.

When transistor Q_1 is fully ON or saturated, it will have approximately 0 Ω of resistance between its collector and emitter. As a result, R_C and R_E determine the value of I_C when Q_1 is saturated.

$$I_{C(Sat.)} = \frac{V_{CC}}{R_C + R_E}$$

$$I_{C(Sat.)} = \frac{V_{CC}}{R_C + R_E} = \frac{20 \text{ V}}{4 \text{ k}\Omega + 5 \text{ k}\Omega} = \frac{20 \text{ V}}{9 \text{ k}\Omega} = 2.2 \text{ mA}$$

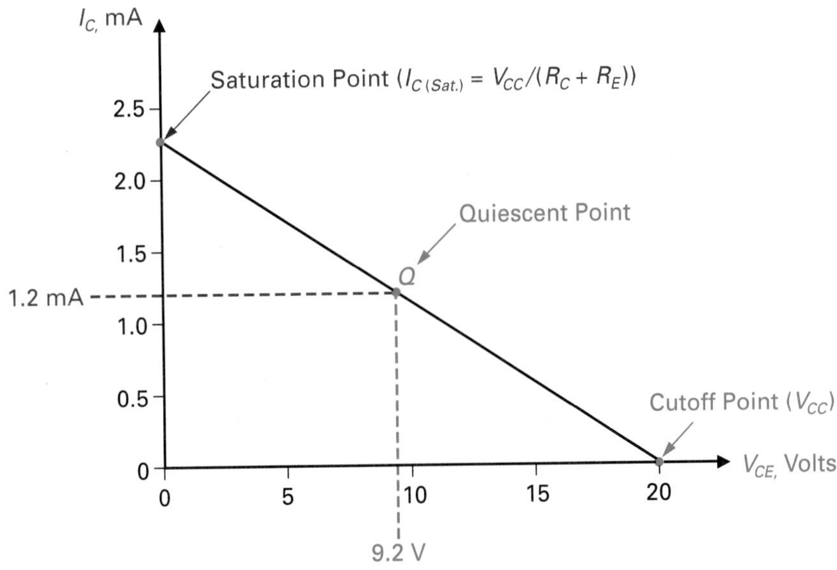

FIGURE 21-31 **The DC Load Line for the Circuit in Figure 21-30.**

As you can see in Figure 21-31, at saturation, I_C is maximum at 2.2 mA, and V_{CE} is 0 V because Q_1 is equivalent to a closed switch (0 Ω) between Q_1's collector and emitter.

$$V_{CE(Sat.)} = 0 \text{ V}$$

At the other end of the dc load line in Figure 21-31, we can see how the transistor's characteristics are plotted when it is cut off. When Q_1 is cut OFF, it is equivalent to an open switch between collector-to-emitter. Therefore all of the V_{CC} supply voltage will appear across the series circuit open.

$$V_{CE(Cutoff)} = V_{CC}$$

$$V_{CE(Cutoff)} = V_{CC} = 20 \text{ V}$$

As you can see in Figure 21-31, when Q_1 is cut OFF, all of the V_{CC} supply voltage will appear across Q_1's collector-to-emitter terminals, and I_C will be blocked and equal to zero.

$$I_{C(Cutoff)} = 0 \text{ mA}$$

Generally, the value of the voltage divider resistors R_1 and R_2 are chosen so that the value of base current (I_B) is near the middle of the dc load line. Referring to Figure 21-31, you can see that by plotting our previously calculated values of I_C (which at rest was 1.2 mA) and V_{CE} ($V_{CE} = 9.2$ V), we obtain a Q point that is near the middle of the dc load line.

EXAMPLE:

Calculate the following for the circuit shown in Figure 21-32.

a. V_B and V_E

b. Determine whether C_E will have any effect on the dc operating voltages.

c. I_C

d. V_C and V_{CE}

e. Sketch the circuit's dc load line and include the saturation, cutoff, and Q points.

Solution:

a.

$$V_B = \frac{R_2}{R_1 + R_2} \times V_{CC} = \frac{2.2 \text{ k}\Omega}{10 \text{ k}\Omega + 2.2 \text{ k}\Omega} \times 12 \text{ V} = 2.16 \text{ V}$$

$$V_E = V_B - 0.7 \text{ V} = 2.16 \text{ V} - 0.7 \text{ V} = 1.46 \text{ V}$$

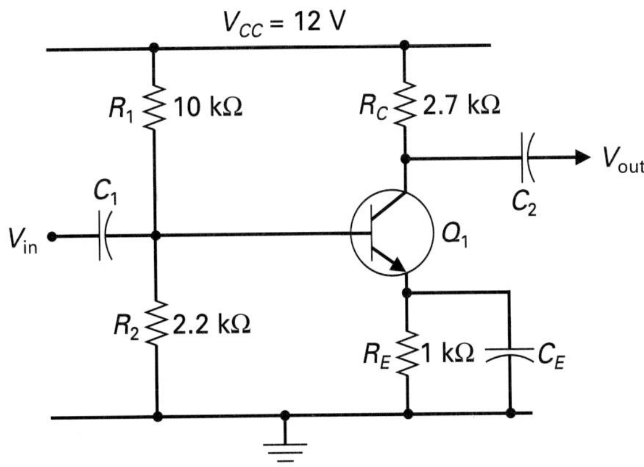

FIGURE 21-32 A Common-Emitter Amplifier Circuit Example.

b. Since all capacitors can be thought of as a dc block, C_E will have no effect on the circuit's dc operating voltages.

c.
$$I_E = \frac{V_E}{R_E} = \frac{1.46\ \text{V}}{1\ \text{k}\Omega} = 1.46\ \text{mA}$$
$$I_C \cong I_E = 1.46\ \text{mA}$$

d.
$$V_{R_C} = I_C \times R_C = 1.46\ \text{mA} \times 2.7\ \text{k}\Omega = 3.9\ \text{V}$$
$$V_{out} \text{ or } V_C = V_{CC} - V_{R_C} = 12\ \text{V} - 3.9\ \text{V} = 8.1\ \text{V}$$
$$V_{CE} = V_{CC} - (V_{R_C} + V_E) = 12\ \text{V} - (3.9\ \text{V} + 1.46\ \text{V}) = 12\ \text{V} - 5.36\ \text{V} = 6.64\ \text{V}$$

e.
$$I_{C(Sat.)} = \frac{V_{CC}}{R_C + R_E} = \frac{12\ \text{V}}{2.7\ \text{k}\Omega + 1\ \text{k}\Omega} = \frac{12\ \text{V}}{3.7\ \text{k}\Omega} = 3.24\ \text{mA}$$
$$V_{CE(Cutoff)} = V_{CC} = 12\ \text{V}$$
$$Q\ Point,\ I_C = 1.46\ \text{mA}\ and\ V_{CE} = 6.64\ \text{V}$$

(This information is plotted on the graph in Figure 21-33)

Voltage-divider bias applications. Voltage-divider bias circuits are used in analog or linear circuit applications such as the amplifier circuit discussed in the previous example. In these circuits, the bipolar transistor is equivalent to a variable-resistor and is controlled by an alternating input signal voltage.

Unlike the base-biased circuit, the voltage-divider biased circuit has very good temperature stability due to the emitter resistor R_E. To explain this in more detail, let us assume that there is an increase in the temperature surrounding a voltage-divider circuit, such as the example circuit in Figure 21-32. As temperature increases, it causes an increase in the transistor's internal currents ($I_B\uparrow$, $I_E\uparrow$, $I_C\uparrow$) because all semiconductor devices have a negative temperature coefficient of resistance (temperature $\uparrow$, $R\downarrow$, $I\uparrow$). An increase in $I_E\uparrow$ will cause an increase in the voltage drop across $R_E\uparrow$, which will decrease the voltage dif-

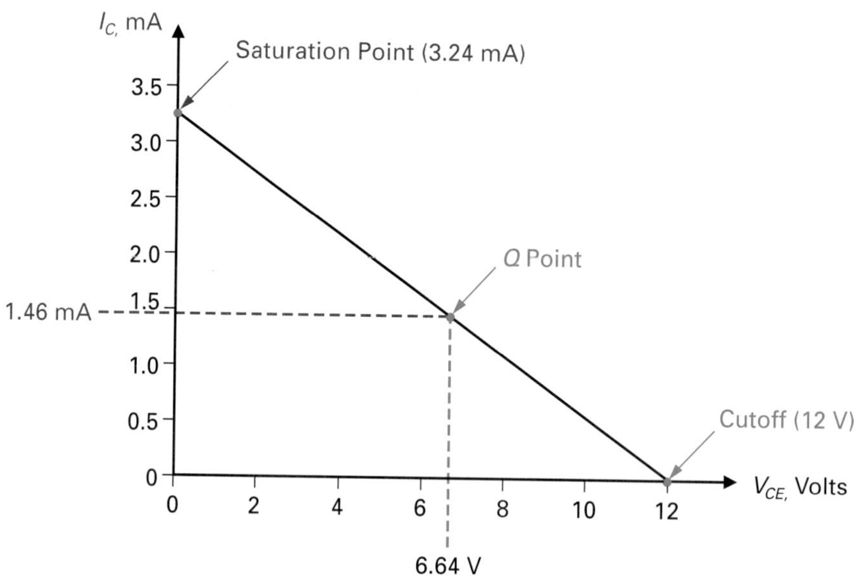

FIGURE 21-33 The DC Load Line for the Circuit in Figure 21-32.

ference between the transistor's base and emitter ($V_{BE}\downarrow$). Decreasing the forward bias applied to the transistor's emitter diode will decrease all of the transistor's internal currents ($I_B\downarrow$, $I_E\downarrow$, $I_C\downarrow$) and return them to their original values. Therefore, a change in output current (I_C) due to temperature will effectively be fed back to the input and change the input current (I_B), which is why a circuit containing an emitter resistor is said to have **emitter feedback** for temperature stability.

Emitter Feedback The coupling from the emitter output to the base input in a transistor amplifier.

Voltage-divider bias troubleshooting. Figure 21-34 repeats our original voltage-divider biased circuit example and the previously calculated values of circuit voltage and current under normal operating conditions. As in any troubleshooting exercise, remember to apply the three step troubleshooting procedure of "diagnose, isolate, and repair."

Once again, begin by looking for the obvious errors:

a. Is the power supplied to the circuit (+20 V) present and of the correct value?

b. Is ground connected to the circuit?

c. If this is a newly constructed circuit, are all of the components connected correctly, especially the three terminals of the transistor?

Once you have determined that the problem is, in fact, within the transistor circuit, perform a visual check of the circuit to look for anything out of the normal.

The next step would be to use test equipment to determine what "effect" the problem is causing and what could be the "cause." To give you some exam-

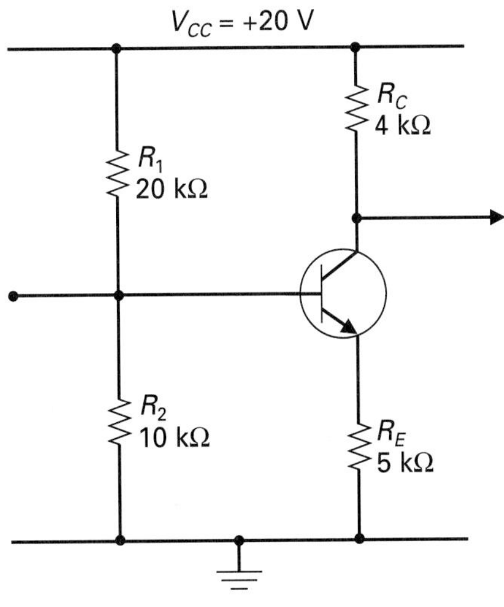

$V_B = 6.7$ V
$V_E = 6.0$ V
$V_{R_C} = 4.8$ V
$V_C = V_{out} = 15.2$ V
$V_{CE} = 9.2$ V

$I_E = 1.2$ mA
$I_C = 1.2$ mA
$I_{C(Sat.)} = 2.2$ mA

FIGURE 21-34 Troubleshooting a Voltage-Divider Biased Circuit.

ples, some of the more common problems that can occur within a voltage-divider biased common-emitter bipolar transistor circuit are listed below. Keep in mind that if a component problem turns the transistor fully OFF or ON, you may have to disconnect power and perform the ohmmeter test on the transistor to determine whether the problem is the transistor or another component in the circuit affecting the transistor.

CAUSE	EFFECT
R_1 Open	With R_1 open, there is no path for base current to $+V_{CC}$, no voltage developed across R_2, and therefore $V_B = 0$ V. Transistor is OFF, and equivalent to open switch between collector and emitter, therefore; $V_E = 0$ V, $V_C = +V_{CC}$.
R_2 Open	If R_2 opens, base of transistor will be a large positive voltage and so it will turn heavily ON (saturate). Transistor is equivalent to a closed switch between collector and emitter, and therefore; $V_{CE} = 0$ V, $V_E = R_E \times I_{C(Sat)}$, $V_C = V_E$, $V_B = V_E + 0.7$ V.
R_E Open	With R_E open, there can be no emitter current, and therefore there can be no input base current or output collector current. Transistor will be OFF, therefore V_C will equal $+V_{CC}$.
R_C Open	With R_C open, there can be no path for the output collector current, $I_C = 0$. Since B-E junction is forward biased I_E will equal I_B, and so; V_E will be a small voltage, $V_B = V_E + 0.7$ V, $V_C = V_E$ (all three transistor terminal voltages will be low).

SELF-TEST REVIEW QUESTIONS FOR SECTION 21-2

1. The bipolar transistor is a ____________ (voltage/current) controlled device.
2. When a bipolar transistor is being operated in the active region, its emitter diode is ____________ biased and its collector diode is ____________ biased.
3. Which of the following is correct:
   a. $I_E = I_C - I_B$
   b. $I_C = I_E - I_B$
   c. $I_B = I_C - I_E$
4. When a transistor is in cutoff, it is equivalent to a/an ____________ between its collector and emitter.
5. When a transistor is in saturation, it is equivalent to a/an ____________ between its collector and emitter.
6. Which of the bipolar transistor circuit configurations has the best
   a. Voltage gain
   b. Current gain
   c. Power gain
7. Which biasing method makes use of two series connected resistors across the V_{CC} supply voltage?
8. Which biasing technique has a single resistor connected in series with the base of the transistor?

SUMMARY

Bipolar Junction Transistors

1. A transistor is a three-element device made of semiconductor materials used to control electron flow, the amount of which can be controlled by varying the voltages applied to its three elements. Having the ability to control the amount of current through the transistor allows us to achieve two very important applications: switching and amplification.
2. Transistors are generally classified as being either bipolar or unipolar. The bipolar junction transistor (BJT) has two P-N junctions whereas the unipolar transistor has only one P-N junction.
3. The bipolar transistor has three alternately doped semiconductor regions called the emitter (E), base (B), and collector (C).
4. The NPN transistor type has a thin, lightly doped p-type semiconductor layer sandwiched between two n-type semiconductor regions.
5. The PNP transistor type has a thin, lightly doped n-type region sandwiched between two p-type regions.
6. As a memory aid for the NPN transistor's schematic symbol, you may want to remember that when the emitter arrow is "**N**ot **P**ointing i**N**" to the base, the transistor is an **"NPN."**

7. The package will protect the transistor from humidity and dust, provide a means for electrical connection between the three semiconductor regions and the three transistor terminals, and serve as a heat sink to conduct away any heat generated by the transistor.
8. Transistor package types are normally given a reference number. These designations begin with the letters "TO," which stands for transistor outline and are followed by a number.
9. A transistor can be thought of as containing two diodes: a base-to-collector diode and a base-to-emitter diode. With an NPN transistor, both diodes will be back-to-back and pointing away from the base. For a PNP transistor, the base-collector and base-emitter diode will both be pointing into the base.
10. Transistors are basically controlled to operate as a switch, or as a variable resistor:
    - **a.** The ON/OFF switching action of the transistor is controlled by the transistor's base-to-emitter (*B-E*) diode. If the *B-E* diode of the transistor is forward biased, the transistor will turn ON and be equivalent to a closed switch between its collector and emitter. If the *B-E* diode of the transistor is reverse biased, the transistor will turn OFF and be equivalent to an open switch between its collector and emitter.
    - **b.** The variable resistor action of the transistor is also controlled by the transistor's base-to-emitter (*B-E*) diode. When the base input control voltage is between the OFF input voltage and the fully ON input voltage, the transistor is partially ON, and therefore the transistor's emitter-to-collector resistance is somewhere between maximum ohms and zero ohms.
11. The name transistor was derived from the fact that through base control we can "transfer" different values of "resistance" between the emitter and collector. This effect of "transferring resistance" is known as transistance, and the component that functions in this manner is called the transistor.
12. The transistor initiated the multibillion dollar semiconductor industry and was the key element behind many other inventions, such as integrated circuits (ICs), optoelectronic devices, and digital computer electronics.
13. A digital logic gate circuit makes use of the transistor's ON/OFF switching action. Transistors are used to construct logic gate circuits; gates are used to construct flip-flop circuits; flip-flops are used to construct register and counter circuits, and these circuits are used to construct microprocessor, memory and input/output circuits—the three basic blocks of a digital computer.
14. When used as a variable resistor, the transistor is the controlling element in many analog or linear circuits applications such as amplifiers, oscillators, modulators, detectors, regulators, and so on. The most important of these applications is amplification, which is the boosting in strength or increasing in amplitude of electronic signals. A voltage gain of 10 means that the output voltage is ten times larger than the input voltage. The transistor does not produce this gain magically within its NPN semiconductor structure. The gain or amplification is achieved by the input signal controlling the conduction of the transistor, which takes energy from the collector supply

voltage and develops this energy across the load resistor. Amplification is achieved by having a small input voltage control a transistor and its large collector supply voltage so that a small input voltage change results in a similar but larger output voltage change.

15. The current at the transistor's emitter terminal is called the emitter current (I_E). The forward biased emitter diode will cause the steady stream of electrons entering the emitter to head toward the base region. The base is a very thin, lightly doped region with very few holes in relation to the number of electrons entering the transistor from the emitter. Consequently, only a few electrons combine with the holes in the base region and flow out of the base region. This relatively small current at the transistor's base terminal is called the base current (I_B). Because only a few electrons combine with holes in the base region, there is an accumulation of electrons in the base's *p* region. These free electrons, feeling the attraction of the large positive collector supply voltage, will travel through the *n*-type collector junction and out of the transistor to the positive external collector supply voltage. The current emerging out of the transistor's collector is called the collector current (I_C). Because both the collector current and base current are derived from the emitter current, we can state that the sum of base current and the collector current will equal the emitter current.

16. An increase in the emitter current ($I_E\uparrow$) will cause a corresponding increase in collector and base current ($I_C\uparrow$, $I_B\uparrow$). Looking at this from a different angle, an increase in the applied base voltage will increase the forward bias applied to the emitter diode of the transistor, which will draw more electrons up from the emitter and cause an increase in I_E, I_B, and I_C. Similarly, a decrease in the applied base voltage will decrease the forward bias applied to the emitter diode of the transistor, which will decrease the number of electrons being drawn up from the emitter and cause a decrease in I_E, I_B, and I_C. The applied input base voltage will control the amount of base current, which will in turn control the amount of emitter and collector current and the conduction of the transistor. This is why the bipolar transistor is known as a current-controlled device.

17. A transistor is said to be in the active region when its base-emitter junction is forward biased (emitter diode is ON), and the base-collector junction is reverse biased (collector diode is OFF). In this mode, the transistor is equivalent to a variable-resistor between the collector and emitter.

18. A transistor is in cutoff when the bias voltage is reduced to a point that it stops current in the transistor. In cutoff, both the emitter and collector diode of the transistor will be biased OFF, the transistor is equivalent to an open switch between the collector and emitter, and transistor current is zero.

19. A transistor is in saturation when the bias voltage is increased to such a point that any further increase in bias voltage will not cause any further increase in current through the transistor. In saturation, both the emitter and collector diodes are said to be forward biased, and the transistor is equivalent to a closed switch.

20. Generally the PNP transistor is not employed as much as the NPN transistor in most circuit applications. The only difference that occurs with PNP transistor circuits is that the polarity of V_{CC} and the base bias voltage (V_B) need to be reversed to a negative voltage.

BJT Circuit Configuration Characteristics (Figure 21-35)

21. Although there are many thousands of different bipolar transistor circuits, all of these circuits can be classified into one of three groups based on which of the transistor's leads is used as the "common reference." These three different circuit configurations are the:
    - **a.** Common-emitter (*C-E*), in which the input signal is applied between the base and emitter, while the output signal appears between the transistor's collector and emitter. With this circuit arrangement, the input signal controls the transistor's base current, which in turn controls the transistor's output collector current, and the emitter lead is common to both the input and output.
    - **b.** Common-base (*C-B*), in which the input signal is applied between the transistor's emitter and base, the output signal is developed across the transistor's collector and base, and the base is common to both input and output.
    - **c.** Common-collector (*C-C*), in which the input is applied between the base and the collector, the output is developed across the emitter and the collector, and the collector is common to both the input and output.
22. The current gain in a common-emitter configuration is called the transistor's beta (symbolized β) and is the ratio of output collector current to input base current. The term *hfe* is often used instead of dc current gain (β_{DC}) in manufacturer data sheets.
23. The common-emitter circuit is not only used to increase the level of current between input and output. It can also be used to provide an increase in the amplitude of the input signal voltage or produce a voltage gain (A_V).
24. The common-emitter circuit provides both current gain and voltage gain. Because power is equal to the product of current and voltage ($P = V \times I$), it is not surprising that the *C-E* circuit configuration also provides a high value of power gain (A_P).
25. The set of collector characteristic curves graph shows the relationship between a transistor's input base current, output collector current, and collector-to-emitter voltage drop.
26. The input resistance (R_{in}), or input impedance (Z_{in}), of a transistor is the amount of opposition offered to an input signal by the transistor's input junction.
27. The output resistance (R_{out}), or output impedance (Z_{out}), of a transistor is the amount of opposition offered to an output signal by the transistor's output junction.

FIGURE 21-35 BJT Circuit Configuration Characteristics.

Common-Emitter (C-E) Circuit

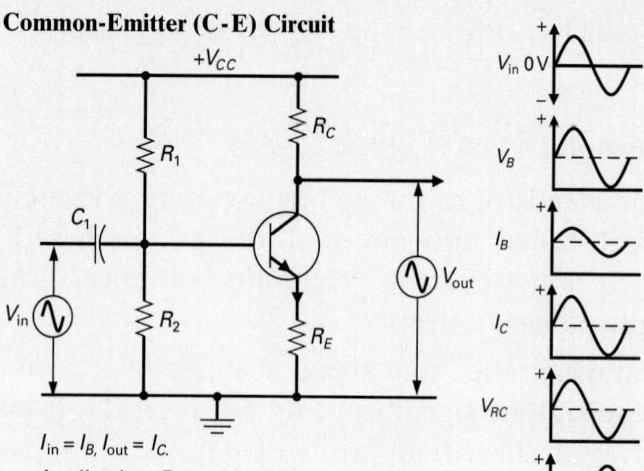

$I_{in} = I_B$, $I_{out} = I_C$.

Application: Power Amplifier or Switch

CHARACTERISTICS

- Current Gain: Medium, 50
- Voltage Gain: High, 100 to 500
- Power Gain: Highest, 200 to 20,000
- Input Resistance: Medium, 1 kΩ to 5 kΩ
- Output Resistance: High, 40 kΩ to 60 kΩ
- Phase Shift: 180°

Current Gain: (DC)$\beta_{DC} = \frac{I_C}{I_B}$, (AC)$\beta_{AC} = \frac{\Delta I_C}{\Delta I_B}$

Voltage Gain: $A_V = \frac{\Delta V_{out}}{\Delta V_{in}}$

Power Gain:

$$A_P = \frac{P_{out}}{P_{in}} = \frac{\Delta V_{out} \times \Delta I_{out}}{\Delta V_{in} \times \Delta I_{in}} = A_V \times \beta_{AC}$$

Common-Base (C-B) Circuit

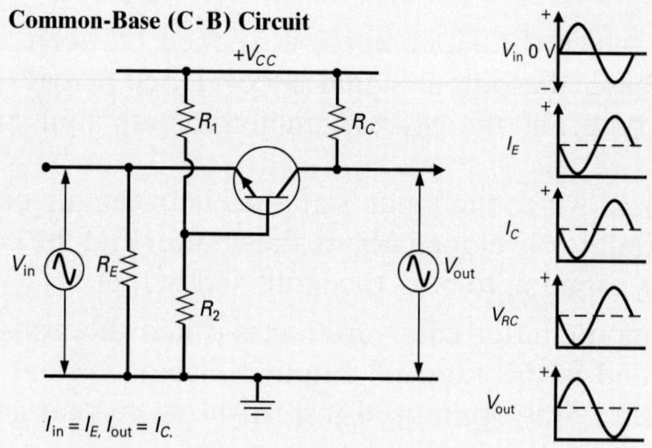

$I_{in} = I_E$, $I_{out} = I_C$.

Application: Voltage Amplifier or Switch

CHARACTERISTICS

- Current Gain: Low, 0.99
- Voltage Gain: Highest, 200 to 2000
- Power Gain: Medium, 200 to 1000
- Input Resistance: Very low, 15 to 150 Ω
- Output Resistance: Very high, 250 kΩ to 1 MΩ
- Phase Shift: 0°

Current Gain: (DC)$\alpha_{DC} = \frac{I_C}{I_E}$, (AC)$\alpha_{AC} = \frac{\Delta I_C}{\Delta I_E}$

Voltage Gain: $A_V = \frac{\Delta V_{out}}{\Delta V_{in}}$

Power Gain:

$$A_P = \frac{P_{out}}{P_{in}} = \frac{\Delta V_{out} \times \Delta I_{out}}{\Delta V_{in} \times \Delta I_{in}} = A_V \times \alpha_{AC}$$

Common-Collector (C-C) Circuit

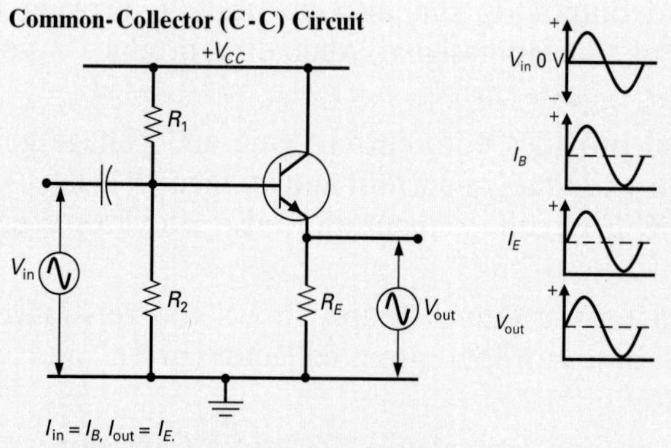

$I_{in} = I_B$, $I_{out} = I_E$.

Application: Current Amplifier or Switch, and Impedance or Resistance Matching Device

CHARACTERISTICS

- Current Gain: Highest, 60
- Voltage Gain: Low, less than 1
- Power Gain: Low, 20 to 80
- Input Resistance: High, 2 kΩ to 500 kΩ
- Output Resistance: Very low, 25 Ω to 1 kΩ
- Phase Shift: 0°

Current Gain: (DC) $= \frac{I_E}{I_B} = 1 + \beta$, AC $= \frac{\Delta I_E}{\Delta I_B}$

Voltage Gain: $A_V = \frac{\Delta V_{out}}{\Delta V_{in}}$

Power Gain:

$$A_P = \frac{P_{out}}{P_{in}} = \frac{\Delta V_{out} \times \Delta I_{out}}{\Delta V_{in} \times \Delta I_{in}}$$

$= A_V \times$ AC Current Gain

Input Resistance: $R_E \times \beta_{AC}$

28. The current gain in a common-base configuration is called the transistor's alpha (symbolized α) and is the ratio of output collector current to input emitter current.
29. Although the common-base circuit configuration does not achieve any current gain, it does have a very high voltage gain.
30. The common-collector circuit provides a slightly higher gain than the *C-E* circuit because the common-collector's output current (I_E) is slightly higher than the *C-E*'s output current (I_C).
31. The common-collector circuit is often referred to as an emitter-follower or voltage-follower because the emitter output voltage seems to track or follow the phase and amplitude of the input voltage.
32. The very high input resistance and low output resistance of the common-collector configuration can be used as a resistance or impedance matching device that can match, or isolate, a high resistance (low current) source to a low resistance (high current) load. By acting as a buffer current amplifier, the *C-C* circuit can ensure that power is efficiently transferred from source to load.

BJT Testing

33. The transistor tester is a special test instrument that can be used to test both NPN and PNP bipolar transistors. This special meter can be used to determine whether an open or short exists between any of the transistor's three terminals, the transistor's dc current gain (β_{DC}), and whether an undesirable value of leakage current is present through one of the transistor's junctions.
34. If the transistor tester is not available, the ohmmeter can be used to detect open or shorted junctions, which are the most common transistor failures.
35. Reverse biasing either the collector or emitter diode of any good transistor should cause the ohmmeter to display a relatively high resistance (several hundred thousand ohms or more). If a reverse biased emitter or collector diode has a low ohms reading, the respective transistor junction can be assumed to be shorted.
36. Forward biasing either the collector or emitter diode of any good transistor should cause the ohmmeter to exhibit a low resistance (several hundred ohms or less). If a forward biased emitter or collector diode has a high ohms reading, the respective transistor junction can be assumed to be open.

BJT Biasing (Figure 21-36)

37. The ac operation of a transistor is determined by the "dc bias level," or "no-input signal level." This steady-state or dc operating level is set by the value of the circuit's dc supply voltage (V_{CC}) and the value of the circuit's biasing resistors. This single supply voltage, and the one or more biasing resistors, set up the initial dc values of transistor current (I_B, I_E and I_C) and transistor voltage (V_{BE}, V_{CE} and V_{BC}).

FIGURE 21-36 BJT Biasing.

Base-Biasing

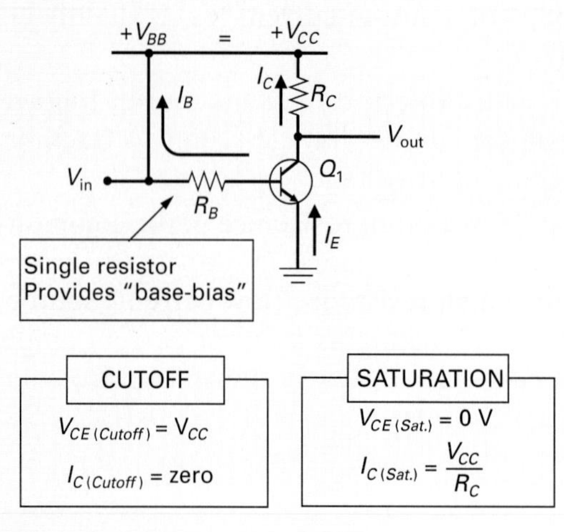

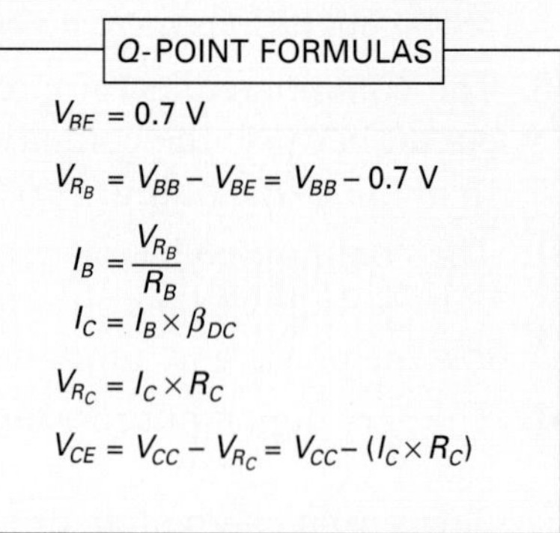

CUTOFF

$V_{CE\,(Cutoff)} = V_{CC}$

$I_{C\,(Cutoff)} = \text{zero}$

SATURATION

$V_{CE\,(Sat.)} = 0\text{ V}$

$I_{C\,(Sat.)} = \frac{V_{CC}}{R_C}$

Advantage: Circuit Simplicity

Disadvantage: Temperature instability causes Q-point shift

Application: Switching (ON/OFF) Circuits.

Voltage-Divider Biasing

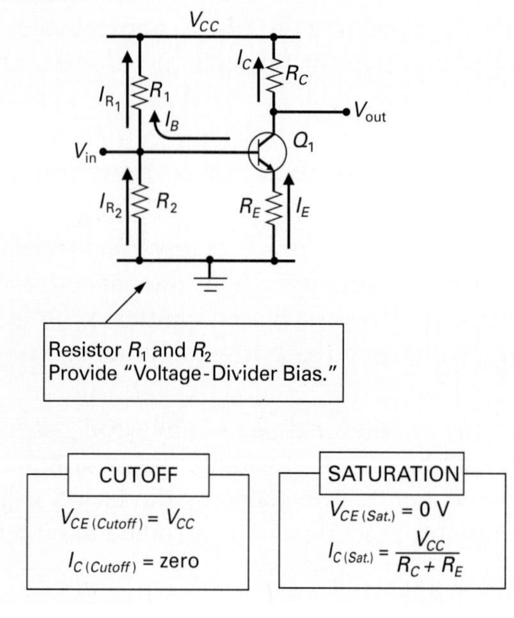

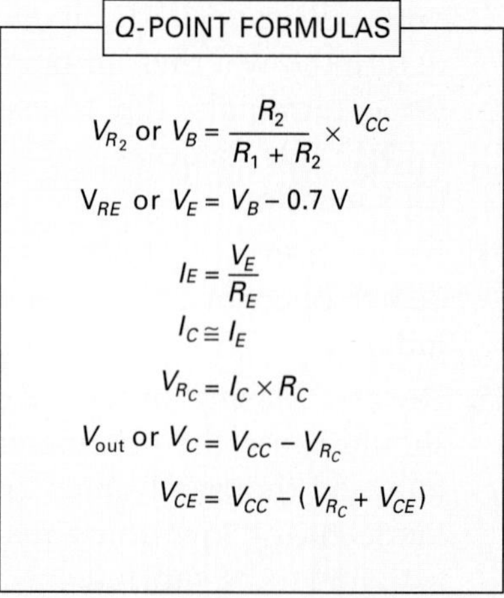

CUTOFF

$V_{CE\,(Cutoff)} = V_{CC}$

$I_{C\,(Cutoff)} = \text{zero}$

SATURATION

$V_{CE\,(Sat.)} = 0\text{ V}$

$I_{C\,(Sat.)} = \frac{V_{CC}}{R_C + R_E}$

Advantage: Good temperature stability

Disadvantage: Needs additional bias resistor

Application: Analog or linear circuits, such as the amplifier

38. With base biasing, the emitter diode of the transistor is forward biased by applying a positive base-bias voltage ($+V_{BB}$) via a current-limiting resistor (R_B) to the base of Q_1.

39. The dc load line is a line plotted on a graph representing all the dc operating points of the transistor for a given load resistance.

40. Base-bias circuits are used in switching circuit applications like the two-state ON/OFF lamp circuit discussed earlier. In these circuits, the bipolar transistor is equivalent to a switch and controlled by a HIGH/LOW input voltage that drives the transistor between the two extremes of cutoff and saturation.
41. The advantage of this biasing technique is circuit simplicity because only one resistor is needed to set the base bias voltage. The disadvantage of the base-biased circuit is that it cannot compensate for changes in its dc bias current due to changes in temperature.
42. Voltage-divider biasing is the most widely used biasing technique. The name of this biasing method comes from the two resistor series voltage-divider (R_1 and R_2) connected to the transistor's base.
43. Voltage-divider bias circuits are used in analog or linear circuit applications like the amplifier. In these circuits, the bipolar transistor is equivalent to a variable-resistor and is controlled by an alternating input signal voltage.
44. Unlike the base-biased circuit, the voltage-divider biased circuit has very good temperature stability due to the emitter resistor R_E which provides emitter feedback.

NEW TERMS

AC Alpha
AC Beta
Active Region
Amplification
Amplifier Circuit
Base
Base Biasing
Base-Collector Diode
Base Current
Base-Emitter Diode
Beta
Bias Voltages
Bipolar Junction Transistor
Bipolar Transistor
BJT
Breakdown Region
Buffer Current Amplifier
Collector
Collector Characteristic Curves
Collector Current
Collector Diode
Common-Base Circuit
Common-Collector Circuit
Common-Emitter Circuit
Configuration
Current Controlled Device
Current Gain
Cutoff Point
Cutoff Region
DC Alpha
DC Beta
DC Load Line
Degenerative Effect
Emitter
Emitter Current
Emitter Diode
Emitter Feedback
Emitter Follower
General Purpose Transistor
High-Current Transistor
High-Power Transistor
High-Voltage Transistor
Hybrid Parameters
Impedance Matching
Input Impedance
Input Resistance
Input Signal Voltage Change
Inverter Gate
NOT Gate
NPN Transistor
Output Impedance
Output Resistance
Output Signal Voltage Change
PNP Transistor
Power Gain
Quiescent Point
Saturation Point
Saturation Region
Transistance

Transistor
Transistor Outline Package
Transistor Tester
Voltage Controlled Switch
Voltage Controlled Variable Resistor
Voltage-Divider Biasing
Voltage Follower
Voltage Gain

REVIEW QUESTIONS

Multiple-Choice Questions

1. The bipolar junction transistor has three terminals called the:

a. Drain, source, gate
b. Anode, cathode, gate
c. Main terminal 1, main terminal 2, gate
d. Emitter, base, collector

2. The term bipolar junction transistor was given to the device because it has:

a. Two P-N junctions **c.** One *p* region and one *n* region
b. Two magnetic poles **d.** Two magnetic junctions

3. An NPN transistor is normally biased so that its base is ____________.

a. Positive **b.** Negative

4. Which is considered the most common bipolar junction transistor configuration?

a. Common base **c.** Common emitter
b. Common collector **d.** None of the above

5. A common-collector circuit is often called a/an ____________.

a. Base follower **c.** Collector follower
b. Emitter follower **d.** None of the above

6. With the NPN transistor schematic symbol, the emitter arrow will point ____________ the base, whereas with the PNP transistor schematic symbol, the emitter arrow will point ____________ the base.

a. Towards, away from **b.** Away from, towards

7. The transistor's ON/OFF switching action is made use of in ____________ circuits.

a. Analog **b.** Digital **c.** Linear **d.** Both (a) and (c) are true

8. The transistor's variable resistor action is made use of in ____________ circuits.

a. Analog **b.** Digital **c.** Linear **d.** Both (a) and (c) are true

9. Approximately 98 percent of the electrons entering the ____________ of a bipolar transistor will arrive at the ____________, and the remainder will flow out of the ____________.

a. Emitter, collector, base **c.** Collector, emitter, base
b. Base, collector, emitter **d.** Emitter, base, collector

10. The common-base circuit configuration achieves the highest ____________ gain, the common-emitter achieves the highest

______________ gain, and the common-collector achieves the highest ______________ gain.

a. Voltage, current, power **c.** Voltage, power, current
b. Current, power, voltage **d.** Power, voltage, current

11. Which of the following abbreviations is used to denote the voltage drop between a transistor's base and emitter?

a. I_{BE} **b.** V_{CC} **c.** V_{CE} **d.** V_{BE}

12. Which of the following abbreviations is used to denote the voltage drop between a transistor's collector and emitter?

a. V_C **b.** V_{CE} **c.** V_E **d.** V_{CC}

13. A transistor's ______________ specification indicates the gain in dc current between the input and output of a common emitter circuit.

a. α_{AC} **b.** α_{DC} **c.** β_{AC} **d.** β_{DC}

14. Consider the following for a base-biased bipolar transistor circuit: R_B = 33 kΩ, R_C = 560 Ω, Q_1 (β_{DC}) = 25, V_{CC} = +10 V. What is V_{BE}?

a. 1.43 mV **c.** 0.7 V
b. 25 × 33 kΩ **d.** Not enough information given to calculate.

15. Which point on the dc load line results in an $I_C = V_{CC}/R_C$ and a V_{CE} = 0 V?

a. Saturation point **b.** Cutoff point **c.** *Q* point **d.** None of the above

16. Which point on the dc load line results in a $V_{CE} = V_{CC}$, and an I_C = 0?

a. Saturation point **b.** Cutoff point **c.** *Q* point **d.** None of the above

17. The midway point on the dc load line at which a transistor is biased with dc voltages when no signal input is applied is called the:

a. Saturation point **b.** Cutoff point **c.** *Q* point **d.** None of the above

18. Which transistor biasing method makes use of one current limiting resistor in the base circuit?

a. Base-Bias **c.** Emitter follower bias
b. Voltage-Divider bias **d.** Current divider bias

19. A forward biased transistor emitter or collector diode should have a ______________ resistance, while a reverse biased emitter and collector diode should have a ______________ resistance.

a. Low, low **b.** High, low **c.** High, high **d.** Low, high

20. A transistor tester will check a transistor's:

a. Opens or shorts between any of the terminals
b. Gain
c. Reverse leakage current value
d. All of the above

Essay Questions

21. What are the two basic transistor types? (21-1-1)

22. Give the full names of the following abbreviations. (Chapter 21)

a. BJT **c.** SMT **e.** V_{CE} **g.** I_B **i.** *C-E* **k.** A_P **m.** R_{in} **o.** ΔV_{out}
b. TO Package **d.** A_V **f.** I_C **h.** V_C **j.** β_{DC} **l.** *Q* point **n.** α_{AC} **p.** $I_{C(Sat.)}$

23. Briefly describe how transistors are constructed and packaged. (21-1-2)
24. Define transistor and transistance. (21-1-3)
25. Briefly describe how the bipolar transistor can be made to operate as a: (21-1-3)

    a. Switch **b.** Variable resistor

26. Briefly describe how the BJT is used in: (21-1-4)

    a. Digital circuit applications **b.** Analog circuit applications

27. Sketch and briefly describe the operation of (21-1-4)

    a. A transistor logic gate **b.** A transistor amplifier

28. Briefly describe how a transistor achieves a gain in voltage between input and output. (21-1-4)
29. What is the relationship between a bipolar transistor's emitter current, base current, and collector current? (21-2-1)
30. Why is the bipolar transistor known as a current controlled device? (21-2-1)
31. In normal operation, which of the bipolar transistor's P-N junctions or diodes is forward biased, and which is reverse biased? (21-2-1)
32. What is the bipolar transistor equivalent to when it is operated in: (21-2-1)

    a. Saturation **b.** The active region **c.** Cutoff

33. What are the three bipolar transistor circuit configurations? (21-2-2)
34. Briefly describe the following terms: (21-2-2)

    a. DC beta **c.** DC alpha
    b. AC beta **d.** AC alpha

35. What is a collector characteristic curve? (21-2-2)
36. List the key characteristics of each of the three transistor configurations. (21-2-2)
37. Why is the common-emitter circuit configuration the most widely used? (21-2-2)
38. Why is the common-collector circuit configuration well suited as an impedance matching device? (21-2-2)
39. Why is impedance matching a desirable condition? (21-2-2)
40. What characteristics of a transistor will a transistor tester check? (21-2-4)
41. Use a sketch to show how an ohmmeter can be used to check a transistor's P-N junctions for opens and shorts. (21-2-4)
42. Sketch a base-biased common-emitter circuit. (21-2-5)
43. What is the disadvantage of a base-biased circuit? (21-2-5)
44. Sketch a voltage-divider biased: (21-2-5)

    a. Common-emitter circuit
    b. Common-base circuit
    c. Common-collector circuit

45. How does an emitter resistor prevent Q point shift due to temperature change? (21-2-5)

Practice Problems

46. Identify the type and terminals of the transistors shown in Figure 21-37.
47. A bipolar transistor is correctly biased for operation in the active region when its emitter diode is forward biased and its collector diode is reverse biased. Referring to Figure 21-38, which of the bipolar transistor circuits is correctly biased?
48. Identify which of the leads of the bipolar transistor packages in Figure 21-39 is the emitter.

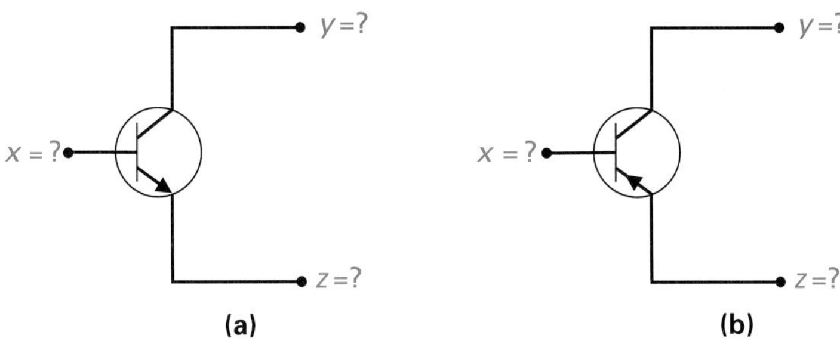

FIGURE 21-37 **Identify the Transistor Type and Terminals.**

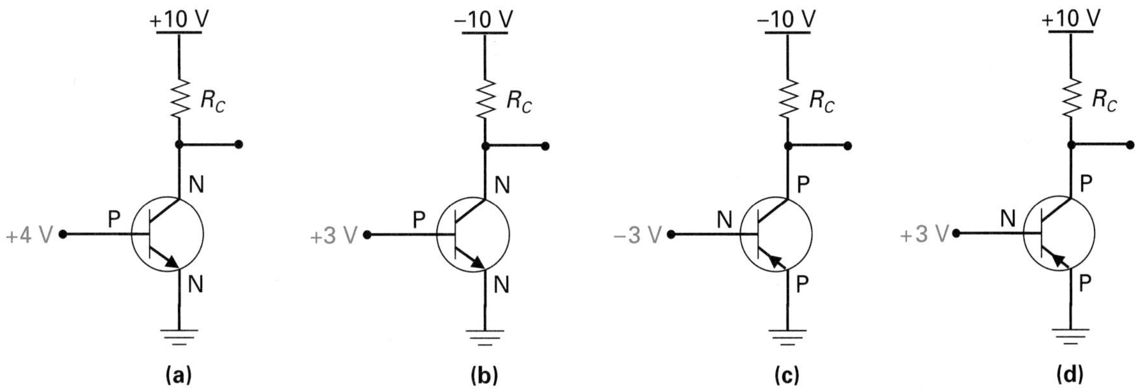

FIGURE 21-38 **Identifying the Correctly Biased (Active Region) Bipolar Transistors.**

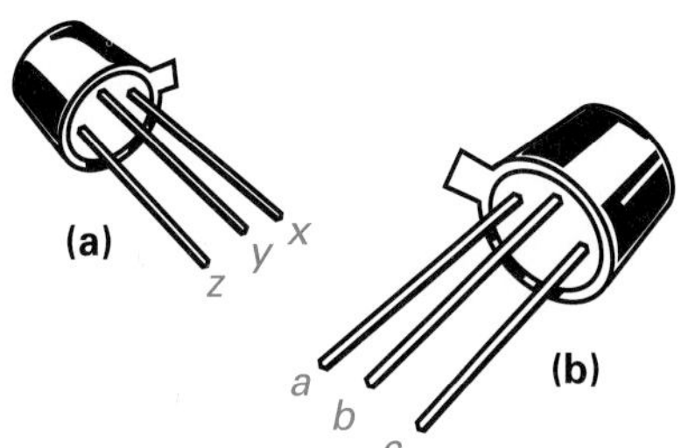

FIGURE 21-39 **Identifying the Bipolar Transistor's Emitter Lead.**

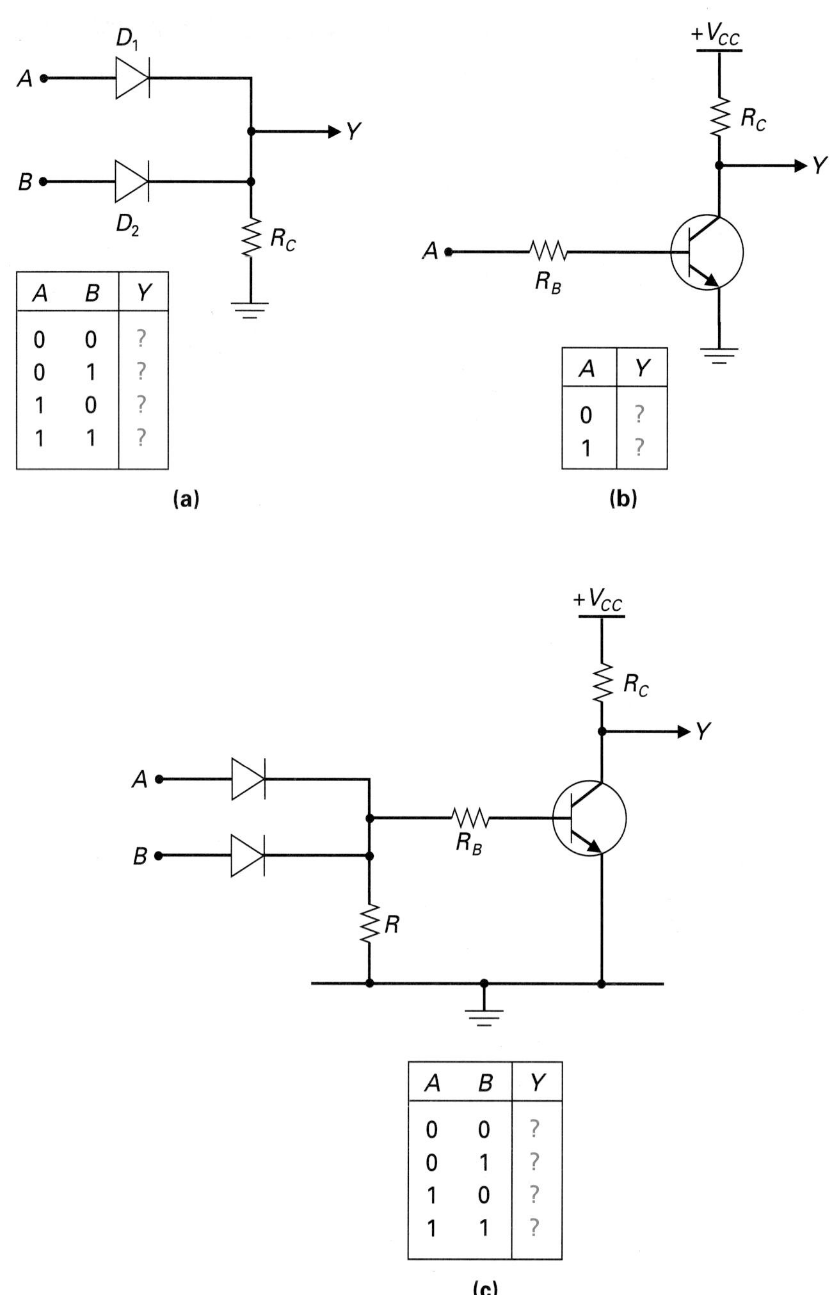

FIGURE 21-40 **Digital Logic Gate Circuits (Logic 0 = 0 V, Logic 1 = +5 V).**

49. Digital circuits are often referred to as switching circuits because their control devices (diodes or transistors) are switched between the two extremes of ON and OFF. These digital circuits are also called two-state circuits because their control devices are driven into either the saturation state (ON) or cutoff state (OFF). Figure 21-40 illustrates three digital logic gates. Complete their function tables by indicating what logic level will be at the output *Y* for each of the input combinations.

50. Calculate the value of the missing current in the following examples:

a. $I_E = 25$ mA, $I_C = 24.6$ mA, $I_B = ?$
b. $I_B = 600$ µA, $I_C = 14$ mA, $I_E = ?$
c. $I_E = 4.1$ mA, $I_B = 56.7$ µA, $I_C = ?$

51. Calculate the value of the missing current in the examples in Figure 21-41.

52. Calculate the dc beta for the transistor circuits in Figure 21-41.

53. Calculate the voltage gain (A_V) of the transistor amplifier whose input/output waveforms are shown in Figure 21-42.

54. Identify the configuration of the actual bipolar transistor electronic system circuits shown in Figure 21-43.

55. Identify the bipolar transistor type, and the biasing technique used in Figure 21-43.

56. Calculate the following for the base biased transistor circuit shown in Figure 21-44:

a. I_B **b.** I_C **c.** V_{CE}

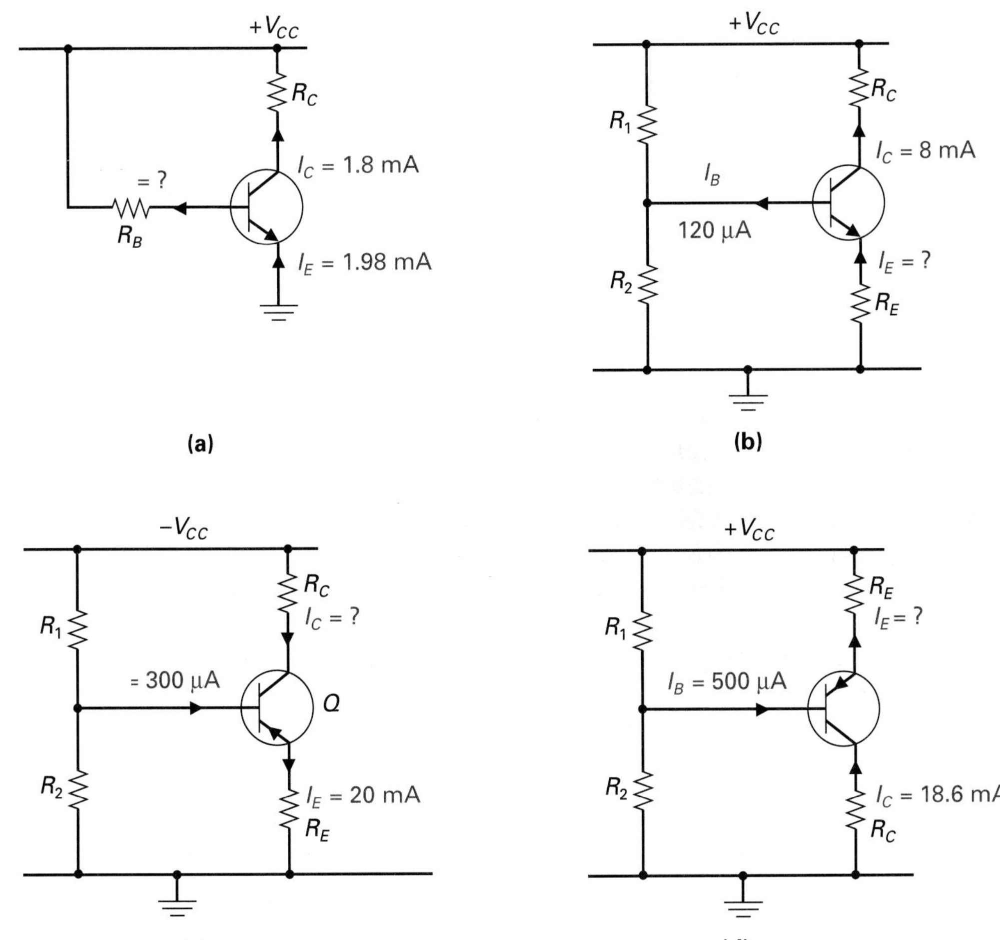

FIGURE 21-41 **Calculating Transistor Current Values.**

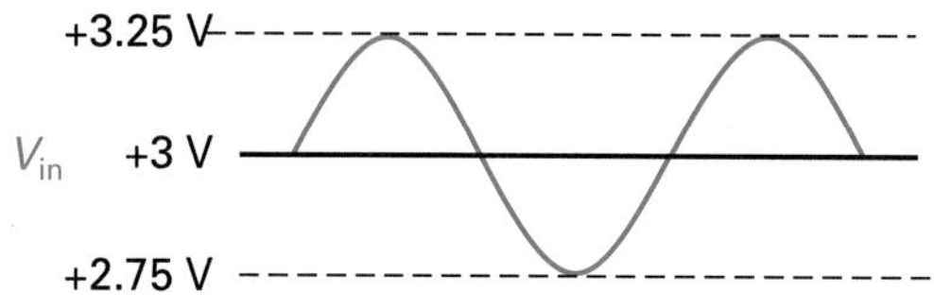

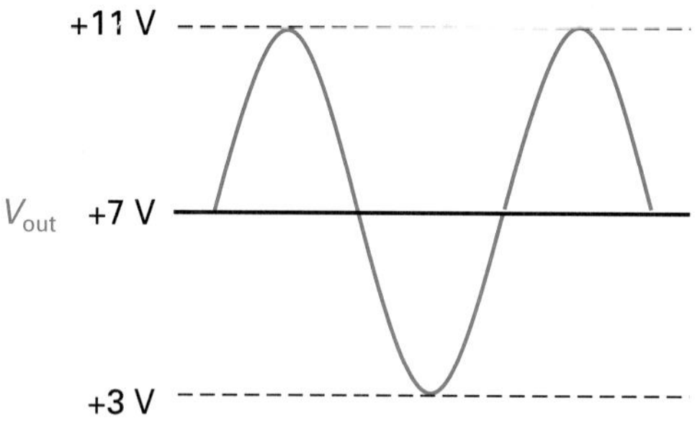

FIGURE 21-42 **Transistor Amplifier Input/Output Waveforms.**

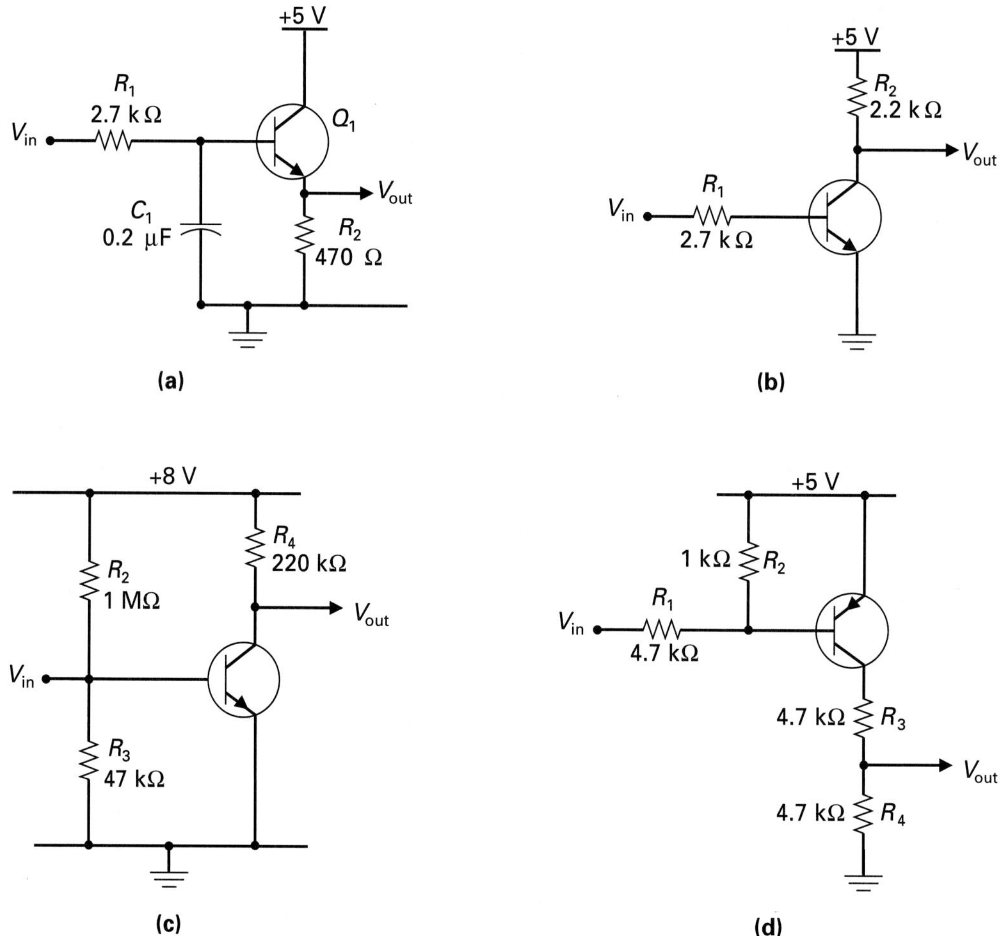

FIGURE 21-43 **Identifying the Configuration of Actual BJT Circuits.**

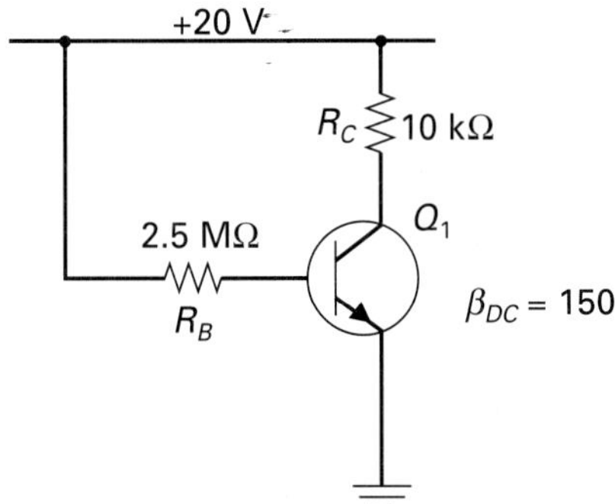

FIGURE 21-44 **Base Biased Transistor Circuit.**

57. Sketch the dc load line for the circuit in Figure 21-44, showing the saturation, cutoff, and Q points.

58. Calculate the following for the voltage-divider biased transistor circuit shown in Figure 21-45:

a. V_B and V_E **b.** I_C **c.** V_C **d.** V_{CE}

59. Sketch the dc load line for the circuit in Figure 21-45, showing the saturation, cutoff, and Q points.

60. In Figure 21-45, does Kirchhoff's voltage law apply to the voltage-divider made up of R_C, R_{CE}, and R_E?

Troubleshooting Questions

61. Briefly describe what characteristics a transistor tester can check.

62. How can an ohmmeter be used to test transistors?

63. Which of the transistors being tested in Figure 21-46 are good or bad? If bad, state the suspected problem.

64. Figure 21-47 shows a base biased circuit with its normal circuit values. Determine what "effect" the two faults in Figure 21-47(a) and (b) will have

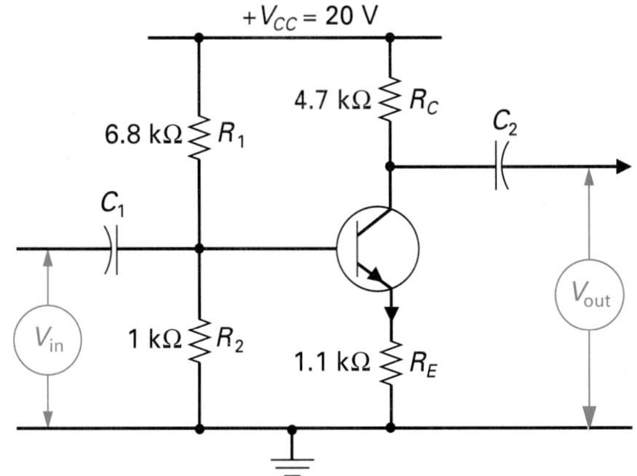

FIGURE 21-45 **Voltage-Divider Biased Transistor Circuit.**

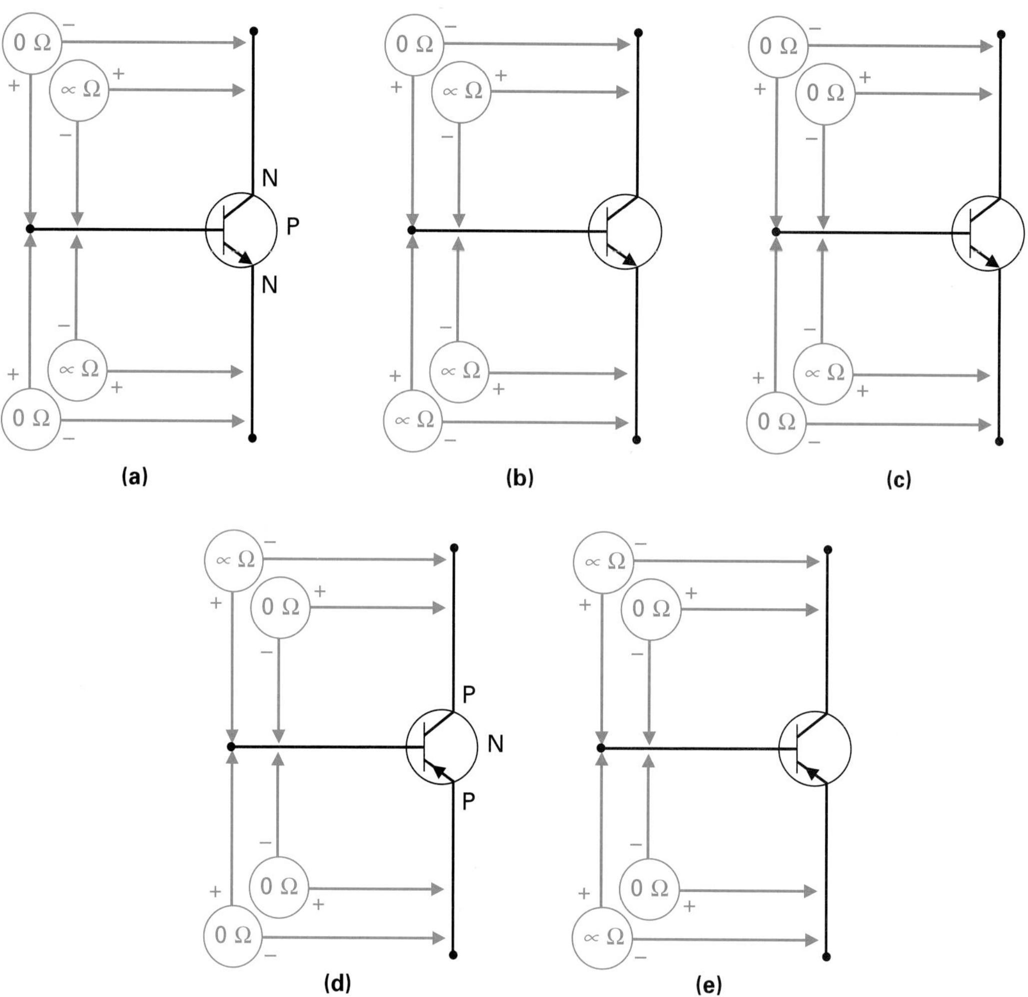

FIGURE 21-46 **Testing Transistors with the Ohmmeter.**

on this circuit. Describe the logic behind the cause and effect, and list other circuit faults that could possibly have the same effect.

65. Figure 21-48 shows a voltage-divider biased circuit with its normal circuit values. Determine what you think is the "root cause" of the three faults in Figure 21-48 (a), (b), and (c), based on the voltage readings given. Describe the logic behind the cause and effect, and list other circuit faults that could possibly have the same effect.

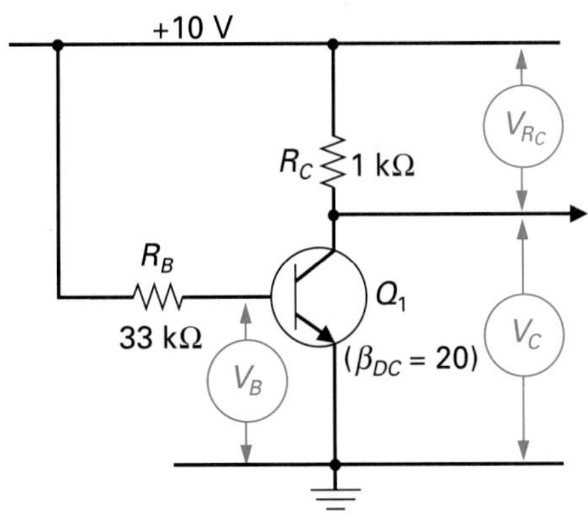

Under Normal Operating Conditions:

$V_{BE} = 0.7$ V $I_B = 282$ μA
$V_{R_B} = 9.3$ V $I_C = 5.6$ mA
$V_{R_C} = 5.6$ V $I_{C(Sat.)} = 10$ mA
$V_{CE} = 4.4$ V $I_{B(Sat.)} = 500$ μA

Cause—Q_1 Base resistor is open

Effect— Q_1 =
V_B =
V_C =
V_{R_C} =

(a)

Cause—Emitter diode is open

Effect— Q_1 =
V_B =
V_C =
V_{R_C} =

(b)

FIGURE 21-47 Troubleshooting Base Biased Circuit Problems.

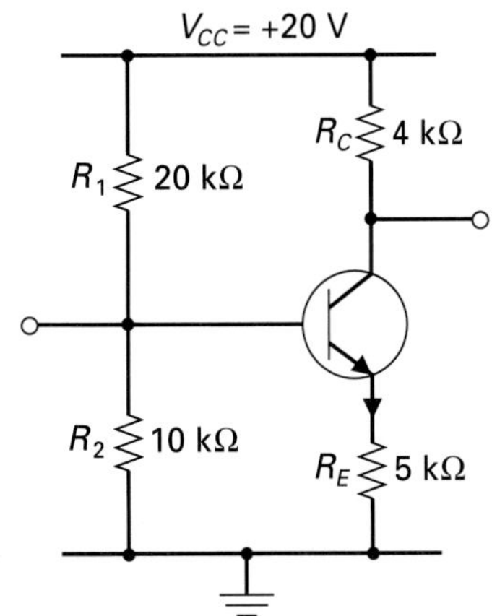

Under Normal Operating Conditions:

$V_B = 6.7$ V $I_E = 1.2$ mA
$V_E = 6.0$ V $I_C = 1.2$ mA
$V_{R_C} = 4.8$ V $I_{C(Sat.)} = 2.2$ mA
$V_C = V_{out} = 15.2$ V
$V_{CE} = 9.2$ V

Effect— $V_B = 0$ V
$V_E = 0$ V
$V_C = +20$ V

Root Cause—?

(a)

Effect— $V_B = 11.7$ V
$V_E = 11.0$ V
$V_C = 11.0$ V

Root Cause—?

(b)

Effect— $V_B = 6.7$ V
$V_E = 0$ V
$V_C = +20$ V

Root Cause—?

(c)

FIGURE 21-48 Troubleshooting Voltage-Divider Bias Circuits.

CHAPTER

22

OBJECTIVES

After completing this chapter, you will be able to:

1. Name the two different types of field effect transistors.

2. Describe the physical construction and operation of the junction FET (JFET), and name and identify its three terminals.

3. Describe why the term "field effect" is used in the name of an FET and why the term "junction" is used in JFET.

4. Explain the JFET operation, and the following characteristics:
   a. V_P, V_{BR}, I_{DSS}, $V_{GS(OFF)}$
   b. Transconductance
   c. High-input impedance

5. Interpret the JFET specifications given in a typical manufacturer's data sheet.

6. Calculate the different values of circuit voltage and current for the following transistor biasing methods:
   a. Gate Biasing
   b. Self Biasing
   c. Voltage-Divider Biasing

7. Identify and list the characteristics of the following three FET circuit configurations:
   a. Common Source
   b. Common Gate
   c. Common Drain

8. Explain how the JFET's characteristics can be used in digital and analog circuit applications.

9. Describe how to test a JFET and how to troubleshoot JFET circuits.

10. Explain the meaning of the term "metal oxide semiconductor field effect transistor."

11. Define the terms "depletion-mode operation" and "enhancement-mode operation."

12. Describe how the D-MOSFET is constructed, how it operates, and its characteristics.

13. Explain D-MOSFET biasing and typical applications.

14. Discuss the operation, characteristics, and application of the dual-gate D-MOSFET.

15. Describe how the E-MOSFET is constructed, how it operates, and its characteristics, and biasing.

16. List the typical applications of the E-MOSFET.

Field Effect Transistors (FETs)

OUTLINE

VIGNETTE

Spitting Lighting Bolts

Nikola Tesla was born in Yugoslavia in 1856. He studied mathematics and physics in Prague, and in 1884 he emigrated to the United States.

In New York he met Thomas Edison, the self-educated inventor who is best known for his development of the phonograph and the incandescent light bulb. Both men were gifted and eccentric and, due to their common interest in "invention," they got along famously. Because Tesla was unemployed, Edison offered him a job. In his lifetime, Edison would go on to take out 1,033 patents and become one of the most prolific inventors of all time. Tesla would go on to invent many different types of motors, generators, and transformers, one of which is named the "Tesla coil" and produces five-foot lighting bolts. With this coil, Tesla investigated "wireless power transmission," the only one of his theories that has not come into being.

Both Tesla and Edison had very strong views on different aspects of electricity and, as time passed, the two men began to engage in very long, loud, and angry arguments. One such discussion concerned whether power should be distributed as alternating current or direct current. Eventually, the world would side with Tesla and choose ac. At the time, this topic, like many others, would cause a hatred to develop between the two men. Eventually Tesla left Edison and started his own company. However, the anger remained, and on one occasion when they were both asked to attend a party for their friend Mark Twain, both refused to come because the other had been invited.

In 1912, Tesla and Edison were both nominated for the Nobel prize in physics, but because neither one would have anything to do with the other, the prize went to a third party—proving that bitterness really will cause a person to cut off his nose to spite his face.

When angry count up to four; when very angry, swear.
Mark Twain

INTRODUCTION

In the previous chapters, we have concentrated on all aspects of the bipolar junction transistor or BJT. We have examined its construction, operation, characteristics, testing, and basic circuit applications. In this chapter we will examine another type of transistor called the *field effect transistor,* which is more commonly called an FET (pronounced "eff-ee-tee"). Like the BJT, the FET has three terminals and can operate as a switch and can be used in digital circuit applications. It can also operate as a variable resistor and be used in analog or linear circuit applications. In fact, as we step through this chapter, you will see many similarities between the BJT and FET. You will also notice a few distinct differences between these two transistor types, and these differences are what make the BJT ideal in some applications and the FET ideal in other applications.

There are two types of field effect transistors or FETs. One type is the *junction field effect transistor,* which is more typically called a JFET (pronounced "jay-fet"). The other type is the *metal oxide semiconductor field effect transistor,* which is more commonly called a MOSFET (pronounced "moss-fet"). In this chapter we will examine the operation, characteristics, applications, and testing of these two types of field effect transistors..

22-1 JUNCTION FIELD EFFECT TRANSISTOR (JFET)

Like the bipolar junction transistor, the **junction field effect transistor** or **JFET** is constructed from *n*-type and *p*-type semiconductor materials. However, the JFET's construction is very different from the BJT's construction, and therefore we will need to first see how the JFET device is built before we can understand how it operates.

Junction Field Effect Transistor (JFET)
A field-effect transistor made up of a gate region diffused into a channel region. When a control voltage is applied to the gate, the channel is depleted or enhanced, and the current between source and drain is thereby controlled.

22-1-1 JFET Construction

Just as the bipolar junction transistor has two basic types (NPN BJT or PNP BJT), there are two types of junction field effect transistor called the ***n*-channel JFET** and ***p*-channel JFET.** The construction and schematic symbol for these two JFET types are shown in Figure 22-1.

To begin with, let us examine the construction of the more frequently used *n*-channel JFET, shown in Figure 22-1(a). This type of JFET basically consists of an *n*-type block of semiconductor material on top of a *p*-type substrate, with a "U" shaped *p*-type section attached to the surface of a *p*-type substrate. Like the BJT, the JFET has three terminals called the **gate, source,** and **drain.** The gate lead is attached to the *p*-type substrate, and the source and drain leads are attached to either end of an *n*-type **channel** that runs through the middle of the "U" shaped *p*-type section. In the simplified two-dimensional view in the inset in Figure 22-1(a), you can see the *n*-type channel that exists between the *n*-channel JFET's source and drain. The schematic symbol for the *n*-channel JFET is shown in Figure 22-1(b), and, as you can see, the gate lead's arrowhead points into the device. To aid your memory, you can imagine this arrowhead as a P-N junction diode as shown in the inset in Figure 22-1(b). The gate lead is connected to the diode's anode, which is a *p*-type material, and the source and drain leads are connected to either end of the diode's cathode, which is an *n*-type material. Because the source-to-drain channel is made from an *n*-type material, this is an *n*-channel JFET. The *p*-type gate and *n*-type source and drain makes this *n*-channel JFET equivalent to an NPN BJT, which has a *p*-type base and *n*-type emitter and collector, as shown in the inset in Figure 22-1(b).

***n*-Channel JFET**
A junction field effect transistor having an *n*-type channel between source and drain.

***p*-Channel JFET**
A junction field effect transistor having a *p*-type channel between source and drain.

Gate
One of the field effect transistor's electrodes (also used for thyristor devices).

Source
One of the field effect transistor's electrodes.

Drain
One of the field effect transistor's electrodes.

Channel
A path for a signal.

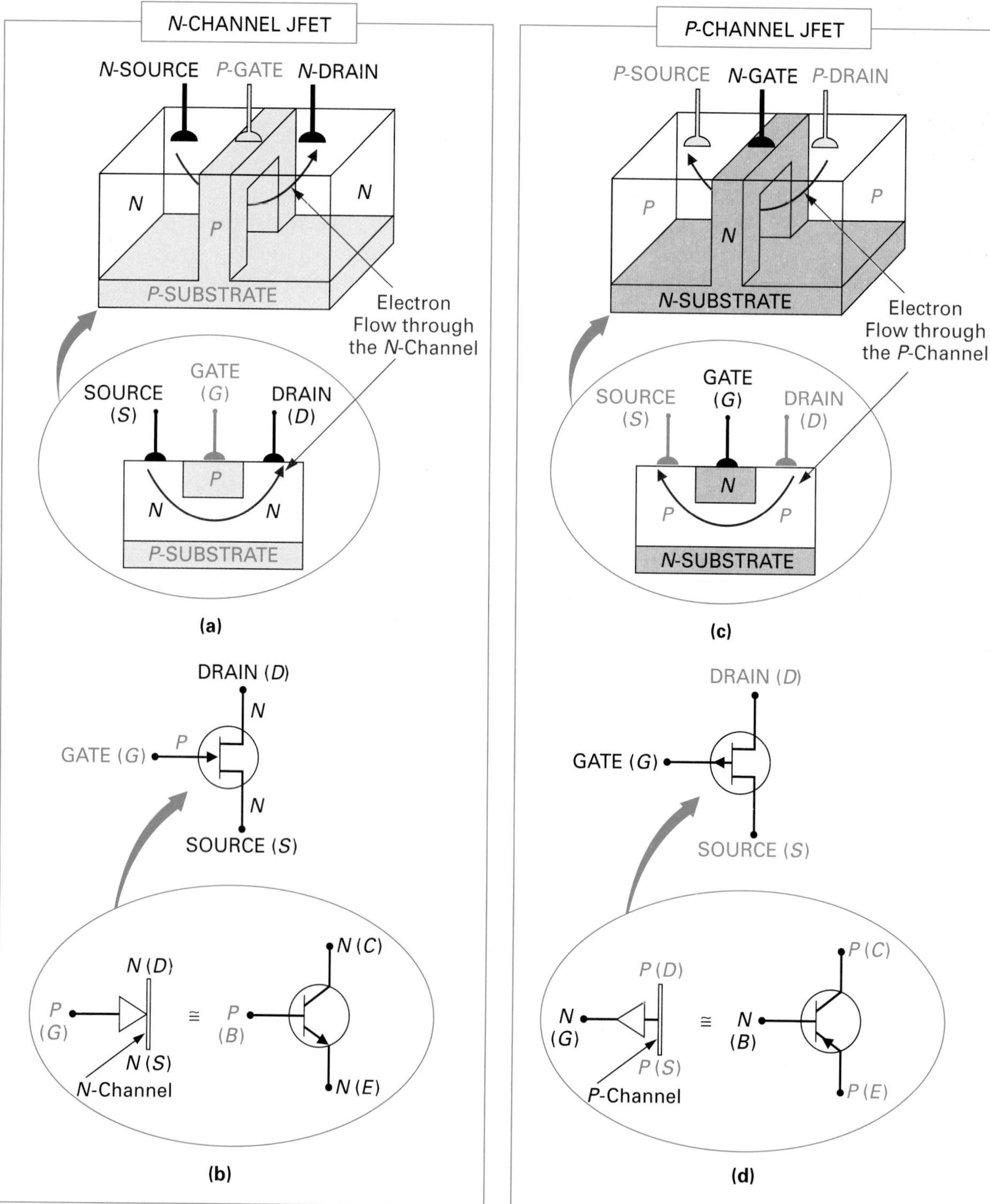

FIGURE 22-1 **The Junction Field Effect Transistor (JFET) Types.**

Figure 22-1(c) shows the construction of the *p*-channel JFET. This JFET type is constructed in exactly the same way as the *n*-channel JFET except that the gate lead is attached to an *n*-type substrate, and the source and drain leads are attached to either end of a *p*-type channel. Looking at the schematic symbol for the *p*-channel JFET in Figure 22-1(d), you can see that the gate's arrowhead points out of the device. To help you distinguish the symbols used for the *n*-channel JFET from the *p*-channel JFET, once again imagine the arrowhead as a P-N junction diode as shown in the inset in Figure 22-1(d). Because the gate lead is connected to the diode's cathode, the gate must be an *n*-type material. Because the source and drain leads are connected to either end of the diode's cathode, the source-to-drain channel is therefore made from a *p*-type material, and this is a *p*-channel JFET. The *n*-type gate and *p*-type source and drain makes this *p*-channel JFET equivalent to a PNP BJT, which has an *n*-type base and *p*-type emitter and collector, as shown in the inset in Figure 22-1(d).

22-1-2 JFET Operation

As you know, an NPN bipolar transistor needs both a collector supply voltage ($+V_{CC}$) and a base-emitter bias voltage (V_{BE}) in order to operate correctly. The same is true for the JFET, which requires both a **drain supply voltage ($+V_{DD}$)** and a **gate-source bias voltage (V_{GS})**, as shown in Figure 22-2(a). The $+V_{DD}$ bias voltage is connected between the drain and source of the *n*-channel JFET and will cause a current to flow through the *n*-channel. This source-to-drain current—which is made up of electrons because they are the majority carriers within an n-type material—is called the JFET's **drain current (I_D)**. The value of drain current passing through a JFET's channel is dependent on two elements: the value of $+V_{DD}$ applied between the drain and source, and the value of V_{GS} applied between gate and source. Let us examine in more detail why these applied voltages control the value of drain current passing through the JFET's channel.

Drain-Supply Voltage ($+V_{DD}$) The bias voltage connected between the drain and source of the JFET, which causes current to flow.

Gate-Source Bias Voltage (V_{GS}) The bias voltage applied between the gate and source of a field effect transistor.

Drain Current (I_D) A JFET's source-to-drain current.

The Relationship between + V_{DD} and I_D (Figure 22-2)

The value of $+V_{DD}$ controls the amount of drain current between source and drain because it is this supply voltage that controls the potential difference applied across the channel, as seen in Figure 22-2(a). Therefore, an increase in the voltage applied across the JFET's drain and source ($+V_{DD}\uparrow$) will increase the amount of drain current ($I_D\uparrow$) passing through the channel. Similarly, a decrease in the voltage applied across the JFET's drain and source ($+V_{DD}\downarrow$) will decrease the amount of drain current ($I_D\downarrow$) passing through the channel. In most cases, a schematic diagram will show the V_{DD} supply voltage connection to the JFET as a source connection to ground and a drain connection up to $+V_{DD}$, as shown in Figure 22-2(b).

The Relationship between V_{GS} and I_D (Figure 22-3)

The value of V_{GS} controls the amount of drain current between source and drain because it is this voltage that controls the resistance of the channel. Figure 22-3(a) shows that when the V_{GS} bias voltage is 0 volts (which is the same as

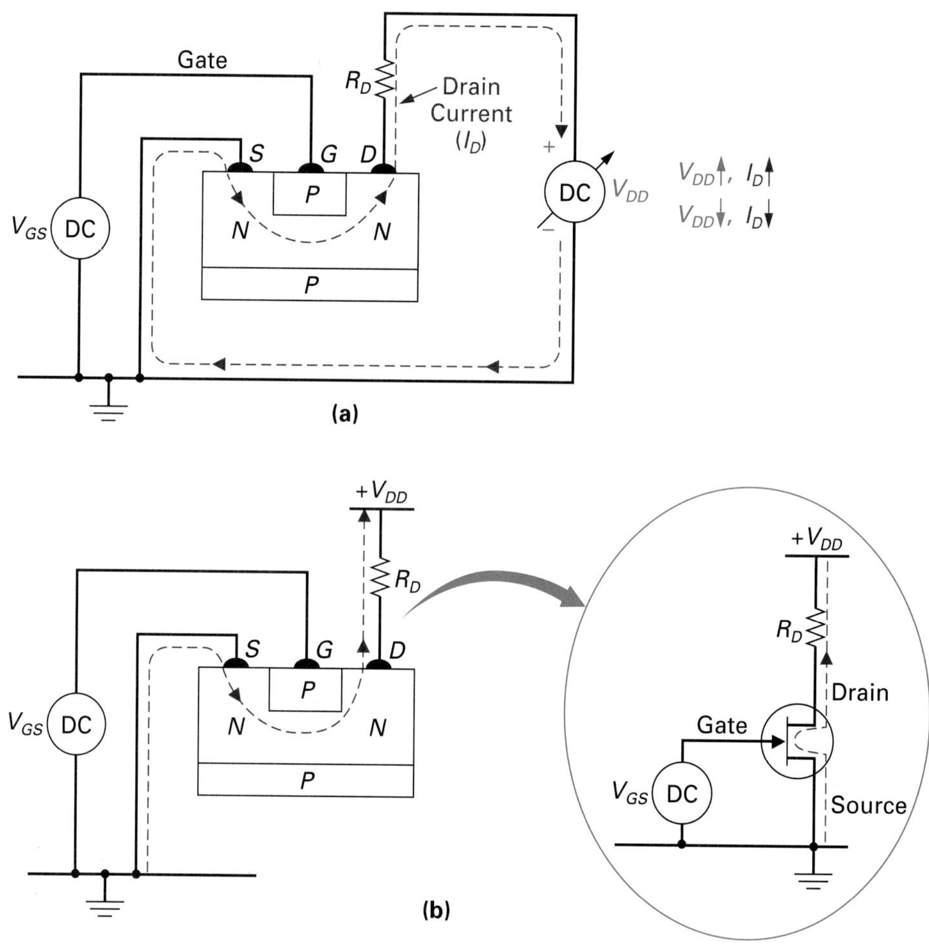

FIGURE 22-2 The Relationship Between a JFET's DC Supply Voltage ($+V_{DD}$) and Output Current (I_D).

connecting the gate to ground), there is no potential difference between the gate and source, and so the gate-to-source P-N junction will be reverse biased. As with all reverse-biased junctions, a small depletion layer will form and spread into the channel. Although it appears as though two depletion regions exist, in fact they are both part of the same depletion region that extends around the wall of the *n*-channel. This extremely small depletion region will offer very little opposition to I_D, and so drain current will be large. In Figure 22-3(b), V_{GS} is increased to −2 V (made more negative), and therefore the gate-to-source P-N junction will be further reverse biased. This causes an increase in the depletion region, decrease in the channel's width, and a decrease in drain current. In Figure 22-3(c), V_{GS} is increased to −4 V, and therefore the gate-to-source P-N junction will be further reverse biased, causing an increase in the depletion region, decrease in the channel's width, and a further decrease in drain current.

In most circuit applications, the $+V_{DD}$ supply voltage is maintained constant and the V_{GS} input voltage is used to control the resistance of the channel and the value of the output current, I_D. This can be seen more clearly in the

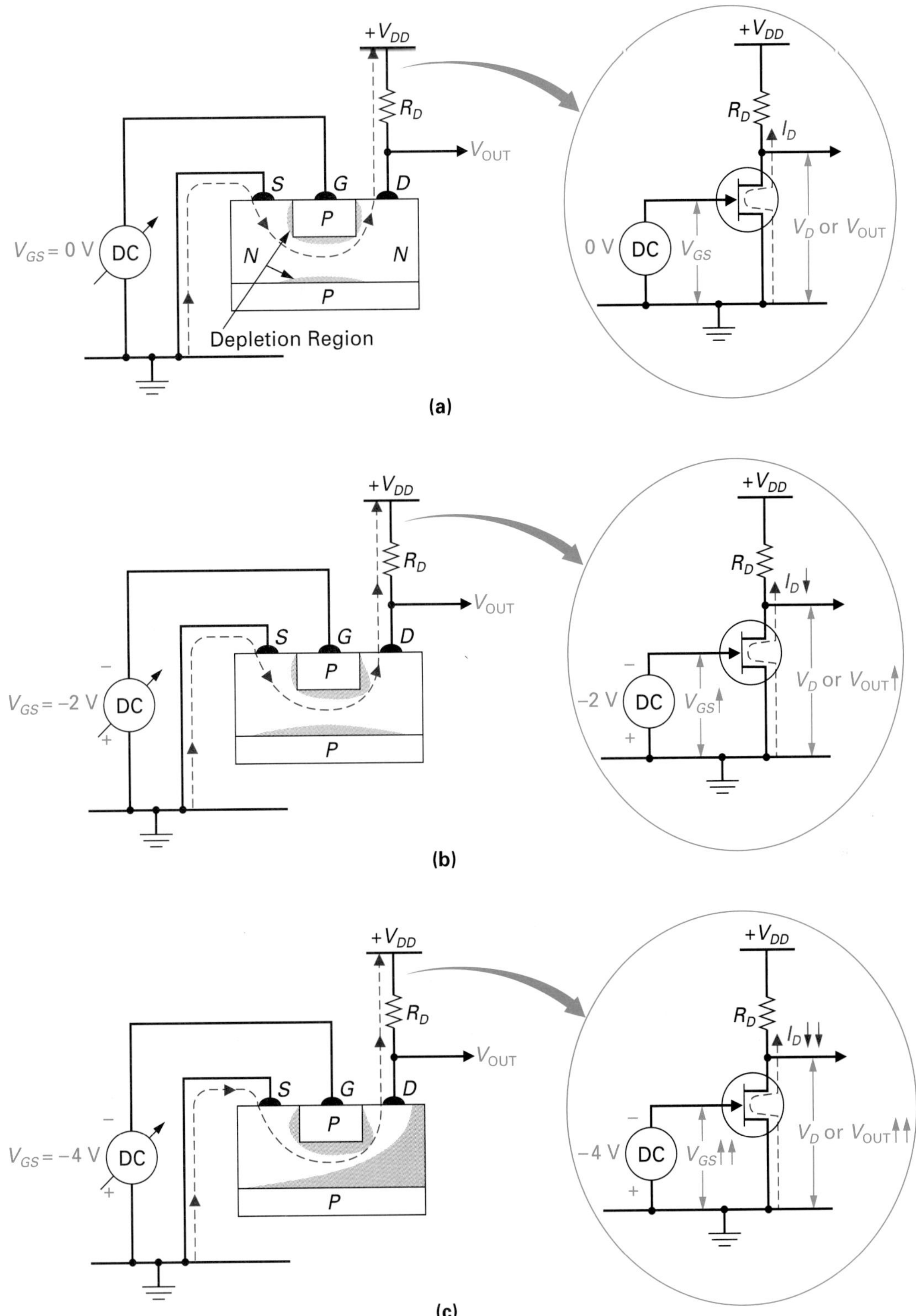

FIGURE 22-3 The Relationship Between a JFET's Input Voltage (V_{GS}) and Output Current (I_D).

insets in Figure 22-3. Because the output voltage (V_D or V_{OUT}) is dependent on the resistance between the JFET's source and drain, by controlling the value of I_D, we can control the output voltage. For instance, if the input voltage V_{GS} is made more negative, the resistance of the channel will be increased, causing the output current I_D to decrease, and therefore the voltage developed between drain and ground (V_D or V_{OUT}) to increase. The gate-to-source junction of an FET is normally always reverse biased by the input voltage (V_{GS}), and it is this input voltage that controls the output current (I_D) and output voltage (V_{OUT}). Because the gate-to-source junction of an FET is normally always reverse biased by the input voltage (V_{GS}), there will be no input current. This characteristic accounts for the FET's naturally high input impedance. The operation of an FET is very different from the BJT, which normally uses an input voltage to forward bias the base-to-emitter junction and vary the input current (which varies the output current), and therefore the output voltage. This is the distinct difference between an FET and a BJT. An FET's input junction is normally reverse biased, and therefore the input voltage controls the output current. A BJT's input junction is normally forward biased and therefore the input current controls the output current. This difference is why BJTs are known as **current-controlled devices** and FETs are known as **voltage-controlled devices.** In fact, the name "field-effect transistor" is derived from this voltage control action because the applied input voltage will generate an electric field. It is this electric field that varies the size of the depletion region, and therefore the resistance of the channel between the FET's drain and source output terminals. In other words, the "effect" of the electric "field" causes "transistance," which is the transferring of different values of resistance between the output terminals.

Current-Controlled Devices
A device in which the input junction is normally forward biased and the input current controls the output current.

Voltage-Controlled Devices
A device in which the input junction is normally reverse biased and the input voltage controls the output current.

The term "junction" is attached to this type of FET because of the single P-N junction formed between the gate and the source-to-drain channel. Therefore, an *n*-channel JFET has a single P-N junction between gate to channel, and a *p*-channel JFET has a single N-P junction between gate to channel.

The field effect transistor is also often referred to as a **unipolar device** because only one type of semiconductor material exists between the output terminals (*n*-type or *p*-type channel between source and drain), and therefore the charge carriers have only one polarity (unipolar). Compare this to a BJT, which is a **bipolar device** because there is a change in semiconductor material between the output terminals (NPN or PNP between emitter and collector), and the charge carriers can be one of two polarities (bipolar, because both majority and minority carriers are used).

Unipolar Device
A device in which only one type of semiconductor material exists between the output terminals and therefore the charge carriers have only one polarity (unipolar).

Bipolar Device
A device in which there is a change in semiconductor material between the output terminals (NPN or PNP between emitter and collector), so the charge carriers can be one of two polarities (bipolar).

22-1-3 JFET Characteristics

Like the BJT, the JFET's response to certain variables is best described by using a graph. Figure 22-4(a) shows a graph plotting drain current (I_D) against drain-to-source voltage (V_{DS}). As you have probably already observed, this **drain characteristic curve** is very similar to a bipolar transistor's collector characteristic curve. Starting at 0 V and moving right along the horizontal axis, you can see that an increase in the drain supply voltage ($+V_{DD}$), and therefore an increase in V_{DS}, will result in a continual increase in I_D. At a certain V_{DS} voltage (in this

Drain Characteristic Curve
A plot of the drain current (I_D) versus the drain-to-source voltage (V_{DS}).

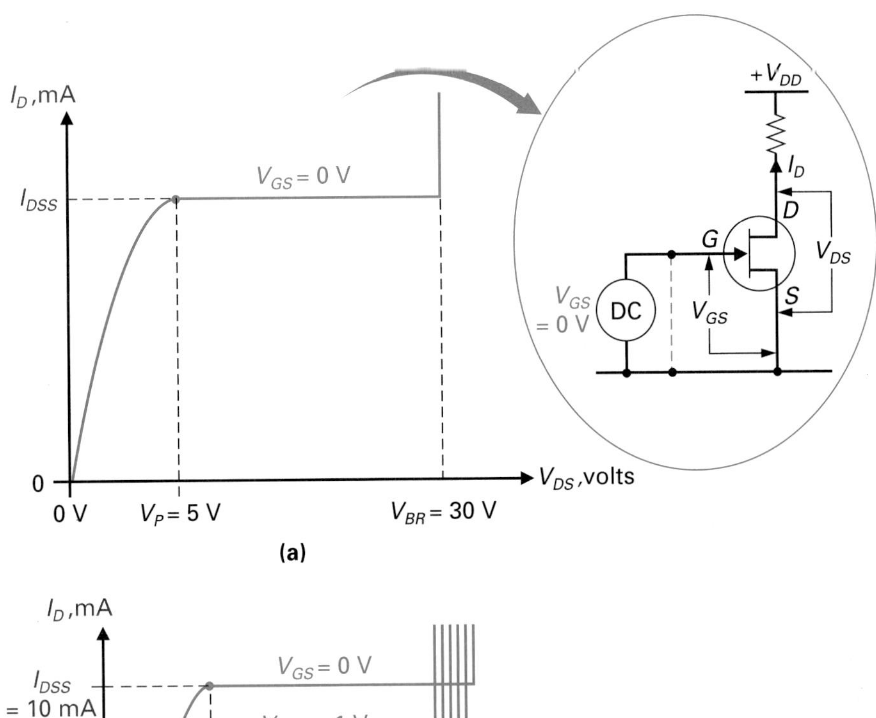

(a)

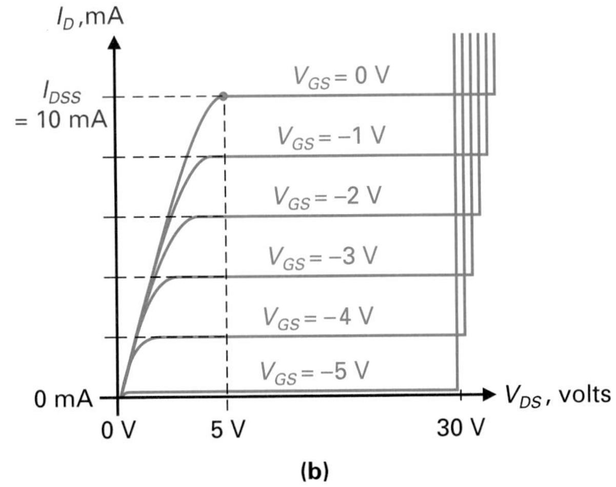

(b)

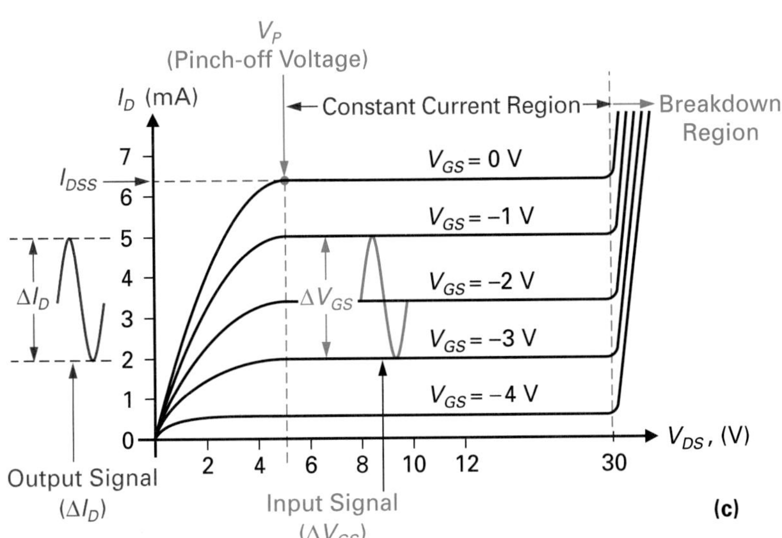

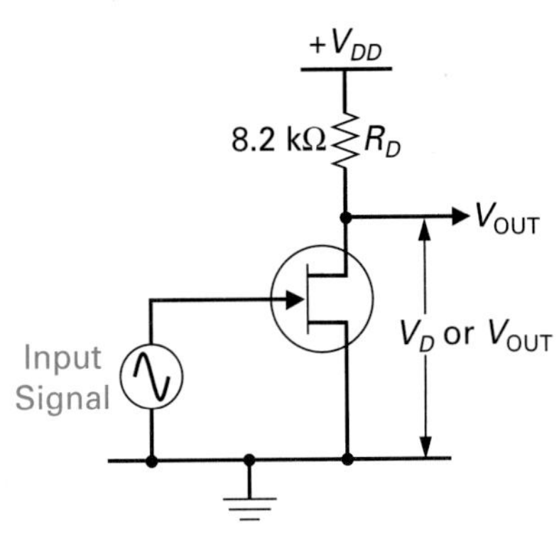

(c)

FIGURE 22-4 **JFET Characteristics.**

Pinch-Off Voltage (V_P)
The value of V_{DS} at which further increases in V_{DS} will cause no further increase in I_D.

Constant-Current Region
The flat portion of the drain characteristic curve. In this region I_D remains constant despite changes in V_{DS}

Breakdown Voltage (V_{BR})
The voltage at which a damaging value of I_D will pass through the JFET.

Drain-to-Source Current with Shorted Gate (I_{DSS})
The maximum value of drain current, achieved by holding V_{GS} at 0 V.

Gate-to-Source Cutoff Voltage or $V_{GS(OFF)}$.
The negative V_{GS} bias voltage that causes I_D to drop to approximately zero.

example 5 V), further increases in V_{DS} will cause no further increase in I_D. This value of V_{DS} is called the **pinch-off voltage (V_P)** because it is the point at which the bias voltage has caused the depletion region to pinch off or restrict drain current. From this point on, further increases in V_{DS} are counteracted by increases in the resistance of the channel, and therefore I_D remains constant. This is shown by the flat portion of the graph in Figure 22-4(a) and is called the **constant-current region** because I_D remains constant despite changes in V_{DS}. If V_{DS} is further increased (by increasing $+V_{DD}$), the JFET will eventually reach its **breakdown voltage (V_{BR}),** at which time a damaging value of I_D will pass through the JFET.

In the example graph in Figure 22-4(a), we plotted what would happen to I_D as V_{DS} increased with V_{GS} at 0 V. In Figure 22-4(b) we will examine what will happen to an *n*-channel JFET when the gate-source junction is reverse biased by several negative voltages. As previously described in the JFET operation section, a negative voltage is normally applied to reverse-bias the gate and set up a depletion region. As V_{GS} is made more negative, the gate will be further reverse biased, and the corresponding I_D value will be smaller. Therefore, when V_{GS} is at 0 V, a maximum value of drain current is passing through the JFET's channel. This maximum value of drain current is called the **drain-to-source current with shorted gate (I_{DSS}).** This name is derived from the fact that when $V_{GS} = 0$ V, as shown in the inset in Figure 22-4(a), the gate and source terminals of the JFET are at the same potential of zero volts, and therefore the gate is effectively shorted to the source as shown by the dashed line. The drain-to-source current with shorted gate (I_{DSS}) rating is therefore the maximum current that can pass through the channel of a given JFET. When given on a specification sheet, this rating is equivalent to a bipolar transistor's $I_{C(SAT)}$ rating.

Returning to Figure 22-4(b), you can see that if V_{GS} is made more negative, the depletion regions within the JFET will get closer and closer and eventually touch, cutting off drain current. This negative V_{GS} bias voltage that causes I_D to drop to approximately zero is called the **gate-to-source cutoff voltage or $V_{GS(OFF)}$.** In the example in Figure 22-4(b), when $V_{GS} = -5$ V, I_D is almost zero and therefore $V_{GS(OFF)} = -5$ V. When cut OFF, the JFET will be equivalent to an open circuit between drain and source, and subsequently all of the drain supply voltage (V_{DD}) will appear across the open JFET ($V_{DS} = V_{DD}$).

To summarize the specifications in Figure 22-4(b),

When $V_{GS} = 0$ V, $I_D = I_{DSS} = 10$ mA

$V_P = 5$ V

$V_{BR} = 30$ V

$V_{GS(OFF)} = -5$ V

Constant Current Region = V_P to $V_{BR} = 5$ V to 30 V

22-1-4 *Transconductance*

Figure 22-4(c) illustrates a JFET test circuit and its associated characteristic graph. Before we see how a JFET can be made to amplify, let's first summarize the details given in this graph. If we first consider the curve when $V_{GS} = 0$ V, you can see that up to V_P, I_D increases in almost direct proportion to V_{DS}. This is because the depletion region is not sufficiently large enough to affect I_D, so the channel is simply behaving as a semiconductor with a fixed resistance value between source and drain.

When V_{DS} is equal to V_P, the drain current (I_D) will be pinched into an extremely narrow channel between the wedge-shaped depletion region. Any further increase in V_{DS} will have two effects:

1. Increase the pinching effect on the channel, which will resist current flow, and
2. Increase the potential between the drain and source, which will encourage current flow.

The net result is that channel resistance increases in direct proportion with V_{DS} and consequently I_D remains constant, as shown by the flat portion of the characteristic curve.

Assuming a fixed value of V_{DD}, any increase in the negative voltage of V_{GS} will cause a corresponding decrease in I_D. Therefore, beyond V_P, I_D is controlled by small-signal changes (such as the input signal) in V_{GS} and is independent of changes in V_{DS}. This section of the curve between V_P and V_{BR} is called the constant-current region.

Like the bipolar transistor, an FET can be used to amplify a signal, as shown in Figure 22-4(c). As before, the amount of amplification achieved is a ratio between output and input. For a bipolar transistor, the amount of gain is equal to the ratio of input current to output current (beta). For an FET, there is no input current, and therefore an FET's gain is equal to the ratio of output current change (ΔI_D) to input voltage change (ΔV_{GS}). This ratio is called the FET's **transconductance** (symbolized δ_m).

Transconductance Also called mutual conductance, it is the ratio of a change in output current to the initiating change in input voltage.

$$\delta_m = \frac{\Delta I_D}{\Delta V_{GS}}$$

δ_m = transconductance in siemens (S)
ΔI_D = change in drain current
ΔV_{GS} = change in gate-source voltage

EXAMPLE:

Calculate the transconductance of the FET for the example shown in Figure 22-4(c).

Solution:

$$\delta_m = \frac{\Delta I_D}{\Delta V_{GS}} = \frac{5\text{ mA} - 2\text{ mA}}{-1\text{ V} - (-3\text{ V})} = \frac{3\text{ mA}}{2\text{ V}} = 1.5\text{ millisiemens}$$

A high-gain FET will produce a large change in I_D for a small change in V_{GS}, resulting in a high-transconductance figure ($\delta_m \uparrow$).

22-1-5 *Voltage Gain*

Because transconductance is the ratio of output current change (ΔI_D) to input voltage change (ΔV_{GS}), it is no surprise that this ratio is used to determine a JFET's voltage gain. The voltage gain formula is as follows

$$A_V = \delta_m \times R_D$$

EXAMPLE:

Calculate the voltage gain for the circuit example shown in Figure 22-4(c).

Solution:

$$A_V = \delta_m \times R_D = 1.5 \text{ mS} \times 8.2 \text{ k}\Omega = 12.3$$

This means that the output voltage will be 12.3 times greater than the input voltage.

22-1-6 *JFET Data Sheets*

Throughout this section, we have used certain JFET specifications in our calculations, such as $V_{GS(\text{OFF})}$ and I_{DSS}. As an example, Figure 22-5 shows the data sheet for a typical *n*-channel JFET. As before, notes have been inserted within the data sheet to describe any confusing ratings; however, most of these ratings are self-explanatory.

22-1-7 *JFET Biasing*

The biasing methods used in FET circuits are very similar to those employed in BJT circuits. In this section we will examine the circuit calculations for the three most frequently used JFET biasing methods: gate-biasing, self-biasing and voltage-divider biasing.

Gate Biasing

Figure 22-6(a) shows a gate-biased JFET circuit. The gate supply voltage ($-V_{GG}$) is used to reverse bias the gate-source junction of the JFET. With no gate current, there can be no voltage drop across R_G and therefore the voltage at the gate of the JFET will equal the dc gate-supply voltage.

$$V_{GS} = V_{GG}$$

In the example in Figure 22-6,

$$V_{GS} = V_{GG} = -1.5 \text{ V}$$

DEVICE: 2N5484 Through 2N5486—N-Channel JFET

Most of these specifications are self-explanatory. $V_{GS\,(off)}$ which has been given throughout is listed in the OFF characteristics, while I_{DSS} is listed in the ON characteristics. The only confusing maximum rating is "Forward Gate Current" because the gate is never normally forward biased (V_{GS} = a negative voltage or zero volts). This rating indicates that if the gate accidentally becomes forward biased, gate current must not exceed 10 mA dc or the JFET will be destroyed.

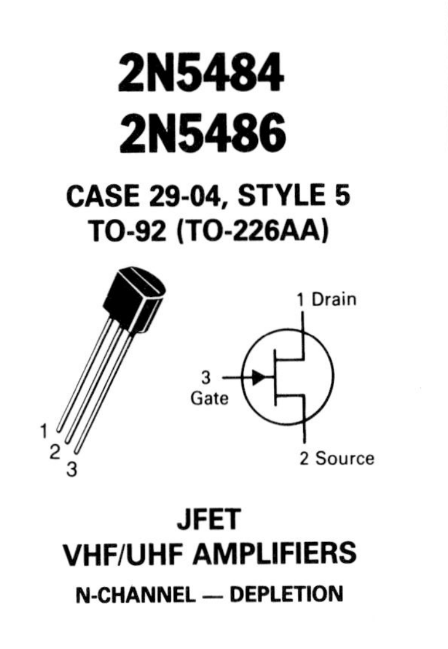

MAXIMUM RATINGS

Rating	Symbol	Value	Unit
Drain-Gate Voltage	V_{DG}	25	Vdc
Reverse Gate-Source Voltage	V_{GSR}	25	Vdc
Drain Current	I_D	30	mAdc
Forward Gate Current	$I_{G(f)}$		mAdc
Total Device Dissipation @ T_C = 25°C Derate above 25°C	P_D	310 2.82	mW mW/°C
Operating and Storage Junction Temperature Range	T_J, T_{stg}	−65 to +150	°C

ELECTRICAL CHARACTERISTICS (T_A = 25°C unless otherwise noted.)

Characteristic		Symbol	Min	Typ	Max	Unit
OFF CHARACTERISTICS						
Gate-Source Breakdown Voltage (I_G = −1.0 μAdc, V_{DS} = 0)		$V_{(BR)GSS}$	−25	—	—	Vdc
Gate Reverse Current (V_{GS} = −20 Vdc, V_{DS} = 0)		I_{GSS}	—	—	−1.0	nAdc
(V_{GS} = −20 Vdc, V_{DS} = 0, T_A = 100°C)			—	—	−0.2	μAdc
Gate Source Cutoff Voltage (V_{DS} = 15 Vdc, I_D = 10 nAdc)	2N5484	$V_{GS(off)}$	−0.3	—	−3.0	Vdc
	2N5485		−0.5	—	−4.0	
	2N5486		−2.0	—	−6.0	
ON CHARACTERISTICS						
Zero-Gate-Voltage Drain Current (V_{DS} = 15 Vdc, V_{GS} = 0)	2N5484	I_{DSS}	1.0	—	5.0	mAdc
	2N5485		4.0	—	10	
	2N5486		8.0	—	20	

$V_{GS\,(off)}$

I_{DSS}

FIGURE 22-5 Data Sheet for an *n*-Channel JFET. (Copyright of Motorola. Used by permission.)

Knowing V_{GS}, we can calculate I_D if the JFET's current (I_{DSS}) and voltage ($V_{GS(OFF)}$) specification limits are known by using the following formula.

$$I_D = I_{DSS}\left(1 - \frac{V_{GS}}{V_{GS(OFF)}}\right)^2$$

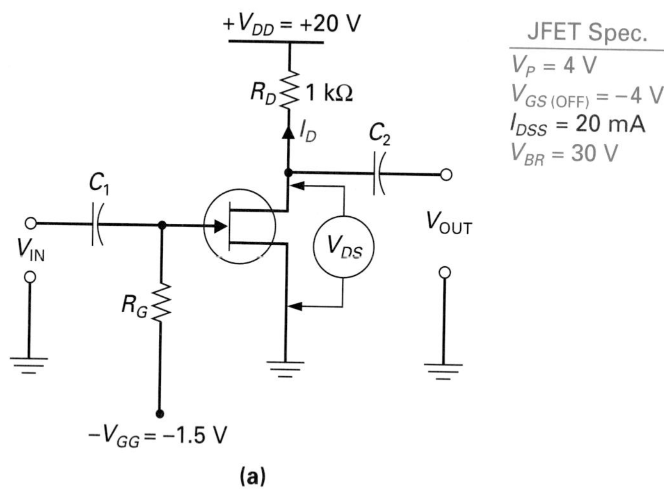

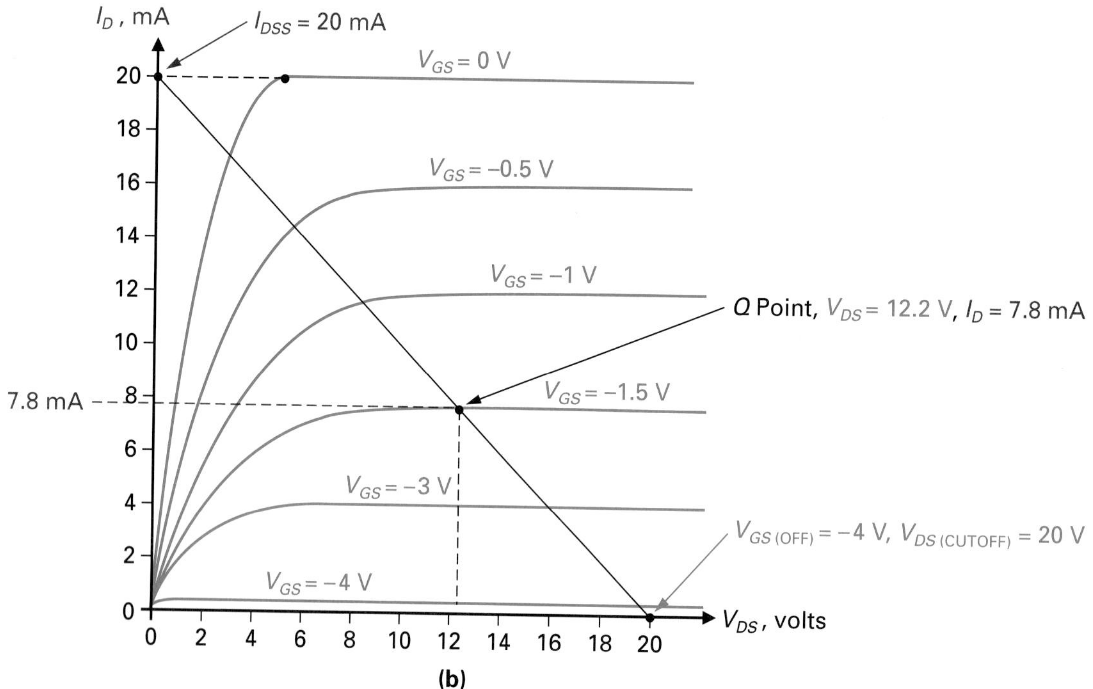

FIGURE 22-6 A Gate-Biased JFET Circuit. (a) Basic Circuit. (b) Drain Characteristic Curves and DC Load Line.

In the example in Figure 22-6,

$$\begin{aligned} I_D &= 20\text{ mA}\left(1 - \frac{-1.5\text{ V}}{-4\text{ V}}\right)^2 \\ &= 20\text{ mA}\,(1 - 0.375)^2 \\ &= 20\text{ mA} \times 0.625^2 \\ &= 20\text{ mA} \times 0.39 = 7.8\text{ mA} \end{aligned}$$

Now that I_D is known, we can calculate the voltage drop across R_D using Ohm's law.

$$V_{RD} = I_D \times R_D$$

In the example in Figure 22-6,

$$V_{RD} = I_D \times R_D = 7.8 \text{ mA} \times 1 \text{ k}\Omega = 7.8 \text{ V}$$

Because V_{RD} plus V_{DS} will equal V_{DD}, we calculate V_{DS} once V_{RD} is known with the following formula.

$$V_{DS} = V_{DD} - V_{RD}$$

In the example in Figure 22-6,

$$V_{DS} = V_{DD} - V_{RD} = 20 \text{ V} - 7.8 \text{ V} = 12.2 \text{ V}$$

Figure 22-6(b) shows the drain characteristic curve and dc load line for the example JFET circuit in Figure 22-6(a). Like the bipolar transistor, the JFET's dc load line extends between the maximum output current point, or saturation point (when the JFET is fully ON, $I_{DSS} = 20$ mA), to the maximum output voltage point (when the JFET is cut OFF, $V_{DS} = 20$ V). The dc operating point, or Q point, which was determined with the previous calculations, is also plotted on the dc load line in Figure 22-6(b).

EXAMPLE:

Calculate the following for the circuit shown in Figure 22-7.

1. V_{GS}
2. I_D
3. V_{DS}
4. Maximum value of I_D
5. V_{DS} when $V_{GS} = V_{GS(\text{OFF})}$
6. Q point

Solution:

1. $V_{GS} = V_{GG} = -3 \text{ V}$
2. $I_D = I_{DSS}\left(1 - \dfrac{V_{GS}}{V_{GS(\text{OFF})}}\right)^2 = 15 \text{ mA}\left(1 - \dfrac{-3 \text{ V}}{-6 \text{ V}}\right)^2 = 15 \text{ mA}\,(1 - 0.5)^2 = 3.75 \text{ mA}$
3. $V_{DS} = V_{DD} - V_{RD}$ (since $V_{RD} = I_D \times R_D$, we can substitute)
   $V_{DS} = V_{DD} - (I_D \times R_D) = 15 \text{ V} - (3.75 \text{ mA} \times 1.2 \text{ k}\Omega) = 15 \text{ V} - 4.5 \text{ V} = 10.5 \text{ V}$
4. Maximum value of $I_D = I_{DSS} = 15$ mA
5. When $V_{GS} = V_{GS(\text{OFF})}$, the JFET is cut off and equivalent to an open switch between drain and source. In this condition all of the drain supply voltage will appear across the open JFET.

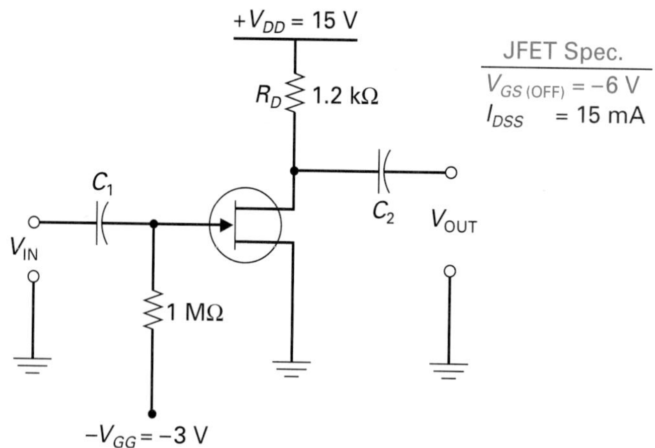

FIGURE 22-7 **A Gate-Biased JFET Circuit Example.**

$$V_{DS(\text{CUTOFF})} = V_{DD} = 15 \text{ V}$$

6. The dc operating or Q point is

$$V_{GS} = -3 \text{ V}$$

$$I_D = 3.75 \text{ mA}$$

$$V_{DS} = 10.5 \text{ V}$$

Self Biasing

Figure 22-8(a) shows how to self-bias a JFET circuit. One advantage of this biasing method over gate biasing is that only a single drain supply voltage is needed (V_{DD}) instead of both V_{DD} and a negative gate supply voltage ($-V_{GG}$). The other difference you may have noticed is that a source resistor (R_S) has been included, and R_G has been connected to ground. Although this arrangement seems completely different to the gate bias circuit, the inclusion of R_S and the grounding of R_G will achieve the same result, which is to reverse bias the JFET's gate-source junction. Figure 22-8(b) illustrates how this is achieved. Since there is no gate current in a JFET circuit ($I_G = 0$), all of the current flowing into the source will travel through the channel and flow out of the drain. Therefore

$$I_S = I_D$$

For the example in Figure 22-8,

$$I_S = I_D = 7 \text{ mA}$$

Now that I_S is known, we can calculate the voltage drop across the source resistor (V_{RS}), and therefore the voltage at the JFET's source (V_S).

$$V_{RS} = V_S = I_S \times R_S$$

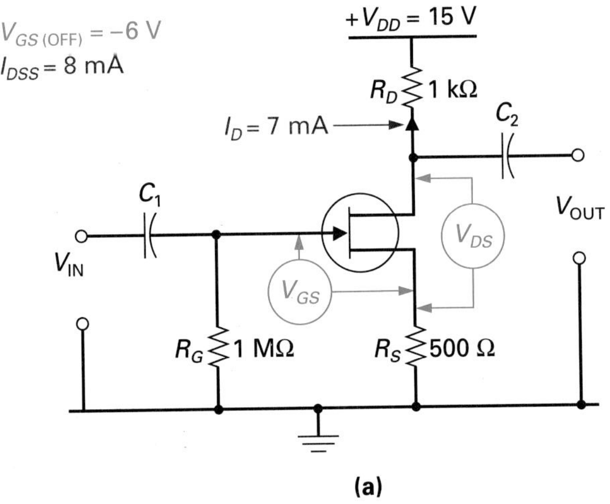

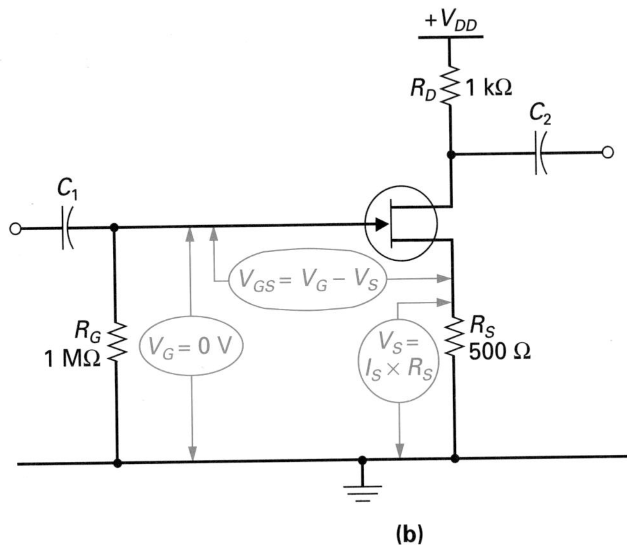

FIGURE 22-8 **A Self-Biased JFET Circuit. (a) Basic Circuit. (b) How R_S Develops a $-V_{GS}$.**

For the example in Figure 22-8,

$$V_{RS} = V_S = I_S \times R_S = 7 \text{ mA} \times 500\ \Omega = +3.5 \text{ V}$$

Because $I_G = 0$ A, there will be no voltage drop across R_G, and so the voltage at the gate of the JFET will be 0 V.

$$V_G = 0 \text{ V}$$

Now that we know that $V_S = +3.5$ V and $V_G = 0$ V, we can see how the JFET's gate-source junction is reverse biased. To reverse bias a gate-biased JFET, we simply made the gate voltage negative with respect to the source that is at 0 V. With a self-biased JFET, we achieve the same result by making the source voltage positive with respect to the gate that is at 0 V. This makes the gate of the JFET negative with respect to the source. This potential difference from gate-to-source (V_{GS}) is therefore equal to

$$V_{GS} = V_G - V_S$$

Because $V_S = I_S \times R_S$ and $V_G = 0$ V, we can substitute the previous formula to obtain

$$V_{GS} = 0 \text{ V} - (I_S \times R_S)$$

or

$$V_{GS} = -(I_S \times R_S)$$

or because $I_S = I_D$

$$V_{GS} = -(I_D \times R_S)$$

In the example in Figure 22-8,

$$\begin{aligned} V_{GS} &= -(I_S \text{ or } I_D \times R_S) \\ &= -(7 \text{ mA} \times 500 \ \Omega) \\ &= -3.5 \text{ V} \end{aligned}$$

If $-V_{GS}$ and R_S are known, we could transpose the above equation to calculate I_D.

$$I_D = \frac{V_{GS}}{R_S}$$

In the example in Figure 22-8,

$$I_D = \frac{V_{GS}}{R_S} = \frac{3.5 \text{ V}}{500 \ \Omega} = 7 \text{ mA}$$

The final calculation is to determine the voltage at the JFET's drain with respect to ground (V_D) and the drain-to-source voltage drop across the JFET.

$$V_D = V_{DD} - V_{RD}$$

Because $V_{RD} = I_D \times R_D$,

$$V_D = V_{DD} - (I_D \times R_D)$$

In the example in Figure 22-8,

$$V_D = V_{DD} - (I_D \times R_D) = 15 \text{ V} - (7 \text{ mA} \times 1 \text{ k}\Omega) = 15 \text{ V} - 7 \text{ V} = 8 \text{ V}$$

Now that the voltage drops across R_D (V_{RD}) and R_S (V_{RS}) are known, we can calculate the voltage drop across the JFET's drain to source (V_{DS}).

$$V_{DS} = V_{DD} - (V_{RD} + V_{RS})$$

In the example in Figure 22-8,

$$V_{DS} = V_{DD} - (V_{RD} + V_{RS}) = 15 \text{ V} - (7 \text{ V} + 3.5 \text{ V}) = 15 \text{ V} - 10.5 \text{ V} = 4.5 \text{ V}$$

EXAMPLE:

Calculate the following for the circuit shown in Figure 22-9.

1. V_S
2. V_{GS}
3. V_{DS}
4. I_D maximum
5. V_{DS} when the JFET is OFF
6. V_D

Solution:

1. Because $I_S = I_D$, $V_S = I_S \times R_S = 4 \text{ mA} \times 500 \ \Omega = 2 \text{ V}$
2. $V_{GS} = V_G - V_S = 0 \text{ V} - 2 \text{ V} = -2 \text{ V}$
3. $V_{DS} = V_{DD} - (V_{RD} + V_{RS}) = 10 \text{ V} - [(I_D R_D) + 2 \text{ V}]$
   $= 10 \text{ V} - [(4 \text{ mA} \times 1.2 \text{ k}\Omega) + 2 \text{ V}] = 10 \text{ V} - (4.8 + 2 \text{ V})$
   $= 10 \text{ V} - 6.8 \text{ V} = 3.2 \text{ V}$
4. I_D maximum $= I_{DSS} = 8 \text{ mA}$
5. $V_{DS(\text{CUTOFF})} = V_{DD} = 10 \text{ V}$
6. $V_D = V_{DS} + V_{RS} = 3.2 \text{ V} + 2 \text{ V} = 5.2 \text{ V}$

As previously mentioned, one advantage of this self-biased JFET method is that only a drain supply voltage is needed (V_{DD}). The gate supply voltage (V_{GG}) is not needed due to the inclusion of a source resistor that reverse biases the JFET's gate-source junction by applying a positive voltage to the source with respect to the 0 V on the gate. This method of effectively sending back a negative voltage from the source to the gate is known as "negative feedback." It not only enables us to bias a JFET with one supply voltage, it also provides temperature stability. Any change in the ambient temperature will cause a change in the semiconductor JFET's conduction, which would move the JFET's Q point away

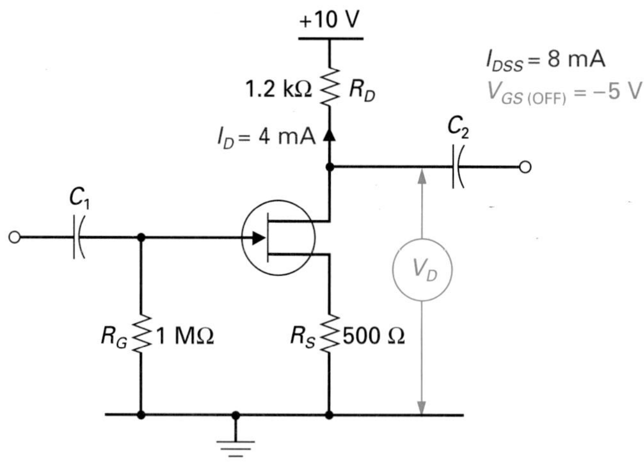

FIGURE 22-9 **A Self-Biased JFET Circuit Example.**

from its desired setting. The inclusion of R_S will prevent the Q point from shifting due to temperature in the same way as a BJT's emitter resistor. If temperature were to increase, for instance ($T\uparrow$), the resistance of the semiconductor would decrease ($R\downarrow$) because all semiconductor materials have a negative temperature coefficient of resistance ($T\uparrow$, $R\downarrow$), and this will cause the channel current to increase. If the drain current increases ($I_D\uparrow$), the voltage drop across R_S will increase (V_{RS} or $V_S\uparrow = I_D\uparrow \times R_S$). This increase in V_S will increase the gate-source reverse voltage ($-V_{GS}\uparrow$), causing the JFET's channel to get narrower and the drain current to decrease ($I_D\downarrow$) and counteract the original increase. Similarly, a decrease in temperature will cause a decrease in I_D which will decrease the gate-source reverse bias, resulting in an increase in I_D. The Q point will remain relatively stable despite changes in temperature when a JFET circuit has a source resistor included.

Voltage-Divider Biasing

Referring to the voltage-divider biased JFET circuit shown in Figure 22-10, you will probably notice that it is very similar to the voltage-divider biased BJT circuit discussed previously. Like the self-biased circuit, the inclusion of a source resistor stabilizes the Q point despite ambient temperature changes. In addition, using a voltage divider to determine the gate-bias voltage ensures that V_{GS}, and therefore the circuit, has increased stability.

The gate voltage (V_G) is calculated using the following voltage divider formula

$$V_{R_2} \text{ or } V_G = \frac{R_2}{R_1 + R_2} \times V_{DD}$$

For the example in Figure 22-10,

$$V_{R_2} \text{ or } V_G = \frac{R_2}{R_1 + R_2} \times V_{DD} = \frac{5\ \text{M}\Omega}{10\ \text{M}\Omega + 5\ \text{M}\Omega} \times 15\ \text{V} = 5\ \text{V}$$

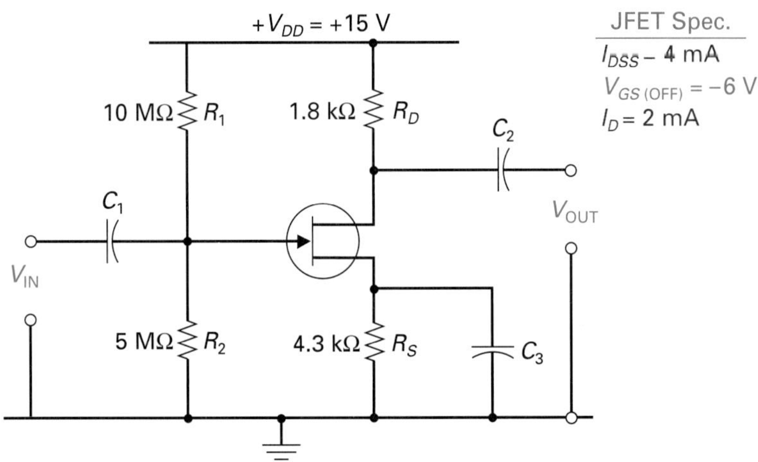

FIGURE 22-10 A Voltage Divider Biased JFET Circuit.

Because $I_D = I_S$ ($I_G = 0$) and the drain resistance and current are known, we can next calculate the voltage drop across the source resistor (V_{RS}), drain resistor (V_{RD}), and JFET's source-drain junction (V_{DS}).

$$I_S = I_D$$

$$I_S = I_D = 2 \text{ mA}$$

$$V_{RS} = I_S \times R_S$$

$$V_{RS} = 2 \text{ mA} \times 4.3 \text{ k}\Omega = 8.6 \text{ V}$$

$$V_{RD} = I_D \times R_D$$

$$V_{RD} = 2 \text{ mA} \times 1.8 \text{ k}\Omega = 3.6 \text{ V}$$

$$V_{DS} = V_{DD} - (V_{RS} + V_{RD})$$

$$V_{DS} = 15 \text{ V} - (8.6 \text{ V} + 3.6 \text{ V}) = 2.8 \text{ V}$$

Now that the JFET's gate and source voltages are known (V_G and V_S), we can calculate the value of gate-source reverse bias ($-V_{GS}$).

$$V_{GS} = V_G - V_S$$

$$(V_G = V_{R_2}, V_S = V_{RS})$$

For the example in Figure 22-10,

$$V_{GS} = 5 \text{ V} - 8.6 \text{ V} = -3.6 \text{ V}$$

EXAMPLE:

Calculate the following for the voltage-divider biased JFET circuit shown in Figure 22-11:

1. V_G
2. I_S
3. V_S
4. V_{DS}
5. V_{GS}
6. V_D, when $V_{GS} = V_{GS(\text{OFF})}$
7. I_D, when $V_{GS} = 0$ V

Solution:

1. $V_G = \dfrac{R_2}{R_1 + R_2} \times V_{DD} = \dfrac{10\ \text{M}\Omega}{100\ \text{M}\Omega + 10\ \text{M}\Omega} \times 30\ \text{V} = 2.7\ \text{V}$
2. $I_S = I_D = 3.6$ mA
3. $V_S = V_{RS} = I_S \times R_S = 3.6\ \text{mA} \times 2.7\ \text{k}\Omega = 9.7\ \text{V}$
4. $V_{DS} = V_{DD} - (V_{RS} + V_{RD}) = 30\ \text{V} - [9.7\ \text{V} + (I_S \times R_D)]$
   $= 30\ \text{V} - [9.7\ \text{V} + (3.6\ \text{mA} \times 5\ \text{k}\Omega)] = 30\ \text{V} - (9.7\ \text{V} + 18\ \text{V}) = 2.3\ \text{V}$
5. $V_{GS} = V_G - V_S = 2.7\ \text{V} - 9.7\ \text{V} = -7\ \text{V}$
6. When $V_{GS} = V_{GS(\text{OFF})}$, JFET is OFF and $V_D = V_{DD} = 30$ V
7. When $V_{GS} = 0$ V, $I_D = \text{maximum} = I_{DSS} = 6$ mA

22-1-8 *JFET Circuit Configurations*

The three JFET circuit configurations are illustrated in Figure 22-12 along with their typical circuit characteristics. Like the bipolar transistor configurations, the term "common" is used to indicate which of the JFET's leads is common to both

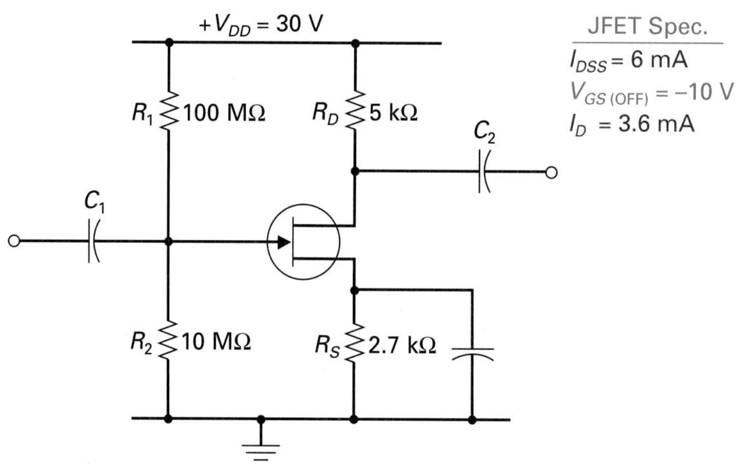

FIGURE 22-11 A Voltage Divider Biased Circuit Example.

	Voltage Gain	Input Impedance	Output Impedance	Circuit Appearance and Application	Waveforms
Common Source (a)	5–10 (Voltage Amp)	Very High 1–15 MΩ	Low 2–10 kΩ	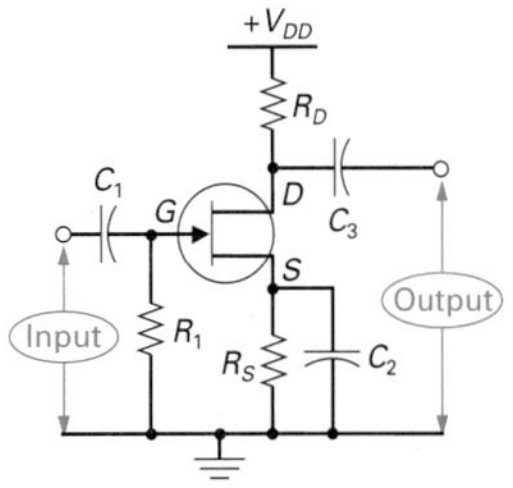 $+V_{DD}$, R_D, C_1, G, D, C_3, S, Input, R_1, R_S, C_2, Output. Most widely used FET configuration. It is mainly used as a voltage amplifier, however it is also used as an impedance matching device and can handle the high radio frequency signals.	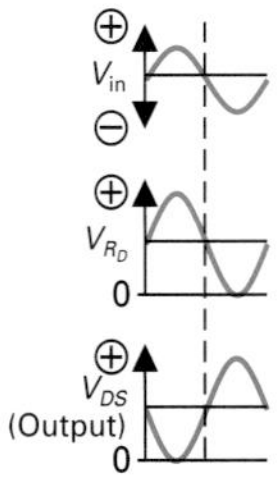 V_{in}, V_{R_D}, V_{DS} (Output). V_{in} and V_{out} are out of phase (180° phase shift)
Common Gate (b)	2–5	Very Low 200–1500 Ω	Medium 5–15 kΩ	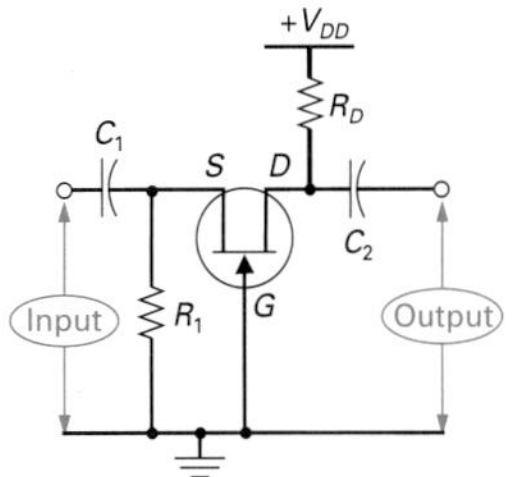 $+V_{DD}$, R_D, C_1, S, D, C_2, Input, R_1, G, Output. This configuration is used to amplify radio frequency signals due to its very stable nature at high frequencies. It is also used as a buffer to match a low impedance source to a high impedance load.	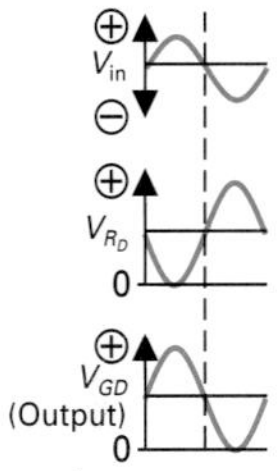 V_{in}, V_{R_D}, V_{GD} (Output). V_{in} and V_{out} are in phase (0° phase shift)
Common Drain (c)	0.98	Very High 1500 MΩ	Low 10 kΩ	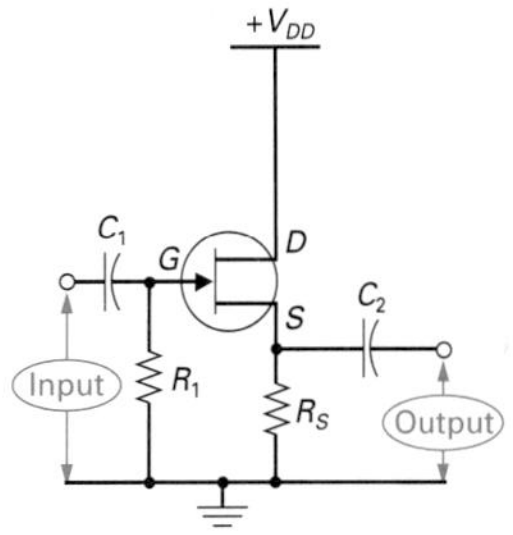 $+V_{DD}$, C_1, G, D, S, C_2, Input, R_1, R_S, Output. This amplifier is commonly called a source follower as the source follows whatever is applied to the gate. Its very high input impedance will not load down (and therefore not distort) signals from high impedance signal sources, such as a microphone, and its low output impedance is ideal to drive a low impedance load such as an audio amplifier.	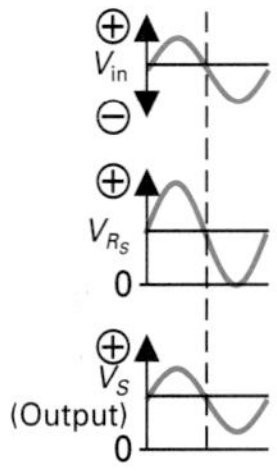 V_{in}, V_{R_S}, V_S (Output). V_{in} and V_{out} are in phase (0° phase shift)

FIGURE 22-12 **JFET Circuit Configurations.**

the input and output. In this section we will examine the characteristics of these three configurations: common-source, common-gate, and common-drain.

Common-Source (*C-S*) Circuits

Similar to its bipolar counterpart, the common-emitter configuration, the **common-source configuration** is the most widely used JFET circuit and is detailed in Figure 22-12(a). The input is applied between the gate and source and the output is taken between the drain and source, with the source being common to both input and output. The ac input will pass through the coupling capacitor C_1 and be

Common-Source Configuration
An FET configuration in which the source is grounded and common to the input and output signal.

superimposed on the dc gate-bias voltage provided by resistor R_1, which sets up the dc operating or Q point. As the signal input changes, it will cause a change in gate voltage, which will cause a corresponding change in the output drain current. The output voltage developed between the FET's drain and ground is 180° out of phase with the input because an increase in $V_{IN}\uparrow$ and, therefore $V_{GS}\uparrow$, will cause an increase in $I_D\uparrow$, a decrease in the voltage drop across the FET ($V_{DS}\downarrow$), and a decrease in the output voltage $V_{OUT}\downarrow$. Resistor R_S is included to provide temperature stability and, as with the bipolar transistor, the source decoupling capacitor C_2 is included to prevent degenerative feedback.

When a small ac input signal is applied to the gate of a common-source amplifier, the variations in voltage at the gate control the JFET, which effectively acts as a variable resistor, varying the output drain current. These changes in drain current will vary the voltage drop across R_D and the drain-to-source voltage drop, which, with R_S, determines the output voltage. Referring to the characteristics listed in Figure 22-12(a), you can see that the output voltage (V_{OUT}) of the common-source JFET configuration can be five to ten times larger than the gate control input voltage (V_{IN}). If a high amount of voltage gain is desired, R_D is made relatively large (typically greater than 20 kΩ) and the JFET is biased so that its drain-to-source resistance is also high. A larger resistance will develop a larger voltage.

Also listed in the common-source characteristics in Figure 22-12(a) is the very high input resistance and the relatively low output resistance of this circuit. The high input resistance is due to the JFET's reverse biased gate-source junction, which permits no gate input current, and therefore has a very large resistance. This key characteristic means that the common-source JFET circuit is ideal in applications where we need to provide voltage amplification but do not want to load down a source that can only generate a small input signal. Such applications include

1. Digital circuits in which the outputs of many circuits are connected to one another, and therefore the output resistances of all the circuits load one another. As a result, the signals generated by these circuits are small, and a circuit is needed that will not load the signal source but will still provide voltage gain.
2. Analog circuits in which it can amplify both dc and low- and high-frequency ac input signal voltages. The *C-S* circuit's high input impedance makes it ideal at the front end of systems such as the first RF amplifier stage following the antenna and the first stage in a voltmeter, in which it will not load the source yet will amplify a wide range of input signal voltages.

Common-Gate (*C-G*) Circuits

Common-Gate Configuration An FET configuration in which the gate is grounded and common to the input and output signal.

The **common-gate circuit configuration** shown in Figure 22-12(b) is very similar to its bipolar counterpart, the common-base circuit. The input is applied between the source and gate, while the output appears across the drain and gate. Self-bias resistor R_1 sets up the static Q point, and the input is applied through the coupling capacitor C_1 and will cause a change in the JFET's source voltage. An increase in source voltage will cause a decrease in the V_{GS} forward bias (n-type source is driven positive), a decrease in I_D, a decrease in the voltage drop across R_D, and therefore an increase in the voltage dropped between the FET's drain and gate. Because the voltage developed across the JFET's drain and gate is applied to the output, an increase in the input produces an increase in the output, and so the input and output voltage are in-phase with one another. Similarly, as the input voltage decreases, the gate-source forward bias will increase. Therefore, I_D will increase, and there will be more voltage developed across R_D and less voltage developed at the output.

Referring to the common-gate characteristics listed in Figure 22-12(b), you can see that this circuit can be used to provide a small voltage gain. Because the input is applied to the JFET's high-current source terminal, the input resistance is very low. This low input resistance and relatively high output resistance makes the circuit ideal in applications where we need to efficiently transfer power between a low-resistance source and a high-resistance load.

Common-Drain (*C-D*) Circuits

Comparable to the bipolar transistor's common-collector or emitter follower, the **common-drain configuration** shown in Figure 22-12(c) is sometimes called a **source follower** because the source output voltage follows in polarity and amplitude the input voltage at the gate. Once again, self-bias resistor R_1 sets up the quiescent operating point, and an ac gate input voltage will cause a variation in I_D. When the input voltage at the gate swings positive, the FET will conduct more current, less voltage will be developed across the FET drain to source, and therefore more voltage will be developed across R_S and the output. Similarly, a decrease in the input voltage will cause the resistance of the JFET's drain-source junction to increase. Therefore, V_{DS} will increase and V_{RL}, or V_{OUT}, will decrease.

Common-Drain Configuration
An FET configuration in which the drain is grounded and common to the input and output signal.

Source Follower
Another name used for a common-drain circuit configuration.

Referring to the common-drain characteristics listed in Figure 22-12(c), you can see that the output voltage is slightly less than the input voltage (circuit does not provide any voltage gain). The input resistance of the common-drain circuit configuration is extremely high due to the JFET's reverse biased gate and R_S connection, and the output resistance is relatively very low. Inserting a common-drain circuit between a high-resistance source and a low-resistance load will ensure that the two opposite resistances are matched and power is efficiently transferred.

22-1-9 JFET Applications

It is the high input impedance of the JFET, and therefore its ability not to load a source, and the voltage amplification ability that are mainly made use of in circuit application. Like the bipolar junction transistor, the JFET can be made to function as a switch or as a variable resistor. Let us begin by examining how the JFET's switching ability can be made use of in digital or two-state circuits.

Digital (Two-State) JFET Circuits

As a switch, the JFET makes use of only two points on the load line: saturation (in which it is equivalent to a closed switch between source and drain) and cutoff (in which it is equivalent to an open switch between source and

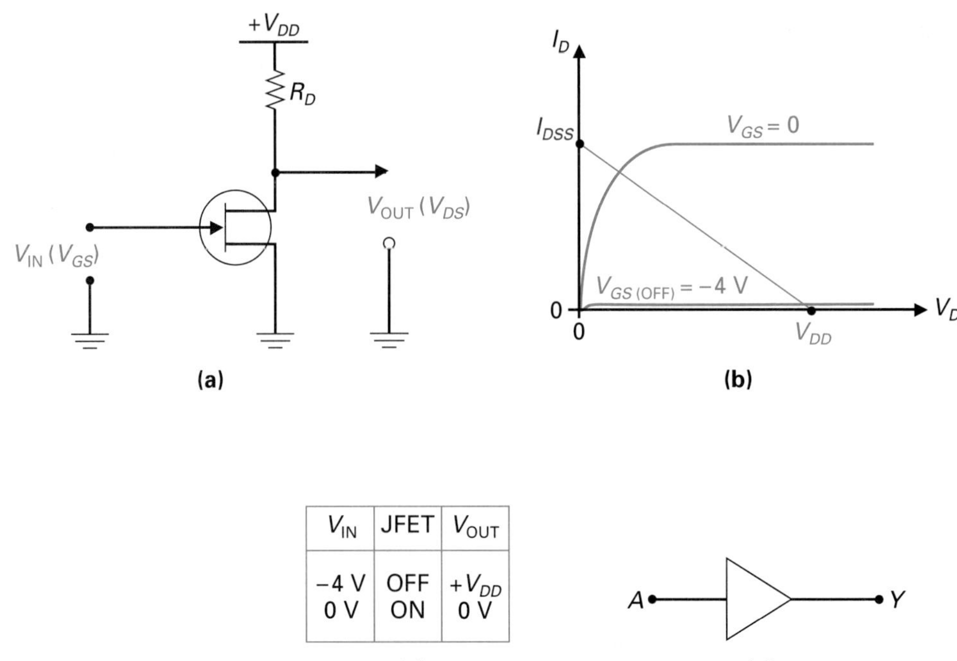

V_{IN}	JFET	V_{OUT}
−4 V	OFF	+V_{DD}
0 V	ON	0 V

(c)

FIGURE 22-13 The JFET's Switching Action—Digital Circuit Applications. (a) Basic Switching Circuit. (b) Load Line. (c) Input/Output Voltages. (d) Digital Buffer/Driver Schematic Symbol.

drain). Figure 22-13(a) shows an ON/OFF JFET switch circuit and its associated load line in Figure 22-13(b). Figure 22-13(c) shows the input/output voltages for each of the circuit's two operating states. When $V_{GS} = V_{GS(\text{OFF})}$ (−4 V), the JFET is cut OFF (lower end of the load line) and is equivalent to an open switch between source and drain. With the JFET's drain-source open, $I_D = 0$ mA, and the drain supply voltage will be applied to the output ($V_{DS} = V_{\text{OUT}} = +V_{DD}$). On the other hand, when $V_{GS} = 0$ V, the JFET is saturated (upper end of the load line) and is equivalent to a closed switch between source and drain. With the JFET's drain-source closed, I_D = max. = I_{DSS}, and the 0 V at the source will be applied to the output ($V_{DS} = V_{\text{OUT}} = 0$ V).

A typical FET application in digital circuits would be a buffer circuit, which is used to isolate one device from another. The high input impedance of the FET does not load the input circuit or circuits, while the low output impedance of the FET provides a high output current to the output circuit. The high output current and buffering or isolating characteristics of these circuits account for why they are also called buffer-drivers. The schematic symbol of the buffer-driver is shown in Figure 22-13(d).

The high input impedance ($Z_{\text{IN}}\uparrow$) of the FET is also made use of in other FET integrated circuits (ICs). When Z_{IN} is high, circuit current is low ($I\downarrow$), and therefore power dissipation is low ($P_D\downarrow$). This condition is ideal for digital integrated circuits (ICs) which contain thousands of transistors all formed onto one small piece of silicon. The low power dissipation of the JFET enables us to densely pack many more components into a very small area.

Analog (Linear) JFET Circuits

In two-state applications, the JFET is made to operate between the two extreme points of saturation (0 Ω) and cutoff (max. Ω). By controlling the gate-source bias voltage (V_{GS}), the resistance between the JFET's drain and source (R_{DS}) can be changed to be any value between 0 Ω and maximum Ω. The JFET can therefore be made to act as a variable resistor, with an increase in negative V_{GS} causing a larger R_{DS}. In contrast, a decrease in negative V_{GS} causes a smaller R_{DS}. This is illustrated in Figure 22-14(a).

It is the reverse biased gate-source junction of a JFET that gives the JFET its key advantage: *an extremely high input impedance* (typically in the high-MΩ range). In addition, the JFET can provide a small voltage gain and has been found to be a very low noise component. All these characteristics make it an ideal choice as an amplifier.

In previous chapters, we have seen how a light load (large resistance $R_L\uparrow$ and small $I\downarrow$) does not pull down the source voltage by any large amount, whereas a heavy load (small resistance $R_L\downarrow$ and large $I\uparrow$) will pull down the source voltage. A heavy load results in less output voltage ($V_{RL}\downarrow$) and an increased current and heat loss at the source. The overall effect is that a small load resistance or impedance causes less power to be delivered to the load.

The circuit in Figure 22-14(b) shows how a common-source JFET has been connected to function as a preamplifier, which is a circuit that provides gain for a very weak input signal. In this example, the JFET preamplifier matches the high-impedance (small signal) crystal microphone to a low-impedance power amplifier. The reverse biased gate-source junction of a JFET pre-amplifier will offer a large load impedance to the source or microphone. This light load ($R_L\uparrow$) input resistance of the JFET will therefore permit most of the signal voltage being generated by the microphone to be applied to the JFET's gate and then be amplified. In other words, the high input impedance of the JFET amplifier circuit will not pull down the voltage signal being generated by the microphone. Therefore, maximum power will be transferred from source to load.

Figure 22-14(c) shows how a JFET at the front end of a voltmeter or oscilloscope will provide a very high input impedance and therefore not load the circuit under test. In this example, the meter will measure 10 V across R_2, and because the ohms per volt (Ω/V) rating of the voltmeter is 125 kΩ/V, the meter input impedance is

$$Z_{\text{IN}} = \Omega/\text{V} \times V_{\text{Measured}}$$
$$= 125\ \text{k}\Omega/\text{V} \times 10\ \text{V} = 1.25\ \text{M}\Omega$$

A 1.25 MΩ meter resistance in parallel with the 5 kΩ resistance of R_2 will have very little effect (1.25 MΩ in parallel with 5 kΩ = about 5 kΩ), so an accurate reading will be obtained.

Figure 22-14(d) shows how the JFET can be used as a radio frequency (RF) amplifier. Studying this circuit, you can see that both the gate and drain contain tuned circuits in the same way as the previously discussed BJT RF amplifier cir-

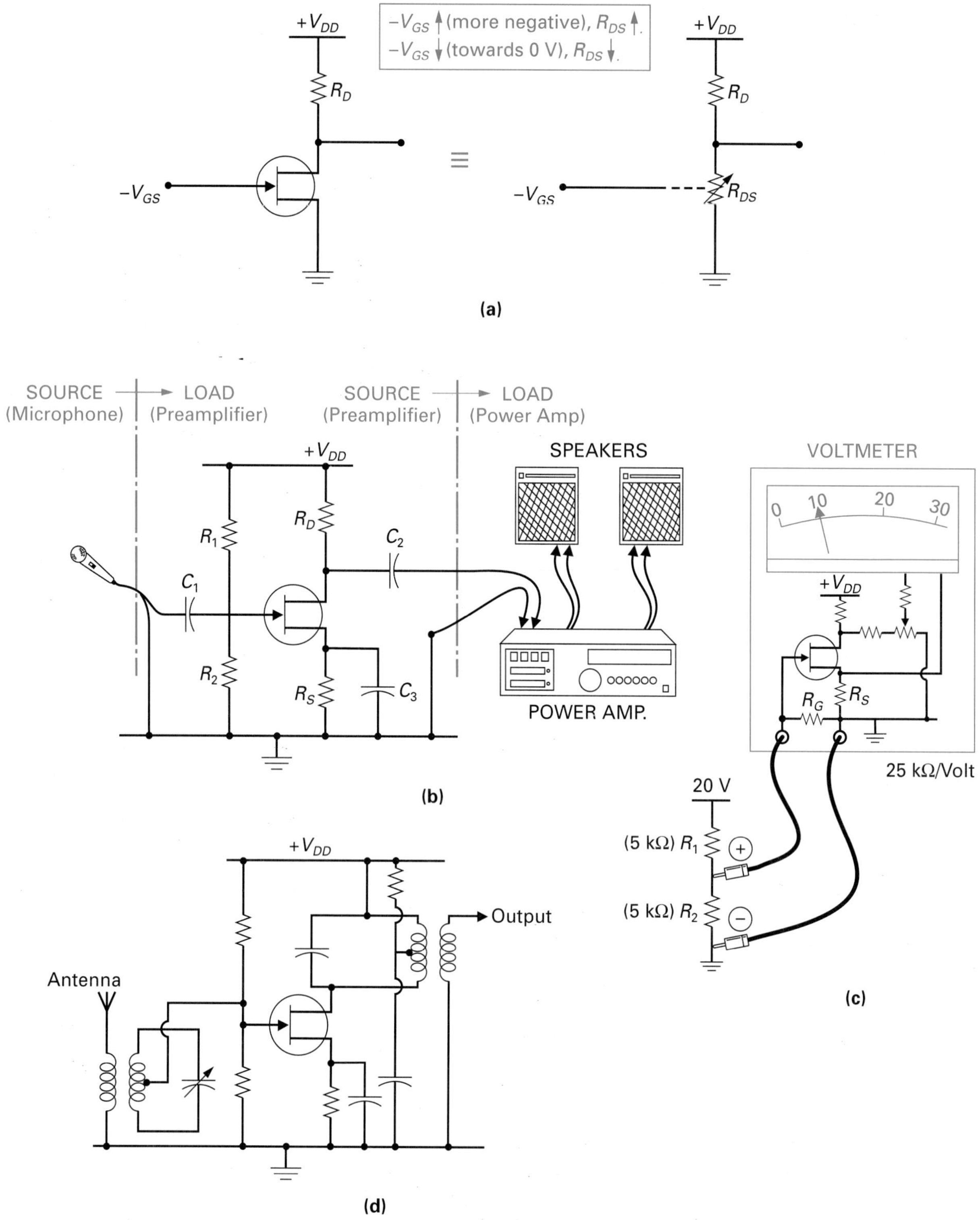

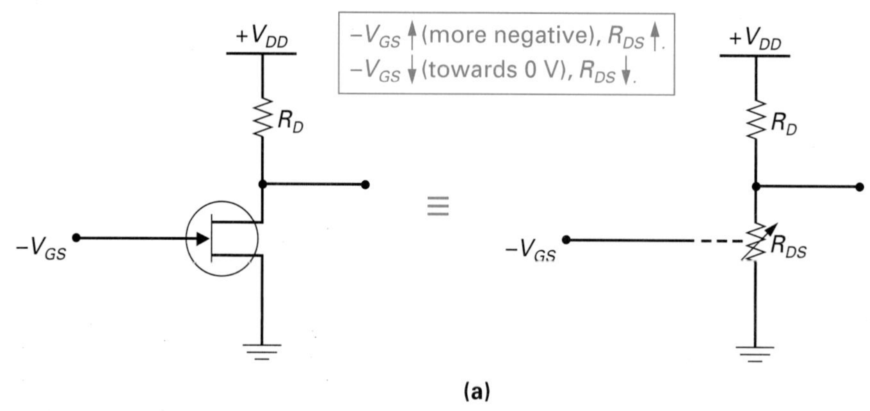

FIGURE 22-14 The JFET's Variable-Resistor Action—Analog Circuit Applications. (a) Equivalent Circuit. (b) Application 1: An Audio Preamplifier Circuit. (c) Application 2: A Voltmeter High Input Impedance Circuit. (d) Application 3: An RF Amplifier Circuit.

cuits. However, there are two advantages that the JFET has over the bipolar transistor as a front end RF amp.

1. The very weak signals injected into the antenna will have a very small value of current. Because the JFET is a voltage operated device, it requires no input current, and it will respond well to the small voltage signal variations picked up by the antenna.
2. The JFET is a very low noise component. Because any noise generated at the front end will be amplified along with the signal at each of the following amplifier stages, this JFET characteristic is ideal in this application.

22-1-10 *Testing JFETs*

Testing JFETs

The transistor tester shown in Figure 22-15(a) can be used to test both BJTs and FETs. This tester can be used to determine

1. Whether an open or short exists between any of the terminals,
2. The FET's transconductance/gain, and
3. The FET's value of I_{DSS} and leakage current.

If a transistor tester is not available, the ohmmeter can be used to detect the most common failures: opens and shorts. Figure 22-15(b) indicates what resistance values should be obtained between the terminals of a good *n*-channel and *p*-channel JFET. Looking at these ohmmeter readings, you can see that because the JFET has only one P-N junction (gate-to-channel), it is relatively simple to test with an ohmmeter for an open or shorted junction.

Troubleshooting JFET Circuits

Let us now examine how to troubleshoot JFET circuits. As with all other circuit troubleshooting, the first step is to isolate whether the problem is within the JFET circuit or external to the JFET circuit. Figure 22-15(c) shows the "normal condition" for a typical voltage divider biased JFET circuit. Because I_G is normally zero, I_1 should equal I_2, and, as we already know, $I_S = I_D$. Because the JFET has only one P-N junction, the two basic circuit problems that can develop are as follows.

1. A shorted gate-source junction will have the effect shown in Figure 22-15(d). With a short between gate and source, there will be no voltage difference between gate and source (V_{GS} will be 0 V), V_G will be a positive voltage, and I_D will be a maximum ($I_D = I_{DSS}$). The gate-source short will allow current to flow out of the gate, and therefore I_S will not equal I_D. Because I_G will combine with I_1, I_1 will not equal I_2. More than likely, the gate current will cause the gate junction to open.
2. An open-gate junction will have the effect shown in Figure 22-15(e). With the gate open, the applied V_{GS} voltage will be present at the gate, and therefore everything will seem normal. However, because V_{GS} is 0 V due to the open gate, I_D will equal maximum ($I_D = I_{DSS}$) and be constant despite changes in the input V_{GS}.

(a)

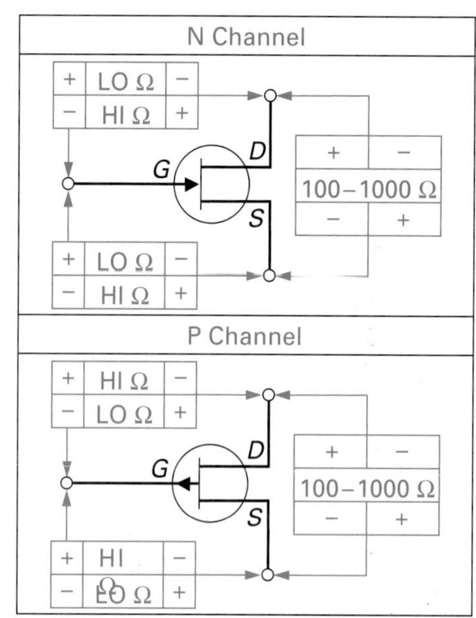

(b)

$+V_{DD}$

I_1 R_1 R_D I_D I_G I_2 R_2 R_S I_S

$I_1 = I_2$ $I_S = I_D$

$I_G = 0$

NORMAL CONDITION

(c)

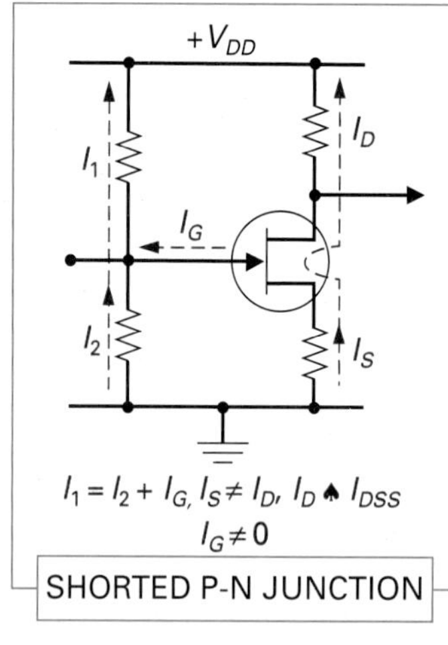

(d)

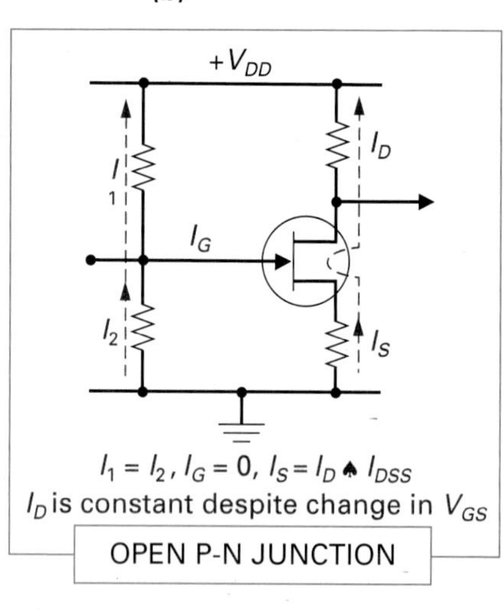

(e)

Cause	Effect
R_2 Open	With R_2 open, $V_2 = +V_{DD}$, and JFET is heavily forward biased and therefore destroyed. Replacement JFET will also be destroyed unless R_2 open is first fixed.
R_1 Open	R_2 provides self-bias. Gate-source bias will increase and output signal may be clipped.
R_D Open	With no drain supply voltage, I_D and I_S will be zero and the JFET will be OFF.
R_S Open	With R_S open, the supply voltage will appear across R_S ($V_S = +V_{DD}$).

(f)

FIGURE 22-15 **Testing JFETs and Troubleshooting JFET Circuits. (Photo courtesy of Sencore, Inc.— Test Equipment for the Professional Servicer. 1-800-SENCORE.)**

If the JFET circuit component at fault is not the JFET itself, we will need to isolate the effect we are getting, and then try to determine the cause. Figure 22-15(f) lists the symptoms you should get from different JFET circuit component failures.

SELF-TEST REVIEW QUESTIONS FOR SECTION 22-1

1. The BJT is a ____________ operated device while the FET is a ______ ______ operated device.
2. The gate-source junction of a JFET is always ____________ biased.
3. When $V_{GS} = 0$ V, $I_D = I_{DSS}$. (True/False)
4. When $V_{GS} = V_{GS(OFF)}$, $I_D = \text{max}$. (True/False)
5. ____________ is a ratio of an FET's output current change to input voltage change.
6. A JFET has a ____________ input impedance due to its ____________ biased gate-source junction.
7. What component in a JFET circuit provides temperature stability?
8. Like self bias, ____________ bias has negative feedback and therefore maintains the Q point stable.
9. Which FET circuit configuration is most widely used like its BJT common-emitter counterpart?
10. Which JFET circuit configuration could provide a high input impedance and a good value of voltage gain?
11. Which JFET circuit configuration is best suited for providing a very high input impedance and low output impedance?
12. Which JFET characteristic is made use of in most circuit applications?
13. Why is the JFET ideal as an RF preamplifier?
14. Using a transistor tester to test a 2N5484 JFET, what typical readings should we obtain for the following:
    a. $V_{GS(OFF)}$
    b. $I_{DSS(MAXIMUM)}$
15. Will the gate-to-source and gate-to-drain of a JFET test with an ohmmeter like any other P-N junction?

22-2 THE METAL OXIDE SEMICONDUCTOR FIELD EFFECT TRANSISTOR (MOSFET)

With the JFET, an input voltage of zero volts would reverse bias the P-N junction, resulting in a maximum channel size and a maximum value of source-to-drain current. To decrease the size of the channel, the input voltage was made negative to further reverse bias the gate-source junction. This action would deplete the channel of free carriers, reducing the size of the channel and therefore the source-to-drain current. This type of action is actually called "depletion-mode operation," since an input voltage is used to deplete the channel and, therefore, reduce the channel's size and current. The MOSFET does not have a P-N gate-channel junction like the JFET. It has a "metal gate" that is insulated from the "semiconductor channel" by a layer of "silicon dioxide," hence the name "metal oxide semiconductor." Like all "field effect transistors" (FETs), the input voltage will generate an "electric field" which will have the "effect" of changing the channel's size.

The key difference between the JFET and MOSFET is that the JFET's input voltage would always have to be zero or a negative voltage in order to reverse bias the gate-source junction. With the MOSFET, the input voltage can be either a positive or negative voltage since gate current will always be zero because the gate is insulated from the channel. To examine each of these input voltage possibilities:

a. If the input voltage is negative, the resulting electric field depletes the channel, reducing its size, and the MOSFET is said to be operating in the depletion mode.

b. If, on the other hand, the input voltage is positive, the resulting electric field enhances the channel, increasing its size, and the MOSFET is said to be operating in the enhancement mode.

The MOSFET can therefore be operated in either the depletion or enhancement mode due to its insulated gate. The two different types of MOSFET are given names based on their normal mode of operation. For instance, the *depletion-type MOSFET (D-type MOSFET* or *D-MOSFET)*, should actually be called a DE-MOSFET because it can be operated in both the depletion mode and the enhancement mode, whereas, the *enhancement-type MOSFET (E-type MOSFET* or *E-MOSFET*) is correctly named since it can only be operated in the enhancement mode. In this chapter, we will examine the construction, operation, characteristics, circuit biasing, applications, and testing of these two MOSFET types.

Depletion-Type MOSFET
A field effect transistor with an insulated gate (MOSFET) that can be operated in either the depletion or enhancement mode.

***n*-Channel D-Type MOSFET**
A depletion type MOSFET having an *n*-type channel between its source and drain terminals.

***p*-Channel D-Type MOSFET**
A depletion type MOSFET having a *p*-type channel between its source and drain terminal.

22-2-1 The Depletion-Type (D-Type) MOSFET

The **depletion-type MOSFET** construction is slightly different from the JFET, and therefore we will need to first see how the D-MOSFET device is built before we can understand how it operates.

D-MOSFET Construction

Like the JFET and BJT, the D-MOSFET has two basic transistor types called the ***n*-channel D-type MOSFET** and ***p*-channel D-type MOSFET.** The construction and schematic symbol for these two D-MOSFET types are shown in Figure 22-16.

To begin with, let us examine the construction of the more frequently used *n*-channel D-type MOSFET, shown in Figure 22-16(a). This type of MOSFET

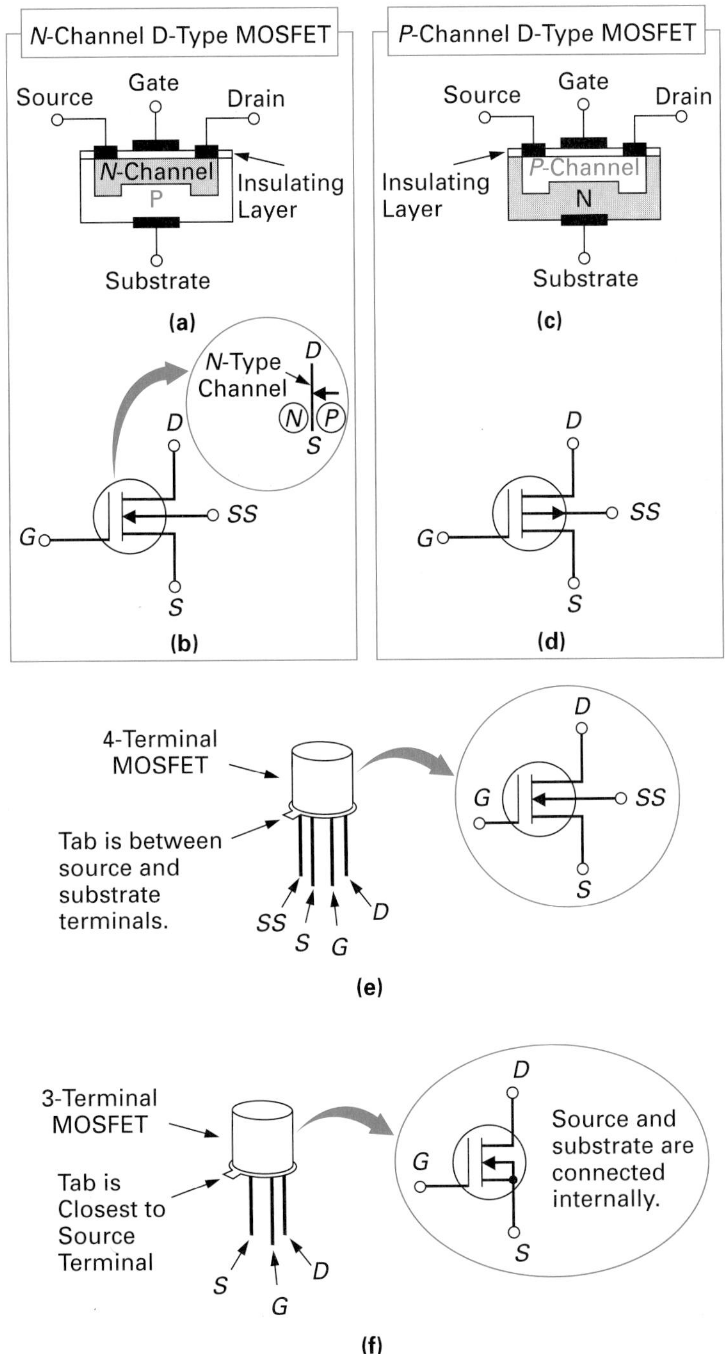

FIGURE 22-16 **D-Type MOSFET Construction and Types.**

basically consists of an *n*-type channel formed on a *p*-type substrate. A source and drain lead are connected to either end of the *n*-channel, and an additional lead is attached to the substrate. In addition, a thin insulating (silicon dioxide) layer is placed on top of the *n*-channel, and a metal plated area with a gate lead attached is formed on top of this insulating layer. Figure 22-16(b) shows the schematic symbol for an *n*-channel D-MOSFET, and, as you can see, the arrow on the substrate (*SS*) or base (*B*) lead points into the device. As a memory aid, imagine this arrowhead as a P-N junction diode as shown in the inset. The source and drain leads are connected to either end of the diode's *n*-type cathode. Therefore, this device must be an *n*-channel D-MOSFET. The basic difference in the construction and schematic symbol of the *p*-channel D-MOSFET can be seen in Figure 22-16(c) and (d).

Figure 22-16(e) and (f) show how the MOSFET is available as a four-terminal or three-terminal device. In some applications, a separate bias voltage will be applied to the substrate terminal for added control of drain current and the four-terminal device will be used. In most circuit applications, however, the three-terminal device, which has its source and substrate lead internally connected, is all that is needed.

D-MOSFET Operation

Figure 22-17(a) shows the typical drain characteristic curve for an *n*-channel depletion-type MOSFET. As you can see, this set of curves has the same general shape as the JFET's set of drain curves and the BJT's set of collector curves. The key difference is that V_{GS} is plotted for both positive and negative values. This is because the D-MOSFET should actually be called a DE-MOSFET because it can be operated in both the depletion mode (in which V_{GS} is a negative value) and the enhancement mode (in which V_{GS} is a positive value). To best understand the operation of the D-MOSFET, let us examine the three operation diagrams shown in Figure 22-17(b), (c), and (d).

Zero-Volt Operation: The center operation diagram, Figure 22-17(b), shows how the *n*-channel D-MOSFET will respond to a V_{GS} input of zero volts. When $V_{GS} = 0$ V, the gate and source terminals are at the same zero volt potential, and therefore the gate is effectively shorted to the source. The value of drain current passing through the channel is called the I_{DSS} value (I_{DSS} is the drain-to-source current passing through the channel when the gate is shorted to the source). Therefore, when

$$V_{GS} = 0 \text{ V}, I_D = I_{DSS}$$

When zero volts is applied to the input of a D-type MOSFET, therefore, it will conduct a value of drain current. With no input, this device is ON, which is why the D-type MOSFET is known as a "normally ON" device.

Enhancement-Mode: The upper operation diagram, Figure 22-17(c), shows how the *n*-channel D-MOSFET will respond when V_{GS} is made positive. In this condition, the channel is enhanced or widened, and the value of I_D is increased above I_{DSS}. Therefore, when

$$V_{GS} = +V, I_D > I_{DSS}$$

Let us examine in more detail why the channel is widened by a positive gate voltage. Because the valence-band holes in the *p*-type material (majority carriers) will be repelled by a positive gate voltage, and the conduction band electrons in the *p*-type material (minority carriers) will be attracted to the channel by the positive gate voltage, there will be a build-up of elec-

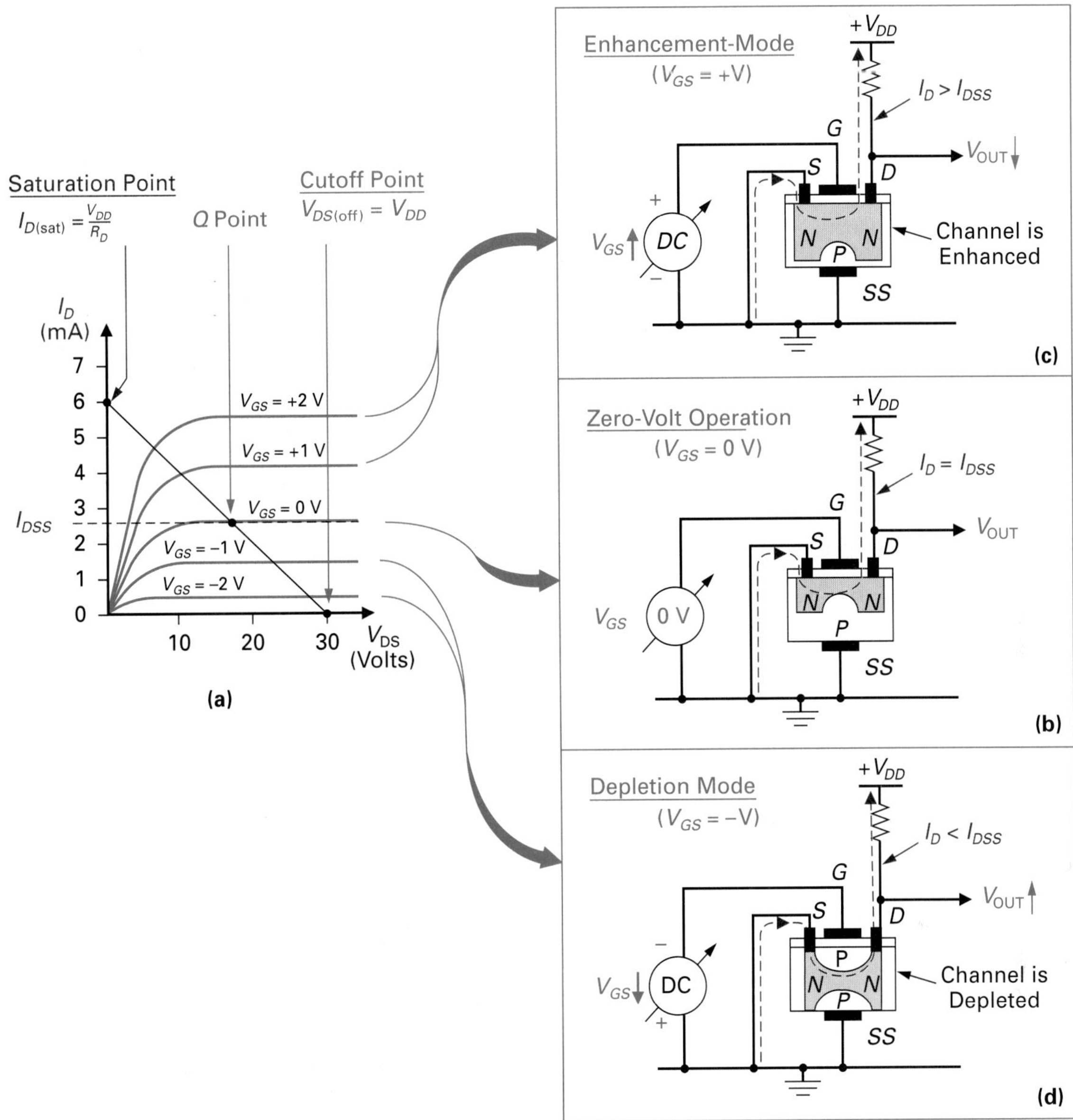

FIGURE 22-17 D-Type MOSFET Operation and Characteristics.

trons in the *p*-type material near the channel. This build-up of electrons in the *p*-type material below the channel will effectively widen the size of the channel, reducing its resistance, and therefore increasing I_D to a value greater than I_{DSS}.

Depletion-Mode: The lower operation diagram, Figure 22-17(d), shows how the *n*-channel D-MOSFET will respond when V_{GS} is made negative. In this condition, the channel is depleted of free carriers, and therefore the value of I_D is decreased below I_{DSS}. Therefore, when

$$V_{GS} = -V,\ I_D < I_{DSS}$$

To summarize the n-channel MOSFET's operation, when V_{GS} was either zero volts or a negative voltage, the n-channel D-MOSFET acted in almost exactly the same way as an n-channel JFET. However, unlike the JFET, the D-MOSFET can have a forward-biased gate-to-source P-N junction because the silicon dioxide insulating layer prevents any current from passing through the gate and will still maintain a high input resistance. This dual operating ability is why the depletion-type MOSFET or D-MOSFET should actually be called a depletion-enhancement or DE-MOSFET.

The drain characteristic curves of the D-MOSFET can be used to plot the device's dc load line, as shown in Figure 22-17(a), with

$$I_{D(\text{sat})} = \frac{V_{DD}}{R_D} \text{ at saturation, and}$$

$$V_{DS(\text{OFF})} = V_{DD} \text{ at cutoff.}$$

As with the JFET, the D-MOSFET's transconductance is equal to the ratio of output current change (ΔI_D) to input voltage change (ΔV_{GS}),

$$\delta_m = \frac{\Delta I_D}{\Delta V_{GS}}$$

and the D-MOSFET's voltage gain is equal to

$$A_V = \delta_m \times R_D$$

EXAMPLE:

A D-MOSFET circuit has the following specifications:

$$I_{DSS} = 2 \text{ mA}, V_{GS(\text{OFF})} = -6 \text{ V}, R_D = 3 \text{ k}\Omega, V_{DD} = 12 \text{ V}$$

Calculate the following two extremes on the D-MOSFET's load line:

1. $I_{D(\text{sat})}$
2. $V_{DS(\text{OFF})}$

Solution:

1. When the D-MOSFET is saturated, it is equivalent to a closed switch and therefore the only resistance is that of R_D.

$$I_{D(\text{sat})} = \frac{V_{DD}}{R_D} = \frac{12 \text{ V}}{3 \text{ k}\Omega} = 4 \text{ mA}$$

2. When the D-MOSFET is cut off, it is equivalent to an open switch, and therefore the full drain supply voltage will appear across the open between drain and source.

$$V_{DS(\text{OFF})} = V_{DD} = 12 \text{ V}$$

D-MOSFET Biasing

Like the JFET, the D-MOSFET can be configured in the same way as a common-drain, common-gate, or common-source circuit, with all of the dc and ac configuration characteristics being the same. As far as biasing, the D-MOS-

FET is easier to bias than the JFET because of its ability to operate in either the depletion mode ($-V_{GS}$) or the enhancement mode ($+V_{GS}$). In fact, one of the most frequently used D-MOSFET biasing methods is to simply have no biasing at all. This biasing method is called **zero biasing** because the Q point is set at zero volts ($V_{GS} = 0$ V), as seen in Figure 22-18. This makes biasing the D-MOSFET very simple because no gate or source bias voltages are needed. The ac input signal developed across R_G is therefore applied to the extremely high input impedance of the D-MOSFET, causing an increase and decrease in the conduction of the MOSFET above and below the $V_{GS} = 0$ V, Q point.

Zero Biasing
A configuration in which no bias voltage is applied at all.

D-MOSFET Applications

The D-MOSFET is most frequently used in analog or linear circuit applications. This is because the D-MOSFET can be very simply biased at a midpoint in the load line and then have its output current varied above and below this natural Q point in a linear fashion. This, coupled with the D-MOSFET's almost infinite input impedance and low noise properties, makes it ideal as a preamplifier at the front end of a system. Figure 22-19 shows how two D-MOSFETs can be used to construct a typical front-end **cascode amplifier circuit.** This cascode amplifier circuit consists of a self-biased common-source amplifier (Q_1) in series with a voltage-divider biased common-gate amplifier (Q_2). The input signal (V_{IN}) is applied to Q_1's gate, and the amplified output at Q_1's drain is then passed to Q_2's source, where it is further amplified by Q_2 before appearing at Q_2's drain, and therefore at the output (V_{OUT}).

Cascode Amplifier Circuit
An amplifier circuit consisting of a self-biased common-source amplifier in series with a voltage-divider biased common-gate amplifier.

The FET's only limiting factor is that its high input impedance starts to decrease as the input signal's frequency increases. Refer to the inset in Figure 22-19, which shows how the gate, insulator, and channel of a D-MOSFET form a capacitor. This input capacitance of typically 5 pF has very little effect at low input signal frequencies ($X_C\uparrow = 1/2\pi f\downarrow C$) because the input impedance is high ($X_C\uparrow$ therefore $Z_{IN}\uparrow$) and the loading effect is negligible. At higher radio frequency, however, ($X_C\downarrow = 1/2\pi f\uparrow C$) the input impedance is lowered ($X_C\downarrow$ therefore $Z_{IN}\downarrow$) and the D-MOSFET loses its high input impedance advantage. To compensate for this disadvantage, FETs are often connected in series, as in Fig-

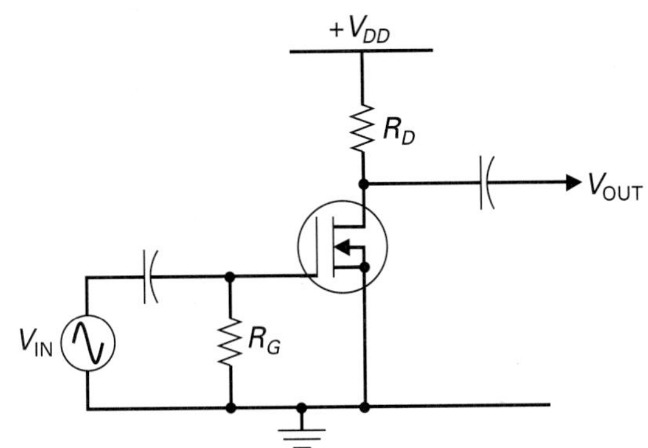

FIGURE 22-18 **Zero Biasing a D-MOSFET.**

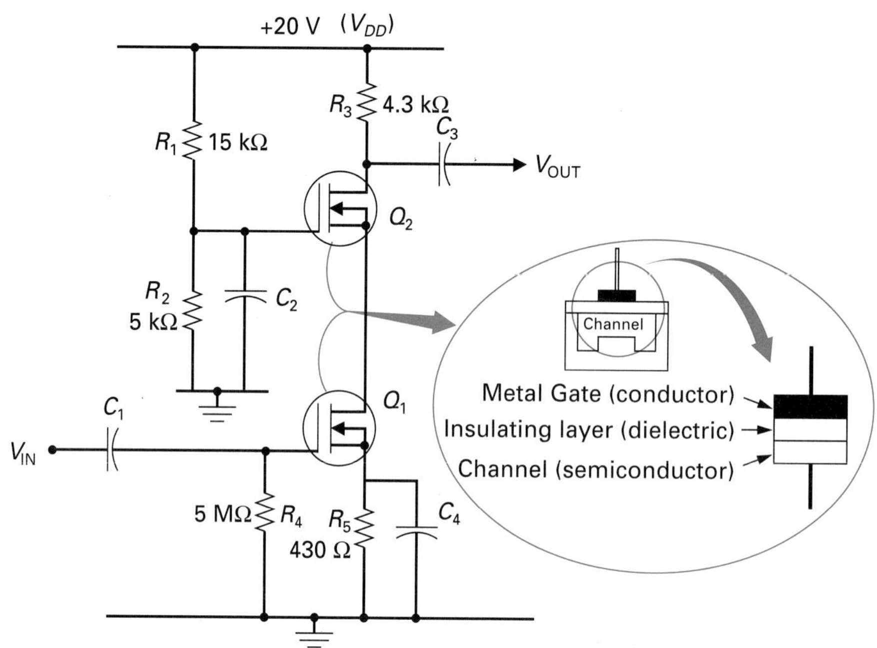

FIGURE 22-19 **A D-MOSFET Analog Circuit Application—Cascode Amplifier.**

ure 22-19, so that their input capacitances are also in series. You may recall from your introductory dc/ac electronics theory that series connected capacitors have a lower total capacitance than either of the individual capacitance values. Therefore, the overall input capacitance of two series-connected D-MOSFETs will be less than that of a single D-MOSFET, making this cascode amplifier ideal as a high radio frequency (RF) amplifier: a low input capacitance ($C_{IN}\downarrow$) means a high input reactance ($X_{C(IN)}\uparrow$), and therefore a high input impedance ($Z_{IN}\uparrow$) at high frequencies.

Dual-Gate D-MOSFET

Dual-Gate D-MOSFET
A metal oxide semiconductor FET having two separate gate electrodes.

To compensate for the D-MOSFET's input capacitance problem, the **dual-gate D-MOSFET** was developed. The construction and schematic symbol for the dual-gate D-MOSFET is shown in Figure 22-20(a). In most applications, the dual-gate D-MOSFET is connected so it acts as two series-connected D-MOSFETs, as shown in the cascode amplifier circuit in Figure 22-20(b). With this amplifier, the ac input signal drives the lower gate, which acts like a common-source amplifier. The output of the common-source lower section of the dual-gate D-MOSFET drives the upper half, which acts like a common-gate amplifier. The inset in Figure 22-20(b) shows how the dual-gate D-MOSFET is equivalent to two series-connected D-MOSFETs. As with the previous cascode amplifier, the overall input capacitance of a dual-gate D-MOSFET is less than that of a standard D-MOSFET, and if capacitance is low, X_C, and therefore Z_{IN}, are high.

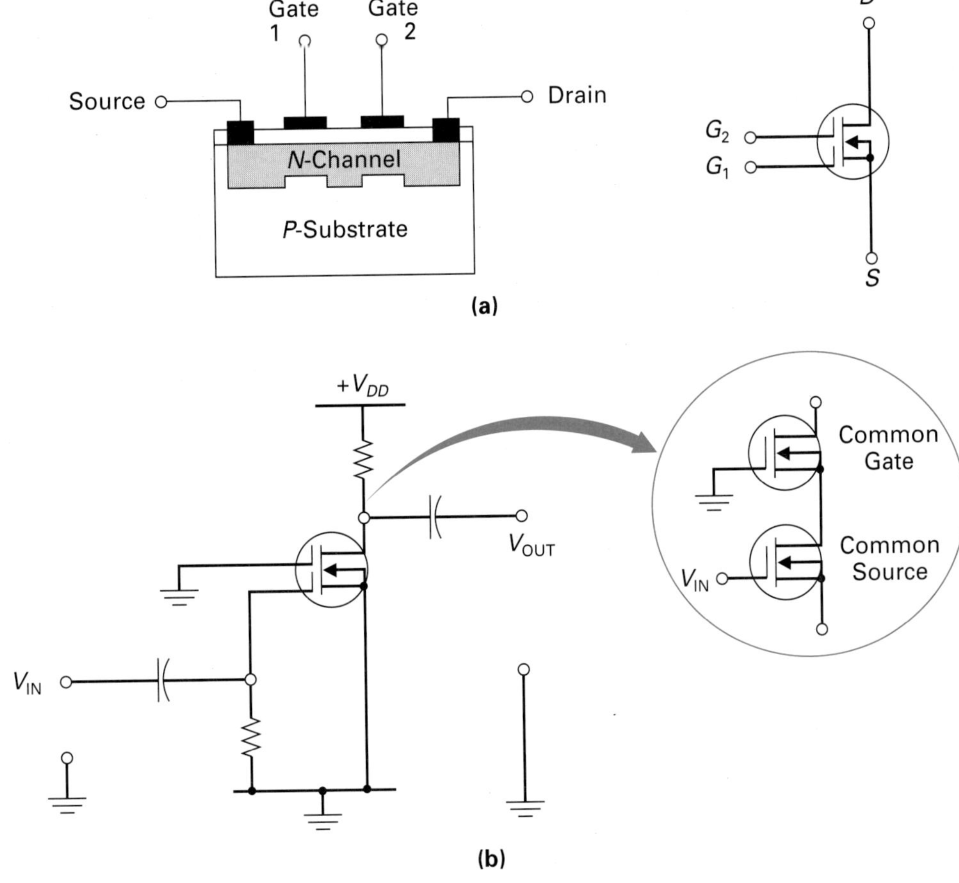

FIGURE 22-20 The Dual-Gate D-MOSFET. (a) Dual-Gate D-MOSFET Construction and Schematic Symbol. (b) Dual-Gate D-MOSFET Application—Cascode Amplifier.

SELF-TEST REVIEW QUESTIONS FOR SECTION 22-2-1

1. The two different types of MOSFETs are called the ____________ type MOSFET and ____________ type MOSFET.
2. The D-type MOSFET can be operated in both the depletion and enhancement mode. (True/False)
3. The D-type MOSFET is a normally ____________ (ON/OFF) device.
4. Which FET has a higher input impedance: JFET or MOSFET?
5. Why is the D-MOSFET ideal as a preamplifier?
   a. It can be mid-load-line biased when 0 V is applied.
   b. It has a high input impedance.
   c. It has low noise properties.
   d. All of the above
6. The ____________ MOSFET was developed to lower input capacitance so that it can handle high-frequency signals.

22-2-2 *The Enhancement-Type (E-Type) MOSFET*

With an input of zero volts, a D-MOSFET will be ON, and a certain value of current will pass through the channel between source and drain. If the input to the D-MOSFET is made positive, the channel is enhanced, causing the source-to-drain current to increase. If the input is made negative, the channel is depleted, causing the source-to-drain current to decrease. The D-MOSFET can therefore operate in either the enhancement or depletion mode and is called a "normally ON" device because it is ON when nothing (0 V) is applied.

Enhancement-Type MOSFET or E-MOSFET
A field effect transistor with an insulated gate (MOSFET) that can only be turned ON if the channel is enhanced.

The **enhancement-type MOSFET** or **E-MOSFET** can only operate in the enhancement-mode. In other words, when the input is either zero volts or a negative voltage, the transistor is OFF and there is no source-to-drain current. However, when the input is made positive, the E-MOSFET will turn ON, resulting in a source-to-drain channel current. The E-MOSFET is therefore a "normally OFF" device because it is OFF when nothing (0 V) is applied.

E-MOSFET Construction

n-Channel E-Type MOSFET
An enhancement type MOSFET having an n-type channel between its source and drain terminals.

p-Channel E-Type MOSFET
An enhancement type MOSFET having a p-type channel between its source and drain terminals.

As with all of the other transistor types, it is easier to understand the operation and characteristics of a device once we have seen how the component is constructed. The E-MOSFET has two basic transistor types called the ***n*-channel E-type MOSFET** and ***p*-channel E-type MOSFET.** The construction and schematic symbol for these two E-MOSFET types are shown in Figure 22-21.

To begin with, let us examine the construction of the more frequently used n-channel E-type MOSFET, shown in Figure 22-21(a). Studying the construction of this E-MOSFET, notice that no channel exists between the source and drain. Consequently, with no gate bias voltage, the device will be OFF. When the gate is made positive, however, electrons will be attracted from the substrate, causing a channel to be induced between the source and drain. This enhanced channel will permit drain current to flow, and any further increase in gate voltage will cause a corresponding increase in the size of the channel and therefore the value of I_D.

The schematic symbol for the n-channel E-type MOSFET can be seen in Figure 22-21(b). The construction and schematic symbol for the p-channel E-MOSFET can be seen in Figure 22-21(c) and (d). As with the D-MOSFET, both 4-terminal and 3-terminal devices are available, with the 3-terminal device having a common connection between source and substrate as shown in Figure 22-21(e).

The only difference between the E-type and D-type MOSFET symbols is the three dashed lines representing the drain, substrate, and source regions. The dashed line is used instead of the solid line to indicate that an E-MOSFET has a normally broken path, or channel, between drain, substrate, and source. To aid memory, the three dashed lines used in the "E"-MOSFET symbol could be thought of as the three horizontal prongs in the capital letter "E," as shown in Figure 22-21(f).

E-MOSFET Operation

Figure 22-22(a) shows the typical drain characteristic curves for an n-channel enhancement-type MOSFET. As you can see, this set of drain curves is very similar to the D-MOSFETs set of drain curves; however, in this case only posi-

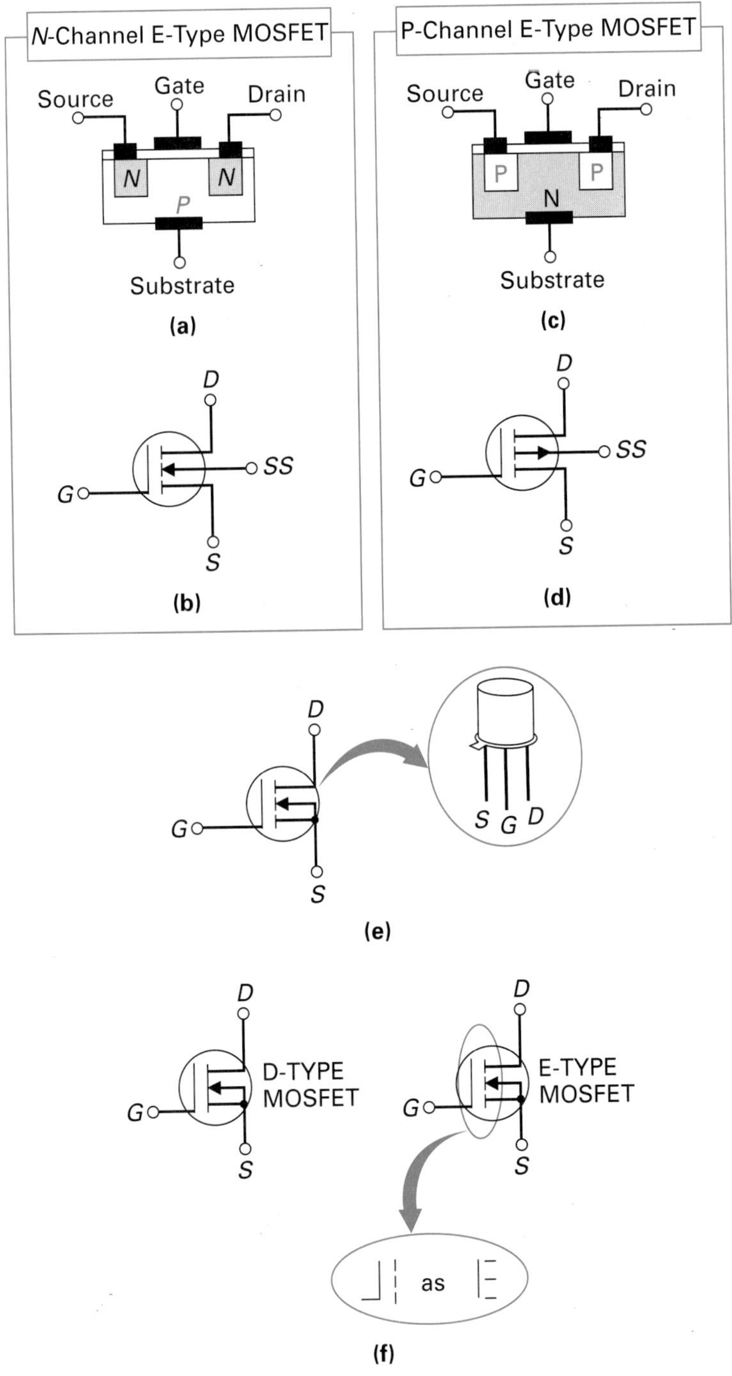

FIGURE 22-21 **E-Type MOSFET Construction and Type.**

tive values of V_{GS} are plotted. Looking at the relationship between V_{GS} and I_D, you may have noticed that any increase in V_{GS} will cause a corresponding increase in I_D.

To best understand the operation of the E-MOSFET, let us examine the three operation diagrams shown in Figure 22-22(b), (c), and (d).

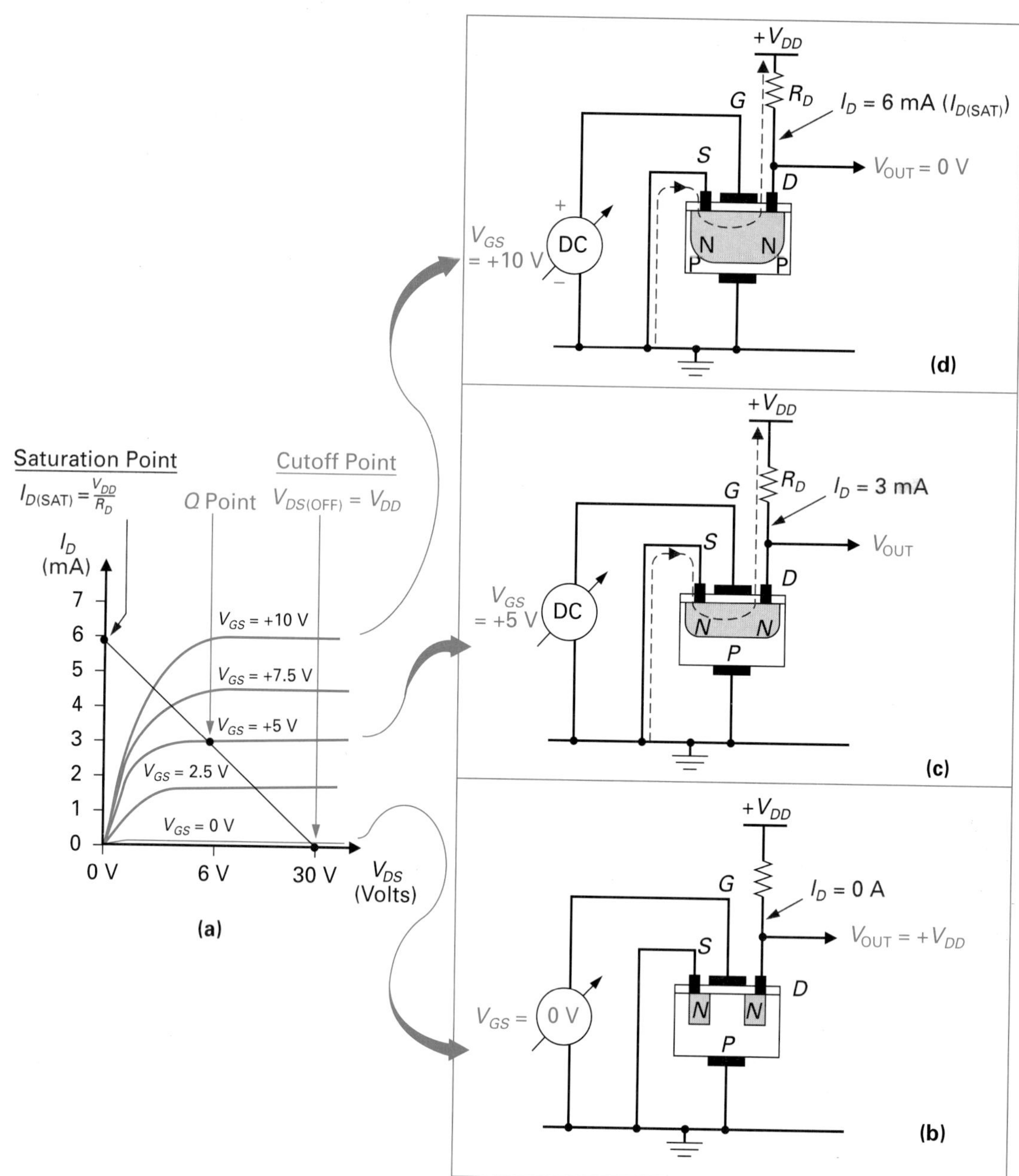

FIGURE 22-22 E-Type MOSFET Operation and Characteristics.

V_{GS} = 0 V Curve: The lower operation diagram, Figure 22-22(b), shows how the n-channel E-MOSFET will respond to a V_{GS} input of zero volts. When $V_{GS} = 0$ V, there is no channel connecting the source and drain, and therefore the drain current will be zero. As a result, $+V_{DD}$ will be present at the output because the E-MOSFET is equivalent to an open switch between drain and source.

V_{GS} = +5 V Curve: The center operation diagram, Figure 22-22(c), shows how the n-channel E-MOSFET will respond to a V_{GS} input of +5 volts. When V_{GS} = +5 V, the E-MOSFET will act in almost exactly the same way as an enhanced D-MOSFET. The positive gate voltage will repel the p-type material's majority carriers (holes) away from the gate, while attracting the p-type material's minority carriers (electrons) towards the gate. This action will form an n-type bridge between the source and drain, and therefore a value of I_D will flow between source and drain, as shown.

V_{GS} = +10 V Curve: The upper operation diagram, Figure 22-22(d), shows how the n-channel E-MOSFET will respond if the V_{GS} input is further increased to +10 volts. When V_{GS} = +10 V, the attraction of electrons and repulsion of holes within the p-type material is increased, causing the channel's width to increase and I_D to also increase. As a result, 0 V will be present at the output because the E-MOSFET is equivalent to a closed switch between drain and source.

In summary, when V_{GS} is zero volts, drain current is also zero. As the value of V_{GS} is increased (made more positive), the channel becomes wider, causing I_D to increase. On the other hand, as the value of V_{GS} is decreased (made less positive), the channel becomes narrower, causing I_D to decrease. In other words, when the input is either zero volts or a negative voltage, the transistor is OFF, and there is no source-to-drain current. However, when the input is made positive, the E-MOSFET will turn ON, resulting in a source-to-drain channel current. The E-MOSFET is therefore a "normally OFF" device because it is OFF when nothing (0 V) is applied.

Although it cannot be seen in the set of drain curves in Figure 22-22(a), V_{GS} will have to increase to a positive threshold voltage of about +1 V before a channel will be induced and a small value of drain current will flow. This "threshold level" is a highly desirable characteristic because it prevents noise or any low-level input signal voltage from turning ON the device. This advantage makes the E-type MOSFET ideally suited as a switch because it can be turned ON by an input voltage and turned OFF once the input voltage falls below the threshold level.

The p-channel E-type MOSFET operates in much the same way as the n-channel device except that holes are attracted from the substrate to form a p-channel and the V_{GS} and V_{DS} bias voltages are reversed.

E-MOSFET Biasing

Like the D-MOSFET, the E-MOSFET can be configured as a common-drain, common-gate, or common-source circuit. Unlike the D-MOSFET and JFET, the E-MOSFET cannot be biased using self bias or zero bias because V_{GS} must be a positive voltage. As a result, gate bias and voltage-divider bias can be used; however, more frequently E-MOSFETs are **drain-feedback biased,** as shown in Figure 22-23(a). In this example, R_D equals 8 kΩ, and R_G (which feeds back a positive voltage from the drain, hence the name drain-feedback bias) equals 100 MΩ. Because an E-MOSFET has an extremely high input impedance (due to the insulated gate), no current will flow in the gate circuit. With no gate

Drain-Feedback Biased
A configuration in which the gate receives a bias voltage fed back from the drain.

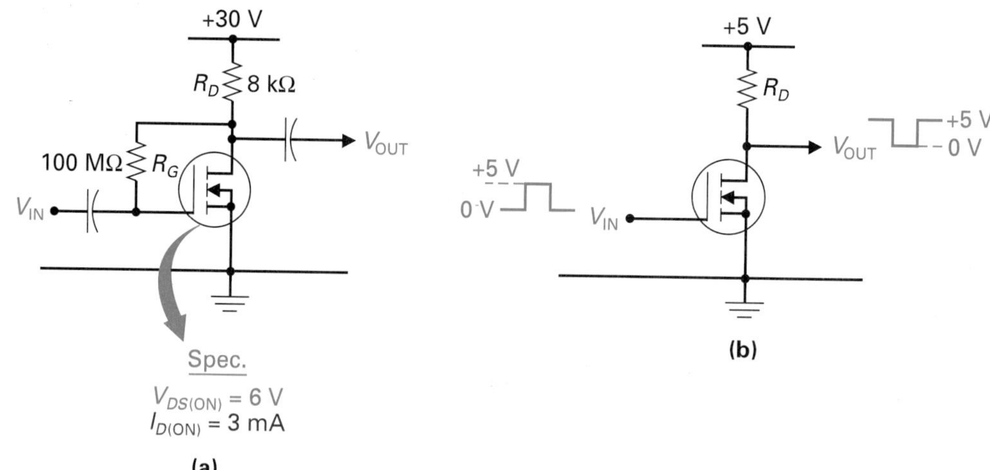

FIGURE 22-23 Biasing E-Type MOSFETs. (a) Drain-Feedback Biasing. (b) Voltage Controlled Switching.

current, there will be no voltage drop across the gate resistor ($V_{RG} = 0$ V), and therefore the voltage at the gate will be at the same potential as the voltage at the drain.

$$V_{GS} = V_{DS}$$

To help set up the Q point, most manufacturer's data sheets specify a load line midpoint drain current $I_{D(\mathrm{ON})}$ and drain voltage $V_{DS(\mathrm{ON})}$. In the example in Figure 22-23(a), when the E-MOSFET is ON, or conducting, and $I_D = 3$ mA, the E-MOSFET's drain-to-source voltage drop ($V_{DS(\mathrm{ON})}$) is 6 V. The value of R_D in Figure 22-23(a) has been chosen so that this E-MOSFET circuit will be biased at its specified Q point. To check, we can use the formula

$$V_{DS} = V_{DD} - (I_{D(\mathrm{ON})} \times R_D)$$

In the example in Figure 22-23(a)

$$V_{DS} = V_{DD} - (I_{D(\mathrm{ON})} \times R_D) = 30\text{ V} - (3\text{ mA} \times 8\text{ k}\Omega)$$
$$= 30\text{ V} - 24\text{ V} = 6\text{ V}$$

EXAMPLE:

A drain-feedback biased E-MOSFET circuit has the following specifications: $R_D = 1$ kΩ, $I_{D(\mathrm{ON})} = 10$ mA, $V_{DD} = 20$ V. Calculate V_{DS} and V_{RD}.

Solution:

$$V_{DS} = V_{DD} - (I_{D(\mathrm{ON})} \times R_D)$$
$$= 20\text{ V} - (10\text{ mA} \times 1\text{ k}\Omega)$$
$$= 20\text{ V} - 10\text{ V} = 10\text{ V}$$

The constant drain-supply voltage (V_{DD}) will be evenly divided across R_D and the E-MOSFET's drain-to source ($V_{DS} = 10$ V, $V_{RD} = 10$ V).

In most instances, the E-MOSFET will be used in digital two-state switching circuit applications. In this case, there will be no need for a gate resistor (R_G) because the input voltage (V_{IN}) will either turn ON or OFF the E-MOSFET, as shown in Figure 22-23(b). For example, when $V_{IN} = 0$ V, the E-MOSFET is OFF, therefore $V_{OUT} = +V_{DD} = +5$ V, whereas when $V_{IN} = +5$ V, the E-MOSFET is ON and therefore $V_{OUT} = 0$ V.

E-MOSFET Applications

The E-MOSFET is more frequently used in digital or two-state circuit applications. One reason is that it naturally operates as a "normally OFF-voltage controlled switch" because it can be turned ON when the gate voltage is positive and turned OFF when the gate voltage falls below a threshold level. This threshold level is a highly desirable characteristic because it prevents noise from false triggering, or accidentally turning ON, the device. The other E-MOSFET advantage is its extremely high input impedance, which means that the device's circuit current, and therefore power dissipation, are low. This enables us to densely pack or integrate many thousands of E-MOSFETs onto one small piece of silicon, forming a high component density integrated circuit (IC). These low-power and high-density advantages make the E-MOSFET ideal in battery-powered small-size (portable) applications such as calculators, wristwatches, notebook computers, hand-held video games, digital cellular phones, and so on.

Vertical-Channel E-MOSFET (VMOS FET)

As just mentioned, the E-MOSFET is generally used in two-state or digital circuit applications, where it acts as a normally OFF switch. In most digital circuit applications, the channel current is small, and therefore a standard E-MOSFET can be used. However, if a larger current-carrying capability is needed, the **vertical-channel E-MOSFET** or VMOS FET can be used. Figure 22-24(a) shows the construction of a VMOS FET. The gate at the top of the device is insulated from the source (which is also at the top), and as with all E-MOSFETs, no channel exists between the source terminal and the drain terminal with no bias voltage applied. The VMOS FET's semiconductor materials are labeled *P*, *N*+, and *N*– and indicate different levels of doping.

Vertical-Channel E-MOSFET
An enhancement type MOSFET that, when turned ON, forms a vertical channel between source and drain.

When this *n*-channel VMOS FET is biased ON (gate is made positive with respect to source), as seen in Figure 22-24(b), a vertical *n*-type channel is formed between source and drain. This channel is much wider than a standard E-MOSFET's horizontal channel, which is why VMOS FETs can handle a much higher drain current.

Figure 22-24(c) shows how the high-current capability of a VMOS FET can be made use of in an interfacing circuit application. In this circuit, a VMOS FET is used to interface a low-power source input signal to a high-power load. Referring to the current specifications of the VMOS FET, relay, and motor, you can see that each device is used to step up current. The standard E-MOSFET sup-

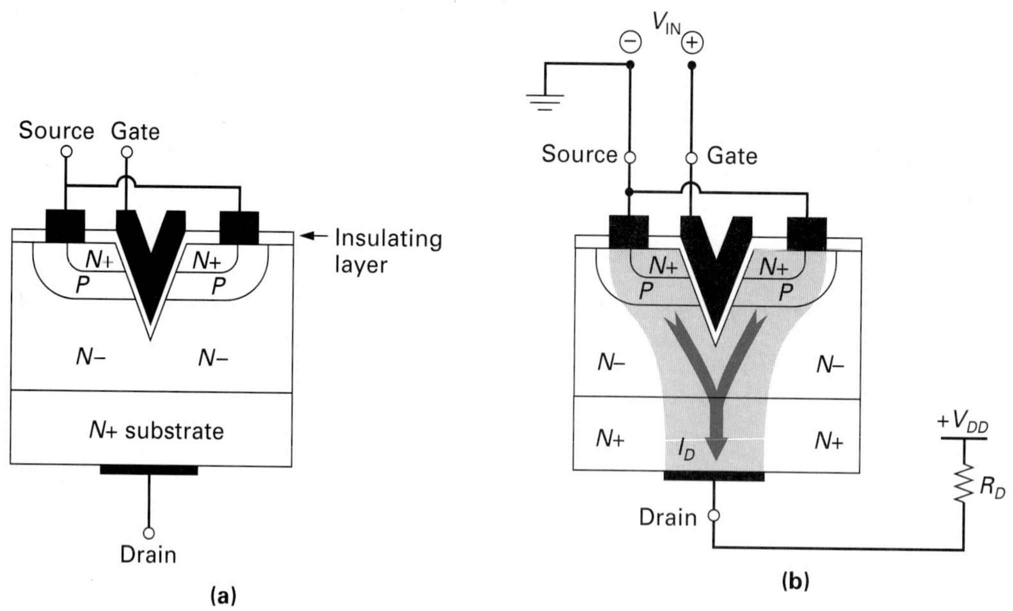

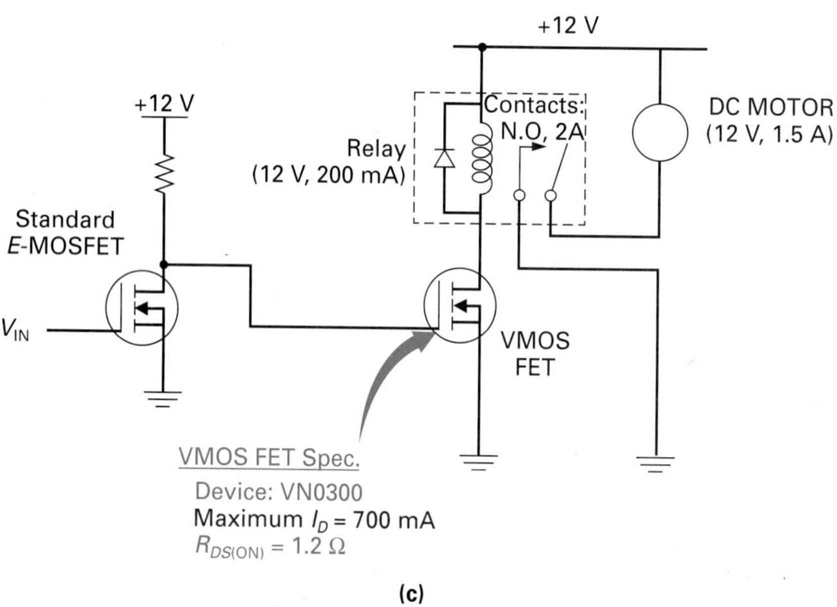

FIGURE 22-24 The Vertical Channel E-MOSFET (VMOS FET). (a) VMOS FET OFF. (b) VMOS FET ON. (c) VMOS FET Application Circuit—Low-Power Signal to High-Power Load Circuit.

plies a low-power input signal to the VMOS FET, which can handle enough current to actuate a relay whose contacts can handle enough current to switch power to the dc motor. To be more specific, if V_{IN} is LOW, the standard E-MOSFET will turn OFF, producing a HIGH output to the gate of the VMOS FET, turning it ON. When the VMOS FET turns ON, it effectively switches ground through to the lower end of the relay coil, energizing the relay and closing its normally open contacts. The closed relay contacts switch ground through to the lower end of the motor, which turns ON because it now has the full +12 V supply across its terminals. The motor will stay ON as long as V_{IN} stays LOW. If V_{IN} were to go HIGH, the standard E-MOSFET inverter would produce a LOW output, which would turn OFF the VMOS FET, relay, and motor.

Other than its high-current capability, the VMOS FET also has a positive temperature coefficient of resistance, which means an increase in temperature will cause a decrease in drain current ($T\uparrow$, $R\uparrow$, $I\downarrow$), and this will prevent thermal runaway. This gives the VMOS FET a distinct advantage over the BJT power amplifier, which has a negative temperature coefficient of resistance ($T\uparrow$, $R\downarrow$, $I\uparrow$). This means that if maximum ratings are exceeded, an increase in temperature will cause an increase in current, which will generate a further increase in heat (temperature) and current, and so on.

MOSFET Data Sheets

Throughout this chapter, we have used certain MOSFET specifications in our calculations, such as I_{DSS}, $V_{DS(ON)}$, $I_{D(ON)}$, and so on. As an example, Figure 22-25 shows the data sheet for a typical E-type MOSFET, and, as you can see, these specifications are listed.

MOSFET Handling Precautions

Certain precautions must be taken when handling any MOSFET devices. The very thin insulating layer between the gate and the substrate of a MOSFET can easily be punctured if an excessive voltage is applied. Your body can build up extremely large electrostatic charges due to friction. If this charge came in contact with the pins of a MOSFET device, an electrostatic-discharge (ESD) would occur, resulting in a possible arc across the thin insulating layer causing permanent damage. Most MOSFETs presently manufactured have zeners internally connected between gate and source to bypass high voltage static or in-circuit potentials and protect the MOSFET. However, it is important to remember the following.

1. All MOS devices are shipped and stored in a "conductive foam" or "protective foil" so that all of the IC pins are kept at the same potential, and therefore electrostatic voltages cannot build up between terminals.
2. When MOS devices are removed from the conductive foam, be sure not to touch the pins because your body may have built up an electrostatic charge.
3. When MOS devices are removed from the conductive foam, always place them on a grounded surface such as a metal tray.
4. When continually working with MOS devices, use a "wrist grounding strap," which is a length of cable with a 1 MΩ resistor in series. This prevents electrical shock if you come in contact with a voltage source.
5. All test equipment, soldering irons, and work benches should be properly grounded.
6. All power in equipment should be off before MOS devices are removed or inserted into printed circuit boards.

DEVICE: 2N7002LT1—N-Channel Silicon E-MOSFET

MAXIMUM RATINGS

Rating	Symbol	Value	Unit
Drain-Source Voltage	V_{DSS}	60	Vdc
Drain-Gate Voltage (R_{GS} = 1 MΩ)	V_{DGR}	60	Vdc
Drain Current – Continuous TC = 25°C(1) T_C = 100°C(1) – Pulsed (2)	I_D I_D I_{DM}	±115 ±75 ±800	mA
Gate-Source Voltage – Continuous – Non-repetitive (tp ≤50 μs)	V_{GS} V_{GSM}	±20 ±40	Vdc Vpk
Total Power Dissipation T_C = 25°C T_C = 100°C Derate above 25°C ambient	P_D	200 80 1.6	mW mW/°C

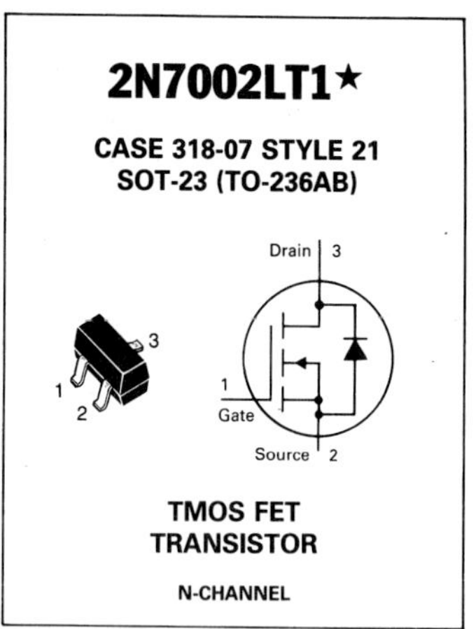

ELECTRICAL CHARACTERISTICS (T_A = 25°C unless otherwise noted.)

Characteristic	Symbol	Min	Typ	Max	Unit
OFF CHARACTERISTICS					
Drain-Source Breakdown Voltage (V_{GS} = 0, I_D = 10 μA)	$V_{(BR)DSS}$	60	—	—	Vdc
Zero Gate Voltage Drain Current (V_{GS} = 0, V_{DS} = 60 V) T_J = 25°C T_J = 125°C	I_{DSS}	— —	— —	1.0 500	μAdc
Gate-Body Leakage Current Forward (V_{GS} = 20 Vdc)	I_{GSSF}	—	—	100	nAdc
Gate-Body Leakage Current Reverse (V_{GS} = −20 Vdc)	I_{GSSR}	—	—	−100	nAdc

(1) The Power Dissipation of the package may result in a lower continuous drain current.
(2) Pulse Width ≤ 300 μs, Duty Cycle ≤ 2.0%.

Characteristic	Symbol	Min	Typ	Max	Unit
ON CHARACTERISTICS*					
Gate Threshold Voltage (V_{DS} = V_{GS}, I_D = 250 μA)	$V_{GS(th)}$	1.0	—	2.5	Vdc
I_D (on) → On-State Drain Current (V_{DS} ≥ 2.0 $V_{DS(on)}$, V_{GS} = 10 V)	$I_{D(on)}$	500	—	—	mA
V_{DS} (on) → Static Drain-Source On-State Voltage (V_{GS} = 10 V, I_D = 500 mA) (V_{GS} = 5.0 V, I_D = 50 mA)	$V_{DS(on)}$	— —	— —	3.75 .375	Vdc
Static Drain-Source On-State Resistance (V_{GS} = 10 V, I_D = 500 mA) T_C = 25°C T_C = 125°C (V_{GS} = 5.0 V, I_D = 50 mA) T_C = 25°C T_C = 125°C	$r_{DS(on)}$	— — — —	— — — —	7.5 13.5 7.5 13.5	Ohms
Forward Transconductance (V_{DS} ≥ 2.0 $V_{DS(on)}$, I_D = 200 mA)	g_{FS}	80	—	—	mmhos
DYNAMIC CHARACTERISTICS					
Input Capacitance (V_{DS} = 25 V, V_{GS} = 0, f = 1.0 MHz)	C_{iss}	—	—	50	pF
Output Capacitance (V_{DS} = 25 V, V_{GS} = 0, f = 1.0 MHz)	C_{oss}	—	—	25	pF
Reverse Transfer Capacitance (V_{DS} = 25 V, V_{GS} = 0, f = 1.0 MHz)	C_{rss}	—	—	5.0	pF
SWITCHING CHARACTERISTICS*					
Turn-On Delay Time (V_{DD} = 25 V, I_D ≅ 500 mA, R_G = 25 Ω, R_L = 50 Ω)	$t_{d(on)}$	—	—	30	ns
Turn-Off Delay Time (V_{DD} = 25 V, I_D ≅ 500 mA, R_G = 25 Ω, R_L = 50 Ω)	$t_{d(off)}$	—	—	40	ns
BODY-DRAIN DIODE RATINGS					
Diode Forward On-Voltage (I_S = 11.5 mA, V_{GS} = 0 V)	V_{SD}	—	—	−1.5	V
Source Current Continuous (Body Diode)	I_S	—	—	−115	mA
Source Current Pulsed	I_{SM}	—	—	−800	mA

*Pulse Test: Pulse Width ≤ 300 μs, Duty Cycle ≤ 2.0%.

FIGURE 22-25 An E-Type MOSFET Data Sheet. (Courtesy of Motorola. Used by permission.)

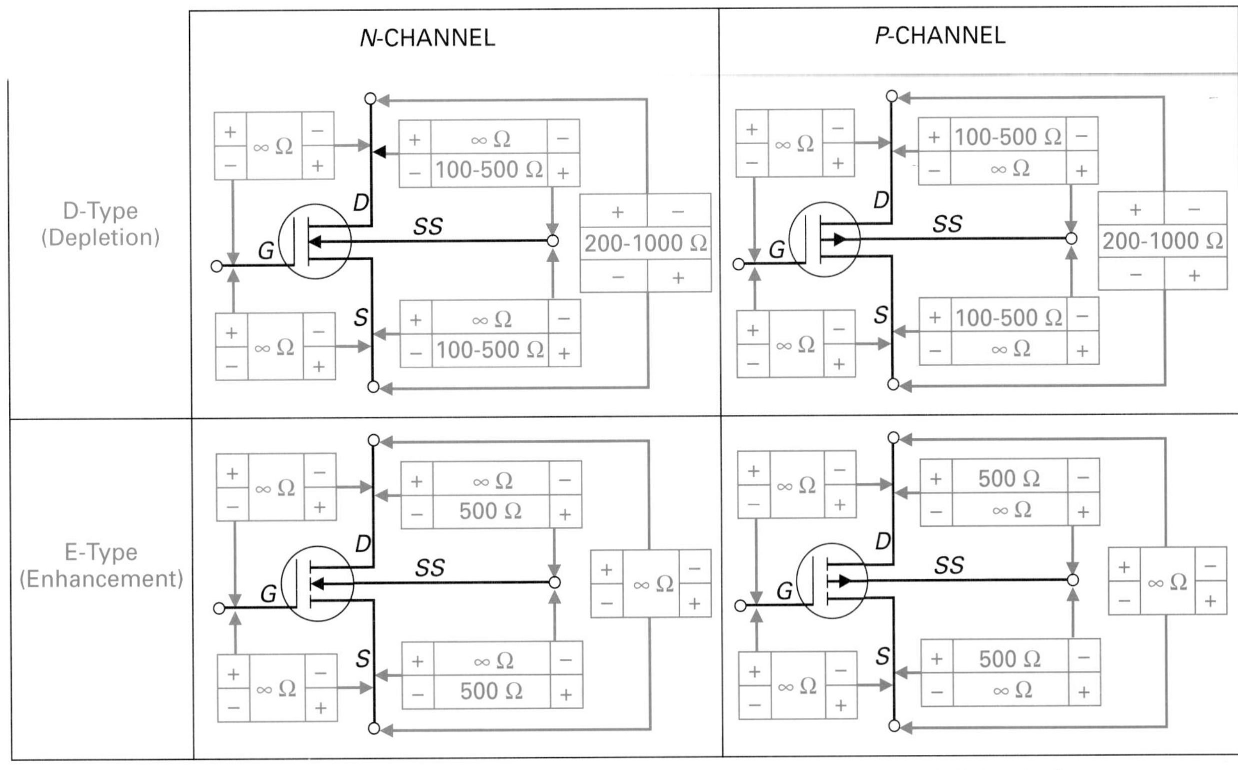

[+] = Positive Lead of Ohmmeter [−] = Negative Lead of Ohmmeter ∞ = Infinite Resistance

FIGURE 22-26 **Testing MOSFETs with the Ohmmeter.**

7. Any unused MOSFET terminals must be connected because an unused input left open can build up an electrostatic charge and float to high voltage levels.
8. Any boards containing MOS devices should be shipped or stored with the connection side of the board in conductive foam.

MOSFET Testing

Like the BJT and JFET, MOSFETs can be tested with the transistor tester to determine: first, whether an open or short exists between any of the terminals; second, the transistor's transconductance/gain; and third, the value of I_{DSS} and leakage current.

If a transistor tester is not available, the ohmmeter can be used to determine the most common failures: opens and shorts. Figure 22-26 shows what resistance values should be obtained between the terminals of a good *n*-channel or *p*-channel D-MOSFET or E-MOSFET when testing with an ohmmeter. Remember that the gate-to-channel resistance of a MOSFET is always an open due to the insulated gate.

SELF-TEST REVIEW QUESTIONS FOR SECTION 22-2-2

1. The E-Type MOSFET can be operated in both the depletion and enhancement mode. (True/False)
2. The E-Type MOSFET is a normally ____________ (ON/OFF) device.
3. The E-type MOSFET is generally
   a. Zero biased **c.** Drain-Feedback biased
   b. Base biased **d.** None of the above
4. The E-MOSFET naturally operates as a voltage controlled switch. (True/False)
5. What are the $I_{D(ON)}$ and $V_{DS(ON)}$ values for a 2N4351?
6. If the 2N4351 E-MOSFET was used as a voltage controlled switch, how long would it take
   a. To switch from OFF to ON
   b. To switch from ON to OFF
7. What general precautions should be observed when handling MOSFETs?
8. The gate-to-source or gate-to-drain resistance of a good E-MOSFET should always measure ____________ ohms.

SUMMARY

1. The field effect transistor, more commonly called an FET (pronounced "eff-ee-tee"), has three terminals and can operate as a switch—and therefore be used in digital circuit applications—or as a variable resistor—and therefore be used in analog or linear circuit applications.
2. There are two types of field effect transistors or FETs. One type is the junction field effect transistor, which is more typically called a JFET (pronounced "jay-fet"). The other type is the metal oxide semiconductor field effect transistor, which is more commonly called a MOSFET (pronounced "moss-fet").

The Junction Field Effect Transistor—JFET (Figure 22-27)

3. Like the bipolar junction transistor, the junction field effect transistor or JFET is constructed from *n*-type and *p*-type semiconductor materials.
4. Just as the bipolar junction transistor has two basic types (NPN BJT or PNP BJT), there are two types of junction field effect transistor called the *n*-channel JFET and *p*-channel JFET.
5. The more frequently used *n*-channel JFET consists of an *n*-type block of semiconductor material on top of a *p*-type substrate with a "U" shaped *p*-type section attached to the surface of a *p*-type substrate.
6. Like the BJT, the JFET has three terminals called the gate, source, and drain. With the *n*-channel JFET, the gate lead is attached to the *p*-type substrate, and the source and drain leads are attached to either end of an *n*-type channel that runs through the middle of the "U" shaped *p*-type section.

QUICK REFERENCE SUMMARY SHEET

FIGURE 22-27 The Junction Field Effect Transistor (JFET).

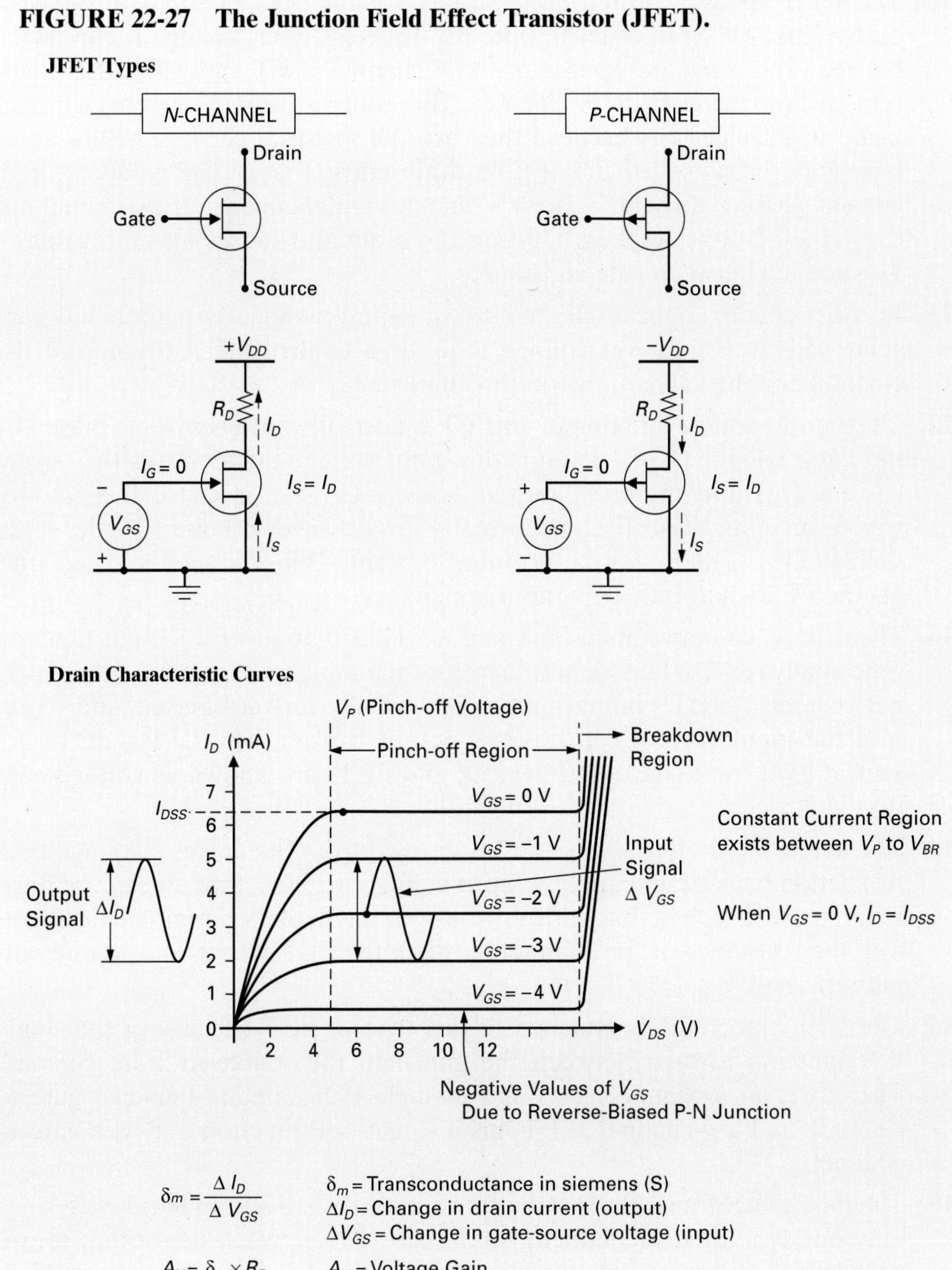

7. With the schematic symbol of the *n*-channel JFET, the gate lead's arrowhead points into the device.

8. The *p*-channel JFET is constructed in exactly the same way as the *n*-channel JFET except that the gate lead is attached to an *n*-type substrate, and the source and drain leads are attached to either end of a *p*-type channel.

9. With the schematic symbol of the *p*-channel JFET, the gate's arrowhead points out of the device.
10. The JFET requires both a drain supply voltage ($+V_{DD}$) and a gate-source bias voltage (V_{GS}) in order to operate. The $+V_{DD}$ bias voltage is connected between the drain and source of the *n*-channel JFET and will cause a current to flow through the *n*-channel. This source-to-drain current, which is made up of electrons because they are the majority carriers within an *n*-type material, is called the JFET's drain current (I_D). The value of drain current passing through a JFET's channel is dependent on two elements: the value of $+V_{DD}$ applied between the drain and source and the value of V_{GS} applied between gate and source.
11. In most circuit applications, the $+V_{DD}$ supply voltage is maintained constant and the V_{GS} input voltage is used to control the resistance of the channel and the value of the output current, I_D.
12. The gate-to-source junction of an FET is normally always reverse biased by the input voltage (V_{GS}), and it is this input voltage that controls the output current (I_D) and therefore output voltage (V_{OUT}). Because the gate-to-source junction of an FET is normally always reverse biased by the input voltage (V_{GS}), there will be no input current. This characteristic accounts for the FET's naturally high input impedance.
13. The difference between an FET and a BJT is that an FET's input junction is normally reverse biased, and therefore the input voltage controls the output current. A BJT's input junction is normally forward biased, and therefore the input current controls the output current. This is why BJTs are known as current-controlled devices, and FETs are known as voltage-controlled devices.
14. The name "field-effect transistor" is derived from the device's voltage control action because the applied input voltage will generate an electric field. It is this electric field that varies the size of the depletion region, and therefore the resistance of the channel between the FET's drain and source output terminals.
15. The term "junction" is attached to this type of FET because of the single P-N junction formed between the gate and the source-to-drain channel. Therefore, an *n*-channel JFET has a single P-N junction between gate to channel, and a *p*-channel JFET has a single N-P junction between gate to channel.
16. The field effect transistor is also often referred to as a unipolar device since only one type of semiconductor material exists between the output terminals, and therefore the charged carriers have only one polarity (unipolar). Compare this to a BJT, which is a bipolar device because there is a change in semiconductor material between the output terminals, and the charged carriers can be one of two polarities (bipolar, since both majority and minority carriers are used).
17. The drain characteristic curve is a graph plotting drain current (I_D) against drain-to-source voltage (V_{DS}).
18. At a certain V_{DS} voltage, further increases in V_{DS} will cause no further

increase in I_D. This value of V_{DS} is called the pinch-off voltage (V_P) because it is the point at which the bias voltage has caused the depletion region to pinch off, or restrict, drain current. From this point on, further increases in V_{DS} are counteracted by increases in the resistance of the channel, and therefore I_D remains constant. This flat portion in the drain characteristic curve is called the constant-current region because I_D remains constant despite changes in V_{DS}. If V_{DS} is further increased (by increasing $+V_{DD}$), the JFET will eventually reach its breakdown voltage (V_{BR}), at which time a damaging value of I_D will pass through the JFET.

19. When V_{GS} is at 0 V, a maximum value of drain current is passing through the JFET's channel. This maximum value of drain current is called the drain-to-source current with shorted gate (I_{DSS}).

20. If V_{GS} is made more and more negative, a point will be reached where the depletion regions within the JFET will get closer and closer—then eventually touch—cutting off drain current. This negative V_{GS} bias voltage that causes I_D to drop to approximately zero is called the gate-to-source cut-off voltage or $V_{GS(\mathrm{OFF})}$.

21. Like the bipolar transistor, an FET can be used to amplify a signal. As before, the amount of amplification achieved is a ratio between output and input. For a bipolar transistor, the amount of gain is equal to the ratio of input current to output current (beta). For an FET, there is no input current, and therefore an FET's gain is equal to the ratio of output current change (ΔI_D) to input voltage change (ΔV_{GS}). This ratio is called the FET's transconductance (symbolized δ_m).

22. Because transconductance is the ratio of output current change (ΔI_D) to input voltage change (ΔV_{GS}), it is no surprise that this ratio is used to determine a JFET's voltage gain.

JFET Biasing Methods (Figure 22-28)

23. The three most frequently used JFET biasing methods are gate biasing, self biasing, and voltage-divider biasing.

24. Gate biasing is the most simple of the three biasing methods.

25. With self biasing, only a single drain supply voltage is needed (V_{DD}), whereas gate-biasing requires both V_{DD} and a negative gate supply voltage ($-V_{GG}$).

26. With a self-biased JFET, we achieve the same result as gate-biasing by making the source voltage positive with respect to the gate, which is at 0 V. This makes the gate of the JFET negative with respect to the source.

27. The Q point will remain relatively stable despite changes in temperature when a JFET circuit has a source resistor included.

28. Like the self-biased circuit, the voltage-divider biased circuit includes a source resistor to stabilize the Q point despite ambient temperature changes. In addition, using a voltage divider to determine the gate-bias voltage ensures that V_{GS}, and therefore the circuit, has increased stability.

QUICK REFERENCE SUMMARY SHEET

FIGURE 22-28 JFET Biasing Methods.

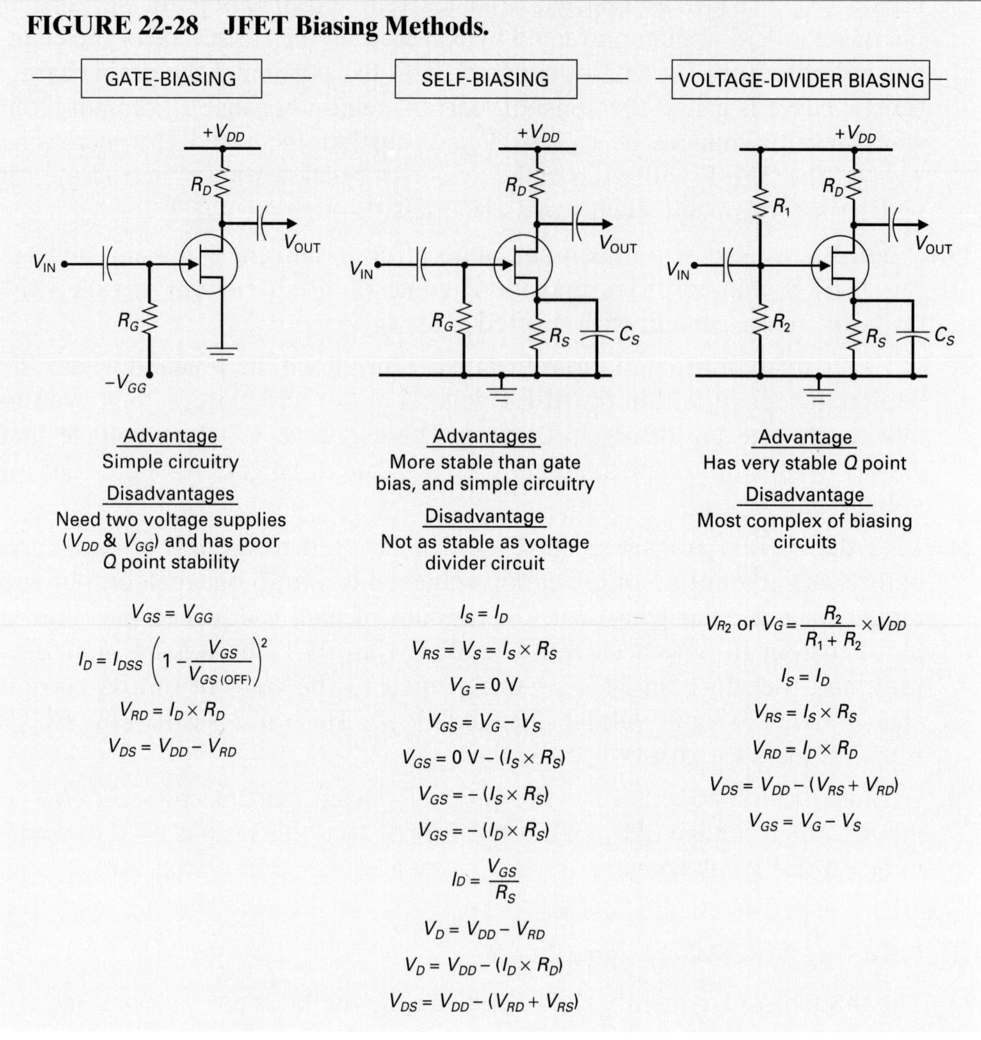

JFET Circuit Configurations (Figure 22-29)

29. The three JFET circuit configurations are called common-source, common-gate, and common-drain.

30. Similar to its bipolar counterpart, the common-emitter configuration, the common-source configuration is the most widely used JFET circuit. The input is applied between the gate and source and the output is taken between the drain and source, with the source being common to both input and output. Resistor R_S is included to provide temperature stability, and, as with the bipolar transistor, the source decoupling capacitor C_2 is included to prevent degenerative feedback.

31. When a small ac input signal is applied to the gate of a common-source amplifier, the variations in voltage at the gate control the JFET, which effectively acts as a variable resistor, varying the output drain current.

QUICK REFERENCE SUMMARY SHEET

FIGURE 22-29 JFET Circuit Configurations.

	Voltage Gain	Input Impedance	Output Impedance	Circuit Appearance and Application	Waveforms
Common Source (a)	5–10 (Voltage Amp)	Very High 1–15 MΩ	Low 2–10 kΩ	Most widely used FET configuration. It is mainly used as a voltage amplifier, however it is also used as an impedance matching device and can handle the high radio frequency signals.	V_{in} and V_{out} are out of phase (180° phase shift)
Common Gate (b)	2–5	Very Low 200–1500 Ω	Medium 5–15 kΩ	This configuration is used to amplify radio frequency signals due to its very stable nature at high frequencies. It is also used as a buffer to match a low impedance source to a high impedance load.	V_{in} and V_{out} are in phase (0° phase shift)
Common Drain (c)	0.98	Very High 1500 MΩ	Low 10 kΩ	This amplifier is commonly called a source follower as the source follows whatever is applied to the gate. Its very high input impedance will not load down (and therefore not distort) signals from high impedance signal sources, such as a microphone, and its low output impedance is ideal to drive a low impedance load such as an audio amplifier.	V_{in} and V_{out} are in phase (0° phase shift)

These changes in drain current will vary the voltage drop across R_D and the drain-to-source voltage drop that, with R_S, determines the output voltage.

32. The output voltage (V_{OUT}) of the common-source JFET configuration can be five to ten times larger than the gate-control input voltage (V_{IN}).

33. The common-source circuit has a very high input resistance and relatively low output resistance. The high input resistance is due to the JFET's reverse biased gate-source junction, which permits no gate input current, and therefore has a very large resistance. This key characteristic means that the common-source JFET circuit is ideal in applications where we need to provide voltage amplification but do not want to load down a source that can only generate a small input signal.

34. The common-gate circuit configuration is very similar to its bipolar counterpart, the common-base circuit. The input is applied between the source and gate, while the output appears across the drain and gate.
35. The common-gate circuit can be used to provide a small voltage gain. Because the input is applied to the JFET's high current source terminal, the input resistance is very low. This low input resistance and relatively high output resistance makes the circuit ideal in applications where we need to efficiently transfer power between a low-resistance source and a high-resistance load.
36. Comparable to the bipolar transistor's common-collector, or emitter follower, the common-drain configuration is sometimes called a source-follower because the source output voltage follows in polarity and amplitude the input voltage at the gate.
37. With the common-drain circuit, the output voltage is slightly less than the input voltage (circuit does not provide any voltage gain). The input resistance of the common-drain circuit configuration is extremely high due to the JFET's reverse biased gate and R_L connection, and the output resistance is relatively very low. Inserting a common-drain circuit between a high-resistance source and a low-resistance load will ensure that the two opposite resistances are matched, and therefore power is efficiently transferred.

JFET Applications

38. It is the high input impedance of the JFET, and therefore its ability not to load a source, and the voltage amplification ability that are mainly made use of in circuit application. Like the bipolar junction transistor, the JFET can be made to function as a switch or as a variable resistor.
39. As a switch, the JFET makes use of only two points on the load line: saturation (in which it is equivalent to a closed switch between source and drain) and cutoff (in which it is equivalent to an open switch between source and drain).
40. A typical FET application in digital circuits would be a buffer circuit, which is used to isolate one device from another. The high input impedance of the FET does not load the input circuit or circuits, while the low output impedance of the FET provides a high output current to the output circuit. The high output current and buffering, or isolating, characteristics of these circuits account for why they are also called buffer-drivers.
41. The high input impedance ($Z_{IN}\uparrow$) of the FET is also made use of in other FET integrated circuits (ICs). When Z_{IN} is high, circuit current is low ($I\downarrow$) and therefore power dissipation is low ($P_D\downarrow$). This condition is ideal for digital integrated circuits (ICs) which contain thousands of transistors all formed onto one small piece of silicon. The low power dissipation of the JFET enables us to densely pack many more components into a very small area.
42. In two-state applications, the JFET is made to operate between the two extreme points of saturation (0 Ω) and cutoff (max. Ω). By controlling the

gate-source bias voltage (V_{GS}), the resistance between the JFET's drain and source (R_{DS}) can be changed to be any value between 0 Ω and maximum Ω. The JFET can therefore be made to act as a variable resistor with an increase in negative V_{GS} causing a larger R_{DS} and, in contrast, a decrease in negative V_{GS} causing a smaller R_{DS}.

43. It is the reverse-biased gate-source junction of a JFET that gives the JFET its key advantage: an extremely high input impedance (typically in the high MΩ range). In addition, the JFET can provide a small voltage gain and has been found to be a very low noise component. All these characteristics make it an ideal choice as an amplifier.
44. The *C-S* circuit is typically used at the front end of systems such as the first RF amplifier stage following the antenna and the first stage in a voltmeter. It will not load the source, yet it will amplify a wide range of input signal voltages.

Testing JFETs and Troubleshooting JFET Circuits

45. The transistor tester can be used to test both BJTs and FETs. This tester can be used to determine
    a. Whether an open or short exists between any of the terminals,
    b. The FET's transconductance/gain, and
    c. The FET's value of I_{DSS} and leakage current.
46. The ohmmeter can be used to detect the most common JFET failures: opens and shorts.
47. As with all other circuit troubleshooting, the first step is to isolate whether the problem is within the JFET circuit or external to the JFET circuit.
48. Because the JFET has only one P-N junction, the two basic circuit problems that can develop are as follows.
    a. With a shorted gate-source junction there will be no voltage difference between gate and source (V_{GS} will be 0 V), and so V_G will be a positive voltage, and I_D will be a maximum ($I_D = I_{DSS}$). The gate-source short will allow current to flow out of the gate, and therefore I_S will not equal I_D. Because I_G will combine with I_1, I_1 will not equal I_2. More than likely, the gate current will cause the gate junction to open.
    b. With an open gate junction, the applied V_{GS} voltage will be present at the gate, and therefore everything will seem normal. However, because V_{GS} is 0 V due to the open gate, I_D will equal maximum ($I_D = I_{DSS}$) and be constant despite changes in the input V_{GS}.
49. The MOSFET does not have a P-N gate-channel junction like the JFET. It has a "metal gate" that is insulated from the "semiconductor channel" by a layer of "silicon dioxide," hence the name "metal oxide semiconductor."
50. Like all "field effect transistors (FETs)" the MOSFET's input voltage will generate an "electric field" that will have the "effect" of changing the channel's size.
51. The key difference between the JFET and MOSFET is that the JFET's input voltage would always have to be zero or a negative voltage in order

to reverse bias the gate-source junction. With the MOSFET, the input voltage can be either a positive or negative voltage because gate current will always be zero (the gate is insulated from the channel).

52. The MOSFET can be operated in either the depletion or enhancement mode due to its insulated gate. The two different types of MOSFET are given names based on their normal mode of operation. The depletion-type MOSFET (D-type MOSFET or D-MOSFET) should actually be called a DE-MOSFET because it can be operated in both the depletion mode and the enhancement mode. The enhancement-type MOSFET (E-type MOSFET or E-MOSFET) is correctly named because it can only be operated in the enhancement mode.

The Depletion-Mode MOSFET (Figure 22-30)

53. The D-MOSFET has two basic transistor types called the *n*-channel D-type MOSFET and *p*-channel D-type MOSFET.
54. This type of MOSFET basically consists of an *n*-type channel formed on a *p*-type substrate. A source and drain lead are connected to either end of the *n*-channel, and an additional lead is attached to the substrate. In addition, a thin insulating (silicon dioxide) layer is placed on top of the *n*-channel, and a metal plated area with a gate lead attached is formed on top of this insulating layer.
55. In some applications, a separate bias voltage will be applied to the substrate terminal for added control of drain current, and the four-terminal device will be used. In most circuit applications, however, the three-terminal device, which has its source and substrate lead internally connected, is all that is needed.
56. The D-MOSFET's set of drain curves have the same general shape as the JFET's set of drain curves and the BJT's set of collector curves. The key difference is that V_{GS} is plotted for both positive and negative values. This is because the D-MOSFET should actually be called a DE-MOSFET because it can be operated in both the depletion mode (in which V_{GS} is a negative value) and the enhancement mode (in which V_{GS} is a positive value).
57. When $V_{GS} = 0$ V, the gate and source terminals are at the same zero-volt potential, and therefore the gate is effectively shorted to the source. The value of drain current passing through the channel in this mode is called the I_{DSS} value (I_{DSS} is the drain-to-source current passing through the channel when the gate is shorted to the source).
58. When zero volts is applied to the input of a D-type MOSFET, it will conduct a value of drain current. Therefore with no input this device is ON, which is why the D-type MOSFET is known as a "normally ON" device.
59. When V_{GS} is made positive, the channel is enhanced or widened, and therefore the value of I_D is increased above I_{DSS}.
60. When V_{GS} is made negative, the channel is depleted of free carriers, and therefore the value of I_D is decreased below I_{DSS}.

QUICK REFERENCE SUMMARY SHEET

FIGURE 22-30 The Depletion-Mode MOSFET (D-MOSFET).

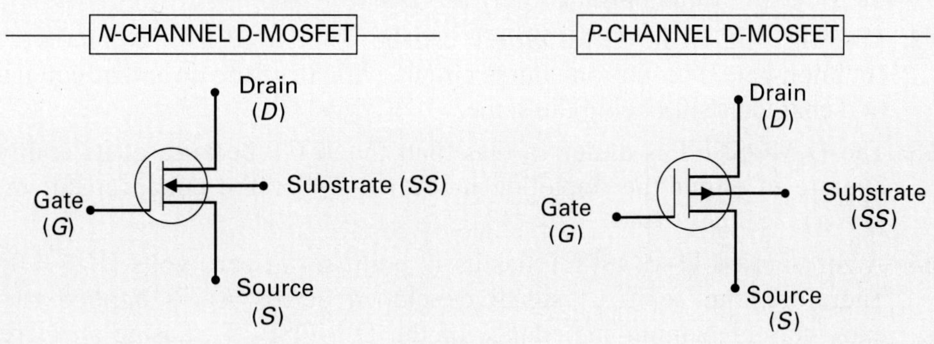

Drain Characteristic Curves

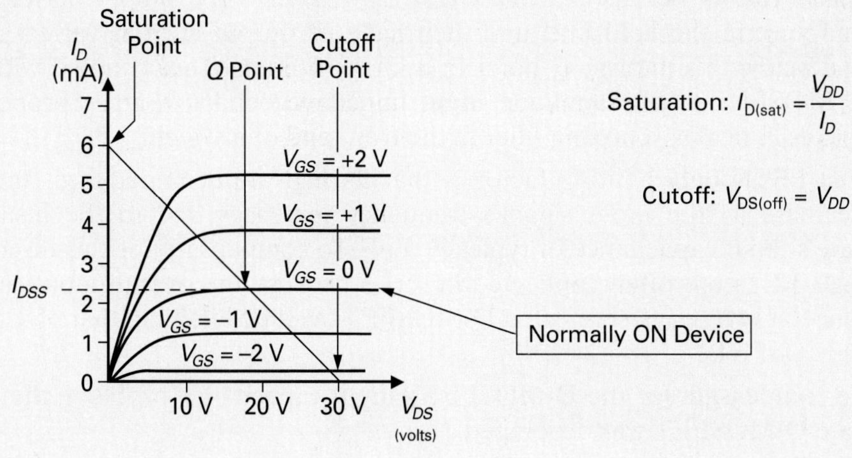

Saturation: $I_{D(sat)} = \frac{V_{DD}}{I_D}$

Cutoff: $V_{DS(off)} = V_{DD}$

Zero Biasing

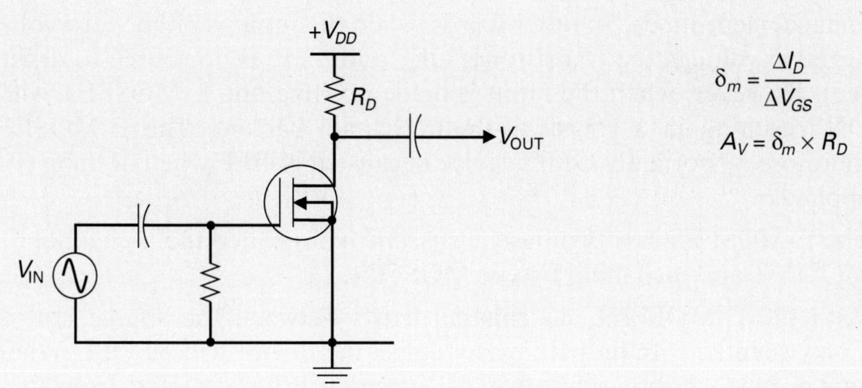

$$\delta_m = \frac{\Delta I_D}{\Delta V_{GS}}$$

$$A_V = \delta_m \times R_D$$

61. As with the JFET, the D-MOSFET's transconductance is equal to the ratio of output current change (ΔI_D) to input voltage change (ΔV_{GS}), and the D-MOSFET's voltage gain is equal to the product of transconductance and the value of drain resistance.
62. Like the JFET, the D-MOSFET can be configured as a common-drain, common-gate or common-source circuit, with all of the dc and ac configuration characteristics being the same.
63. The D-MOSFET is easier to bias than the JFET because of its ability to operate in either the depletion mode ($-V_{GS}$) or the enhancement mode ($+V_{GS}$).
64. A zero-biased D-MOSFET has its Q point set at zero volts ($V_{GS} = 0$ V). Therefore, the ac input signal developed across R_G is applied to the extremely high input impedance of the D-MOSFET, causing an increase and decrease in the conduction of the MOSFET.
65. The D-MOSFET is most frequently used in analog or linear circuit applications. This is because the D-MOSFET can be very simply biased at a midpoint in the load line, and then have its output current varied above and below this natural Q point in a linear fashion. This, coupled with the D-MOSFET's almost infinite input impedance and low-noise properties, makes it ideal as a preamplifier at the front end of a system.
66. The FET's only limiting factor is that its high input impedance starts to decrease as the input signal's frequency increases due to the insulated gate's input capacitance of typically 5 pF. To compensate for this disadvantage, FETs are often connected in series because the overall input capacitance of two series-connected D-MOSFETs will be less than that of a single D-MOSFET.
67. To compensate for the D-MOSFET's input capacitance problem, the dual-gate D-MOSFET was developed.

The Enhancement-Mode MOSFET (Figure 22-31)

68. The enhancement-type MOSFET or E-MOSFET can only operate in the enhancement-mode. In other words, when the input is either zero volts or a negative voltage, the transistor is OFF and there is no source-to-drain current. However, when the input is made positive, the E-MOSFET will turn ON, resulting in a source-to-drain channel current. The E-MOSFET is therefore a "normally OFF" device because it is OFF when nothing (0 V) is applied.
69. The E-MOSFET has two basic transistor types called the *n*-channel E-type MOSFET and *p*-channel E-type MOSFET.
70. With the E-MOSFET, no channel exists between the source and drain. Consequently, with no gate bias voltage, the device will be OFF. When the gate is made positive, however, electrons will be attracted from the substrate, causing a channel to be induced between the source and drain. This enhanced channel will permit drain current to flow, and any further increase in gate voltage will cause a corresponding increase in the size of the channel and therefore the value of I_D.

FIGURE 22-31 The Enhancement-Mode MOSFET (E-MOSFET).

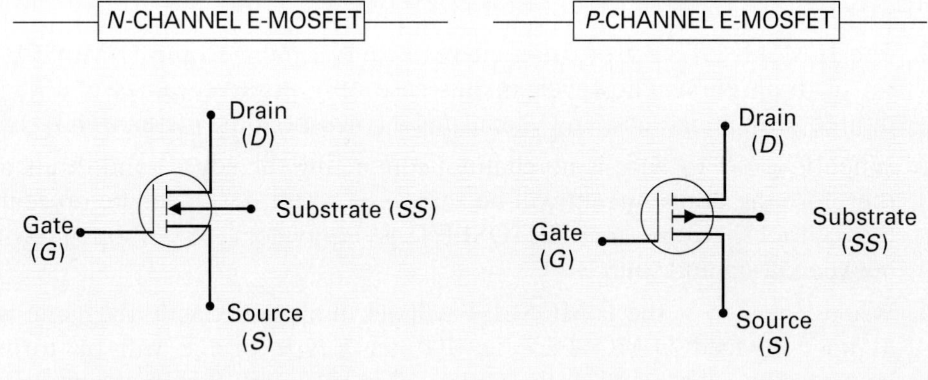

Drain Characteristic Curves

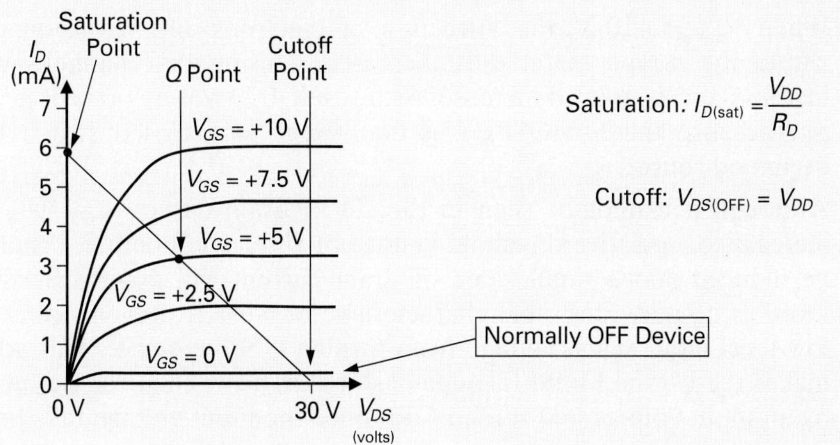

Saturation: $I_{D(\text{sat})} = \frac{V_{DD}}{R_D}$

Cutoff: $V_{DS(\text{OFF})} = V_{DD}$

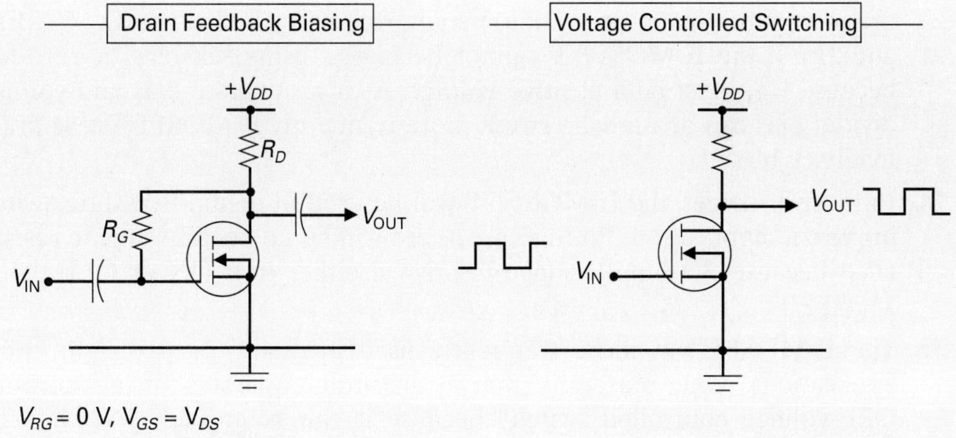

$V_{RG} = 0$ V, $V_{GS} = V_{DS}$

$V_{DS} = V_{DD} - (I_{D(\text{ON})} \times R_D)$

71. The only difference between the E-type and D-type MOSFET symbols is the three dashed lines representing the drain, substrate, and source regions. The dashed line is used instead of the solid line to indicate that an E-MOSFET has a normally broken path or channel between drain, substrate, and source.

72. The E-MOSFET's set of drain curves is very similar to the D-MOSFET's set of drain curves; however, in this case only positive values of V_{GS} are plotted, with an increase in V_{GS} causing a corresponding increase in I_D.

73. When $V_{GS} = 0$ V, there is no channel connecting the source and drain, and therefore the drain current will be zero. As a result, $+V_{DD}$ will be present at the output because the E-MOSFET is equivalent to an open switch between drain and source.

74. When $V_{GS} = +5$ V, the E-MOSFET will act in almost exactly the same way as an enhanced D-MOSFET in that an *n*-type bridge will be formed between the source and drain, and a value of I_D will flow between source and drain.

75. When $V_{GS} = +10$ V, the attraction of electrons and repulsion of holes within the *p*-type material is increased, causing the channel's width to increase and I_D to also increase. As a result, 0 V will be present at the output because the E-MOSFET is equivalent to a closed switch between drain and source.

76. Although it cannot be seen in the set of drain curves, V_{GS} will have to increase to a positive threshold voltage of about +1 V before a channel will be induced and a small value of drain current will flow. This "threshold level" is a highly desirable characteristic because it prevents noise or any low-level input signal voltage from turning ON the device. This advantage makes the E-type MOSFET ideally suited as a switch: it can be turned ON by an input voltage and turned OFF once the input voltage falls below the threshold level.

77. Like the D-MOSFET, the E-MOSFET can be configured as a common-drain, common-gate, or common-source circuit. Unlike the D-MOSFET and JFET, the E-MOSFET cannot be biased using self-bias or zero-bias because V_{GS} must be a positive voltage. As a result, gate-bias and voltage-divider bias can be used; however, more frequently E-MOSFETs are drain-feedback biased.

78. In most instances, the E-MOSFET will be used in digital two-state switching circuit applications. In this case, there will be no need for a gate resistor (R_G) because the input voltage (V_{IN}) will either turn ON or OFF the E-MOSFET.

79. The E-MOSFET is more frequently used in digital or two-state circuit applications. One reason is that it naturally operates as a "normally OFF-voltage controlled switch" because it can be turned ON when the gate voltage is positive and turned OFF when the gate voltage falls below a threshold level. This threshold level is a highly desirable characteristic because it prevents noise from false triggering, or accidentally turning ON, the device. The other E-MOSFET advantage is its extremely

high input impedance, which means that the device's circuit current, and therefore power dissipation, are low. This enables us to densely pack or integrate many thousands of E-MOSFETs onto one small piece of silicon, forming a high component density integrated circuit (IC). These low-power and high-density advantages make the E-MOSFET ideal in battery-powered small-size (portable) applications such as calculators, wristwatches, notebook computers, hand-held video games, digital cellular phones, and so on.

The Vertical Channel MOSFET (VMOS FET)

80. If a larger current-carrying capability than the standard E-MOSFET is needed, the vertical-channel E-MOSFET or VMOS FET can be used.

81. When this *n*-channel VMOS FET is biased ON (gate is made positive with respect to source), a vertical *n*-type channel is formed between source and drain. This channel is much wider than a standard E-MOSFET's horizontal channel, which is why VMOS FETs can handle a much higher drain current.

82. Other than its high current capability, the VMOS FET also has a positive temperature coefficient of resistance, which means an increase in temperature will cause a decrease in drain current ($T\uparrow$, $R\uparrow$, $I\downarrow$), and this will prevent thermal runaway.

MOSFET Handling Precautions

83. The very thin insulating layer between the gate and the substrate of a MOSFET can easily be punctured if an excessive voltage is applied. Your body can build up extremely large electrostatic charges due to friction. If this charge comes in contact with the pins of a MOSFET device, an electrostatic discharge (ESD) would occur, resulting in a possible arc across the thin insulating layer causing permanent damage.

84. Most MOSFETs presently manufactured have zeners internally connected between gate and source to bypass high voltage static or in-circuit potentials and protect the MOSFET.

85. When handling MOSFETs it is important to remember the following.

- **a.** All MOS devices are shipped and stored in a "conductive foam" so that all of the IC pins are kept at the same potential, and therefore electrostatic voltages cannot build up between terminals.
- **b.** When MOS devices are removed from the conductive foam, be sure not to touch the pins because your body may have built up an electrostatic charge.
- **c.** When MOS devices are removed from the conductive foam, always place them on a grounded surface such as a metal tray.
- **d.** When continually working with MOS devices, use a "wrist grounding strap," which is a length of cable with a 1 MΩ resistor in series. This prevents electrical shock if you come in contact with a voltage source.
- **e.** All test equipment, soldering irons, and work benches should be properly grounded.

f. All power in equipment should be off before MOS devices are removed or inserted into printed circuit boards.

g. Any unused MOSFET terminals must be connected because an unused input left open can build up an electrostatic charge and float to high voltage levels.

h. Any boards containing MOS devices should be shipped or stored with the connection side of the board in conductive foam.

MOSFET Testing

86. Like the BJT and JFET, MOSFETs can be tested with the transistor tester to determine: first, whether an open or short exists between any of the terminals; second, the transistor's transconductance/gain; and third, the value of I_{DSS} and leakage current.

87. If a transistor tester is not available, the ohmmeter can be used to determine the most common *n*-channel and *p*-channel D-MOSFET and E-MOSFET failures: opens and shorts. Remember that the gate-to-channel resistance of a MOSFET is always an open due to the insulated gate.

NEW TERMS

Bipolar Family
Buffer Driver
Cascode Amplifier
Common-Drain Circuit Configuration
Common-Gate Circuit Configuration
Common-Source Circuit Configuration
Constant Current Region
Depletion-Mode Operation
Depletion-type MOSFET (D-MOSFET)
Drain
Drain Characteristic Curve
Drain Feedback Biasing
Drain-to-Source Current with Shorted Gate (I_{DSS})
Dual Gate D-MOSFET
Enhancement-mode Operation
Enhancement-type MOSFET (E-MOSFET)
Field Effect Transistor (FET)
Gate
Gate Biasing
Gate-to-Source Cut-off Voltage ($V_{GS(OFF)}$)
Junction Field Effect Transistor (JFET)
Metal Oxide Semiconductor FET (MOSFET)
Metal Oxide Semiconductor Field Effect Transistor (MOSFET)
n-Channel JFET
Negative Feedback
p-Channel JFET
Pinch-off Voltage
Preamplifier
Self Biasing
Source
Source Follower
Temperature Stability
Transconductance
Vertical Channel E-MOSFET
VMOS
Zero Biasing

Multiple-Choice Questions

REVIEW QUESTIONS

1. A (an) ____________ has three terminals called the gate, source, and drain.

   a. BJT **c.** Bipolar Transistor
   b. Zener **d.** JFET

2. The ____________ channel JFET schematic symbol has the arrow pointing out while the ____________ channel JFET schematic symbol has the arrow pointing in.

   a. *D-S, p* **c.** *n, p*
   b. *p, G-S* **d.** *p, n*

3. The BJT is a ____________ controlled device whereas the FET is a ____________ controlled device.

   a. voltage, current **b.** current, voltage

4. The gate-source junction of a JFET is always ____________ biased since the input voltage is normally ____________ or some ____________ voltage.

   a. reverse, 0 V, negative **c.** reverse, –4 V, positive
   b. forward, 0 V, negative **d.** forward, –4 V, negative

5. When $V_{GS} = V_{GS(\text{OFF})}$, $I_D = ?$

   a. I_{DSS} **c.** Maximum
   b. Zero **d.** V_P

6. Which JFET circuit configuration is also known as a source follower?

   a. Common-Drain **c.** Common-Gate
   b. Common-Source **d.** Both (a) and (b) are true

7. Which JFET circuit configuration provides a high input impedance and a good voltage gain?

   a. Common-Drain **c.** Common-Gate
   b. Common-Source **d.** Both (a) and (c) are true

8. Which biasing method makes use of a $-V_{GG}$ supply voltage?

   a. Self Biasing **c.** Base Biasing
   b. Gate Biasing **d.** Voltage-Divider Biasing

9. Which biasing method uses a source resistor?

   a. Self Biasing **c.** Voltage-Divider Biasing
   b. Gate Biasing **d.** Both (a) and (c)

10. Transconductance is a ratio of

    a. ΔI_D to ΔV_{DS} **c.** ΔI_D to ΔV_{GS}
    b. ΔV_{GD} to ΔI_D **d.** ΔV_{GS} to ΔV_{DS}

11. The input resistance of an FET is much higher than the input resistance of a BJT.

    a. True
    b. False

12. With a junction FET, as V_{GS} is made more negative the depletion region will ____________, the channel size will get ____________, and therefore I_D will ____________.

    a. decrease, larger, increase
    b. increase, smaller, decrease
    c. decrease, smaller, increase
    d. increase, larger, decrease

13. Which JFET circuit configuration has phase inversion between input and output?

    a. Common-Drain
    b. Common-Source
    c. Common-Gate
    d. Both (b) and (c)

14. With a *p*-channel JFET, the gate should be ____________ with respect to the source, and the drain should be ____________ with respect to the source.

    a. negative, negative
    b. negative, positive
    c. positive, positive
    d. positive, negative

15. The current between ____________ leads of a JFET is controlled by varying the reverse bias voltage applied to the ____________ leads.

    a. gate and source, source and drain
    b. source and drain, drain and gate
    c. source and drain, gate and source
    d. drain and gate, gate and source

16. The ____________ has an insulated gate that allows us to use either a positive or negative gate input voltage.

    a. JFET
    b. MOSFET
    c. BJT
    d. VJT

17. Which type of FET operates in the depletion mode?

    a. JFET
    b. E-MOSFET
    c. D-MOSFET
    d. Both (a) and (c)

18. Which type of FET makes use of an electric field to change the channel's size?

    a. JFET
    b. E-MOSFET
    c. D-MOSFET
    d. All of the above

19. The ____________ has drain current when the gate voltage is zero and is therefore known as a normally ____________ device.

    a. D-MOSFET, OFF
    b. D-MOSFET, ON
    c. E-MOSFET, OFF
    d. E-MOSFET, ON

20. The ____________ has no drain current when gate voltage is zero and is therefore known as a normally ____________ device.

    a. D-MOSFET, OFF
    b. D-MOSFET, ON
    c. E-MOSFET, OFF
    d. E-MOSFET, ON

21. Which type of FET operates in both the depletion and enhancement mode?

    a. JFET
    b. D-MOSFET
    c. E-MOSFET
    d. Both (b) and (c)

22. Which type of MOSFET can use zero biasing to bias it to a mid-load-line point?

a. D-MOSFET **c.** VMOS FET
b. E-MOSFET **d.** Both (b) and (c)

23. The dual-gate MOSFET is ideal for interfacing low-voltage devices to high-power loads.

a. True **b.** False

24. With the ____________, the gate voltage must be greater than the device's threshold voltage to produce drain current.

a. JFET **c.** E-MOSFET
b. D-MOSFET **d.** All of the above

25. The vertical channel MOSFET is ideal for interfacing low-voltage devices to high-power loads.

a. True **b.** False

26. Which of the following transistors has the highest input impedance?

a. BJT **c.** MOSFET
b. JFET **d.** Both (a) and (c) are true

27. D-MOSFETs are more frequently used in ____________ circuit applications whereas E-MOSFETs are more often used in ____________ circuit applications.

a. analog, digital **c.** digital, linear
b. linear, analog **d.** two-state, digital

28. A digital code consists of a series of HIGH and LOW voltages called ____________.

a. NOT gates **c.** Logic levels
b. CMOS **d.** Both (a) and (b)

29. The ____________ has made its mark in two-state circuits in which it uses only two points on the load line and acts like a voltage controlled ____________.

a. E-MOSFET, variable resistor **c.** D-MOSFET, variable resistor
b. E-MOSFET, switch **d.** D-MOSFET, switch

30. To avoid possible damage to MOS devices while handling, testing, or in operation, the following precaution should be taken.

a. All leads should be connected except when being tested or in actual operation.
b. Pick up devices by plastic case instead of leads.
c. Do not insert or remove devices when power is applied.
d. All of the above

Essay Questions

31. What are the basic differences between a BJT and an FET? (22-1)

32. Identify the schematic symbols shown in Figure 22-32. (22-1-1)

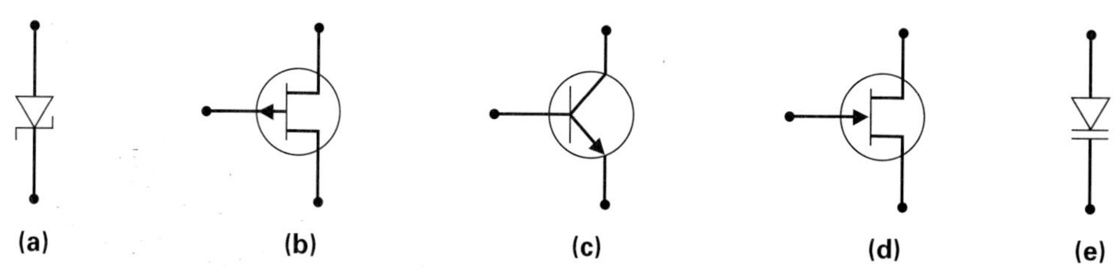

FIGURE 22-32 Schematic Symbols.

33. Give the full names of the following abbreviations
    a. JFET (22-1)
    b. MOSFET (Intro.)
    c. BJT (Intro.)
    d. V_{GS} (22-1-2)
    e. I_{DSS} (22-1-3)
    f. $V_{GS(OFF)}$ (22-1-3)
34. Name the two different types of JFETs. (22-1-1)
35. Briefly describe the operation of a JFET. (22-1-2)
36. Why is an FET referred to as a voltage-operated device, and the BJT referred to as a current-operated device? (22-1-2)
37. Name the two different types of FETs. (Intro.)
38. In relation to the drain characteristic curve of a JFET, define the following points: (22-1-3)
    a. Pinch-Off Voltage (V_P)
    b. Constant Current Region
    c. Breakdown Voltage (V_{BR})
    d. Drain-to-Source Current with Shorted Gate (I_{DSS})
    e. Gate-to-Source Cut-Off Voltage ($V_{GS(OFF)}$)
39. What is transconductance? (22-1-4)
40. What key JFET characteristic is made use of in application circuits? (22-1-9)
41. List the three different JFET circuit configurations, and briefly describe their characteristics. (22-1-8)
42. Briefly describe the following JFET biasing methods: (22-2)
    a. Gate Biasing
    b. Self Biasing
    c. Voltage-Divider Biasing
43. Why is the FET better suited to miniaturization than the BJT? (22-1-9)
44. Describe some of the typical digital and analog circuit applications of the JFET. (22-1-9)
45. What two specifications need to be obtained from a manufacturer's data sheet in order to determine JFET circuit current and voltage values? (22-1-6)
46. Define the following terms: (22-2)
    a. Depletion-mode operation
    b. Enhancement-mode operation
47. How does a MOSFET differ from a JFET? (22-2)
48. Why should a D-MOSFET really be called a DE-MOSFET? (22-2-1)

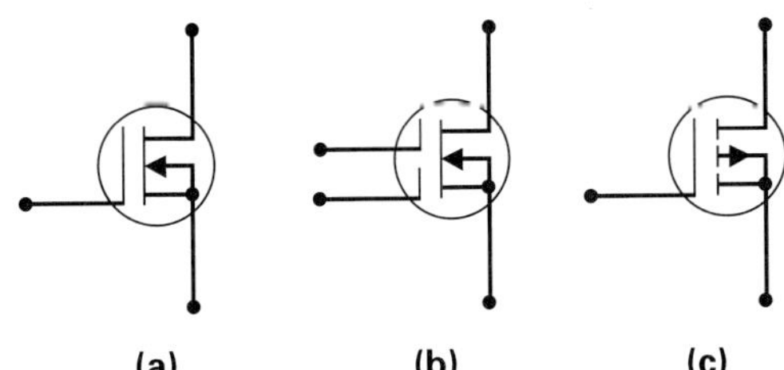

FIGURE 22-33 MOSFET Schematic Symbols.

49. Identify the device schematic symbols shown in Figure 22-33. (22-2-1)
50. Briefly describe the operation of the D-MOSFET. (22-2-1)
51. Which MOSFET is called a "normally ON" device and which is called a "normally OFF" device? Describe why. (22-2-1 and 22-2-2)
52. How would a D-MOSFET typically be biased to a midpoint on its load line? (22-2-1)
53. Briefly describe the operation and application of the dual-gate MOSFET. (22-2-1)
54. Briefly describe the operation of the E-MOSFET. (22-2-2)
55. How would an E-MOSFET typically be biased to a midpoint on its load line? (22-2-2)
56. Why are MOSFETs ideally suited for digital circuit applications? (22-2-2)
57. Briefly describe the operation and application of the vertical channel MOSFET. (22-2-2)
58. Briefly describe the circuit applications for the D-MOSFET and E-MOSFET. (22-2-1 and 22-2-2)
59. Referring to the data sheet in Figure 22-25, determine the following for 2N7002 MOSFET: (22-2-1)
    - **a.** Maximum value of drain current
    - **b.** Drain-to-source reverse breakdown voltage
    - **c.** Input capacitance
60. Why should MOSFETs be handled carefully, and what are some of the procedures to follow? (22-2-1)

Practice Problems

61. Referring to Figure 22-5, you can see that when the gate of a 2N5484 JFET is reverse biased (normal operation), the gate reverse current (I_{GSS}) = 1.0 nA when $V_{GS} = -20$ V ($V_{DS} = 0$ V). Calculate the input or gate-source impedance of the JFET.
62. Calculate the following for the amplifier circuit in Figure 22-34:
    - **a.** Transconductance **b.** Voltage Gain
63. Calculate the following for the circuit in Figure 22-35:
    - **a.** V_{GS} **d.** $I_{D(\text{MAXIMUM})}$
    - **b.** I_D **e.** Q point
    - **c.** V_{DS}

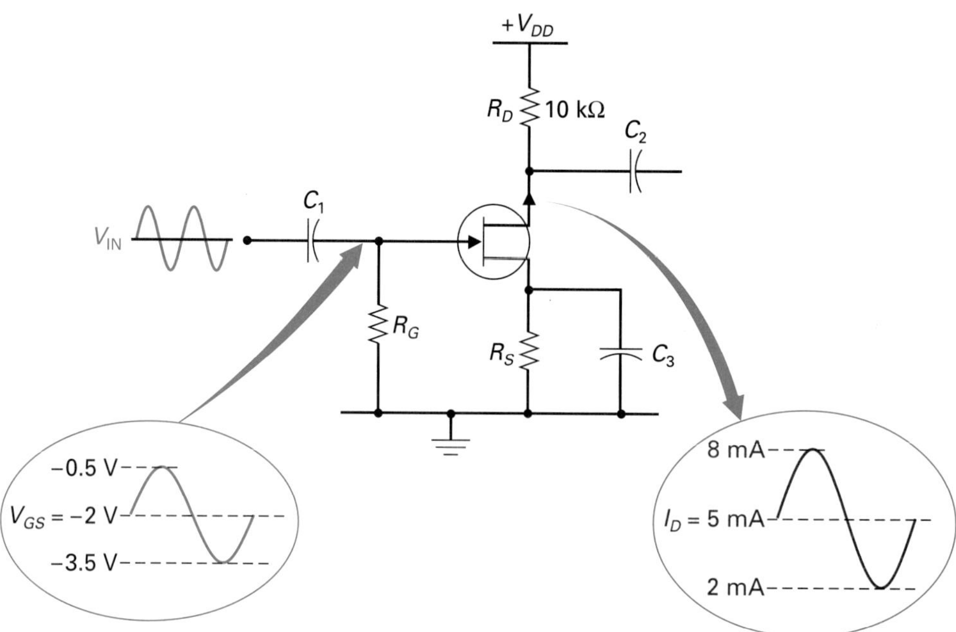

FIGURE 22-34 A Common-Source Amplifier.

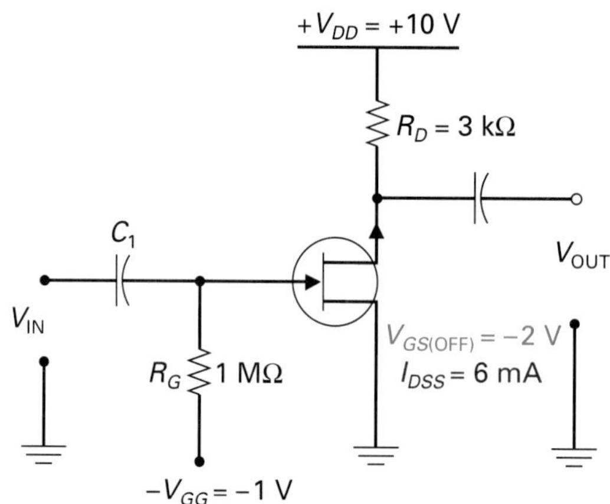

FIGURE 22-35 A Gate-Biased JFET Circuit.

64. Calculate the following for the circuit in Figure 22-36:

a. V_S **b.** V_{GS} **c.** V_{DS}

65. Calculate the following for the circuit in Figure 22-37:

a. V_G **b.** V_S **c.** V_{GS} **d.** V_{DS}

66. Referring to Figure 22-38,

a. What is the circuit configuration?
b. What is the circuit's voltage gain?
c. What is a circuit like this typically used for?

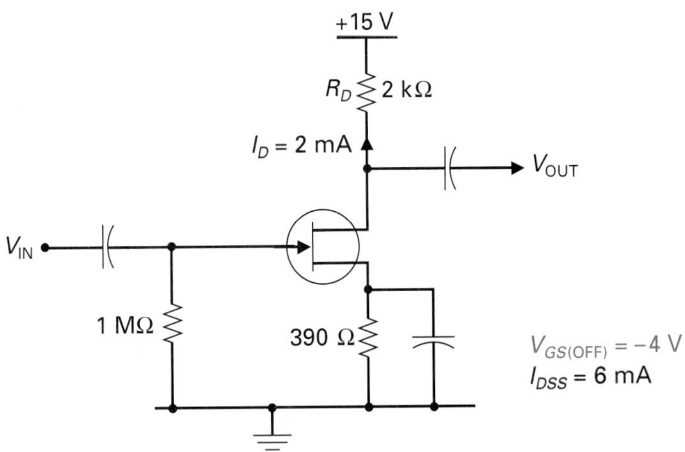

FIGURE 22-36 A Self-Biased JFET Circuit.

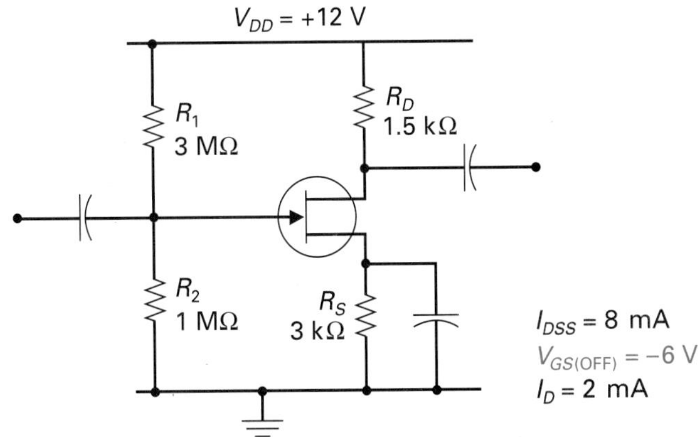

FIGURE 22-37 A Voltage Divider Biased JFET Circuit.

67. Referring to the circuit in Figure 22-39,

a. What is the circuit configuration?
b. What will be the output when the input is 0 V?
c. What will be the output when the input is –6 V?
d. Is there phase inversion between input and output?

68. Calculate the following for the circuit shown in Figure 22-40:

a. $I_{D(\text{sat})}$ **b.** $V_{DS(\text{OFF})}$

69. Calculate the following for the circuit shown in Figure 22-41:

a. I_D **b.** V_{DS}

70. What biasing method was used in Figure 22-40 and 22-41?

71. Identify the circuit in Figure 22-42. What will the approximate output voltage be if the input equals

a. $V_{\text{IN}} = 0$ V **b.** $V_{\text{IN}} = +5$ V

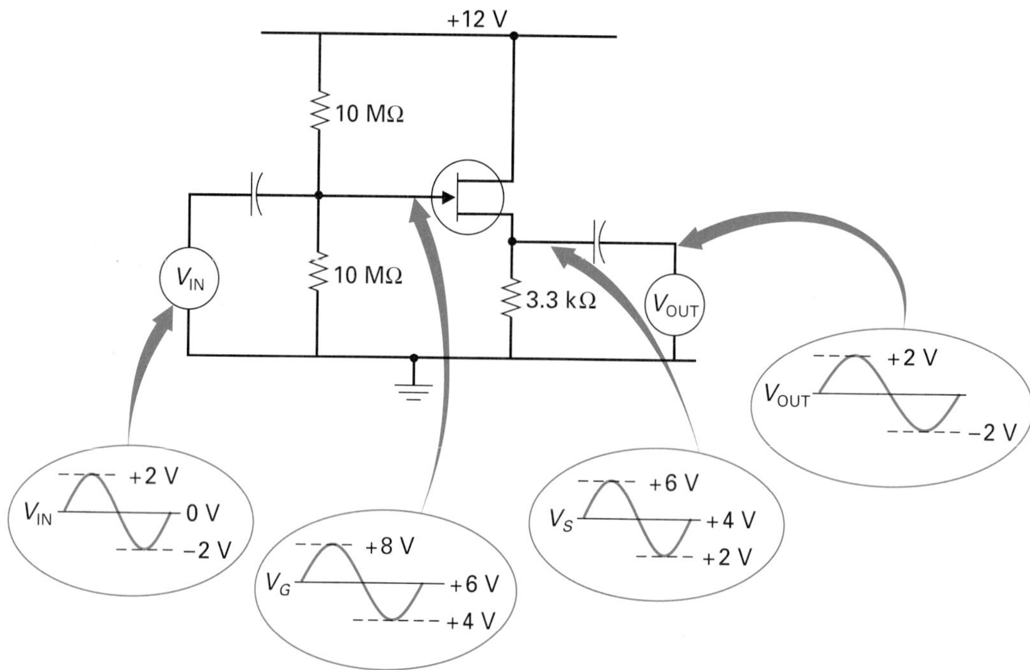

FIGURE 22-38 **An Analog JFET Circuit.**

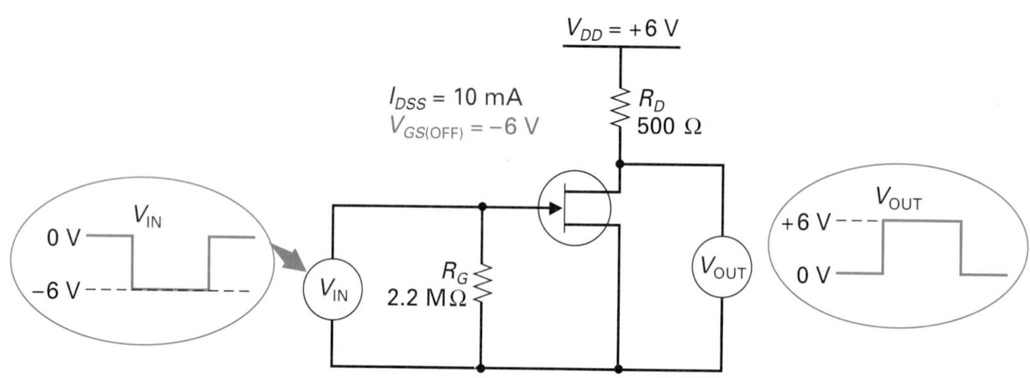

FIGURE 22-39 **A Digital JFET Circuit.**

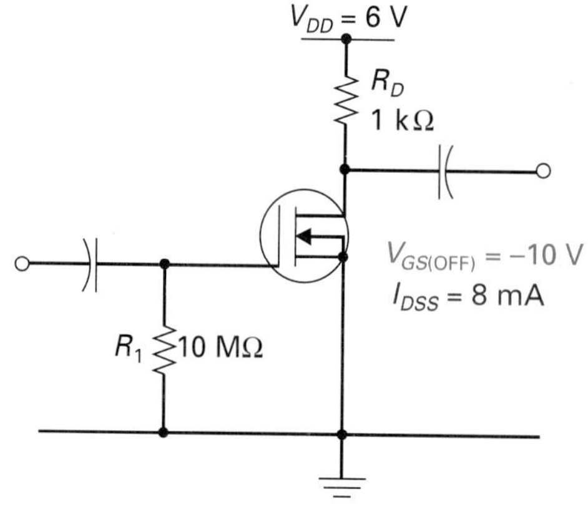

FIGURE 22-40 **A D-MOSFET Amplifier Circuit.**

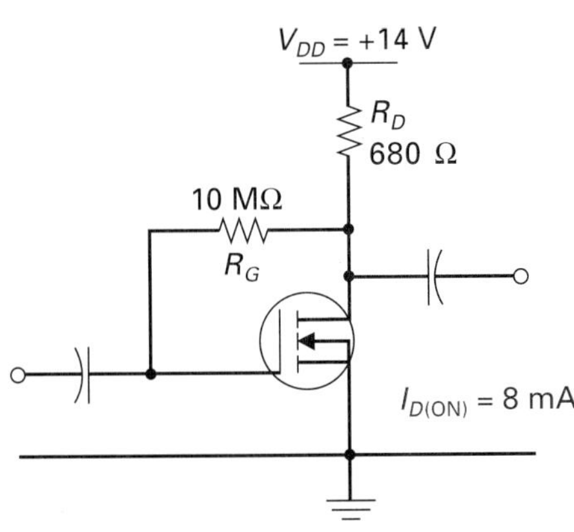

FIGURE 22-41 An E-MOSFET Amplifier Circuit.

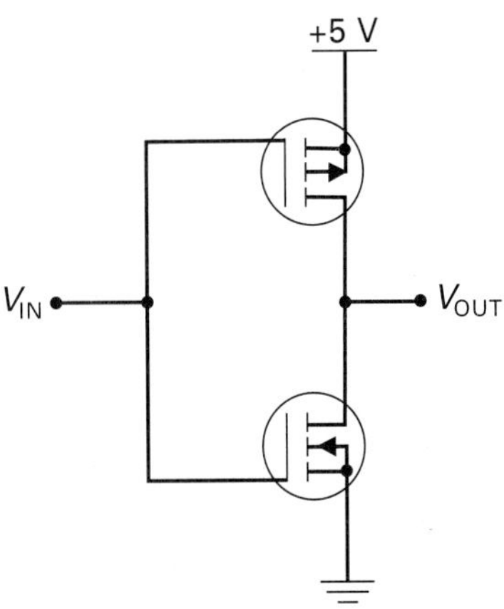

FIGURE 22-42 MOSFET Circuit.

72. Referring to the circuit in Figure 22-43, answer the following questions:

a. Is Q_1 ON or OFF, and why?

b. Calculate V_{OUT} when $V_{IN} = +5$ V.

c. Calculate V_{OUT} when $V_{IN} = 0$ V.

Troubleshooting Questions

73. Briefly list what characteristics of a JFET can be tested with a transistor tester.

74. How can an ohmmeter test JFETs?

75. Which of the JFETs in Figure 22-44 are good, and which are bad? If bad, state the suspected problem.

76. If you had a choice of placing the pins of a good MOSFET transistor on a plastic or metal tray, which would you choose?

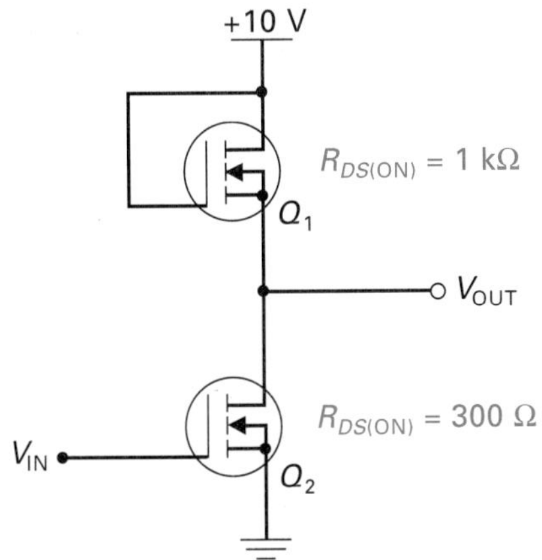

FIGURE 22-43 A MOSFET Acting as a Load Resistor.

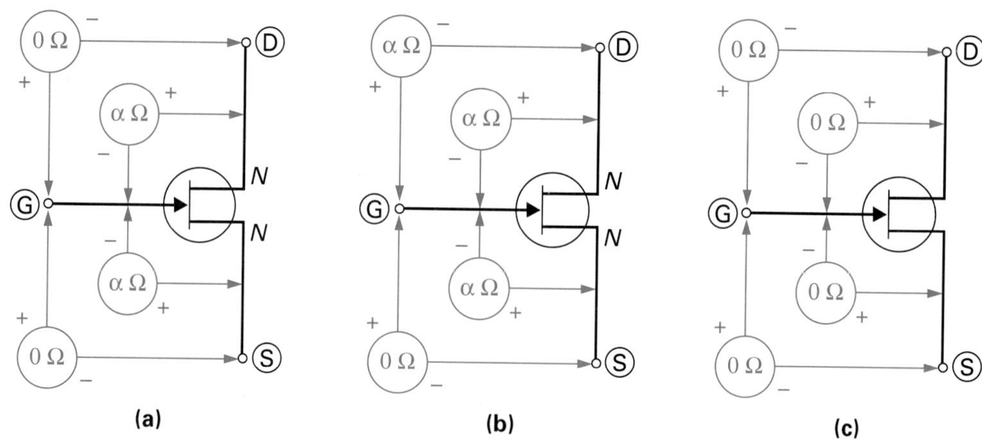

FIGURE 22-44 **Testing JFETs with the Ohmmeter.**

77. What resistance would you expect the ohmmeter to show in Figure 22-45?

78. Are the MOSFETs in Figure 22-46 good or bad?

79. Is there a problem with the circuit in Figure 22-47? If so, what do you think it could be?

80. Is there a problem with the circuit in Figure 22-48? If so, what do you think it could be?

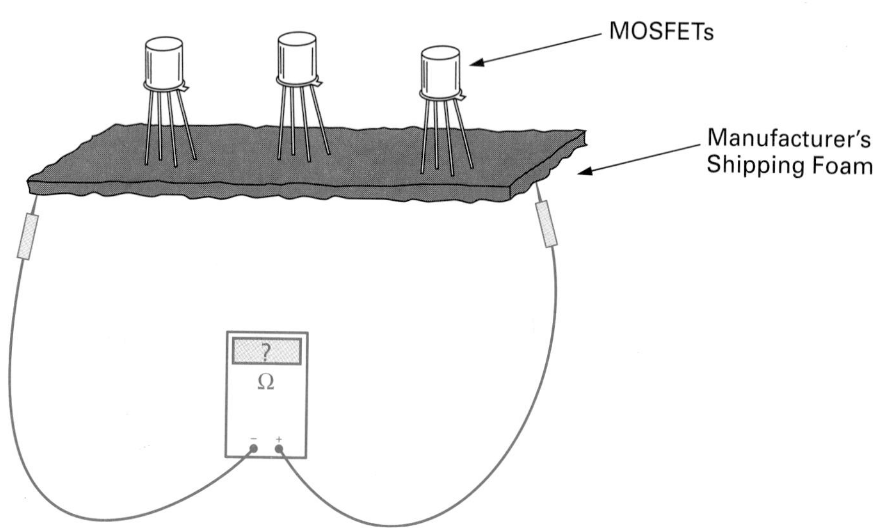

FIGURE 22-45 **MOSFETs in Manufacturer's Shipping Foam.**

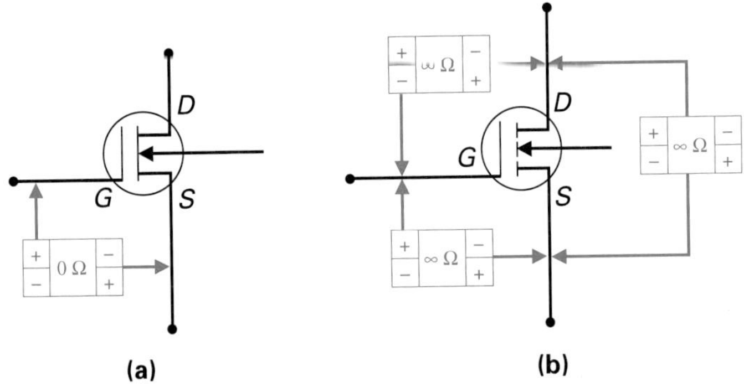

FIGURE 22-46 **Testing MOSFETs with the Ohmmeter.**

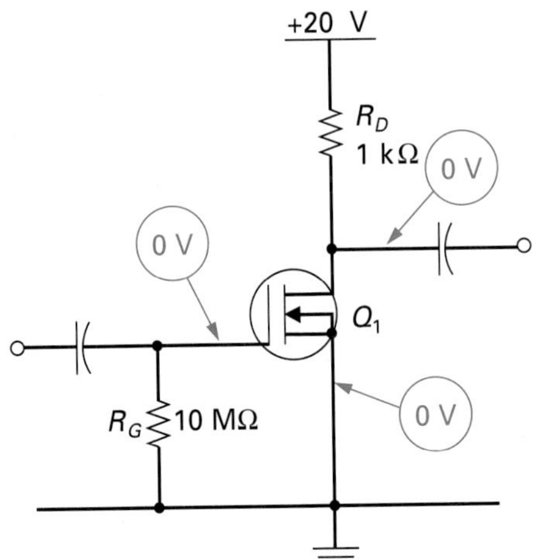

FIGURE 22-47 **D-MOSFET Circuit Troubleshooting.**

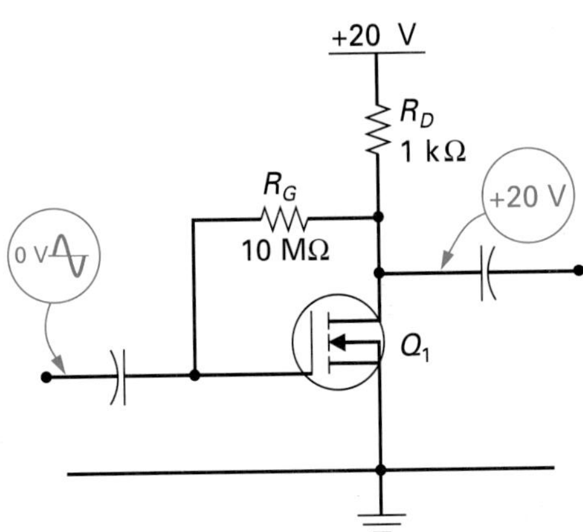

FIGURE 22-48 **E-MOSFET Circuit Troubleshooting.**

CHAPTER

23

OBJECTIVES

After completing this chapter, you will be able to:

1. Describe the symbol, package types, and internal block diagram of the operational amplifier or op-amp.
2. Explain how the op-amp's first differential amplifier stage gives the op-amp its high input impedance, high differential gain, and low common mode gain characteristics.
3. Define common-mode rejection ratio.
4. Describe how the internal second stage voltage amplifier gives the op-amp its very high gain characteristic.
5. Describe how the internal third stage emitter-follower amplifier gives the op-amp its low output impedance and current gain characteristics.
6. Explain how the op-amp operates in the following three basic circuit applications:
   - **a.** Open-loop comparator circuit
   - **b.** Closed-loop inverting amplifier circuit
   - **c.** Closed-loop noninverting amplifier circuit
7. Interpret a typical op-amp data sheet.
8. Compare the characteristics of several different op-amp types.
9. Describe some of the more typical op-amp circuit malfunctions.
10. Identify and describe the operation of the following op-amp circuit applications:
    - **a.** Voltage follower circuit
    - **b.** Summing amplifier circuit
    - **c.** Difference amplifier circuit
    - **d.** Differentiator circuit
    - **e.** Integrator circuit
    - **f.** Signal generator circuits
    - **g.** Active filter circuits

Operational Amplifiers

OUTLINE

VIGNETTE

I See

Standing six feet six inches tall, Jack St. Clair Kilby was a quiet introverted man from Kansas. Excited about the prospect of joining the very well-respected Massachusetts Institute of Technology (MIT) to further his education, he was thoroughly disappointed when he failed the mathematics entrance exam by three points. For the next ten years he worked for a manufacturer of radio and television parts, paying particularly close attention to the new component on the block, the transistor.

In May of 1958, Texas Instruments, a new fast-growing company who developed the first commercially available silicon transistor just four years earlier, offered him a job in their development lab. A project was underway to print electronic components on ceramic wafers and then wire them and stack them together to make a circuit. The more Kilby became involved in the project, the more he realized how complicated and ridiculous the method was. The idea to miniaturize was a good one, but a different solution was needed.

Two months later in July, the company shut down for summer vacation, but Kilby was forced to work because he had not accrued vacation time. This proved to be a blessing in disguise for Kilby who found himself in the lab with a lot of time and resources available to him to develop his idea. His idea was to build resistors and capacitors from the same semiconductor material that was being used to manufacture transistors. This would mean that all of the components that make up a circuit could be manufactured simultaneously on a single slice of semiconductor material.

A few months later Kilby presented his prototype to a very skeptical boss. It contained five components all connected by tiny wires with the complete assembly held together by large blobs of wax. Kilby suddenly found himself the owner of a patent and the richly deserved acclaim that always goes along with being first. The first integrated circuit or IC was born on a thin wafer of germanium just two-fifths of an inch long. Texas Instruments demonstrated its miniaturization advantage by building a computer for the air force using their newly developed technique in a 587 IC. Its rewards were immediately apparent when a 78 cubic foot monster computer was replaced with a more powerful unit measuring only 6.5 cubic inches.

INTRODUCTION

The operational amplifier was initially a vacuum tube circuit used in the early 1940s in analog computers. The name "operational amplifier" or "op-amp" was chosen because the circuit was used as a high-gain dc "amplifier" performing mathematical "operations." These early circuits were expensive and bulky, and they found very little application until the semiconductor integrated circuit was developed in 1958 by Jack Kilby at Texas Instruments. Circuits that once needed hundreds of discrete or individual components can now be integrated into a single IC, making equipment smaller, more energy efficient, cheaper, and easier to design and troubleshoot.

Today's IC op-amp is a very high-gain dc amplifier that can have its operating characteristics changed by connecting different external components. This makes the op-amp very versatile, and it is this versatility that has made the op-amp the most widely used linear IC.

In this chapter, we will be examining the op-amp's operation, characteristics, typical circuit applications, and troubleshooting.

23-1 OPERATIONAL AMPLIFIER BASICS

To begin with, Figure 23-1 introduces the **operational amplifier,** or **op-amp,** by showing its schematic symbol in Figure 23-1(a) and internal circuit in Figure 23-1(b). It would be safe to say that you will not really be learning anything that has not already been covered because the op-amp's internal circuit is simply a combination of three previously covered amplifier circuits. These three circuits are all interconnected and contained within a single IC, and together they function as a "high-gain, high input impedance, low output impedance amplifier."

Operational Amplifier (Op-Amp)
Special type of high-gain amplifier.

23-1-1 *Op-Amp Symbol*

Referring again to Figure 23-1(a), you can sec that the triangle shaped amplifier symbol is used to represent the op-amp in an electronic schematic diagram. Comparing the two symbols, you may have noticed that in some cases the two power supply connections are not shown, even though power is obviously applied.

Let us now examine the op-amp's input and output terminals shown in Figure 23-1. The two op-amp inputs are labeled "–" and "+." The "–" or negative input is called the **inverting input** because any signal applied to this input will be amplified and inverted between input and output (output is 180° out of phase with input). On the other hand, the "+" or positive input is called the **noninverting input** because any signal applied to this input will be amplified but not inverted between input and output (output is in phase with input). An input signal will normally be applied to only one of these inputs, while the other input is used to control the op-amp's operating characteristics.

Inverting Input
The inverting or negative input of an op-amp.

Noninverting Input
The noninverting or positive input of an op-amp.

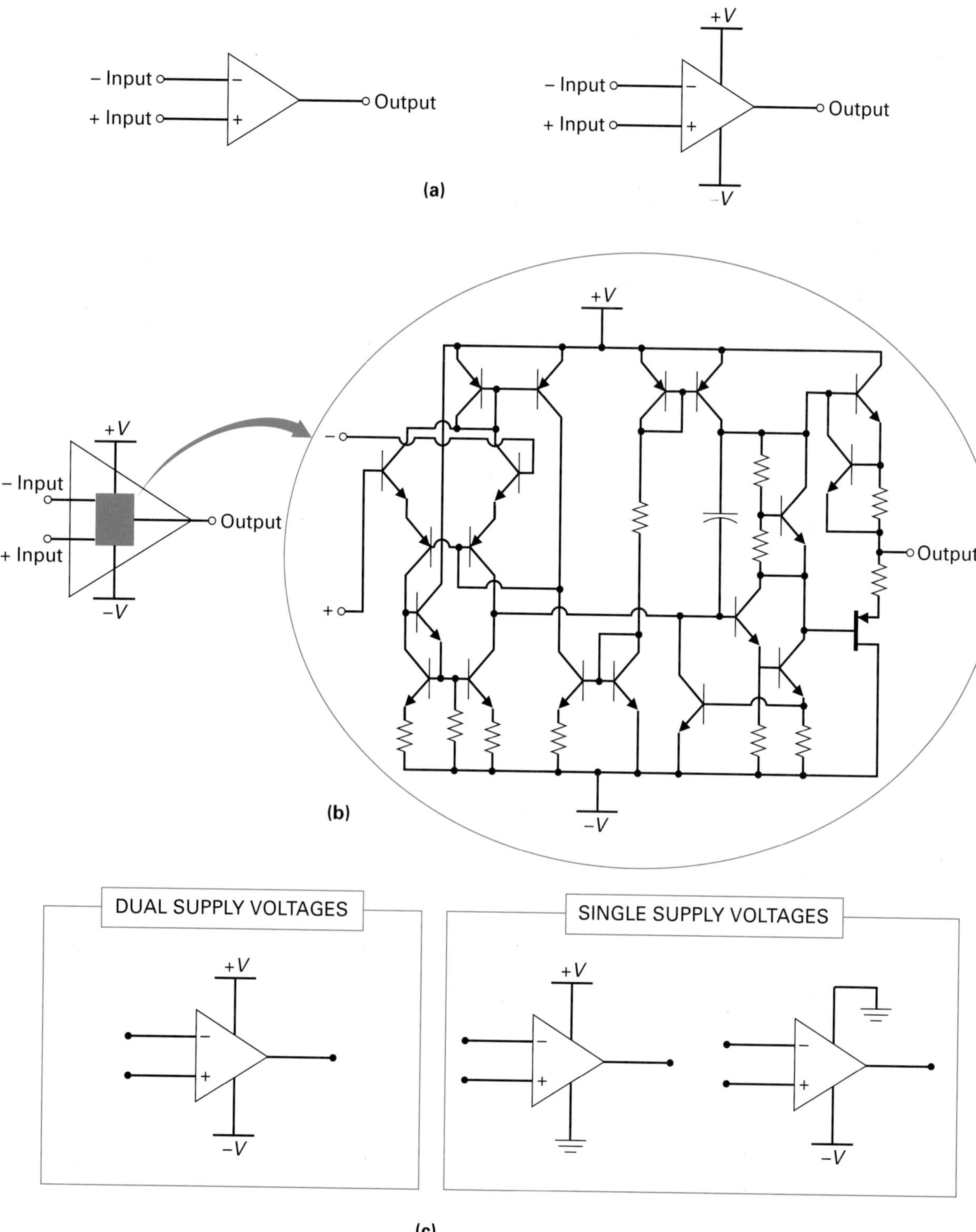

FIGURE 23-1 The Operational Amplifier. (a) Schematic Symbols. (b) Internal Circuit. (c) Power Supply Connections.

The two power supply connections to the op-amp are labeled "+V" and "−V." Figure 23-1(c) shows how power to the op-amp can be supplied by dual supply voltages or by a single supply voltage. When two supply voltages are used (dual supply voltages), the voltage values are of the same value but of opposite polarity (for example, +12 V and −12 V). On the other hand, when only one supply voltage is used (single supply voltage), a positive or negative voltage is applied to its respective terminal while the other terminal is grounded (for example, +5 V and ground or −5 V and ground). Having both a positive and negative power supply voltage will allow the output signal to swing positive and negative, above and below zero. As with all high gain amplifiers, however, the output voltage can never exceed the value of the +V and −V supply voltages.

23-1-2 Op-Amp Packages

The entire op-amp circuit is placed within one of two basic packages, shown in Figure 23-2(a) and (b). The TO-5 metal can package is available with 8, 10, or 12 leads, while the dual in-line through-hole and surface-mount packages typically have 8 or 14 pins.

Like all ICs, an identification code is used to indicate the device manufacturer, device type, and key characteristics. Figure 23-2(c) lists some of the more common manufacturer prefix codes, operating temperature codes, and package codes. In this example, the "MC 741C N" code indicates that the 741 op-amp is made by Motorola, it is designed for commercial application in which the temperature range is between 0 to 70°C, and the package is a through-hole DIP with longer leads.

Referring back to the IC packages in Figure 23-2(a) and (b), you can see that in addition to the two inputs, single output, and two power supply terminals, there are two additional leads labeled balance. These two inputs will normally be connected to a potentiometer that can be adjusted to set the output at zero volts when both inverting and noninverting inputs are at zero volts. **Balancing** the op-amp in this way is generally needed due to imbalances within the op-amp's internal circuit.

Balancing
Setting the output of an op-amp to zero volts when both inverting and noninverting inputs are at zero volts.

SELF-TEST REVIEW QUESTIONS FOR SECTION 23-1

1. Is the operational amplifier a discrete component or an integrated circuit?
2. List the names of the op-amp's 5 terminals.
3. Name the two basic op-amp package types.
4. Briefly describe the meaning of each of the three parts in an op-amp's identification code.

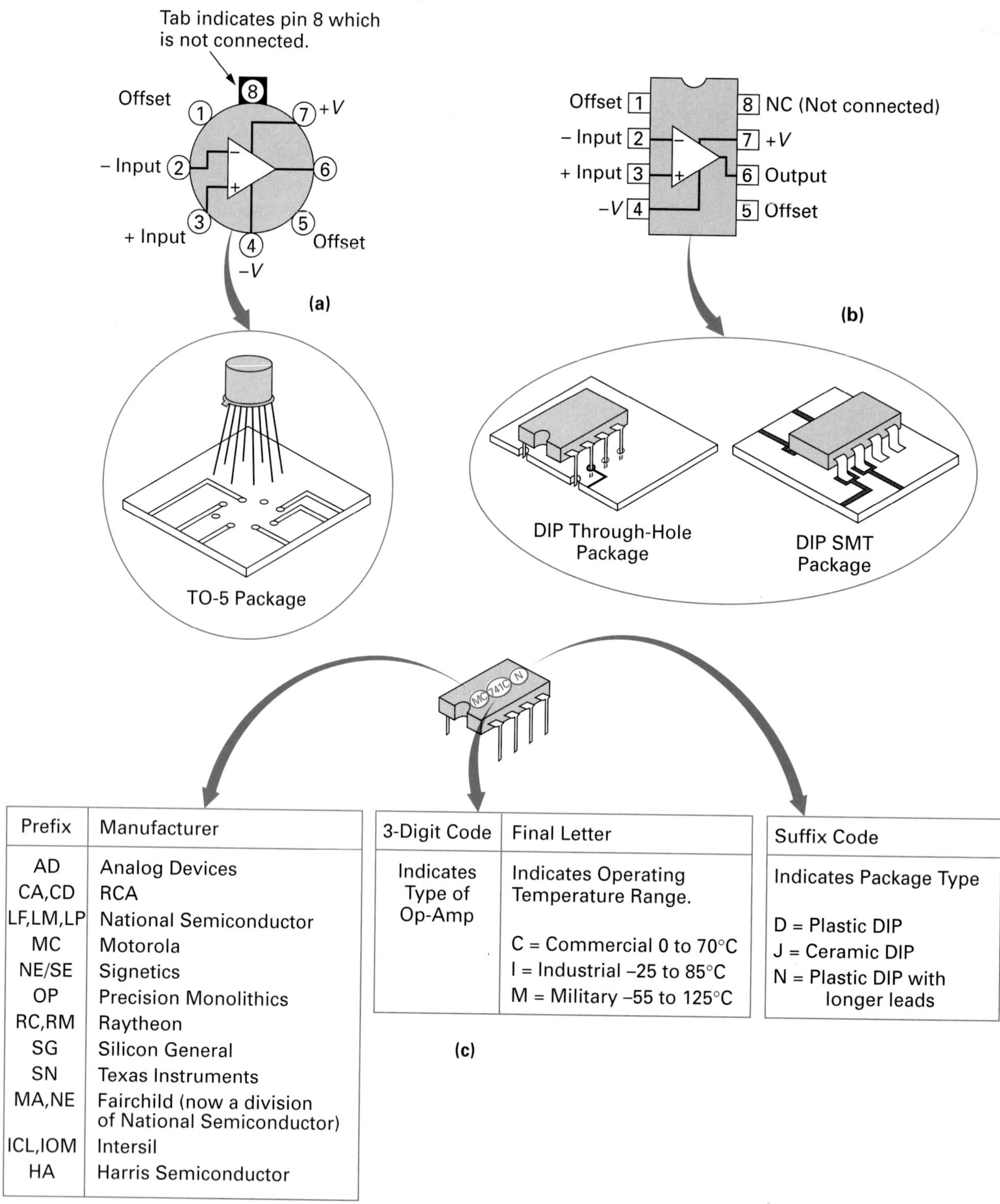

Prefix	Manufacturer
AD	Analog Devices
CA,CD	RCA
LF,LM,LP	National Semiconductor
MC	Motorola
NE/SE	Signetics
OP	Precision Monolithics
RC,RM	Raytheon
SG	Silicon General
SN	Texas Instruments
MA,NE	Fairchild (now a division of National Semiconductor)
ICL,IOM	Intersil
HA	Harris Semiconductor

3-Digit Code	Final Letter
Indicates Type of Op-Amp	Indicates Operating Temperature Range. C = Commercial 0 to 70°C I = Industrial –25 to 85°C M = Military –55 to 125°C

Suffix Code
Indicates Package Type D = Plastic DIP J = Ceramic DIP N = Plastic DIP with longer leads

FIGURE 23-2 Op-Amp Package Types and Identification Codes.

23-2 OP-AMP OPERATION AND CHARACTERISTICS

As mentioned previously, the operational amplifier contains three amplifier circuits, and these three circuits are all interconnected and contained within a single IC. Referring to the block diagram of the op-amp in Figure 23-3(a) you can see that these three circuits are *a differential amplifier, a voltage amplifier,* and *an output amplifier.* Combined, these three circuits give the op-amp its key characteristics, which are *high-gain, high-input impedance,* and *low-output impedance.* We will briefly review their characteristics because combined they determine the characteristics of the op-amp.

23-2-1 The Differential Amplifier within the Op-Amp

The differential amplifier within the op-amp is connected to operate in its "differential-input, single-output mode." The operation of the differential amplifier in this mode is reviewed in Figure 23-3(b). When both input signals are equal in amplitude and in phase with one another, they are referred to as **common mode input signals,** as seen in the waveforms in Figure 23-3(b). On the other hand, if the input signals are out of phase with one another, they are referred to as **differential mode input signals,** as seen in the other set of waveforms in Figure 23-3(b). The differential amplifier will amplify differential input signals while rejecting common mode input signals. The questions you may be asking at this stage are what are common mode input signals and why should we want to reject them? The answers are as follows: temperature changes and noise are common-mode input signals, and they are unwanted signals. Let us examine these common-mode signals in more detail.

Common Mode Input Signals
Input signals to an op-amp that are in phase with one another.

Differential Mode Input Signals
Input signals to an op-amp that are out of phase with one another.

1. Temperature variations within electronic equipment affect the operation of semiconductor materials and, therefore, the operation of semiconductor devices. These temperature variations can cause the dc output voltage of the first stage to drift away from its normal Q point. The second-stage amplifier will amplify this voltage change in the same way as it would amplify any dc input signal and so will all of the following amplifier stages. The increase or decrease in the normal Q-point bias for all of the amplifier stages will get progressively worse due to this thermal instability, and the final stage may have a Q point that is so far off of its mid position that an input signal may drive it into saturation or cutoff, causing signal distortion. With the difference amplifier, any change that occurs due to temperature changes will affect both stages and so will not appear at the output of the differential amplifier, due to its **common mode rejection.**
2. The second common-mode input signal that the differential amplifier removes is noise. It is often necessary to amplify low-level signals from low-sensitivity sources such as microphones, light detectors, and other transducers. High-gain amplifiers are used to increase the amplitude of these small input signals up to a more usable level that is large enough to drive or con-

Common Mode Rejection
The ability of a device to reject a voltage signal applied simultaneously to both input terminals.

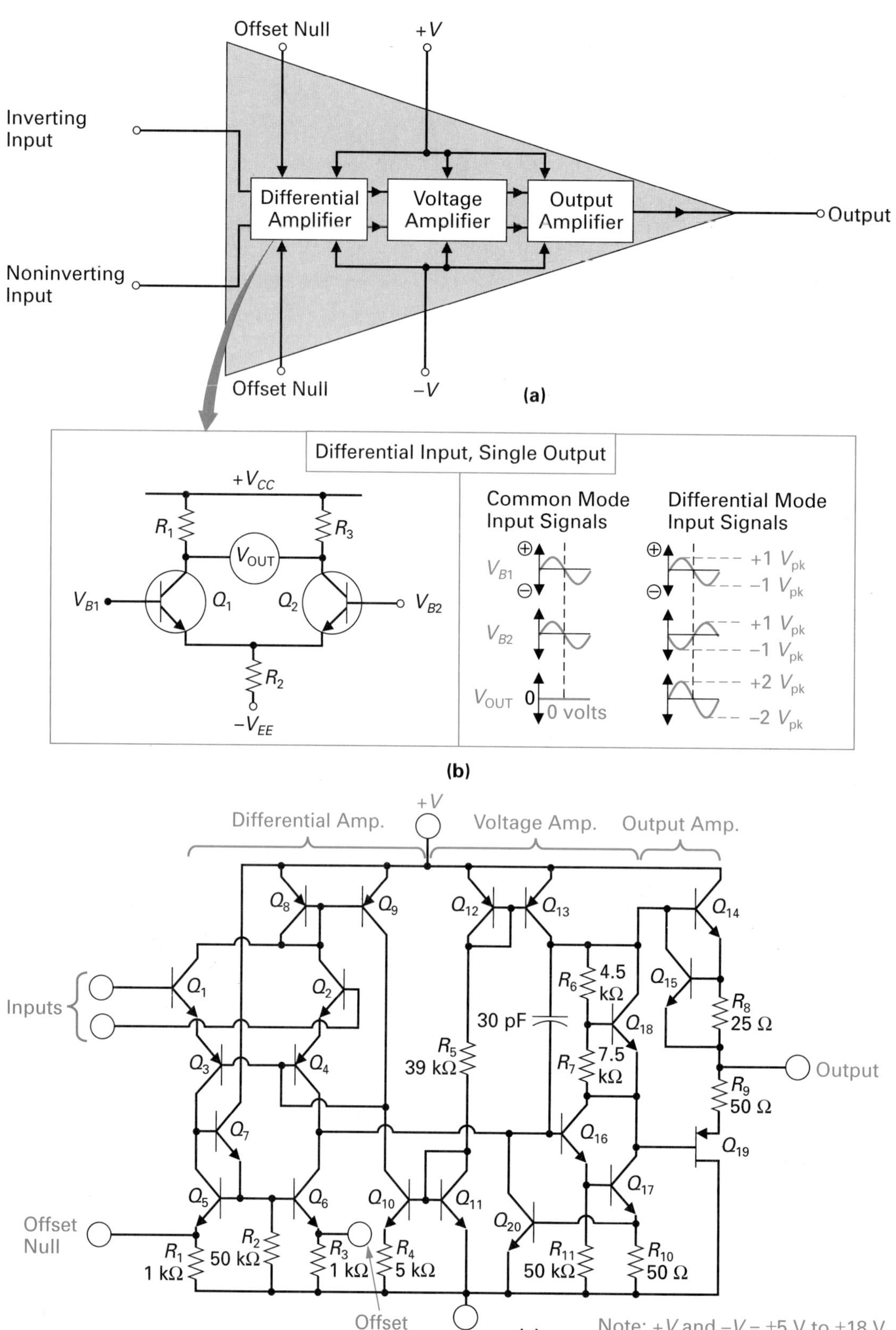

FIGURE 23-3 The Operational Amplifier's Internal Circuit. (a) Block Diagram. (b) The Differential Amplifier's Operation. (c) Circuit Diagram.

trol a load, such as a loudspeaker. The 60 Hz ac power line, or any other electrical variation, can induce a noise signal along with the input signal at the input of this high-gain amplifier. Since these noise signals will be induced at all points in the circuit, and be identical in amplitude and phase, the differential amplifier will block these unwanted signals because they will be present at both inputs of the differential amplifier (noise will be a common-mode input). A true input signal on the other hand, will appear at the two inputs of the differential amplifier as a differential input signal, and therefore be amplified.

23-2-2 *Common-Mode Rejection Ratio*

To summarize, a differential input will be amplified by the op-amp's differential amplifier and passed to the output, while unwanted signals caused by temperature variations or noise will appear as common-mode input signals and therefore be rejected. An op-amp's ability to provide a high **differential gain (A_{VD})** and a low **common-mode gain (A_{CM})** is directly dependent on its internal differential amplifier and is a measure of an op-amp's performance. This ratio is called the **common-mode rejection ratio (CMRR)** and is calculated with the following formula:

Differential Gain (A_{VD})
The amplification of differential-mode input.

Common-Mode Gain (A_{CM})
The amplification of common-mode input.

Common-Mode Rejection Ratio (CMRR)
The ratio of an operational amplifier's differential gain to the common-mode gain.

$$\text{CMRR} = \frac{A_{VD}}{A_{CM}}$$

Looking at this formula, you can see that the higher the A_{VD} (differential gain), or the smaller the A_{CM} (common-mode gain), the higher the CMRR value, and therefore the better the operational amplifier. This ratio can also be expressed in dBs by using the following formula:

$$\text{CMRR} = 20 \times \log \frac{A_{VD}}{A_{CM}}$$

EXAMPLE:

If an op-amp's differential amplifier has a differential gain of 5,000 and a common-mode gain of 0.5, what is the operational amplifier's CMRR? Express the answer in standard gain and dBs.

Solution:

$$\text{CMRR} = \frac{A_{VD}}{A_{CM}} = \frac{5000}{0.5} = 10{,}000$$

$$\text{CMRR} = 20 \times \log \frac{A_{VD}}{A_{CM}} = 20 \times \log 10{,}000 = 20 \times 4 = 80 \text{ dB}$$

A CMRR of 10,000 or 80 dB means that the op-amp's desired input signals will be amplified 10,000 times more than the unwanted common-mode input signals.

23-2-3 *The Op-Amp Block Diagram*

Now that we have reviewed the differential amplifier's circuit characteristics, let us return to the op-amp block diagram in Figure 23-3(a). It is the op-amp's differential-amplifier stage that provides the good common-mode rejection and high differential gain. Because the op-amp's "–" and "+" inputs are applied to either base of the diff-amp, we know that input current will be very small. It is this circuit characteristic that provides the op-amp with another key feature, which is a high input impedance ($I\downarrow$, $Z\uparrow$). The voltage-amplifier stage following the diff-amp usually consists of several darlington-pair stages that provide an overall op-amp voltage gain of typically 50,000 to 200,000. The final output stage consists of a complementary emitter-follower stage to provide a low output impedance and high current gain, so that the op-amp can deliver up to several milliamps, depending on the value of the load.

23-2-4 *The Op-Amp Circuit Diagram*

The complete internal circuit of a typical op-amp can be seen in Figure 23-3(c). With integrated circuits, it is better to have transistors function as resistors wherever possible because they occupy less chip space than actual resistors. This accounts for why the circuit seems to contain many transistors that have their base and collector leads connected. You may also have noticed that no coupling capacitors have been used so that the op-amp can amplify both ac and dc input signals. As with most schematics, the inputs are shown on the left, output on the right, and power is above and below. As discussed previously, the two balancing, or **offset null, inputs** will normally be connected to an external potentiometer that can be adjusted to set the output at zero volts when both the inverting and noninverting inputs are at zero volts. Balancing the op-amp to find the zero-volt output point, or null, in this way is generally needed due to slight imbalances within the op-amp's internal circuit.

Offset Null Inputs
The two balancing inputs used to balance an op-amp.

An important point to realize at this time is that the op-amp is a single component, and up until this time we have concentrated on an understanding of the op-amp's internal circuitry because this helps us to better understand the circuit's normal input/output relationships and characteristics. These operational characteristics are important if we are going to be able to isolate whether a circuit malfunction is internal or external to the op-amp. However, because it is impossible to repair any internal op-amp failures, we will not concentrate on every detail of the op-amp's internal circuit.

23-2-5 *Basic Op-Amp Circuit Applications*

Now that we have an understanding of the op-amp's characteristics, let us now put it to use in some basic circuit applications. To begin with, we will examine the comparator circuit.

The Open-Loop Comparator Circuit

Figure 23-4 shows how the op-amp can be used to function as a **comparator,** which is a circuit that is used to detect changes in voltage level. In Figure 23-4(a), the inverting input (–) of the op-amp is grounded and the input signal is applied to the op-amp's noninverting input (+). Referring to the associated waveforms, you can see that when the input swings positive relative to the negative input (which is 0), the output of the amplifier goes into immediate saturation due to the very large gain of the op-amp. For example, if the op-amp had a voltage gain of 25,000 ($A_V = 25{,}000$), even a small input of +25 mV would cause the op-amp to try and drive its output to 625 V ($V_{OUT}) = V_{IN} \times A_V = 25\text{ mV} \times 25{,}000 = 625\text{ V}$). Since the maximum possible positive output voltage cannot exceed the positive supply volt-

Comparator
An op-amp used without feedback to detect changes in voltage level.

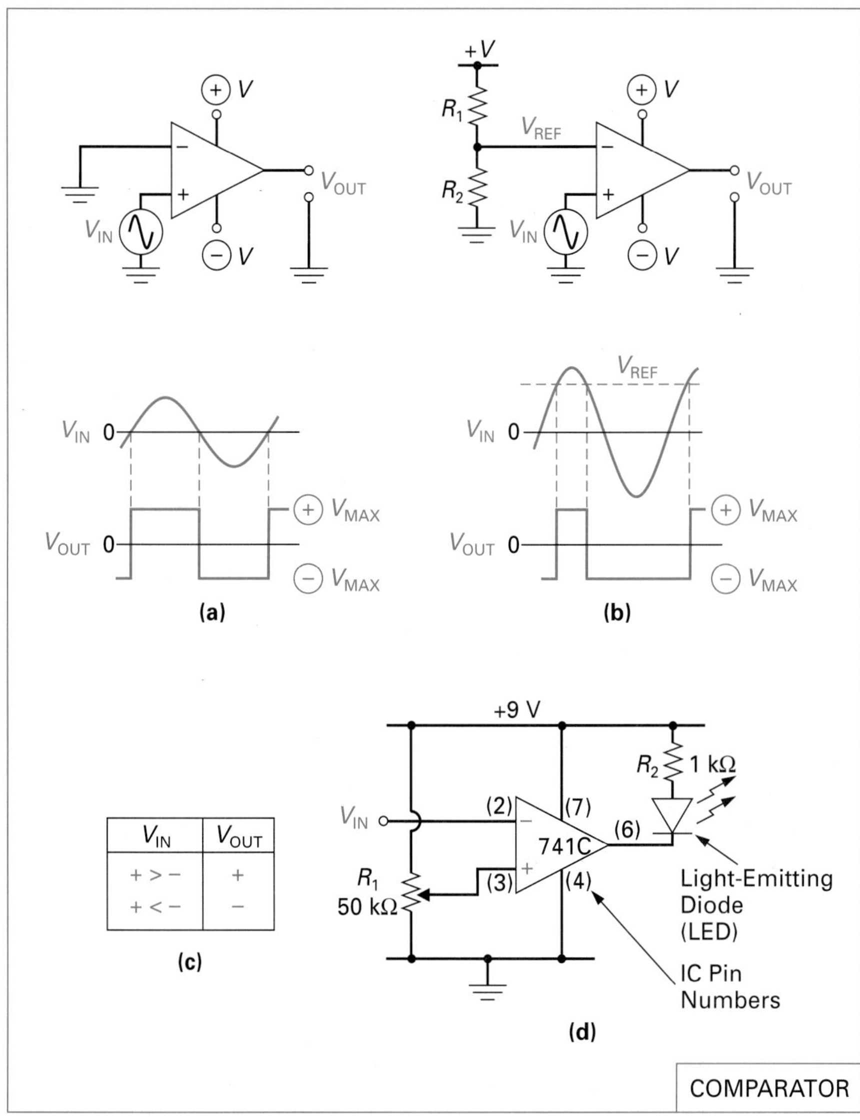

V_{IN}	V_{OUT}
+ > −	+
+ < −	−

FIGURE 23-4 The Open-Loop Comparator Circuit.

age (+V), the output goes to its maximum positive limit, which is equal to the +V supply voltage. When the input swings negative, the amplifier is driven immediately into its opposite state (cut-off), and the output goes to its maximum negative limit, which is equal to the $-V$ supply voltage.

Figure 23-4(b) shows how a voltage divider made up of R_1 and R_2 can be used to supply the inverting input (–) of the op-amp with a reference voltage (V_{REF}) that can be determined by using the voltage divider formula.

$$V_{REF} = \frac{R_2}{R_1 + R_2} \times (+V)$$

Referring to the associated waveforms in Figure 23-4(b), you can see that whenever the ac input signal is more positive than the reference voltage, the output is positive. On the other hand, whenever the ac input signal is less than the reference voltage, the output is negative.

The table in Figure 23-4(c) summarizes the operation of the comparator circuit. In the first line of the table you can see that when the negative input is negative relative to the positive input, the output will go to its positive limit ($V_{OUT} = +V$). In the second line of the table you can see that when the opposite occurs (the negative input is positive, or the positive input is negative), the output will go to its negative limit ($V_{OUT} = -V$).

EXAMPLE:

Briefly describe the operation of the 741C op-amp comparator circuit shown in Figure 23-4(d).

Solution:

Potentiometer R_1 sets up the reference voltage (V_{REF}), which can be anywhere between 0 V and +9 V. When the input voltage (V_{IN}) to the negative input of the op-amp is greater than the positive reference voltage, the op-amp's output will be equal to the $-V$ voltage supply (which is ground), and so the LED will turn ON.

Open-Loop Mode
A control system that has no means of comparing the output with the input for control purposes.

Closed-Loop Mode
A control system containing one or more feedback control loops in which functions of the controlled signals are combined with functions of the commands to tend to maintain prescribed relationships between the commands and the controlled signals.

The Closed-Loop Inverting Amplifier Circuit

The op-amp is usually operated in either the **open-loop mode** or **closed-loop mode.** With the previously discussed comparator circuit, the op-amp was operating in its open-loop mode because there was no signal feedback from output to input. In most instances, the op-amp is operated in the closed-loop mode, in which there is signal feedback from output back to input. This feedback signal will always be out of phase with the input signal and therefore oppose the original signal, which is why it is called "degenerative or negative feedback." Negative feedback, however, is necessary in nearly all op-amp circuits for the following reasons:

1. Because the op-amp has such an extremely high gain, even a very small input signal will be amplified to a very large signal, which will drive the op-amp out of its linear region and into saturation and cutoff. Negative feedback will lower the op-amp's gain, and therefore control the op-amp to prevent output waveform distortion.

2. Having such a high gain can cause the amplifier to go into oscillation due to positive feedback. Negative feedback prevents an amplifier from going into oscillation by reducing the op-amp's gain.
3. The open-loop gain of an op-amp can have a very large range of value for the same device. For example, the 741's open-loop gain can be anywhere from a minimum of 25,000 to 200,000. Including negative feedback in the op-amp circuit will reduce the gain to a consistent value so that the same part can be relied on to provide the same response.

Inverting Amplifier Circuit
An op-amp circuit that produces an amplified output signal that is 180° out of phase with the input signal.

Figure 23-5(a) shows how an op-amp can be connected as an **inverting amplifier circuit,** which produces an amplified output signal that is 180° out of phase with the input signal. Looking at the output voltage label ($-V_{OUT}$), notice that the negative symbol preceding V_{OUT} is being used to indicate the 180° phase inversion between input and output. In this circuit arrangement, the input signal (V_{IN}) is applied through an input resistor (R_{IN}) to the inverting input (–) of the op-amp, while the noninverting input (+) is connected to ground. A feedback loop is connected from the output back to the inverting input via the feedback resistor R_F.

Let us now take a closer look at the closed-loop feedback system that occurs within this amplifier circuit. If the applied input voltage was zero volts ($V_{IN} = 0$ V), the differential input signal (which is the difference between the op-amp's "+" and "–" inputs) will be 0 V, because both the inverting and noninverting inputs will now be at 0 V. A differential input of zero volts therefore will generate an output of zero volts. If the input signal was now to swing positive toward +5 V, the

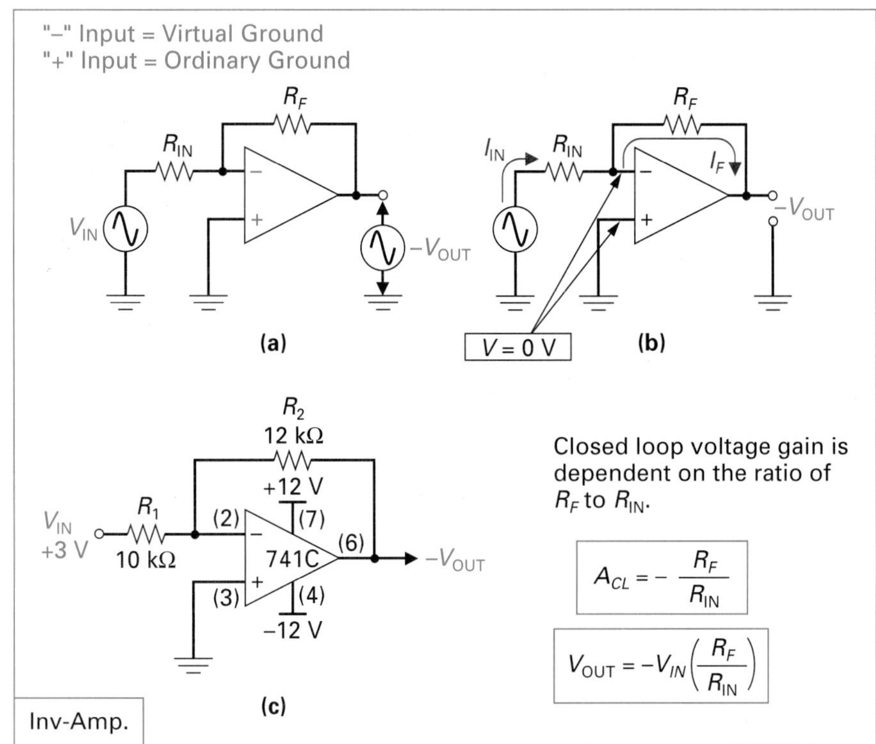

FIGURE 23-5 **The Closed-Loop Inverting Amplifier Circuit.**

output (V_{OUT}) would swing negative due to the internal op-amp circuit phase inversions. This negative output voltage swing would be applied back to the inverting input via R_F to counteract the original positive input change. The feedback path is designed so that it cannot completely cancel the input signal, for if it did, there would be no input, and therefore no output or feedback. In most instances, the feedback voltage (V_F) will greatly restrain the input voltage change to the point that a +5 V input change at V_{IN} will only be felt as a +5 micro volt change at the op-amp's inverting input. Therefore, even though V_{IN} seems to change in values measured in volts, the inverting input of the op-amp will only change in values measured in micro volts. In fact, if the voltage at the inverting input of the op-amp is measured with a voltmeter as shown in Figure 23-5(b), the "–" input appears to remain at 0 V due to the very minute change at the "–" input. The inverting input of the op-amp in this case would be defined as a **virtual ground,** which is different from an **ordinary ground.** A virtual ground is a voltage ground because this point is at zero volts; however, it is not a current ground because it cannot sink or conduct away any current. An ordinary ground on the other hand, is at zero volts and can sink any amount of current.

Virtual Ground
A ground for voltage but not for current.

Ordinary Ground
A connection in the circuit that is said to be at ground potential or zero volts. Because of its connection to earth it has the ability to conduct electrical current to and from earth.

Returning to the inverting amplifier circuit in Figure 23-5(b), we can now analyze this circuit in a little more detail now that we know its basic operation. To begin with, if the "–" input of the op-amp is at 0 V, then all of the input voltage (V_{IN}) will be dropped across the input resistor (R_{IN}). Therefore, the input current can be calculated if we know the value of V_{IN} and R_{IN}, with the following formula:

$$I_{IN} = \frac{V_{IN}}{R_{IN}}$$

Knowing that the left side of R_F is at 0 V means that all of the output voltage will be developed across R_F, and therefore the value of feedback current (I_F) can be calculated with the formula:

$$I_F = \frac{-V_{OUT}}{R_F}$$

The extremely high input impedance of the op-amp means that only a very small fraction of the input current will enter the inverting input. In fact, nearly all of the current flowing through R_{IN} and reaching the "–" op-amp input will leave this virtual ground point via the easiest path which is through R_F. Therefore, it can be said that the feedback current is equal to the input current, or

$$I_{IN} = I_F$$

If I_{IN} (which equals V_{IN}/R_{IN}) and I_F (which equals $-V_{OUT}/R_F$) are equal, then

$$\frac{V_{IN}}{R_{IN}} = \frac{-V_{OUT}}{R_F}$$

If this is rearranged, we arrive at the following:

$$\frac{V_{\text{OUT}}}{V_{\text{IN}}} = -\frac{R_F}{R_{\text{IN}}}$$

Because the voltage gain of an amplifier is equal to

$$A_V = \frac{V_{\text{OUT}}}{V_{\text{IN}}}$$

and because $V_{\text{OUT}}/V_{\text{IN}} = -R_F/R_{\text{IN}}$, the **closed-loop voltage gain (A_{CL})** of the inverting operational amplifier is equal to the ratio of R_F to R_{IN}.

Closed-Loop Voltage Gain (A_{CL})
The voltage gain of an amplifier when it is operated in the closed-loop mode.

$$A_{CL} = -\frac{R_F}{R_{\text{IN}}}$$

(Negative symbol preceding R_F/R_{IN} indicates signal inversion)

By rearranging the previous equation $V_{\text{OUT}}/V_{\text{IN}} = -R_F/R_{\text{IN}}$, we can arrive at the following formula for calculating the output voltage of this circuit.

Since

$$\frac{V_{\text{OUT}}}{V_{\text{IN}}} = -\frac{R_F}{R_{\text{IN}}}$$

$$V_{\text{OUT}} = -V_{\text{IN}}\left(\frac{R_F}{R_{\text{IN}}}\right)$$

The input impedance of this inverting op-amp is equal to the value of R_{IN} because the input voltage (V_{IN}) is developed across the input resistor (R_{IN}).

EXAMPLE:

Calculate the output voltage of the inverting amplifier shown in Figure 23-5(c).

Solution:

$$V_{\text{OUT}} = -V_{\text{IN}}\left(\frac{R_F}{R_{\text{IN}}}\right) = -3\text{ V} \times \frac{12\text{ k}\Omega}{10\text{ k}\Omega} = -3\text{ V} \times 1.2 = -3.6\text{ V}$$

The Closed-Loop Noninverting Amplifier Circuit

Figure 23-6(a) shows how an op-amp can be configured to operate as a **noninverting amplifier circuit.** The input voltage (V_{IN}) is applied to the op-amp's noninverting input (+), and therefore the output voltage (V_{OUT}) will be in phase with the input. To achieve negative feedback, the output is applied back to the inverting input (–) of the op-amp via the feedback network formed by R_F and R_1.

Noninverting Amplifier Circuit
An op-amp circuit that produces an amplified output signal that is in phase with the input signal.

Let us now take a closer look at the closed-loop negative feedback system that occurs within this amplifier circuit. The output voltage (V_{OUT}) is propor-

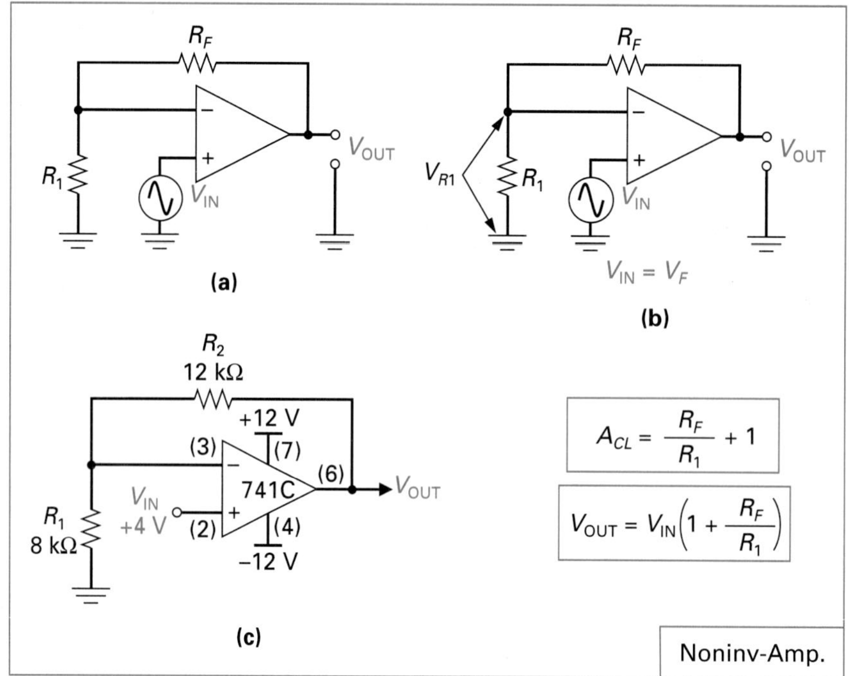

FIGURE 23-6 **The Closed-Loop Noninverting Amplifier Circuit.**

tionally divided across R_F and R_1, with the feedback voltage (V_F) developed across R_1 being applied to the inverting input (−) of the op-amp, as shown in Figure 23-6(b). Because V_{OUT} is in phase with V_{IN}, the feedback voltage (V_F) will also be in phase with V_{IN}, and therefore these two in-phase inputs to the op-amp will be common-mode input signals. As a result, feedback will be degenerative. However, because V_{OUT} is slightly larger than V_{IN}, there will be a small difference between V_{IN} and V_F, and this differential input will be amplified. To summarize, the noninverting op-amp provides negative feedback by feeding back an in-phase common-mode signal, and this degenerative feedback will lower the op-amp's gain to

1. prevent output waveform distortion,
2. prevent the amplifier from going into oscillation, and
3. reduce the gain of the op-amp to a consistent value.

The very small micro volt difference between V_{IN} and V_F will be amplified; however, since there is such a very small difference between these two input signals, it can be said that

$$V_{IN} = V_F$$

Because the closed-loop voltage gain of any op-amp is equal to

$$A_{CL} = \frac{V_{OUT}}{V_{IN}}$$

the gain of the noninverting amplifier could also be calculated with

$$A_{CL} = \frac{V_{OUT}}{V_F}$$

By using the voltage divider formula, we can develop a formula for calculating V_F.

$$V_F = \frac{R_1}{R_1 + R_F} \times V_{\text{OUT}}$$

By rearranging the above formula as follows

$$\frac{V_{\text{OUT}}}{V_F} = \frac{R_1 + R_F}{R_1} \quad \text{or} \quad \frac{V_{\text{OUT}}}{V_F} = \frac{R_1}{R_1} + \frac{R_F}{R_1} = \frac{R_F}{R_1} + 1$$

and because $V_{\text{OUT}}/V_F = A_{CL}$, the noninverting op-amp's gain can also be calculated with the formula

$$A_{CL} = \frac{R_F}{R_1} + 1$$

Because the output voltage of an amplifier is equal to the product of input voltage and gain ($V_{\text{OUT}} = V_{\text{IN}} \times A_{CL}$), we can add V_{IN} to the previous closed-loop gain formula in order to calculate output voltage.

$$V_{\text{OUT}} = V_{\text{IN}}\left(1 + \frac{R_F}{R_1}\right)$$

With the inverting op-amp, the input impedance is determined by the input resistor (R_{IN}). With the noninverting amplifier, there is no resistor connected because V_{IN} is applied directly into the very high input impedance of the op-amp. As a result, the noninverting amplifier circuit has an extremely high input impedance.

EXAMPLE:

Calculate the output voltage from the noninverting amplifier shown in Figure 23-6(c).

Solution:

$$V_{\text{OUT}} = V_{\text{IN}}\left(1 + \frac{R_F}{R_1}\right) = 4\text{ V}\left(1 + \frac{12\text{ k}\Omega}{8\text{ k}\Omega}\right) = 4\text{ V} \times (1 + 1.5) = 4\text{ V} \times 2.5 = +10\text{ V}$$

The open-loop comparator and closed-loop inverting and noninverting amplifier circuits are just three of many op-amp application circuits. In the final section of this chapter, we will examine many other typical op-amp circuit appli-

DEVICE: μA 747—Dual Linear Op-Amp IC

DESCRIPTION

The 747 is a pair of high-performance monolithic operational amplifiers constructed on a single silicon chip. High common-mode voltage range and absence of "latch-up" make the 747 ideal for use as a voltage-follower. The high gain and wide range of operating voltage provides superior performance in integrator, summing amplifier, and general feedback applications. The 747 is short-circuit protected and requires no external components for frequency compensation. The internal 6dB/octave roll-off insures stability in closed-loop applications. For single amplifier performance, see μA741 data sheet.

FEATURES

- No frequency compensation required
- Short-circuit protection
- Offset voltage null capability
- Large common-mode and differential voltage ranges
- Low power consumption
- No latch-up

PIN CONFIGURATION

N Package

INV. INPUT A 1
NON-INVERTING INPUT A 2
OFFSET NULL A 3
V– 4
OFFSET NULL B 5
NON-INVERTING INPUT B 6
INVERTING INPUT B 7
14 OFFSET NULL A
13 V + A
12 OUTPUT A
11 NO CONNECT
10 OUTPUT D
9 V + B
8 OFFSET NULL B

TOP VIEW

SL00100

ABSOLUTE MAXIMUM RATINGS

SYMBOL	PARAMETER	RATING	UNIT
V_S	Supply voltage	±18	V
$P_{D\ MAX}$	Maximum power dissipation T_A=25°C (still air)[1]	1500	mW
V_{IN}	Differential input voltage	±30	V
V_{IN}	Input voltage[2]	±15	V
	Voltage between offset null and V-	±0.5	V
T_{STG}	Storage temperature range	-65 to +150	°C
T_A	Operating temperature range	0 to +70	°C
T_{SOLD}	Lead temperature (soldering, 10sec)	300	°C
I_{SC}	Output short-circuit duration	Indefinite	

Explanation of Key Maximum Ratings

Supply Voltage: This is the maximum voltage that can be used to power the op-amp.

Maximum Power Dissipation: This is the maximum power the op-amp can dissipate.

Differential Input Voltage: This is the maximum voltage that can be applied across the + and – inputs.

Input Voltage: This is the maximum voltage that can be applied between an input and ground.

Operating Temperature Range: The temperature range in which the op-amp will operate within the manufacturer's specifications.

Open Short Circuit Duration: This is the amount of time that the op-amp's output can be short circuited to ground or to a supply voltage. This op-amp has an "indefinite" rating since it has an internal circuit that will turn OFF the op-amp's output and protect the internal circuitry if an output short occurs.

FIGURE 23-7 **An Op-Amp Data Sheet. (Courtesy of Philips Semiconductors.)**

cations, but for now let us review our understanding of the op-amp by examining a typical manufacturer's data sheet.

23-2-6 *An Op-Amp Data Sheet*

To better understand the characteristics of the op-amp, Figure 23-7 shows the data sheet for a "747," which is a 14-pin IC containing two 741 op-amps. Like most data sheets, this one contains a general description of the device, a pin-configuration diagram, a listing of maximum ratings, an internal-circuit diagram, and a listing of input/output characteristics. Notes have been included so that you will be able to understand the meaning of most key terms.

DEVICE: μA 747—Dual Linear Op-Amp IC

EQUIVALENT SCHEMATIC

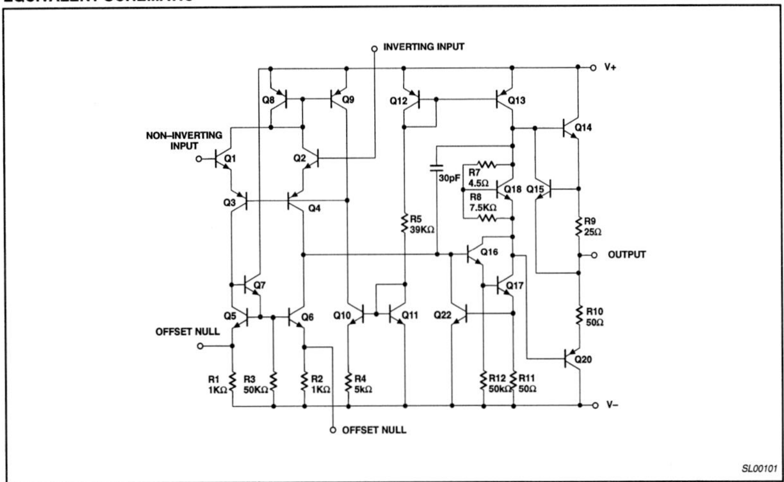

DC ELECTRICAL CHARACTERISTICS

T_A=25°C, V_{CC} = ±15V unless otherwise specified.

SYMBOL	PARAMETER	TEST CONDITIONS	μA747C Min	μA747C Typ	μA747C Max	UNIT
V_{OS}	Offset voltage	R_S≤10kΩ		2.0	6.0	mV
		R_S≤10kΩ, over temp.		3.0	7.5	mV
$\Delta V_{OS}/\Delta T$				10		μV/°C
I_{OS}	Offset current			20	200	nA
		Over temperature		7.0	300	nA
$\Delta I_{OS}/\Delta T$				200		pA/°C
I_{BIAS}	Input current			80	500	nA
		Over temperature		30	800	nA
$\Delta I_B/\Delta T$				1		nA/°C
V_{OUT}	Output voltage swing	R_L≥2kΩ, over temp.	±10	±13		V
		R_L≥10kΩ, over temp.	±12	±14		V
I_{CC}	Supply current each side			1.7	2.8	mA
		Over temperature		2.0	3.3	mA
P_d	Power consumption			50	85	mW
		Over temperature		60	100	mW
C_{IN}	Input capacitance			1.4		pF
	Offset voltage adjustment range			±15		mV
R_{OUT}	Output resistance			75		Ω
	Channel separation			120		dB
PSRR	Supply voltage rejection ratio	R_S≤10kΩ, over temp.		30	150	μV/V
A_{VOL}	Large-signal voltage gain (DC)	R_L≥2kΩ, V_{OUT}=±10V	25,000			V/V
		Over temperature	15,000			V/V
CMRR	Common-mode rejection ratio	R_S≤10kΩ, V_{CM}=±12V Over temperature	70			dB

Explanation of Key Ratings

Input Offset Voltage: The voltage that must be applied to one input for the output voltage to be zero.
Input Offset Current: The difference of the two input bias currents when the output voltage is zero.
Input Bias Current: The average of the currents flowing into both inputs (ideally input bias currents are equal).
Input Resistance: This is the resistance of either input when the other input is grounded.

Output Resistance: This is the resistance of the op-amp's output.
Output Short-Circuit Current: Maximum output current that the op-amp can deliver to a load.

Supply Current: The current that the op-amp circuit will draw from the power supply.
Slew rate: The maximum rate of change of output voltage under large signal conditions.
Channel Separation: When two op-amps are within one package, there will be a certain amount of interference or "crosstalk" between op-amps.

FIGURE 23-7 (continued)

OP-AMP TYPE	INPUT IMPEDANCE	OUTPUT IMPEDANCE	OPEN-LOOP GAIN (Min.)	CMRR (dB)
741 C	2 MΩ	75Ω	25,000	70
101	800 kΩ	Low	25,000	70
108 A	70 MΩ	Low	80,000	96
351	High	Low	25,000	70
318	High	Low	25,000	70
357	High	Low	50,000	80
363	High	Low	1,000,000	94
356	High	Low	25,000	80

FIGURE 23-8 **Comparing the Characteristics of Several Op-Amps.**

Figure 23-8 compares the characteristics and cost of several different op-amp types to the 741, which is very popular due to its good performance and low price. Referring to the cost-factor column, you can see, for example, that the LF351 is twice the price of the 741. Studying the key characteristics in this figure, you can see that the 741, for example, will typically have an input impedance of 2 MΩ, an output impedance of 75Ω, an open-loop gain of 25,000, and a common-mode rejection ratio of 70 dB for any input signal from 0 Hz up to 1 MHz. The gain bandwidth column lists only the upper frequency limit because the op-amp has no internal coupling capacitors, and therefore the lower frequency limit extends down to dc signals, or 0 Hz.

23-2-7 *Troubleshooting Op-Amp Circuits*

Now that we know how operational amplifier circuits are supposed to work, let us see how we can troubleshoot a problem when they do not work. First, let us review our basic troubleshooting procedure.

Step 1: DIAGNOSE

The first step is to determine whether a problem really exists. To carry out this step, a technician must collect as much information as possible about the system, circuit, and components used, and then diagnose the problem.

Step 2: ISOLATE

The second step is to apply a logical and sequential reasoning process to isolate the problem. In this step, a technician will operate, observe, test, and apply troubleshooting techniques in order to isolate the malfunction.

Step 3: REPAIR

The third and final step is to make the actual repair, then test the circuit.

Isolating the Problem to the Op-Amp Circuit

Once you have diagnosed that an op-amp circuit is malfunctioning, the next step is to isolate whether the problem is within the op-amp's internal circuit, or external to the op-amp IC. For example, Figure 23-9(a) shows a typical closed-

COST FACTOR	SLEW RATE (V/μs)	GAIN BANDWIDTH	FEATURES
1	0.5	1 MHz	Low cost
1	0.5	1 MHz	Low cost
—	0.3	1 MHz	Precision low drift
2	13	4 MHz	Low bias current
5	70	15 MHz	High slew rate
4	30	20 MHz	High CMRR
45	—	2 MHz	Low noise; high rejection
3	10	5 MHz	Improved 741

FIGURE 23-8 (continued)

loop inverting op-amp circuit. As with all amplifiers, the circuit receives power from a dc power supply, an input signal from the input-signal source, and supplies an output signal to a load. Therefore, our first step is to isolate whether the problem is within the op-amp circuit or within one of the external circuits. To achieve this, the following checks should be made.

1. Power Supply or Amplifier? If the amplifier is not functioning as it should, first check the power (+12 V and −12 V) and ground connection to the amplifier circuit. If the dc power supply voltage to the circuit is not correct, you will next need to determine whether the problem is in the source (the dc power supply circuit), or in the load (the amplifier circuit). If the amplifier is disconnected from the dc power supply, and the voltage returns to its normal potential, the amplifier has developed some short within its circuit and this is pulling down the $+V_{CC}$ supply voltage. If, on the other hand, the amplifier is disconnected from the dc power supply, and the voltage supply voltage still remains at its incorrect value, the problem is within the dc power supply circuit.
2. Amplifier Source or Output Load? If the output signal from the op-amp is incorrect, remember that a bad load can pull down a voltage signal and make it appear as though the amplifier is faulty. When you have no output, or a bad output, you will first need to disconnect the output of the amplifier from the load to determine whether the problem is in the source (amplifier) or the load. If the load is disconnected and the signal is good, the impedance of the load is pulling down the output of the amplifier. On the other hand, if the load is disconnected and the signal output of the amplifier is still bad, the problem is in the amplifier.
3. Signal Source or Amplifier Load? If the output signal from an op-amp is incorrect, do not backtrack through the amplifier circuit. Go immediately to the amplifier's input and check the input signal to the first stage. Once again, if the input signal to the amplifier is incorrect, we will first have to isolate whether the problem exists in the signal source, or in the amplifier. If the input signal source is disconnected from the amplifier, and the signal generated by V_{IN} returns to its normal level, a fault in the amplifier input is

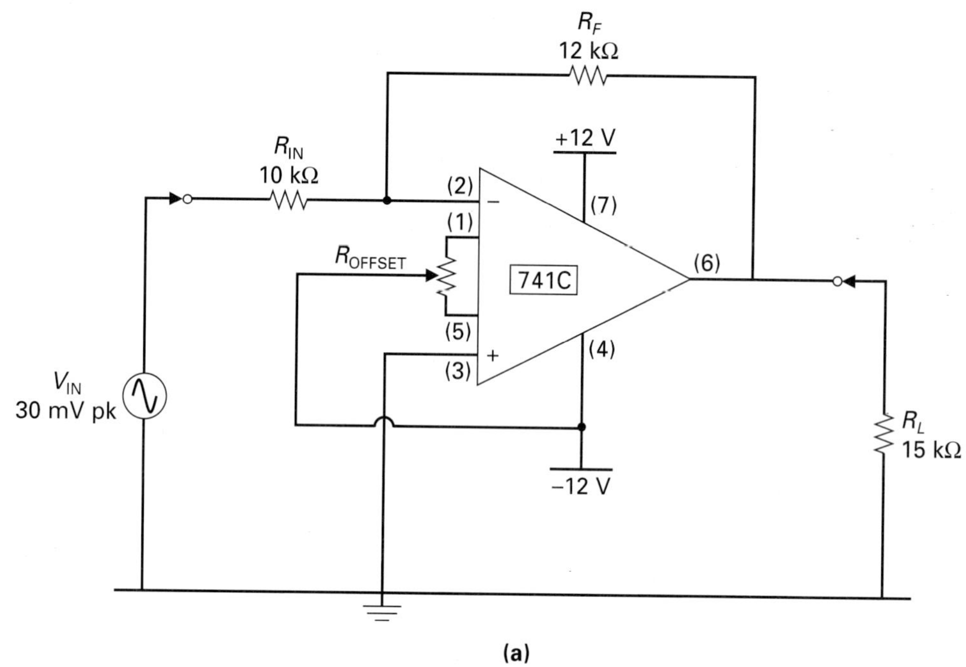

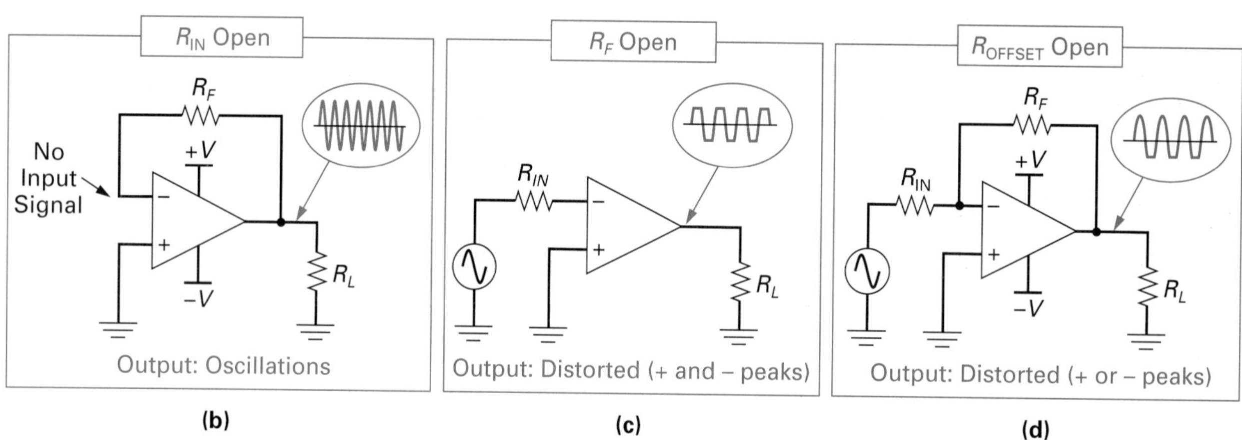

FIGURE 23-9 Troubleshooting Op-Amp Circuits.

pulling down the input signal. On the other hand, if the input signal source is disconnected from the amplifier, and the signal does not return to its normal level, a fault exists in the input signal source circuit.

Troubleshooting an Op-Amp Circuit

Once you have isolated the problem to the op-amp circuit, you will next need to isolate the faulty component. Because op-amp circuits only contain a handful of components, the isolating process is relatively easy, especially because each component's fault produces its own distinct effect. Figure 23-9(b), (c), and (d) show the effect you should get if each of the circuit resistors were to open.

1. If R_{IN} were to open as shown in Figure 23-9(b), the input signal would be disconnected from the amplifier. Although this condition should result in no output, in most cases it sends the op-amp circuit into oscillation: if the output is positive when R_{IN} opens, a positive voltage will be fed back to the inverting input via R_F. This positive input will be inverted to a negative output, which will be fed back via R_F producing a positive output, which will be fed back via R_F producing a negative output, and so on. The result will be a continuous millivolt oscillation at the output.
2. If R_F were to open as shown in Figure 23-9(c), the feedback path would be disconnected and the gain of the op-amp circuit will increase to its maximum open loop value. For example, if a 30 mV pk input signal were applied to a 741 with a minimum open-loop gain of 25,000, the op-amp would try to produce an output of 750 V pk. Because the output voltage can never exceed the $+V$ and $-V$ supply voltages, the result will be a heavily distorted waveform that has its positive and negative peaks clipped when the op-amp is driven into saturation and cutoff.
3. If the offset resistor were to open, the output of the op-amp will not be zero when the input is zero. In fact, the output will be offset by an amount that is equal to the offset voltage times the op-amp's closed loop gain. As a result, the output waveform will have either its positive or negative peaks clipped, depending on whether the op-amp is unbalanced in the positive or negative direction.
4. As shown previously, the op-amp has an internal circuit that contains a large number of components. If you have isolated the problem to the op-amp's internal circuit, and you have determined that all of the op-amp's externally connected resistors are okay, the next step will be to replace the op-amp IC itself because it is impossible to repair any internal op-amp failures. If only one of the op-amps within a dual op-amp package such as the 747 has gone bad, the whole IC will have to be replaced. When removing an op-amp IC, be sure to note the orientation of the package because a replacement put in backwards will more often than not be destroyed.

SELF-TEST REVIEW QUESTIONS FOR SECTION 23-2

1. Can the op-amp be used to amplify dc as well as ac signal inputs?
2. What is the difference between an open-loop and closed-loop op-amp circuit?
3. List the key characteristics of an op-amp.
4. What are the three basic amplifier types within an op-amp?
5. Which of the following circuits is connected in an open-loop mode?
   a. Comparator **b.** Inverter amplifier **c.** Noninverting amplifier
6. Why is it important for an op-amp circuit to have negative feedback?

23-3 ADDITIONAL OP-AMP CIRCUIT APPLICATIONS

The operational amplifier's flexibility and characteristics make it the ideal choice for a wide variety of circuits applications. In fact, because the op-amp is the most frequently used linear IC, it is safe to say that you will find several op-amps in almost every electronic system. Although it is impossible to cover all of these applications, many circuits predominate, and others are merely variations on the same basic theme. In this section we will concentrate on the operation and characteristics of all of the most frequently used op-amp circuit applications.

23-3-1 *The Voltage Follower Circuit*

Voltage Follower Circuit
An op-amp circuit that has a direct feedback to give unity gain so the output voltage follows the input voltage. Used in applications where a very high input impedance and very low output impedance are desired.

Figure 23-10 shows how an op-amp can be connected to form a noninverting **voltage follower circuit.** Using the noninverting closed-loop gain formula discussed previously, we can calculate the voltage gain of this circuit.

$$A_{CL} = \frac{R_F}{R_1} + 1 = \frac{0\ \Omega}{0\ \Omega} + 1 = 0 + 1 = 1$$

With a gain of 1, the output voltage will be equal to the input voltage—so what is the advantage of this circuit? The answer is the op-amp characteristics of a high-input impedance and a low-output impedance. Similar to the BJT's emitter-follower and the FET's source-follower, the op-amp voltage-follower circuit derives its name from the fact that the output voltage follows the input voltage in both polarity and amplitude. This circuit is therefore ideal as a buffer, interfacing a high-impedance source to a low-impedance load.

23-3-2 *The Summing Amplifier Circuit*

Summing Amplifier Circuit (or Adder Circuit)
An op-amp circuit that will sum or add all of the input voltages.

The **summing amplifier circuit,** or adder amplifier, consists of two or more input resistors connected to the inverting input of an op-amp as shown in Figure 23-11(a). This circuit will sum or add all of the input voltages, and therefore the output voltage will be

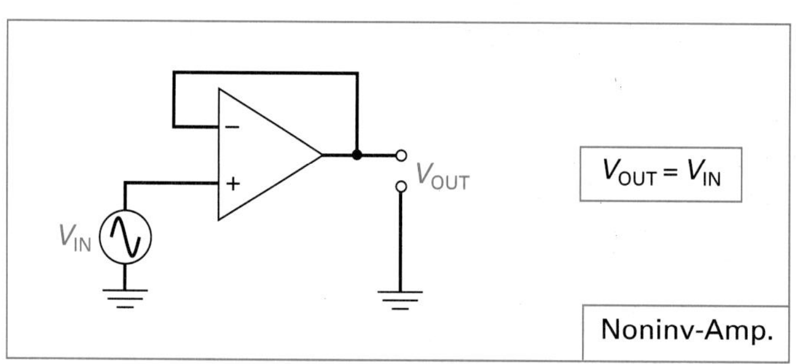

FIGURE 23-10 Voltage Follower Circuit.

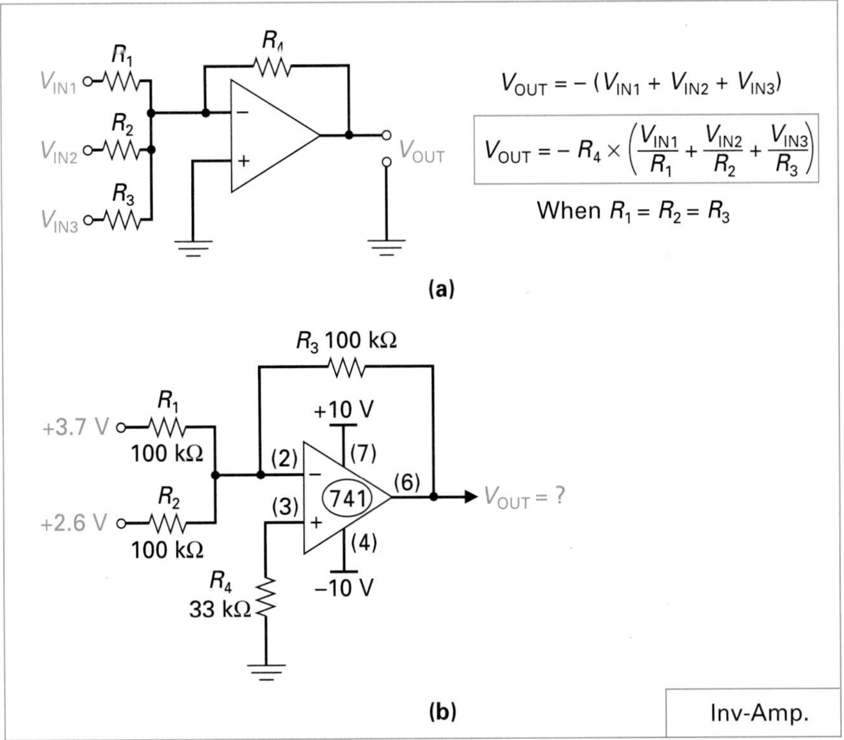

FIGURE 23-11 **Summing Amplifier Circuit.**

$$V_{OUT} = -(V_{IN1} + V_{IN2} + V_{IN3})$$

Using Ohms law ($V = R \times I$), we can also calculate the output voltage with the formula

$$V_{OUT} = -R_4 \times \left(\frac{V_{IN1}}{R_1} + \frac{V_{IN2}}{R_2} + \frac{V_{IN3}}{R_3}\right)$$

Once again, the negative sign preceding the formula indicates that the output signal will be opposite in polarity to the two or three input signals.

EXAMPLE:

Calculate the output voltage of the summing amplifier circuit shown in Figure 23-11(b).

Solution:

$$V_{OUT} = -(V_{IN1} + V_{IN2}) = -(3.7\text{ V} + 2.6\text{ V}) = -6.3\text{ V}$$

23-3-3 *The Difference Amplifier Circuit*

Differential Amplifier Circuit
An op-amp circuit in which the output voltage is equal to the difference between the two input voltages.

Figure 23-12(a) shows how the op-amp can be connected to operate as a difference or **differential amplifier circuit.** In this circuit application, the op-amp will simply be making use of its first internal amplifier stage, which (as mentioned earlier) is a diff-amp. In this circuit, all four resistors are normally of the same value, and the output voltage is equal to the difference between the two input voltages.

$$V_{OUT} = V_{IN2} - V_{IN1}$$

EXAMPLE:

Calculate the output voltage of the difference amplifier circuit shown in Figure 23-12(b).

Solution:

$$V_{OUT} = V_{IN2} - V_{IN1} = 6\text{ V} - 3.2\text{ V} = 2.8\text{ V}$$

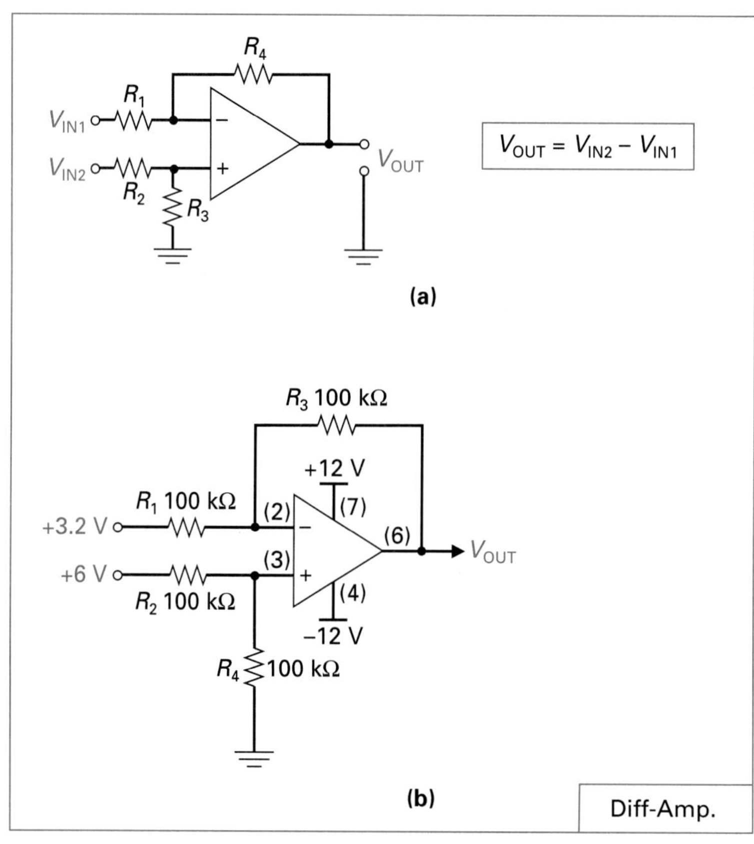

FIGURE 23-12 **The Difference Amplifier Circuit.**

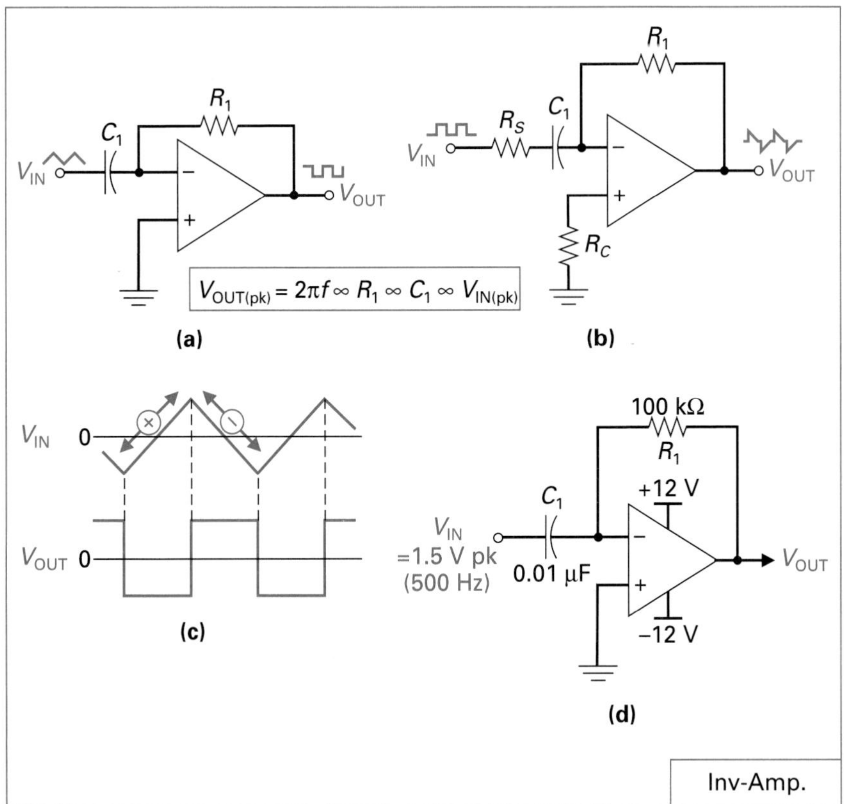

FIGURE 23-13 **Differentiator Circuit.**

23-3-4 *The Differentiator Circuit*

The op-amp **differentiator circuit** (not to be confused with the previous differential circuit) is similar to the basic inverting amplifier except that R_{IN} is replaced by a capacitor, as shown in Figure 23-13(a). Including a capacitor in any circuit means that we will develop problems as the frequency of the input signal increases because capacitive reactance is inversely proportional to frequency. This means that the reactance of the input capacitor will decrease for input signals that are higher in frequency, and therefore the input voltage applied to the op-amp and output voltage from the op-amp will increase with frequency. Including an additional resistor (R_S) in series with the input capacitor, as shown in Figure 23-13(b), will decrease the high-frequency gain because gain will now be a ratio of R_F/R_S.

Differentiator Circuit
A circuit whose output voltage is proportional to the rate of change of the input voltage. The output waveform is the time derivative of the input waveform.

Figure 23-13(c) shows the differentiator's input/output waveforms, with the peak of the output square wave being equal to

$$V_{OUT(pk)} = 2\,\pi f \times R_1 \times C_1 \times V_{IN(pk)}$$

EXAMPLE:

Calculate the peak output voltage of the differentiator circuit shown in Figure 23-13(d).

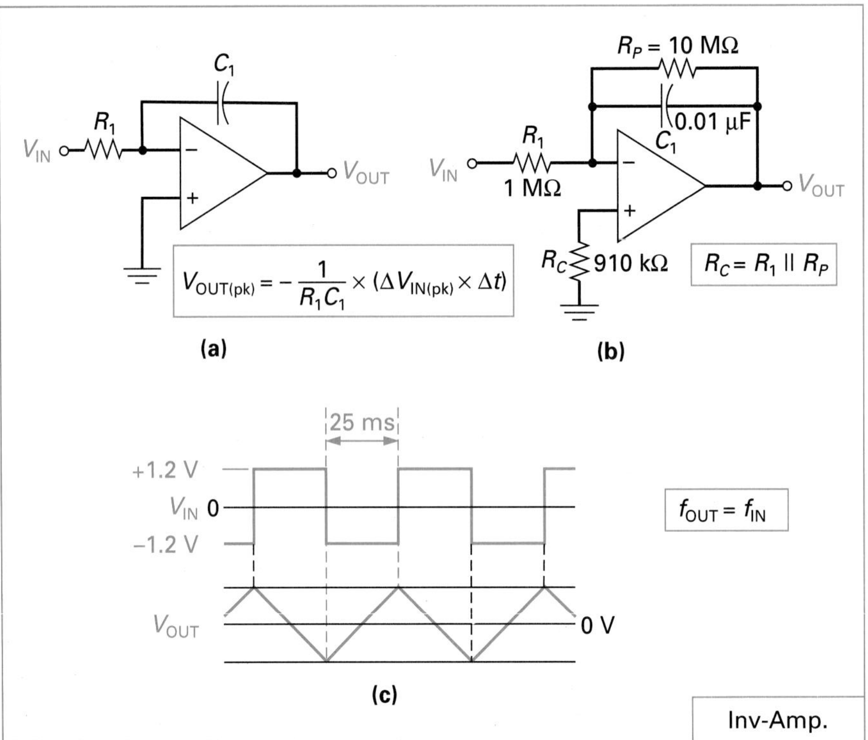

FIGURE 23-14 **Integrator Circuit.**

Solution:

$$V_{OUT(pk)} = 2\,\pi f \times R_1 \times C_1 \times V_{IN(pk)}$$

$$= (2 \times \pi \times 500\text{ Hz}) \times 100\text{ k}\Omega \times 0.01\ \mu\text{F} \times (1.5\text{ V pk}) = 4.7\text{ V}$$

23-3-5 *The Integrator Circuit*

Integrator Circuit A circuit with an output which is the integral of its input with respect to time.

Figure 23-14(a) shows how the position of the resistor and capacitor in the differentiator circuit can be reversed to construct an op-amp **integrator circuit.** As in the differentiator circuit, the capacitor will alter the gain of the op-amp because its capacitive reactance changes with frequency. To compensate for this effect, a parallel resistor (R_P) is included in shunt with the capacitor, as shown in Figure 23-14(b), to decrease the low-frequency gain: gain will now be a ratio of R_P/R_1.

Figure 23-14(c) shows the integrator's input/output waveforms, with the peak of the output triangular wave being equal to

$$V_{OUT(pk)} = \frac{1}{R_1C_1} \times (\Delta V_{IN(pk)} \times \Delta t)$$

EXAMPLE:

Calculate the peak output voltage of the integrator circuit shown in Figure 23-14(b) considering the input/output waveforms given in Figure 23-14(c).

Solution:

$$V_{OUT(pk)} = \frac{1}{R_1C_1} \times (\Delta V_{IN(pk)} \times \Delta t) = \frac{1}{1\ M\Omega \times 0.01\ \mu F} \times (0\ V\ to\ 1.2\ V\ pk \times 25\ ms)$$

$$= 100 \times (1.2\ V\ pk \times 25\ ms) = 100 \times 0.03 = 3\ V\ pk$$

23-3-6 Signal Generator Circuits

Figure 23-15 shows how the op-amp can be connected to act as a signal generator, which is a circuit that will convert a dc supply voltage into a repeating output signal.

In Figure 23-15(a), the **twin-T sine-wave oscillator,** which has two T-shaped feedback networks, will generate a repeating sine wave output at a frequency equal to

Twin-T Sine-Wave Oscillator
An oscillator circuit that makes use of two T-shaped feedback networks.

Square-Wave Generator
A circuit that generates a continuously repeating square wave.

Relaxation Oscillator
An oscillator circuit whose frequency is determined by an *RL* or *RC* network, producing a rectangular or sawtooth output waveform.

$$f_0 = \frac{1}{2\pi RC}$$

EXAMPLE:

Calculate the frequency of the oscillator shown in Figure 23-15(a).

Solution:

$$f_0 = \frac{1}{2\pi RC} = \frac{1}{2 \times \pi \times 6.8\ k\Omega \times 0.033\ \mu F} = 709\ Hz$$

In Figure 23-15(b), the op-amp has been connected to a **square-wave generator,** or as it is more frequently called a **relaxation oscillator.** The output signal is fed back to both the inverting and noninverting inputs of the op-amp. The capacitor will charge and discharge through R controlling the frequency of the output square wave, which is equal to

$$f_0 = \frac{1}{2\ RC \log\left(\frac{2R_1}{R_2} + 1\right)}$$

Triangular-Wave Generator
A signal generator circuit that produces a continuously repeating triangular wave output.

Connecting the output of the square-wave generator in Figure 23-15(b) into the inputs of the circuits in Figure 23-15(c) and (d), we can generate a triangular or staircase output waveform. The **triangular-wave generator** in Figure 23-15(c) is simply the integrator circuit discussed previously, with its output frequency equal to the square wave input frequency. When the switch is closed in

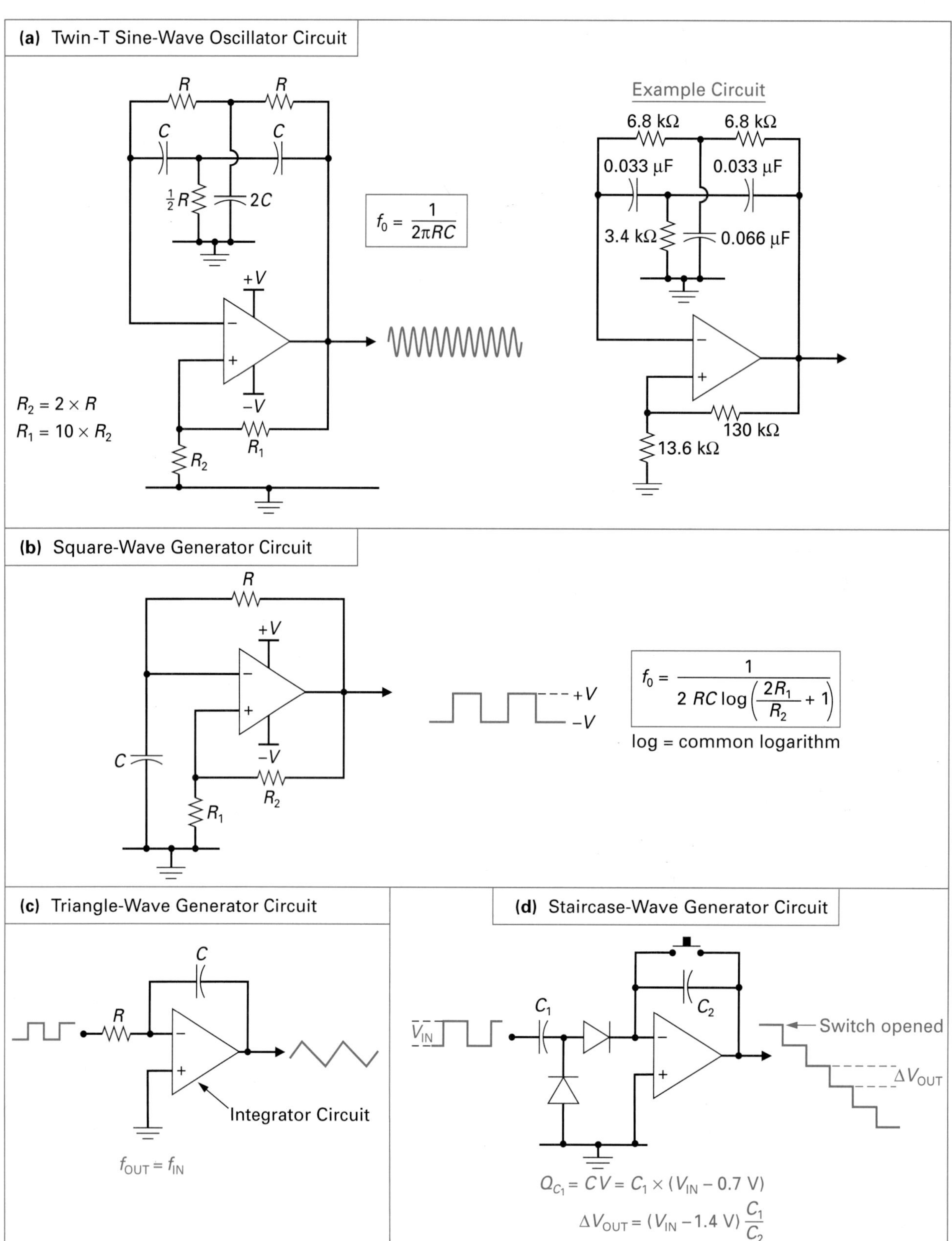

FIGURE 23-15 Signal Generator Circuits.

the **staircase-wave generator** in Figure 23-15(d) the capacitor is bypassed and will therefore not charge. On the other hand, when the switch is open, C_2 will be charged by each input cycle, producing equal output steps that have the following voltage change

Staircase-Wave Generator
A signal generator circuit that produces an output signal voltage that increases in steps.

$$\Delta V_{OUT} = (V_{IN} - 1.4\text{ V})\,\frac{C_1}{C_2}$$

23-3-7 *Active Filter Circuits*

Passive filters are circuits that contain passive or nonamplifying components (resistors, capacitors, and inductors) connected in such a way that they will pass certain frequencies while rejecting others. An **active filter** on the other hand, is a circuit that uses an amplifier with passive filter elements to provide frequency paths with rejection characteristics. Active filters, like the op-amp circuits seen in Figure 23-16, have several advantages over passive filters.

Active Filter
A circuit that uses an amplifier with passive filter elements to provide frequency paths with rejection characteristics.

1. Because the op-amp provides gain, the input signal passed to the output will not be attenuated, and therefore better response curves can be obtained.
2. The high input impedance and low output impedance of the op-amp means that the filter circuit does not interfere with the signal source or load.
3. Because active filters provide gain, resistors can be used instead of inductors, and therefore active filters are generally less expensive.

Figure 23-16 illustrates how the op-amp can be connected to form the four basic active-filter types.

Active High-Pass Filter

Figure 23-16(a) illustrates the simple op-amp circuit, frequency response, and relevant formulas for an **active high-pass filter.** As before, the gain of this inverting amplifier is dependent on the ratio of R_F to R_{IN}. When capacitors are included in any circuit, impedance (Z) must be considered instead of simply resistance, and gain is now equal to the ratio of feedback impedance to input impedance.

Active High-Pass Filter
A circuit that uses an amplifier with passive filter elements to pass all frequencies above a cut-off frequency.

$$A_{CL} = -\frac{Z_F}{Z_{IN}}$$

The input RC network will offer a high impedance to low frequencies, resulting in a low voltage gain. At high frequencies, the RC network will have a low impedance, causing a high voltage gain. The cutoff frequency for this circuit can be calculated with the following formula when $C_1 = C_2$.

$$f_C = \frac{1}{2\pi RC}$$

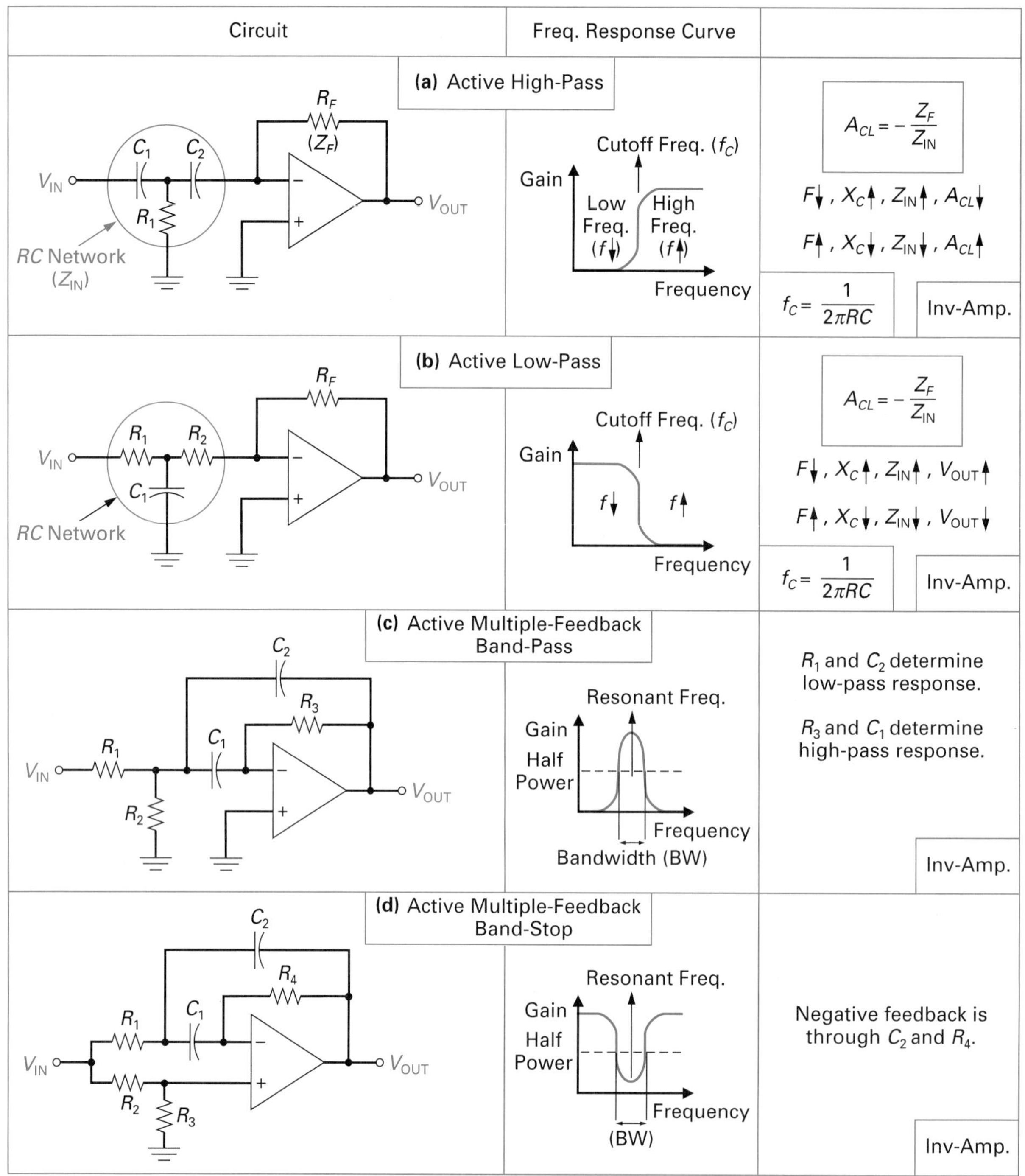

FIGURE 23-16 Active-Filter Circuits.

Active Low-Pass Filter
Amplifier circuit with passive filter elements to pass all frequencies below a cut-off frequency.

Active Low-Pass Filter

Figure 23-16(b) illustrates the op-amp circuit, frequency response curve, and relevant formulas for an **active low-pass filter.** At low frequencies, the capacitor's reactance is high, and low-frequency signals will be passed to the op-amp's input to be amplified and passed to the output. As frequency increases, the capacitive reactance of C_1 will decrease; more of the signal will be shunted away from the op-amp and will not appear at the output. The cutoff frequency for this circuit can be calculated with the following formula when $R_1 = R_2$.

$$f_C = \frac{1}{2\pi RC}$$

Active Band-Pass Filter

Figure 23-16(c) illustrates how the op-amp can be connected to form an **active band-pass filter.** At frequencies outside of the band, V_{OUT} is fed back to the input without being attenuated, and therefore the input signal amplitude is almost equal to the feedback signal amplitude. This results in almost complete cancellation of the signal and therefore a very small output voltage. On the other hand, for the narrow band of frequencies within the band, the feedback network will increase its amount of attenuation. This increase of attenuation means that a very small feedback signal will appear back at the negative input of the op-amp and will have a very small degenerative effect. As a result, the change at the input of the op-amp will be larger when the input signal frequencies are within this band, and the voltage out will also be larger.

Active Band-Pass Filter
A circuit that uses an amplifier with passive filter elements to pass only a band of input frequencies.

Active Band-Stop Filter

Figure 23-16(d) illustrates how the op-amp can be connected to form an **active band-stop filter,** also known as a band-reject or notch filter. The basic operation of this circuit is opposite to that of the previously discussed band-pass filter. At frequencies outside of the band, the feedback signal will be heavily attenuated, and therefore the degenerative effect will be small and the output voltage large. On the other hand, at frequencies within the band, the feedback signal will not be heavily attenuated, and therefore the degenerative effect will be large and the output voltage small.

Active Band-Stop Filter
A circuit that uses an amplifier with passive filter elements to block a band of input frequencies.

SELF-TEST REVIEW QUESTIONS FOR SECTION 23-3

1. Which op-amp circuit provides a voltage gain of 1 and is used as a buffer?
2. Which op-amp circuit will sum all of the input voltages?
3. What is the basic circuit difference and input/output waveform difference between the integrator and differentiator circuit?
4. Which op-amp circuit will generate an output that is equal to the difference between the two inputs?
5. Sketch a circuit showing how the op-amp can be connected to generate a repeating square wave output.
6. What is the difference between an active filter and a passive filter?

SUMMARY

The Operational Amplifier (Figure 23-17)

1. The operational amplifier was initially a vacuum tube circuit used in the early 1940s in analog computers.
2. The name "operational amplifier" or "op-amp" was chosen because the circuit was used as a high-gain dc "amplifier" performing mathematical "operations."
3. The early op-amp circuits were expensive and bulky, and they found very little application until the semiconductor integrated circuit was developed in 1958 by Jack Kilby at Texas Instruments. Circuits that once needed hundreds of discrete or individual components can now be integrated into a single IC, making equipment smaller, more energy efficient, cheaper, and easier to design and troubleshoot.
4. Today's IC op-amp is a very high-gain dc amplifier that can have its operating characteristics changed by connecting different external components. This makes the op-amp very versatile, and it is this versatility that has made the op-amp the most widely used linear IC.
5. The op-amp's internal circuit is simply a combination of three previously covered amplifier circuits. These three circuits are all interconnected and contained within a single IC, and together they function as a "high-gain, high input impedance, low output impedance amplifier."
6. The triangle shaped amplifier symbol is used to represent the op-amp in an electronic schematic diagram.
7. The two op-amp inputs are labeled "–" and "+." The "–" or negative input is called the inverting input because any signal applied to this input will be amplified and inverted between input and output (output is 180° out of phase with input). On the other hand, the "+" or positive input is called the noninverting input because any signal applied to this input will be amplified but not inverted between input and output (output is in phase with input). An input signal will normally be applied to only one of these inputs, while the other input is used to control the op-amp's operating characteristics.
8. The two power supply connections to the op-amp are labeled "$+V$" and "$-V$." When two supply voltages are used (dual supply voltages), the voltage values are of the same value, but of opposite polarity (for example, +12 V and –12 V). On the other hand, when only one supply voltage is used (single supply voltage), a positive or negative voltage is applied to its respective terminal while the other terminal is grounded (for example, +5 V and ground or –5 V and ground). Having both a positive and negative power supply voltage will allow the output signal to swing positive and negative, above and below zero. As with all high-gain amplifiers, however, the output voltage can never exceed the value of the $+V$ and $-V$ supply voltages.
9. The entire op-amp circuit is placed within a TO-5 metal can package or a dual-in-line through-hole or surface-mount package.
10. Like all ICs, an identification code is used to indicate the device manufacturer, device type, and key characteristics.

QUICK REFERENCE SUMMARY SHEET

FIGURE 23-17 The Operational Amplifier (Op-Amp).

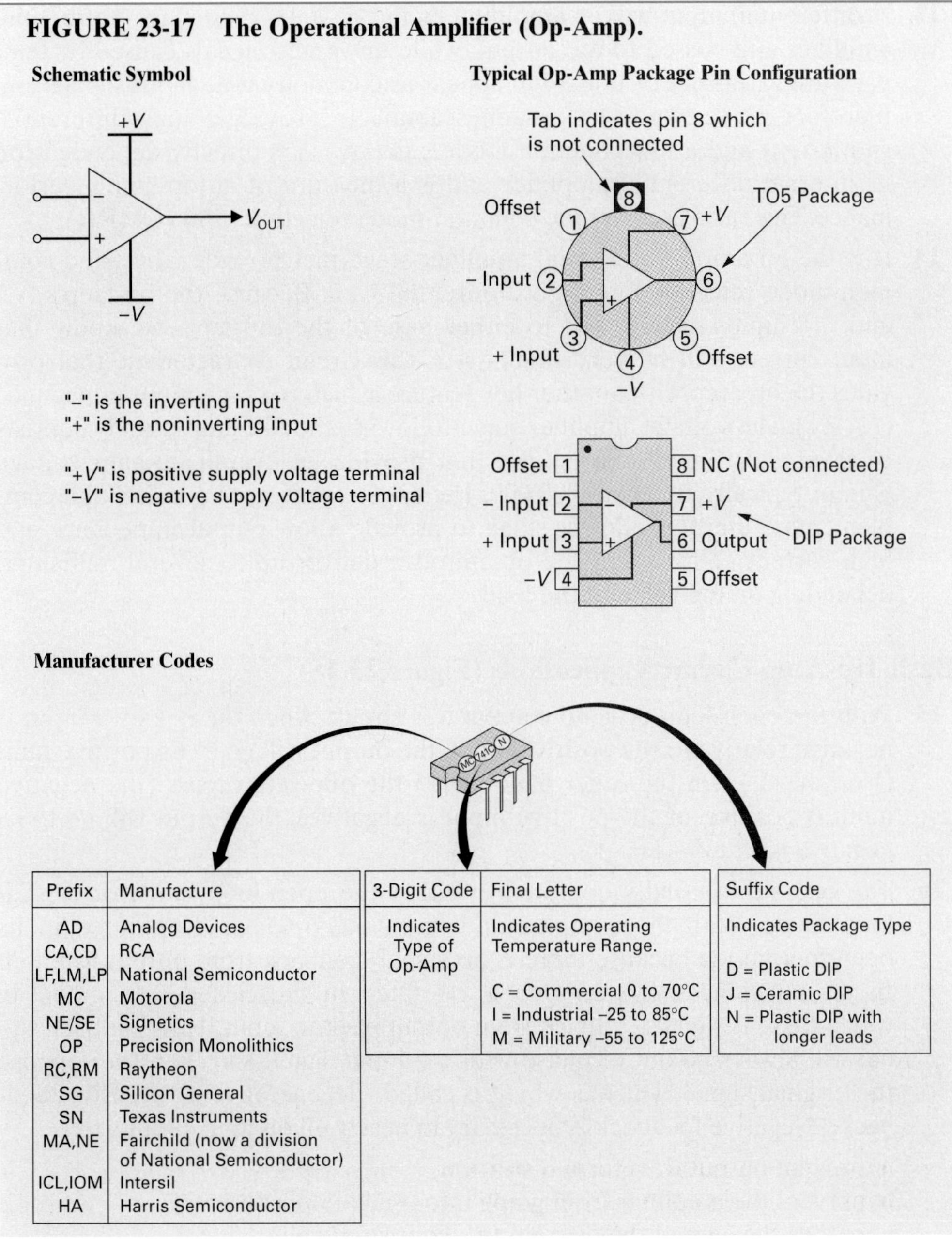

Prefix	Manufacture
AD	Analog Devices
CA,CD	RCA
LF,LM,LP	National Semiconductor
MC	Motorola
NE/SE	Signetics
OP	Precision Monolithics
RC,RM	Raytheon
SG	Silicon General
SN	Texas Instruments
MA,NE	Fairchild (now a division of National Semiconductor)
ICL,IOM	Intersil
HA	Harris Semiconductor

3-Digit Code	Final Letter
Indicates Type of Op-Amp	Indicates Operating Temperature Range. C = Commercial 0 to 70°C I = Industrial –25 to 85°C M = Military –55 to 125°C

Suffix Code
Indicates Package Type D = Plastic DIP J = Ceramic DIP N = Plastic DIP with longer leads

11. The two offset or balancing inputs will normally be connected to a potentiometer that can be adjusted to set the output at zero volts when both the inverting and noninverting inputs are at zero volts. Balancing the op-amp in this way is generally needed due to imbalances within the op-amp's internal circuit.

12. The three circuits all interconnected and contained within the single op-amp IC are a differential amplifier, a voltage amplifier, and an output amplifier.

Combined, these three circuits give the op-amp its key characteristics, which are high gain, high input impedance, and low output impedance.

13. A differential input will be amplified by the op-amp's first stage differential amplifier and passed to the output, while unwanted signals caused by temperature variations or noise will appear as common-mode input signals and therefore be rejected. An op-amp's ability to provide a high differential gain (A_{VD}) and a low common-mode gain (A_{CM}) is directly dependent on its internal differential amplifier and is a measure of an op-amp's performance. This ratio is called the common-mode rejection ratio (CMRR).

14. It is the op-amp's differential amplifier stage that provides the good common-mode rejection and high differential gain. Because the op-amp's "–" and "+" inputs are applied to either base of the diff-amp, we know that input current will be very small—it is this circuit characteristic that provides the op-amp with another key feature, which is a high input impedance ($I\downarrow$, $Z\uparrow$). The voltage amplifier stage following the diff-amp usually consists of several darlington-pair stages that provide an overall op-amp voltage gain of typically 50,000 to 200,000. The final output stage consists of a complementary emitter-follower stage to provide a low output impedance and high current gain so that the op-amp can deliver up to several milliamps, depending on the value of the load.

Basic Op-Amp Circuit Applications (Figure 23-18)

15. With the open-loop op-amp comparator circuit, when the negative input is negative relative to the positive input, the output will go to its positive limit ($V_{OUT} = +V$). On the other hand, when the opposite occurs (the negative input is positive, or the positive input is negative), the output will go to its negative limit ($V_{OUT} = -V$).

16. The op-amp is usually operated in either the open-loop mode or closed-loop mode. With the comparator circuit, the op-amp is operated in its open-loop mode because there is no signal feedback from output to input. In most instances, the op-amp is operated in the closed-loop mode, in which there is signal feedback from output back to input. This feedback signal will always be out of phase with the input signal and therefore oppose the original signal, which is why it is called "degenerative or negative feedback." Negative feedback is necessary in nearly all op-amp circuits to

a. prevent output waveform distortion,
b. prevent the amplifier from going into oscillation, and
c. reduce the gain of the op-amp to a consistent value.

17. The inverting operational amplifier circuit produces an amplified output signal that is 180° out of phase with the input signal. In this circuit arrangement, the input signal (V_{IN}) is applied through an input resistor to the inverting input (–) of the op-amp, while the noninverting input (+) is connected to ground. A feedback loop is connected from the output back to the inverting input via the feedback resistor.

18. The feedback voltage (V_F) will greatly restrain the input voltage change to the point that a +5 V input change at V_{IN} will only be felt as a +5 micro volt

FIGURE 23-18 Basic Op-Amp Circuit Applications.

Open-Loop Comparator Circuit

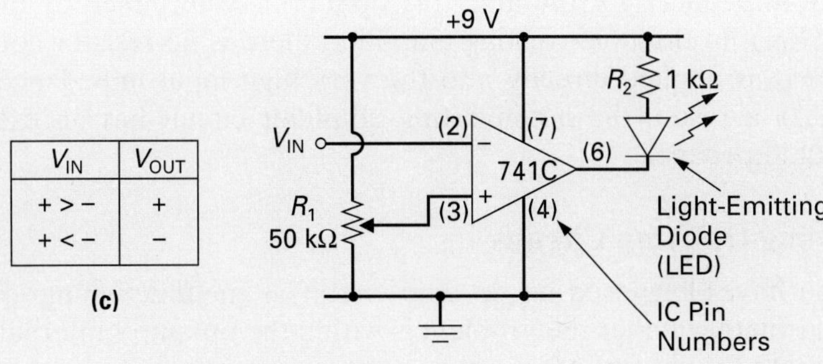

Closed-Loop Inverting Amplifier Circuit

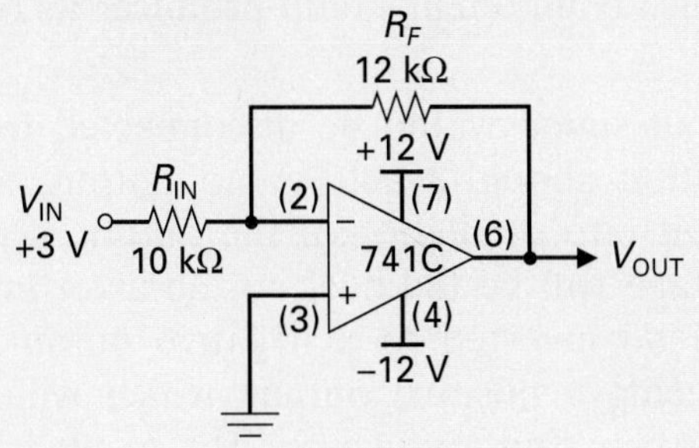

Closed loop voltage gain is dependent on the ratio of R_F to R_{IN}.

$$A_{CL} = -\frac{R_F}{R_{IN}}$$

$$V_{OUT} = -V_{IN}\left(\frac{R_F}{R_{IN}}\right)$$

Closed-Loop Noninverting Amplifier Circuit

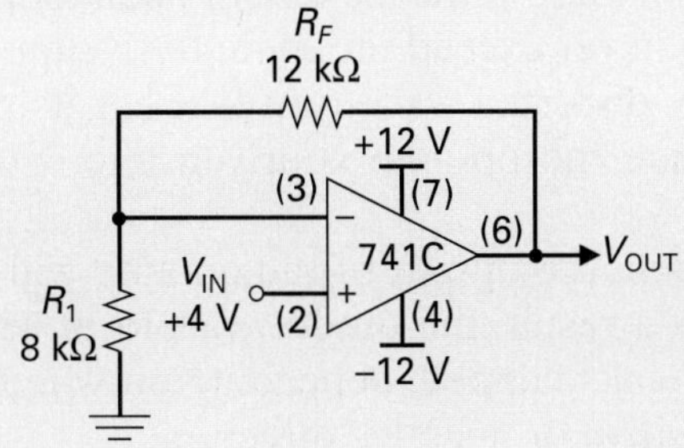

$$A_{CL} = \frac{R_F}{R_1} + 1$$

$$V_{OUT} = V_{IN}\left(1 + \frac{R_F}{R_1}\right)$$

change at the op-amp's inverting input. In fact, if the voltage at the inverting input of the op-amp is measured with a voltmeter, the "−" input appears to remain at 0 V due to the very minute change at the "−" input. The inverting input of the op-amp in this case would be defined as a virtual ground, which is different from an ordinary ground. A virtual ground is a voltage ground because this point is at zero volts; however, it is not a current ground because it cannot sink or conduct away any current. An ordinary ground, on the other hand, is at zero volts and can sink any amount of current.

19. With the noninverting amplifier circuit, the input voltage (V_{IN}) is applied to the op-amp's noninverting input (+), and the output voltage (V_{OUT}) will be in phase with the input. To achieve negative feedback, the output is applied back to the inverting input (–) of the op-amp via the feedback network formed by R_2 and R_1.
20. The input impedance for the inverting op-amp is determined by the input resistor (R_{IN}). In the noninverting amplifier, there is no resistor connected because V_{IN} is applied directly into the very high input impedance of the op-amp. As a result, the noninverting amplifier circuit has an extremely high input impedance.

Troubleshooting Op-Amp Circuits

21. Once you have diagnosed an op-amp circuit as malfunctioning, the next step is to isolate whether the problem is within the op-amp's internal circuit or external to the op-amp IC.
22. The isolating process is relatively easy because op-amp circuits only contain a handful of components and each component's fault produces its own distinct effect.

    a. If R_{IN} were to open, the input signal would be disconnected from the amplifier. Although this condition should result in no output, in most cases it sends the op-amp circuit into oscillation: if the output is positive when R_{IN} opens, a positive voltage will be fed back to the inverting input via R_F, this positive input will be inverted to a negative output, which will be fed back via R_F producing a positive output, which will be fed back via R_F producing a negative output, and so on. The result will be a continuous millivolt oscillation at the output.

    b. If R_F were to open, the feedback path would be disconnected and the gain of the op-amp circuit will increase to its maximum open loop value. Because the output voltage can never exceed the $+V$ and $-V$ supply voltages, the result will be a heavily distorted waveform that has its positive and negative peaks clipped when the op-amp is driven into saturation and cutoff.

    c. If the offset resistor were to open, the output of the op-amp will not be zero when the input is zero. As a result, the output waveform will have either its positive or negative peaks clipped, depending on whether the op-amp is unbalanced in the positive or negative direction.
23. If you have isolated the problem to the op-amp's internal circuit, and you have determined that all of the op-amp's externally connected resistors are okay, the next step will be to replace the op-amp itself because it is impossible to repair any internal op-amp failures.

Additional Op-Amp Circuit Applications

24. The noninverting voltage follower circuit has a voltage gain of 1. Similar to the BJT's emitter-follower and the FET's source-follower, the op-amp voltage-follower circuit derives its name from the fact that the output voltage follows the input voltage in both polarity and amplitude. This circuit is

ideal as a buffer for interfacing a high-impedance source to a low-impedance load.

25. The inverting summing amplifier circuit, or adder amplifier, will sum or add all of the input voltages.
26. With the difference or differential amplifier circuit, the output voltage is equal to the difference between the two input voltages.
27. The op-amp differentiator circuit (not to be confused with the previous differential circuit) is similar to the basic inverting amplifier except that R_{IN} is replaced by a capacitor. The reactance of the input capacitor will decrease for input signals that are higher in frequency, and therefore the input voltage applied to the op-amp and output voltage from the op-amp will increase with frequency. Including an additional resistor (R_S) in series with the input capacitor will decrease the high-frequency gain, and therefore gain will be a ratio of R_F/R_S.
28. With the op-amp integrator circuit, a resistor is connected in the input and a capacitor is connected in the feedback path. Like the differentiator circuit, the capacitor will alter the gain of the op-amp because its capacitive reactance changes with frequency. To compensate for this effect, a parallel resistor (R_P) is included in shunt with the capacitor to decrease the low-frequency gain because gain will now be a ratio of R_P/R_1.

Op-Amp Signal Generator Circuits

29. A signal generator is a circuit that will convert a dc supply voltage into a repeating output signal.
30. The twin-T sine-wave oscillator, which has two T-shaped feedback networks, will generate a repeating sine wave output.
31. With the square-wave op-amp generator, or relaxation oscillator, the output signal is fed back to both the inverting and noninverting inputs of the op-amp. The capacitor will charge and discharge through R, and this cycle will control the frequency of the output square wave.
32. The triangular-wave generator is simply an integrator circuit with its output frequency equal to the square wave input frequency.
33. When the switch is closed in the staircase-wave generator, the capacitor is bypassed and will not charge. On the other hand, when the switch is open, C_2 will be charged by each input cycle, producing equal output voltage steps.

Op-Amp Active Filter Circuits

34. Passive filters are circuits that contain passive or nonamplifying components (resistors, capacitors, and inductors) connected in such a way that they will pass certain frequencies while rejecting others.
35. An active filter is a circuit that uses an amplifier with passive filter elements to provide frequency paths with rejection characteristics.

36. Active filters have several advantages over passive filters.

   a. Because the op-amp provides gain, the input signal passed to the output will not be attenuated, and therefore better response curves can be obtained.

   b. The high input impedance and low output impedance of the op-amp means that the filter circuit does not interfere with the signal source or load.

   c. Because active filters provide gain, resistors can be used instead of inductors, and therefore active filters are generally less expensive.

37. With the active high-pass filter, the input *RC* network will offer a high impedance to low frequencies resulting in a low voltage gain, while at high frequencies the *RC* network will have a low impedance causing a high voltage gain.

38. With the active low-pass filter, the capacitor's reactance is high at low frequencies, and therefore low-frequency signals will be passed to the op-amp's input to be amplified and passed to the output. As frequency increases, the capacitive reactance of C_1 will decrease, and therefore more of the signal will be shunted away from the op-amp and consequently not appear at the output.

39. At frequencies outside of the band, V_{OUT} is fed back to the input without being attenuated, and therefore the input signal amplitude of the active band-pass filter is almost equal to the feedback signal amplitude. This results in almost complete cancellation of the signal and therefore a very small output voltage. On the other hand, for the narrow band of frequencies within the band, the feedback network will increase its amount of attenuation. This increase of attenuation means that a very small feedback signal will appear back at the negative input of the op-amp and will have a very small degenerative effect.

40. At frequencies outside of the band, the feedback signal will be heavily attenuated, and therefore the degenerative effect of an active band-stop filter will be small and the output voltage large. On the other hand, at frequencies within the band, the feedback signal will not be heavily attenuated, and therefore the degenerative effect will be large and the output voltage small.

NEW TERMS

Active Filter
Active High-Pass Filter
Active Low-Pass Filter
Active Multiple Feedback Band-Pass Filter
Active Multiple Feedback Band-Stop Filter
Balancing
Closed-Loop Circuit
Closed-Loop Mode
Closed-Loop Voltage Gain
Common-Mode Gain
Common-Mode Input Signals
Common-Mode Rejection
Common-Mode Rejection Ratio
Comparator Circuit
DC Offsets
Difference Amplifier
Differential Amplifier
Differential Gain
Differential Mode Input Signals
Differentiator Circuit
Input Bias Current
Input Offset Current

Input Offset Voltage
Integrator Circuit
Inverting Amplifier
Inverting Input
Noninverting Amplifier
Noninverting Input
Offset Null Inputs
Open-Loop Circuit
Open-Loop Gain
Open-Loop Mode
Operational Amplifier
Ordinary Ground
Passive Filter
Relaxation Oscillator
Signal Generator
Slew Rate
Square-Wave Generator
Staircase Wave Generator
Summing Amplifier
Triangular Wave Generator
Twin-T Oscillator
Virtual Ground
Voltage Follower

REVIEW QUESTIONS

Multiple-Choice Questions

1. When a differential amplifier is used in the differential-input, single-output mode, it has a ____________ differential gain and a ____________ common-mode gain.
   a. High, high **c.** Low, low
   b. High, low **d.** Low, high
2. The op-amp's internal circuit contains a ____________, ____________, and ____________ amplifier stage.
   a. Differentiator, current, power
   b. Integrator, voltage, output
   c. Darlington-pair, emitter follower, summing
   d. A differential, darlington-pair, emitter follower
3. The op-amp's differential amplifier stage provides the op-amp with a
   a. Low common-mode gain **c.** High input impedance
   b. High differential gain **d.** All of the above
4. Which transistor circuit is used in the op-amp's final output stage to provide a low output impedance and high current gain?
   a. Common emitter **c.** Common collector
   b. Common base **d.** Both (a) and (c)
5. Could an op-amp circuit be constructed using discrete components?
   a. Yes **b.** No
6. The comparator is considered a/an ____________ loop op-amp circuit.
   a. Common
   b. Open
   c. Differential
   d. Closed
7. What is the lower frequency limit of an op-amp?
   a. 20 Hz **b.** 6 Hz **c.** DC **d.** 7.34 Hz
8. A virtual ground is a ground to ____________ but not to ____________.
   a. Current, voltage **b.** Voltage, current

9. The feedback loop in a closed-loop op-amp circuit provides
   a. Positive feedback
   b. Negative feedback
   c. Degenerative feedback
   d. Both (a) and (b)
   e. Both (b) and (c)
10. The ____________ input(s) of an op-amp is used to compensate for slight differences in the transistors in the differential amplifier stage.
   a. Inverting
   b. DC offset
   c. $+V$
   d. V_{OUT}

Essay Questions

11. Why is the term "operational" used to describe the op-amp? (Intro.)
12. Sketch the op-amp schematic symbol. (23-1-1)
13. Describe some of the different op-amp package types and the meaning of the manufacturer codes. (23-1-2)
14. What are the three basic amplifier blocks within an op-amp? (23-2)
15. In what mode is the differential amplifier used within the op-amp? (23-2-1)
16. What is the difference between common-mode input signals and differential-mode input signals? (23-2-1)
17. Define "common-mode rejection ratio" in relation to the op-amp. (23-2-2)
18. Sketch the basic block diagram of an op-amp's internal circuit, and list the characteristics of each block. (23-2-3)
19. What is the difference between an open-loop and closed-loop op-amp circuit? (23-2-5)
20. Sketch a simple op-amp comparator circuit and briefly describe its operation. (23-2-5)
21. Why is it necessary for an op-amp to have negative feedback? (23-2-5)
22. Sketch a simple inverting operational amplifier circuit, and briefly describe its operation. (23-2-5)
23. How is the gain, and therefore output voltage, of an inverting op-amp circuit determined? (23-2-5)
24. Sketch a simple noninverting operational amplifier circuit, and briefly describe its operation. (23-2-5)
25. How is the gain, and therefore output voltage, of a noninverting op-amp circuit determined? (23-2-5)
26. What is the difference between a virtual ground and an ordinary ground? (23-2-5)
27. What is the difference between a 741 and 747 op-amp IC? (23-2-6)
28. Sketch and describe the operation of the following op-amp application circuits:
   a. Voltage follower circuit (23-3-1)
   b. Summing amplifier circuit (23-2-2)
   c. Difference amplifier circuit (23-3-3)
   d. Differentiator circuit (23-3-4)
   e. Integrator circuit (23-3-5)

29. Sketch a circuit showing how an op-amp can be connected to operate as a sine-wave, square-wave, and triangular-wave generator. (23-3-6)

30. What is a staircase generator circuit? (23-3-6)

31. What is the difference between a passive filter and an active filter, and what are the advantages of active filters? (23-3-7)

32. Show how the op-amp can be used as an active filter by sketching an example circuit and explaining its operation. (23-3-7)

Practice Problems

33. Explain the meaning of the following op-amp manufacturer codes:

    a. LM 318C N **b.** NE 101C D

34. Calculate the common-mode rejection ratio of an op-amp if it has a common-mode gain of 0.8 and a differential gain of 27,000. Also give the answer in dBs.

35. Identify the circuit shown in Figure 23-19. Why must the input to this circuit always be a negative voltage?

36. What would be the output voltage from the circuit in Figure 23-19 if the input voltage were –1.6 V?

37. Identify the circuits shown in Figure 23-20(a) and (b), and then sketch the shape of the output waveform if a square wave were applied to the inputs.

38. Referring to the circuit shown in Figure 23-21, what would be the voltage out if a +7.3 V input were applied?

39. Identify the circuits shown in Figure 23-22(a) and (b), and then calculate the output voltages for the given input voltages.

40. Which of the circuits shown in Figure 23-23 is an active high-pass filter and which is an active low-pass filter?

41. Calculate the cutoff frequency of the circuit in Figure 23-23(a) if $C_1 = C_2 = 0.1\ \mu F$, $R_1 = 10\ k\Omega$.

42. Calculate the cutoff frequency of the circuit in Figure 23-23(b), if $R_1 = R_2 = 33\ k\Omega$, $C_1 = 0.33\ \mu F$.

43. Identify the circuit shown in Figure 23-24 and calculate its closed-loop voltage gain.

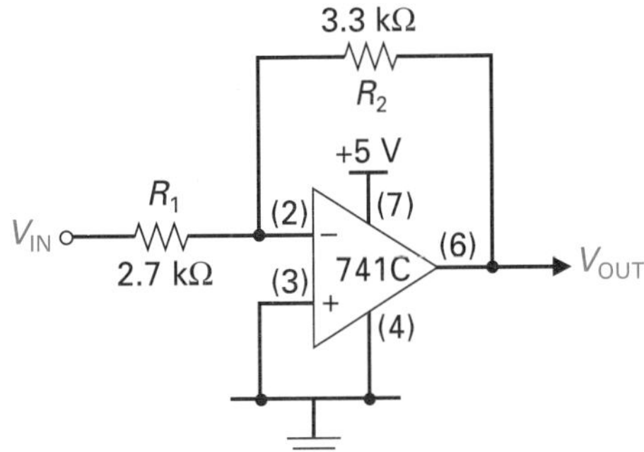

FIGURE 23-19 **A 741C Op-Amp Circuit.**

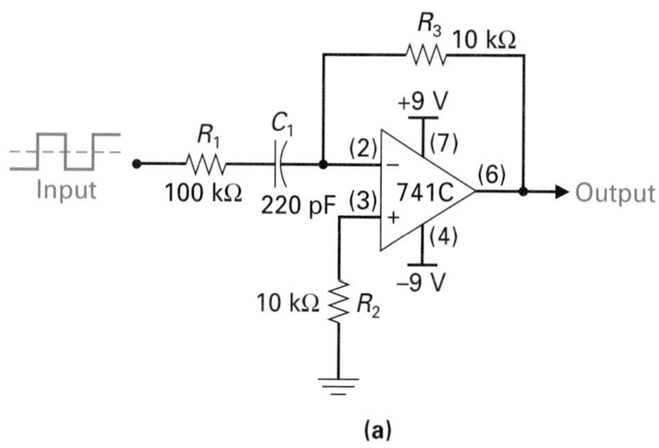

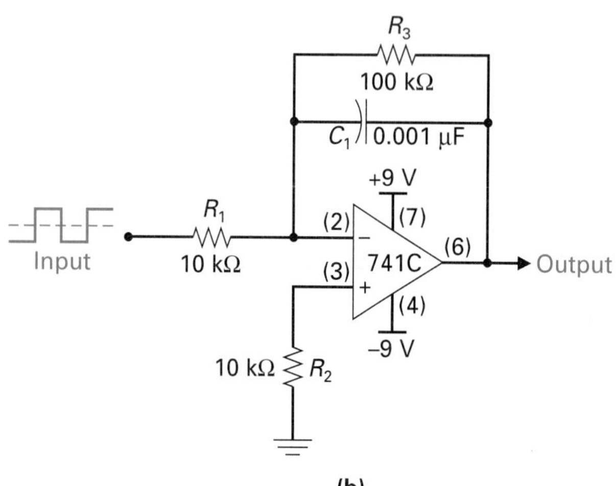

FIGURE 23-20 Two Applications for the 741C Op-Amp.

44. What would be the output voltage from the circuit in Figure 23-24 for the input voltage given?

Troubleshooting Questions

45. If the circuit in Figure 23-24 were not functioning as it should, how would you determine whether the problem is (23-2-7)

a. In the source

b. In the load

c. In the circuit supply voltages

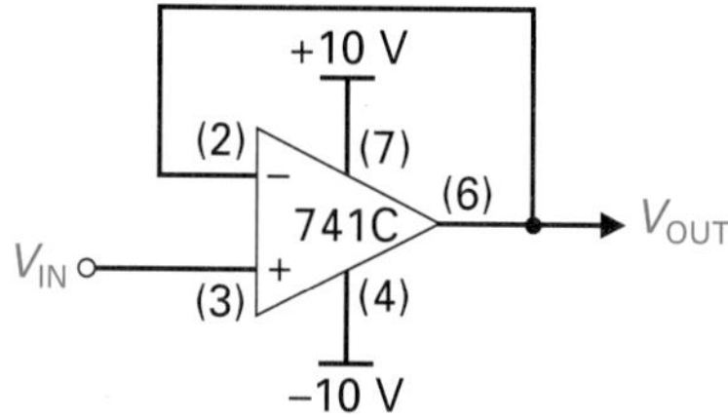

FIGURE 23-21 An Op-Amp Circuit.

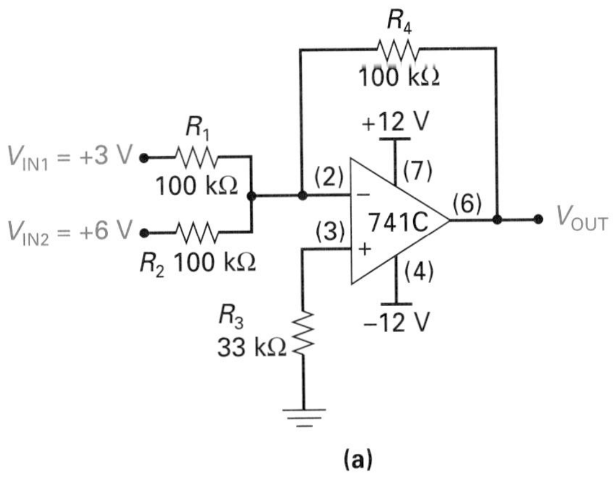

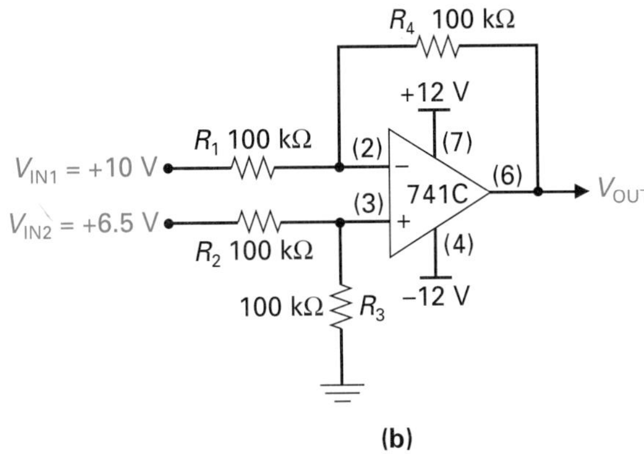

FIGURE 23-22 Op-Amp Circuit Examples.

46. If the following faults were introduced to the circuit in Figure 23-24, what would be the circuit symptoms? (23-2-7)

a. R_{IN} open
b. R_F open
c. R_{OFFSET} open

47. As an exercise in troubleshooting, construct the circuit shown in Figure 23-24. Then introduce some of the following circuit problems, note the symptoms, and logically explain why a cause has a certain effect.

a. Disconnect one of the inputs to the op-amp.
b. Disconnect one of the supply voltages.
c. Adjust the offset resistor to one extreme end of its track.
d. Disconnect the load.

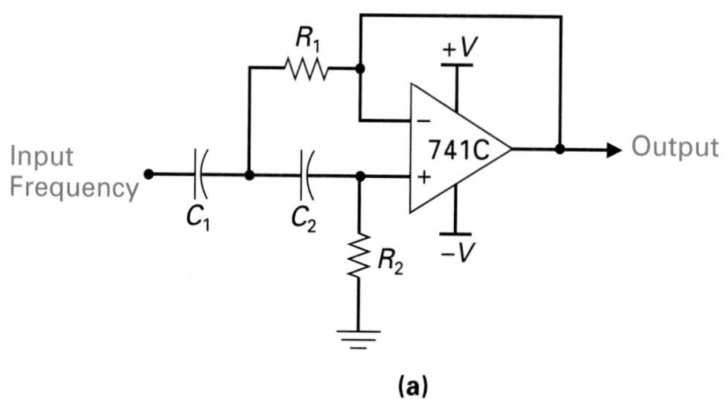

(a)

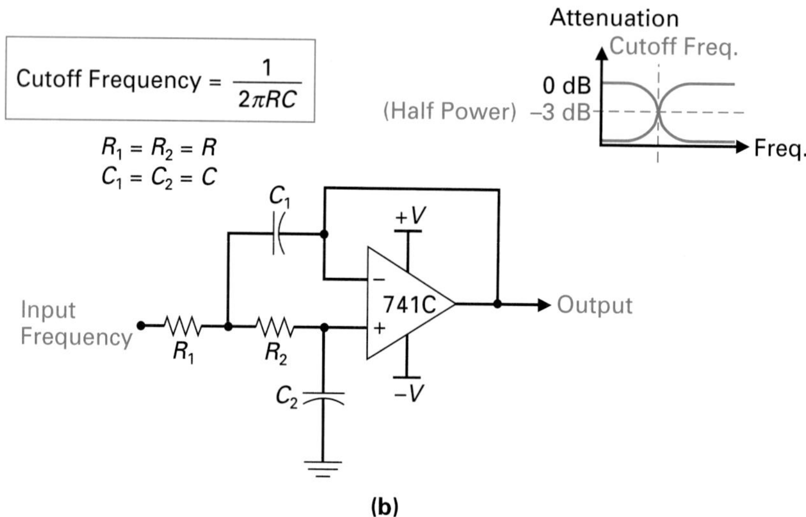

(b)

FIGURE 23-23 **Op-Amp Active Filter Circuits.**

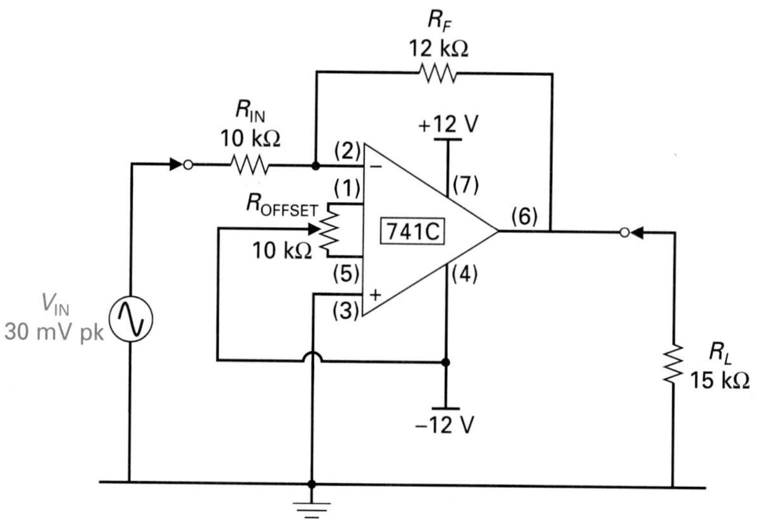

FIGURE 23-24 **Op-Amp Amplifier Circuit.**

Index

C

D

E

F

J

K

L

M

N

Q

R

S

T

U

V

W

Z